Lecture Notes in Artificial Intelligence 11606

Subseries of Lecture Notes in Computer Science

More information about this series at http://www.springer.com/series/1244

Franz Wotawa · Gerhard Friedrich ·
Ingo Pill · Roxane Koitz-Hristov ·
Moonis Ali (Eds.)

Advances and Trends in Artificial Intelligence

From Theory to Practice

32nd International Conference
on Industrial, Engineering and Other Applications
of Applied Intelligent Systems, IEA/AIE 2019
Graz, Austria, July 9–11, 2019
Proceedings

Editors
Franz Wotawa
Institute for Software Technology
Graz University of Technology
Graz, Austria

Gerhard Friedrich
Department of Applied Informatics
University of Klagenfurt
Klagenfurt, Austria

Ingo Pill
Institute for Software Technology
Graz University of Technology
Graz, Austria

Roxane Koitz-Hristov
Institute for Software Technology
Graz University of Technology
Graz, Austria

Moonis Ali
Department of Computer Science
Texas State University
San Marcos, TX, USA

ISSN 0302-9743 ISSN 1611-3349 (electronic)
Lecture Notes in Artificial Intelligence
ISBN 978-3-030-22998-6 ISBN 978-3-030-22999-3 (eBook)
https://doi.org/10.1007/978-3-030-22999-3

LNCS Sublibrary: SL7 – Artificial Intelligence

This Springer imprint is published by the registered company Springer Nature Switzerland AG
The registered company address is: Gewerbestrasse 11, 6330 Cham, Switzerland

Preface

This volume contains the conference proceedings of the 32nd edition of the International Conference on Industrial, Engineering and Other Applications of Applied Intelligent Systems (IEA/AIE), which was held during July 9–11, 2019, in Graz, Austria.

This annual event always tries to bring together academic and industrial researchers from all areas of intelligent systems research in order to discuss theoretical foundations and their application in practice. Without doubt, today many smart, intelligent systems assist us in our everyday live. Be it the more or less successful smart behavior of our mobile phone that aims to save battery power so that we have it available for a full day, autonomous trains connecting airport terminals, or in public transport, recommender systems drawing our attention to products we might need or might have forgotten to re-order, or autonomous rovers exploring outer space or hazard sites. The IEA/AIE conference series has always been a frontier platform from which to disseminate new technology for implementing intelligent systems, new possible applications, and also experience reports regarding best practices for the last mile to finally deploy intelligent systems. Also this year, our international Program Committee with reviewers from 25 countries selected 73 papers out of the 151 registered abstracts for presentation at the conference and publication in these proceedings. Selecting the 41 full papers and 32 short papers was indeed not an easy process. We want to thank all the reviewers for their efforts with the reviews and the lively discussions that led to the final acceptance decisions. Without the authors' hard work on their research and also the corresponding submitted manuscripts, there would be no program. Thus, our special thanks go these authors, both those of the accepted papers and also those we might see at a future edition of IEA/AIE.

The IEA/AIE 2019 program included three special tracks and three keynote presentations given by distinguished scientists and practitioners. Our heartfelt thanks go to Reiner John, Dietmar Jannach, and Louise Travé-Massuyès for their keynotes. We also want to thank the organizers of the three special tracks on "Automated Driving and Autonomous Systems," "Mobile and Autonomous Robotics," and "AI for Tourism" for their support. Concluding these acknowledgments, we would like to thank everyone who helped to make IEA/AIE 2019 a success. This includes all authors, Program Committee members, reviewers, and keynote speakers, as well as the organizers, and all the participants of IEA/AIE 2019—the heart and soul of any conference.

May 2019

Franz Wotawa
Gerhard Friedrich
Ingo Pill
Roxane Koitz-Hristov
Moonis Ali

Preface

Organization

General Co-chairs

Moonis Ali	Texas State University, USA
Franz Wotawa	Graz University of Technology, Austria

Local Chair

Ingo Pill	Graz University of Technology, Austria

Program Committee Co-chairs

Gerhard Friedrich	University of Klagenfurt, Austria
Franz Wotawa	Graz University of Technology, Austria

Special Topics Co-chairs

Mihai Nica	AVL List GmbH, Austria
Paul Plöger	Bonn-Rhein-Sieg University of Applied Science, Germany
Gerald Steinbauer	Graz University of Technology, Austria
Franz Wotawa	Graz University of Technology, Austria

Publication Chair

Roxane Koitz-Hristov	Graz University of Technology, Austria

Publicity Chair

Patrick Rodler	University of Klagenfurt, Austria

Web and Local Committee

Jörg Baumann	Graz University of Technology, Austria
Elisabeth Orthofer	Graz University of Technology, Austria
Petra Pichler	Graz University of Technology, Austria

Program Committee

Rui Abreu	University of Lisbon and INESC-ID, Portugal
Otmane Ait Mohamed	Concordia University, Canada
Moonis Ali	Texas State University, USA

Carlos Alonso Gonzáles	Universidad de Valladolid, Spain
Abdelmalek Amine	Tahar Moulay University of Saida, Algeria
Artur Andrzejak	Universität Heidelberg, Germany
Farshad Badie	Aalborg University, Denmark
Suzanne Barber	The University of Texas at Austin, USA
Fevzi Belli	University of Paderborn, Germany
Salem Benferhat	University of Artois, France
Jiang Bian	University of Florida, USA
Mathias Brandstötter	Joanneum Research Forschungsgesellschaft mbH, Austria
Ivan Bratko	University of Ljubljana, Slovenia
Shyi-Ming Chen	National Taiwan University of Science and Technology, Taiwan
Sung-Bae Cho	Yonsei University, South Korea
Alessandro Cimatti	Fondazione Bruno Kessler, Italy
Flavio S. Correa da Silva	University of Sao Paulo, Brazil
Georgios Dounias	University of the Aegean, Greece
Alexander Feldman	Palo Alto Research Center, USA
Alexander Felfernig	Graz University of Technology, Austria
Alexander Ferrein	Fachhochschule Aachen University of Applied Sciences, Germany
Philippe Fournier-Viger	Harbin Institute of Technology, China
Gerhard Friedrich	University Klagenfurt, Austria
Erik Frisk	Linköping University, Sweden
Hamido Fujita	Iwate Prefectural University, Japan
Matjaz Gams	Jozef Stefan Institute, Slovenia
Christophe Gonzales	LIP6, Sorbonne Universite, France
Morten Goodwin	University of Agder, Norway
Ole-Christoffer Granmo	University of Agder, Norway
Alban Grastien	Data61, Australia
Maciej Grzenda	Warsaw University of Technology, Poland
Hans W. Guesgen	Massey University, New Zealand
Abdelwahb Hamou-Lhadj	Concordia University, Canada
Enrique Herrera-Viedma	University of Granada, Spain
Rattikorn Hewett	Texas Tech University, USA
Michael Hofbaur	Joanneum Research Forschungsgesellschaft mbH, Austria
Birgit Hofer	Graz University of Technology, Austria
Takayuki Ito	Nagoya Institute of Technology, Japan
Dietmar Jannach	University Klagenfurt, Austria
Meir Kalech	Ben-Gurion University, Israel
Johan de Kleer	Palo Alto Research Center, USA
Roxane Koitz-Hristov	Graz University of Technology, Austria
Philippe Leray	Nantes University, France
Mark Levin	Russian Academy of Sciences, Russia
Wolfgang Mayer	University of South Australia, Australia

Ian Miguel	University of St. Andrews, UK
Silvia Miksch	Vienna University of Technology, Austria
Maryam Sadat Mirzaei	RIKEN-Kyoto University, Japan
Malek Mouhoub	University of Regina, Canada
Nysret Musliu	Vienna University of Technology, Austria
Ngoc-Thanh Nguyen	Wroclaw University of Technology, Poland
Oliver Niggeman	Institute for Industrial IT, Germany
Jae Oh	Syracuse University, USA
Sigeru Omatu	Osaka Institute of Technology, Japan
Beatrice Ombuki-Berman	Brock University, Canada
Alberto Ortiz	University of the Balearic Islands, Spain
Ingo Pill	Graz University of Technology, Austria
Paul Plöger	Bonn-Rhein-Sieg University of Applied Science, Germany
Gregory Provan	University College Cork, Ireland
Belarmino Pulido	Universidad de Valladolid, Spain
Srini Ramaswamy	ABB Inc., USA
Wolfgang Reif	University of Augsburg, Germany
Patrick Rodler	University Klagenfurt, Austria
Indranil Roychoudhury	Schlumberger Software Technology Innovation Center, USA
Samira Sadaoui	University of Regina, Canada
Aleksander Sadikov	University of Ljubljana, Slovenia
Miguel Sànchez-Marrè	Polytechnic University of Catalonia, Spain
Konstantin Schekotihin	University Klagenfurt, Austria
Ali Selamat	Universiti Teknologi Malaysia, Malaysia
Sabrina Senatore	University of Salerno, Italy
Neal Snooke	Aberystwyth University, UK
Michael Spranger	Sony Computer Science Laboratories Inc., Japan
Gerald Steinbauer	Graz University of Technology, Austria
Roni Stern	Ben-Gurion University, Israel
Peter Struss	Technical University of Munich, Germany
Anna Sztyber	Warsaw University of Technology, Poland
Karim Tabia	University of Artois, France
Oliver Tazl	Graz University of Technology, Austria
Choh Man Teng	Institute for Human and Machine Cognition, USA
Louise Travé-Massuyès	LAAS-CNRS, France
Barbara Vantaggi	Sapienza University of Rome, Italy
Toby Walsh	University of New South Wales, Australia
Yutaka Watanobe	University of Aizu, Japan
Stefan Woltran	Vienna University of Technology, Austria
Franz Wotawa	Graz University of Technology, Austria
Marina Zanella	University of Brescia, Italy
Markus Zanker	Free University of Bolzano-Bozen, Italy
Jernej Zupancic	Jozef Stefan Institute, Slovenia

Additional Reviewers

Darshana Abeyrathna
Ahmed Abouzeid
Marwan Ammar
Medina Andresel
Ayman A. Atallah
Thomas Bach
Lutz Büch
Ludovik Coba
Diego E. D. Costa
Michael Eichholzer
Dejan Georgiev
Mohammad Ghanavati
Saeed Gorji
Manuel Herold
Raheleh Jafari
Raoul Jetley
Zilong Jiao
Anastassia Kuestenmacher
Nicolas Limpert
Jan Maly
Victor Matarè
Martin Mozina
Deebul Nair
Sowmith Nethula
Van Du Nguyen
Christopher Sanford
Jivitesh Sharma
Will Snipes
Gabriele Sottocornola
Matthias Straka
Sithu Sudarsan
Panagiotis Symeonidis
Van Cuong Tran
Johannes P. Wallner
Razieh N. Zaeem

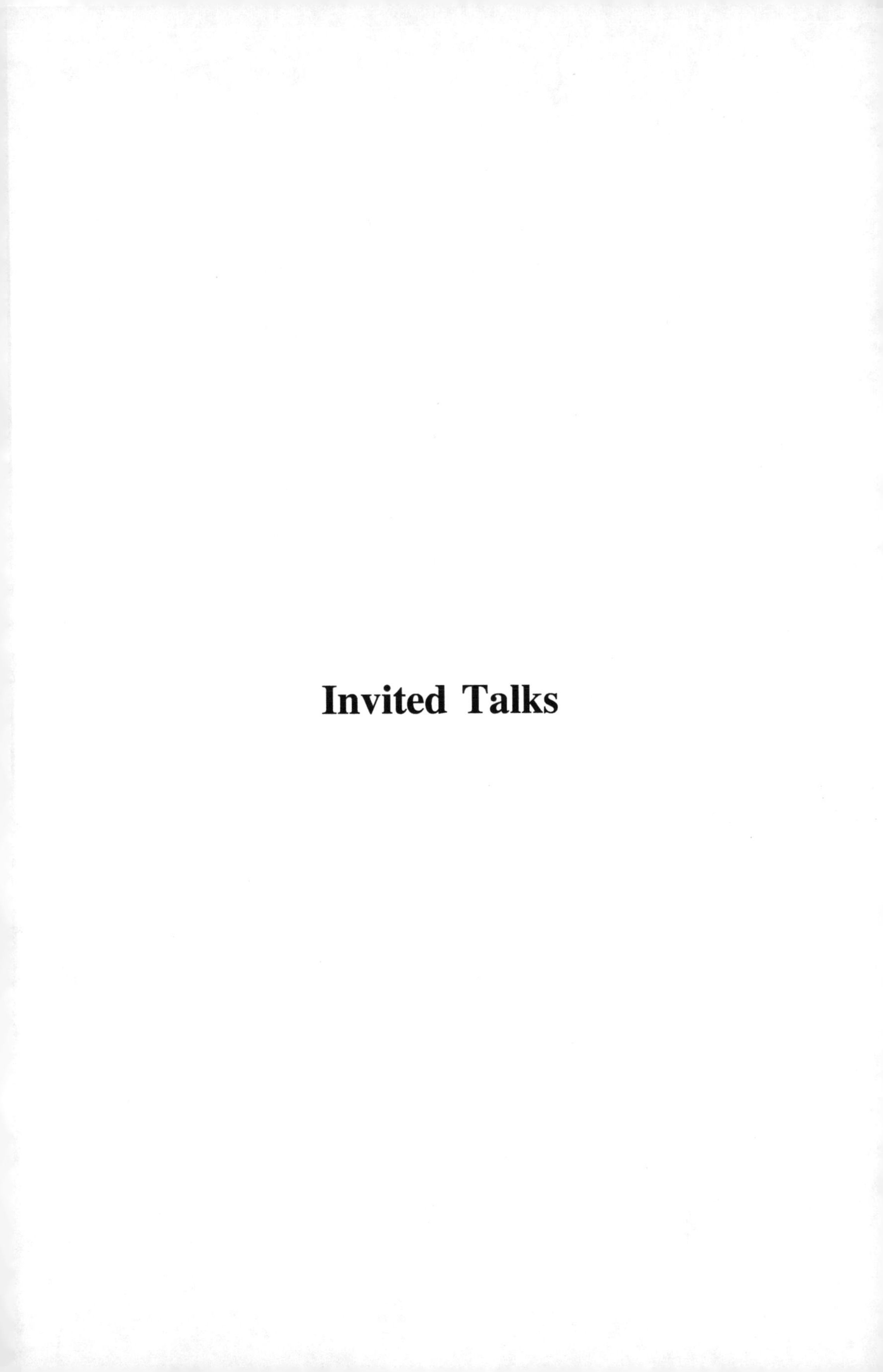

Invited Talks

Session-Based Recommendation: Challenges and Recent Advances

Dietmar Jannach

University of Klagenfurt, Austria

Abstract. In many applications of recommender systems, a larger fraction of the user population are first-time users or are not logged in when they use the service. In these cases, the item suggestions by the recommender cannot be based on individual long-term preference profiles. Instead, the recommendations have to be determined based on the observed short-term behavior of the users. Due to the high practical relevance of session-based recommendation, different proposals were made in recent years to deal with the particular challenges of the problem setting. In this talk we will review some of these challenges and provide a survey on recent advances in the field. A specific focus on the talk will be on the particularities of the e-commerce domain.

Bio: Dietmar Jannach is a full professor of Information Systems at the University of Klagenfurt, Austria. Before joining this university in 2017, he was a professor of Computer Science at TU Dortmund, Germany. In his research, he focuses on the application of intelligent system technology to practical problems and the development of methods for building knowledge-intensive software applications. In the last years, Dietmar Jannach worked on various practical aspects of recommender systems. He is the main author of the first textbook on the topic published by Cambridge University Press in 2010 and was the co-founder of a tech startup that created an award-winning product for interactive advisory solutions.

The 2nd Wave of AI – Thesis for Success of AI in Trustworthy, Safety Critical Mobility Systems

Reiner John

Infineon Technologies AG, Germany

Abstract. In the era of digital transformation, when flexibility and deep understanding in the operation of complex products becomes the key competitive advantage, Artificial Intelligence (AI) is the accepted method to drive the digitalization for the transformation of the industry and their industrial products. These products with highest complexity are based on multi-dimensional requirements as well as novel components, e.g. dedicated CPUs which support AI operations in the cloud and at the edge, as well as dedicated sensors with specialized AI capabilities. One of the most prominent examples is the automotive industry and the products based on high semiconductor content for functional integration, such as highly automated cars including the related industrial and manufacturing itself. The change towards more AI driven applications is present and it is faster and faster emerging in nearly all areas of the industry and will enable new innovative industrial and manufacturing models. The key enabler to the certification and uses of the 3rd Generation of AI methods in safety critical systems is the understanding and the transparence of the decision-making process. The key hurdle to implement this is the certification process which is mandatory for safety critical systems in mobility.

Bio: Reiner John received a diploma degree in Electrical Engineering from the University of Metz/Perpignan, France, in 1984. He started his career at the Siemens Semiconductor Group in Munich in the test system development. From 1989 to 1991 he was responsible for the training of customers in the Siemens Automation Group. From 1991 to 1996 he joined the Siemens Automotive Division in Regensburg, and was responsible for software development processes of µ-controllers. In 1996 he joined the Siemens Semiconductors Division and has been working in several quality- and production management positions. From 2000 to 2006 he was responsible for the Infineon Silicon Foundry Taiwan office. Currently he is responsible for the coordination of public funded projects at Infineon's R&D Funding Department, including several large projects related to automated and assisted driving.

Contributions of Diagnostic Reasoning to the General Demand for AI in the Industry

Louise Travé-Massuyès

Centre National de la Recherche Scientifique (LAAS-CNRS), France

Abstract. AI applications have never been as popular as today. The enthusiasm of all branches of the industry is in tune and agrees to say that AI technologies can lift many industrial locks for customer value creation, productivity improvement, and insight discovery. Huge business opportunities are expected in this process. Numerous applications are also foreseen in medicine and heath care, agriculture and environment, transport/mobility, and energy domains. Faced with this expectation, where do we situate ourselves? In this talk, I will focus on engineering and process applications and will identify the main requests and the needs in these domains. I will then focus on my area of expertise, which is diagnostic reasoning, and explain how existing diagnosis theories can bring their contribution based on the presentation of some applications that address specific needs in these domains. I will conclude my talk by drawing my picture of what is still missing to satisfy the current expectation.

Bio: Louise Travé-Massuyès holds a position of Directrice de Rechercheat Laboratoire d'Analyse et d'Architecture des Systèmes, Centre National de la Recherche Scientifique (LAAS-CNRS), Toulouse, France; head of the Diagnosis and Supervisory Control Team (DISCO) from 1994 to 2015. Her research interests are all related to diagnosis reasoning, tackled by model-based and data-driven approaches. This theme, which she developed throughout her career, led her to consider various formalisms to address the family of problems covered by the diagnosis field. She has been particularly active in establishing bridges between the diagnosis communities of Artificial Intelligence and Automatic Control. She is among the coordinators of the "USER" Strategic Field, assigned to diagnosis and health monitoring topics, within the French Aerospace Valley World Competitiveness Cluster, and serves as the contact evaluator for the French Research Funding Agency. She serves as Associate Editor for the well-known Artificial Intelligence Journal. She is member of the International Federation of Automatic Control IFAC Safeprocess Technical Committee.

Contents

AI for Estimation and Prediction

"It Could Be Worse, It Could Be Raining": Reliable Automatic Meteorological Forecasting for Holiday Planning

Matteo Cristani[1], Francesco Domenichini[2], Claudio Tomazzoli[1], and Margherita Zorzi[1](✉)

[1] Dipartimento di Informatica, Università di Verona, Verona, Italy
margherita.zorzi@univr.it
[2] ARPAV, Padova, Italy

Abstract. Weather forecasting is a logical process that consists in evaluating the predictions provided by a set of stochastic models, compare these and take a conclusion about the weather in a given area and a given interval of time. Meteorological forecasting provides reliable predictions about the weather within a given interval of time. The automation of the forecasting process would be helpful in a number of contexts. For instance, when forecasting about underpopulated or small geographic areas is out of the human forecasters' tasks but is central, e.g., for tourism. In this paper, we start to deal with these challenging tasks by developing a defeasible reasoner for meteorological forecasting, which we evaluate against a real-world example with applications to tourism and holiday planning.

Keywords: Automatic reasoning · Weather forecasting · Labelled logical framework · Spindle defeasible automatic reasoner

1 Introduction

Meteorological forecasting has become a commonly required web service and weather forecasting websites are nowadays among the most visited ones. Producing a meteorological forecast is, however, an complex task to perform, that involves expertise and experience. Typical pipeline for a forecasting starts with the production of weather computational models. The evolution of the weather in a given geographic area are provided, and further evaluated with a blend of specific and general criteria, including accuracy and stability estimates. These models are not directly accessible online because of their size, often larger than terabytes. The forecaster relies on the execution of a sophisticated reasoning on data, comparing the models and evaluating the confidence degree in a range of possibilities compatible with the models themselves with the above mentioned

F. Wotawa et al. (Eds.): IEA/AIE 2019, LNAI 11606, pp. 3–11, 2019.
https://doi.org/10.1007/978-3-030-22999-3_1

criteria as well. In this paper we propose an implementation method for a decision support technology that uses non monotonic reasoning to accommodate conflicting rules in the system, so that the process of decision making not only results fuzzy, but is also subject to revisions and constrained by confidence. This reflects the method of production of weather forecasting suggester by experts, that we involved in the research.

The rest of the paper is organised as follows. In Sect. 2 we formalise a defeasible reasoner for meteorological forecasting. Section 3 is devoted to a real-world example. In Sect. 4 we discuss related work, and in Sect. 5 we draw conclusions[1].

2 A Defeasible Reasoner for Meteorological Forecasting

What is commonly intended as "weather forecasting" can be logically model as a *conclusion* the forecaster derives from a set of *premises*, by the application of some rules.

In this section, we refer a logical framework called MeteoLOG, that formalizes the hybrid reasoning at the basis of meteorological forecasting. Real world meteorological forecasts usually make extensive use of "smoothening" expressions to represent uncertainty in future meteorological conditions; MeteoLOG, initially introduced in [6], benefits from three standard logical approaches: defeasible logic [13], labeled deduction systems [7,16,21,22] and fuzzy frameworks [2,10].

In [6] we introduce the *syntax of formulae* and of *labels*, along with a notion of *prevalence*, which imports a defeasible flavour into the system. We also provide an intuitive description of the label-elimination algorithm Tournament, which represents the basis of the reasoning process that we develop below.

In this paper, we only deal with ground formulas modeling meteorological forecasting values, which we call *Assertional Maps (Assertional Maps (AMs))*. AMs provide quantitative information and they represent the basic piece of knowledge used for forecasting. In the real world, they are collected worldwide, from different forecasting sites and through a number of different technologies. The internationally accepted set of numerical weather conditions revealed in AMs concerns: *Temperature*, *Pressure*, *Humidity*, *Snowfalls*, *Wind*, *Precipitations*, *Visibility*.

From an abstract viewpoint, AMs express rough assertions about weather to be processed and evaluated. They are simply represented by suitable predicates on space-time coordinates pointing out a numerical weather condition, expressed in a suitable measuring system.

Formally, an AM is a five-ary predicate $Q(x, y, z, \tau^r, q)$, where Q is a numerical weather condition, x, y and z represent geographic coordinates, τ^r represents the forecasting time (the interval of the validity of the assertion) and q is the effective measured value, represented by a 2-dimensional vector (v, d), where v is the numerical value and d is the *direction*. For instance,

[1] An extended version of this work is available online at https://arxiv.org/abs/1901.09867.

$Rain(45.43, 11.80, 06/04/2018, 14{:}05{:}00\ CET, 5\,\text{mm})$ represents that the assertion *Rain* on ground level, point of measure $(45.43, 11.80)$ on GPS coordinates, on 06/04/2018 at 14:05:00 *CET* was 5 mm.

Since we are interested in the reasoning process behind the forecasting, we now focus on *models* experts apply to derive information from AM. We formalize such methods and related notions by means of labels, and import into the formulas additional information such as the precision of the method and the detection time (the instant in which the method has been applied to generate the map). This information is crucial for the forecaster's work, since the choice of the (as much as possible) correct maps is mainly based on methodological information.

Labeled Assertional Maps (LAMs) are obtained by labelling AMs. This formally models the additional information the forecaster have to evaluate and decide if a rough AM expressing a prediction is admissible for forecasting or not.

Labels represent *contextualised methods*, i.e., a forecasting method applied to a data gathering sample, performed in a given instant of time, weighted with an some accuracy information; they are pairs of the kind $\langle \lambda, \tau^t \rangle$, where λ represents a model and τ represents the instant in which the map has been generated.

Each method can be associated with an accuracy value $\lambda.a$, a function that extracts the accuracy information from the method λ. An LAM is then a labeled formula $\langle \lambda, \tau^t \rangle : Q(x, y, z, \tau^r, q)$. Having defined the syntax of MeteoLOG, we are currently working at the definition of a suitable semantics and a natural deduction system [3,19] for MeteoLOG.

Tournament is an algorithm that returns a defeasible theory, given an ordered set of *Metarules*, a set of *accuracies* and actual time. Informally, the algorithm maps *assertions* into defeasible rules and facts; when it finds possible conflicts it generates a set of defeasible conflicting rules and then, using the accuracy information, it generates the priority rules to solve the conflicts so that the method with best accuracy prevails; in case of even accuracy, the latest LAM prevails. Once a set of LAMs has been collected and an *accuracy* set has been acknowledged, Tournament algorithm starts with a sifting action on the set of labeled assertions. First, it discharges LAMs that are out of date. Second, it orders LAMs on the basis of priorities, obtaining some AM for each numerical weather condition we are interested in. This operation corresponds to a label-elimination: once priorities have been derived, the majority of information about the forecasting method became useless. As an output of this step, we obtain a set of *defeasible rules* to be given as input to the *Reasoner*, which derives then a set of numerical weather condition also called a weather *scenario*. Priorities of these generated defeasible conflicting rules are given by a *function* (named "supremacy") that takes into account two conflicting rules so that, depending on this function definition, the resulting output can differ from both source rules.

Forecasting reasoning can be divided into three steps: (i) the quantitative forecasting (invisible to the final user) the reasoner generates; (ii) the qualitative forecasting (also invisible to the final user) called in the following *sharp forecasting*; (iii) the qualitative, natural language based forecasting destined to the final user, called in the following *smoothed forecasting*. Between the first two phases, a mapping between data and a suitable *forecasting lexicon* occurs (in other words, a natural language processing phase [23,24]).

Once the final set of reliable assertional maps has been collected, the forecaster can proceed with data analysis and the releasing of the weather bulletin. Every one has a wide experience in weather forecasting as a final user. It is well known that weather bulletins are offered in a friendly form. For example, if the forecaster infers that the probability that tomorrow it will rain is very small, she doesn't release the assertion "$Rain(45.43, 11.80, 06/04/2018, 14{:}05{:}00\ CET, 5\,\mathrm{mm})$ with probability 15%" but the understandable natural language sentence "partially cloudy, possible scattered rains". This final step provides a "smoothing" phase to the output of the previous one in which some adjectives can be added to give evidence to the uncertainty of the event. We don't fully model this final step of weather forecasting; nonetheless, we propose an example of a possible automatisation of the human task, leaving the full development for future work (see Sect. 5).

3 Reference Implementation

To illustrate concretely how the proposed approach can fit a real-life scenario, let us consider a weather forecast considering the seaside part of Veneto, which is located in the north-east of Italy and albeit being not a remote area it exploits several neighbor small touristic places. For the sake of space, but without loss of generality, we limit the weather forecast to cloud, wind and sea conditions and to only three points; we label these points North, South and Center, the latter representing roughly the position of the famous city of Venice (see Fig. 1). Sea Conditions have only one point, representing the sea in the area. We use only two forecasting maps and we limit the time-frame to only two values, representing two and one days after the present: respectively t_2, t_1, t_0. We have as input two forecasting sources, coming from different forecasting models such as IFS (also known as ECMWF for European Center Medium Weather Forecast) and GFS (Global Forecast System), plus the map of observations.

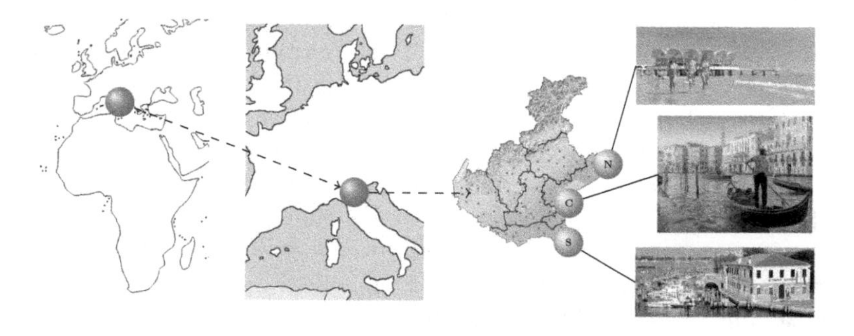

Fig. 1. Some touristic places in Veneto

In the following, for the sake of space, we will shorten $\{cloudiness : \alpha\%, Wind : \beta\ knots\ from\ \gamma\}$ using a form like $\{\alpha\%, \beta\ \gamma\}$ where N, S, C stands for *North*, *South*, *Center* respectively. We will also denote "Sea: δ cm wave" as "Sea: δ". The first source obtained with the GFS prevision model asserts, using this shorten form,

at time t_0*: North:*$\{90\%, 18\ N\}$*, Center:*$\{90\%, 18\ N\}$*, South:*$\{90\%, 10\ N\}$ *Sea: 190,*
at time t_1*: North:*$\{90\%,\ 8\ N\}$*, Center:*$\{90\%,\ 8\ E\}$*, South:*$\{90\%,\ 5\ E\}$ *Sea: 100,*
at time t_2*: North:*$\{90\%, 18\ N\}$*, Center:*$\{90\%,\ 8\ E\}$*, South:*$\{90\%,\ 5\ E\}$ *Sea: 100.*

The second source obtained with ECMWF asserts,

at time t_0*: North:*$\{90\%, 15\ NE\}$*, Center:*$\{90\%, 15\ NE\}$*, South:*$\{90\%, 15\ NE\}$ *Sea: 160,*
at time t_1*: North:*$\{75\%,\ 5\ NE\}$*, Center:*$\{75\%,\ 5\ NE\}$*, South:*$\{75\%,\ 5\ N\}$ *Sea: 90,*
at time t_2*: North:*$\{30\%, 5\ N\}$*, Center:*$\{30\%,\ 5\ N\}$*, South:*$\{30\%,\ 5\ N\}$ *Sea: 50.*

The observation map, which only relates data at t_0 states that

at time t_0*: North:*$\{90\%, 15\ NE\}$*, Center:*$\{90\%, 15\ NE\}$*, South:*$\{90\%, 15\ NE\}$ *Sea: 190.*

We know from knowledge experts that ECMWF has a better accuracy than GFS: numerically $a(ECMWF, t_1) = 0.85$, $a(ECMWF, t_2) = 0.80$, $a(GFS, t_1) = 0.45$, $a(GFS, t_2) = 0.40$. These assertions, using "E" for ECMWF, "G" for GFS, "O" for observation and "C" for "cloudiness", "W" for "wind" and "S" for "sea conditions" can be represented in the proposed formalism as

$\langle G, t_0\rangle : C(North, t_0, 90)$	$\langle G, t_0\rangle : C(Center, t_0, 90)$	$\langle G, t_0\rangle : C(South, t_0, 90)$
$\langle G, t_0\rangle : C(North, t_1, 90)$	$\langle G, t_0\rangle : C(Center, t_1, 90)$	$\langle G, t_0\rangle : C(South, t_1, 90)$
$\langle G, t_0\rangle : C(North, t_2, 90)$	$\langle G, t_0\rangle : C(Center, t_2, 90)$	$\langle G, t_0\rangle : C(South, t_2, 90)$
$\langle E, t_0\rangle : C(North, t_0, 90)$	$\langle E, t_0\rangle : C(Center, t_0, 90)$	$\langle E, t_0\rangle : C(South, t_0, 90)$
$\langle E, t_0\rangle : C(North, t_1, 75)$	$\langle E, t_0\rangle : C(Center, t_1, 75)$	$\langle E, t_0\rangle : C(South, t_1, 75)$
$\langle E, t_0\rangle : C(North, t_2, 50)$	$\langle E, t_0\rangle : C(Center, t_2, 50)$	$\langle E, t_0\rangle : C(South, t_2, 50)$
$\langle O, t_0\rangle : C(North, t_0, 90)$	$\langle O, t_0\rangle : C(Center, t_0, 90)$	$\langle O, t_0\rangle : C(South, t_0, 90)$
$\langle G, t_0\rangle : W(North, t_0, [N, 18])$	$\langle G, t_0\rangle : W(Center, t_0, [N, 18])$	$\langle G, t_0\rangle : W(South, t_0, [N, 10])$
$\langle G, t_0\rangle : W(North, t_1, [N, 8])$	$\langle G, t_0\rangle : W(Center, t_1, [E, 8])$	$\langle G, t_0\rangle : W(South, t_1, [E, 5])$
$\langle G, t_0\rangle : W(North, t_2, [N, 8])$	$\langle G, t_0\rangle : W(Center, t_2, [E, 8])$	$\langle G, t_0\rangle : W(South, t_2, [E, 5])$
$\langle E, t_0\rangle : W(North, t_0, [NE, 15])$	$\langle E, t_0\rangle : W(Center, t_0, [NE, 15])$	$\langle E, t_0\rangle : W(South, t_0, [NE, 15])$
$\langle E, t_0\rangle : W(North, t_1, [NE, 5])$	$\langle E, t_0\rangle : W(Center, t_1, [NE, 5])$	$\langle E, t_0\rangle : W(South, t_1, [NE, 5])$
$\langle E, t_0\rangle : W(North, t_2, [N, 5])$	$\langle E, t_0\rangle : W(Center, t_2, [N, 5])$	$\langle E, t_0\rangle : W(South, t_2, [N, 5])$
$\langle O, t_0\rangle : W(North, t_0, [NE, 15])$	$\langle O, t_0\rangle : W(Center, t_0, [NE, 15])$	$\langle O, t_0\rangle : W(South, t_0, [NE, 15])$
$\langle G, t_0\rangle : S(Sea, t_0, 190)$	$\langle G, t_0\rangle : S(Sea, t_1, 100)$	$\langle G, t_0\rangle : S(Sea, t_2, 100)$
$\langle E, t_0\rangle : S(Sea, t_0, 160)$	$\langle E, t_0\rangle : S(Sea, t_1, 50)$	$\langle E, t_0\rangle : S(Sea, t_2, 10)$
$\langle O, t_0\rangle : S(Sea, t_0, 190)$		

This is therefore a set of metarules, so after the *Translator* has done its elaboration the resulting defeasible rules (see [20] for defeasible rules notation)

$r_{fcg_{11}}: \Rightarrow CN_g t_0 90$ $\quad r_{fcg_{21}}: \Rightarrow CN_g t_1 90$ $\quad r_{fcg_{31}}: \Rightarrow CN_g t_2 90$ $\quad r_{co_{11}}: \rightarrow CN t_0 90$
$r_{fcg_{12}}: \Rightarrow CC_g t_0 90$ $\quad r_{fcg_{22}}: \Rightarrow CC_g t_1 90$ $\quad r_{fcg_{32}}: \Rightarrow CC_g t_2 90$ $\quad r_{co_{12}}: \rightarrow CE t_0 90$
$r_{fcg_{13}}: \Rightarrow CS_g t_0 90$ $\quad r_{fcg_{23}}: \Rightarrow CS_g t_1 90$ $\quad r_{fcg_{33}}: \Rightarrow CS_g t_2 90$ $\quad r_{co_{13}}: \rightarrow CS t_0 90$

$r_{fce_{11}}: \Rightarrow CN_e t_0 90$ $\quad r_{fce_{21}}: \Rightarrow CN_e t_1 75$ $\quad r_{fce_{31}}: \Rightarrow CN_e t_2 30$
$r_{fce_{12}}: \Rightarrow CC_e t_0 90$ $\quad r_{fce_{22}}: \Rightarrow CC_e t_1 75$ $\quad r_{fce_{32}}: \Rightarrow CC_e t_2 30$
$r_{fce_{13}}: \Rightarrow CS_e t_0 90$ $\quad r_{fce_{23}}: \Rightarrow CS_e t_1 75$ $\quad r_{fce_{33}}: \Rightarrow CS_e t_2 30$

$r_{wg_{11}}: \Rightarrow WN_g t_0 N18$ $\quad r_{wg_{21}}: \Rightarrow WN_g t_1 N8$ $\quad r_{wg_{31}}: \Rightarrow WN_g t_2 N8$ $\quad r_{wo_{11}}: \rightarrow WN t_0 NE15$
$r_{wg_{12}}: \Rightarrow WC_g t_0 N18$ $\quad r_{wg_{22}}: \Rightarrow WC_g t_1 E8$ $\quad r_{wg_{32}}: \Rightarrow WC_g t_2 E8$ $\quad r_{wo_{12}}: \rightarrow WC t_0 NE15$
$r_{wg_{13}}: \Rightarrow WS_g t_0 N10$ $\quad r_{wg_{23}}: \Rightarrow WS_g t_1 E5$ $\quad r_{wg_{33}}: \Rightarrow WS_g t_2 E5$ $\quad r_{wo_{13}}: \rightarrow WS t_0 NE15$

$r_{we_{11}}: \Rightarrow WN_e t_1 NE15$ $\quad r_{we_{21}}: \Rightarrow WN_e t_1 NE5$ $\quad r_{we_{31}}: \Rightarrow WN_e t_2 N5$
$r_{we_{12}}: \Rightarrow WC_e t_1 NE5$ $\quad r_{we_{22}}: \Rightarrow WC_e t_1 NE5$ $\quad r_{we_{32}}: \Rightarrow WC_e t_2 N5$
$r_{we_{13}}: \Rightarrow WS_e t_1 N5$ $\quad r_{we_{23}}: \Rightarrow WS_e t_1 NE5$ $\quad r_{we_{33}}: \Rightarrow WS_e t_2 N5$

$r_{sg_{11}}: \Rightarrow Sea_g t_0 190$ $\quad r_{sg_{21}}: \Rightarrow Sea_g t_1 100$ $\quad r_{sg_{31}}: \Rightarrow Sea_g t_2 100$ $\quad r_{so_{11}}: \rightarrow Sea_o t_0 190$
$r_{se_{11}}: \Rightarrow Sea_e t_0 160$ $\quad r_{se_{21}}: \Rightarrow Sea_e t_1 50$ $\quad r_{se_{31}}: \Rightarrow Sea_e t_2 10$

$r_{cg_{11}}: CN_g t_1 90, CN_e t_1 75 \Rightarrow CN t_1 88$ $\quad r_{ce_{11}}: CN_g t_1 90, CN_e t_1 75 \Rightarrow CN t_1 78$
$r_{cg_{12}}: CC_g t_1 90, CC_e t_1 75 \Rightarrow CC t_1 88$ $\quad r_{ce_{12}}: CC_g t_1 90, CC_e t_1 75 \Rightarrow CC t_1 78$
$r_{cg_{13}}: CS_g t_1 90, CS_e t_1 75 \Rightarrow CS t_1 88$ $\quad r_{ce_{13}}: CS_g t_1 90, CS_e t_1 75 \Rightarrow CS t_1 78$
$r_{wg_{11}}: WN_g t_1 N8, WN_e t_1 NE5 \Rightarrow WN t_1 N7$ $\quad r_{we_{11}}: WN_g t_1 N8, WN_e t_1 NE5 \Rightarrow WN t_1 NE6$
$r_{wg_{12}}: WC_g t_1 E8, WC_e t_1 NE5 \Rightarrow WC t_1 E7$ $\quad r_{we_{12}}: WC_g t_1 E8, WC_e t_1 NE5 \Rightarrow WC t_1 NE6$

$r_{wg_{13}}: WS_g t_1 E5, WS_e t_1 N5 \Rightarrow WS t_1 E5$ $\quad r_{we_{13}}: WS_g t_1 E5, WS_e t_1 N5 \Rightarrow WS t_1 N5$
$r_{sg_{11}}: Sea_g t_1 100, Sea_e t_1 50 \Rightarrow Sea_{t_1} 95$ $\quad r_{se_{11}}: Sea_g t_1 100, Sea_e t_1 50 \Rightarrow Sea_{t_1} 65$

$v_{c_{11}}: CN t_1 88 \Rightarrow \neg CN t_1 78$ $\quad v_{c_{12}}: CC t_1 88 \Rightarrow \neg CC t_1 78$ $\quad v_{c_{11}}: CS t_1 88 \Rightarrow \neg CS t_1 78$
$v_{c_{21}}: CN t_1 78 \Rightarrow \neg CN t_1 88$ $\quad v_{c_{22}}: CC t_1 78 \Rightarrow \neg CC t_1 88$ $\quad v_{c_{23}}: CS t_1 78 \Rightarrow \neg CS t_1 78$
$v_{w_{11}}: WN t_1 N7 \Rightarrow \neg WN t_1 NE6$ $\quad v_{w_{12}}: WC t_1 E7 \Rightarrow \neg WC t_1 NE6$ $\quad v_{w_{11}}: WS t_1 E5 \Rightarrow \neg WS t_1 NE5$
$v_{w_{21}}: WN t_1 NE6 \Rightarrow \neg WN t_1 N7$ $\quad v_{w_{22}}: WC t_1 NE6 \Rightarrow \neg WC t_1 E7$ $\quad v_{w_{23}}: WS t_1 NE5 \Rightarrow \neg WS t_1 E5$
$v_{s_{11}}: Sea_{t_1} 95 \Rightarrow \neg Sea_{t_2} 75$ $\quad v_{s_{21}}: Sea_{t_1} 75 \Rightarrow \neg Sea_{t_2} 95$

$r_{cg_{21}}: CN_g t_2 90, CN_e t_2 30 \Rightarrow CN t_2 68$ $\quad r_{ce_{21}}: CN_g t_2 90, CN_e t_2 30 \Rightarrow CN t_2 38$
$r_{cg_{22}}: CC_g t_2 90, CC_e t_2 30 \Rightarrow CC t_2 68$ $\quad r_{ce_{22}}: CC_g t_2 90, CC_e t_2 30 \Rightarrow CC t_2 38$
$r_{cg_{23}}: CE_g t_2 90, CE_e t_2 30 \Rightarrow CE t_2 68$ $\quad r_{ce_{23}}: CE_g t_2 90, CE_e t_2 30 \Rightarrow CE t_2 38$
$r_{wg_{21}}: WN_g t_2 N8, WN_e t_2 N5 \Rightarrow WN t_2 NE7$ $\quad r_{we_{21}}: WN_g t_2 N8, WN_e t_2 N5 \Rightarrow WN t_2 N6$
$r_{wg_{22}}: WC_g t_2 E8, WC_e t_2 N5 \Rightarrow WC t_2 NE7$ $\quad r_{we_{22}}: WC_g t_2 E8, WC_e t_2 N5 \Rightarrow WC t_2 N6$
$r_{wg_{23}}: WS_g t_2 E5, WS_e t_2 N5 \Rightarrow WS t_2 NE5$ $\quad r_{we_{23}}: WS_g t_2 E5, WS_e t_2 N5 \Rightarrow WS t_2 N5$
$r_{sg_{21}}: Sea_g t_2 100, Sea_e t_2 10 \Rightarrow Sea_{t_2} 80$ $\quad r_{se_{21}}: Sea_g t_2 100, Sea_e t_2 10 \Rightarrow Sea_{t_2} 20$

$v_{c_{31}}: CN t_2 68 \Rightarrow \neg CN t_2 38$ $\quad v_{c_{32}}: CC t_2 68 \Rightarrow \neg CC t_2 38$ $\quad v_{c_{31}}: CS t_2 68 \Rightarrow \neg CS t_2 38$
$v_{c_{41}}: CN t_2 38 \Rightarrow \neg CN t_2 68$ $\quad v_{c_{42}}: CC t_2 38 \Rightarrow \neg CC t_2 68$ $\quad v_{c_{43}}: CS t_2 38 \Rightarrow \neg CS t_2 68$
$v_{w_{31}}: WN t_2 NE7 \Rightarrow \neg WN t_2 N6$ $\quad v_{w_{32}}: WC t_2 NE7 \Rightarrow \neg WC t_2 N6$ $\quad v_{w_{31}}: WS t_2 NE5 \Rightarrow \neg WS t_2 N5$
$v_{w_{41}}: WN t_2 N6 \Rightarrow \neg WN t_2 NE7$ $\quad v_{w_{42}}: WC t_2 N6 \Rightarrow \neg WC t_2 NE7$ $\quad v_{w_{43}}: WS t_2 N5 \Rightarrow \neg WS t_2 NE5$
$v_{s_{31}}: Sea_{t_2} 80 \Rightarrow \neg Sea_{t_2} 20$ $\quad v_{s_{41}}: Sea_{t_2} 20 \Rightarrow \neg Sea_{t_2} 80$

$p_{11}: v_{c_{21}} \rangle r_{cg_{11}}$ $\quad p_{12}: r_{ce_{11}} \rangle v_{c_{11}}$ $\quad p_{13}: v_{c_{41}} \rangle r_{cg_{21}}$ $\quad p_{14}: r_{ce_{21}} \rangle v_{c_{31}}$
$p_{21}: v_{w_{21}} \rangle r_{wg_{11}}$ $\quad p_{22}: r_{we_{11}} \rangle v_{w_{11}}$ $\quad p_{23}: v_{w_{41}} \rangle r_{wg_{21}}$ $\quad p_{24}: r_{we_{21}} \rangle v_{w_{31}}$
$p_{31}: v_{c_{22}} \rangle r_{cg_{12}}$ $\quad p_{32}: r_{ce_{12}} \rangle v_{c_{12}}$ $\quad p_{33}: v_{c_{42}} \rangle r_{cg_{22}}$ $\quad p_{34}: r_{ce_{22}} \rangle v_{c_{32}}$
$p_{41}: v_{w_{22}} \rangle r_{wg_{12}}$ $\quad p_{42}: r_{we_{12}} \rangle v_{w_{12}}$ $\quad p_{43}: v_{w_{42}} \rangle r_{wg_{22}}$ $\quad p_{44}: r_{we_{22}} \rangle v_{w_{32}}$
$p_{51}: v_{c_{23}} \rangle r_{cg_{13}}$ $\quad p_{52}: r_{ce_{13}} \rangle v_{c_{13}}$ $\quad p_{53}: v_{c_{43}} \rangle r_{cg_{23}}$ $\quad p_{54}: r_{ce_{23}} \rangle v_{c_{33}}$
$p_{61}: v_{w_{23}} \rangle r_{wg_{13}}$ $\quad p_{62}: r_{we_{13}} \rangle v_{w_{13}}$ $\quad p_{63}: v_{w_{43}} \rangle r_{wg_{23}}$ $\quad p_{64}: r_{we_{23}} \rangle v_{w_{33}}$
$p_{71}: v_{s_{21}} \rangle r_{sg_{11}}$ $\quad p_{72}: r_{se_{11}} \rangle v_{s_{11}}$ $\quad p_{73}: v_{s_{41}} \rangle r_{sg_{21}}$ $\quad p_{74}: r_{se_{21}} \rangle v_{s_{31}}$

Given this theory, a *defeasible Reasoner* (see [8] for defeasible conclusion notation) concludes $+\partial CNt_1 78$, $+\partial CCt_1 78$, $+\partial CSt_1 78$, $+\partial WNt_1 NE6$, $+\partial WCt_1 NE6$, $+\partial WSt_1 N5$, $+\partial Sea_{t_1} 65$, $+\partial CNt_2 38$, $+\partial CCt_2 38$, $+\partial CSt_2 38$, $+\partial WNt_2 N6$, $+\partial WCt_2 N6$, $+\partial WSt_2 N5$, $+\partial Sea_{t_2} 20$. Therefore, translating numerical value into words we have the results expressed in Fig. 2.

at time t_1 North : Mostly Cloudy, Light Winds from North East
Center : Mostly Cloudy, Light Winds from North East
South : Mostly Cloudy, Light Winds from North
Sea : Slight

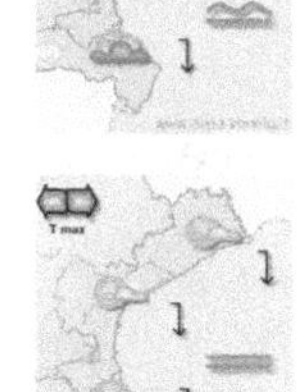

at time t_2 North : Partly Cloudy, Light Winds from North
Center : Partly Cloudy, Light Winds from North
South : Partly Cloudy, Light Winds from North
Sea : Calm

Fig. 2. Weather forecast both in words (left) and iconographically (right)

4 Related Work

After the pioneering studies [5,9] and further engineering investigations on the commercial solutions [17], a first attempt going in the same direction that we following in this paper appeared in the 1990s [11] and inspired many specialized studies [4]. The ontological approach and the usage of the Internet of Things have been applied to forecasting quite recently [1,15] and we acknowledge that the main technical inspirations of the framework discussed here trace back these works. Nevertheless the main influences come from non-monotonic reasoning [12–14,18] and the usage of non-monotonic deduction systems for sensor-based applications (clearly related to the initial part of the forecasting process) [8,20].

5 Conclusions

In this paper, we propose an approach to support meteoroligists in producing weather forecasts. The basic work is a reasoning framework able to simulate in a quite refined way the decision process made by the forecasters in producing weather bulletins. There are several ways to take this study further.

First of all, though the formalism has been shown to be adequate to represent knowledge used by forecasters, it still lacks of a general formalisation for semantics and the proof theory, however we can prove it to be sound and complete against the standard semantics, is not complete with respect to canonical models, that indeed lack in semantics as well. Regarding the algorithm named Tournament we will formalize a proof of correctness and an evaluation of the complexity after the exploitation of the metarules. Defeasible methods are known to be generally linear in rules but not in metarules, for they explode combinatorially the controlled priorities; in the specific case, the set of metarules might be treatable in a lighter way. These aspects are still under investigation. The research team includes a forecaster of the ARPA Veneto weather forecasting service, one of the most valuable forecasting service in Italy, who will lead the development of both the definition of the *supremacy* function and the Tournament algorithm.

We also plan to include more specific features in order to improve the precision of the automatic bulletin, aiming to a completely automatic and (potentially) unsupervised bulletin generator.

References

1. Agresta, A., et al.: An ontology framework for flooding forecasting. In: Murgante, B., et al. (eds.) ICCSA 2014. LNCS, vol. 8582, pp. 417–428. Springer, Cham (2014). https://doi.org/10.1007/978-3-319-09147-1_30
2. Aschieri, F., Zorzi, M.: Non-determinism, non-termination and the strong normalization of system T. In: Hasegawa, M. (ed.) TLCA 2013. LNCS, vol. 7941, pp. 31–47. Springer, Heidelberg (2013). https://doi.org/10.1007/978-3-642-38946-7_5
3. Aschieri, F., Zorzi, M.: On natural deduction in classical first-order logic: Curry-Howard correspondence, strong normalization and Herbrand's theorem. Theor. Comput. Sci. **625**, 125–146 (2016)
4. Carr III, L., Elsberry, R., Peak, J.: Beta test of the systematic approach expert system prototype as a tropical cyclone track forecasting aid. Weather Forecast. **16**(3), 355–368 (2001)
5. Conway, B.: Expert systems and weather forecasting. Meteorol. Mag. **118**(1399), 23–30 (1989)
6. Cristani, M., Domenichini, F., Olivieri, F., Tomazzoli, C., Zorzi, M.: It could rain: weather forecasting as a reasoning process. In: Knowledge-Based and Intelligent Information & Engineering Systems: Proceedings of the 22nd International Conference KES-2018, Procedia Computer Science, vol. 126, pp. 850–859 (2018)
7. Cristani, M., Olivieri, F., Tomazzoli, C., Zorzi, M.: Towards a logical framework for diagnostic reasoning. In: Jezic, G., Chen-Burger, Y.-H.J., Howlett, R.J., Jain, L.C., Vlacic, L., Šperka, R. (eds.) KES-AMSTA-18 2018. SIST, vol. 96, pp. 144–155. Springer, Cham (2019). https://doi.org/10.1007/978-3-319-92031-3_14
8. Cristani, M., Tomazzoli, C., Karafili, E., Olivieri, F.: Defeasible reasoning about electric consumptions. In: 30th IEEE International Conference on Advanced Information Networking and Applications, AINA 2016, pp. 885–892 (2016)
9. Desmarais, M.C., de Verteuil, F., Zwack, P., Jacob, D.: Stratus: a prototype expert advisory system for terminal weather forecasting, pp. 150–154 (1990)
10. Domańska, D., Wojtylak, M.: Application of fuzzy time series models for forecasting pollution concentrations. Expert Syst. Appl. **39**(9), 7673–7679 (2012)
11. Goldberg, E.: FoG. Synthesizing forecast text directly from weather maps, pp. 156–162 (1993)
12. Governatori, G., Olivieri, F., Calardo, E., Rotolo, A., Cristani, M.: Sequence semantics for normative agents. In: Baldoni, M., Chopra, A.K., Son, T.C., Hirayama, K., Torroni, P. (eds.) PRIMA 2016. LNCS (LNAI), vol. 9862, pp. 230–246. Springer, Cham (2016). https://doi.org/10.1007/978-3-319-44832-9_14
13. Governatori, G., Olivieri, F., Scannapieco, S., Cristani, M.: The hardness of revising defeasible preferences. In: Bikakis, A., Fodor, P., Roman, D. (eds.) RuleML 2014. LNCS, vol. 8620, pp. 168–177. Springer, Cham (2014). https://doi.org/10.1007/978-3-319-09870-8_12
14. Governatori, G., Olivieri, F., Scannapieco, S., Rotolo, A., Cristani, M.: The rationale behind the concept of goal. Theory Pract. Log. Program. **16**(3), 296–324 (2016)

15. Kulkarni, A., Mukhopadhyay, D.: Internet of things based weather forecast monitori ng system. IJEEI **9**(3), 555–557 (2018)
16. Masini, A., Viganò, L., Zorzi, M.: Modal deduction systems for quantum state transformations. J. Mult.-valued Log. Soft Comput. **17**(5–6), 475–519 (2011)
17. Moninger, W.: Shootout-89, an evaluation of knowledge-based weather forecasting systems. Mach. Intell. Pattern Recogn. **10**(C), 457–458 (1990)
18. Olivieri, F., Governatori, G., Scannapieco, S., Cristani, M.: Compliant business process design by declarative specifications. In: Boella, G., Elkind, E., Savarimuthu, B.T.R., Dignum, F., Purvis, M.K. (eds.) PRIMA 2013. LNCS (LNAI), vol. 8291, pp. 213–228. Springer, Heidelberg (2013). https://doi.org/10.1007/978-3-642-44927-7_15
19. Prawitz, D.: Ideas and Results in Proof Theory. Studies in Logic and the Foundations of Mathematics, vol. 63, pp. 235–307. Elsevier, Amsterdam (1971)
20. Tomazzoli, C., Cristani, M., Karafili, E., Olivieri, F.: Non-monotonic reasoning rules for energy efficiency. J. Ambient Intell. Smart Environ. **9**(3), 345–360 (2017)
21. Viganò, L., Volpe, M., Zorzi, M.: Quantum state transformations and branching distributed temporal logic. In: Kohlenbach, U., Barceló, P., de Queiroz, R. (eds.) WoLLIC 2014. LNCS, vol. 8652, pp. 1–19. Springer, Heidelberg (2014). https://doi.org/10.1007/978-3-662-44145-9_1
22. Viganò, L., Volpe, M., Zorzi, M.: A branching distributed temporal logic for reasoning about entanglement-free quantum state transformations. Inf. Comput. **255**, 311–333 (2017)
23. Cristani, M., Bertolaso, A., Scannapieco, S., Tomazzoli, C.: Future paradigms of automated processing of business documents. Int. J. Inf. Manag. **40**, 47–65 (2019)
24. Zorzi, M., Combi, C., Lora, R., Pagliarini, M., Moretti, U.: Automagically encoding adverse drug reactions in MedDRA. In: 2015 International Conference on Healthcare Informatics, ICHI 2015, Dallas, TX, USA, 21–23 October 2015, pp. 90–99. IEEE Computer Society (2015)

A Taxonomy of Event Prediction Methods

Fatma Ezzahra Gmati[1], Salem Chakhar[2,3](✉), Wided Lejouad Chaari[1], and Mark Xu[2]

[1] COSMOS, National School of Computer Sciences, University of Manouba, Manouba, Tunisia
fatma.ezzahra.gmati@gmail.com, wided.chaari@ensi-uma.tn
[2] Portsmouth Business School, University of Portsmouth, Portsmouth, UK
{salem.chakhar,mark.xu}@port.ac.uk
[3] CORL, University of Portsmouth, Portsmouth, UK

Abstract. Most of existing event prediction approaches consider event prediction problems within a specific application domain while event prediction is naturally a cross-disciplinary problem. This paper introduces a generic taxonomy of event prediction approaches. The proposed taxonomy, which oversteps the application domain, enables a better understanding of event prediction problems and allows conceiving and developing advanced and context-independent event prediction techniques.

Keywords: Time series · Event prediction · Taxonomy · Data mining

1 Introduction

An event is defined as a timestamped element in a temporal sequence [39]. Examples of events include earthquakes, flooding, and business failure. Event prediction aims to assess the future aspects of event features, e.g. occurrence time and probability, frequency, intensity, duration and spatial occurrence. Event prediction problem is encountered in different research and practical domains, and a large number of event prediction approaches have been proposed in the literature [46, 51, 60]. However, most of existing event prediction approaches have been initially designed and used within a specific application domain [16, 17, 59] while event prediction is naturally a cross-disciplinary problem.

Events can be categorized into either simple or complex [24]. A complex event is a collection of simple or complex events that can be linearly ordered in event streams or partially ordered in event clouds [24]. In this paper, we distinguish between two types of complex events. Type 1 of complex events represents a collection of events where only the collection characteristics are accessible and measurable. This is due to the fact that the access to the characteristics of simple events is costly, difficult or non-relevant. Earthquakes are good examples of this type of complex events. The predicative analysis of Type 1 complex events can be handled using the classical event prediction approaches. Type 2 complex events represents a collection of events where the characteristics of simple events as well as events collection are accessible and measurable. Examples of Type 2

F. Wotawa et al. (Eds.): IEA/AIE 2019, LNAI 11606, pp. 12–26, 2019.
https://doi.org/10.1007/978-3-030-22999-3_2

complex events include computer system and Internet of Things failures. The predicative analysis of Type 2 complex events is essentially based on Complex Event Processing (CEP) techniques [24,58].

The objective of this paper is to identify and classify the main mature and classical approaches of event prediction. It introduces a generic taxonomy of event prediction approaches that oversteps application domain. This generic cross-disciplinary view enables a better understanding of event prediction problems and opens road for the design and development of advanced and context-independent techniques. The proposed taxonomy distinguishes first three main categories of event prediction approaches, namely generative, inferential and hybrid. Each of these categories contains several event prediction methods, whose characteristics are presented in this paper.

The paper is organized as follows. Section 2 introduces the taxonomy. Sections 3–5 detail the main categories of event prediction approaches. Section 6 discusses some existing approaches. Section 7 concludes the paper.

2 General View of the Taxonomy

The taxonomy in Fig. 1 presents a generic classification of event prediction approaches in time series. This taxonomy includes only classical and mature approaches that are well established in the literature. Furthermore, this taxonomy has been constructed based on some commonly studied event types from several fields, namely finance, geology, hydrology, medicine and computer science. Three main categories of event prediction approaches can be distinguished in Fig. 1:

- **Generative approaches.** These approaches build theoretical models of the system generating the target event and predict future events through simulation. The term generative refers to the strategy adopted by generative science [18] consisting in the modelling of natural phenomena and social behavior through mathematical equations [12] or computational agents [21]. They are adapted to predict events where specific simulation frameworks are accessible, for instance flood modeling and simulation frameworks [12,14]. Generative approaches are mature and well proven. These approaches require a strong expertise in the target event field.
- **Inferential approaches.** Real world is complex and even though physics and mathematics have greatly evolved, our knowledge of rules that control observed phenomena is still superficial [38]. Thus, generative approaches still deficient in cases where the knowledge of the system generating the event is insufficient. Inferential approaches fill this gap. These approaches literally learn and infer patterns from past data.
- **Hybrid approaches.** These approaches combine models constructed from observed data with models based on physics laws. Hence, they employ generative and inferential approaches. The authors in [30] design hybrid approaches by 'conceptual approaches'. The basic idea of hybrid approaches is to use inferential methods to prepare the considerable amount of historical data required as input to generative methods.

These categories will be further detailed in the rest of this paper:

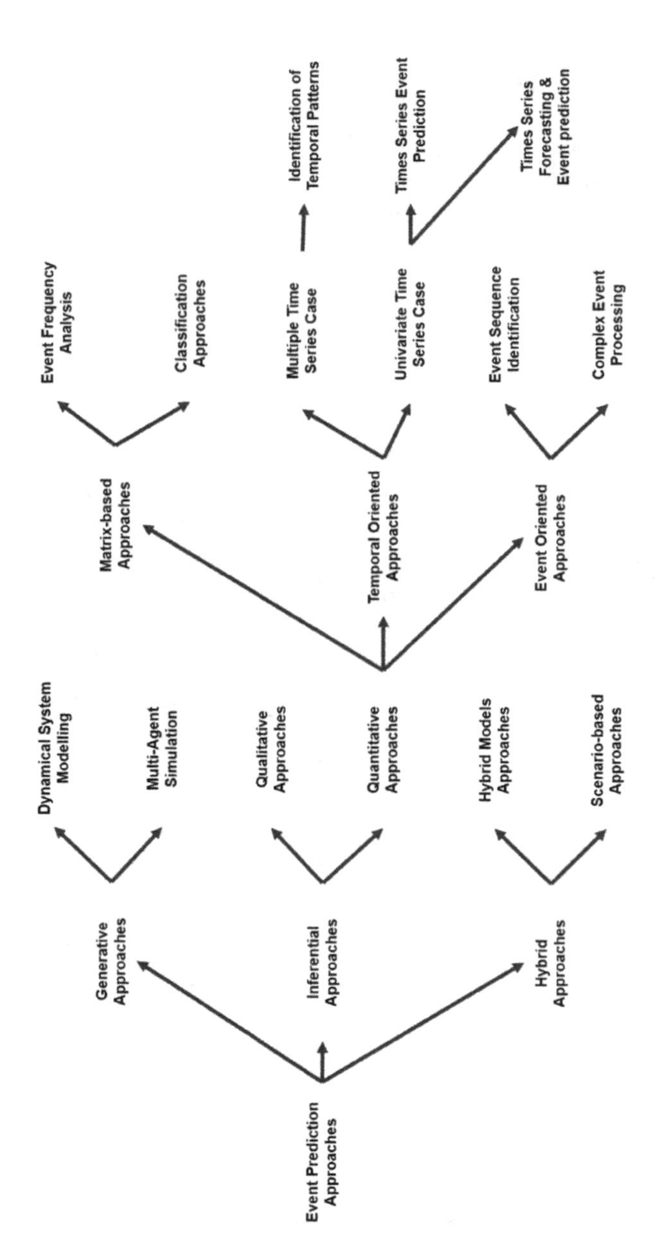

Fig. 1. Taxonomy of events prediction approaches

3 Generative Approaches to Event Prediction

The flowchart in Fig. 2 illustrates graphically the working principle of generative approaches. Three main steps can be distinguished. First, the theoretical structure of the system generating the event is modelled. Second, the obtained model is calibrated and validated using real-world datasets. This step consists in the estimation of the model parameters that fit at best the available data. Finally, simulation is performed, the future states of the system are generated and future event characteristics are deduced.

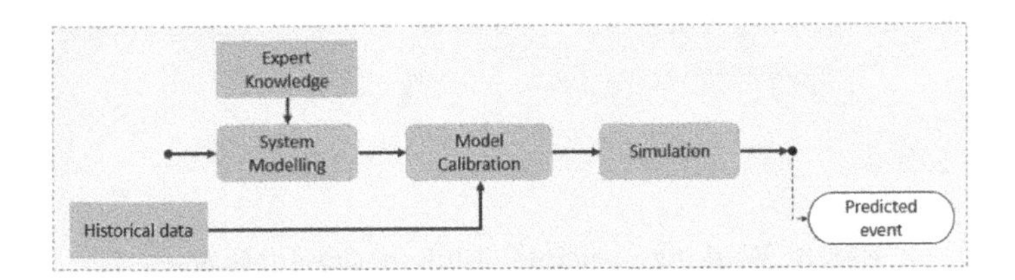

Fig. 2. Working principle of generative approaches

There are two main sub-categories of generative approaches:

- **Dynamical system modeling.** A dynamical system can be described as a set of states S and a rule of change R that determines the future state of the system over time T. In other words, the rule of change $R : S \times T \longrightarrow S$ gives the consequent states of the system for each $s \subset S$. These approaches build the theoretical model, then construct a computational model that implements the theoretical mathematical model [12]. In hydrology field, these models are called hydrodynamic models [37]. Dynamical system modeling approaches depend on the model robustness. They are mainly applied in weather forecast and flood prediction.
- **Agent-based simulation.** These approaches consist in the modelling of the system components behavior as interacting agents. They are effective when human social behavior need to be considered [21].

The main difference between these two sub-categories concerns the model conception foundation. In the first case, differential equations govern the system evolution, whereas, in the second case, logical statements establish the rules and interaction between agents [11].

4 Inferential Approaches to Event Prediction

The working principle of inferential approaches is shown in Fig. 3, where three main steps are involved. First, data is created and analyzed. Second, predictive modelling (i.e. inference) is conducted. Inference may be based on expert opinion or on a quantitative predictive model, as detailed in what follows. Finally, the model is tested over unseen datasets. The event characteristics are deduced from obtained results.

There are two main trends within inferential approaches: qualitative and quantitative. The first case is conducted through human experts while the second relies on statistical or machine learning techniques.

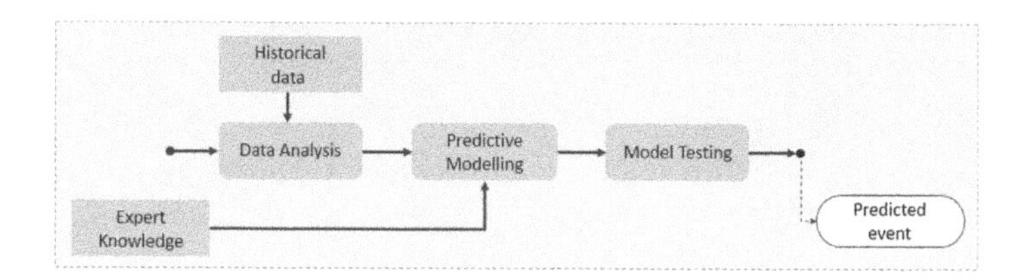

Fig. 3. Working principle of inferential approaches

4.1 Qualitative Approaches

Qualitative approaches relies on Human expertise. Experts in the target event field analyze the data in order to deduce common patterns. Then, they construct mathematical or logical relations between studied variables and event probable occurrence. These relations are commonly called indexes in finance context. Unlike generative approaches where each model must have sound theoretical foundation, qualitative approaches allow subjectivity in the constructed model. According to [2], subjectivity is accepted when human behavior is under study.

4.2 Quantitative Approaches

Within quantitative approaches, data is processed through algorithms and statistical techniques. There is a large number of quantitative approaches. The main difference between them concerns the format of data used to carry out the study. Hence, quantitative approaches are further subdivided according to data format into three subgroups, which are detailed in the following paragraphs. We design by e^t a target event and by e_o^t the occurrence o of the event e^t and t_o its time of occurrence with $o \in [1..n]$; n is the number of past events considered in the study.

Matrix Data Structure-Based Approaches. In this case, data has the format of a matrix. This format is commonly used in statistics. Cases (i.e. observations or learning set) representing the matrix rows are event instances e_o^t. Matrix rows can also be control-cases c_z^t representing random situations that take place at time t_z such that $t_o \neq t_z$ with $z \in [1..m]$; m is the total number of control-cases. The matrix columns are variables (i.e. features, biomarkers, attributes) X_k with $k \in [1..K]$; K is the total number of variables. Finally, data used has the following format: $M = (x_{ij})$ with $i \in [1..n+m]$ and $j \in [1..K]$; and x_{ij} is the value of variable X_j for each observation. Approaches dealing with such format are classification approaches and event frequency analysis approaches.

Classification Approaches. The prediction process involves predictor variables referenced above as X_k. Classification can be supervised or unsupervised. For supervised classification, a decision variable D such that $D \in \{X_k\}$ specifies the predicted outcome. The decision variable can be the event magnitude or simply a binary valued variable specifying the actual occurrence of the event or not [32]. In unsupervised classification, clusters are deduced and interpreted as prediction outcomes [32]. Classification techniques for prediction purpose can be applied as single classifiers [3,47] or as hybrid classifiers [13,15] which is the recent trend in this area. The authors in [32] give a summary of hybrid classifiers for business failure prediction.

Event Frequency Analysis Approaches. The event occurrences are described by a unique random variable or a set of variables X_o. This can be the event intensity (i.e. magnitude for earthquakes) or other characteristics such as the volume and duration for floods [61]. In this category of approaches, the variable outcomes are estimated by analysing frequency distribution of event occurrences. The estimation of outcomes relies on descriptive statistics and consists practically in approximating the variables distribution then deducing their statistical descriptions.

There are two cases for this type of approaches: (i) rare events with a focus on maximum values for X_o (i.e. extreme events) [25]; and (ii) frequent events. For the first case, extreme value theory has become a reference. It involves the analysis of the tail of the distribution. For the second case, known distributions such as Poisson, Gamma or Weibull are considered. Studies extending the extreme value theory for the multivariate case exist but are rather difficult to apply for non-statisticians [17].

Temporal Approaches. In temporal approaches, the time dimension is explicitly considered. Here, e_o^t will be identified on a set of K time series, each time series represents a variable X measured at equal intervals over a time period T such that the time of occurrence of e_o^t, namely t_o, is included in T (see Fig. 4). The set of time series which actually represent the studied data is denoted by $S_T = \{X_k(t); t \in T\}$, $k \in [1..K]$. Temporal approaches may consider unique time series (i.e. univariate time series) or several time series. Two types of methodologies are possible.

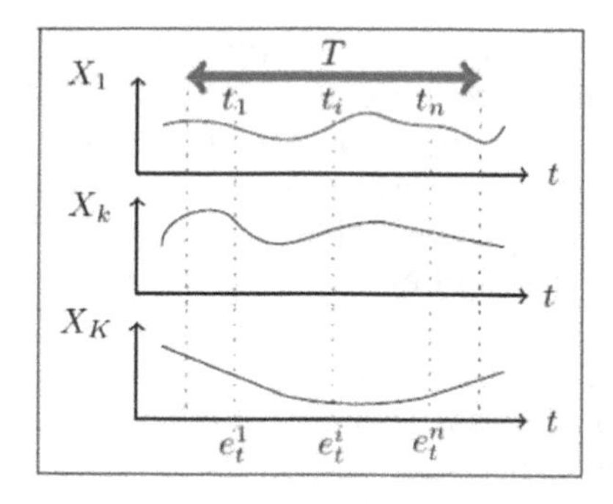

Fig. 4. Target events identification on time series over time interval

Approaches Dealing with Univariate Time Series. In this special case, a unique time series is under consideration. The event prediction can follow two patterns:

- ***Time series forecasting and event detection.*** For this case, time series values are forecasted. Then the target event is detected. It is important to note that event prediction on the basis of time series data is different from time series forecasting. The difference consists in the nature of the predicted outcome. For event prediction, the outcome is an event, hence the goal is to identify the time of occurrence of the event through the analysis of the effect the precursor factors have on time series data. For time series forecasting, the outcome consists in the future values of the time series. The authors in [54] applied this approach for computer systems failure prediction.
- ***Time series event prediction.*** We refer to time series event [43,45] as a notable variation in the time series values that characterizes the occurrence of the target event under study. In this special case, researchers analyze variations, mainly trends, in time series data, preceding the time series event and deduce temporal patterns that can be used for prediction.

Approaches Dealing with Multivariate Time Series. Most works under this category deduce temporal patterns from multiple time series followed by clustering or classification of these patterns in order to deduce future events [8,41]. These approaches adopt the same strategy as with time series event prediction but they are more adapted to the complex aspect of multivariate time series.

Event Oriented Approaches. When the available data is a collection of events, event prediction strategy follows a different path, where the central focus becomes the chronological interrelations between events data and a special target event, the latter can a simple or complex event. In what follows, we will detail two cases: the first is event sequence identification, which is adapted for simple events, and the second is complex event processing which is adapted to complex events.

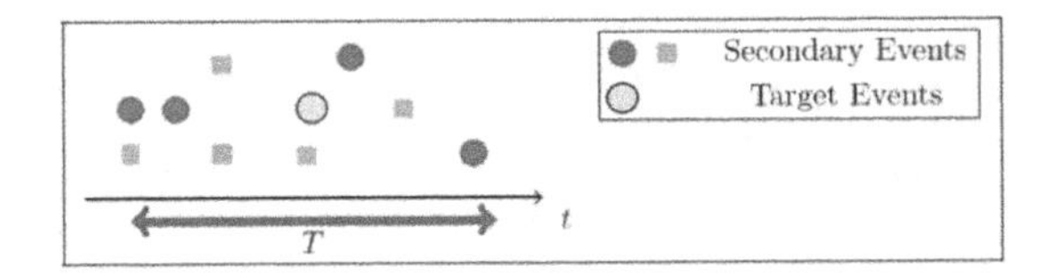

Fig. 5. Graphical illustration of target and secondary events over time

Event Sequence Identification. Within event sequence identification, we consider a set L of secondary events $\{e_l^s\}$ with $l \in [1..L]$ (see Fig. 5). These events are events occurring around the target event and can be used to predict the target event. The secondary event e_l^s occurrences are denoted by $\{e_{lz}^s\}$with $z \in [1..Z_l]$; Z_l the total number of occurrences for the secondary event with index l. The secondary event e_l^s is also described by a set of K variables $\{X_k\}$ with $k \in [1..K]$ (the authors in [59] described these variables as a set feature value pairs). All events set $\{e_{lz}^s\}$ (with $l \in [1..L]$ and $z \in [1..Z_l]$), in addition to $\{e_i^t\}$ (with $i \in [1..n]$), are considered over a common time interval T and temporal sequences are deduced. These events and the corresponding variables describing each event occurrence represent the data format for event sequence identification approaches.

This category of approaches is mainly used in online system failure prediction [34, 49] where secondary events are identified from computer log files.

Complex Event Processing. A complex event is a sequence $<e_1, \ldots, e_m>$ of different simple events chronologically related. The CEP aim at detecting complex events on the basis of an event space as a dataset. The CEP solutions has been applied successfully to predict heart failures [36], where simple events like symptoms are detected and hence an alert predicting the heart stroke (complex event) is enabled. Other applications include computer system failure prediction [6], Internet of things failure prediction [56] and bad traffic prediction [1].

5 Hybrid Approaches to Event Prediction

The working principle of hybrid approaches is given in Fig. 6. As shown in this figure, hybrid approaches combine steps from generative and inferential approaches. The starting point is both available historical data (like inferential approaches) and knowledge about system generating the event (like generative approaches). The outputs of these two parallel steps are combined into a general model. The next step consists in model calibration and validation against real-world datasets (similarly to generative approaches). Finally, simulation of the future states of the system is performed and the predicted outcome is deduced.

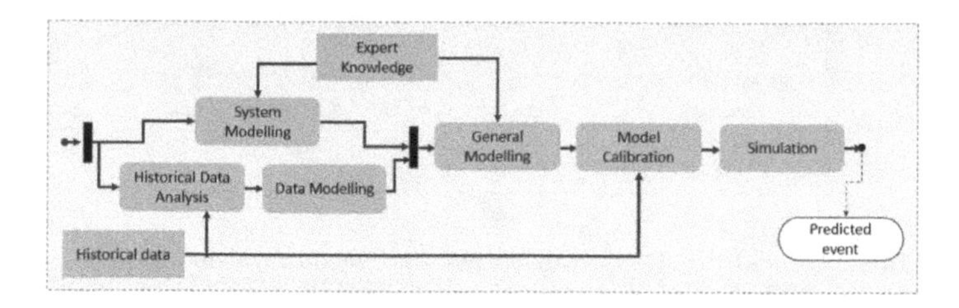

Fig. 6. Working principle of hybrid approaches

This category can be further subdivided into two sub-groups:

- **Scenario based approaches.** These approaches construct a mathematical model of the system generating the target event. Then, they vary the model input data according to different possible scenarios extracted from the historical data records. The various outputs of the model represent all possible results. Scenario based approaches are often seen as solution to uncertainty issues [28]. A classic example of scenario based approaches is ensemble streamflow prediction [23], which is mainly used in flood prediction.
- **Mixed data models approaches.** These approaches combine models constructed from observed data with models based on physics laws. Hence they employ generative and inferential modeling techniques. The authors in [30] design this type of approaches by 'conceptual models'. Generally, generative models involve a considerable amount of historical data, so inferential models, such as time series modelling techniques, are used to generate the required input data.

6 Discussion

Table 1 provides some examples illustrating the application of discussed categories of methods in different application domains. This table shows that some approaches are devoted to some specific event types. For example, hybrid approaches are widely applied in hydrology, especially for flood prediction. Event sequence identification is mainly applied for computer system failure prediction. This is due to the availability of secondary events through system logs. Multi-agent simulation requires strong knowledge in computer science and may be too complex for non-specialists. This explains its application for restrained fields. Qualitative prediction approaches are well adapted to predict low risk related events such as in financial context. However, they can be unreliable for major events such as floods and earthquakes.

The approaches depicted in the taxonomy have several drawbacks. For instance, generative and hybrid approaches fail to produce a model that generates exactly the real-world outcomes of the studied systems [38]. Dynamical system modeling approaches perform the prediction under the assumption that the system generating the event is deterministic. However errors due to the incomplete modeling of the system make this assumption very strong in some cases. For instance, earthquake prediction studies until now fail to model the dynamics of tectonic plaques accurately [39].

Within inferential approaches, the authors in [25] argue that in event frequency analysis, fitting event characteristic variables to a known probability law can lead to inaccurate results [25]. The predictive ability of event frequency analysis approaches is relatively limited but they can be used to assist the prediction process by analysing the studied phenomenon. In addition, the authors in [2] remark that inferential approaches fail to analyze the data holistically and they mainly focus on a truncated aspect of the data. In addition, they fail to take into account the system dynamics and interactions between variables [2]. At this level, one should observe that classification approaches can take into account interactions between variables. Furthermore stream mining models, prepared to deal with concept drift, can address system dynamics by evolving the machine learning model.

The authors in [2] advocate that qualitative approaches are more effective if they are combined with quantitative analysis, since a holistic consideration of event context, its dynamics but also temporal evolution by experts may overcome the restrictive view of data by quantitative approaches. The advantage of classification based approaches as a prediction technique is the availability of a range of proven tools and computing packages, making its application open to large public. However, classification based approaches fail to handle uncertainty in data and do not take into account preferences in input variables. In addition, classification based approaches neglect time dimension, as stated in [5]. Furthermore, the approximate time to event occurrence defined by [59] as lead time can be inaccurately estimated, which is a drawback for risk prevention procedures, especially concerning major events such as floods and earthquakes. At this level, we should mention that the proposal of [26] presents some solutions to address this issue by using composed labels having the form (Event, Time) for predicating event occurrence time and (Event,Intensity,Time) for predicating event occurrence time and intensity.

An interesting approach to reduce the effect of these shortcomings is to combine classification and pattern identification techniques, as suggested by [5]. In this respect, the authors in [19,20] combine self-organising maps and temporal patterns to predict firms failure. More specifically, the authors in [20] use the term 'failure trajectory' to refer to temporal patterns representing the firm health over time while the author in [19] uses the term 'failure process' as a temporal pattern, and refers to it as a typology of firm behaviour over time. In both cases the patterns are used to classify firms; hence predict business failure event. More recently, the authors in [26] introduce rough set based classification techniques with an explicit support of temporal patterns identification.

Table 1. Examples of event prediction methods and applications

Category	Business failure	Stock market variation	Earthquakes	Floods	Heart stroke/Health events	Computer System failure	Internet of Things failure
Dynamical system modelling	[42]		[7]	[12]		[55]	
Multi agent simulation					[31]		
Qualitative prediction	[52]						
Event frequency analysis				[48]			
Classification	[13, 15, 29]		[47]		[3, 50]	[27]	
Temporal patterns (Univariate)		[44]	[4, 22, 35, 40]	[16]			
Temporal patterns (Multiple)					[8, 57]		
Forecasting/Event detection						[54]	
Event sequence identification						[34, 49]	
Complex event processing					[36]	[6]	[56]
Hybrid models approaches				[9, 10]			
Scenario based approaches				[33, 53]			

7 Conclusion

This paper introduces a taxonomy of event prediction approaches. It represents a generic view of event prediction approaches that oversteps the problem considered and application domain. The proposed taxonomy has several practical and theoretical benefits. First, it extends the application domain of existing and new event prediction approaches. Second, opens road for designing and developing more advanced and context-independent techniques. Third, it helps users in selecting the appropriate approach to use in a given problem.

Several points need to be investigated in the future. First, the proposed taxonomy is far from exhaustive. We then intend to extend the present work by considering additional application domains and event types. Second, several event prediction approaches can be used for the same event type. Then, it would be interesting to design a generic guideline or some rules permitting to select the event prediction method to be used in a given problem, which will reduce the cognitive effort required from the expert.

References

1. Akbar, A., Khan, A., Carrez, F., Moessner, K.: Predictive analytics for complex IoT data streams. IEEE Internet Things J. **4**(5), 1571–1582 (2017)
2. Alaka, H., Oyedele, L., Owolabi, H., Ajayi, S., Bilal, M., Akinade, O.: Methodological approach of construction business failure prediction studies: a review. Constr. Manag. Econ. **34**(11), 808–842 (2016)
3. Austin, P., Lee, D., Steyerberg, E., Tu, J.: Regression trees for predicting mortality in patients with cardiovascular disease: what improvement is achieved by using ensemble-based methods? Biometrical J. **54**(5), 657–673 (2012)

4. Aydin, I., Karakose, M., Akin, E.: The prediction algorithm based on fuzzy logic using time series data mining method. World Acad. Sci. Eng. Technol. **51**(27), 91–98 (2009)
5. Balcaen, S., Ooghe, H.: 35 years of studies on business failure: an overview of the classic statistical methodologies and their related problems. Br. Acc. Rev. **38**, 63–93 (2006)
6. Baldoni, R., Montanari, L., Rizzuto, M.: On-line failure prediction in safety-critical systems. Future Gener. Comput. Syst. **45**, 123–132 (2015)
7. Barbot, S., Lapusta, N., Avouac, J.P.: Under the hood of the earthquake machine: toward predictive modeling of the seismic cycle. Science **336**(6082), 707–710 (2012)
8. Batal, I., Cooper, G., Fradkin, D., Harrison Jr., J., Moerchen, F., Hauskrecht, M.: An efficient pattern mining approach for event detection in multivariate temporal data. Knowl. Inf. Syst. **46**(1), 115–150 (2015)
9. Bergstrom, S.: Development and application of a conceptual runoff model for Scandinavian catchments. Techncial report, SMHI RHO 7 (1976)
10. Blazkov, S., Beven, K.: Flood frequency prediction for data limited catchments in the Czech Republic using a stochastic rainfall model and topmodel. J. Hydrol. **195**(1–4), 256–278 (1997)
11. Bosse, T., Sharpanskykh, A., Treur, J.: Integrating agent models and dynamical systems. In: Baldoni, M., Son, T.C., van Riemsdijk, M.B., Winikoff, M. (eds.) DALT 2007. LNCS (LNAI), vol. 4897, pp. 50–68. Springer, Heidelberg (2008). https://doi.org/10.1007/978-3-540-77564-5_4
12. Brunner, G.: HEC-RAS river analysis system hydraulic reference manual. version 5.0. Technical report, Hydrologic Engineering Center, Davis, CA (2016)
13. Cabedo, J., Tirado, J.: Rough sets and discriminant analysis techniques for business default forecasting. Fuzzy Econ. Rev. **20**(1), 3–37 (2015)
14. Casulli, V., Stelling, G.: Numerical simulation of 3D quasi-hydrostatic, free-surface flows. J. Hydraul. Eng. **124**(7), 678–686 (1998)
15. Cheng, M.Y., Hoang, N.D.: Evaluating contractor financial status using a hybrid fuzzy instance based classifier: case study in the construction industry. IEEE Trans. Eng. Manag. **62**(2), 184–192 (2015)
16. Damle, C., Yalcin, A.: Flood prediction using time series data mining. J. Hydrol. **333**, 305–316 (2006)
17. Denny, M., Hunt, L., Miller, L., Harley, C.: On the prediction of extreme ecological events. Ecol. Monogr. **93**(3), 397–421 (2009)
18. Dodig-Crnkovic, G., Giovagnoli, R.: Computing nature-a network of networks of concurrent information processes. In: Dodig-Crnkovic, G., Giovagnoli, R. (eds.) Computing Nature, vol. 7, pp. 1–22. Springer, Heidelberg (2013). https://doi.org/10.1007/978-3-642-37225-4_1
19. du Jardin, P.: Bankruptcy prediction using terminal failure processes. Eur. J. Oper. Res. **242**(1), 286–303 (2015)
20. du Jardin, P., Séverin, E.: Predicting corporate bankruptcy using a self-organizing map: an empirical study to improve the forecasting horizon of a financial failure model. Decis. Support Syst. **51**(3), 701–711 (2011)
21. Epstein, J.: Generative Social Science: Studies in Agent-Based Computational Modeling. Princeton University Press, Princeton (2006)
22. Florido, E., Martínez-Álvarez, F., Morales-Esteban, A., Reyes, J., Aznarte-Mellado, J.: Detecting precursory patterns to enhance earthquake prediction in Chile. Comput. Geosci. **76**, 112–120 (2015)

23. Franz, K., Hartmann, H., Sorooshian, S., Bales, R.: Verification of national weather service ensemble streamflow predictions for water supply forecasting in the colorado river basin. J. Hydrometeorol. **4**(6), 1105–1118 (2003)
24. Fülöp, L., Beszédes, A., Tóth, G., Demeter, H., Vidács, L., Farkas, L.: Predictive complex event processing: a conceptual framework for combining complex event processing and predictive analytics. In: Proceedings of the Fifth Balkan Conference in Informatics, BCI 2012, pp. 26–31. ACM, New York (2012)
25. Ghil, M., et al.: Extreme events: dynamics, statistics and prediction. Nonlinear Process. Geophys. **18**, 295–350 (2011)
26. Gmati, F.E., Chakhar, S., Lajoued Chaari, W., Chen, H.: A rough set approach to events prediction in multiple time series. In: Mouhoub, M., Sadaoui, S., Ait Mohamed, O., Ali, M. (eds.) IEA/AIE 2018. LNCS (LNAI), vol. 10868, pp. 796–807. Springer, Cham (2018). https://doi.org/10.1007/978-3-319-92058-0_77
27. Hamerly, G., Elkan, C.: Bayesian approaches to failure prediction for disk drives. In: Proceedings of the Eighteenth International Conference on Machine Learning, ICML 2001, pp. 202–209. Morgan Kaufmann Publishers Inc., San Francisco (2001)
28. Hull, T.: A deterministic scenario approach to risk management. In: Enterprise Risk Management Symposium, Society of Actuaries, April edn, Chicago, IL, pp. 1–7 (2010)
29. Iturriaga, F., Sanz, I.: Bankruptcy visualization and prediction using neural networks: a study of U.S. commercial banks. Expert Syst. Appl. **42**(6), 2857–2869 (2015)
30. Devia, G.K., Ganasri, B., Dwarakish, G.: A review on hydrological models. Aquatic Procedia **4**, 1001–1007 (2015)
31. Li, Y., Lawley, M.A., Siscovick, D.S., Zhang, D., Pagán, J.A.: Agent-based modeling of chronic diseases: a narrative review and future research directions. Preventing Chronic Dis. **13** (2016). https://doi.org/10.5888/pcd13.150561
32. Lin, W.Y., Hu, Y., Tsai, C.F.: Machine learning in financial crisis prediction: a survey. IEEE Trans. Syst. Man Cybern. **42**(4), 421–436 (2012)
33. Hopson, T.M., Webster, P.: A 1–10-day ensemble forecasting scheme for the major river basins of Bangladesh: forecasting severe floods of 2003–07. J. Hydrometeorol. **11**(3), 618–641 (2010)
34. Mannila, H., Toivonen, H., Verkamo, A.I.: Discovery of frequent episodes in event sequences. Data Min. Knowl. Discov. **1**(3), 259–289 (1997)
35. Martínez-Álvarez, F., Troncoso, A., Morales-Esteban, A., Riquelme, J.: Computational intelligence techniques for predicting earthquakes. In: International Conference on Hybrid Artificial Intelligence Systems, pp. 287–294 (2011)
36. Mdhaffar, A., Rodriguez, I., Charfi, K., Abid, L., Freisleben, B.: CEP4HFP: complex event processing for heart failure prediction. IEEE Trans. NanoBiosci. **16**(8), 708–717 (2017)
37. Merkuryeva, G., Merkuryev, Y., Sokolov, B., Potryasaev, S., Zelentsov, V., Lektauers, A.: Advanced river flood monitoring, modelling and forecasting. J. Comput. Sci. **10**, 77–85 (2014)
38. Meyers, R.: Extreme Environmental Events: Complexity in Forecasting and Early Warning. Springer, New York (2010)
39. Mitsa, T.: Temporal Data Mining. CRC Press, Boca Raton (2010)
40. Morales-Esteban, A., Martínez-Álvarez, F., Troncoso, A., Justo, J., Rubio-Escudero, C.: Pattern recognition to forecast seismic time series. Expert Syst. Appl. **37**, 8333–8342 (2010)

41. Morchen, F., Ultsch, A.: Discovering temporal knowledge in multivariate time series. In: Weihs, C., Gaul, W. (eds.) Classification - The Ubiquitous Challenge, pp. 272–279. Springer, Heidelberg (2005). https://doi.org/10.1007/3-540-28084-7_30
42. Nwogugu, M.: Decision-making, risk and corporate governance: new dynamic models/algorithms and optimization for bankruptcy decisions. Appl. Math. Comput. **179**(1), 386–401 (2006)
43. Povinelli, R.: Time series data mining: identifying temporal patterns for characterization and prediction of time series events. Ph.D. thesis, Marquette University, Milwaukee, WI (1999)
44. Povinelli, R.J.: Identifying temporal patterns for characterization and prediction of financial time series events. In: Roddick, J.F., Hornsby, K. (eds.) TSDM 2000. LNCS (LNAI), vol. 2007, pp. 46–61. Springer, Heidelberg (2001). https://doi.org/10.1007/3-540-45244-3_5
45. Povinelli, R., Feng, X.: A new temporal pattern identification method for characterization and prediction of complex time series events. IEEE Trans. Knowl. Data Eng. **15**(2), 339–352 (2003)
46. Preston, D., Protopapas, P., Brodley, C.: Event discovery in time series. In: Apte, C., Park, H., Wang, K., Zaki, M. (eds.) Proceedings of the 2009 SIAM International Conference on Data Mining, pp. 61–72. SIAM (2009). https://doi.org/10.1137/1.9781611972795.6
47. Rafiei, M., Adeli, H.: NEEWS: a novel earthquake early warning model using neural dynamic classification and neural dynamic optimization. Soil Dyn. Earthq. Eng. **100**, 417–427 (2017)
48. Razmi, A., Golian, S., Zahmatkesh, Z.: Non-stationary frequency analysis of extreme water level: application of annual maximum series and peak-over threshold approaches. Water Resour. Manag. **31**(7), 2065–2083 (2017)
49. Sahoo, R., et al.: Critical event prediction for proactive management in large-scale computer clusters. In: Proceedings of the Ninth ACM SIGKDD International Conference on Knowledge Discovery and Data Mining, pp. 426–435. ACM, New York (2003)
50. Samuel, O., Grace, G., Sangaiah, A., Fang, P., Li, G.: An integrated decision support system based on ANN and Fuzzy_AHP for heart failure risk prediction. Expert Syst. Appl. **68**, 163–172 (2017)
51. Tak-chung, F.: A review on time series data mining. Eng. Appl. Artif. Intell. **24**(1), 164–181 (2011)
52. Tamari, M.: Financial ratios as a means of forecasting bankruptcy. Manag. Int. Rev. **6**(4), 15–21 (1966)
53. Thielen, J., Bartholmes, J., Ramos, M.H., de Roo, A.: The European flood alert system-part 1: concept and development. Hydrol. Earth Syst. Sci. **13**(2), 125–140 (2009)
54. Vilalta, R., Apte, C., Hellerstein, J., Ma, S., Weiss, S.: Predictive algorithms in the management of computer systems. IBM Syst. J. **41**(3), 461–474 (2002)
55. Vrugt, J., ter Braak, C., Clark, M., Hyman, J.M., Robinson, B.: Treatment of input uncertainty in hydrologic modeling: doing hydrology backward with Markov chain Monte Carlo simulation. Water Resour. Res. **44**(12) (2008). https://doi.org/10.1029/2007WR006720
56. Wang, C., Vo, H., Ni, P.: An IoT application for fault diagnosis and prediction. In: 2015 IEEE International Conference on Data Science and Data Intensive Systems, pp. 726–731. IEEE (2015)

57. Wang, S.: Online monitoring and prediction of complex time series events from nonstationary time series data. Ph.D. thesis, Rutgers University-Graduate School-New Brunswick (2012)
58. Wang, Y., Gao, H., Chen, G.: Predictive complex event processing based on evolving Bayesian networks. Pattern Recogn. Lett. **105**, 207–216 (2018)
59. Weiss, G., Hirsh, H.: Learning to predict rare events in categorical time-series data. Techncal report, AAAI (1998). www.aaai.org
60. Yan, X.B., Lu, T., Li, Y.J., Cui, G.B.: Research on event prediction in time-series data. In: Proceedings of International Conference on Machine Learning and Cybernetics, Shanghai, vol. 5, pp. 2874–2878, August 2004
61. Yue, S., Ouarda, T., Bobee, B., Legendre, P., Bruneau, P.: The Gumbel mixed model for flood frequency analysis. J. Hydrol. **226**, 88–100 (1999)

Infilling Missing Rainfall and Runoff Data for Sarawak, Malaysia Using Gaussian Mixture Model Based K-Nearest Neighbor Imputation

Po Chan Chiu[1,2,3](✉), Ali Selamat[1,2,4,5], and Ondrej Krejcar[5]

[1] School of Computing, Faculty of Engineering, Universiti Teknologi Malaysia, 81310 Johor Bahru, Johor, Malaysia
pcchiu@unimas.my
[2] MagicX (Media and Games Center of Excellence), Universiti Teknologi Malaysia, 81310 Johor Bahru, Johor, Malaysia
[3] Faculty of Computer Science and Information Technology, Universiti Malaysia Sarawak, 94300 Kota Samarahan, Sarawak, Malaysia
[4] Malaysia Japan International Institute of Technology (MJIIT), Universiti Teknologi Malaysia Kuala Lumpur, Jalan Sultan Yahya Petra, 54100 Kuala Lumpur, Malaysia
[5] Faculty of Informatics and Management, University of Hradec Kralove, Rokitanského 62, 500 03 Hradec Kralove, Czech Republic

Abstract. Hydrologists are often encountered problem of missing values in a rainfall and runoff database. They tend to use the normal ratio or distance power method to deal with the problem of missing data in the rainfall and runoff database. However, this method is time consuming and most of the time, it is less accurate. In this paper, two neighbor-based imputation methods namely K-nearest neighbor (KNN) and Gaussian mixture model based KNN imputation (GMM-KNN) were explored for gap filling the missing rainfall and runoff database. Different percentage of missing data entries were inserted randomly into the database such as 2%, 5%, 10%, 15% and 20% of missing data. Pros and cons of these two methods were compared and discussed. The selected study area is Bedup Basin, located at Samarahan Division, Sarawak, East Malaysia. It is observed that the GMM-KNN imputation method results in the best estimation accuracy for the missing rainfall and runoff database.

Keywords: GMM-KNN · KNN · Imputation · Missing rainfall · Runoff data

1 Introduction

Hydrological missing data poses a challenge in hydrological and environmental modelling. Hydrologists are often encountered problem of missing values in a rainfall and runoff database. Rainfall is the quantity of rain that falls in a location over a period of time [1]. Meanwhile runoff refers as amount of water that discharged in surface streams. Missing data occurs when data values are not available or incomplete in the database. The incompleteness of rainfall and runoff data may due to equipment

F. Wotawa et al. (Eds.): IEA/AIE 2019, LNAI 11606, pp. 27–38, 2019.
https://doi.org/10.1007/978-3-030-22999-3_3

malfunctioned, measurement errors and changes to instrumentation over time [2]. If the missing data is left untreated, it reduces the power and the precision of hydrological research. These missing data would result in uncertainty particularly water flow information and it affects the plan ahead of time to deal with extremes such as flood and climate change.

As floods become increasingly more frequent in Malaysia, the analysis of rainfall and runoff plays a significant role in the field of climatology and hydrological studies [3–6]. However, rainfall and runoff data analysis is always challenged by the shortage of consecutive data at Sarawak rivers. In many instances, while analyzing the hydrological data for Sarawak rivers, there is a shortage of rainfall records of several gauging stations at Bedup Basin, and most of these records are incomplete. In addition to that, a study by Ismail *et al.* [7] concluded that the best data treatment method of a target station is different from the other target stations in Peninsular Malaysia. Therefore, this paper explores a suitable technique for handling missing rainfall and runoff data at Bedup Basin, Sarawak.

The remaining of this work is organized as follows. Section 2 reviews the existing handling missing data techniques; Sect. 3 presents a case study; Sect. 4 describes the imputation methods and Sect. 5 reports the experimental results both on KNN and GMM-KNN imputation methods. Finally, Sect. 6 summarizes the main conclusions.

2 Related Work

Numerous techniques have been proposed to estimate missing values [7–9]. Imputation is a procedure that is used to fill in missing values with substitutes [9]. The normal ratio method (NR) and the inverse distance weighting method (IDW) are the two types of the traditional missing data handling methods. The methods are the most popular approach for estimation of missing rainfall records. Suhaila *et al.* [8] adapted the inverse distance weighting (IDW) method for estimation of missing rainfall data. The study reported that the target station could be affected most by the nearest stations. Kamaruzaman *et al.* [10] have compared different methods such as inverse distance weighted (IDW), modified correlation weighted (MCW), combination correlation with inverse distance (CCID) and averaging correlation and inverse distance (ACID) to examine the best imputation methods for treating daily rainfall at 104 stations in Peninsular Malaysia. Meanwhile interpolation techniques such as arithmetic average (AA) method, inverse distance (IDW) method, normal ratio (NR) method and coefficient of correlation (CC) method were compared in a study by Ismail *et al.* [7]. There are several shortcomings, such as the overestimation or underestimation of association among variables and lack of information available from the neighbor stations. If there is no information could be used from the neighbor stations, the mean on the same day and month but at different year will be taken as the estimation of the missing values at the missing entries. Hence, this method is less accurate and time consuming as compared to other missing data imputation techniques.

The nearest neighbor stations are progressively being used to estimate the missing values in the database. Ferrari and Ozaki [11] used the nearest neighbor station to estimate the missing data based on the statistical imputation and quality control procedures to model the drought period. Furthermore, Teegavarapu and Chandramouli [12] used the inverse distance weighting method (IDW) to estimate missing rainfall values which is based on the values recorded at all other available nearby stations.

Other than that, artificial neural networks (ANNs) have become one of the most promising tools for treating missing data problem. In a study by Dastorani *et al.* [13], artificial neural networks (ANNs) and adaptive neuro-fuzzy inference system (ANFIS) methods were proposed to predict the missing flow data using the data from neighboring sites. The study revealed that the ANFIS technique demonstrated a superior prediction of missing flow data in arid land stations. Besides that, Mispan *et al.* [14] employed Levenberg-Marquadt back propagation algorithm in predicting missing stream flow data in Langat River Basin, Malaysia. The training and validation results are satisfactory; which r values range from 0.91 to 0.97 for flow parameters.

Another approach to treat the missing data problem is using Gaussian mixture model based KNN (GMM-KNN) method. Ding and Ross [15] have proposed Gaussian mixture model based KNN (GMM-KNN) imputation method for treating missing scores in biometric fusion. In the study, Ding and Ross [15] reported that GMM-KNN method performs better than the other imputation methods such as K-nearest neighbor (KNN) method, likelihood-based method, Bayesian-based method and multiple imputation (MI) method at multiple training set sizes and missing rates because it retains the natural structure of the original dataset. On the other hands, the other imputation methods such as KNN method did not capture the shape of the original scores very well. Therefore, the study indicates that the GMM-KNN imputation method results in the best recognition accuracy in the context of multibiometric fusion. However, the GMM-KNN imputation method has not been explored in the context of rainfall and runoff in Malaysia. Therefore, this study intends to explore the estimation of missing data using hydrological data from neighboring gauging sites in Sarawak and GMM-KNN imputation method.

3 Material and Method

3.1 Study Area

The study area is located in Sungai Bedup Basin, an upstream of Sadong Basin in Sarawak as shown in Fig. 1. This basin has a maximum stream length of 10 km and is situated approximately 80 km from Kuching city.

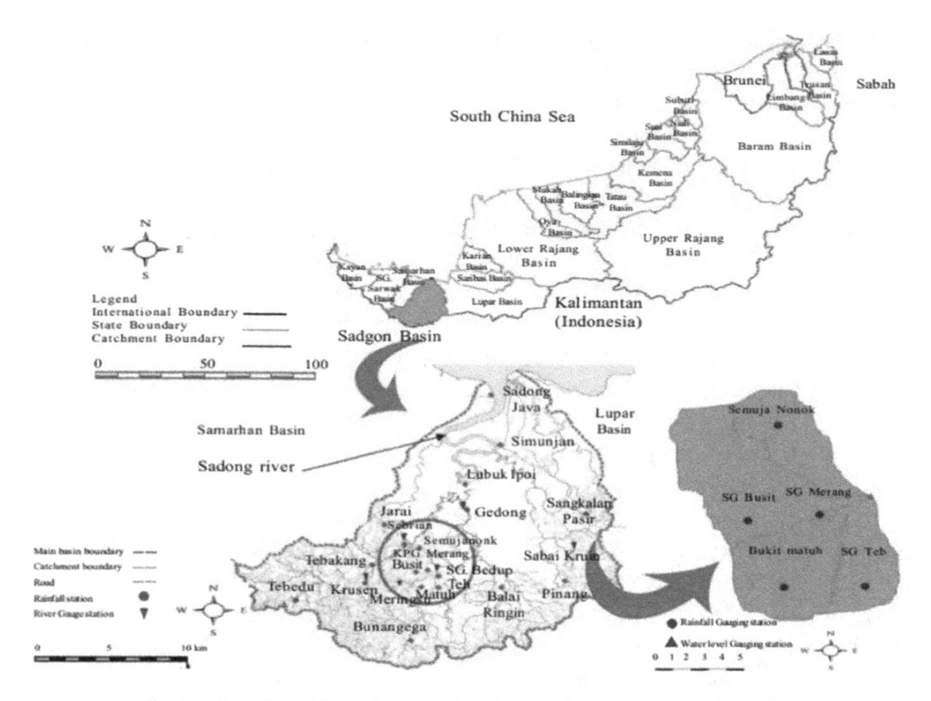

Fig. 1. Locality Map of Bedup Basin and the gauging stations [16]

Sungai Bedup Basin consists of five rainfall gauging stations and one river stage gauging station. The details of the gauging stations are presented in the following Table 1.

Table 1. Gauging stations of Bedup Basin, Sarawak.

Station name	Station number	Latitude	Longitude	Data collected
Bukit Matuh	1005079	001 03 50	110 35 35	Rainfall
Semuja Nonok	1105035	001 06 25	110 35 50	Rainfall
Sungai Busit	1005080	001 05 25	110 34 40	Rainfall
Sungai Merang	1006033	001 05 40	110 36 25	Rainfall
Sungai Teb	1006037	001 03 15	110 37 00	Rainfall
Sungai Bedup	1006028	001 05 10	110 37 50	Runoff

3.2 Data

The dataset used in this study consists of five rainfalls and one runoff data that were collected from Department of Irrigation and Drainage, Sarawak. A daily rainfall and runoff dataset consisting of 24-month records has been selected to evaluate the performance of imputation methods, as shown in Table 2. The selected dataset is prepared with some data's missing. In this work, rate-based schema has been used to randomly select a specific proportion of the entries and then removed from the complete dataset

[17]. Different percentages of missing data are inserted randomly into the dataset. The missing percentage varies as 2%, 5%, 10%, 15% and 20% missing of the total data entries. For each missing entry, a proportion of the rainfall and runoff entries will be randomly selected and removed from the dataset. According to Little and Rubin's [18] missing data mechanism, the missing value in this study has been classified as missing completely at random (MCAR). The reason is because of the occurrence of missingness in the rainfall and runoff data of the area at Bedup basin is not affected by the data in that area or any area.

Table 2. Fragment of the data from the gauging stations of Bedup Basin.

Bukit Matuh (mm)	Semuja Nonok (mm)	Sungai Busit (mm)	Sungai Merang (mm)	Sungai Teb (mm)	Sungai Bedup (m^3)
1	53.5	35.5	53.5	36	1.28
40	4.5	11	2	1	1.546
41	40	?	39.5	55	1.433
0	23	26.5	22.5	46.5	1.556
34.5	0.5	0	0	?	2.23
4	?	34	44.5	30.5	?
116.5	?	7.5	5.5	2	1.764
0	148.5	?	119.5	112.5	1.783
0	0.5	0.5	0.5	0.5	2.789
0	0.5	0	0	0	2.796

Missing data

3.3 Data Correlation Between the Investigated Stations

The correlation of daily rainfall and runoff at different stations is important to calculate the strength of a relationship between the data values. Hence, the correlation between the daily rainfall and runoff at different nearby stations was investigated (Table 3).

Table 3. Correlation matrix of investigated stations

	Bukit Matuh	Semuja Nonok	Sungai Busit	Sungai Merang	Sungai Teb	Sungai Bedup
Bukit Matuh	1.0000	0.0086	0.0043	0.0187	0.0088	0.0545
Semuja Nonok	0.0086	1.0000	0.8139	0.8538	0.7348	0.0956
Sungai Busit	0.0043	0.8130	1.0000	0.8455	0.7897	0.0874
Sungai Merang	0.0187	0.8544	0.8450	1.0000	0.8191	0.0759
Sungai Teb	0.0088	0.7366	0.7876	0.8166	1.0000	0.1042
Sg Bedup	0.0545	0.0970	0.0865	0.0931	0.1170	1.0000

As seen in Table 3, the rainfall at the Semuja Nonok is most correlated with the Sungai Busit, Sungai Merang and Sungai Teb stations respectively. The correlation of the rainfall at the Bukit Matuh station are relatively lower as compared to the rest of the rainfall stations. In addition to that, the runoff at the Sg Bedup has low correlation with all the other rainfall stations, within the range of 0.0545 to 0.1042. Generally, the rainfall and runoff correlation values in different stations are positively correlated with their respective stations.

3.4 Performance Measures

In this study, the root mean square error (RMSE) and the mean absolute error (MAE) are used to evaluate the performance of GMM-KNN and KNN imputation methods. The root mean square error (RMSE) calculates the average square errors of the treated datasets and the error is measured by Eq. (1).

$$RMSE = \sqrt{\frac{\sum_{i=1}^{N}(O_i - T_i)^2}{N}} \tag{1}$$

The mean absolute error (MAE) provides the average error in the treated datasets. The error is calculated based on Eq. (2).

$$MAE = \frac{1}{N}\sum_{i=1}^{N}|O_i - T_i| \tag{2}$$

N = total number of observations
O = actual values of observation
T = imputed values.

4 Imputation Methods

4.1 K-Nearest Neighbor Imputation

In context of limited data availability such as time series of rainfall and runoff at Sungai Bedup Basin, this study uses the nearest neighbor stations to estimate the missing values of the basin's datasets. Among the nearest neighbor imputation algorithms, K-Nearest Neighbor (KNN) imputation is one of the easiest and efficient methods used to fill in the missing values in the datasets [19].

In this work, the built-in KNN imputation from Matlab is adopted in this study. KNN imputes missing data using nearest-neighbor method. KNN replaces the missing values with a weighted mean of the k nearest-neighbor columns. The weights are inversely proportional to the distances from the neighboring columns in terms of

Euclidean distance. In this study, k = 5 is found to provide the best imputation accuracy in the dataset (not shown here).

4.2 Gaussian Mixture Model Based KNN Imputation (GMM-KNN)

Another efficient nearest neighbor imputation is Gaussian mixture model based KNN imputation (GMM-KNN). The GMM-KNN is proposed by Ding and Ross [15] in their study on handling missing scores in biometric fusion. Two main steps are essential in GMM imputation, that are the density estimation using the GMM assumption and the imputation itself based on this estimated density. For the density estimation, a simulated dataset (s), D_{sim} is generated from Gaussian mixture distribution. The dataset, D_{sim} is simulated from a multivariate normal distribution and then fit a GMM to the data using Matlab. If the $D_{sim} = 10$, the density estimation will return a GMM with a number of estimated parameters such as ten distinct means, covariances matrices and component proportions to the data. Then KNN imputation process can be used based on the generated Gaussian mixture distribution. The key idea of GMM-KNN is to find the most similar vectors as "donors" in the training set. The Euclidean distance measurement d is employed to find the "nearest" donors for the incomplete score vectors. The GMM-KNN scheme can be summarized in the following steps, as shown in Algorithm 1 [15]. According to our analysis (not shown here), generally, k = 5 and $D_{sim} = 1$ provide the best imputation accuracy on the dataset in terms of RMSE, MAE and computational time.

Algorithm 1 Gaussian Mixture Model KNN imputation (GMM-KNN)

1. Use the estimated parameters of GMM, to simulate a dataset D_{sim}, having a similar or larger size than D_{ori}
2. For each observation x, apply the distance function d to find k = 5 nearest neighbours in the simulated set D_{sim}
3. The missing variables x, are imputed by the average of corresponding variables from the nearest neighbours taken from D_{sim}

where D_{sim} = simulated dataset (s) that generated from Gaussian mixture distribution
D_{ori} = original dataset

5 Results and Discussion

This study uses the GMM-KNN and KNN methods to impute the missing values in a rainfall and runoff database from East Malaysia. The root mean square error (RMSE) and mean absolute error (MAE) are used to evaluate the performance of GMM-KNN and KNN imputation methods.

Figures 2 and 3 illustrates the imputation performances of GMM-KNN and KNN models at different percentages of missing data.

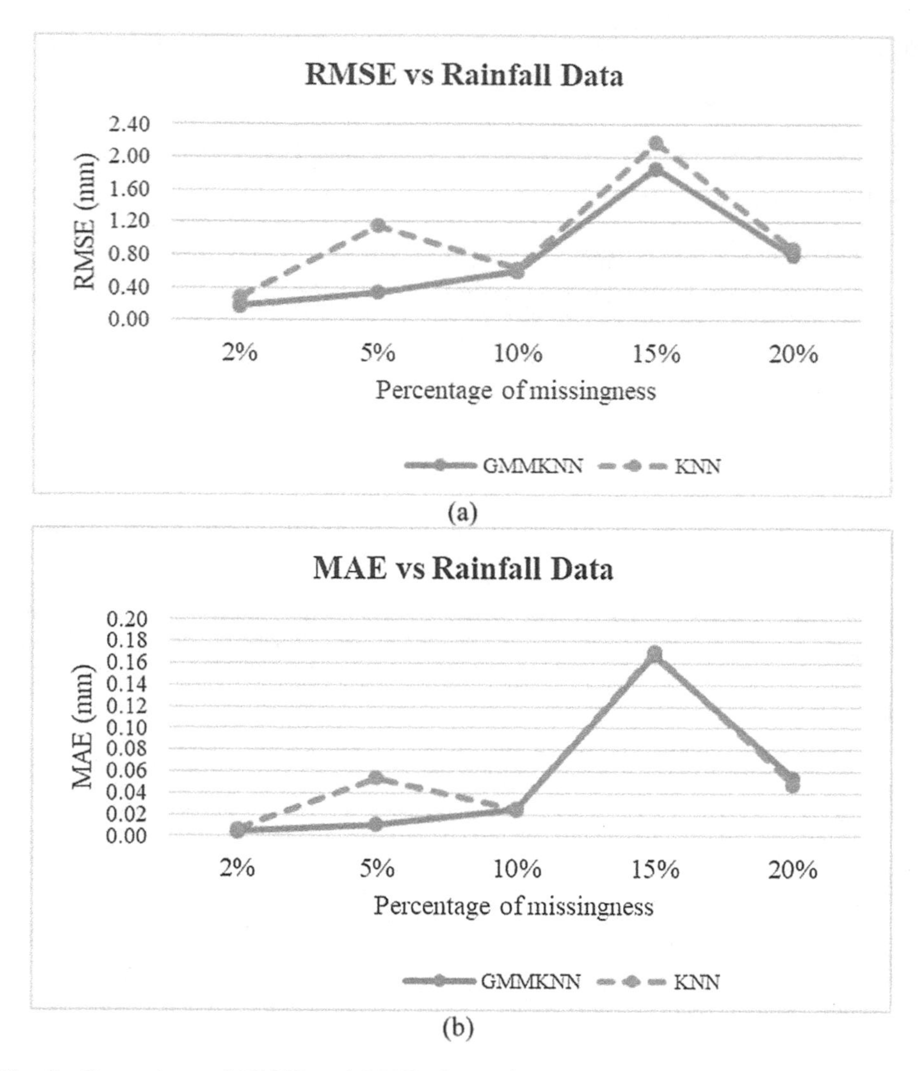

Fig. 2. Comparison of RMSE and MAE plots using GMM-KNN and KNN imputation at various percentage of missing rainfall data

Generally, the GMM-KNN models perform better than KNN model at different missing rainfall and runoff entries, which support the findings by Ding and Ross [15]. As seen in Fig. 2(a), GMM-KNN imputation provides quite accurate predictions at the missing entry of 2%, 5%, 10%, 15% and 20% with the low root mean square error (RMSE), ranges between 0 to 1.8 mm. In addition to this, the observed data points are close to the GMM-KNN model's predicted values at the missing entries between 2% to 10% of missing values in the datasets. However, there are slightly increase in the rainfall value of RMSE for KNN imputation method compared to GMM-KNN.

In Fig. 2(b), the comparison of mean absolute error (MAE) at various percentage of missing data demonstrated that KNN imputation generates higher MAE in rainfall

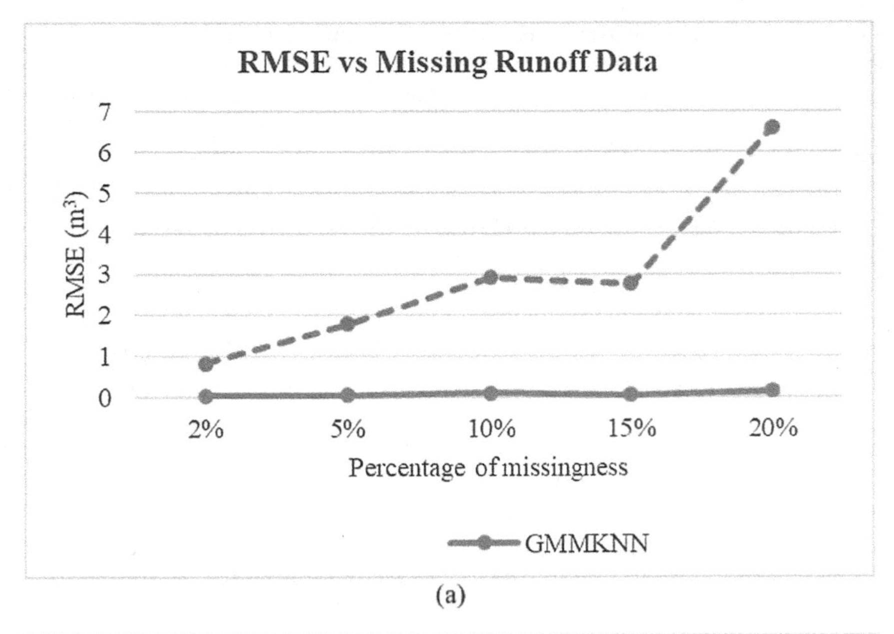

(a)

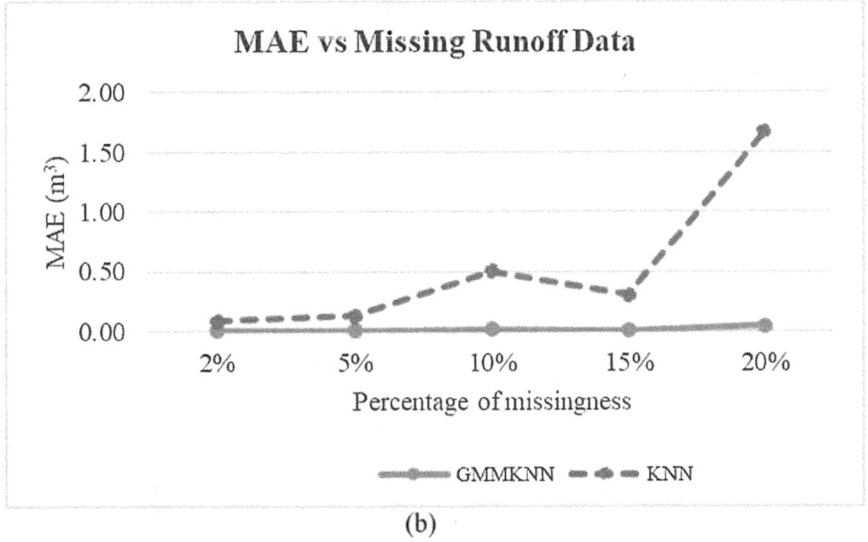

(b)

Fig. 3. Comparison of RMSE and MAE plots using GMM-KNN and KNN imputation at various percentage of missing runoff data

values at the missing entries of 2%, 5%, 10%, 15% and 20% as compared to GMM-KNN imputation. In contrast, GMM-KNN offers the best performances at multiple missing entries, with the MAE values range between 0.004 mm to 0.167 mm. Since the GMM-KNN imputation uses a simulated dataset D_{sim} that is synthetically generated for the donor imputation pool, GMM-KNN has a better chance of finding the closest from the nearest neighbor. However, it is revealed that the values of RMSE and MAE of both methods are not linearly increased with the increasing of the amount of missingness. This could be due to the reason that the methods are not sensitive to the proportion of the missing values in the datasets, as reported by Suhaila et al. in the estimation of missing rainfall data [8].

In addition to that, a study has been conducted by Kamaruzaman *et al.* [10] using weighting methods for treating missing daily rainfall data at the missing percentage of 10% within the context of Peninsular Malaysia. Kamaruzaman *et al.* [10] reported that ACID method obtained good results in the test statistic for treating missing daily rainfall data in terms of RMSE and MAE with an average of 12.78 and 6.28 respectively. As compared to the findings by Kamaruzaman *et al.* [10], the performance of GMM-KNN imputation has improved in terms of RMSE and MAE, where the error values decreased by an average of 11 mm and 6 mm respectively. Furthermore, Ismail *et al.* [7] revealed that IDW method is to be the best missing data estimation method among the interpolation techniques for most of the rainfall stations located at Terengganu, Malaysia. For example, Ismail *et al.* [7] reported the values of RMSE and MAE for station TR_b range between 12.765 mm to 14.76 mm and between 6.042 mm to 6.704 mm respectively. Comparison of this study to the study by Ismail *et al.* [7] shows that GMM-KNN method has lower values of RMSE and MAE, where the error values decreased by an average of 6.97 mm and 3.78 mm respectively at multiple missing rainfall entries. Hence, it is clearly shown that GMM-KNN imputation is the best method for finding the missing rainfall entries at Bedup Basin station.

Meanwhile the example of plots of RMSE and MAE for the runoff missing data are shown in Fig. 3. In Fig. 3(a) and (b), it is observed that the GMM-KNN imputation results in a low value of RMSE, with the range of 0.04 to 0.15 m^3 and MAE, with the range of 0.004 to 0.04 m^3 at all the missing entries. One possible reason is that a good density GMM model positively increases the accuracy of the neighbor-based imputation method.

However, KNN imputation has an increased runoff value of RMSE and MAE when the percentages of missing entries increase. Besides that, KNN imputation has high values of RMSE and MAE, between the range of 0.8 to 6.6 m^3 and the range of 0.08 to 1.7 m^3 respectively. This may be due to the low correlation between the runoff data and the rainfall data of the target stations. Since KNN imputation is based on nearest neighbor method, the accuracy of KNN imputation could highly affected by the correlation of the nearby target stations. As a result, the accuracy of the KNN decreased gradually with the decreased value of correlation between the runoff and the rainfall data. A closer inspection revealed that the decreased value of the correlation between the data could be due to the proportions of missing data. The proportions of missing values that contain more relevance to the target station could lead to overestimate and underestimate the missingness. Therefore, GMM-KNN imputation provides a much better performance for filling rainfall and runoff missing entries than the KNN imputation at Bedup Basin station.

6 Conclusion

In this study, the daily rainfall and runoff data at six gauging stations located in Bedup Basin was considered. The GMM-KNN and KNN methods were applied to fill the missing entries at different percentage of rainfall and runoff missing entries. The results demonstrated that GMM-KNN method performs better than KNN method and it is suitable to be applied for finding the missing rainfall and runoff database.

However, the drawback of GMM-KNN imputation is GMM may fail to work if the dimensionality of the problem is too high. For future work, it is recommended to consider hybrid GMM with other missing data estimation techniques on real world missing datasets.

Acknowledgments. The authors sincerely acknowledge the Department of Irrigation and Drainage (DID), Sarawak, Malaysia for providing the rainfall and runoff data in this study. The authors wish to thank Universiti Teknologi Malaysia (UTM) under Research University Grant Vot-20H04, Malaysia Research University Network (MRUN) Vot 4L876 and the Fundamental Research Grant Scheme (FRGS) Vot 5F073 supported under Ministry of Education Malaysia for the completion of the research. The works were also supported by the SPEV project, University of Hradec Kralove, FIM, Czech Republic (ID: 2102–2019). We are also grateful for the support of Ph.D. student Sebastien Mambou in consultations regarding application aspects.

References

1. Selase, A.E., Agyimpomaa, D.E., Selasi, D.D., Hakii, D.M.: Precipitation and rainfall types with their characteristic features. J. Nat. Sci. Res. **5**(20), 1–3 (2015). www.iiste.org
2. Sattari, M.T., Rezazadeh-Joudi, A., Kusiak, A.: Assessment of different methods for estimation of missing data in precipitation studies. Hydrol. Res. **48**(4), 1032–1044 (2017)
3. Kuok, K.K., Harun, S., Shamsudin, S.M.: Global optimization methods for calibration and optimization of the hydrologic Tank model's parameters. Can. J. Civ. Eng. **1**(1), 2–14 (2010)
4. Kuok, K.K., Kueh, S.M., Chiu, P.C.: Bat optimisation neural networks for rainfall forecasting: case study for Kuching city. J. Water Clim. Change (2018)
5. Valizadeh, N., El-Shafie, A., Mirzaei, M., Galavi, H., Mukhlisin, M., Jaafar, O.: Accuracy enhancement for forecasting water levels of reservoirs and river streams using a multiple-input-pattern fuzzification approach. Sci. World J. **2014** (2014)
6. Yaseen, Z.M., El-Shafie, A., Afan, H.A., Hameed, M., Mohtar, W.H., Hussain, A.: RBFNN versus FFNN for daily river flow forecasting at Johor River, Malaysia. Neural Comput. Appl. **27**(6), 1533–1542 (2016)
7. Ismail, W.N., Zin, W.Z., Ibrahim, W.: Estimation of rainfall and stream flow missing data for Terengganu, Malaysia by using interpolation technique methods. Malay. J. Fundam. Appl. Sci. **13**(3), 213–217 (2017)
8. Suhaila, J., Sayang, M.D., Jemain, A.A.: Revised spatial weighting methods for estimation of missing rainfall data. Asia-Pac. J. Atmos. Sci. **44**(2), 93–104 (2008)
9. Eskelson, B.N., Temesgen, H., Lemay, V., Barrett, T.M., Crookston, N.L., Hudak, A.T.: The roles of nearest neighbor methods in imputing missing data in forest inventory and monitoring databases. Scand. J. For. Res. **24**(3), 235–246 (2009)

10. Kamaruzaman, I.F., Zin, W.Z., Ariff, N.M.: A comparison of method for treating missing daily rainfall data in Peninsular Malaysia. Malay. J. Fundam. Appl. Sci. **13**(4–1), 375–380 (2017)
11. Ferrari, G.T., Ozaki, V.: Missing data imputation of climate datasets: implications to modeling extreme drought events. Revista Brasileira de Meteorologia **29**(1), 21–28 (2014)
12. Teegavarapu, R.S., Chandramouli, V.: Improved weighting methods, deterministic and stochastic data-driven models for estimation of missing precipitation records. J. Hydrol. **312** (1–4), 191–206 (2005)
13. Dastorani, M.T., Moghadamnia, A., Piri, J., Rico-Ramirez, M.A.: Application of ANN and ANFIS models for reconstructing missing flow data. Environ. Monit. Assess. **166**, 421–434 (2010)
14. Mispan, M.R., Rahman, N.F., Ali, M.F., Khalid, K., Bakar, M.H., Haron, S.: Missing river discharge data imputation approach using artificial neural network. J. Eng. Appl. Sci. **10**(22) (2015)
15. Ding, Y., Ross, A.: A comparison of imputation methods for handling missing scores in biometric fusion. Pattern Recogn. **45**(3), 919–933 (2012)
16. Kuok, K.K., Harun, S., Shamsuddin, S.M., Chiu, P.C.: Evaluation of daily rainfall-runoff model using multilayer perceptron and particle swarm optimization feed forward neural networks. J. Environ. Hydrol. **18**(10), 1–6 (2010)
17. Oba, S., Sato, M.A., Takemasa, I., Monden, M., Matsubara, K.I., Ishii, S.: A Bayesian missing value estimation method for gene expression profile data. Bioinformatics **19**(16), 2088–2096 (2003)
18. Little, R.J., Rubin, D.B.: Statistical Analysis with Missing Data. Wiley, Hoboken (2014)
19. Zainuri, N.A., Jemain, A.A., Muda, N.: A comparison of various imputation methods for missing values in air quality data. Sains Malaysiana **44**(3), 449–456 (2015)

On Using "Stochastic Learning on the Line" to Design Novel Distance Estimation Methods for *Three-Dimensional Environments*

Jessica Havelock[1], B. John Oommen[1,2(✉)], and Ole-Christoffer Granmo[2]

[1] School of Computer Science, Carleton University, Ottawa, Canada
oommen@scs.carleton.ca

[2] Centre for Artificial Intelligence Research, University of Agder, Grimstad, Norway

Abstract. We consider the unsolved problem of Distance Estimation (DE) when the inputs are the x and y coordinates (i.e., the latitudinal and longitudinal positions) of the points under consideration, *and* the elevation/altitudes of the points specified, for example, in terms of their z coordinates (3DDE). The aim of the problem is to yield an accurate value for the real (road) distance between the points specified by *all the three* coordinates of the cities in question (This is a typical problem encountered in a GISs and GPSs.). In our setting, the distance between *any pair of cities* is assumed to be *computed* by merely having access to the coordinates and *known* inter-city distances of a *small subset* of the cities, where these are also specified in terms of their 3D coordinates. The 2D variant of the problem has, typically, been tackled by utilizing parametric functions called "Distance Estimation Functions" (DEFs). To solve the 3D problem, we resort to the Adaptive Tertiary Search (ATS) strategy, proposed by Oommen *et al.*, to affect the learning. By utilizing the information provided in the 3D coordinates of the nodes and the true road distances from this *subset*, we propose a scheme to estimate the inter-nodal distances. In this regard, we use the ATS strategy to calculate the best parameters for the DEF. While "Goodness-of-Fit" (GoF) functions can be used to show that the results are competitive, we show that *they are rather not necessary to compute the parameters*. Our results demonstrate the power of the scheme, even though we completely move away from the traditional GoF-based paradigm that has been used for four decades. Our results conclude that the 3DDE yields results that are far superior to those obtained by the corresponding 2DDE.

Keywords: Road distance estimation · Estimating real-life distances · Learning Automata · Adaptive Tertiary Search · Stochastic Point Location

F. Wotawa et al. (Eds.): IEA/AIE 2019, LNAI 11606, pp. 39–49, 2019.
https://doi.org/10.1007/978-3-030-22999-3_4

1 Introduction

In this paper, we consider the problem of estimating the real (road) distances between points (cities), where the inputs are the x and y coordinates of the points under consideration, and their z coordinates. We refer to this problem as 3DDE. This problem is much more complicated than the corresponding 2D problem that has been studied for a few decades (the 2DDE), because in hilly terrains, estimating the road distances is more complex than on flat domains.

Historically, the input to Distance Estimation (DE) problems are, typically, the start and end locations in the form of x and y co-ordinates of the locations in the Cartesian plain, or the latitude and longitude in the geographic region. However, determining the actual "road distances" (the physical distance to be traveled on the "roads" built in the community) for an area, is more challenging. These road distances, also synonymously known as traveling distances or "true" distances, can depend on the network, the terrain, the geographical impediments like rivers or canyons, and of course, the direct distance between the respective points – which serves as a lower bound for the "true" distances. The 2DDE and 3DDE problems involve finding the best estimator for these true distances.

This problem has been studied for over four decades, and its solutions have been used in many practical applications, such as in developing vehicle scheduling software, vehicle routing, and in partitioning districts for fire-fighters [1,2,7].

Legacy Methods: Distance Estimation Functions. To historically do DE, one typically resorts to Distance Estimating *Functions* (DEFs) [5] which are simultaneously good estimators, characterized by low computations. All these DEFs involved parameters whose values are obtained by a "training" phase to best fit the data of the system being characterized, with some "true" known *a priori* road distances. The accuracy of the estimations depends on the DEF, the system and the available data.

Our Proposed Approach. In this paper, we apply a new method for determining the DEF, earlier pioneered in [3], namely the Adaptive Tertiary Search (ATS) [6]. The ATS uses Learning Automata (LA) to perform a stochastic search "on a line" to determine the parameter sought for. Our most "daring" step is to *completely move away from invoking Goodness-of-Fit (GoF) criteria for the DEFs* as has been done for more than four decades [8,9].

2 Distance Estimation: Core Concepts

DE is a topic that has been extensively studied for more than four decades. Due to space limitations, it is impossible to survey the field here. But we refer the reader to [3] and [4], where it has been surveyed in greater detail.

DE is, typically, done by determining the appropriate DEF, which is a mapping from $R^d \times R^d$ to R. It returns the estimate of the true distance. The inputs to the DEF are the locations of the two points, and it produces an estimate of the distance between them by incorporating the set of parameters into the DEF. This set of parameters is *learnt* so that the DEF best represents the space.

Definition 1. *A* ***Distance Estimation Function (DEF)*** *is defined as a function* $\pi(P_1, P_2|\Lambda) : R^d \times R^d \longrightarrow R$*, in which* $P_1 = \langle x_1, x_2, ..., x_d \rangle$ *and* $P_2 = \langle y_1, y_2, ..., y_d \rangle$ *are points in* R^d*, and* Λ *is a set of parameters characterizing* π*, learnt using a set of training points with known true inter-point distances.*

This set, Λ is, typically, learnt by minimizing a GoF function which is used to measure how well a region is represented by the DEF. Central to DE is the concept of GoF functions which are measures of how good a DEF estimates the true (but unknown) distances. Several GoF functions have been utilized in the literature, and the most common one is the sum of Square Deviation (SD).

The most common types of DEFs are those based on the family of L^p norms, traditionally used for computing distances:

$$L_p(X) = \left(\sum_{i=1}^{n} (|x_i|^p \right)^{1/p} . \tag{1}$$

The various well-known L^p norms have been used as stepping stones to design DEFs as seen in [5]. The input to these functions are the co-ordinates of the input vectors, X and Y. These DEFs are first trained on the subset of the co-ordinates of the cities and *their* known (region-based) true distances, and this yields the "best" parameters for the DEF given the training data. Thereafter, the DEF can be used for DE for other unknown inter-city distances.

3 The Adaptive Tertiary Search and Its Use in DE

The solution that we propose for DE is based on a scheme relevant to the Stochastic Point Location (SPL) problem. To formulate the SPL, we assume that there is a Learning Mechanism (LM) whose task is to determine the optimal value of some variable (or parameter), λ. We assume that there is an optimal choice for λ - an unknown value, say $\lambda^* \in [0, 1]$. As in [3], in this paper, we shall use the ATS [6] to solve the DE problem, although we could have used other solutions.

Table 1. The decision table for the ATS scheme.

O^1	O^2	O^3	New sub-interval
Inside	Left	Left	Δ^1
Left	Left	Left	Δ^1
Right	Inside	Left	Δ^2
Right	Left	Left	$\Delta^1 \cup \Delta^2$
Right	Right	Inside	Δ^3
Right	Right	Left	$\Delta^2 \cup \Delta^3$
Right	Right	Right	Δ^3

To determine λ^* within the resolution of accuracy, the original search interval is divided into three equal and disjoint subintervals, Δ^i, where $i = 1...3$. The subintervals are searched using a two-action LA. The LA returns the $\lambda(n)$, the estimated position of λ^* from that subinterval $O^i \in \{Left, Right, Inside\}$. From these, a new search interval is obtained based on the table given in Table 1. This is repeated until the search interval is smaller than the resolution of accuracy. The search interval will yield the required resolution within a finite number of epochs because the size of the search interval is decreasing [6]. After the interval is small enough, the midpoint of the final interval is the estimate for λ^*.

3.1 Updating Search Intervals

Let us first consider the case where the DEF has two parameters, say k and p. The strategy for our search will be to use the ATS to determine the best value for k and p, say k^* and p^*, respectively. It is crucial that the *order* of updating the search intervals in the k and p spaces is considered when determining these multiple parameters. If this is not done correctly, it may result in the premature reduction of a search interval. The order of executing the searching, and the pruning of the intervals must also be maintained while searching for the two parameters, k and p, simultaneously. Thus, all the subintervals must be searched before the intervals are updated simultaneously, as shown in [3] and [4].

The set of LA operate in the same manner as in [6], except for how it deals with the additional parameters. When the LA is learning information about how it should update the value for k, it uses values of p from within its *current* search interval and vice versa. As a result, each LA operates with the knowledge of the *current* search interval of *all* the other parameters, as explained in [3] and [4].

This process of searching for multiple parameters can be done in parallel by assuming that for each learning loop, the other parameter's value is either the maximum or the minimum of its current search interval. This is a consequence of the monotonicity of the DEFs, as discussed in Sect. 3.3.

3.2 The Corresponding LA

Each LA is provided with two inputs, namely the parameter that it is searching for, and all the search intervals, and yields as its output the relative location of the parameter in question. It does this by producing a decision (Left, Right or Inside) based on *its* final belief after communicating with *its* Environment.

The LA starts out with a uniform belief, 50%, for both "Left" and "Right". It then makes a decision based on its current belief. If the decision is "Left", then the LA picks a point in the left half of the interval at random; otherwise (i.e., the decision is "Right") the point is chosen from the right half of the interval. Once the decision is made, the LA asks the Environment for a response. The LA uses the Linear Reward-Inaction (L_{RI}) update scheme, and so the current belief is only updated if the Environment's response is positive.

The LA and the Environment repeat this loop for a large number, say N_∞, of iterations. After they are done communicating, the LA produces its output as per the LA algorithm briefly described below, but omitted here in the interest of space. It is found in [3] and [4]. If the LA's belief of "Right" is greater than $1-\epsilon$, the parameter in question is to the right side of the current search interval, and so its output is "Right". Conversely, if the belief of "Left" is greater than $1-\epsilon$, the LA's final decision is "Left". If neither of these cases emerge, the LA does not have a belief greater than $1-\epsilon$ that the parameter is to the "Right" or "Left", and in this case, the LA decides that the parameter's optimal value is "Inside" the present interval. Again, the entire algorithm is formally given in [3] and [4] (omitted here in the interest of space).

3.3 The Corresponding Environment

Each LA requires feedback from a specific Environment. This feedback informs the LA if it has made the correct decision, i.e., choosing the right or left half of the subinterval. It is easy to obtain this answer because it only involves a single parameter at a time. To further explain this, consider the DEFs below:

$$F(k,p) = k \cdot F_1(X_1, X_2, p), \text{and where,} \tag{2}$$

$$F_1(X_1, X_2, p) = \left(\sum_{i=1}^{d} |x_{1i} - x_{2i}|^p \right)^{1/p}. \tag{3}$$

Although nothing specific can be said about the monotonicity characteristics of $F(k,p)$, we see from Eq. (2) that by virtue of the fact that it is always positive and that it can be factored, it is monotonically *increasing* with k for any fixed value, p. Similarly, from Eq. (3), since $F_1(X_1, X_2, p)$ is not a function of k, it is monotonically *decreasing* with p for any fixed value of k. These properties allow the Oracle to respond accordingly when finding k, and for the corresponding LA to move in the desired direction (i.e., "Left" or "Right") in the space that only involves the single parameter k. The contrary monotonicity properties allow the Oracle to respond according to a corresponding algorithm (*EnvironmentResponseP*, explained in[1] in [3] and [4]) when determining p, and for the corresponding LA to move in the desired direction in the space that involves only p.

4 Testing and Results: 3-Dimensional Environments

As opposed to the theory and results presented in the predecessor paper [3], in this paper, we have considered the problem of DE in three dimensions, i.e., 3DDE. To permit us to do testing of 3DDE, since real-life data is currently not available[2], we have resorted to utilizing "realistic" artificially-constructed data. Data of this type will give us a better understanding of how the DE method

[1] The proofs of the relevant claims are also included in these publications.

[2] The use of a third dimension (the altitude) could be especially beneficial if the region in question has a predominantly hilly or mountainous terrain, as in Perugia, Italy, or Zermatt and Switzerland.

behaves. Using the artificially-constructed data, we have compared the results from the 3DDE using the two dimensional method presented in [3]. From these comparisons, we are able to show which method is superior in these situations.

4.1 Experimental Setup

To compare the 3DDE with the typical two dimensional DE (2DDE), we used the ATS in combination with the weighted L^p DEF chosen because it has been well studied for DE and is very versatile. It can also out-perform many other DEFs in networks of differing structures. Finally, it possesses a simplicity that generally performs well when one considers that it is a DEF with only two variables. The metrics we used were the Normalized Absolute-value Difference (NAD), the sum of Square Deviation (SD), the Relative Absolute-value Difference (RAD) and Expected Percent (EP) errors which are recommended in the literature.

All of the data sets consisted of 100 points, representing the cities. This value was chosen based on the common data set sizes in the literature. The distance between the intermediate points, D, was another parameter that had to be defined. We used the value $D = 0.05$, which was based on the largest rate of change of the terrain and the desired accuracy for the inter-point distances. This inter-point distance was small enough to account for the features on all the surfaces considered in this paper. Further, both the 2D and 3DDE methods used the same data sets. The only difference for the 2DDE method was that it did not incorporate the third dimension.

For both the noiseless and the noisy data, there were three types of surfaces (given below) used to construct the data sets. The first type was a single hill, an example of which is shown in Fig. 1 (left). Five different single Gaussian hill surfaces, $N(0.5, \sigma)$, were constructed to determine the accuracy of the DE of both "easy" and "hard" terrains. The parameters for these surfaces are shown in Table 2. The second type of surface used was a valley. This consisted of two Gaussian hills, $N(0, \sigma_1) + N(1, \sigma_2)$ located at opposite corners of the terrain, which had a 1×1 unit base. An example of this is shown in Fig. 1 (center). Again, the steepness of the valley was varied according to the values in Table 2. The last surface consisted of two Gaussian hills located side-by-side, $N(\mu_1, \sigma_1) + N(\mu_2, \sigma_2)$. This is, clearly, the most complete surface and is shown in Fig. 1

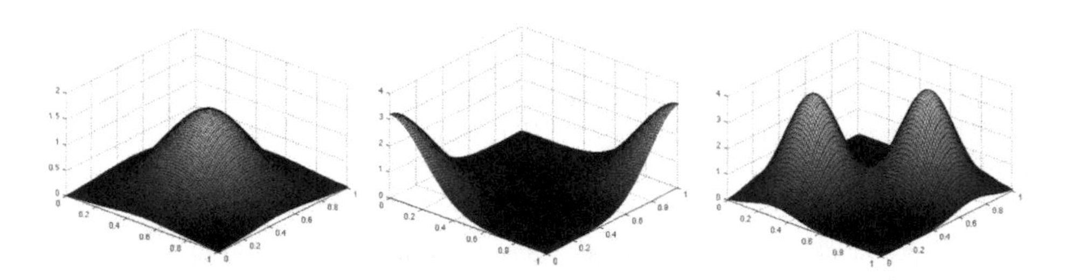

Fig. 1. Example of 3D Surfaces used in our experiments.

(right). The differences between the two hill surfaces are outlined in Table 2. For this scenario, the steepness was also changed.

4.2 Results for Noisy Surfaces

The results of the 3DDE was compared to the 2DDE over 20 runs. Further, the level of noise used was different for each run.

Table 2. Parameters of the 3D testing surfaces used in our experiments.

Surface:	Single hill		Valley		Two hills	
Number	Mean (μ)	Variance (σ^2)	Means (μ_1, μ_2)	Variances (σ_1^2, σ_2^2)	Means (μ_1, μ_2)	Variances (σ_1^2, σ_2^2)
1	(0.5)	(0.3)	(0.0, 1.0)	(0.02, 0.02)	(0.2, 0.8)	(0.09, 0.09)
2	(0.5)	(0.2)	(0.0, 1.0)	(0.05, 0.05)	(0.2, 0.8)	(0.09, 0.07)
3	(0.5)	(0.1)	(0.0, 1.0)	(0.09, 0.09)	(0.2, 0.8)	(0.07, 0.07)
4	(0.5)	(0.09)	(0.0, 1.0)	(0.12, 0.12)	(0.2, 0.8)	(0.07, 0.05)
5	(0.5)	(0.08)	(0.0, 1.0)	(0.15, 0.15)	(0.2, 0.8)	(0.05, 0.05)

Single Hill Surfaces. Table 3 shows the errors for the surfaces containing a single hill. These surfaces are in order from the flattest to the steepest. The first single hill surface had a value of $\sigma^2 = 0.3$. This was a relatively flat hill where the z-component was irrelevant, and as a result, the 2DDE marginally outperformed the 3DDE. The 3DDE method had a slight improvement to about 8.8% when $\sigma^2 = 0.2$, while the 2DDE error increased to over 9% error. When σ^2 was increased to values greater than or equal to 0.1, the 2DDE produced an

Table 3. Results for 20 runs of 2D ATS and 3D ATS on noisy single hill surfaces.

Name σ^2	Hill-1 0.3	Hill-2 0.2	Hill-3 0.1	Hill-4 0.09	Hill-5 0.08
	3D	3D	3D	3D	3D
SD	3.05	3.13	8.01	12.26	23.16
NAD	40.09	38.46	56.36	63.70	77.81
RAD	0.0950	0.0871	0.1369	0.1726	0.2135
EP	0.0922	0.0884	0.1296	0.1464	0.1789
	2D	2D	2D	2D	2D
SD	2.24	3.78	32.89	30.91	79.61
NAD	36.05	44.28	145.42	147.94	209.79
RAD	0.0787	0.0933	0.2722	0.2608	0.3627
EP	0.0829	0.1018	0.3343	0.3401	0.4823

error greater than 26%. For the 3DDE, all the errors were less than 21%. One observes that errors obtained increased with the steepness of the hill.

Valley Surfaces: The results of the 2DDE and 3DDE on the valley surfaces with noisy distances are summarized in Table 4. For the valley surfaces, the 2DDE and the 3DDE have the closest performance on the valley with the steepest sidewalls. The 2DDE had errors of 13%, and the 3DDE had an expected error for an estimated distance and the overall error of 11% and 15% respectively. As the valleys flattened out, the RAD and EP errors for the 2DDE increased to 35% and 40% respectively. Thereafter, the 2DDE improved as the valleys continued to flatten. For the flattest valley, the 2DDE had errors of 18% and 20%. The 3DDE decreased to a maximum of 5%. When $\sigma^2 > 0.02$, the errors were all between 12% and 9%. The reader must observe the superiority of the 3DDE.

Table 4. Results for 20 runs of 2D ATS and 3D ATS on various noisy valley surfaces.

Name σ^2	Valley-1 0.02	Valley-2 0.05	Valley-3 0.09	Valley-4 0.12	Valley-5 0.15
	3D	3D	3D	3D	3D
SD	21.63	5.93	7.39	4.29	3.56
NAD	49.40	46.24	48.27	45.99	42.40
RAD	0.1469	0.1128	0.1196	0.1012	0.0915
EP	0.1136	0.1063	0.1110	0.1057	0.0975
	2D	2D	2D	2D	2D
SD	17.23	31.42	53.60	26.22	12.80
NAD	57.95	146.90	174.09	136.31	86.96
RAD	0.1318	0.3036	0.3593	0.2863	0.1784
EP	0.1332	0.3377	0.4002	0.3133	0.1999

Table 5. Results for 20 runs of 2D ATS and 3D ATS on noisy surfaces with two hills.

Name (σ_1^2, σ_2^2)	Two Hills-1 $(0.09, 0.09)$	Two Hills-2 $(0.09, 0.07)$	Two Hills-3 $(0.07, 0.07)$	Two Hills-4 $(0.07, 0.05)$	Two Hills-5 $(0.05, 0.05)$
	3D	3D	3D	3D	3D
RAD	0.1728	0.1635	0.1637	0.2238	0.2546
EP	0.1553	0.1552	0.1529	0.1957	0.2239
	2D	2D	2D	2D	2D
RAD	0.3236	0.3810	0.6673	0.7219	0.8441
EP	0.4029	0.4738	0.7499	0.8736	1.0960

Two Hill Surfaces: Table 5 shows the results for the noisy distances on the surface with two hills. The surfaces are arranged, based on the mean height of the two hills from the shortest to the highest.

The 3DDE had RAD and EP errors ranging from 25% to 15%. The terrains with the smallest error contained hills where $\sigma_1^2 = 0.09$, $\sigma_2^2 = 0.07$ and $\sigma_1^2 = 0.07$, $\sigma_2^2 = 0.07$. The largest error of 25% for the 3DDE was caused by estimating distances for the surface with the two steepest hills, $\sigma_1^2 = 0.05$ and $\sigma_2^2 = 0.05$. This surface also forced the 2DDE to produce its largest error of 109%, which is actually unacceptable, and which clearly advocates the use of the 3DDE domain. In fact, the smallest error that the 2DDE produced for the surfaces with two hills was 32%, which was for the flattest hills.

4.3 Discussion

Results for Noisy Surfaces: The discussion, for results for noisy surfaces, will be focused on the RAD errors. These errors, for both the 2DDE and the 3DDE, are plotted against the steepness of the surface in Fig. 2(a)–(c). The flatter surfaces are on the right of each plot while the steeper surfaces are on the left.

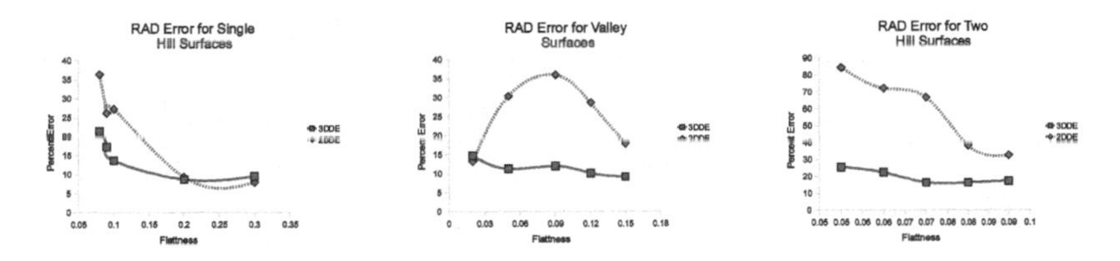

Fig. 2. Plot of RAD errors versus the steepness of the terrain for (a) the noisy single hill data sets, (b) the noisy valley data sets, and (c) the noisy two hill data sets.

Single Hill Surfaces: Both the 3DDE and the 2DDE had the same trend for the single hill surfaces. They both perform better for the flatter surfaces.

Valley Surfaces: The 3DDE had little variance in the errors. They ranged from 15% to 9%, with the steepest valley having the largest errors, and the flattest valley having the lowest errors. All of the noisy 3DDE errors were larger than their noiseless counterparts. The 2DDE did not, however, follow a linear trend. It performed best on the steepest valley, and the flattest valley.

Two Hill Surfaces: Just as for the noiseless data, the surfaces with two hills were the hardest for both methods to produce accurate estimates. Both performed best on the flatter hills, and worst on the steepest.

To get a closer look at where the errors were coming from, Fig. 4 shows the distribution of errors on noiseless and noisy data sets respectively. Each figure contains 500 points. The errors are represented by how red each point is. The black points contain the smallest errors while the red, the highest (Fig. 3).

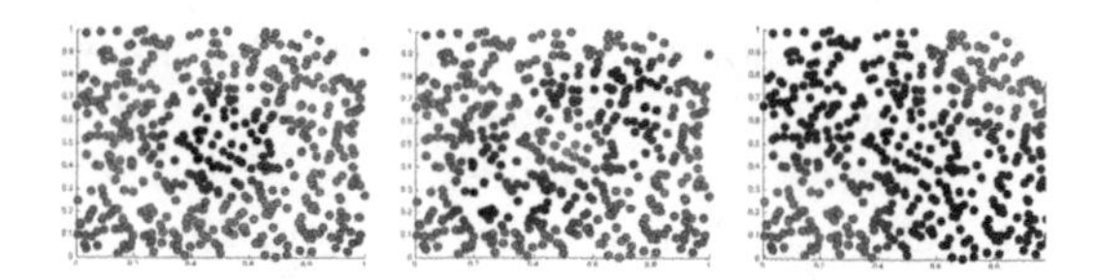

Fig. 3. The distribution of the errors for the noiseless 3D data sets. In 3a (left) the error distribution for a noiseless hill surface is shown, 3b (center) shows the error distribution for two hills, and 3c (right) shows the distribution for a valley. (Color figure online)

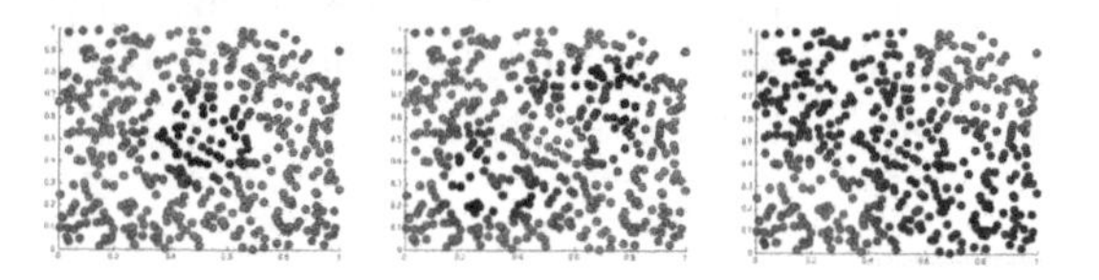

Fig. 4. The distribution of the errors for the noisy 3D data sets. In 4a (left) the error distribution for a noisy hill surface is shown, 4b (center) shows the error distribution for two hills, and 4c (right) shows the distribution for a valley. (Color figure online)

The trends for both the noiseless and noisy data sets were the same. For surfaces with one and two hills, the ATS favored the tops of the hills. As opposed to this, for the valley surfaces, the ATS produces the best estimates at the bottom of the valley.

5 Conclusions

In this paper, we have considered the Distance Estimation (DE) problem that has been studied for almost four decades. This paper has generalized the problem studied to-date to also permit the input to be the elevation/altitudes of the points – their z coordinates (3DDE). To the best of our knowledge, this is the first and pioneering attempt to solving 3DDE, the DE problem in which the inputs are all the three coordinates of the cities. Further, we have assumed that the distance between *any pair of cities* is to be *computed* by merely having access to the coordinates and *known* inter-city distances of a *small subset* of the cities.

Our solution departs from the legacy methods in that we ignore the use of so-called "Goodness-of-Fit" (GoF) functions. Rather, we have used the field of Learning Automata (LA) and in particular, the Adaptive Teriary Search (ATS) used to solve the Stochastic Point Location (SPL) problem. The results obtained also lead us to conclude that the pioneering application of the ATS to the 3DDE yields results that are far superior to those obtained by the corresponding 2DDE.

References

1. Brimberg, J., Love, R.F., Walker, J.H.: The effect of axis rotation on estimation. Eur. J. Oper. Res. **80**, 357–364 (1995)
2. Erkut, H., Polat, S.: A simulation model for an urban fire fighting system. OMEGA - Int. J. Manag. Sci. **20**(4), 535–542 (1992)
3. Havelock, J., Oommen, B.J., Granmo, O.-C.: Novel distance estimation methods using "stochastic learning on the line" strategies. IEEE Access **6**, 48438–48454 (2018). https://doi.org/10.1109/ACCESS.2018.2868233
4. Havelock, J., Oommen, B.J., Granmo, O.-C.: Novel distance estimation methods for 3D spaces using "stochastic learning on the line" strategies. Unabridged version of this paper (2019)
5. Love, R.F., Morris, J.G.: Modelling inter-city road distances by mathematical functions. Oper. Res. Q. **23**(1), 61–71 (1972)
6. Oommen, B.J., Raghunath, G.: Automata learning and intelligent tertiary searching for stochastic point location. IEEE Trans. Syst. Man Cybern. **28**(6), 947–954 (1998)
7. Oommen, J., Altınel, I.K., Aras, N.: Discrete vector quantization for arbitrary distance function estimation. IEEE Trans. Syst. Man Cybern. **28**(4), 496–510 (1998)
8. Ortega, F.A., Mesa, J.A.: A methodology for modelling travel distances by bias estimation. Sociedad de Estadistica e lnvestigacion Operativa **6**(2), 287–311 (1998)
9. Uster, H., Love, R.F.: Application of a weighted sum of order p to distance estimation. IIE Trans. **33**(8), 675–684 (2001)

Predicting the Listing Status of Chinese Listed Companies Using Twin Multi-class Classification Support Vector Machine

Sining Zhao and Hamido Fujita(✉)

Faculty of Software and Information Science, Iwate Prefectural University, Iwate 0200693, Japan
g231p201@s.iwate-pu.ac.jp, HFujita-799@acm.org

Abstract. Multi-class classification problem is research challenge in many applications. Listing companies' statuses are signals on different risk levels in China's stock markets. The prediction of the listing statuses is complex problem due to imbalance in the data, due to different values and features. In the literature when the list status is divided into two categories for simple measurements using binary classification model, accurate risk management cannot achieved correctly. In this work, we have used SMOTE and wrapper feature selection to reprocess data. Accordingly, we have proposed an algorithm named as Twin-KSVC (twin multi-class support vector machine) which is used for multi-class classification problem by "1-versus-1-versus-rest" structure. Our experiments tested on large sample of data set; show that we could achieve better performance, in comparison with other approach. We have tested our algorithm on different strategies of feature selection for comparison purposes.

Keywords: Multi-class · Classification · Listing-status prediction · Twin-KSVC · SMOTE · Wrapper feature selection

1 Introduction

The Chinese stock market was reopened in 1990. It is a non-profit organization directly administered by the China Securities Regulatory Commission (CSRC). Much of the regulatory framework for the listing of equity issues is administrative, instead of market-based. When a listed company has unusual financial or other specific conditions, the company's stock will receive special treatment (ST) label as a risk warning label [1]. In our study, there are four listing-status groups which are denoted by A, B, D, and X, respectively. A is considered as a normal company; B is considered as a company with risk warning; D is considered as a company with another risk warning; X is considered as a delisted company. Correctly predicting the listing status of a listed company is very important for the company's stakeholders. Our goal is to predict changes in the company's listing status to help investors manage their stock portfolio risk and help creditors, suppliers, and customers accurately assess the company's credit risk [2–4].

F. Wotawa et al. (Eds.): IEA/AIE 2019, LNAI 11606, pp. 50–62, 2019.
https://doi.org/10.1007/978-3-030-22999-3_5

For correctly predicting the listing status of a listed company, we need to establish classification models in machine learning. SVM [5] is aa supervised learning method that is often used for binary classification problems. SVM solves quadratic programming problems (QPPs), ensuring that an optimal solution is obtained. It is well known that the complexity of the usual SVM is O(n^3), where n is the total size of training data. To reduce the computational cost of SVM, Jayadeva et al. proposed a twin support vector machine (TWSVM) [6], which solves two smaller-sized QPPs instead of a complex QPP in a traditional SVM. Thus, the running time of TWSVM is four times faster than the usual SVM. As the data used in our paper, there are four listing-status groups need to classify. The twin multi-class support vector machine (Twin-KSVC) is a novel machine learning algorithm based in SVM, which aims at finding two non-parallel planes for multi-class. Since the data used is imbalanced data which has many features, we need to preprocess the data to reduce the feature space and time cost [7–13]. In this study, SMOTE [14, 15] and wrapper feature selection [16–19] are used for data preprocessing.

The paper is organized as follows: Sect. 1 is the introduction. Section 2 briefs on SMOTE, feature selection, support vector machine, twin support vector machine, and twin support vector machine, and give the corresponding algorithm. Section 3 introduces the proposed method with an experimental framework and corresponding algorithm. Section 4 is the experiment of actual data. Conclusion and future work are given in Sect. 5.

2 Related Research

In this section, we give a review for SMOTE, feature selection, support vector machine, twin support vector machine, and twin support vector machine formulations.

2.1 SMOTE

Preprocess data to reduce the feature space and time cost is important task in data analytics. SMOTE and wrapper feature selection are used for data preprocessing, in this paper. Over-sampling in data analysis are techniques used to adjust the class distribution of data set. And the most common technique is known as Synthetic Minority Over-sampling Technique (SMOTE) which is an improved scheme based on the random oversampling algorithm. The minority class is over-sampled by creating synthetic examples rather than by simple copy [20].

The algorithm flow is as follow:

Algorithm 1. SMOTE

Input: Training set X, the number of nearest neighbors k.
Output: Set of synthetic samples x_{new}

1. For each minority sample x, the distance from all samples in the minority sample set is calculated by Euclidean distance to obtain its k-nearest neighbors.
2. Amount of synthetic samples N% is set according to the sample imbalance ratio.
3. For each minority sample x, $\overline{x}$ neighbor samples are randomly selected from k-nearest neighbors.
4. Create synthetic samples x_{new} randomly by (1)

$$x_{new} = \mathrm{x} + \mathrm{rand}(0, 1) \cdot (\bar{x} - \mathrm{x}) \tag{1}$$

2.2 Feature Selection

Feature selection is a process by selecting a subset of original attributes to reduce the feature space, optimally. Feature selection includes the three main categories: filters, wrappers and embedded methods. Filter methods can process high-dimensional data sets easily and fast. However, these methods ignore the interaction with the classifier and feature dependencies. Wrapper methods include the interaction between features and the classifier. Wrapper methods have high computational costs. Embedded methods combine the advantages of the both previous methods. However, not all classifiers can be easily adjusted to embed the search for the optimal feature subset. In our study, we utilize sequential forward selection (SFS) which is a simple searching approach of wrapper to search for the optimal feature subset with the minimum error on the training set from the classification model.

Sequential feature selection algorithms are a family of greedy search algorithms that are used to reduce an initial d-dimensional feature space to a k-dimensional feature subspace where k < d. The algorithm flow is as follow:

Algorithm 2. SFS

Input: The empty set N_0, d-dimensional feature x
Output: New selected features N_k

1. Start with the empty set $\mathrm{N}_0 = \{\emptyset\}$
2. Select the next best feature $x^* = \arg\max_{x \notin \mathrm{N}_i} J(\mathrm{N}_i + x)$
3. Update $\mathrm{N}_{i+1} = \mathrm{N}_i + x^*; i = i + 1$
4. Go to 2 end for i = k is met.

2.3 Support Vector Machine

SVM is a new machine learning technique which solves QPPs, ensuring that an optimal solution is obtained. SVM is primarily a classier method that performs classification tasks by constructing hyper-planes in a multidimensional space that separates cases of different class labels.

Consider a binary classification problem, given a training set T = $\{(x_1, y_1), (x_2, y_2), \cdots, (x_l, y_l)\}$, where $x_i \in R^n$ and $y_i \in \{-1, 1\}$. The hyper-plane is denoted as $x^T w + b = 0$, in order to get high accuracy and maximum classification margin, it can be rewritten as: $y_i(w \cdot x_i - b) \geq 1,\ i = 1, 2 \cdots l$. The distance between these two hyper-planes is $2/\|w\|$, so to maximize the distance between the planes we need to minimize $\|w\|$ [21], to reach optimal solution as shown in Eq. (2):

$$\begin{gathered}\min_{w,b} \varphi(w) = \tfrac{1}{2}(w^T w) \\ \text{s.t.} \forall_i : y_i(x_i w + b) \geq 1.\end{gathered} \tag{2}$$

Lagrange function is to achieve the minimum cost function:

$$\mathrm{L}(\mathrm{w}, \mathrm{b}, \lambda) = \frac{1}{2}(w^T w) - \sum_{i=1}^{k} \lambda_i (y_i(x_i w + b) - 1) \tag{3}$$

2.4 Twin Support Vector Machine

Twin support vector machine (TWSVM) generates two nonparallel planes for binary classification. The goal of TWSVM is to optimize a pair of QPPs by generating two nonparallel planes that each hyper-plane is nearer to one class as far as possible from the other class. Therefore, TWSVM solves a pair of smaller QPPs instead of a complex QPP in a traditional SVM. The complexity of the usual SVM is no more than O(n^3), TWSVM solves two problems, if each of data is approximately equal to n/2, and the complexity of TWSVM is $\mathrm{O}(2 \cdot (n/2)^\wedge 3) = \mathrm{O}(\mathrm{n}^\wedge 3/4)$. Thus, the running time of TWSVM is four times faster than the usual SVM.

Consider a binary classification problem, training set is redefine as $\mathrm{A} = (\mathrm{A}_+, e_1)$ and $\mathrm{B} = (\mathrm{B}_-, e_2)$,which are expressed as positive and negative class data samples, respectively. The core is to construct two nonparallel hyper-planes in n-dimension input space: $x^T w_1 + b_1 = 0$ and $x^T w_2 + b_2 = 0$.

The two nonparallel hyper-planes are obtained by solving the following pair of QPPs:

$$\begin{aligned}&\min_{w_1, b_1, \varepsilon} \tfrac{1}{2}\|Aw_1 + b_1\|^2 + c_1 e_2^T \varepsilon \\ \text{s.t.}\quad & -(Bw_1 + e_2 b_1) + \varepsilon \geq e_2,\ \varepsilon \geq 0\end{aligned} \tag{4}$$

and

$$\begin{aligned} \min_{w_2,b_2,\eta} & \ \frac{1}{2}\|Bw_2+b_2\|^2+c_2e_1^T\eta \\ \text{s.t.} & \ -(Aw_2+e_1b_2)+\eta\geq e_1,\ \eta\geq 0 \end{aligned} \tag{5}$$

where c_1, c_2 are penalty parameters. ε, η are slack variables. e_1, e_2 are vectors of ones of appropriate dimensions.

$$\text{Class i} = \arg\min_{k=1,2}\left|x^T w_i + b_i\right| \tag{6}$$

The test data sample is calculated from two nonparallel hyper-plane by (6) and is assigned to the class from which its distance is closer.

2.5 Twin Multi-class Support Vector Machine

Compared to TWSVM, the twin multi-class support vector machine (Twin-KSVC) can handle multi-class by generating two nonparallel planes for two certain classes of multi-class data. Each nonparallel planes is nearer to one class and is as far as possible from the remaining class.

Training set is redefine as $\mathrm{A} = (\mathrm{A}_+, e_1)$, $\mathrm{B} = (\mathrm{B}_-, e_2)$ and $\mathrm{C} = (\mathrm{C}, e_3)$. which are expressed as positive and negative class data samples and remaining class. The two nonparallel hyper-planes are: $x^T w_1 + b_1 = 0$ and $x^T w_2 + b_2 = 0$.

The two nonparallel hyper-planes are obtained by solving the following pair of QPPs:

$$\begin{aligned} \min_{w_1,b_1,\varepsilon,\eta} & \ \frac{1}{2}\|Aw_1+b_1\|^2+c_1e_2^T\varepsilon+c_2e_3^T\eta \\ \text{s.t.} & \ -(Bw_1+e_2b_1)+\varepsilon\geq e_2, \\ & -(Cw_1+e_3b_1)+\eta\geq e_3(1-\epsilon), \\ & \varepsilon\geq 0,\ \eta\geq 0 \end{aligned} \tag{7}$$

and

$$\begin{aligned} \min_{w_2,b_2,\varepsilon^*,\eta^*} & \ \frac{1}{2}\|Bw_2+b_2\|^2+c_3e_1^T\varepsilon^*+c_4e_3^T\eta^* \\ \text{s.t.} & \ (Aw_2+e_1b_2)+\varepsilon^*\geq e_1, \\ & (Cw_2+e_3b_2)+\eta^*\geq e_3(1-\epsilon), \\ & \varepsilon^*\geq 0,\ \eta^*\geq 0 \end{aligned} \tag{8}$$

where c_1, c_2, c_3, c_4 are penalty parameters. ε, η, ε^*, η^* are slack variables. e_1, e_2, e_3 are vectors of ones of appropriate dimensions.

To solve this problem by introducing the Lagrange function:

$$\max_{\alpha} \; e_4^T \alpha^T R_r (S^T S)^{-1} R_r^T \alpha \quad \text{s.t.} \quad 0 \leq \alpha \leq H, \tag{9}$$

and

$$\max_{\beta} \; e_5^T \beta^T S_r (R^T R)^{-1} S_r^T \beta \quad \text{s.t.} \quad 0 \leq \beta \leq M, \tag{10}$$

where $S = [A\ e_1]$, $R = [B\ e_2], D = [C\ e_3]$, $R_r = [R; D]$, $S_r = [S; D]$,

$H = [c_1 e_2; c_2 e_3]$, $M = [c_3 e_1; c_4 e_3]$, $e_4 = [e_2; e_3(1-\epsilon)]$, $e_5 = [e_1; e_3(1-\epsilon)]$.

We can obtain the augmented vector by (11) and (12):

$$\begin{bmatrix} w_1 \\ b_1 \end{bmatrix} = -(S^T S)^{-1} R_r^T \alpha \tag{11}$$

and

$$\begin{bmatrix} w_2 \\ b_2 \end{bmatrix} = (R^T R)^{-1} S_r^T \beta \tag{12}$$

For a new test data x_i, we label it by the following decision function:

$$f(x_i) = \begin{cases} 1, & w_1^T x_i + b_1 > -1 + \epsilon \\ -1, & w_2^T x_i + b_2 < 1 - \epsilon \\ 0, & else \end{cases}$$

3 Proposed Method

3.1 Experimental Framework

Our framework is shown in Fig. 1. The row data is separated into training data and testing data.

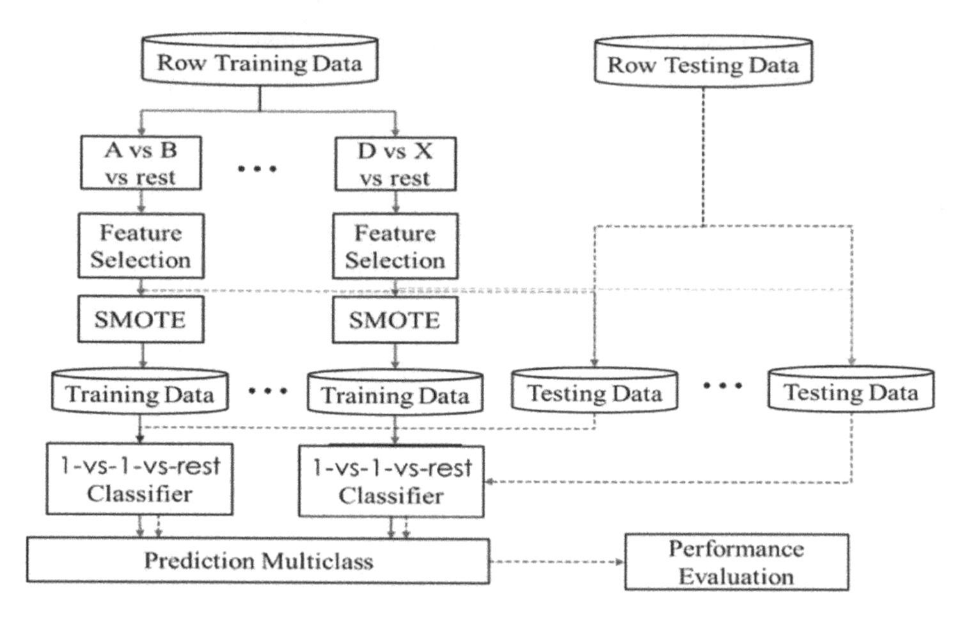

Fig. 1. The framework of training and testing on multi-class classification models.

3.2 Algorithm

We obtain the training data and testing data algorithm based on the framework of training and testing the multi-class classification models.

The algorithm flow is as follow:

Algorithm 3. Training data with 1-vs-1-vs-rest Twin-KSVC

Input: Training data set X_i with N features, labels y_i for i $\in (1,2,\cdots,k)$
Output: k(k-1)/2 classifiers

1. Transform the training data to k(k-1)/2 1-vs-1-vs-rest new classes.
2. Rank the N feature using feature selection in each class, and select the top-ranked N^* features based on ranking.
3. Use SMOTE to balance data in each class.
4. Compute the k(k-1)/2 classifiers using Twin-KSVC.

Algorithm 4. Predict testing data

Input: Testing data set X_i^* with N features, labels y_i^* for i $\in (1,2,\cdots,k)$, k(k-1)/2 N^*rankings
Output: predict data $y_i^{'}$ and performance evaluation

1. Generate k(k-1)/2 testing data with different N^*rankings.
2. Compute the k(k-1)/2 predicted testing data and get total votes for each class.
3. The given testing point xi will be assigned to the label that gets the most votes.
4. Compare predicted results $y_i^{'}$ with actual results y_i^* to obtain performance evaluation

---

4 Experiment

4.1 Dataset

We use data related to China Stock Market and Accounting Research Database (CSMARD) from the GTA database. There are a total of 18619 company-year observations dating from 1999 to 2011. The number of observations of different listing statuses in each observed years is shown in Fig. 1. And each observation retains 208 features and one class label.

There are four different listing statuses in the dataset which include "A", "B", "D"

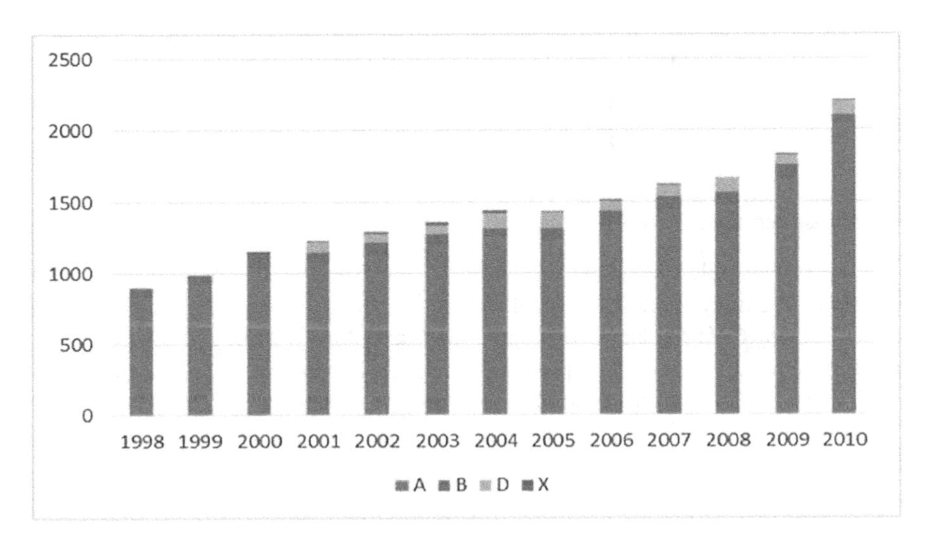

Fig. 2. The number of observations of different listing statuses in each observed years.

and "X". Most of the companies which are in listing status "A" generally tend to continue to be in the same status instead of being changed to another status. And most of the companies in listing status "B" also generally continue to be in the same status but some of them rise to "A" due to improved business conditions. The companies in listing status "D" may continue to be in the same status or rise to "A" or "B" but few of them may slip to "X". Through Fig. 2, we know that the data is imbalanced multiclass data. And "X" status is only a small percentage of the overall data. Therefore, it is extremely difficult to correctly predict "X".

4.2 Experimental Settings

The dataset is separated into a training set and a testing set. The training set includes the data from 1998 to 2006, and the testing set includes the data from 2007 to 2010. Because the data utilized is a multiclass dataset, so we transform the training data with four different classes to six 1-vs-1-vs-rest new classes. Each class is handled by SFS and SMOTE. In our study, we select the number of features taking each value in set {4, 6, 8, 10}. The average accuracy of feature selection is shown in Table 1. The linear

Twin-KSVC and the nonlinear Twin-KSVC both are used respectively for prediction. We utilize Radial basis function kernel (RBF) as the kernel of nonlinear Twin-KSVC. Radial basis function kernel is shown by (13),

$$\mathrm{K}(x, x^*) = \exp(-\frac{\|x - x^*\|^2}{2\sigma^2}) \tag{13}$$

where $\|x - x^*\|^2$ is recognized as the squared Euclidean distance between the two feature vectors where σ is a free parameter.

Table 1. The average accuracy of feature selection

The number of features	4	6	8	10
Average-score	0.8976	0.9048	0.9067	0.9076

4.3 Measurements of the Evaluation

Because we use multi-class data, in addition to the traditional measurements of the evaluation (Accuracy, Precision, Recall & F1), the related formula is shown in (14)–(17). We utilize Cohen's kappa (CK) measures and Macro-averaged F1 (MF1) measures that are used on multi-class problems. Cohen's kappa measure is a statistic which is consistency measurement of the evaluator with qualitative projects. Macro-averaged F1 is arithmetic mean of each class F1 value. CK and MF1 are shown in (18)–(19). The confusion matrix is shown in Table 2.

$$\text{Accuracy} = \frac{TP + TN}{TP + FP + TN + FN} \tag{14}$$

$$\text{Precision} = \frac{TP}{TP + FP} \tag{15}$$

$$\text{Recall} = \frac{TP}{TP + FN} \tag{16}$$

$$F1 = 2 \cdot \frac{\text{Precision} \cdot \text{Recall}}{\text{Precision} + \text{Recall}} \tag{17}$$

$$\begin{gathered} p_e = \frac{(TP + \mathrm{FP}) \cdot (TP + \mathrm{FN}) + (\mathrm{FN} + \mathrm{TN}) \cdot (\mathrm{FP} + \mathrm{TN})}{(TP + FP + TN + FN)^2} \\ \mathrm{CK} = \frac{\text{Accuracy} - p_{e\cdot}}{1 - p_e} \end{gathered} \tag{18}$$

$$\mathrm{MF1} = \frac{1}{k}\sum_{i=1}^{k} F1_i \tag{19}$$

Table 2. Confusion matrix

	Actual positive	Actual negative
Predict positive	TP	FP
Predict negative	FN	TN

4.4 Experiment Result

Table 3 shows the results of Average accuracy, MF1, CK using Twin-KSVC and RBF Twin-KSVC method. Figure 3 shows the comparisons among these computed values. According to these results, we can realize, when the number of selected features is set to six using Twin-KSVC, the average accuracy is maximum with 0.90. Although the minimum value is 0.82, the divergence between the maximum value and the minimum value are not so big. Although the accuracy is somehow satisfactory, it cannot indicate the performance between these frameworks. So we still need to refer to MF1 and CK values because of imbalanced data. The maximum result of MF1 is 0.38 when the number of selected features is set to eight and ten using RBF Twin-KSVC. The maximum result of CK is 0.35 when the number of selected features is set to four using Twin-KSVC. Although the maximum values of MF1 and CK are not in same number of selected features, the divergence between the maximum value and the minimum value are not so big, like the average accuracy shown in Table 3 and Fig. 3.

Table 3. Average accuracy, MF1, CK of Twin-KSVC and RBF Twin-KSVC

The number of features		4	6	8	10
Twin-KSVC	Accuracy	0.88	0.90	0.89	0.89
	MF1	0.33	0.36	0.32	0.33
	CK	0.35	0.34	0.30	0.32
RBF Twin-KSVC	Accuracy	0.88	0.83	0.84	0.82
	MF1	0.37	0.35	0.38	0.38
	CK	0.31	0.32	0.33	0.33

Although the divergence of average accuracy, MF1 and CK are not so big, for the prediction of the results, it is important that prediction of the listing status “D” and “X” of a listed company, is correct. As previously explained, the few of “D” may slip to “X”. So when we get total votes for each class after computing the predicted testing data, the four listing-status groups are predicted in the same fashion.

As we give priority to the sequence in “X”, “D”, “B”, “A”. In Table 4 and Fig. 4, the maximum result of the accuracy of “D” is 77% when the number of selected features is set to six using RBF Twin-KSVC. The maximum result of the accuracy of “X” is 11% when the number of selected features is set to eight and ten using Twin-KSVC. The maximum result of the accuracy of “B” is 23% when the number of selected features is set to four using RBF Twin-KSVC. The maximum result of the accuracy of “A” is 97% when the number of selected features is set to eight and ten using Twin-KSVC.

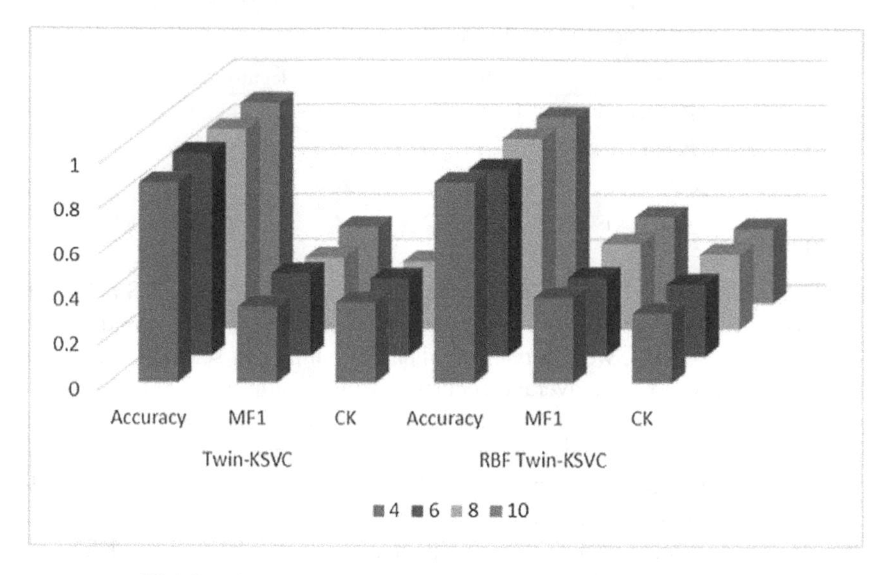

Fig. 3. The Accuracy, MF1, CK, of the four different classes.

This study is based on the work reported in [2], which has used three different types of multiclass classification models with LSSVM to predict listing status in order to achieve better performance. In [2], the maximum result of accuracy of "X" is 44.8% in OVO, the maximum result of accuracy of "D" is 49.8%, the maximum result of accuracy of "B" is 68.6% in OVAPES, and the maximum result of accuracy of "A" is 84.6% in OVAPES. In comparison to [2] we can say our results could achieve better performance using Twin-KSVC.

Table 4. Classification accuracy on four different classes

The number of features		4	6	8	10
Twin-KSVC	A	0.95	0.97	0.97	0.97
	B	0	0.07	0.08	0.16
	D	0.55	0.34	0.1	0.08
	X	0.04	0.04	0.11	0.11
RBF Twin-KSVC	A	0.95	0.87	0.89	0.87
	B	0.23	0.06	0.14	0.15
	D	0.23	0.77	0.62	0.7
	X	0	0.04	0.04	0.04

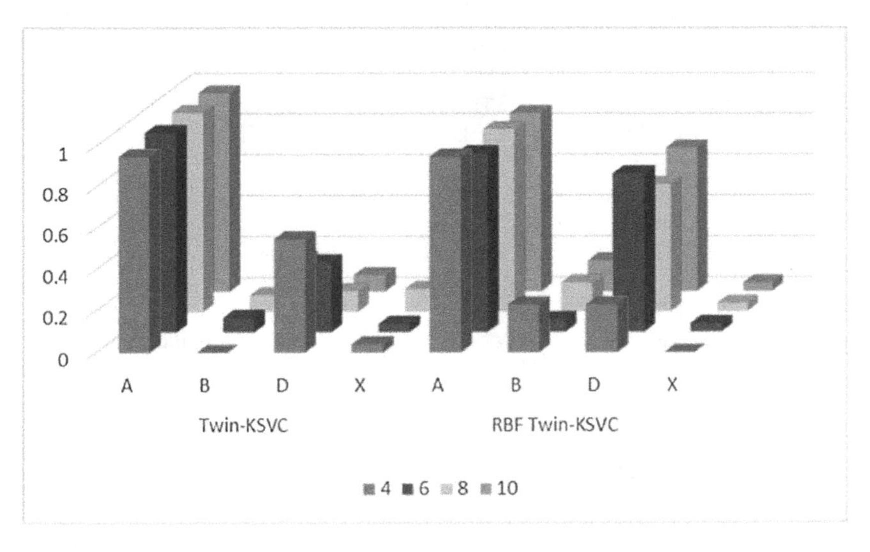

Fig. 4. Classification accuracy on four different classes

5 Conclusion and Future Work

In this study, we consider listing status transitions of Chinese listed companies as a multi-class classification problem. To solve this problem, we use a new algorithm Twin-KSVC which is used for the multi-class classification problem by "1-versus-1-versus-rest" structure. We also use SMOTE and wrapper feature selection to reprocess data. This is because the good quality of the algorithm of Twin-KSVC. In addition, the running time of Twin-KSVC is faster than the LSSVM.

From Fig. 4, it is shown that no framework and setting can achieve maximum classification accuracy for all four listing statuses. According to the experimental results, in comparison to [2], the accuracy of "D" has improved dramatically from 49.8% to 77%, and the accuracy of "A" also has improved slightly from 84.6% to 97%. Because of the difficulty of distinguishing between the characteristics of companies in class "B" and class "D", and few number of "D" may slip to "X". We give priority to "D" when results are predicted to be the same amount to help investors avoid the greater risks of their stock portfolios. In comparison to [2], there are two reasons why the accuracy of the listing status "X" is extremely few. One possible reason is that there are few observations of delisted companies. Another possible reason is that the listed companies may not be given a delisting risk warning on China's stock due to their financial performance but rather because of operations risk and negative audits from accounting agencies. So the prediction of "X" is difficult. In this study, the accuracy of "B" and "X" is not very satisfactory.

In future work, we will use more new data to perfect the classifier on other method like a KNN-weighted multi-class twin support vector machine (KMTSVM) to increase the accuracy of "B" and "D". Also, we think another voting method can be investigated to increase the accuracy.

References

1. Cheng, E., Xia, F., Wang, G.Y.: The special treatment designation and information transmission in the Chinese stock market. Math. Methods Financ. Bus. Adm. 139–151 (2014)
2. Zhou, L., Tama, K.P., Fujita, H.: Predicting the listing status of Chinese listed companies with multi-class classification models. Inf. Sci. **328**, 222–236 (2016)
3. Zhou, L., Si, Y.-W., Fujita, H.: Predicting the listing statuses of Chinese-listed companies using decision trees combined with an improved filter feature selection method. Knowl.-Based Syst. **128**, 93–101 (2017)
4. Zhou, L., Wang, Q., Fujita, H.: One versus one multi-class classification fusion using optimizing decision directed acyclic graph for predicting listing status of companies. Inf. Fusion **36**, 80–89 (2017)
5. Hsu, C.-W., Lin, C.-J.: A comparison of methods for multiclass support vector machines. IEEE Trans. Neural Netw. **13**(2), 415–425 (2002)
6. Jayadeva, R., Khemchandani, S.C.: Twin support vector machines for pattern classification. IEEE Comput. Soc. **29**(5), 905–910 (2007)
7. Panga, X., Xub, C., Xu, Y.: Scaling KNN multi-class twin support vector machine via safe instance reduction. Knowl.-Based Syst. **148**, 17–30 (2018)
8. Angulo, C., Parra, X., Català, A.: K-SVCR. A support vector machine for multi-class classification. Neurocomputing **55**(1–2), 57–77 (2003)
9. Xu, Y., Guo, R., Wang, L.: A twin multi-class classification support vector machine. Cogn. Comput. **5**(4), 580–588 (2013)
10. Ding, S., Yu, J., Zhao, H.: Twin support vector machines based on particle swarm optimization. J. Comput. **8**(9), 2296–2303 (2013)
11. Xu, Y., Wang, L.: A weighted twin support vector regression. Knowl. Based Syst. **33**, 92–101 (2012)
12. Muller, K.R., Mika, S., Ratsch, G., Tsuda, K., Scholkopf, B.: An Introduction to kernel-based learning algorithms. IEEE Trans. Neural Netw. **12**(2), 199–222 (2001)
13. Tomar, D., Agarwal, S.: Twin support vector machine: a review from 2007 to 2014. Egypt. Inform. J. **16**(1), 55–69 (2015)
14. Chawla, N.V., Bowyer, K.W., Hall, L.O., Kegelmeyer, W.P.: SMOTE: synthetic minority over-sampling technique. J. Artif. Intell. Res. **16**, 321–357 (2002)
15. Gao, T.: Hybrid classification approach of SMOTE and instance selection for imbalanced datasets. Iowa State University (2015). https://lib.dr.iastate.edu/etd/14331
16. Panthonga, R., Srivihokb, A.: Wrapper feature subset selection for dimension reduction based on ensemble learning algorithm. Procedia Comput. Sci. **72**, 162–169 (2015)
17. Chandrashekar, G., Sahin, C.G.: A survey on feature selection methods. Comput. Electr. Eng. **40**, 16–28 (2014)
18. Last, M., Kandel, A., Maimon, O.: Information-theoretic algorithm for feature selection. Pattern Recogn. Lett. **22**, 799–811 (2001)
19. Kohavi, R., John, G.H.: Wrappers for feature subset selection. Artif. Intell. **97**, 273–324 (1997)
20. Zhaoa, J., Xu, Y., Fujita, H.: An improved non-parallel Universum support vector machine and its safe sample screening rule. Knowl.-Based Syst. **170**, 79–88 (2019). https://doi.org/10.1016/j.knosys.2019.01.031
21. Zhang, C., et al.: Multi-imbalance: an open-source software for multi-class imbalance learning. Knowl.-Based Syst. https://doi.org/10.1016/j.knosys.2019.03.001

Robust Query Execution Time Prediction for Concurrent Workloads on Massive Parallel Processing Databases

Zhihao Zheng[1], Yuanzhe Bei[2], Hongyan Sun[3], and Pengyu Hong[1](✉)

[1] Brandeis University, Waltham, MA 02453, USA
{zhihaozh,hongpeng}@brandeis.edu
[2] MicroFocus Vertica, Cambridge, MA 02140, USA
yuanzhe.bei@microfocus.com
[3] YinTech Innovation Labs, Waltham, MA 02451, USA
hongyan.sun@yintechlabs.com

Abstract. Reliable query execution time prediction is a desirable feature for modern databases because it can greatly help ease the database administration work and is the foundation of various database management/automation tools. Most exiting studies on modeling query execution time assume that each individual query is executed as serialized steps. However, with the increasing data volume and the demand for low query latency, large-scale databases have been adopting the massive parallel processing (MPP) architecture. In this paper, we present a novel machine learning based approach for building a robust model to estimate query execution time by considering both query-based statistics and real-time system attributes. The experiment results demonstrate our approach is able to reliably predict query execution time in both idle and noisy environments at random levels of concurrency. In addition, we found that both query and system factors are crucial in making stable predictions.

Keywords: Query execution · Machine learning · Concurrent

1 Introduction

Commercial databases hold companies' most critical information and need to be maintained at high-availability with stable-latency at all times. They need to be installed and tuned very carefully (e.g., fault tolerance, knob setting, resource pool setting, etc.). Nevertheless, no matter how comprehensive a database has been tuned, it is still challenging to maintain stable-latency [2]. In real world scenarios, databases receive various queries with a wide range of complexities at any given time. Some of those queries are sub-optimal and even do not make much sense, and may cause a database to execute with unexpected long latency and fail to guarantee service quality [4]. In those cases, database administrators

F. Wotawa et al. (Eds.): IEA/AIE 2019, LNAI 11606, pp. 63–70, 2019.
https://doi.org/10.1007/978-3-030-22999-3_6

usually do not have much control and have no clue how to adjust databases in time to deal with the problem.

The situation becomes worse when a database is loaded with concurrent workloads [11], in which we can frequently observe significant degradation of query performance resulting in serious consequences. For example, a sub-second dashboard query may be slowed down to dozens of seconds or even minutes resulting in poor user experience [16]; a system with periodical ETL process can break if the report generation query failed to finish within the interval of two consecutive ETL jobs [9]; with Software as a Service (SaaS) providers the service-level agreement (SLA) can break and result in severe revenue loss [13]; etc. The degradation can be due to resource contention, as concurrent queries may compete on the same resources - disk I/O, network, memory, threads, etc [17]; the degradation can also be caused by less available resources that lead to lower execution parallelism or spills [6]. The following is a classic disastrous case. When a poorly written query (usually generated by dashboard/business intelligence (BI) tool) is sent to the database unintentionally, the query can consume most resources in the system, especially modern analytical databases with extensive distributed/parallelized query execution engines [18]. The whole database performance can be significantly impacted. Even worse. The query may take very long time to run or "never" finish. A typical solution is to allocate dedicated/cascaded resource pools, where short queries and long-run queries have different dedicated pools [10], while short queries can be evicted to the long-run pool if they does not finish within a predefined time threshold. Nevertheless, such solution can only work well if the workload is to some extent known or predictable. In addition, significant resources can be wasted on those queries that need to be evicted and cascaded to a different pool.

The above problem can be much better solved if query execution time can be somewhat predicted. Many researches have been done in estimating area [1,5, 15,20,21]. Existing works on estimating the running time of concurrent queries either limit the concurrent query workload to a given set of queries or assume that the execution of each query is serialized. However, in modern analytical database platforms built on massively parallel processing (MPP) architecture.

This study takes a first step to explore the possibility to estimate dynamic workload on Vertica analytic platform [12], a column-oriented relational database system built on the MPP architecture, and is commercialized from the C-Store project [19]. Our goal is to achieve reliable estimations of the running time of any arbitrary query at various levels of concurrency. We adopted a data driven approach and applied machine learning to automatically construct a query execution time estimator.

2 Method

In this section, we explain our machine learning approach for building a model to infer query execution time under mixing workloads with high concurrency. We used Random Forest [7,8,14] to build an ensemble regressor whose inputs include the features describing the states of the operating system and the features extracted from the query plan generated by the Vertica Database for each query.

Constructing appropriate features is essential to building a good prediction model. The Vertica database uses a data flow based cost model, due to its simplicity and robustness. A query is decomposed into a set of operators, which are arranged into a operation tree. Each operator is responsible for running a certain algorithm to perform a sub-task. The Vertica database can estimate the amount resources needed by an operator to process a data flow, The resources are classified into four categories: CPU, Memory, Disk, Network. The Vertica database chooses the best query plan based on the previous mentioned cost model. The execution engine of Vertica is multi-threaded and streamed into a pipeline based on the chosen query plan. More than one operator can run simultaneously at any given time. By using the information from the cost model, we are able to learn the relationship between the required resources and its corresponding execution time. However, this approach only works in single-thread settings, since such features only describes the query itself. When multiple queries are run concurrently, the execution of each query will be significantly affected by others. Thus, we introduced the following system level features to characterize the system wise workload:

1. **CPU circles:** the number of CPU circles requested by an operator.
2. **Memory Data:** the memory size required by an operate.
3. **Network Data:** the amount of data need to be transferred via network by an operator.
4. **Disk Data:** the amount of data to be split out to disk by an operator.
5. **Optimizer Cost:** the cost estimated by the Vertica Optimizer.
6. **Input Rows:** the row count of the related tables to be scanned.
7. **Estimated Output Rows:** the estimated row count of output rows, which is calculated based on Cardinality.
8. **CPU utilization:** the percentage of CPU utilization. If there are multiple CPUs, we take the average.
9. **Query Concurrency:** the number of queries running concurrently.

3 Experiments

The experiments were carried out using the Vertica database. The Vertica database ran on a single cluster of three independent nodes. We used all 22 TPC-H query templates [3] to randomly generate 22000 queries (1000 for each query template) with 10 GB standard TPC-H data. The data was collected by running those 22000 queries at each of the 20 concurrency levels. For a given concurrency level of K ($1 \leq K \leq 20$), we concurrently run K sessions and each session continuously executes queries randomly sampled from the 22000 queries generated above. We conducted the experiments in two environments. One of them is a "stand-alone" environment where only our experiments are allowed to run, and the other is a "noisy" environment where there are other unknown applications running simultaneously with our experiments. The running time variations are high for queries generated across all query templates and are also high within queries generated from the same query template. Hence, it is challenging to accurately and robustly estimate query execution time. For all the experiments reported below, we split the data into two parts: 70 % of queries as the training data and the rest as the test data. Below we first discuss the results of using individual features in making predictions (Sect. 3.1), which give us some intuitions about their contributions, and then report the results of using all features (Sect. 3.2).

3.1 Query Execution Time Prediction Using Individual Features

This experiment was carried out in a "stand-alone" environment. Table 1 lists the mean relative errors of using individual features in query execution time prediction. The relative error of a prediction is calculated as $|T_{predict} - T_{truth}|/T_{truth}$ where $T_{predict}$ is the predicted time and T_{truth} is the ground truth. The log relative error is calculated as $log(T_{predict}/T_{truth})$. A zero log relative error means $T_{predict} = T_{truth}$. In Table 1, we divide the test data into five tiers based on their real execution time (e.g., the first tier is the fastest 20% queries, and the last tier is the slowest 20% queries). More specifically, the execution time ranges of the queries are (0, 3.4 s) in Tier 1, [3.4 s, 7.1 s] in Tier 2, [7.1 s, 12.4 s] in Tier 3, [12.4 s, 23.2 s] in Tier 4, and (23.2 s, $+\infty$) in Tier 5. Table 1 show that individual features perform poorly under mixing workloads. The overall relative error across tiers ranges from 0.86 to 2.84, which means the predicted execution time of a 10-minute query can range from 1 min to 38 min. Since we used the Mean Square Error as the cost function to optimize the parameters of the query execution prediction model, the trained model focuses more on slow queries, which contribute more significantly to the overall error than short queries. This leads to extremely unstable predictions on short-run queries. We also observe that the model actually performs best for middle-run-time queries and its performance drops more for queries that need longer time to run.

3.2 Results of Integrating Multiple Features

To deal with the unstable problem with using individual features to make predictions, we explore the approach of integrating multiple features, which takes slightly more computational resources. This experiment was carried out in a "stand-alone" environment. Shown in Table 2, the mean relative error of stand-alone is 0.156, which is about five times better than the best results achievable by using individual features. The variation between different tiers is also significantly smaller, which indicating the model performs robustly for a variety of queries. Figure 1(a) shows that most of the log relative errors are close to zero (zero means a prediction is identical to the ground truth), and few percentages of them are larger than 0.5. Higher concurrency level also makes it more challenging to accurately predict query execution time (see Table 3). To investigate the benefits of using the "system-level" information, we performed an experiment that used only the intrinsic features (i.e., features derived from a

Table 1. The rest results of the models using individual feature in a "stand-alone" environment. The queries are grouped into 5 tiers based on their execution time. See the main text for detailed explanations.

Tier	Optimizer cost	CPU cycles	Memory data	Network data	CPU utilization
1	2.02	2.16	2.07	10.23	5.74
2	1.06	1.05	1.07	2.37	1.47
3	0.36	0.38	0.39	0.80	1.00
4	0.41	0.39	0.41	0.16	0.42
5	0.43	0.43	0.42	0.57	0.49
Overall	0.86	0.88	0.87	2.82	1.82
Tier	Disk data	Input rows	Output rows	Query concurrency	
1	10.35	3.03	3.42	5.35	
2	2.34	1.00	0.99	1.39	
3	0.77	0.37	0.34	0.97	
4	0.15	0.42	0.48	0.43	
5	0.57	0.43	0.45	0.47	
Overall	2.84	1.05	1.13	1.72	

query, excluding Query Concurrency and CPU Utilization). As shown in Table 2, the results are a little better than those of using individual features, however, are significantly worse those those of integrating all features. Thus, including the "system-level" features is essential to the accurate estimation. The above results clearly show that integrating multiple features, as we designed, can significantly improve robustness and accuracy of predictions across a wide spectrum of queries.

Table 2. The test results of the model using all features in a "stand-alone" environment. The queries were grouped into tiers in the same way to Table 1.

Tier	Mean relative error all features	Mean relative error intrinsic only
1	0.169	1.911
2	0.158	0.994
3	0.161	0.379
4	0.156	0.438
5	0.137	0.439
Overall	0.156	0.832

3.3 Noisy Environment

Most previous works conducted their experiments in "stand-alone" environments. Neglecting the effects of system status can lead to inferior performances.

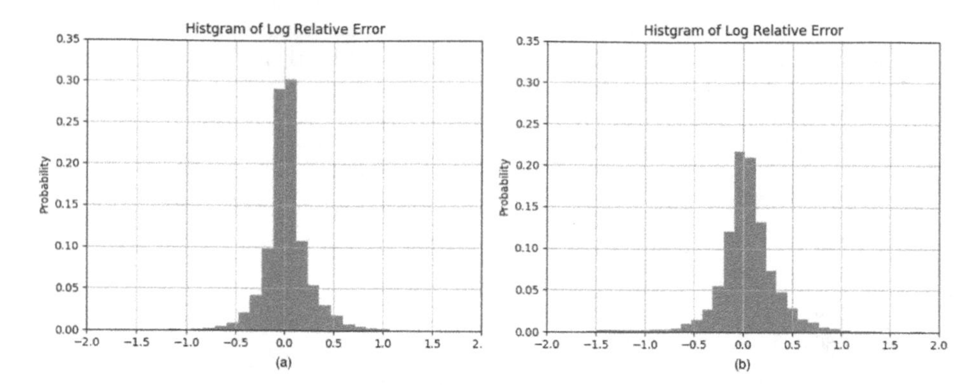

Fig. 1. Histogram of the log relative errors of the model integrating all features in a (a) “stand-alone” and (b) “noisy“environment.

Table 3. The test results of the model integrating all features in a “stand-alone” environment. The results are grouped into their corresponding concurrency levels.

Concurrency	Mean relative error	Concurrency level	Mean relative error level
1	0.024	11	0.181
3	0.119	13	0.212
5	0.094	15	0.560
7	0.229	17	0.331
9	0.203	19	0.260
10	0.186	20	0.307

When a system is busy due to whatever reasons, it will slow down the execution of database. By considering the “system-level” features, our approach can perform more robust than others. To test the robustness of our approach, we did another experiment on a public server where other users may run unknown applications on it. We call this a “noisy environment” because the data is more noisy. Shown in Fig. 1(b), the distribution of the relative error in a “noisy” environment spreads out more than that in a “stand-alone” environment, but still concentrating around 0. One encouraging observation is that the mean relative errors of short-run queries is under 0.4 even in a “noisy” condition, and the performance of our approach on long-run queries is only slightly affected by unknown “noise” in the environment.

4 Conclusion

In this paper, we present a data-driven method that applies Random Forest to build model for predicting query execution time under mixing workloads. Tested on a commercial database Vertica, we demonstrate that our approach is able to robustly and accurately estimate the running time under various levels of workloads and concurrency. We designed a set of features include the intrinsic ones, which can be extracted from each query (or the execution plan of each query), and the system-level ones (e.g., CPU Utilization and Query Concurrency). Our experiments show that each individual feature is not enough for accurately estimation the query execution time because each feature alone does not provide enough information about actual multi-thread query execution. Our experiments also show that the system-level features can be used to significantly stabilize and improve the performance of our approach. Combining the intrinsic and system-level features together in a Random Forest fashion, we are able to model the query execution characteristics well even in a noisy environment. In future, we plan to incorporate our approach into real database systems to help allocate queries into different resource pools based on the estimated query execution time, and detect poorly written queries before execution.

References

1. Agarwal, S., Mozafari, B., Panda, A., Milner, H., Madden, S., Stoica, I.: Blinkdb: queries with bounded errors and bounded response times on very large data. In: Proceedings of the 8th ACM European Conference on Computer Systems, EuroSys 2013, pp. 29–42. ACM, New York (2013). https://doi.org/10.1145/2465351.2465355
2. Chaudhuri, S., Weikum, G.: Rethinking database system architecture: towards a self-tuning RISC-style database system. In: Proceedings of the 26th International Conference on Very Large Data Bases, VLDB 2000, pp. 1–10. Morgan Kaufmann Publishers Inc., San Francisco (2000). http://dl.acm.org/citation.cfm?id=645926.671696
3. Council, T.P.P.: TPC-H benchmark specification, **21**, 592–603 (2008). http://www.tcp.org/hspec.html
4. Dageville, B., Das, D., Dias, K., Yagoub, K., Zait, M., Ziauddin, M.: Automatic SQL tuning in oracle 10G. In: Proceedings of the Thirtieth International Conference on Very Large Data Bases, VLDB 2004, vol. 30, pp. 1098–1109. VLDB Endowment (2004). http://dl.acm.org/citation.cfm?id=1316689.1316784
5. Duggan, J., Cetintemel, U., Papaemmanouil, O., Upfal, E.: Performance prediction for concurrent database workloads. In: Proceedings of the 2011 ACM SIGMOD International Conference on Management of Data, SIGMOD 2011, pp. 337–348. ACM, New York (2011). https://doi.org/10.1145/1989323.1989359
6. Golov, N., Rönnbäck, L.: Big data normalization for massively parallel processing databases. In: Jeusfeld, M.A., Karlapalem, K. (eds.) ER 2015. LNCS, vol. 9382, pp. 154–163. Springer, Cham (2015). https://doi.org/10.1007/978-3-319-25747-1_16
7. Ho, T.K.: The random subspace method for constructing decision forests. IEEE Trans. Pattern Anal. Mach. Intell. **20**(8), 832–844 (1998)
8. Ho, T.K.: Random decision forests. In: Proceedings of the Third International Conference on Document Analysis and Recognition, vol. 1, pp. 278–282. IEEE (1995)
9. Jörg, T., Deßloch, S.: Towards generating ETL processes for incremental loading. In: Proceedings of the 2008 International Symposium on Database Engineering & Applications, pp. 101–110. ACM (2008)

10. Krompass, S., Kuno, H., Wiener, J.L., Wilkinson, K., Dayal, U., Kemper, A.: Managing long-running queries. In: Proceedings of the 12th International Conference on Extending Database Technology: Advances in Database Technology. EDBT 2009, pp. 132–143. ACM, New York (2009). https://doi.org/10.1145/1516360.1516377
11. Kuno, H., Dayal, U., Wiener, J.L., Wilkinson, K., Ganapathi, A., Krompass, S.: Managing dynamic mixed workloads for operational business intelligence. In: Kikuchi, S., Sachdeva, S., Bhalla, S. (eds.) DNIS 2010. LNCS, vol. 5999, pp. 11–26. Springer, Heidelberg (2010). https://doi.org/10.1007/978-3-642-12038-1_2
12. Lamb, A., et al.: The vertica analytic database: C-store 7 years later. Proc. VLDB Endow. **5**(12), 1790–1801 (2012). https://doi.org/10.14778/2367502.2367518
13. Lehner, W., Sattler, K.: Database as a service (DBaaS). In: 2010 IEEE 26th International Conference on Data Engineering (ICDE 2010), pp. 1216–1217 (2010). https://doi.org/10.1109/ICDE.2010.5447723
14. Liaw, A., Wiener, M., et al.: Classification and regression by randomforest. R News **2**(3), 18–22 (2002)
15. Macdonald, C., Tonellotto, N., Ounis, I.: Learning to predict response times for online query scheduling. In: Proceedings of the 35th International ACM SIGIR Conference on Research and Development in Information Retrieval, SIGIR 2012, pp. 621–630. ACM, New York (2012). https://doi.org/10.1145/2348283.2348367
16. Pelkonen, T., et al.: Gorilla: a fast, scalable, in-memory time series database. Proc. VLDB Endow. **8**(12), 1816–1827 (2015). https://doi.org/10.14778/2824032.2824078
17. Rahm, E., Marek, R.: Dynamic multi-resource load balancing in parallel database systems. In: Proceedings of the 21st International Conference on Very Large Data Bases, VLDB 1995, pp. 395–406. Morgan Kaufmann Publishers Inc., San Francisco (1995). http://dl.acm.org/citation.cfm?id=645921.673163
18. Stonebraker, M., et al.: MapReduce and parallel DBMSs: friends or foes? Commun. ACM **53**(1), 64–71 (2010). https://doi.org/10.1145/1629175.1629197
19. Stonebraker, M., et al.: C-store: a column-oriented DBMS. In: Proceedings of the 31st International Conference on Very Large Data Bases. VLDB 2005, pp. 553–564. VLDB Endowment (2005). http://dl.acm.org/citation.cfm?id=1083592.1083658
20. Wu, W., Chi, Y., Zhu, S., Tatemura, J., Hacigümüs, H., Naughton, J.F.: Predicting query execution time: are optimizer cost models really unusable? In: 2013 IEEE 29th International Conference on Data Engineering (ICDE), pp. 1081–1092, April 2013. https://doi.org/10.1109/ICDE.2013.6544899
21. Wu, W., Chi, Y., Hacígümüş, H., Naughton, J.F.: Towards predicting query execution time for concurrent and dynamic database workloads. Proc. VLDB Endow. **6**(10), 925–936 (2013). https://doi.org/10.14778/2536206.2536219

Thompson Sampling Based Active Learning in Probabilistic Programs with Application to Travel Time Estimation

Sondre Glimsdal(✉) and Ole-Christoffer Granmo(✉)

Centre for Artificial Intelligence Research, University of Agder, Grimstad, Norway
{sondre.glimsdal,ole.granmo}@uia.no

Abstract. The pertinent problem of Traveling Time Estimation (TTE) is to estimate the travel time, given a start location and a destination, solely based on the coordinates of the points under consideration. This is typically solved by fitting a function based on a sequence of observations. However, it can be expensive or slow to obtain labeled data or measurements to calibrate the estimation function. Active Learning tries to alleviate this problem by actively selecting samples that minimize the total number of samples needed to do accurate inference. Probabilistic Programming Languages (PPL) give us the opportunities to apply powerful Bayesian inference to model problems that involve uncertainties. In this paper we combine Thompson Sampling with Probabilistic Programming to perform Active Learning in the Travel Time Estimation setting, outperforming traditional active learning methods.

Keywords: Thompson Sampling · Distance Estimation · Probabilistic Programming · Active Learning · Travel Time Estimation

1 Introduction

An important part of traveling is estimating travel time. Without knowing the time it will take to travel between two locations it can be difficult to plan ahead and ensure that things go according to plan. Many services already provide good estimates for well-known scenarios such as car travel and public transportation. For these services, estimates are typically based on first determining a route and then adding up the individual components of that route to obtain the total travel time.

However, in many situations, the navigation system may fail to provide adequate information to form a route, leaving it unable to provide travel time estimates. These situations could occur for instance when hiking cross country or when traveling in areas where shortcuts and obstacles that do not appear on maps are frequent, such as in urban city centers. An alternative approach is

F. Wotawa et al. (Eds.): IEA/AIE 2019, LNAI 11606, pp. 71–78, 2019.
https://doi.org/10.1007/978-3-030-22999-3_7

to focus on estimating the *true road distance*. While this is an interesting approach, the data required for estimation is significantly harder to gather. Not only does one need a timekeeping device, one also need some way to accurately track velocity. We avoid this by instead focusing on the actual time it takes to travel between two points.

Definition 1. *A Travel Time Estimation Function (TTEF) is a function* $R^d \times R^d \rightarrow R$, *where* a *is the start location,* b *is the destination, and* θ *is the parameters learned from the observations.*

The challenge is thus to determine θ by observing the actual travel time between locations and generalize from those observations. To minimize this calibration cost, the number of observations should be kept at a minimum. An important task is therefore to gather information in such a manner that each observation maximizes the gain in estimation accuracy. The objective of this paper is to address the calibration of TTEFs using as few data points as possible. We achieve this by formulating the problem as an Active Learning problem.

1.1 Active Learning

Active Learning (AL) has emerged as an effective tool for bridging supervised learning and unsupervised learning [2,12]. The settings where supervised learning thrive are those abundant with labeled data, for instance sentiment analysis for movie reviews where the reviews typically have been assigned e.g. a star-rating by the reviewer, allowing the collection of large amounts of labeled data [9].

This is in contrast to other fields such as medical imaging, where one often need human experts to manually label the data. In such cases, it becomes pertinent with learning algorithms that maximize the information gained from each labeled example. An active learner operates by carefully selecting the most beneficial example to be labeled, with the result that fewer examples have to be labeled in total, while simultaneously performing as well as a passive learner, i.e., a learner that simply observes the labeled examples.

The active learning paradigm can roughly be divided into two parts based on the nature of the unsupervised examples, it is either *pool-based* where all the examples are available without a label, or *stream-based* where the examples are given as a stream, feeding one example at a time. In our novel variant of the TTE, the data is neither stream based nor pool based, it is instead a hybrid between the two types of AL. In TTE the learner is faced with a stream of pools, where the learner may only select one example from each pool, discarding the rest, as clarified below.

1.2 Active Learning in Travel Time Estimation

We define the data generating process of TTE as follows. An observer is standing on a location $a_{t=1}$ and then has to select a destination from a set or pool of n distinct locations, $D_{t=2} = \{d_1, d_2, \ldots, d_n\}$. Once a destination $a_{t=2} \in D_{t=2}$ is selected the observer travel from $a_{t=1}$ to $a_{t=2}$ and record the travel time δ_1. This process is then repeated with $a_{t=2}$ as the new starting location. A new destination $a_{t=3}$ needs to be selected, now from $D_{t=3}$, and we obtain δ_2 – the travel time between $a_{t=2}$ and $a_{t=3}$. An important factor that makes TTE more difficult is that the observation δ_t does not only depend on a_t but on a_{t-1} as well.

1.3 Probabilistic Programming

Probabilistic Programming (PP) is an attempt to close the representation gap between the much celebrated probabilistic graphical models (PGM) such as Bayesian Networks and Markov Networks and the more specialized algorithms that are typically represented as a mixture of pseudo code, natural language, and mathematics. The idea is to formulate the entire model, from sample generation to the joint distribution in a unified representation framework, and let the underlying architecture handle the inference. This alleviates the need for highly specialized algorithms and lets the designer focus on designing a correct model, rather than focusing on models that are easy to do inference on. With the advances in computational power, a wide array of PPL have appeared in the literature. In this paper we employ PyMC3 [11], which is built on top of the Theano framework [15].

1.4 Paper Contributions

In this paper, we demonstrate the effectiveness of using Probabilistic Programming to solve the TTE problem, while simultaneously applying Thompson Sampling based Active Learning to minimize the number of observations required. To further investigate the effectiveness of this approach we also show that it performs comparable to traditional baselines for active learning on a well know regression problem [4].

2 Related Work

2.1 Active Learning

The highly effective Query By Committee (QBC) [4,13] algorithm is based on the premise that a committee of unique learners label each potential data point. That is, in a pool-based setting each data point in the pool is labeled by each learner. The next data point to obtain a label for is simply the data point where the learners disagree the most. For the simple case with binary labeled points and two learners, any point where the two learners disagree is considered as the next query point. In cases where the labels are real-valued, an alternative approach is to select the point that is expected to reduce prediction error the most [4]. For real-valued regression problems, the data point that maximizes the variance of the training set after being added is selected [3].

A critical aspect of the QBC algorithm is the disagreement between the learners. In the original work [13] a randomized algorithm was used. However, a more general approach is to train the same algorithm on different subsets of the data, as in query by bagging and query by boosting [8].

Bandit based active learning is a well-explored area of research [1,5,10], but this class of approaches is ill-suited to the TTE problems due to the simple fact that they require a pool based approach where one can track the uncertainty for each possible query-point as part of the active learning.

2.2 Distance Estimation

The field of Distance Estimation (DE) has primarily been dominated by the use of parameterized functions of a simple yet effective form. These functions are then calibrated using a set of inter-connected points and their distances [7] by maximizing the Goodness of Fit (GoF) between the observed values and the underlying function. Recently, an Adaptive Tertiary Search (ATS) based method that does not explicitly depend on GoF was proposed [6]. Instead, this method depends on the sign between the estimated distance and the actual, observed distance, and can thus be seen as a form of gradient descent.

To be consistent with previous work we will restrict ourself to the family of Weighted LP functions:

$$\text{W-L}_p(X) = k\left(\sum |x_i|^p\right)^{1/p}$$

where k is the linear weight, and $p \in R^+$ denotes the p-norm.

3 Active Learning with Thompson Sampling for Travel Time Estimation

The principle of Thompson Sampling (TS) can be summarized as follows. Given a distribution $\pi(\theta)$ over a parameter θ to be estimated, we sample an instance s from $\pi(\theta)$. We then assume that s is, in fact, the correct underlying value for θ. Thus, we explore by assuming that s is optimal and gather information I_s that we use to update the distribution over $\pi_{t+1}(\theta \mid I_s)$. Consequently, we also exploit our previous knowledge as the distribution over θ gets sharper towards the optimal value as t increases.

In the context of active learning in a probabilistic program, the objective of TS is to *convince* the maximum a-posteriori (MAP) model M_{map} that the TS sampled model M_{ts} is optimal by selecting the observation $o \in O$ such that the difference between $M_{\text{map}}(o)$ and $M_{\text{ts}}(o)$ is minimized after observing o.

In contrast, QBC is based on generating a committee $M_{\text{map}}^{(i)}, i = 1, 2, \ldots, m$ where each MAP estimate is based on a different subset of the data. This inherently means that the quality of an individual committee $M_{\mathrm{map}}^{(i)}$ is worse than the MAP estimate of a TS model M_{map} that employs all available data.

Algorithm 1. The TS-PPL algorithm

Result: A program $P(T)$ based on T observations.
Set a prior program $P(0)$;
for $t = 1, 2, \ldots, T$ **do**
  $M_{\mathrm{map}}(t) = \mathrm{MAP}(P(t-1))$;
  $M_{\mathrm{ts}}(t) \sim P(t-1))$;
  $o(t) = \mathrm{argmax}_{o_i \in O(t)}[M_{\mathrm{map}}(o_i; t) - M_{\mathrm{ts}}(o_i; t)]^2$;
  $y(t) = \mathrm{Query}(o(t))$
  $P(t) = P(t-1) \leftarrow o(t), y(t)$
end

4 Experiments

To demonstrate the efficiency of TS-PPL we apply it to two different problems. First, we investigate the performance for learning real-valued functions as done in [2]. Second, we investigate how it perform in the Travel Time Estimation problem. The metric of interest for the experiments will be the head-to-head results generated from identical experimental data and model. That is, the trials will be identical except the choice of observations to label. The objective function is to minimize the error on a separate hold-out set and thus, the cumulative error E_T is the sum of errors from $t = 0$ to $t = T$. The head-to-head metric between A and B is therefore the fraction of trials where scheme A have a lower cumulative error than B at the reported time-step t, i.e. $\sum_{i=0}^{N} \mathbb{1}[E_t^{(i)}(A) < E_t^{(i)}(B)]/N$.

4.1 Active Learning of Real-Valued Functions

The objective in a real-valued function is to minimize the generalization error between the learned function and the underlying true function, e.g. the difference in the area under the curve. We will now test TS-PPL with a standard function learning experimental setup [2], that have an underlying true function as shown in Eq. 1, with $z = \frac{x-0.2}{0.4}$, $a = 1, b = -1, c = 0$ and $\epsilon \sim N(0, 0.1^2)$.

$$f(x) = ax^2 + bx + c + \delta\frac{z^3 - 3z}{\sqrt{6}} + \epsilon \qquad (1)$$

Algorithm 2. The PP for the function regression problem

$a \sim N(\mu = 1, \sigma = 2)$
$b \sim N(\mu = 1, \sigma = 2)$
$c \sim N(\mu = 1, \sigma = 2)$
$s \sim |N(\mu = 0, \sigma = 10)|$
$\mu_t = ax^2 + bx + c$
$y_{\mathrm{obs}} \sim N(\mu = \mu_t, \sigma = s \mid y)$

The available observations are drawn from $N(0.2, 0.4^2)$ and presents a serious challenge for QBC to outperform due to the Signal-To-Noise Ratio (SNR) of $0.4^2/0.3^2 = 1.8$ shift between the underlying function and the test distribution that the candidate points are drawn from. In Table 1 we observe that the standard QBC outperform the TS-PPL algorithm when the assumed function type is approximately correct ($\delta = 0.005$). However, it is outperformed when the difference between the assumed model and the underlying model is large ($\delta = 0.05$).

Table 1. The result of head-to-head comparisons between the different methods based on 5k trials in the function approximation scenario. The data is given in the format X/Y where X is the fractions of wins in head-to-head matches from $t = 20$ and Y is the fraction of wins from $t = 40$.

	δ	arms = 5	arms = 10	arms = 20	arms = 100
TS vs QBC	0.05	0.56/0.50	0.48/0.43	0.55/0.42	0.56/0.53
TS vs Passive	0.05	0.58/0.57	0.69/0.61	0.66/0.64	0.84/0.83
QBC vs Passive	0.05	0.57/0.61	0.65/0.64	0.62/0.65	0.76/0.75
TS vs QBC	0.005	0.49/0.48	0.46/0.45	0.41/0.41	0.51/0.51
TS vs Passive	0.005	0.55/0.64	0.71/0.67	0.72/0.69	0.71/0.72
QBC vs Passive	0.005	0.58/0.62	0.68/0.69	0.77/0.73	0.68/0.71

4.2 Travel Time Estimation

Similar to [6], we conduct the TTE experiments on publicly available data from the TSPLIB Symmetric Traveling Salesman Problem Instances (MP-TESTDATA) [14] with N=29. The $O(t) = \{(x_i, y_i)\}^n$ pairs available for observations at time t is drawn from $x_i \sim U(0, x_{\max}), y_i \sim U(0, y_{\max})$ where $x_{\max} = 2300$ and $y_{\max} = 1900$. The purpose is to draw the observations uniformly from the entire dataset.

The oracle computes the travel time from $\boldsymbol{a}$ to $\boldsymbol{b}$, denoted $Q(\boldsymbol{a}, \boldsymbol{b})$, as

$$||\boldsymbol{a} \rightarrow \boldsymbol{p}||_{L_1} + \text{TravelTime}(\boldsymbol{p}, \boldsymbol{q}) + ||\boldsymbol{q} \rightarrow \boldsymbol{b}||_{L_1} \tag{2}$$

where $\boldsymbol{p}, \boldsymbol{q}$ is the closest points in the dataset to $\boldsymbol{a}$ and $\boldsymbol{b}$ respectively and TravelTime$(\boldsymbol{p}, \boldsymbol{q})$ is provided by the dataset.

The PP used is defined as a Bayesian prior over the W-L$_p$ model from [6]. and is as follows:

Algorithm 3. The PP for the Travel Time problem

$k \sim N(\mu = 1, \sigma = 2)$
$p \sim N(\mu = 1, \sigma = 2)$
$s \sim |N(\mu = 0, \sigma = 50)|$
$\mu_t = k \; ||l_i - l_{i-1}||_p$
$y \sim N(\mu = \mu_t, \sigma = s \mid t_1, \ldots, t_n)$

Table 2. The result of head-to-head comparisons between the different methods based on 5k trials. The data is given in the format X/Y where X is the fractions of wins in head-to-head matches from $t = 20$ and Y is the fraction of wins from $t = 40$.

	arms = 2	arms = 5	arms = 10
TS-PPL vs QBC	0.65/0.64	0.54/0.58	0.61/0.61
TS-PPL vs Passive	0.55/0.58	0.70/0.71	0.72/0.72
QBC vs Passive	0.45/0.46	0.64/0.57	0.72/0.65
	arms = 20	arms = 50	arms = 100
TS-PPL vs QBC	0.64/0.62	0.66/0.64	0.50/0.57
TS-PPL vs Passive	0.70/0.70	0.63/0.70	0.69/0.66
QBC vs Passive	0.60/0.60	0.57/0.62	0.69/0.60

The results for comparing between OBC and TS-PPL is found in Table 2. From the results, it is quite clear that TS-PPL outperforms QBC as well as Passive for the TTE problem achieving near 10% better results. This indicates that when the problem is not a simple regression problem, anchoring the selection process in the MAP estimate, as done in TS, gives a better trade-off than anchoring in the variance over a committee.

5 Conclusion

We have proposed TS-PPL an effective scheme for performing Active Learning in Probabilistic Programs. We have shown that TS-PPL can be applied to both a standard regression problem and a more complex problem in the Travel Time Estimation problem. Our method significantly outperforms the strong baseline of Query by Committee as well as passive learning for Travel Time Estimation. TS-PPL further gives competitive results in the case of regression.

References

1. Bouneffouf, D., Laroche, R., Urvoy, T., Feraud, R., Allesiardo, R.: Contextual bandit for active learning: active Thompson sampling. In: Loo, C.K., Yap, K.S., Wong, K.W., Teoh, A., Huang, K. (eds.) ICONIP 2014. LNCS, vol. 8834, pp. 405–412. Springer, Cham (2014). https://doi.org/10.1007/978-3-319-12637-1_51
2. Burbidge, R., Rowland, J.J., King, R.D.: Active learning for regression based on query by committee. In: Yin, H., Tino, P., Corchado, E., Byrne, W., Yao, X. (eds.) IDEAL 2007. LNCS, vol. 4881, pp. 209–218. Springer, Heidelberg (2007). https://doi.org/10.1007/978-3-540-77226-2_22
3. Cohn, D.A., Ghahramani, Z., Jordan, M.I.: Active learning with statistical models. J. Artif. Intell. Res. **4**, 129–145 (1996)
4. Freund, Y., Seung, H.S., Shamir, E., Tishby, N.: Selective sampling using the query by committee algorithm. Mach. Learn. **28**(2–3), 133–168 (1997)

5. Ganti, R., Gray, A.G.: Building bridges: viewing active learning from the multi-armed bandit lens. arXiv preprint arXiv:1309.6830 (2013)
6. Havelock, J., Oommen, B.J., Granmo, O.C.: Novel distance estimation methods using "stochastic learning on the line" strategies. IEEE Access **6**, 48438–48454 (2018)
7. Love, R.F., Morris, J.G.: Modelling inter-city road distances by mathematical functions. J. Oper. Res. Soc. **23**(1), 61–71 (1972)
8. Mamitsuka, N.A.H., et al.: Query learning strategies using boosting and bagging. In: Machine Learning: Proceedings of the Fifteenth International Conference (ICML 1998), vol. 1 (1998)
9. McAuley, J., Yang, A.: Addressing complex and subjective product-related queries with customer reviews. In: Proceedings of the 25th International Conference on World Wide Web, pp. 625–635. International World Wide Web Conferences Steering Committee (2016)
10. Osugi, T., Kim, D., Scott, S.: Balancing exploration and exploitation: a new algorithm for active machine learning. In: Fifth IEEE International Conference on Data Mining, 8-p. IEEE (2005)
11. Salvatier, J., Wiecki, T.V., Fonnesbeck, C.: Probabilistic programming in python using PYMC3. PeerJ Comput. Sci. **2**, e55 (2016)
12. Settles, B.: Active learning literature survey. Computer Sciences Technical report 1648, University of Wisconsin-Madison (2009)
13. Seung, H.S., Opper, M., Sompolinsky, H.: Query by committee. In: Proceedings of the Fifth Annual Workshop on Computational Learning Theory, pp. 287–294. ACM (1992)
14. Skorobohatyj, G.: MP-TESTDATA—the TSPLIB symmetric traveling salesman problem instances. http://elib.zib.de/pub/mptestdata/tsp/tsplib/tsp/index.html. Accessed 16 Nov 2015
15. Theano Development Team: Theano: a Python framework for fast computation of mathematical expressions. arXiv e-prints abs/1605.02688, May 2016. http://arxiv.org/abs/1605.02688

Towards Analyzing the Impact of Diversity and Cardinality on the Quality of Collective Prediction Using Interval Estimates

Van Du Nguyen[1(✉)], Hai Bang Truong[2], and Ngoc Thanh Nguyen[3,4]

[1] Faculty of Information Technology, Nong Lam University, Ho Chi Minh City, Vietnam
nvdu@hmcuaf.edu.vn

[2] Faculty of Computer Science, University of Information Technology, Vietnam National University Ho Chi Minh City (VNU-HCM), Ho Chi Minh City, Vietnam
bangth@uit.edu.vn

[3] Department of Information Systems, Faculty of Computer Science and Management, Wroclaw University of Science and Technology, Wrocław, Poland
Ngoc-Thanh.Nguyen@pwr.edu.pl

[4] Faculty of Information Technology, Nguyen Tat Thanh University, Ho Chi Minh City, Vietnam

Abstract. Recently, many research results have indicated that diversity is the most important characteristic of crowd-based applications. However, it is a point estimates-based finding in which single values are used as the representation of individual predictions on a real-life cognition task. This paper presents a study on how cardinality and diversity influence the quality of collective prediction using interval estimates. By means of computational experiments, we have found that these factors positively influence the quality of collective prediction. Besides, the results also indicate that the hypothesis *"the higher the diversity, the better the quality of collective prediction"* is true. Furthermore, the findings also reveal a cardinality threshold in which its increase does not significantly influence the quality of collective prediction.

Keywords: Wisdom of crowds · Collective intelligence · Interval estimates

1 Introduction

Recently, many research results have indicated that using the wisdom of crowds is an efficient approach to solving some difficult cognition tasks in the real world [1, 11, 14]. Even the results in [5] have shown that the crowd-based predictions could outperform those produced by traditional forecasting methods. However, they are point estimates-based findings in which single values are used as representations of individual prediction on a given task. Moreover, in practice, it is hard for members to provide individual predictions in the form of single values. For example, a sale expert can find it is easy to give a prediction on the growth of sales in the form of an interval value

F. Wotawa et al. (Eds.): IEA/AIE 2019, LNAI 11606, pp. 79–86, 2019.
https://doi.org/10.1007/978-3-030-22999-3_8

such as from 3% to 5%. From this fact, in order to differentiate this case from the former one, we called it *interval estimates*. Notice that we do not take into consideration the case in which the proper value is an interval as in the task of temperature forecasting for a specific region on the next day.

The general research problem of the wisdom of crowds is as follows: for a given task T - cognition task, a number of autonomous members, who are invited for providing individual predictions on T. These members can be humans or even agent systems. In addition, they are often diverse in backgrounds, knowledge bases, etc. Moreover, we consider the unified representations of individual predictions are interval values. The proper value, which is a single value, will be treated as an interval whose lower and upper values are identical. From a set of interval values (individual predictions), some aggregation methods have been invoked to determine a consistent one, (called *collective prediction*). This consistent prediction is often considered as a representative of the collective as a whole. Referring to [7], many methods have been proposed for integrating individual predictions. They aim at aggregating interval predictions into a single value (similar to the case of point estimates) based on the unweighted or weighted average of their midpoints. Meanwhile, in this work, the output of the aggregation method is still an interval value and the criteria proposed in Nguyen [16] will be taken into account for such a task.

Recently, many publications have revealed the positive role of diversity in the emergence of Collective Intelligence [1, 4, 11, 13]. In fact, it can be seen that diverse collective members will have new perspectives, new information, etc. about the given problem and collective knowledge is often larger than the *"normal sum"* of individual knowledge [8]. In addition, the correlated errors among individual predictions can be reduced if collective members are diverse [6]. However, these findings are based on *point estimates*. This positive effectiveness leads to a natural question: how diversity influences collective prediction in case of interval estimates? For this aim, the computational experiments will be conducted to determine the relationship between diversity and the quality of collective prediction. Concretely, we suppose collectives having the same cardinality but varying in diversity level. In other words, we will try to prove the following hypothesis *"the higher the diversity, the better the quality of collective prediction."* Beyond the impact of diversity, the problem of maintaining diversity in a collective is also an important issue. Recent studies on *point estimates* have shown that the common approach that can be used to enhance the diversity of a collective is expanding cardinality [2, 3, 15]. Taking into account this characteristic, we also focus on how cardinality influences the quality of collective prediction using interval estimates. In the current literature, these problems have not been widely devoted.

The structure of the paper is as follows: Some basic notions are described in Sect. 2. Section 3 presents the research model and hypotheses. Computational experiments on the impact of cardinality and diversity on the quality of collective prediction will be reported in Sect. 4. The last section concludes the paper and discusses some further research.

2 Background

2.1 Collective Prediction Determination

In this work, by *collective* we denote a set of interval values (called *individual predictions*) provided by a number of collective members who are often autonomous members on the same cognition task. The representation of individual predictions in a collective is described as follows: $X = \{x_1, x_2, \ldots, x_n\}$ where n represents the size of collective X (called *cardinality*). In this paper, we assume the structure of predictions is an interval. In particular, each prediction in a collective is described as follows: $\forall x_i \in X : x_i = \langle x_{i*}, x_i^* \rangle$ where x_{i*} and x_i^* present the lower and the upper bounds of individual prediction x_i. The distance between predictions is measured using the following formula [16]:

$$d(x_i, x_j) = \frac{\left|x_i^* - x_j^*\right| + \left|x_{i*} - x_{j*}\right|}{2}$$

where $d(x_i, x_j) \in \Re_+$.

In general, the aggregation methods worked out previously for interval estimates are mainly based on the midpoints of interval predictions [7]. These methods aim at aggregating interval predictions to a single value based on the unweighted or weighted average of their midpoints. Meanwhile, in this work, the output of the aggregation method is still an interval prediction. For such a task, the criteria mentioned in [16] (1-Optimality, 2-Optimality) are taken into consideration.

2.2 Collective Prediction Measures

With estimating tasks, the most important measure that can be used to reflect the quality of collective prediction is the accuracy of collective prediction [9, 10]. However, with applications of the Wisdom of Crowds, it is well-known the outperformance of a collective as a whole over collective members in finding solutions to a given problem. Therefore, the best collective prediction is the one that outperforms all individual predictions in a collective [15]. The measures of the quality of collective prediction used in this work are as follows:

Definition 1. *The quality of collective prediction can be measured based on:*

- *The accuracy of collective prediction* [9]:

$$Diff(X) = 1 - d(r, x)$$

- *The relationship between accuracies of collective prediction and individual predictions*

$$WR(X) = \frac{1}{n}\sum_{i=1}^{n} f(r, x, x_i)$$

where $f(r, x, x_i) = \begin{cases} 1, & \text{if } d(r,x) < d(r,x_i) \\ 0, & \text{otherwise} \end{cases}$

3 Research Model and Hypotheses

In group decision making, it can be seen that the more consistent the individual opinions (or preferences), the higher the quality of collective opinion. However, in Collective Intelligence, diversity has been proven to be useful in improving to better collective predictions [14]. Besides, by enlarging the cardinality, one can maintain the diversity of a collective because the diversity of a collective will potentially enhance as the cardinality increases. As mentioned earlier, however, these findings are based on *point estimates*. Therefore, in this work, we investigate how these factors influence collective prediction using interval estimates. Notice that we only take into consideration diversity of individual predictions in a collective and the pairwise distances between individual predictions in a collective are taken into account for its measure [9]. The definition is described in the following formula:

$$c(X) = \begin{cases} \frac{1}{n(n-1)} \sum_{i=1}^{n} d(x_i, X), & \text{for } n > 1 \\ 0, & \text{otherwise} \end{cases}$$

where $d(x_i, x_j)$ represents the distance between x_i and x_j and $d(x_i, X) = \sum_{j=1}^{n} d(x_i, x_j)$ represents the sum of distances from prediction x_i to predictions of collective X.

According to the findings in the case of point estimates, in the remaining parts of the paper we will try to prove the following hypotheses:

H1: *The increase in the quality of collective prediction is a consequence of an increase in the cardinality.*

H2: *The higher the diversity, the better the quality of collective prediction.*

In the remaining part of the paper, we will conduct computational experiments to prove these hypotheses.

4 The Impact of Cardinality and Diversity on Collective Prediction

4.1 Settings

This section presents computational experiments to study the influence of factors such as cardinality and diversity on collective prediction. For this aim, the experiments are conducted with the following assumptions:

- $\forall x_i \in U : d(r, x_i) \leq \partial$
- $(r = \langle r_*, r^* \rangle)$, where $(r_* = r^*)$
- $\forall x_i \in U : (x_{i*} \geq |r_* - \partial|) \wedge (x_i^* \leq |r^* + \partial|)$

According to the first assumption, we assume a predefined threshold (∂) in which the distance from the proper value to potential predictions in set U does not exceed. Meanwhile, the second one means that the proper value will be treated as an interval in which the lower and upper bounds are identical. Meanwhile, the third assumption describes the potential lower and upper bounds of each interval prediction. In the following experiments, we assume that the value of ∂ is equal to 500 and the proper value is 1000. The proper value and the lowest and highest predictions can be treated as interval values as follows: $\langle 1000, 1000 \rangle$, $\langle 500, 500 \rangle$, and $\langle 1500, 1500 \rangle$ respectively. Moreover, we also assume that the length of each interval prediction does not exceed 10% of ∂ (that is $\forall x_i \in U : |x_{i*} - x_i^*| \leq 50$). By this assumption, we will ignore the phenomenon of which collective members have little information about the proper value of the given problem, can lead to suboptimal collective predictions. For example, a sale expert reports that the sales will grow by 4.5%–4.7%. In this case, it is likely she has much evidence about the growth of sales. However, if someone reports that the sales will grow by 0%–30%, then she seems to know very little about the growth of sales.

4.2 The Relationship Between Diversity and the Quality of Collective Prediction

This section presents an influence analysis of diversity on the quality of collective prediction using computational experiments. We assume that collectives are identical cardinality but are various diversity levels. Similar to the previous settings, the cardinalities 29, 129, and 229 will be used. For each cardinality value, we performed some prior computational experiments to determine the corresponding diversity levels. In this section, we use five levels of diversity as follows: Div1: $(c(X) \geq 0.0) \wedge (c(X) < 0.1)$; Div2: $(c(X) \geq 0.1) \wedge (c(X) < 0.2)$; Div3: $(c(X) \geq 0.2) \wedge (c(X) < 0.3)$; Div4: $(c(X) \geq 0.3) \wedge (c(X) < 0.4)$; Div5: $(c(X) \geq 0.4) \wedge (c(X) < 0.5)$.

This subsection will try to show the positive influence of diversity on the quality of collective prediction. The experimental results are described in Fig. 1.

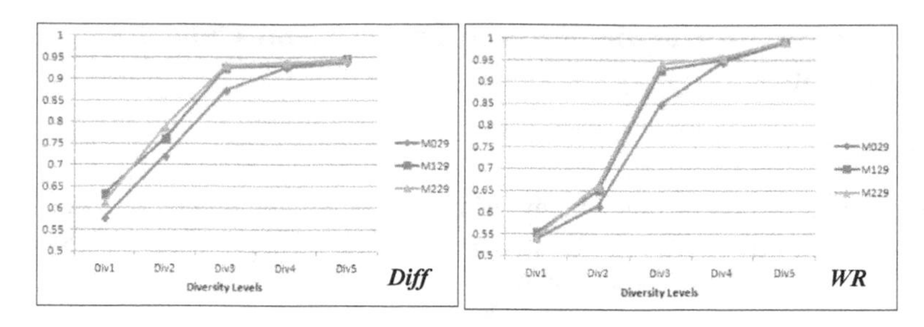

Fig. 1. *Diff* and *WR* of collectives with cardinalities 29, 129 and 229

As we expected, diversity positively influences the quality of collective prediction. The difference between the qualities of collective predictions in the case collectives have the same cardinality but vary in diversity is statistically significant (according to the Mann-Whitney U test). In all cases, the p values are less than the significance level 0.05. The statistical analysis also did not find a significant difference on between the qualities of collective predictions for collectives having the same diversity level but having different cardinality values. For *WR* measure, we have also found the similar result. From these findings, it can be seen that to achieve a better collective prediction, instead of being to be large, a collective should be enough diverse. From this fact, it can be stated that maintaining diversity in a collective is an important issue. In the next section, we focus on determining the impact of cardinality on the diversity of a collective. In addition, in the previous section, we have found that cardinality is positively associated with the quality of collective prediction.

4.3 The Relationship Between Cardinality and the Quality of Collective Prediction

As aforementioned, in order to maintain the diversity in a collective, cardinality is considered as an effective factor for such a task. It is because when the cardinality increases, the diversity of individual predictions and the variety of backgrounds will be able to enhance. Accordingly, we will study the relationship between the cardinality and the quality of collective prediction. For this aim, the cardinality will be increased from 9 to 1009 (each step by 10 elements).

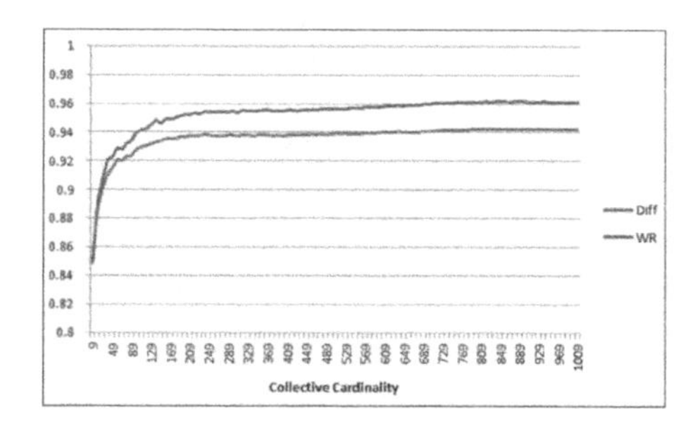

Fig. 2. *Diff* and *WR* of collectives with cardinalities increased from 9 to 1009

Notice that, with each setting, we run 100-repetition and the results reported in Fig. 2 are the average of 100 corresponding values. As we expected, when we increase the cardinality, the quality of collective prediction will be improved. Moreover, the increase of cardinality does not significantly affect the quality of collective prediction, if the cardinality is *"large enough."* According to the previous settings, the potential individual predictions will belong to [500, 1500]. Based on the computational result for point estimates, the cardinality needed to achieve a reliable prediction with 95% confidence interval and ± 50 is about 129. Taking into account this cardinality as a *"large enough"* value, we perform a statistical test to check the significant difference between the qualities of collective predictions with some selected cardinalities. For such a task, the Mann-Whitney test, a nonparametric test, is used with cardinalities of 29, 129, 229 and 329. Regarding *Diff* measure, we have found significant differences between the qualities of collectives of 29 and collectives of 129, 229, and 329. However, the differences between collectives of 129 and collectives of 229, and 329 are not statistically significant (*p-values* are much higher than 0.05). For *WR* measure, we have also found the similar result.

Furthermore, we computed the values of R^2 in the relationship between the log of cardinality and the quality of collective prediction. Accordingly, the values of R^2 increase up to the cardinality of 129 (i.e., 0.94 for *Diff* measure and 0.95 for *WR* measure). For cardinalities of 229, 329, etc. the values of R^2 are decreased (for both *Diff* and *WR* measures). From these results, it can be stated that with some restrictions, the hypothesis *"The increase in the quality of collective prediction is caused by an increase in the cardinality"* is true. However, the results also show that when the cardinality is large enough, its increase does not cause any significant impact on the quality of collective prediction.

5 Conclusion

From the prominent role of diversity in Collective Intelligence in case of point estimates, this paper has presented an influence analysis of diversity and cardinality on the quality of collective prediction using interval estimates. Regarding the impact of cardinality, the experimental results have shown that expanding cardinality is an effective approach to improving the quality of collective prediction. Besides, they also revealed a cardinality threshold in which the quality of collective prediction is not significant increase as the cardinality increases. Regarding the impact of diversity, the computational experiment has been conducted by considering collectives are identical cardinality but are various diversity levels. The findings have indicated that diversity plays a major role in improving the quality of collective predictions. In conclusion, with some restrictions, it can be noted that the hypothesis *"the higher the diversity, the better the quality of collective prediction"* is true.

In the future work, we intend to evaluate the performance of different aggregation methods used for collective prediction determination. The impact of other criteria such as independence, decentralization will also be investigated. For such a task, methods for solving conflicts proposed in [12] will be taken into consideration. We also concentrate on the diversity of members that can be considered as an efficient approach to fostering the diversity of individual predictions.

Acknowledgment. This paper is partially funded by Vietnam National University Ho Chi Minh City (VNU-HCM) under grant number C2018-26-09.

References

1. Armstrong, J.S.: Combining forecasts. In: Armstrong, J.S. (ed.) Principles of Forecasting, pp. 417–439. Springer, Boston (2001). https://doi.org/10.1007/978-0-306-47630-3_19
2. Conradt, L.: Collective behaviour: when it pays to share decisions. Nature **471**, 40–41 (2011)
3. Cui, R., Gallino, S., Moreno, A., Zhang, D.J.: The operational value of social media information (2015). SSRN: https://doi.org/10.2139/ssrn.2702151
4. Hong, L., Page, S.E.: Groups of diverse problem solvers can outperform groups of high-ability problem solvers. Proc. Natl. Acad. Sci. USA **101**, 16385–16389 (2004)
5. Lang, M., Bharadwaj, N., Di Benedetto, C.A.: How crowdsourcing improves prediction of market-oriented outcomes. J. Bus. Res. **69**, 4168–4176 (2016)
6. Larrick, R.P., Soll, J.B.: Intuitions about combining opinions: misappreciation of the averaging principle. Manag. Sci. **52**, 111–127 (2006)
7. Lyon, A., Wintle, B.C., Burgman, M.: Collective wisdom: methods of confidence interval aggregation. J. Bus. Res. **68**, 1759–1767 (2015)
8. Maleszka, M., Nguyen, N.T.: Integration computing and collective intelligence. Expert Syst. Appl. **42**, 332–340 (2015)
9. Nguyen, N.T.: Advanced Methods for Inconsistent Knowledge Management. Springer, London (2008). https://doi.org/10.1007/978-1-84628-889-0
10. Nguyen, N.T.: Inconsistency of knowledge and collective intelligence. Cybern. Syst. **39**, 542–562 (2008)
11. Page, S.E.: The Difference: How the Power of Diversity Creates Better Groups, Firms, Schools, and Societies. Princeton University Press, Princeton (2007)
12. Pietranik, M., Nguyen, N.T.: A multi-attribute based framework for ontology aligning. Neurocomputing **146**, 276–290 (2014)
13. Robert, L., Romero, D.M.: Crowd size, diversity and performance. In: Proceedings of the 33rd Annual ACM Conference on Human Factors in Computing Systems, pp. 1379–1382. ACM, Seoul (2015)
14. Surowiecki, J.: The Wisdom of Crowds. Doubleday/Anchor, New York (2005)
15. Wagner, C., Suh, A.: The wisdom of crowds: impact of collective size and expertise transfer on collective performance. In: 47th Hawaii International Conference on System Sciences, pp. 594–603 (2014)
16. Zgrzywa, M.: Consensus determining with dependencies of attributes with interval values. J. Univers. Comput. Sci. **13**, 329–344 (2007)

Applied Neural Networks

Biometric Fish Classification of Temperate Species Using Convolutional Neural Network with Squeeze-and-Excitation

Erlend Olsvik[1], Christian M. D. Trinh[1], Kristian Muri Knausgård[2](✉), Arne Wiklund[1], Tonje Knutsen Sørdalen[3,4], Alf Ring Kleiven[3], Lei Jiao[1], and Morten Goodwin[1]

[1] Centre for Artificial Intelligence Research, University of Agder, 4879 Grimstad, Norway

[2] Department of Engineering Sciences, University of Agder, 4879 Grimstad, Norway
kristianmk@ieee.org

[3] Institute of Marine Research (IMR), His, Norway

[4] Department of Natural Sciences, Centre for Coastal Research (CCR), University of Agder, Kristiansand, Norway

Abstract. Our understanding and ability to effectively monitor and manage coastal ecosystems are severely limited by observation methods. Automatic recognition of species in natural environment is a promising tool which would revolutionize video and image analysis for a wide range of applications in marine ecology. However, classifying fish from images captured by underwater cameras is in general very challenging due to noise and illumination variations in water. Previous classification methods in the literature relies on filtering the images to separate the fish from the background or sharpening the images by removing background noise. This pre-filtering process may negatively impact the classification accuracy. In this work, we propose a Convolutional Neural Network (CNN) using the Squeeze-and-Excitation (SE) architecture for classifying images of fish without pre-filtering. Different from conventional schemes, this scheme is divided into two steps. The first step is to train the fish classifier via a public data set, i.e., Fish4Knowledge, without using image augmentation, named as pre-training. The second step is to train the classifier based on a new data set consisting of species that we are interested in for classification, named as post-training. The weights obtained from pre-training are applied to post-training as a priori. This is also known as transfer learning. Our solution achieves the state-of-the-art accuracy of 99.27% accuracy on the pre-training. The accuracy on the post-training is 83.68%. Experiments on the post-training with image augmentation yields an accuracy of 87.74%, indicating that the solution is viable with a larger data set.

Keywords: Biometric fish classification · CNN · Squeeze-and-Excitation · Temperate species · Natural environment

F. Wotawa et al. (Eds.): IEA/AIE 2019, LNAI 11606, pp. 89–101, 2019.
https://doi.org/10.1007/978-3-030-22999-3_9

1 Introduction

Coastal marine ecosystems are highly complex, productive, and important spawning, nursing and feeding areas for numerous of fish species, but studying such biodiversity is often logistically challenging and time-consuming [13,16]. With the recent advancement in cost-effective high definition underwater camera technologies, large volumes of observations from remote areas are now allowing us to test predictions about species' cryptic behaviour, fundamental ecological processes and environmental changes [12]. Yet, video data analysis is extremely labour intensive and only a fraction of the available recordings can be analyzed manually, greatly limiting the utility of the data. In addition, accuracy of visual-based assessments is highly dependent on conditions in the underwater environment (depth, light, background noise) and taxonomical expertise in interpreting the videos [1].

Computer vision solutions have been increasingly applied to marine ecology to tackle these problems [6,9,14]. One such solution, the commercial product CatchMeter [17], consists of a light box with a camera that photographs and classify the fish as well as provide a length estimate. Fish are recognized by utilizing a threshold for detecting the outline of fish in the images and has a very high classification accuracy of 98.8%. However, the fish are photographed in a relatively structured environment, which has limited applicability in studies of natural behaviour in the wild.

A specific Convolutional Neural Network (CNN) called Fast R-CNN stands out as it applies object detection to extract only the fish from images taken from natural environment, actively ignoring background noise [9]. The approach starts by pre-training an AlexNet [8] on the ImageNet database. The AlexNet is then modified to train on a subset of the Fish4Knowledge data set [5]. As the final step, the Fast R-CNN takes the pre-trained weights and region proposals made by AlexNet as input, and achieves a mean average precision of 81.4%. In another solution [6], pre-training is applied to a CNN similar to AlexNet. The network consists of five convolutional layers and three fully-connected layers. Pre-training is performed using 1000 images from 1000 categories from the ImageNet data set, and the learned weights are then applied to a CNN after adapting it to the Fish4Knowledge data set. Post-training is then performed using as few as 50 images per category and 10 categories from the Fish4Knowledge data set. The images from the Fish4Knowledge data set are pre-processed using image de-noising. The accuracy achieved on the 1420 test images is 85.08% using very small amounts of data.

The highest reported accuracy for the Fish4Knowledge data set so far is 98.64%. The result was achieved by first applying filters to the original images to extract the shape of the fish and remove the background, and then use a CNN with a Support Vector Machine (SVM) classifier function [14]. The network is named DeepFish, which consists of three standard convolution layers and three fully-connected layers. In addition, previous solutions usually apply a preprocessing of the images in order to remove the noise in the targeted image as much as possible, and to outline the area where fish are located [6,14]. Although this process can indeed improve the system performance, the set of filters must be chosen carefully, as it may result in a negative performance impact in a live and dynamic scenario. Useful object background information may unintentionally be removed during the segmentation pre-processing, such as indicated by comparison of background discarding Fast R-CNN and background encoding YOLO in [15]. Considering the noise tolerant nature of CNN with Squeeze-and-Excitation (SE) architecture, it could be an advantage to use the original image to maintain maximum information content.

In this paper, we further explore CNN using the most recent SE architecture, which, to the best of our knowledge, has not previously been utilized in fish classification. In addition to the learning algorithm, we also collect a new data set of temperate fish species in this work. Clearly, the Fish4Knowledge data set is currently limited to tropical fish species. If a CNN is trained on this data set alone, it may not be able to classify fish species in other ecosystems. Therefore, the trained model based on the Fish4Knowledge needs to be further tuned and validated to fit specific ecosystems of interest. Our approach is to first pre-train the network on the Fish4Knowledge data set to learn generic fish features, and then the learned weights from pre-training are adopted as a starting point for further training on the new data set containing images of temperate fish species, which is called post-training. This two-step process is known as transfer learning [18]. The solution based on SE-architecture requires no pre-processing of images, except re-sizing to the appropriate CNN input size.

The remainder of the paper is structured as follows. Section 2 describes the data sets used to train the neural network, and then the detailed network structure and configurations are presented. Section 3 discusses the experimental results for the CNN approach before the work is summarized in the last section.

2 Data Sets and Deep Learning Approaches

2.1 The Data Sets

Two data sets were used in the test, the Fish4Knowledge data set [9] and a Norwegian data set with temperate species collected by the research team. Fish4Knowledge is used in pre-training of the neural network, while the temperate data set is used in the post-training. Some differences between the data sets are: (1) The Fish4Knowledge has in addition to the fish images categorized images in trajectories, e.g. a sequence of images taken from the same video sequence or stream. (2) The temperate data set has in addition to the other species a separate folder for male and female Symphodus melops (S. melops). Some individuals of male S. melops have also been tracked and captured by camera multiple times.

Fish4Knowledge. The Fish4Knowledge data set is a collection of images, extracted from underwater videos of fish, off the coast of Taiwan. There is a total of 27230 images cataloged into 23 different species. The top 15 species accounts for 97% of the images, and the single top species accounts for around 44% of the images. The number of images for each species range from 25 to 12112 between the species. This creates a very imbalanced data set. Further, the images size ranges from approximately 30×30 pixels to approximately 250×250 pixels. Another observation in the data set, is that most of the images are taken from a viewpoint along the anteroposterior axis, or slightly tilted from that axis. In that subset of images, most of these images are from the left or right lateral side, exposing the whole dorsoventral body plan in the image. There are some images from the anterior view, but few from the posterior end. Among all the images there were not many images from the true dorsal viewpoint. Most of the selected species have a compressed body plan, e.g. dorsoventral elongate. This creates a very distinct shape when the images are taken from a lateral viewpoint. Hence, images taken from the dorsal view creates a thin, short shape. The images also have a background that is relatively light, enhancing the silhouette of the fish.

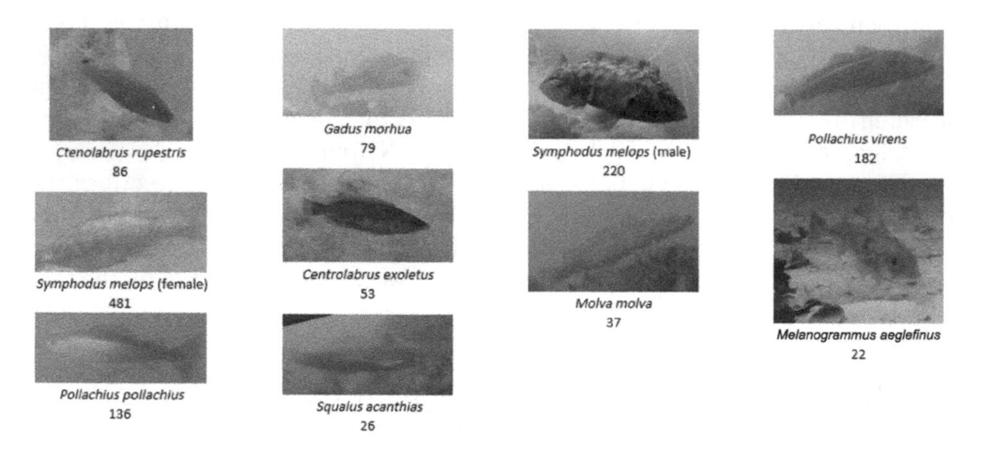

Fig. 1. Distribution of the temperate species data set.

Temperate Fish Species. The temperate data set consists of an image collection of some of the most abundant fish species in Northern Europe. The video recordings were sampled by scientists at the Institute of Marine Research (IMR) in Norway in two different occasions. One part is from video recordings taken from May to June 2015 in a remote and shallow bay on the Austevoll archipelago (Norway, North Sea). GoPro Hero4+ (black) cameras were deployed at 2–5 m of depth around small reef sites to record the nesting behaviour of *Symphodus melops*. Recording conditions varied between sites and days, especially in sun exposure and background noise. All videos were recorded in full HD resolution of 1920×1080 pixels with default settings. Colourful males of *S. melops* build nests to attract females who lay their eggs for the males to care for until they hatch [2]. The females are brown in colour and easily distinguished from nesting males. Some males employ a strategy to look indistinguishable from females and do not build nests, but instead sneak' on other males' nest [3]. Because of the morphological appearances of the different sexes, nest-building males are labelled as "males" in the data set (accounting for approximately 17% of the images in the data set), while females and sneaker males are labelled as "females" (accounting for about 36%). Two other wrasse species from these videos were also categorized. The second part of the data set was collected with stereo baited remote underwater video (stereo-BRUV). The stereo-BRUV consists of two calibrated GoPro Hero4+ (black) cameras. The cameras were deployed between 8 and 35 m in 2 coastal areas of Norway: south-estern coast (county of Aust-Agder) and mid-western coast (county of Trøndelag). The stereo-BRUV data sampling is normally used for monitoring marine biodiversity [10] and temporal trends in fish assemblages [11]. A single video frame often contains more than one fish (of same species and/or different species). Differences in depth, visibility, habitat, distance from camera and angle of the fish secured a high variability in pictures of each species. Except from the spiny dogfish (*Squalus acanthias*), the five other species were from the family Gadidae (*Gadus morhua, Pollachius virens, Pollachius pollachius, Molva molva and Malanogrammus aeglefinus*). Overall, the Norwegian temperate data set has a higher image noise (visibility, background, resolution) and variability of angle of the fish compared with the Fish4Knowledge data set. This is expected to reduce the classification accuracy, but be more realistic for analysis of observations in the natural environment. Figure 1 illustrates a snapshot of the data set.

2.2 CNN-SENet Structure

A CNN-SENet, is a Convolutional Neural Network with an added squeeze and excitation (SE) architectural element, that re-calibrates channel wise-feature responses adaptively [4]. The architecture of the CNN-SENet, depicted in Fig. 2, is configured with the following parameters. Image size in height (H), width (W) and depth channels; the number of learnable filters (F); the batch size (B) (default 16), the filter size (S), and reduction ratio (r) as described in [4]. Lastly the number of fish species classifications needs to be added, as parameter C. The input layer takes image of size 200×200 with a depth of 3 color channels, R, G, and B. The output is batch normalized before entering the Squeeze-and-Excitation function, called SE block, depicted in Fig. 3. The SE block performs a feature re-calibration through the (1) squeeze operation preventing the network from becoming channel-dependent. This exploits contextual information outside the receptive field and is achieved by doing global average pooling on each input channel before reshaping, and (2) the excitation operation that utilizes the output from the squeeze function by fully capture channel-wise dependencies. This is achieved by the two fully-connected (FC) layers sandwiching the reduction layer, and finally a sigmoid activation layer. Before exiting the SE block, the output from the excitation function is multiplied with the original batch normalized output. This multiplied output is then added to a ReLU layer performing an element-wise activation function, rendering the dimension size unchanged. The output is then sent to a Max Pooling layer, that uses a 2×2 filter to reduce and re-size the height and width spatially, rendering an output of $98 \times 98 \times 32$. This core portion of the network is stacked to the size of the kernel size, in this case the size of five. The first iteration has a convolutional layer of 32 filters in 5×5. The second and third has 64 filters in 3×3, the forth 128 filter in 2×2, and the fifth 256 filters in 2×2, with all layers applying a horizontal and vertical stride of 1.

Furthermore, the network has 3 FC layers. The first, with 256 neurons, takes the output from the last convolutional layer that is first flattened. The output is then batch normalized before sent to the second FC layer, with 256 neurons. A reduction function is applied after the output from the FC layer is batch normalized. Before entering the last FC layer, with C neurons, a dropout layer of 50% is applied. The final layer, softmax, applies a classifier function to obtain the probability distribution for each class per input image, using a categorical cross-entropy with the Adam optimizer [7].

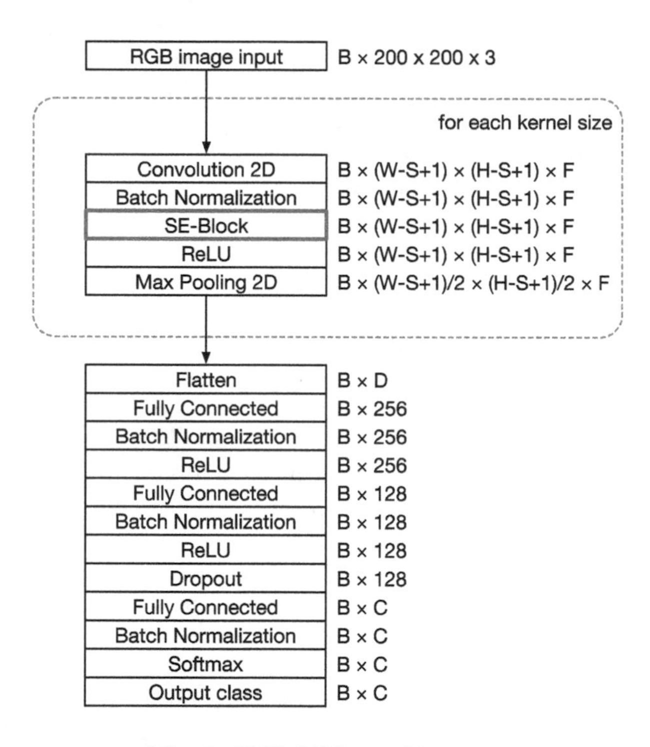

Fig. 2. CNN-SENet architecture.

In CNN-SENet, there are certain parameters that need to be configured, including dropout percentage, learning rate, and batch normalization, that are

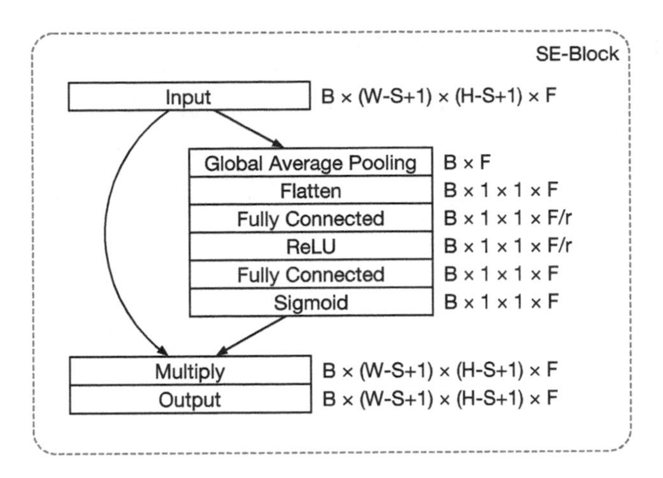

Fig. 3. Squeeze-and-Excitation block.

discussed presently. The parameters are configured based on trial-and-error method. For the dropout percentage, clearly, the higher the dropout, the more the information is lost during training because forward- and back-propagation are carried out only on the remaining neurons after dropout is applied. Different percentages of the dropout are tested, and 50% is configured in this study due to the better overall performance achieved. The learning rates when using the Adam optimizer should be tuned to further optimize the network. After numerous trials, the learning rate is configured as 0.001 without decay. For batch normalization, it has been tested and the results with batch normalization is slightly better than without it. In more details, the accuracy on the testing set without batch normalization is 98.35%, while the accuracy with batch normalization is 99.27%. With the above parameters, the model trains faster and has a higher validation accuracy, that concludes the architecture of CNN-SENet.

To compare CNN-SENet with DeepFish, Table 1 illustrates the main differences between the two. Clearly, CNN-SENet has a more sophisticated structure than DeepFish.

Table 1. Differences between CNN-SENet and DeepFish.

	CNN-SENet	DeepFish
Image size	200×200	47×47
Testing samples	4126	3098
Network architecture	Basic with SE blocks	Basic
Classifier	Softmax	SVM
Convolutional layers	5	3

3 Experiments and Results

Accuracy and performance of the new fish classification CNN-SENet is quantified and compared with the state-of-the-art networks represented by Inception-V3, ResNet-50 and Inception-ResNet-V2. Additionally, a simplified version of the CNN-SENet, without the Squeeze-and-Excitation blocks, is included to explore how the spatial relationship between fish image colors and other feature layers affect results [4].

3.1 Experiments

Three different experiments were performed. Pre-training with Fish4Knowledge, post-training with the new temperate Fish Species data set described in Subsect. 2.1 and post-training with an extended version of the new data set using image augmentation techniques. For all three experiments, the applicable data set was divided into 70% training images, 15% validation images and 15% testing images. Both training and validation images are integral parts of the training process, while the testing images were kept out-of-the-loop for independent verification of the "end product".

All benchmarked networks are trained for 50 epochs with images adapted to their input image size of 200 × 200 RGB pixels, with the notable exception of the 299 × 299 RGB pixels required by Inception-ResNet-V2.

Pre-training. Pre-training was performed using a data set consisting of 19149 Fish4Knowledge images, with an additional 4126 images for verification and 4126 images reserved for testing. The selected training configuration consists of a single run with 50 training epochs and a batch size of 16. Results from pre-training are evaluated using weights from the epoch with highest validation accuracy, and not necessarily the final epoch.

Table 2. Testing accuracy and time per epoch on pre-training.

Network	Testing accuracy	Time one epoch
Inception-V3	99.18%	923 s
ResNet-50	98.86%	646 s
Inception-ResNet-V2	98.59%	2221 s
CNN-SENet	99.27%	197 s
CNN-SENet without Squeeze-and-Excitation	99.15%	159 s

Post-training. Post-training was performed using 712 images of four fish classes from the temperate fish species data set described in Sect. 2.1. An additional 155 images was used for verification during training, and a subset of 155 images of the

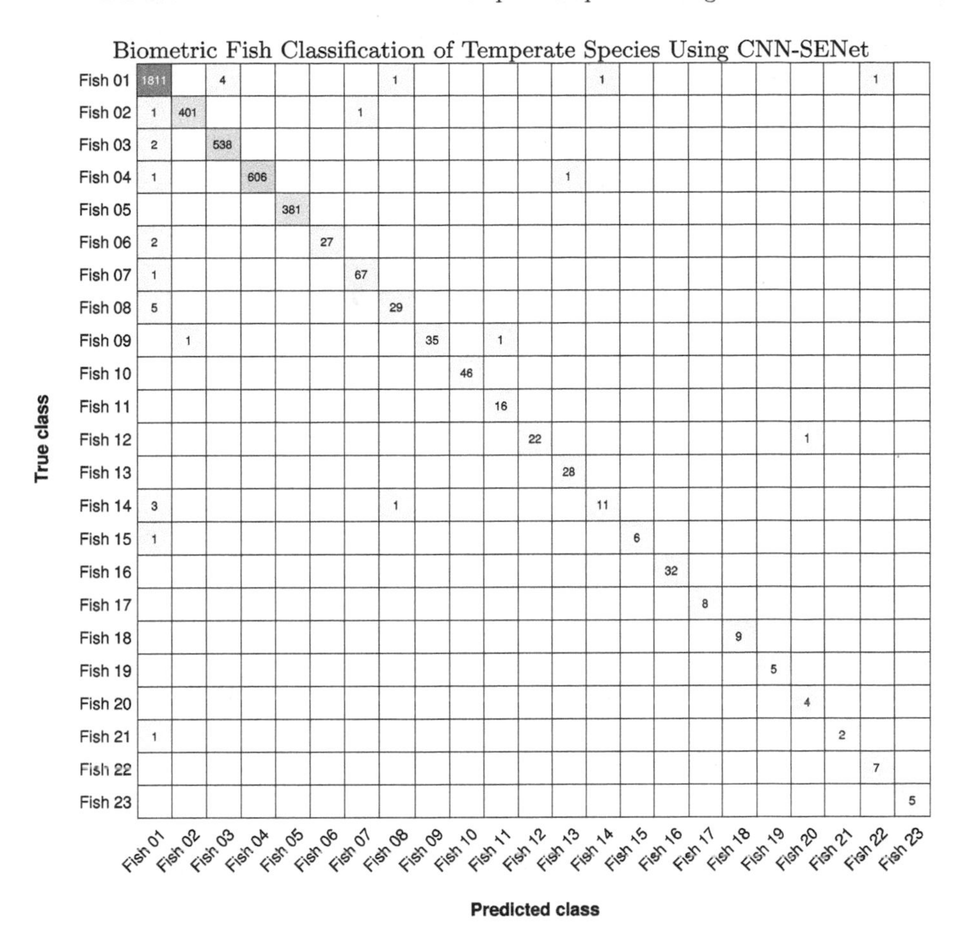

Fig. 4. Confusion matrix for Fish4Knowledge data set pre-training with CNN-SENet.

same classes were reserved for testing. Corkwing wrasse (male), Corkwing wrasse (female), Pollach and Coalfish were selected for the experiment as a reasonable number of images of different individuals under varying conditions was available for these species.

The post-training process consists of 50 epochs and a batch size of 8. The batch size was reduced, compared to pre-training, to compensate for the relatively small number of available temperate fish images. Weights from the pre-training step are loaded before initiating post-training, and post-training accuracy is evaluated using the weights from the final epoch.

The rationale for this post-training method is to make use of the more or less generic fish identification features learned from the large Fish4Knowledge data set. Post-training will then start with the network in a "fish-class-sensitive" state and proceed by learning specific features of the temperate species on top of this.

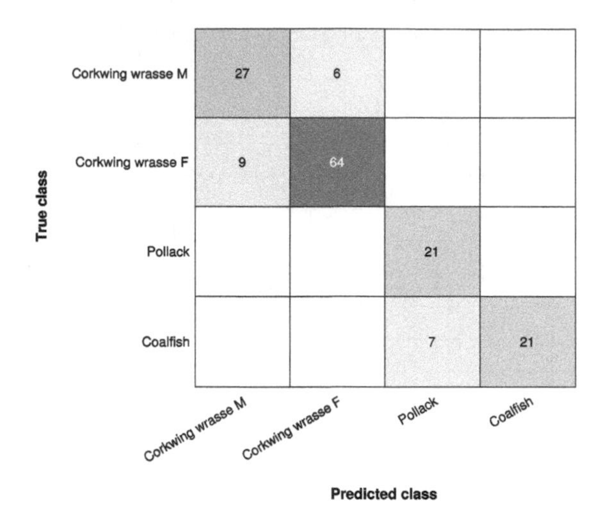

Fig. 5. Confusion matrix for temperate data set post-training with CNN-SENet.

Fish4Knowledge consists of images of 23 different classes. The selected subset of the temperate data set consists of 4 classes. To prepare the loaded pre-trained model for post-training, the last fully connected (FC) layer with 23 output neurons, suitable for 23 fish classes, is replaced with a similar layer with 4 output neurons.

Table 3. Average testing accuracy over 10 runs and time per epoch on post-training.

Network	Testing accuracy	Time one epoch
Inception-V3	85.42%	33 s
ResNet-50	82.39%	47 s
Inception-ResNet-V2	78.84%	91 s
CNN-SENet	83.68%	9 s
CNN-SENet without Squeeze-and-Excitation	82.32%	7 s

Post-training with Image Augmentation. Data augmentation techniques in machine learning aims at reducing overfitting problems by expanding a data set (base set) by introducing label-preserving transformations. For an image data set, this means that transformed copies of the original images in the base set are produced. These additional training data enables a network under training to learn more generic features, by reducing sensitivity to augmentation operations that transforms the image but not severely the characterizing visual features of for example a fish [8].

The main algorithm flow is the same as for the post-training version, but the data set was expanded by using the following transformation operations. Images are rotated randomly within a specific range, according to an uniform distribution. Images are vertically and horizontally shifted a random fraction of the image size. Scaling and shearing transformations are applied randomly, and lastly half of the images are flipped horizontally.

3.2 Results

Pre-training. Results from pre-training on Fish4Knowledge are presented in Table 2. The testing accuracy is on par with or exceeds the level of accuracy achieved with previous state-of-art solutions described in Sect. 1.

CNN-SENet with Squeeze-and-Excitation achieves 99.15% test accuracy, almost identical results as the Inception-V3 algorithm when it comes to accuracy. However, the run time for each epoch is roughly three times larger for Inception-V3. The training-runtime is expected to be reflected in prediction. CNN-SENet without Squeeze-and-Excitation is faster than the SE-version, but also slightly less accurate during these tests.

Inception-ResNet-V2 achieves the lowest test accuracy and also the highest time consumed for each epoch during training. The required input image size is 299×299, compared to 200×200 for the other networks under test. As the required resolution is higher than the resolution of most Fish4Knowledge images, the necessary upscaling process may negatively affect accuracy. Additionally, the larger input size also dramatically increases the computational complexity and leads to longer time on each epoch.

A confusion matrix for the CNN-SENet pre-training run is included as shown in Fig. 4. Fish 01 seems to attract more wrong predictions than the other species. The reason for this is unknown, but the imbalance in the data set could explain some of the behavior, as the ability to learn Fish 01 will be more rewarding during training as it occurs more frequently.

Post-training with and Without Image Augmentation. Results from the post-training experiment indicates that this is a more challenging image recognition task. Without image augmentation, the highest average testing accuracy achieved was 85.42% using the Inception-V3 CNN algorithm as listed in Table 3. CNN-SENet performance is few percent below, but with a significantly better training time for each epoch. All bench-marked algorithms show significantly reduced accuracy compared to the results from pre-training. The temperate species data set used for post-training is challenging, in the sense that it contains few images overall. The data set also consists of pictures of fish under low visibility conditions, and situations where the fish silhouette is not always prominent. A confusion matrix for the CNN-SENet post-training run is included as shown in Fig. 5.

Image augmentation, as described in Sect. 3.1, improves the results for post-training for all benchmarked algorithms, as shown in Table 4. The ResNet-50

Table 4. Average testing accuracy over 10 runs on post-training with image augmentation.

Network	Testing accuracy
Inception-V3	88.45%
ResNet-50	90.20%
Inception-ResNet-V2	82.39%
CNN-SENet	87.74%
CNN-SENet without Squeeze-and-Excitation	83.55%

network reaches just above 90% testing accuracy. CNN-SENet accuracy increases approximately four percentage points compared to post-training without image augmentation. The training time for each epoch does not change notably using image augmentation, so the metric was omitted from Table 4.

4 Conclusions

We propose a Convolutional Neural Network implementing the Squeeze-and-Excitation (CNN-SE) architecture, which is specifically tuned and trained for biometric classification of fish. The experimental results show that CNN-SENet achieves the state-of-the-art accuracy of 99.27% on the Fish4Knowledge data set without any data augmentation or image pre-processing. For post-training, where the CNN-SENet is specialized for recognizing temperate fish species, the achieved average accuracy is 83.68%. The lower accuracy can be explained by the small size of the new temperate species data set combined with high variation in image data. For both approaches, CNN-SENet with SE blocks has a higher accuracy than without the SE blocks, indicating that SE has a positive effect on accuracy. In conclusion, we show that CNN with SE architecture is a powerful and effective tool for automatic analysis of fish images taken in the wild, but future work should make use of much larger and well-labelled data sets.

References

1. Francour, P., Liret, C., Harvey, E.: Comparison of fish abundance estimates made by remote underwater video and visual census. Naturalista Siciliano **23**, 155–168 (1999)
2. Halvorsen, K.T., et al.: Male-biased sexual size dimorphism in the nest building corkwing wrasse (symphodus melops): implications for a size regulated fishery. ICES J. Mar. Sci. **73**(10), 2586–2594 (2016)
3. Halvorsen, K.T., Sørdalen, T.K., Vøllestad, L.A., Skiftesvik, A.B., Espeland, S.H., Olsen, E.M.: Sex- and size-selective harvesting of corkwing wrasse (Symphodus melops)-a cleaner fish used in salmonid aquaculture. ICES J. Mar. Sci. **74**(3), 660–669 (2017). Jonathan Grabowski, H. (ed.)

4. Hu, J., Shen, L., Sun, G.: Squeeze-and-excitation networks. CoRR abs/1709.01507 (2017)
5. Huang, P.X., Boom, B.B., Fisher, R.B.: Fish recognition ground-truth data (2013). Accessed 30 Jan 2018
6. Jin, L., Liang, H.: Deep learning for underwater image recognition in small sample size situations. In: OCEANS 2017-Aberdeen, pp. 1–4. IEEE (2017)
7. Kingma, D.P., Ba, J.: Adam: a method for stochastic optimization. CoRR abs/1412.6980 (2014)
8. Krizhevsky, A., Sutskever, I., Hinton, G.E.: ImageNet classification with deep convolutional neural networks. In: Advances in Neural Information Processing Systems, pp. 1097–1105 (2012)
9. Li, X., Shang, M., Qin, H., Chen, L.: Fast accurate fish detection and recognition of underwater images with fast R-CNN. In: OCEANS 2015 MTS/IEEE Washington, pp. 1–5. IEEE (2015)
10. Mallet, D., Pelletier, D.: Underwater video techniques for observing coastal marine biodiversity: a review of sixty years of publications (1952–2012). Fish. Res. **154**, 44–62 (2014)
11. Mclean, D.L., Harvey, E.S., Meeuwig, J.J.: Declines in the abundance of coral trout (Plectropomus leopardus) in areas closed to fishing at the Houtman Abrolhos Islands, Western Australia. J. Exp. Mar. Biol. Ecol. **406**(1), 71–78 (2011)
12. Pelletier, D., Leleu, K., Mou-Tham, G., Guillemot, N., Chabanet, P.: Comparison of visual census and high definition video transects for monitoring coral reef fish assemblages. Fish. Res. **107**(1), 84–93 (2011)
13. Perry, D., Staveley, T.A.B., Gullström, M.: Habitat connectivity of fish in temperate shallow-water seascapes. Front. Mar. Sci. **4**, 440 (2018)
14. Qin, H., Li, X., Liang, J., Peng, Y., Zhang, C.: DeepFish: accurate underwater live fish recognition with a deep architecture. Neurocomputing **187**, 49–58 (2016)
15. Redmon, J., Divvala, S.K., Girshick, R.B., Farhadi, A.: You only look once: unified, real-time object detection. CoRR abs/1506.02640 (2015)
16. Weinstein, B.G.: A computer vision for animal ecology. J. Anim. Ecol. **87**(3), 533–545 (2017)
17. White, D., Svellingen, C., Strachan, N.: Automated measurement of species and length of fish by computer vision. Fish. Res. **80**(2–3), 203–210 (2006)
18. Yosinski, J., Clune, J., Bengio, Y., Lipson, H.: How transferable are features in deep neural networks? CoRR abs/1411.1792 (2014)

Distance Metrics in Open-Set Classification of Text Documents by Local Outlier Factor and Doc2Vec

Tomasz Walkowiak(✉), Szymon Datko, and Henryk Maciejewski

Faculty of Electronics, Wrocław University of Science and Technology, Wrocław, Poland
{tomasz.walkowiak,szymon.datko,henryk.maciejewski}@pwr.edu.pl

Abstract. In this paper, we investigate the influence of distance metrics on the results of open-set subject classification of text documents. We utilize the Local Outlier Factor (LOF) algorithm to extend a closed-set classifier (i.e. multilayer perceptron) with an additional class that identifies outliers. The analyzed text documents are represented by averaged word embeddings calculated using the fastText method on training data. Conducting the experiment on two different text corpora we show how the distance metric chosen for LOF (Euclidean or cosine) and a transformation of the feature space (vector representation of documents) both influence the open-set classification results. The general conclusion seems to be that the cosine distance outperforms the Euclidean distance in terms of performance of open-set classification of text documents.

Keywords: Text mining · Subject classification · Open-set classification · Word embedding · fastText · Local Outlier Factor · Cosine distance · Standarization

1 Introduction

The classification of texts, i.e. automatically assigning a text to one of predefined subject groups, becomes a tool useful in many areas (like digital libraries, newspaper repositories, categorization of scientific papers, questions, and answering systems, selection of tourist offers). However, practical usage of such a tool requires an extension of classical closed-set classifiers to open-set ones. Since a classical approach associates a new document to one of the trained classes, even if the document is actually not related to any of them, it can lead to very spectacular mistakes made by close-set text classification tools, for example assigning a random text to some class.

Therefore, we propose an extension to the standard closed-set classification schema. First, we build a standard classification model using the available training dataset. Next, we utilize the Local Outlier Factor (LOF) [1] algorithm to extend the result with an additional class that identifies outliers. LOF provides a measure of dissimilarity (outlierness factor), which proves useful for

F. Wotawa et al. (Eds.): IEA/AIE 2019, LNAI 11606, pp. 102–109, 2019.
https://doi.org/10.1007/978-3-030-22999-3_10

high-dimensional data. Other approaches to open-set classification of text documents involve the utilization of statistical-based concepts, like inter-quartile range-based criteria, [13], or similarity estimation with simple threshold-based decision mechanism applied for the *aposteriori* probability [2,3]. There are also approaches based on the usage of convolutional neural networks [11].

Moreover, we analyze how the distance metric chosen for LOF and a transformation of the feature space (vector representation of documents) influence the open-set classification results.

The paper is organized as follows. Section 2 describes the doc2vec representation of documents, the method of open-set classification and Local Outlier Factor algorithm used by this method. In Sect. 4 we discuss the used distance metrics and methods of feature space transformation. Next section presents the used corpora, experiments, and results of the comparative study.

2 Open Set Classification

2.1 Doc2vec

Several approaches to representing documents by feature vectors are available like classical bag-of-words and a number of its modifications. Recently, word embedding methods gained large popularity. They mostly act as a lookup table that maps each word into a continuous multidimensional vector space [8]. Word2Vec allows constructing a feature vector of an entire document (doc2vec) by simple average of word embeddings [5]. Within this paper, we have used a recent deep learning method – fastText [5].

The main idea behind fastText is to perform word embedding and classifier learning in parallel. FastText forms the linear model, since it consists of word embeddings, simple averaging and linear soft-max classifier. Therefore it is very effective to train and use.

2.2 Classification

Having doc2vec vectors we can train a typical (closed-set) classifier using standard machine learning algorithms based on the training set. Next, it can be used to assign any new document (described by doc2vec values) to one of classes occurring in the training set. In other words, the classifier splits the feature space into areas related to the trained classes. Hence, they associate a new document with one of the trained classes (winning class), even if the document is actually not related to any of the classes (subject categories) known to the classifier.

To overcome this problem we propose post-processing of typical classifier results. It includes calculation of dissimilarity between the feature vector and the winning class feature vectors (of all documents from the training set that belongs to the winning class). It could be done by measuring the outlierness factor [13]. If the factor is above given threshold we reject the decision made by the classifier and assign the document to the 'outlier' class.

2.3 Outlierness Factor

As an outlierness factor is required in the procedure described above, we propose to use the Local Outlier Factor (LOF) [1]. It is a measure based on a weighted Euclidean metric, aiming to find outliers by comparing vectors to their local neighborhoods. It works by calculating an average distance between a given point, its neighbors and their (neighbors) neighbors to determine the local density of points in the given point's surrounding.

The open-set classification procedure proposed in Sect. 2.2 requires thresholds for each class. We could set it up assuming that the training data sets are contaminated by outliers, i.e. they include a given proportion of vectors with LOF values larger than the threshold. This proportion is called contamination[1].

3 Analyzed Distance Metrics and Transformations of Feature Space

Original LOF [1] is based on Euclidean distance (often called L_2 norm). However, Beyer et al. [6] suggest that L_2 norm fails in high dimensions. Whereas, the *cosine distance* (mostly in the form of cosine similarity) is widely and successfully used in the analysis of word2vec data [7], as well as in doc2vec [10]. The cosine distance is not vulnerable to any scaling of the given vector's size and it is assumed that it acts much better in high dimensional space.

The *standardization*, i.e. removing the mean and scaling to the unit variance, is a widely used transformation of data in the machine learning. It allows matching the requirement of normal data distribution that is assumed in many classification algorithms. It is known in the statistics also as the z-score. It could be seen as moving and linear scaling of the input data. A mean and a standard deviation are calculated on training data and later on used during open-set classification.

The *normalization* scales the vector to the unit norm. In contrary to the standardization, it operates on an individual object and does not require any parameters estimated on other data. Normalization maps each data point onto unit n-sphere. It has some interesting properties. The cosine distance in original and normalized space are equal. Moreover, the cosine distance between vectors is equal to half of the square of Euclidean distance between them. Therefore, the results of algorithms based on the nearest neighbor (as LOF, for example) should be almost the same when someone uses cosine distance or Euclidean one on normalized vectors [12].

4 Evaluation

4.1 Data Sets

To evaluate the performance of the proposed open set classification method and analyze the influence of distance metrics and feature space transformation in

[1] https://scikit-learn.org/0.19/modules/generated/sklearn.neighbors.LocalOutlierFactor.html.

a real task the text corpora of different subject classes are needed. We have used two data sets: texts from English newsgroups (20*newsgroups*) and Polish Wikipedia (*Wiki*) articles.

The first corpus (20*newsgroups*) is a commonly used collection of nearly 20,000 forum posts[2] divided into 20 subject categories. The data were divided into training and testing sets. However, for the purpose of open set classification we need also an outlier data. For this purpose, we have selected following categories: *misc.forsale*, *talk.politics.misc*, *talk.politics.guns*, *talk.politics.mideast*, *talk.religion.misc* and *alt.atheism*, then we removed them from the training data.

The second corpus (*Wiki*) consists of ca. 10,000 Polish language Wikipedia articles [9], coming from 34 subject areas. In this case, the outlier data consists of randomly selected articles from Polish press news [14]. The number of outliers was equal to the size of a test partition.

4.2 Experiment Overview

The proposed method was evaluated on corpora described in Sect. 4.1. Firstly (for each corpus), the word2vec model was built using the fastText algorithm (Sect. 2.1) on the training set. Next, doc2vec feature vectors, as an average of word2vec values for each word in a document were calculated, forming the feature vector space. For closed-set classification, the multilayer perceptron (MLP) [4] with Broyden–Fletcher–Goldfarb–Shanno (BFGS) nonlinear optimization learning algorithm was used. The MLP model was built on the training set. Then, a constructed model was examined on both testing sets, labeling all documents to trained categories (closed-set classification). Later the Local Outlier Factor measure was used to verify if the assignment to categories was correct and to catch incorrect labels, marking mismatched data as outliers (open-set classification).

Finally, knowing all true labels from the original data-set, the evaluation of classification was performed. We measured a number of correct decisions from all assignments made to a specific class (precision) and a number of correct decisions from all assignments expected to a specific class (recall). Then, a harmonic mean of these values, called **f1-score** was calculated. The results, reported later, are given as the average of f1-scores for each class weighted by support (the number of instances in each class). It is important to notice that 50% of the testing data consists of outliers (for open-set data) so they have important influence on the final results.

4.3 Results

In a Table 1 we report the f1-score calculated for closed-set and open-set tasks for 20*newsgroups* and *Wiki* corpora. The experiments were performed for 100 dimensional word embeddings and contamination parameter equal to 0.1. The second column is presented for a reference showing how the closed-set classifier performs in task dedicated for it (only the closed-set test data were used). Next

[2] http://qwone.com/~jason/20Newsgroups/.

columns show the results for open-set tasks (the closed-set test data and outliers are used). It can be noticed (second and third column) that introducing outlier data (50% of all documents) to a classical classification method (closed-set one) results in almost 2.5 time degradation of f1-score. However, the proposed by authors method (Sect. 2.2) is capable of improving the outcome (the fourth column) and achieve almost 63% or 54% of f1-score, depending on the outlier data set.

Next, we have analyzed the influence of feature vectors' transformation (none, standardization and normalization) and distance metric (Euclidean and cosine) on proposed method's performance (Table 2). It can be noticed that replacing a standard for LOF Euclidean distance with cosine one leads to the improvement of performance. Additional improvement, but only in case of *Wiki* corpus, is achieved when standardization is used. As it was mentioned in Sect. 3, the results for cosine distance with and without normalization are equal to the results obtained for Euclidean distance with normalization.

We have also tested other distance metrics available in SciPy[3] package. The results are not shown since they were never better than for cosine one and many times even worse than for Euclidean metric.

Figure 1 shows the relation between contamination parameter of Local Outlier Factor (Sect. 2.3) and f1-score for 20*newsgroups* and *Wiki* corpus respectively. We have shown (blue lines) the results for cosine and Euclidean distance, as well as cosine with standardization of vector space. For a reference, which

Table 1. Classification results: f1-score (word2vec dimension: 100, LOF contamination: 0.1)

Dataset	Closed-set	Open-set	
Method	Closed-set (MLP)		Open-set (MLP + LOF)
20 newsgroups	0.7949	0.3073	0.6292
Wiki	0.8333	0.3119	0.5361

Table 2. Open set classification results: f1-score (word2vec dimension: 100, LOF contamination: 0.1)

Distance	Transformation	20*newsgroups*	*Wiki*
Euclidean	-	0.6292	0.5361
Euclidean	Standardization	0.6248	0.5358
Euclidean	Normalization	**0.6512**	0.7339
Cosine	-	**0.6512**	0.7335
Cosine	Standardization	0.6438	**0.7902**
Cosine	Normalization	**0.6512**	0.7335

[3] https://docs.scipy.org/doc/scipy/reference/spatial.distance.html.

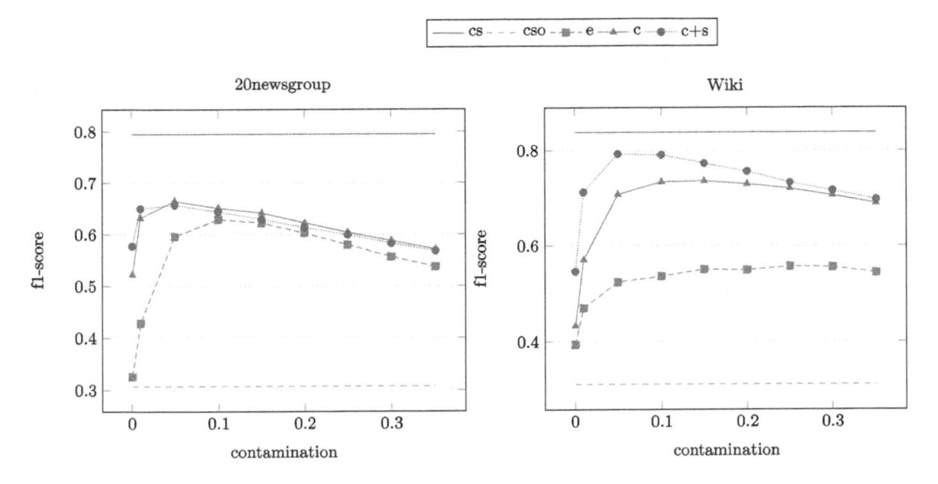

Fig. 1. f1-score for different values of contamination (cs - closed-set classifier without outliers, cso - closed-set classifier with open set data, e - open-set classifier with euclidean distance, c - open-set classifier with cosine distance, c+s - open-set classifier with cosine distance and standardization) (Color figure online)

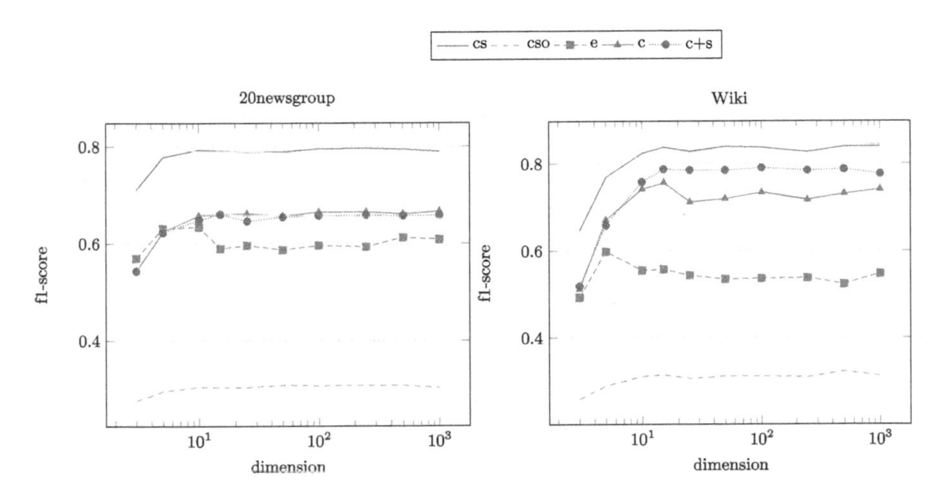

Fig. 2. f1-score in function of feature vector dimension (cs - closed-set classifier without outliers, cso - closed-set classifier with open set data, e - open-set classifier with euclidean distance, c - open-set classifier with cosine distance, c+s - open-set classifier with cosine distance and standardization)

could be interpreted as the bottom and top limit (red lines), we have also presented f1-scores for the closed-set task performed by the MLP on data with and without outliers. Obviously, these two do not depend on the contamination value (LOF is not involved). For the remaining three (open-set classification), when the values of contamination are rising, the f1-score is improving. Achieving its maximum at contamination value equal to 0.05 and 0.1 (only for 20*newsgroups* and pure Euclidean based LOF). After that, there is a decrease. Moreover, it could be noticed how cosine based LOF outperforms the Euclidean one, regardless of the data set and contamination value.

Next, we have analyzed the influence of word embedding dimensionality (range 3 – 1000) on the open-set classification performance. The results are presented in Fig. 2 for contamination parameter equal to 0.05 for 20*newsgroups* and 0.1 for *Wiki*. The impact of doc2vec dimension for values larger than 10 appears not very significant. This may suggest that fastText algorithm is so effective in finding well-distinguishing features (word embeddings) that after some specific dimension there is only insignificant redundancy introduced.

Again, we can notice that cosine distance outperforms the Euclidean one. However, it is much more significant for *Wiki* corpus.

5 Conclusion

In this work, we showed how to extend the standard closed-set classifier (MLP was used during reported experiments) for the open-set classification of text documents described by doc2vec feature vectors. It is done by utilizing the Local Outlier Factor on document embeddings. In the experiment, we evaluated the proposed method on a collection of nearly 20,000 forum posts in English and Wikipedia articles in Polish (with 34 subject areas). The results show that the proposed extension is capable to work effectively in an open set environment.

Moreover, we researched various distance metrics and measured their performance in the task of open-set classification of text documents. Results show that using the cosine distance metric in LOF procedure we reach highest overall score on both examined datasets.

We have also studied the effect of two common transformations of feature vectors - standardization and normalization. In case of one of data sets, standardization allowed to boost results, whereas normalization gives the same results as a usage of cosine distance.

Acknowledgement. This work was sponsored by National Science Centre, Poland (grant 2016/21/B/ST6/02159).

References

1. Breunig, M.M., Kriegel, H.P., Ng, R.T., Sander, J.: LOF: Identifying density-based local outliers. SIGMOD Rec. **29**(2), 93–104 (2000). https://doi.org/10.1145/335191.335388
2. Doan, T., Kalita, J.: Overcoming the challenge for text classification in the open world. In: 2017 IEEE 7th Annual Computing and Communication Workshop and Conference (CCWC), pp. 1–7. IEEE (2017)
3. Fei, G., Liu, B.: Breaking the closed world assumption in text classification. In: Proceedings of the 2016 Conference of the North American Chapter of the Association for Computational Linguistics: Human Language Technologies, pp. 506–514 (2016)

4. Hastie, T., Tibshirani, R., Friedman, J.: The Elements of Statistical Learning. SSS. Springer, New York (2009). https://doi.org/10.1007/978-0-387-84858-7. Autres impressions : 2011 (corr.), 2013 (7e corr.)
5. Joulin, A., Grave, E., Bojanowski, P., Mikolov, T.: Bag of tricks for efficient text classification. In: Proceedings of the 15th Conference of the European Chapter of the Association for Computational Linguistics: Volume 2, Short Papers, pp. 427–431. Association for Computational Linguistics (2017)
6. Beyer, K., Goldstein, J., Ramakrishnan, R., Shaft, U.: When is "nearest neighbor" meaningful? In: ICDT 1999 Proceedings of the 7th International Conference on Database Theory, pp. 217–235 (1999)
7. Mikolov, T., Chen, K., Corrado, G., Dean, J.: Efficient estimation of word representations in vector space. CoRR abs/1301.3781 (2013). http://arxiv.org/abs/1301.3781
8. Mikolov, T., Sutskever, I., Chen, K., Corrado, G.S., Dean, J.: Distributed representations of words and phrases and their compositionality. In: Advances in Neural Information Processing Systems, pp. 3111–3119 (2013)
9. Młynarczyk, K., Piasecki, M.: Wiki train - 34 categories, CLARIN-PL digital repository (2015). http://hdl.handle.net/11321/222
10. Pandey, N.: Density based clustering for cricket world cup tweets using cosine similarity and time parameter. In: 2015 Annual IEEE India Conference (INDICON), pp. 1–6 (2015). https://doi.org/10.1109/INDICON.2015.7443520
11. Prakhya, S., Venkataram, V., Kalita, J.: Open set text classification using convolutional neural networks. In: Proceedings of the 14th International Conference on Natural Language Processing, pp. 466–475. NLP Association of India, Kolkata (2017)
12. Qian, G., Sural, S., Gu, Y., Pramanik, S.: Similarity between Euclidean and cosine angle distance for nearest neighbor queries. In: Proceedings of the 2004 ACM Symposium on Applied Computing, SAC 2004, pp. 1232–1237. ACM, New York (2004). https://doi.org/10.1145/967900.968151
13. Walkowiak, T., Datko, S., Maciejewski, H.: Algorithm based on modified angle-based outlier factor for open-set classification of text documents. Appl. Stochast. Models Bus. Ind. **34**(5), 718–729 (2018)
14. Walkowiak, T., Malak, P.: Polish texts topic classification evaluation. In: Proceedings of the 10th International Conference on Agents and Artificial Intelligence - ICAART, vol. 2, pp. 515–522. INSTICC, SciTePress (2018)

Hydropower Optimization Using Deep Learning

Bernt Viggo Matheussen[1(✉)], Ole-Christoffer Granmo[2], and Jivitesh Sharma[2]

[1] Agder Energi AS, University of Agder, Grimstad, Norway
bernt.viggo.matheussen@ae.no
[2] Centre for Artificial Intelligence Research, University of Agder, Grimstad, Norway
{ole.granmo,jivitesh.sharma}@uia.no
https://www.ae.no, https://www.uia.no, https://cair.uia.no

Abstract. This paper demonstrates how deep learning can be used to find optimal reservoir operating policies in hydropower river systems. The method that we propose is based on the implicit stochastic optimization (ISO) framework, using direct policy search methods combined with deep neural networks (DNN). The findings from a real-world two-reservoir hydropower system in southern Norway suggest that DNNs can learn how to map input (price, inflow, starting reservoir levels) to the optimal production pattern directly. Due to the speed of evaluating the DNN, this approach is from an operational standpoint computationally inexpensive and may potentially address the long-standing problem of high dimensionality in hydropower optimization. Further on, our method may be used as an input for decision-theoretic planning, suggesting the policy that will give the highest expected profit. The approach also permits for a broader use of pre-trained neural networks in historical reanalysis of production patterns and studies of climate change effects.

Keywords: Scheduling · Markov · Hydrology · Reservoir · Energy · Water

1 Introduction

Operations of a multi-reservoir hydropower river system are challenging in many aspects. Depending on the system configuration, production planners must consider factors such as price developments, flooding risks, environmental constraints, and hydrological information, before they can decide upon the schedule of the hydropower production. Further on, forecasts of future inflow and price are highly uncertain. Additionally, the dynamics of a power system is non-linear. These properties make it difficult to optimize reservoir management. Indeed, the core of the problem involves a complex high-dimensional non-convex state space search. To cope with this complexity, researchers have for many decades developed optimization algorithms that can be used for decision support in daily hydropower operations [1]. Among the many techniques available, various linear

F. Wotawa et al. (Eds.): IEA/AIE 2019, LNAI 11606, pp. 110–122, 2019.
https://doi.org/10.1007/978-3-030-22999-3_11

and dynamic programming based methods seem to have gained a momentum compared to other methods. Some examples of these methods are linear programming (LP) [2,3], non-linear programming (NLP) [4,5] and dynamic programming (DP) [6,7].

Unfortunately, traditional methods have severe weaknesses. Most notably are the well know problems referred to as the "curse of dimensionality and modeling" [13]. Linear programming schemes, for instance, face difficulties with uncertain information, requiring extensive Monte Carlo simulation. Furthermore, it is difficult to capture non-linear relationships, e.g., the hydraulic head may not be represented correctly. Further, computational complexity multiplies with both the number of system components and the extent of the time horizon. Dynamic programming, on the other hand, handles input uncertainty. They scale linearly in the time dimension and capture non-linear and uncertain relationships. Such schemes generally require that models and system dynamics are fully known, at least at a probabilistic level. Further, computational complexity grows exponentially with the number of system components, making this class of schemes incapable of modeling larger systems in full complexity and detail. Temporal resolutions are often coarse (weekly, monthly), which results in an underestimation of flooding in smaller and steep catchments, resulting in sub-optimal solutions. Finally, in an operational setting, computation time is crucial for decision making, since the allocation of production units are synchronized with the physical markets. Due to their high computational demand, the classical techniques are slow to run, which is unpractical in time-constrained operations.

Several authors have applied meta-heuristic methods to mitigate some of these challenges. A recent paper [8] applies the so-called Firefly algorithm (FA) to the hydropower optimization problem. The algorithm is motivated by the grouping behavior of fireflies and has strong similarities to the particle swarm method [13]. The firefly technique is applied to the reservoir operation problem, and the authors claim that the Firefly algorithm is superior compared to genetic algorithms (GA). Another modern meta-heuristic method is the invasive weed optimization (IWO) algorithm [15]. IWO is a novel evolutionary algorithm inspired by colonizing weeds. Other methods include search algorithms such as gradient ascent/descent and simulated annealing. For more on this see [1,15,16] and citations therein.

Machine learning (ML) has further been used in water resources management. Labadie [1] refers to work starting in the 1970s shaping ideas on how to incorporate various ML concepts into the reservoir management problem. After that, a wide range of methodologies have appeared in the literature trying to improve either deterministic or stochastic methods. For example, DeRigo tried to use neural networks to estimate the Bellman function in stochastic dynamic programming [9]. Lee [10] demonstrated how the Q-Learning method in reinforcement learning was used on a two-reservoir river system. They claim that the method outperformed more classical techniques. Such an approach was further investigated and modified by [11] who used a tree-based reinforcement learning algorithm trying to identify optimal water reservoir operation for the Lake Como river system in Italy. The results showed improved performance compared to traditional (stochastic) DP methods. More recently Dariane used neural networks and reinforcement learning in combination with a meta-heuristic method (particle swarm) to optimize a river-system in Iran [13]. Lately, Sangiorgo used a NN-based ISO technique to optimize the Nile multi-reservoir system [14]. The results demonstrated among other things that NNs can reduce computational demand, facilitating real-time operation.

From the literature cited, we can see that different ML techniques have been applied to reservoir management problems around the world. Nevertheless, the application of these techniques to river systems located in the Nordic region is absent in the literature. The reason for this is not apparent, but in the Nordics, all power stations are connected to a competitive electricity market; thus both price and inflow must be accounted for in the optimization procedure. To our knowledge, the inclusion of price and inflow in an ML-based optimization algorithm has not been investigated before, for the Nordic region. It should be noted that the industry has used decision support tools based on the well-known LP and DP methods for more than a decade [3].

In this research we propose a modified ISO framework that has similarities to the work of [1,12–14]. Our method combines the use of historical data, synthetic time series, meta-heuristic optimization based on multi-start gradient ascent, and neural networks (NN), in an intricate interplay, explained in the next section. The ISO framework was chosen for several reasons. First of all, the hydraulics of a hydropower river system is strongly non-linear, so it was desirable with a method that could handle non-linearities in the optimization step. It is known that LP can find global optima for linear problems, but if the problem is not linear in the first place, then it may be questioned whether LP methods find "real-life" global optima. One of the advantages NNs provide in this setting is their low computational demand in operation after training has been completed. With proper tuning, they can run sufficiently fast on almost any device. Lastly, due to inflow and price forecasts being highly uncertain, our method must be able to deal with stochastic input. Our unique combination of methods supports the inclusion of price and inflow in the optimization and the use of neural networks in a theoretical decision planning concept. The method handles uncertainty (ensemble of price and inflow) and is based on a continuous model (as opposed to a discrete representation of states and actions). Such an approach has to our knowledge not been tested on a multi-reservoir system previously.

The rest of this paper is organized as follows. We first describe, in detail, the mathematical methods we propose using to resolve the hydropower optimization problem. Then we present the study area and available data used for evaluating the method. Finally, we discuss our results and provide conclusions and pointers for further research.

2 Method

The objective of this research is to develop a stochastic hydropower optimization algorithm that can maximize expected profit given an ensemble of forecasts of future reservoir inflow and market price. All relevant constraints and initial conditions must be taken into account. To do this we apply a modified version of the implicit stochastic optimization (ISO) method described in [1,12–14]. Figure 1 illustrates the components of our proposed architecture, inspired by the ISO framework.

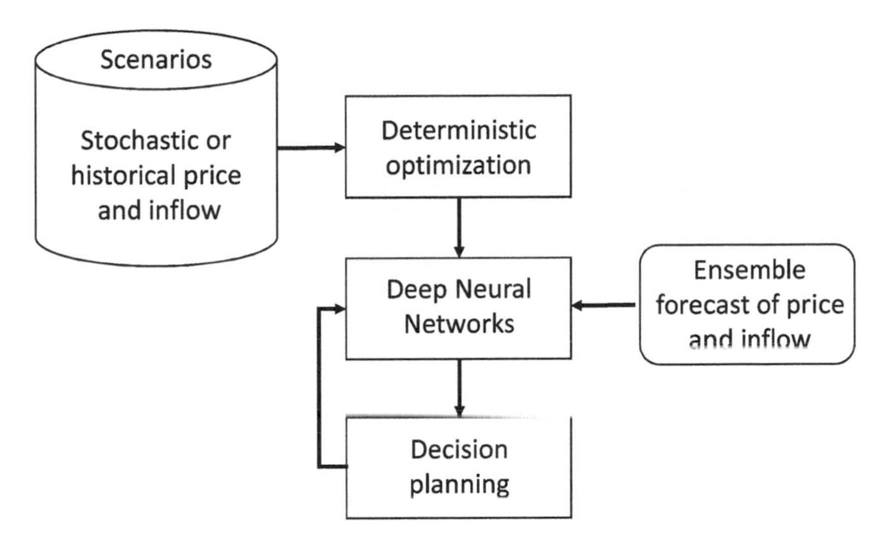

Fig. 1. Schematics of our implicit stochastic optimization (ISO) architecture, drawing upon previous work on ISO [1,12–14].

The first step in the method is to build Markov chains (MC) from the historical data of price and inflow. After that, we can sample from the Markov chains to generate an unlimited amount of scenarios with a given temporal resolution (e.g., hourly or daily) and planning horizon (e.g., 40 days). For each of these scenarios of price and inflow, we use a multi-start gradient ascent method to find the optimal operating policies. The reason for choosing this method was that it is a commonly used method in ML, it can find close to near-optimal solutions for many different problems, and it is easy to implement. It should be noted that any other deterministic optimization method can be used for this purpose (e.g., meta-heuristic methods). After a large number of scenarios have been generated and optimal operating policies for these scenarios have been found, we can use this input to train deep neural networks. In this work, we use dense multi-layer perceptron networks to make a mapping between the input scenario and the optimal operating policies. This mapping can then be used in a two-step procedure we refer to as "Decision planning". In all brevity, the first step of the decision planning is to use the DNNs to get a first estimate of the optimal policy for each of the ensemble of scenarios. Second, for each of the decisions found, we simulate the effect of this decision in one time step. Then we find the expected profit for the rest of the time horizon using another DNN. This allows us to effectively identify robust near-optimal strategies, as detailed in the following.

2.1 Time Series Generation with Markov Chains

For our time series generation module, we assume that there exist historical data of price and inflow for the hydropower system under consideration. The data are reported with a specified temporal resolution (e.g., daily time steps), but are continuous-valued. Based on the spread (min, max) and from visual inspection, the discrete time continuous-valued data may be turned into a discrete-valued sequence. From this we can compute the one-step state-transition probability matrix, defining a Markov Chain (MC) [16]. We use this matrix to sample an arbitrary amount of time sequences for training the DNN. To make the training data more diverse, we further added uniform noise to the discrete values, rendering the sampled sequences continuous-valued.

We define a scenario to be a discrete-time-continuous-valued sequence of a specified time horizon (T, e.g., 40 days) with price and inflow data. Besides this, a scenario also consists of initial reservoir levels as well as the expected power price at the end of the planning horizon (named rest price throughout the rest of this paper). An ensemble of such scenarios is illustrated in Fig. 3.

2.2 Deterministic Optimization of Operation Policies

For each of the scenarios sampled from the MCs, the optimal deterministic operating policy (i.e., water release decisions) must be identified. In this work, we decided to use a multi-start, one-at-a-time empirical gradient ascent method [16].

$$\nabla F = \left(\frac{\partial F}{\partial d_{1,0}}, \frac{\partial F}{\partial d_{1,1}}, \frac{\partial F}{\partial d_{2,0}}, \frac{\partial F}{\partial d_{2,1}}, \dots \frac{\partial F}{\partial d_{T,0}}, \frac{\partial F}{\partial d_{T,1}} \right) \tag{1}$$

The method involves finding the empirical gradient illustrated in Eq. 1 changing one parameter (hatch release or production level) for one time step at a time. After that, we check the change in profit and then resetting it back to original before checking next time step. After all time steps have been tested we find the time step with the highest change in total profit. After this, we repeat it all over. This procedure is then started with several different initial values of the policies (hatch release and production level).

Equations 2 to 7 defines the deterministic optimization problem that must be resolved for each of the scenarios (s). For each hydraulic node in the system (n), profit (π) is calculated as the income ($e_t * w_t$), then subtracting the operational costs (c). We also model the effect of an operation policy that violates operational constraints such as minimum reservoir levels, or start-stop costs of machinery. In practice, this would also include minimum flow requirements and other environmental constraints, but this is not included in this work. The value of the remaining water (I) in the reservoirs, at the end of the planning horizon, is estimated assuming a rest price (e_{T+1}), and an average reservoir height. The value is added to the total profit for each of the hydraulic nodes (power station or reservoir) in the system. This optimization (maximization) is carried out for each scenario (s) in the ensemble.

$$Max(F), \qquad F = \sum_{n=1}^{N} \pi_{n,s}(d) \tag{2}$$

$$\pi_{n,s}(d) = \begin{cases} \sum_{t=1}^{T} \left(e_t\, w_t - c_t\right) + I(x_{T+1}, w_{T+1}, e_{T+1}) & \text{if } powerstation \\ \sum_{t=1}^{T} \left(-c_t\right) + I(x_{T+1}, w_{T+1}, e_{T+1}) & \text{if } reservoir \end{cases} \tag{3}$$

Subject to:

$$w_t = \kappa(\eta \rho g(h(x_t) - h_{ps} - \gamma d_{t,n}^2) d_{t,n}) \tag{4}$$

$$d_{min,n} \leq d_t \leq d_{max,n} \tag{5}$$

$$x_{n,min} \leq x_{n,t} \leq x_{n,max} \tag{6}$$

$$x_{n,t} = x_{n,t-1} - \delta d_{t,n} + \delta \sum_{j=1}^{N_u} (o_{j,t} + q_{j,t} + d_{j,(t-\tau)}) \tag{7}$$

where:

e_t = Price for energy [Euro/MWh]
e_{T+1} = Rest price: expected price after the planning horizon [Euro/MWh]
w_t = Power production [MWh]
c_t = Costs for violating reservoir levels, or start-stop [Euro]
η = Turbine and generator efficiency [fraction, unit free]
ρ = Density of water [1000 kg/m^2]
g = Gravity [9.81 m/s^2]
h = Water elevation height at reservoir or power station [m]
γ = Friction loss coefficient [s^2/m^5]
$d_{t,n}$ = Water release decision from powerplant or reservoir [m^3/s]
$x_{n,t}$ = Filling in reservoir n, [m^3]
I = Expected profit of residual water [Euro]
q = Natural inflow to node [m^3/s]
o = Overflow from upstream node (reservoir) [m^3/s]
T = Number of time stages
N = Number of nodes, (N_u number of upstream nodes)
κ = Daily time step conversion factor (24*3600)/(1000000*3600)
δ = Time step length (24*3600) [s]

2.3 Decision Planning

In this work, we assume that production planners are using an ensemble of price and inflow scenarios to identify the optimal water release decisions for a given hydropower system. Despite the availability of an ensemble of scenarios, enhanced with optimal decisions, making the correct decision is still difficult. The reason is that we do not know in advance which scenario actually will play out. Instead, we get a possibly wide range of plausible decisions, dependent on the input scenario. To resolve the problem of singling out the best decision to make, given an uncertain future, we apply a two-step procedure. Firstly, for all the scenarios in the ensemble ($s \in S$), we use the previously trained neural networks to find the associated optimal release decisions ($d \in D$). This provides a distribution over plausible optimal decisions conditioned on the ensemble data. Only the release decisions in the first time step are considered, even the deterministic optimization algorithm have considered the whole planning horizon when generating training data. Secondly, for each of the release decisions, we realize the first time step, update the reservoir content (Eq. 7), and then expected (mean) profit for all scenarios is calculated with another trained neural network. In general, this means finding the release decisions for the first time step that gives the highest expected stochastic profit after the first time step has been realized. This is shown in Eq. 8. Since we are using neural networks that have already been trained in previous steps, the computational requirements are relatively low.

$$\underset{d_{t_1} \in \mathcal{D}}{\operatorname{argmax}}\, E[\pi \mid d_{t_1}] = \{d_{t_1} \mid d_{t_1} \in D \wedge \forall d^* \in D : E[\pi \mid d^*] \leq E[\pi \mid d_{t_1}]\} \tag{8}$$

To resolve Eq. 8 efficiently, the NNs (Fig. 1) must be trained to provide a mapping between inflow-price and the various release decisions. Also needed is a network that can provide the maximum profit for a given scenario. The reason for this is that Eq. 8 requires that each scenario's profit is estimated. From an operational point of view, the use of neural networks replaces the need for deterministic optimization using time-consuming heuristic methods. In this work, we study a simple "two-node" system, which requires three neural networks to be trained (hatch release, production, profit) – more on this in the results section.

3 Study Area and Available Data

In this research the hydropower system of Kvinesdal, located in southern-Norway, is used as a case study. A location map, and schematics of the system is shown in Fig. 2. The system consists of two reservoirs and one hydropower station. The uppermost reservoir, named Tjeldaasvatn, is laying at an average elevation of 312 m above sea level (m.a.s.l). The reservoir is connected to the downstream reservoir Stampetjonn through an open channel. The water release out of Tjeldaasvatn is controlled by a hatch that can release up to $1\,\text{m}^3/\text{s}$ of

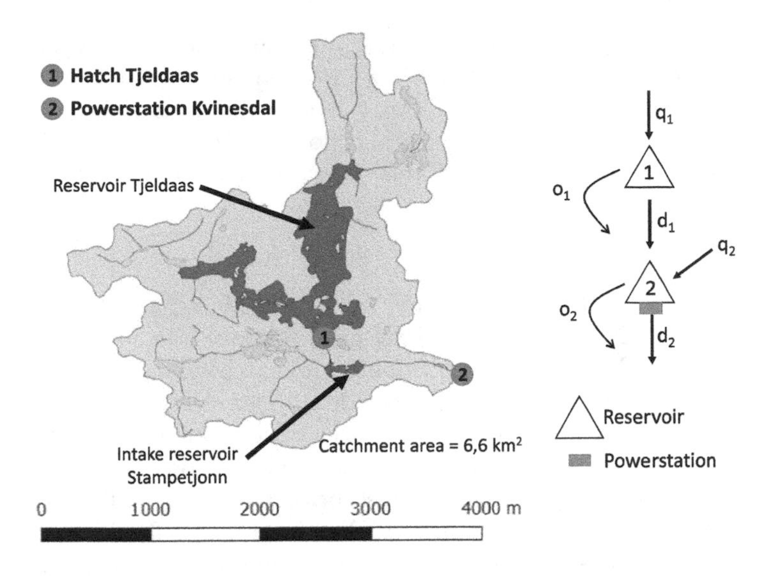

Fig. 2. Schematic of the Kvinesdal hydropower river system (southern-Norway).

water. The lowermost reservoir Stampetjonn serves as an intake reservoir to the Kvinesdal power plant. It is located at an average elevation of 302 m.a.s.l. Both the upper and lower reservoirs are constructed with overflow safety spillways that transports water into the old river bed during high water levels (above upper regulated water heights). During operations, water from the intake reservoir is transported through a $\sim$ 1 km circular pipe down to the power station at 38 m.a.s.l. Kvinesdal power plant has an installed maximum power capacity of 1.4 MW at a water usage of around 0.69 m^3/s. In Norway, this is considered to be a small station. During the period from 2007 until 2018 the power station produced energy for more than 90% of the time. This indicates relatively high water availability for the system. The powerplant is owned and operated by Agder Energi AS - a Norwegian provider of renewable energy. The plant is connected to the Nordic electricity grid and is part of the Nordic physical electricity market.

In this research, time series of price and inflow to the Kvinesdal system was provided from the historical archive of Agder Energi. The historical records covered the time period from January 1, 1996, until August 31, 2017, and had a temporal resolution of 24 h. The Pearson correlation (PC) between price and inflow was calculated to be −0.042, for the daily data over the time period 2007–2018. Data treated with a moving average filter (180 days) had a PC of 0.041. Based on this, it was assumed that there is neither dependence nor correlation between the market price and the inflow data. Thus, they can be treated separately for the rest of this work. It should be noted that the inflow data used in this work represents the inflow to the whole river system (catchment area 6.6 km^2). In the hydraulic calculations of the river system, local inflow to the system must be provided for both the upper and the lower reservoirs. This was resolved by splitting the inflow data into two time series scaled after the contributing area to the reservoirs. An effect of this is that both the upper and lower reservoirs receive local natural inflow at the exact time. Due to this research being a proof-of-concept, it was chosen to neglect this effect.

4 Results

The methodology used in this research involved training and testing of three neural networks (NN). These were all designed to make a mapping between the input (price and inflow of scenario), and output (hatch release, production level, and total profit) for a scenario. It was decided to use classical multi-perceptron fully connected (dense) feedforward networks. The input to the networks was chosen to be forty (thirty-nine for profit) days with inflow, price and the rank of the price. Besides the time series, rest price and initial fillings of the two reservoirs were also used as input to the NNs. A total of 116400 input scenarios with price, inflow, and so on, were prepared by sampling from the Markov chains fitted to the historical time-series data. All the scenarios were optimized with the multi-start gradient ascent method and the resulting policies (water release from the hatch, production level, profit) were used as output values (supervisory signal) to the neural networks. For hatch release and production level, data only for the first day of the planning horizon was used as the supervisory signal.

All the input data were normalized to have values between zero and one. Each neural network used five hidden layers in addition to the input and output layers. Hyper-parameters, width and depth and choice of activation functions were found by trial and error. A combination of hyperbolic tangents (tanh), rectified linear units (relu) and sigmoid activation functions showed the highest score on the objective criteria. It was decided to use 90 % of the available data as training for the neural networks and the remaining part for testing. The networks were trained using the RMSProb gradient algorithm [17] updated with the back-propagation method, optimizing for Mean Square Error (MSE) between the predictions and the supervisory data.

Table 1. Values of the objective criteria obtained during training and testing of the neural networks.

	Training				Testing			
Obj.crit	MSE	P.corr	MAE	Bias	MSE	P.corr	MAE	Bias
Prod	0.014	0.937	0.066	-0.001	0.025	0.889	0.089	-0.001
Hatch	0.012	0.954	0.057	-0.004	0.036	0.863	0.096	-0.001
Profit	0.000	0.999	0.005	0.002	0.000	0.999	0.005	0.002

Table 1 shows the results from the training and testing of the three deep neural networks (Prod., Hatch, Profit). In addition to MSE, three other objective criteria were used to quantify the performance of the NNs, i.e., Pearson correlation coefficient (P.corr), mean absolute error (MAE) and bias (difference in average). In general, it can be seen that the MSE is below 0.014 for the training period and below 0.036 for all the three networks in the testing period. The Pearson correlations are all around 0.94 and above. The MAE is around 0.06, and lower and bias is under 0.07. It can also be seen that the objective criteria are in general lower for the testing period. This is as expected since we test with data that has not been seen by the training algorithm. The results show that it is possible to make a mapping between inflow and price information and optimum production patterns for a hydropower system.

Figure 3, (a) and (b), illustrates an ensemble of price and inflow forecasts (scenarios) representative for the Kvinesdal hydropower system. The ensemble forecast has a planning horizon of forty days and was made available through the internal forecasting systems used by Agder Energi. The data were generated by the use of meteorological, hydrological and physically based power system models (price models). Due to time constraints and page limitations in this paper, it was chosen to neglect further details on how the forecasts were made. It was decided to use these data as external test data, assuming that they represent an actual stochastic forecast.

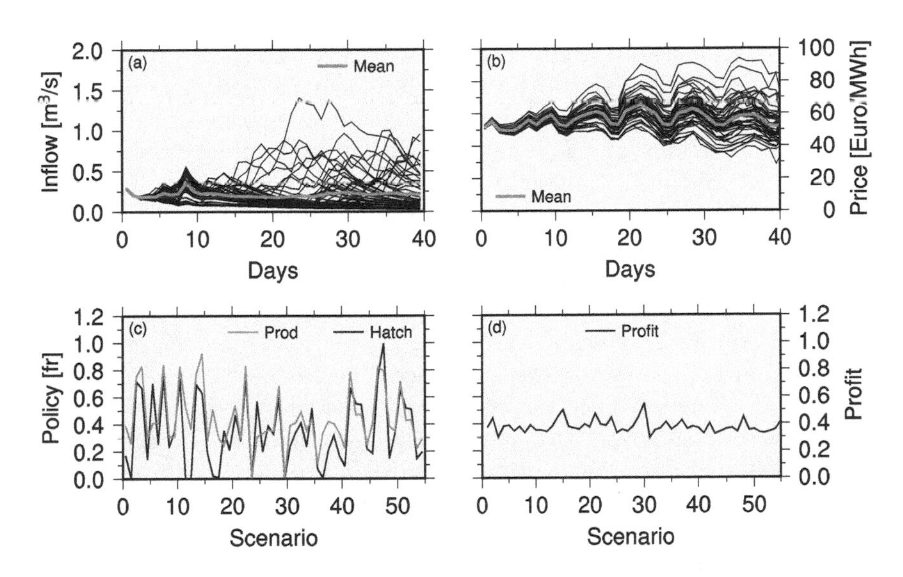

Fig. 3. Example of a real world ensemble forecast of inflow (a) and powerprice (b). Results from the decision planning method is shown in (c) and (d).

The ensemble forecast shown in Fig. 3, (a) and (b), were used as input to the decision theoretic planning approach described in earlier sections, and mathematically shown in Eq. 8. The method first calculates optimum hatch and production policies for each scenario using the Prod. and Hatch NNs. The results from this calculation is shown in Fig. 3(c). Secondly each of the optimum policies for the first day is then realized and new values of reservoirs levels are calculated after the first day. After that, the profit for all scenarios are computed for each of the policies, and the expectation (average profit) is calculated. The scenario with the highest profit corresponds to the policy that should be chosen as decision. In Fig. 3(d), the profit expressed as a fraction between zero and one is shown. It can be seen that it is for scenario 30 that we find the highest expected profit. In Fig. 3(c), we can see that this scenario has zero policy for both hatch and production. So the decision for the input ensemble shown in Fig. 3(a) and Fig. 3(b), would be 0.0 release of water from the upper to the lower reservoir. At the same time we should not produce since the policy for Prod is zero.

5 Discussion and Conclusions

This paper demonstrates how deep learning can be used together with relatively simple search algorithms to find optimal reservoir operating policies in hydropower river systems. The method is based on the ISO framework which uses ensemble input to make a stochastic optimization for a given system. The findings suggest that deep NNs can learn how to map input (price, inflow, starting reservoir levels) to the optimal production pattern directly. This approach is from an operational standpoint computationally in-expensive and may be utilized in several ways in the future. First, the NNs may be used to provide starting policies for metaheuristic optimization of longer time horizons or scenarios with higher temporal resolutions. This may potentially disseminate the problem of "curse of dimensionality". Although, as pointed out by Dariane [13], if the neural networks are trained indirectly through the use of metaheuristic algorithms, they may be slow to run for complex systems with long time horizons. Secondly, historical re-analysis of optimum production patterns is hard to do with classical methods, due to extreme computational demands, but this is not an issue when using NNs. Such abilities can be useful when studying the effects of climate change on hydropower operations since long time series must be used in climate studies. Further, since the proposed method has a distinction between the hydraulic simulator and the neural networks, there are no limitations in what type of constraints that can be included in the model. Such constraints may be ramping restrictions on how fast water release are allowed to change, it could also be minimum flow requirements or even identification of optimal water release for salmon habitat.

One weakness in this work is that the proposed optimization algorithm has only been tested for one example data set. In the future, the method must be tested for a larger number of cases and be benchmarked against other stochastic optimization algorithms. It is a paradox that the real global optimum for many situations may never be found although we run multiple algorithms on powerful computers. The methods we apply to the reservoir problem may only be tested and benchmarked against each other. Further on, the proposed method should also be tested on more complex hydropower system, with several more reservoirs and power stations, to see if it scales. A coarser temporal resolution should also be used in the analysis.

Acknowledgements. This research has been supported by Agder Energi, the Norwegian Research Council (ENERGIX program), and the University of Agder. We also thank Jarand Roeynstrand at Agder Energi for input on an early version of this article.

References

1. Labadie, J.W.: Optimal operation of multi-reservoir systems: state-of-the-art review. J. Water Resour. Plann. Manag., 93 (2004). https://doi.org/10.1061/(ASCE)0733-9496(2004)130:2(93)
2. Yoo, J.-H.: Maximization of hydropower generation through the application of a linear programming model. J. Hydrol. **376**, 182–187. https://doi.org/10.1016/j.jhydrol.2009.07.026
3. Belsnes, M.M., Wolfgang, O., Follestad, T., Aasgård, E.K.: Applying successive linear programming for stochastic short-term hydropower optimization. Electr. Power Syst. Res. **130**, 167–180 (2016)
4. Moosavian, S.A.A., Ghaffari, A., Salimi, A.: Sequential quadratic programming and analytic hierarchy process for nonlinear multiobjective optimization of a hydropower network. In: Optimal Control Applications and Methods. https://doi.org/10.1002/oca.909
5. Jothiprakash, V., Arunkumar, R.: Multi-reservoir optimization for hydropower production using NLP technique. KSCE J. Civ. Eng. **18**(1), 344–354 (2014)
6. Allen, B., Bridgeman, R.G.: Dynamic programming in hydropower scheduling. J. Water Resour. Plann. Manag., ASCE. https://doi.org/10.1061/(ASCE)0733-9496(1986)112:3(339)
7. Zhao, T., Zhao, J., Yang, D.: Improved dynamic programming for hydropower reservoir operation. J. Water Resour. Plann. Manag. **140**(3), 365–374 (2012)
8. Garousi-Nejad, I., Bozorg-Haddad, O., Loáiciga, H.: Modified firefly algorithm for solving multi-reservoir operation in continuous and discrete domains. J. Water Resour. Plann. Manag. 142(9) (2016). https://doi.org/10.1061/(ASCE)WR.1943-5452.0000644
9. DeRigo, D., Rizzoli, A.E., Soncini-Sessa, R., Weber, E., Zenesi, P.: Neuro-dynamic programming for the efficient management of reservoir networks. In: Proceedings of MODSIM 2001 International Congress on Modelling and Simulation, Vol. 4, pp. 1949–1954. Modelling and Simulation Society of Australia and New Zealand, December 2001. ISBN 0-867405252
10. Lee, J.H., Labadie, J.W.: Stochastic optimization of multi-reservoir systems via reinforcement learning. Water Resour. Res. **43**, W11408 (2007). https://doi.org/10.1029/2006WR005627
11. Castelletti, A., Galelli, S., Restelli, M., Soncini-Sessa, R.: Tree-based reinforcement learning for optimal water reservoir operation. Water Resour. Res. **46**, W09507 (2010). https://doi.org/10.1029/2009WR008898
12. Loucks, D.P.: Implicit stochastic optimization and simulation. In: Marco, J.B., Harboe, R., Salas, J.D. (eds.) Stochastic Hydrology and its Use in Water Resources Systems Simulation and Optimization. NATO ASI Series (Series E: Applied Sciences), vol. 237. Springer, Dordrecht (1993). https://doi.org/10.1007/978-94-011-1697-8_19

13. Dariane, A.B., Moradi, A.M.: Comparative analysis of evolving artificial neural network and reinforcement learning in stochastic optimization of multi-reservoir systems. Hydrol. Sci. J. **61**(6), 1141–1156 (2016). https://doi.org/10.1080/02626667.2014.986485
14. Sangiorgio, M., Guariso, G.: NN-based implicit stochastic optimization of multi-reservoir systems management. Water **10**, 303 (2018)
15. Azizipour, M., Ghalenoei, V., Afshar, M.H., Solis, S.S.: Optimal operation of hydropower reservoir systems using weed optimization algorithm. Water Resour. Manag. **30**, 3995–4009 (2016). https://doi.org/10.1007/s11269-016-1407-6
16. Russel, S., Norvig, P.: Artificial Intelligence: A modern Approach, 3rd edn., pp. 131 or 589. Prentice Hall. ISBN 13 978-0-13-604259-4
17. Keras documentation. https://keras.io/optimizers/

Towards Real-Time Head Pose Estimation: Exploring Parameter-Reduced Residual Networks on In-the-wild Datasets

Ines Rieger(✉), Thomas Hauenstein, Sebastian Hettenkofer, and Jens-Uwe Garbas

Fraunhofer-Institute for Integrated Circuits IIS, Am Wolfsmantel 33, 91058 Erlangen, Germany
ines.rieger@iis.fraunhofer.de

Abstract. Head poses are a key component of human bodily communication and thus a decisive element of human-computer interaction. Real-time head pose estimation is crucial in the context of human-robot interaction or driver assistance systems. The most promising approaches for head pose estimation are based on Convolutional Neural Networks (CNNs). However, CNN models are often too complex to achieve real-time performance. To face this challenge, we explore a popular subgroup of CNNs, the Residual Networks (ResNets) and modify them in order to reduce their number of parameters. The ResNets are modified for different image sizes including low-resolution images and combined with a varying number of layers. They are trained on in-the-wild datasets to ensure real-world applicability. As a result, we demonstrate that the performance of the ResNets can be maintained while reducing the number of parameters. The modified ResNets achieve state-of-the-art accuracy and provide fast inference for real-time applicability.

Keywords: Head pose estimation · Residual Network · Real-time

1 Introduction

Head poses are a key aspect of human non-verbal communication. As a consequence, automatic head pose estimation plays an important role in human-computer interaction. Several use cases in real-world scenarios include head pose estimation: In autonomous driving, head poses are used to estimate the driver's level of attention. Inattentive drivers can then be encouraged to focus on the road again [16]. In order to diminish the risk of collisions, the attention level of the surrounding pedestrians can also be estimated from their head poses [1,9]. Head pose estimation is further used as one of the key aspects in real-time human-robot interaction, e.g. in domestic environments [26] to provide a natural

F. Wotawa et al. (Eds.): IEA/AIE 2019, LNAI 11606, pp. 123–134, 2019.
https://doi.org/10.1007/978-3-030-22999-3_12

interaction mode with its users. In behavioural studies, head poses can be used to identify social groups [17] or a person's target of interest [21]. Head poses are also part of the Facial Action Coding System (FACS) [8] to decode emotions and therefore contribute to the interpretation of facial expressions [13]. Thus, real-time head pose estimation is crucial for several real-world applications, but still faces challenges such as slow inference time or robustness for these settings. This paper aims at providing a real-time solution by exploring small Convolutional Neural Network (CNN) models for low resolution input images trained on in-the-wild datasets.

CNNs, a specialized kind of feed-forward neural network, have proven to be advantageous for various image and video processing tasks such as object detection or object recognition [18]. One particular successful CNN architecture is the Residual Network (ResNet) architecture [10], which provides effective training with very deep networks through shortcut connections. The shortcut connections enclose blocks of stacks of convolutional layers and enable a second way to propagate information forward and backward through the network. Veit et al. [31] proved that not all blocks contribute equally to the learning process by investigating the gradient flow. Presumably, a more shallow ResNet architecture could learn the same representations and perform as well as a deeper ResNet architecture. Based on this assumption and the overall success of ResNets, this paper contributes to the field of real-time head pose estimation by exploring various modified ResNets. We can summarize our key contributions as follows:

- We start to reduce the model parameters by adapting the original ResNet architecture for training with images of 112×112 pixels instead of 224×224 pixels. We further reduce the parameters by adapting the 18-layer ResNet for low-resolution images of 64×64 pixels. This 18-layer ResNet contains less parameters than the 18-layer ResNet originally proposed by He et al. [10].
- The modified ResNets are evaluated on two in-the-wild datasets: The *Annotated Facial Landmarks in the Wild* (AFLW) dataset [14] and the *Annotated Faces in the Wild* (AFW) benchmark dataset [34]. In-the-wild datasets ensure real-world applicability, which is important for use cases such as driver assistance systems or human-robot interaction. These datasets include no depth information.
- The performance of the implemented ResNets is evaluated with a five-fold cross-validation on the AFLW dataset and with a five time training-testing cycle on the AFW dataset. Multiple training cycles not only contribute to the robustness of the results, but also mitigate the non-deterministic behaviour of multi-thread training on GPUs. The results are measured in mean absolute error and accuracy.
- We compute the number of parameters and measure the inference time on a CPU and on a GPU. Low model complexity and the corresponding fast inference time is important for real-time applications.

The ResNet models are trained to estimate the head poses represented by Euler Angles, which measure the orientation of a rigid body in a fixed coordinate system [4].

2 Related Work

Head pose estimation approaches can be grouped in appearance-based methods, model-based methods and nonlinear regression methods.

Appearance-based methods compare new head images with a set of exemplary, annotated heads and pick the most similar one [2,24]. Despite the advantage of a simple implementation and an easy extension for new heads, there is a huge disadvantage: The method is based on the premise that similar images also have similar head poses, and thus ignore the impact of identity.

In contrast to appearance-based methods, model-based methods follow a geometric approach by not taking the whole face into account, but only certain facial key-points. One approach uses the POSIT algorithm [3] to fit an averaged 3-dimensional facial model onto a 2-dimensional face image annotated with facial key-points, and then computes the head pose [14]. Another approach is to measure the distance of the facial key-points of the 2-dimensional image to a reference coordinate system [22]. The drawback of the model-based approaches is the need of high accuracy in the facial key-point detection. Estimating head poses from images with occluded face regions is therefore difficult.

To cover the complex feature space required for head pose estimation in images in in-the-wild settings, nonlinear regression methods can present a solution. The first nonlinear regression methods for head pose estimation were support vector regression [23], random forests [5,6] and multilayer perceptrons (MLP) [28,29]. With the rise of computational power, CNNs emerged around 2007 in the field of image based head pose estimation. In contrast to MLPs, CNNs display a high tolerance to shift and distortion variance. There are several recent approaches that employ the in-the-wild datasets AFLW and AFW for training and use the AFW dataset as a testing benchmark: Patacchiola and Cangelosi [25] compare various LeNet-5 [19] variants trained with different gradient and adaptive gradient methods. Ruiz et al. [27] train a 50-layer ResNet with a combined loss function of mean squared loss and cross entropy loss for all three angles. They achieve good results and outperform Patacchiola and Cangelosi. Kumar et al. [15] use a Heatmap-CNN (H-CNN) that learns local and global structural dependencies for detecting facial landmarks and estimating the head pose. The H-CNN includes Inception modules [30] that consist of parallel threads of stacked convolutional layers and therefore display an architecture similar to ResNets. Hsu et al. [12] train their multi-loss CNN based on a combined L2 loss regression and ordinal regression loss. To counteract the *gimbal lock* [20], an ambiguity problem in the Euler angle representation, they use quaternions as head pose representation. Hsu et al. [12] find that their pretrained Quaternion Net outperforms their network using Euler Angles for head pose representation. Wu et al. [32] train their combined face detection network on an augmented AFLW dataset combined with an unreleased own head pose dataset. An evaluation on the AFW dataset achieves state-of-the-art performance. Zhang et al. [33] use a cross-cascading regression network with two submodules, one for facial landmark detection and one for head pose estimation and achieve state-of-the-art performance on the AFLW dataset.

3 Residual Networks (ResNets)

The ResNet is one of the most popular architectures for image processing with very deep neural networks. The architecture was proposed by He et al. [10], who won benchmark competitions like the ImageNet Large Scale Visual Recognition Competition (ILSVRC) 2015[1] and the Common Objects in Context (COCO)[2] competition with an ensemble of ResNets. They have also proven the successful training with over one thousand layers [10].

3.1 Original ResNets

ResNets use the concept of shortcut connections with the effect that the input of the subsequent layers does not only contain the information of the immediate preceding layer, but from all preceding layers. This is contradictory to hierarchical CNNs, where the input of the layers does only contain information of the directly preceding layer. The shortcut connections can resolve the problem of vanishing or exploding gradients and the degradation of the training error in hierarchical CNNs. The degradation of the training error describes that the training and test error increases, when the network depth grows.

A shortcut connection is called an *identity shortcut*, when the input and output dimension stays the same within a block. Identity shortcuts do not introduce additional parameters in the network. When the dimension increases, He et al. [10] consider two options: (1) Identity mapping with extra zero entries or (2) projection shortcuts using 1×1 convolutions. As ResNets do not use as many filters as classic CNNs to achieve the same depth, they are parameter-reduced.

The residual block (see Fig. 1 (right side)) described by He et al. [10] uses the Rectified Linear Unit (ReLU) activation function with a preceding batch normalization layer. The shortcut connection encloses a block of two convolutional layers $F(x_l)$, where the input x_l is added to the result of the two convolutional layers, resulting in x_{l+1} in the forward propagation (see Eq. 1). This equation does not hold for blocks using the projection shortcut, but since the likelihood for such blocks is low, He et al. [10] do not expect this to have a great impact.

$$x_{l+1} = x_l + F(x_l) \tag{1}$$

In the backpropagation, the incoming gradient is split into two additive terms: The error is propagated through the shortcut connection and through the residual function, where the weights of the convolutional layers are adjusted. Since any gradient is a summation, they are not likely to vanish.

To enhance the performance of the ResNet, He et al. [11] propose a pre-activated residual block, where the Batch Normalization layer is placed before instead of after the convolutional layer. The presented experiments in this paper use the pre-activated form of the residual block.

---

[1] http://image-net.org/challenges/LSVRC/2015/, accessed 14.12.2018.

[2] http://cocodataset.org/#detections-challenge2015, accessed 14.12.2018.

3.2 Modified ResNets

The challenging aspect of real-time applications is the complexity of the trained models and the resulting computing time. To address this challenge, we explore three parameter-reduced ResNets of different depths and for different image sizes.

These ResNets (see Table 1) are modified to process images with a lower resolution of 112 × 112 pixels and 64 × 64 pixels instead of 224 × 224 pixels as in the original ResNet [11]. As a first step, the stride of the first convolutional layer is changed to one, so the size of the feature map is not reduced. This is different to the original ResNet, where a stride of two is used. Consequently, as the convolutional layer does not reduce the size of the feature map, the ResNet can process input images with a size of 112 × 112 pixels instead of 224 × 224 pixels. The ResNet34-112 and ResNet18-112 are modified to take 112×112 pixels as input. Their layers are divided in four stacks similar to [10]. The smallest proposed ResNet of He et al. [10] is a ResNet with 18 layers divided in four stacks. To further reduce the parameters, we propose the ResNet18-64, which uses only three stacks instead of four stacks. This allows low resolution inputs of 64 × 64 pixels while significantly decreasing the number of parameters.

Table 1. Overview of modified ResNets

ResNet model	Input size	Stacks	Layers	Parameters
ResNet34-112	112 × 112 pixels	[3,4,6,3]	34	21.27×10^6
ResNet18-112	112 × 112 pixels	[2,2,2,2]	18	11.17×10^6
ResNet18-64	64 × 64 pixels	[2,3,3,0]	18	4.25×10^6

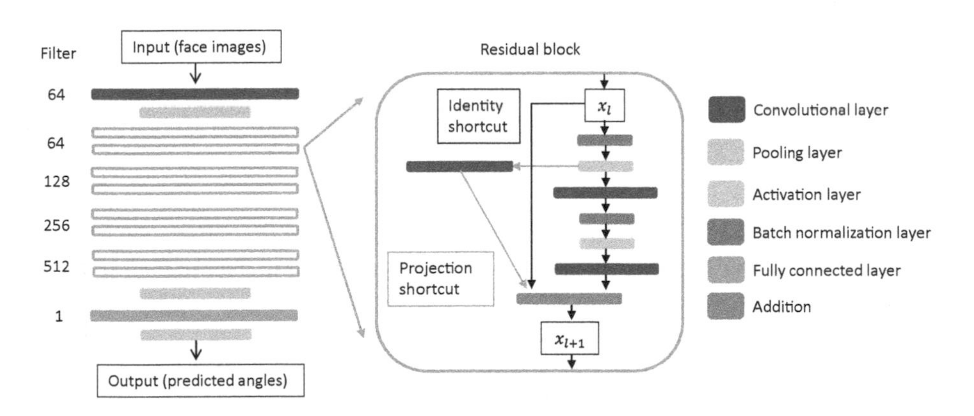

Fig. 1. Modified ResNet architecture with 18 layers divided in four stacks for inputs of 112 × 112 pixels (ResNet18-112).

The modified ResNets (see Fig. 1) use pre-activated residual blocks with projection shortcuts for an increase in dimensionality and identity shortcuts, when the dimension stays the same. In contrast to the original ResNet [11], the hyperbolic tangent function is used as the activation function instead of the ReLU function. The tangent function computes values in the range of [−1, 1] and the labels and image pixel values are normalized to values in this range as well. Furthermore, the projection shortcuts for increasing the dimensions are placed after instead of before the first set of batch normalization and tangent activation layer, as the batch normalization provides a regularization effect of the image pixel values.

4 Datasets

To train the models for real-life settings, two in-the-wild datasets are used for training and testing.

4.1 Annotated Facial Landmarks in the Wild (AFLW)

The AFLW dataset [14] provides a large variety of different faces with regard to ethnicity, pose, expression, age, gender and occlusion. The faces are in front of natural background under varying lighting conditions. The dataset contains 25,993 annotated faces in 21,997 images. The license agreement does not allow publication of the AFLW database.[3] All images are annotated with face coordinates and the three angles yaw, pitch and roll. 56% of the faces are tagged as female and 44% are tagged as male. Koestinger et al. [14] state that the rate of non-frontal faces of 66% is higher than in any other dataset. The distribution of poses of the AFLW dataset is not uniform, showing fewer images with a strong head rotation. The yaw angle has a range from $-125.1°$ to $168.0°$, the pitch angle from $-90.0°$ to $90.0°$ and the roll angle from $-178.2°$ to $179.0°$. The head poses were computed with the POSIT algorithm using manually annotated facial key-points [3]. However, it is worth noting that the resulting head poses were not manually verified. The AFLW dataset has extremely wide ranges for all angles, which supersede realistic head movements by far [7].

4.2 Annotated Faces in the Wild (AFW)

The AFW dataset [34] shows a wide variety of ethnicity, pose, expression, age, gender and occlusion. The license agreement does not allow publication of the AFW database. The faces are positioned in front of natural cluttered backgrounds. There are 468 faces in 205 images. An annotation of the angles yaw, pitch and roll as well as face coordinates are provided. The yaw angle has a range from $-105°$ to $90°$, the pitch angle from $-45°$ to $30°$ and the roll angle from $-15°$ to $15°$, all annotated manually in steps of $15°$. Since the yaw angle has the widest range, this angle is normally used when testing with this dataset.

[3] https://www.tugraz.at/institute/icg/research/team-bischof/lrs/downloads/aflw/, accessed 26.03.2019.

5 Experiments

In this section, we describe the pre-processing, training parameters, evaluation methods and results of the trained models including a comparison to other state-of-the-art approaches regarding the performance and number of parameters.

5.1 Pre-processing

As explained in Sect. 4.1, some samples in the AFLW dataset are annotated with unrealistic values. Furthermore, few images are provided for extreme angles. Following the approach of Patacchiola and Cangelosi [25], we filter the dataset and only keep images in the following label ranges: $\pm 100°$ for the yaw angle, $\pm 45°$ for the pitch angle and $\pm 25°$ for the roll angle. The yaw angle of the AFW dataset is also restricted to $\pm 100°$, as this angle is used for testing the trained networks. Both datasets are converted to greyscale. We crop the images using the annotated face coordinates. Each image is scaled down to 112×112 pixels and 64×64 pixels respectively. Face images smaller than the required size are left out. For the AFW dataset, the use of face images greater than 150 pixels is an additional constraint, following the protocol in [34]. To normalize the values, the labels are rescaled from $[-100, 100]$ to $[-1, 1]$ and the pixel values are rescaled from $[0, 255]$ to $[-1, 1]$.

In total, four datasets are prepared for our training: AFLW-112, AFLW-64, AFW-112 and AFW-64. The total amount of face images is 16,931 in AFLW-112, 20,872 in AFLW-64, 325 in AFW-112 and 352 in AFW-64.

5.2 Methods

The proposed ResNet architectures were implemented using TensorFlow and trained on a Nvidia Tesla P100 GPU. The pre-processing, training and evaluation is implemented in one pipeline. The convolutional weights are initialized with the variance scaling initializer and trained with an initial learning rate of 0.1, which is decreased by the factor 10 after 30, 60, 80 and 90 epochs. The weight decay λ of the L2 regularization excludes the loss of the batch normalization layers and has a value of 0.0002. We use a batch size of 256. All modified ResNets are trained separately for each angle.

There are two training and testing procedures: (1) Five-fold cross-validation with the AFLW-112 and AFLW-64 dataset and (2) Five training-testing cycles with training on the whole AFLW-112 or AFLW-64 dataset and testing on the AFW-112 or AFW-64 dataset (see Fig. 2). The ResNet18-112 and ResNet34-112 are trained for 200 epochs in both cases and the ResNet18-64 is trained for 120 epochs in case (1) and for 150 epochs in case (2). The epoch number was determined empirically.

The results are measured in mean absolute error (MAE) (see Eq. 2), where $\hat{y}$ describes the predicted values and y the true values in degrees. The number of testing examples is n.

$$\frac{1}{n}\sum_{i=1}^{n}|\hat{y}_i - y_i| \tag{2}$$

As in other approaches, the predicted and true values are mapped on discrete categories with a size of 15° (i.e. ...,]−7.5,7.5],]7.5,22.5],...) to predict the accuracy. If the predicted value is in the same category as the true value, the predicted value is classified as correct, otherwise as incorrect. A further applied evaluation method considers the mapped predicted and true values as true, if it matches the true category or the adjoining categories. This gives a range of 45°, where the predicted value can be classified as correct.

The mapping of the true and the predicted values on categories is problematic, because the cases where these values are located near the borders of the categories can distort the result. Furthermore, it is questionable, if the evaluation method with mapped categories ±15° error has high significance, as a wide range of degrees is considered as a correct prediction. The MAE on the other hand provides a clear interpretation of the results.

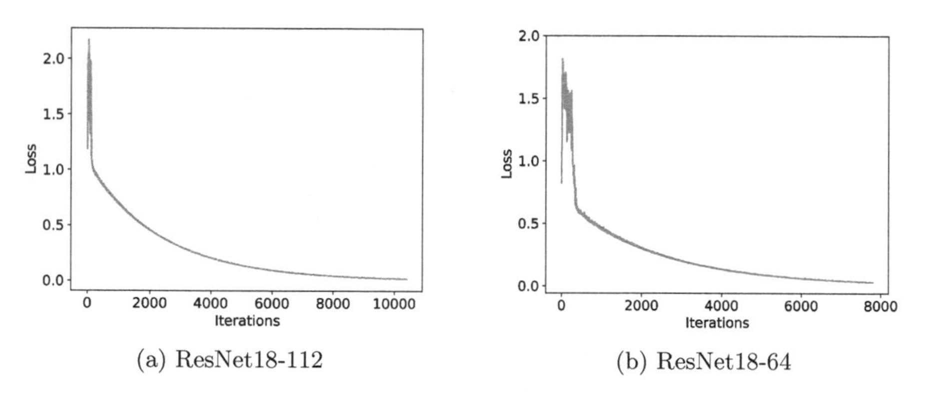

(a) ResNet18-112 (b) ResNet18-64

Fig. 2. Average training loss of five-fold cross-validation on the AFLW-112 and AFLW-64 dataset

5.3 Results

Table 2 shows the average training results for the three modified ResNets, evaluated on the pre-processed AFLW and AFW datasets with the methods explained in Sect. 5.2. As presumed in the introduction, the comparison between the evaluated ResNets shows that their results for the three angles are quite similar. The results are similar, when tested on the AFLW dataset and when tested on the AFW dataset, only the ResNet34-112 shows worse results than the ResNet18-64 and ResNet18-112.

Since the distribution of correctly classified images across label ranges is important for applications using head pose estimation, heatmaps are also considered as an evaluation tool. The heatmaps (see Fig. 3) show that the ResNet18-64 displays a more uniform distribution over the categories than the ResNet18-112.

In comparison to the ResNet18-112, the ResNet18-64 shows a higher percentage of correctly classified images in categories closer to ±100° and a similar percentage of correctly classified images in categories closer to 0°.

Table 2. Average results of the modified ResNets: (1) tested with a five-fold cross-validation on the AFLW dataset and (2) tested on the AFW dataset in five training-testing cycles (marked with (AFW))

Angle	MAE	Std. Dev.	Category	Category ±15°
ResNet34-112, CPU: 8 fps, GPU: 100 fps, 21.27×10^6 parameters				
Yaw	8.1°	±9.3°	56.0%	93.8%
Pitch	6.2°	±5.4°	61.8%	97.8%
Roll	3.8°	±3.8°	78.5%	99.8%
Yaw (AFW)	15.6°	±15.8°	38.0%	77.9%
ResNet18-112, CPU: 17 fps, GPU: 142 fps, 11.17×10^6 parameters				
Yaw	8.4°	±9.4°	54.0%	93.3%
Pitch	6.0°	±5.3°	62.4%	98.1%
Roll	3.8°	±3.7°	78.4%	99.8%
Yaw (AFW)	13.5°	±15.3°	42.6%	83.8%
ResNet18-64, CPU: 50 fps, GPU: 250 fps, 4.25×10^6 parameters				
Yaw	8.5°	±8.9°	53.4%	93.4%
Pitch	6.5°	±5.5°	59.6%	97.7%
Roll	3.9°	±3.8°	77.8%	99.7%
Yaw (AFW)	13.2°	±13.3°	41.1%	83.3%

The parameter number decreases with the reduction of the model's complexity (see Table 2). As expected, the ResNet18-64 with 18 layers and an input image size of 64 × 64 pixels has the lowest number of parameters. The inference time was measured once on an Intel Core i7-6700 CPU running at 3.40 GHz and once on this CPU equipped with a NVIDIA GeForce GTX 1060 6 GB GPU. The frames per second (fps) rates of the different ResNets show a significant speed-up with the reduction of the model complexity and resulting decrease of parameters. The ResNet18-64 achieves 50 fps on the CPU, which is suitable for most real-time applications.

Other approaches evaluated on the AFLW and AFW datasets are summarized in Table 3. The number of parameters of [27] is based on their provided open source implementation, which is executable on a GPU based system.[4] In order to compare the frame rate, we reimplemented the LeNet-5 variant of [25]. In comparison, our ResNet18-64 has the lowest number of parameters while predicting more accurately than the LeNet-5 variant [25] and nearly as accurate as

[4] https://github.com/natanielruiz/deep-head-pose, accessed 09.01.2019.

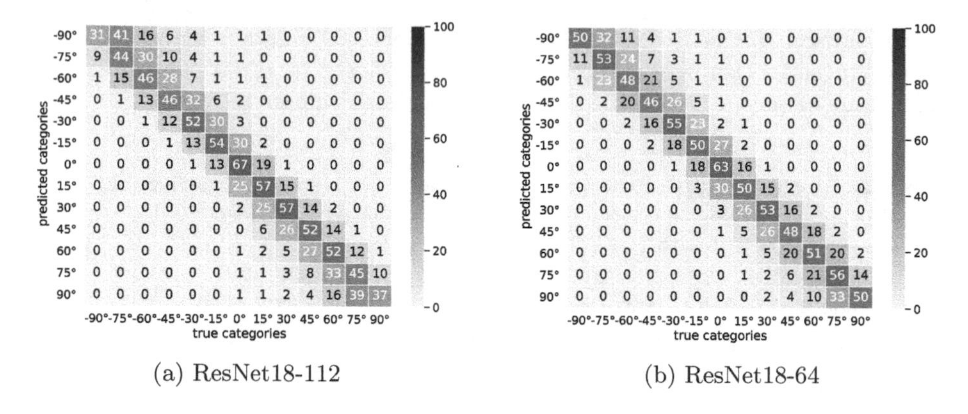

(a) ResNet18-112 (b) ResNet18-64

Fig. 3. These figures depict the average yaw angle heatmaps of a five-fold cross-validation on the AFLW-112 and AFLW-64 dataset. The distribution of images of each predicted category is in percentage.

Table 3. Results of the modified networks on the AFLW dataset in MAE and on the AFW dataset with ±15° error tolerance

Approach	AFLW			AFW	Parameters	CPU
	Yaw	Pitch	Roll	Yaw		
ResNet50 [27]	**6.3°**	**5.9°**	**3.8°**	**96.2%**	24.0×10^6	-
LeNet-5 variant [25]	9.5°	6.8°	4.2°	75.3%	4.6×10^6	17 fps
ResNet18-64 (ours)	8.5°	6.5°	3.9°	83.3%	$\mathbf{4.25 \times 10^6}$	**50 fps**

the ResNet50 [27]. Patacchiola and Cangelosi [25] also use low-resolution images with 64×64 pixels, while Ruiz et al. [27] take larger images with 224×224 pixels. To improve the computational efficiency, we believe that low-resolution images are better suited for real-world applications. Compared to our reimplementation of the LeNet-5 variant, our ResNet18-64 achieves a significantly higher frame rate on our CPU setup. Overall, our parameter-reduced ResNet18-64 achieves state-of-the-art precision and at the same time real-world applicability, even on CPUs.

6 Conclusion and Future Work

In this paper, we explored parameter-reduced Residual Networks (ResNets) of varying complexity for head pose estimation in order to achieve real-time performance. Based on the presumption that not all residual blocks contribute equally to the learning process, we showed that it is possible to reduce the number of parameters of the ResNet architecture while maintaining the performance. We proposed two new ResNet architectures for inputs of 112×112 pixels, one with 18 layers and one with 34 layers. To reduce the number of parameters even further, we proposed the ResNet18-64 with 18 layers for low resolution inputs of

64 × 64 pixels. The ResNet18-64 achieves real-time capability even on a CPU based system with a performance close to state-of-the-art results. To ensure real-world applicability, we evaluated the modified ResNets on the two in-the-wild datasets AFLW and AFW. In the future, it is possible to extend this approach to a model evaluating all three angles at once.

References

1. Benenson, R., Omran, M., Hosang, J., Schiele, B.: Ten years of pedestrian detection, what have we learned? In: Agapito, L., Bronstein, M.M., Rother, C. (eds.) ECCV 2014. LNCS, vol. 8926, pp. 613–627. Springer, Cham (2015). https://doi.org/10.1007/978-3-319-16181-5_47
2. Beymer, D.: Face recognition under varying pose. In: CVPR, vol. 94, p. 137. Citeseer (1994)
3. Dementhon, D.F., Davis, L.S.: Model-based object pose in 25 lines of code. Int. J. Comput. Vision **15**(1–2), 123–141 (1995)
4. Diebel, J.: Representing attitude: Euler angles, unit quaternions, and rotation vectors. Matrix **58**(15–16), 1–35 (2006)
5. Fanelli, G., Dantone, M., Gall, J., Fossati, A., Van Gool, L.: Random forests for real time 3D face analysis. Int. J. Comput. Vision **101**(3), 437–458 (2013)
6. Fanelli, G., Gall, J., Van Gool, L.: Real time head pose estimation with random regression forests. In: CVPR 2011, pp. 617–624. IEEE (2011)
7. Ferrario, V.F., Sforza, C., Serrao, G., Grassi, G., Mossi, E.: Active range of motion of the head and cervical spine: a three dimensional investigation in healthy young adults. J. Orthop. Res. **20**(1), 122–129 (2002)
8. Friesen, E., Ekman, P.: Facial Action Coding System: A Technique for the Measurement of Facial Movement. Consulting Psychologist Press, Palo Alto (1978)
9. Geronimo, D., Lopez, A.M., Sappa, A.D., Graf, T.: Survey of pedestrian detection for advanced driver assistance systems. IEEE Trans. Pattern Anal. Mach. Intell. **32**(7), 1239–1258 (2010)
10. He, K., Zhang, X., Ren, S., Sun, J.: Deep residual learning for image recognition. In: Proceedings of the IEEE Conference on Computer Vision and Pattern Recognition, pp. 770–778 (2016)
11. He, K., Zhang, X., Ren, S., Sun, J.: Identity mappings in deep residual networks. In: Leibe, B., Matas, J., Sebe, N., Welling, M. (eds.) ECCV 2016. LNCS, vol. 9908, pp. 630–645. Springer, Cham (2016). https://doi.org/10.1007/978-3-319-46493-0_38
12. Hsu, H.W., Wu, T.Y., Wan, S., Wong, W.H., Lee, C.Y.: QuatNet: quaternion-based head pose estimation with multi-regression loss. IEEE Trans. Multimedia **21**(4), 1035–1046 (2018)
13. Izard, C.E.: Human Emotions. Springer, Heidelberg (2013)
14. Koestinger, M., Wohlhart, P., Roth, P.M., Bischof, H.: Annotated facial landmarks in the wild: a large-scale, real-world database for facial landmark localization. In: 2011 IEEE International Conference on Computer Vision Workshops (ICCV Workshops), pp. 2144–2151. IEEE (2011)
15. Kumar, A., Alavi, A., Chellappa, R.: KEPLER: keypoint and pose estimation of unconstrained faces by learning efficient H-CNN regressors. In: 2017 12th IEEE International Conference on Automatic Face and Gesture Recognition (FG 2017), pp. 258–265. IEEE (2017)
16. Kuwahara, J., Nakazato, H.: Driving assistance system, US Patent 9,855,892, 2 January 2018

17. Leach, M.J., Baxter, R., Robertson, N.M., Sparks, E.P.: Detecting social groups in crowded surveillance videos using visual attention. In: Proceedings of the IEEE Conference on Computer Vision and Pattern Recognition Workshops, pp. 461–467 (2014)
18. LeCun, Y., Bengio, Y., Hinton, G.: Deep learning. Nature **521**(7553), 436 (2015)
19. LeCun, Y., Bottou, L., Bengio, Y., Haffner, P.: Gradient-based learning applied to document recognition. Proc. IEEE **86**(11), 2278–2324 (1998)
20. Lepetit, V., Fua, P., et al.: Monocular model-based 3D tracking of rigid objects: a survey. Found. Trends® Comput. Graph. Vis. **1**(1), 1–89 (2005)
21. Leroy, J., Rocca, F., Mancas, M., Gosselin, B.: Second screen interaction: an approach to infer TV watcher's interest using 3D head pose estimation. In: Proceedings of the 22nd International Conference on World Wide Web, pp. 465–468. ACM (2013)
22. Li, D., Pedrycz, W.: A central profile-based 3D face pose estimation. Pattern Recogn. **47**(2), 525–534 (2014)
23. Li, Y., Gong, S., Liddell, H.: Support vector regression and classification based multi-view face detection and recognition. In: Proceedings Fourth IEEE International Conference on Automatic Face and Gesture Recognition (Cat. No. PR00580), pp. 300–305. IEEE (2000)
24. Niyogi, S., Freeman, W.T.: Example-based head tracking. In: Proceedings of the Second International Conference on Automatic Face and Gesture Recognition, pp. 374–378. IEEE (1996)
25. Patacchiola, M., Cangelosi, A.: Head pose estimation in the wild using convolutional neural networks and adaptive gradient methods. Pattern Recogn. **71**, 132–143 (2017)
26. van der Pol, D., Cuijpers, R.H., Juola, J.F.: Head pose estimation for a domestic robot. In: Proceedings of the 6th Conference on Human-Robot Interaction, pp. 277–278. ACM (2011)
27. Ruiz, N., Chong, E., Rehg, J.M.: Fine-grained head pose estimation without keypoints. In: Proceedings of the IEEE Conference on Computer Vision and Pattern Recognition Workshops, pp. 2074–2083 (2018)
28. Schiele, B., Waibel, A.: Gaze tracking based on face-color. In: International Workshop on Automatic Face and Gesture Recognition, vol. 476. University of Zurich Department of Computer Science Multimedia Laboratory (1995)
29. Stiefelhagen, R.: Estimating head pose with neural networks-results on the Pointing04 ICPR workshop evaluation data. In: Proceedings of Pointing 2004 Workshop: Visual Observation of Deictic Gestures, vol. 1 (2004)
30. Szegedy, C., et al.: Going deeper with convolutions. In: Proceedings of the IEEE Conference on Computer Vision and Pattern Recognition, pp. 1–9 (2015)
31. Veit, A., Wilber, M.J., Belongie, S.: Residual networks behave like ensembles of relatively shallow networks. In: Advances in Neural Information Processing Systems, pp. 550–558 (2016)
32. Wu, H., Zhang, K., Tian, G.: Simultaneous face detection and pose estimation using convolutional neural network cascade. IEEE Access **6**, 49563–49575 (2018)
33. Zhang, W., et al.: Cross-cascading regression for simultaneous head pose estimation and facial landmark detection. In: Zhou, J., et al. (eds.) CCBR 2018. LNCS, vol. 10996, pp. 148–156. Springer, Cham (2018). https://doi.org/10.1007/978-3-319-97909-0_16
34. Zhu, X., Ramanan, D.: Face detection, pose estimation, and landmark localization in the wild. In: 2012 IEEE Conference on Computer Vision and Pattern Recognition, pp. 2879–2886. IEEE (2012)

Autonomous Systems and Automated Driving

A Rule-Based Smart Control for Fail-Operational Systems

Georg Engel, Gerald Schweiger, Franz Wotawa, and Martin Zimmermann(✉)

Institute for Software Technology,
CD Laboratory for Quality Assurance Methodologies for Autonomous Cyber-Physical Systems, Technische Universität Graz, Inffeldgasse 16b/2, 8010 Graz, Austria
{engel,gerald.schweiger,wotawa,martin.zimmermann}@ist.tugraz.at

Abstract. When systems become smarter they have to cope with faults occurring during operation in an intelligent way. For example, an autonomous vehicle has to react appropriately in case of a fault occurring during driving on a highway in order to assure safety for passengers and other humans in its surrounding. Hence, there is a need for fail-operational systems that extend the concept of fail-safety. In this paper, we introduce a method that relies on rules for controlling a system. The rules specify the behavior of the system including behavioral redundancies. In addition, the method provides a runtime execution engine that selects the rules accordingly to reach a certain goal. In addition, we present a language and an implementation of the method and discuss its capabilities using a case study from the mobile robotics domain. In particular, we show how the rule-based fail-operational system can adapt to a fault occurring at runtime.

Keywords: Fail-operational systems · Adaptive behavior · Self-healing systems · Autonomous and mobile systems

1 Introduction

Autonomous systems must have the capabilities to make decisions during operation in an independent fashion without considering external control. Ideally such systems must not only be able to react to external stimuli coming from the system's environment in a smart way, but also in case of internal faults or misinterpretations of sensor inputs causing deviations from ordinary behavior. Reacting on internal faults is especially relevant for safety critical systems like autonomous vehicles, where the vehicle itself has not only to detect the fault, but also to react in a smart way in order to reach a safe state autonomously. Such behavior requires that the system is capable of compensating faults at least for a certain amount of time. For example, if the autonomous vehicle detects a fault in its powertrain, stopping operation in a blind bend might not be safe for the passengers. It might be a wiser choice to go to the closest parking space and

F. Wotawa et al. (Eds.): IEA/AIE 2019, LNAI 11606, pp. 137–145, 2019.
https://doi.org/10.1007/978-3-030-22999-3_13

keep operation even under degraded conditions. Hence, such autonomous vehicles have to have build in capabilities for implementing fail-operational behavior. Obviously, fail-operational behavior can only be provided in case of redundancy. Redundancy can be achieved either via using spare parts that are used to replace broken parts autonomously, or re-configuring the system during operation to achieve the desired functionality. The latter deals with either using components in an undesired way to replace a broken function, or to go to degraded mode that can still be achieved with the available functionality. Regardless of the underlying redundancy, a fail-operational system requires a control program that enables the use of redundancy in case of a fault. In this paper, we contribute to the implementation of fail-operational systems, and introduce a rule-based programming language that is capable of deciding which redundant behavior to use during operation. In addition, we discuss a case study from the autonomous mobile robot's domain showing that the control system relying on the rule-based program is capable of dealing with faults in the powertrain by going to a degraded mode.

The idea behind the programming language RBL is to provide means for specifying control rules that select redundant behavior during runtime and to have an interface between the selected rules and the rest of the system. In particular, our implementation allows to call external Java methods used for communicating with the rest of the system. RBL also offers means for computing weights for rules that should be selected. These weights are not fixed, but vary during runtime based on the success of actions executed previously. Hence, actions that are not successful when carried out are less often executed. For example, if we have two rules implementing the same functionality but using redundant components that are triggered via their Java interface, the rule that most often lead to a successful execution will also be more likely used in the future. Whenever there is a fault in the corresponding component, the action will fail, the rule weight will decrease, and the other alternative rule will be more likely selected for execution. Besides the underlying foundations behind the RBL language, which is based on previous work [9,20], we introduce an extension that supports fine tuning of the weight calculation of the rules.

2 The RBL Programming Language

The RBL language is a programming language that uses rules to model fault-tolerant systems. RBL is an extension to a previously proposed rule-based language [20]. Every rule in RBL has a set of preconditions and post conditions. If all preconditions are satisfied the rule can be executed. If the rule was successfully executed, the post conditions will be true and can enable new rules to be executed afterwards. A simple example for a rule is: *"If the water bottle is full and open, you can pour out the water, after that the bottle is empty"*. In this section, we give a brief overview about the syntax and semantic of RBL and describe the interface between RBL, Java and a Modelica Model. For more details about RBL and a more formal definition of the underlying semantics, we refer the interested reader to a previous paper [20].

Every RBL program comprises a set of rules, where the basic syntax of a rule is: `<preconditions> -> <postconditions> <action> [<weight modifiers>].` where `<preconditions>` and `<postconditions>` are sets of propositions that can also be empty. It is worth noting that in the postcondition we are able to add proposition and to remove proposition using `+` and `-` respectively before writing the proposition. A rule might be selected if all the preconditions are satisfied. The `<action>` is a Java class that implements a specific Java interface. A method of this class is called whenever the rule is executed. Before we discuss the `<weight modifiers>`, which are a new extension to the language, we will first briefly explain the fault-tolerant behavior of RBL.

For each execution RBL generates a list of rules, which when executed in succession, satisfies all precondition of the rules and the last rule in the list is a goal rule (denoted by "#" as postcondition). If there is more than one of such lists, meaning there are redundancies in the rules, RBL takes the list with the smallest number of rules and highest weight. The original rule-based language calculates the weight as follows:

$$w = (1 - current\_act) * (1 - damp). \tag{1}$$

After each run *current_act* and *damp* are updated. Where *current_act* is increased if the rule was selected in the path and decreased if it was not selected, meaning *current_act* is a measure of how often the rule was selected in the past. *damp* on the other hand is decreased if the rule was executed and was successful and increased when it failed, meaning it is a measure of how successful the rule was in the past. Formula (Eq. 1) together with the update enables the fault-tolerant behavior.

To fine-tune the fault-tolerant behavior to different situations, we introduce two new extensions to the language: (i) **Aging** is another update step that occurs after each run. *damp* will be decreased or increased by the aging value based on whether the aging target is currently smaller or greater than *damp*. (ii) **Activity and damping scaling** can be used to disable, or lessen, the effect of *damp* and *current_act* on the weight calculation. This leads to following new weight calculation:

$$w = (1 - (current\_act * act\_scaling)) * (1 - (damp * damp\_scaling)). \tag{2}$$

The `<weight modifiers>`, with which the two new concepts can be configured, are 5 values separated by a comma, if a value is left blank a default value is used. For example `[0.2,,,0,1]` is a valid `<weight modifiers>`. The different values have following meaning accordingly to their order:

1. **Damping value** is the value by which *damp* is increased or decreased
2. **Aging value** is the value by which *damp* ages
3. **Aging target** is the value towards *damp* ages
4. **Activity scaling** is the value of *act_scaling* in Equation (Eq. 2)
5. **Damping scaling** is the value of *damp_scaling* in Equation (Eq. 2)

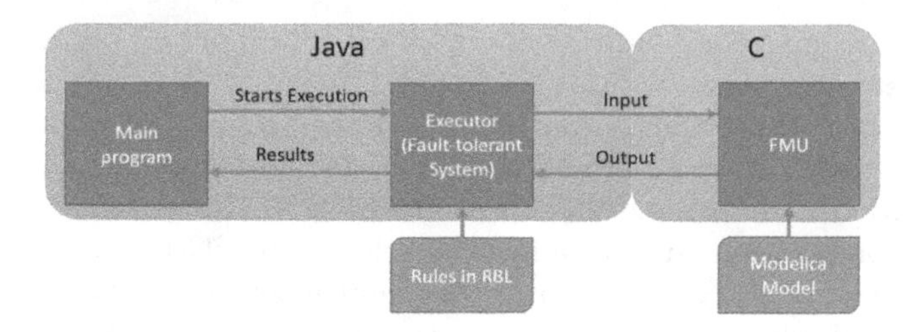

Fig. 1. Architecture of the RBL and FMU interface.

Table 1. Faults over time of the different fault modes.

	0 to 10 s	10 to 20 s	20 to 30 s
Normal	No fault	No fault	No fault
Fault A	Left wheel faulty	No fault	No fault
Fault B	No fault	Left wheel faulty	Right wheel faulty
Fault C	No fault	Left and right wheel faulty	No fault

For example, the rule *"If the water bottle is full and open, you can pour out the water, after that the bottle is empty."* can be expressed in RBL as follows:

```
b_open, b_full -> +b_empty -b_full action.emptyBottle [0.1,,,0.5,1].
```

3 The Mobile Robot Case Study

In this section, we discuss a case study where we outline the use of RBL to implement fail operational behavior for a mobile robot. The mobile robot includes a differential drive, where each wheel is connected with a motor, and a wheel encoder attached for measuring the rotational speed during operation. For details of the mathematical modelling of the powertrain we refer to [3]. In the case study we assume a controller that is able to set the voltage for the motors such that they start rotating. The speed of rotation is proportional to the voltage provided. The task of the controller is to set the voltage such that the robot moves straight. The controller itself only receives the rotational speed measured using the wheel encoders as inputs. We further assume that in case of a fault of the motor the rotational speed does not follow the given voltage anymore. For simplicity, we assume that in the fault case the rotational speed is half the expected speed. For the experiment based on the case study, we implemented the kinetics of the robot drive using Modelica [5].

In order to couple the Java-based control program with the Modelica model capturing the kinematics of the robot, we made use of co-simulation. Co-simulation is a simulation method involving a collaboration of various solvers and often tools [6]. Usually, co-simulation is chosen when different specialized tools are employed to model different subsystems of a heterogeneous complex system.

Various co-simulation interfaces are available [4], a prominent one is the Functional Mockup Interface (FMI) [1]. It has been developed as tool-independent standard in the ITEA2 European Advancement project MODELISAR, and is by now widely supported by many tools. FMI supports both model exchange and co-simulation of dynamic models, providing a zip of an executable and xml-files describing metadata for the subsystem, for a detailed discussion see e.g. [16]. For our experiments, we used the open-source library javaFMI [7] as an interface between a controller implemented in Java and a dynamical simulation model implemented in Modelica.

We briefly discuss the underlying co-simulation framework used for the experimental evaluation. The framework comprises three main components: the main program, the executor, and the Functional Mock-up Unit (FMU). In Fig. 1 we depict the general underlying architecture. The main program starts the execution and evaluates the results. The executor is generated out of the RBL rules and is responsible for the fault-tolerant behavior. The executor communicates through javaFMI with the FMU and evaluates the results from the FMU. The FMU is generated directly from the Modelica model of the robot.

4 Experimental Evaluation

In this section we will describe our experimental evaluation. First, we will give an overview of our experimental setup, then we will present and discuss our results.

Experimental Setup: For our experimental evaluation we coupled the former mentioned Modelica Model with the RBL language, as we described in Sect. 3. The goal of the robot was to drive in a straight line and if faults occur going in a degraded mode.

The only inputs the fault-tolerant system can give to the robot was that either both motors of the wheels get the same voltage, or the left or right motor gets half as much voltage as the other motor. In case of a fault this would mean that the healthy wheel should match the speed of the faulty wheel, by reducing the voltage of the healthy wheel to half. As a feedback the fault-tolerant system only knows if both wheels turn at the same speed, or not.

The ruleset of RBL for our experiment was very simple as we can see in the following listing. To note is that we disabled that *current_act* has any influence on our system, this means the only relevant measure for choosing a rule for our experiment was how often the rules succeed.

```
-> #drive actions.leftHalfDrive [0.2,,,0,1].
-> #drive actions.rightHalfDrive [0.2,,,0,1].
-> #drive actions.normalDrive [0.2,,,0,1].
```

To see the fail-tolerant behavior in action we tested 4 different configurations of the model, where each was executed for 30 s. The concrete faults can be seen in Table 1.

Table 2. Results of the different fault modes for runs with and without RBL rules.

	Without RBL		With RBL	
	% of straight	Distance from Y	% of straight	Distance from Y
Normal	100.0%	0.0	99.8%	3
Fault A	66.7%	15,264.6	99.7%	695.2
Fault B	33.4%	1,762.3	99%	1,473.8
Fault C	100.0%	0.0	99.8%	3
Average	75.0%	4,256.7	99.6	543.7

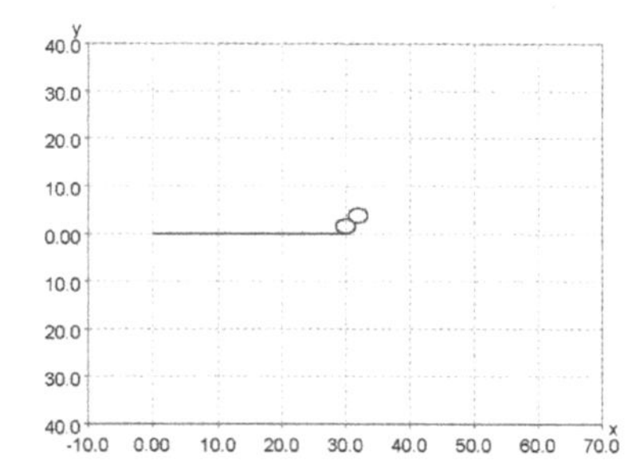

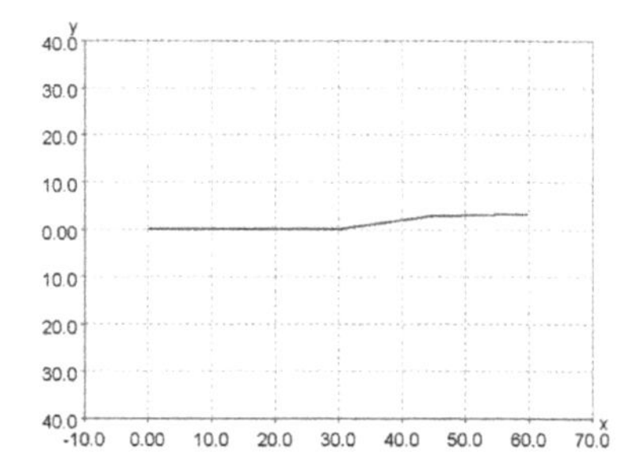

Fig. 2. Visual comparison of the test without RBL (left) and with RBL (right) with fault B. We can clearly see that the robot with RBL drives a more straight line towards infinity than the robot without RBL.

Results: To understand our result, we used 3 metrics: (i) The percentage of the time the robot actually drove straight, meaning both wheels turned at the same speed. (ii) the sum of the distances from the robot to the y axis per time step. (iii) we plotted the path of the robot in absolute coordinates to compared them visually. We compared runs without the RBL rules , and runs with RBL, i.e., where the fault-tolerant system could change the supplied voltage to the wheels in case of a fault.

We conclude from Table 2 that in the case no fault occurs (Normal), or both wheels are faulty (Fault Mode C), the robot without the RBL rules performance a bit better than the robot with RBL rules. This is due to the nature of the fault-tolerant system that first has to find the correct voltage supply for the robot. However, the other tests show that the robot with RBL rules perform much better. It has always a success rate of over 99%. Interesting to note is that the % of driving straight does not correlate to the distance from the y axis. In Fig. 2, we can see why. When we look at the Figure of the path without the RBL rules (left), we can see that even though the robot drives wrong most of the time it always circles back to the y axis which leads to a value comparable to the distance to the Y axis from the robot with RBL rules. On the figure however we can clearly see that the robot did not fulfil its target to drive in a straight line. All tests run with the RBL rules however fulfil this target within a small margin.

5 Related Research and Conclusions

Early work in the domain of adaptive and self-healing systems include [15] and [10] where the authors focus on reconfiguration and self-repair of hardware. Musliner et al. [11] introduced an approach for self-adaptive software to be used in real-time environments. The use of self-healing in the context of autonomous systems and automotive is not novel. Seebach and colleagues [17] introduced an approach for implementing such a behavior.

Pell and colleagues [12] were one of the first making use of model-based reasoning [2,14] for allowing systems gaining additional autonomy, which finally led to a model-based control system for space probes [13]. In the latter papers, models are used to allow the system to react to internal faults occurring during operation in a smart way. Steinbauer and Wotawa [18] used this idea and implemented a system for mobile robots that was able to detect software issues and to restart software modules causing these issues at runtime to bring the robot back to operation. Hofbaur et al. [8] introduced a system that is able to change control models for mobile robots at runtime in case of faults in the robot's drive. For a summary, of model-based approaches for self-adaptive systems we refer to [19]. See [21] for more information and details regarding the formal background of model-based reasoning for self-adaptive systems.

In this paper we presented a case study that uses RBL to combine a control-system with a physical model of a robot that captures the kinematics of the robot. Hence, we are able to demonstrate that the proposed approach for self-adaptive systems also works in a simulated environment that is very close to the real world. We tested our approach by modeling a robot and introducing faults in the system. The experimental results indicate that RBL can be effectively used for controlling such a robot in case of faults. In future research, we will extend the case study capturing different fault scenarios and also different drives. Moreover, we will compare the outcome with other means for implementing fail-operational behavior like model-based reasoning.

Acknowledgments. The research was supported by ECSEL JU under the project H2020 737469 AutoDrive - Advancing fail-aware, fail-safe, and fail-operational electronic components, systems, and architectures for fully automated driving to make future mobility safer, affordable, and end-user acceptable. AutoDrive is funded by the Austrian Federal Ministry of Transport, Innovation and Technology (BMVIT) under the program "ICT of the Future" between May 2017 and April 2020. More information https://iktderzukunft.at/en/bmvit. The financial support by the Austrian Federal Ministry for Digital and Economic Affairs and the National Foundation for Research, Technology and Development is gratefully acknowledged.

References

1. Blochwitz, T., et al.: The functional mockup interface for tool independent exchange of simulation models. In: 8th International Modelica Conference 2011, pp. 173–184 (2009)
2. Davis, R.: Diagnostic reasoning based on structure and behavior. Artif. Intell. **24**, 347–410 (1984)
3. Dudek, G., Jenkin, M.: Computational Principles of Mobile Robotics. Cambridge University Press, Cambridge (2010)
4. Engel, G., Chakkaravarthy, A.S., Schweiger, G.: A general method to compare different co-simulation interfaces: demonstration on a case study. In: Obaidat, M.S., Ören, T., Rango, F.D. (eds.) SIMULTECH 2017. AISC, vol. 873, pp. 351–365. Springer, Cham (2019). https://doi.org/10.1007/978-3-030-01470-4_19
5. Fritzson, P.: Object-Oriented Modeling and Simulation with Modelica 3.3 - A Cyber-Physical Approach, 2nd edn. Wiley-IEEE Press, New York (2014)
6. Gomes, C., Thule, C., Broman, D., Larsen, P.G., Vangheluwe, H.: Co-simulation: state of the art. CoRR abs/1702.0, February 2017
7. Hernández-Cabrera, J.J., Évora-Gómez, J., Roncal-Andrés, O.: javaFMI. https://bitbucket.org/siani/javafmi/wiki/Home. Accessed 25 Jan 2019
8. Hofbaur, M.W., Köb, J., Steinbauer, G., Wotawa, F.: Improving robustness of mobile robots using model-based reasoning. J. Intell. Rob. Syst. **48**(1), 37–54 (2007)
9. Krenn, W., Wotawa, F.: Intelligent, fault adaptive control of autonomous systems. In: Madrid, N.M., Seepold, R.E.D. (eds.) Intelligent Technical Systems. Lecture Notes in Electrical Engineering, vol. 38, pp. 175–188. Springer, Dordrecht (2009). https://doi.org/10.1007/978-1-4020-9823-9_13
10. Moreno, J.M., Madrenas, J., Faura, J., Cantó, E., Cabestany, J., Insenser, J.M.: Feasible evolutionary and self-repairing hardware by means of the dynamic reconfiguration capabilities of the FIPSOC devices. In: Sipper, M., Mange, D., Pérez-Uribe, A. (eds.) ICES 1998. LNCS, vol. 1478, pp. 345–355. Springer, Heidelberg (1998). https://doi.org/10.1007/BFb0057636
11. Musliner, D., Goldman, R., Pelican, M., Krebsbach, K.: Self-adaptive software for hard real-time environments. Intell. Syst. Appl. **14**(4), 23–29 (1999)
12. Pell, B., et al.: A remote-agent prototype for spacecraft autonomy. In: Proceedings of the SPIE Conference on Optical Science, Engineering, and Instrumentation, Volume on Space Sciencecraft Control and Tracking in the New Millennium, Bellingham, Waschington, U.S.A., Society of Professional Image Engineers (1996)
13. Rajan, K., et al.: Remote agent: an autonomous control system for the new millennium. In: Proceedings of the 14th European Conference on Artificial Intelligence (ECAI), Berlin, Germany, August 2000
14. Reiter, R.: A theory of diagnosis from first principles. Artif. Intell. **32**(1), 57–95 (1987)
15. Rincon, F., Teres, L.: Reconfigurable hardware systems. In: Proceedings of the International Semiconductor Conference, New York, NY, USA, vol. 1, pp. 45–54 (1998)
16. Schweiger, G., Gomes, C., Engel, G., Hafner, I., Schoeggl, J.P., Nouidui, T.S.: Functional mockup-interface : an empirical survey identifies research challenges and current barriers. In: American Modelica Conference 2018 (2018)

17. Seebach, H., et al.: Designing self-healing in automotive systems. In: Xie, B., Branke, J., Sadjadi, S.M., Zhang, D., Zhou, X. (eds.) ATC 2010. LNCS, vol. 6407, pp. 47–61. Springer, Heidelberg (2010). https://doi.org/10.1007/978-3-642-16576-4_4
18. Steinbauer, G., Mörth, M., Wotawa, F.: Real-time diagnosis and repair of faults of robot control software. In: Bredenfeld, A., Jacoff, A., Noda, I., Takahashi, Y. (eds.) RoboCup 2005. LNCS (LNAI), vol. 4020, pp. 13–23. Springer, Heidelberg (2006). https://doi.org/10.1007/11780519_2
19. Steinbauer, G., Wotawa, F.: Model-based reasoning for self-adaptive systems – theory and practice. In: Cámara, J., de Lemos, R., Ghezzi, C., Lopes, A. (eds.) Assurances for Self-Adaptive Systems. LNCS, vol. 7740, pp. 187–213. Springer, Heidelberg (2013). https://doi.org/10.1007/978-3-642-36249-1_7
20. Wotawa, F., Zimmermann, M.: Adaptive system for autonomous driving. In: 2018 IEEE International Conference on Software Quality, Reliability and Security Companion (QRS-C), pp. 519–525, July 2018
21. Wotawa, F.: Reasoning from first principles for self-adaptive and autonomous systems. In: Lughofer, E., Sayed-Mouchaweh, M. (eds.) Predictive Maintenance in Dynamic Systems, pp. 427–460. Springer, Cham (2019). https://doi.org/10.1007/978-3-030-05645-2_15

Autonomous Monitoring of Air Quality Through an Unmanned Aerial Vehicle

Víctor H. Andaluz[1(✉)], Fernando A. Chicaiza[1(✉)], Geovanny Cuzco[2(✉)], Christian P. Carvajal[1(✉)], Jessica S. Ortiz[1(✉)], José Morales[3(✉)], Vicente Morales[4(✉)], Darwin S. Sarzosa[1(✉)], Jorge Mora-Aguilar[1(✉)], and Gabriela M. Andaluz[5(✉)]

[1] Universidad de las Fuerzas Armadas ESPE, Sangolquí, Ecuador
{vhandaluz1, jsortiz4, dssarzosa1, jlmora2}@espe.edu.ec
[2] Universidad Nacional de Chimborazo, Riobamba, Ecuador
gcuzco@unach.edu.ec
[3] Escuela Superior Politécnica de Chimborazo, Riobamba, Ecuador
jose.morales@espoch.edu.ec
[4] Universidad Técnica de Ambato, Ambato, Ecuador
jvmorales99@gmail.com
[5] Universidad Internacional del Ecuador, Quito, Ecuador
gaandaluzor@uide.edu.ec

Abstract. The monitoring of air quality allows to evaluate the amount of harmful particles for health that are being released. Under this paradigm and knowing the current methods to monitor these parameters, this work proposes the use of a UAV for commercial use and the construction of a card for gas measurement. Additionally and with the objective of having complete control over the vehicle, the article proposes the development of a library for the control and monitoring of the instrumentation of a commercial drone, through which the validation of control algorithms is proposed. As a result of this work, two real experiments on a rural environment and an urban environment are carried out to validate both the library created and the method of acquiring information on air quality.

Keywords: Air quality · UAV · Linear algebra · Advanced controller

1 Introduction

Air quality is a major concern in several cities around the world; prolonged exposure to elements such as carbon monoxide (CO), nitrogen oxides (NOx), sulfur dioxide (SO_2), ozone (O_3) and particulate material (PM) [1], significantly affect human health and are responsible for a variety of respiratory diseases, cardiovascular, diabetes [2], anemia [3], cognitive neuro [4], psychological [5], cancer [6], among others. In addition, the air pollution is responsible for the environmental problems such as acidification and eutrophication [7] of ecosystems. Currently, the measurement of air quality has been developed focused in two areas: ***(i) the legislative,*** for compliance with environmental regulations [8]; ***(ii) scientific research***, related to studies of impact on the environment,

F. Wotawa et al. (Eds.): IEA/AIE 2019, LNAI 11606, pp. 146–157, 2019.
https://doi.org/10.1007/978-3-030-22999-3_14

the biology and medicine [2, 4, 9]. In both cases the measurement is performed using networks of fixed and mobile certified reference equipment stations with high operation costs. Networks of stations for monitoring provide accurate information but in a given region, useful for compliance with regulations environmental but insufficient to influence health studies [8]. It is of great scientific interest to increase data collection in order to generate high resolution air quality maps, so that low cost sensor platforms on mobile devices provide the sector with great potential and, by incorporating sensors to Unmanned Aerial Vehicles (UAVs) provides the researcher with real-time monitoring possibilities at different heights.

UAVs can be used in several applications whether they are civilian or military due to its versatility and continuous evolution, considered alternative tool for low cost for gathering information in large areas and inaccessible, its found in applications such as: ***(i) surveillance***, mapping and 3D modeling [10], ***(ii) Precision agriculture***, as agricultural monitoring [11], weed detection, pest control; ***(iii) Monitoring*** of rivers and lakes, during the natural disaster information collection, natural reserves, among others [12]. These aircraft can: ***(a) Fly autonomously*** to the on-board processors and sensors from global positioning (GPS), inertial sensors for navigation as gyroscopes, accelerometers, electromechanical systems [13], altitude, and; ***(b) Tele-Operate,*** controlled from a ground station under human supervision [13]. In both cases the UAVs are exposed to multiple disturbances requiring a greater complexity in the system of navigation and control [14].

The control of a UAV can be organized into three large groups: ***(i) Based on learning,*** or free of models where algorithms have been developed with *Fuzzy Logic* as proposed by [15] applied to trajectory tracking. [16] applies *neural networks* to the prediction of collisions that unlike [17] employs *learning reinforcement*; ***(ii) Lineal Controllers***, its great advantage is the ease of implementation in a real platform and employs a linearized model of the quadrotor, applications have been found where PID drivers are used, LQG/LQR, as well as **Robust Mixed H_∞/H_2** [18], who perform feedback control with noise for attitude control and tracking of an unmanned aerial vehicle, other applications have focused on trajectory tracking using gain scheduling [19] and; ***(iii) Non-Linear Controllers***, which can be completely non-linear, where applying backstepping provides good results according [20] and [21], instead linearization processes where jobs have been found using sliding mode [22], predictive models, adaptive and linearization in the feedback [23], a different approach that allows the tracking of continuous trajectories to pieces using linear algebra can be seen in [24, 25] Linear algebra control is an advanced control technique that does not require complex calculations and the computational requirement is low, which allows it to be implemented in low-performance processors and maintain an adequate performance for the execution of tasks such as**: position control, speed control and, road tracking.** For the described, in this article, we present a control technique employed by the linear algebra method for the autonomous flight of the unmanned aerial vehicle UAV, for the monitoring of air quality, obtaining gas measurements such as carbon monoxide CO, dioxide of sulfur SOx, Ozone Ox and carbon monoxide CO_2.

2 Problem Formulation

The measurement of air pollution parameters is characterized by temporal and spatial environments because most of these measurements are made in situ using monitoring networks with; fixed stations, which are delineated to a specific area; and mobile stations, which is limited to uses in time and place, to the two systems currently used, it is impossible to measure gases in a stratified form by height and in trajectories that the investigation requires. The need to analyze environmental information at different heights involves the use of emerging technologies for the collection of environmental data using UAVs that are able to track the road continuously and with low impact due to disturbances to which they are exposed.

Figure 1 shows the multilayer control scheme proposed for the present work, where each module has been considered as a fully functional entity that operates independently where: *(i) Off-line planning layer* is responsible for generating the route planning considering height, position and speed; *(ii)* On-line planning layer allows to modify trajectory references during a work cycle in order to modify the routes during the development of a mission; *(iii) Non-linear controller*, is responsible for generating the maneuvering signals to meet the objective of the tasks of position control, trajectory or path monitoring, at desired speeds, considering the state of the variables coming from; *(iv) The UAV layer* represents the kinematics and general dynamics of the quadrocopters and, finally; *(v) The environment layer* that is responsible for relating the environmental variables to be collected as CO, SO_2, O_3, CO_2, T, H and considers additionally the variables to be taken into account for the evasion of obstacles and completing the task to collect environmental data.

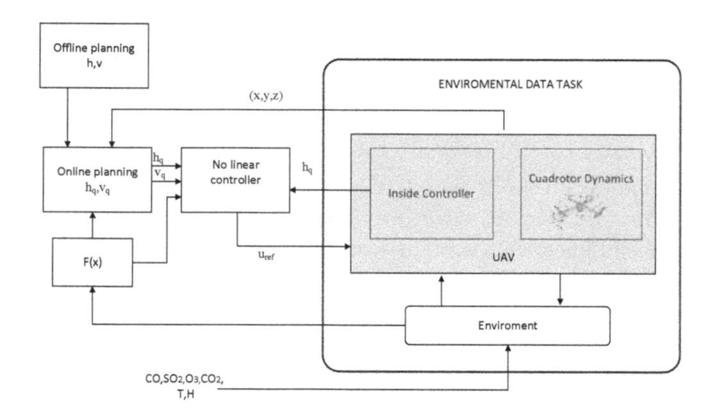

Fig. 1. Multi-layered scheme of the autonomous quadrocopters

3 Modeling and Control

The kinematic model of an UAV results in the location of the point of interest based on the location of the aerial platform. The *instantaneous kinematic model of an UAV gives* the derivative of its interest point location as a function of the location of the aerial mobile platform, $\dot{\chi} = \partial f / \partial \mathbf{q}(\mathbf{q})\mathbf{v}$, where $\dot{\boldsymbol{\chi}} = [\dot{\chi}_X \; \dot{\chi}_Y \; \dot{\chi}_Z]$ is the vector of the point interest velocity, $\mathbf{v}$ is the control vector of mobility of the aerial mobile manipulator, in this case the dimension depend of the maneuverability velocities of UAV.

For this case of study, then the kinematic model of UAV is conform by a set of four velocities represented at the spatial frame $<\chi>$. The displacement of the UAV is guided by the three linear velocities v_1, v_m and v_n defined in a rotating right-handed spatial frame $<\chi>$, and the angular velocity ω, as shown in Fig. 2.

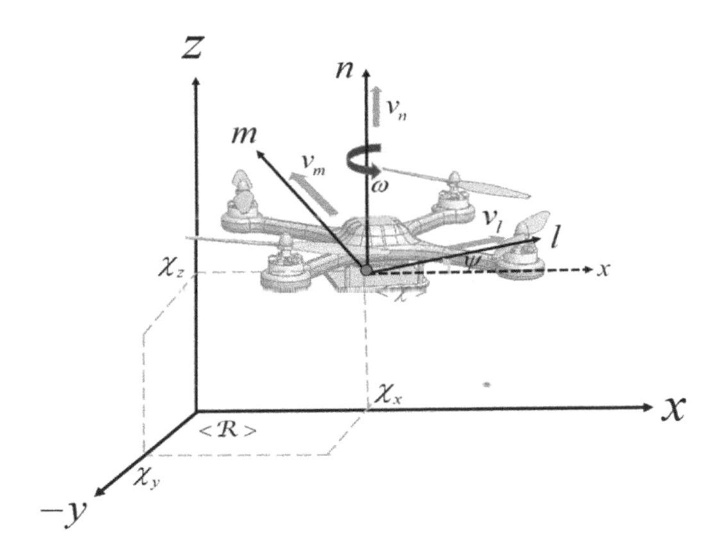

Fig. 2. Kinematic scheme of UAV

In other words, the cartesian motion of the quadcopter at the inertial frame $<R>$ is defined as,

$$\begin{cases} \dot{\chi}_x = v_1 \cos \psi - v_m \sin \psi \\ \dot{\chi}_x = v_1 \sin \psi - v_m \cos \psi \\ \dot{\chi}_z = v_n \\ \dot{\psi} = \omega \end{cases} \tag{1}$$

where $\dot{\chi}_x$, $\dot{\chi}_y$ $\dot{\chi}_z$ and ψ are all measured with respect to the inertial frame $<R>$. The point of interest (whose position is being controlled) is the center of gravity of the UAV. Also the equation system (1) can be written in compact form as,

$$\begin{bmatrix} \dot{\chi}_x \\ \dot{\chi}_y \\ \dot{\chi}_z \\ \dot{\psi} \end{bmatrix} = \begin{bmatrix} \cos\psi & -\sin\psi & 0 & 0 \\ \sin\psi & \cos\psi & 0 & 0 \\ 0 & 0 & 1 & 0 \\ 0 & 0 & 0 & 1 \end{bmatrix} \begin{bmatrix} v_l \\ v_m \\ u_{vn} \\ \omega \end{bmatrix}$$

$$\dot{\chi}(t) = \mathbf{J}(\psi)\mathbf{v}(t) \tag{2}$$

where $\dot{\chi} \in \Re^n$ with $n = 4$ represents the vector of axis velocities of the $<R>$ system and the angular velocity around the axis Z; Jacobian matrix that defines a linear mapping between the vector of the UAV velocities $\mathrm{v}(t)$; and the control of maneuverability of the quadcopter is defined $v \in \Re^n$.

Kinematic Controller

The proposed controller is based on calculations with numerical methods, which is considered the kinematic model of the UAV (4) defined at the moment of time k.

$$\dot{\chi}(k) = \mathbf{J}(\psi(k))\mathbf{v}(k) \tag{3}$$

In (5) using a discrete-time transformation by means of the Euler method and taking into account that the discrete time $t = kT_0$, where T_0 represents the sampling time y $k \in \{1, 2, 3, 4, 5\ldots\}$ we obtain:

$$\frac{\chi(k) - \chi(k-1)}{T_0} = \mathbf{J}(\psi(k))\mathbf{v}(k) \tag{4}$$

For the design of the kinematic control of the UAV, takes into account the cinematic model (2). In order to achieve the proposed control task, the following expression is considered:

$$\frac{1}{T_0}(\chi(k) - \chi(k-1))\upsilon_d(k) + \frac{1}{T_0}(\mathbf{W}(\mathrm{P_d}(k-1) - \chi(k-1))) \tag{5}$$

where, $\mathrm{P_d}$ is the desired path, $\mathbf{W}(\tilde{\chi}(k-1))$ is a diagonal matrix that control error weights, defined as $\mathbf{W}(\tilde{\chi}_m(k-1)) = w_m/(1 + |\tilde{\chi}_m(k-1)|)$, where m represents the operational coordinates quadcopter robot.

Now, to generate the system equations consider (5) and (4), the system can be rewritten as $\mathbf{Au} = \mathbf{b}$

$$\underbrace{\mathbf{J}(\psi(k))}_{\mathrm{A}}\underbrace{\mathbf{v}(k)}_{\mathrm{u}} = \underbrace{\mathbf{v}_\mathrm{d}(k) + \frac{1}{T_0}(\mathbf{W}(\mathrm{P_d}(k-1) - \chi(k-1)))}_{\mathrm{b}} \tag{6}$$

Then, the following control law is proposed as:

$$\mathbf{v}_{c}(k) = \mathbf{J}^{-1}(\psi(k))\left(\mathbf{v}_{d}(k) + \frac{1}{T_0}(\mathbf{W}(\mathrm{P}_{d}(k-1) - \chi(k-1)))\right) \tag{7}$$

4 Experimental Results

The validation of the proposed controller and the gas acquisition system is presented in this section through the execution of two experiments. Taking into account that toxic gas emissions vary depending on factors such as: presence of vehicular traffic, number of industries, density of flora, population density, and so on, the first experiment is carried out in a location far from this type of factors, while the second is executed on a totally urban space to contrast results. Physically, the UAV-Acquisition Board includes the drone with propeller protectors, on which the card is placed with the gas sensors. The card is embedded in the vehicle through a coupling that does not alter the dynamics of the UAV and whose maximum weight is 125 g, considering that additional batteries are not included in addition to the integrated in the Phantom, see Fig. 3.

Fig. 3. UAV – Acquisition Board

4.1 First Experiment

The first experiment shows the execution of a circular path of 5 m radius and 25 m high in the environment with the presence of apparently reduced contamination. The stroboscopic movement of the experimental execution is shown in Fig. 4, where the actual execution of the vehicle on a relatively remote rural location is reconstructed, but with the consideration of wind disturbances of more than 25 km/h. Also, the performance of the proposed controller is demonstrated through Fig. 5, where a clear tendency to zero of the control errors is observed given the control actions (Fig. 6).

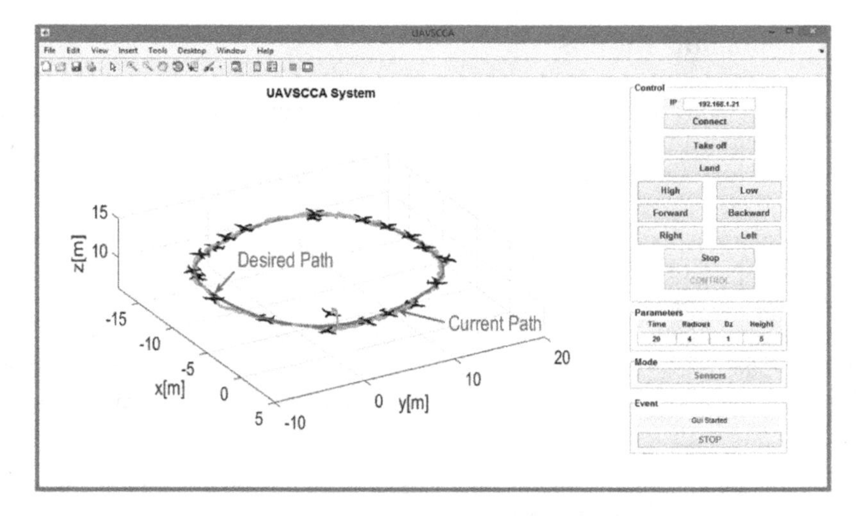

Fig. 4. Stroboscopic movement of the first experiment

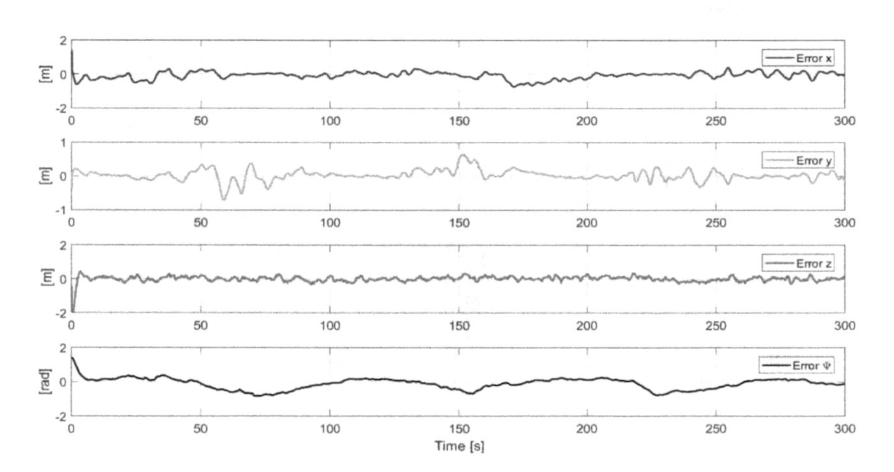

Fig. 5. Control errors in each of the coordinate axes

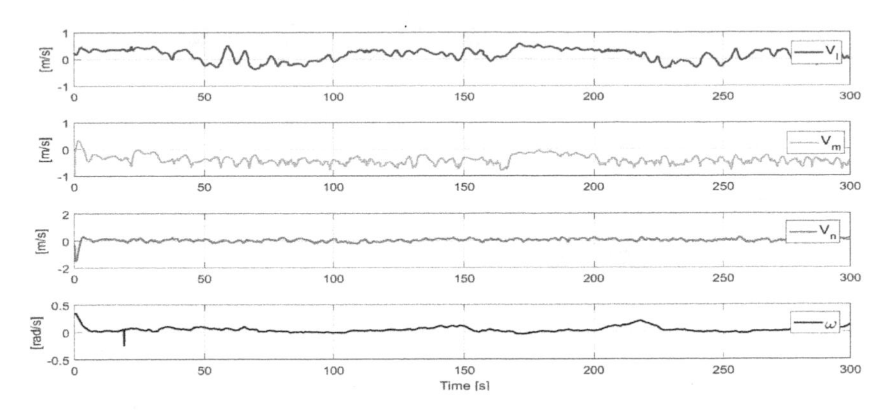

Fig. 6. Control actions to correct errors in each coordinate

On the other hand, the information acquired from the gas sensors is shown in Fig. 7. The variation shown in the location away from sources of pollution shows a low rate of presence of gases.

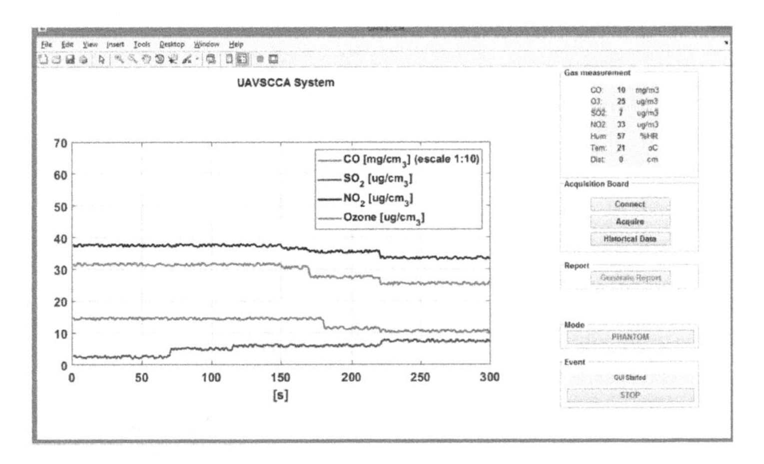

Fig. 7. Graphical plotting of wind parameters measurement

Second Experiment

The second experiment considers an urban space to perform gas monitoring, where the path to be followed is described as an ascending circle on the Z axis. Similar to the first experiment, the stroboscopic movement of the UAV execution is shown in Fig. 8, where the controller is validated for path following taking into account the robustness that it presents in the face of wind disturbances. As a result, Figs. 9 and 10, respectively, show the tendency of errors to zero, as well as the control actions to correct the tracking errors.

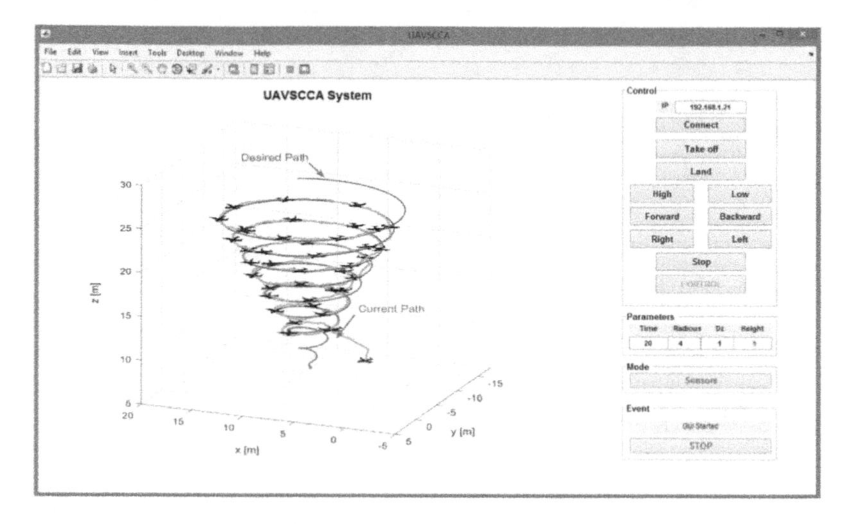

Fig. 8. Stroboscopic movement of the second experiment

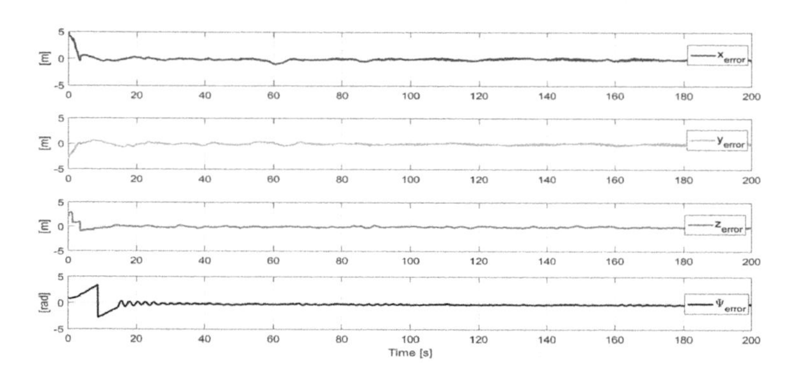

Fig. 9. Control errors in each of the coordinate axes

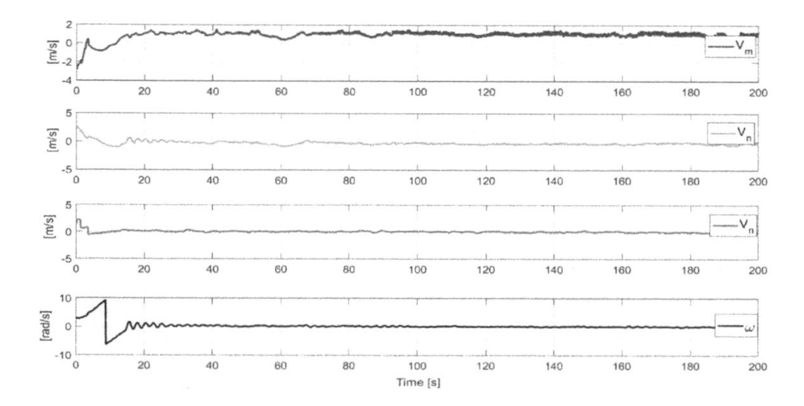

Fig. 10. Control actions to correct errors in each coordinate

Finally, Fig. 11 presents the measurement of gases in an urban place, in an hour considered as the one with the highest vehicular traffic and industrial production. The programmed trajectory allows to have a complete view of gas dispersion, where by wind actions the decrease in the presence of the measured variables is noticed as the vehicle increases the flight height.

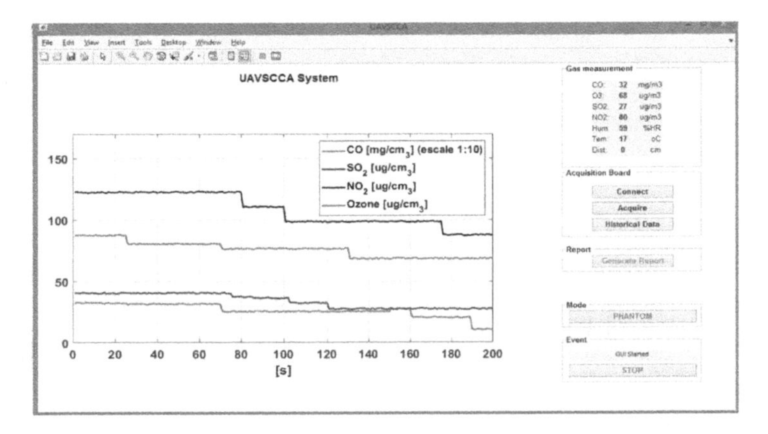

Fig. 11. Graphical plotting of wind parameters measurement, second experiment

5 Conclusions

The work proposes a system to monitor the amount of toxic gases scattered in the air through the use of a UAV. By means of an unmanned aerial vehicle, it is proposed to include an air information acquisition card, so that the movement over unstructured spaces is not difficult. For this, the construction of an acquisition card that is embedded in the vehicle is considered, considering that the dynamics and weight do not influence the correct execution of flights. Additionally, the programming of libraries based on the SDK of the commercial robot Phantom 3 PRO is proposed in order to verify the control algorithms pro-posed for road tracking. Finally, and through two real experiments, both the construction of the card and the manipulation of the drone through a mathematical software are shown, considering two scenarios where the variation of air parameters differ from each other.

Acknowledgements. The authors would like to thanks to the Corporación Ecuatoriana para el Desarrollo de la Investigación y Academia–CEDIA for the financing given to research, development, and innovation, through the CEPRA projects, especially the project CEPRA-XI-2017-06; Control Coordinado Multi-operador aplicado a un robot Manipulador Aéreo; also to Universidad de las Fuerzas Armadas ESPE, Universidad Técnica de Ambato, Escuela Superior Politécnica de Chimborazo, Universidad Nacional de Chimborazo, and Grupo de Investigación ARSI, for the support to develop this work.

References

1. Kumar, P., et al.: The rise of low-cost sensing for managing air pollution in cities. Environ. Int. **75**, 199–205 (2015)
2. Strak, M., et al.: Long-term exposure to particulate matter, NO_2 and the oxidative potential of particulates and diabetes prevalence in a large national health survey. Environ. Int. **108**(2), 228–236 (2017)
3. Honda, T., Pun, V.C., Manjourides, J., Suh, H.: Anemia prevalence and hemoglobin levels are associated with long-term exposure to air pollution in an older population. Environ. Int. **101**, 125–132 (2017)
4. Tziviana, L., et al.: Effect of long-term outdoor air pollution and noise on cognitive and psychological functions in adults. Int. J. Hyg. Environ. Health **216**, 11 (2014)
5. Liu, L., et al.: Influence of exposure to coarse, fine and ultrafine urban particulate matter and their biological constituents on neural biomarkers in a randomized controlled crossover study. Environ. Int. **101**, 89–95 (2017)
6. Ortega-García, J.A., López-Hernández, F.A., Cárceles-Álvarez, A., Fuster-Soler, J.L., Sotomayor, D.I., Ramis, R.: Childhood cancer in small geographical areas and proximity to air-polluting industries. Environ. Res. **156**(March), 63–73 (2017)
7. Britton, A.J., Hewison, R.L., Mitchell, R.J., Riach, D.: Pollution and climate change drive long-term change in Scottish wetland vegetation composition. Biol. Conserv. **210**(April), 72–79 (2017)
8. Castell, N., et al.: Can commercial low-cost sensor platforms contribute to air quality monitoring and exposure estimates? Environ. Int. **99**, 293–302 (2017)
9. Honda, T., Pun, V.C., Manjourides, J., Suh, H.: Associations between long-term exposure to air pollution, glycosylated hemoglobin and diabetes. Int. J. Hyg. Environ. Health **220**(7), 1124–1132 (2017)
10. Ortiz, J.S., Zapata, C.F., Vega, A.D., Andaluz, V.H.: Path planning based on visual feedback between terrestrial and aerial robots cooperation. In: Zeghloul, S., Romdhane, L., Laribi, M. A. (eds.) Computational Kinematics. MMS, vol. 50, pp. 96–105. Springer, Cham (2018). https://doi.org/10.1007/978-3-319-60867-9_12
11. Andaluz, Víctor H., et al.: Nonlinear controller of quadcopters for agricultural monitoring. In: Bebis, G., et al. (eds.) ISVC 2015. LNCS, vol. 9474, pp. 476–487. Springer, Cham (2015). https://doi.org/10.1007/978-3-319-27857-5_43
12. Siebert, S., Teizer, D.J.: Mobile 3D mapping for surveying earthwork using an unmanned aerial vehicle (UAV). J. Chem. Inf. Model. **53**(9), 1689–1699 (2013)
13. Pajares, G.: Overview and current status of remote sensing applications based on unmanned aerial vehicles (UAVs). Photogram. Eng. Remote Sens. **81**(4), 281–330 (2015)
14. Szafranski, G., Czyba, R.: Different approaches of PID control UAV type quadrotor, pp. 70–75 (2011)
15. Kayacan, E., Maslim, R.: Type-2 fuzzy logic trajectory tracking control of quadrotor VTOL aircraft with elliptic membership functions. IEEE/ASME Trans. Mechatron. **22**(1), 339–348 (2017)
16. Chakravarty, P., Kelchtermans, K., Roussel, T., Wellens, S., Tuytelaars, T., Van Eycken, L.: CNN-based single image obstacle avoidance on a quadrotor, pp. 6369–6374 (2017)
17. Kahn, G., Villaflor, A., Pong, V., Abbeel, P., Levine, S.: Uncertainty-aware reinforcement learning for collision avoidance (2017)
18. Emam, M., Fakharian, A.: M. Emam 1, 4993

19. Abdurrohman, M.Q., et al.: A modified gain schedulling controller by considering the sparseness property of UAV quadrotors. J. Mechatron. Electr. Power Veh. Technol. **6**(1), 9 (2015)
20. Parhizkar, N.: Experimental investigation of rotational control of a constrained quadrotor using backstepping method, no. 1
21. Chen, F., Lei, W., Zhang, K., Tao, G., Jiang, B.: A novel nonlinear resilient control for a quadrotor UAV via backstepping control and nonlinear disturbance observer. Nonlinear Dyn. **85**, 1281–1295 (2016)
22. Xu, J., Wang, M., Qiao, L.: Dynamical sliding mode control for the trajectory tracking of underactuated unmanned underwater vehicles. Ocean Eng. **105**, 54–63 (2015)
23. Bonna, R., Camino, J.: Trajectory tracking control of a quadrotor using feedback linearization. In: Proceedings of XVII International Symposium on Dynamic Problems of Mechanics, no. 2013 (2015)
24. Andaluz, V.H., Carvajal, C.P., Pérez, J.A., Proaño, L.E.: Kinematic nonlinear control of aerial mobile manipulators. In: Huang, Y., Wu, H., Liu, H., Yin, Z. (eds.) ICIRA 2017. LNCS (LNAI), vol. 10464, pp. 740–749. Springer, Cham (2017). https://doi.org/10.1007/978-3-319-65298-6_66
25. Andaluz, V.H., et al.: Robot nonlinear control for unmanned aerial vehicles' multitasking. Assem. Autom. **38**, 645–660 (2018)

Intelligent Parking Management by Means of Capability Oriented Requirements Engineering

Mohamed Salah Hamdi[1(✉)], Adnane Ghannem[1], Pericles Loucopoulos[2], Evangelia Kavakli[3], and Hany Ammar[4]

[1] Information Systems Department, Ahmed Bin Mohammed Military College, Doha, Qatar
mshamdi@abmmc.edu.qa
[2] Manchester Business School, The University of Manchester, Manchester, UK
[3] Cultural Technology and Communication Department, University of the Aegean, Mytilene, Greece
[4] Lane Department of Computer Science and Electrical Engineering, West Virginia University, Morgantown, WV, USA

Abstract. Capability Oriented Requirements Engineering (CORE) is an emerging research area where designers are faced with the challenge of analyzing changes in the business domain, capturing user requirements, and developing adequate IT solutions taking into consideration these changes and answering user needs. CORE aims at providing continuously a certain level of quality (business, service, security, etc.) in dynamically changing circumstances such as smart city operations. Intelligent management of transportation is a complex smart city operation, and hence, an optimal application domain for CORE. A specific application within the domain of intelligent management of transportation is the smart management of parking. In this paper, we propose dealing with the intelligent management of parking spaces by using CORE. In an I-Parking system that offers personalized advice to users, we deal with change and introduce smartness by means of "Goal Models", "Informational Models", "Capability Models" and "Actor Dependency Models".

Keywords: Smart cities · Smart city operations · Intelligent management of transportation · I-Parking · CORE

1 Introduction

The high degree of urbanization, at a worldwide level, presents numerous and serious challenges to municipalities as well as citizens. Cities are increasingly held accountable for the wellbeing of their residents while at the same time risk losing control because of the sheer size of the ecosystem that comprises a city. Citizens, enabled with modern technology, are in a position to take on more responsibility for the quality of life in their communities. Information and Communication Technology (ICT) is becoming a strong

F. Wotawa et al. (Eds.): IEA/AIE 2019, LNAI 11606, pp. 158–172, 2019.
https://doi.org/10.1007/978-3-030-22999-3_15

enabler for cities to turn "smarter". Smarter cities of the future will drive sustainable economic growth, through offering smarter solutions to various everyday problems. What's more, Smart City Operations (SCOs) will have the potential to analyze data for better decisions, anticipate problems to resolve them proactively, and coordinate resources to operate effectively. In the context of SCOs, Intelligent Transportation Systems (ITSs) aim to achieve traffic efficiency by minimizing traffic problems. ITSs may include traffic assessment and management; in-vehicle and on-road safety management, either active or passive; emergency management; driver modeling; Parking management; infotainment; environmental effects of transportation; and application of technologies like sensor networks or network entities' control techniques.

In this paper, we focus on the intelligent management of parking spaces. Large cities are overcrowded with vehicles, which puts pressure on the parking infrastructure, and hence the need for effective management. Especially during peak hours, the search for a parking spot becomes a time consuming and challenging process. Dealing efficiently with such scenarios, will have positive impact on economic activities' efficiency, social interactions, and cost [1]. Intelligent Parking Systems (IPSs) can support parking activities in many different ways such as providing parking availability information, and offering automated navigation to help the driver to find the most suitable parking spot based on his profile. One of the challenges when dealing with intelligent parking is managing change, especially in the case of big events (e.g. World Cup QATAR 2022). This paper addresses the challenge of developing an IPS in a systematic, generic, and repeatable manner. Capability Oriented Requirements Engineering (CORE), a methodology that encompasses all activities from requirements to system design and implementation, is presented. This methodology is motivated by the notion of 'capability' which is defined as "the confluence of capacity and ability to achieve a desired goal under specified standards and conditions through a combination of ways and means to perform a set of tasks" [2]. Capability modeling is emerging as a paradigm that facilitates the alignment of technical systems and organizational systems [3] taking into consideration aspects of modeling, patterns, contextualization, and adaptation. There are distinct types of capability depending on the impact and the increasing importance. These range from (a) the deployment of resources, through (b) the combination of use of resources and know-how to (c) dynamic capabilities where the focus is on re-configuration, and re-creation of resources to address environmental changes [4]. In the context of big events (e.g., organization of the football world cup or Olympic games), capability modeling can bring originality and innovation into the idea of IPSs, help deal with the aforementioned challenges, and make parking management more efficient. To demonstrate the applicability and value of the proposed approach, the methodology is applied on a system called "I-Parking" which is part of a larger project dealing with delivering different prototypes for safety management, for customized recommendations of city features, and for security management.

2 Background

Important requirements need to be addressed from a CORE point of view taking into consideration social, technological, and economic requirements [5]. SCOs are characterized by their dynamic change. The development of a software system for such an application encompasses three challenges. First, ***managing the issue of complexity*** of software design and development which arises from the continuous changes to contexts, constraints, and functionality [6]. Second, ***understanding the requirements of the organization*** [5] that owns the individual system, the interaction between the components that collaborate together, and the dynamics of the whole software system in order to understand the orchestration and configuration of different individual systems. Third, ***evolving the software system and the users' needs simultaneously*** [5] in order to maintain a high level of quality and utility of the software system.

To deal with these challenges, the notion of capability has emerged complementing ***traditional enterprise modeling approaches*** by representing organizational knowledge from a result-based perspective [7]. In the literature, Barney [8] has distinguished between two views: (1) the Resource Based View (RBV) and (2) the Dynamic Capability View (DCV). In the RBV, the focus of the researchers is on identifying the possession of valuable, rare, inimitable, and non-substitutable resources of the enterprise as a source of sustainable advantage. In the DCV, the focus of the researchers is on the dynamic aspect of capabilities aiming at the "ability to integrate, build, and reconfigure internal and external competences to address rapidly changing environments". This is also called *dynamic capability*.

Approaches for modeling enterprise capabilities have been proposed by the academia [9, 10] as well as by the industry [11] with the aim of linking strategic objectives and high-level organizational requirements to technological artifacts. Danesh and Yu [12] argue that a *business capability* (represents the what) is at a higher level than a *business process* (represents the how). The business capability is representing a conceptual service performed by a set of people and processes. The people and processes are supported by the relevant application, information, and underlying technology. Business processes describe the methods an organization employs in order to provide and leverage business capabilities [3]. According to Loucopoulos and Kavakli [13] enterprise capability reveals the following features: (i) A capability is associated with a certain owner (a business entity such as a department, an organization, a person, a system, etc.); (ii) A capability denotes the fitness of its owner for achieving a certain result (business goal, customer need, project objective, etc.); (iii) A capability encapsulates the resources (processes, people, technology, assets, etc.) required by the capability owner for possessing this capability; (iv) A capability is context-specific. Its application depends on specific parameters within the enterprise environment (social context, economic context, cultural context, etc.).

The introduction of capability in software development, analysis, and especially design is still in early stages. For example, Iacob et al. [9] extended the meta-model of ARCHIMATE using the notion of capabilities, resources, and values to enable strategic alignment of technical projects. Bērziša et al. [14] have proposed a meta-model relating enterprise capability to the domain context, business processes, and enterprise objectives. The need for operating in changing environments is addressed by integrating organizational development with information system development taking into account changes in the application context of the solution. The design of adjustable services that can adapt to changes in parameters of the capability context is enabled by modeling the context of the capabilities of the enterprise. Based on the proposal of Iacob et al. [9], Azevedo et al. [15] claim that the notion of capability is subjective and subject to multiple interpretations of the dependencies between capability-related concepts and other elements of the enterprise architecture. This increases the need for more rigor in the conceptualization of capability. The authors proposed the use of capability semantics for some relevant concepts [9] based on the Unified Foundation Ontology (UFO) [16] that represents a synthesis of a selection of foundational ontologies. A large number of recent works [17–20] has shown that foundational ontologies can be used to evaluate conceptual modeling languages and to develop guidelines for their use. One of the main application domains of conceptual modeling and methods is the business modeling called also in the Model-Driven Architecture (MDA) approach of the Object Management Group (OMG), a "computation-independent model" because it must not be expressed in terms of IT concepts, but solely in terms of business language. The business domain, since it contains so many different kinds of things, poses many challenges to foundational ontologies. For this purpose, we have to capture the ontological categories underlying natural language and human cognition. These are also reflected in conceptual modelling languages such as ER diagrams or UML class diagrams called by Gangemi et al. [21] a 'descriptive ontology' as opposed to 'prescriptive ontology', which claims to be 'realistic' and robust against the state of the art in scientific knowledge. The use of UFO has revealed some additional relationships between capability and the structural and behavioral elements of the enterprise architecture.

Several approaches were proposed in the literature to support smart parking systems using many techniques such as agent-based systems, fuzzy-based systems, GPS-based systems, wireless-based systems, vision-based systems, and vehicular communication. Some studies considered street parking [22–25]. Some other studies targeted private parking spaces such as those dedicated for stadiums or for shopping malls. To the best of our knowledge, none of these studies has tackled the problem of designing an intelligent parking from a capability-driven perspective.

In the adopted CORE conceptual framework, capabilities focus on assets, these being capacities and abilities. There are passive assets, the resources, and dynamic assets, the agents, where such agents collaborate with each other. Collaboration is defined as dependencies between the agents. These dependencies may involve the exchange of passive assets (i.e. resources), the execution of some task, or the achievement of a goal. Requirements engineering involves both current capabilities (related to current goals) and desired capabilities (related to change goals), in order to model the transformations from the former to the latter. Quality requirements, referred to as soft goals, are related to business goals. Internal capabilities are fully owned by the considered software system; they can cooperate with external capabilities (for instance, in order to get up-to-date information about the context where the system is operating). Requirements engineering within the CORE framework involves three main activities, which are carried out iteratively, thus incrementally refining the results being produced: information elicitation, business requirements modeling, and system requirements modeling.

The intelligent management of a parking lot, especially in the case of a large-scale sports event, requires the coordination between many actors (stakeholders) and the development of solutions involving different technological systems and human experts. The current work is done in the context of a larger project aiming at delivering innovative models, techniques, and tools for the advancement of theories in both Requirements Engineering (RE) and SCOs and concerned with the development of a prototype for SCOs with a focus on the intelligent management of transportation for Doha during the World Cup 2022. Emphasis is on the design and implementation of novel methodologies and algorithms for traffic/emergency/parking management. Different prototypes including a prototype for safety management, a prototype for customized recommendations of city features, a prototype for security, and a prototype for intelligent parking (I-Parking) are being developed.

In order to offer a good quality of service to all stakeholders (i.e., persons, organizations) involved in this event, there is a need for optimizing the use of a variety of resources (e.g., human, technological, physical infrastructure). In a parking lot area, for example, drivers usually want to get the adequate parking spot as soon as possible while taking into consideration their special needs (e.g., spot close to the targeted store or gate of the stadium). The design of an I-Parking application that helps to overcome the challenges of alignment, agility, and sustainability in relationship with dynamically changing requirements is the main motivation for this work. The CORE approach seems appropriate because it integrates organizational development with information system development taking into account changes in the application context of the solution. The concepts appearing in the CORE models, especially the informational model, are extracted from an ontology model developed in previous work [26].

3 The CORE Approach

3.1 The Conceptual Framework

The CORE approach encourages the consideration of capabilities as a motivating factor in RE activities. The conceptual modelling framework applied in CORE is based on recent work [7, 27, 28]. It employs a set of complimentary and intertwined modelling paradigms based on enterprise capabilities, goals, actors, and information objects.

In SCOs applications, a key consideration is to focus on the management of a city's assets and on the way these assets are transformed into on-line networks of collaborating social objects in a similar way to social networks. In this context, the CORE approach is particularly suitable since it focuses its attention on assets and their collaboration for achieving a certain enterprise goal. This notion is shown in the meta-model of Fig. 1. CAPABILITIES focus on ASSETS (capacities and abilities). For ASSETS, we distinguish between PASSIVE (resources) and DYNAMIC (agents) assets. DYNAMIC ASSETS represent the social dimension focusing on the COLLABORATION between these agents. COLLABORATION is defined as dependencies between the agents. These dependencies may involve the exchange of

PASSIVE ASSETS (resources), the execution of some TASK, or the achievement of a GOAL. In a RE setting we are interested in both CURRENT CAPABILITIES and DESIRED CAPABILITIES in order to model the necessary transformations from the former to the latter. Capabilities are related to GOALS. For the goals, we distinguish between CURRENT GOALS and CHANGE GOALS, which are related to CURRENT CAPABILITIES and DESIRED CAPABILITIES, respectively. Goal Oriented Requirements Engineering (GORE) has long been established as a strong method for identifying and analyzing the intentionality of stakeholders [29, 30]. Goal-oriented approaches such as the Knowledge Acquisition in autOmated Specification (KAOS) method [31], i* [32], and Enterprise Knowledge Development (EKD) [33] describe the 'causal transformation' of strategic goals into one or more sub-goals that constitute the means of achieving desired ends. There is a number of quality requirements (referred to as SOFT GOALS) that are related to BUSINESS GOALS. These soft goals motivate the analysis for discovering the Key Performance Indicators (KPIs) that affect the operationalization of related goals and provide the basis for evaluating or revising current enterprise behavior.

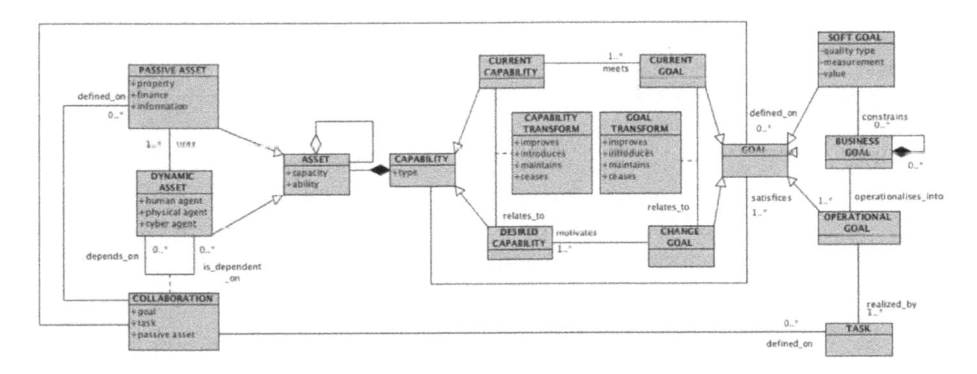

Fig. 1. Main CORE concepts and their relationships.

3.2 Capability-Driven RE

Capability-driven RE using the CORE framework involves three main activities: 'information elicitation', 'business requirements modeling', and 'system requirements modeling'. These activities are carried out in an iterative manner resulting in a stepwise refinement of the results being produced:

- **Information Elicitation:** refers to the collection of information related to the user case using a number of instruments (online forms, structures elicitation forms, collaborative workshops, onsite visits). It results in textual descriptions of the business requirements in terms of the user needs and aspirations with respect to the foreseen functionality and quality of the new system under development. The use of natural language has the advantage of ease of transferability but is prone to ambiguity and inconsistencies. Defining those concepts that are relevant to the

SCOs in a clear and consistent manner is done through textual analysis, which results in the list of concepts that describe the application domain. However, it does not define their structure, nor the interrelationships that exist between these concepts, which hinders any potential analysis. The use of conceptual modelling overcomes this shortcoming.

- **Business Requirements Modeling:** the capability-driven framework applied in CORE [7, 27, 28], is used to model business requirements in terms of enterprise capabilities, goals, actors, and information objects following the rationale discussed in Sub-sect. 3.1. The output of this step is a set of models that can be used by all stakeholders to ascertain whether the conceptualization of the business requirements is accurate and relevant.
- **System Requirements Modeling:** in this step the business requirements should be transformed into system requirements shifting from user needs to the system behavior that satisfies these needs. This behavior can be described using well accepted standards, such as UML use case diagrams. Additional, UML diagrams (e.g. activity diagrams or sequence diagrams) can be used to identify the sequence of activities for realizing each use case, as well as the system components involved and the input/output information flows between these components.

4 Application of CORE on Intelligent Parking Management

Using the capability framework presented in Sect. 3, this section provides a 'walk-through' of the approach showing how it has been used to model the requirements of the I-Parking application and enable the intelligent management of a parking lot. We focus mainly on the conceptual models developed during the capability and requirements modelling steps.

4.1 Modelling Business Requirements

Capability Model. The capability model focuses on the capacities and abilities necessary for a particular application. Capability is a higher-level concept that gives us the opportunity to consider the essential functioning of the application, without having to consider how this functioning comes about. Based on the analysis of the information provided by the stakeholders, Intelligent Parking Management requires three main capabilities denoted in the model shown in Fig. 2 as *Parking Monitoring*, *Guidance Provision*, and *Information Management*. These capabilities entail the existence of certain capacities (in the form of resources) that the enterprise possesses and the ability (in the form of means or skills inherent in the resources) for the resources to be of functional use. For example, Parking Monitoring is a capability of I-Parking that deploys resources such as IoT Devices (sensors, cameras, etc.) in collaboration with the Parking Attendant (human agents). This capability has abilities such as monitoring the parking lot entrance and exit gates, classifying incoming vehicles based on some features such as type (Truck, Emergency car, Personal car, etc.) and size (big, medium, and small), monitoring the parking spaces, and detecting vacant spaces. In the model of Fig. 2, the capacities and the abilities are shown for each capability. Intelligent Parking Management is possible because of the collaboration between these capabilities. For example, Guidance Provision, deploys a GPS system for providing vehicle routing and collaborates both with Information Management, in order to identify the driver's profile and parking reservations, and with Parking Monitoring for detecting parking vacancies. Similarly, Information Management collaborates with Parking Monitoring for updating information regarding the status of the parking lot. Capabilities CAP1... CAP3 are considered as internal capabilities, i.e., they are fully owned by the I-Parking system. In addition to these internal capabilities, there are two external capabilities with which the I-Parking system collaborates, but these capabilities are not owned, not controlled, and not subject to any influence by the I-Parking system. The external capabilities are *Transportation Management* (CAP4) and *Traffic Management* (CAP5). These capabilities can collaborate with the internal capabilities in order to keep the I-Parking system connected and up-to-date with the external events such as traffic jams and accidents. This is particularly important in the case of big events, such as sporting events.

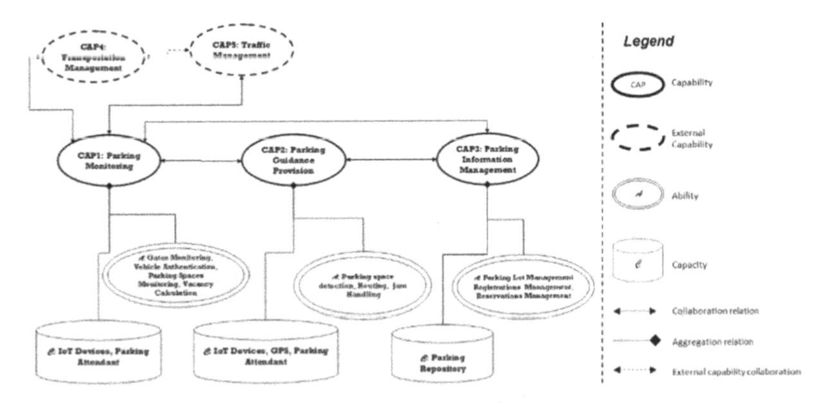

Fig. 2. Capability model for the I-Parking application.

Goal Model. The goal model focuses on enterprise's objectives for retaining, acquiring, or developing the necessary capabilities for the application. It describes the 'causal transformation' of strategic goals into one or more sub-goals that constitute the means of achieving the desired ends. Each step can result in the identification of new goals that are linked to the original one through causal relations, thus forming a hierarchy of goals.As can be seen in the goal model for the I-parking application shown in Fig. 3, the main objective is to manage a parking intelligently.

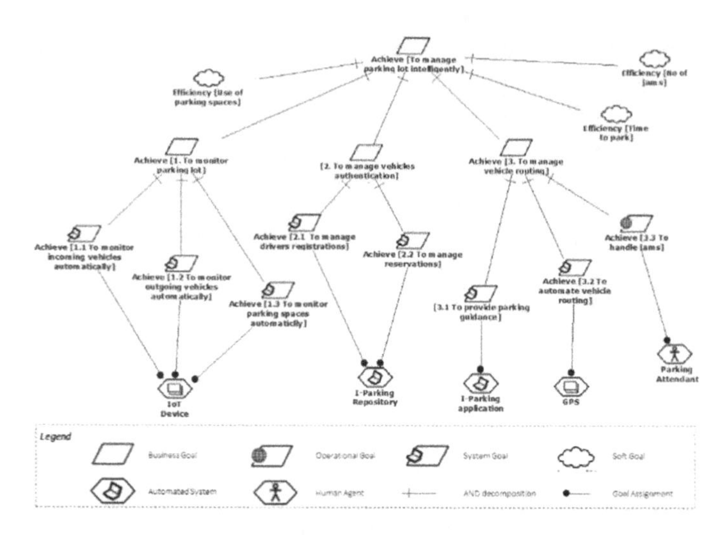

Fig. 3. Goal model for the I-Parking application.

Three sub-goals are derived from this goal: (1) *To monitor parking lot*; (2) *To manage vehicle authentication*; and (3) *To manage vehicle routing*. These are in turn decomposed into a number of operational goals, which are finally assigned to specific human and automated agents. For example, goal *1.3 To monitor parking spaces automatically* is assigned to the *IoT Device*, whilst the goal *3.3 To handle jams* is assigned to the *Parking Attendant*. It should be noted that these actors had already been identified as resources (capacities) and documented in the capability model. There is a number of quality requirements (referred to as soft goals) that are related to these goals, e.g., the efficiency concerning the use of parking spaces. These soft goals provide the basis for evaluating or revising the I-Parking application behavior.

Actor Dependency Model. The goal model described in the previous section gives an intentional description of the I-Parking application. The actor-dependency model provides the socio-technical context of the identified capabilities. There are three types of intentional dependencies: (1) goal-dependency; (2) resource-dependency; and (3) task-dependency. In a goal-dependency, the depender actor depends on the dependee actor to create a certain state in the world without influencing the dependee's decisions to achieve the goal. In resource-dependency, the depender actor depends on the dependee actor for the availability of an entity which can be physical (e.g., equipment) or informational (e.g., jam or feedback). Finally, in task-dependency, the depender depends on the dependee to create an activity by making decisions without disclosing them to the dependee. The I-Parking actor dependency model is illustrated in Fig. 4. There is a complex network of relationships between the actors. For example, the *Driver* depends on *I-Parking* for the achievement of goal *3.1 To provide parking guidance*. *I-Parking* in turn, depends on the *GPS* for the achievement of goal *3.2 To automate vehicle routing*. At the same time, *I-Parking* depends on the *I-Parking Repository* for the provision of specific informational resources namely *Sensor data* and Parking data.

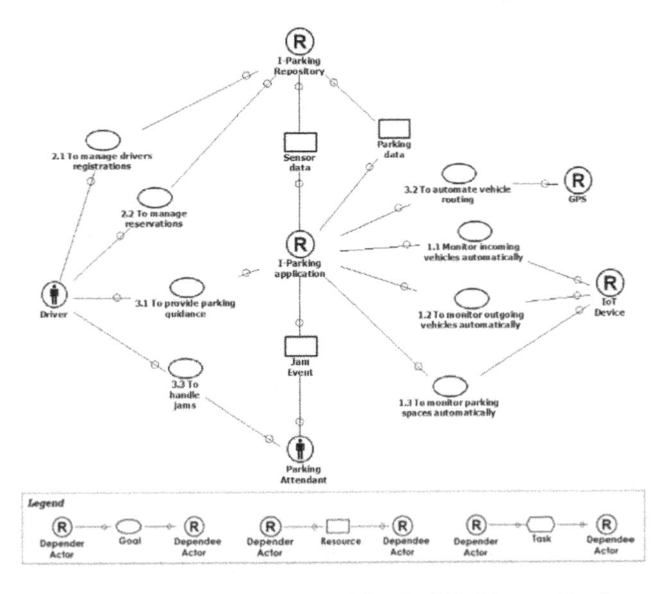

Fig. 4. Actor dependency model for the I-Parking application.

Informational Model. Figure 5 represents a part of the informational model of the I-Parking application. The informational model represents the concepts, relationships, attributes, operations, constraints, and rules used to specify the semantics of the data for the chosen domain. We represented the informational model for the I-Parking application as a UML class diagram. The developed ontology is a generic example that represents the concepts involved in most intelligent parking scenarios. The link between two concepts is an association that is defined as an "Object Property". Roles are specified in each extremity of the association.

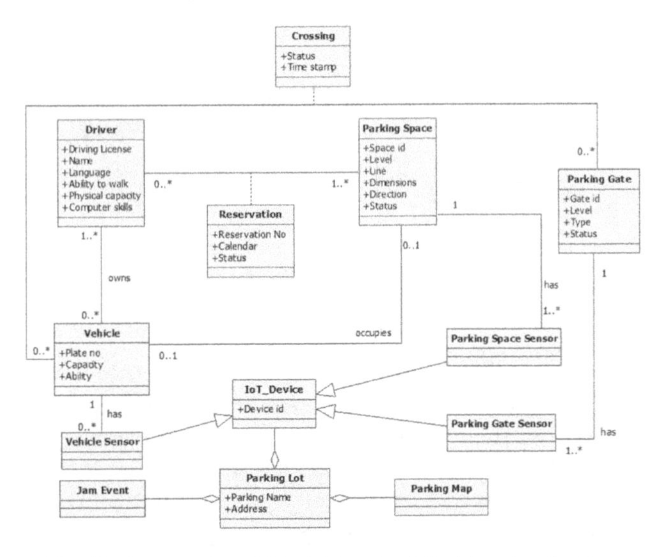

Fig. 5. Informational Model for the I-Parking Application.

Inter-model Relationships. As mentioned in Sect. 3.1, the four models are intertwined. In particular, the required capabilities model acts as an anchor point for the rest of the models, which will use this model for answering: "why does the enterprise need these capabilities?" (answered by the goal model), "what socio-technical actors are involved and how do they co-operate in order to meet these enterprise goals?" (answered by the actor dependency model), and "what kind of information is used in this co-operation?" (answered by the informational object model). This is illustrated in Fig. 6, which, using small fractions of the complete models, shows the intertwining between these models. As shown in Fig. 6, the collaboration between the capabilities "*CAP2: Parking Guidance Provision*" and "*CAP3: Parking Information Management*" gives rise to the *Parking data* exchange of resources between the *I-Parking application* and the *I-Parking Repository*. This is identified as an informational resource and is modeled in the informational model as the aggregate object class *Parking Lot*. Similarly, the existence of the *GPS* in the actor dependency model is due to the business goal *3.2 To automate vehicle routing*. This example illustrates the synergy that can be created between the four types of models. The designer has the advantage to iterate and refine the models by examining the influences of the concepts in one model on the others.

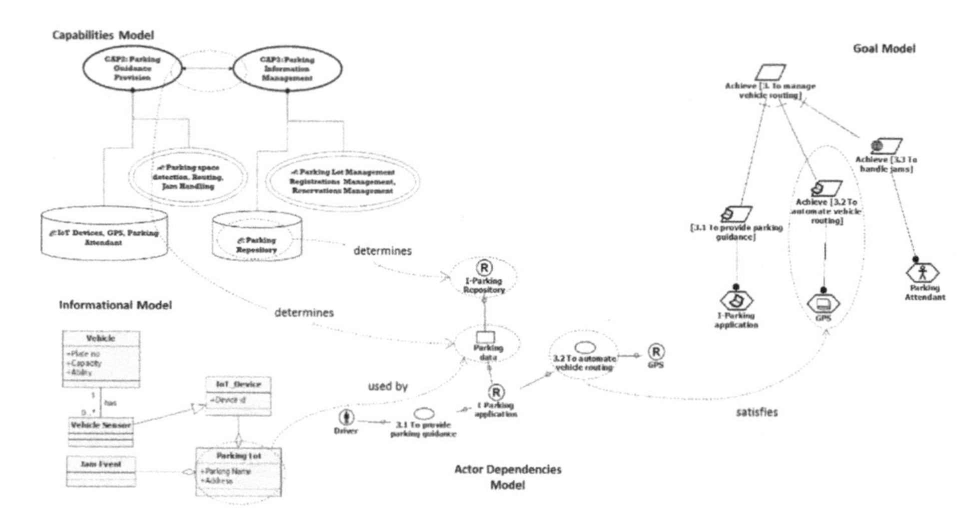

Fig. 6. Inter-model relationships for the I-parking application.

4.2 Modeling System Requirements

The previous models express the I-Parking requirements from a business perspective. They describe the desired state that the user wishes to achieve and define the system boundaries in terms of the actors that collaborate with the system in order for these goals to be achieved. However, they do not describe how these collaborations are realized in terms of specific interactions between the system and its users. To describe these interactions, we propose to use UML use case diagrams. For example the use case diagram of Fig. 7 (left), describes the interaction between the vehicle *Driver* and the *I-Parking* system that emanates from the goal dependency related to the achievement of goal *3.1 To provide parking guidance*. By logically grouping all the use cases, we can determine the functional components of the I-Parking system, as shown in Fig. 7 (right).

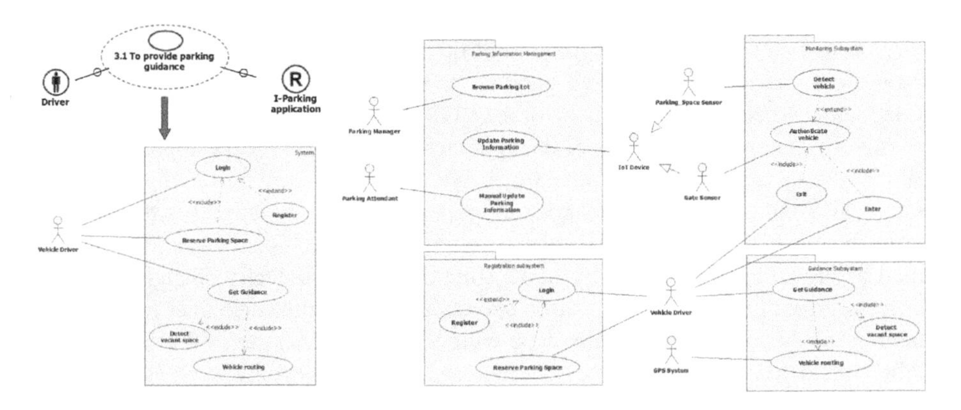

Fig. 7. (Left) Interaction between the vehicle Driver and the I-Parking system. (Right) Use case diagram of the I-Parking application.

5 Implementation Scenario

The implementation scenario of intelligent parking management (I-Parking) highlights the effectiveness of the quest for sustainable mobility in smart cities and exemplifies the provision of smart mobility services in urban environments in the framework of intelligent management of transportation as a complex smart city operation. The I-Parking system would consist of a smart phone application that communicates with a server to obtain information about vacant parking spaces and to issue guidance to help the driver reach the parking space. The server runs a parking management system that maintains a database providing information about vacant parking spaces. The information is constantly updated using appropriate sensors. Figure 8 sketches the way the

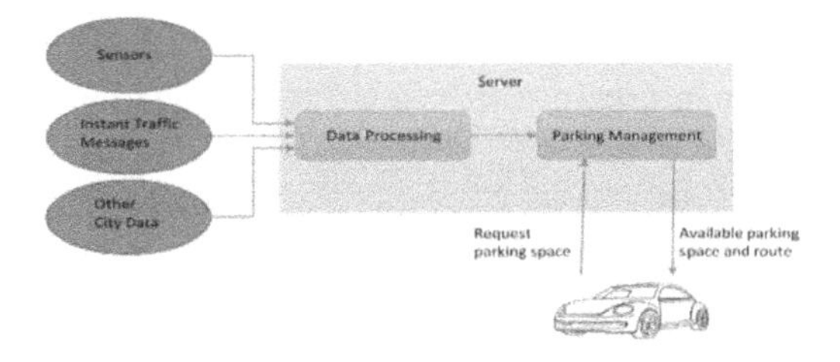

Fig. 8. I-Parking system.

I-Parking system works. Such a system will make parking more effective by allowing high parking space utilization and fast parking spot finding time, which will improve the mobility efficiency and may reduce energy consumption and save the environment.

6 Conclusions and Future Work

In this paper, we presented a CORE based approach aimed at facilitating enterprise agility in terms of dynamic configuration of the enterprise behavior within the context of transportation management, and parking management as a special case. We highlighted some aspects of the adequacy of the CORE based development process on an I-Parking application. The impacts of the approach in terms of process improvements lie in its ability to allow the development of a system (in this case an Intelligent Parking System (IPS)) in a systematic, generic, and repeatable way. What makes the proposed approach specifically applicable for the development of intelligent systems is that it is aimed at providing a good quality of IT solutions in dynamically changing environments, as typically needed by smart city operations. One of the challenges to be faced is enabling the system to dynamically cope with continuous changes in the environment. In the considered scenario changes are represented by big events, such as large-scale sport events (e.g. football world cup or Olympic games). Although the methodology described in this paper seems hard to follow without the support of an automated tool, it is potentially interesting as no study in the literature has so far tackled the problem of designing an IPS from a capability-driven perspective. In future work, we are planning to have our investigations go into two directions. First, we will go further with the assessment of the performance of I-Parking prototype to test the adequacy of the CORE based approach. Second, we will apply this approach to other more complex smart city operations such as the management of the entrance/exit of stadiums in the framework of the Doha world cup games 2022.

Acknowledgment. This work was made possible by NPRP grant # [7-662-2-247] from Qatar Research Fund (a member of Qatar Foundation). The findings achieved herein are solely the responsibility of the authors.

References

1. Li, C.-C., Chou, S.-Y., Lin, S.-W.: An agent-based platform for drivers and car parks negotiation. In: IEEE International Conference on Networking, Sensing and Control (2004)
2. DoD: Systems Engineering Guide for Systems of Systems. https://www.acq.osd.mil/se/docs/se-guide-for-sos.pdf. Accessed 30 Mar 2019
3. Ulrich, W., Rosen, M., Rosen, D., Rosen, B., Rosen, T.: The business capability map: building a foundation for business/IT alignment. In: Business & Enterprise Architecture (2012)
4. Wang, C.L., Ahmed, P.K.: Dynamic capabilities: a review and research agenda. Int. J. Manag. Rev. **9**(1), 31–51 (2007)
5. Jarke, M., et al.: The brave new world of design requirements. Inf. Syst. **36**(7), 992–1008 (2011)

6. Lyytinen, K., Loucopoulos, P., Mylopoulos, J., Robinson, W.N. (eds.): Design Requirements Engineering: A Ten-Year Perspective: Design Requirements Workshop, Cleveland, OH, USA, 3–6 June 2007, Revised and Invited Papers, vol. 14. Springer, Heidelberg (2009). https://doi.org/10.1007/978-3-540-92966-6
7. Loucopoulos, P., Kavakli, E.: Capability modeling with application on large-scale sports events. In: AMCIS 2016, San Diego, USA (2016)
8. Barney, J.B.: Firm resources and sustained competitive advantage. In: Economics Meets Sociology in Strategic Management, pp. 203–227. Emerald Group Publishing Limited (2000)
9. Iacob, M.E., Quartel, D., Jonkers, H.: Capturing business strategy and value in enterprise architecture to support portfolio valuation. In: 2012 IEEE 16th International Enterprise Distributed Object Computing Conference, pp. 11–20 (2012)
10. Stirna, J., et al.: Capability driven development – an approach to support evolving organizations. In: Sandkuhl, K., Seigerroth, U., Stirna, J. (eds.) The Practice of Enterprise Modeling, vol. 134, pp. 117–131. Springer, Heidelberg (2012). https://doi.org/10.1007/978-3-642-34549-4_9
11. Ulrich, W., Rosen, M.: The business capability map: the "rosetta stone" of business/IT alignment. Cut. Consort. Enterp. Architect. **14**(2), 1–23 (2011)
12. Danesh, M.H., Yu, E.: Modeling enterprise capabilities with i*: reasoning on alternatives. In: Iliadis, L., Papazoglou, M., Pohl, K. (eds.) CAiSE 2014. LNBIP, vol. 178, pp. 112–123. Springer, Cham (2014). https://doi.org/10.1007/978-3-319-07869-4_10
13. Loucopoulos, P., Kavakli, E.: Capability modeling with application on large-scale sports events. In: Twenty-Second Americas Conference on Information Systems (AMCIS 2016), San Diego, USA (2016)
14. Bērziša, S., et al.: Capability driven development: an approach to designing digital enterprises. Bus. Inf. Syst. Eng. **57**(1), 15–25 (2015)
15. Azevedo, C.L.B., et al.: Modeling resources and capabilities in enterprise architecture: a well-founded ontology-based proposal for ArchiMate. Inf. Syst. **54**, 235–262 (2015)
16. Guizzardi, G., Wagner, G., de Almeida Falbo, R., Guizzardi, R.S.S., Almeida, J.P.A.: Towards ontological foundations for the conceptual modeling of events. In: Ng, W., Storey, V.C., Trujillo, J.C. (eds.) ER 2013. LNCS, vol. 8217, pp. 327–341. Springer, Heidelberg (2013). https://doi.org/10.1007/978-3-642-41924-9_27
17. Green, P., Rosemann, M.: Integrated process modeling: an ontological evaluation. Inf. Syst. **25**(2), 73–87 (2000)
18. Evermann, J., Wand, Y.: Towards ontologically based semantics for UML constructs. In: S.Kunii, H., Jajodia, S., Sølvberg, A. (eds.) ER 2001. LNCS, vol. 2224, pp. 354–367. Springer, Heidelberg (2001). https://doi.org/10.1007/3-540-45581-7_27
19. Guizzardi, G., Herre, H., Wagner, G.: On the general ontological foundations of conceptual modeling. In: Spaccapietra, S., March, S.T., Kambayashi, Y. (eds.) ER 2002. LNCS, vol. 2503, pp. 65–78. Springer, Heidelberg (2002). https://doi.org/10.1007/3-540-45816-6_15
20. Opdahl, A.L., Henderson-Sellers, B.: Ontological evaluation of the UML Using the Bunge–Wand–Weber model. Softw. Syst. Model. **1**(1), 43–67 (2002)
21. Gangemi, A., Guarino, N., Masolo, C., Oltramari, A., Schneider, L.: Sweetening ontologies with DOLCE. In: Gómez-Pérez, A., Benjamins, V.R. (eds.) EKAW 2002. LNCS (LNAI), vol. 2473, pp. 166–181. Springer, Heidelberg (2002). https://doi.org/10.1007/3-540-45810-7_18
22. Pazos, N., et al.: Dynamic street-parking optimisation. In: 2016 IEEE 30th International Conference on Advanced Information Networking and Applications (AINA) (2016)

23. Tasseron, G., Martens, K., van der Heijden, R.: The potential impact of vehicle-to-vehicle communication on on-street parking under heterogeneous conditions. IEEE Intell. Transp. Syst. Mag. **8**(2), 33–42 (2016)
24. Bock, F., Eggert, D., Sester, M.: On-street parking statistics using LiDAR mobile mapping. In: 2015 IEEE 18th International Conference on Intelligent Transportation Systems (2015)
25. Rajabioun, T., Ioannou, P.A.: On-street and off-street parking availability prediction using multivariate spatiotemporal models. IEEE Trans. Intell. Transp. Syst. **16**(5), 2913–2924 (2015)
26. Ghannem, A., Hamdi, M.S., Abdelmoez, W., Ammar, H.H.: A context model development process for smart city operations. In: 2015 IEEE International Conference on Service Operations and Logistics, and Informatics (SOLI), pp. 122–127 (2015)
27. Loucopoulos, P., Kavakli, E.: Capability oriented enterprise knowledge modeling: the CODEK approach. Domain-Specific Conceptual Modeling, pp. 197–215. Springer, Cham (2016). https://doi.org/10.1007/978-3-319-39417-6_9
28. Loucopoulos, P.: Capability modeling as a strategic analysis tool - keynote extended abstract. In: IEEE Conference on Requirements Engineering: RePa Workshop, Beijing, China. IEEE Computer Society (2016)
29. Yu, E., Mylopoulos, J.: Why goal-oriented requirements engineering. In: Fourth International Workshop on Requirements Engineering: Foundation for Software Quality (REFSQ 1998), Pisa, Italy (1998)
30. Horkoff, J., et al.: Strategic business modeling: representation and reasoning. Softw. Syst. Model. (2012). https://doi.org/10.1007/s10270-012-0290-8
31. Lamsweerde, A.V.: Goal-oriented requirements engineering: a guided tour. In: Proceedings of the Fifth IEEE International Symposium on Requirements Engineering, pp. 249–262. IEEE Computer Society (2001)
32. Yu, E.S.K.: Modelling organizations for information systems requirements engineering. In: IEEE International Symposium on Requirements Engineering, San Diego, California. IEEE Computer Society Press (1993)
33. Kavakli, V., Loucopoulos, P.: Focus issue on legacy information systems and business process change: modelling of organisational change using the EKD framework. Commun. Assoc. Inf. Syst. **2**(1), 6 (1999)

Neural Network Control System of Motion of the Robot in the Environment with Obstacles

Viacheslav Pshikhopov, Mikhail Medvedev, and Maria Vasileva(✉)

JSC «SDB of Robotics and Control Systems», Southern Federal University, Taganrog, Russia
marv@sfedu.ru

Abstract. The article deals with the combined motion control system which provides an autonomous movement of the robot in an uncertain environment. The motion planning level is implemented on a cascade neural network of deep learning. The proposed structure of the network allows decomposing the task of planning a path to the task of deciding whether to maneuver and the task of selecting a direction to bypass an obstacle. The motion control level is implemented in the form of a hybrid system that includes the neural network correction of the path, and the algorithm for avoiding collisions, built on the basis of unstable modes. The control system was modeled and as the result of modeling the quality of control system was obtained. The results of experiments confirming the performance of the control system are presented. It is proposed to classify the environment of operation of the robot according to the complexity of the current situation, depending on the need for maneuver. The environment is classified into complexity classes, the number of which depends on the number of active network cascades.

Keywords: Movement planning · Two-dimensional environment · Neural networks · Combined control

1 Introduction

The convolutional neural network is the first deep learning network that was used to solve the symbol recognition problem [1, 13]. At present, deep learning networks are widely used in a text [2] and object [3] recognition tasks, in navigation [4], for scene understanding [5] and using learned patterns in other areas [6]. Promising directions of neural network systems are researches related to deep learning technologies and self-learning with reinforcements and using of knowledge bases (ontologies) and programs of logical inference, posteriori training.

In automatic control systems, neural networks are used as adaptive regulators, identifiers and path planners. Sustainability and training of neural networks are very special and the most important things at these systems.

Article [7] considers the adaptive neural network control of helicopter when there are parametric and functional uncertainties. Neural network control was synthesized using Lyapunov's method. It gives an opportunity to tracking reference signal with

F. Wotawa et al. (Eds.): IEA/AIE 2019, LNAI 11606, pp. 173–181, 2019.
https://doi.org/10.1007/978-3-030-22999-3_16

small errors. Paper [8] presents a learning algorithm for dynamic recurrent Elman's neural networks based on swarm particle optimization. Developed a new method of control where dynamic identifier makes identification of ultrasonic engine's speed. Monograph [9] presents a neural network algorithm of path planning in uncertain 2D environments. As a result of the comparative analysis showed neural network's high efficiency when solving path planning tasks. Paper [10] proposed the neural network for identification of reverse dynamics of a discrete object. It allows giving predictive properties to the control system.

Using neural networks as motion controllers require a high degree of reliability. It is achieved by increasing of training set volume to dozens of millions of images. This article research the robot control system which was created by using cascades of deep neural networks. This structure of the system allows forming the number of cascades of a neural network that corresponds to the current situation assessment.

2 Formulation of the Problem

We consider a two-dimensional region with dimensions Lx × Ly, see Fig. 1. The current position (x, y) of the robot (see Fig. 1) is marked by rectangle and the desired position (xc, yc) is marked by circle. Obstacles are shown by shaded circles.

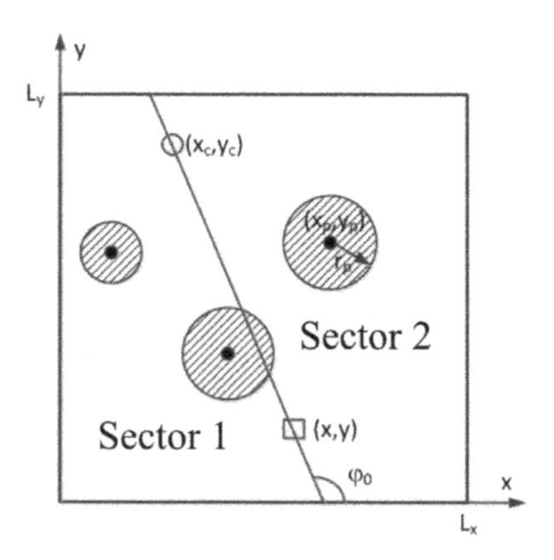

Fig. 1. Area of operation of the robot

Each obstacle is characterized by the coordinates of the center of the circle (xp, yp) and the radius rp. Obstacles are quasistationary i.e. speed of their movement is significantly lower than the maximum speed of the robot.

The task is to develop the learning system of robot motion planning, which automatically implements the path planning and path tracking from the current position (x, y) to the desired position (xc, yc) in the way that to avoid of collision with obstacles. Wherein the training system must develop a sign of the complexity of the situation and change the structure of the planner depending on the current situation.

3 Structure of the Learning System of Motion Planning

The solution of the task is carried out based on a cascade neural network which is trained within the structure shown in Fig. 2. Situation Si in the form of an image comes into a neural network consisting.

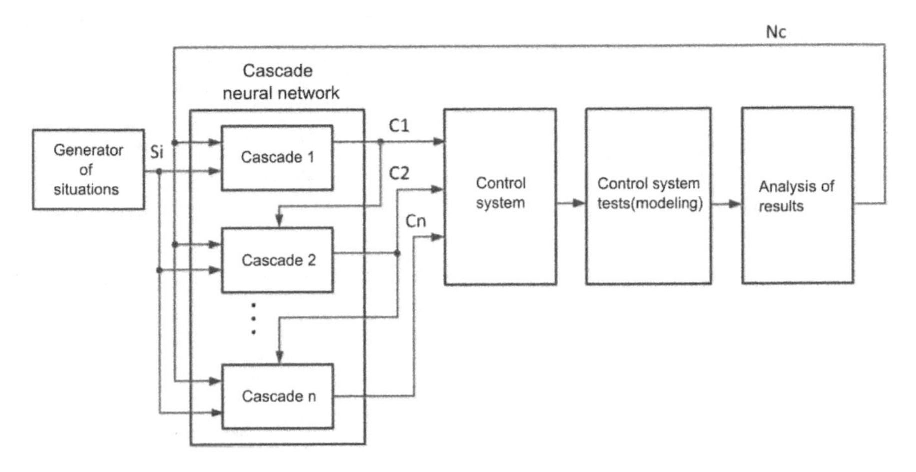

Fig. 2. Structure of the learning control system

The task of the first cascade is to determine the situation in which a maneuver is required. The first sign of a neural network cascade C_1 is equal to zero in the case when the robot must move directly to the target, without performing maneuvers. When $C_1 = 1$, then the robot must perform the maneuver. The task of cascade 2 is to determine the sector where the maneuver is performed. Sign C_2 is 1 if a maneuver is required in sector 1, and C_2 is −1 if a maneuver is required in sector 2 (see Fig. 1). In this way, the second cascade of the neural network limits the maneuver in the yaw angle in the range from $\varphi 0$ to $\varphi 0+ \pi$ for sector 1, and from $\varphi 0$ to $\varphi 0 - \pi$ for sector 2. The third cascade of the neural network produces a sign C_3, which selects from the sector defined by the second cascade a subsector. For example, if the cascade 2 in the situation shown in Fig. 1 chose sector 2, then cascade 3 limits maneuvering in the yaw angle in the range from 2 $\varphi 0$ to $\varphi 0 - \pi/2$, or from $\varphi 0 - \pi/2$ to $\varphi 0 - \pi$.

The cascade structure allows you to increase the reliability of the neural network operating.

The neural network forms signs $C_1, C_2, \ldots, C_n$. These signs are inputs of the motion control system. In the control system, the path of the robot is planned in the selected neural network sector. This is performed by one of the known methods [11]. A similar neural network is built to control the maneuvers and speed of the robot.

The evaluation of the success of reaching the target point (x_c, y_c) is making based on the modeling or experiment results. If the robot reached the target point and did not allow the collision with the obstacles, then the goal was achieved successfully. In another way, the goal wasn't achieved.

The cascaded neural network consists of n/2 levels. Each level consists of two cascades. The first cascade of each level forms a sign that determines the need for maneuver in a given sector. The second cascade divides the sector into two equal parts and chooses one of the parts for maneuver. Wherein, as a result of training for each situation is created the number of active levels is equal to that required for successfully obstacles avoiding. Depending on the number of cascades involved, a classification of the current situation by the level of complexity can be proposed. If $C_1 = 0$, then the maneuver is not required, and the situation corresponds to the level of complexity 0. If $C_1 = 1$, and $C_3 = 0$, then a maneuver is required without specifying a sector in the chosen direction. This situation corresponds to the level of complexity 1. If $C_i = 1$ ($i = 3, 5, \ldots$), then the level of complexity is (ni + 1)/2. Parameter n_i is the index of the last non-zero odd parameter C_i. On the one hand, the classification is related to the distance between obstacles, which characterize the complexity of the environment in terms of the possibility of the desired position achievement. On the other hand, this approach to assessing complexity takes into account the need and accuracy of maneuver.

The complexity of the situation can be used to select the most appropriate method of motion planning [9], as well as to control the speed of the robot.

4 Developing of a Learning Path Planning System

We consider a rectangular area with parameters $L_x = 10$ m, $L_y = 10$ m. The maximum radius of the obstacle is 2.5 m. It is assumed that the scene is pre-mapped, i.e. the position and radii of all obstacles are known.

Here are equations of the kinematics of the mobile robot with differential drive

$$\dot{x} = r_k(\omega_L + \omega_R)\cos\varphi \quad \dot{y} = r_k(\omega_L + \omega_R)\sin\varphi, \tag{1}$$

$$\dot{\varphi} = r_k(\omega_R - \omega_L)/a, \tag{2}$$

where (x, y) are coordinates of the robot; φ is an angle of the robot's yaw; ω_L, ω_R are rotation speeds of the left and right wheels; r_k, a are geometrical parameters.

The motion control is based on the method [12] that solving path and position control problems for mobile robot described by nonlinear equations.

The desired angle of yaw φ^* and the desired speed are:

$$\varphi^* = \arctan\frac{y - y_c}{x - x_c} + C_1 C_2 f(\xi) \quad V^* = \gamma V_k \|p_n\|, \tag{3}$$

where C_1, C_2 are parameters produced by the neural network; V_k is desired speed of the robot when moving towards the target; pn = $[x - x_c \; y - y_c]^T$ is directing vector; $\|p_n\|$ is the Euclidean norm of the directing vector; ξ is a solution of the equation

$$\dot{\xi} = -(T_\xi - \beta)\xi + \beta, \tag{4}$$

T_ξ is the constant positive parameter that determines the value of the characteristic number of the system (4) when the robot moves in a stable mode; β is bifurcation parameter [12], that defined by the expression:

$$\beta = (T_\xi + \beta_0)\mathrm{sgn}\left(\sum_{i=1}^{n} [|r_i - r_s| - (r_i - r_s)]\right), \tag{5}$$

r_i is a distance to the i-th obstacle; r_s is safety distance to the obstacle; n is the number of obstacles; sgn – the sign function; β_0 is a positive parameter.

If a robot maneuver is required then $C_1 = 1$, in a different way $C_1 = 0$. If $C_1 = 1$, then the second cascade participates in the decision-making process. Parameter C_2 takes the values +1 or −1 and is the output of the cascade 2 of the neural network. The following pairs of cascades work in a similar way to cascades 1 and 2. The function $f(\xi)$ determines bounds of the sectors for the robot.

Thus, the first pair of cascades of the neural network decides whether to maneuver (cascade 1) and select the sector for maneuver (cascade 2). The sector size is π radians. The second pair of cascades of the neural network decides if the maneuver is necessary within the selected segment (cascade 3) and selects a subsector within the sector (cascade 4). The subsector size is $\pi/2$ radians. The subsequent pairs of cascades of the neural network work similarly.

Parameter β_0 defines the value of the eigenvalue of Eq. (4) when the robot moves in an unstable mode. As follows from (5), for $\beta = 0$, the eigenvalue of Eq. (4) is $-T_\xi$. When approaching an obstacle, $r_i < r_s$, the bifurcation parameter changes to the value $T_\xi + \beta_0$. As a consequence, the characteristic number of Eq. (4) changes to the value β_0.

The speeds of the robot wheels are calculated on the basis of the desired orientation angle and of the speed of movement in accordance with the expressions:

$$\begin{bmatrix} \omega_R \\ \omega_L \end{bmatrix} = r_k \begin{bmatrix} 1 & 1 \\ 1/a & 1/a_k \end{bmatrix}^{-1} \begin{bmatrix} V^* \\ \varphi^* \end{bmatrix}. \tag{6}$$

To learn the first cascade, the following procedure is applied. At the first stage, scenes are generated with randomly located obstacles and random initial and target positions of the robot. Obstacles may overlap, and the initial and final position of the robot are at a distance not less than $r_s = 1$ m. The resulting scenes are fed into the control system at $C_1 = 0$, without using the neural network. The control system generates the effects of providing linear motion to the target. Next, the working of control system is simulated. Based on the simulation results, we determine the minimum distance r_{min} from the robot to obstacles when moving toward the target. If the distance r_{min} is greater than the permissible value, then the conclusion is made that in this situation it is not necessary to maneuver. If the distance r_{min} is smaller than the permissible value, then the conclusion is made that in this situation it is necessary to maneuver.

The first cascade is the convolutional neural network with 5 hidden layers. The learning set size for the training was 1000 images. Size of image is 131 × 175 pixels. The layer 'convolution2dLayer' is a sliding filter with a 3 × 3 pixel. After convolution we use a layer 'batchNormalizationLayer'. After each normalization layer, the image is processed by a filter 'maxPooling2dLayer' with a 2 × 2 dimension which selects the maximum values from the matrix of pixel values. The layer 'SoftmaxLayer' is applied so that the network can be used for classification to more than two categories. In the output layer 'classificationLayer' the desired class is finally determined.

During the training, the size of the packets was optimized. All data that was divided into packets are going through the neural network. The highest accuracy of recognition of the verification data set was 90.02%.

To learn the second cascade scenes of $C_1 = 1$ are selected. Since the resulting sample of images is smaller than the initial set, it is complemented by generating situations under the initial conditions of the robot, which are approximately at a distance r_s. to any of the obstacles. In this case, each generated situation belongs to the "Maneuver needed" class and can be used to train a 2nd cascade of the network. Training of the 2nd cascade of the neural network is carried out in the same way. The obtained accuracy of training on a sample of 614 situations for each class ("Maneuver to the left" and "Maneuver to the right") is equal to 93.3%.

The trained two cascades determine the minimal configuration of the neural network, which plans the path of the robot.

5 The Results of Numerical Studies

Consider the simulation results. For this, 394 additional scenes are generated, in 160 cases maneuver is required. The simulation results of an intelligent robot control system with a neural network scheduler are presented in Fig. 3. Numerical parameters characterizing the quality of the control system are presented in Table 1.

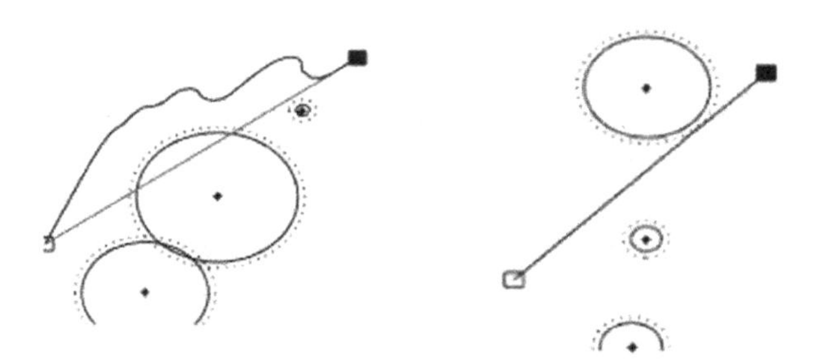

Fig. 3. Results of control system simulation

Table 1. Numerical evaluation of the control system quality

Parameter	Cascade of neural networks	Single neural network
	Parameter value	
The total number of experiments	394	387
The percentage of successful tests, %	92.9	83.2
The percentage of correct decisions 1th cascade, %	91.2	–
The percentage of correct decisions 2nd cascade, %	94.2	–

Table 1 also shows the results of the operation of the control system using a neural network consisting of one cascade, which classifies the current situation into three classes: maneuver to the left; no maneuver needed; maneuver to the right. The structure of the neural network corresponds to Fig. 2. The training sample is the same as the sample used to train the cascade network. As can be seen from Table 1, the proposed cascade network has the accuracy of making the right decisions 10% higher than the network with one cascade.

6 The Results of the Experiment

Consider the results of the experiment, its purpose is to test the performance of the neural network control system in practice. Experimental studies were carried out using a wheeled robot operating on the basis of the Raspberry Pi 3 microcontroller with the ROS operating system installed. The robot was coordinated in space using an RP lidar. On a stationary PC, in the Matlab program, using a trained neural network, the trajectory was calculated and divided into points. The set of points was transmitted via the Wi-Fi radio channel to the robot's OS, which worked them out one by one. The interaction of ROS with Matlab was implemented using the Robotics System Toolbox software package. The result of the route is shown in Fig. 4 as a combination of robot positions at different times.

Fig. 4. Results of experimental modeling of the control system

7 Conclusion

The article proposes algorithms for the intelligent control system of a ground robot. A cascade neural network is proposed, which allows decomposing the problem of multidimensional environmental classification into separate classification tasks into two classes. This approach allows increasing the percentage of correct decisions made by the neural network. On the base of simulation results, it was shown that a cascade neural network, compared to a single neural network, is 10% more accurate in deciding how to bypass obstacles. It was proposed to estimate the complexity of the situation (in terms of the robot's movement towards the target) by the number of cascades required to bypass obstacles. This approach to estimating the complexity of a situation allows for two factors to be taken into account. They are needed for maneuver and the accuracy of maneuvering. The proposed algorithms are experimentally investigated using a wheeled robot. The experiments confirmed the efficiency of the control system and the advantages of the proposed approach.

The developed neural network system is trained in a certain environment. However, it does not plan all the way to the target point, and only generates the direction of movement at the current time. This allows it to be used in an uncertain dynamic environment. Traditional planning methods in such a situation require recalculation of the entire trajectory of movement at each step. On the other hand, the use of a specific map for training allows you to not apply training with reinforcements, which is effective in the case when it is impossible to work out in advance the optimal plan of movements.

Acknowledgment. The study is supported by RSCF grant 18-19-00621, at Joint stock Company «Scientific-Design bureau of Robotics and Control Systems».

References

1. LeCun, Y., et al.: Backpropagation Applied to Handwritten Zip Code Recognition. Neural Comput. **1**(4), 541–551 (1989)
2. Kim, Y.: Convolutional neural networks for sentence classification. In: Proceedings of the 2014 Conference on Empirical Methods in Natural Language Processing (EMNLP 2014), pp. 1746–1751 (2014)
3. Angelova, A., Krizhevsky, A., Vanhoucke, V.: Pedestrian detection with a Large-Field-Of-View deep network. In: IEEE International Conference on Robotics and Automation (2015)
4. Engel, J., Schöps, T., Cremers, D.: LSD-SLAM: large-scale direct monocular SLAM. In: Fleet, D., Pajdla, T., Schiele, B., Tuytelaars, T. (eds.) ECCV 2014. LNCS, vol. 8690, pp. 834–849. Springer, Cham (2014). https://doi.org/10.1007/978-3-319-10605-2_54
5. Zhu, J., Park, T., Isola, P., Efros, A.A.: Unpaired image-to-image translation using cycle-consistent adversarial networks. In: IEEE International Conference on Computer Vision (ICCV) (2017)
6. Gholami, B., Rudovic, O., Pavlovic, V.: Probabilistic unsupervised domain adaptation for knowledge transfer across visual categories. In: The IEEE International Conference on Computer Vision (ICCV), pp. 3581–3590 (2017)

7. Ge, S., Ren, B., Tee, K.: Adaptive neural network control of helicopters with unknown dynamics. In: 45th IEEE Conference on Decision and Control, pp. 3022–3027 (2006)
8. Hong-Wei, G., Wen-Li, D., Feng, Q., Lu, W.: A dissimilation particle swarm optimization-based Elman network and applications for identifying and controlling ultrasonic motors. In: Melin, P., Castillo, O., Ramírez, E.G., Kacprzyk, J., Pedrycz, W. (eds.) Analysis and Design of Intelligent Systems using Soft Computing Techniques. Advances in Soft Computing, vol. 41, pp. 1345–1360. Springer, Heidelberg (2007). https://doi.org/10.1007/978-3-540-72432-2_40
9. Guzik, V., Pereverzev, V., Pyavchenko, A., Saprykin R., Pshikhopov, V.: Neural networking path planning based on neural-like structures. In: Path Planning for Vehicles Operating in Uncertain 2D Environments, pp. 25–96. Elsevier, Butterworth-Heinemann (2017)
10. Notkin, B.: Neural network identification of the inverse dynamics of an object for design of predictive control system. In Proceedings of the 4-th International Conference SICPRO 2005, Moscow (2005)
11. LaValle, S.: Planning Algorithms, 842 p. Cambridge University Press, Cambridge (2006)
12. Pshikhopov, V., Medvedev, M., Gaiduk, A., Belyaev, V., Fedorenko, R., Krukhmalev, V.: Position-trajectory control system for robot on base of airship. In: 2013 Proceedings of the IEEE Conference on Decision and Control, pp. 3590–3595 (2013)
13. LeCun, Ya., Yoshua, B., Geoffrey, H.: Deep learning. Nature **521**(7553), 436–444 (2015)

On Board Autonomy Operations for OPS-SAT Experiment

Simone Fratini[1(✉)], Julian Gorfer[2], and Nicola Policella[1]

[1] European Space Agency - ESA/ESOC,
Robert-Bosch-Strasse 5, 64293 Darmstadt, Germany
{simone.fratini,nicola.policella}@esa.int

[2] Department of Data Science and Knowledge Engineering,
Maastricht University, Bouillonstraat 8-10,
6211 LH Maastricht, The Netherlands
j.gorfer@student.maastrichtuniversity.nl

Abstract. Upcoming space missions are requiring a higher degree of on-board autonomy operations to increase quality science return, to minimize close-loop space-ground decision making, and to enable new scenarios. Artificial Intelligence technologies like Machine Learning and Automated Planning are becoming more and more popular as they can support data analytics conducted directly on-board as input for the on-board decision making system that generates plans or updates them while being executed. This paper describes the planning and execution architecture under development at the European Space Agency to target this need of autonomy for the OPS-SAT mission to be launched in 2019.

Keywords: Autonomous systems · Planning and scheduling

1 Introduction

The mission of scientific satellites in space consists, at a very general level, in making scientific observations and in downloading collected data to Earth trough ground stations. Current common practice for most of these missions, is to generate plans in the ground mission control facilities, and then to upload them as static sequences of time-tagged telecommands to be executed by the satellite at some point in the future. This approach has obvious limitations because of the intrinsic uncertainty at execution time: for instance in case of Earth observing satellites, meteorological conditions can significantly impact observations quality and the amount of data actually generated by the experiments is rarely known in advance with enough precision to generate accurate dump plans. Moreover, temporal communication delays, particularly for deep space missions, make even more critical the problem: local and spontaneous phenomena cannot be observed properly, because when they are detected through an analysis of dumped data

J. Gorfer—Work performed at ESA under traineeship activity.

F. Wotawa et al. (Eds.): IEA/AIE 2019, LNAI 11606, pp. 182–195, 2019.
https://doi.org/10.1007/978-3-030-22999-3_17

on ground, it is usually too late to organize a proper observation plan or to update the plan on board to follow them.

As a result, the efficiency of these systems is reduced by contingent conditions difficult to predict at planning time and difficult to accommodate in a rigid process that keeps planning and execution strictly distinct: a lot of low quality data are being collected and dumped for instance, the amount and effectiveness of planned experiments is reduced because of the use of very conservative models for memory management and because of the impossibility of supporting opportunistic science.

Upcoming missions instead, either targeting Earth, planets or deep space, are requiring a higher degree of on-board autonomy operations to increase quality science return, to minimize close-loop space-ground decision making, and to enable new operational scenarios. Machine Learning and Automated Planning are enabling technology to support data analytic directly on-board and to provide the input for the on-board planner. This paper describes an on-board planning and execution architecture deployed at the European Space Agency (ESA) to target this need of autonomy for the OPS-SAT mission, a 3-Unit CubeSat structure. The mission aims at testing and validating new techniques in mission control and on-board systems [1]. It consists of a satellite that is only 30 cm high but that contains an experimental computer that is ten times more powerful than any current ESA spacecraft.[1]

The architecture presented is based on an automated, domain independent, planner, a platform-agnostic executive, and an on-board image classification system. The project here described constitutes a first attempt to the application of Artificial Intelligence in ESA for on board autonomy on a flying mission. Initial steps were taken in the past to validate, on the ground segment, building blocks technologies such as diagnostics and automated planning. The proposal aims at enhancing the state of the art of the Agency taking inspiration from previous NASA-JPL missions that have pioneered and proven the value of AI planning and scheduling for injecting autonomy in space applications: Remote Agent Experiment (RAX) on Deep Space 1 (DS-1) [2] and the Autonomous Sciencecraft Experiment (ASE) on Earth Observing 1 (EO-1) [3]. Lately, NASA launched the IPEX CubeSat [4] to validate new technologies for on-board image processing and autonomous operations.

The paper first introduces the mission scenario and the autonomy experiment. Then it presents the architecture implemented and describes one of the testing scenarios used to validate on-board autonomous capability on OPS-SAT.

2 OPS-SAT

The OPS-SAT mission has been conceived as an opportunity for testing new concepts and technology in an in-flight environment. The idea is that the validation through OPS-SAT, it would make easier the acceptance and adoption of these

[1] OPS-SAT launch date is currently scheduled for November 2019.

solutions in future missions. This is mainly because of the difficulty to perform live testing in the domain of mission control systems. No-one wants to take any risk with an existing, valuable satellite unless strictly necessary. This makes very complex to introduce new concepts, techniques or systems in orbit. OPS-SAT is based on a low-cost satellite that is rock-solid safe and robust even if there are any malfunctions due to testing. The robustness of the basic satellite itself will give ESA flight control teams the confidence they need to upload and try out new, innovative control software submitted by experimenters; the satellite can always be recovered if something goes wrong during an experiment execution. Achieving this level of performance and safety at a low cost is a challenge. To do this, OPS-SAT combines off-the-shelf subsystems as typically used with cubesats, the latest terrestrial microelectronics for the on-board computer, and the experience ESA-ESOC has gained in safely operating satellites for the last 40 years. OPS-SAT can be seen as an open, flying "laboratory" for in-orbit demonstration of (r)evolutionary new control concepts and software systems that would be otherwise too risky to trial on a "real" satellite.

2.1 On-Board Autonomy Experiment

Our on-board autonomy experiment is based on two pillars: (1) the possibility of analyzing images directly on-board and (2) the presence of an intelligent planner to promptly react to the results of the analysis by accommodating the current plans and/or to generate new plans for achieving new goals. In particular, the goal of our experiment is not limited to validate the adoption of on-board autonomy capabilities in the case of planning satellite tasks, but it also considers to evaluate the necessary changes to be introduced in the ground system to properly control the satellite without limiting the on-board intelligence.

A superficial analysis could assume that functions usually performed on ground will disappear from the ground segment once performed on-board. However a closer look shows that autonomy induces three evolutions in the role of the Ground Segment [5]:

1. Commanding and monitoring of on-board autonomous (decisional) processes. This can be quite different from Monitor & Control (M&C) of classical processes (i.e. how to M&C on-board orbit control, on-board mission planning or advanced Fault Detection, Isolation and Recovery (FDIR), etc.);
2. On ground automation, that participates to the global system autonomy (i.e. automation of Telemetry/Telecommands loop, mission plan update, detailed telemetry downloading);
3. Support to on-board autonomous processes. Enhanced autonomy induces a new re-partition of processes between Ground and Space, but even if some decisions are taken on-board, it is still valuable that part of it is performed on ground (i.e. for FDIR, platform management, orbit management).

The latter is crucial for mitigating the risks associated with on-board autonomy. All possible situations cannot be tested before launch, and on-board constraints usually do not allow embarking the most capable software. In that case, the ground segment shall provide functions to handle extreme situations.

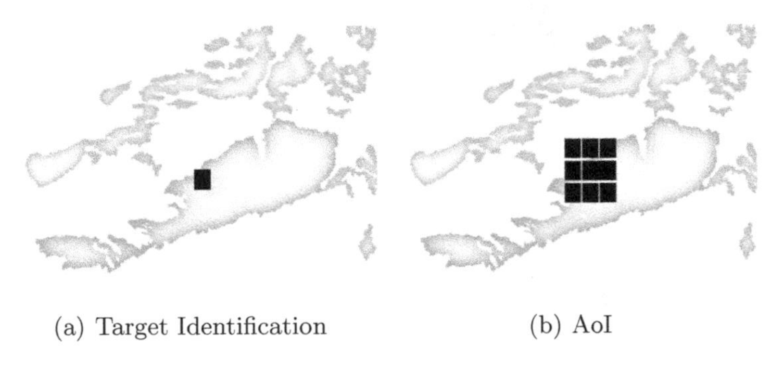

(a) Target Identification (b) AoI

Fig. 1. Survey of the area of interest

Testing Scenario. The potentialities of the architecture for on-board autonomy presented in this paper are demonstrated in a testing scenario which is composed by two phases. In the first phase the satellite is continuously scouting for an interesting phenomena (e.g., the presence of a volcano) by using an on-board image classification algorithms. In practice the satellite will repeatedly taking pictures and analyze them until the particular *target* picture is found.[2]

The finding of the target (see example in Fig. 1(a)) stops the first phase and triggers a second one in which a survey of the *Area of Interest* (AoI) is planned. This consists in planning a series of pictures in order to cover a larger area around the target picture, Fig. 1(b). Each picture of the survey will have latitude and longitude values such that the area around the identified target is consistently covered. More precisely, given a target identified by the couple of latitude and longitude values (x, y), the survey of the area consists of a set of actions $TakePicture(x', y')$ to take pictures at latitudes (x', y') around (x, y)[3].

In the allocation of the pictures several aspects shall be considered. First, the satellite orbit: in fact, given the geographic location of each picture, the orbit is used to determine the time windows when the satellite has direct line-of-sight to the picture's location. Second, the limited on-board memory: each picture has to be stored in the on-board memory before it can be downloaded to the ground (no continuous communication to earth is possible). Last, the limited availability of ground stations to download the data. This can be seen as a set of downloading activities which routinely empty the on-board memory (completely or partially depending of their duration).This

[2] It is worth noting that, during this scouting phase, the pictures will not be stored in the memory nor downloaded to the ground.

[3] Considering the distance between two adjacent images, δ_{lat} and δ_{long}, and *level* the number of 'circles' around the target that should be analyzed, we have, $\forall i, j = -level, ..., level$:

$$x' = x + i * \delta_{lat}$$
$$y' = y + j * \delta_{long}$$

.

resulting planning problem requires to find a time allocation of a set of pictures, $TP = \{TakePicture(x^0, y^0), .., TakePicture(x^n, y^n)\}$, considering that:

- each picture has a set of windows of opportunity, O_i;
- the capacity of the on-board memory, C_j;
- a set of pre-defined downloading opportunities $D = \{dl_0, .., dl_m\}$.

It is important to notice that the problem does not imply any temporal order in which to take the different pictures composing the survey. Furthermore, during the execution of the plan, further adjustment could be required. In particular we will consider to analyze directly on-board the quality of the single pictures taken from the AoI. For instance cloudy pictures could be of no use, therefore the plan should be updated to re-take the picture.

The experiment follows the path first investigated in [3]: where in [3] a scheduler is used on-board to allocate set of flexible activities, in our case a more complex automated planner is used to also generate on-board the activities to achieve high level survey goals.

3 System Architecture

Figure 2 illustrates the system architecture. This is composed by (1) an automated planner that provides the deliberative capabilities (in green); (2) an executive module to instantiate plans into commands for the platform, providing controlling and execution monitoring capabilities (in pink); (3) a set of data analyzer and anomaly detection modules (in blue); (4) an overall controller that masters the interactions among the planner, the executive and the data analyzers.

The Command and Telemetry Dispatcher or CTD (in black in Fig. 2) is responsible for connecting the system to the platform. This is done by interfacing to a middleware layer, the Nanosat Mission Operation Framework, or NMF [6]. The framework, highlighted in yellow in the figure, wraps the different OPS-SAT's hardware components and modules such as the GPS receiver, the HD-Camera Module, or the Magnetometer. The autonomy system can directly manipulate a subset of these components and read sensory values with simple Java API method calls. Changing the attitude[4] of the spacecraft or taking pictures are examples of the functionality provided by the NMF.

The CTD translates control statements received from the Executor into NMF procedure calls, and returns parameter values to telemetry queries from the Executor and/or the Controller. For instance when a picture is taken, the Controller uses the CTD to send it to the Image Classification module. The latter performs a different tasks depending on the phase of the experiment. During scouting, the image classifier check the presence of a particular event (e.g., a volcano). During survey instead, the image classifier focuses only on the quality of the picture taken. In both phases the results from the image classification analysis are sent to the controller that can take different decisions: (1) save the

[4] The attitude defines configuration and orientation of the satellite.

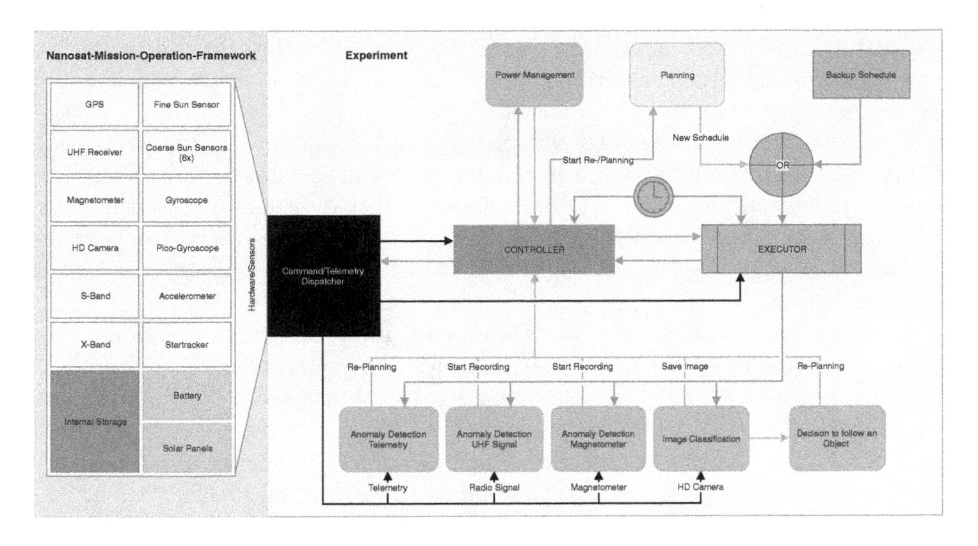

Fig. 2. System overview. (Color figure online)

image and plan a survey if a target is found; (2) do nothing if the target is not in the image; (3) don't save the image if cloudy and plan a new acquisition for the picture.

3.1 Deliberative Technology

The deliberative capabilities are provided by a planning technology based on constraint based temporal planning with timelines. The planner has been developed on top of APSI (Advanced Planning and Scheduling Architecture) [7], an ESA software framework designed to improve the cost-effectiveness and flexibility of AI planning and scheduling tools deployment.

Timeline-based temporal planning is inspired by classical Control Theory, where a problem is modeled as a set of entities, or *timelines*, whose properties vary in time, such as one or more physical subsystems. The timelines evolve over time concurrently while their behaviors can be affected by control decisions[5].

Specifically the planning problems is defined using two classes of modeling components, *state variables* and *resources*, and their valid interactions are specified by means of *synchronizations*.

State variables represent components that can take sequences of symbolic states subject to various (possibly temporal) transition constraints. This primitive allows the definition of *timed automata* representing the constraints that specify the logical and temporal allowed transitions of a timeline. A timeline for a state variable is valid if and only if it represents a *timed word* accepted by the

[5] A detailed description of the timeline based approach, state of the art of the technologies in use and basic concepts is out of the scope of this paper. More information can be found, for example, in [8–11].

automaton. The timed automaton (or in the APSI case, the state variable) is a very powerful modeling primitive, widely studied [12], and for which different algorithms exist to find valid timelines.

Resources are used to model any physical or virtual entity of limited availability, like those commonly used in constraint-based schedulers [13]. A resource timeline represents the resource profile of availability over time as resulting from the quantitative use/production/consumption over a time interval or at time instants. A resource timeline is valid if and only if the profile never exceeds the resource capacity.

The physical and technical constraints that influence the interaction of the sub-systems (modeled either as state variables or resources) are represented by temporal and logical synchronizations among the values taken by the automata and/or resource allocations on the timelines. These constructs define valid schema of values allowed on timelines and link the values of the timelines with resource allocations. In particular they allow the definition of Allen's relations [14] like quantitative temporal relations among time points and time intervals as well as constraints on the parameters of the related values. From a planning perspective, the synchronizations define the cause-effect relations among the states of the model, describing how a given status can be achieved.

In this context, problem solving consists of generating feasible timelines to model a desired system behavior, simulating the temporal evolution of relevant subsystems status and of relevant parameters values. Timelines are then used to control the platform in a closed-loop control schema: an *executor* translates timeline's values into commands and synchronizes actual values read from the platform with the simulated values into the timelines. When discrepancies are detected, a new plan is generated starting from the actual status of the platform.

One of the reasons for the use of this type of planning is the capability to enable, in a *flexible* way, the integration of planning and scheduling. A plan is constituted by a set of timelines and temporal/value constraints among them. A timeline is a sequence of values, a set of ordered transition points between the values and a set of distance constraints between transition points. When the transition points are time-bounded by the planning process instead of being exactly specified we refer to the timeline as *time flexible* and to the plan resulting from a set of flexible timeline as a *flexible plan*.

A flexible plan is defined over temporal variables usually named *time points*. Temporal reasoning is framed as a constraint satisfaction problem over time points bounding the possible occurrence of time points to achieve relations stated among them. Hence the events in the plan have a possible lower and upper bound of temporal occurrence instead of a fix occurrence, and these bounds can be dynamically recomputed, by propagating new constraints, when the plan is executed and an actual temporal occurrence is stated for the time points. Simple Temporal Networks (STN) [15] are the most popular formalism for constraint based temporal reasoning. The temporal information is encoded as a distance graph. The vertices correspond to the (temporal) events and the edges to the temporal constraints between events. The edges are labeled with the lower and upper temporal bounds for the distance constraint between events. An STN is said to be consistent iff there exists a schedule (an assignment to the time points) that satisfies all the temporal constraints.

In situations where everything is under the control of the agent driven by the plan, a consistent plan is enough to ensure a reliable control. But in dynamic and partially unpredictable environments where some occurrences depend on others observable only at execution time, a formalism more suitable is STNUs (Simple Temporal Network with Uncertainty) [16], an extension of STNs where time points are classified into *executable*, which actual occurrence can be chosen by the agent executing the plan, and *contingent*, where the occurrence is chosen by the environment external to the executor. In the STNU case, the consistency definition is not directly applied due to the uncertain assignment of contingent time points at planning time. For a STNU, instead than consistency checking, a *controllability* [17] property is defined, which verifies whether an agent can generate a valid/consistent schedule to any situation that may arise in the external world (*strong* controllability) or, if this is not possible, if the agent can have a strategy, at execution time, to choose future controllable events given the contingent occurrences observed in the past (*dynamic* controllability).

In the OPS-SAT experiment some events presents uncertainty in their temporal duration (see Sect. 4 for some examples), hence the Planner will generate temporally flexible plans and leave to the Executor the task of instantiating the schedule at run time (see Sect. 3.3).

3.2 Controller

The Controller is responsible for integrating the planner with the executive and the analyzer. Figure 3 shows the control workflow. The current testing scenario is limited to two main tasks are considered. To take pictures searching for a specific target, and to survey an Area of Interest. Once the task is selected, the controller monitors the current status of the platform, via the CTD, and defines a planning problem with goals and initial conditions for the planner. The Planner creates a flexible plan that is then passed to the Executor.

The Executor works with the CTD to dispatch commands and to inject values into the plan. Three possible situations can occur: (1) the execution goes as planned, i.e. the flexibility provided by the plan is sufficient to achieve the goals. In this case the controller is notified with success and the cycle restart from the pool of tasks; (2) during execution an exception triggers a re-planning, i.e. some temporal occurrences goes out of the planned bounds or values not planned are injected into the timelines. In this case the execution aborts and the control is given back to the Controller, that is in charge of generating a new sets of goals. In the testing scenario a no-nominal situation occurs after taking an image when it is classified as cloudy, hence a replanning occurs after discarding the image; (3) an interrupt from one of the analyzers triggers a re-planning. In this case the plan is being executed as nominal, but an external trigger fosters the autonomous generation of new tasks. In this case the Controller can either interrupt the

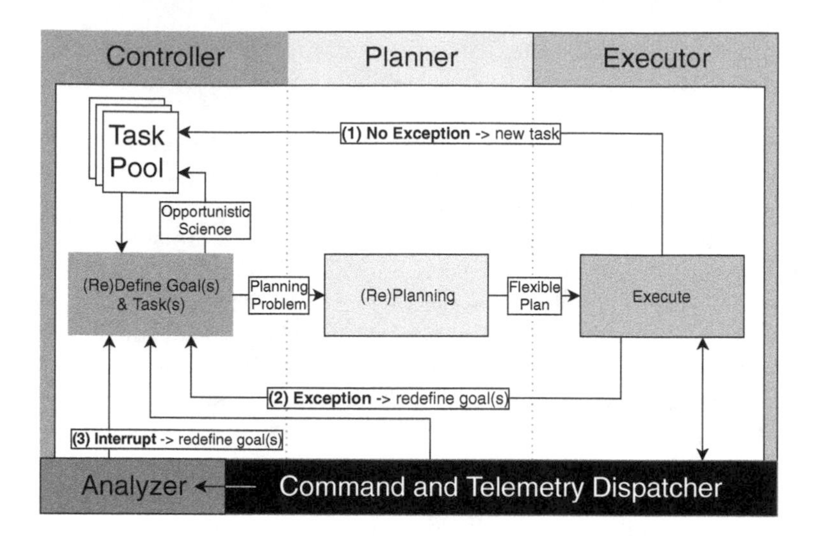

Fig. 3. Control schema

execution and generate a new plan or decide to generate only the tasks and wait for them to be selected in following iterations.

3.3 Executive

The executive is organized into three core services: the Clock, the Plan Executor and the Timeline Executor. The Clock (shown as a clock icon in Fig. 2) generates ticks in a periodic interval and sends notifications to synchronize the system real time with the plan virtual time in the threads under concurrent execution. A tick contains an increasing id number and the timestamp of its release, triggering the Controller and the Executor. In this way it can be determined how much time passed since the last trigger and if a tick was accidentally skipped because a thread were unable to process it.

The Executor process is initiated by the Controller that passes a flexible plan in which each timeline provides a set of transitions between values to be scheduled with respect to the transitions of the other timelines. While controllable transitions are under the control of the Executor, the uncontrollable ones are translated into queries for the CTD that will inform the Executor on their occurrence. For each timeline, the Executor creates a separate thread and monitors the execution ensuring controllability, propagating events as constraints on the temporal database or on the values in the timelines. Once all the threads have reached the end, the Executor sends a notification to the Controller that the plan has been successfully completed. Conversely, if an exception arises, the Executor raises an exception for the Controller and aborts the process.

Table 1. Precision, Recall and F1-Score.

Accuracy	0.937
Precision	0.939
Recall	0.932
F1-Score	0.935

Table 2. Confusion matrix.

		Prediction	
		Clouds	**Clear**
True Label	**Clouds**	80	3
	Clear	6	54

3.4 Data Analyzers

The analyzers implemented are based on Convolutional Neural Network for image classification (see for instance [18] for a survey). In particular several neural networks have been created to analyze information coming from the NMF platform.

A first model has been introduced to analyze images and determine the level of cloudiness. A small Convolutional Neural Network, which has two Convolutional Layers followed by Maxpooling Layers and two fully connected Dense Layers, can classify images to be cloudy/clear with an accuracy of 0.937 and an F1-Score of 0.935 (see Table 1). However, the model misclassifies an image as cloudy in case of presence of ice or snow. Table 2 shows the confusion matrix of the validation set. The six misclassified cloud images are all images that contain a high percentage of white pixels due to snow and ice. A possible explanation could be that the image gets downscaled from 2000 × 2000 pixels to 128 × 128 pixels which leads to an information loss. Also OPS-SAT camera can only take RGB images and has no other channels available to distinguish between white clouds and white snow. The three misclassified pictures with no clouds on the other hand could be misclassified because of inconsistent labelling. To solve the problem of misclassifying snow and ice images as cloudy, the model needs more data. It would also help to have other channels available to distinguish clouds from snow and ice.

The second model classifies an image to identify interesting objects to track. The same small Convolutional Neural Network can be used to classify points of interest in images. The only difference to the previous model is the bigger output layer. Each node of the output layer refers to a class. A preliminary concept of this model classifies following labels: (1) Volcano, (2) Island, (3) Human-made, (4) Land, and (5) Water. One image can have multiple classification labels. For example an image containing an island also contains water.

4 Model

For the case of the Testing Scenario described in Sect. 2.1, we need to control the camera payload, the memory storage sub-system and the attitude of the satellite. Relevant information out of control of the planning system but needed at planning time is the GPS coordinate covered by the satellite in a given time interval. Besides that we model also the image classification sub-system. We model then the planning problem starting with a *Mission Timeline* MT where we post the goals describing the objectives to be achieved with the plan. The mission timeline state variable will take the following values:

- SEARCH TARGET($?t, ?x, ?y, ?file\_id$) - the goal of searching for a target of type $?t$, storing the picture in a file with the id $?file\_id$;, returning the GPS coordinates $\langle ?x, ?y \rangle$;
- TAKEPICTURE($?x, ?y, ?file\_id$) - the goal of taking a picture in $\langle ?x, ?y \rangle$ (GPS coordinates) and store the picture in a file with the id $?file\_id$;
- IDLE() - the idle status.

The camera is modeled as a state variable CAM taking the following values:

- CAMIDLE(), when the camera is not taking pictures;
- TAKEPIC($?file\_id$) when the payload is taking a picture that will be stored in a file with id $= ?file\_id$.

The attitude is modeled by a state variable ATT taking three possible values:

- LOCKED($?x, ?y$), when the satellite is actively pointing the target in $\langle ?x, ?y \rangle$ and following it along the orbit;
- UNLOCKED($?x, ?y$), when the satellite is pointing towards $\langle ?x, ?y \rangle$ but not actively tracking the target;
- LOCKING($?x, ?y$) when the satellite is locking onto $\langle ?x, ?y \rangle$. The attitude is actively maintained only during the lock status.

Once the images have been acquired the lock is released and re-acquired when a different area has to be observed. Hence transitions of the state variable ATT follows the loop $\rightarrow$ LOCKED(...) $\rightarrow$ UNLOCKED(...) $\rightarrow$ LOCKING(...) $\rightarrow$ LOCKED(...) $\rightarrow$ and so on.

The *Memory* MEM is modeled as resource timeline that keeps track of the spacecraft memory allocation.

The GPS coordinates are stored in the timeline of a state variable GPS. This timeline is not under the control of the planner as it depends on the orbit of the satellite which cannot be changed by the experiments. It is generated in advance by querying an on-board service that gives information on when OPS-SAT will be covering a range of GPS coordinates and provided to the planner in the planning problem with goals to achieve. Since we are not interested in all the coordinates covered by the satellite along its orbit, we model a state variable taking only the following values: NOTINPOSITION(), denoting that the spacecraft is currently not covering any area of interest for the current planning session, and POSITION($?x, ?y$) when the satellite is above an area covering the $\langle ?x, ?y \rangle$ GPS coordinates.

The image classification sub-system is modeled by means of a state variable CLASS taking the following values:

- CLASSIDLE(), when no classification algorithm is classifying at the moment;
- CLASSIFYCLOUDS($?file\_id$), when the image taken by the camera with id $file\_id$ is currently being classified as cloudy or clear;

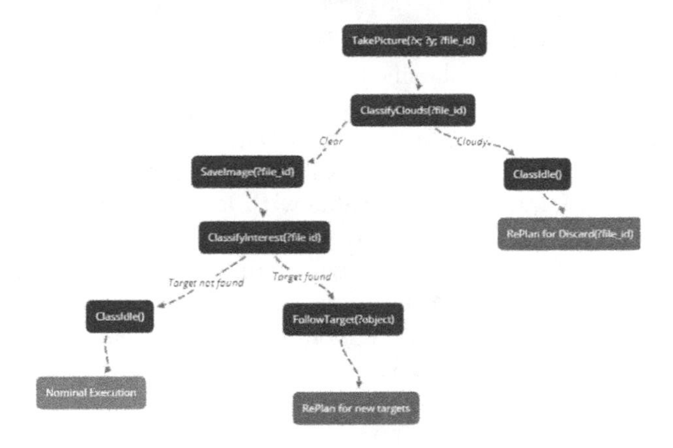

Fig. 4. Model execution logic

- SAVEIMAGE(?*file_id*) denotes the procedure to save the image once its been classified as clear;
- CLASSIFYINTEREST(?*file_id*) denotes that the image gets classified again to analyze the level of interest in the picture;
- FOLLOWTARGET(?*object*) when an interesting object needs to be followed has been identified.

The system status switches from CLASSIDLE() to CLASSIFYCLOUDS(?*file_id*) just after an image has been taken. At this point we plan assuming that an image is clear and no interesting target to follow are identified in the image (see Fig. 4). This is the nominal plan, then after having classified the image the plan switches the system into a status SAVEIMAGE(?*file_id*) followed by CLASSIFYINTEREST(?*file_id*) and back to CLASSIDLE(). If during execution an image is classified as cloudy, a value CLASSIDLE() is injected into the timeline and a replanning occurs (in this case to recalculate the memory occupation that results different than planned since the image is being discarded). Similarly if after CLASSIFYINTEREST(?*file_id*) a status FOLLOWTARGET(?*object*) is injected in place of the idle status planned, the re-planning procedure is fired to generate a new plan to follow the interesting object found (in the following orbits). If everything goes as planned, the image is classified, saved, analyzed and then the execution moves to the next observation planned.

The goal TAKEPICTURE($?x, ?y, ?file\_id$) on the mission's timeline MT can be achieved by (a) having the camera taking a picture with id = $?file\_id$, (b) with the attitude locked in $\langle ?x, ?y \rangle$ and (c) classifying the picture once taken. Hence we have synchronizations in the model stating that (a-c) a value TAKEPICTURE($?x, ?y, ?file\_id$) CONTAINS[6] both a value TAKEPIC(?*file_id*) on the timeline CAM and a value CLASSIFYCLOUDS(?*file_id*) on the timeline CLASS;

[6] CONTAINS, DURING and BEFORE are temporal constraints in Allen's temporal logic [14] among the interval where the values occur.

(b) occurs DURING a value LOCKED($?x, ?y$) on the timeline ATT. In addition to that, since an image can be classified once the acquisition procedure has terminated, we specify that the value CLASSIFYCLOUDS($?file\_id$) on the timeline textscclass occurs AFTER the value TAKEPIC($?file\_id$) on the timeline CAM. Similar statements synchronize the locking values during the GPS visibility of the target and the SAVEIMAGE($?file\_id$) value with a memory resource allocation.

Regarding controllability, most of the states defined above can be executed under the control of the planner but have an uncertain duration. During planning only bounds of acceptable duration can be considered, hence the states are modeled with controllable starting point and contingent ending points. For instance the already mentioned locking value for the attitude subsystem or the take image value, whose actual duration depends on the acquisition camera procedures. Also the actual footprint in memory of an image can only be evaluated with some uncertainty. For this reason, the flexible planning (Sect. 3.1) and the execution strategies (Sect. 3.3) are necessary to support this experiment.

5 Conclusions

As its main contribution, the paper presents an on-board autonomous control architecture which combines a domain independent planner, a platform-agnostic executive, and an on-board image classification system. The project here described constitutes a first attempt to the application of Artificial Intelligence in ESA for on board autonomy on a flying mission. Initial steps were taken in the past to validate, on the ground segment, building blocks technologies such as diagnostics and automated planning.

In order to validate the architecture, a testing scenario is currently under the System Validation Test. The paper presents and discuss in detail a part of this test. The experiment uses the Nanosat Mission Operation Framework ([6]). The NMF and the Experiment are implemented in Java 8. To run the experiment on the computer, we use the Netbeans IDE 8.2 and to simulate the satellite we use the Nanosat-MO-Simulator (Developer Version 2.0) of the NMF SDK. The Dell Latitude computer we use for this experiment is equipped with Windows 10 64-Bit, 16 GB DDR4 SDRAM and an Intel Core i7-7600U with 2 cores and 2.89 GHz processing speed. To test the experiment on real hardware, we use a MitySOM 5CSx System on Module dual-core 800 MHz ARM Processor running with a Linux distribution named Ångström 32-Bit.

Future work foresees to support the on-board system with a ground system that could complement and support the autonomous capability. The ground system will have the role of validating the plan generated on-board with respect to the overall mission objectives. For instance, the satellite can run multiple experiments in parallel and another experiment could require the use of the same payload (the on-board camera). For what concerns the extension of the on board capabilities, analysis of radio frequencies and magnetic forces as well as anomaly detection of telemetry values [19] is currently work in progress.

References

1. OPS-SAT: OPS-SAT Web Site (2018). https://www.esa.int/Our_Activities/Operations/OPS-SAT
2. Muscettola, N., Nayak, P.P., Pell, B., Williams, B.C.: Remote agent: to boldly go where no AI system has gone before. Artif. Intell. **103**, 5–47 (1998)
3. Chien, S., et al.: Using autonomy flight software to improve science return on earth observing one. J. Aerosp. Comput. Inf. Commun. **2**, 196–216 (2005)
4. Chien, S., et al.: Onboard autonomy on the intelligent payload experiment CubeSat mission. J. Aerosp. Inf. Syst. **14**(6), 307–315 (2017). https://doi.org/10.2514/1.I010386
5. Grandjean, P., Pesquet, T., Muxi, A.M.M., Charmeau, M.C.: What on-board autonomy means for ground operations: an autonomy demonstrator conceptual design. In: SpaceOps 2004 (2004)
6. NMF: Nanosat Mission Operation Framework Web Site (2018). https://nanosat-mo-framework.github.io/
7. APSI:APSI Software Distribution Web Site (2017). https://essr.esa.int/project/apsi-advanced-planning-and-scheduling-initiative
8. Muscettola, N.: HSTS: integrating planning and scheduling. In: Zweben, M., Fox, M.S. (eds.) Intelligent Scheduling. Morgan Kauffmann, Burlington (1994)
9. Frank, J., Jonsson, A.: Constraint based attribute and interval planning. J. Constraints **8**, 339–364 (2003)
10. Fratini, S., Pecora, F., Cesta, A.: Unifying planning and scheduling as timelines in a component-based perspective. Arch. Control Sci. **18**, 231–271 (2008)
11. Chien, S., et al.: A generalized timeline representation, services, and interface for automating space mission operations. In: Proceedings of the 12th International Conference on Space Operations, SpaceOps, AIAA (2012)
12. Alur, R., Dill, D.L.: A theory of timed automata. Theor. Comput. Sci. **126**, 183–235 (1994)
13. Baptiste, P., Pape, C.L., Nuijten, W.: Constraint-Based Scheduling. Kluwer Academic Publishers, Norwell (2001)
14. Allen, J.: Maintaining knowledge about temporal intervals. Commun. ACM **26**, 832–843 (1983)
15. Dechter, R., Meiri, I., Pearl, J.: Temporal constraint networks. Artif. Intell. **49**, 61–95 (1991)
16. Vidal, T., Ghallab, M.: Dealing with uncertain durations in temporal constraint networks dedicated to planning. In: ECAI, pp. 48–54. Wiley, Chichester (1996)
17. Morris, P., Muscettola, N., Vidal, T.: Dynamic control of plans with temporal uncertainty. In: Proceedings of the 17th International Joint Conference on Artificial Intelligence. IJCAI 2001, San Francisco, CA, USA, vol. 1, pp. 494–499. Morgan Kaufmann Publishers Inc. (2001)
18. Lu, D., Weng, Q.: A survey of image classification methods and techniques for improving classification performance. Int. J. Remote Sens. **28**, 823–870 (2007)
19. Martinez Heras, J., Donati, A.: Enhanced telemetry monitoring with novelty detection. AI Mag. **35**, 37–46 (2014)

Practical Obstacle Avoidance Path Planning for Agriculture UAVs

Kaipeng Wang, Zhijun Meng(✉), Lifeng Wang, Zhenping Wu, and Zhe Wu

School of Aeronautic Science and Engineering, Beihang University, Beijing, China
mengzhijun@buaa.edu.cn

Abstract. This research deals with the coverage path problem (CPP) in a given area with several known obstacles for agriculture Unmanned Aerial Vehicles (UAVs). The work takes the geometry characteristics of the field and obstacles into consideration. A practical method of the coverage path planning process is established. An obstacle avoidance path planning is used to find a coverage path for agriculture UAVs. The method has been tested with an Android application and is already applied in reality. The results turn out that the method is complete for this kind of coverage path planning problem.

Keywords: Coverage path problem · Obstacle avoidance · Agriculture UAVs

1 Introduction

In recent years, Unmanned Aerial Vehicles (UAVs) are widely used in agriculture activities. Most UAVs have the ability to take autonomous flight. However, in an area with obstacles such as trees, poles or cabins which are usually located within the working fields, the agriculture UAVs have to be manipulated by remote control pilots to keep safety. This could be very inconvenient, and even dangerous in large scale field where human eyesight is limited. In order to make agriculture UAVs capable to work in fields of obstacles automatically, we give out this practical obstacle avoidance path planning method. Besides obstacle avoidance, it's also a coverage path planning method that makes the UAV able to cover the whole working area.

The coverage path problem (CPP) was firstly investigated by Zelinsky et al. [1] on a ground robot in 1993, to minimize the length, energy consumption, and travel time. Later, Carvalho et al. [2] proposed an algorithm for a cleaning robot in an industrial environment with unknown obstacles. The path planning problem has attracted a lot of attention with the development of UAVs. Li et al. [3] studied an exact cellular decomposition method for UAV path planning in a polygon region. For the purpose of precision agriculture mapping, Barrientos

This work is supported by National Natural Science Foundation (NNSF) of China under Grant 61702023 and 91538204.

F. Wotawa et al. (Eds.): IEA/AIE 2019, LNAI 11606, pp. 196–203, 2019.
https://doi.org/10.1007/978-3-030-22999-3_18

et al. [4] divided the polygon area and conducted path planning for each subarea for multi-UAVs. Torres et al. [5] presented a path planning algorithm for a single UAV with the less turns and coping with both convex and non-convex regions. Li et al. [6] used an algorithm to minimize the consuming of energy by the UAV for covering 3D terrain. And a study in [7] introduced a multi-robot boundary coverage problem with the application of inspection of blade surfaces inside a turbine. Moreover, the survey given by Galceran and Carreras [8] conducted a comprehensive explanation on CPP.

In this study, we adopt multirotor UAVs as the agent, which are mostly used in agriculture activities. The problem is described in Sect. 2. Then we use a cut and search strategy to get an obstacle avoidance path for the mission in Sect. 3. Finally we give out the results of this method with an Android application in Sect. 4. A conclusion is made in Sect. 5.

2 Problem Description

In practice, the field and obstacles data are measured with the GPS position of their corners. The workers measure the GPS position of the field or obstacles one by one to form polygons to represent the areas. Any GPS device or a precise map can be easily used to take the samples.

Input Definition:

- The point class is defined as $\mathcal{PT} = \{(Lat, Lon, Xn, Yn) \in \mathbb{R}^4\}$. The original Lat and Lon can be measured with a GPS device like smart phone, UAV onboard GPS etc. And X and Y can be calculated with the simplified equations.
- The corner class is defined as $\mathcal{CR} = \{(pre, next, data), data \in \mathcal{PT}\}$, pre points to the previous corner and $next$ points to the next point. It's a double linked list. The corners of a district polygon can be easily visited with these pointers.
- The district class is defined as $\mathcal{DT} = \{(firstCorner, next)\}$, $firstCorner$ points to the first corner of this certain district and $next$ points to the next district. Both field and obstacles can be described with the district class.
- The environment class is defined as $\mathcal{ENV} = \{field, obstacle\}$, $field \in \mathcal{DT}$, $obstacle$ points to the head of the obstacles, each of them is an instance of class $\mathcal{DT}$. There should be only one instance of the environment class in a certain obstacle avoidance coverage path planning problem.

Intermediate Definition:

- The couple point class is defined as $\mathcal{CP} = \{(pre, next, couple, data, distr, edge), data \in \mathcal{PT}\}$, pre points to the previous couple point in the chain, $next$ points to the next couple point in the chain, $couple$ points to the couple point with which it coupled, it must be either its pre or its $next$. An instance of the couple point should be located just on the edge of one district. So *distr* points to the district the instance locates. And *edge* points to the first point of the edge it locates.

Output Definition:

- The set of mission $\mathcal{M} = \{No, Agri, ...,\}$ is the misson for UAV to do after flying over the current point. In the agriculture application case the element No means do nothing, the element $Agri$ means the UAV should turn on the pump and start or keep on spraying.
- The path point class is defined as $\mathcal{PPT} = \{(pre, next, data, mission), data \in \mathcal{PT}, mission \in \mathcal{M}\}$, *pre* points to the previous path point, *next* points to the next path point.

The output of the method is a pointer of $\mathcal{PPT}$ which points to the head of a list of path points, so that it represents the finally planned path.

3 Path Planning Method

This path planning method is geometry based. Firstly we cut the environment with the working internal. Secondly the couples of the cutting points are made to compose the working segments. Each couple is made up of two couple points that locate in one cutting line. The segment between every two couple points is totally located in the field area. Thirdly we search the couple points one by one to find the working path of the field with obstacles. The details of the method is discussed in the following subsections.

3.1 Environment Cutting

Imagine a huge knife that has many parallel blades. The internal of the blades are equal to the working internal. The field with obstacles can be cut into slices. The range of the blades is from the most left side of the environment to the most right side of the environment. You can see it in Fig. 1.

In the practice, we should let top and bottom range of the blades a little more than the environment. So that any edge in the environment will be cut by the blades.

3.2 Make Couples to Form Working Segments

As demonstrated in Fig. 1, a couple is made up of two points that lies on one cutting line. Each point of a couple is on one edge of either the field polygon or obstacle polygons, so that the segment formed by the points of a couple must be located within the field area and without collision with any obstacle.

For all the cut lines, we could find out all the couples on it as Algorithm 1.

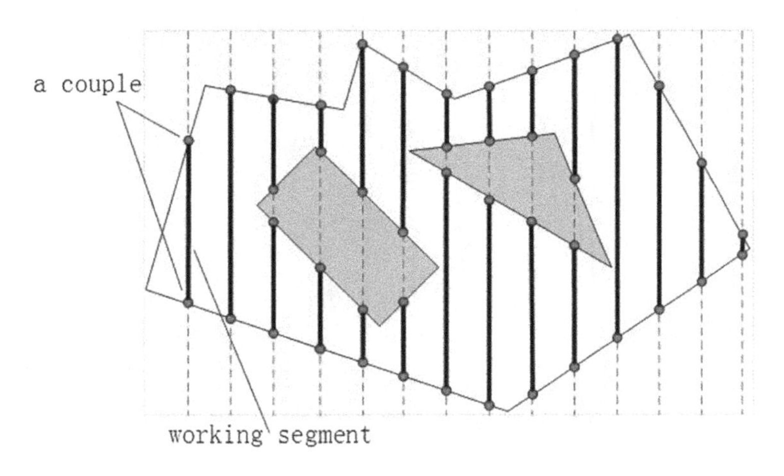

Fig. 1. Make couples to form working segments

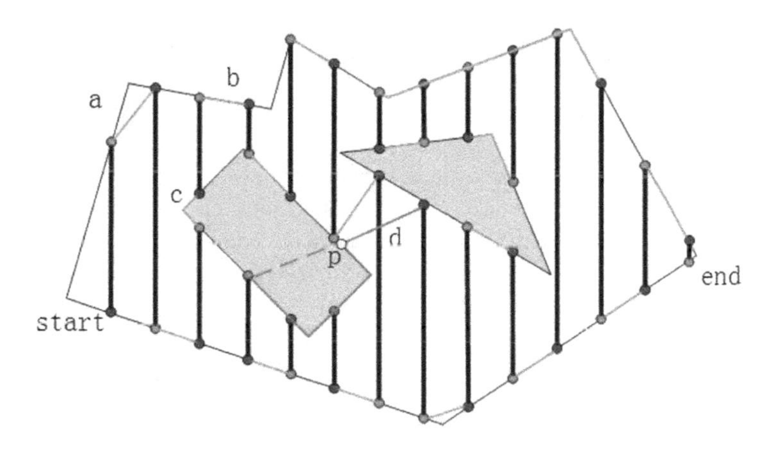

Fig. 2. Find the path through all coupled points

Algorithm 1. Find Couples

1: **for all** cut line L **do**
2: Calculate the cross points of L with $env.field$ and all the obstacles
3: Sort the cross points from bottom to top into a sequence
4: Every two points in the sequence is a couple
5: **end for**

You can see that the segment between the two points in a couple is in the working field. It's called a working segment that the agriculture UAVs should keep working during the flight through this segment. All the working segments will cover the field without any collision. Then we should find a path to connect all the working segments together. It's called the transfer segment that the agriculture UAVs should stop working during the flight through the transfer segments.

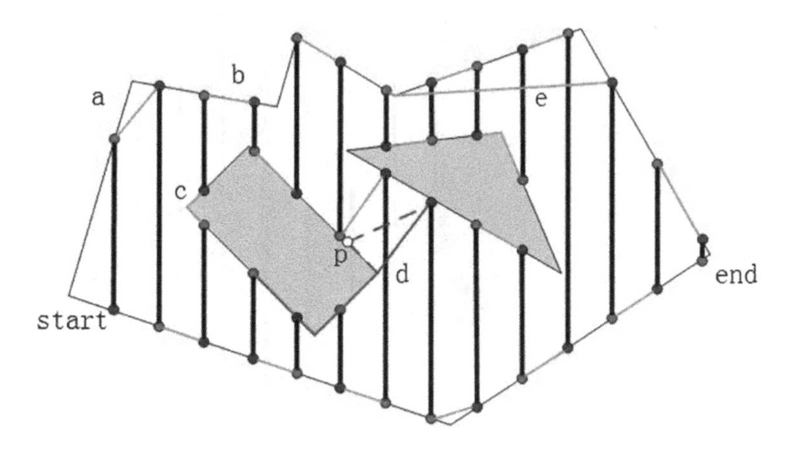

Fig. 3. Refine the coverage path

3.3 Form the Obstacle Avoidance Coverage Path

All the coupled points are put in a sequence marked unused firstly. If the couple point of the current point is marked as unused, the path should go through the working segment between the couple to the couple point. And mark the current point as used and make its couple point as the new current point. But if the couple point of current point is used, we should find the nearest unused point in the sequence and find the path to it in the field area without collision with the obstacles. It's described in Algorithm 2.

Algorithm 2. Find Path to a Target Point

1: Set *cur* to the head of couple point link
2: **while** $cur \neq NULL$ or used **do**
3: Put *cur* in the output chain
4: Mark *cur* used
5: **if** *cur.couple* is unused **then**
6: $cur \leftarrow cur.couple$
7: **else**
8: Find the nearest couple point in unused set
9: Find a path from *cur* to nearest with Alg. 3
10: Put the path in the output chain
11: $cur \leftarrow$ this nearest couple point
12: **end if**
13: **end while**

3.4 Find Path to a Target Point

In order to find the most convenient path to the nearest unused point, we make the nearest unused point as the target point. The following method is used to find the path to the target point. A general algorithm to find path from current point to a target point is described as Algorithm 3. And it's used recursively in its procedure.

Algorithm 3. Find Path to a Target Point

1: Input: current point p_0, set target point p_t
2: **if** $\overline{p_0 p_t}$ is totally in the field without collision with obstacles **then**
3:     Connect p_0 and p_t
4:     **return**
5: **else**
6:     **if** p_0 and p_t are on the same edge of the same polygon **then**
7:         Connect p_0 and p_t
8:         **return**
9:     **else**
10:         **if** p_0 and p_t are on the different edges of the same polygon **then**
11:             Find the nearest path through the edges of this polygon
12:             **return**
13:         **else**
14:             Cut the polygons with $\overline{p_0 p_t}$
15:             Sort the cross points fom p_0 to p_t
16:             Find path one by one from p_0 to p_t recursively with this algorithm
17:         **end if**
18:     **end if**
19: **end if**

There are some principles below to make up this method. The front ones are prior than the behind ones.

(1) If the segment between the target point and current point is totally in the field without collision, connect the two points directly.
(2) If the target point and current point are on the same edge of the same polygon, connect the two points directly.
(3) If the target point and current point are on the different edges of the same polygon, find the nearest path through the edges of the polygon.
(4) Else, make a line between the two points. Sort all the cutting points from current point to the target point. Find path one by one using the 4 principle recursively.

In Fig. 2, we mark some examples of the given principles. In case a, the segment between the two points is in the field without collision, so they are connected directly. In case b, the two points are on the same edge of the same polygon, they are connected directly as well. In case c, the two points are on different edges of the same polygon, they are connected through the edges of the polygon. In case d, the two points are on different polygons. The line between the two points cut the polygon with a new point p. We find a path to point p with principle (3) and then find a path from p to the target with principle (1). So the path could finally be find recursively using the four given principles.

3.5 Refine the Coverage Path

The working segments are fixed. So we can just refine the transfer segments. We will judge if a point in the transfer segment is suitable to delete to make it shorter.

The principle is that if the line between the point in front and point behind in three continuous points is totally in the field without any collision, the middle point could be deleted.

As shown in Fig. 3, d and e are refined with the principle given above.

With this given procedure we can get the coverage path in any polygon shaped field with any kind of polygon shaped obstacles. The method is based on geometry characteristics. Every step of the method is described in detail with clear illustration. It's easy for practical using in the agriculture UAV systems.

4 Results and Application

This study is part of a project for agriculture UAV system. The method described in the paper has been coded with C language and been used in an Android application. As it's showed in Fig. 4, the field is a concave polygon and there exists three obstacles in the environment. One obstacle is crossed with the field and the other two are in the region of the field. The result shows that the field

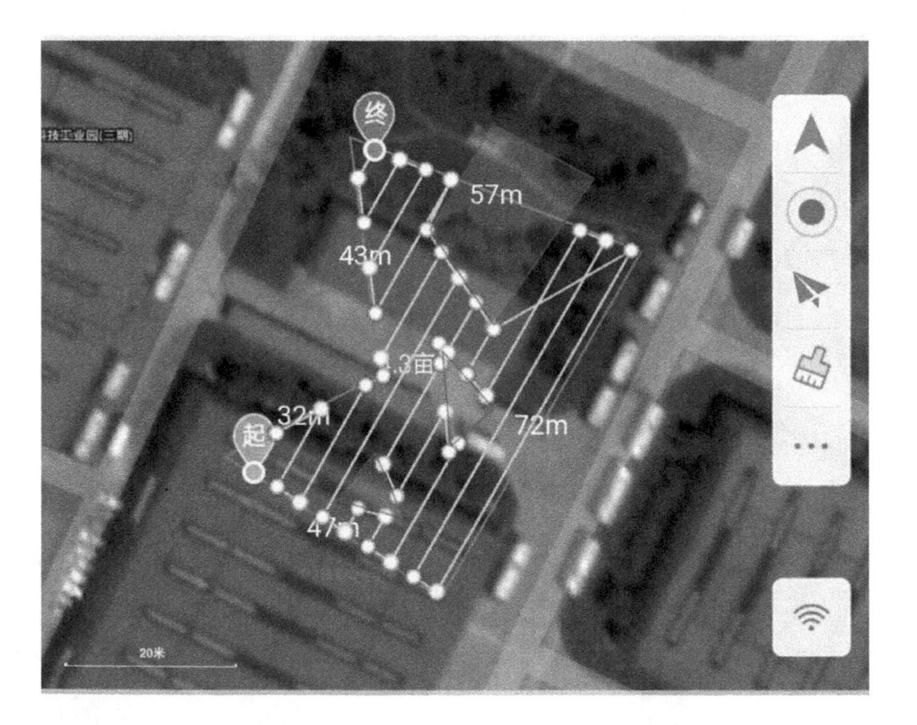

Fig. 4. An android application of this method

polygon is trimmed by the crossed obstacle polygon. The final result will cover the field finally. It is proved that the method is fast and practical.

5 Conclusion

We take the geometry characteristics into consideration to make out this practical obstacle avoidance path planning method. A pipeline of the coverage path planning process was established. An obstacle avoidance path planning method is used to find a coverage path for agriculture UAVs. The method has been tested with an Android application. The results turn out that the method is complete and efficient for this kind of coverage path planning problem. And it could be used in many similar situations.

References

1. Zelinsky, A., Jarvis, R.-A., Byrne, J.-C., Yutta, S.: Planning paths of complete coverage of an unstructured environment by a mobile robot. In: 2006 Proceedings of the International Conference on Advanced Robotics, Tokyo, Japan, pp. 533–538 (1993)
2. Dc Carvalho, R.-N., Vidal, H.-A., Vieira, P., Ribeiro, M.-I.: Complete coverage path planning and guidance for cleaning robots. In: Proceedings of the 1997 IEEE International Symposium on Industrial Electronics, Guimaraes, Portugal, pp. 677–682 (1997)
3. Li, Y., Chen, H., Joo Er, M., Wang, X.: Coverage path planning for UAVs based on enhanced exact cellular decomposition method. Mechatronics **21**(5), 876–885 (2011)
4. Barrientos, A., et al.: Aerial remote sensing in agriculture: A practical approach to area coverage and path planning for fleets of mini aerial robots. J. Field Robot **28**(5), 667–689 (2011)
5. Torres, M., Pelta, D.-A., Verdegay, J.-L., Torres, J.-C.: Coverage path planning with unmanned aerial vehicles for 3D terrain reconstruction. Expert Syst. Appl. **55**, 441–451 (2016)
6. Li, D., Wang, X., Sun, T.: Energy-optimal coverage path planning on topographic map for environment survey with unmanned aerial vehicles. Electron. Lett. **52**(9), 699–701 (2016)
7. Easton, K., Burdick, J.: A coverage algorithm for multi-robot boundary inspection. In: 2005 Proceedings of the IEEE International Conference on Robotics and Automation, Barcelona, Spain, pp. 727–734 (2005)
8. Galceran, E., Carreras, M.: A survey on coverage path planning for robotics. Robot. Auton. Syst. **61**(12), 1258–1276 (2013)

Data Science and Security

A Fault-Driven Combinatorial Process for Model Evolution in XSS Vulnerability Detection

Bernhard Garn[1], Marco Radavelli[2](✉), Angelo Gargantini[2], Manuel Leithner[1], and Dimitris E. Simos[1]

[1] SBA Research, 1040 Vienna, Austria
{bgarn,mleithner,dsimos}@sba-research.org
[2] University of Bergamo, Bergamo, Italy
{marco.radavelli,angelo.gargantini}@unibg.it

Abstract. We consider the case where a knowledge base consists of interactions among parameter values in an input parameter model for web application security testing. The input model gives rise to attack strings to be used for exploiting XSS vulnerabilities, a critical threat towards the security of web applications. Testing results are then annotated with a vulnerability triggering or non-triggering classification, and such security knowledge findings are added back to the knowledge base, making the resulting attack capabilities superior for newly requested input models. We present our approach as an iterative process that evolves an input model for security testing. Empirical evaluation on six real-world web application shows that the process effectively evolves a knowledge base for XSS vulnerability detection, achieving on average 78.8% accuracy.

Keywords: Combinatorial testing · XSS vulnerability · Security testing · Model evolution

1 Introduction

Computer users of today often interact via web-applications with services offered by various parties. This new way of connecting the client and server side with each other has brought about a multitude of novel security challenges, both for the client and the server [3]. One major threat to web applications is posed by Cross-Site Scripting (XSS), which continues to be included in the OWASP Top 10 most critical web application security risks [5]. Security testing is a vital and expensive part of the software development lifecycle. An effective testing technique applied to security testing is *Combinatorial Testing* (CT), for the capability of detecting failures and fail conditions with a small amount of tests that need to be executed compared to the whole input space [14]. Given a discrete finite model of the *system under test* (SUT), made of parameters with a

F. Wotawa et al. (Eds.): IEA/AIE 2019, LNAI 11606, pp. 207–215, 2019.
https://doi.org/10.1007/978-3-030-22999-3_19

```
Model wavsep_xss

Parameters:
  JSO: { P1 P2 P3 P4 P5 P6 P7 P8}           PAY: { P1 P2 P3 P4 P5 }
  INT: { P1 P2 P3 P4 P5 P6 P7 P8 P9 P10}    EVH: { P1 P2 P3 }
  PAS: { P1 P2 P3 P4 P5 P6 P7 P8 P9 P10}    WS: Boolean
  JSE: { P1 P2 P3 P4 P5 P6 P7 P8 P9}

Constraints:
  # ! JSE==JSE.P2 #    # ! JSE==JSE.P3 #          # ! JSE==JSE.P4 #
  # ! INT==INT.P9 #    # ! (JSO==JSO.P2 && WS==false && PAS==PAS.P7) #
  ...
```

Fig. 1. Knowledge base K_3 for NavigateCMS: abstract attack model (initially it had no constraints), with detected XSS vulnerability constraints, in CTWedge

finite list of possible values, called *input parameter model* (IPM), and given an interaction strength t, CT creates a test suite guaranteeing the appearance of all t-way interactions of parameter values, for any selection of t parameters [11]. In the case for testing for exploiting XSS vulnerabilities, the IPM specifies an attack grammar and is also called *(abstract) attack model.* The aim of this paper is to present a way to evolve knowledge bases for security testing. In our approach, the evolution of a knowledge base consists in the integration of learned constraints, using BEN [9], into the IPM. In particular, the contribution of this paper consists in an automated technique to detect all the conditions under which vulnerabilities are triggered, by using combinatorial testing. Evaluation shows that the process is able to evolve the knowledge base to achieve, on average over all the benchmarks, 78.8% accuracy, in 14 min computation time.

The rest of the paper is structured as follows. In Sect. 2 we give basic definitions, in Sect. 3 we discuss our proposed process, and Sect. 4 presents the results of our case study experiments. We provide a brief overview over related work in Sect. 5, and we conclude the paper with further research directions in Sect. 6.

2 Preliminaries

In the course of *combinatorial security testing* (cf. [14]), attack models have appeared in the form of a BNF grammar, e.g. [2,13]. In this paper, we will follow this established terminology for designing XSS attack models to be used in conjunction with combinatorial methods. We denote with K_i a knowledge base at time i, encoded as an abstract attack model (IPM, see Fig. 1). Given an IPM, an abstract test case f is a particular assignment of values for its parameters. An abstract test suite is used to derive a concrete test suite, where abstract test cases are being translated into concrete XSS attack strings via a translation function τ. For example, given the following abstract test case:

$$(\mathsf{JSO} = 2, \mathsf{WS} = 1, \mathsf{INT} = 3, \mathsf{EVH} = 2, \mathsf{PAY} = 2, \mathsf{PAS} = 5, \mathsf{JSE} = 7)$$

for each parameter, the respective integer value corresponds to a concrete parameter value (i.e., a string), and the translated concrete test case is obtained by concatenating all these strings together in the order given by the IPM:

```
<script> onError= alert(1) ') '\>
```

The resulting string can be submitted against the SUT, and a boolean function `orac` decides if the outcome of the execution of the translated test case $\tau(f)$ against the SUT triggers an XSS vulnerability or not. We denote also the function $eval(f) := oracle(\tau(f))$, that is *true* if the test case f triggered an XSS vulnerability (i.e., f is a *vulnerability triggering* test case). The generated test vectors aim at producing valid JavaScript code when these are executed against SUTs. A description of parameters that appear in the attack model is mentioned in [2,8,13]. At any time point i, the knowledge base K_i may be used to create test cases, and to classify an abstract test case as either vulnerability-triggering or not, depending on whether the *constraints* are satisfied. This capability of the knowledge base is denoted as a *model* function, which takes as input an abstract test case, and gives as output a "best guess" that may or may not be correct w.r.t. the actual result of the function *eval*. The attack model initially contains only combinatorial parameters and no constraints. During the process, the knowledge base is enriched by the conditions used to identify the vulnerabilities.

To put this problem into a formal setting, a *knowledge base fault* occurs when a test f is classified as non-vulnerable in the model ($\neg model(f)$), despite it actually triggers a vulnerability in the SUT ($eval(f)$) (**False Negative**: it entails a loss of potentially valuable information for fixing the vulnerability); or when a test is being marked as vulnerability-triggering in the model ($model(f)$), despite it does not trigger a vulnerability ($\neg eval(f)$) (**False Positive**: it triggers a false alarm, and the programmers may consequently waste effort in fixing pieces of code that did not trigger any vulnerability). When such a discrepancy is fixed by updating the model function, we say that the knowledge base evolves.

Since in our experiments it yielded better results, we decided to consider the convention for which the initial model function considers any test to be non-vulnerability triggering; and during the process constraints are added in order to identify and subsequently exclude all the tests that do not trigger any vulnerability. We call this convention *pessimistic* approach, in contrast with the *optimistic* one in which initially any test is vulnerability-triggering, and constraints are added to isolate the tests that actually trigger some vulnerability.

Let us complete some notation for combinatorial analysis. A combination c is an assignment (i.e. configuration) on a subset $Dom(c)$ of all the possible parameters P in the attack model, such that $Dom(c) \subseteq P$. We call *size* of the combination the cardinality of $Dom(c)$. A combination c identifies a set of tests: *c represents* a test f if all the parameters in c are also present in f, associated to the same values. Formally, $c \subseteq f : \forall p \in Dom(c), f(p) = c(p)$. A combination c is *suspicious* in a test set $F \subseteq \Gamma$ if c represents *only failed tests in F*. Formally, $\forall f \in F : c \subseteq f \rightarrow model(f) \neq eval(f)$. For the purposes of this work, we assume that the constraints are in conjunctive relation among each other.

3 Process for Model Evolution

Figure 2 shows an overview of our process to automatically evolve an abstract attack model initialized without constraints, to detect conditions that trigger XSS vulnerabilities. The process proceeds according to the following steps:

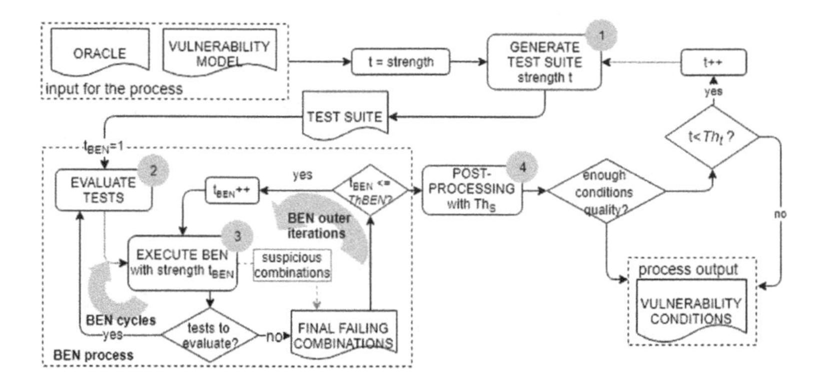

Fig. 2. Condition detection meta-process

1. From a defined initial interaction strength t, derive a t-way test suite.
2. Mark the test cases as failing or passing according to the current model and the evaluator. If all the tests pass and Th_t is not yet reached, increment the strength $t \leftarrow t+1$, and go to point 1. Otherwise, evaluate the tests. Internally, the evaluator executes the translation function τ to obtain the concrete test string to insert in the *reflection* URL to query the SUT. PhantomJS[1] then analyzes the HTML code returned by the SUT for a specific *target function* (the `alert()` function in our case[2]) that was included as payload. If any of the target functions were executed, the injection was successful; if the page loads normally and does not produce any errors, the injection is deemed unsuccessful. Lastly, if JavaScript errors are observed on the page, it is likely that some content was injected, but not in a form that constitutes a usable injection (thus resulting in incorrect syntax). Our current approach regards these test cases as failing, but future evolutionary approaches might take advantage of this particular classification.
3. Pass the evaluated test suite to BEN [9] to derive suspicious combinations, together with their *suspiciousness* level. We call BEN multiple times specifying the size t_{BEN} of the suspicious combinations to detect[3], from 1 to Th_{BEN}.

[1] PhantomJS (http://phantomjs.org/) is a headless browser environment enabling introspection of events such as network requests, document edits and JavaScript errors.

[2] In theory, any valid JavaScript functions that will not be called during the normal operation of the SUT can be chosen instead.

[3] Note that t_{BEN} is different from the strength t for generating the initial test suite.

BEN may also ask for a few additional tests (up to 10 tests at a time: *inner BEN cycles*) to reduce the amount of suspicious combinations and improve the accuracy of the computed *suspiciousness* levels.

4. The suspicious combinations from BEN are then translated into a set of constraints for the current knowledge base K_i, by negating the corresponding boolean expression (obtained by putting the assignments in conjunction) of every combination whose *suspiciousness* value is above the threshold Th_S. Due to low accuracy in the detected suspicious combinations, we noticed that K_i often results to be a contradiction, which is normally not the case in real-world systems. Therefore, we *post-process* the constraints by computing the *unsat-cores* of K_i and removing all such clauses starting from the *least-suspicious* constraints (according to BEN), until K_i is not a contradiction any longer. The process eventually quits if either the user is *satisfied* with the quality of K_i[4], or the threshold Th_t is reached. Otherwise, increase t and go to point 1.

Table 1. XSS reflection sites on WAVSEP benchmarks

SUT ID	SUT name	Reflection site	vulnerability ratio (t=5)
1	Tag2HtmlPageScope	<body>$input</body>	17.08 %
2	Tag2TagStructure	<input type="text" value="$input">	4.06 %
3	Event2TagScope	<img src="$input">	4.63 %
4	Event2DoubleQuotePropertyScope	<img src="$input">	3.45 %
5	MiniCMS	/mc-admin/page.php?date=$input	80.2 %
6	NavigateCMS	/navigate.php?fid=$input	60.0 %

4 Experiments

The process has been implemented in Java using `CTWedge` [7] to represent and update attack models, `ACTS` [16] to generate combinatorial test suites of a defined strength, and `BEN` [9] as a tool to detect suspicious combinations and compute suspiciousness. Experiments were executed on a PC with Intel i7 3.40 GHz processor and 16 GB RAM. We run the process on six real web-applications: four are part of the WAVSEP[5] project, and two are open source content management systems: MiniCMS and NavigateCMS. Each SUT receives over HTTP one GET parameter which is rejected on the page in different contexts, and might optionally be altered by a specific sanitization function. Table 1 shows, for each SUT, the respective *vulnerability ratio*, i.e., the ratio of tests that triggered an XSS vulnerability ($eval(f)$) out of the total number of tests executed, that, given the

[4] In this case, we believe that the suspiciousness average and standard deviation could be useful indicators of the F_1 score that the currently inferred model may have.

[5] WAVSEP: Web Application Vulnerability Scanner Evaluation Project, https://github.com/sectooladdict/wavsep.

Table 2. Quality metrics for the inferred models ($Th_{BEN} = 3$, and $Th_S = 0$)

	sut	Time (s)	Constraints	Suspiciousness avg. ± s.d	Accuracy	Precision	Recall (TPR)	Specificity (TNR)	F_1 score
$Th_t = 4$	1	246	146	0.254 ± 0.0288	72.6	32.5	55.7	76.1	41.0
	2	141	38	0.362 ± 0.00753	93.3	32.6	59.5	94.8	42.1
	3	1437	794	0.137 ± 0.0456	87.4	15.7	39.7	89.7	22.5
	4	1088	551	0.136 ± 0.0483	89.0	13.7	42.3	90.6	20.7
	5	1445	623	0.342 ± 0.0287	78.3	87.0	85.8	48.3	86.4
	6	679	80	0.344 ± 0.0147	52.0	66.6	40.0	70.0	50.0
	avg	839	372	0.263 ± 0.0289	78.8	41.4	53.8	78.3	43.8
$Th_t = 3$	1	9.3	644	0.196 ± 0.0423	41.4	19.8	79.8	33.5	31.8
	2	4.2	164	0.224 ± 0.0874	78.9	14.2	83	78.7	24.2
	3	28.5	764	0.207 ± 0.0628	75.4	12.9	75.7	75.3	22
	4	9.3	123	0.125 ± 0.0379	72.3	8.53	73.7	72.2	15.3
	5	58.5	1340	0.318 ± 0.0268	80.1	80.2	99.9	0.306	89
	6	37.6	2460	0.296 ± 0.0245	62.2	61.8	97.1	9.91	75.5
	avg	24.6	915	0.228 ± 0.0470	68.4	32.9	84.9	45.0	43.0

practical infeasibility of the exhaustive test suite, we compute on all the tests generated up to strength t = 5 (42830 tests).[6]

To assess the quality of the evolved knowledge base from our method, we use the typical metrics of information retrieval: in particular, *precision* ($\frac{TP}{TP+FP}$), and *recall* ($\frac{TP}{TP+FN}$) give a measure of how the process isolates true positives, *accuracy* gives an overall ratio of correctly classified tests, and the F_1 score is considered to be a good candidate synthesis index of the inferred model's quality. For the experiments, we set the parameters of the process as follows:

- $Th_{BEN} = 3$. We limited the size of detected suspicious combinations as the BEN process becomes too slow when computing suspiciousness of the too many combinations of size 4 (or larger) on these attack models.
- t, the initial strength of test suite, is set to 2.
- $Th_t = 4$. We limit the maximum strength of the initial test suite since even $t = 5$ (about 36000 tests) would make the BEN process too slow.
- $Th_S = 0$, as we want all the suspicious combinations to be considered.

Test suites for interaction strengths $t \in \{3, 4\}$ had 900 and 7200 test cases, respectively. For each SUT, Table 2 reports the number of constraints included in the inferred model, the average suspiciousness with its standard deviation, and the accuracy, precision, recall, specificity, and F_1 score of the final model, computed over all tests up to strength 5, as for the *vulnerability ratio* in Table 1.

RQ1: *What is the quality of the model obtained by the approach?* We observe that the inferred model achieves an average accuracy of 78.8%, with a maximum of 93.3%. Precision has an average of 41.4%, ranging from 13.7% to 87%, and recall (54% on average) is higher than precision. F_1 score is on average 43.8%, with a maximum of 86.4% on SUT5. With relatively few tests ($t = 4$ out of 7

[6] The tests suites were generated using the IpoF algorithm, implemented in ACTS.

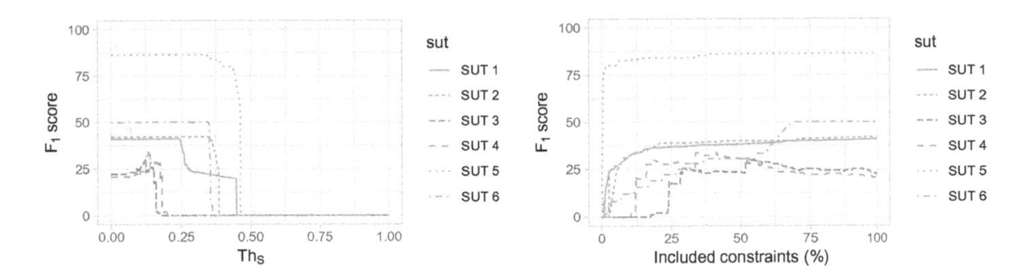

Fig. 3. Achieved F_1 score of final model by varying Th_S, when $Th_t = 4$

parameters), the final model is of good quality, but not completely accurate. We can also observe that F_1 is proportional to the vulnerability ratio of the SUT.

RQ2: *How does the quality of the inferred model vary depending on* Th_t? By increasing Th_t from 3 to 4, the time taken, as expected, increases, while the number of constraints, in most cases, decreases, meaning that with more tests our process is able to describe the vulnerability conditions with fewer constraints. On average, although recall decreases, both precision, accuracy and specificity increase, and F_1 slightly increases. This means that the classification improves.

RQ3: *Which is the computational effort of the proposed process?* Table 2 also reports the total execution time, excluding the actual test execution, as test results are cached, except for the few tests (30 at most) that BEN may ask during the process. For the first two SUTs, the process takes less than 250 s to complete, but up to 24 min are needed for SUT5 with $Th_t = 4$. Most of the computation time is used internally by BEN; by limiting to 3 the strength of the initial test suite, the total time is always below 1 min for every SUT.

RQ4: *How does variations of* Th_S *affect model quality?* The highest F_1 score is achieved with low values of Th_S (see Fig. 3), except for SUT3 and SUT4, for which a $\overline{Th_S} \simeq 0.13$ achieves the maximum F_1 score. However, we can notice that at least around 25% of the constraints (starting from the least suspicious ones) can be removed with negligible impact on the final F_1 score.

5 Related Work

XSS vulnerability detection is not a novel topic in computer science research. Duchene et al. [4] used model based testing and fuzzing to discover XSS vulnerabilities; Melicher et al. [12] proposed improvements on using the DOM model to generate and detect XSS attacks; Simos et al. [13] proposed a combinatorial approach to find attack vectors that trigger XSS vulnerabilities; Jia et al. [10] used machine learning and hyper-heuristic search to improve combinatorial tests; Temple et al. [15] proposed a machine-learning approach to infer constraints among parameters that, although not sound, achieves high precision (about 90%) and recall (80%). Although these works use model-based testing, the usage of combinatorial testing for XSS vulnerability detection to *classify* vulnerabilities based on the input and describe *completely* the vulnerability space of a part of a web application, evolving a *knowledge base*, is the main novelty of our approach. The first phase of BEN [9] as a failure-inducing combination detection and ranking tool has been used by Gargantini et al. [6] to repair constraints in combinatorial models, evaluating different test generation policies.

6 Conclusion and Future Work

We presented an automated iterative process based on combinatorial testing to evolve an attack model to include conditions among input parameters that trigger XSS vulnerabilities in web applications. Our approach is based on the notion of suspicious combination, i.e., whose appearance in a test vector would trigger a discrepancy between the *best-guess* of the current model, and the actual outcome when executed against the SUT. Identification of constraints among XSS attack parameters helps to better understand the root cause of an XSS vulnerability and provides insights about how to fix a flawed sanitization function. As future work, we plan to improve the process by reducing the required tests, using information from previous step, and evaluating alternatives to BEN, such as MixTgTe [1]. We believe that this approach can be extended to other security vulnerabilities related to sanitization functions, and to detect discrepancies between a functional system specification and its implementation. Another direction is to further simplify the detected constraints, to reduce them in number and present them to the user in a more readable way.

References

1. Arcaini, P., Gargantini, A., Radavelli, M.: Efficient and guaranteed detection of t-way failure-inducing combinations. In: IEEE International Conference on Software Testing, Verification and Validation ICST Workshops (2019)
2. Bozic, J., Garn, B., Kapsalis, I., Simos, D., Winkler, S., Wotawa, F.: Attack pattern-based combinatorial testing with constraints for web security testing. In: IEEE International Conference on Software Quality, Reliability and Security (2015)
3. Catteddu, D.: Cloud computing: benefits, risks and recommendations for information security. In: Serrão, C., Aguilera Díaz, V., Cerullo, F. (eds.) IBWAS 2009. CCIS, vol. 72, pp. 17–17. Springer, Heidelberg (2010). https://doi.org/10.1007/978-3-642-16120-9_9
4. Duchene, F., Groz, R., Rawat, S., Richier, J.-L.: XSS vulnerability detection using model inference assisted evolutionary fuzzing. In IEEE International Conference on Software Testing, Verification and Validation (ICST), pp. 815–817 (2012)
5. O. Foundation: OWASP Top 10 2017. https://www.owasp.org/index.php/Top_10-2017_A7-Cross-Site_Scripting_(XSS). Accessed 19 Apr 2018
6. Gargantini, A., Petke, J., Radavelli, M.: Combinatorial interaction testing for automated constraint repair. In: IEEE International Confernce on Software Testing, Verification and Validation ICST Workshops, pp. 239–248 (2017)

7. Gargantini, A., Radavelli, M.: Migrating combinatorial interaction test modeling and generation to the web. In: IEEE International Conference on Software Testing, Verification and Validation ICST Workshops, pp. 308–317 (2018)
8. Garn, B., Kapsalis, I., Simos, D.E., Winkler, S.: On the applicability of combinatorial testing to web application security testing: a case study. In: Proceedings of the 2014 Workshop on Joining AcadeMiA and Industry Contributions to Test Automation and Model-Based Testing, pp. 16–21. ACM (2014)
9. Ghandehari, L.S., Lei, Y., Kacker, R., Kuhn, D.R.R., Kung, D., Xie, T.: A combinatorial testing-based approach to fault localization. IEEE Trans. Softw. Eng. (2018)
10. Jia, Y., Cohen, M.B., Harman, M., Petke, J.: Learning combinatorial interaction test generation strategies using hyperheuristic search. In: Proceedings of the International Conference on Software Engineering - Volume 1, ICSE 2015, pp. 540–550 (2015)
11. Kuhn, D., Kacker, R., Lei, Y.: Introduction to Combinatorial Testing. Chapman & Hall/CRC, London (2013)
12. Melicher, W., Das, A., Sharif, M., Bauer, L., Jia, L.: Riding out DOMsday: towards detecting and preventing DOM cross-site scripting. In: Proceedings of Network and Distributed System Security Symposium. Internet Society (2018)
13. Simos, D.E., Kleine, K., Ghandehari, L.S.G., Garn, B., Lei, Y.: A combinatorial approach to analyzing cross-site scripting (XSS) vulnerabilities in web application security testing. In: Wotawa, F., Nica, M., Kushik, N. (eds.) ICTSS 2016. LNCS, vol. 9976, pp. 70–85. Springer, Cham (2016). https://doi.org/10.1007/978-3-319-47443-4_5
14. Simos, D.E., Kuhn, R., Voyiatzis, A.G., Kacker, R.: Combinatorial methods in security testing. IEEE Comput. **49**, 40–43 (2016)
15. Temple, P., Galindo, J.A., Acher, M., Jézéquel, J.-M.: Using machine learning to infer constraints for product lines. In: Proceedings of the 20th International Systems and Software Product Line Conference, pp. 209–218. ACM (2016)
16. Yu, L., Lei, Y., Kacker, R.N., Kuhn, D.R.: ACTS: a combinatorial test generation tool. In: IEEE International Conference on Software Testing, Verification and Validation (2013)

An Efficient Algorithm for Deriving Frequent Itemsets from Lossless Condensed Representation

JianTao Huang, Yi-Pei Lai, Chieh Lo, and Cheng-Wei Wu(✉)

Department of Computer Science and Information Engineering, National Ilan University, Yilan City, Taiwan, ROC
wucw@niu.edu.tw

Abstract. Mining *frequent itemsets* (abbr. *FIs*) from dense databases usually generates a large amount of itemsets, causing the mining algorithms to suffer from long execution time and high memory usage. *Frequent closed itemset* (abbr. *FCI*) is a *lossless condensed representation* of FI. Mining only the FCIs allows to reducing the execution time and memory usage. Moreover, with correct methods, the complete information of FIs can be derived from FCIs. Although many studies have presented various efficient approaches for mining FCIs, few of them have developed efficient algorithms for deriving FIs from FCIs. In view of this, we propose a novel algorithm called *DFI-Growth* for efficiently deriving FIs from FCIs. Moreover, we propose two strategies, named *maximum support selection* and *maximum support replacement* to guarantee that all the FIs and their supports can be correctly derived by DFI-Growth. To the best of our knowledge, the proposed DFI-Growth is the first kind of tree-based and pattern growth algorithm for deriving FIs from FCIs. Experiments show that DFI-Growth is superior to the most advanced deriving algorithm [12] in terms of both execution time and memory consumption.

Keywords: Frequent itemset mining · Frequent closed itemset mining · Lossless condensed representation · Deriving algorithm

1 Introduction

The purpose of *frequent itemset* (*abbr. FI*) *mining* [1, 7, 13] is to discover sets of items appearing frequently in databases. This technology has been widely applied to various real-life applications, such as *association analysis* [5], *web mining* [20], *text mining* [9], and *bioinformatics* [16]. Many efficient methods have been developed for mining FIs, such as *Apriori* [1], *FP-Growth* [7] and *Eclat* [18]. When mining FIs from *sparse databases*, these methods usually have good performance. This is because the correlations between items in sparse databases are relatively weak, thereby the lengths of FIs in the databases are also relatively short. In this situation, traditional FI mining methods can take good use of the *downward closure property* [1, 7] to effectively prune the search space. *Transaction database* is a typical example of sparse database.

However, in real-life scenarios, many kinds of datasets are *dense datasets*, such as plant feature data (e.g., the *Mushrooms* dataset [4]), records of game steps (e.g., the

F. Wotawa et al. (Eds.): IEA/AIE 2019, LNAI 11606, pp. 216–229, 2019.
https://doi.org/10.1007/978-3-030-22999-3_20

Connect and *Chess* datasets [4]), and census statistical data (e.g., the *Pumsb* dataset [4]). When mining FIs from dense datasets, traditional FI mining methods usually generate a large amount of FIs, causing then to suffer from long execution time and huge memory consumption. Intuitively, the more itemsets an approach needs to process, the more processing resources it consumes. When dealing with dense datasets, the performance of FI mining methods may decrease greatly. This is because that, in dense datasets, the correlations between items are relatively strong, and the lengths of FIs are relatively long. In this situation, the downward closure property [1, 7] could not work well for effectively pruning the search space.

To resolve the above problems, many studies have devoted to developing *condensed representations of* FIs. The experimental results of previous studies [2, 6, 8, 11] have showed that mining condensed representations from *dense dataset* can greatly reduce the execution time and memory usage. Many types of condensed representations of FI have been proposed, like *free itemset* [2], *maximal itemset* [6], *generator itemset* [8], and *closed itemset* [11]. Among these representations, *closed itemset* is the most popular one. *Frequent closed itemset* (abbr. *FCI*) is a *lossless condensed representation* [8] of FIs. Through the correct algorithm, all the FIs and their supports can be completely derived from all the FCIs. Therefore, the complete set of FCIs retains the complete information of FIs without any loss. Though many studies [8, 17, 19] have proposed for efficiently mining FCIs, few of them consider to develop algorithms for efficiently deriving FIs from FCIs (called *deriving algorithms* for simplicity). However, developing efficient deriving algorithms is a very important work. When all the FIs cannot be successfully mined from the dense datasets, an alternative solution is to mine all the FCIs from the datasets first, and then applying efficient deriving algorithms to recover the complete set of FIs and their supports from FCIs.

However, *traditional deriving algorithm* [12] adopts *breadth-first search* and *top-down* strategy to derive FIs. It derives FIs of length $k - 1$ from FCIs and FIs of length k. For sake of simplicity, the traditional deriving algorithm is called *LevelWise* in this work. A drawback of LevelWise is that it needs to maintain the complete set of FIs of length k and $(k - 1)$ in the memory during the deriving process. Therefore, when there are many FIs of length k and $(k - 1)$, LevelWise will consume a large amount of memory, or even cannot complete the whole deriving task due to running out of memory. Moreover, the deriving process of LevelWise involves a large number of search operations, which may result in the long execution time problem.

In view of the above, this paper proposes a novel algorithm, named *DFI-Growth* (*Deriving Frequent Itemsets based on Pattern Growth*), for efficient deriving FIs and their supports from FCIs. The design idea of DFI-Growth is to first construct a FP-Tree to maintain the information of FCIs, and then uses FP-Growth to generate FIs from the FP-Tree. However, the traditional FP-Tree and FP-Growth are not designed for the deriving task and directly applying them will result in incorrect results. Therefore, we propose two new strategies and modify the algorithmic processes of FP-Tree and FP-Growth, making them be able to deriving the correct information of FIs from FCIs. The proposed two strategies are respectively called *maximum support replacement* and *maximum support selection*. The former is applied during the construction of FP-Tree, while the latter is the pattern generation. We perform experiments on both real-life and synthetic datasets to evaluate the performance of the proposed DFI-Growth algorithm. Extensive experiments show that DFI-Growth is quite efficient and able to correctly deriving all the FIs and their supports. Moreover, its performance is significantly better than the current best method LevelWise [12] in terms of both execution time and memory usage. For instance, on the Chess dataset, when the minimum support threshold is set to 40%, LevelWise cannot complete the whole deriving task within 12 h (i.e., 43,200 s). However, the proposed DFI-Growth only takes 20 s to complete the task.

Table 1. An example of a transaction database.

Tid	Transaction
1	ACTW
2	CDW
3	ACTW
4	ACDW
5	ACDTW
6	CDT

2 Basic Concept and Definitions

Let $I^*= \{I_1, I_2,\ldots, I_M\}$ be a finite set of distinct *items*. A *transaction database* $D = \{t_1, t_2, \ldots, t_H\}$ is a set of *transactions*, where each transaction $t_R \in D\,(1 \leq R \leq H)$ is a subset of I^* and has a unique *transaction identifier R*. An *itemset* is a set of items. If all the items of an itemset X are contained in a transaction T, X is said to *be contained in T*, which is denoted by $X \subseteq T$.

Definition 1 (Length of an itemset). The *length* of an itemset $X =\{I_1, I_2,\ldots, I_K\}$ is defined as K, where K is the total number of distinct items in X.

Definition 2 (Tidset of an itemset). The *Tidset* of an itemset X is denoted as $r(X)$ and defined as the set of transaction identifiers of all the transactions containing X in D.

Definition 3 (Support of an itemset). The *support count of an itemset X* is denoted as $SC(X)$ and defined as $|r(X)|$. Besides, the *support of X* is defined as $SC(X)/|D|$, where $|D|$ is the total number of transactions in D.

Definition 4 (Frequent itemset). An itemset X is called *frequent itemset* iff the support of X is no less than a user-specified *minimum support threshold* θ $(0 < \theta \leq 1)$.

Definition 5 (Super-itemset and sub-itemset). If an itemset X is a *subset* of another itemset Y, then X is called *sub-itemset* of Y. Besides, Y is called *super-itemset* of X.

Definition 6 (Closure of an itemset). The *closure of an itemset X* is denoted as $\rho(X)$ and defined as

$$\rho(X) \bigcap\nolimits_{R \in |r(X)|} t_R,$$

where $\rho(X)$ returns the largest super-itemset Y in D such that $X \subseteq Y$ and $SC(X) = SC(Y)$. For example, the closure of the itemset {W} is $\rho(\{W\}) = t_1 \cap t_2 \cap t_3 \cap t_4 \cap t_5$ = {ACTW} ∩ {CDW} ∩ {ACTW} ∩ {ACDW} ∩ {ACDTW} = {CW}.

Property 1. For any itemset X, $SC(X) = SC(\rho(X)) \Leftrightarrow r(X) = r(\rho(X))$.

Definition 7 (Closed itemset). An *itemset X* is called *closed itemset* iff there exists no other itemset Y such that $X \subset Y$ *and* $SC(X) = SC(Y)$. Otherwise, X is a *non-closed itemset*. Moreover, the closure of a closed itemset X is equal to X itself (i.e., $\rho(X) = X$).

For example, in the database of Table 1, the itemset {W} is non-closed because $\rho(\{W\}) = \{CW\}$. However, the itemset {C} is closed because $\rho(\{C\}) = \{C\}$.

Definition 8 (Frequent closed itemset). A *closed itemset X* is called *frequent closed itemset* iff the support of X is not less than a user-specified minimum support threshold θ $(0 < \theta \leq 1)$.

Problem Definition. Let F and C be the complete sets of FIs and FCIs in a transaction database D, where each itemset in F or C is associated with its support count. Given C, the problem to be solved in this work is to develop an algorithm for efficiently and correctly deriving F from C.

3 Related Work

3.1 Frequent Itemset Mining

Extensive studies have been proposed for mining FIs. The *Apriori* algorithm [1] is one of the well-known algorithms for FI mining. Apriori adopts a breadth-first search strategy and candidate generation-and-test scheme to discover FIs. Although Apriori has been applied to various applications and domains, it needs to scan the original database several times and may generate a large amount of candidates during the mining process, which degrades the performance for discovering FIs. Another famous algorithm is *FP-Growth* [7], which is usually used as the benchmarking algorithm for performance comparison in pattern mining field. *FP-Growth* uses a compact *FP-Tree* as the internal data structure of the algorithm and adopts a pattern growth methodology to generate FIs from the FP-Tree. The most impressive contribution of FP-Growth is that it only needs to scan the original database twice and discovers FIs without candidate generation. It is widely recognized that the pattern growth algorithms generally are superior to Apriori-like algorithms. In addition to Apriori and FP-Growth, there are many other efficient FI mining algorithms that have been proposed, such as *H-mine* [14], *DHP* [13], *Eclat* [18], and *COFI-Mine* [10]. Although FI mining algorithms may have good performance when mining FIs from sparse databases, while mining FIs from dense databases, a large number of FIs in dense databases may seriously degrade their performance. This is because that the more FIs need to be processed, the more calculations are needed for the algorithms.

3.2 Frequent Closed Itemset Mining

Instead of mining all the FIs, some studies propose to mine the *condensed representations of FIs*. In general, a condensed representation of FIs is a subset of FIs, thereby the total number of itemsets in the representation is generally less than that of FIs. Since the number of itemsets to be mined is reduced, the execution time and memory usage spent by the mining task are also reduced. Different types of condensed representations of FIs have been proposed, including *maximal itemset* [6], *closed itemset* [11], *non-derivable itemset* [3] and *generator itemset* [8]. Among these representations, *frequent closed itemset* is a *lossless condensed representation of FIs*. Mining only the FCIs will not lose information of any FIs. Due to this special property, FCI mining has become a popular and widely researched technology and many studies have been proposed for efficiently mining FCIs, such as *A-Close* [11], *CLOSET+* [17], *CHARM* [19] and *DCI_Closed* [8].

Nevertheless, few of these studies focus on developing algorithms for deriving FIs from FCIs. However, developing efficient deriving algorithms is an important task. In many applications, users may need to discover FIs from dense datasets instead of FCIs. In this case, discovering only the FCIs cannot fulfill the need of users, while discovering FIs by traditional FI mining algorithms may face the inefficiency problem. A solution to this problem is to discover FCIs first, and then generate the complete information of FIs through a correct and efficient deriving algorithm. To the best of our knowledge, only one study [12] had proposed the deriving algorithm (called LevelWise in this work). However, this it was proposed in 1999, which means that over the past nearly 20 years, no any other studies re-raise this research problem and develop new deriving algorithms.

3.3 Traditional Deriving Method

This subsection introduces the process of the LevelWise algorithm [12], which consists of two main steps below.

Step 1. The algorithm sorts all the FCIs in length descending order. Assuming that the length of the longest FCI is M and let L_M denotes the set of all the itemsets of length M collected so far. Set a variable K to M.

Step 2. For each itemset X in L_K, the algorithm generates all its *sub-itemsets* of length $(K - 1)$. Then, for each generated sub-itemset Y, the algorithm searches from L_{K-1} to see whether there exists an itemset Y' in L_{K-1} such that $Y = Y'$. If it cannot find such itemset, it puts a key-value pair (Y, s) into L_{K-1}, where s is called *the current count* of Y and s is initially set to the support count of X. Otherwise, if such itemset Y' can be found in L_{K-1} and $SC(X)$ is greater than the current count of Y', then the algorithm replaces the current count of Y' by $SC(X)$. Finally, the algorithm decreases the variable K by 1 and repeats the process of Step 2 until $K = 1$.

Although LevelWise is the first algorithm for deriving FIs from FCIs. However, when L_K and L_{K-1} store too many itemsets, LevelWise will suffer from the huge memory consumption and spend long execution time for searching and matching items. In the worst case, it is even unable to complete the whole process due to running out of memory space.

4 The Proposed Method: DFI-Growth

This section introduces the proposed algorithm *DFI-Growth* (*Deriving Frequent Itemsets based on Pattern Growth*). DFI-Growth is developed based on the adjustment of FP-Tree and FP-Growth. We briefly describe the FP-Tree tree structure adopted in DFI-Growth. In DFI-Growth, the FP-Tree is used to maintain the information of FCIs instead of transactions. In a FP-Tree, each node N represents an item $I \in I^*$ and consists of five fields: *N.name*, *N.count*, *N.parent*, *N.child* and *N.link*. The *N.name* and *N.count* fields store the item name of I and a count value of I, respectively. The *N.parent* and *N.child* fields store a parent node of N and all the children nodes of N, respectively. The *N.link* field stores a node link, which points to another node having the same item name as N. Each FP-Tree is associated with a *header table*. A header table records the information of items and these items are all sub-itemsets of FCIs. In a header table H, each item $I \in I^*$ in H consists of three fields: $H[I]$*.name*, $H[I]$*.count* and $H[I]$*.link*, which respectively records the name of item I, a count of I, and a node link pointing to a FP-Tree node having the same item name as I. To learn more information about FP-Tree and header table, readers can refer to [7].

Frequent Closed Itemset	Support Count
{CWA}	4
{CW}	5
{CD}	4
{C}	6
{CT}	4
{CWD}	3
{CWAT}	3

Itemset	Maximum Support Selection	Support Count
{C}	max{ *SC*({CWA}), *SC*({CW}), *SC*({CD}), *SC*({C}), *SC*({CT}), *SC*({CWD}), *SC*({CWAT}) }	6
{W}	max{ *SC*({CWA}), *SC*({CW}), *SC*({CWD}), *SC*({CWAT}) }	5
{A}	max{ *SC*({CWA}, *SC*({CWAT} }	4
{D}	max{ *SC*({CD}, *SC*({CWD} }	4
{T}	max{ *SC*({CT}, *SC*({CWAT} }	4

Fig. 1. An example of the MSS-I strategy.

Subroutine: *InsertFCI*(N, I_w)
01 **If**(N has a child N' such that N'*.name* $== I_w$)
02 **then** N'*.count* $\leftarrow max\{N'.count,\ SC(X)\}$;
03 **else**
04 **Create** a new child node N' and
05 **Set** N'*.name* $\leftarrow I_w$, N'*.count* $\leftarrow SC(X)$, N'*.parent* $\leftarrow N$;
06 **Set** N'*.link* to the lastly created node having the same item name as I_w;
07 $w \leftarrow w + 1$, **If**($w \leq k$), **call** *InsertFCI*(N', I_w);

Fig. 2. The pseudo code of the *InsertFCI* subroutine.

4.1 Construction of a FP-Tree for Deriving Task

This subsection introduces the construction process of a FP-Tree for the deriving task, which consists of the following four main steps.

Step 1. Let C be the complete set of FCIs discovered from a database D. DFI-Growth scans C once and derives the support count of each item appearing in C according to the proposed *Maximum Support Selection for Item* (abbr. *MSS-I*) strategy.

Strategy 1 (MSS-I). Let I be an item and $S(I) = \{c_1, c_2, \ldots, c_z\}$ be the set of all the FCIs containing I. The support count of I is equal to $max\{SC(c_1), SC(c_2), \ldots, SC(c_z)\}$.

For example, if the minimum support count is set to 3, the left side of Fig. 1 shows the complete set of FCIs in the database of Table 1. Therefore, C = {{CWA}:4, {CWD}:3, {CWAT}:3, {CW}:5, {CD}:4, {CT}:4, {C}:6}, where each number beside each itemset is its support count. The set of all the FCIs containing the item {W} is $S(\{W\})$ = {{CWA}:4, {CWD}:3, {CWAT}:3, {CW}:5}. Therefore, the support count of {W} is $SC(\{W\}) = max\{4, 3, 3, 5\} = 5$. The readers may wonder "*why the support count of {W} is neither 3 nor 4, but it is 5?*". This is because that the support count of {CW} is 5, which means that there are totally 5 transactions containing {CW} and all of these transactions also contain {W}. Therefore, the support count of {W} is 5. Then, the readers may raise another question "*Is it possible that the support count of {W} be higher than 5, for example be 6 or other value?*" The answer to this question is "*impossible*". The reason is that if an itemset X is non-closed (Definition 6), X must exists only one closure $\rho(X) = Y$ ($Y \neq X$ and $Y \in C$) such that $SC(X) = SC(\rho(X))$ (Definition 7 and Property 1). Otherwise, if an itemset X is closed, its closure must be equal to X itself (Definition 7). After applying the MSS-I strategy, the correct support counts of items are shown in the right side of Fig. 1.

Step 2. Creating a header table H and putting items into H with a fixed sorting order f. Then, creating a node R as the root of FP-Tree T. Scanning C again and sorting items in each FCI by the sorting order f.

Step 3. Inserting each sorted FCI into the FP-Tree T by calling the subroutine *InsertFCI*. The pseudo code of the subroutine *InsertFCI* is shown in Fig. 2. Each time when a frequent closed itemset $X = \{I_1, I_2, \ldots, I_K\}$ is retrieved, the two variables N and w are respectively set to the root node R and 1. Besides, setting the variable I_w to I_1, where I_1 is the first item of the closed itemset X. Then, calling the subroutine $InsertFCI(N, I_w)$ to insert X into the FP-Tree T.

The process of the subroutine $InsertFCI(N, I_t)$ works as follows. If the node N has a child node N' such that $N'.name = I_w$, then the algorithm applies the *maximum support replacement* (abbr. *MSR*) strategy for setting $N'.count$ (Line 1–Line 2).

Strategy 2 (MSR). If $SC(X) > N'.count$, then replacing $N'.count$ by $SC(X)$.

Otherwise, if $SC(X)$ is not greater than $N'.count$, then the algorithm creates a new node N' and sets it as the child node of N. Besides, setting $N'.count$ and $N'.link$ to the support count of X and the lastly created node having the same item name as I_w, respectively (Line 3–Line 6). Next, increasing the variable w by 1. After that, if w is

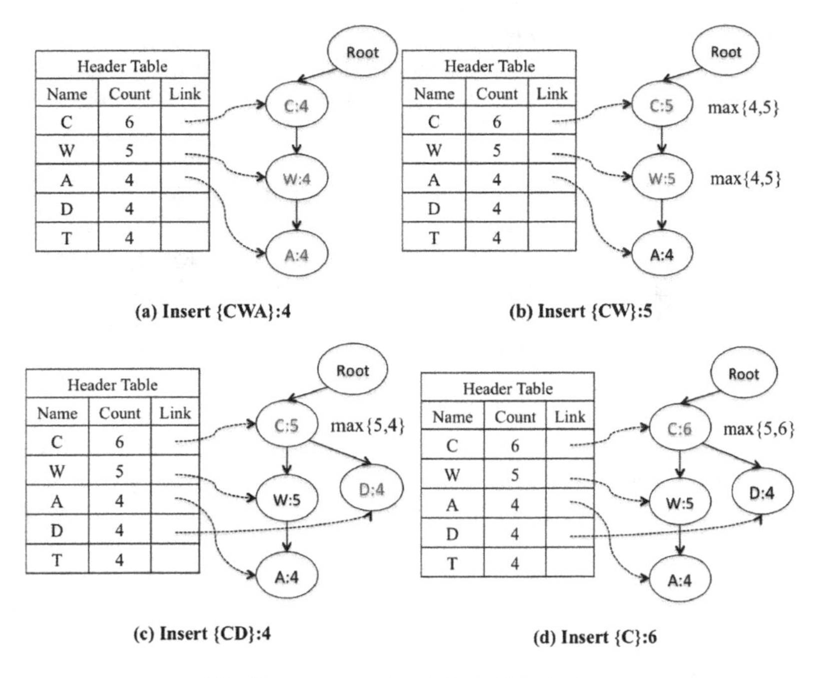

Fig. 3. An example of the MSR strategy.

not higher than K, then recursive calling the subroutine *InsertFCI* (N', I_w) for processing the w-th item of X (Line 7).

For instance, Fig. 3 shows an example of the MSR strategy. Figure 3(a) shows the resulting FP-Tree after inserting the FCI {CWA}:4. Figure 3(b) shows the process for inserting the frequent closed itemset {CW}:5. When inserting the first item of {CW} into the FP-Tree, there exists a node $N' = \{C\}$ already under the root of the FP-Tree. Besides, the item name of N' is equal to that of the first item of X (i.e., {C}). Therefore, DFI-Growth applies MSR strategy to setting $N'.count$. Because the support count of {CW} (i.e., 5) is greater than $N'.count$ (i.e., 4), $N'.count$ is replaced by $SC(\{CW\}) = 5$. After that, the second item of {CW} is processed by the same way. Figure 3(c) and (d) show the insertion process of the frequent closed itemsets {CD}:4 and {C}:6, respectively.

Step 4. During the process of FP-Tree construction, all the nodes having the same item name will be linked together by the node link of the header table H and related nodes in FP-Tree T. Through these node links, DFI-Growth allows to traversing nodes having the same item name efficiently.

4.2 Deriving FIs by DFI-Growth

After constructing the FP-Tree T and the header table H, the algorithm calls the subroutine *DFI-Growth*(H, F, α) to derive FIs from T, where the variable α is initialize to ϕ. The process of the subroutine *DFI-Growth*(H, F, α) works as follows.

Step 1. The algorithm visits every item I in the header table H. For each visited item, the algorithm outputs $\{\alpha \cup I\}$ and $H[I].count$ as a FI and its support count. Through the node link in $H[I].link$, DFI-Growth traverses all the nodes having the same item name as I in the FP-Tree T.
Step 2. Let $\beta = \{\beta_1, \beta_2, \ldots, \beta_m\}$ be a set of nodes having the same item name as I. For each node $\beta_i \in \beta$ $(1 \leq i \leq m)$, the algorithm visits from parent node of β_i to the root node R of T. Then, the algorithm collects the items of each visited nodes. These items form a sequence P_i $(1 \leq i \leq m)$ and the sequence is called *conditional itemset*. Each *conditional itemset* P_i is associated with a count called *conditional itemset count* and is denoted as $CIC(P_i)$. The conditional itemset count of P_i is initially set to $\beta_i.count$. After processing all the nodes in β in the same way, a conditional database of the itemset $\{\alpha \cup I\}$ is obtained. Let $D_{\{\alpha \cup I\}} = \{P_1{:}CIC(P_1), P_2{:}CIC(P_2), \ldots, P_m{:}CIC(P_m)\}$ denotes the conditional database of $\{\alpha \cup I\}$.
Step 3. Scanning the conditional database $D_{\{\alpha \cup I\}}$ once to find all the distinct items in conditional itemsets. For each item x, the algorithm calculates the support count of the frequent itemset $\{\alpha \cup I \cup x\}$ by the proposed *Maximum Support Selection for itemSet* (abbr. *MSS-S*) strategy.
Strategy 3 (MSS-S). Let x be an item appearing in $D_{\{\alpha \cup I\}}$ and $P^* = \{P'_1 : CIC(P'_1), P'_2 : CIC(P'_2), \ldots, P'_h : CIC(P'_h)\}$ be the set of all the conditional itemsets containing x in $D_{\{\alpha \cup I\}}$. The support count of the frequent itemset $\{\alpha \cup I \cup x\}$ is equal to $max\{CIC(P'_1), CIC(P'_2), \ldots, CIC(P'_h)\}$, where $CIC(P'_i)$ is the conditional itemset count of P'_i $(1 \leq i \leq n)$.

Conditional Database of {T}	
Conditional Itemset	Conditional Itemset Count
{CWA}	3
{C}	4

FI	Maximum Support Selection	Support Count
{TC}	*max*{*CIC*({CWA}), *CIC*({C})}	4
{TW}	*max*{*CIC*({CWA})}	3
{TA}	*max*{*CIC*({CWA})}	3

Fig. 4. An example of the MSS-S strategy.

Figure 4 shows an example of the MSS-S strategy. Considering the conditional database of the itemset {T}, there are two conditional itemsets {CWA}:3 and {C}:4. By the step of the algorithm, the value of the variable α currently is ϕ. The set of all the conditional itemsets containing the item {C} is {{CWA}:3, {C}:4}. The support count of $\{\phi \cup T \cup C\} = \{TC\}$ is $max\{CIC(\{CWA\}), CIC(\{C\})\} = max\{3, 4\} = 4$. Therefore, the algorithm outputs {TC} as a FI and its support count is 4.
Step 4. The algorithm puts items in $D_{\{\alpha \cup I\}}$ into a new header table H' in the sorting order f. Then, the algorithm scans the conditional database $D_{\{\alpha \cup I\}}$ again and sorts items in each conditional itemset in $D_{\{\alpha \cup J\}}$ according to f.
Step 5. Creating a new node R' as the root of a new FP-Tree T' for the itemset $\{\alpha \cup I\}$. Inserting every sorted conditional itemset in $D_{\{\alpha \cup I\}}$ into T' to construct the conditional FP-Tree of the itemset $\{\alpha \cup I\}$. Each time when a sorted conditional itemset J is inserted, set the variables N and I_w to R and I_1, and calling the

function *InsertFCI*(N, I_w). After constructing T', all the nodes in T' having the same item name are linked together by the node links.
Step 6. If the constructed FP-Tree T' contains nodes other than the root node R', then calling the subroutine *DFI-Growth*(H', T', $\alpha \cup I$) to derive FIs recursively.

5 Experimental Results

In this section, we evaluate the performance of the proposed DFI-Growth algorithm and compare it with the current best algorithm LevelWise [12]. The purpose of the both algorithms is the same, which is deriving FIs and their supports from FCIs. All the experiments are conducted on a computer equipped with Intel® Core™ i7-8700 3.20 GHz CPU with 16 GB of RAM, running on Windows 10 Professional OS. All the compared algorithms are implemented in Java programming language. Five real-life datasets and one synthetic dataset are used for the performance evaluation. The five real-life datasets used in the experiments are *Mushrooms*, *Chess*, *Connect*, *BMS_WebView_1*, and *Chainstore*, while the synthetic dataset is c20d10k. All the datasets are download from SPMF [4]. Table 2 shows characteristics of the datasets. Both execution time and memory usage of the algorithms are evaluated. We measure the maximum memory usage of the algorithms by using the Java API.

Table 2. Characteristics of different datasets used in the experiments.

Datasets	Number of transactions	Number of distinct items	Type
Mushrooms	8,124	119	Dense
Chess	3,196	75	Dense
Connect	67,557	129	Dense
BMS_WebView_1	59,602	497	Sparse
Chainstore	1,112,949	461	Sparse
c20d10k	10,000	192	Sparse

Figures 5 and 6 respectively show the execution times and memory usages of the compared algorithms. The execution times of DFI-Growth and LevelWise on dense datasets Mushrooms, Chess and Connect are shown in Fig. 5(a), (b) and (c), respectively. As shown in these figures, the performance of DFI-Growth is superior to LevelWise. For example, on Mushrooms dataset, when the minimum support threshold is lower than 1%, LevelWise even cannot complete the whole deriving task due to running out of memory. Besides, on the Chess and Connect datasets, LevelWise even cannot complete the deriving task within 12 h (i.e., 43200 s) under low minimum support thresholds. However, our proposed DFI-Growth only takes around 60 s to derive the complete FIs and their supports from FCIs. This is because that LevelWise adopts breadth-first search and top-down method to derive FIs from FCIs. When there are many FCIs and FIs, LevelWise needs to spend a lot of time for searching and matching items. Moreover, it always keeps two levels of itemsets in the memory, which

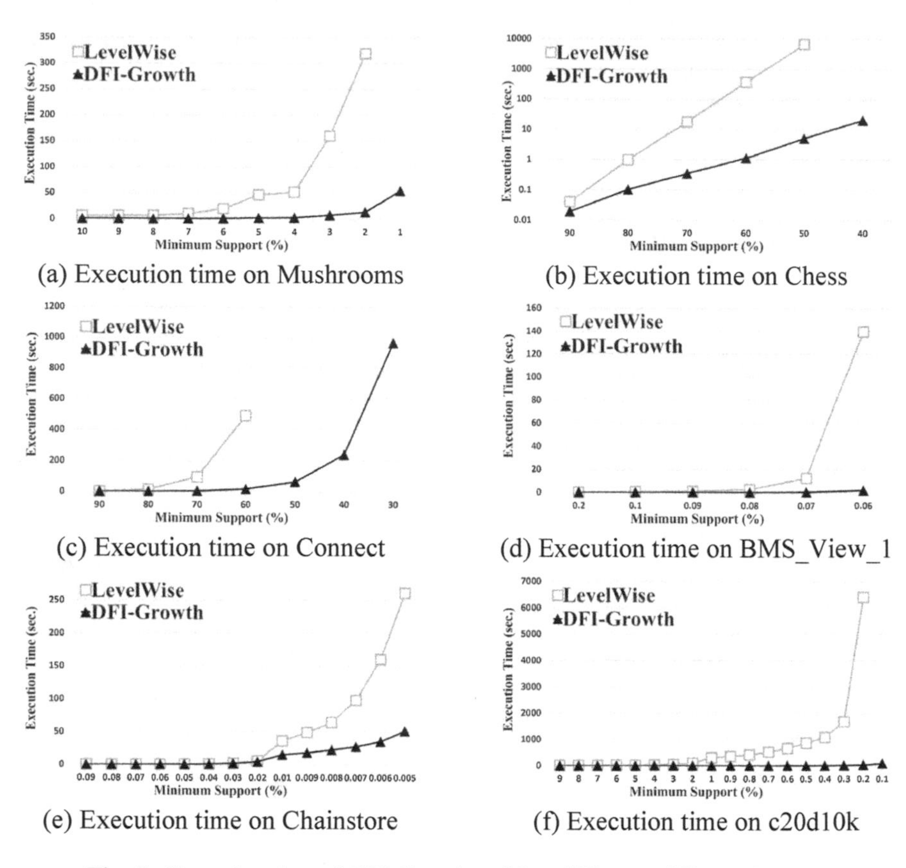

(a) Execution time on Mushrooms (b) Execution time on Chess

(c) Execution time on Connect (d) Execution time on BMS_View_1

(e) Execution time on Chainstore (f) Execution time on c20d10k

Fig. 5. Execution time of DFI-Growth and LevelWise on different datasets.

occupies considerable memory spaces during the deriving process. However, DFI-Growth adopts pattern growth with divide-and-conquer scheme to derive FIs, which partitions the whole deriving tasks into smaller independent one and thus generally takes less time than LevelWise.

Figure 5(d), (e) and (f) show the execution times of DFI-Growth and LevelWise on all the sparse datasets. From these figures, we can observe that the performance gap between of LevelWise and DFI-Growth is small when the threshold is high. This is because that when the threshold is high, the number of FCIs is small and the lengths of FCIs are relative short. Therefore, both of the algorithms can work well when the threshold is high. However, with decreasing thresholds, the performance gap between LevelWise and DFI-Growth increases significantly. As shown in the experiments, no matter the execution time or memory consumption, the proposed DFI-Growth generally has much better performance than LevelWise.

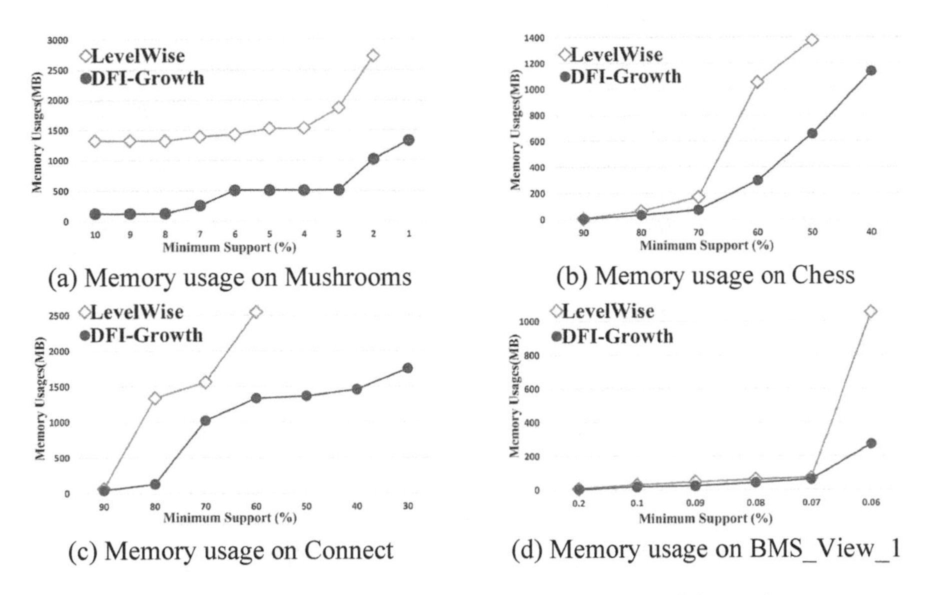

Fig. 6. Memory usage of LevelWise and DFI-Growth on different datasets.

6 Conclusion

Given a complete set of FCIs and their supports, the interesting problem addressed in this paper is "*how to efficiently, correctly, and completely derive FIs from FCIs?*". To resolve this problem, we propose a novel algorithm, named *DFI-Growth* (*Deriving Frequent Itemsets based on Pattern Growth*), for the deriving task. Moreover, we propose two novel strategies, named *maximum support selection* (abbr. *MSS*) and *maximum support replacement* (abbr. *MSR*), and incorporate them into DFI-Growth for efficiently deriving the complete FIs with correct supports. The MSS strategy is proposed for calculating the correct supports of FIs using information of FCIs. The MSR strategy is applied during the construction of FP-Tree, and it is proposed to set the correct information of FP-Tree nodes. Experiments show that the execution time and memory usage of DFI-Growth are significantly better than that of the current best algorithm LevelWise [12]. For example, on the Chess dataset, LevelWise cannot successfully complete the deriving task within 12 h, while DFI-Growth only takes 20 s to finish. The implementations of LevelWise [21] and DFI-Growth [22] to the well-known open-source data mining library SPMF[4], for benefiting related research works and real-life applications.

Acknowledgment. This work is supported in part by Ministry of Science and Technology, Taiwan, ROC under grant no. 107-2218-E-197-002 and 108-2634-F-009-010.

References

1. Agrawal, R., Srikant, R.: Fast algorithms for mining association rules. In: Proceedings of International Conference on Very Large Data Bases, pp. 487–499 (1994)
2. Boulicaut, J.-F., Bykowski, A., Rigotti, C.: Approximation of frequency queries by means of free-sets. In: Zighed, D.A., Komorowski, J., Żytkow, J. (eds.) PKDD 2000. LNCS (LNAI), vol. 1910, pp. 75–85. Springer, Heidelberg (2000). https://doi.org/10.1007/3-540-45372-5_8
3. Calders, T., Goethals, B.: Mining all non-derivable frequent itemsets. In: Elomaa, T., Mannila, H., Toivonen, H. (eds.) PKDD 2002. LNCS, vol. 2431, pp. 74–86. Springer, Heidelberg (2002). https://doi.org/10.1007/3-540-45681-3_7
4. Fournier-Viger, P., Gomariz, A., Gueniche, T., Soltani, A., Wu, C., Tseng, V.S.: SPMF: a Java open-source pattern mining library. J. Mach. Learn. Res. **15**, 3569–3573 (2014)
5. Gupta, S., Mamtora, R.: A survey on association rule mining in market basket analysis. Int. J. Inf. Comput. Technol. **4**(4), 409–414 (2014)
6. Gouda, K., Zaki, M.: GenMax: an efficient algorithm for mining maximal frequent itemsets. Data Min. Knowl. Discov. **11**(3), 223–242 (2005)
7. Han, J., Pei, J., Yin, Y.: Mining frequent patterns without candidate generation. In: Proceedings of ACM SIGMOD International Conference on Management of Data, pp. 1–12 (2000)
8. Lucchese, C., Orlando, S., Perego, R.: Fast and memory efficient mining of frequent closed itemsets. IEEE Trans. Knowl. Data Eng. **18**(1), 21–36 (2006)
9. Liu, J., Shang, J., Wang, C., Ren, X., Han, J.: Mining quality phrases from massive text corpora. In: Proceedings of ACM SIGMOD International Conference on Management of Data, pp. 1729–1744 (2015)
10. Mohammed, E.H., Zaiane, O.R.: COFI-tree mining: a new approach to pattern growth with reduced candidacy generation. In: Proceedings of IEEE International Conference on Data Mining (2003)
11. Pasquier, N., Bastide, Y., Taouil, R., Lakhal, L.: Discovering frequent closed itemsets for association rules. In: Proceedings of 7th International Conference on Database Theory, pp. 398–416 (1999)
12. Pasquier, N., Bastide, Y., Taouil, R., Lakhal, L.: Efficient mining of association rules using closed itemset lattices. Inf. Syst. **24**(1), 25–46 (1999)
13. Park, J.S., Chen, M.S., Yu, P.S.: An effective hash-based algorithm for mining association rules. In: Proceedings of ACM SIGMOD International Conference on Management of Data, pp. 175–186 (1995)
14. Pei, J., Han, J., Lu, H., Nishio, S., Tang, D., Yang, S.: H-Mine: hyper-structure mining of frequent patterns in large databases. In: Proceedings of IEEE International Conference on Data Mining, pp. 441–448 (2001)
15. Prabha, S., Shanmugapriya, S., Duraiswamy, K.: A survey on closed frequent pattern mining. Int. J. Comput. Appl. **63**(14), 47–52 (2013)
16. Ting, S.L., Shum, C.C., Kwok, S.K., Tsang, A.H.C., Lee, W.B.: Data mining in biomedicine: current applications and further directions for research. J. Softw. Eng. Appl. 150–159 (2009)
17. Wang, J., Han, J., Pei, J.: CLOSET+: searching for the best strategies for mining frequent closed itemsets. In: Proceedings of 9th ACM SIGKDD International Conference on Knowledge Discovery and Data Mining, pp. 236–245 (2003)
18. Zaki, M.J.: Scalable algorithms for association mining. IEEE Trans. Knowl. Data Eng. **12**(3), 372–390 (2000)

19. Zaki, M.J., Hsiao, C.J.: CHARM: an efficient algorithm for closed itemset mining. In: Proceedings of SIAM International Conference on Data Mining, pp. 457–473 (2002)
20. Zhang, Q., Segall, R.: Web mining: a survey of current research, techniques, and software. Int. J. Inf. Technol. Decis. Making **7**(4), 683–720 (2008)
21. Source code of the implemented LevelWise algorithm released in SPMF. http://www.philippe-fournier-viger.com/spmf/LevelWise
22. Source code of the proposed DFI-Growth algorithm released in SPMF. http://www.philippe-fournier-viger.com/spmf/DFI-Growth

Discovering Stable Periodic-Frequent Patterns in Transactional Data

Philippe Fournier-Viger[1(✉)], Peng Yang[2], Jerry Chun-Wei Lin[3], and Rage Uday Kiran[4]

[1] School of Humanities and Social Sciences, Harbin Institute of Technology (Shenzhen), Shenzhen, China
philfv8@yahoo.com

[2] School of Computer Science and Technology, Harbin Institute of Technology (Shenzhen), Shenzhen, China
pengyeung@163.com

[3] Department of Computing, Mathematics and Physics, Western Norway University of Applied Sciences (HVL), Bergen, Norway
jerrylin@ieee.org

[4] Institute of Industrial Science, University of Tokyo, Tokyo, Japan
uday.rage@gmail.com

Abstract. Periodic-frequent patterns are sets of items (values) that periodically appear in a sequence of transactions. The periodicity of a pattern is measured by counting the number of times that its periods (the interval between two successive occurrences of the patterns) are greater than a user-defined $maxPer$ threshold. However, an important limitation of this model is that it can find many patterns having a periodicity that vary widely due to the strict $maxPer$ constraint. But finding stable patterns is desirable for many applications as they are more predictable than unstable patterns. This paper addresses this limitation by proposing to discover a novel type of periodic-frequent patterns in transactional databases, called Stable Periodic-frequent Pattern (SPP), which are patterns having a stable periodicity, and a pattern-growth algorithm named SPP-growth to discover all SPP. An experimental evaluation on four datasets shows that SPP-growth is efficient and can find insightful patterns that are not found by traditional algorithms.

Keywords: Pattern mining · Periodic pattern · Stable periodicity · Lability

1 Introduction

Frequent itemset mining (FIM) [1–4] is a popular data analysis task. The goal is to identify all patterns that frequently appear in records of a transactional database. A pattern is said to be frequent if its support (occurrence frequency) is no less than a user-defined minimum support threshold. Although discovering frequent patterns is useful, too many frequent patterns are often found, and

F. Wotawa et al. (Eds.): IEA/AIE 2019, LNAI 11606, pp. 230–244, 2019.
https://doi.org/10.1007/978-3-030-22999-3_21

many of them are uninteresting to users. To address this issue, several variations of FIM have been developed to select small sets of patterns that are interesting to users based on various constraints. This includes discovering maximal frequent patterns [5], closed frequent patterns [6], high-utility patterns [7], and periodic frequent patterns [8,9,11,12,14,15]. Frequent patterns that periodically occur in a database are called *periodic-frequent patterns* (PFP). Finding these patterns has many practical applications such as for analyzing the behavior of website users and the performance of recommender systems [10]. Finding periodic patterns can also help to understand the purchase behavior of customers by discovering sets of products that are periodically bought. For instance, one may find that a customer buys bread every week. Such pattern can then be used for marketing [8,12].

Several algorithms have been proposed to discover PFP in a transaction database (a sequence of transactions). Most of them measure the periodicity of a pattern by counting the number of times that its periods (number of events between two consecutive occurrences) are greater than a user-specified $maxPer$ threshold. In *full* PFP mining [8,11], a pattern is called periodic if none of its periods are greater than $maxPer$. A drawback of this model is that it is too strict as a pattern is discarded if it has only one period exceeding $maxPer$. For example, a pattern indicating that a customer buys milk every day would be discarded just because the customer skipped one day. An alternative called *partial* PFP mining [14] was then proposed, which relaxed the $maxPer$ constraint to allow a certain number of periods to exceed $maxPer$. But this model is not strict enough, as a pattern may be considered periodic even if it has some very long periods. For example, buying coffee may be considered as periodic if a customer buys it on many consecutive days even if he then did not buy it for a year. Thus, traditional models for discovering PFP patterns are inadequate because the size of periods for some patterns may vary widely in a real-life database but traditional periodicity measures do not take this into account. For several applications such as market basket analysis, it is desirable to identify stable periodic-frequent patterns, that is patterns that have periods that are more or less stable in terms of size over time. Such patterns can be useful to better forecast product demand and improve inventory management strategies.

In this paper, we propose a solution for discovering stable periodic-frequent patterns using a novel measure of stability. This paper has three main contributions. First, a novel measure named *lability* is proposed to assess the stability of patterns. Second, a pattern-growth algorithm, called Stable Periodic-frequent Pattern-growth (SPP-growth), is proposed to efficiently discover the complete set of stable periodic-frequent itemsets. Third, several experiments were conducted on synthetic and real-life datasets to evaluate the efficiency of SPP-growth, and patterns found using the proposed stability measure. Experimental results show that the proposed approach is efficient for finding stable periodic-frequent patterns and that insightful patterns are discovered, which are not discovered using traditional approaches.

The rest of the paper is organized as follows. Section 2 describes related work. Section 3 describes the proposed model of SPPs. Section 4 introduces our algorithm to find all SPPs in transactional databases. Section 5 reports experimental results. Finally, Sect. 6 draws a conclusion.

2 Related Work

The problem of frequent itemset mining is defined as follows [1,2]. Let I be a set of items or symbols. Each subset $X \subseteq I$ is said to be an itemset. The length of an itemset containing k items is said to be k. Furthermore, all itemsets of a length k are called k-itemsets. A *transactional database* $D = \{T_1, T_2, ..., T_n\}$ is a set of transactions where each transaction T_c is an itemset with a unique Transaction IDentifier (TID) c, and where the TID can also represent the transaction time (or timestamp). The support of an itemset X in a database D is denoted as $sup(X)$ and defined as $|\{T|T \in D \wedge X \subseteq T\}|$. In other words, $sup(X) = |g(X)|$, where $g(X)$ is the set of transactions containing X.

Table 1. A transactional database

TID	Itemset	TID	Itemset
1	a, b, c, e	6	b, c, e
2	a, b, c, d	7	b, c, d, e
3	a, b, e	8	a, c
4	c, e	9	a, b, d
5	b, d, e	10	b

For example, consider the database of Table 1, which will be used as running example. This database contains ten transactions $(T_1, T_2, ..., T_{10})$. Transaction T_2 has TID 2 and is a 4-itemset $\{a, b, c, d\}$. The set of transactions containing the itemset $\{a, b\}$ is $g(\{a, b\}) = \{T_1, T_2, T_3, T_9\}$. Hence, the support of $\{a, b\}$ is $sup(\{a, b\}) = |g(\{a, b\})| = 4$.

Definition 1 (Frequent itemset mining). *The problem of frequent itemset mining consists of discovering the frequent itemsets. An itemset X is frequent if sup(X) $\geq$ minSup, where minSup is a user-specified minimum support threshold.*

For instance, if $minSup = 5$, there are five frequent itemsets: {a} : 5, {b} : 8, {c} : 6, {e} : 6 and {b,e} : 5, where each itemset X is annotated with $sup(x)$.

Various algorithms have been proposed to discover frequent itemsets, such as Apriori [1] and FP-Growth [2]. However, these algorithms cannot be applied to mine PFPs as they do not evaluate the periodic behavior of patterns. Inspired by studies on frequent itemset mining, researchers have designed algorithms to discover periodic-frequent patterns in transaction databases [2,11–14]. Several applications of mining periodic-frequent patterns were presented in previous studies [8,10]. The traditional (*full*) periodic-frequent pattern mining model is defined as follows [8].

Definition 2 (Periods of an itemset). *Let there be an itemset* X *and a database* $D = \{T_1, T_2, ..., T_n\}$. *The set of transactions containing* X *is denoted as* $g(X) = \{T_{gX(1)}, T_{gX(2)}, ..., T_{gX(sup(X))}\}$. *Let* $T_{gX(i+1)}$ *and* $T_{gX(i)}$, $i \in [0, sup(X)]$ *be two consecutive TIDs where* X *appears. For an integer* $i \geq 0$, *the number of transactions or the time difference between* $T_{gX(i+1)}$ *and* $T_{gX(i)}$ *is said to be a period of* X, *defined as* $per(X, i) = gX(i+1) - gX(i)$. *For simplicity of computation, it is considered that* X *appears in two additional transactions where* $gX(0) = 0$ *and* $gX(sup(X)+1) = |D|$. *The periods of an itemset* X *is a list of periods defined as* $per(X) = \{gX(1) - gX(0), gX(2) - gX(1), ..., gX(sup(X) + 1) - gX(sup(X))\}$. *Thus,* $per(X) = \bigcup_{0 \leq z \leq sup(X)} (gX(z+1) - gX(z))$ *and* $|per(X)| = |g(X)| + 1 = sup(X) + 1$.

For example, consider the itemset $\{b, e\}$. This itemset appears in transactions T_1, T_3, T_5, T_6 and T_7, and thus $g(\{b, e\}) = \{T_1, T_3, T_5, T_6, T_7\}$. The periods of this itemset are $per(\{b, e\}) = \{1,2,2,1,1,3\}$.

Definition 3 (Periodic-frequent pattern). *Let* $per(X)$ *be the set of all periods of* X. *Then, the periodicity of* X *can be defined as* $maxper(X) = max(per(X))$. *An itemset* X *is a periodic-frequent pattern if* $sup(X) \geq minSup$ *and* $maxper(X) \leq maxPer$, *where* $minSup$ *and* $maxPer$ *are user-defined thresholds.*

For example, if $minSup = 5$ and $maxPer = 2$, the complete set of (*full*) PFPs is {b}: (8, 2) and {c}: (6, 2), where each PFP X is annotated with a pair $(sup(X), maxper(X))$.

Tanbeer et al. [8] proposed the problem of mining PFPs and the PF-growth algorithm. Then, the MTKPP [11] algorithm was designed, which relies on a depth-first search and a vertical database representation. But these two algorithms have the drawback that a pattern is discarded if only one of its periods exceeds $maxPer$ (i.e., the item {e} has a periodic behavior, but it is regarded as non periodic since it has a period of $3 > maxPer = 2$). Several variations of the above definition were proposed to address some of its limitations. Surana et al. [13] proposed to associate a minimum support threshold and a maximum periodicity to each item, to evaluate each item in a different way. But the number of parameters becomes equal to the number of items. Kiran et al. [14] relaxed the maximum periodicity constraint by considering that a pattern X is (*partial*) periodic if its *periodic-frequency* ($\frac{|\{i | per(X,i) \leq maxPer\}|}{|per(X)|}$) is no less than a user-defined threshold. However, a major drawback of that definition is that a pattern that has some very long periods will still be considered as periodic (i.e., the item {a} is periodic even if it is periodic only in a very short time-interval). In summary, these studies measure the periodicity of a pattern by counting the number of times that its periods are less than $maxPer$ but ignore by how much these periods deviates from $maxPer$. To our best knowledge, this study first considers the problem of finding stable periodic-frequent patterns by taking into account by how much the periods of each pattern deviate from $maxPer$.

3 The Proposed Model

The proposed model defines the concept of stable periodic-frequent patterns. In this model, the definition of periodic-frequent patterns is extended to capture the frequent patterns having a stable periodic behavior. The basic idea of the proposed model is to assess the periodic stability of a pattern by calculating the cumulative sum of the difference between each period length and $maxPer$. The proposed model is defined as follows.

Definition 4 (Lability of an itemset). *Let $T_{gX(i+1)}$ and $T_{gX(i)}$, $i \in [0, sup(X)]$ be two consecutive TIDs where X appears. The i-th lability of X is defined as $la(X,i) = max(0, la(X,i-1) + per(X,i) - maxPer)$, where $la(X,-1) = 0$. Moreover, it can be concisely written as*

$$la(X,i) = max(0, la(X,i-1) + gX(i+1) - gX(i) - maxPer)$$

The lability of an itemset X is a list of periods defined as $la(X) = \{la(X,0), la(X,1), ..., la(X, sup(X))\}$, and $|la(X)| = |per(X)| = sup(X) + 1$.

For example, consider the item {d}. The terms for computing its lability are $la(\{d\},0) = max(0, la(\{d\},-1) + per(\{d\},0) - maxPer) = max(0, 0+1-2) = 0$, $la(\{d\},1) = 1$, $la(\{d\},2) = 1$, $la(\{d\},3) = 1$ and $la(\{d\},4) = 0$. Hence, the lability of $\{d\}$ is $la(\{d\}) = \{0,1,1,1,0\}$.

Based on Definition 4, it can be observed that the periodic behavior of a pattern is stable (lability is zero) if its periods are always no greater than $maxPer$. If a pattern has periods larger than $maxPer$, its lability will increase, and these exceeding values will be accumulated by the lability measure. And if there exists periods of a pattern that are smaller than $maxPer$, its lability will decrease for these periods until it reaches a minimum of zero. Thus, the *lability* of a pattern changes over time depending on its periodic behavior, and each value exceeding $maxPer$ is accumulated. A low lability value (close to zero) means a stable periodic behavior while a high value means an unstable one. Hence, this measure can be used to find stable patterns by limiting the maximum lability.

Definition 5 (Stable periodic-frequent pattern). *Let $la(X)$ be the set of all i-th lability of a pattern X. The stability of X is defined as $maxla(X) = max(la(X))$. An itemset X is a stable periodic-frequent pattern (SPP) if $sup(X) \geq minSup$ and $maxla(X) \leq maxLa$, where $minSup$ and $maxLa$ are thresholds.*

For example, by continuing the previous example, if the user specifies that $maxLa = 1$, the complete set of SPPs is {b}: (8, 0), {c}: (6, 0), {e}: (6, 1) and {b,e}: (5, 1), where each SPP X is annotated with ($sup(X)$, $maxla(X)$).

It is interesting to note that if $maxLa = 0$, SPPs are the traditional PFPs. Thus the proposed SPPs is a generalization of the traditional definition of PFPs.

Definition 6 (Problem definition). *Given a transaction database (D), set of items (I), user-defined minimum support threshold ($minSup$), user-defined maximum periodicity threshold ($maxPer$) and maximum lability threshold ($maxLa$),* *the problem of finding stable periodic-frequent patterns is to discover each pattern X in D such that sup(X) $\geq minSup$ and $maxla(X) \leq maxLa$.*

To develop an efficient algorithm for mining SPPs, it is important to design efficient pruning strategies. To use the stability measure for pruning the search space, the following theorem is proposed.

Lemma 1 (Monotonicity of the maximum lability). *Let X and Y be itemsets such that $X \subset Y$. It follows that $maxla(Y) \geq maxla(X)$.*

Proof. Since $X \subset Y$, $g(Y) \subseteq g(X)$. If $g(Y) = g(X)$, then X and Y have the same periods, thus $la(Y) = la(X)$ and $maxla(Y) = maxla(X)$. If $g(Y) \subset g(X)$, then for each transaction $\{T_z | T_z \in g(X) \wedge T_z \notin g(Y)\}$, the corresponding period $per(X, z)$ will be replaced by a larger period $per(Y, z)$. Thus, any period in $per(Y)$ cannot be smaller than a period in $per(X)$. Hence, $maxla(Y) \geq maxla(X)$.

Theorem 1 (Maximum lability pruning). *Let X be an itemset appearing in a database D. X and its supersets are not SPPs if $maxla(X) > maxLa$. Thus, if this condition is met, the search space consisting of X and all its supersets can be discarded.*

Proof. By definition, if $maxla(X) > maxLa$, X is not a SPP. By Lemma 1, supersets of X are also not SPPs.

4 The SPP-Growth Algorithm

This subsection introduces the proposed *SPP-growth* algorithm. It performs two steps: (i) compressing the database into a stable periodic-frequent tree (SPP-tree) and (ii) mining the SPP-tree to find all stable periodic-frequent patterns.

4.1 The SPP-Tree Structure

The SPP-tree structure consists of a prefix-tree and a SPP-list. The SPP-list consists of entries having three fields: item name (i), support (S) and maximum lability (ML). The prefix-tree structure of the SPP-tree is similar to that of the FP-tree [2]. However, to calculate both the support and maximum lability of patterns, the SPP-tree nodes explicitly maintain occurrence information for each transaction by maintaining an occurrence TID (or timestamp) list, called *TID-list* at the last node of every transaction. Hence, two types of nodes are maintained in a SPP-tree: **ordinary** nodes and **tail** nodes. Ordinary nodes are similar to FP-tree nodes [2], whereas tail nodes are the last items of any sorted transaction. The structure of a tail node is $i[t_a, t_b, \cdots, t_c]$, where i is the node's item name and t_j ($j \in [1, n]$) is a TID where item i is the last item. To facilitate tree traversal, each node in the prefix-tree maintains parent, children and node traversal pointers. Besides, unlike FP-tree nodes, SPP-tree nodes do not maintain a support count value. To increase the likelihood of obtaining a compact tree, items in the prefix-tree are arranged in descending order of their **support** of SPP. The next paragraphs explain how a SPP-tree is constructed and mined to extract SPPs.

(a)

i	*S*	*ML*	t_{cur}
a	1	0	1
b	1	0	1
c	1	0	1
e	1	0	1

(b)

i	*S*	*ML*	t_{cur}
a	2	0	2
b	2	0	2
c	2	0	2
e	1	0	1
d	1	0	2

(c)

i	*S*	*ML*	t_{cur}
a	5	3	9
b	8	0	10
c	6	0	8
e	6	0	7
d	4	1	9

(d)

i	*S*	*ML*
a	5	3
b	8	0
c	6	0
e	6	1
d	4	1

(e)

i	*S*	*ML*
b	8	0
c	6	0
e	6	1

Fig. 1. The SPP-list of of Table 1 after scanning (a) the first transaction, (b) the second transaction, (c) the entire database, (d) after adding $t = gX(sup(X) + 1)$ to each item, and (e) the final SPP-list containing the sorted list of items.

4.2 Constructing an SPP-tree

To construct a SPP-list, a temporary array t is used to record the current TID (or timestamp) of an item. Algorithm 1 describes the steps for constructing a SPP-list. Consider the database of Table 1 and that $minSup$, $maxPer$ and $maxLa$ values are set to 5, 2 and 1, respectively. Figures 1(a)–(c) show the construction of the SPP-list after scanning the first, second, and all transactions of the database, respectively (line 2 to 9 of Algorithm 1). Figures 1(d) shows the result of adding $t = g(sup(X) + 1)$ to every item (line 10 of Algorithm 1). Figures 1(e) shows the SPP-list containing stable periodic-frequent items, sorted by descending order of *support* (line 10 of Algorithm 1).

After finding stable periodic-frequent items, the database is scanned again to construct the prefix-tree of the SPP-tree (Algorithm 2). The construction of the prefix-tree is similar to how a FP-tree is constructed [2]. But it has to be noted that only **tail** nodes of a SPP-tree maintain TIDs (or timestamps). Figure 2(a)–(e) show the construction of the SPP-tree after scanning the first, second, eighth, and all transactions of the database. In a SPP-tree, an item header table is built so that each item points to its occurrences in the tree via a chain of node-links to facilitate tree traversal. For simplicity, we do not show these node-links in the illustrations. They are created in the same way as for the FP-tree.

4.3 Mining an SPP-tree

The SPP-tree is then mined as follows. A bottom-up scan is done to browse each stable periodic-frequent item of the header table of the SPP-tree. The conditional pattern base of each item is constructed (a projected database, consisting of the set of prefix paths of the SPP-tree co-occurring with the suffix itemset) to collect the TIDs of its ancestors and calculate its $SPP\text{-}list$. Then, the item's conditional SPP-tree is constructed (where ancestors that have a support less than $minSup$ or a $maxla$ larger than $maxSup$ are pruned), and mining is pursued recursively on the resulting tree. The algorithm is said to be of type pattern-growth because it recursively grows each SPP by appending single items from

Algorithm 1. Construction of an SPP-list

input : D: a transactional database,
$minSup$, $maxPer$, $maxLa$: the user-specified thresholds
output: the SPP-list containing the sorted list of items

1 Record the TID $t_{last}(i)$ of the last transaction containing each item i in a temporary array. Set the maximum lability $ML(i)$, the lability $la(i)$ and support $S(i)$ of each item i to 0;
2 **foreach** *transaction* $T \in D$ *with TID* t_{cur} **do**
3   **foreach** *item* $i \in T$ **do**
4         $S(i) = S(i) + 1;$
5         $la(i) = max(0, la(i) + t_{cur} - t_{last}(i) - maxPer);$
6         $ML(i) = max(ML(i), la(i));$
7         $t_{last}(i) = t_{cur};$
8   **end**
9 **end**
10 Set $t_{cur} = |D|$ and update each item in the SPP-list. Remove each item i such that $S(i) < minSup$ or $ML(i) > maxLa$ from the SPP-list. Consider the remaining items of the SPP-list as stable periodic-frequent items and sort them by descending order of support.

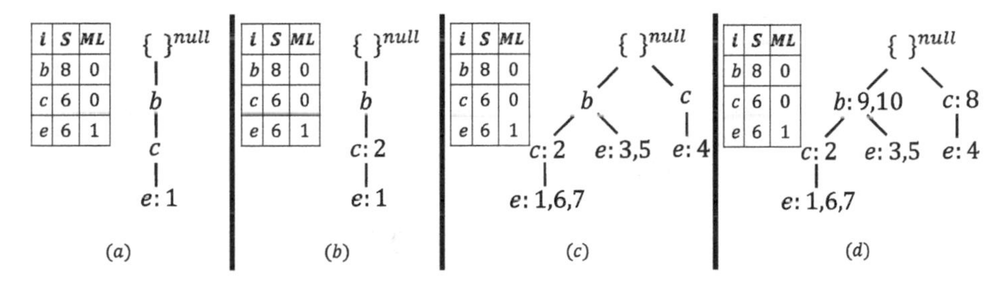

Fig. 2. Construction of an SPP-tree after scanning the (a) first transaction, (b) second transaction, (c) eighth transaction, and (d) the entire database.

its generated conditional SPP-tree. The procedure to discover SPPs from the SPP-tree is shown in Algorithm 3.

To illustrate how the SPP-tree is processed during mining, Fig. 3(a) shows the SPP-tree of Fig. 2(d) after removing the bottommost item e. Note that the *TID*-list of e has been pushed-up to its parent node since the node e was a *tail*-node in the original SPP-tree. Besides, Fig. 3(b) shows the prefix-tree of suffix item e. A *conditional pattern base* is a path from the root node to a leaf node. The conditional pattern bases of e are thus {b:3,5}, {bc:1,6,7} and {c:4}. Using the prefix-tree of e, it is rather simple to compute the *maxLa* and *support* of each item that is co-occurring with the suffix itemset. The pruning conditions are checked and nodes not respecting the thresholds are deleted from the SPP-tree, while stable periodic patterns are output. The resulting tree is shown in Fig. 3(c). The item c was deleted because its support is less than 5,

Algorithm 2. Construction of an SPP-tree

input : D: a transactional database,
SPP-list: contains stable periodic-frequent items, their S and ML

1 Create the root of the SPP-tree, T, and label it with "null" ;
2 **foreach** *transaction* $T \in D$ *with TID* t_{cur} **do**
3 Sort stable periodic-frequent items in T according to the order of *SPP-list*. Let the sorted candidate item list be $[p|P]$, where p is the first item and P is the remaining list. Call $insert\_tree([p|P], ts_{cur}, T)$, which is performed as follows. If T has a child N such that $N.item\text{-}name \neq p.item\text{-}name$, then create a new node N. Link its parent to T. Let its node-link be linked to nodes with the same *item-name* via the node-link structure. Remove p from $[p|P]$. If P is empty, add t_{cur} to the leaf node; else, call $insert\_tree(P, ts_{cur}, N)$ recursively.
4 **end**

Algorithm 3. The SPP-Growth algorithm

input : T: a SPP-tree, α: suffix itemset (initial value is $\emptyset$)
output: the set of SPPs

1 **while** *T's header table* $T^{ht} \neq \emptyset$ **do**
2 $i = T^{ht}[T^{ht}.size - 1]$;
3 $\beta = \alpha \cup i$, output β;
4 Traverse the node-link of i to construct β's conditional pattern base and collect its TIDs where β has appeared in D and calculate $SPP\text{-}list_\beta$;
5 **if** $SPP\text{-}list_\beta \neq \emptyset$ **then**
6 Construct β's conditional tree T_β and call $SPP\text{-}Growth(T_\beta, \beta)$;
7 **end**
8 Remove i from T and push i's TIDs to its parent nodes.
9 **end**

while *be* is output as a stable periodic pattern. Then same process of creating a prefix-tree and its corresponding conditional tree is repeated in the same way to consider other pattern extensions. The whole process of mining patterns using each item is repeated if the header table of the SPP-tree is empty. Because the SPP-growth algorithm starts from SPP-tree of single items and recursively explores the conditional SPP-tree of patterns and only prunes the search space using Theorem 1, it can be seen that this procedure is correct and complete to discover all SPPs.

5 Experimental Evaluation

Since SPP-growth is the first algorithm for mining SPPs, its performance is not compared with other algorithms. SPP-growth is implemented in Java. Experiments were performed on a computer having a 64 bit Xeon E3-1270 3.6 GHz

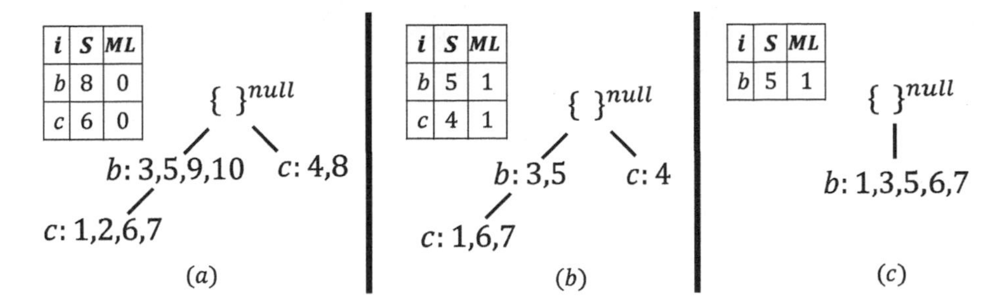

Fig. 3. Mining stable periodic-frequent patterns using suffix item e. (a) The SPP-tree after removing the item e, (b) prefix-tree of suffix item e, (c) conditional tree of suffix item e.

CPU, running Windows 10 and having 64 GB of RAM. The algorithm is evaluated in terms of performance on both synthetic (*T10I4D100K*) and real-world (*mushroom*, *OnlineRetail* and *kosarak*) datasets, obtained from the SPMF website [16]. Characteristics of the datasets are presented in Table 2, where $|D|$, $|I|$, T_{min}, T_{max} and T_{avg} denote the number of transactions, distinct items, minimum transaction length, maximum transaction length and average transaction length, respectively. Datasets were chosen because they have different characteristics (dense/sparse datasets, long/short transactions, few/many items).

Table 2. Characteristics of the datasets

Dataset	$\|D\|$	$\|I\|$	T_{min}	T_{max}	T_{avg}
T10I4D100K	100,000	870	1	29	10
mushroom	8,124	119	23	23	23
kosarak	990,002	41,270	1	2,498	8
OnlineRetail	541,909	2,603	1	1,108	1

Figure 4 shows the runtime requirements of the SPP-growth algorithm for different $minSup$, $maxPer$ and $maxLa$ values on $mushroom$ and $T10I4D100K$, respectively. In these charts, values for the maximum lability thresholds ($maxLa$) are shown on the x axis, while the y axis denotes execution times. The notation S-P denotes the SPP-growth algorithm with $minSup = S$ and $maxPer = P$. The following two observations are drawn from Fig. 4:

- Increasing $maxLa$ often increase the runtime. The reason is that increasing $maxLa$ increases the range of lability values accepted for SPPs. Thus, more patterns must be considered in the search space.
- The SPP-growth algorithm has better performance on the sparse dataset than on the dense dataset. The reason is that patterns in sparse datasets are

more likely to be unstable. Hence, SPP-growth algorithm can eliminate many candidate patterns on such datasets.

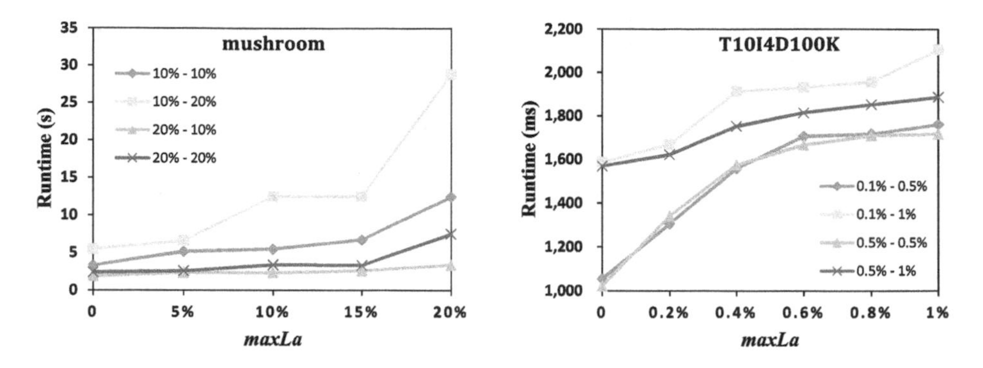

Fig. 4. Execution times for different parameter values

Figure 5 shows the number of SPPs generated for different $minSup$, $maxPer$ and $maxLa$ values for *mushroom* and $T10I4D100K$, respectively. The following observations can be drawn:

- For fixed $maxLa$ and $maxPer$ values, increasing $minSup$ may decrease the number of stable periodic-frequent patterns. The reason is that some patterns will then fail to satisfy the higher $minSup$ threshold value.
- Similarly, for fixed $maxLa$ and $minSup$ values, increasing $maxPer$ may increase the number of SPPs. The reason is that as $maxPer$ is increased, frequent patterns having longer periods may become stable periodic-frequent patterns.
- For the $T10I4D100K$ dataset, the pattern count increases rapidly as the *support* and $maxPer$ threshold are increased at the same time. This is

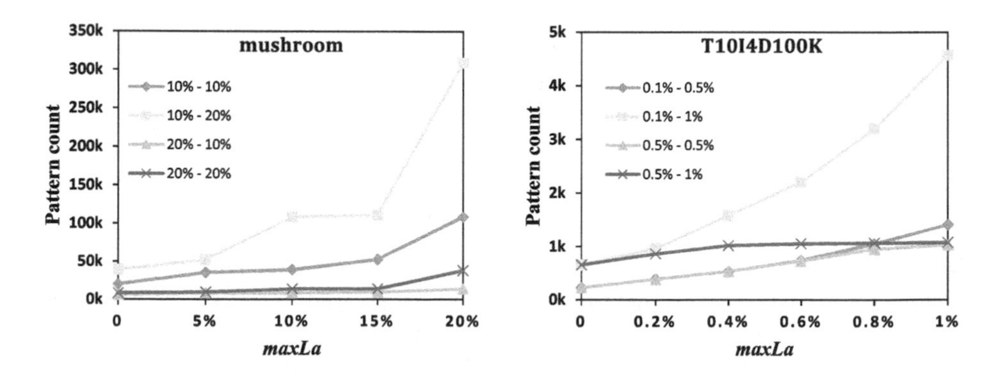

Fig. 5. Number of SPPs found for different parameter values

because patterns in sparse datasets are more likely to be candidate patterns. Hence, increasing the *support* and *maxPer* thresholds at same time results in more candidate patterns that need to be considered as potential SPPs.

We also evaluated the proposed algorithm's scalability in terms of execution time, number of SPPs found and number of tree nodes when the number of transactions is varied. For this experiment, the real *kosarak* dataset is used, since it has a large number of distinct items and transactions. The dataset was divided into five parts and then the performance of the algorithm was measured after adding each part to the previous ones. Figure 6 shows the experiment's results for $minSup = 0.1\%$, $maxPer = 2\%$ and $maxLa = 1\%$. It is clear that the runtime, number of patterns and number of nodes increase along with the database size. This is reasonable because the *maxLa* and *maxPer* of patterns also increase with the database size. Hence, the algorithm finds more patterns and spend time to build additional tree nodes when size is increased.

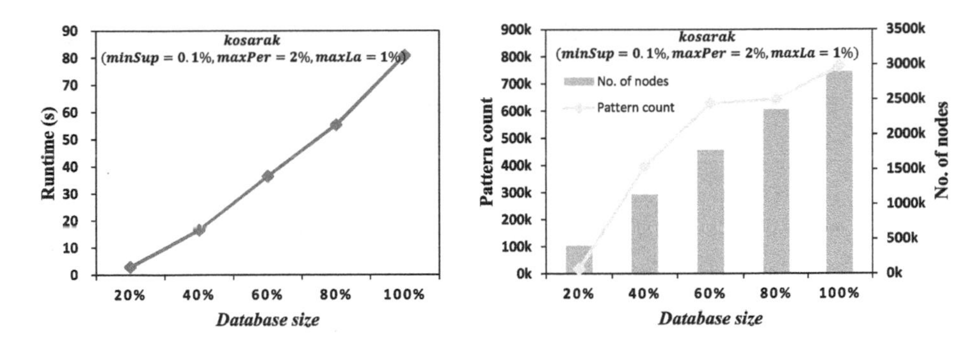

Fig. 6. Scalability of SPP-growth when varying the database size

Table 3. Comparison of peak memory usage.

Algorithms			*T10I4D100K* (MB)	No. of nodes
minSup	*maxPer*	*maxLa*		
0.1%	0.5%	0%	263	334,153
0.1%	0.5%	0.4%	296	575,854
0.1%	1%	0%	324	596,766
0.1%	1%	0.4%	421	652,840
0.2%	0.5%	0%	263	334,153
0.2%	0.5%	0.4%	296	575,854
0.2%	1%	0%	323	596,351
0.2%	1%	0.4%	389	638,054

In another experiment, the peak memory usage of SPP-growth was recorded for different parameter values on $T10I4D100K$. Results (Table 3) show that SPP-growth consumes less memory for high $minSup$, low $maxPer$ and low $maxLa$ values. This is reasonable as more patterns can be pruned.

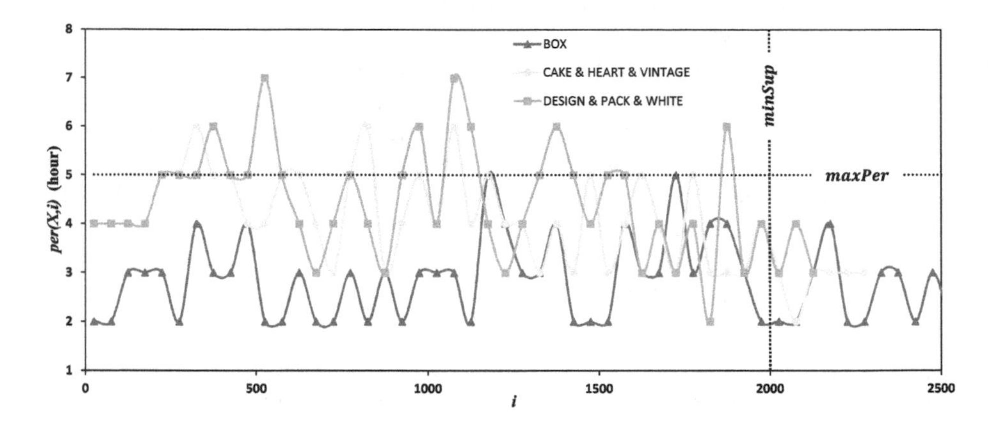

Fig. 7. Periods of some interesting SPPs found in *OnlineRetail*

We also analyzed the SPPs found in the real *OnlineRetail* sale dataset to assess their usefulness. *OnlineRetail* contains transactions of a UK-based online store from 01/12/2010 to 09/12/2011. Data was segmented into hours to obtain 2,975 non empty transactions. For $minSup = 2,000$, $maxPer = 5$ h and $maxLa = 2$ h, 32,284 SPPs are found, and only 197 PFPs for $maxLa = 0$ h. This shows that many stable patterns can be saved using the proposed algorithm. Figure 7 shows some found SPPs, which are {Box}: (2553, 0h), {Cake, Heart, Vintage}: (2284, 1 h) and {Design, Pack, White}: (2131, 2 h), where each SPP X is annotated with $(sup(X), maxla(X))$. The X-axis indicates period numbers of patterns, and the Y-axis indicates the value $per(X, i)$ for the i-th period. Note that to reduce the number of points on that chart, only the maximum value for each group of 50 periods is shown in Fig. 7. It can be observed that frequent patterns exceeding $maxPer$ having a stable periodic behavior are obtained, while such patterns would be ignored by traditional PFP mining algorithms due to the $maxPer$ constraint. The stable patterns found about the sale of products are also deemed interesting as they indicate stable sale trends. Such information could be used to forecast product sales.

6 Conclusion

This paper proposed a novel problem of mining stable periodic-frequent patterns. A new $maxLa$ measure has been designed to assess the stability of an itemset's periodic behavior in a database. A pattern-growth algorithm has also been proposed to find SPPs. An experimental evaluation on both synthetic and real datasets shows that SPP-growth is efficient and can find useful patterns. For future work, we plan to adapt the concepts of stability to mine other types of patterns such as stable periodic sequential patterns.

References

1. Bodon, F., Schmidt-Thieme, L.: The relation of closed itemset mining, complete pruning strategies and item ordering in apriori-based FIM algorithms. In: Jorge, A.M., Torgo, L., Brazdil, P., Camacho, R., Gama, J. (eds.) PKDD 2005. LNCS (LNAI), vol. 3721, pp. 437–444. Springer, Heidelberg (2005). https://doi.org/10.1007/11564126_43
2. Han, J., Pei, J., Yin, Y.: Mining frequent patterns without candidate generation. In: Proceedings of 26th ACM SIGMOD International Conference on Management of Data, pp. 1–12 (2000)
3. Zaki, M.J., Gouda, K.: Fast vertical mining using diffsets. In: Proceedings of 9th ACM SIGKDD International Conference on Knowledge Discovery and Data Mining, pp. 326–335 (2003)
4. Fournier-Viger, P., Lin, J.C.-W., Vo, B., Truong, T.C., Zhang, J., Le, H.B.: A survey of itemset mining. Wiley Interdisc. Rev. Data Min. Knowl. Discov. **7**(4), e1207 (2017)
5. Gouda, K., Zaki, M.J.: Efficiently mining maximal frequent itemsets. In: Proceedings of 17th IEEE International Conference on Data Mining, pp. 163–170 (2001)
6. Pasquier, N., Bastide, Y., Taouil, R., Lakhal, L.: Discovering frequent closed itemsets for association rules. In: Beeri, C., Buneman, P. (eds.) ICDT 1999. LNCS, vol. 1540, pp. 398–416. Springer, Heidelberg (1999). https://doi.org/10.1007/3-540-49257-7_25
7. Fournier-Viger, P., Wu, C.-W., Zida, S., Tseng, V.S.: FHM: faster high-utility itemset mining using estimated utility co-occurrence pruning. In: Andreasen, T., Christiansen, H., Cubero, J.-C., Raś, Z.W. (eds.) ISMIS 2014. LNCS (LNAI), vol. 8502, pp. 83–92. Springer, Cham (2014). https://doi.org/10.1007/978-3-319-08326-1_9
8. Tanbeer, S.K., Ahmed, C.F., Jeong, B.S., Lee, Y.K.: Discovering periodic-frequent patterns in transactional databases. In: 13th Proceedings of Pacific-Asia Conference on Knowledge Discovery and Data Mining, pp. 242–253 (2009)
9. Kiran, R.U., Kitsuregawa, M., Reddy, P.K.: Efficient discovery of periodic-frequent patterns in very large databases. J. Syst. Softw. **112**, 110–121 (2016)
10. Fong, A.C.M., Zhou, B., Hui, S.C., Hong, G.Y., Do, T.: Web content recommender system based on consumer behavior modeling. IEEE Trans. Consum. Electron. **57**(2), 962–969 (2011)
11. Amphawan, K., Lenca, P., Surarerks, A.: Mining top-K periodic-frequent pattern from transactional databases without support threshold. In: Papasratorn, B., Chutimaskul, W., Porkaew, K., Vanijja, V. (eds.) IAIT 2009. CCIS, vol. 55, pp. 18–29. Springer, Heidelberg (2009). https://doi.org/10.1007/978-3-642-10392-6_3
12. Fournier-Viger, P., Lin, J.C.-W., Duong, Q.-H., Dam, T.-L.: PHM: mining periodic high-utility itemsets. In: Perner, P. (ed.) ICDM 2016. LNCS (LNAI), vol. 9728, pp. 64–79. Springer, Cham (2016). https://doi.org/10.1007/978-3-319-41561-1_6
13. Surana, A., Kiran, R.U., Reddy, P.K.: An efficient approach to mine periodic-frequent patterns in transactional databases. In: Cao, L., Huang, J.Z., Bailey, J., Koh, Y.S., Luo, J. (eds.) PAKDD 2011. LNCS (LNAI), vol. 7104, pp. 254–266. Springer, Heidelberg (2012). https://doi.org/10.1007/978-3-642-28320-8_22

14. Kiran, R.U., Venkatesh, J.N., Fournier-Viger, P., Toyoda, M., Reddy, P.K., Kitsuregawa, M.: Discovering periodic patterns in non-uniform temporal databases. In: Kim, J., Shim, K., Cao, L., Lee, J.-G., Lin, X., Moon, Y.-S. (eds.) PAKDD 2017. LNCS (LNAI), vol. 10235, pp. 604–617. Springer, Cham (2017). https://doi.org/10.1007/978-3-319-57529-2_47
15. Fournier-Viger, P., Li, Z., Lin, J.C.-W., Kiran, R.U., Fujita, H.: Discovering periodic patterns common to multiple sequences. In: Ordonez, C., Bellatreche, L. (eds.) DaWaK 2018. LNCS, vol. 11031, pp. 231–246. Springer, Cham (2018). https://doi.org/10.1007/978-3-319-98539-8_18
16. Fournier-Viger, P., Gomariz, A., Gueniche, T., Soltani, A., Wu, C., Tseng, V.S.: SPMF: a Java open-source pattern mining library. J. Mach. Learn. Res. (JMLR) **15**, 3389–3393 (2014)

Graphical Event Model Learning and Verification for Security Assessment

Dimitri Antakly[1,2(✉)], Benoît Delahaye[2], and Philippe Leray[2]

[1] GFI informatique, Paris, France
dimitri.antakly@gfi.fr
[2] Université de Nantes/LS2N UMR CNRS, 6004 Nantes, France
{benoit.delahaye,philippe.leray}@univ-nantes.fr

Abstract. The main objective of our work is to assess the *security* of a given real world system by verifying whether this system satisfies given properties and, if not, how far it is from satisfying them. We are interested in performing formal verification of this system based on event sequences collected from its execution. In this paper, we propose a preliminary model-based approach where a Graphical Event Model (GEM), learned from the event streams, is considered to be representative of the underlying system. This model is then used to check a certain *security property*. If the property is not verified, we also propose a search methodology to find another *close* model that satisfies it. Our approach is generic with respect to the verification procedure and the notion of distance between models. For the sake of completeness, we propose a distance measure between GEMs that allows to give an insight on how far our real system is from verifying the given property. The interest of this approach is illustrated with a toy example.

Keywords: Model-based learning · Formal verification · Graphical Event Models (GEMs) · Event streams

1 Introduction

In order to build a secure access to data in a real world system and to ensure its safeness from any upcoming potential threat one should learn the dependencies and behavior of the different components of the system, identify malicious behaviors and act at the right moment to intercept them. Some of the existing modeling formalisms are better tailored for the verification of given properties or hypothesis, others for learning behaviors and dependencies. Probabilistic finite automaton, for instance, were used in modeling and verification of known or desired behaviors [5]. Petri Nets were used in modeling and verification of several parallel tasks as well as in Process Mining [8]. Probabilistic graphical models were used in Machine Learning for the representation of dependencies between the different variables of a system.

Each of these cited formalisms has its own advantages and disadvantages. Nonetheless, all of the formalisms cited above and the ones that are in the same

F. Wotawa et al. (Eds.): IEA/AIE 2019, LNAI 11606, pp. 245–252, 2019.
https://doi.org/10.1007/978-3-030-22999-3_22

family have a common flaw, the discretisation of time, which can be described as a representational bias in the learning of these formalisms. Thus, from a security point of view, it is better to use continuous time modeling formalisms that allow knowing exactly when to act and not only what action to take; for example when predicting a system failure or forecasting future user tendencies.

To explore the dynamics of a wide variety of systems behavior based on collected event streams, there exist many advanced continuous time modeling formalisms: for instance, continuous time Bayesian networks, Markov jump processes [6], Poisson networks and graphical event models (GEMs) [2]. In this work we are interested in *Recursive Timescale Graphical Models* (or RTGEMs) [2] a sub-family of GEMs, that present advantages compared to the other formalisms.

Appropriate learning and verification techniques should be adapted for the type of formalism that we wish to use. Standard model checking, for example, is used as an verification method [1]. It has been applied to many formalisms, but to the best of our knowledge, never adapted to RTGEMs. Another valid solution for verification are approximation methods, such as Statistical Model Checking (SMC) [4], which is an efficient technique based on simulations and statistical results. SMC has been successfully applied to probabilistic graphical models such as dynamic Bayesian networks (DBNs) in [3]. In the same way, SMC could be easily adapted to RTGEMs.

The main objective of this work is to learn a model (if one exists) that is at the same time representative of the real world system and secure. We are not only interested in evaluating the fitness of the model using standard scoring techniques but also in its suitability from a security point of view. Hence, we propose a strategy where we choose to learn the "optimal" RTGEM (the one that most fits the data). If this model does not satisfy a specific *security* property we seek to find another RTGEM, in its *close* neighborhood, that does. To do so, an appropriate model-based strategy is proposed and a distance measure is introduced in order to compare two RTGEMs. The strategy we propose contains three main steps, the learning of the model, the space exploration and model verification phase, and finally the distance calculation.

This paper is divided into four sections, Sect. 2 consists in definitions and some background context that will be useful further on. Section 3 contains the proposed strategy, that is illustrated by a toy example in Sect. 4. Section 5 is reserved for the conclusion and perspectives.

2 Background

The data we are using consists in timed sequences with strictly increasing timestamps (we use $t_0 = 0$ and $t^* = t_{n+1}$ as conventions). Thus, our data is written x_{t^*} for the sequence of events $(t_1, l_1), ..., (t_n, l_n)$, with $0 < t_i < t_{i+1} < t^*$ for all $1 \leq i \leq n-1$ and where l_i are labels chosen from a finite label vocabulary $\mathcal{L}$. We write $|x_{t^*}|$ for the size of our data x_{t^*} (the number of events in the sequence). The history at time t is the set of all the events that occurred before t, h_i denotes the ith history $h_i = (t_1, l_1), ..., (t_{i-1}, l_{i-1})$.

2.1 Graphical Event Models

A *Graphical Event Model* (GEM) is defined as a directed graph $\mathcal{G} = (\mathcal{L}, E)$ that can represent data of the type x_{t^*} as given above, as well as the dependencies between the different labels (or events) in time. In this work we are only interested in *Markov* GEMs, where the *conditional intensity functions* $\lambda_l(t \mid h)$ satisfy the following property:

$$\lambda_l(t \mid h) = \lambda_l(t \mid [h]_{Pa(l)})$$

where $Pa(l)$ are the parents of l in $\mathcal{G}$. This means that the conditional intensity of a certain label l at time t only depends on the history of the parents of l and not the entire history of the process. We also note that conditional intensity functions are *piecewise-constant*, which means that they take constant values for a certain period of time based on the observed history. More details about GEMs can be found in [2].

2.2 Recursive Timescale Graphical Event Models

Recursive Timescale Graphical Event Models as described in [2] are a class of GEMs where each dependency between two events is defined for a given finite *timescale* which specifies the temporal horizon and the granularity of the dependency represented by that edge. Formally, a timescale is a set T of half-open intervals $(a, b]$ (with $a \geq 0$ and $b > a$) that form a partition of some interval $(0, t_h]$, where t_h is the highest value of T and is called the *horizon* of an edge e. An RTGEM $M = (\mathcal{G}, \mathcal{T})$ consists of a GEM $\mathcal{G} = (\mathcal{L}, E)$ and a set of timescales $\mathcal{T} = T_{e(e \in E)}$ corresponding to the edges E of the graph $\mathcal{G}$. The "recursive" form of this formalism comes from the fact that it is constructed using a forward search algorithm, usually starting from an empty model (only containing nodes that are not connected). The set of elementary operators, allowed in the learning of RTGEMs in a forward search algorithm, is the following $\mathcal{O}_F = \{add, split, extend\}$. The "add" operator adds a non-existing edge to the model and its corresponding timescale $T = (0, c]$, with c a constant. The "split" operator splits one interval $(a, b]$ in the timescale of a chosen edge into two intervals $(a, \frac{a+b}{2}], (\frac{a+b}{2}, b]$. The extend operator extends the horizon of a chosen edge by adding the interval $(t_h, 2t_h]$, with t_h being the previous horizon.

The conditional intensity functions now have parameters (they are also piecewise-constant), i.e. $\lambda_l(t \mid h) = \lambda_{l, c_l(h,t)}$ where the index $c_l(h, t)$ is the *parent count vector* of bounded counts over the intervals in the timescales of the parents of l. For the following example, we consider that all RTGEMs are bounded by 1, thus only the fact that a parent has occurred (or not) within the corresponding timescale is important.

Example 1. Consider the TGEM illustrated in Fig. 1. We have $\mathcal{L} = \{A, B, C, D\}$; for the event B for example, $c_B(h, t) = [0, 1, 1]$ means that there was no A in $[t - 3, t)$, there was an A in $[t - 6, t - 3)$ and there was a D in $[t - 5, t)$. Hence, the conditional intensity functions for the variable B are of the form:

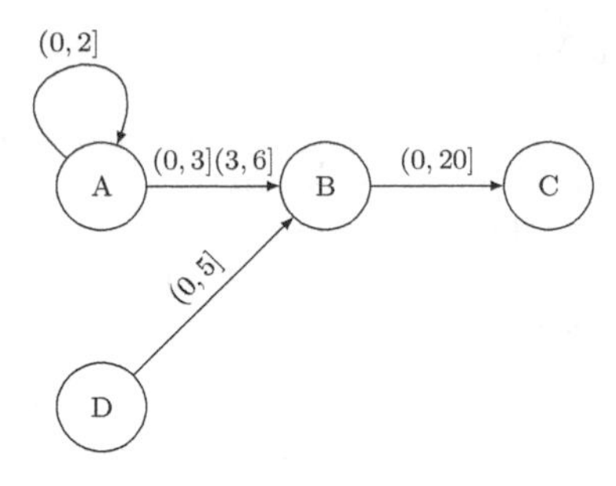

Fig. 1. Example of a four variables RTGEM

$\lambda_{B,000}$, $\lambda_{B,001}$, etc. All conditional intensity functions are equal to constants making them piecewise-constant depending on the corresponding combination of parents.

In the learning algorithm proposed in [2], a backward search follows a forward search. This means that, implicitly, the authors use symmetric operators in their backward search. For the sake of convenience we write: $\mathcal{O}_F^{-1} = \{remove\_edge, fusion, remove\_interval\}$ for these symmetric operators, that do the inverse of the ones cited beforehand. In order to learn the fittest RTGEM, a Greedy Forward-Backward search is applied, based on an adapted Bayesian Information Criterion (BIC) score to select the RTGEM that most represents the data.

3 Proposed Approach

3.1 Problem Statement

In the learning phase of a model, where we want to learn the "fittest" model that best represents reality, we tend to adapt scores and metrics that evaluate the complexity and resemblance of the different learned models compared to the real data ($\mathcal{D}$), in order to chose the optimal one. From a *security* point of view, it is also important to verify if our model (that represents reality to a certain degree) satisfies certain security rules or properties, and if not, how far the current model is from verifying them. The probability that a model M verifies a security property ϕ is written $P(\phi \mid M)$. The problem we want to solve can be written as follows:

$$\exists M^*, M^* = argmaxP(D \mid M) \; with \; P(\phi \mid M) > c \tag{1}$$

with $c \in [0, 1]$ a given constant. The security properties we are looking to verify are *qualitative* and generally address a limited number of events in our graph. We denote $l_\phi \in \mathcal{L}_\phi$ the labels of the events concerned by the security property ϕ. The problem as stated in equation 1 cannot be solved using classical multi-objective optimization heuristics, because of the fact that a qualitative property cannot be optimized, it is either true or false (it cannot become "truer").

3.2 Proposed Strategy

The strategy we propose to solve (1) is described in the following generic algorithm consisting in three main steps. The first step is the learning phase, the second step is the model space exploration phase and model verification, and the last step is the distance calculation between two models.

Algorithm 1. Proposed Strategy

input: $\mathcal{D}, \phi$
output: M^*, Δ
1: $M^o = \underset{M \in RTGEM}{\text{argmax}} \; P(\mathcal{D} \mid M)$
2: $\mathcal{N} = \mathcal{N}_c(\mathcal{N}_{l_\phi}(M^o))$
3: $M^* = \text{find}\{M \in \mathcal{N}, P(\phi \mid M) > c\}$
4: $\Delta = \text{SHD}(M^o, M^*)$

The first line of Algorithm 1 corresponds to the learning phase of the fittest RTGEM M^o (Sect. 2.2). Lines 2 and 3 correspond to the model space exploration phase and model verification, where we try to find a model M^*, in the "close" neighborhood of M^o, that verifies the security property. This step will be explained in details in Sects. 3.3 and 3.4. The last line of the algorithm consists in calculating the distance between the optimal and the selected model (if one exists). The *distance* measure we propose will be defined and explained in Sect. 3.5.

3.3 Model Space Exploration

The security properties we would like to verify address a number of particular variables l_ϕ in our model. The notation $\mathcal{N}_{l_\phi}(M^o)$, on line 2 of Algorithm 1, defines the neighborhood of M^o limited by the labels l_ϕ that are concerned by the security property ϕ. The neighborhood $\mathcal{N}$ that we consider, is the transitive closure ($\mathcal{N}_c$) of the previous neighborhood, limited by the number of allowed operators that is fixed beforehand. We check if the initial model verifies the security property in the first step of our find function before doing any space exploration. The idea of the model space exploration (in line 3) consists in doing a finite number of operations on the concerned labels l_ϕ of the model M^o, while staying in $\mathcal{N}$, in order to find a model that verifies the property. We check after each operation, if the obtained model satisfies the property or not. The search stops immediately when we find a model that verifies the property. The operations that are allowed are the ones in the sets $\mathcal{O}_F$ and $\mathcal{O}_F^{-1}$.

The find function can be defined using any search technique: an exhaustive technique like DFS (Depth First Search) or BFS (Breadth First Search) for example. It can also be random, like the random walk technique or a greedy search with an objective to improve $P(\phi \mid M)$ in order to make it higher than c.

3.4 Model Verification

In practice, we are interested in two main types of queries that can be verified on continuous-time graphical models. The first type of queries targets the order or number of occurrences of given events. The second type of queries addresses time or the timing of given events. By adapting these queries (or a conjunction of them) to the system's security standards we obtain our *security properties* that allow us to classify a model as normal (or dangerous) from a security point of view. Certain types of properties can be formalized using an extended version of LTL (Linear Time Logic) [1], with the addition of past time intervals over the variables. For example we can write $\Box^{100}(C \Rightarrow A_{(0,5]} \wedge B_{(0,10]})$, meaning that all the occurrences of C within the next 100 time units (if it ever occurs) must imply the occurrence of A and B in the past within their respective timescales.

These types of properties could be verified using exact verification methods like standard model checking techniques [1], but standard model checking is subject to state space explosion and to the best of our knowledge was never adapted to Graphical Event Models. In practice we could also use an approximation method, like Statistical Model Checking (SMC) [4] that consists in simulating the model and verifying on each sampled data sequence if the given property is verified.

3.5 Distance Between Models

To the best of our knowledge, there is no existing metric of distance between RTGEMs. In the literature, the popular Hamming distance has been adapted for some probabilistic graphical models such as Bayesian networks [7]. In the following, we propose an extension of the Structural Hamming Distance (SHD), adapted to RTGEMs, where we evaluate the amount of differing information on two different edges. Consider two RTGEMs with the same set of labels $\mathcal{L}$, $G_1 = (\mathcal{L}, E_1)$ and $G_2 = (\mathcal{L}, E_2)$, we define:

$$\text{SHD}(G_1, G_2) = \sum_{e \in E_{sd}} 1 + \sum_{e \in E_{inter}} d(\mathcal{T}(e, G_1), \mathcal{T}(e, G_2)) \tag{2}$$

Where $E_{sd} = \{E_1 \setminus E_2\} \cup \{E_2 \setminus E_1\}$ are the edges of each model that are not present in the other one and $E_{inter} = E_1 \cap E_2$ is the set of edges that are present in both models. $\mathcal{T}(e, G_1)$ and $\mathcal{T}(e, G_2)$ are the lists of endpoints of the intervals on the timescales of the corresponding edge e in graph G_1 and G_2 respectively. A timescale in an RTGEM can be represented by a vector $v = [0, a, b, c, ...]$ where the values are the successive timestamp values. We write v_1 and v_2 for the values of a timescale (on a given edge that is present in both graphs) of G_1 and G_2 respectively. We write $v_{id} = |v_1 \cap v_2|$, for the identical endpoints in the two vectors; and $v_{nid} = |v_1 \setminus v_2| + |v_2 \setminus v_1|$, for the endpoints that are not identical in the two vectors. Thus, we define the elementary distance as follows:

$$d(\mathcal{T}(e, G_1), \mathcal{T}(e, G_2)) = \frac{v_{nid}}{v_{nid} + v_{id}} \tag{3}$$

Equation (2) corresponds to adding 1 to the global distance when the edge (or the dependency between two nodes) exists in a graph but not the other, and adding a value d in $[0, 1)$ corresponding to the difference between the timescales when an edge exists in both graphs.

4 Toy Example

The purpose of the following example is to illustrate the interest of the proposed strategy on a real life application. We consider a prepaid card online service, where the possible actions are *Recharge*, *Check account*, *Transfer money* (to transfer money to his or to another bank account) and *Log out*. We suppose that the optimal RTGEM (M^o) that best fits the real behavior of users is as shown in Fig. 2. A security query ϕ that we can verify on this example can be of the form $\square^{1000}(\text{Transfer Money} \Rightarrow \text{Recharge}_{(0,20]} \vee \text{Check account}_{(0,5]})$, and for instance we want the model to satisfy the property $P(\phi \mid M^o) > 0.8$. In other words we would like to ensure a behavior for users that we consider safe: every time a user wants to transfer money, an action where he checks his account must have occurred right beforehand or a recharging of his account must have occurred not long ago (because he may have made some purchases very recently after the recharge and is aware of his balance). If our system does not verify this property we would say that the average user should be more "careful" while using the service and that the global behavior of the service is not secure.

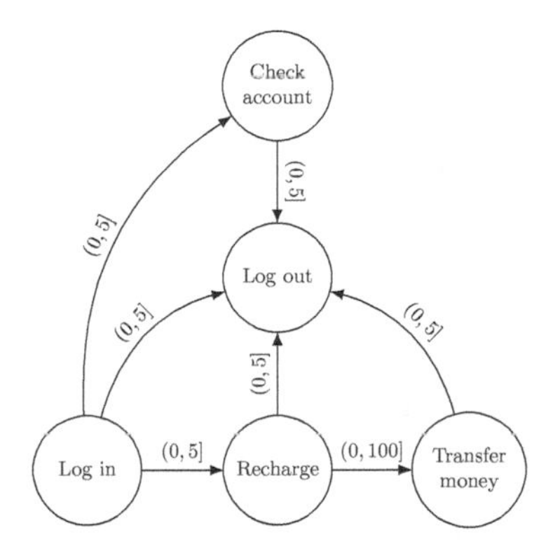

Fig. 2. The learned model

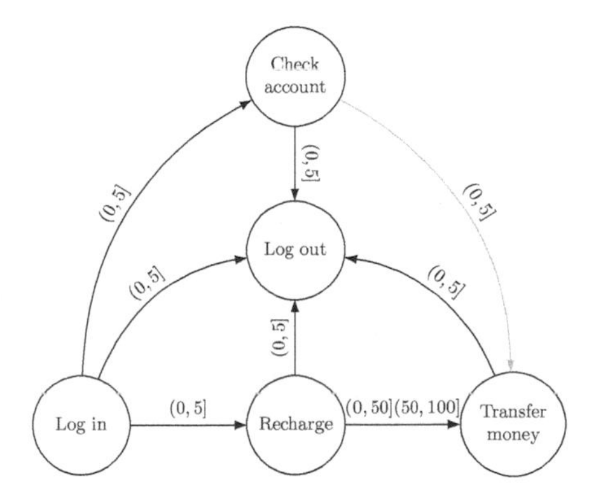

Fig. 3. A modified more secure model (Color figure online)

We notice, only from looking at the model, that $P(\phi \mid M^o)$ is low and that the learned behavior does not verify the security property mainly because of the missing dependency between "*Check account*" and "*Transfer money*". Hence, users are transferring money without checking their accounts first. Furthermore, after recharging and using their card they never directly check their accounts but they sometimes directly transfer money.

By doing a limited number of allowed operations on the labels that are addressed by ϕ (l_ϕ = {Transfer Money, Recharge, Check account}), we can obtain the RTGEM (M^*) of Fig. 3, where the modifications are in red. Clearly, this obtained model is more *secure* considering the security property ϕ, because we now have a smaller interval between "recharge" and "transfer money" that is taken into consideration and we have added the edge between "check account" and "transfer money".

The distance is SHD(M^o, M^*) = 1.333 in this case, because of the added edge and the split on the interval.

5 Conclusion and Perspectives

In conclusion, we have proposed a preliminary model-based strategy in learning and verification for security assessments, as well as a distance measure between graphical models. Our strategy consists in learning the model that best represents the real data, in checking if a *close* model exists that verifies a certain security property and in computing a distance, that we defined, between the two models to see how far the *fittest* model is from verifying the property.

In the future we plan to apply our algorithm and start the experimentation on a real world case study to evaluate its complexity and advantages, especially in systems verification, in order to compare it with other verification approaches. Finally, we are also planning on studying the correlation between the relative distance measure we compute and the change in the satisfaction probability of a certain property.

References

1. Baier, C., Katoen, J.P.: Principles of Model Checking. MIT Press, Cambridge (2008)
2. Gunawardana, A., Meek, C.: Universal models of multivariate temporal point processes. In: Proceedings of the 19th International Conference on Artificial Intelligence and Statistics, pp. 556–563 (2016)
3. Langmead, C.J.: Generalized queries and Bayesian statistical model checking in dynamic Bayesian networks: application to personalized medicine. In: Proceedings of the 8th Annual International Conference on Computational Systems Bioinformatics, pp. 201–212. Life Sciences Society (2009)
4. Legay, A., Delahaye, B., Bensalem, S.: Statistical model checking: an overview. In: Barringer, H., et al. (eds.) RV 2010. LNCS, vol. 6418, pp. 122–135. Springer, Heidelberg (2010). https://doi.org/10.1007/978-3-642-16612-9_11
5. Mao, H., Chen, Y., Jaeger, M., Nielsen, T.D., Larsen, K.G., Nielsen, B.: Learning probabilistic automata for model checking. In: 2011 Eighth International Conference on Quantitative Evaluation of Systems (QEST), pp. 111–120. IEEE (2011)
6. Rao, V., Teh, Y.W.: Fast MCMC sampling for Markov jump processes and extensions. J. Mach. Learn. Res. **14**(1), 3295–3320 (2013)
7. Tsamardinos, I., Brown, L.E., Aliferis, C.F.: The max-min hill-climbing Bayesian network structure learning algorithm. Mach. Learn. **65**(1), 31–78 (2006)
8. Van Der Aalst, W.: Process Mining: Discovery, Conformance and Enhancement of Business Processes. Springer, Heidelberg (2014). https://doi.org/10.1007/978-3-642-19345-3

Predicting User Preference in Pairwise Comparisons Based on Emotions and Gaze

S. Angelastro(✉), B. Nadja De Carolis(✉), and S. Ferilli(✉)

Dipartimento di Informatica, Università di Bari, Bari, Italy
{sergio.angelastro,berardina.decarolis,stefano.ferilli}@uniba.it

Abstract. Emotions have an impact to almost all decisions. They affect our choices and are activated as feedback during the decision process. This work aims at investigating whether behavior patterns can be learned and used to predict the user's choice. Specifically, we focused on pairwise image comparisons in a preference elicitation experiment, and exploited a Process Mining approach to learn preferences. We proposed and evaluated a strategy based on experienced emotions and gaze behaviour, whose results show promising prediction performance.

Keywords: User modeling · Emotion analysis · Gaze behavior · Process mining

1 Introduction

In human decision-making both rational thinking and emotions have an important role. Although decisions are supposed to be a consequence of an intentional and rational behavior, emotions strongly impact choices as they may affect the logical reasoning. Loewenstein and Lerner [11] proposed a decision-making model in which emotions are divided into two types: those anticipating future emotions and those immediately experienced while deliberating and deciding. Therefore, when making a choice, emotions experienced during the decision process can be used as feedback about one's preferences. In the field of Recommender Systems (RSs), preference learning is a necessary step to produce good recommendations. Usually this process is based on explicit feedback, however, recent trends use approaches based on implicit feedback inferred by analyzing user's behavior during the interaction [18]. Nowadays, it has been recognized that successful recommendations need to take into account user perceptions of what is being recommended that in some way elicits interest, perplexity, curiosity, or, more generally, an emotional response in the user. Early studies have shown the potential for improving RSs by analyzing clicking behavior [1] or incorporating eye tracking data [19]. Moreover, behavioral data enables the analysis on user high-level decision-making processes [8]. Also emotions experienced during the interaction with a RSs reflect liking/disliking of content and, therefore, can be used as an

F. Wotawa et al. (Eds.): IEA/AIE 2019, LNAI 11606, pp. 253–261, 2019.
https://doi.org/10.1007/978-3-030-22999-3_23

implicit relevance feedback [15]. Following this approach, our hypothesis is that by detecting behavior patterns typical of the decision-making process it is possible to learn a model to predict future choices.

To address this issue, we investigated how both gaze behavior and emotions, detected from facial expressions, could be used to this aim. Therefore, we performed experiments in which pairwise items' image comparisons was used as a preference elicitation method. Then, to learn the model, a process mining approach has been used and, specifically, the WoMan framework. It includes prediction features that have proven to be effective in several domains, and that may support the aims of this paper. Experiments show that the prediction task, based on both emotions and gaze, yields good results. Specifically, in the 81% of the cases the system correctly classifies the final user's choice before half of the process span, and in just 15% it is unable to provide a suggestion since, in the current phase of the research, the user indecision modeling is missing.

This paper is structured as follows. The next two sections present two frameworks for cognitive emotions recognition and gazes detection, and one for learning and predict preferences. Then, Sect. 4 reports the details of the proposed approach and comments on the experimental outcomes. Finally, in the last section, some conclusions and future work issues are discussed.

2 Emotions Recognition and Gaze Detection

Emotions are part of our everyday living and influence decision-making. The work described in [14] emphasizes that one of the four roles played by emotions in decision-making is to provide information through the display of both positive and negative emotions that arise directly from the options being considered by the decision maker [16]. In particular, according to [3], we are interested in secondary emotions, since decision-making is a cognitive process.

In the context of RSs, the interaction is mainly performed through a display in front of which the user looks at items to make a decision about what to select. Then it is feasible to recognize emotions from facial expressions. To this aim, we developed a Facial Expressions Recognition (FER) system specifically trained on a suitable dataset. The system is called **FEAtuREs** (Facial Expression Analysis for Recognition of Emotions), able to analyze facial expressions both from recorded video and in real time [5]. It follows a commonly used pipeline in FER systems [4]. It performs facial expression recognition on a single image considering as a region of interest the whole face. The set of descriptors used in FEAtuREs is based on the Histogram of Oriented Gradients (HOG) [2]. The classification of the facial expression is done by a multi-class Support Vector Machine (SVM) adopting the "one-against-one" strategy. The final prediction is returned by a voting system among all the classifiers.

The system has been trained and validated on a dataset that integrates three different ones: (i) "EU-Emotion Stimulus Set (EESS)" [12]; (ii) "The Cambridge Mindreading (CAM) Face-Voice Battery" [9]; (iii) "The Cambridge Mindreading Face-Voice Battery for Children (CAM-C)" [10]. In particular, we selected eleven

Preference: a		
timestamp	gaze	emotion
201811062104048	left	boredom
201811062104054	left	interest
201811062104059	left	interest
201811062104065	left	interest
20181106210407	right	boredom
201811062104076	right	boredom
201811062104082	right	boredom

Fig. 1. (a) FEAtuREs interface (left) - (b) User's behavior example: a is chosen (right)

cognitive emotions that were mentioned in literature as relevant to the decision-making process. The output of this selection is a set of 4184 images whose distribution is the following: enthusiasm (498), interest (340), surprise (295), curiosity (453), concentration (495), attention (374), disappointment (370), boredom (270), perplexity (369), discomfort (461), frustration (259). The average accuracy is of 92% calculated using k-fold evaluation with k = 10. The interface for cognitive emotions recognition is illustrated in Fig. 1(a). Besides emotion recognition, FEAtuREs performs also gaze detection and tracking to detect, in the pairwise comparison, the side of the screen (left or right) the user is looking at. It has been implemented using Dlib and OpenCV functions[1]. Then, a specific instance of a decision-making process is bounded between a starting point, when the system shows for the first time the item's pairs, and an ending point, when a choice is made. While the user is engaged in this process, the flow of emotions and gazes is gathered and collected by FEAtuREs. The user preference choice process is labeled as a or b depending on the selected item. In Fig. 1(b) is shown an example of the user behavior, gazes and emotions with the timestamp in which they occur, collected before the item a is chosen.

3 Process Mining: The WoMan Framework

While Process Mining and Management (PMM) techniques [17] have been typically motivated by, and exploited in, business and industrial domains, they have been recently used for other application fields, as well. In this paper we aim at checking whether they can be effective for user preference prediction, where the user behavior is seen as a process of emotions and gazes. So, a quick recall of PMM basics may be helpful here. A *process* consists of actions performed by agents, formally specified by a *workflow*, which defines their allowed compositions[2]. A process execution is described in terms of *events* associated to the performed activities. A *case* is a particular execution of activities compliant to a given workflow. Case *traces* consist of lists of events associated to time points. A *task* is a generic piece of work, and an *activity* is its actual execution.

[1] Head pose estimation using OpenCV and Dlib: https://www.learnopencv.com/.
[2] Sequential, parallel, conditional, or iterative composition.

Specifically, we adopted WoMan (Workflow Management), an incremental, declarative [13], and logic-based PMM framework [6]. In addition to more typical PMM tasks, such as process mining and supervision, it proved able to support the prediction task[3] in many application domains (see [7] for more technical details). WoMan takes as input trace elements consisting of 6-tuples $\langle T, E, W, P, A, O \rangle$, where T is the event timestamp, E is the type of the event (one of 'begin' or 'end' of 'activity' or 'process'), W is the name of the reference workflow, P is the case identifier, A is the activity name, and O is the progressive occurrence number of A. WoMan models describe the structure of workflows using two elements: **task** (a kind of activities that is allowed in the process) and **transition** (the allowed connections between set of tasks). WoMan consists of several modules. The learning module, **WIND** (Workflow INDucer), learns or refines a process model according to a case. The supervision module, **WEST** (Workflow Enactment Supervisor and Trainer), takes the case events as long as they are available, and returns information about their compliance with the currently available model for the process they refer to. While in supervision mode, WoMan can make several kinds of predictions. Specifically, when the enacted process is unknown, **WOGUE** (Workflow Guesser) returns a ranking (by confidence) of a set of candidate process models.

4 Process Mining for Preference Learning and Prediction

Approach. We used WoMan to face two problems: *preferences learning*, by modeling user's behavioral aspects on a pairwise comparison, and *preference prediction*, guessing future user's choices during the decision-making process.

User's behavior can be defined as a flow of emotions and gazes, framed in the context of a decision-making process. This flow can be seen as a process consisting of activities, i.e., emotions and/or gazes, performed (implicitly) by agents, i.e., users[4], and a process concerning the preference's choice.

We devised and tested several strategies to deal with these kinds of information. Due to space constraints, we report and evaluate only one, namely a strategy based on a combination of gazes, proved to be a powerful source of information in RSs area [19], and emotions. For convenience, we will refer with D_g = {*left,right*} to the domain of gazes, and with D_e = {*annoyance, attention, boredom, concentration, curiosity, disappointment, enthusiasm, frustration, interest, perplexity, surprise*} to the domain of emotions. The idea is to consider gazes and emotions together, in order to give them equal importance, by transforming D_a and D_g into a unique domain $D_{a,g}$ defined as their Cartesian product, and containing all the possible syntactic concatenations. The number of possible tasks will be higher than the ones which consider domains individually, carrying more informative power at the expense of the process complexity.

[3] Given a workflow and an intermediate status of a process execution, the goal is predicting how the execution might proceed, or what kind of process is being enacted, among a set of candidates.

[4] Users will be ignored since the interest is not on user's profiling.

Example 1. Consider the user's behavior in Fig. 1(b). The event trace of the case, named c_1, of the process a (user preference) obtained by applying our strategy is:

$$\langle 201811062104048, begin\_of\_process, a, c_1, start, 1\rangle$$
$$\langle 201811062104048, begin\_of\_activity, a, c_1, boredom\_left, 1\rangle$$
$$\langle 201811062104054, end\_of\_activity, a, c_1, boredom\_left, 1\rangle$$
$$\langle 201811062104054, begin\_of\_activity, a, c_1, interest\_left, 1\rangle$$
$$\langle 201811062104059, end\_of\_activity, a, c_1, interest\_left, 1\rangle$$
$$\ldots$$
$$\langle 201811062104082, begin\_of\_activity, a, c_1, boredom\_right, 3\rangle$$
$$\langle 201811062104082, end\_of\_activity, a, c_1, boredom\_right, 3\rangle$$
$$\langle 201811062104082, end\_of\_process, a, c_1, stop, 1\rangle$$

Note that the name of the activity carries both gaze and emotion information. Whenever a new activity X is detected a *begin_of_activity* event is stored, just after being terminated the previous activity, if any, by recording an *end_of_activity* event. Several occurrences of the same activity are progressive numbered.

In order to learn preferences from user's behavior, the module **WIND** is applied, discovering one model for each preference, containing tasks and transitions. Since there is no concurrency relation between activities, a transition happen whenever a new activity in the event trace is detected, and consists in stepping from the current to the next one.

The experiments concern the evaluation of the proposed strategy by assessing whether performance on the task of preference prediction in the decision-making process at least overcomes a total random classifier.

Datasets and Models Description. Datasets have been created by collecting the user's behaviors, while they made a choice on each image pairs sourced from:

- **IAPS**[5] (International Affective Picture System) domain, an images database designed to provide a standard for studies in emotions domain; 34 users (15 females and 19 males) for 18 pairs of images, paired by opposite valence (positive or negative), and randomly placed to the left or to the right. They produced 280 cases of choice, 145 for a and 135 for item b;
- **ComPro** (COMmercial PROducts) domain, images selected in the domain of commercial products taken from several areas, e.g., clothing, food and art; 42 users (18 females and 24 males) for 28 pairs of images, paired by category and not fully conflicting. They produced 578 cases, where 275 concern the item a and 303 the item b.

We applied the proposed strategy to transform user's cases into event logs, and discover the process models they refer to. Table 1 reports some statistics on each experimental event log and model: number of cases (#cases) and events (#events) (also on average per case), tasks (#task) and transitions (#trans). ComPro involves more cases than IAPS, allowing to discover many more different

[5] http://csea.phhp.ufl.edu/Media.html.

transitions, while #task is almost the same since few cases are sufficient to mine the most of them. The overall #events in ComPro is higher than the one in IAPS, as it involves more cases, while the average length is almost the same. The number of discovered tasks and transitions between the 2 processes, for both datasets, is quite comparable.

Table 1. Datasets and models statistics

Domain	Process *a*					Process *b*				
	#cases	#events		#task	#trans	#cases	#events		#task	#trans
		Overall	Avg	Overall	Overall		Overall	Avg	Overall	Overall
IAPS	145	2842	19.6	15	109	135	2872	21.27	17	141
ComPro	275	5920	21.53	22	184	303	6434	21.23	22	185

Table 2. Preference prediction statistics

Dataset	Performance						
	Acc		P	R	F-score	Indecision	Avg sequence
	I	D					
IAPS	**81%**	**95%**	**95%**	**96%**	**95%**	**15%**	19,56
ComPro	62%	80%	79%	81%	80%	22%	21,38

Performance Evaluation. The experimental procedure was as follows. A 10-fold cross validation procedure was run on each event log (translated dataset). Process models was learned from each training set by applying the WoMan's module WIND [6]. Finally they was used as a reference to call WOGUE on each event in the test sets to predict which kind of process was in execution. As long as case events were tested, WOGUE returned a ranking of the set of candidate processes, using the one with highest Mean Reciprocal Rank as case label.

Table 2 reports datasets on the row headings and corresponding average performance on the columns. Column *Acc* (for *Accuracy*) reports the ratio of cases that WOGUE has correctly classified, distinguishing between the one under indecision, *Acc-I*, i.e., even when it is unable to assign a label, and the one when a decision was made, *Acc-D*. The column *Indecision* reports the ratio of cases in which WOGUE didn't assigned a label. Columns *P* (for *Precision*), *R* (for *Recall*), and *F-score* report classical predictive measures (only on labeled cases), and finally, column *Avg Sequence* reports the average length, in terms of number of events, of the tested cases. The winning strategies are reported in bold.

Our strategy, given its powerful representation formalism, outperforms, in all datasets, a total random classifier, being *Acc* greater than 50%. Moreover, predictive measures, summarized by the *F-score*, yielded a 95% for IAPS, and a 80% for ComPro, which means that it can fairly predict preferences.

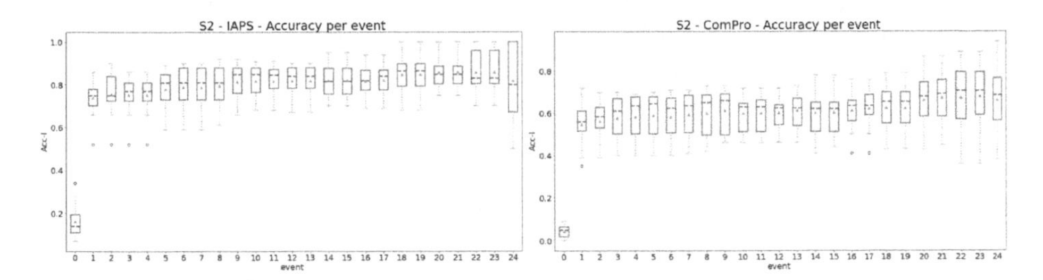

Fig. 2. IAPS - (left) and **ComPro** - (right) for predictions.

Specifically, the prediction task in IAPS is 95% accurate in assigning labels, and it is still high (81%) even under indecision (15% of *Indecision*). It is 95% precise (P) in making predictions, correctly covering the 96% (R) of the cases. As regards the ComPro dataset, performance are not as good as on IAPS, nevertheless it yielded about 80% of *Acc-D*, P, and R, when it decided a label, while *Acc-I* is 62% (being *Indecision* value equal to 22%). This may be due to both challenging elicitation skill of ComPro and lack of a user's indecision model.

Graphs in Fig. 2 show the 10-folds *Acc-I* trend, in form of box plots' sequence, per *event*. Since the *Avg Sequence* in Table 2 is between 19–21, and since the interest is on guessing as early as possible the user's choice, only the first 25 events are shown. The *Acc-I* value converges rapidly after few events to the one in Table 2, for both datasets, therefore it is quite reliable at the beginning. Specifically, the accuracy on IAPS is about 80% after 6 events (a third of *Avg Sequence*), and the accuracy on ComPro is about 60%–70% within 10 events (a half of *Avg Sequence*), which are noteworthy results. Low accuracy values are outliers and occur especially at the beginning, while good ones are in boxes[6], in general distributed around the mean. Since boxes in graphs are skewed to the maximum up to a half of the sequence, it is very likely to have a correct prediction in this time span. These results confirmed that both emotions and gazes are important in modeling the user's behavior for preference prediction. Considering trends in Fig. 2 and the average sequence length in Table 2, the system can provide reliable predictions within a half of the sequence. However a user's indecision behavior model would help to increase the overall accuracy.

5 Conclusions and Future Directions

Emotions are fundamental to almost all decisions. This work investigated the use of the WoMan Process Mining framework for preference learning and prediction. We focused on pairwise image comparisons in a preference elicitation experiment, assessing the effectiveness of a strategy which takes into account gaze behaviour and emotions. Results confirm that it significantly outperforms prediction of a total random classifier. In future work, we plan to exploit other

[6] The range between the first and third quartile.

strategies and to perform new experiments with more users. We also plan to ask a marketing expert to suggest how to select suitable items to be shown in the pairwise comparison. Then we will integrate our approach in a game in which the system will guess as early as possible the user's choice.

References

1. Agichtein, E., Brill, E., Dumais, S.: Improving web search ranking by incorporating user behavior information. In: Proceedings of the 29th Annual International ACM SIGIR Conference on Research and Development in Information Retrieval, pp. 19–26. ACM (2006)
2. Dalal, N., Triggs, B.: Histograms of oriented gradients for human detection. In: 2005 IEEE Computer Society Conference on Computer Vision and Pattern Recognition. CVPR 2005, vol. 1, pp. 886–893. IEEE (2005)
3. Damasio, A., et al.: Descartes' Error: Emotion, Reason, and the Human Brain. GP Putnam's Sons, New York (1994)
4. Del Coco, M., Carcagnì, P., Palestra, G., Leo, M., Distante, C.: Analysis of HOG suitability for facial traits description in FER problems. In: Murino, V., Puppo, E. (eds.) ICIAP 2015. LNCS, vol. 9280, pp. 460–471. Springer, Cham (2015). https://doi.org/10.1007/978-3-319-23234-8_43
5. D'Errico, F., Paciello, M., De Carolis, B., Vattanid, A., Palestra, G., Anzivino, G.: Cognitive emotions in e-learning processes and their potential relationship with students' academic adjustment. Int. J. Emot. Educ. **10**(1), 89–111 (2018)
6. Ferilli, S.: WoMan: logic-based workflow learning and management. IEEE Trans. Syst. Man Cybern.: Syst. **44**, 744–756 (2014)
7. Ferilli, S., Esposito, F., Redavid, D., Angelastro, S.: Predicting process behavior in WoMan. In: Adorni, G., Cagnoni, S., Gori, M., Maratea, M. (eds.) AI*IA 2016. LNCS (LNAI), vol. 10037, pp. 308–320. Springer, Cham (2016). https://doi.org/10.1007/978-3-319-49130-1_23
8. Glaholt, M.G., Reingold, E.M.: Eye movement monitoring as a process tracing methodology in decision making research. J. Neurosci. Psychol. Econ. **4**(2), 125 (2011)
9. Golan, O., Baron-Cohen, S., Hill, J.: The Cambridge Mindreading (CAM) face-voice battery: testing complex emotion recognition in adults with and without asperger syndrome. J. Autism Dev. Disord. **36**(2), 169–183 (2006)
10. Golan, O., Sinai-Gavrilov, Y., Baron-Cohen, S.: The Cambridge Mindreading Face-voice Battery for Children (CAM-C): complex emotion recognition in children with and without autism spectrum conditions. Mol. Autism **6**(1), 22 (2015)
11. Loewenstein, G., Lerner, J.S.: The role of affect in decision making. Handb. Affect. Sci. **619**(642), 3 (2003)
12. O'Reilly, H., et al.: The EU-emotion stimulus set: a validation study. Behav. Res. Methods **48**(2), 567–576 (2016)
13. Pesic, M., van der Aalst, W.M.P.: A declarative approach for flexible business processes management. In: Eder, J., Dustdar, S. (eds.) BPM 2006. LNCS, vol. 4103, pp. 169–180. Springer, Heidelberg (2006). https://doi.org/10.1007/11837862_18
14. Pfister, H., Böhm, G.: The multiplicity of emotions: a framework of emotional functions in decision making. Judgm. Decis. Mak. **3**(1), 5 (2008)
15. Tkalčič, M., De Carolis, B., de Gemmis, M., Odić, A., Košir, A. (eds.): Emotions and Personality in Personalized Services: Models, Evaluation and Applications. HIS. Springer, Cham (2016). https://doi.org/10.1007/978-3-319-31413-6

16. Tkalčič, M., Maleki, N., Pesek, M., Elahi, M., Ricci, F., Marolt, M.: A research tool for user preferences elicitation with facial expressions. In: Proceedings of the Eleventh ACM Conference on Recommender Systems, pp. 353–354. ACM (2017)
17. van der Aalst, W., et al.: Process mining manifesto. In: Daniel, F., Barkaoui, K., Dustdar, S. (eds.) BPM 2011. LNBIP, vol. 99, pp. 169–194. Springer, Heidelberg (2012). https://doi.org/10.1007/978-3-642-28108-2_19
18. White, R.W., Jose, J.M., Ruthven, I.: Comparing explicit and implicit feedback techniques for web retrieval: TREC-10 interactive track report. In: Proceedings of the Tenth Text Retrieval Conference (TREC-10), pp. 534–538 (2002)
19. Xu, S., Jiang, H., Lau, F.: Personalized online document, image and video recommendation via commodity eye-tracking. In: Proceedings of the 2008 ACM Conference on Recommender Systems, pp. 83–90. ACM (2008)

Decision Support Systems and Recommender Systems

A Classification Method of Photos in a Tourism Website by Color Analysis

Jun Sasaki[1(✉)], Shuang Li[1], and Enrique Herrera-Viedma[2]

[1] Iwate Prefectural University, Takizawa, Iwate 152-52, Japan
jsasaki@iwate-pu.ac.jp, g236q005@s.iwate-pu.ac.jp
[2] University of Granada, 18071 Granada, Spain
viedma@decsai.ugr.es

Abstract. The number of Foreign Independent Tour (FIT) is increasing in the world. This research aims to develop a personal adaptive tourism recommendation system (PATRS) for FIT. This paper describes the concept of PATRS and related researches. In order to develop the PATRS, an easy feature extraction method from a tourism website is required. The classification of photos of tourism spots is an important technology to realize the feature extraction from numerous information in the website. This paper proposes a classification method of photos in a major tourism website by color analysis. From the results on the experiments, we confirmed that the photos in a tourism website can be classified into four classes by the proposed method.

Keywords: Classification method · Tourism recommendation system · Color analysis · Feature extraction

1 Introduction

Recently, the number of Foreign Independent Tour (FIT) is beyond of the number of package type tourists in Japan [1]. The travelers of FIT are finding their point-of-interest (POI) by using tourism websites in their visiting areas. On the other hand, it is difficult to find personally adaptive POI because there are too many tourism information in major tourism areas. In order to easily find a personally adaptive POI, an easy feature extraction of POI from too many information on a tourism website is required. The classification of photos of tourism spots is an important technology to realize the feature extraction. This paper proposes a classification method of photos in a major tourism website by color analysis.

This paper introduces the concept of PATRS, issues on the development and related works in Sects. 2, 3 and 4. In order to develop the PATRS, it is necessary to classify the POI to find a personally adaptive POI easily. In Sect. 5, we describe the purpose of the classification of photos in a tourism website. Sect. 6 proposes the classification method and shows experimental results. Finally, in Sect. 7, the paper describes the conclusion and the future issues in this research.

F. Wotawa et al. (Eds.): IEA/AIE 2019, LNAI 11606, pp. 265–278, 2019.
https://doi.org/10.1007/978-3-030-22999-3_24

2 Issues in FIT

TripAdvisor [2] is the most popular website (for getting tourism information) for Japanese travellers who go abroad. This site has many information on hotels, tourism spots, restaurants and flight schedule, etc. in Japanese. In the case of major tourism spots, it is difficult to find personally adaptive POI because there are too many tourism information. For example, St. Petersburg, which is popular tourism area in Russia, has 1173 tourism spots, 475 museums, 447 tours, 183 concerts and shows, 405 entertainment facilities and 266 shops etc. in the website of the TripAdvisor. The website recommends some tourism spots and tours, but it is not for one user but for every user. It is difficult to confirm the recommendation is satisfied or not for the user.

The aim of this research is to develop a new tourism recommendation system in order to solve above-mentioned problems. In the first step, we describes the required technologies to develop the system and the related works on the technologies. The classification and the feature extraction are also the required technologies for solving the problem. Then, as one of the classification of POI, the paper proposes a classification method of photos from a tourism website by color analysis. In the following sections, this paper describes our research target and study area, related works, our proposal and experiments.

3 Research Target and Study Area

The Fig. 1 shows the concept of the personal adaptive tourism recommendation system (PATRS). Travelers input the travel area, their purpose and experience to the system, the Input Processing Module creates some keywords and conditions from the input data and retrieve the POI database (DB). The POI DB stores the characteristics data, which the POI Characteristics Analyzing Module extracts by analyses of comment and photo data in tourism websites and Social Network Services (SNSs) for each POI in the user traveling area. The Characteristics Analyzing Module sends some candidates of POI to the Adaptation Calculation Module, where the priority level of recommendation on each candidate of POI is calculated. The Visualization Module and shows the results for the users with a good human interface such as a Map style.

As the first step, the research aims to design of the POI Characteristics Analyzing Module. It should have an easy finding function of personally adaptive POI for a FIT by using the comment and photo data in a tourism websites. In the design of the POI Characteristics Analyzing Module, a classification and feature extraction function of POI are important technologies. We discuss on the classification method of photos in a tourism website in the paper.

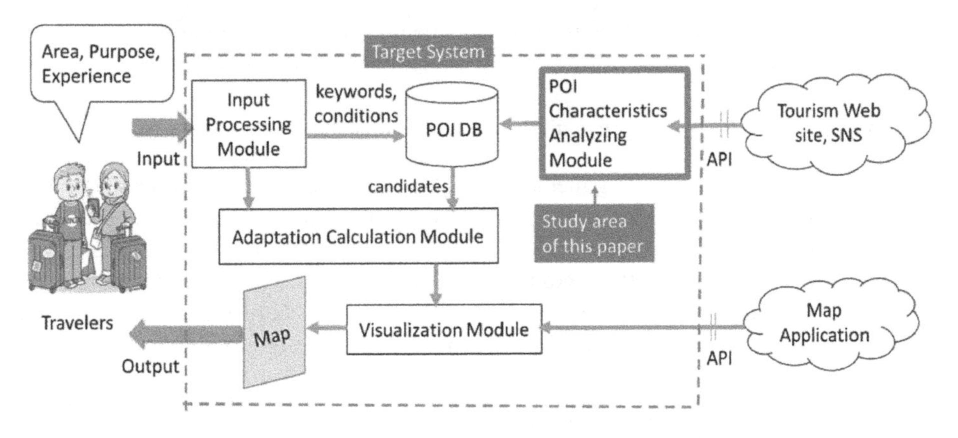

Fig. 1. The concept of PATRS and study area of this paper.

4 Related Works

In order to develop the PATRS, many theories and technologies are required. We propose a development flow of PATRS shown in Fig. 2. Based on the flow, the authors have been researching some related issues.

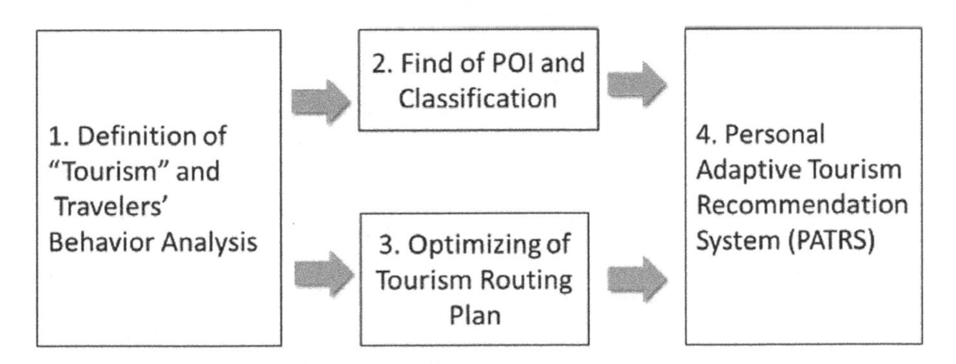

Fig. 2. Development flow of PATRS.

4.1 Definition of "Tourism" and Travelers' Behavior Analysis

The reference [3] analyzed the behavior of inbound tourists (who were coming to Japan from other country) in Tokyo metropolitan area from the data of IC transit cards. The reference [4] proposes a note on modeling for changes of tourist purposes using fuzzy systems. The authors proposed a finding method of target users who were interested in regional areas using online advertising and Social Network Services (SNSs) [5]. Kasahara et al. constructed a tourist behavior model by using a tourists tracking application in a regional environment [6]. Li et al. defined the concept of "smart tourism", which is a travel by using information systems and information environment in China [7]. Recently, semantic analysis as well as data analysis has been trying. For example, there is a sentiment analysis on Twitter for tourism based on applications in reference [8].

4.2 Find of POI and Classification

There are many works on finding method of POI and the classification. Some researcher have been used mainly, Twitter or Flickr data to find a POI and classification as shown in [9, 10] etc. The authors also have been used Flicker to find POIs of inbound tourists in rural area in Japan [11]. The references [12] and [13] used the "similarity" among tourists and tourism spots to find and recommend POIs. Further, there are some works on extraction and classification method of the name or official account of POIs [14, 15] etc.

4.3 Optimizing Tourism Routing Plan

Though some papers such as [16] discussed on the optimal routing for sightseeing and its solution theoretically from old time, there is no practical method yet. As, actual tourism plans are usually complex because of depending on travellers' condition such as the time schedule, its budget, weather and traffic of the transfer, it is a difficult problem.

4.4 Personal Adaptive Tourism Recommendation System (PATRS)

There are many proposals on the recommendation of tourism spots. For example, there are reports on the method such as using "blog", tourism website, Wikipedia, Google Maps, web photos Collaborative Filtering, contents analysis and ontology etc. [17–21]. Recently, recommendation systems and decision making support systems for a group that with considering privacy and finding method of POI and using location based service are attractive research issues [22, 23]. Further, Kinoshita et al. proposed a personal navigation system to assist sightseeing across multiple days [24]. Kinoshia and Yokokishizawa proposed a tour route planning support system using Kansei evaluation and multi-objective gas [25]. Further, Kurata developed a dialogue based travel-plan creation support tool for walking plan around a city [26]. The authors have been developing tourism recommendation systems using data from SNS [27]. We also proposed a personal tour planning system for use in urban and rural areas [28].

Though there are many tourism recommendation systems proposed, nobody developed a system for recommending an appropriate sightseeing plan for individual travelers to foreign countries. This research aims to develop the PATRS that is useful especially when a person makes a personal foreign trip. The classification of POI is one of the required technologies to develop the PATRS. As a first step of this thesis, we discuss on a classification method of photos in a tourism website.

5 Purpose of Classification of Photos in a Tourism Website

Generally, travelers obtain the information of POIs by tourism magazines and websites. The information from such media of POIs is limited and it cannot be completely correct and unfair, because the media biased the information by writers and editors in the media companies. Therefore, we propose use the comments and photos collected from tourism websites or SNSs and try to extract the features of POIs. We thought that the comments and photos of general tourists could be correct and fair written by the authors who actually went to POIs. Regarding to the analysis of comments, we reported other papers. In this paper, we describe a classification method of photos in a tourism website. In the future study, we will combine the analysis of comments and the classification method of photos to develop the personally adaptive recommendation system.

Generally, tourism websites and SNSs consist of photos taken by travellers and comments (text) posted by them. We analyse these data to extract features of POI and classify them. Objective information can be obtained from the photo data and subjective information can be obtained from comment data. This paper describes the photo data analysis as a first step. It is possible to estimate an object from photo data by using a recent AI technology. However, in order to improve estimation accuracy using AI, a large amount of teacher data is required. This paper challenges to clarify the possibility of obtaining objective information from colour analysis of photo data without using AI.

6 Classification Method and Experiments

6.1 Classification Method

The purpose of this paper is to investigate the possibility of the classification of photos in a tourism website. Application of image recognition by Artificial Intelligent (AI) is also conceivable, but enormous teacher data is required. Therefore, we thought that some kind of classification not by AI but by color analysis of the photograph, and conducted an experiment showing the possibility.

We selected the TripAdvisor Japan as the most popular tourism website for Japanese tourists. Three tourism areas are selected, the first is St. Petersburg (area “S”) in Russia as a major sightseeing area, the second is Graz (area “G”) in Austria as a middle class famous sightseeing area and the third is Lappeenranta (area “L”) in Finland as a not so famous area. We selected 15 photos from the first to 15th recommended POI in each tourism area, and the total number of photos is 45. We will call the first recommended POI in the area “S” as “S1” and the second POI as “S2”, so we have 45 POIs of S1, S2, …, S15, G1, G2, …, G15, L1, L2, …, and L15 (Table 1).

Table 1. Experimental conditions.

No.	Item	Content	Reference
1	Tourism website	TripAdvisor Japan	https://www.tripadvisor.jp/
2	Tourism area	St. Petersburg (Russia): area “S”	A famous sightseeing area
		Graz (Austria):area “G”	A middle class famous sightseeing area
		Lappeenranta (Finland): area “L”	Not so famous sightseeing area

(continued)

Table 1. (*continued*)

No.	Item	Content	Reference
3	Photos of POIs	Top 15 POIs in each tourism area recommended by the tourism website. Total number of phots is 45.	Date: December 31 in 2018
4	Color analysis tool	Color analysis tool "Irotoridori" provided by Ironodata. Inc.	https://ironodata.info/extraction/irotoridori.php

To analyze the photos, we used a color analysis tool which is a free website provided by a Japanese company named "Irotoridori" [29]. When we input a photo data to this color analysis tool, it presents the analysis results of color components. It shows the color component of the photograph and its proportion. Figure 3 shows an example of the result on POI "S1". It shows the photo includes 12 color components of color code from #2040C0 to #606060, and the ratio of the color components is from 95% to 13%, respectively.

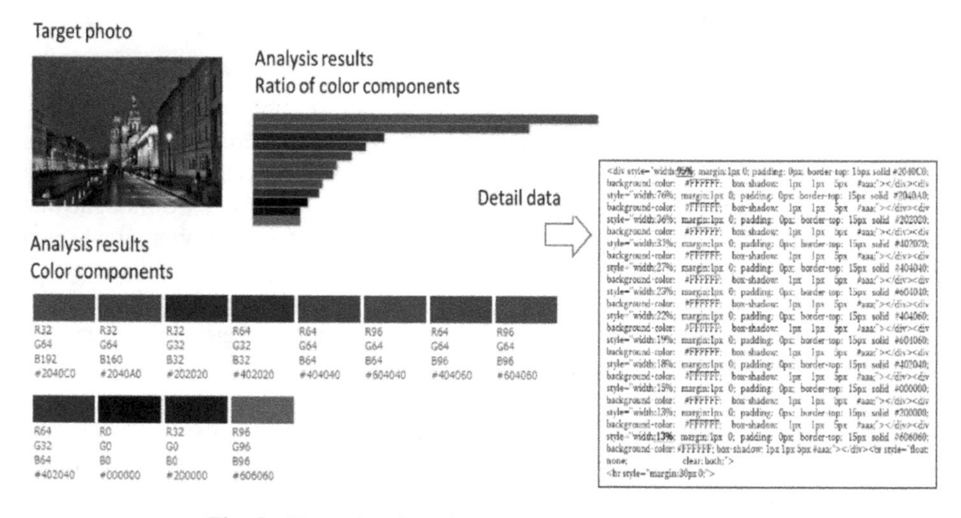

Fig. 3. Example of a color analysis result on POI "S1".

In the color analysis, the three of hue, brightness and contrast are important items. We decided to use hue to discriminate whether the composition of the photo is artificial or natural. We changed the combination of *R (Red), G (Green)* and *B (Blue)* as shown in Table 2 and judged whether the color produced was artificial or natural. By a try and error works, we could find the following formulas on hue, *Hu* to distinguish artificial or natural,

$$Hu = 3R - (G + 2B) \geq 0, \text{ the color component is artificial,} \quad (1)$$

$$Hu = 3R - (G + 2B) < 0, \text{ the color component is natural.} \quad (2)$$

We calculated the brightness, Br, and the contrast, Co as follows.

$$Br = MAX(R, G, B). \tag{3}$$

Where, $MAX(R, G, B)$ indicates the maximum value of *R, G, B* of the average of the photo, and,

$$Co = \frac{MAX(R, G, B) - MIN(R, G, B)}{MAX(R, G, B)} \tag{4}$$

Where, $MIN(R, G, B)$ indicates the minimum value of *R, G, B* of the average of the photo.

Table 2. Judgment of natural or artificial color.

No.	R	G	B	Color	3*R-(G+2*B)	Category	No.	R	G	B	Color	3*R-(G+2*B)	Category
1	0	0	1	Black	-2	Natural	33	128	0	0	Dark red	384	Artificial
2	0	0	64	Dark blue	-128		34	128	0	64	Dark red	256	
3	0	0	128	Navy blue	-256		35	128	0	128	Purple	128	
4	0	0	249	Blue	-498		36	128	0	249	Blue	-114	Natural
5	0	64	0	Dark green	-64		37	128	64	0	Brown	320	Artificial
6	0	64	64	Black	-192		38	128	64	64	Engine color	192	
7	0	64	128	Deep blue	-320		39	128	64	128	Purple	64	
8	0	64	249	Light blue	-562		40	128	64	249	Blue	-178	Natural
9	0	128	0	Green	-128		41	128	128	0	Military color	256	Artificial
10	0	128	64	Dark green	-256		42	128	128	64	Military color	128	
11	0	128	128	Blue green	-384		43	128	128	129	Gray	-2	Natural
12	0	128	249	Bright light blue	-626		44	128	128	249	Blue	-242	
13	0	249	0	Yellow green	-249		45	128	249	0	Yellow green	135	Artificial
14	0	249	64	Light green	-377		46	128	249	64	Yellow green	7	
15	0	249	128	Negai color	-505		47	128	249	128	Negai color	-121	Natural
16	0	249	249	Bright light blue	-747		48	128	249	249	Bright light blue	-363	
17	64	0	0	Dark red	192	Artificial	49	249	0	0	Red	747	Artificial
18	64	0	64	Dark purple	64		50	249	0	64	Vermilion	619	
19	64	1	128	Purple	-65	Natural	51	249	0	128	Rose color	491	
20	64	0	249	Light blue	-306		52	249	0	249	Wisteria	249	
21	64	64	0	Military color	128	Artificial	53	249	64	0	Vermilion	683	
22	64	64	65	Dark gray	-2	Natural	54	249	64	64	Vermilion	555	
23	64	64	128	Deep purple	-128		55	249	64	128	Rose color	427	
24	64	64	249	Blue	-370		56	249	64	249	Wisteria	185	
25	64	128	1	Green	62	Artificial	57	249	128	0	Orange	619	
26	64	128	64	Deep green	-64	Natural	58	249	128	64	Orange	491	
27	64	128	128	Gray	-192		59	249	128	128	Pink	363	
28	64	128	249	Blue	-434		60	249	128	249	Wisteria	121	
29	64	249	0	Yellow green	-57		61	249	249	0	Yellow	498	
30	64	249	64	Yellow green	-185		62	249	249	64	Yellow	370	
31	64	249	128	Negai color	-313		63	249	249	128	Light Yellow	242	
32	64	249	249	Light blue	-555		64	249	249	250	White	-2	Natural

Table 3 shows an example of the calculation results on *Hu*, *Br* and *Co* for the photo of POI "S1". In this table, the 50% of "Judg." means 50% of this photo is artificial. We call the value as a ratio of artificial, *Ra*. Then we calculated all of the *Ra* for every photo data and compared with the visual value as shown in Fig. 4. The visual value means the average of artificial ratio on each photo evaluated visually by the two authors of the paper. From the comparison shown in Fig. 4, we could confirm the calculation results expressed the tendency of the change artificial ratio. As the *Ra* decreases, the error is increasing. We found the $Ra = 100\%$ expressed the photo showed an inside of building or almost near a building.

Table 3. Example of the calculation results on hue, brightness and contrast for POI "S1".

%	Color Code	R	G	B	3*R-(G+ 2 *B)	Judg.	Brightness	Contrast
95	#2040C0	32	64	192	-352	0		
76	#2040A0	32	64	160	-288	0		
36	#202020	32	32	32	0	1		
31	#402020	64	32	32	96	1		
27	#404040	64	64	64	0	1		
23	#604040	96	64	64	96	1		
22	#404060	64	64	96	-64	0		
19	#604060	96	64	96	32	1		
18	#402040	64	32	64	32	1		
15	#000000	0	0	0	0	1		
13	#200000	32	0	0	96	1		
13	#606060	96	96	96	0	1		
388	Average	47.9	53.4	108.5		50%	108.5	0.6

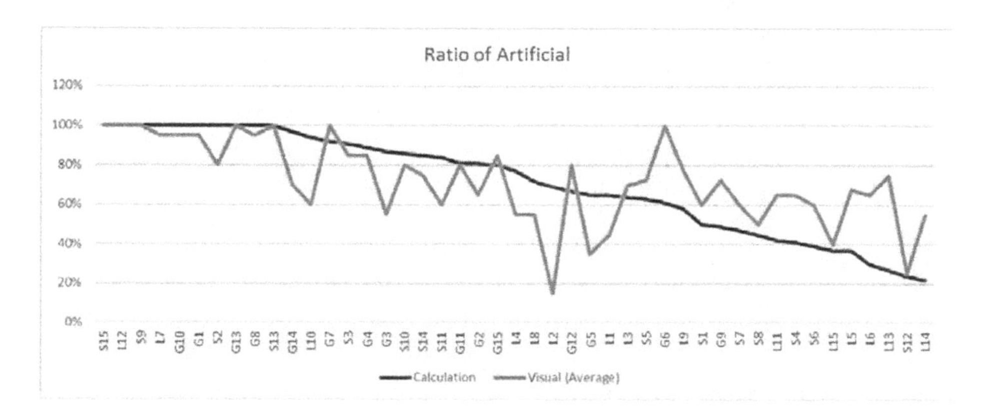

Fig. 4. Comparison between calculation and visual (average) on ratio of artificial (*Ra*) on all photos.

Next, the brightness (*Br*) and the contrast (*Co*) are important elements to know the ratio of artificial and natural and weather condition in a photo. When both of *Br* and *Co* are high, the photo will have artificial things (buildings) in clear weather. When only *Br* is high, the photo will have artificial things (buildings) in cloudy weather. When both of *Br* and *Co* are low, the photo will be natural things.

By the method mentioned above, we can classify all targeted photos into the next four classes and each judgement method is as follows.

Class A: a photo in a building, $Ra = 100\%$

Class B: a photo of mainly a building in a clear weather day, both of *Br* and *Co* are over average, except Class A,

Class C: a photo of mainly a building in a cloudy weather day, only *Br* is over average, except Class A,

Class D: a photo of mainly a natural park, others.

6.2 Experiments

Table 4, 5, 6 and 7 show the experimental results on Class A, B, C and D, respectively. The evaluation was as follows,

Class A: total ten photos, five photos were in buildings, four photos were almost buildings and one photo (S2) was not almost building, the classification successful ratio was 90%,
Class B: total eleven photos, nine photos were mainly buildings, two photos (L15 and S12) were not mainly buildings and all photos were in a clear weather day, the classification successful ratio was 81.8%,
Class C: total twelve photos, nine photos were mainly buildings, three photos (G3, S10 and S5) were not mainly buildings, the weather was cloudy or clear, and the classification successful ratio was 75%,
Class D: total twelve photos, four photos were mainly natural parks, five photos were sightseeing spots, three photos were museums, and the classification successful ratio was 33.3%.

As a result, in the proposed classification method, we confirmed as follows,

(1) The artificial ratio calculated by *Hu* is useful to find and classify indoor photos ($Ra = 100\%$),
(2) If both of *Br* and *Co* are over average, the photos show almost building in clear weather,
(3) If *Br* is over average, the photos show almost building but it is difficult to judge if the weather is clear or cloudy,
(4) If a both of *Br* and *Co* are under average, the photos show natural parks but it is not deterministic.

Table 4. Classification result (**Class A:** *Ra = 100%*).

POI *Ra* *Br* *Co*	Photo *Features*	POI *Ra* *Br* *Co*	Photo *Features*	POI *Ra* *Br* *Co*	Photo *Features*
S15 *100%* *155.9* *0.162*	 *In Museum*	L12 *100%* *90.4* *0.839*	 *In Shopping Center*	S9 *100%* *98.7* *0.137*	 *In Museum*
L7 *100%* *106.0* *0.073*	 *In front of Zoo*	G10 *100%* *143.7* *0.020*	 *Government Building*	G1 *100%* *127.6* *0.078*	 *Sightseeing Spot*
S2 *100%* *141.9* *0.069*	 *Museum*	G13 *100%* *94.9* *0.145*	 *In Church*	G8 *100%* *104.6* *0.071*	 *Shopping Street*
S13 *100%* *92.0* *0.137*	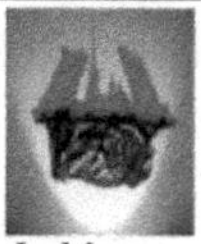 *In Museum*				

Table 5. Classification result (**Class B:** *Both Br and Co are over average, except Class A, Average Br = 125.1, Average Co = 0.185*).

POI *Ra* *Br* *Co*	Photo *Features*	POI *Ra* *Br* *Co*	Photo *Features*	POI *Ra* *Br* *Co*	Photo *Features*
G9 *49%* *135.2* *0.270*	 *Building*	S7 *47%* *141.1* *0.181*	 *Concert Hall*	S8 *45%* *153.3* *0.331*	 *Museum*
L11 *42%* *157.7* *0.210*	 *Museum*	S6 *39%* *138.2* *0.274*	 *Church*	L15 *37%* *172.0* *0.344*	 *Sightseeing Spot*
L5 *37%* *129.4* *0.323*	 *Amusement Park*	L6 *30%* *126.6* *0.369*	 *Church*	L13 *27%* *171.9* *0.228*	 *Station*
S12 *24%* *133.1* *0.352*	 *Museum*	L14 *22%* *164.7* *0.413*	 *Museum*		

Table 6. Classification result (**Class C:** *only Br is over average, except Class A, Average Br = 125.1*).

POI *Ra* *Br* *Co*	Photo *Features*	POI *Ra* *Br* *Co*	Photo *Features*	POI *Ra* *Br* *Co*	Photo *Features*
G14 *97%* *133.6* *0.151*	 *Church*	S3 *91%* *150.1* *0.012*	 *Museum*	G3 *87%* *125.4* *0.107*	 *Sightseeing Spot*
S10 *85%* *147.3* *0.078*	 *Museum*	S14 *85%* *152.9* *0.016*	 *Museum*	G11 *81%* *147.2* *0.069*	 *Sightseeing Spot*
L4 *77%* *132.0* *0.050*	 *Sightseeing Spot*	L8 *72%* *151.8* *0.040*	 *Sightseeing Spot*	L3 *64%* *149.8* *0.020*	 *Sightseeing Spot*
S5 *63%* *149.9* *0.139*	 *Natural Park*	L9 *58%* *125.3* *0.107*	 *Shopping Center*	S4 *41%* *125.3* *0.122*	 *Church*

Table 7. Classification result (**Class D:** *others*).

POI *Ra* *Br* *Co*	Photo *Features*	POI *Ra* *Br* *Co*	Photo *Features*	POI *Ra* *Br* *Co*	Photo *Features*
G4 *89%* *89.9* *0.287*	 *Sightseeing Spot*	S11 *84%* *79.6* *0.306*	 *Museum*	G5 *65%* *86.3* *0.246*	 *Natural Park*
G6 *61%* *59.1* *0.181*	 *In Museum*	S61 *50%* *108.5* *0.558*	 *Sightseeing Spot*	L10 *94%* *105.4* *0.096*	 *Museum*
G7 *92%* *88.3* *0.013*	 *Sightseeing Spot*	G2 *81%* *118.8* *0.121*	 *Natural Park*	G15 *80%* *118.2* *0.170*	 *Sightseeing Spot*
L2 *69%* *114.9* *0.101*	 *Natural Park*	G12 *67%* *94.6* *0.139*	 *Sightseeing Spot*	L1 *65%* *97.9* *0.163*	 *Natural Park*

7 Conclusion

This paper described the concept of PATRS and related researches. In order to develop the PATRS, classification technologies of POIs an easy feature extraction method from a tourism website is required. The classification of photos of tourism spots is an important technology to realize the feature extraction from numerous information in the website. This paper proposed a classification method of photos in a major tourism website by color analysis.

The feature of the method is to extract color elements and calculate the ratio of artificial colors in the photo from *Hu* of each color elements. By the experiment using 45 photos from a tourism website, we confirmed the method is useful to find and classify indoor photos. Further, the brightness and the contrast of photos are useful to classify photos of mainly buildings or natural parks, and clear weather or cloudy weather. As it is not deterministic results, we have to study additional classification method.

We are reporting in other paper an analytical method for extracting feature words of POI from comments of tourism website. In the future, we plan to combine the comment analysis and the photo analysis mentioned in this paper to classify and to extract features of POI more precisely. After, we will develop the module for finding POI adapted to personal preferences. Finally, we would like to connect those researches to the development of the personal adaptive tourism recommendation system (PATRS).

Acknowledgements. We would like to thank I-O DATA Foundation and the financing of project TIN2016-75850-R from the FEDER Funds for support of this research.

References

1. Japanese Travel Trade News [Official]. https://www.travelvoice.jp/english/jtb-report-2017-shows-fit-outnumbers-package-tours-in-the-overseas-travel-market-of-japan/
2. TripAdvisor. https://www.tripadvisor.jp/
3. Yabe, N., Kurata, Y.: An analysis of the behavior of inbound tourists in the Tokyo Metropolitan Area from the data of IC transit cards. Theor. Appl. GIS **21**(1), 35–46 (2013). (in Japanese)
4. Maeda, Y., Goto, F., Masui, H., Masui, F., Kamemaru, S., Suzuki, M.: A note on modeling for changes of tourist purposes. Biomedical Fuzzy Syst. Assoc **16**(2), 29–36 (2014). (in Japanese)
5. Sasaki, J., Takahashi, S., Shuang, L., Komatsu, I., Yamada, K., Takagi, M.: Finding target users interested in regional areas using online advertising and social network services. In: Fujita, H., Guizzi, G. (eds.) SoMeT 2015. CCIS, vol. 532, pp. 203–215. Springer, Cham (2015). https://doi.org/10.1007/978-3-319-22689-7_15
6. Kasahara, H., Tamura, K., Iiyama, M., Mukunoki, M., Minoh, M.: Tourist behavior model construction based on tracks of tourists using regional environmental factors and application. J. Inf. Process. Soc. Japan **57**(5), 1411–1420 (2016). (in Japanese)
7. Li, Y., Clerk, H., Huang, C., Duan, L.: The concept of smart tourism in the context of tourism information services. Tourism Manag. **58**(2017), 293–300 (2017)

8. De Úñez, X.M.G., Núñez-Valdez, E.R., Espada, J.P., Crespo, R.G., Garacía-Díaza, V.: Proposal for sentiment analysis on Twitter for tourism-based applications. In: New Trends in Intelligent Software Methodologies, Tools and Techniques, (SOMET 2018), pp. 713–721. IOS Press (2018)
9. Cano, E., Varga, A., Ciravegna, F.: Volatile classification of point of interests based on social activity streams. Research Gate (2011). https://vdocuments.mx/
10. Kurata, Y., Ai, H., Sanada, F.: Creation of innovative tourist maps based on the user-posted data of a photo-sharing site. Int. J. Tourism Sci. **8**, 151–154 (2015)
11. Li, S., Takahashi, S., Yamada, K., Takagi, M., Sasaki, J.: Analysis of SNS photo data taken by foreign tourists to Japan and a proposed adaptive tourism recommendation system. In: Progress in Informatics and Computing (PIC2017), pp. 323–327. IEEE Press, December 2017
12. Watanabe, S., Yoshino, T.: Tourist information visualization system for improvement discovery based on the similarity among tourist spots. In: DICOMO2016 Symposium, pp. 1357–1362, July 2016. (in Japanese)
13. Sato, N., Nanba, E., Ishino, A., Takezawa, T.: Construction of charm discovery system by comparison of similar tourist spots. In: Conference of Tourism Information of Japan (2016). (in Japanese)
14. Ochiai, K., Yamada, W., Fukazawa, Y., Kikuchi, H., Matsuo, Y.: POI official account classification method using twitter posts and profile information. J. Inf. Process. Soc. Jpn. (Database) **9**(2), 11–12 (2016). (in Japanese)
15. Nakamura, M., Ototake, H., Yoshimura, K.: Classification method of sightseeing spots using Word2Vec. In: The 69th Conference of Kyushu branch of the institute of Electrical and Information Engineers, 07-2A-08, p. 290 (2016). (in Japanese)
16. Matsuda, Y., Nakamura, M., Kang, D., Miyagi, H.: A fuzzy optimal routing problem for sightseeing. IEE J. Trans. EIS **125**(8), 1350–1357 (2005). (in Japanese)
17. Yamashita, A., Kawamura, H., Ohuchi, A.: Development of user adaptive tourism website with weblog information gathering. In: 22nd Fuzzy System Symposium, vol. 8C2-4, pp. 827–830, September 2006. (in Japanese)
18. Togashi, J., Sugimoto, T.: An interactive sightseeing spot recommendation system using wikipedia and google maps. In: IPSJ General Conference, no. 71, pp. 321–322 (2009)
19. Cao, L., Luo, J., Gallagher, A., Jin, X., Han, J., Hauang, T.S.: A worldwide tourism recommendation system based on geotagged web photos. In: 2010 IEEE International Conference on Acoustics Speech and Signal Processing (ICASSP) (2010)
20. Tarui, Y.: Recommendation system of tourist site using collaborative filtering method and contents analysis method. Jobu Univ. Bull. **36**, 1–14 (2011). (in Japanese)
21. Moreno, A., Valls, A., Isern, D., Marin, L., Borràs, J.: Ontology-based personalized recommendation of Tourism and Leisure Activities. Eng. Appl. Artif. Intell. (Elsevier) **26**(1), 633–651 (2013)
22. Okuzono, M., Hokota, M., Hirano, H., Masuko, S., Hoshino, J.: Recommendation system of sightseeing area for groups. IPSJ SIG Technical report, vol. 2015-HCI-162, no. 19 (2015)
23. Mihara, Y., et al.: Visit recommendation systems for Smartphones without privacy overhead. J. Inf. Process. Soc. Jpn. **7**(2), 87–96 (2017). (in Japanese)
24. Kinoshita, T., Nagata, M., Murata, Y., Shiata, N., Yasumoto, K., Ito, M.: Personal navigation system to assist sightseeing across multiple days. J. Inf. Process. Soc. Jpn. **47**(12), 3179–3187 (2006). (in Japanese)
25. Kinoshita, Y., Yokokishizawa, H.: A proposal of a tour route planning support system using Kansei evaluation and multi-objective gas. In: 24th Fuzzy System Symposium, vol. FC2-1, pp. 864–869(2008). (in Japanese)

26. Kurata, Y.: Walking plan around a city for you - development of a dialogue base travel plan creation support tool, Institute of Systems, Control and Information Engineers, vol. 75, no. 8, pp. 348–353 (2013). (in Japanese)
27. Takahashi, S., Li, S., Yamada, K., Takagi, M., Sasaki, J.: Case Study of tourism course recommendation system using data from social network services. In: New Trends in Intelligent Software Methodologies, Tools and Techniques (SoMeT2017), pp. 339–342. IOS Press, September 2017
28. Li, S., Komatsu, I., Yamada, K., Takagi, M., Sasaki, J.: Personal tour planning system (PTPS) for use in urban and rural areas. In: New Trends in Intelligent Software Methodologies, Tools and Techniques (SOMET 2018), pp. 259–269. IOS Press (2018)
29. Irotoridori. https://ironodata.info/extraction/irotoridori.php

Improving Customer's Flow Through Data Analytics

Nang Laik Ma[1](✉) and Murphy Choy[2]

[1] School of Business, Singapore University of Social Sciences, Singapore 599491, Singapore
nlma@suss.edu.sg
[2] Edinburgh Business School, Heriot Watt University, Edinburgh EH14 4AS, UK

Abstract. In this paper, we focus on improving the customer's flow by harnessing the power of analytics and focusing on the arrival process of passengers at one of the busiest airports in Asia. As there is a recent growth in travelers, the airport is undergoing expansion and is thus under tremendous pressure to utilise its resources effectively and efficiently. We first leverage the historical data of the arrival flights, passenger load, and on-time performance flag indicator in order to predict the arriving passenger' load for the immigration counters and taxi queues. We then build a decision support system using simulation to estimate the optimal number of immigration counter requirements so as to minimize the waiting time at the queues. This is also done to predict the number of taxis required to meet the service level agreement and to ensure the seamless flow of customers at various touch points to improve customer' satisfaction. The tool developed has benefited the manager in his daily operations, and advanced his decision making process supported by data rather than personal experience or "gut" feeling.

Keywords: Airport operations · Decision support system · Passenger load · Customers flow · Arrival process · Simulation · Manage queue

1 Introduction

As one of the major airports in Asia, our client undertakes the key functions of focusing on airport operations and management, air hub development, commercial activities and airport emergency services. Over the past 10 years, the number of travelers has been increasing and in year 2017, it handled more than 100 airlines with routes heading to 400 cities around the world handling 62.2 million passenger a year. The airport is ranked as one of the world's top ten airports with passenger service a priority in maintaining the position, thus the airport operator needs to plan its resources and capacity effectively and efficiently to remain competitive.

The focus of the study will be on arriving passenger ("the passenger", known better as the customer) experience at the airport under the Airport Operations. The scope of the business process covered in this study starts from the time the aircraft lands and

F. Wotawa et al. (Eds.): IEA/AIE 2019, LNAI 11606, pp. 279–286, 2019.
https://doi.org/10.1007/978-3-030-22999-3_25

ends at the point when the passenger leaves the airport using one of the many transport options available at the airport.

Most of the airlines do not submit the arrival passenger load information until the last hour, therefore the airport usage of the average passenger load of X% to do the daily planning is a very rough estimate. Furthermore, airport operators also need to work closely with its partners such as the Immigration Controller for their manpower allocation and various taxi operators. Based on the current practice, the operator will use the average passenger load X% for each flight and share the information with the Immigration Controller on a weekly basis. Immigration officers are full-time staff and the scheduling of staff is done weekly according to the average person load shared by the airport operators. Due to security reason, neither of them can deploy any additional manpower at a short notice or part-time basis to cover any additional workload.

Since the airport is located at the outskirt of the city areas, most taxi drivers will not come to the airport unless they are in the vicinity, or they are dropping off another passengers at the airport. There is a sign-board near the terminal which displays the number of arrival fight and estimated number of passenger arriving for the taxi driver to view. This information is refreshed in an interval of every 30 min and it is based on the average passenger load of X% and the Scheduled Time of Arrival (STA) of the flight. However in reality, all these numbers are subjected to changes due to unforeseen circumstances on the ground. At the peak hour, passengers may need to wait for more than 30 min at the immigration counters with an additional 30 min to queue for taxi, which makes the passenger very unhappy and lowers the customer satisfaction index for the airport. Thus, estimating the accurate passenger load is very important in resource planning to maintain the airport reputation and customers' satisfaction.

The business objective of the project is to satisfy its arrival passenger' experience from the point their flight lands at the airport, walking through immigration and finally leaving from the airport via various transport options. Using the available historical data, we do our preliminary research and analysis based on the following:

a. Segmentation study of the arrival passengers
b. On time performance of the arrival aircrafts
c. Build a simulation model to provide the airport operator an indication on the number of immigration counters and taxis required to optimize available resources.

In Sect. 2, we are going to do a literature review of the industry and related topics of resource planning and simulation. Section 3 will focus on data analysis, and in subsequent sections, we build the model for decision making and present the results. Finally, we indicate the limitations of the model and the future direction for the research.

2 Literature Review

For the departure process, check-in counters are one of the important resources at the airport as the delays and queues at the check-in counters are perceived as poor service and may result in flight delays. In this article [1], the author modelled the check-in scheduling problem using integer programming model to identify the minimum number of counters required under the realistic constraints. The simulation model also developed to study the stochastic nature of various parameters such as number of passengers on the flights and check-in counter opening and closing time to evaluate and improve the operational efficiency at the airport.

Queues are very common problems at most of the airports around the world. There are also numerous researchers who look at the queues. [2] is based on queues management at the service counters at Vancouver International airport using simulation and mathematical programming approach. LP was used to determine the minimum number of staff required and the optimal schedules for the staff.

[3] developed a methodology to define the service level standards at the airport terminal based on the passenger perception. It is based on the processing time, waiting time and space available for each person at one of the international airports in Brazil. In [4–7], the authors focused on the use of simulation model to evaluate the terminal performance and utilisation. In [8], the authors proposed an Eulerian model of air traffic flows in NAS (National airspace system) and present a distributed feedback control approach to manage the flows. [9] focused on runway resource allocation model for strategic planning which select the best runway for landing and takeoff based on a list of safety constraints. In reference [10], the authors used hierarchical forecasting method to predict the monthly departure passenger movement over the next twelve month for macro level planning. They have developed the forecasting model using SAS Forecast Studio and the mean absolute percentages error is less than 3%, which indicate the reliability of the model to be used for real-world application.

Based on the literature review, we have identified that most of the researchers focus on check-in counters utilisation, terminal performance and the simulation of check-in counters queues. There is very limited research that is based on the arrival passengers flow at the airport which contributed up to 50% of the total annual passengers load at this particular airport. With the customers who have various choices to fly to various destination, the arrival passenger's perception of the airport is also important so that they will have a seamless flow from the time the planes touch down at the airport, through immigration and to the time they leave the airport using one of the transport choice (taxi). This experience should be an interest of the airport operators and is the main focus of our research in this paper.

3 Data Analysis and Model Development

After the aircraft lands at the airport and is on-chocks to the designated gate, passengers will leave the aircraft and proceed to the immigration for clearance. We need to predict the passenger load based on the historical data, on-time performance and estimated time of arrival for each flight (Fig. 1).

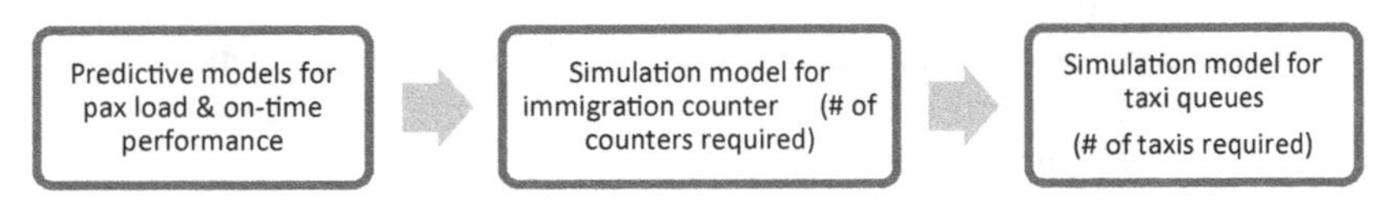

Fig. 1. Overview of our research focus

3.1 Predictive Model for the Passenger Load

We have decided to use most recent one year of historical data for our analysis as it will include winter and summer season. There are hundreds of thousands 162,356 of records with 24 fields. Some of the data fields are the STA of the flight, actual arrival time, the aircraft type, airline, flight number, origin of the flight whether it is a transfer flight or enroute flight, gate assigned and the actual passenger count. There is no missing data and we use 100% of the data for our predictive model of passenger load and on-time performance.

We analysed that the average pax load for the arrival passengers is 75%, however the average is not useful for this case, as the pax load is much higher for budget airlines. The day of the week also plays an important part as there are more travellers arriving the country on weekends than weekdays.

Refer to previous work [10], we have applied different data mining models such as multiple linear regression (MLR), decision tree (DT) and neural network (NN) and derived the business rule to determine the Passenger load for each flight. Passenger load (define as 'pax load' thereafter) is the ratio of total pax on board to maximum capacity based on airline and aircraft type. The decision tree (DT) performs better than the other predictive models such as MLR and NN with mean absolute percentage error (MAPE) of 13% as compared to 15% and 16% respectively. The error also varies significantly among different airline due to the number of records and past historical performance. We are only using the internal data available and not including data that are related to airline promotion or activities. Some of the important factors to predict the pax load are airline, original of the flight and day of the week. We have deployed the decision tree model to predict the pax load for the airport operators to be used on a daily basis.

3.2 Predictive Model for the On-Time Performance

On-time performance of flight is one of the key performance indicators for the airlines industry around the globe. Flights arriving within 15 min from STA are considered on-time, those that arrive before/after are considered not on-time (early or late).

$$\text{Arrival Time Difference for each flight} = \text{actual arrival time} - \text{STA} \tag{1}$$

We derive arrival time difference for each flight by subtracting STA from actual arrival time. A flight is late if the arrival time difference is positive and early if the time difference is negative. The flight is on time if the absolute value of arrival time difference is less than or equal to 15 min, otherwise it is flagged as not on time.

We also analyzed same data stated above and derive a target flag called on-time performance. Based on the initial data analysis, the top airline has 86% on time performance and absolute time difference is 18 min. However, for some airlines the average time difference can be as high as 40 min. Thus, different airlines having huge gap on their on-time performance will make the planning more challenging for the airport terminal manager.

We use NN and DT to predict the on-time performance flag. NN and DT were evaluated and DT performs better than NN with a higher accuracy rate of 85.45% as opposed to 84.36% by NN. We have decided to deploy DT model to improve our decision making. Finally, we have combined the predicted pax load and on-time performance indicator to estimate the number of arriving passengers for resource planning at the immigration and at the taxi queue.

3.3 Simulation Models for Immigration and Taxi Queues

Gate assignment, on-time performance, estimated time of arrival and passenger count are some critical inputs for the building of the simulation model for the immigration counter and taxi transport option. For simplicity, we use the gate pre-assigned by the terminal in the data provided.

We use the predicted pax load model with on-time performance and compute the total arriving passengers for each fight and consolidate the total number of passenger which will be arriving for every half-an-hour period. If the on-time performance is true, we predict that the flight will arrive within 15 min of STA; however, if it is not on-time, we assume that it will be delayed for another 15 min and will arrive within the next 30 min.

We use normal distribution with mean average walking time and standard deviation to account for the variability of walking time from the gate to the immigration counters. As mentioned earlier, the number of passengers will affect the waiting time and queue lengths at the immigration, we need to estimate the right number of counters to be opened at the immigration to meet the service level agreement, which is 80% of the passengers need to be service within Y minutes at the immigration.

From the time the aircraft taxi in to the gate assigned, passengers will leave the aircraft. From there we will derive when the first passenger arrival time to immigration counter using the equation below. This will give us an indication when the first passenger can be serviced at the immigration counter which will help to identify the number of passengers' arrival at each half hour interval in the immigration hall.

Time first passenger arrive immigration hall = $T_{on\_choke} + T_{exitplane} + T_{Walking}$ We used an estimate based on our previous work, where the service time at the immigration counters is assumed to follow an exponential distribution with a meanservice time of 1 min per passenger. We set an initial number of immigration counters to be opened at the particular time, it is based on the initial pax load estimation. Monte Carlo simulation is performed, we keep track of the queuing time of the passenger, average queue length, and total time spent at the immigration is recorded. We vary the number of immigration counters until the agreed service level is achieved or the maximum counters is opened. Most of the inputs are parameterized so that the managers can change these inputs for various scenarios such as peak hour demand or shortage of immigration officers.

The taxi management system that we have developed here aim to achieve a win-win situations for the taxi companies and airport operator. The model will estimate the number of taxi required based on more accurate passenger arrival information and this can be shared with the taxi operator to strengthen collaboration between both parties. If there are enough taxi coming to the airport, this will minimize the passengers waiting time for taxi and the taxi drivers will be happy if they can quickly pick up the passengers after they arrive.

The proportion of passengers who are taking taxi will be higher during mid-night to early morning when the public transports are not operating. The average number of passengers on a taxi is between 1 to 4 persons with an average of 2.5 per taxi based on some information collected on the ground. Using these input parameters, we built another simulation model which is used to provide an indication on the number of taxi required. For this taxi queue model, most passengers will join the bank queue – single queue and they will be served based on first come first served principal. The correct number of taxi arriving will help to clear the taxi queue quickly and reduce the waiting time for each passenger.

4 Model Output

4.1 Immigration Counters Management System

The predicted pax load from model above is used to compute the number of immigration counter required, number of passengers who are not able to clear immigration in the original half hour interval and overflow to the next half hour and number of taxi required are some of the key outputs from the model.

The maximum capacity indication could give the airport operator that there is capacity issues based on the current clearance rate at the immigration counter and the number of available counters. In a half hour interval where result give an indicator of >1, the operator may need to consider deploying more manpower at the immigration counters or consider changing the arrival gate, so that the passengers will walk longer and arrive at different time to lessen the load.

With different input parameters, we could predict the number immigration counters required for a 24 h time window, as seen in Fig. 2. This provided a range of results for the airport so that they can better anticipate any possible changes in event for the day.

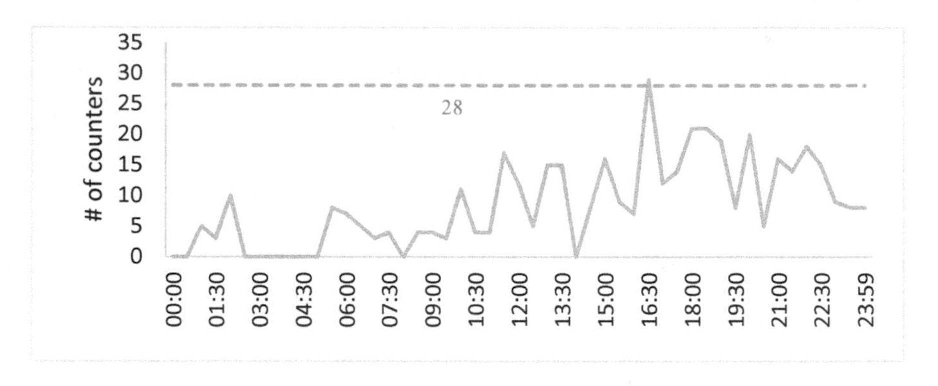

Fig. 2. Number of immigration counter required using predicted pax load

4.2 Taxi Management System

Once the passengers clear the immigration and collect their baggage from the baggage claim area, they will pass through custom clearance before they can exit the restricted areas. Once they are out of the restricted areas and arrive at the main meeting hall, there are various transport options available for them to proceed to their next destination. Public transport like train, buses, taxi, as well as private transport options like arranged transfer, etc.

The taxi management system provided the estimated number of taxi required at each half hour interval. The proportion of passengers who will be taking the taxi and the average number of passengers per taxi will determine the number of taxi required. We use average 2.5 passengers per taxi with 30% of passengers taking taxi and the proportion will be higher at the wee hours where the public transport is not operating (from mid night to 6:00 am). Since this is a model for the airport to better plan and manage their resource so as to improve their passengers' arrival process, they will need to have the data earlier so that there is ample time for their planning. Figure 3 shows the output given by the taxi management system using predicted load. The number derived in the model will be a more accurate figure from the current practice of using the average pax load of X%.

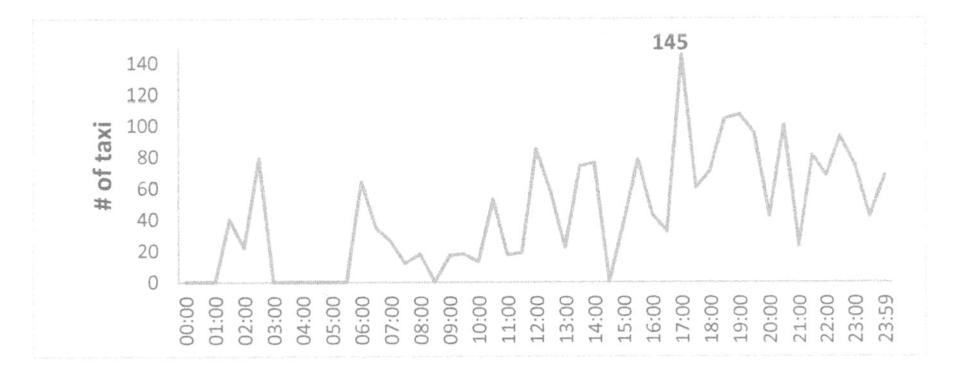

Fig. 3. Number of taxi required based on predicted load

5 Model Output

Improving passenger arrival process is an important aspect of the airport operations. We have analyzed one year of historical data to predict the passenger load and on-time performance of the flights and build simulations model to predicted the number of immigrations counters and taxi required. Using the developed system, the terminal managers can change various inputs to run different scenarios in simulation in order to plan for the unexpected which need proper management. For future research, we may use real-time data from sensors on the ground to track the number of passengers arriving at the airport and provide more accuracy data for planning and monitoring. Harnessing the power of analytics, we can expect the arrival passenger to spend minimum time at the queues, and enjoy the best customer experience.

References

1. Al-Sultan, A.T.: Airport check-in optimization by IP and simulation in combination. Int. J. Math. Comput. Sci. **9**(1), 403–406 (2015)
2. Atkins, D., Begen, M.A., Luczny, B., Parkinson, A., Puterman M.L.: Right on queue. OR/MS Today. **30**(2) (2003)
3. Correia, A.R., Wirasinghe, S.C.: Development of level of service standards for airport facilities: Application to São Paulo International Airport. J. Air Transp. Manag. **13**(2), 97–103 (2007)
4. Chang, W., Lee, C.D., Mangano, F.: Computer simulation of terminal utilization. Airport Forum **8**(3), 63–67 (1978a)
5. Chun, H.W., Tak Mak, R.W.: Intelligent resource simulation for an airport check-in counter allocation system. IEEE Trans. Syst. Man Cybern. Part C Appl. Rev. **29**(3), 325–335 (1999)
6. Jim, H.K., Chang, Z.Y.: An airport passenger terminal simulator: a planning and design tool. Simul. Pract. Theor. **6**(4), 387–396 (1998)
7. Joustra, P.E., Van Dijk, N.M.: Simulation of check-in at airports. In: Simulation Conference, 2001, Proceedings of the Winter, vol. 2, pp. 1023–1028 (2001)
8. Le Ny, J., Balakrishnan, H.: Distributed feedback control for an Eulerian model of the National Airspace System. In: 2009 American Control Conference, ACC 2009, pp. 2891–2897, 10–12 June 2009
9. Li, L., Clarke, J.-P.: Airport configuration planning with uncertain weather and noise abatement procedures. In: 2010 IEEE/AIAA 29th Conference on Digital Avionics Systems Conference (DASC), pp. 2.B.5-1–2.B.5-9, 3–7 October 2010
10. Ma, N.L.: Forecasting passenger flows using data analytics. In: Benferhat, S., Tabia, K., Ali, M. (eds.) IEA/AIE 2017. LNCS (LNAI), vol. 10350, pp. 211–220. Springer, Cham (2017). https://doi.org/10.1007/978-3-319-60042-0_24

Towards Similarity-Aware Constraint-Based Recommendation

Muesluem Atas(✉), Thi Ngoc Trang Tran, Alexander Felfernig, Seda Polat Erdeniz, Ralph Samer, and Martin Stettinger

Institute of Software Technology, Graz University of Technology, Inffeldgasse 16b/II, 8010 Graz, Austria
{muesluem.atas,ttrang,alexander.felfernig,spolater,rsamer, martin.stettinger}@ist.tugraz.at
http://ase.ist.tugraz.at/

Abstract. Constraint-based recommender systems help users to identify useful objects and services based on a given set of constraints. These decision support systems are often applied in complex domains where millions of possible recommendations exist. One major challenge of constraint-based recommenders is the identification of recommendations which are similar to the user's requirements. Especially, in cases where the user requirements are inconsistent with the underlying constraint set, constraint-based recommender systems have to identify and apply the most suitable diagnosis in order to identify a recommendation and to increase the user's satisfaction with the recommendation. Given this motivation, we developed two different approaches which provide similar recommendations to users based on their requirements even when the user's preferences are inconsistent with the underlying constraint set. We tested our approaches with two real-world datasets and evaluated them with respect to the runtime performance and the degree of similarity between the original requirements and the identified recommendation. The results of our evaluation show that both approaches are able to identify recommendations of similar solutions in a highly efficient manner.

Keywords: Decision support systems · Constraint-based recommender systems · Similarity measures · Recommendation similarity

1 Introduction

Recommender Systems (RS) have become an essential means for guiding users in a personalized way to interesting or useful objects and services (often referred to as *items*) [9,19]. These decision support systems help users to identify useful items matching their wishes and needs, such as movies, books, songs, web sites, financial services, travel destinations, and restaurants [5,7,9,14]. In contrast to traditional recommendation approaches such as collaborative [11] and

F. Wotawa et al. (Eds.): IEA/AIE 2019, LNAI 11606, pp. 287–299, 2019.
https://doi.org/10.1007/978-3-030-22999-3_26

content-based filtering [15], constraint-based RS [1,9] recommend products and services based on a given constraint set. These systems are usually applied in complex domains such as cars, personal computers (PC), and financial services. They allow individual customization of complex industrial products and services in order to satisfy individual customer needs (i.e, user requirements) [18]. When interacting with constraint-based RS, users articulate their requirements (e.g., when interacting with a PC recommender, a user specifies different properties such as *memory type, memory size, processor type, price,* and *brand of PC*). In this context, inconsistent requirements will be automatically or manually adapted [1,3]. Finally, a recommendation will be suggested to the user. A complex product such as a personal computer can have millions of recommendations and the RS has to deal with difficult problems such as system maintainability, consistency maintenance, and efficient response times. One major challenge of constraint-based recommenders is to identify the most suitable recommendation for a user based on his/her articulated requirements. Especially, in cases where user requirements are inconsistent with the underlying constraint set, users have to be supported in finding a way out from the *no recommendation could be found* [18] dilemma (i.e., identifying a diagnosis). Besides, after identifying and applying a diagnosis, the proposed recommendation should be similar to the user's defined requirements in order to increase the user's satisfaction and be computed with a performance acceptable for interactive settings. The identification of a recommendation which is similar to a user's requirements is a challenging task if millions of recommendations exist. The most naive solution is the comparison of each possible recommendation with a given set of user requirements in order to identify the most similar recommendation for the user which is not possible due to an unacceptable runtime performance.

To the best of our knowledge, such similarity-aware constraint-based RS do not exist. A related work is presented in Eiter et al. [4] where the authors analyse several decision/optimization versions of identifying similar and diverse solutions in the context of *Answer Set Programming* (ASP). The authors introduce offline and online methods to determine the computational complexity of similar/diverse solutions. Hebrard et al. [8] present a number of practical approaches to identify the distance of similar and diverse solutions in constraint programming and focus on determining the computational complexity of distance functions for similar and diverse solutions. The approach suggested by [8] calculates the whole set of possible solutions at once and then identifies similar and diverse solutions. In our approaches, we do not calculate the whole set of possible solutions, since this is not feasible due to the high complexity of item domains. Given this motivation, we developed two different approaches that suggest similar recommendations to the users based on their requirements even if the user preferences are inconsistent with the underlying constraint set.[1] We tested our

[1] The work presented in this paper has been partially conducted within the scope of the research projects *WeWant* (basic research project funded by the Austrian Research Promotion Agency - 850702) and *OpenReq* (Horizon 2020 project funded by the European Union - 732463).

approaches with two different datasets (PC and bike recommendations) and evaluated them with regard to their runtime performance and the degree of similarity between the original requirements and identified recommendations. The results of our evaluation indicate that our approaches are able to identify similar recommendations with a high similarity degree in a highly efficient manner.

The remainder of this paper is structured as follows. In Sect. 2, we introduce a working example from the *personal computer* domain. Section 3 presents the identification and application of a diagnosis based on users' requirements and the determination of candidate recommendations (i.e., possible solutions). In Sect. 4, we introduce similarity metrics to calculate the similarity between the original requirements and the possible solutions in order to identify the solution with the highest similarity. Section 5 introduces our developed approaches for the identification of similar recommendations. Section 6 provides the evaluation results of both approaches with two different real-world datasets. Finally, we conclude the paper with a discussion of some ideas for future work in Sect. 7.

2 Working Example

For demonstration purposes, we introduce a constraint-based recommendation scenario from the domain of personal computers (PC). The example presented in this section introduces the KB (variable definitions and constraints) and user requirements regarding a PC. For simplicity reasons, we used only some of the PC variables as a constraint satisfaction problem (CSP) which is often used for the definition of a constraint-based recommendation task [20].

Definition 1: Constraint-Based Recommendation Task. A constraint-based recommendation task expressed in CSP representation is defined as a triple (V,D,C) where $V = \{v_1, v_2, ..., v_n\}$ is a set of finite domain variables, $D = \{dom(v_1), dom(v_2), ..., dom(v_n)\}$ refers to the set of variable domains and $C = C_{KB} \cup C_R$ corresponds to the set of constraints representing *product-specific constraints* (C_{KB}) and *requirement constraints defined by a user* (C_R).

A simplified example of a constraint-based recommendation task in the PC domain is the following. In this context, the variable *max-price* represents the maximal price of a PC in Euro, *min-hd-cap* corresponds to the minimum hard-disc capacity in GB, *price* refers to the price of a PC in Euro, *pro-freq* represents the clock-rate of a processor in GHz, *mb-ram-cap* refers to the capacity of the motherboard RAM in GB, and *hd-cap* represents the capacity of the hard-disc in GB. Additionally, there are variables which indicate the importance of PC variables from the user's point of view.[2] For instance, the expression *imp-price = 5* implies that the defined price limit is very important for the user whereas the expression *imp-price = 1* indicates a price limit which is not important from the user's point of view. The product knowledge is represented as $C_{KB} = \{c_1 - c_5\}$ and the user requirements are expressed as $C_R = \{c_6 - c_{13}\}$.

[2] If this information is not provided, equal importance of all variables is assumed.

- $V = \{$**max-price**, **min-hd-cap**, **price**, **pro-freq**, **mb-ram-cap**, **hd-cap**, **imp-price**, **imp-hd-cap**, **imp-pro-freq**, **imp-mb-ram-cap**$\}$
- $D = \{dom($**max-price**$) = \{1000, 2000, 3000, 3500\}, dom($**min-hd-cap**$) = \{512, 1024\}, dom($**price**$) = [400 ... 3500], dom($**pro-freq**$) = \{2, 2.2, 2.6, 3.15\}, dom($**mb-ram-cap**$) = \{8, 16\}, dom($**hd-cap**$) = \{64, 128, 256, 512, 1024\}, dom($**imp-price**$) = dom($**imp-hd-cap**$) = dom($**imp-mb-ram-cap**$) = dom($**imp-pro-freq**$) = [1 ... 5]\}$
- C_{KB}=$\{$ c_1: (**hd-cap**≥512 && **mb-ram-cap**≥8) ⇒ **price**≥2250, c_2: (**hd-cap**≥1024 && **mb-ram-cap**≥16) ⇒ **price**≥3250, c_3: **mb-ram-cap**=16 ⇒ **pro-freq**≥3.15, c_4: **price**≤**max-price**, c_5: **min-hd-cap**≥**hd-cap**$\}$
- $C_R = \{c_6$: **max-price** = 2000, c_7 : **min-hd-cap** = 1024, c_8 : **pro-freq** = 2.6, c_9 : **mb-ram-cap** = 16, c_{10} : **imp-price** =3, c_{11} : **imp-hd-cap** =2, c_{12} : **imp-mb-ram-cap** =5, c_{13} : **imp-pro-freq** =1$\}$

A constraint-based recommendation can be defined based on the given constraint-based recommendation task.

Definition 2: Constraint-based recommendation. A constraint-based recommendation for a recommendation task is defined as an instantiation I $= \{v_1 = ins_1, v_2 = ins_2, ..., v_n = ins_n\}$ where $ins_i \in dom(v_i)$. A constraint-based recommendation is *consistent* if the instantiations in I are consistent with the $\bigcup c_i \in C$. Furthermore, a recommendation for a constraint-based recommendation task is *complete* if all variables in V are instantiated and *valid* if the recommendation is consistent and complete. In this paper, we ranked our solutions and always recommend the first solution (i.e., recommendation) to the user. For a detailed discussion at ranking approaches for solutions we refer to [21].

3 Identification of Personalized Diagnoses

For the recommendation task introduced in Sect. 2, it is not possible to find a solution due to some inconsistencies between the user requirements C_R and the product-specific constraints C_{KB}. For instance, user's constraints regarding the *clock-rate of the processor* and *RAM capacity of the motherboard* (c_8, c_9) contradict each other, because the third constraint in the knowledge base (KB) indicates that the *clock-rate of the processor* must be greater or equal to the value of 3.15 GHz if the *RAM capacity of the motherboard* is 16 GB. Consequently, we have to identify a minimal set of user constraints which has to be adapted or deleted in order to get rid of the *no recommendation could be found* dilemma. In some certain cases where the user requirements C_R are inconsistent with the underlying constraint set C_{KB}, the users have to be supported in identifying constraints which trigger an inconsistency. Such constraints can be determined on the basis of the minimal conflict detection principle [10]. On the basis of minimal conflict sets, diagnoses (i.e., hitting sets) can be determined thereof [17]. Such diagnoses are proposals of requirements which should be changed such that the system is able to find a solution.

Definition 3: Conflict Set. A conflict set is a set $CS \subseteq C_R$ such that $CS \cup C_{KB}$ is inconsistent. A conflict set CS is minimal if and only if there does not exist a conflict set $CS' \subset CS$. In the working example defined in Sect. 2, there exist two minimal conflict sets: $CS_1 = \{c_6, c_7\}$ and $CS_2 = \{c_8, c_9\}$. CS_1 and CS_2 are conflict sets since each individual conflict set is in conflict with C_{KB}. A basic approach to determine minimal diagnoses from minimal conflict sets is the so-called *hitting set directed acyclic graph* (HSDAG) [17].

Definition 4: Diagnosis. A diagnosis defines a set of constraints $\Delta \subseteq C_R$ such that $C_{KB} \cup (C_R - \Delta)$ is consistent. A diagnosis Δ is defined as minimal if and only if there does not exist a diagnosis $\Delta' \subset \Delta$ such that $C_{KB} \cup (C_R - \Delta')$ is consistent. Based on the identified minimal conflict sets, the following diagnoses can be identified by using the HSDAG approach: $\Delta_1 = \{c_6, c_9\}$, $\Delta_2 = \{c_6, c_8\}$, $\Delta_3 = \{c_7, c_8\}$, and $\Delta_4 = \{c_7, c_9\}$. A basic approach for resolving conflicts is to adapt or to delete (see Sect. 5) the constraints contained in a diagnosis set. In order to identify the most suitable diagnosis for the user, the importance of diagnosed constraints (i.e, diagnosed variables) from the user's point of view has to be taken into account. For instance, selecting the first diagnosis ($\Delta_1 = \{c_6, c_9\} = \{$*max-price* $= 2000$ €, *mb-ram-cap* $= 16$ $GB\}$) would require the deletion or adaptation of the constraints regarding *max-price of PC* and the *RAM capacity of motherboard* variables, but as indicated in Sect. 2, the *price of PC* is *moderately important* (imp-price=3) and the *RAM capacity of motherboard* is very important (imp-mb-ram-cap=5) for the user. The deletion or adaptation of important user constraints decreases the user satisfaction. In such cases where several diagnoses exist, one should select the diagnosis which contains unimportant user constraints such that users are still satisfied with the proposed constraint adaptation. Moreover, this strategy helps to identify recommendations which are similar to users' requirements, because the similarity calculation (between user requirements and identified recommendations) applied in our approach depends on the importance of the variables (see Sect. 4). This means, the adaptation of important variables will often deteriorate the degree of similarity compared to the adaptation of less important variables. For an automated minimal diagnosis detection, we apply the FastDiag [6] algorithm, which allows an efficient calculation of one diagnosis at a time. Furthermore, FastDiag enables to identify a minimal diagnosis which consists of unimportant user requirements. If the user requirements provided to FastDiag are already sorted based on their importance, then the algorithm tries to identify a minimal diagnosis which consists of unimportant constraints (i.e., user requirements) in the first half of the constraint list. In our working example, Δ_3 contains less important constraints compared to Δ_1, Δ_2, and Δ_4. The application of $\Delta_3 = \{c_7, c_8\} = \{$*min-hd-cap*$=$ 1024 GB, *pro-freq* $= 2.6$ $GHz\}$ leads to six different solutions. One of the six possible solutions is the following:

price = 1000 €, **pro-freq** = 3.15 GHz, **hd-cap**=64 GB, **mb-ram-cap**= 16 GB

Our approaches (see Sect. 5), take the first 10 solutions and recommend the solution to the user which is most similar to the user's requirements. The similarity between the user's requirements and the identified solutions can be calculated using similarity metrics presented in Sect. 4.

4 Determination of Similarity Degree Using Similarity Metrics

Similarity metrics [13] are applied for the similarity calculation between the user requirements and a recommendation (see Formulae 1–5). The metrics are denoted as *more-is-better* (MIB; e.g., hard-disc capacity of a PC), *less-is-better* (LIB; e.g., price of a PC), *nearer-is-better* (NIB; e.g., clock-rate of the processor should be as near as possible to 2.6 GHz), and *equal-is-better* (EIB; color of a PC) [12]. The term *sim(r,u)* indicates the similarity between a recommendation r from a set of recommendations and requirements of a user u. The notation $s(r_i, u_i)$ represents the similarity between the requirement of user u and the recommendation r with respect to the variable i (attribute-level similarity). In addition, *imp(i)* denotes the importance of a variable i from the user's point of view and *val(i)* represents the *value* of variable i. The terms $minval(r_i)/maxval(r_i)$ are minimum/maximum values of a variable i taken from the KB definition.

$$sim(r,u) = \frac{\sum_{i \in variables} s(r_i, u_i) * imp(i)}{\sum_{i \in variables} imp(i)} \tag{1}$$

$$MIB : s(r_i, u_i) = \frac{val(u_i) - minval(r_i)}{maxval(r_i) - minval(r_i)} \tag{2}$$

$$LIB : s(r_i, u_i) = \frac{maxval(r_i) - val(u_i)}{maxval(r_i) - minval(r_i)} \tag{3}$$

$$NIB : s(r_i, u_i) = 1 - \frac{|val(u_i) - val(r_i)|}{maxval(r_i) - minval(r_i)} \tag{4}$$

$$EIB : s(r_i, u_i) = \begin{cases} 1 & \text{if } r_i = u_i \\ 0 & \text{otherwise} \end{cases} \tag{5}$$

5 Approaches for the Identification of Similar Recommendations

We developed two approaches for the identification of recommendations which are similar to the user's requirements. The *soft relaxation-based approach* is based on a soft relaxation of inconsistent user requirements. A soft relaxation in this context makes a strictly specified user requirement less strict. For instance, a soft relaxation for a user constraint *price* = 100 € can be represented as a deviation of 10 €: {*price* ≥ 90 € && *price* ≤ 110 €}. This relaxation strategy is used to identify items which are similar to the users' specifications. In contrast to a soft relaxation, a hard relaxation does not use any deviation, it simply deletes the specified value instead. The *hard relaxation-based approach* identifies similar recommendations by deleting the diagnosed user requirements (= hard relaxation) and using search heuristics from CHOCO constraint solver.

Soft Relaxation-Based Approach: The first developed approach for the identification of similar recommendations is based on a soft relaxation of inconsistent user requirements. This approach uses following lines of the Algorithm 1: 1–4, 10, and 12–17. It takes at first user's requirements and checks whether the user's requirements are consistent with the underlying KB by using a constraint solver (see lines 1–2 in Algorithm 1). If they are consistent, the recommender will only consider the first 10 solutions and recommends the solution which has the highest similarity (see line 13). As a KB of complex items such as cars, PCs, and smart homes consisting of hundreds of constraints, the user's requirements are often inconsistent with the underlying KB. In the case of an inconsistency, a suitable diagnosis has to be identified. For identifying a suitable diagnosis, the user's requirements will be sorted by their importance. Thereafter, the FASTDIAG algorithm analyzes the already sorted user constraints and identifies a diagnosis (see line 3). After the identification of a suitable diagnosis, all variables in the diagnosis set will be relaxed (soft relaxation) in order to find solutions similar to the specified user requirements (see line 4). The goal of the relaxation is to avoid empty search results. Such strategies try to identify solutions which are similar to the user's requirements. The presented relaxation strategy is a basic approach for identifying similar numerical values [2]. For the relaxation of non-numerical variables (e.g., *color* of the PC), there also exist some strategies. Wilson and Martinez [22] present improved versions of heterogeneous distance functions for nominal variable values by representing the variables as vectors. This means that, variables are represented by different aspects (e.g., the colors are presented in a RGB color model). Another approach to relax non-numerical variables is the relaxation based on the popularity (i.e., the most popular variable value will be used). This strategy can help to identify recommendations where the user's requirements are inconsistent with the underlying KB, but it is not able to identify recommendations similar to the user's requirements. For instance, if the user specifies that the color of the PC should be *white* ($\{c_1{:}color{=}white\}$), a relaxation of *color=black* does not make sense, even if the *black* color is popular. We apply another simple approach for non-numerical variables. If a non-numerical user requirement was inconsistent with the underlying KB, then its neighbor values from the set of variable domains were used for the relaxation. For instance, assuming that a user specifies that the color of the PC should be *yellow* which would be inconsistent and that the domain of the color-variable is defined as follows: dom(**color**) = {black, brown, red, orange, yellow, green, blue, gray, white}. In such cases, we are choosing the neighbors of the user's specified value and relax the non-numerical user requirement as follows: $\{color = yellow \;\|\; color = orange \;\|\; color = green\}$. After the relaxation of the inconsistent variables, a recommendation will be suggested to the user (see line 10). Thereafter, the similarity metrics mentioned in Sect. 4 are applied in order to calculate the similarity between the user requirements and the recommended item. Finally, the average of all similarities for all users is taken into account in order to determine the quality of the *soft relaxation-based approach.*

Hard Relaxation-Based Approach: This approach identifies recommendations similar to the user's requirements by deleting the diagnosed user requirements (i.e., hard relaxation) and by using constraint solver heuristics. This approach uses the following lines of the Algorithm 1: 1–3, 6–8, and 12–17. First, the approach takes the user's requirements and checks whether the user requirements C_R are consistent with the underlying KB C_{KB}. In the case of a consistency, the recommender will only consider the first 10 solutions and recommend the solution which has the highest similarity with the original user requirements (see line 1–2 in Algorithm 1). Otherwise, FastDiag will only consider all those user requirements which are already ordered with respect to their importance and identify a diagnosis with minimal cardinality. The identified diagnosis will most probably contain less important user requirements which is a strategy to prevent a deterioration of the user's satisfaction. Thereafter, all diagnosed user requirements will be deleted from the user constraint set which guarantees that at least one item can be recommended based on the user's remaining constraints. After that, the variable- and value-ordering heuristics of the Choco [16] constraint solver are applied in order to identify similar recommendations.[3] In Choco, the user defines constraints and tries to identify solutions which satisfy his/her requirements by using alternating constraint filtering algorithms with a search mechanism. The following Choco heuristics are applied in our approach:

- **Choco value-ordering heuristics:** *IntDomainMax, IntDomainMin, IntDomainMedian, IntDomainRandom, IntDomainRandomBound, IntDomainMiddle*
- **Choco variable-ordering heuristics:** *FirstFail, Largest, Smallest, Random, AntiFirstFail, MaxRegret*

At the beginning, a variable-ordering heuristic has to be selected for the application of the Choco heuristics, to determine the ordering of the variables. Then, a value-ordering heuristic can be applied for each variable to determine the ordering of the values. There are 36 different heuristic combinations (6xVariable- and 6xValue-Ordering heuristics). The application of a heuristic combination will not affect the recommendation list, but its application will lead to a different ranking of the list. Our goal is to identify the heuristic combination which leads to a recommendation list where the most similar recommendations are located on the top of the list. However, we did not apply Choco value-ordering heuristics for each variable, because for some variables the used value-ordering heuristics should not change. For instance, for the *price of the PC*, only the *less-is-better (IntDomainMin Value ordering heuristic)* metric makes sense (i.e., the cheaper the PC is, the higher the user's satisfaction would be). Therefore, for some variables, the applied value-ordering heuristic will not change and for the remaining variables, we tried all the heuristic combinations. The heuristic combination which leads to the most similar recommendations is later used to test the *hard relaxation-based approach* (see line 8). Finally, the rest of the algorithm will be executed as explained before.

[3] Choco [16] is a free open-source constraint solver library for the Java programming language. http://www.choco-solver.org/.

Algorithm 1. Identification of similar requirements based on soft- and hard-relaxation

```
1: for user u : users do
2:    if checkConsistency(u.reqs) == false then
3:       diagVars = FASTDIAG(orderReqsBasedOnImp(u.reqs))
4:       relaxReqs = reqsRelaxation(diagVars)
5:       if checkConsistency(u.reqs + relaxReqs) == false then
6:          deleteDiagConstraints(diagVars)
7:          applyChocoHeuristics()
8:          rec = getRecommendation(u.reqs - diagVars, heuristic)
9:       else
10:         rec = getRecommendation(u.reqs + relaxReqs)
11:      end if
12:   else
13:      rec = getRecommendation(u.reqs)
14:   end if
15:   similarityPerUser += calculateSim(rec, u.reqs)
16: end for
17: similarity = similarityPerUser / users.size()
```

6 Evaluation

The evaluation of similarity-aware constraint-based recommendation based on both approaches is presented in this section. The training and testing of our approaches are based on two different knowledge bases (from *personal computer* and *bike* domains) defined by the *Configuration Benchmarks Library* (CLib).[4]

6.1 Personal Computer Dataset

The first dataset represents the KB of a *personal computer* which consists of 45 variables with different domain values and of more than 200 KB constraints. Such knowledge bases from complex domains have usually millions of solutions and the similarity calculation between the user requirements and all possible solutions in order to identify the most similar recommendation is not possible due to the poor runtime performance. For testing our approaches, we artificially generated 500 random user requirements.[5] Additionally, we also randomly generated the importance of variables from the user's point of view.

Soft Relaxation-Based Approach: The result of this approach achieves a very high similarity on average, but it does not always guarantee to find recommendations for all users (see Table 1). However, in certain cases where the recommender can identify a solution, the similarity degree will be very high since all the variable values of the identified recommendation will be close to the requirements defined by the user. The non-cumulative normal distribution of the

[4] https://www.itu.dk/research/cla/externals/clib/, Maintained by CLA group. KB definition in CSP representation: https://github.com/CSPHeuristix/CDBC/.

[5] All user requirements were inconsistent with the underlying KB.

similarities is depicted in Fig. 1. As shown in Table 1 and Fig. 1, the average of similarities based on the *soft relaxation-based approach* is very high and data points are close to the mean ($\mu = 94{,}11$ % and $\sigma = 2{,}66$).

Hard Relaxation-Based Approach: In this approach, we train the system with 500 automatically generated user requirements in order to identify the most suitable CHOCO heuristic. We figured out that the *Largest* variable-ordering as well as the *IntDomainMedian* value-ordering heuristic combination achieve the highest similarity on average. This means that the CHOCO constraint solver orders variables based on the largest values in its domain and then selects the median value from the variable domain. After the identification of the most suitable heuristic combination, we tested the same approach with another 500 user requirements which were not used in the training phase.

The results show (see Table 1 and Fig. 1) that the mean and the standard deviation of the *soft relaxation-based approach* are significantly better than the mean and the standard deviation of the *hard relaxation-based approach*. However, an average similarity of 84,68 % and a standard deviation of 9,91 % is also an acceptable result. Moreover, this approach is able to identify recommendations for all 500 users, whereby the *soft relaxation-based approach* identifies recommendations only for 82 out of 500 users. The time consumption of both approaches is about 20 seconds which means that a recommendation per user can be calculated in ~40 ms which is quite acceptable.[6] The main reason for the huge time consumption in both approaches is the CHOCO constraint solver which creates a new CHOCO model for each user. A new CHOCO model will be generated for each user which takes all the user's requirements and KB constraints into the account.

6.2 Bike Dataset

The second dataset represents the KB of a *bike recommendation* which consists of 34 variables with different domain values and more than 350 KB constraints.[7] Examples for constraints can be *frame-type, color,* or *tire-height* of the bike. Furthermore, there are also constraints regarding the customers (i.e., users) such as *gender, height,* and *weight* of a customer.

Soft Relaxation-Based Approach: The evaluation on both datasets shows similar results (see Table 1 and Fig. 1). The results show that the average similarity of the *soft relaxation-based approach* is very high and the data points are very close to the mean. Furthermore, the time consumption of the *soft relaxation-based approach* using the *bike* dataset is much higher than the *personal computer* dataset. The reason for this is the high number of KB constraints in the *bike* dataset (120 ms vs. 40 ms per recommendation).

[6] Our approaches were implemented in programming language Java and were executed on a computer with following properties: Windows 10 Enterprise; 64-bit operating system; Intel(R) Core(TM) i5-5200 CPU @ 2,20 GHz processor; 8,00 GB RAM.

[7] For training and testing our approaches, we automatically generated again 500 user requirements. All user requirements were inconsistent with the underlying KB.

Table 1. Similarity results of the *soft-* and *hard relaxation-based approaches* on both datasets. μ indicates the mean and σ the standard deviation.

Dataset	Relaxation type	μ	σ	Margin of error	Confidence interval	Number of recommendations	Time
PC	Soft	94,11%	2,66%	94,11 ± 0,08%	95%	82/500	23,2 s
	Hard	84,68%	9,91%	84,68 ± 0,04%	95%	500/500	19,4 s
Bike	Soft	91,24%	7,02%	91,24 ± 0,08%	95%	190/500	60,2 s
	Hard	81,31%	16,17%	81,31 ± 0,06%	95%	500/500	42,1 s

Hard Relaxation-Based Approach: We can from the results (see Table 1 and Fig. 1) observe that the *hard relaxation-based approach* is able to work independently from the domain. The application of this approach on both datasets leads to similar results. As already discussed, the time consumption of the *hard relaxation-based approach* using the *bike* dataset takes longer than using the *personal computer* dataset (84 ms vs. 40 ms per user recommendation).

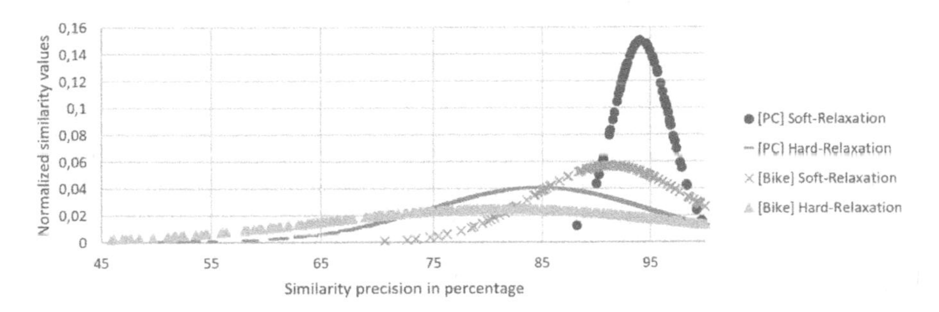

Fig. 1. Normal distribution of similarities on *PC* and *bike* datasets using *soft-* and *hard relaxation-based* approach.

Our observation of the characteristics of each approach using datasets from different domains leads to the conclusion that both approaches are able to identify similar recommendations independently from the domain even when the user's requirements are inconsistent with the underlying KB. The *soft relaxation-based approach* is able to identify similar recommendations with a high similarity (>91%), but it is not able to identify recommendations for all the users. The *hard relaxation-based approach* can identify similar recommendations for all users, but its average similarity (>81%) is lower than the average similarity of the *soft relaxation-based approach.* To properly counteract these undesired issues, we propose a hybrid approach which combines the advantages of both approaches (see Algorithm 1). The hybrid approach simply tries to identify a recommendation using the *soft relaxation-based approach.* Whenever a recommendation can be identified, the similarity will be very high. Otherwise, the *hard relaxation-based approach* will be applied which guarantees that at least one recommendation can be suggested to each user.

7 Conclusion and Future Work

This paper presented the analyzation of two different constraint-based RS which can recommend items similar to the user's requirements. Both recommendation approaches are able to identify similar recommendations even when the user's requirements are inconsistent with the underlying KB. We evaluated both RS in terms of the similarity degree and the runtime performance with KB from different domains and figured out that both approaches are able to recommend similar items in an effective and efficient way. Finally, we propose a hybrid recommender system which can identify similar recommendations with a high degree of similarity by combining the advantages of both approaches.

With regard to *similarity-aware constraint-based recommendations*, we want to evaluate our approaches in other domains such as *cars*, *round trips*, and *smart homes* in order to determine their performance. Moreover, we plan to develop our own heuristic based on the idea of the Choco heuristics in order to identify recommendations which are similar to the user's requirements. Another idea regarding future work is to develop intelligent relaxation strategies for numerical and non-numerical variables in order to recommend similar items to the users.

Acknowledgments. The work presented in this paper has been conducted within the scope of the research projects WeWant (basic research project funded by the Austrian Research Promotion Agency) and OpenReq (Horizon 2020 project funded by the European Union - 732463).

References

1. Burke, R.: Knowledge-based recommender systems. Encycl. Libr. Inf. Syst. **32**(2000), 175–185 (2000)
2. Dabrowski, M., Acton, T.: Beyond similarity-based recommenders: preference relaxation and product awareness. In: Huemer, C., Setzer, T. (eds.) EC-Web 2011. LNBIP, vol. 85, pp. 296–307. Springer, Heidelberg (2011). https://doi.org/10.1007/978-3-642-23014-1_25
3. de Kleer, J., Mackworth, A.K., Reiter, R.: Readings in model-based diagnosis. In: Characterizing Diagnoses and Systems, pp. 54–65. Morgan Kaufmann Publishers Inc., San Francisco (1992)
4. Eiter, T., Erdem, E., Erdoğan, H., Fink, M.: Finding similar or diverse solutions in answer set programming. In: Hill, P.M., Warren, D.S. (eds.) ICLP 2009. LNCS, vol. 5649, pp. 342–356. Springer, Heidelberg (2009). https://doi.org/10.1007/978-3-642-02846-5_29
5. Felfernig, A., Atas, M., Tran, T.N.T., Stettinger, M., Erdeniz, S.P., Leitner, G.: An analysis of group recommendation heuristics for high- and low-involvement items. In: Benferhat, S., Tabia, K., Ali, M. (eds.) IEA/AIE 2017. LNCS (LNAI), vol. 10350, pp. 335–344. Springer, Cham (2017). https://doi.org/10.1007/978-3-319-60042-0_39

6. Felfernig, A., Schubert, M., Reiterer, S.: Personalized diagnosis for over-constrained problems. In: Proceedings of the Twenty-Third International Joint Conference on AI, IJCAI 2013, pp. 1990–1996. AAAI Press (2013)
7. Gasparic, M., Janes, A.: What recommendation systems for software engineering recommend. J. Syst. Softw. **113**(C), 101–113 (2016)
8. Hebrard, E., Hnich, B., O'Sullivan, B., Walsh, T.: Finding diverse and similar solutions in constraint programming. In: Proceedings of the 20th National Conference on Artificial Intelligence, AAAI 2005, vol. 1, pp. 372–377. AAAI Press (2005)
9. Jannach, D., Zanker, M., Felfernig, A., Friedrich, G.: Recommender Systems: An Introduction, 1st edn. Cambridge University Press, New York (2010)
10. Junker, U.: Quickxplain: preferred explanations and relaxations for over-constrained problems. In: Proceedings of the 19th National Conference on Artifical Intelligence, AAAI 2004, pp. 167–172. AAAI Press (2004)
11. Konstan, J.A., Miller, B.N., Maltz, D., Herlocker, J.L., Gordon, L.R., Riedl, J.: Grouplens: applying collaborative filtering to usenet news. Commun. ACM **40**(3), 77–87 (1997)
12. McSherry, D.: Similarity and compromise. In: Ashley, K.D., Bridge, D.G. (eds.) ICCBR 2003. LNCS (LNAI), vol. 2689, pp. 291–305. Springer, Heidelberg (2003). https://doi.org/10.1007/3-540-45006-8_24
13. McSherry, D.: Maximally successful relaxations of unsuccessful queries. In: 15th Conference on AI and Cognitive Science, pp. 127–136. AAAI Press (2004)
14. Paraschakis, D.: Recommender systems from an industrial and ethical perspective. In: Proceedings of the 10th ACM Conference on Recommender Systems, RecSys 2016, pp. 463–466. ACM, New York (2016)
15. Pazzani, M., Billsus, D.: Learning and revising user profiles: the identification of interesting web sites. Mach. Learn. **27**(3), 313–331 (1997)
16. Prud'homme, C., Fages, J.-G., Lorca, X.: Choco Documentation. TASC - LS2N CNRS UMR 6241, COSLING S.A.S. (2017)
17. Reiter, R.: A theory of diagnosis from first principles. Artif. Intell. **32**(1), 57–95 (1987)
18. Reiterer, S., Felfernig, A., Jeran, M., Stettinger, M., Wundara, M., Eixelsberger, W.: A wiki-based environment for constraint-based recommender systems applied in the e-government domain. In: Posters, Demos, Late-breaking Results and Workshop Proceedings of the 23rd Conference on UMAP, Dublin, Ireland, 29 June–3 July 2015
19. Ricci, F., Rokach, L., Shapira, B.: Introduction to Recommender Systems Handbook. In: Ricci, F., Rokach, L., Shapira, B., Kantor, P.B. (eds.) Recommender Systems Handbook, pp. 1–35. Springer, Boston (2011). https://doi.org/10.1007/978-0-387-85820-3_1
20. Tsang, E.P.K.: Foundations of Constraint Satisfaction. Computation in Cognitive Science. Academic Press, Cambridge (1993)
21. Von Winterfeldt, D.: Decision analysis and behavioral research (1986)
22. Wilson, R.D., Martinez, T.R.: Improved heterogeneous distance functions. J. Artif. Int. Res. **6**(1), 1–34 (1997)

Understand the Buying Behavior of E-Shop Customers Through Appropriate Analytical Methods

Jaroslav Olejár, František Babič(✉), and Ľudmila Pusztová

Department of Cybernetics and Artificial Intelligence,
Faculty of Electrical Engineering and Informatics,
Technical University of Košice, Letná 9, 042 00 Košice, Slovak Republic
{jaroslav.olejar.2, frantisek.babic,
ludmila.pusztova.2}@tuke.sk

Abstract. Customer satisfaction represents a crucial goal for every seller. In e-commerce, it is possible to increase this factor by a better understanding of customers purchasing behavior based on collected historical data. In a period of a continually growing amount of data, it is not an easy task to effectively pre-process and analyses. Our motivation was to understand the buying behavior of the on-line e-shop customer through appropriate analytical methods. The result is a knowledge set that retailers could use to deliver products to specific customers, to meet their expectations, and to increase his revenues and reputation. For recommendations generation, we used a collaborative filtering method and matrix factorization associated with Singular Value Decomposition (SVD) algorithm. For segmentation, we selected the K-Means algorithm and the RFM method. All methods produced interesting and potentially useful results that will be evaluated and deployed into practice.

Keywords: E-shop · Transactions · Recommendations · Segmentation

1 Introduction

The growing number of internet users generates a large volume of information. However, information interesting for one user may not be attractive to the other users. Using recommendation systems can be a way to offer the user the information that will be useful to him. We can define the recommendation system as a user decision-making strategy in the information environment [1]. The main task is to filter out a wide range of information that will be interesting for the user, useful and in line with his preferences. This system uses user information and the experience of other customers to provide the right alternatives to help navigate in the unknown domain. The basis of the recommendation system represents a database with stored data describing relationships between users and objects. Customer segmentation is an approach of dividing a customer base into groups of individuals that are similar in specific ways relevant to marketing, such as age, gender, interests, and purchasing habits.

F. Wotawa et al. (Eds.): IEA/AIE 2019, LNAI 11606, pp. 300–307, 2019.
https://doi.org/10.1007/978-3-030-22999-3_27

We aimed to create a baseline model (an initial knowledge set) that can serve for further analysis. This model can be used to confirm already expected knowledge and to validate newly acquired knowledge.

The paper is organized as follows. At first, we introduce the topic and our motivation to deal with it. Section 1.1 described briefly existing related works or studies to investigate possible suitable methods, advantages, or disadvantages. Section 2 introduces the used methods and CRISP-DM methodology. Section 3 presents the performed experiments and obtained results. Finally, Sect. 4 presents our conclusions.

1.1 Related Work

Sarwar et al. have analyzed recommendation algorithms in their work [2] to increase the scalability of collaborative filtering methods and to improve the quality of customer recommendations. They used Movie Lens datasets with movie ratings by customers and an E-commerce dataset with product purchases. Authors generated recommendations using the association rules and the most frequent item method. Authors also used the Singular value decomposition technique to reduce the data dimensionality and F1 metric for evaluation. On Movie Lens dataset, the most frequent item had F1 value 0.22, and the association rule method 0.21. On e-commerce dataset, F1 was lower, 0.16 for both approaches.

Shao-Lun Lee focused on supporting commodity recommendations based on customer preferences in the retail business [3]. He worked with the customer transaction database and made a selection of target customers using RFM scores and ensured that recommendations would be directed only to crucial customers.For these customers, the author derived NRS (Normalized Relative Spending) value to identify their preferences to divide them into clusters. After that, he applied the C4.5 algorithm to assigned particular commodities to the clusters likely to be purchased by customers in them. Finally, he generated a list of recommendations for each customer with accurate precision of 85%.

Jamali and Ester focused their research on utilizing social networking information to improve recommendations and reduce the impact of user's cold start on the expected quality [4]. They used different approaches how to solve this problem: the Social Matrix Factorization algorithm (SocialMF), the STE algorithm similar to SocialMF, Basic Matrix Factorization algorithm (BaseMF) and the most frequent item algorithm. The authors applied these algorithms on the Flixster dataset containing movie ratings and Epinions dataset with different types of object's ratings. The RMSE metric was used to evaluate the generated recommendations. The SocialMF algorithm achieved the best results on both data sets, specifically 0.815 on Flixster dataset and 1.075 on Epinions dataset.

Sari et al. performed a customer segmentation research [5]. Tavakoli et al. combined the RFM model with customers clustering using K-Means. They applied this model to the Digikala Company, the biggest E-Commerce in the Middle East, and the results showed improvements in the number of the purchase and average monetary of the baskets [6].

By analyzing these works, we better understand the issue of recommendation systems. They inspired us by using algorithms of collaborative filtering and matrix factorization for generating recommendations. We also used the K-Means algorithm and the RFM method for customer segmentation. The mentioned works dealt with various available data different from our sample but provided important knowledge framework to solve this topic.

2 Methods

The most popular methods for recommendation generation are content-based, collaborative, and hybrid filtering [7]. In our experiments, we used collaborative filtering. This method uses a database providing a comprehensive matrix of evaluations. This matrix includes users, objects, and object ratings by users representing by numerical values. Its main advantage is the better extraction of recommendations for those objects that are difficult to describe through metadata, such as videos and music. This method assumes that users with similar interests in the past will have similar interests in the future. Recommendations are not generated solely based on the user's ratings but based on other user ratings.

The K-Nearest Neighbor algorithm is a type of instance-based learning, or lazy learning, where the function is only approximated locally, and all computation is deferred until classification [8]. The output is a class membership. The algorithm classifies an object by a plurality vote of its neighbors, with the objective being assigned to the class most common among its k nearest neighbors.

The K-means algorithm is a type of partitioning algorithm, widely used for its smooth implementation and fast execution [9]. It takes numerical parameters and partitions a set of (n) objects into (k) clusters so that the resulting intra-cluster similarity is high while inter-cluster similarity is low. Cluster similarity is measured with the mean value of distances between objects in a cluster, which can be considered as the cluster's center of gravity. The performance of clusters may be affected by the chosen value of (k), as searching for the appropriate number for a given data set is generally "a trial and error" procedure.

RFM is a method used for analyzing customer value [10]. It divides the customers into groups by Recency (time since last order), Frequency (number of orders), and Monetary (total cost of the order). We have to calculate for each customer value of 1, 2, 3, or 4 for each of these three metrics. Value 1 means customer group with spending the most money, made the highest number of orders, and their last order was recent. Value 4 represents the customers who spend the least money, made the least orders, and their last order was a long time ago. The best customers for sellers are who have the value 1 in all these metrics.

Exploratory data analysis (EDA) is an approach to analyzing data sets to summarize their main characteristics, often with visual methods.

2.1 CRISP-DM Methodology

We managed our analytical process in line with the CRISP-DM (Cross-Industry Standard Process for Data Mining), representing the most popular methodology for data analytics and data science. This methodology defines six main phases [11, 12]:

Business understanding deals with a specification of business goal followed with its transformation to a specific analytical task(s). Based on this specification, we can select relevant mining methods and the necessary resources.

Data understanding starts with a collection of necessary data for the specified task and ends with a detailed description, including some statistical characteristics.

Data preparation is usually the most complex and most time-consuming phase, generally taking 60 to 70% of the overall time. It contains data aggregation, cleaning, reduction, or transformation. The result covers prepared data for modeling phase.

Modeling deals with the application of suitable data mining algorithms on the pre-processed data. Also, it is necessary to specify the correct metrics for results evaluation, e.g., accuracy, ROC, precision, recall, etc.

The evaluation phase is oriented towards the evaluation of generated models and obtained results based on specified goals in business understanding.

The deployment contains the exploitation of created mining models in real cases, their adaptation, maintenance, and collection of acquired experiences and knowledge.

3 Experiments and Results

In this section, we summarized the main points of our analytical process performed following the CRISP-DM methodology.

3.1 Business Understanding

The business goal of our analytical process was to understand the buying behavior of the on-line e-shop customer better. The result will be a knowledge set that retailers could use to deliver products to specific customers, to meet their expectations, and to increase his revenues and reputation. This set should be presented in a simple and understandable form for people without more in-depth knowledge from the data analytics domain. Responsible staff will evaluate all the results from their business perspectives. From the analytical point of view, we would extract recommendations and possible interesting hidden patterns in the data. For this purpose, we used collaborative recommendation techniques, clustering, and exploratory data analytics.

3.2 Data Understanding and Preparation

For ease of understanding, we have decided to combine the phase of data understanding and preparation into one chapter.

The anonymized customer's number data sample was provided by the e-shop offering nutritional supplements for athletes. The data contained more than 3 million records (orders) described by 13 relevant variables (Table 1); more than 70 thousand unique customers who bought at least one item of goods.

Table 1. Attributes description.

Name	Description
order_id	Unique identification of the order
store_name	Name of the store by country
created_at	Time and date of order in the format: year-month-day hour:min: sec
shipping_description	The way of delivering the products to the customer (personal, post, courier)
method	Payment methods for products (cash, card, transfer, etc.)
customer_id	Anonymized customer's number
country_id	Acronym of the customer country
city	Customer village
item_id	Unique identification of product
name	Name of the products
sku	Unique identification + short name of the same products group
gty_ordered	The number of ordered products
price	Price of products in €
is_wholesale	Information, if the products are sold to the wholesaler

The data sample contained more than 70 thousand unique customers who bought at least one item of goods. At the same time, there were almost 5 thousand individual items of products that have been purchased by at least one customer. We started with EDA to provide easy to understand characteristics of the historical customers' behavior like the ratio between the number of created orders and the related hour of the day. The graphs showed that customers most buy between nine and ten o'clock in the evening. From the morning, the numbers have increased to lunchtime. A slight decrease occurred in the afternoon. The lowest amount of orders was created late in the night. The most preferred payment method was cash on delivery, almost 80%. Most often order that cost more than 9€ and less than 25€. We can notice that after crossing the limit 20€, most orders had the round price.

Some records were duplicated, for example, the correct product name, but wrong product price, or in reverse. We solved this issue by associating three variables item_id, parent_item_id, and product_type into a new variable. This operation reduced the number of records to around 1.7 million. Also, we solve the inconsistencies related to the countries and the right currency.

In the end, we chose four different samples for modeling phase. The first sample contained 30 thousand customers. These customers have the highest number of various products in their orders. The second sample included only products with the price higher than 9.95€; the third one higher than 19.95€. The last sample consisted only of products that were first added to the cart to prevent the cases that a product was added just for completion.

3.3 Modeling and Evaluation

We used two different approaches to generating recommendations, namely collaborative filtering based on object similarity using the k-Nearest Neighbor (k-NN) algorithm and method for matrix factorization associated with Singular Value Decomposition (SVD) algorithm.

For building the k-NN model, we used the binary matrix M × N, M - the number of individual products, N -the number of unique customers. The matrix values were 0 (customer did not buy this product) or 1 (customer buy this product). For calculation, we have used the cosine similarity. Based on several experimental attempts, we set up the k parameter to 5. The result was a list of recommendations for each product with a calculated cosine distance (the cosine of an angle between two vectors, $1 - (u*v)/\|u\|_2 * \|v\|_2$). The lower value means more significant similarity of products. A perfect example is represented by the protein bars that had a low value of cosine similarity (0.364). It means that it is interesting for the e-shop to recommend other flavors of the bars in the case that customer put one of them into a shopping cart. A similar approach can be applied to the peanut butter with cashew butter (0.752), non-calorie syrup (0.746) or multivitamin complex (0.734). The lower distance, in this case, also have the other types of peanut butter (0.602). The recommendations for Joint Nutrition do not have similar low cosine similarity but shows customer habits including other types of vitamins (0.747 - multivitamin complex, 0.793 - Vitamin C) or nutritional supplements (BCAA - 0.828). The results are different depending on the data sample. When we used all products, we often had the same types of products in our recommendations. If we filtered products by price, we did not get more interesting products in our recommendations. A good alternative was to filter products by order of addition to the cart, but the lowest value of cosine distance is usually on the first sample with 30 thousand customers.

Using the factorization method, we wanted to find the products that the customer has not bought yet, but are likely to be interested in them. We also created the matrix M × N, but in this case, M was the number of customers, and N - the number of products. The matrix values represented some purchases for each product by the customer. We normalized all values on a scale from −1 to 1 and applied the SVD method with the number of dimension 20. The result was a multiplication of three matrices U, VT, and Σ. Based on this multiplication, we were able to predict the products for each customer. For example, we identified a customer buying various types of gainers, and the recommendations included vitamins, proteins, or sports bottle (did not include in his previous orders). The second customer bought many products in the form of tablets, so the first recommendation is a pill box. Another customer bought mainly vitamins and protein bars, so the recommendations include other products from these categories like joint nutrition, Omega 3, etc.

Each customer can be assigned to a group (cluster) based on his or her buying behavior. We call this approach customer segmentation, and a clustering algorithm can do it. We used the K-Means algorithm and focused on the number of purchased products and an average price paid for one product. For each customer, we calculated how much products he bought, the total cost of the order, based on it the average price for one product. We removed extreme values, i.e., the customers who have purchased more than 300 products or whose average cost per one product was over 100€. Finally, we normalized the values. We used the Elbow method [13] to determine the optimal number of obtained clusters (k value = 4). We expected the following four clusters: few or many purchased products/low or high price. In the results, we saw that the last cluster could not be created, probably because of previous processing operations (Fig. 1).

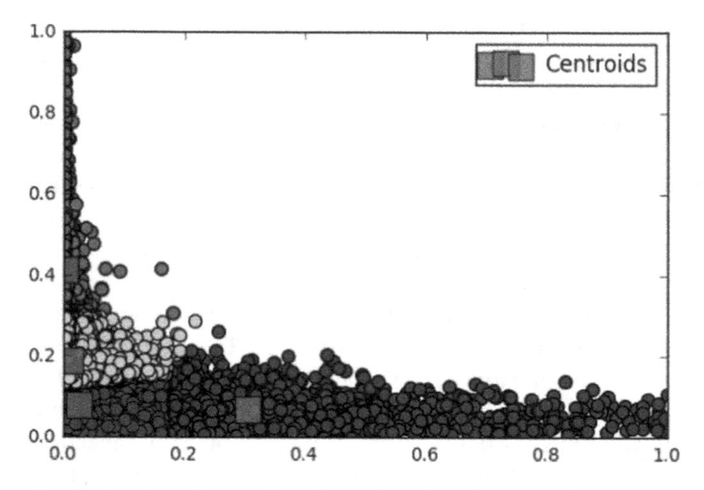

Fig. 1. Price distribution (x-axis: number of purchased products, y-axis: an average price paid, both normalized).

Before we applied the RFM method, we calculated for each customer how many orders he made, a total price, and how many days had elapsed since the last purchase (our boundary was 1/10/2018). After removing the extreme values (customers with more than 300 orders or more than 10 thousand paid costs), we divided each of the three metrics into quarters and assigned customers to one of them. We can characterize the top extracted customers as follows: at least 10 orders, their last purchase was no more than 70 days ago, and their total price was at least 90€ (RFM = 111).

4 Conclusion

This paper describes several methods to understand the buying behavior of the e-shop customer. The collaborative filtering using k-Nearest Neighbor algorithm resulted in a list of recommendations for each product with a calculated cosine metric, such as different flavors of the protein bars. The factorization method identified for each customer the products that have not bought yet, but are likely to be interested, e.g., the sports bottle for the customer with the higher frequency of the gainers buying. The last experiments dealt with customers' segmentation. For this purpose, we used two methods like K-Means and RFM. All results were plausible and step by step evaluated by the cooperating company. The company plans to implement a more advanced product recommendation mechanism for customers in the short term. Tested approaches such as collaborative filtering and matrix factoring represent an option with high deployment potential. They already use the RFM method to segment customers but with different settings. Our results will make it possible to compare both approaches and optimize the use of this method

Acknowledgments. The work was partially supported by the Slovak Grant Agency of the Ministry of Education and Academy of Science of the Slovak Republic under grant no. 1/0493/16 and The Slovak Research and Development Agency under grant no. APVV-16-0213.

References

1. Isinkaye, G.O., Folajimi, Y.O., Ojokoh, B.A.: Recommendation systems: principles, methods and evaluation. Egypt. Inf. J. **16**, 261–273 (2015)
2. Sarwar, B., Kapyris, G., Konstan, J., Riedl, J.: Analysis of recommendation algorithms for e-commerce. In: Proceedings of the 2nd ACM conference on Electronic commerce - EC 2000, pp. 158–167. ACM Press, New York (2000)
3. Lee, S.: Commodity recommendations of retail business based on decision tree induction. Expert Syst. Appl. **37**(5), 3685–3694 (2010)
4. Jamali, M., Ester, M.: A matrix factorization technique with trust propagation for recommendation in social networks. In: Proceedings of the Fourth ACM Conference on Recommender Systems - RecSys 2010, pp. 135–142. ACM Press, New York (2010)
5. Sari, J.N., Nugroho, L., Ferdiana, R., Santosa, P.I.: Review on customer segmentation technique on ecommerce. J. Comput. Theor. Nanosci. **22**(10), 3018–3022 (2016)
6. Tavakoli, M., Molavi, M., Masoumi, V., Mobini, M., Etemad, S., Rahmani, R.: Customer segmentation and strategy development based on user behavior analysis, RFM model and data mining techniques: a case study. In: 2018 IEEE 15th International Conference on e-Business Engineering (ICEBE), Xi'an, China, pp. 119–126 (2018)
7. Bobadilla, J., Ortega, F., Hernando, A., Gutiérrez, A.: Recommender systems survey. Knowl.-Based Syst. **45**, 109–132 (2013)
8. Altman, N.S.: An introduction to kernel and nearest-neighbor nonparametric regression. Am. Stat. **46**(3), 175–185 (1992)
9. Hothor, T., Everitt, B.S.: A Handbook of Statistical Analyses Using R, 2nd edn. Chapman and Hall/CRC, Boca Ratoyn (2009)
10. Fader, P.S., Hardie, B.G., Lee, K.L.: RFM and CLV: Using iso-value curves for customer base analysis. J. Mark. Res. **42**(4), 415–430 (2005)
11. Chapman, P., et al.: CRISP-DM 1.0 Step-by-Step Data Mining Guide (2000)
12. Shearer, C.: The CRISP-DM Model: the new blueprint for data mining. J. Data Ware-Hous. **5**(4), 13–22 (2000)
13. Ketchen, D.J., Shook, C.L.: The application of cluster analysis in strategic management research: an analysis and critique. Strateg. Manag. J. **17**(6), 441–458 (1996)

Using Conformal Prediction for Multi-label Document Classification in e-Mail Support Systems

Anton Borg[1(✉)], Martin Boldt[1], and Johan Svensson[2]

[1] Department of Computer Science and Engineering, Blekinge Institute of Technology, 371 79 Karlskrona, Sweden
{anton.borg,martin.boldt}@bth.se
[2] Telenor Sverige AB, 371 80 Karlskrona, Sweden

Abstract. For any corporation the interaction with its customers is an important business process. This is especially the case for resolving various business-related issues that customers encounter. Classifying the type of such customer service e-mails to provide improved customer service is thus important. The classification of e-mails makes it possible to direct them to the most suitable handler within customer service. We have investigated the following two aspects of customer e-mail classification within a large Swedish corporation. First, whether a multi-label classifier can be introduced that performs similarly to an already existing multi-class classifier. Second, whether conformal prediction can be used to quantify the certainty of the predictions without loss in classification performance. Experiments were used to investigate these aspects using several evaluation metrics. The results show that for most evaluation metrics, there is no significant difference between multi-class and multi-label classifiers, except for Hamming loss where the multi-label approach performed with a lower loss. Further, the use of conformal prediction did not introduce any significant difference in classification performance for neither the multi-class nor the multi-label approach. As such, the results indicate that conformal prediction is a useful addition that quantifies the certainty of predictions without negative effects on the classification performance, which in turn allows detection of statistically significant predictions.

Keywords: Conformal prediction · Multi-label classification · Customer support e-mail

1 Introduction

An important part of any corporation is the various interaction processes with the customers. This is especially important for resolving various business-related issues that customers encounter, since failing to resolve such issues in an efficient manner risk negatively affect both the image and the reputation of the corporation. In highly competitive markets a single negative customer service experience

F. Wotawa et al. (Eds.): IEA/AIE 2019, LNAI 11606, pp. 308–322, 2019.
https://doi.org/10.1007/978-3-030-22999-3_28

can deter potential new customers from a company or increase the risk of existing customers to drop out [14], both negatively affecting the sales. Although recent years have shown a shift in the means of communication between customers and customer service divisions within corporations, e.g. using autonomous chatbots or social network-based communication solutions, traditional e-mails still account for an important means of communication due to both its ease and widespread use within almost all customer age groups. Thus, implementing efficient customer service processes that target customer e-mail communication is a necessity for larger corporations as they receive large numbers of such customer service e-mails each day.

In this study we investigate improvements of customer service e-mail management using a supervised learning paradigm with a multi-label classifier. The semi-automated customer service e-mail management system studied exists within one of the bigger telecom operators in Europe with over 200 million customers worldwide, and some 2.5 million in Sweden. When these customers experience problems they often turn to e-mail as their means of communication with the company, by submitting an e-mail to a generic customer service e-mail address. Consequently, such customer service e-mails might be assigned to random customer service personnel. However, experience and knowledge concerning the different areas requiring support might differ. A person with knowledge in the economic aspects of the business does not have the same knowledge in the technical aspects. To address this problem, an intelligent model that classifies the type of issue in an e-mail makes it is easier to direct e-mails to the most suitable handlers.

While traditional classification approaches often use a binary approach, e.g. in spam classification an e-mail is classified as either of two labels (spam or ham), this might not be true for customer service e-mails. In such a setting, the label range might not be binary, but multiple labels can be present in each e-mail, e.g., an e-mail can contain multiple topics. Such classification problems are called multi-label classification [24]. While multi-label classification is not new, using conformal prediction in this setting is not really investigated. Conformal prediction provides not only the predicted label for each instance tested, but also provides the prediction within a certain probability [19,22]. As such, an instance can be predicted with a certain confidence. Inductive conformal prediction is a conformal prediction approach that splits the training set into two parts, the proper training set and the calibration set. As in any supervised learning approach, a model is trained by an underlying algorithm using the training set, and that model is then used to classify instances in the test set. However, for each classification the instances in the calibration set together with the instance that is about to be classified is used to compute the level of certainty in the prediction, represented as a p-value [19]. In the case of e-mail classification, this makes conformal prediction interesting, especially regarding the quantification of confidence in each prediction.

Each customer service e-mail might have different labels, and each label might be of different importance to the customer. For example, a customer service e-mail might contain two paragraphs of text, one detailing a technical issue, and the other one an order errand. As such, the e-mail should be classified with for instance the *Invoice*, *TechicalIssue*, and the *Order* labels. By leveraging the confidence output of the conformal predictor, the labels can be ranked according to the confidence of each prediction. Thus, it is possible to determine the certainty of each subject in the e-mail and process the e-mail in the customer service system accordingly. In our example, the *Invoice* label might have a high confidence, but the *TechicalIssue* and *Order* labels have lower confidences. Consequently, the e-mail is first sent to a customer service representative that is experienced in invoice cases. After that, the same representative might be able to deal with the other aspects or forward the case where relevant.

1.1 Aims and Objectives

In this study we investigate to what extent conformal prediction can be used in multi-label classification of customer service messages based on the content in e-mails. Further, since the state of the current solution used by the studied company is a rule-based multi-class classifier it is considered to be the baseline. Thus, the studied multi-label classifier is compared to the multi-class classifier for determining the primary label of e-mails.

1.2 Scope and Limitation

The scope of this study is within a Swedish setting, involving e-mail messages written in Swedish sent to the customer service branch of the studied telecom company. However, the problem studied is general enough to be of interest for other organizations as well. It is important to stress that classification confidence is not equal to classification importance. That is, the study does not take into account that different sections of an e-mail could have different importance to either the customer or the customer service persons.

2 Background and Related Work

There exist various machine learning-based classification approaches for e-mail classification and the solutions differ with the wide range of classification problems. The most common approach is the binary classification, i.e. a problem with two outcomes ($C = c_0, c_1$). An example of binary classification problem in the e-mail domain is spam classification, where an instance (i.e. e-mail) can be classified as either spam or ham [4]. Other research have investigated separating e-mails into corporate and personal, to enable users to focus on corporate e-mails during work hours [25]. The classification is done based on extracted social network features rather than the e-mail content itself, something that has also been investigated with regard to spam classification as well [4,27]. Other research have investigated automatic e-mail classification for separating complaints from non-complaints [7], improving customer service efficiency by detecting e-mails that require responses comparing the automated approach with a rule-based system [8], or determining the part of an e-mail in a conversation thread [18].

When the number of potential classes are more than two ($C = c_0, ..., c_m$), the problem is transformed into a multi-class problem. As such, the classifier have m potential classes available, instead of two. However, the classes are still mutually exclusive [10]. Research into multi-class classification have investigated for instance classifying e-mail content into specific e-mail folders/clusters based on the content, a.k.a. foldering [1,17]. Other research have focused on the detection of cues indicating emotional aspects within e-mail content [12], or determining the most relevant user-actions (reply, read, and delete) based on e-mail content [13].

However, e-mails and other types of documents often include multiple topics. As such, such e-mails might be more appropriately classified with multiple labels. Techniques for doing so are known as multi-label classification [24], and although similar to multi-class classification there is an important distinction in that the classes are not mutually exclusive [10]. Thus, each instance have a minimum of one label and a maximum of m labels [23]. Multi-label classification has been applied to not only content annotation, automated tagging, but also in other domains such as medicine and marketing [24].

When it comes to studies investigating conformal prediction within multi-label learning only two prior studies have been found within the literature. First, Wang et al. compare Random Forest, Naïve Bayes and a k-Nearest Neighbor (KNN) classifiers (all three using conformal prediction) with a baseline KNN without conformal prediction [15]. The application area is medical classification and the problem is multi-label-oriented, more specifically diagnosing 736 chronic fatigue medical cases using 95 symptoms as labels. The authors further investigate how classification performance is affected by different significance levels, ranging from 0.8–0.99. The results show that the proposed Random Forest implementation outperforms the other candidate models, including the traditional KNN implementation, over the five evaluation metrics used. Random Forest also shows best classification performance over the different significance levels used.

Secondly, Harris Papadopoulos studies the use of conformal prediction using two popular multi-label image datasets [20]. The study investigates how two Neural Network-based models that make use of conformal prediction compares against traditional models learned by both the Neural Network, KNN and Naïve Bayes algorithms. The performance is evaluated over different significance levels as well as label-sets and is measured over four evaluation metrics, including F_1-score (both micro and macro). The results show that the conformal-based Neural Network models outperform the traditional methods, and that the associated confidence p-value are found to be both informative and reliable.

While there are some prior research investigating using conformal prediction for multi-label classification, the research is limited. In particular, to knowledge of the authors, the approach has not been investigated in the context of e-mail-

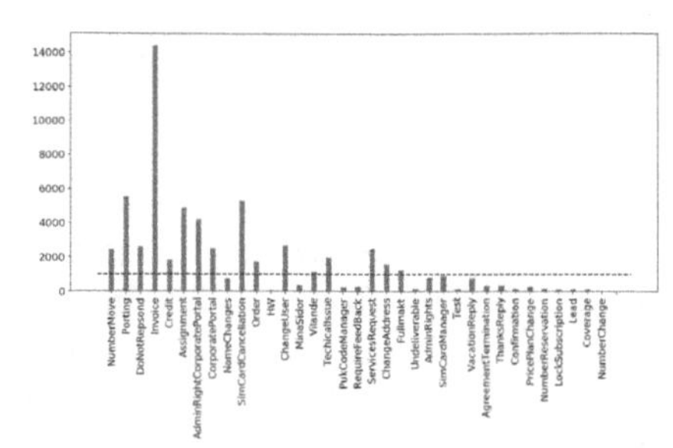

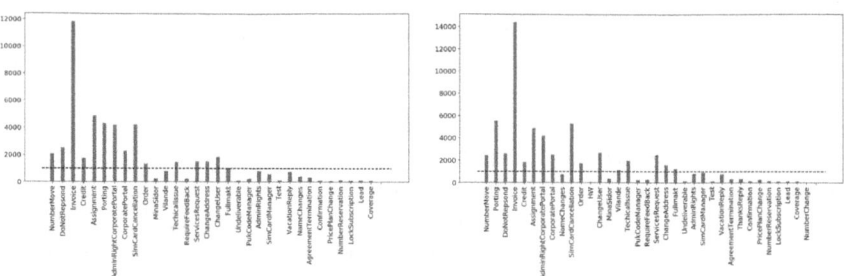

(a) Label frequency for the multi-class dataset, also showing the pruning threshold at 1000 e-mails.

(b) Label frequency in multi-label dataset, also showing the pruning threshold at 1000 e-mails.

Fig. 1. Label frequencies per dataset.

based customer support. As such, it is of interest to investigate this application as this form the identified research gap.

3 Method

In this section, we describe the data used, present our experimental setup, the algorithms evaluated, as well as the metrics and the statistical tests chosen for the study.

3.1 Data

The dataset consists of 51,682 e-mails from the customer service department from a Swedish branch of a major telecom corporation. Each e-mail consists of the subject line, to address, and the content and it is labeled with at least one label. In total there exists 36 distinct labels, each independent from the others, where several of these might be present in any given e-mail. The labels have been set by a rule-based system that was manually developed, configured and fine-tuned over several years by domain expertise within the company. A *DoNotUnderstand* label acts as the last resort for any e-mail that the current labeling system is unable to classify. Those e-mails have been excluded from the dataset, as there exists no real labels to evaluate against. Each e-mail has been anonymized and in addition the *To* and *From* address was completely removed.

Two datasets were constructed, d_{mc} and d_{ml}, where d_{mc} is the data modeled as a multi-class problem, with the primary label for each instance being used. d_{ml} is the data modeled as a multi-label problem, with every label for each instance being used. Due to the large number of labels and the small number of e-mails per label, rarely occurring labels were marked to be removed. A minimum threshold of 1,000 were chosen. The label frequency for both datasets can be observed in Fig. 1a and b for the multi-class dataset and the multi-label datasets respectively, as well as the threshold.

The datasets where randomly subsampled to 20% of their original size (i.e. of the original 51,682 instances, 10,336 were kept). The labels below the threshold is removed from the datasets. Any e-mail instance for which all labels were excluded was also removed from the datasets, since there did not exist any ground truth to evaluate against for those e-mails. The result is that the d_{ml} dataset consists of 9,637 instances and 16 labels, and the d_{mc} dataset consists of 9,138 instances and 14 labels.

3.2 Preprocessing

The e-mail messages were transformed into a bag-of-word representation as both uni-grams and bi-grams. Uni-grams are when sentences are divided into single words, and bi-grams are when sentences are divided into pairs of word. In the second case, words are no longer independent of each other and algorithms can take into context word combinations. An initial pre-study indicated that tri-grams and above did not show an increase in performance [5].

Once a bag-of-words representation using uni- or bi-grams were finished, any stop-words in the datasets were removed. As the primary language in this data set is Swedish, a list of Swedish stop-words was used[1]. However, the Swedish stop-words were extended by English stop-words, as a fair amount of English also occurs due to the corporate environment. The remaining words (after removing stop-words) were then transformed into unigrams and bigrams ranked using a term-frequency and inverse document frequency (TF-IDF) vectorizer and the 3,000 words with highest ranks were kept [16]. The parameter values were chosen after initial tests. The TF-IDF algorithm weights each word based on the term frequency, i.e. how frequent each word is in each document, and the inversed document frequency, i.e. the inverse fraction of documents that contain the word. The term frequency indicates if a word is indicative of a document and the inverse document frequency normalizes each word according to how frequent it is occurring in all documents.

It needs to be stressed that no specific feature selection algorithm was used since this is an initial study for testing the approach of applying multi-label classification and conformal prediction on e-mail content classification. Thus, investigating suitable feature selection algorithms, including more extensive parameter tuning, could be an interesting avenue for future studies in order to improve classification performance beyond the results presented in this paper.

3.3 Experiment Setup

The support vector machine (SVM) classification algorithm was used in the experiments, more specifically the scikit-learn SVC algorithm was used. The motivation for this choice is that SVM has been shown to perform well in other text classification tasks [21]. However, since the SVM implementation in the scikit-learn package does not support multi-label classification, the multi-label

[1] https://gist.github.com/peterdalle/8865eb918a824a475b7ac5561f2f88e9.

and multi-class problems were transformed into one-vs-rest problems, i.e. for each label a binary classifier was trained [11]. As such, the problem was transformed into m binary problems, which is possible since there is no inter-dependencies between labels.

Inductive conformal prediction (ICP) was used to calculate the prediction levels for the models. ICP was used in favor of transductive conformal prediction, since the latter is much more computational demanding [2,19]. This study uses the *nonconformist*[2] conformal prediction package for Python.

The nonconformist package and the one-vs-rest scikit-learn package were extended to enable using the nonconformist package between the one-vs-rest class and the underlying classification algorithm. The conformal predictions were made with both a significance level of 95% and no significance level at all. As such, conformal prediction is compared against a standard approach.

Finally, the experimental setup used a 10 times 10-fold cross-validation approach based on the preprocessed labeled data presented in Subsect. 3.1. The evaluation metrics are detailed in Subsect. 3.4. T-tests were used to detect possible differences between the two approaches [10]. The magnitude of difference between the standard approach and the conformal prediction approach for each metric were calculated using the objective Cohen's d measure of effect sizes [6].

3.4 Evaluation Metrics

The experiments were evaluated using standard evaluation metrics calculated based on the True Positives (TP), False Positives (FP), True Negatives (TN), and False Negatives (FN). The evaluation metrics consists of the F_1-score (micro average), Jaccard index, Accuracy, and area under roc-curve (AUC) (micro average) [10]. It should be noted that for the multi-class case, Accuracy, Jaccard index, and F_1-score will be equivalent.

Jaccard index compares two sets and calculates the similarity by dividing the size of the intersection with the size of the union of the two sets [3], i.e. as in Eq. 1:

$$Jacc = \frac{|A \cap B|}{|A \cup B|} \tag{1}$$

The Jaccard index is calculated between the prediction and the correct labels. The metric score is between $0 - 1$, where 1 is a perfect match. Accuracy is in general defined as in Eq. 2 [26]:

$$Acc = \frac{TP + TN}{TP + TN + FP + FN} \tag{2}$$

It is a measurement of how well the model is capable of predicting TP and TN compared to the total number of instances. There exist two aspects for the accuracy metric that should be highlighted. First, for the multi-class case, the accuracy is equivalent to the Jaccard index. Second, for the multi-label setting,

[2] https://github.com/donlnz/nonconformist.

Table 1. The measured metrics' means from the experiment with the multi-label approach (standard deviations are shown within parenthesis). Cohen's d quantifies the size of the difference between both groups per metric.

Metric	Multi-label	Multi-label*	Cohen's d
F_1-score	0.7504 (0.0117)	0.7190 (0.0126)	2.5826
Accuracy	0.6109 (0.0157)	0.5780 (0.0153)	0.8417
Jaccard index	0.6722 (0.0134)	0.6285 (0.0145)	3.1302
Hamming loss	0.0328 (0.0015)	0.0350 (0.0015)	1.4667
AUC	0.8309 (0.0075)	0.8017 (0.0080)	3.7658

*: predicted using conformal prediction with a significance level of 0.05

each instance prediction must match exactly the true labels, i.e. very conservative comparison. Accuracy ranges between $0 - 1$, where 1 is a perfect score.

$$F_1 = 2 * \frac{Precison * Recall}{Precision + Recall} \tag{3}$$

$$Precision = \frac{TP}{TP + FP} \tag{4}$$

$$Recall = \frac{TP}{TP + FN} \tag{5}$$

However, in the case of micro-averaging, precision and recall are calculated according to Eqs. 6 and 7 respectively, where n is the number of classes. Micro-averaging is used as the number of labels vary between classes [26]. Similar to the previous metrics, F_1-score ranges between $0 - 1$, where 1 is a perfect score.

$$Precision_{\mu} = \frac{TP_1 + ... + TP_n}{TP_1 + ... + TP_n + FP_1 + ... + FP_n} \tag{6}$$

$$Recall_{\mu} = \frac{TP_1 + ... + TP_n}{TP_1 + ... + TP_n + FN_1 + ... + FN_n} \tag{7}$$

Hamming loss measures the fraction of labels that are incorrect compared to the total number of labels [23]. A score of 0 indicates no incorrect label predictions.

The AUC calculates the area under a curve (which in this case is the ROC). The AUC, a.k.a. Area under ROC curve (AUROC), is a standard performance measure in data mining applications. Further, it does not depend on an equal class distribution and misclassification cost [9].

Table 2. The measured metrics' means from the experiment with the multi-class approach (standard deviations are shown within parenthesis). Cohen's d quantifies the size of the difference between both groups per metric.

Metric	Multi-class	Multi-class*	Cohen's d
F_1-score	0.6297 (0.0144)	0.6291 (0.0145)	0.0415
Accuracy	0.6297 (0.0144)	0.6291 (0.0145)	0.0415
Jaccard index	0.6297 (0.0144)	0.6291 (0.0145)	0.0415
Hamming loss	0.3703 (0.0144)	0.3709 (0.0145)	0.0415
AUC	0.8006 (0.0077)	0.8003 (0.0078)	0.0387

*: predicted using conformal prediction with a significance level of 0.05

4 Results

When observing the results in Tables 1 and 2 the results suggest the feasibility of using SVM for e-mail and document classification. It should be noted that the accuracy of a random guesser should be $1/|C|$, where $|C|$ is the number of classes. Consequently, the accuracy baseline for the multi-class setup in this study is 0.0714 (i.e. 1/14) and for the multi-label setup it is 0.0625 (i.e. 1/16). Given this, an accuracy of 0.6109 (std 0.0157) and 0.6297 (std 0.0144) should be considered quite good.

Comparing significant classifications in the multi-label approach with the case where no significance level was used show slightly lower evaluation scores, as shown in Table 1. T-tests were performed between the two approaches for the different evaluation metrics. All of the evaluation metrics show significant difference; AUC (t(198) = 26.6193, $p < 0.05$), Jaccard index (t(198) = 22.0849, $p < 0.05$), Hamming loss (t(198) = −10.4520, $p < 0.05$), Accuracy (t(198) = 14.9141, $p < 0.05$), F_1-score (t(198) = 18.1770, $p < 0.05$). The difference per metric between both approaches are quantified using the Cohen's d measure, which ranged 0.8417 – 3.7658. Such effect size values indicate a large difference between the conformal approach and the standard approach. Thus, that there is decreased classification performance when using conformal prediction at significance level of 0.05.

Comparing significant classifications with results where no significance level was used also show very similar results for the multi-class approach, see Table 2. Similar to the multi-label results, T-tests were performed between the two approaches for the different evaluation metrics. However, no statistical significant difference were detected between the two approaches; AUC (t(198) = 0.2717, $p = 0.7861$), Jaccard index (t(198) = 0.2717, $p = 0.7861$, Hamming loss (t(198) = 0.2717, $p = 0.7861$), Accuracy (t(198) = 0.2717, $p = 0.7861$, F_1-score (t(198) = 0.2717, $p = 0.7861$). It should be pointed out that although the test-statistica and p-value is the same, this is only due to the precision used. A higher precision shows differences between the measurements used. Further, the effect sizes

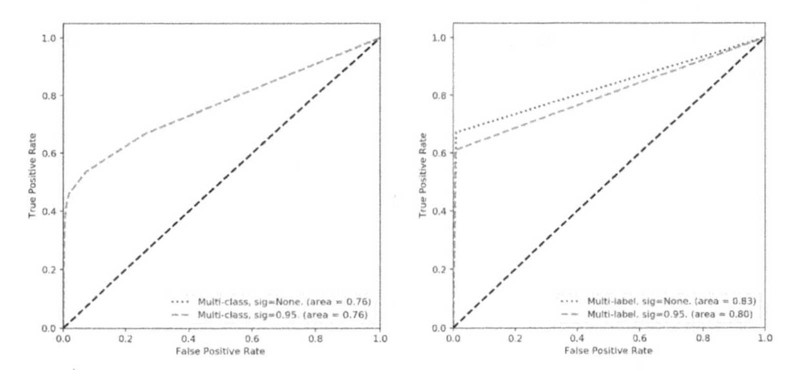

(a) Results for the multi-class dataset. (b) Results for the multi-label dataset.

Fig. 2. Micro-averaged AUC.

quantified by the Cohen's d measure ranged 0.0387 and 0.0415, which means that only negligible differences exist between the approaches.

Comparing AUC for multi-label and multi-class indicates a statistical significant difference when using no significance level ($t(198) = 28.0141$, $p < 0.05$), but not when using a significance level ($t(198) = 1.2184$, $p = 0.2244$). The Accuracy, Jaccard index, and F_1-score are difficult to compare between the two approaches. Given the conservative accuracy measurement in the multi-label context the metric is not really comparable to the multi-class context but can give an indication of how exact the multi-label performed. Between the multi-class and multi-label approach, T-tests for accuracy found significant differences between both the significant predictions ($t(198) = 8.2172$, $p < 0.05$) and the non-significant predictions($t(198) = 22.4172$, $p < 0.05$). In this case the Jaccard index and F_1-score is similar between multi-label and multi-class. The F_1-score and the Jaccard index are slightly better for multi-label, the T-test for both evaluation metrics indicate that significant difference was detected in all but one case. T-test for F_1-score were $t(198) = -60.9200$, $p < 0.05$ and $t(198) = -43.0720$, $p < 0.05$, for significant and non-significant predictions respectively. Similarly, the T-test for Jaccard index were $t(198) = -20.1286$, $p < 0.05$ and $t(198) = 0.2979$, $p = 0.7664$, for significant and non-significant predictions respectively.

For the Hamming loss, however, the results for the multi-label classifier is lower, indicating a lower number of instances incorrectly classified compared to the multi-class approach. In this case the hamming loss for multi-label is between $0.0313 - 0.0343$ compared to multi-class which is between $0.3559 - 0.3874$, indicating significantly lower mislabeling for the previous approach. T-test comparing the Hamming loss between the two approaches confirm this, both without a significance level ($t(198) = 228.9471$, $p < 0.05$) and using a significance level ($t(198) = 224.4472$, $p < 0.05$).

The micro-averaged AUC between the two approaches does indicate a difference, see Fig. 2. Further, the AUC for individual labels indicates that even though instances can have multiple labels, and as such increase the complexity of the training, the results were not that much worse per individual label, as per Fig. 3. For multi-class, no difference between statistical significant and non-significant predictions where observable.

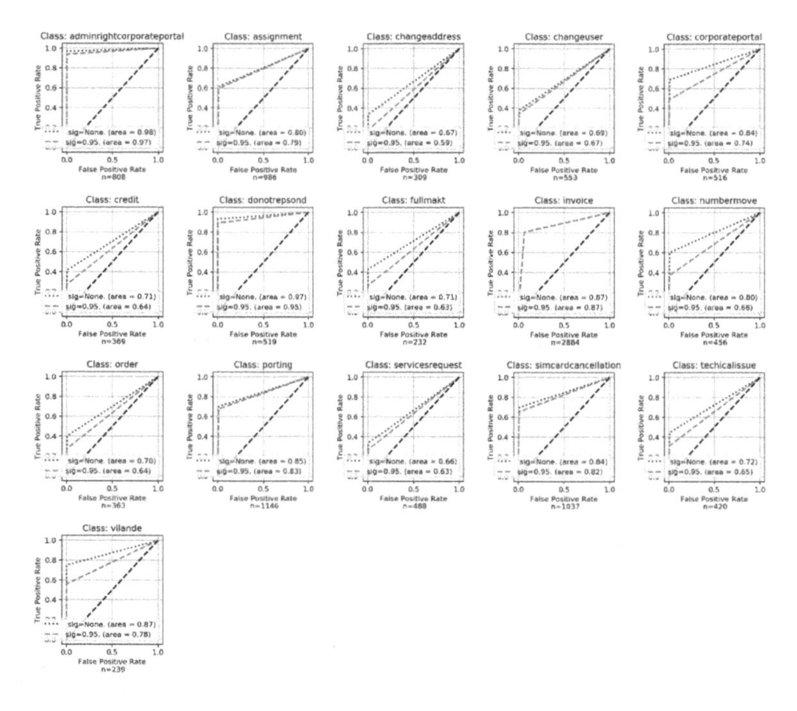

Fig. 3. Multi-label dataset

5 Discussion

A multi-label approach allows the use of multiple labels per document, which in the context of customer support e-mail classification allows improved modeling of real-world e-mail communication patterns. For instance, by attaching several labels to e-mails where customers include multiple customer service errands in the same message. This can also take into account the shift of subjects throughout e-mail conversations, e.g. where an initial e-mail regarding a technical issue later is shifted to an invoice errand. Supporting these real-world e-mail communication patterns is an important aspect in customer service e-mail management. Thus, overall multi-label approaches seem more suitable (compared to multi-class) in this context.

Another important aspect in customer support e-mail classification, as in most classification domains, is to provide a measure of certainty together with the classification [22]. This is one of the advantages that the use of conformal prediction provides, since the framework provides p-values for each classification. These p-values indicate whether the classifications are statistically significant or not given the chosen significance level. Taking the p-values into account allow domain experts to differentiate between instances that are predicted with statistic significance and instances that are not. The former are instances where the model is relatively more certain of the predictions, and the latter are instances where the model is more likely to be mistaken. Thus, the latter are instances that are more likely in need of manual verification by domain expertise and p-values can be used to identify those instances. The p-values could further be used for visualization purposes in order to assist the domain experts in maintaining, for instance, an overview of the classification operation in a system by using aggregated descriptive statistics of the p-values.

In addition to the p-values the conformal prediction framework also provides a credibility score for each instance that indicates how well the model recognizes the instance [22]. That is how novel the instance is compared to the training data which the model has been generated on. Therefore, the credibility score can be used to detect when instances have changed significantly since the model was generated, which means that the model needs to be retrained on an updated sample of data. However, in this context this could possible also be used in combination with new product launches or campaigns to detect if there is a change in the e-mail content for specific labels.

Further, any manually verified predictions, especially predictions without statistic significance, are useful to include when retraining models as they can include new knowledge about the class. This might be cases where the original training data didn't include similar cases (i.e. low credibility of predictions), but where domain experts can provide label-information.

However, all of these possibilities that conformal prediction enable come at a cost of decreased classification performance. For the investigated multi-label approach, the classification performance decrease quite notably, as shown in Table 1 where the effect sizes indicate large notable differences for all metrics. One possible explanation for this is that an e-mail shares the content between labels and as such it might be more difficult for the model to distinguish between the labels, especially when only significant predictions are produced. For the multi-class approach there is a smaller decrease in performance, which is indicated by the negligible effect sizes shown in Table 2. So, the possibilities provided by using conformal prediction is to various extent a trade-off against the classification performance. In many applications it is acceptable to have slightly lower classification performance in order to get the benefits from conformal prediction.

Finally, by using multi-label learning, e-mails that are labeled as *DoNotUnderstand* might also have been given several other possible labels. Using conformal prediction, the labels for each instance can be ranked based on e.g. their confidence score, and as such provide an insight into possible correct labels.

5.1 Case Implications

As stated before, the telecommunication company labels each e-mail using a rule-based system. In the current setting, only the main label is used without any probability information. These labels are then used in the customer support organization, where manager can set up support queues. A support queue consists of a combination of labels decided by a manager, e.g. $queue_1$ consists of e-mails that can be labeled with either *ChangeUser, Invoice, Assignment* and $queue_2$ might consist of e-mails that can be labeled with either *Order, TechnicalIssue*. The different customer support teams then subscribe to queue decided by their manager. Throughout their workday, customer support personnel picks e-mails from their queue to work with.

By adding conformal prediction around the document classifier, be it a rule-based learner or a LSTM classifier, the confidence of each classification can be utilized. Given that two labels have been assigned to an e-mail using multi-label classification, it is possible to use confidence to indicate if the labels have been assigned with a similar confidence or if the model indicates that there is a large disparity between the labels. This would allow a more fine-grained division of e-mails into queues, and by extension the development of specialized customer support teams. Regarding the former, when adding e-mails to queues, the certainty can be used as a threshold for inclusion in the queue, e.g. a label is only included into a queue if the model has an adequate certainty. Further, in the case of the manual rules setup and used by the company, the confidence score can also indicate when a rule has erroneously classified an instance.

Using the credibility score can indicate when the contents for a label has changed to such an extent that it is no longer clearly recognizable, e.g. technical issues might be changed by the introduction of new devices or services. This might indicate that the label contains subtopics or that customer behavior has altered by other factors.

6 Conclusion and Future Work

This work investigates the use of multi-class and multi-label approaches for document classification in a Swedish e-mail customer service system based on the SVM learning algorithm. More specifically the use of conformal prediction within this problem domain is investigated. One of the advantages of conformal predictions is that it indicates which classifications can be considered statistically significant [22]. Both multi-class and multi-label approaches performed had high AUC scores, 0.8006 (std 0.0077) and 0.8309 (std 0.0075) for multi-class and multi-label respectively. Further, the experimental results show that performance loss for using conformal prediction was negligible for both the multi-class approaches, but not for the multi-label approaches. However, the Hamming loss was significantly better for the multi-label approach compared to the multi-class approach ($t(198) = 224.4472$, $p < 0.05$), which indicates significantly less mislabels for the multi-label approach.

Since the use of conformal predictions to provide significant predictions did not have any significant performance impact for either multi-label or multi-class approach, it is the authors opinion that the benefit of indicating statistically significant classifications outweighs the added complexity by introducing the conformal predictions framework. This is especially the case for multi-label predictions, where an extra label might indicate which domain experts are asked to handle an instance.

However, in this study a one-vs-rest approach has been used, but it can be beneficial for companies to have a prioritized list of labels. Therefore, it would be motivated to investigate conformal prediction for multi-label classification when taking this into account.

References

1. Alsmadi, I., Alhami, I.: Clustering and classification of email contents. J. King Saud Univ. - Comput. Inf. Sci. **27**(1), 46–57 (2015). https://doi.org/10.1016/j.jksuci.2014.03.014
2. Balasubramanian, V., Ho, S.S., Vovk, V.: Conformal prediction for reliable machine learning: Theory, Adaptations and applications. Newnes (2014)
3. Borg, A., Boldt, M.: Clustering residential burglaries using modus operandi and spatiotemporal information. Int. J. Inf. Technol. Decis. Mak. **15**(01), 23–42 (2016). https://doi.org/10.1142/S0219622015500339
4. Borg, A., Lavesson, N.: E-mail classification using social network information. In: 2012 Seventh International Conference on Availability, Reliability and Security, pp. 168–173, August 2012. https://doi.org/10.1109/ARES.2012.84
5. Cavnar, W.B., Trenkle, J.M.: N-gram-based text categorization. In: Proceedings of the Third Symposium on Document Analysis and Information Retrieval, pp. 161–175 (1994)
6. Cohen, J.: Stat. Power Anal. Behav. Sci., 2nd edn. Lawrence Earlbaum Associates, Hillsdale (1988)
7. Coussement, K., den Poel, D.V.: Improving customer complaint management by automatic email classification using linguistic style features as predictors. Decis. Support Syst. **44**(4), 870–882 (2008). https://doi.org/10.1016/j.dss.2007.10.010
8. Dredze, M., Brooks, T., Carroll, J., Magarick, J., Blitzer, J., Pereira, F.: Intelligent email: reply and attachment prediction. In: Proceedings of the 13th International Conference on Intelligent User Interfaces, IUI 2008, pp. 321–324. ACM, New York (2008). https://doi.org/10.1145/1378773.1378820
9. Fawcett, T.: ROC graphs: notes and practical considerations for researchers. Mach. Learn. **31**(1), 1–38 (2004)
10. Flach, P.: Machine Learning: The Art and Science of Algorithms that Make Sense of Data. Cambridge University Press, Cambridge (2012)
11. Galar, M., Fernández, A., Barrenechea, E., Bustince, H., Herrera, F.: An overview of ensemble methods for binary classifiers in multi-class problems: experimental study on one-vs-one and one-vs-all schemes. Pattern Recogn. **44**(8), 1761–1776 (2011)
12. Gupta, N., Gilbert, M., Fabbrizio, G.D.: Emotion detection in email customer care. Comput. Intell. **29**(3), 489–505 (2013). https://doi.org/10.1111/j.1467-8640.2012.00454.x

13. Ha, Q.M., Tran, Q.A., Luyen, T.T.: Personalized email recommender system based on user actions. In: Bui, L.T., Ong, Y.S., Hoai, N.X., Ishibuchi, H., Suganthan, P.N. (eds.) SEAL 2012. LNCS, vol. 7673, pp. 280–289. Springer, Heidelberg (2012). https://doi.org/10.1007/978-3-642-34859-4_28
14. Halpin, N.: The customer service report: why great customer service matters even more in the age of e-commerce and the channels that perform best (2016). http://www.businessinsider.com/customer-service-experiences-are-more-important-than-ever-in-the-age-of-e-commerce-2016-3?r=US&IR=T&IR=T
15. Wang, H., Liu, X., Lv, B., Yang, F., Hong, Y.: Reliable multi-label learning via conformal predictor and random forest for syndrome differentiation of chronic fatigue in traditional Chinese medicine. PLOS One **9**(6), 1–14 (2014). https://doi.org/10.1371/journal.pone.0099565
16. Witten, I.H., Frank, E., Hall, M.: Data Mining - Practical Machine Learning Tools and Techniques, 3rd edn. Elsevier, Amsterdam (2011)
17. Koren, Y., Liberty, E., Maarek, Y., Sandler, R.: Automatically tagging email by leveraging other users' folders. In: In Proceedings of KDD 2011, pp. 913–921. ACM (2011)
18. Nenkova, A., Bagga, A.: Email classification for contact centers. In: Proceedings of the 2003 ACM Symposium on Applied Computing, SAC 2003, pp. 789–792. ACM, New York (2003). https://doi.org/10.1145/952532.952689
19. Papadopoulos, H.: Inductive conformal prediction: theory and application to neural networks. In: Tools in Artificial Intelligence. InTech (2008)
20. Papadopoulos, H.: A cross-conformal predictor for multi-label classification. In: Iliadis, L., Maglogiannis, I., Papadopoulos, H., Sioutas, S., Makris, C. (eds.) AIAI 2014. IAICT, vol. 437, pp. 241–250. Springer, Heidelberg (2014). https://doi.org/10.1007/978-3-662-44722-2_26
21. Sebastiani, F.: Machine learning in automated text categorization. ACM Comput. Surv. (CSUR) (2002). http://portal.acm.org/citation.cfm?id=505283
22. Shafer, G., Vovk, V.: A tutorial on conformal prediction. J. Mach. Learn. Res. **9**(Mar), 371–421 (2008)
23. Tsoumakas, G., Katakis, I., Vlahavas, I.: Mining multi-label data. In: Maimon, O., Rokach, L. (eds.) Data Mining and Knowledge Discovery Handbook, pp. 667–685. Springer, Boston (2009). https://doi.org/10.1007/978-0-387-09823-4_34
24. Tsoumakas, G., Zhang, M.L., Zhou, Z.H.: Introduction to the special issue on learning from multi-label data. Mach. Learn. **88**(1), 1–4 (2012). https://doi.org/10.1007/s10994-012-5292-9
25. Wang, M.F., Jheng, S.L., Tsai, M.F., Tang, C.H.: Enterprise email classification based on social network features. In: 2011 International Conference on Advances in Social Networks Analysis and Mining, pp. 532–536, July 2011. https://doi.org/10.1109/ASONAM.2011.89
26. Yang, Y.: An evaluation of statistical approaches to text categorization. Inf. Retrieval **1**(1), 69–90 (1999). https://doi.org/10.1023/A:1009982220290
27. Yelupula, K., Ramaswamy, S.: Social network analysis for email classification. In: Proceedings of the 46th Annual Southeast Regional Conference on XX, ACM-SE 46, pp. 469–474. ACM, New York (2008). https://doi.org/10.1145/1593105.1593229

Fault Detection and Diagnosis

A Posteriori Diagnosis of Discrete-Event Systems with Symptom Dictionary and Scenarios

Nicola Bertoglio, Gianfranco Lamperti, and Marina Zanella(✉)

Department of Information Engineering, University of Brescia, Brescia, Italy
{n.bertoglio001,gianfranco.lamperti,marina.zanella}@unibs.it

Abstract. Offline knowledge compilation enables an online diagnosis process that can manage in a linear time any sequence of observables. In *a posteriori* diagnosis, this sequence, called a *symptom*, is the input, and the corresponding collection of sets of faults, each set being a *candidate*, is the output. Since the compilation is computationally hard, we propose to compile only the knowledge chunks that are relevant to some phenomena of interest, each described as a *scenario*. If, on the one hand, a partial knowledge compilation does not ensure the completeness of the resulting collection of candidates, on the other, it allows attention to be focused on the most important of them. Moreover, the compiled structure, called *symptom dictionary*, can incrementally be extended over time.

Keywords: Diagnosis · Model-based reasoning · Discrete-event systems · Finite automata · Symptom dictionary · Scenarios · Preprocessing

1 Introduction

A discrete-event system (DES) [2] is an abstraction of a (real) dynamic system that can be exploited for model-based reasoning. A conceptual model of a DES is either a finite automaton, where each transition is either normal or faulty, or a network of communicating finite automata [1,3], each representing the behavior of a component within a distributed topology. An *a posteriori* diagnosis problem consists of a DES model and a *symptom*, namely a temporally-ordered sequence of discrete sensor measurements called *observations*. Solving a problem means drawing a set of faults, called a *candidate*, from each sequence of the DES state transitions that can explain the symptom. Since diagnosis is notoriously NP-hard also for static systems, knowledge compilation techniques have been adopted by research on model-based diagnosis of DESs since the mid '90s. The *diagnoser approach* [8] proposed to compile (offline) the knowledge about the whole DES into a *diagnoser*, to be exploited (online) in order to process any symptom in a time that is linear in the symptom length. However, the diagnoser generation is impractical because of a combinatorial explosion of states even for distributed

F. Wotawa et al. (Eds.): IEA/AIE 2019, LNAI 11606, pp. 325–333, 2019.
https://doi.org/10.1007/978-3-030-22999-3_29

DESs including few components. To make knowledge compilation viable, this paper proposes to build a structure, called an *open symptom dictionary*, that does not account for all the possible symptoms, but only for those whose length is limited by an upper bound, called *distance* and/or for those that are relevant to a restricted set of (either critical or most probable) behaviors of interest.

2 Discrete-Event Systems

A DES [5–7] is a network of *components*, each modeled as a communicating automaton [1] and endowed with input and output terminals. Each output terminal of a component is connected with the input terminal of another by a *link*. A component transition is triggered either (1) spontaneously (by the empty event ε), or (2) by an (external) event coming from the extern of the DES, or (3) by an (internal) event coming from another component. Initially, the DES is *quiescent*, with no internal event being present. When a component performs a transition, it possibly generates new events on its output terminals; such events are bound to be consumed to trigger the transitions of other components, on so on until the DES becomes quiescent anew. A transition placing an output event on a link can occur only if this link is empty. A DES moves from the initial to the final quiescent state through a sequence of component transitions, called a *trajectory*. Each state x of a DES $\mathcal{X}$ is a pair (S, E), where S is the array of the current states of the components and E is the array of the (possibly empty) events currently placed on the links. The (possibly infinite) set of trajectories can be specified by a DFA $\mathcal{X}^*$, called the *space* of $\mathcal{X}$, $\mathcal{X}^* = (\Sigma, X, \tau, x_0, X_f)$ where Σ (the alphabet) is the set of component transitions, X is the set of states, $\tau : X \times \Sigma \mapsto X$ is the transition function, x_0 is the initial (quiescent) state, and X_f is the set of final (quiescent) states.

Example 1. The DES, called $\mathcal{V}$, in Fig. 1, which will be considered in all the examples in the paper, represents a (simplified) autonomous vacuum cleaner. It includes two components, a battery b and a cleaning device d, and two links. The models of b and d are automata, each endowed with states (circles) and

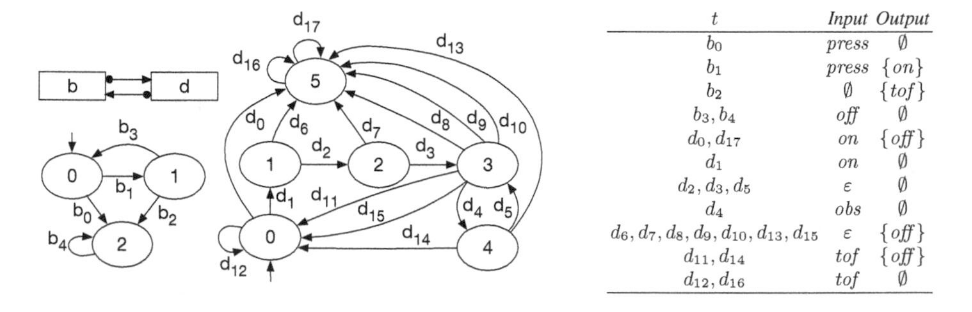

t	*Input*	*Output*
b_0	*press*	$\emptyset$
b_1	*press*	$\{on\}$
b_2	$\emptyset$	$\{tof\}$
b_3, b_4	*off*	$\emptyset$
d_0, d_{17}	*on*	$\{off\}$
d_1	*on*	$\emptyset$
d_2, d_3, d_5	ε	$\emptyset$
d_4	*obs*	$\emptyset$
$d_6, d_7, d_8, d_9, d_{10}, d_{13}, d_{15}$	ε	$\{off\}$
d_{11}, d_{14}	*tof*	$\{off\}$
d_{12}, d_{16}	*tof*	$\emptyset$

Fig. 1. DES $\mathcal{V}$ with models of battery and device (left) and transitions details (right).

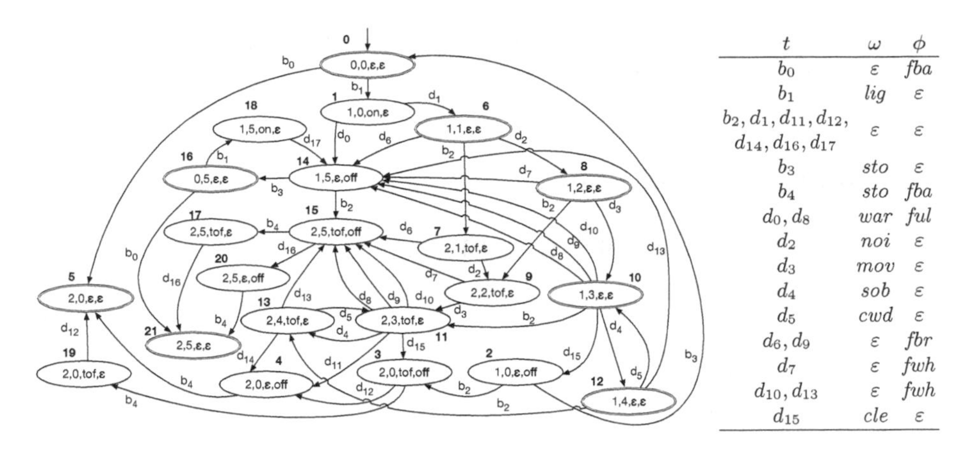

t	ω	ϕ
b_0	ε	*fba*
b_1	*lig*	ε
$b_2, d_1, d_{11}, d_{12}, d_{14}, d_{16}, d_{17}$	ε	ε
b_3	*sto*	ε
b_4	*sto*	*fba*
d_0, d_8	*war*	*ful*
d_2	*noi*	ε
d_3	*mov*	ε
d_4	*sob*	ε
d_5	*cwd*	ε
d_6, d_9	ε	*fbr*
d_7	ε	*fwh*
d_{10}, d_{13}	ε	*fwh*
d_{15}	*cle*	ε

Fig. 2. Space $\mathcal{V}^*$ (left) and mapping table $\mu(\mathcal{V})$ (right).

transitions (arcs). On the right, each transition t is associated with the input (triggering) event and the set of output events (incidentally, every transition of $\mathcal{V}$ has an output event at most). In the space $\mathcal{V}^*$, depicted in Fig. 2, each state is a quadruple (s_b, s_d, e_1, e_2), with s_b being the state of b, s_d the state of d, e_1 the internal event placed at the input terminal of d, and e_2 the internal event placed at the input terminal of b (ε means no event). The initial state is $(0, 0, \varepsilon, \varepsilon)$; final states are double circled (here as well as in all the next figures). For subsequent referencing, the states are renamed $0 \cdots 21$. Owing to cycles, the set of trajectories in $\mathcal{V}^*$ is infinite. One of them is $[b_1, d_1, d_6, b_3]$, ending in state 16.

Let $\mu(\mathcal{X})$ be a finite set of triples (t, ω, ϕ), one for each component transition t, with $\omega \in \Omega \cup \{\varepsilon\}$, $\phi \in \Phi \cup \{\varepsilon\}$, where ε is the empty symbol, Ω a finite set of *observations*, and Φ a finite set of *faults*. A triple (t, ω, ϕ) has the following meaning: if $\omega \neq \varepsilon$, then t is *observable*, else t is *unobservable*; if $\phi \neq \varepsilon$, then t is *faulty*, else t is *normal*.

The *symptom* $\mathcal{O}$ of a trajectory $\mathcal{T} \in \mathcal{X}^*$ is the finite *sequence* of observations involved in $\mathcal{T}$, and the *diagnosis* δ of $\mathcal{T}$ is the finite *set* of faults involved in $\mathcal{T}$: we say that $\mathcal{T}$ *implies* both $\mathcal{O}$ and δ, denoted $\mathcal{T} \Rightarrow \mathcal{O}$ and $\mathcal{T} \Rightarrow \delta$, respectively. The solution, called *explanation*, of an a posteriori diagnosis problem $(\mathcal{X}, \mathcal{O})$ is the (finite) set of diagnoses (called *candidates*) implied by the trajectories of $\mathcal{X}$ that imply $\mathcal{O}$: $\Delta(\mathcal{O}) = \{\, \delta \mid \mathcal{T} \in \mathcal{X}^*, \mathcal{T} \Rightarrow \mathcal{O}, \mathcal{T} \Rightarrow \delta \,\}$.

Example 2. Figure 2 shows $\mu(\mathcal{V})$. Let $\mathcal{O} = [lig, sto]$ be a symptom. Based on $\mathcal{V}^*$, six trajectories imply $\mathcal{O}$, namely $\mathcal{T}_1 = [b_1, d_1, d_6, b_3]$, $\mathcal{T}_2 = [b_1, d_1, d_6, b_3, b_0]$, $\mathcal{T}_3 = [b_1, d_1, d_6, b_2, b_4, d_{16}]$, $\mathcal{T}_4 = [b_1, d_1, d_6, b_2, d_{16}, b_4]$, $\mathcal{T}_5 = [b_1, d_1, b_2, d_6, b_4, d_{16}]$, and $\mathcal{T}_6 = [b_1, d_1, b_2, d_6, d_{16}, b_4]$, where $\mathcal{T}_1$ involves the faulty transition d_6, $\mathcal{T}_2$ involves the faulty transitions d_6 and b_0, and $\mathcal{T}_3 \cdots \mathcal{T}_6$ involve the faulty transitions d_6 and b_4. The explanation of $\mathcal{O}$ is $\Delta(\mathcal{O}) = \{\{fbr\}, \{fba, fbr\}\}$.

$\Delta(\mathcal{O})$ can be generated online by *abducing* the subspace of $\mathcal{X}^*$ involving all and only the trajectories implying $\mathcal{O}$. However, this process is too time consuming to be practical.

3 Symptom Dictionary

The *extended space* $\mathcal{X}^+$ of a DES $\mathcal{X}$ is a DFA $\mathcal{X}^+ = (\Sigma, X^+, \tau^+, x_0^+, X_{\mathrm{f}}^+)$ where: X^+ is the set of states (x, δ), $x \in X$, $\delta \subseteq \Phi$; $x_0^+ = (x_0, \emptyset)$; X_{f}^+ is the set of final states (x, δ), where $x \in X_{\mathrm{f}}$; $\tau^+ : X^+ \times \Sigma \mapsto X^+$ is the transition function, where $\tau^+((x, \delta), t) = (x', \delta')$ iff $\tau(x, t) = x'$ and $\delta' = \delta \cup \{\phi\}$ if $(t, \omega, \phi) \in \mu(\mathcal{X}), \phi \neq \varepsilon$, otherwise $\delta' = \delta$. In practice, $\mathcal{X}^+$ can be produced directly by processing $\mathcal{X}$, without generating $\mathcal{X}^*$.

Proposition 1. *The language of $\mathcal{X}^+$ equals the language of $\mathcal{X}^*$. Besides, if $x^+ = (x, \delta)$ is a final state in $\mathcal{X}^+$, then, for each trajectory $\mathcal{T}$ ending in x^+, $\mathcal{T} \Rightarrow \delta$.*

Example 3. Figure 3 displays $\mathcal{V}^+$ (states are renamed $0 \cdots 35$). Considering in $\mathcal{V}^+$ the six trajectories detailed in Example 2, where $\mathcal{T}_1$ ends in state $13 = (16, \{\mathit{fbr}\})$ and the other five trajectories end in state $33 = (21, \{\mathit{fbr}, \mathit{fba}\})$, respectively, we have $\mathcal{T}_1 \Rightarrow \{\mathit{fbr}\}$, while the other five trajectories imply $\{fbr, fba\}$, as claimed in Proposition 1.

Let $\mathcal{X}_{\mathrm{n}}^+$ be the NFA obtained by substituting the symbol t, marking each transition in $\mathcal{X}^+$, with ω, where $(t, \omega, \phi) \in \mu(\mathcal{X})$. The *symptom dictionary* of $\mathcal{X}$ is the DFA $\mathcal{X}^\oplus$ obtained by determinization [3] of $\mathcal{X}_{\mathrm{n}}^+$, where each final state $x^\oplus$ of $\mathcal{X}^\oplus$ is marked with $\Delta(x^\oplus)$, the set of diagnoses associated with the final states of $\mathcal{X}_{\mathrm{n}}^+$ included in $x^\oplus$.

Proposition 2. *The language of $\mathcal{X}^\oplus$ equals the set of symptoms of $\mathcal{X}$. Besides, if $\mathcal{O}$ is a symptom with accepting state $x^\oplus$, then $\Delta(x^\oplus) = \Delta(\mathcal{O})$.*

Example 4. Figure 4 outlines $\mathcal{V}^\oplus$; on the right, each state $v^\oplus$ of $\mathcal{V}^\oplus$ is described in terms of the $\mathcal{V}^+$ states it includes[1] (where final states are in bold) and the associated set of diagnoses. Considering the symptom $\mathcal{O} = [\mathit{lig}, \mathit{sto}]$, defined in Example 2, with accepting state 4 in $\mathcal{V}^\oplus$, we have $\Delta(4) = \{\{\mathit{fbr}\}, \{\mathit{fbr}, \mathit{fba}\}\} = \Delta(\mathcal{O})$.

The generation of $\mathcal{X}^\oplus$ is beyond dispute for large DESs since in the worst case it is at least exponential with the number of components and links. Instead of the whole $\mathcal{X}^\oplus$, we might generate a prefix of it, denoted $\mathcal{X}_{[d]}^\oplus$, comprehending only the states at distance $\leq d$ and the transitions exiting the states at distance $< d$, where the *distance* of a state is the minimum number of transitions connecting the initial state with this state.

[1] Each DFA state here includes only the significant NFA states. A state is *significant* when it is either final or it is exited by a transition marked with a (non null) observation.

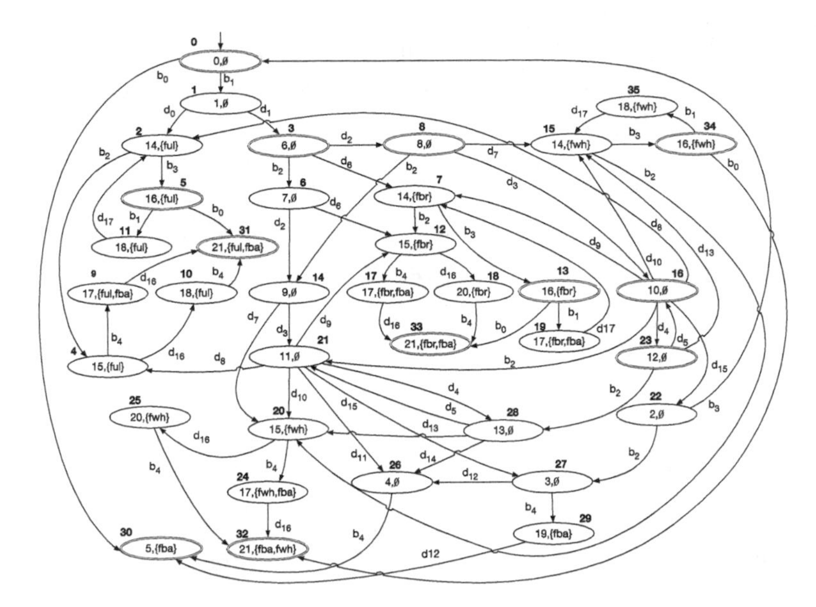

Fig. 3. Extended space $\mathcal{V}^{+}$.

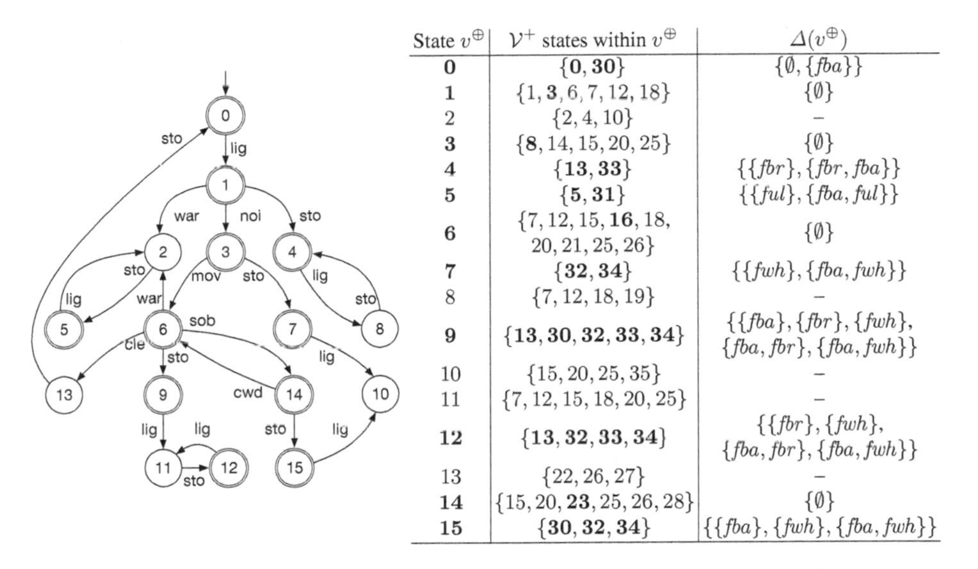

State $v^{\oplus}$	$\mathcal{V}^{+}$ states within $v^{\oplus}$	$\Delta(v^{\oplus})$
0	$\{\mathbf{0}, \mathbf{30}\}$	$\{\emptyset, \{fba\}\}$
1	$\{1, \mathbf{3}, 6, 7, 12, 18\}$	$\{\emptyset\}$
2	$\{2, 4, 10\}$	–
3	$\{\mathbf{8}, 14, 15, 20, 25\}$	$\{\emptyset\}$
4	$\{\mathbf{13}, \mathbf{33}\}$	$\{\{fbr\}, \{fbr, fba\}\}$
5	$\{\mathbf{5}, \mathbf{31}\}$	$\{\{ful\}, \{fba, ful\}\}$
6	$\{7, 12, 15, \mathbf{16}, 18, 20, 21, 25, 26\}$	$\{\emptyset\}$
7	$\{\mathbf{32}, \mathbf{34}\}$	$\{\{fwh\}, \{fba, fwh\}\}$
8	$\{7, 12, 18, 19\}$	–
9	$\{\mathbf{13}, \mathbf{30}, \mathbf{32}, \mathbf{33}, \mathbf{34}\}$	$\{\{fba\}, \{fbr\}, \{fwh\}, \{fba, fbr\}, \{fba, fwh\}\}$
10	$\{15, 20, 25, 35\}$	–
11	$\{7, 12, 15, 18, 20, 25\}$	–
12	$\{\mathbf{13}, \mathbf{32}, \mathbf{33}, \mathbf{34}\}$	$\{\{fbr\}, \{fwh\}, \{fba, fbr\}, \{fba, fwh\}\}$
13	$\{22, 26, 27\}$	–
14	$\{15, 20, \mathbf{23}, 25, 26, 28\}$	$\{\emptyset\}$
15	$\{\mathbf{30}, \mathbf{32}, \mathbf{34}\}$	$\{\{fba\}, \{fwh\}, \{fba, fwh\}\}$

Fig. 4. Symptom dictionary $\mathcal{V}^{\oplus}$ (left) and relevant state details (right).

Example 5. With reference to Fig. 4, $\mathcal{V}^{\oplus}_{[3]}$ includes the states $0 \cdots 8$, and the transitions exiting the states $0 \cdots 4$ (thus excluding the states at distance 3, namely 5, 6, 7 and 8).

$\mathcal{X}^{\oplus}_{[d]}$ provides the explanation of every symptom that either is not longer than d or encompasses some cycle(s) of $\mathcal{X}^{\oplus}_{[d]}$. If a symptom does not belong to the language of $\mathcal{X}^{\oplus}_{[d]}$, such as $\mathcal{O} = [lig, noi, mov, sto]$ for $\mathcal{V}^{\oplus}_{[3]}$, $\Delta(\mathcal{O})$ needs to be generated online by reconstructing the portion of $\mathcal{V}^{+}$ including all the trajectories implying $\mathcal{O}$, a costly technique.

4 Scenarios and Open Dictionary

To solve a diagnosis problem relevant to a symptom longer than d in a lighter way, a DES can be accompanied by a set of behaviors, the *scenarios*, that are required to be explained efficiently. Each scenario is a regular language on the set of the component transitions and it is described by the DFA recognizing it, denoted $\mathcal{S} = (\hat{\Sigma}, \hat{S}, \hat{\tau}, \hat{s}_0, \hat{S}_{\rm f})$.

Example 6. For $\mathcal{V}$, we define a scenario, depicted in Fig. 5 (for simplicity, several transitions are factorized into a single arc), which is interested in the fault-free trajectories.

The *space* of $\mathcal{X}$ *constrained* by $\mathcal{S}$ is a DFA $\mathcal{X}^{*}_{\mathcal{S}} = (\Sigma, X_{\mathcal{S}}, \tau_{\mathcal{S}}, x_{\mathcal{S}_0}, X_{\mathcal{S}_{\rm f}})$ where: $X_{\mathcal{S}}$ is the set of states $(x, \hat{s})$, where $x \in X$ and $\hat{s} \in \hat{S}$; $x_{\mathcal{S}_0} = (x_0, \hat{s}_0)$; $X_{\mathcal{S}_{\rm f}}$ is the set of final states $(x, \hat{s})$, where $x \in X_{\rm f}$ and $\hat{s} \in \hat{S}_{\rm f}$; $\tau_{\mathcal{S}}$: $X_{\mathcal{S}} \times \Sigma \mapsto X_{\mathcal{S}}$, where $\tau_{\mathcal{S}}((x, \hat{s}), t) = (x', \hat{s}')$ iff $\tau(x, t) = x'$ and $\hat{s}' = \hat{\tau}(\hat{s}, t)$.

Proposition 3. *If $\mathcal{T}$ is a trajectory in $\mathcal{X}^{*}_{\mathcal{S}}$, then $\mathcal{T}$ is also a trajectory in both $\mathcal{X}^{*}$ and $\hat{\mathcal{S}}$.*

The notion of an extended space applies to a *constrained extended space* $\mathcal{X}^{+}_{\mathcal{S}}$ too.

Example 7. Displayed in the center of Fig. 5 is the constrained extended space $\mathcal{V}^{+}_{\mathcal{S}_1}$, where each $\mathcal{V}^{+}_{\mathcal{S}_1}$ state includes just the relevant $\mathcal{V}^{+}$ state (both the scenario state 0 and the empty diagnosis are omitted since they are shared by all $\mathcal{V}^{+}_{\mathcal{S}_1}$ states).

Proposition 4. *If $\mathcal{T}$ is a trajectory in $\mathcal{X}^{+}_{\mathcal{S}}$, then $\mathcal{T}$ is a trajectory in $\mathcal{X}^{+}$. Besides, if $x^{+}_{\mathcal{S}} = (x_{\mathcal{S}}, \delta)$ is the accepting state of $\mathcal{T}$ in $\mathcal{X}^{+}_{\mathcal{S}}$, where $x_{\mathcal{S}} = (x, \hat{s})$, then (x, δ) is the accepting state of $\mathcal{T}$ in $\mathcal{X}^{+}$.*

Let δ be the diagnosis of $\mathcal{X}$ relevant to $\mathcal{T}$, this being a trajectory of all the scenarios in a set σ. The pair (δ, σ) is a *complex diagnosis* of $\mathcal{X}$, where σ is called the *scope* of δ. By definition, the *empty scenario* of $\mathcal{X}$ is $\varepsilon = \Sigma^{*}$, where Σ is the whole set of component transitions. Any diagnosis δ involved in a symptom dictionary $\mathcal{X}^{\oplus}$ can be represented as a complex diagnosis $(\delta, \{\varepsilon\})$. Also, we have $\mathcal{X}^{*} = \mathcal{X}^{*}_{\varepsilon}$, $\mathcal{X}^{+} = \mathcal{X}^{+}_{\varepsilon}$, and $\mathcal{X}^{\oplus} = \mathcal{X}^{\oplus}_{\varepsilon}$.

Let $\mathbf{S}$ be a set of scenarios for a DES $\mathcal{X}$ and $\mathcal{O}$ a symptom of $\mathcal{X}$. The *complex explanation* of $\mathcal{O}$ is the set of complex diagnoses $\mathbf{\Delta}(\mathcal{O}) = \{(\delta, \sigma) \mid \sigma = \{\mathcal{S} \mid \mathcal{S} \in \mathbf{S}, \mathcal{T} \in \mathcal{X}^{*}_{\mathcal{S}}, \mathcal{T} \Rightarrow \mathcal{O}, \mathcal{T} \Rightarrow \delta\}\}$. Let $\mathcal{X}^{\rm n}_{\mathcal{S}}$ be the NFA obtained

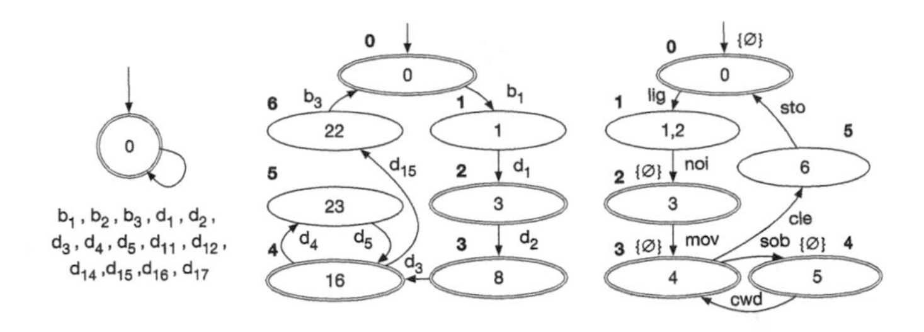

Fig. 5. Scenario $\mathcal{S}_1$ (left), $\mathcal{V}^{+}_{\mathcal{S}_1}$ (center), and constrained dictionary $\mathcal{V}^{\oplus}_{\mathcal{S}_1}$ (right).

from a constrained extended space $\mathcal{X}^{+}_{\mathcal{S}}$ by replacing the symbol t, marking each transition, with ω, where $(t, \omega, \phi) \in \mu(\mathcal{X})$. The *dictionary* of $\mathcal{X}$ *constrained by* scenario $\mathcal{S}$ is the DFA $\mathcal{X}^{\oplus}_{\mathcal{S}}$ obtained by determinization of $\mathcal{X}^{\mathrm{n}}_{\mathcal{S}}$, where each final state $x^{\oplus}_{\mathcal{S}}$ of $\mathcal{X}^{\oplus}_{\mathcal{S}}$ is marked with the set of complex diagnoses $\mathbf{\Delta}(x^{\oplus}_{\mathcal{S}}) = \{(\delta, \{\mathcal{S}\}) \mid (x^{+}, \delta) \in x^{\oplus}_{\mathcal{S}}, x^{+} \in X^{+}_{\mathrm{f}}\}$.

Example 8. Shown in Fig. 5 is the constrained dictionary $\mathcal{V}^{\oplus}_{\mathcal{S}_1}$[2], which is also outlined (after renaming) in Fig. 6, along with $\mathcal{V}^{\oplus}_{[3]}$.

Let $\{\mathcal{X}^{\oplus}_{\mathcal{S}_1}, \ldots, \mathcal{X}^{\oplus}_{\mathcal{S}_k}\}$ be a set of constrained dictionaries, with $X^{\oplus}_{\mathrm{f}_i}$ being the set of final states of $\mathcal{X}^{\oplus}_{\mathcal{S}_i}$, $i \in [1 \mathinner{..} k]$. Let $\hat{\mathcal{X}}^{\oplus}_{\mathrm{n}}$ be the NFA obtained by creating an initial state $x^{\oplus}_0$ and k ε-transitions, each transition exiting $x^{\oplus}_0$ and entering the initial state of $\mathcal{X}^{\oplus}_{\mathcal{S}_i}$. The *open dictionary* of $\mathcal{X}$ is the DFA $\hat{\mathcal{X}}^{\oplus}$ obtained by determinization of $\hat{\mathcal{X}}^{\oplus}_{\mathrm{n}}$, where each final state $\hat{x}^{\oplus}$ is marked with the set of complex diagnoses $\mathbf{\Delta}(\hat{x}^{\oplus}) = \{(\delta, \sigma) \mid x^{\oplus}_i \in \hat{x}^{\oplus}, x^{\oplus}_i \in X^{\oplus}_{\mathrm{f}_i}, i \in [1 \mathinner{..} k], (\delta, \sigma_i) \in \mathbf{\Delta}(x^{\oplus}_i), \sigma \supseteq \sigma_i\}$.

Example 9. Figure 7 shows the open dictionary that merges $\mathcal{V}^{\oplus}_{[3]}$ and $\mathcal{V}^{\oplus}_{\mathcal{S}_1}$, and (on the right) the complex diagnoses relevant to its final states. Some remarks apply to $\hat{\mathcal{V}}^{\oplus}$. We do not know whether all the trajectories whose symptoms

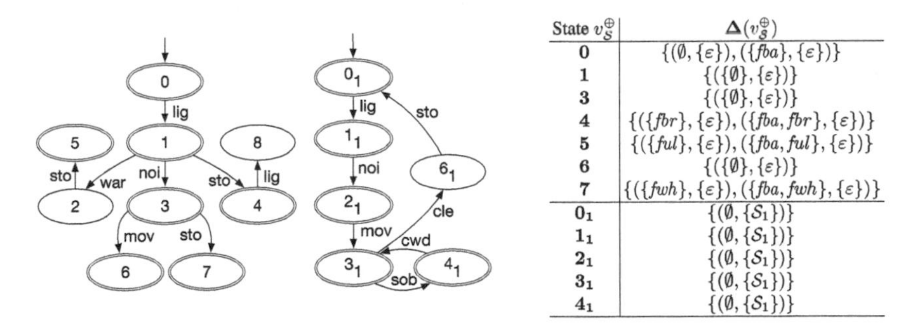

State $v^{\oplus}_{\mathcal{S}}$	$\mathbf{\Delta}(v^{\oplus}_{\mathcal{S}})$
0	$\{(\emptyset, \{\varepsilon\}), (\{fba\}, \{\varepsilon\})\}$
1	$\{(\{\emptyset\}, \{\varepsilon\})\}$
3	$\{(\{\emptyset\}, \{\varepsilon\})\}$
4	$\{(\{fbr\}, \{\varepsilon\}), (\{fba, fbr\}, \{\varepsilon\})\}$
5	$\{(\{ful\}, \{\varepsilon\}), (\{fba, ful\}, \{\varepsilon\})\}$
6	$\{(\{\emptyset\}, \{\varepsilon\})\}$
7	$\{(\{fwh\}, \{\varepsilon\}), (\{fba, fwh\}, \{\varepsilon\})\}$
0_1	$\{(\emptyset, \{\mathcal{S}_1\})\}$
1_1	$\{(\emptyset, \{\mathcal{S}_1\})\}$
2_1	$\{(\emptyset, \{\mathcal{S}_1\})\}$
3_1	$\{(\emptyset, \{\mathcal{S}_1\})\}$
4_1	$\{(\emptyset, \{\mathcal{S}_1\})\}$

Fig. 6. $\mathcal{V}^{\oplus}_{[3]}$, $\mathcal{V}^{\oplus}_{\mathcal{S}_1}$ (left) and explanations for final states (right).

[2] Each state of the DFA includes only the significant states of the NFA (cf. Footnote 1).

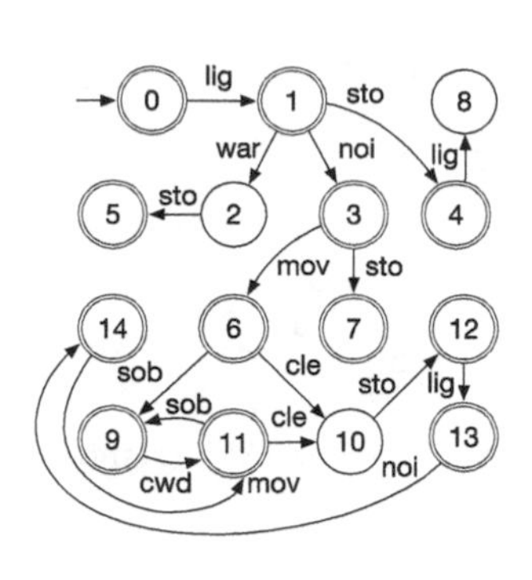

State $\hat{p}^{\oplus}$	States within $\hat{p}^{\oplus}$	$\Delta(\hat{p}^{\oplus})$
0	$\{\mathbf{0}, \mathbf{0_1}\}$	$\{(\emptyset, \{\varepsilon, \mathcal{S}_1\}), (\{fba\}, \{\varepsilon\})\}$
1	$\{\mathbf{1}, \mathbf{1_1}\}$	$\{(\emptyset, \{\varepsilon, \mathcal{S}_1\})\}$
2	$\{2\}$	–
3	$\{\mathbf{3}, \mathbf{2_1}\}$	$\{(\emptyset, \{\varepsilon, \mathcal{S}_1\})\}$
4	$\{\mathbf{4}\}$	$\{(\{fbr\}, \{\varepsilon\}), (\{fba, fbr\}, \{\varepsilon\})\}$
5	$\{\mathbf{5}\}$	$\{(\{ful\}, \{\varepsilon\}\}), (\{fba, ful\}, \{\varepsilon\})\}$
6	$\{\mathbf{6}, \mathbf{3_1}\}$	$\{(\{\emptyset\}, \{\varepsilon, \mathcal{S}_1\})\}$
7	$\{\mathbf{7}\}$	$\{(\{fwh\}, \{\varepsilon\}), (\{fba, fwh\}, \{\varepsilon\})\}$
8	$\{8\}$	–
9	$\{\mathbf{4_1}\}$	$\{(\{\emptyset\}, \{\mathcal{S}_1\})\}$
10	$\{6_1\}$	–
11	$\{\mathbf{3_1}\}$	$\{(\{\emptyset\}, \{\mathcal{S}_1\})\}$
12	$\{\mathbf{0_1}\}$	$\{(\{\emptyset\}, \{\mathcal{S}_1\})\}$
13	$\{\mathbf{1_1}\}$	$\{(\{\emptyset\}, \{\mathcal{S}_1\})\}$
14	$\{\mathbf{2_1}\}$	$\{(\{\emptyset\}, \{\mathcal{S}_1\})\}$

Fig. 7. Open dictionary $\hat{\mathcal{V}}^{\oplus}$ (left) and relevant state details (right).

are compatible with a normal behavior have been diagnosed, as states 9 and $11 \ldots 14$ are marked with $\{(\emptyset, \{\mathcal{S}_1\})\}$, not with $\{(\emptyset, \{\varepsilon, \mathcal{S}_1\})\}$. The sets of candidates associated with the states 0, 1, and $3 \cdots 7$ equal the explanations of the corresponding symptoms as every scope includes ε.

5 Conclusion

This paper proposes to produce offline an open dictionary. If the symptom given online is a word of the open dictionary, either every diagnosis relevant to the symptom is associated with the empty scenario or no diagnosis is associated with the empty scenario. In the former case, the set of diagnoses is sound and complete, in the latter it is sound (and possibly incomplete). If the symptom is not a word of the open dictionary (or it is a word whose set of candidates is possibly incomplete), then the dictionary can be extended with such a word.

Scenarios resemble *fault supervision patterns* [4]. However, such patterns are meant to generalize the concept of a fault from one transition to a whole paradigmatic faulty evolution. Scenarios are instead the means to reduce the size of the compiled knowledge, while at the same time focusing attention on specific evolutions of the DES. Moreover, a supervision pattern is synchronized with the whole space of the DES, whose generation is computationally prohibitive. The current approach does not generate the whole space, instead a constrained space is built for each specific scenario.

A dictionary resembles the diagnoser [8]. The set of diagnoses computed by the diagnoser approach, given whichever (physically possible) symptom, is sound and complete if the DES is diagnosable. The set of diagnoses computed by an unconstrained symptom dictionary, instead, is sound and complete independently of the diagnosability of the DES. Finally, the *scope* of a diagnosis, as introduced in this paper, enables the diagnostician to focus attention on a set of phenomena that are reckoned as interesting, whereas no such information is provided by the diagnoser approach and only a single phenomenon of interest can be taken into account by the supervision pattern approach.

Acknowledgments. This work was supported in part by Lombardy Region (Italy), project *Smart4CPPS, Linea Accordi per Ricerca, Sviluppo e Innovazione, POR-FESR 2014-2020 Asse I.*

References

1. Brand, D., Zafiropulo, P.: On communicating finite-state machines. J. ACM **30**(2), 323–342 (1983). https://doi.org/10.1145/322374.322380
2. Cassandras, C., Lafortune, S.: Introduction to Discrete Event Systems, 2nd edn. Springer, New York (2008)
3. Hopcroft, J., Motwani, R., Ullman, J.: Introduction to Automata Theory, Languages, and Computation, 3rd edn. Addison-Wesley, Reading (2006)
4. Jéron, T., Marchand, H., Pinchinat, S., Cordier, M.: Supervision patterns in discrete event systems diagnosis. In: Workshop on Discrete Event Systems (WODES 2006), pp. 262–268. IEEE Computer Society, Ann Arbor (2006)
5. Lamperti, G., Zanella, M.: Context-sensitive diagnosis of discrete-event systems. In: Walsh, T. (ed.) Twenty-Second International Joint Conference on Artificial Intelligence (IJCAI 2011), vol. 2, pp. 969–975. AAAI Press, Barcelona (2011)
6. Lamperti, G., Zanella, M., Zhao, X.: Introduction to Diagnosis of Active Systems. Springer, Cham (2018). https://doi.org/10.1007/978-3-319-92733-6
7. Lamperti, G., Zhao, X.: Diagnosis of active systems by semantic patterns. IEEE Trans. Syst. Man Cybern.: Syst. **44**(8), 1028–1043 (2014). https://doi.org/10.1109/TSMC.2013.2296277
8. Sampath, M., Sengupta, R., Lafortune, S., Sinnamohideen, K., Teneketzis, D.: Diagnosability of discrete-event systems. IEEE Trans. Autom. Control **40**(9), 1555–1575 (1995)

Detecting Fraudulent Bookings of Online Travel Agencies with Unsupervised Machine Learning

Caleb Mensah[1], Jan Klein[2](✉), Sandjai Bhulai[1], Mark Hoogendoorn[1], and Rob van der Mei[2]

[1] Vrije Universiteit, De Boelelaan 1105, 1081 HV Amsterdam, The Netherlands
{s.bhulai,m.hoogendoorn}@vu.nl

[2] Centrum Wiskunde & Informatica, Science Park 123, 1098 XG Amsterdam, The Netherlands
{j.g.klein,r.d.van.der.mei}@cwi.nl

Abstract. Online fraud poses a relatively new threat to the revenues of companies. A way to detect and prevent fraudulent behavior is with the use of specific machine learning (ML) techniques. These anomaly detection techniques have been thoroughly studied, but the level of employment is not as high. The airline industry suffers from fraud by parties such as online travel agencies (OTAs). These agencies are commissioned by an airline carrier to sell its travel tickets. Through policy violations, they can illegitimately claim some of the airline's revenue by offering cheaper fares to customers.

This research applies several anomaly detection techniques to detect fraudulent behavior by OTAs and assesses their strengths and weaknesses. Since the data is not labeled, it is not known whether fraud has actually occurred. Therefore, unsupervised ML is used. The contributions of this paper are, firstly, to show how to shape the online booking data and how to engineer new and relevant features. Secondly, this research includes a case study in which domain experts evaluate the detection performance of the considered ML methods by classifying a set of 75 bookings. According to the experts' analysis, the techniques are able to discover previously unknown fraudulent bookings, which will not have been found otherwise. This demonstrates that anomaly detection is a valuable tool for the airline industry to discover fraudulent behavior.

Keywords: Fraud · Anomaly detection · Unsupervised learning · Airline · Online travel agent

1 Introduction

Since industries have expanded their services to the internet to reach more customers, new ways have evolved to claim part of a company's revenue. Aviation faces a considerable problem with these malpractices. In 2008, airline industries

F. Wotawa et al. (Eds.): IEA/AIE 2019, LNAI 11606, pp. 334–346, 2019.
https://doi.org/10.1007/978-3-030-22999-3_30

all over the world missed out on 1.4 billion US dollars due to fraud. This was around 1.3% of their total revenue, although the rates were up to 4% in parts such as the Middle East and Latin America. Nowadays, these figures are expected to be even higher [1]. One of the conductors of fraud in the airline industry are online travel agencies (OTAs). Such an agency specializes in selling travel products including flights, hotels and rental cars to customers online. There is a wide variety of different kinds of OTAs, but they share at least one similarity: they all have an agency agreement with the supplier to resell its products [2]. In this case, the airline carrier allows the OTA access to its booking system to sell airplane seats. This expands the reach of the carrier, and therefore, increases its revenue. However, some OTAs violate the policies conducted by the airline organization in order to get access to cheaper ticket fares. This is possible, because an airplane seat can have a different price depending on several well-known factors. These include the seat's class (economy or business), the flight destination and the remaining time until departure. More specifically, when a flight consists of multiple flight segments, the price of a single segment can differ depending on the other segments in the complete flight. Here, a flight segment can be seen as the part between the departure and arrival of an airplane. If it lands more than once, there are multiple flight segments. An OTA can add one or more artificial segments to a flight to possibly get access to relatively lower prices. Later on, it can cancel these segments, which leads to revenue loss for the airline company. Therefore, the airline carrier desires to discover these malpractices to avoid losing profit.

In general, fraudulent behavior is assumed to be unusual, and hence, (largely) deviates from the expected, normal behavior. A way to discover such anomalous behavior is with the use of outlier detection techniques. Usually, the data with potentially fraudulent behavior is unlabeled, suggesting the use of unsupervised machine learning (unsupervised ML). This can be applied in a wide variety of domains, such as insurance, health care and cyber-security, with the same goal of finding malicious activities in data [3]. However, most of the applications are to discover and prevent bank fraud. For example, Bolton et al. propose the use of unsupervised profiling methods to detect credit card fraud in financial transactions on a customer-based level [4], while Ferdousi et al. examine the occurrence of fraud in stock market data as anomalous behavior in an evolving time series [5].

In the airline industry, the data consists of flight bookings, which can be seen as customer-based data changing through time. However, there are some important differences between bookings and financial transactions. First of all, customers are usually not aware of an OTA conducting fraud and are not directly affected by it. Fraud can even be advantageous to the customer who can purchase a cheaper flight ticket. Furthermore, a booking can be fraudulent because of how it changes through time, in contrast with fraud in a single financial transaction. Lastly, OTAs are part of the business model and are necessary for the airline carrier to make a profit. Of course, the majority of them act sincerely.

Since the airline industry has some characteristics which set it apart from other fields in which fraud occurs, it is interesting to examine how anomaly detection methods perform. More importantly, we were not able to find literature on the detection of fraudulent behavior of OTAs. This paper addresses that research gap. The contributions of our research are, firstly, to show how three different algorithms are applied to the booking data of OTAs to discover violations of the policies conducted by the airline carrier. This allows us not only to eventually block fraudulent bookings, but this can also enrich domain experts with new knowledge on how to avoid malicious behavior from happening. Before the techniques are applied, practically usable data is constructed from raw booking datasets. To this end, existing features are modified and new variables are added. Secondly, we show the importance of the engineered features in discovering fraudulent bookings. An evaluation set of 90 bookings is constructed for domain experts to classify as normal or fraudulent. We assess how well the anomaly detection methods are able to find these fraudulent observations.

2 Data

The data used in this research was obtained from an airline company. It has several kinds of features. The first type is based on the passengers' travel requirements information, summarized as a passenger booking. It consists of features such as travel dates, travel routes, ticket information and associated OTAs for all flights planned for the coming 360 days. There were some observations with missing values for some of the features. It was decided not to remove all of them, since having missing values in certain fields could be related to fraudulent behavior of OTAs. Missing information could be due to an error in the reservation system, which could have been exploited by an OTA. The second type of features contains information about revenue for each created booking. Actual revenue data is only available for ticketed (paid) reservations, while it is estimated for non-ticketed reservations using historical revenue data. The third type of variables is directed at the OTAs themselves. It provides characteristics such as an unique identifier and their location (or market).

The goal is to find fraudulent bookings and the corresponding OTAs which violate the policies of the airline carrier. The observations in this raw dataset were given on a flight segment level, but the data needs to be booking-based, i.e., each observation should indicate a booking. Therefore, the flight segments corresponding to the same booking had to be merged.

3 Methods

In this section, we discuss how the segment-based raw data was merged and how new variables were constructed from the raw features. Furthermore, we provide a preliminary analysis of the data and introduce which ML techniques were used for the experiments. Lastly, we discuss some transformations that were applied to the dataset with the goal to improve the results.

3.1 Feature Engineering

Before the experiments were carried out, new variables which better represent the underlying characteristics of the data were extracted from the raw features which were described in Sect. 2. They can be categorized into two classes: *(i)* revenue-based features, and *(ii)* booking-based features.

Revenue-Based Features. The first category of features was derived from the variables containing revenue information. These new features were introduced to describe the relative amount of revenue generated per booking and to compare the expected revenue with the ticketed revenue received. The predictions in revenue were based on a historical horizon of fifteen days, which was advised by domain experts. They expected the majority of the changes to occur during this time window. Moreover, the *predicted minimum* and *maximum* revenue were added as features and a feature describing the *changes in revenue* over the time horizon was included. A relatively large difference between the predicted maximum and actual revenue could indicate malicious booking behavior. Since these new features were obtained per flight segment, the records corresponding to the same booking were aggregated (by taking both the sum and average) to obtain one observation for each feature per booking. Furthermore, the *ticketing time* and a feature describing the *variation in ticketing times* for the flight segment were included. The ticketing time is the time it takes before a booking has been paid for. When a flight is legitimately booked online, the payment is expected to be done directly for the whole flight, and hence, there should be no or only a small variation in the ticketing times of the flight segments. A relatively large variation could indicate fraudulent behavior.

Booking-Based Features. The second class of features was composed from the raw booking features. These new features do not only describe the important characteristics of the booking, but they also represent the OTA providing the flight ticket. As mentioned before, flight segments corresponding to the same booking were aggregated to obtain one observation per booking. A new feature of interest is *point of commencement (PoC) circumvention.* This feature checks whether the effective PoC is equal to the true PoC. Here, the effective PoC is the starting point the passenger is expected to depart from, while the true PoC is the actual starting point. PoC circumvention occurs when a fake flight segment is added to a booking to get access to a cheaper fare. Before the airplane departs, this flight segment is canceled while the lower flight price is retained. A difference between the effective and true PoC of a booking coincides with one or more cancellations or additions of flight segments, so features were added which explicitly indicate this behavior. This was done by comparing booking data on successive days and by calculating the differences in the number of flight segments in each booking. Furthermore, variables which indicate whether an OTA chronologically books flight segments (from first departure until last arrival) were included. These features are directly linked to policy violations.

Lastly, several other features were composed which capture other booking related data, such as the *number of passengers* in a booking, the *length of stay*, *number of days between cancellations*, and so on.

Final Dataset. After the feature engineering process, the final dataset consists of $P = 84$ numerical features on $N = 17{,}886$ unique bookings. The total number of unique OTAs is 158. Now, anomalies can be found on a booking level. Each booking is connected to an OTA, making it possible to find the agencies which were potentially conducting fraud.

3.2 Feature Analysis

Before the ML algorithms were applied, a preliminary feature analysis on fifteen days of data was performed. The purpose of this study is to give an insight into the booking data in the considered airline market and to examine what fraudulent behavior of an OTA could be. The features of interest in this exploratory study were those which are concerned with PoC circumvention.

Table 1. Number of flight segments in six different bookings for the past fifteen days.

Booking	Days from now into the past														
	14	13	12	11	10	9	8	7	6	5	4	3	2	1	0
1	7	7	7	7	7	7	7	7	7	6	6	6	6	6	6
2	6	6	7	7	7	7	7	7	7	7	7	6	6	6	6
3	5	5	5	5	5	5	5	5	5	5	5	5	5	5	6
4	7	7	7	7	7	7	7	7	7	7	7	7	7	6	6
5	2	2	2	2	2	2	2	2	2	2	4	4	2	2	2
6	6	6	7	7	6	6	6	6	6	7	7	6	6	6	6

To this end, the cancellations made in the bookings were examined. Table 1 shows the adjustments made in several bookings. These modifications were not just caused by cancellations, but also by the addition of new flight segments. This occurs in, for example, the last row. It shows an increase in the number of flight segments on day 12 and a decrease (cancellation) two days later on day 10, which is odd. This process repeats itself on day 5. It is unlikely that a passenger made such adjustments. A deeper analysis of a booking with canceled flight segments is shown in Table 2. Here, the two tables represent the same booking, but on different days. Note that the first two rows on the first day are not present on the second day anymore: these flight segments have been canceled. It is interesting to note that the values of the second column, the segment identifier, were not adjusted when flight segments were deleted. After examining this for several bookings with cancellations, it was concluded that the segment identifier

Table 2. Description of a particular booking on the first and second day. The columns indicate the departure date, segment identifier, departure location, arrival location, effective PoC, true PoC, and PoC circumvention, respectively.

Dep. day	Seg. ID	Dep. loc.	Arr. loc.	Eff. PoC	True PoC	PoC circumv.
Booking properties on day 1						
0	1	L_1	L_2	NA	PoC_1	NA
0	2	L_2	L_3	NA	PoC_1	NA
5	3	L_3	L_4	NA	PoC_1	NA
5	4	L_4	L_1	PoC_1	PoC_1	0
Booking properties on day 2						
5	3	L_3	L_4	NA	PoC_2	NA
5	4	L_4	L_1	PoC_1	PoC_2	1

Table 3. Overview of the descriptive statistics in the PoC features.

Description	Value
Percentage of PoC circumvented flight segments	7.27%
Number of unique effective PoCs	115
Number of unique true PoCs	138

was never modified. Hence, unexpected behavior in that variable could indicate this kind of fraud.

An overview of descriptive PoC characteristics is given in Table 3. Here, the percentage of flight segments with PoC circumvention is around 7%. Moreover, the table shows that the number of unique true POCs is greater than the number of effective PoCs. This difference indicates that there are at least 23 locations being used to circumvent the availability.

3.3 Anomaly Detection Techniques

Three anomaly detection techniques were considered in this research: isolation forest, one-class Support Vector Machine, and k-means clustering, which are explained in this section. These methods were chosen such that a wide variety of anomaly detection techniques was considered. Since no labeled data is available, unsupervised methods were used. They assume that the majority of the observations is normal, while only a small fraction is abnormal. This is the case for fraudulent bookings in the airline industry.

Isolation Forest. The first unsupervised technique was designed by Liu et al. in 2008. In contrast to traditional anomaly detection methods, an isolation forest explicitly separates anomalies rather than determining normal behavior and identifying anomalies as deviations from that behavior. This algorithm is more effective and efficient in detecting anomalies than commonly used distance- and density-based methods [6]. In short, an isolation forest determines how long it takes for each observation to be separated, which is done by continuously splitting features between their minimum and maximum values. Since the splits are performed on a feature level, the importance of each feature can be easily derived. Each isolation tree $t \in \{1, \ldots, T\}$ in an isolation forest of $T \in \mathbb{N}$ trees yields a path length $h_t(\mathbf{x}_i)$ for every observation $\mathbf{x}_i \in \mathbb{R}^P$, $i \in \{1, \ldots, N\}$, with P the number of features and N the total number of observations. Anomalies are the records with the smallest average path lengths, because they can be isolated rapidly.

There are two hyperparameters in an isolation forest: the sub-sampling size $\psi \in \mathbb{N}$, and T. The first parameter controls the training data size per tree, while the second one determines how many isolation trees are constructed during training. The *anomaly score* $s_N(\mathbf{x}_i)$ determines how anomalous observation $\mathbf{x}_i$ is. It is defined as

$$s_N(\mathbf{x}_i) = 2^{-\frac{\overline{h}(\mathbf{x}_i)}{c(N)}} \in (0, 1),$$

where $\overline{h}(\mathbf{x}_i) = (1/T) \sum_{t=1}^{T} h_t(\mathbf{x}_i)$ is the average path length of $\mathbf{x}_i$ in the isolation forest and $c(N) = 2H_{N-1} - 2(N-1)/N$ is the expected path length with H_n the n-th harmonic number. Liu et al. offer some rules of thumb: if $s_N(\mathbf{x}_i) \gg 0.5$, then $\mathbf{x}_i$ can be considered as an anomaly; if $s_N(i) \ll 0.5$, then $\mathbf{x}_i$ can be regarded as normal; and if $s_N(\mathbf{x}_i) \approx 0.5$, then the status of $\mathbf{x}_i$ is vague.

One-Class Support Vector Machine. The second unsupervised method applied in this research was designed by Schölkopf et al. in 1999 [7]. The goal of this Support Vector Machine (SVM) is to identify one specific class amongst all observations. This results in trying to separate the observations belonging to the normal class from the rest of the feature space. Hence, the instances which do not lie within the non-linear normality boundary are considered to be anomalous. Therefore, one-class SVM (ocSVM) is a boundary-based algorithm. It has been shown that such algorithms perform better than density-based techniques, since they solve a fundamentally easier problem [8]. Consequently, ocSVM is widely used in the field of anomaly detection.

The goal of ocSVM is to separate the data from the origin with maximum margin. A mathematical problem (a quadratic program) is solved to determine the normality boundary, yielding an optimal normal vector $\mathbf{w}$ and margin ρ. There is a hyperparameter $\nu \in [0, 1]$ acting as a trade-off between the fraction of anomalies in the data and the number of training examples used as support vectors [9]. The anomaly score $s(\mathbf{x}_i)$ of an observation $\mathbf{x}_i$ is given by

$$s(\mathbf{x}_i) = \operatorname{sgn}((\mathbf{w} \cdot \mathbf{\Phi}(\mathbf{x}_i)) - \rho),$$

where $\mathbf{\Phi}$ is a map into a dot product space related to the chosen kernel function. Now, if $s(\mathbf{x}_i) < 0$, then $\mathbf{x}_i$ can be regarded as anomalous; if $s(\mathbf{x}_i) > 0$, then $\mathbf{x}_i$ can be considered normal; and if $s(\mathbf{x}_i) = 0$, then $\mathbf{x}_i$ is exactly on the boundary and its status is not determined.

k-Means Clustering. The third and final anomaly detection technique considered is a clustering technique. k-means is an unsupervised, iterative algorithm proposed by Stuart Lloyd in 1957. It is one of the most popular clustering methods because of its simplicity [10]. In k-means, N observations have to be clustered into k clusters. Each cluster is represented by the mean (centroid) of the observations it contains. The clustering is performed such that the inter-cluster similarity is minimized, while the intra-cluster similarity is maximized. The similarity is determined by the Euclidean distance of the feature value to the mean value of the observations in the cluster: the smaller the distances, the higher the similarity.

The k-means algorithm converges quickly to a local optimum. Here, $k \in \mathbb{N}$ is a hyperparameter which, for example, can be determined using the elbow method. Here, the proportion of explained variance by the model is plotted as a function of the cluster size k. For small values of k, an increasing k will explain relatively much additional variance, but less additional variance is explained when k gets large. The optimal k is the value such that there is a bend in the plot. Now, to perform anomaly detection, a cluster boundary is introduced for each of the k clusters. This is a hypersphere around the cluster mean such that 95% of all the cluster observations are within the sphere, assuming that 5% of the observations are considered anomalous. For a new observation $\mathbf{x}_i$, first the closest cluster is chosen, and then it is determined whether $\mathbf{x}_i$ is within the boundary. If it is not, it can be considered anomalous.

3.4 Data Transformations

Finally, we investigated whether some data transformations had a positive effect on the anomaly detection performance of the algorithms discussed in Sect. 3.3. The transformations that were considered are normalization and standardization. To normalize the data, the feature values were linearly scaled such that all values lie in the interval $[0, 1]$. An advantage of normalization, or min-max scaling, is that each feature contributes equally, since all values are bounded in the same interval. Consequently, there is no feature overshadowing the other variables because of its large (absolute) values. However, a disadvantage is that the dispersion of the data is lost, possibly making it more difficult to detect anomalies. Standardization ensures that each feature has mean 0 and variance 1. The advantage of standardization over normalization is that the loss of dispersion is smaller.

Since tree-based models can handle varying feature ranges, normalization and standardization are not required in an isolation forest. However, the ocSVM and k-means methods are sensitive to magnitudes, and could therefore benefit from these transformations.

4 Experimental Setup

As mentioned in Sect. 3.3, there were several hyperparameters which had to be determined beforehand. For the isolation forest, these constants were the sub-sampling size $\psi \in \mathbb{N}$ and the number of trees $T \in \mathbb{N}$. Liu et al. [11] argue that $\psi = 256$ and $T = 100$ are large enough to enable convergence of the average path length of each observation. Next, the parameter ν in the one-class Support Vector Machine was chosen to be 0.05, since we assumed that about 5% of the observations were anomalous. The number of clusters k, which was a hyperparameter for k-means, was determined by the elbow method and varied for the different experiments that were performed: $k = 2$ for no modifications, $k = 9$ for normalization and $k = 38$ for standardization. Since there are no labels, there was no ground truth in the data to which the hyperparameters could be optimized.

All described procedures were performed in Python 3.6 with the libraries numPy and Pandas. The results were obtained from an evaluation set. This set was determined by the anomaly scores calculated by the three discussed ML methods. For each technique, the anomaly scores of the bookings were ranked in a descending order. Then, a random subset of 10 bookings was taken from the top 30, one from the 30 scores around the median, and one from the 30 lowest scores, yielding a sample of 30 bookings for each anomaly detection method. The sample observations from the top 30 were predicted to be fraudulent, while the other observations were considered normal. In total, there were 3 samples of 30 bookings each. There was an overlap between the samples, i.e., the algorithms ranked some of the observations in the same regions. Hence, they were selected more than once for the samples. There were 75 unique bookings in the total of 90. The sample bookings were classified by the domain experts as fraudulent (1) or normal (0), making it possible to assess the detection power of the algorithms. In the samples 39 fraudulent bookings were found, while 36 bookings were deemed normal.

5 Results

5.1 Performance of Anomaly Detection Techniques

To determine the quality of the models, the precision, recall, F_1 score and F_2 score were calculated. The latter two are given by the formula

$$F_\beta = \frac{(1+\beta^2) \cdot \text{precision} \cdot \text{recall}}{\beta^2 \cdot \text{precision} + \text{recall}},$$

Table 4. Results with no data transformation on method-specific evaluation samples

Model	Precision	Recall	F_1 score	F_2 score
isolation forest	0.75	0.8	0.774	0.789
ocSVM	0.769	0.588	0.667	0.617
k-means	1	0.444	0.615	0.5

where $\beta = 1$ for the F_1 score and $\beta = 2$ for the F_2 score. The F_2 score weighs recall more than the F_1 score does, i.e., it puts more emphasis on false negatives than false positives. This was done for the unmodified final dataset, which is the data without normalization or standardization. The performance metrics for the method-specific samples of size 30 are shown in Table 4. The construction of these samples is explained in Sect. 4. The isolation forest performed slightly better in finding policy violations (recall = 0.8) than making the distinction between normal and fraudulent observations (precision = 0.75). This was the other way around for the one-class Support Vector Machine (ocSVM) and k-means clustering, since both techniques had a higher precision than recall. Both the F_1 and F_2 scores suggest that the isolation forest performed the best. In fraud detection, reducing false negatives is usually more important than reducing false positives, since missing a fraudulent observation is deemed more harmful than raising a false alarm. False positives only bother domain experts with extra investigation time, while false negatives result in a potentially large loss of revenue. Hence, the F_2 score better represents how desirably the anomaly detection method performed. Note that this evaluation was done on the different method-specific samples. Although there is some overlap between them, a direct comparison of the performance measures is risky.

Table 5. Results with no data transformation on complete evaluation sample.

Model	Precision	Recall	F_1 score	F_2 score
isolation forest	0.706	0.615	0.657	0.632
ocSVM	0.75	0.462	0.571	0.5
k-means	0.8	0.308	0.444	0.351

The three ML methods were also applied to the complete evaluation sample. This set is the combination of the three method-specific samples of 30 observations each. The complete sample consists of 75 unique bookings. The results for the data with no modifications are presented in Table 5. The performance of the methods is comparable to the results on the method-specific samples in terms of F_1 and F_2 score. The isolation forest still ranks the best ($F_2 = 0.632$), followed by the ocSVM ($F_2 = 0.5$) and k-means clustering ($F_2 = 0.351$). This comparison was based on the same sample for each technique, thus strengthening the claims. Since there are 39 actual fraudulent bookings and 36 normal instances, the F_2 score is expected to be approximately 0.503 when the predicted labels are assigned by unbiased coin flips. This means only the isolation forest performed better than this threshold value.

The normalization and standardization procedures had remarkable influences on the results, as can be seen in Tables 6 and 7. For the method-specific samples, the precision and recall are both 0 for the isolation forest and ocSVM, performing severely worse than without data transformations. This was expected to some extent for the isolation forest, since the segregation of the observations is done

Table 6. Results on transformed method-specific evaluation data.

Model	Normalized				Standardized			
	Prec.	Recall	F_1	F_2	Prec.	Recall	F_1	F_2
isolation forest	0	0	0	0	1	0.067	0.125	0.082
ocSVM	0	0	0	0	0	0	0	0
k-means	0.909	0.556	0.690	0.602	0.8	0.444	0.571	0.488

Table 7. Results on transformed complete evaluation data.

	Normalized				Standardized			
Model	Prec.	Recall	F_1	F_2	Prec.	Recall	F_1	F_2
isolation forest	1	0.026	0.05	0.032	0.5	0.077	0.133	0.093
ocSVM	0.5	0.026	0.049	0.032	1	0.026	0.05	0.032
k-means	0.742	0.590	0.657	0.615	0.786	0.564	0.657	0.598

more rapidly with large variations in the data. However, this was not expected for the ocSVM. According to literature, transforming the data should benefit an SVM. This could be due to the Gaussian radial basis function that we used in this research. Nevertheless, the performance of k-means clustering increased from $F_2 = 0.5$ to $F_2 = 0.602$ with normalization. There was a slight decrease for standardization from $F_2 = 0.5$ to $F_2 = 0.488$.

For the combined sample, the performance of all three considered anomaly detection methods moderately increased in terms of F_2 score compared to the method-specific samples. In short, the isolation forest performed the best on its own sample of 30 observations without any data transformations ($F_2 = 0.789$). This was also the case for the one-class SVM ($F_2 = 0.617$). k-Means clustering performed the best on the complete evaluation sample with normalized data ($F_2 = 0.615$). Note that all these values are larger than the threshold value of 0.503.

5.2 Feature Evaluation

The results of the anomaly detection methods and the advise of the domain experts allowed us to construct a set of features which were deemed to be the most likely to identify suspicious behavior of an OTA. The list of the five most important features is given in Table 8. The first feature indicates the sum of the segment identifiers divided by the corresponding triangular number: $\binom{S+1}{2}$, where S is the number of flight segments in the booking. We expect the sum to equal the triangular number (and so the feature value to be 1), since the segments are usually labeled in an ascending order from 1 to S. The order ratio feature is not equal to 1 for the booking shown in Table 2, because flight segments have been canceled. The second variable is related to PoC circumvention.

As discussed in Sect. 3.2, the fact that PoC circumvention has occurred could indicate fraudulent behavior. We also showed in Tables 1 and 2 how the number of cancellations, which is the third most important feature, could be connected to fraud. The fourth feature has not been discussed in the feature analysis, but an unexpected value of this feature also suggests malicious behavior. Finally, the last feature in Table 8 indicates whether the OTA creating the booking is not equal to the OTA owning it.

Table 8. List of features to identify suspicious activity.

List of features that detect suspicious activity
Order ratio
PoC circumvention ratio
Number of cancellations
Number of booking class switches
Number of OTA owners which are unequal to the creator

6 Discussion and Conclusion

The goal of this research was to discover policy violations conducted by OTAs with the use of three anomaly detection methods. To this end, the raw data was analyzed and new variables were constructed to better describe the behavior of the OTAs. We demonstrated that these new features were important in detecting fraudulent bookings. This encourages domain experts to monitor these variables to detect some of the fraudulent behavior and avoid revenue loss. Moreover, this advises an airline organization to update its policy agreement with the OTAs to prevent such malpractices from happening in the future.

Together with the domain experts, we concluded that most of the anomalies were caused by cancellation activity in the bookings, suggesting that the values of the features corresponding to this behavior give a strong indication of fraud. However, there were instances in which normal bookings were detected as fraudulent, which happened because of complex and highly unusual flights. Also, there were instances in which the domain experts marked a booking as fraudulent, but it was based on a gut feeling. Here, the benefit of using unsupervised ML becomes evident: these bookings would never have been found when the bookings were only analyzed on a feature-based level. Moreover, since we were not able to find literature about this research field, we took an important step in understanding fraudulent behavior conducted by OTAs.

One of our suggestions for future research is to broaden the scope to make the results more generalizable. Firstly, we considered the bookings of one airline market. It is possible that the behavior of OTAs is significantly different for another market. Secondly, because of time constraints, only 75 records ($\approx$0.42%)

were analyzed by the domain experts. This means the results could be notably different when a new sample is considered. Another suggestion for future research is to find out at which stage in the booking process the models are able to detect fraud in an online setting.

References

1. Centre for Aviation: Fraud costs airlines USD1.4 billion a year. Regional airlines the fraudsters' "carriers of choice". https://centreforaviation.com/analysis/reports/fraud-costs-airlines-usd14-billion-a-year-regional-airlines-the-fraudsters-carriers-of-choice-48150. Accessed 2 Apr 2019
2. Rezgo Booking Software: What is an OTA? https://www.rezgo.com/glossary/ota. Accessed 2 Apr 2019
3. Chandola, V., Banerjee, A., Kumar, V.: Anomaly detection: a survey. ACM Comput. Surv. **41**(3), Article No. 15 (2009)
4. Bolton, R., Hand, D.: Unsupervised profiling methods for fraud detection. Credit Scoring Credit Control **7**, 235–255 (2001)
5. Ferdousi, Z., Maeda, A.: Unsupervised outlier detection in time series data. In: 22nd International Conference on Data Engineering Workshops, pp. 51–56. IEEE, Atlanta (2006)
6. Liu, F., Ting, K., Zhou, Z.-H.: Isolation forest. In: Proceedings of the 2008 Eighth IEEE International Conference on Data Mining, ICDM 2008, pp. 413–422. IEEE Computer Society, Washington (2008)
7. Schölkopf, B., Williamson, R., Smola, A., Shawe-Taylor, J., Platt, J.: Support vector method for novely detection. In: Advances in Neural Information Processing Systems, pp. 582–588 (2000)
8. Tax, D., Duin, R.: Uniform object generation for optimizing one-class classifiers. J. Mach. Learn. Res. **2**, 155–173 (2001)
9. Heller, K., Svore, K., Keromytis, A., Stolfo, S.: One class support vector machines for detecting anomalous windows registry accesses. In: Proceedings of the workshop on Data Mining for Computer Security, vol. 9. IEEE, Melbourne (2003)
10. Lloyd, S.: Least squares quantization in PCM. IEEE Trans. Inf. Theor. **28**(2), 129–137 (1982)
11. Liu, F., Ting, K., Zhou, Z.-H.: Isolation-based anomaly detection. ACM Trans. Knowl. Discov. Data (TKDD) **6**(1), 4 (2012)

Learned Constraint Ordering for Consistency Based Direct Diagnosis

Seda Polat Erdeniz(✉), Alexander Felfernig, and Muesluem Atas

Graz University of Technology, Inffeldgasse 16B/2, 8010 Graz, Austria
{spolater,alexander.felfernig,muatas}@ist.tugraz.at
http://ase.ist.tugrat.at

Abstract. Configuration systems must be able to deal with inconsistencies which can occur in different contexts. Especially in interactive settings, where users specify requirements and a constraint solver has to identify solutions, inconsistencies may more often arise. In inconsistency situations, there is a need of diagnosis methods that support the identification of minimal sets of constraints that have to be adapted or deleted in order to restore consistency. A diagnosis algorithm's performance can be evaluated in terms of time to find a diagnosis (runtime) and diagnosis quality. Runtime efficiency of diagnosis is especially crucial in real-time scenarios such as production scheduling, robot control, and communication networks. However, there is a trade off between diagnosis quality and the runtime efficiency of diagnostic reasoning. In this paper, we deal with solving *the quality-runtime performance trade off problem* of direct diagnosis. In this context, we propose *a novel learning approach for constraint ordering in direct diagnosis*. We show that our approach improves the runtime performance and diagnosis quality at the same time.

Keywords: Configuration systems · Diagnosis · Matrix factorization

1 Introduction

Configuration systems [11] are used to find solutions for problems which have many variables and constraints. A configuration problem can be defined as a constraint satisfaction problem (*CSP*) [17]. If constraints of a *CSP* are inconsistent, no solution can be found. In this context, diagnosis [1] is required to find at least one solution for an inconsistent *CSP*.

There are several diagnosis approaches. One of them is *direct diagnosis* which employs queries to check the consistency of the constraint set without the need to identify the corresponding conflict sets. When diagnoses have to be provided in real-time, response times should be less than a few seconds [2]. For example, in communication networks, efficient diagnosis is crucial to retain the quality of service. However, in direct diagnosis approaches, there is a clear trade-off between runtime performance of diagnosis calculation and diagnosis quality [4].

F. Wotawa et al. (Eds.): IEA/AIE 2019, LNAI 11606, pp. 347–359, 2019.
https://doi.org/10.1007/978-3-030-22999-3_31

To address this challenge, we propose *Learned Constraint Ordering (LCO)* for direct diagnosis. Our approach learns constraint ordering heuristics from inconsistent historical transactions which include inconsistent user requirements. Using historical inconsistent transactions, we build a sparse matrix and then employ matrix factorization techniques to estimate diagnoses. After this offline learning phase, the most similar transaction to the new inconsistent requirement set is found and the corresponding constraint ordering heuristic (which is calculated in the offline phase) is applied to reorder the inconsistent constraints before direct diagnosis. Thanks to the learned ordering of constraints, direct diagnosis algorithms can solve the diagnosis task with a high quality diagnosis result in a shorter runtime compared to direct diagnosis without constraint ordering. We provide a working example to demonstrate the effects of our approach. Finally, based on experimental evaluations, we show that using our constraint ordering approach with direct diagnosis algorithms is superior to the baseline (direct diagnosis algorithms without constraint ordering) on popular benchmark constraint satisfaction problems.

2 Preliminaries

In this section, we give an overview of the basic definitions in consistency-based configuration and diagnosis, and introduce a running example for this paper. Finally, we state our evaluation criteria for direct diagnosis.

2.1 Configuration Task

The following (simplified) assortment of digital cameras (see Table 1) and a set of inconsistent user requirements (see Table 2) for selecting a digital camera from the camera product table will serve as a working example to demonstrate how our approach works.

Table 1. The camera product table

	$Camera_1$	$Camera_2$	$Camera_3$	$Camera_4$	$Camera_5$
effectiveResolution	20.9	6.1	6.1	6.2	6.2
display	3.5	2.5	2.2	1.8	1.8
touch	yes	yes	no	no	no
wifi	yes	yes	no	no	no
nfc	no	no	no	no	no
gps	yes	yes	no	no	yes
videoResolution	UHD	UHD	No	UHD	4K
zoom	3.0	3.0	7.8	5.8	3.0
weight	475	475	700	860	560
price	659	659	189	2329	469

The working example is formed as a configuration task in Table 2 on the basis of Definition 1. As shown in Table 2, it has a variable set (V) with 10 variables (which are also listed in the first column of Table 1) and only one knowledge base constraint (c_1) which allows to select one camera (from the available cameras in Table 1). Other constraints are defined in the set of user requirements (REQ_{Lisa}).

Definition 1 ***(Configuration Task).*** *A configuration task can be defined as a CSP(V, D, C). V =$\{v_1,v_2, ..., v_n\}$ represents a set of finite domain variables. $D = \{dom(v_1), dom(v_2), ..., dom(v_n)\}$ represents a set of variable domains $dom(v_n)$ where $dom(v_n)$ represents the domain of variable v_n. $C = (C_{KB} \cup REQ)$ where C_{KB}=$\{c_1, c_2, ..., c_q\}$ is a set of domain specific constraints (the configuration knowledge base) that restricts the possible combinations of values assigned to the variables in V. REQ=$\{c_{q+1}, c_{q+2}, ..., c_t\}$ is a set of user requirements, which is also represented as constraints. A configuration (S) for a configuration task is a set of assignments S= $\{v_1 = a_1, v_2 = a_2, ..., v_n = a_n\}$ where $a_i \in dom(v_i)$ which is consistent with C.*

Table 2. An inconsistent configuration task: CSP_{Lisa}

V, D	v_1: **effectiveResolution**: $\{6.1Megapixel, 6.2Megapixel, 20.9Megapixel\}$, v_2: **display**: $\{1.8inches, 2.2\,in., 2.5\,in., 3.5\,in.\}$, v_3: **touch**:$\{no, yes\}$, v_4: **wifi**:$\{no, yes\}$, v_5: **nfc**:$\{no, yes\}$, v_6: **gps**:$\{no, yes\}$, v_7: **videoResolution**: $\{No, UHD, 4K\}$, v_8: **zoom**:$\{3.0x, 5.8x, 7.8x\}$, v_9: **weight**: $\{475g, 560g, 700g, 860g, 1405g\}$, v_{10}: **price**: {€ 189, € 469, € 659, € 2329, € 5219}
C_{KB}	c_1: (Camera$_1\vee$ Camera$_2\vee$ Camera$_3\vee$ Camera$_4\vee$ Camera$_5$)
REQ_{Lisa}	c_2: **effectiveResolution** = 20.9Megapixel c_3: **display** = 2.5 in. c_5: **wifi** = yes c_7: **gps** = yes c_9: **zoom** = 5.8x

2.2 Diagnosis Task

In a configuration system with a set of available products (see Table 1), a configuration task may not satisfy all constraints (see Table 2). Such a "no solution could be found" dilemma is caused by at least one conflict between constraints in the knowledge base C_{KB} and (a) the user requirements REQ or (b) within the set C_{KB} itself.

In such inconsistency cases we can help users to resolve the conflicts with a diagnosis Δ which is a set of constraints. Assuming that C_{KB} is consistent, we can say that the knowledge base always will be consistent if we remove REQ. Removing Δ from REQ leads to a consistent knowledge base (see Definition 2).

Definition 2 ***(REQ Diagnosis Task).*** *A user requirements diagnosis task (RDT) is defined as a tuple (C_{KB}, REQ) where REQ is the set of given user requirements and C_{KB} represents the constraints part of the configuration knowledge base. A diagnosis for a REQ diagnosis task (C_{KB}, REQ) is a set $\Delta \subseteq REQ$, s.t. $C_{KB} \cup (REQ - \Delta)$ is consistent (which means that there is at least one solution). $\Delta = \{c_1, c_2, .., c_n\}$ is minimal if there does not exist a diagnosis $\Delta' \subset \Delta$, s.t. $C_{KB} \cup (REQ - \Delta')$ is consistent.*

In Definition 3, we introduce the term *minimal diagnosis* which helps to reduce the number of constraints within a diagnosis.

Definition 3 ***(Minimal Diagnosis).*** *A minimal diagnosis Δ is a diagnosis (see Definition 2) and there doesn't exist a subset $\Delta' \subset \Delta$ which has the same property of being a diagnosis.*

The *REQ diagnosis task* of *Lisa* (RDT_{Lisa}) can be resolved by two minimal diagnoses. The removal of the set $\Delta_{Lisa\_1} = \{c_2, c_9\}$ or $\Delta_{Lisa\_2} = \{c_3, c_9\}$ leads to a consistent configuration knowledge base.

2.3 Direct Diagnosis

Algorithmic approaches to provide efficient solutions for diagnosis problems are many-fold. Basically, there are two types of approaches *conflict-directed diagnosis* [15] and *direct diagnosis* [4].

Conflict-directed diagnosis algorithms first calculate conflicts then find diagnoses. Therefore, their runtime performance are not sufficient for real-time scenarios. *Direct diagnosis* algorithms determine diagnoses by executing a series of queries. These queries check the consistency of the constraint set without the need of pre-calculated conflict sets.

Quality of diagnoses and runtime performance of direct diagnosis algorithms are based on the ordering of the constraints in the set of user requirements: *the lower the importance of a constraint means the lower the index of the constraint.* The lower the ordering conflicting constraint has the higher the probability that this constraint will be part of the diagnosis [3].

Users typically prefer to keep the important requirements and to change or delete (if needed) the less important ones [6]. The major goal of (model-based) diagnosis tasks is to identify the *preferred (leading) diagnoses* [7]. For the characterization of a preferred diagnosis we will rely on a total ordering of the given set of constraints in REQ. Such a total ordering can be achieved, for example, by directly asking the customer regarding the preferences, by applying multi attribute utility theory where the determined interest dimensions correspond with the attributes of REQ or by applying the orderings determined by conjoint analysis [12].

2.4 Evaluation Criteria

We can evaluate the performance of a direct diagnosis algorithm based on runtime performance, diagnosis quality (in terms of minimality), and combined performance (runtime and minimality).

Runtime. $runtime(\Delta)$ represents the time spent during the diagnostic search to find Δ. This spent time can be measured in milliseconds or in the number of consistency checks (#**CC**) applied till a diagnosis is found. For a more accurate runtime measurement (excluding the operating system's effects on runtime, etc.), the number of consistency checks can be used.

Minimality. Diagnosis quality can be measured in terms of the degree of minimality of the constraints in a diagnosis, the cardinality of Δ compared to the cardinality of Δ_{min}. $|\Delta_{min}|$ represents the cardinality of a minimal diagnosis. The highest (best) minimality can be 1 according to Formula 1.

$$minimality(\Delta) = \frac{|\Delta_{min}|}{|\Delta|} \tag{1}$$

Combined. Since it is important to satisfy both evaluation criteria runtime performance and minimality at the same time, we also evaluate combined performance based on the Formula 2. $combined(\Delta)$ increases when $minimality(\Delta)$ increases and/or $runtime(\Delta)$ decreases. This means, the direct diagnosis algorithm provides a diagnosis with a good quality (in terms of high minimality, which is maximum 1) in a short time (low runtime) when combined performance is high.

$$combined(\Delta) = \frac{minimality(\Delta)}{runtime(\Delta)} \tag{2}$$

3 Related Work

The most widely known algorithm for the identification of minimal diagnoses is *hitting set directed acyclic graph* (HSDAG) [10]. HSDAG is based on conflict-directed hitting set determination and determines diagnoses based on breadth-first search. It computes minimal diagnoses using minimal conflict sets which can be calculated by QuickXPlain [5]. The major disadvantage of applying this approach is the need of predetermining minimal conflicts which can deteriorate diagnostic search performance. Many different approaches to provide efficient solutions for diagnosis problems are proposed. One approach [19] focuses on improvements of HSDAG.

The direct diagnosis algorithm FlexDiag [4] utilizes an inverse version of the QuickXplain and an associated inverse version of HSDAG. Therefore, it finds directly a diagnosis from an inconsistent constraint set. FlexDiag assures diagnosis determination within certain time limits by systematically reducing the number of solver calls needed using the parameter m. Therefore, authors claim that this specific interpretation of anytime diagnosis leads *a trade-off between diagnosis quality (evaluated, e.g., in terms of minimality) and the time* needed for diagnosis determination. Our proposed constraint ordering approach *LCO* improves their direct diagnosis approach in terms of diagnosis quality at the same time with runtime performance.

Another work DIR [14] determines diagnoses by executing a series of queries. Authors reduce the number of consistency checks by avoiding the computation of minimized conflict sets and by computing some set of minimal diagnoses instead of a set of most probable diagnoses or a set of minimum cardinality diagnoses. Their approach is very similar to [3], with two modifications: (i) a depth-first search strategy instead of breadth-first and (ii) a new pruning rule to remove a constraint from the set of inconsistent constraints. They compared their approach only with the standard technique (based on QuickXplain [5] and HSDAG [10]. In their experiments based on a set of knowledge bases created by automatic matching systems, they show that their direct diagnosis approach outperforms the standard diagnosis approach in terms of runtime. In this work, authors gather the constraint ordering directly from users interactively. Our work differs from [14] in that we do not need active user interactions to determine the constraint ordering. Our approach learns the constraint ordering from historical inconsistent user requirements and their preferred diagnoses.

The importance of constraint ordering is already mentioned in related direct diagnosis work [4,14]. In our approach, we predict the most important constraints for the users based on an important collaborative filtering approach *matrix factorization* [8] and employing historical transactions. We learn a constraint ordering where the predicted most important constraints are placed to the highest orderings. This is because, the implemented direct diagnosis algorithm FlexDiag first start searching a diagnosis among the lowest raking constraints. Therefore, the highest ordering constraints have low probability to be in the diagnosis set.

4 Learned Constraint Ordering (LCO)

In this paper, our motivation is to solve *the quality-runtime performance trade off problem* of direct diagnosis. For this purpose, our proposed method learns constraint ordering heuristics based on historical transactions in offline phase and then in online phase it employs a direct diagnosis algorithm on the re-ordered constraints of diagnosis tasks (active transactions). In this paper, we demonstrate and evaluate our approach based on the direct diagnosis algorithm FlexDiag [4].

4.1 Offline Phase: Learning from Historical Transactions

Our proposed method needs an offline phase in which various constraint ordering heuristics are learned based on historical transactions. For the offline learning, matrix factorization techniques and *historical transactions with inconsistent user requirements* are employed (see Table 3).

Table 3. Historical transactions with inconsistent user requirements.

	Alice	Bob	Tom	Ray	Joe
c_2: resolution	-	6.1	20.9	20.9	6.2
c_3: display	3.5	2.2	-	2.5	2.2
c_4: touch screen	-	-	yes	yes	-
c_5: wifi	-	-	yes	yes	-
c_6: nfc	-	yes	-	-	yes
c_7: gps	-	yes	yes	yes	no
c_8: video resolution	UHD	-	UHD	UHD	UHD
c_9: zoom	3.0	5.8	3.0	5.8	7.8
c_{10}: weight	560	700	475	475	-
c_{11}: price	469	189	469	-	189
Purchase	$Camera_1$	-	$Camera_1$	-	$Camera_3$
Δ_{min}	$c10$, c_{11}	-	c_3, c_{11}	-	c_2, c_6, cv_7

In Table 3, for each user, we have an inconsistent set of user requirements (which leads to "no solution"). After their no-solution situation, some of them (e.g. Alice, Tom, and Joe in Table 3) decided to buy a product (*Purchase*) which does not completely satisfy their requirements. Therefore, they had to eliminate a set of initial requirements which is presented as Δ_{min}. These historical transactions which are completed with a purchase are *complete historical transactions.* The rest of historical transactions, in which users did not complete their transactions with a purchase (e.g. Bob and Ray in Table 3), is called *incomplete historical transactions.* Therefore, we estimate diagnoses of *incomplete historical transactions* using matrix factorization.

The Sparse Matrix. Matrix factorization based collaborative filtering algorithms [9] introduce a rating matrix R (a.k.a., user-item matrix) which describes preferences of users for the individual items the users have rated. Thereby, R represents an $m \times n$ matrix, where m denotes the number of users and n the number of items. The respective element $r_{u,i}$ of the matrix R describes the rating of the item i made by user u. Given the complete set of user ratings, the recommendation task is to predict how the users *would* rate the items which they have not yet been rated by these users.

In our approach, we build a sparse matrix R (user-constraint matrix) using inconsistent historical transactions as shown in Table 4-(a) where columns represent constraints. Therefore, each row of the sparse matrix R represents a set of user requirements (the left half) and their corresponding diagnoses (the right half) if available. User requirements are presented in their normalized values in the range of 0–1, and diagnoses are presented with the presence (1)/non-presence (0) of user requirements.

If there are non-numeric domains in the problem, they are enumerated. For example, the domain v_7: **videoResolution**: $\{No, UHD, 4K\}$ is enumerated as v_7: $\{0, 1, 2\}$. Besides, domain ranges of all constraints in REQ are mapped to [0...1] since matrix factorization needs to use the same range for all values in the matrix. For this purpose, we have employed *Min-Max Normalization* [18].

Matrix Factorization. In terms of *matrix factorization*, the sparse matrix R is decomposed into an $m \times k$ *user-feature matrix* P and a $k \times n$ *constraint-feature matrix* Q which both are used to find the estimated dense matrix PQ^T. Thereby, k is a variable parameter which needs to be adapted accordingly depending on the internal structure of the given data.

In our example, we apply matrix factorization to the sparse matrix in Table 4-(a). Then, the estimated matrix is obtained as shown in Table 4-(b) which includes the estimated diagnoses for Bob and Ray.

Table 4. Matrix factorization estimates a dense matrix PQ^T (b) which closely approximates the sparse matrix R (a).

The sparse matrix (R)

	c_2	c_3	c_4	c_5	c_6	c_7	c_8	c_9	c_{10}	c_{11}	p_{c2}	p_{c3}	p_{c4}	p_{c5}	p_{c6}	p_{c7}	p_{c8}	p_{c9}	p_{c10}	p_{c11}
Alice		1					0.5	0	0.09	0.05	0	0	0	0	0	0	0	0	1	1
Bob	0	0.23			1	1		0.58	0.24	0										
Tom	1		1	1		1	0.5	0	0	0.05	0	1	0	0	0	0	0	0	0	1
Ray	1	0.41	1	1		1	0.5	0.58	0											
Joe	0.006	0.23			1	0	0.5	1		0	1	0	0	0	1	1	0	0	0	0

The estimated dense matrix (PQ^T)

	c_2	c_3	c_4	c_5	c_6	c_7	c_8	c_9	c_{10}	c_{11}	p_{c2}	p_{c3}	p_{c4}	p_{c5}	p_{c6}	p_{c7}	p_{c8}	p_{c9}	p_{c10}	p_{c11}
Alice	1.4	0.2	1.3	1.5	0.3	1.2	0.9	0.6	−0.2	−0.3	0.4	0.2	−0.4	−0.4	0.4	0.4	−0.4	0.2	0.4	1.1
Bob	0.9	0.3	1.2	1.3	1.1	1.3	0.9	1	0	−0.3	0.7	0	−0.4	−0.4	1.1	1.1	−0.4	0.6	0.1	0.5
Tom	1.5	0.3	1.5	1.6	0.3	1.4	0.9	0.6	−0.2	−0.3	0.4	0.3	−0.4	−0.4	0.4	0.4	−0.4	0.3	0.4	1.1
Ray	1.4	0.3	1.5	1.6	0.7	1.5	0.9	0.9	−0.1	−0.3	0.8	0.1	−0.4	−0.4	0.6	0.6	−0.4	0.7	0.2	0.6
Joe	0.7	0.2	0.9	1.1	1.2	0.9	0.8	1.1	0.1	−0.3	0.8	−0.1	−0.3	−0.3	1.1	1.1	−0.4	0.4	0.1	0.4

4.2 Online Phase: Diagnosing Active Transactions

After calculating the estimated matrix in the offline phase, in the online phase we diagnose active transactions which includes inconsistencies as in our working example. In active transactions, users still did not leave the configuration system and need real-time help to remove inconsistencies in their configuration to decide on a product to purchase. Therefore, the configuration system should provide a high quality diagnosis in a reasonable time (before users leave the system without a purchase).

The Most Similar Historical Transaction. We find the most similar historical transaction to the new set of inconsistent requirements using Formula 3 where HT represents a historical transaction, AT represents the active transaction, $HT.c_i$ represents the value of each constraint in the estimated dense matrix PQ^T, and $AT.c_i$ represents the value of each constraints in the active transaction. i represent a constraint index value in the REQ of AT.

$$min(\sqrt{\sum_{i \in AT.REQ} \|HT.c_i - AT.c_i\|^2}) \tag{3}$$

In our camera configuration example, the most similar historical transaction to the active transaction of *Lisa* is the transaction of *Ray*. Therefore, to diagnose RDT_{Lisa}, we use LCO_{Ray}: $\{c_2, c_9, c_6, c_7, c_{11}, c_{10}, c_3, c_4, c_5, c_8\}$. When we only consider user requirements of *Lisa*, we obtain the constraint ordering for Lisa LCO_{Lisa}: $\{c_2, c_9, c_7, c_3, c_5\}$.

Direct Diagnosis with LCO. After the most similar historical transaction is found and its constraint ordering is applied to the active transaction's user requirements to be reordered, the direct diagnosis algorithm is employed on the active diagnosis task with the reordered user constraints. In this working example, user constraints of the diagnosis task is reordered using LCO_{Lisa} as $\{c_2, c_9, c_7, c_3, c_5\}$ and a minimal diagnosis $\Delta = \{c_2, c_9\}$ is found by FlexDiag-(m=1) with performance results (on the basis of evaluation criteria given in Sect. 2.4): #CC = 4, minimality = 1, and combined = 0.250. However, when we employ the diagnosis algorithm FlexDiag-(m=1) on the default order of user constraints $\{c_2, c_3, c_5, c_7, c_9\}$, the same diagnosis $\Delta = \{c_2, c_9\}$ is found with performance results: #CC = 8, minimality = 1, and combined = 0.125. Therefore, using LCO with FlexDiag-(m=1), we double the combined performance of diagnosis when diagnosing the working example.

5 Experimental Evaluation

Settings. We have developed our approach in Java and tested on a computer with an Intel Core i5-5200U, 2.20 GHz processor, 8 GB RAM and 64 bit Windows 7 Operating System and Java Run-time Environment 1.8.0. Constraint satisfaction problems have been solved by *Choco3*[1]. which is a java library for constraint satisfaction problems with a FlatZinc (the target language of MiniZinc) parser. For matrix factorization, we have used the *SVDRecommender* of Apache Mahout [13][2].

[1] http://www.choco-solver.org/.

[2] Using the latent factor k=100 and the number of iterations = 1000.

Datasets. We have used Minizinc-2016 benchmark problems [16] where each problem includes five data files with file extension "*.dzn*".[3] In order to obtain historical and active transactions based on these benchmark problems, we randomly generated 5000 sets of inconsistent user requirements (each with N constraints) based on integer variables.

Results. As discussed throughout this paper, our main objective is to improve the combined performance (runtime performance and diagnosis quality at the same time). We have compared our approach LCO with the baseline *no constraint ordering*. In both cases, for diagnostic search, FLEXDIAG is used with three different m values 1, 2, and 4.

As shown in Table 5, based on the averages (in the last row), LCO outperforms the baseline in terms of *runtime* and *minimality* since with each m value (1, 2, and 4), LCO has lower runtime than the baseline whereas its minimality is higher (or equal) compared to the baseline with each m value (1, 2, and 4).

Based on the results in Table 5, we present two comparisons in Fig. 1 where relations between performance indicators and the number of constraints in the set of user requirements are shown.

In Fig. 1-(a), we observe that *minimality* increases when *runtime* (in #CC) increases. As observed, the number of consistency checks (#CC) are at each m ($m = 1$, 2, and 4) lower when LCO is used. Moreover, at each m ($m = 1$, 2, and 4), LCO also provides better of equal minimality results.

We present combined performance results in Fig. 1-(b). Deviations in the results of LCO are more visible than the baseline, because LCO has greater performance values. When we zoom into the results of the baseline, we also observe similar deviations due to the variations in problems. As observed, our approach improves combined performance significantly.

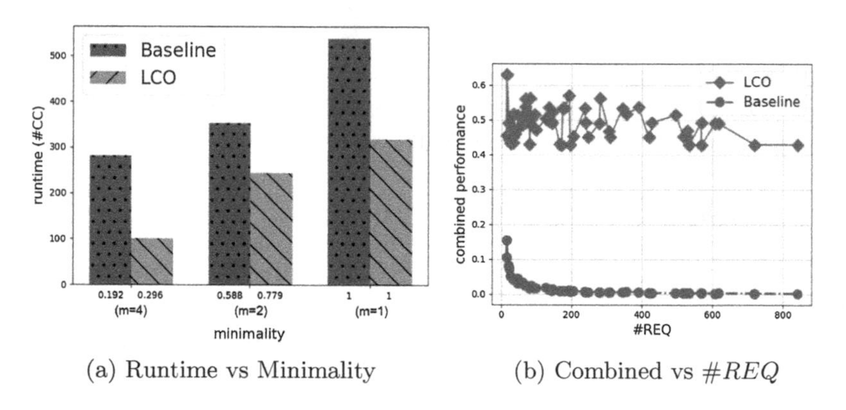

(a) Runtime vs Minimality (b) Combined vs $\#REQ$

Fig. 1. Comparison Graphs based on the Experimental Results in Table 5

[3] http://www.minizinc.org/challenge2016/results2016.html.

Table 5. Experimental results based on Minizinc-2016 Benchmark problems.

EXPERIMENTAL RESULTS																
Minizinc 2016 Benchmark			Inconsistencies		BASELINE						OUR APPROACH					
					runtime			*minimality*			*runtime*			*minimality*		
.mzn	*.dzn*	#vars	#REQs	$\lvert\Delta_{\min}\rvert$	m=1	m=2	m=4	m=1	m=2	m=4	m=1	m=2	m=4	m=1	m=2	m=4
1. cc_base	test.02	136	68	5	27	23	15	1	0.333	0.25	19	15	6	1	0.5	0.333
	test.06	478	239	23	120	80	60	1	0.333	0.166	60	53	22	1	1	0.333
	test.11	159	80	5	32	27	18	1	0.5	0.25	20	16	7	1	1	0.333
	test.13	291	146	11	73	42	36	1	0.5	0.166	36	32	13	1	1	0.333
	test.20	688	344	28	172	98	86	1	0.333	0.166	86	76	29	1	1	0.333
2. celar	CEL.6-S0	557	279	21	139	80	70	1	1	0.143	70	62	23	1	1	0.333
	CEL.6-S4	1136	568	45	284	189	126	1	0.333	0.25	162	114	52	1	1	0.25
	CEL.7-S4	1137	569	45	227	190	126	1	0.333	0.143	162	126	52	1	0.5	0.333
	graph05	2680	1340	99	536	447	335	1	0.333	0.143	383	268	122	1	1	0.25
	scen07	6331	3166	211	1583	1055	703	1	1	0.166	791	703	288	1	1	0.25
3. stepl	kb128_11	17390	8695	790	4348	2484	1932	1	0.5	0.25	2174	1739	725	1	1	0.25
	kb128_14	17390	8695	828	4348	2898	2174	1	0.333	0.166	2174	1932	725	1	0.5	0.333
	kb128_16	17390	8695	870	4348	2484	2174	1	1	0.25	2174	1739	725	1	1	0.25
	kb128_17	17390	8695	870	3478	2898	2174	1	0.333	0.166	2484	1739	725	1	0.5	0.25
	kb192_10	47278	23639	1970	9456	6754	5253	1	0.5	0.166	6754	5253	2149	1	0.5	0.333
4. depot	att48_6	77	39	3	19	11	9	1	1	0.166	10	9	3	1	1	0.25
	rat99_5	69	35	3	14	12	9	1	0.333	0.143	9	8	3	1	1	0.25
	rat99_6	77	39	4	19	11	9	1	0.333	0.25	10	8	4	1	1	0.25
	st70_5	69	35	2	17	10	9	1	0.5	0.166	10	8	3	1	0.5	0.25
	ulysses	69	35	2	17	10	9	1	0.333	0.166	10	7	3	1	1	0.333
5. dcmst	c_v15_d7	494	247	18	124	71	62	1	0.5	0.166	62	49	22	1	0.5	0.25
	c_v20_d5	856	428	39	214	143	107	1	0.333	0.143	122	95	39	1	0.5	0.25
	s_v20_d4	390	195	14	78	65	49	1	1	0.25	49	39	18	1	1	0.333
	s_v20_d5	385	193	14	77	64	48	1	0.333	0.166	48	43	16	1	0.5	0.333
	s_v40_d5	1214	607	45	304	173	135	1	1	0.2	173	121	51	1	0.5	0.25
6. filter	ar_1_3	121	61	6	30	17	13	1	0.333	0.25	17	12	6	1	1	0.25
	dct_1_3	189	95	9	38	32	21	1	0.5	0.143	24	19	9	1	1	0.333
	ewf_1_2	139	70	6	35	20	15	1	1	0.25	17	14	6	1	0.5	0.333
	fir_1_3	92	46	4	18	15	10	1	1	0.166	13	10	4	1	1	0.333
	fir_1_4	92	46	3	18	15	12	1	0.333	0.2	12	10	4	1	0.5	0.333
7. gbac	UD3	1687	844	77	422	241	211	1	0.5	0.25	241	187	70	1	1	0.25
	UD6	561	281	24	140	94	62	1	0.333	0.166	80	62	23	1	1	0.333
	UD10	619	310	24	155	103	77	1	1	0.2	77	62	28	1	1	0.25
	UD3	1938	969	81	388	277	242	1	0.5	0.143	242	194	88	1	1	0.25
	UD5	1439	720	55	360	206	160	1	0.5	0.2	180	160	65	1	1	0.333
8. gfd-sch.	n25f5.	344	172	11	69	57	43	1	0.333	0.166	49	34	14	1	0.5	0.333
	n35f5.	477	239	20	95	80	53	1	0.333	0.25	60	48	20	1	0.5	0.25
	n55f2.	612	306	21	122	102	77	1	1	0.166	87	61	28	1	0.5	0.333
	n60f7.	1236	618	54	247	177	155	1	0.5	0.143	177	137	56	1	0.5	0.333
	n180f.	4950	2475	177	1238	825	550	1	0.5	0.143	619	495	225	1	0.5	0.25
9. map.	m2x2_1	197	99	9	49	33	22	1	0.333	0.25	25	22	8	1	0.5	0.333
	m2x2	358	179	16	90	60	40	1	1	0.2	45	40	16	1	1	0.333
	m3x3	838	419	38	210	120	105	1	0.333	0.166	105	93	35	1	0.5	0.333
	m4x4	841	421	28	168	120	105	1	1	0.25	120	93	38	1	0.5	0.25
	ring	411	206	15	82	59	46	1	0.5	0.2	51	46	19	1	1	0.25
10. m.dag	25_01	136	68	7	34	19	15	1	0.333	0.166	19	15	6	1	0.5	0.333
	25_03	160	80	5	32	27	18	1	0.333	0.143	20	18	7	1	1	0.25
	25_04	133	67	6	27	19	17	1	0.333	0.166	17	15	6	1	1	0.333
	25_06	164	82	8	33	27	18	1	0.5	0.2	23	16	7	1	1	0.333
	31_02	169	85	6	42	24	19	1	0.5	0.2	21	17	8	1	1	0.25
11. mrcp.	j30_1_10	712	356	28	178	102	89	1	1	0.143	89	79	32	1	0.5	0.333
	j30_15_5	333	167	12	67	56	37	1	0.5	0.2	42	37	15	1	0.5	0.333
	j30_17_10	992	496	34	198	142	124	1	1	0.25	142	99	41	1	0.5	0.25
	j30_37_4	780	390	28	195	111	98	1	0.5	0.2	111	78	33	1	0.5	0.25
	j30_53_3	350	175	12	70	58	44	1	0.333	0.25	44	39	15	1	1	0.25
12. nfc	12_2_5	31	16	1	8	5	3	1	0.333	0.2	4	3	1	1	1	0.25
	12_2_10	29	15	1	6	4	4	1	1	0.166	4	3	1	1	1	0.333
	18_3_5	45	23	2	11	6	6	1	1	0.143	6	5	2	1	0.5	0.333
	18_3_10	42	21	2	11	7	5	1	1	0.2	5	5	2	1	0.5	0.333
	24_4_10	54	27	2	11	8	7	1	0.5	0.25	7	5	2	1	1	0.25
13. oocsp	030_e6_cc	1034	517	43	207	148	129	1	1	0.25	129	103	43	1	0.5	0.25
	030_ea4_cc	1068	534	40	267	178	119	1	0.333	0.25	153	119	45	1	0.5	0.333
	030_f7_cc	1058	529	44	265	151	132	1	1	0.166	151	106	44	1	1	0.25
	030_mii8	1053	527	48	263	176	117	1	1	0.25	132	117	44	1	0.5	0.333
	100_r1	3433	1717	149	858	572	429	1	0.333	0.25	429	381	143	1	1	0.333
14. pc	28-4-7-1	252	126	12	50	42	28	1	0.333	0.143	32	25	11	1	1	0.25
	30-5-6-2	270	135	9	54	39	34	1	1	0.143	39	27	11	1	1	0.333
	30-5-6-8	270	135	11	54	45	30	1	1	0.143	34	27	11	1	0.5	0.333
	32-4-8-2	288	144	10	58	41	36	1	0.333	0.2	36	32	13	1	1	0.333
	32-4-8-5	288	144	10	58	41	36	1	0.5	0.2	36	29	13	1	1	0.333
average		2349	1174	101	530	357	276	1	0.588	0.192	315	249	102	1	0.779	0.296

6 Conclusions

In this paper, we proposed a novel learning approach for constraint ordering heuristics to solve *the quality-runtime performance trade off problem* of direct diagnosis. We employed matrix factorization for learning based on historical transactions. Taking the advantage of learning from historical transactions, we calculated possible constraint ordering heuristics in offline phase for solving new diagnosis tasks in online phase.

In particular, we applied our constraint ordering heuristics to reorder user constraints of diagnosis tasks and employed a direct diagnosis algorithm FlexDiag [4] to diagnose the reordered constraints of the diagnosis tasks. The reason to choose this diagnosis algorithm is that *the quality-runtime performance trade off problem* is much more obvious in FlexDiag when its m parameter is increased. However, our approach can be also applicable to other direct diagnosis approaches (e.g. DIR [14]). We compared our approach with a baseline: FlexDiag without heuristics. According to our experimental results, our approach *LCO* solves the *the quality-runtime performance trade off problem* by improving the diagnosis quality (in terms of minimality) and the runtime performance at the same time.

Our approach also improves the diagnosis quality in terms of prediction accuracy by learning the preferred diagnoses of users from historical transactions. However, we could not evaluate this criterion due to the lack of preferred diagnoses information in publicly available real-world configuration benchmarks. As a future work, in order to measure the prediction quality, we can further investigate *LCO*'s prediction accuracy on the basis of a real-world dataset of users preferred diagnoses.

Acknowledgments. The work presented in this paper has been conducted within the scope of the Horizon 2020 projects OpenReq (Grant Nr. 732463) and AGILE (Grant Nr. 688088).

References

1. Bakker, R.R., Dikker, F., Tempelman, F., Wognum, P.M.: Diagnosing and solving over-determined constraint satisfaction problems. IJCAI **93**, 276–281 (1993)
2. Card, S.K., Robertson, G.G., Mackinlay, J.D.: The information visualizer, an information workspace. In: Proceedings of the SIGCHI Conference on Human Factors in Computing Systems, pp. 181–186. ACM (1991)
3. Felfernig, A., Schubert, M., Zehentner, C.: An efficient diagnosis algorithm for inconsistent constraint sets. Artif. Intell. Eng. Des. Anal. Manuf. (AIEDAM) **26**(1), 53–62 (2012)
4. Felfernig, A., et al.: Anytime diagnosis for reconfiguration. J. Intell. Inf. Syst. **51**, 1–22 (2018)
5. Junker, U.: Quickxplain: conflict detection for arbitrary constraint propagation algorithms. In: Workshop on Modelling and Solving Problems with Constraints, IJCAI 2001 (2001)

6. Junker, U.: Preferred explanations and relaxations for over-constrained problems. In: AAAI-2004 (2004)
7. de Kleer, J.: Using crude probability estimates to guide diagnosis. Artif. Intell. **45**(3), 381–391 (1990)
8. Koren, Y.: Collaborative filtering with temporal dynamics. Commun. ACM **53**(4), 89–97 (2010)
9. Koren, Y., Bell, R., Volinsky, C.: Matrix factorization techniques for recommender systems. Computer **42**(8), 30–37 (2009)
10. Reiter, R.: A theory of diagnosis from first principles. Artif. Intell. **32**(1), 57–95 (1987)
11. Sabin, D., Weigel, R.: Product configuration frameworks-a survey. IEEE Intell. Syst. Appl. **13**(4), 42–49 (1998)
12. Schaupp, L.C., Bélanger, F.: A conjoint analysis of online consumer satisfaction1. J. Electron. Commer. Res. **6**(2), 95 (2005)
13. Schelter, S., Owen, S.: Collaborative filtering with apache mahout. Proc. of ACM RecSys Challenge (2012)
14. Shchekotykhin, K.M., Friedrich, G., Rodler, P., Fleiss, P.: Sequential diagnosis of high cardinality faults in knowledge-bases by direct diagnosis generation. ECAI **14**, 813–818 (2014)
15. Stern, R.T., Kalech, M., Feldman, A., Provan, G.M.: Exploring the duality in conflict-directed model-based diagnosis. AAAI **12**, 828–834 (2012)
16. Stuckey, P.J., Feydy, T., Schutt, A., Tack, G., Fischer, J.: The minizinc challenge 2008–2013. AI Mag. **35**(2), 55–60 (2014)
17. Tsang, E.: Foundations of Constraint Satisfaction. Academic Press, Cambridge (1993)
18. Visalakshi, N.K., Thangavel, K.: Impact of normalization in distributed k-means clustering. Int. J. Soft Comput. **4**(4), 168–172 (2009)
19. Wotawa, F.: A variant of Reiter's hitting-set algorithm. Inf. Process. Lett. **79**(1), 45–51 (2001)

On the Usefulness of Different Expert Question Types for Fault Localization in Ontologies

Patrick Rodler(✉) and Michael Eichholzer

Alpen-Adria Universität Klagenfurt, 9020 Klagenfurt, Austria
patrick.rodler@aau.at, michael.eichholzer@aon.at

Abstract. When ontologies reach a certain size and complexity, faults such as inconsistencies or wrong entailments are hardly avoidable. Locating the faulty axioms that cause these faults is a hard and time-consuming task. Addressing this issue, several techniques for semi-automatic fault localization in ontologies have been proposed. Often, these approaches involve a human expert who provides answers to system-generated questions about the intended (correct) ontology in order to reduce the possible fault locations. To suggest as few and as informative questions as possible, existing methods draw on various algorithmic optimizations as well as heuristics. However, these computations are often based on certain assumptions about the interacting user and the metric to be optimized.

In this work, we critically discuss these optimization criteria and suppositions about the user. As a result, we suggest an alternative, arguably more realistic metric to measure the expert's effort and show that existing approaches do not achieve optimal efficiency in terms of this metric. Moreover, we detect that significant differences regarding user interaction costs arise if the assumptions made by existing works do not hold. As a remedy, we suggest a new notion of expert question that does not rely on any assumptions about the user's way of answering. Experiments on faulty real-world ontologies testify that the new querying method minimizes the necessary expert consultations in the majority of cases and reduces the computation time for the best next question by at least 80 % in all scenarios.

Keywords: Ontology debugging · Interactive debugging · Fault localization · Sequential diagnosis · Expert questions · Ontology quality assurance · Ontology repair · Test-driven debugging

1 Introduction

As Semantic Web technologies have become widely adopted in, e.g., government, security and health applications, the quality assurance of the data, information and knowledge used by these applications is a critical requirement. At the core

F. Wotawa et al. (Eds.): IEA/AIE 2019, LNAI 11606, pp. 360–375, 2019.
https://doi.org/10.1007/978-3-030-22999-3_32

of these semantic technologies, ontologies are a means to represent knowledge in a formal, structured and human-readable way, with a well-defined semantics. As ontologies are often developed and cured in a collaborative way by numerous contributors [34,35], are merged by automated alignment tools [11], reach vast sizes and complexities [5], or use expressive logical formalisms such as OWL 2 [6], faults occur regularly during the evolution of ontologies [3,11,14,29]. Since one of the major benefits of ontologies is the capability of using them to perform logical reasoning and thereby solve relevant problems, faults that affect the ontology's semantics are of particular concern for semantic applications. Specifically, such faults may cause the ontology, e.g., to become inconsistent, include unsatisfiable classes, or feature wrong entailments.

One important step towards the repair of such faults is the *localization* of the responsible faulty ontology axioms. To handle nowadays ontologies with often thousands of axioms, several fault localization approaches [9,11,12,30] have been proposed to semi-automatically assist humans in this complex and time-consuming task, amongst them a plug-in, called OntoDebug[1] [27], for the popular ontology editor Protégé. These approaches, which are mainly based on the *model-based diagnosis* framework [10,15], use the faulty ontology along with additional specifications to reason about different fault assumptions. Such fault assumptions are called *diagnoses* if they are consistent with all given specifications. The specifications usually comprehend some requirements to the correct ontology, e.g., in the form of *logical properties* (e.g., consistency, coherency), and/or in terms of necessary and forbidden entailments. The latter are usually referred to as *positive and negative test cases* [4,28,30].

Research on model-based diagnosis has brought up various algorithms [9–11,15,16,32] for computing and ranking diagnoses; however, a frequent problem is that a high number of competing diagnoses might exist where all of them lead to repaired ontologies with necessarily different semantics [16]. Finding the correct diagnosis (pinpointing the actually faulty axioms) is thus crucial for successful and sustainable repair. But, it is a mentally-demanding task for humans since it requires them to reason about and recognize entailments and non-entailments [7] of the ontology under particular fault assumptions. To relieve the user as much as possible, interactive techniques [16,30] have been developed to undertake this task for the most part. What remains to be accomplished by the interacting human—usually an ontology engineer or a domain expert (referred to as *expert* in the sequel)—is the answering of a sequence of system-generated *queries* about the intended ontology. Roughly speaking, this involves the classification of certain axioms as either intended entailments (positive test cases) or non-intended entailments (negative test cases). Several evaluations [21,24,30,32] have shown the feasibility and usefulness of such a query-based approach for fault localization, and its efficiency has been improved by various algorithmic optimizations [8,20,25,31] and the use of heuristics [17,18,23,26,30] for the selection of the most informative questions to ask an expert.

[1] All information about OntoDebug can be found at http://isbi.aau.at/ontodebug/.

However, the used heuristics, algorithms and optimization criteria are based on certain assumptions about the question answering behavior of experts. In this work, we critically discuss existing approaches with regard to these assumptions. Particularly, we characterize different types of experts and show that not all of them are equally well accommodated by current querying approaches. That is, we observe that the necessary expert interaction cost to locate the ontology's faults is significantly influenced by the way queries posed by the debugging system are answered. To overcome this issue, we propose a new way of user interaction that serves all discussed expert types equally well and moreover increases the expected amount of information relevant for fault localization obtained from the expert per asked axiom.

The main idea behind the new approach is to restrict queries—which are, for quite natural reasons, *sets of* axioms in existing methods—only to *single* axioms, as usually done in *sequential diagnosis* applications [10,33], where systems different from ontologies (e.g., digital circuits) are analyzed and such singleton queries are the natural choice. That is, experts are asked single axioms at a time instead of getting batch queries which (possibly) include multiple axioms. Experiments on real-world faulty ontologies manifest the reasonability of the new approach. Specifically, in two thirds of the studied cases, the new querying technique is superior to existing ones in terms of minimizing the number of required expert inputs, regardless of the type of expert. In addition, the time for the determination of the best next query is reduced by at least 80 % in all investigated cases when using singleton queries instead of existing techniques.

The rest of the work is organized as follows. In Sect. 2, we give a short introduction to query-based fault localization in ontologies, before we challenge certain assumptions made by state-of-the-art approaches in the field in Sect. 3. We describe our proposed approach and discuss its pros and cons in Sect. 4. Our experiments and the obtained results are explicated in Sect. 5. We address research limitations and point to relevant future work topics in Sect. 6. In Sect. 7, we summarize the conclusions from this work.

2 Query-Based Fault Localization in Ontologies

We briefly recap basic technical concepts used in works on ontology fault localization, based on [16,30]. As a running example we reuse the example presented in [23].

Fault Localization Problem Instance. We assume a faulty ontology to be given by the finite set of axioms $\mathcal{O} \cup \mathcal{B}$, where $\mathcal{O}$ includes the *possibly faulty* axioms and $\mathcal{B}$ the *correct* (background knowledge) axioms, and $\mathcal{O} \cap \mathcal{B} = \emptyset$ holds. This partitioning of the ontology means that faulty axioms must be sought only in $\mathcal{O}$, whereas $\mathcal{B}$ provides the fault localization context. At this, $\mathcal{B}$ can be useful to achieve a fault search space restriction (if parts of the faulty ontology are marked correct) or a higher fault detection rate (if external approved knowledge is taken into account, which may point at otherwise undetected faults). Besides logical properties such as consistency and coherency[2], requirements to the intended (correct) ontology can be formulated as a set of test cases [4], analogously as it is common practice in software engineering [2]. In particular, we distinguish between two types of test cases, positive (set P) and negative (set N) ones. Each test case is a set (interpreted as conjunction) of axioms; positive ones $p \in P$ *must* be and negative ones $n \in N$ *must not* be entailed by the intended ontology. We call $\langle \mathcal{O}, \mathcal{B}, P, N \rangle$ an *(ontology) fault localization problem instance (FPI)*.

Example 1. Consider the following ontology with the terminology $\mathcal{T}$:[3]

$$\{ \quad ax_1 : \mathit{ActiveResearcher} \sqsubseteq \exists \mathit{writes}.(\mathit{Paper} \sqcup \mathit{Review}) \;, \\ ax_2 : \exists \mathit{writes}.\top \sqsubseteq \mathit{Author} \;, \qquad ax_3 : \mathit{Author} \sqsubseteq \mathit{Employee} \sqcap \mathit{Person} \quad \}$$

and assertions $\mathcal{A} : \{ax_4 : \mathit{ActiveResearcher}(ann)\}$. In natural language, the terminological axioms say that "an active researcher writes something which is a paper, a review or both" (ax_1), that "everybody who writes something is an author" (ax_2), and that "an author is both an employee and a person" (ax_3). To locate faults in the terminology while accepting as correct the assertion (ax_4) and stipulating that Ann is not necessarily an employee (negative test case n_1 : $\{\mathit{Employee}(ann)\}$), one can specify the following FPI: $\mathit{fpi}_{ex} := \langle \mathcal{T}, \mathcal{A}, \emptyset, \{n_1\} \rangle$.□

Fault Hypotheses. Let U_P denote the union of all positive test cases $p \in P$ and $\mathbf{C}_\bot := \{C \sqsubseteq \bot \mid C$ named class in $\mathcal{O}, \mathcal{B}$ or $P\}$. Given that the ontology, along with the positive test cases, is inconsistent or incoherent, i.e., $\mathcal{O} \cup \mathcal{B} \cup U_P \models x$ for some $x \in \{\bot\} \cup \mathbf{C}_\bot$, or some negative test case is entailed, i.e., $\mathcal{O} \cup \mathcal{B} \cup U_P \models n$ for some $n \in N$, some axioms in $\mathcal{O}$ must be accordingly modified or deleted to enable the formulation of the intended ontology. We call such a set of axioms $\mathcal{D} \subseteq \mathcal{O}$ a *diagnosis* for the FPI $\langle \mathcal{O}, \mathcal{B}, P, N \rangle$ iff $(\mathcal{O} \setminus \mathcal{D}) \cup \mathcal{B} \cup U_P \not\models x$ for all $x \in N \cup \{\bot\} \cup \mathbf{C}_\bot$. $\mathcal{D}$ is a *minimal diagnosis* iff there is no diagnosis $\mathcal{D}' \subset \mathcal{D}$. We call $\mathcal{D}^*$ *the actual diagnosis* iff all $ax \in \mathcal{D}^*$ are faulty and all $ax \in \mathcal{O} \setminus \mathcal{D}^*$ are correct. For efficiency and to point to minimally-invasive ontology repairs, fault localization approaches usually restrict their focus to the computation of minimal diagnoses.

Example 2. For $\mathit{fpi}_{ex} = \langle \mathcal{O}, \mathcal{B}, P, N \rangle$ from Example 1, $\mathcal{O} \cup \mathcal{B} \cup U_P$ entails the negative test case $n_1 \in N$, i.e., that Ann is an employee. The reason is that according to $ax_1(\in \mathcal{O})$ and $ax_4(\in \mathcal{B})$, Ann writes some paper or review since she is an active researcher. Due to the additional $ax_2(\in \mathcal{O})$, Ann is also an author because she writes something. Finally, since Ann is an author, she must be both an employee and a person, as postulated by $ax_3(\in \mathcal{O})$. Hence, $\mathcal{D}_1 : [ax_1]$, $\mathcal{D}_2 : [ax_2]$, $\mathcal{D}_3 : [ax_3]$ are (all the) minimal diagnoses for fpi_{ex}, as the deletion of any $ax_i \in \mathcal{O}$ breaks the unwanted entailment n_1. □

---

[2] An ontology $\mathcal{O}$ is *coherent* iff there do not exist any unsatisfiable classes in $\mathcal{O}$. A class C is *unsatisfiable* in $\mathcal{O}$ iff $\mathcal{O} \models C \sqsubseteq \bot$. See also [13, Def. 1 and 2].

[3] Throughout the presented examples, we use Description Logic notation. For details, see [1].

Eliminating Wrong Fault Hypotheses. The main idea model-based diagnosis systems use for fault localization—i.e., to find the actual diagnosis among the set of all (minimal) diagnoses—is that different fault assumptions have (necessarily [16]) different semantic properties in terms of entailments and non-entailments. This fact can be exploited to distinguish between diagnoses by posing queries to an expert. A query is a set of axioms Q which is entailed by some fault assumptions and inconsistent with some other fault assumptions. Asking a query Q corresponds to the question "Is (the conjunction of axioms in) Q an entailment of the intended ontology?". If answered positively, Q is added to the positive test cases P, and otherwise to the negative test cases N. The crucial property which makes a set of axioms Q a query is that at least one diagnosis is ruled out, regardless of whether Q is affirmed or negated. More formally:

Definition 1 (Query). *Given a set of minimal diagnoses* $\mathbf{D}$ *for an FPI* $\langle \mathcal{O}, \mathcal{B}, P, N \rangle$, *a set of axioms* Q *is a* query *(wrt.* $\mathbf{D}$*) iff at least one* $\mathcal{D}_i \in \mathbf{D}$ *is not a diagnosis for* $\langle \mathcal{O}, \mathcal{B}, P \cup \{Q\}, N \rangle$ *and at least one* $\mathcal{D}_j \in \mathbf{D}$ *is not a diagnosis for* $\langle \mathcal{O}, \mathcal{B}, P, N \cup \{Q\} \rangle$.

The expert who answers queries is modeled as a function $\mathsf{expert} : \mathbf{Q} \to \{y, n\}$ where $\mathbf{Q}$ is the query space; $\mathsf{expert}(Q) = y$ iff the answer to Q is positive, else $\mathsf{expert}(Q) = n$.

Example 3. Let the known set of diagnoses for fpi_{ex} be $\mathbf{D} = \{\mathcal{D}_1, \mathcal{D}_2, \mathcal{D}_3\}$ (see Example 2). One query wrt. $\mathbf{D}$ is, e.g., $Q_1 := \{\mathit{ActiveResearcher} \sqsubseteq \mathit{Author}\}$. Because, *(i)* adding Q_1 to P yields that the removal of $\mathcal{D}_1$ or $\mathcal{D}_2$ from $\mathcal{O}$ no longer breaks the unwanted entailment $\mathit{Employee}(\mathit{ann})$, i.e., $\mathcal{D}_1, \mathcal{D}_2$ are no longer minimal diagnoses, *(ii)* moving Q_1 to N means that $\mathcal{D}_3$ is not a minimal diagnosis anymore, as, to prevent the entailment of (the new negative test case) Q_1, at least one of ax_1, ax_2 must be deleted. Note, e.g., $Q_2 := \{\mathit{Author} \sqsubseteq \mathit{Person}\}$ is not a query since no diagnosis in $\mathbf{D}$ is invalidated upon assigning Q_2 to P, i.e., in case of a positive answer no useful information for diagnoses discrimination is gained. This is because Q_2 does not contribute to the violation of n_1 (in fact, the other "part" $\mathit{Author} \sqsubseteq \mathit{Employee}$ of ax_3 does so). □

Problem Definition. The query-based ontology fault localization problem (QFL) is to find for an FPI a series of queries to an expert, the answers of which lead to a single possible remaining fault assumption. The optimization version of the problem includes the additional goal to minimize the effort of the expert. Formally:

Problem 1 ((Optimal) QFL). **Given:** FPI $\langle \mathcal{O}, \mathcal{B}, P, N \rangle$. **Find:** (Minimal-cost) series of queries $Q_1, \dots, Q_k$ s.t. there is only one minimal diagnosis for $\langle \mathcal{O}, \mathcal{B}, P \cup P', N \cup N' \rangle$, where P' (N') is the set of all positively (negatively) answered queries, i.e., $P' := \{Q_i \mid 1 \leq i \leq k, \mathsf{expert}(Q_i) = y\}$ and $N' := \{Q_i \mid 1 \leq i \leq k, \mathsf{expert}(Q_i) = n\}$.

Note, there is no unified definition of the cost of a solution to the QFL problem. Basically, any function mapping the series $Q_1, \dots, Q_k$ to a non-negative real number is possible. We pick up on this discussion again in Sect. 3.

Example 4. Let the actual diagnosis for fpi_{ex} be $\mathcal{D}_3$, i.e., ax_3 is the (only) faulty axiom in $\mathcal{O}$ (intuition: an author is not necessarily employed, but might be, e.g., a freelancer). Then, given fpi_{ex} as an input, solutions to Problem 1, yielding the final diagnosis $\mathcal{D}_3$, are, e.g., $P' = \emptyset, N' = \{\{\exists writes.\top \sqsubseteq Employee\}\}, \{Author \sqsubseteq Employee\}$ or $P' = \{\{ActiveResearcher \sqsubseteq Author\}\}, N' = \emptyset$. Measuring the querying cost by the number of queries, the latter solution (cost: 1) is optimal, the former (cost: 2) not. □

Query-Based Fault Localization. Given an FPI as input, the ontology fault localization process basically consists of four iteratively repeated steps: First, the fault hypotheses computation yielding a sample of diagnoses; second, the determination of the best next query based on the known diagnoses; third, the information acquisition where an expert answers the suggested query; and, fourth, the integration of the gathered information, involving the extension of the FPI's test cases based on the posed query and the given answer. The reiteration of these phases is continued until a stop criterion is met, e.g., a single diagnosis remains. This remaining diagnosis then provably contains only faulty axioms [16].[4] In the following, we will call one execution of this process starting with an input FPI until a single diagnosis is isolated a *fault localization session.*

3 Discussion of Query-Based Fault Localization Approaches

In this section we analyze existing approaches regarding the assumptions they make about (the query answering behavior of) the interacting user, their properties resulting from natural design choices, as well as optimization criteria they consider.

Assumptions about Query Answering. All approaches that draw on the interactive methodology described in Sect. 2 make the assumption *during their computations and optimizations* that the expert evaluates each query as a whole. That is, they perform an assessment of the query effect or (information) gain *based on two possible outcomes* (y and n). However, in fact, since queries might contain multiple axioms, the feedback of an expert to a query might take a multitude of different shapes. Because, the expert might not view the query as an atomic question, but at the axiom level, i.e., inspecting axioms one-by-one. Clearly, to answer the query $Q = \{ax_1, \dots, ax_m\}$ positively—i.e., that the conjunction of the axioms $ax_1, \dots, ax_m$ is an entailment of the intended ontology—one needs to scrutinize and approve the entailment of all single axioms. To negate the query Q, in contrast, it suffices to detect one of the m axioms in Q which is not an entailment of the intended ontology. In this latter case, however, we might

[4] Note, the finally remaining diagnosis does not necessarily contain *all* faulty axioms in the ontology, as, e.g., some existing faults in the ontology might not yet have surfaced in terms of problems such as wrong entailments or unsatisfiable classes. However, the (faultiness of the) axioms in the final diagnosis do(es) explain *all observed problems* in the ontology.

reasonably assume the interacting expert to be able to name (at least this) one *specific* axiom $ax^* \in Q$ that is not an intended entailment. We might think of ax^* as a "witness of the falsehood of the query". This additional information—beyond the mere negative answer n indicating that some *undefined* query-axiom must not be entailed—justifies the addition of $n^* := \{ax^*\}$, instead of Q, to the negative test cases. Please note that n^* provides stronger information than Q, and thus potentially rules out more diagnoses. The reason is that each diagnosis that entails Q (i.e., is invalidated given the negative test case Q) particularly entails ax^* (i.e., is definitely invalidated given the negative test case n^*). Apart from the scenario where experts provide just a falsehood-witness in the negative case, they might give even more information. For instance, an expert could walk through the query-axioms until either a non-entailed one is found or all axioms have been verified as intended entailments. In this case, there might as well be some entailed axioms encountered before the first non-entailed one is detected. The set of these entailed axioms could then be added to the positive test cases—in addition to the negative test case n^*. Alternatively, the expert might also continue evaluating axioms after recognizing the first non-entailed axiom ax^*, in this vein providing the classification of all single query-axioms in Q.

Based on this discussion, we might—besides the *query-based* expert that answers queries as a whole, exactly as specified by the expert function defined in Sect. 2—characterize (at least) three different types of *axiom-based* experts which supply information beyond the mere n label for a query Q in the negative case:[5]

- *Minimalist:* Provides exactly one $ax^* \in Q$ which is not entailed by the intended ontology.
- *Pragmatist:* Provides the first found axiom $ax^* \in Q$ that is not entailed by the intended ontology, and additionally all axioms evaluated as entailments of the intended ontology until ax^* was found.
- *Maximalist:* Provides the classification of each axiom in Q as either an entailment or a non-entailment of the intended ontology.

Consequently: *(i)* Without knowing the answering type of the interacting expert in advance, the binary query evaluation conducted in existing works is generally only an approximation. *(ii)* Even if the expert type is known, it is an open question which form of interaction, i.e., which way of asking the expert, allows to exploit the expert knowledge most beneficially and economically. Our experimental evaluations reported in Sect. 5 shall confirm (i) and bring light to (ii).

Natural Design Choices. As explicated in Sect. 2, the principle behind queries is the comparison of entailments and non-entailments resulting from different fault assumptions (diagnoses). In existing works [26,30], this is often done by

[5] Note, a positive answer (y) implicitly provides *axiom-level* information, i.e., the positive classification of all query-axioms. Thus, the discussed experts differ only in their negation behavior.

computing common entailments for some diagnoses and verifying whether assuming correct these entailed axioms leads to an inconsistency with some other diagnosis. In the light of this strategy, it is quite natural to specify queries as *sets of* axioms. The reasons are the following:

First, it stands to reason to use and further process *all* entailments that a reasoner outputs. Second, the fewer entailments are used, the higher is the chance that these are entailed by all (known) diagnoses and hence do not constitute a query. In fact, it has been shown in [24] that such unsuccessful query verifications can account for a massive query computation time overhead. Third, allowing queries to include a larger number of axioms implies a larger query search space and thus enables to identify a better next query—where "better" applies to the case where a *query-based* expert is assumed and query selection heuristics [23] are used that aim at minimizing the *number of queries*.

Optimization Criteria. The meaning of "minimal-cost" in Problem 1 might be defined in different ways. Most existing works on query-based fault localization, e.g., [16,26,27,30], specify the cost of a solution $Q_1, \dots, Q_k$ to the QFL problem to be *the number of* queries, i.e., k. The underlying assumption in this case is that any two queries mean the same (answering) cost for an expert. Given that queries might include fewer or more axioms of lower or higher (syntactic or semantic) complexity, we argue that this cost measure might be too coarse-grained to capture the effort for an interacting expert in a realistic way. Instead, it might be more suitable to measure the costs at the axiom level.

However, there is a fundamental problem with the optimization criterion that aims at minimizing the number of query-axioms an expert needs to classify during an interactive fault localization session. Because, adopting this criterion, the evaluation and comparison of the goodness of queries while searching for the best next query trivially requires the calculation of the specific query-axioms—for a potentially large number of query candidates. However, the calculation of the specific query-axioms is generally costly in that it involves a high number of calls to expensive reasoning services. A remedy to this problem in terms of a two-staged technique which (i) can assess queries without knowing the specific axioms they contain and (ii) minimizes both the number of queries and the costs at the axiom level is suggested by [24]. However, the expert type taken as a basis for these optimizations is again the *query-based* one, and the number of axioms is only the secondary minimization criterion after the number of queries.

4 New Approach to Expert Interaction

Idea. In the light of the issues pointed out in Sect. 3 and following quite straightforward from the given argumentation, we propose a new way of expert interaction for fault localization in ontologies, namely to abandon "batch-queries" including multiple axioms and to focus on so-called *singleton queries* instead. That is, we suggest to restrict queries to only single-axiom questions. Formally:

Definition 2 (Singleton Query). *Let* $\mathbf{D}$ *be a set of diagnoses for an FPI* $\langle \mathcal{O}, \mathcal{B}, P, N \rangle$. *Then,* Q *is a singleton query (wrt.* $\mathbf{D}$*) iff* Q *is a query (wrt.* $\mathbf{D}$*) and* $|Q| = 1$.[6]

Properties. The *advantages* of singleton queries are the following:

- *Maximally-fine granularity of optimization loop:* Each atomic expert input (i.e., each classified axiom) can be directly taken into account to optimize further computations and expert interactions. Simply put, each axiom the expert is asked to classify is a function of *all* so-far classified axioms.
- *Smaller search space:* There are fewer singleton queries than there are general queries. Therefore, the worst-case search costs for singleton queries are bounded by the worst-case search costs for normal queries.
- *Realistic query assessment:* For singleton queries, the binary-outcome assessment performed by existing approaches is exact, plausible and not just an approximation of the possible real cases, independent of the expert (type). The reason is that there *are* exactly two possible outcomes, namely y (query-axiom is an intended entailment) and n (query-axiom is a non-intended entailment).
- *Direct re-use of existing works:* Concepts (e.g., heuristics) and techniques (e.g., search algorithms) devised for queries can be immediately re-used for singleton queries, because each singleton query *is* a (specific) query.
- *Unequivocal optimization criterion:* Minimization of the number of queries and minimization of the number of query-axioms coincide for singleton queries. This unifies the two competing and arguable views on the query optimization problem.
- *More informative feedback per axiom (assuming query-based expert):* For both singleton and normal queries, the positive assessment of the query implies that all axioms in it are intended entailments. That is, the information acquired per axiom is equal. In case the query is negated, however, singleton queries generally provide more information per axiom. Because, for a normal query a negative answer corresponds to the information that *one of a set of* axioms is not true, whereas we learn from a negated singleton query that *one particular* axiom must not be entailed.
- *Same fault localization efficiency for all expert types:* Singleton queries, by their nature, admit only one style of answering—the answer is positive iff the single comprised axiom must be entailed by the intended ontology, and negative iff it must not be entailed. Thus, all discussed expert types coincide for singleton queries. As an implication of this, it is neither required to ascertain the expert type a priori nor to adapt algorithms to different experts, which makes the query optimization process simpler and the outcome equally suitable for all (discussed) types of users.

On the downside, the smaller search space—apart from the advantage it brings regarding the worst-case query search complexity—can be seen as a *disadvantage*

[6] To stress the difference between singleton queries (Definition 2) and queries in terms of Definition 1, we will henceforth often refer to the latter as *normal queries*.

as well. The reason is that soundness of the query search is more difficult to obtain, i.e., more considerations and computations than for normal queries are required to ensure that the search outcome is indeed a *singleton* query (cf. the discussion on "Natural Design Choices" in Sect. 3). To tackle this, one could try to generate normal queries and post process them by means of query-size minimization techniques similar to those used by existing works [17,30]. The problem is, however, that these techniques do not guarantee the reduction to a single axiom.

Thus, beside all the advantages of singleton queries, an algorithmic and computational challenge towards their efficient generation and optimization remains to be solved.

Computation and Optimization. Despite this open issue regarding general singleton queries, we were able to develop a polynomial time and space algorithm for singleton queries of the form $\{ax\}$ where $ax \in \mathcal{O}$.[7] This algorithm gets an FPI $\langle \mathcal{O}, \mathcal{B}, P, N \rangle$, a set of known minimal diagnoses $\mathbf{D}$ as well as a query selection heuristic h (among those discussed in [23]) as an input, and outputs the globally optimal singleton query of the above-mentioned form. At this, "globally optimal" means optimal in terms of h among *all* queries in the query space. The basis for our algorithm is provided by the theory and strategies for normal queries elaborated in [17], which we extended and adapted accordingly to obtain a method for singleton queries. The full description of the new algorithm is beyond the scope of this work and can be found in [19]. Here, we rather focus on understanding the added value of singleton queries and their comparison with normal queries.

5 Evaluation

Goal. The aim of the following experiments is the analysis of normal queries under different answering conditions (expert types discussed in Sect. 3) and the comparison between normal queries and the proposed singleton queries. Focus of the investigations is the *required effort for the expert* for fault localization and the *query computation time*.

Dataset, Experiment Settings and Measurements. The dataset of faulty (inconsistent and/or incoherent) real-world ontologies used in our experiments is given in Table 1. We used each of these ontologies $\mathcal{O}$ to specify an FPI as $fpi := \langle \mathcal{O}, \emptyset, \emptyset, \emptyset \rangle$, i.e., the background knowledge $\mathcal{B}$ as well as the positive (P) and negative (N) test cases were initially empty. Table 1 also gives an idea of the *diagnostic structure* of the considered FPIs, in terms of the size and logical expressivity[8] of the ontology, as well as the number and minimal/maximal size of

[7] Such (singleton) queries consisting of only axioms explicitly included in the ontology are called *explicit* (singleton) *queries* [17].

[8] The logical expressivity refers to the power of the logical language used in the ontology in terms of how much can be expressed using this language. In general, the higher the expressivity, the higher the cost of reasoning (and thus the cost of computing queries) with the respective logic tends to be. See [1] for more details on the logical expressivity of ontologies.

all minimal diagnoses for the initial problem. As query selection heuristics (h) we used the measures discussed in [26,30]. These are ENT (maximize information gain per query), SPL (maximize worst-case diagnoses elimination rate per query) and RIO (optimize balance between ENT and SPL per query).

For each FPI and each heuristic h we ran 20 fault localization sessions, each time using a different randomly specified actual diagnosis $\mathcal{D}^*$ to be located. To automatically answer queries throughout a session in a way the predefined diagnosis $\mathcal{D}^*$ is finally located, we implemented a function based on $\mathcal{D}^*$ which simulates the interacting user. Specifically, the *query-based* expert was simulated by always outputting an answer to a query Q that does not effectuate the invalidation of $\mathcal{D}^*$; the *axiom-based* experts (*minimalist*, *pragmatist*, *maximalist*) were simulated in a way that, if they classify an axiom ax at all (cf. Sect. 3), then as an entailment if $ax \notin \mathcal{D}^*$ and as a non-entailment else. The size of the diagnoses sample generated before each query computation was set to $|\mathbf{D}| = 10$. Since two of the used heuristics (ENT, RIO) depend on diagnosis probabilities, we sampled and assigned uniform random probabilities to diagnoses for each FPI. For query generation throughout the fault localization sessions, we used the algorithms described in [17] (for normal queries) and [19] (for singleton queries). Note, all (normal and singleton) queries Q computed in our experiments were restricted to include axioms that occur in the ontology $\mathcal{O}$, i.e., $Q \subseteq \mathcal{O}$ for all queries Q.[9]

For each performed fault localization session we measured the number of answered queries ($\#Q$) as well as the number of classified query-axioms ($\#Ax$) required until the predefined $\mathcal{D}^*$ was found with certainty (i.e., until all other diagnoses were ruled out through the answered queries), and the average computation time to find the best next query (*time per Q*).

Table 1. Dataset of faulty ontologies used in the experiments, sorted by the ontology size $|\mathcal{O}|$.

ontology $\mathcal{O}$	$\lvert\mathcal{O}\rvert$	expressivity [1)]	#D/min/max [2)]
Koala (K) [3)]	42	$\mathcal{ALCON}^{(D)}$	10/1/3
University (U) [4)]	50	$\mathcal{SOIN}^{(D)}$	90/3/4
MiniTambis (M) [4)]	173	$\mathcal{ALCN}$	48/3/3
Transportation (T) [4)]	1300	$\mathcal{ALCH}^{(D)}$	1782/6/9
Economy (E) [4)]	1781	$\mathcal{ALCH}^{(D)}$	864/4/8
DBpedia (D) [5)]	7228	$\mathcal{ALCHF}^{(D)}$	7/1/1

Key:
1): Description Logic expressivity [1].
2): #D, min, max denote the number, the minimal as well as the maximal size of minimal diagnoses for the input FPI.
3): Faulty ontology included in the Protégé Project.
4): Sufficiently complex FPIs (#D $\geq$ 40) used in [30].
5): Faulty version of the DB-Pedia ontology, downloaded from https://bit.ly/2RUVbMj.

Experiment Results. First, we observe that, for normal queries, the answering style has a significant impact on the expert's effort, both when using #Ax and #Q as a cost metric. In fact, any axiom-based strategy (pragmatist, maximalist or minimalist) is better than a query-based one (bars in Fig. 1), with savings

[9] This is owed to the fact that the efficient generation of optimal singleton queries including "implicit" axioms, i.e., where $Q \not\subseteq \mathcal{O}$ holds, is still an open research topic (cf. Sect. 4).

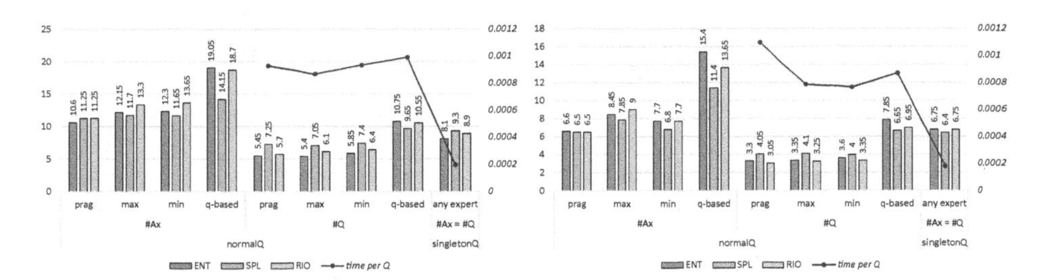

Fig. 1. Overview of observations for ontology E (left) and M (right): The bars show *#Ax* and *#Q* for heuristics ENT (blue), SPL (gray) and RIO (yellow) and for expert types *min*imalist, *prag*matist, *max*imalist, and *q*uery-*based* expert (cf. Sect. 3), for normal queries (*normalQ*) and singleton queries (*singletonQ*). The red line reports *time per Q* (in sec). All plotted values are averages over all 20 fault localization sessions. Bars refer to left y-axis, red line to right y-axis. (Color figure online)

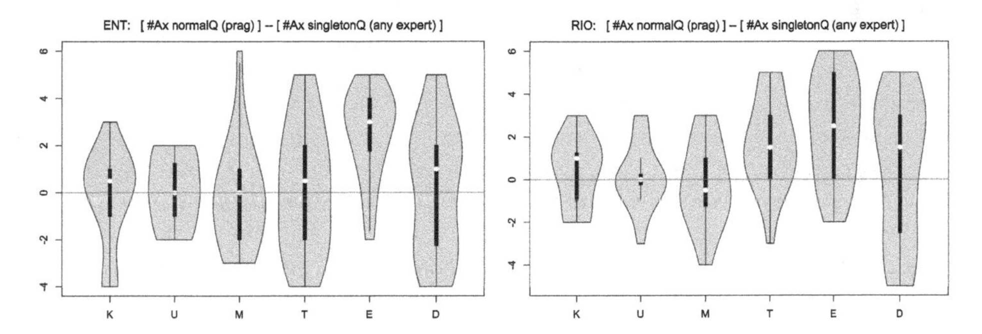

Fig. 2. Comparison between normal and singleton queries for ENT (left) and RIO (right) heuristics: The violin plots show the difference in query answering effort (*#Ax*) between using the best answering strategy (*prag*matist) for normal queries (*normalQ*) and using singleton queries (*singletonQ*), for all ontologies (x-axis) given in Table 1. Each violin plot summarizes the differences per session over all 20 fault localization sessions. White dots in plots indicate the median; if above/below zero (red line), singleton/normal queries are better in the majority of the sessions. (Color figure online)

of up to 57 % wrt. #Ax and up to 58 % wrt. #Q (cf. ENT, pragmatist vs. query-based, M ontology, in Fig. 1). The reason for this is that an axiom-based approach involves strictly more informative answers than a query-based one (cf. Sect. 3).

Second, also among the axiom-based expert types, there are notable cost differences (wrt. #Ax). As it turns out, the pragmatist approach is clearly the best choice to answer normal queries for *all* investigated ontologies.[10] Also, when measuring the cost by #Q (as existing works do), the pragmatist tends to be

[10] Note, the presented figures do not expose all results. However, the observations were greatly consistent over all studied ontologies. See the extended version [19] of this paper for all plots.

the most reasonable type, albeit the differences are just marginal in this case. So far, we conclude that normal queries, for best efficiency and regardless of the adopted query selection heuristic, should *not* be answered simply by y or n, but the interacting expert should evaluate the individual query-axioms, pursuing the pragmatist method (cf. Sect. 3). Note, it is surprising that *one* (axiom-based) answering strategy *always* prevails, as normal queries are optimized based on the assumption of the (fairly different) query-based user.

Third, when comparing singleton with normal queries (answered by the pragmatist strategy), the costs wrt. #Ax are often pretty similar on average (see, e.g., M ontology in Fig. 1), even though with a notable tendency towards a superiority of singleton queries. E.g., for the E ontology and ENT heuristic, we measure an average effort overhead of more than 30 % when relying on normal queries as opposed to singleton ones (Fig. 1). Figure 2 gives a clearer picture of this comparison. E.g., it reveals that, for all ontologies, singleton queries were at least as good as normal ones in the majority of sessions when using ENT as a heuristic. For the RIO heuristic, the results are similar, and in three cases (ontologies K, T, D) even more in favor of singleton queries than for ENT. Over all ontologies and heuristics, singleton queries even led to less expert interactions in more than 66% of the sessions. However, there are scenarios where normal queries outperform singletons on average as well, as evidenced by the RIO and M ontology combination. Moreover, in most scenarios the proportion of sessions where normal queries mean fewer expert consultations (area of violin plots below the red line) is significant. Thus, normal queries *are* a reasonable way of expert interaction, but can match up to singleton queries only if the pragmatist answering behavior is given.

Regarding the computation time per query, we clearly recognize (red lines in Fig. 1) that (optimal) singleton queries are significantly faster determined than (optimal) normal queries. The savings *always* amounted to between 80% and 90%.

6 Research Limitations and Future Work

First, the evaluations in this work are based on simulations of fault localization sessions and objective measures such as computation times or the number of required queries. Although this objective assessment shows a higher average efficiency of the new approach as compared to existing ones, it is important to validate the subjective usefulness of the suggested querying technique, for instance in terms of a user study. This is part of our future work. However, it nevertheless stands to reason that users familiar with normal queries would likewise accept and adopt singleton queries, just because singleton queries represent a particularly *simple subclass of normal queries*.

A second limitation is the restriction to explicit queries—those that are constituted by axioms from the ontology at hand—in our empirical analyses. The reason we did so is because we currently only have an algorithm for the computation and optimization of explicit singleton queries, by drawing on and extending the theory elaborated in [17]. The finding of an *efficient* algorithm that soundly generates implicit singleton queries, in contrast, is an open issue and on our future work agenda. That said, as soon as we have developed an adequate algorithm, we plan to do similar evaluations as done in this work for singleton and normal queries without the restriction to explicit queries.

As a third limitation, it should be noted that the analyzed expert types, as discussed in Sect. 3, provide by no means a complete characterization of all possible cases that could arise. While the discussion in this work bases on the assumption that an expert will provide for each query at the minimum as much information as is necessary to classify the entire query as a positive or negative test case (cf. the expert function in Sect. 2), there are (at least) two further query answering scenarios that are worthwhile considering. First, there is the case where the expert classifies a proper subset (or even none) of the axioms of a normal query positively while not labeling any axiom negatively, e.g., due to laziness or lack of knowledge. Second, there is the case where an expert might misclassify axioms when answering queries. Such "oracle errors" were observed quite commonly in the studies conducted in [21]. Investigating these two scenarios for normal and singleton queries as well as the conception of strategies how to handle these cases is another research avenue we will prospectively pursue.

7 Conclusions

We critically discuss design choices, made assumptions and used optimization criteria of state-of-the-art query-based ontology fault localization approaches. Based on the revealed issues, we propose a new way of asking questions to an expert. Theoretical and empirical analyses using real-world problems demonstrate significant advantages of the novel querying method. Among other things, we learn that the suggested method—as opposed to existing approaches—(1) is simpler, (2) enables exact query optimizations instead of only approximate ones, (3) implies a more than 80 % reduction of the expert's waiting time for the next question, (4) enforces more informative expert inputs, (5) leads to the least fault localization effort for the expert in more than 66% of the cases, and (6) guarantees the same efficiency regardless of the expert's (answering) behavior. Notably, our method is basically applicable to any monotonic knowledge representation language [16], as well as to other model-based diagnosis applications [22].

Acknowledgments. This work was in part supported by the Carinthian Science Fund (KWF), contract KWF-3520/26767/38701. Moreover, we thank Wolfgang Schmid for his technical support during the implementation of our experiments.

References

1. Baader, F., Calvanese, D., McGuinness, D., Nardi, D., Patel-Schneider, P. (eds.): The Description Logic Handbook, 1st edn. Cambridge University Press, Cambridge (2003)
2. Beck, K.: Test-Driven Development: By Example. Addison-Wesley Professional, Boston (2003)
3. Ceusters, W., Smith, B., Goldberg, L.: A terminological and ontological analysis of the NCI thesaurus. Methods Inf. Med. **44**(4), 498 (2005)
4. Felfernig, A., Friedrich, G., Jannach, D., Stumptner, M.: Consistency-based diagnosis of configuration knowledge bases. Artif. Intell. **152**(2), 213–234 (2004)
5. Golbeck, J., Fragoso, G., Hartel, F., Hendler, J., Oberthaler, J., Parsia, B.: The national cancer institute's thesaurus and ontology. JWS **1**(1) (2003). http://dx.doi.org/10.2139/ssrn.3199007
6. Grau, B.C., Horrocks, I., Motik, B., Parsia, B., Patel-Schneider, P., Sattler, U.: OWL 2: the next step for OWL. JWS **6**(4), 309–322 (2008)
7. Horridge, M., Bail, S., Parsia, B., Sattler, U.: The cognitive complexity of OWL justifications. In: Aroyo, L., et al. (eds.) ISWC 2011. LNCS, vol. 7031, pp. 241–256. Springer, Heidelberg (2011). https://doi.org/10.1007/978-3-642-25073-6_16
8. Jannach, D., Schmitz, T., Shchekotykhin, K.: Parallel model-based diagnosis on multi-core computers. JAIR **55**, 835–887 (2016)
9. Kalyanpur, A.: Debugging and repair of OWL ontologies. Ph.D. thesis, University of Maryland (2006)
10. de Kleer, J., Williams, B.C.: Diagnosing multiple faults. Artif. Intell. **32**(1), 97–130 (1987)
11. Meilicke, C.: Alignment incoherence in ontology matching. Ph.D. thesis, University of Mannheim (2011)
12. Nikitina, N., Rudolph, S., Glimm, B.: Interactive ontology revision. JWS **12**(0), 118–130 (2012)
13. Qi, G., Hunter, A.: Measuring incoherence in description logic-based ontologies. In: Aberer, K., et al. (eds.) ASWC/ISWC -2007. LNCS, vol. 4825, pp. 381–394. Springer, Heidelberg (2007). https://doi.org/10.1007/978-3-540-76298-0_28
14. Rector, A.L., Brandt, S., Schneider, T.: Getting the foot out of the pelvis: modeling problems affecting use of SNOMED CT hierarchies in practical applications. JAMIA **18**(4), 432–440 (2011)
15. Reiter, R.: A theory of diagnosis from first principles. Artif. Intell. **32**(1), 57–95 (1987)
16. Rodler, P.: Interactive Debugging of Knowledge Bases. Ph.D. thesis, University of Klagenfurt (2015)
17. Rodler, P.: Towards better response times and higher-quality queries in interactive KB debugging. Technical report, University of Klagenfurt (2016). http://arxiv.org/abs/1609.02584v2
18. Rodler, P.: On active learning strategies for sequential diagnosis. In: DX, pp. 264–283 (2018)
19. Rodler, P., Eichholzer, M.: A new expert questioning approach to more efficient fault localization in ontologies. Technical report, University of Klagenfurt (2019). http://arxiv.org/abs/1904.00317
20. Rodler, P., Herold, M.: StaticHS: a variant of Reiter's hitting set tree for efficient sequential diagnosis. In: SoCS, pp. 72–80 (2018)

21. Rodler, P., Jannach, D., Schekotihin, K., Fleiss, P.: Are query-based ontology debuggers really helping knowledge engineers? Knowledge-Based Systems (2019). https://doi.org/10.1016/j.knosys.2019.05.006
22. Rodler, P., Schekotihin, K.: Reducing model-based diagnosis to knowledge base debugging. In: DX, pp. 284–296 (2018)
23. Rodler, P., Schmid, W.: On the impact and proper use of heuristics in test-driven ontology debugging. In: Benzmüller, C., Ricca, F., Parent, X., Roman, D. (eds.) RuleML+RR 2018. LNCS, vol. 11092, pp. 164–184. Springer, Cham (2018). https://doi.org/10.1007/978-3-319-99906-7_11
24. Rodler, P., Schmid, W., Schekotihin, K.: A generally applicable, highly scalable measurement computation and optimization approach to sequential model-based diagnosis. CoRR abs/1711.05508 http://arxiv.org/abs/1711.05508 (2017)
25. Rodler, P., Schmid, W., Schekotihin, K.: Inexpensive cost-optimized measurement proposal for sequential model-based diagnosis. In: DX, pp. 200–218 (2018)
26. Rodler, P., Shchekotykhin, K., Fleiss, P., Friedrich, G.: RIO: minimizing user interaction in ontology debugging. In: Faber, W., Lembo, D. (eds.) RR 2013. LNCS, vol. 7994, pp. 153–167. Springer, Heidelberg (2013). https://doi.org/10.1007/978-3-642-39666-3_12
27. Schekotihin, K., Rodler, P., Schmid, W.: OntoDebug: interactive ontology debugging plug-in for Protégé. In: Ferrarotti, F., Woltran, S. (eds.) FoIKS 2018. LNCS, vol. 10833, pp. 340–359. Springer, Cham (2018). https://doi.org/10.1007/978-3-319-90050-6_19
28. Schekotihin, K., Rodler, P., Schmid, W., Horridge, M., Tudorache, T.: A Protégé plug-in for test-driven ontology development. In: ICBO (2018)
29. Schulz, S., Schober, D., Tudose, I., Stenzhorn, H.: The pitfalls of thesaurus ontologization-the case of the NCI thesaurus. In: AMIA Annual Symposium (2010)
30. Shchekotykhin, K., Friedrich, G., Fleiss, P., Rodler, P.: Interactive ontology debugging: two query strategies for efficient fault localization. JWS **12–13**, 88–103 (2012)
31. Shchekotykhin, K., Jannach, D., Schmitz, T.: Mergexplain: fast computation of multiple conflicts for diagnosis. In: IJCAI, pp. 3221–3228 (2015)
32. Shchekotykhin, K.M., Friedrich, G., Rodler, P., Fleiss, P.: Sequential diagnosis of high cardinality faults in knowledge-bases by direct diagnosis generation. In: ECAI (2014)
33. Siddiqi, S.A., Huang, J.: Sequential diagnosis by abstraction. JAIR **41**, 329–365 (2011)
34. Smith, B., Ashburner, M., Rosse, C., et al.: The OBO foundry: coordinated evolution of ontologies to support biomedical data integration. Nature Biotechnol. **25**(11), 1251 (2007)
35. Tudorache, T., Noy, N.F., Tu, S., Musen, M.A.: Supporting collaborative ontology development in Protégé. In: Sheth, A., et al. (eds.) ISWC 2008. LNCS, vol. 5318, pp. 17–32. Springer, Heidelberg (2008). https://doi.org/10.1007/978-3-540-88564-1_2

Using Description Logic and Abox Abduction to Capture Medical Diagnosis

Mariam Obeid[1], Zeinab Obeid[2], Asma Moubaiddin[3], and Nadim Obeid[4(✉)]

[1] Royal Surrey County Hospital, Guildford, England
mariam.obeid@nhs.net
[2] School of Medicine, The University of Jordan, Amman, Jordan
zyn0144986@ju.edu.jo
[3] School of Foreign Languages, The University of Jordan, Amman, Jordan
a.mobaiddin@ju.edu.jo
[4] King Abdullah II School for Information Technology, The University of Jordan, Amman, Jordan
obein@ju.edu.jo

Abstract. Medical diagnosis can be defined as the detection of a disease by examining a patient's signs, symptoms and history. Diagnostic reasoning can be viewed as a process of testing hypotheses guided by symptoms and signs. Solutions to diagnostic problems can be found by generating a limited number of hypotheses early in the diagnostic process and using them to guide subsequent collection of data. Each hypothesis, if correct, can be used to pre-dict what additional findings must be present, and the diagnostic process would then be a guided search for these findings. The process depends on the medical knowledge available. Description Logic-based ontologies provide class definitions (i.e., the necessary and sufficient conditions for defining class membership). In medicine, these definitions correspond to diagnostic criteria, i.e., the particular form of diseases should be associated with the relevant disease categories. In this paper, we model medical diagnosis as an (iterative) abductive reasoning process using ALC. ALC is employed to take advantage of its inference services. However, the inference capabilities provided by DL are not sufficient for diagnosis purposes. The contributions of the paper include: (1) arguing for the need for a disease-symptoms ontology, (2) proposing an ontological representation which, beside facilitating abductive reasoning, takes into account the diagnostic criteria such that specific patient conditions can be classified under a specific disease, and (3) employing Abox abduction to capture the process of medical diagnosis (the process of generating and testing hypotheses) on this proposed representation.

Keywords: Medical diagnosis · Description logic · Ontology · Abductive reasoning

1 Introduction

Medical diagnosis can be defined as the detection of a disease by examining a patient's signs, symptoms and history. Symptoms are the most directly observable characteristics of a disease and the very basis of clinical disease classification. Symptoms and

F. Wotawa et al. (Eds.): IEA/AIE 2019, LNAI 11606, pp. 376–388, 2019.
https://doi.org/10.1007/978-3-030-22999-3_33

signs represent the high-level manifestations of a disease that are actually noticed by patients and physicians. Eventually, it is due to certain symptoms that a person will seek professional help. However, the process is challenging because (1) the vast amount and intricacy of medical data hinders the process of a thorough study of the content of the data, (2) symptoms and signs vary widely. Most diseases have their own specific signs and symptoms while many symptoms such as fever, fatigue, and muscle aches are common to a number of diseases. It was emphasized in [6] that diagnostic classification (i.e. classifying a patient as having a particular disease) is an important issue that has to be addressed in medicine.

Ontologies define classes of entities and their interrelations. They are used to organize data according to a theory of the domain. The shared understanding results from the fact that all the agents interpret the concepts with regard to the same ontology. There are many existing biomedical vocabularies and ontologies such as SNOMED-CT which holds a very large number of relationships between medical entities such as diseases, body locations and clinical findings [7, 8]. However, they contain a very small number of symptom-disease relationships needed for clinical diagnoses. Furthermore, ontologies provide class definitions (i.e., the necessary and sufficient conditions for defining class membership). In medicine, these definitions correspond to diagnostic criteria, i.e., the particular form of diseases as described in the records of patients suffering from these disease should be associated with the relevant disease categories. It was shown in [6] that eligibility criteria are often more useful than the Aristotelian definitions traditionally used in ontology for diagnostic classification. They argue that the classificatory principles and properties represented in ontologies may not be suf ficient to classify instances of diseases. They suggest that operational definitions of diseases are more useful. These definitions are mostly based on the association of signs and symptoms. However, as mentioned above, this is difficult to achieve as the symptoms and signs of a disease varies widely and many diseases share a lot of common symptoms and signs.

Description Logic (DL) systems [4, 5] provide some inference capabilities. DL-based ontologies are appropriate for modeling, and reasoning about, knowledge. Some important features of DL systems are that: (1) the core reasoning problems are (usually) decidable and (2) efficient decision procedures have been designed and implemented for these problems. This explains why most ontologies are represented using DL.

Furthermore, the inference capabilities provided by DL are not sufficient for diagnosis purposes. Abduction is considered as the inference process that goes from observations to explanations within a (classical) logical theory (e.g., ontology). Abductive reasoning is a backward chaining inference which involves generating hypotheses and finding the best explanation for some given observations. This may involve the assumption of new knowledge in order to constructively infer the observation. We use both the observation and the available medical knowledge to expressed in a suitable DL (ALC) system using backward chaining to generate the hypotheses. Forward chaining is then used to determine satisfiable and preferred explanation(s).

The objective of this study is to show that formal ontology can adequately represent diagnostic criteria which together with the inference services provided by formal logic (e.g. ALC) and the appropriate employment of abductive reasoning (via Abox abduction) allows the system to propose plausible explanations/diagnoses to observations/symptoms.

The system presented in this paper is not intended to replace the clinician during diagnosis, but to determine the extent to which formal ontology and abductive reasoning can play a supporting role in disease diagnosis.

In this paper, we model medical diagnosis as an (iterative) abductive reasoning process using ALC. ALC is employed to take advantage of its inference services. However, the inference capabilities provided by DL are not sufficient for diagnosis purposes. The contributions of the paper include: (1) making a case for the need for a disease-symptoms ontology, (2) proposing an ontological representation which takes into account the diagnostic criteria such that specific patient conditions can be classified under a specific disease and (3) employing Abox abduction to capture the process of medical diagnosis on this proposed representation.

2 ALC and Tableau Reasoning

ALC has three basic types of entities: individuals/objects represented as constants, concepts which are represented as unary predicates and roles which are represented as binary relations in FOPC [4, 5]. We use P, P_1, … Cp_1, Cp_2, … to denote concept names, R_0, R_1, … to denote role names, a, b, c … for object names, and A, B, C, … for propositional variables. ALC makes available to us the following constructors: negation ($\neg$), conjunction ($\prod$), existential ($\exists$) and universal ($\forall$) restriction.

We shall use T (resp. $\perp$) to denote the universal (resp. bottom) concept. Atomic concepts are validly concepts. The formation of Complex concepts is similar to the FItis a concept. If P is an atomic concept then $\neg$P is a concept. If R is a role, Cp_1 and Cp_2 are concepts then $Cp_1 \sqcap Cp_2$, $Cp_1 \sqcup Cp_2$, $\forall$R.Cp and $\exists$R.Cp are concepts.

ALC has a model theoretic semantics. The semantics are given by an interpretation $\mathbf{I} = (\Delta^{\mathbf{I}}, .^{\mathbf{I}})$ where $\Delta^{\mathbf{I}}$ is a non-empty domain and $.^{\mathbf{I}}$ is a function that maps every individual name to an individual from $\Delta^{\mathbf{I}}$, every concept name to a subset of individual and every role name to a set of pairs of individual from $\Delta^{\mathbf{I}}$. In particular, $\mathbf{T}^{\mathbf{I}} = \Delta^{\mathbf{I}}$ **and** $\perp^{I} = \phi$ (the empty set). The semantic of complex concept constructors is defined inductively as in Table 1 below.

Table 1. The Syntax and semantics of complex concept constructors.

Constructor	Semantics	Syntax
Concept negation	$\Delta^{\mathbf{I}} \backslash Cp^{\mathbf{I}}$	$\neg$Cp
Concept intersection	$Cp1^{\mathbf{I}} \cap Cp2^{\mathbf{I}}$	Cp1 $\sqcap$ Cp2
Concept union	$Cp1^{\mathbf{I}} \cup Cp2^{\mathbf{I}}$	Cp1 $\sqcup$ Cp2
Existential restriction	$\{x: \exists y((x, y) \in R^{I} \wedge y \in Cp^{I}\}$	$\exists$R.Cp
4th-level heading	$\{x: \forall y((x, y) \in R^{I} \rightarrow y \in Cp^{I}\}$	$\forall$R.Cp

An ALC Knowledge Base (KB), KB = (***A, T***) where ***T*** is a TBox and ***A*** is an Abox. ***T*** is the terminological part of KB where relationships between concepts and roles are established. We employ axioms of the form Cp1 $\sqsubseteq$ Cp2 to means that Cp1 is a sub-concept of Cp2, and Cp1$\equiv$ Cp2 to mean that Cp1 and Cp2 are equivalent, i.e., Cp1 $\sqsubseteq$ Cp2 and Cp2 $\sqsubseteq$ Cp1. ***A*** consists of a set of assertions of the form Cp(a) or R(a, b). Cp (a) states that a is an instance of Cp and R(a, b) states that individual a is related to b via role R.

The semantics of ***T*** and ***A*** are defined as shown in Table 2. An interpretation $.^{I}$ satisfies an axiom if, and only if, the semantics of the axiom are respected under $.^{I}$. a model of a KB is an interpretation that satisfies all of its axioms. KB is satisfiable if, and only if, it has, at least, one model.

Table 2. Semantics of DL axioms.

Axiom	Semantics
Cp1 $\sqsubseteq$ Cp2	$Cp1^{I} \subseteq Cp2^{I}$
Cp1 $\equiv$ Cp2	$Cp1^{I} = Cp2^{I}$
C(a)	$a^{I} \in C^{I}$
R(a, b)	$(a^{I}, b^{I}) \in R^{I}$

ALC Inference Services include satisfiability checking, subsumption checking and concepts classification, Instance Checking and KB Consistency checking.

2.1 Tableau Reasoning

Checking satisfiability of concepts in description logics can be performed using a tableau-based algorithm [3]. For instance, to test whether a concept/assertion *Cp* is satisfiable, the algorithm starts with an ABox containing the assertion *Cp*(*x*) where x is a new individual. To test the satisfiability of a subsumption (i.e., Cp1 $\sqsubseteq$ Cp2), the algorithm starts with {(Cp1 $\sqcap$ ¬Cp2)(a)}.

The algorithm requires the assertions to be normalized to Negation Normal Form (NNF), i.e. negations can only appear in front of atomic concepts. This is performed by applying De Morgan's laws and rules for quantifiers. An assertion Cp can be transformed into NNF(Cp) by pushing negation inwards, using the following formulae:

$$\neg(Cp1 \sqcap Cp2) \equiv \neg Cp1 \sqcup \neg Cp2$$
$$\neg(Cp1 \sqcup Cp2) \equiv \neg Cp1 \sqcap \neg Cp2,$$

$$\neg(\exists R.Cp) \equiv (\forall R.\neg Cp)$$
$$\neg(\forall R.Cp) \equiv (\exists R.\neg Cp).$$

One possible way of implementing this method is by employing completion graphs. These are directed graphs in which every node represents a set of assertions. The ABox for a node contains all the assertions of the node, together with the assertions of the nodes on the path to the root.

The notion **completion** is defined as the process where the algorithm continues applying the consistency-preserving transformation rules to the ABox until no more rules productively apply. A rule is productively applicable if its application brings some modification to the Abox ***A***. The Application of the ⊔-rule produces new nodes where each node contains one assertion. The other rules (i.e., ⊓ -rule, ∀-rule and ∃-rule) add assertions to the node where the rule is applied. The ABox for a node contains all the assertions of the node, together with the assertions of the nodes on the path to the root.

Tableau Consistency-Preserving Transformation Rules for ALC

⊓-rule: **if *A contains*** (Cp1 ⊓ Cp2((a) and ***A*** does not contain both Cp1(a) and Cp2 (a) and a is not blocked **then** replace A with A′ = A ∪ {Cp1(a), Cp2(a)}

⊔-rule: **if *A contains*** (Cp1 ⊔ Cp2)(a), and ***A neither contain*** Cp1(a) nor Cp2(a) and a is not blocked, **then** replace ***A*** with ***A′*** = ***A*** ∪ {Cp1(a)} and replace ***A*** with ***A″*** = ***A*** ∪ {Cp2(a)}

∃-rule: **if *A contains*** (∃ r.Cp)(a), a is not blocked, and there is no individual b such that Cp(b) and r(a, b) are in ***A*** **then** create a new individual c which does not occur in ***A*** and replace ***A*** with ***A′*** = ***A*** ∪ {r(a, c), Cp(c)}

∀-rule: if ***A contains*** (∀ r.Cp)(a) and r(a, b), a is not blocked, but ***A*** does not contain b then replace ***A*** with ***A′***= ***A*** ∪ {CP(b)}

Blocking can be used to guarantee terminating proofs even in the presence of concept inclusions. Blocking is used to prevent the application of the same rule again and again i.e., when it is clear that the subtree rooted in some node x is similar to the subtree rooted in some predecessor node y of x. The tableau expansion rules given previously can then be modified so that they apply only to individuals such as a if they are not blocked. In this way, the tableau techniques can be seen to be sound and complete decision procedures for ALC. For more details on blocking cf. [14].

Definition 2.1.

1. An ABox ***A*** is called complete iff none of the transformation rules presented above productively applies to it.
2. ***A*** contains a clash iff there is a concept name Cp and an individual name x such that ***A*** contains both Cp(x) and ¬Cp(x).
3. ***A*** is called closed if it contains a clash, and open otherwise.

Satisfiability is proven if, at least, one of the ABoxes connected to a leaf node does not contain a contradiction. Otherwise unsatisfiability is proven.

To test the satisfiability of a subsumption (i.e., Cp1 ⊑ Cp2), the algorithm starts with {(Cp1 ⊓ ¬Cp2)(a)}. In order to take subsumption axioms and concept definitions in the TBox into account, ABoxes have to be expanded with statements of the form (¬Cp1 ⊔ Cp2)(a) for every individual a in the ABox, for each axiom Cp1 ⊑ Cp2 in the TBox. This is often a costly task, and different methods are used to minimize the need for such expansions.

3 Medical Diagnosis and ABox Abduction

Abduction is basic for medical diagnosis. It is simply the generation of a hypothesis that explains one or more observations (signs). When explaining a given set of observations (signs), the clinician has often to choose among different explanations to choose what is deemed to be the best according to the observations [17]. As not all signs are usually available at the beginning of a diagnostic process, abduction is an iterative process, because new detected signs will be interpreted and integrated to generate one single present explanation for all captured signs so far. The present explanation is not definitive or absolute as it may change in the light of new observations. This process ends when the flow of new signs stops and/or a plausible/ conclusive explanation is reached.

Reasoning involved in abduction goes beyond the information included in the premises. Abduction can generate "plausible" hypotheses or it can be considered as inference "to the best explanation" [17, 18].

Definition 3.1 (ABox abduction problem)**.** Let ***KB*** = (***A, T***) an *ALC* KB and let ***B*** a set of ABox assertions denoted as the abductive query.

We say that (***KB***, ***B***) is an ABox abduction problem iff *K* $\not\models$ ***B*** and ***KB*** $\cup$ ***B*** $\not\models \bot$.

The ABox abduction problem, in DL, is the problem of finding a set of assertions ***H*** that, when added to the knowledge base ***KB*** causes the entailment of a desired set of ABox axioms ***B***. The notion of entailment should be taken as the classical consequence relation, i.e., S entails S' (S |= S') if every model of S is a model of S'.

Definition 3.2 (ABox abduction solution). Let ***H*** a set of ABox assertions. ***H*** is a solution to abductive problem (***KB, B***) iff ***KB*** $\cup$ ***H*** |= ***B***.

Additionally, we say that ***H*** is:

1. consistent iff ***KB*** $\cup$ ***H*** $\not\models \bot$.
2. relevant iff ***H*** $\not\models$ ***B***.
3. minimal iff there is no solution ***H'*** to (***KB***, ***B***) such that ***H'*** |= ***H***.

ABox abduction problem is the problem of finding a set of assertions ***H*** that, when added to the knowledge base ***KB***, causes the entailment of a desired set of ABox axioms ***B***. The notion of entailment should be taken as the classical consequence relation where S entails S' (S |= S') if every model of S is a model of S'.

It has been shown in [10] that computing all abductive solutions, even in the case of propositional logic, is not always practical. Therefore, constraints on the solutions can considerably reduce the search space and allow reasonable benefits of logical-based abduction. Minimality ensures that accepted solutions do not contain unnecessary information.

The consistency requirement discards solutions inconsistent with ***KB***. *In other words, if* ***K*** $\cup$ ***H*** |= $\bot$, then ***H*** is not considered a solution. It is possible to argue that inconsistent solutions could be valuable in a defeasible reasoning setting.

The relevance condition filters out those solutions that entail the query without any contribution to the background knowledge

4 Representational Issues and Examples

In the medical domain, effective Knowledge Representation (KR) requires the use of homogeneous vocabularies to ensure both shared understanding among medical workers and interoperability between medical information systems.

Onology allows us to formally represent and share knowledge. It employs three different types of elements: individuals/constants, concepts/unary relations and roles/binary relations to represent a domain. For instance, a medical ontology Γ that contains medical knowledge about heart problems may employ roles such as *has symptom*, *has disease*, etc., concepts such as ChestPain, ArmPain etc. and individual names John, Mary, etc. It may also employ relations between concepts such as *is-a* provides a hierarchical organization to the concepts. For example, Chest Pain *is-a* Heart Failure symptoms and Cardiac Arrhythmia *is-a* Heart Failure symptom. Another type of relationships can be expressed formally as for example *has-manifestations* relationship between diseases and their manifestations/symptoms such as Heart Failure *has-manifestations* one or more of Heart Failure symptoms. Ontology can also contain entities (individuals) which are domain real objects.

As mentioned in the introduction, ontologies provide class definitions (i.e., the necessary and sufficient conditions for defining class membership). In medicine, these definitions correspond to diagnostic criteria, i.e., the particular form of diseases as described in the records of patients suffering from these disease should be associated with the relevant disease categories. It was shown in [6] that eligibility criteria are often more useful than the Aristotelian definitions traditionally used in ontology for diagnostic classification. They argue that the classificatory principles and properties represented in ontologies may not be sufficient to classify instances of diseases. They suggest that operational definitions of diseases are more useful. These definitions are mostly based on the association of signs and symptoms. However, as mentioned above, this is difficult to achieve as the symptoms and signs of a disease varies widely and many diseases share a lot of common symptoms and signs. For instance, the Heart Failure (HF) symptoms include Edema in lower limb(Ell), Dyspnea (Dy), Hypertension (H), Xerostomia (X), Wheezing (W), Weight Gain (WG), Chest Pain (CP), Cardiac Arrhythmia (CA) and Rapid or irregular heartbeat. Furthermore, it is not required that all the symptoms of a disease are in a patient to be diagnosed as having a heart failure. However, since the heart failure disease has many common symptoms with many other disease such as symptoms such as Angina Pectoris, Mitral Valve Prolapse

Example 1: The medical ontology $\boldsymbol{\Gamma}$ (cf. Fig. 1) contains some medical knowledge about heart problems as described below:

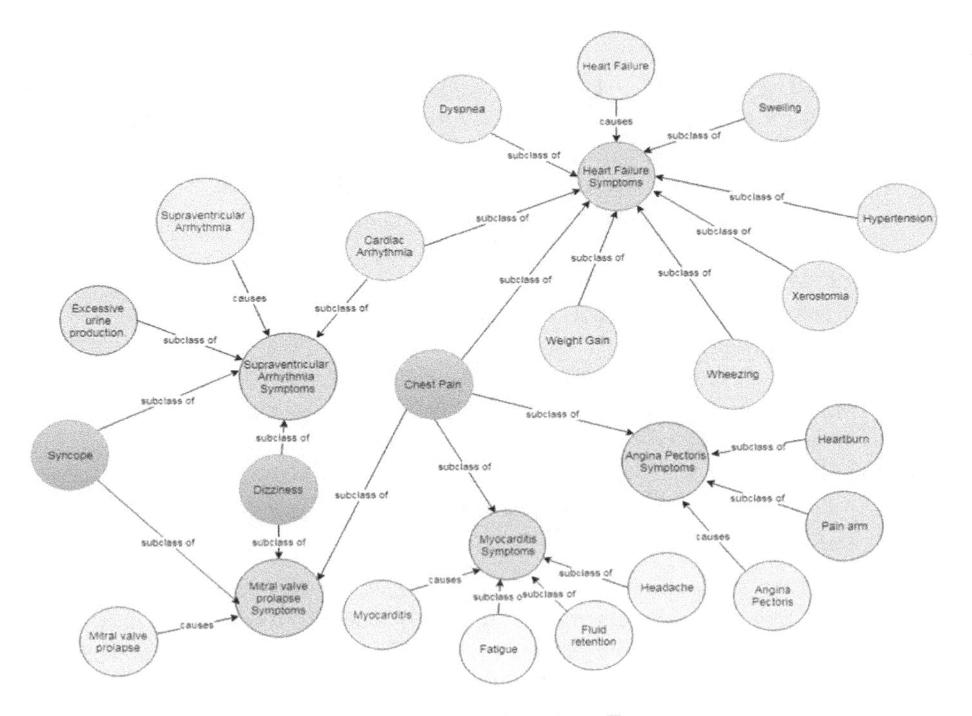

Fig. 1. Medical ontology Γ

The Heart Failure (HF) symptoms include Edema in lower limb(Ell), Dyspnea (Dy), Hypertension (H), Xerostomia (X), Wheezing (W), Weight Gain (WG), Chest Pain (CP), Cardiac Arrhythmia (CA) and Rapid or irregular heartbeat.

The Angina Pectoris (AP) symptoms include Chest Pain (CP), Heartburn (HAP) and Pain in Arm (P).
The Mitral Valve Prolapse (MVP) symptoms include Dizziness (Di), Syncope (Sy) and Chest Pain (CP).
The Myocarditis (M) symptoms include Fatigue (F), Fluid retention (FR), Headache (HM) and Chest Pain (CP).
The Supraventricular Arrhythmia (SA) Symptoms include Excessive Urine Production (EUP), Syncope (Sy) and Dizziness (Di).

Representation of Part of Γ in ALC: The representation of concepts and their relationships are as follows:

Ell $\sqsubseteq$ HF-Sympt, Dy $\sqsubseteq$ HF-Sympt, H $\sqsubseteq$ HF-Sympt, X $\sqsubseteq$ HF-Sympt, W $\sqsubseteq$ HF-Sympt, WG $\sqsubseteq$ HF-Sympt, CP $\sqsubseteq$ HF-Sympt, CA $\sqsubseteq$ HF-Sympt, and so on.

These assertions state that Edema in lower limbs is a heart failure symptom, Dyspnea is a heart failure symptom and so on.

Consider the following rules:

(R1) $\exists$HasDisease.HF $\sqsubseteq$ $\exists$HasSympt.S_1 $\sqcap$ … $\sqcap$ $\exists$HasSympt. S_k
where $S_1 \sqsubseteq$ HF-Sympt, …, $S_k \sqsubseteq$ HF-Sympt and $1 \leq k \leq$ No-of-Atoms (HF-Sympt)

(R2) $\exists$HasDisease.AP $\sqsubseteq$ $\exists$HasSympt.S_1 $\sqcap$ … $\sqcap$ $\exists$ HasSympt.S_k
where $S_1 \sqsubseteq$ AP-Sympt, $S_k \sqsubseteq$ AP-Sympt and $1 \leq k \leq$ No-of-Atoms (AP-Sympt)

(R3) $\exists$HasDisease. MVP $\sqsubseteq$ $\exists$HasSympt.S_1 $\sqcap$ … $\sqcap$ $\exists$HasSympt. S_k
where $S_1 \sqsubseteq$ MVP-Sympt, $S_k \sqsubseteq$ MVP-Sympt and $1 \leq k \leq$ No-of-Atoms (MVP-Sympt)

(R4) $\exists$HasDisease. Myc $\sqsubseteq$ $\exists$HasSympt.S_1 $\sqcap$ $\sqcap$ $\exists$HasSympt.S_k
where $S_1 \sqsubseteq$ Myc-Sympt, $S_k \sqsubseteq$ Myc-Sympt and $1 \leq k \leq$ No-of-Atoms (CP-Sympt)

(R5) $\exists$ HasDisease.SA $\sqsubseteq$ $\exists$HasSympt.S_1 $\sqcap$ … $\sqcap$ $\exists$HasSympt.S_k
where $S_1 \sqsubseteq$ SA-Sympt, $S_k \sqsubseteq$ SA-Sympt and $1 \leq k \leq$ No-of-Atoms (CP-Sympt).

R1 states that a patient suffering from a heart failure will show one or more symptoms, each of which is a heart failure symptom. R2, R3, R4 and R5 can be interpreted in the same way. R2 is concerned with Angina Pectoris, R3 with Mitral Valve Prolapse, R4 with Myocarditis and R6 with Supraventricular Arrhythmia (SA).

Case1. Suppose that we are presented with a 40 years old male, say John, complaining of chest pain. Using ALC inference services (forward chaining) we can infer that:

C11. Chest Pain is a heart failure symptom (CP $\sqsubseteq$ HF-Sympt).
C12. Chest Pain is an Angina Pectoris symptom (CP $\sqsubseteq$ AP-Sympt).
C13. Chest Pain is a Mitral Valve Prolapsed symptom (CP $\sqsubseteq$ MVP-Sympt).
C14. Chest Pain is a Myocarditis Symptom (CP $\sqsubseteq$ Myc-Sympt).

Now, using Abox Abduction (backward chaining), we have the following possible explanations:

C11'. Heart failure, HF using R1,
C12'. Angina Pectoris, AP using R2,
C13'. Mitral Valve Prolapsed, MTV using R3 and
C14'. Myocarditis, Myc using R4.

Case2. Suppose that in addition to chest pain, John complains of dizziness.
Using ALC inference services (forward chaining) we can infer that:

C21. Chest Pain is a heart failure symptom (CP $\sqsubseteq$ HF-Sympt).
C22. Chest Pain is an Angina Pectoris symptom (CP $\sqsubseteq$ AP-Sympt).
C23. Chest Pain is a Mitral Valve Prolapsed symptom (CP $\sqsubseteq$ MVP-Sympt) and Dizziness is a Supraventricular Arrhythmia Symptom (Di $\sqsubseteq$ MVP-Sympt).
C24. Chest Pain is a Myocarditis Symptom (CP $\sqsubseteq$ Myc-Sympt).
C25. Dizziness is a Supraventricular Arrhythmia Symptom (Di $\sqsubseteq$ SA-Sympt).

Now, using Abduction, we have the following possible explanations:

C21’. Heart failure, HF using R1.
C22’. Angina Pectoris, AP using R2.
C23’. Mitral Valve Prolapsed, MTV using R3 and R5.
C14’. Myocarditis, Myc using R4.
C25’. Supraventricular Arrhythmia using R5.
C26’. Heart failure and Supraventricular Arrhythmia using R1+R4.
C27’. Angina Pectoris and Supraventricular Arrhythmia using R2+R5.
C28’. Myocarditis and Supraventricular Arrhythmia using R4+R5.

The most plausible explanation that satisfies the minimality criterion together with the other criteria namely consistency and relevance is C23’.

5 Some Approaches to Abductive Reasoning in DL

A distinction has been made between four different abductive reasoning tasks such as concept abduction, ABox abduction, TBox abduction and KB abduction [11]. Most existing approaches emphasize ABox and concept abduction. Most of these approaches are based on existing proof techniques such as semantic tableau and resolution.

The authors in [12] propose an approach for ABox abduction in ALC. They use semantic tableau in their approach and they perform instance checking on an abductive query. After extracting a full completion graph, the algorithm generates a set of concept assertions for each open branch which would close that branch. Their algorithm is sound but not definitely complete and the solutions may not be semantically minimal.

The authors in [16] propose two approaches for ABox abduction. In one approach, they employ semantic-tableau. In the other, they employ resolution. Both approaches are goal-oriented in the sense that only actions which contribute to the solution are chosen in the proof procedures. These approaches require translation to first order logic (conjunctive normal form) and the solutions are then translated back to description logic. They are both complete and sound for consistent and semantically minimal solutions. However, as the set of possible solutions may contain some inconsistent and non-minimal solutions, there is a need for additional checks in order to ensure consistency and minimality. An approach presented in [9] uses existing abductive logic programming systems. It considers solutions over a finite set of concepts and roles. The solutions are consistent and minimal. However, it does not guarantee completeness since it requires translation to a Datalog program which is approximate and in some cases a solution may not be found.

Other approaches such as [11, 30] to Abox abduction that emphasized medical diagnosis have logical inconsistencies in their examples or formulae. In [11], we have assertions like:

Paul: ∃has symptom.(Laziness Π Pizza Appetite) which can be expressed in First order predicate calculus (FOPC) as:

(∃ y) (has symptom(Paul, y)&Laziness(y)&PizzaAppetite(y))

a: ∃has symptom.(Headache ⊔ Depressed)

which can be expressed in FOPC as:

$(\exists\ y)$ (has symptom(a, y)&(Headache(y) V Depressed(y))

What if a has headache and a is depressed?.
Similarly, in [30], we have assertions like:

$\exists$hasDiag:DM $\sqsubseteq$ $\exists$hasSymp.(S1 $\sqcap$ S2 $\sqcap$ …. $\sqcap$ S_8)

We have similar patterns in formule 3 and 9 in the paper.

6 Conclusions and Future Work

We have, in this paper, made an attempt to model medical diagnosis as an (iterative) abductive reasoning process using ALC. We have an ontological representation which takes into account the diagnostic criteria such that specific patient conditions can be classified under a specific disease. We employed ALC to benefit from its inference services. However, the inference capabilities provided by DL were not sufficient for diagnosis purposes. Therefore, we had to employ Abox abduction to capture the process of medical diagnosis.

It is worthwhile noting that there are issues such as the hierarchy of diseases where a disease could be manifested by some symptoms, signs and the emergence of other diseases. That is, the *Disease-Symptom* concept may include symptoms and diseases. This may require us to use a system such as ALCR+ [15]. Furthermore, time (temporal findings) is essential for the diagnosis of many diseases. This may require us to employ a temporal description logic [2].

Approaches which use classical logic presume complete knowledge of the domain of concern. These approaches cannot deal with inconsistency. When an inconsistency arises in a Knowledge Base (KB), then every conclusion can be derived and the system collapses. However, we only have partial knowledge of any domain. Defeasible rules systems [1, 13, 19–29, 31] are appropriate in those situations as they offer more expressive capabilities and are closer to commonsense reasoning. There are many situations in which conflicting rules may arise on the Web or in other contexts such as: (1) Reasoning with Incomplete Information, (2) Rules with Exceptions, (3) Default Inheritance in Ontologies and (4) Ontology Merging. These issues are not addressed in this paper. However, we hope to be able to elaborate on these issues in a future publication as there will be a need to employ defeasible ontologies and to integrate them with abductive reasoning.

References

1. Al-Shaikh, A., Khattab, H., Moubaiddin, A., Obeid, N.: A defeasible description logic for representing bibliographic data. In: Taha, N., Al-Sayyed, R., Alqatawna, J., Rodan, A. (eds.) Social Media Shaping e-Publishing and Academia, pp. 95–105. Springer, Cham (2017). https://doi.org/10.1007/978-3-319-55354-2_8
2. Artale, A., Kontchakov, R., Wolter, F., Zakharyaschev, M.: Temporal description logic for ontology-based data access. In: IJCAI 2013, pp. 711–717 (2013)

3. Baader, F., Sattler, U.: An overview of tableau algorithms for description logics. Stud. Logica **69**, 5–40 (2001)
4. Baader, F., Nutt, W.: Basic description logics. In: Baader, F., Calvanese, D., McGuinness, D., Nardi, D., Patel-Schneider, P. (eds.) Description Logic Handbook, chapter 2, pp. 43–95 (2003)
5. Baader, F., Calvanese, D., McGuinness, D., Nardi, D., Patel-Schneider, P.F. (eds.): The Description Logic Handbook: Theory, Implementation and Applications. Cambridge University Press, Cambridge, Cambridge (2003)
6. Bertaud-Gounot, V., Duvauferrier, R., Burgun, A.: Ontology and medical diagnosis. Inf. Health Soc. Care **37**(2), 51–61 (2012)
7. Cornet, R., de Keizer, N.: Forty years of SNOMED: a literature review. BMC Med. Inf. Decis. Making **8**(1) (2008)
8. Donnelly, K.: SNOMED-CT: the advanced terminology and coding system for eHealth. Stud. Health Technol. Inf. **121**, 279 (2006)
9. Du, J., Qi, G., Shen, Y., Pan, J.: Towards practical Abox abduction in large OWL DL ontologies. In: Proceedings of the 25th AAAI Conference on Artificial Intelligence, pp. 1160–1165 (2011)
10. Eiter, T., Gottlob, G.: The complexity of logic-based abduction. J. ACM **42**(1), 3–42 (1995)
11. Elsenbroich, C., Kutz, O., Sattler, U., et al.: A case for abductive reasoning over ontologies. In: Proceedings of the 9th OWL: Experiences and Directions, pp. 56–75. IOS Press, Amsterdam (2006)
12. Halland, K., Britz, K.: Naive ABox abduction in ALC using a DL tableau. In: Proceedings of the 25th International Workshop on Description Logics (DL), pp. 443–453 (2012)
13. Hijazi, S., Jabri, R., Obeid, N.: On drug dosage control using description defeasible logic. In: International Conference on Computational Science and Computational Intelligence (CSCI). IEEE (2017)
14. Horrocks, I., Sattler, U., Tobies, S.: Practical reasoning for expressive description logics. In: Ganzinger, H., McAllester, D., Voronkov, A. (eds.) LPAR 1999. LNCS (LNAI), vol. 1705, pp. 161–180. Springer, Heidelberg (1999). https://doi.org/10.1007/3-540-48242-3_11
15. Horrocks, I., Sattler, U.: A description logic with transitive and inverse roles and role hierarchies. J. Log. Comput. **9**(3), 385–410 (1999)
16. Klarman, S., Eendriss, U., Schlobach, S., et al.: ABox abduction in the description logic ALC. J. Autom. Reason. **46**(1), 43–80 (2011)
17. Magnani, L.: Abductive reasoning: philosophical and educational perspectives in medicine. In: Evans, D.A., Patel, V.L. (eds.) Advanced Models of Cognition for Medical Training and Practice. Springer, Berlin (1992). https://doi.org/10.1007/978-3-662-02833-9_2
18. Magnani, L.: Abduction, Reason, and Science: Processes of Discovery and Explanation. Kluwer Academic/Plenum Publishers, New York (2001)
19. Moubaiddin, A., Obeid, N.: The role of dialogue in remote diagnostics. In: Proceedings of the 20th International Conference on Condition Monitoring & Diagnostic Engineering Management (2007)
20. Moubaiddin, A., Obeid, N.: Dialogue and argumentation in multi-agent diagnosis. In: Nguyen, N.T., Katarzyniak, R. (eds.) New Challenges in Applied Intelligence Technologies, Studies in Computational Intelligence, vol. 134, pp. 13–22. Springer, Heidelberg (2008). https://doi.org/10.1007/978-3-540-79355-7_2
21. Moubaiddin, A., Obeid, N.: Partial information basis for agent-based collaborative dialogue. Appl. Intell. **30**(2), 142–167 (2009)
22. Moubaiddin, A., Obeid, N.: On formalizing social commitments in dialogue and argumentation models using temporal defeasible logic. Knowl. Inf. Syst. **37**(2), 417–452 (2013)

23. Moubaiddin, A., Salah, I., Obeid, N.: A temporal modal defeasible logic for formalizing social commitments in dialogue and argumentation models. Appl. Intell. **48**(3), 608–627 (2018)
24. Obeid, N.: Three valued logic and nonmonotonic reasoning. Comput. Artif. Intell. **15**(6), 509–530 (1996)
25. Obeid, N.: Towards a model of learning through communication. Knowl. Inf. Syst. **2**(4), 498–508 (2000)
26. Obeid, N.: A formalism for representing and reasoning with temporal information, event and change. Appl. Intell. **23**(2), 109–119 (2005)
27. Obeid, N., Moubaiddin, A.: Towards a formal model of knowledge sharing in complex systems. In: Szczerbicki, E., Nguyen, N.T. (eds.) Smart Information and Knowledge Management, Studies in Computational Intelligence Series, pp. 53–82. Springer, Heidelberg (2010). https://doi.org/10.1007/978-3-642-04584-4_3
28. Obeid, N., Rao, R.B.: On integrating event definition and event detection. Knowl. Inf. Syst. **22**(2), 129–158 (2010)
29. Obeid, N., Rawashdeh, E., Alduweib, E., Moubaiddin, A.: On ontology-based diagnosis and defeasibility. In: International Conference on Computational Science and Computational Intelligence (CSCI), pp. 57–62. IEEE (2016)
30. Pukancová, J., Martin, H.: Abductive reasoning with description logics: use case in medical diagnosis. In: Proceedings of the 28th International Workshop on Description Logics (2015)
31. Sabri, K.E., Obeid, N.: A temporal defeasible logic for handling access control policies. Appl. Intell. **44**(1), 30–42 (2016)

Intelligent Information Storage and Retrieval

A Lightweight Linked Data Reasoner Using Jena and Axis2

I-Ching Hsu(✉) and Sin-Fong Lyu

Department of Computer Science and Information Engineering,
National Formosa University, 64, Wenhua Rd.,
Huwei Township 632, Yunlin County, Taiwan
hsuic@nfu.edu.tw

Abstract. Semantic Web is rapidly becoming a reality through the development of Linked Data in recent years. Linked Data uses RDF data model to describe statements that link arbitrary data resources on the Internet. It can facilitate to infer new data resources at runtime through the RDF links, and then provide more complete answers as new data resources appear on the Internet. Linked Data provides the means to reach the goal of Semantic Web. At present, Linked Data being used only in the promotion of information sharing or exchange is not a semantic inference due to the lack of an easily shared inference engine. This study addresses the issue developing a Lightweight Linked Data Reasoner (LLDR) which is based on Jena reasoner and is implemented in the apache Axis2. To illustrate the LLDR application, this study developed the Vehicle Ontology to annotate project document from heterogeneous and distributed project resources as Linked Data.

Keywords: Linked Data · Semantic Web · Jena

1 Introduction

The Semantic Web is an extension of the current Web in which information is given well defined meaning, better enabling computers and people to work in cooperation. The term Linked Data is also first introduced by Tim Berners-Lee, which is used to refer to a set of best practices for publishing and connecting structured data on the Web [1]. Linked Data uses RDF data model to describe statements that link arbitrary data resources on the Web. It can facilitate to infer new data resources at runtime through the RDF links, and then provide more complete answers as new data resources appear on the Web. Linked Data provides the means to reach the goal of Semantic Web. Hence, Linked Data can be regarded as an infrastructure to practice the Semantic Web. Semantic Web is rapidly becoming a reality through the development of Linked Data in recent years.

In recent years, more and more Linked Data is created automatically or semi-automatically. D2R Server [2] is a tool for publishing relational databases as Linked Data on the Semantic Web. DBpedia [3] is published as Linked Data that can be regarded as the Semantic Web mirror of Wikipedia. DBpedia extracts structured information from Wikipedia, convert it into RDF data model, and make it freely

F. Wotawa et al. (Eds.): IEA/AIE 2019, LNAI 11606, pp. 391–397, 2019.
https://doi.org/10.1007/978-3-030-22999-3_34

available on the Internet. Linked Data changed our way of sharing resources and information. As well known, Web 2.0 is recognized as the next generation of web applications proposed by O'Reilly [4]. Jena is an inference engine that provides an ontology-based reasoner for semantic Web-based language, including RDF, RDFS, OWL, and rule. Apache Axis2 [1] is a widely use web service engine that helps developers to create, deploy, and run Web Services. Therefore, LLDR hide the detailed Jena programming and Web Services protocols from the developers and make it easier for the developers to use. The existing Linked Data can be explained by LLDR to infer new data resources at runtime.

This paper is organized as follows. The next section presents some related works. Section 3 develops the Vehicle ontology to provide common knowledge and reusable resources for the LLDR. The LLDR is presented in Sect. 4. Section 5 presents an applicable demonstration and experimental results are presented. Finally, summary and concluding remarks are included.

2 Related Works

The Semantic Web is an extension of the current Web in which information is given well defined meaning, better enabling computers and people to work in cooperation. Many studies [5, 6] adopt Semantic Web to build intelligent applications in various domains. Linked Data can be regarded as an infrastructure to practice the Semantic Web. One major feature of existing Web is to adopt Linked Data to build a more maintainable and cooperative Web. Linked Data is a kind of metadata that can be considered as resources over the Internet. Therefore, Linked Data can be used to enhance the intelligence, reusability, and interoperability of Web applications. In [7], the authors shows how linked data sets can be exploited to build rich Web applications with little effort. In order to make it as easy as possible for Web applications to process data, system developer should reuse existing RDF-based ontology from well-known vocabularies wherever possible. Developers should only define new vocabularies if they can not find required vocabularies in existing RDF-based ontology. An application that combines resources from different websites to produce a new Web application is called a Web 2.0 Mashup [8]. The growing availability of Linked Data does not keep pace with the rich semantic descriptions to facilitate the direct deployment of user-tailored services.

3 Project Domain Ontology

The core ingredients of an RDF-based ontology include a set of concepts, a set of properties, and the relationships between the elements of these two sets. The Vehicle Ontology offers the vehicle classification in a high abstraction level and is used to describe the semantic-based relation between classes, such as Vehicle, Department, Car, Bus etc., involved in the traffic domain. Figure 1 shows the semantic structure of Vehicle Ontology as a UML class diagram. Vehicle Ontology is defined based on RDF Schema and a set of well-known vocabularies, including FOAF and DC, that makes it easy for program to process some basic facts about the terms in the Vehicle Ontology. The Vehicle Ontology introduces the following classes and properties.

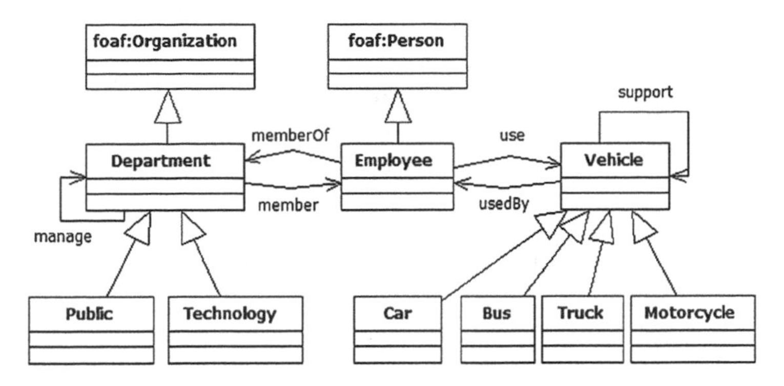

Fig. 1. The UML diagram for Vehicle Ontology

The Department class is a subclass of the foaf:Organization class, which can inherit semantics from the foaf:Organization. The major properties of foaf:Organization include foaf:mbox, foaf:weblog, foaf:made, foaf:holdsAccount, etc. The Department class of Vehicle Ontology is used to describe the agent for traffic domain. The Employee class is a subclass of the foaf:Person class. The major properties of foaf: Person include foaf:firstName, foaf:surname, foaf:family_name, foaf:knows, foaf:interest, foaf:topic_interest, etc. The Employee class can inherit above properties to describe employees of a department. The Department class is composed of two subclasses: Public and Technology. In the Vehicle Ontology, the Vehicle can be divided into four categories: Car, Bus, Truck, and Motorcycle. There is an inverse relation between use property and used By property. The support property is symmetric. The manage property is transitive.

The partial instances and relationships of Vehicle are summarized in Tables 1 and 2, respectively. For example, the entity ID "E3939889" is an instance of Employee class. Therefore, it can inherit semantics from Employee and foaf:Person classes. The relation addresses the relationship (such as, domain "E3939889", range "C9278120") is an instance of use property. There is a use relationship from " E3939889" to "C9278120". Therefore, the relationship can inherit semantic from the use property. In the following, this study illustrates how Vehicle Ontology can be combined with FOAF, and DC to annotate relationships of project resources using the concrete examples.

4 Lightweight Linked Data Reasoner

The Lightweight Linked Data Reasoner (LLDR) can be associated with various domain ontologies and Linked Data. The basic function of Linked Data is to provide RDF-based metadata for Web resources on the Internet. The LLDR is a semantic reasoner

Table 1. The partial instances in Vehicle Ontology

Entity URI	Class
E3939889	Employee
C9278120	Car
P5562109	Public
T6527182	Technology
B6785421	Bus
E3939888	Employee

Table 2. The partial relationships in the instances

Property	Domain	Range
Use	E3939889	C9278120
Use	E3939888	B6785421
Member	T6527182	E3939888
Member	P5562109	E3939889
Manage	P5562109	T6527182
Support	C9278120	C9275512
Support	C9278120	C9275532

that serves as a Web service to support RDF-based reasoning. The LLDR application environment includes the LLDR, Web Server, Knowledge Base, Linked Data Base, and Client. The flow-oriented LLDR architecture is depicted in Fig. 2.

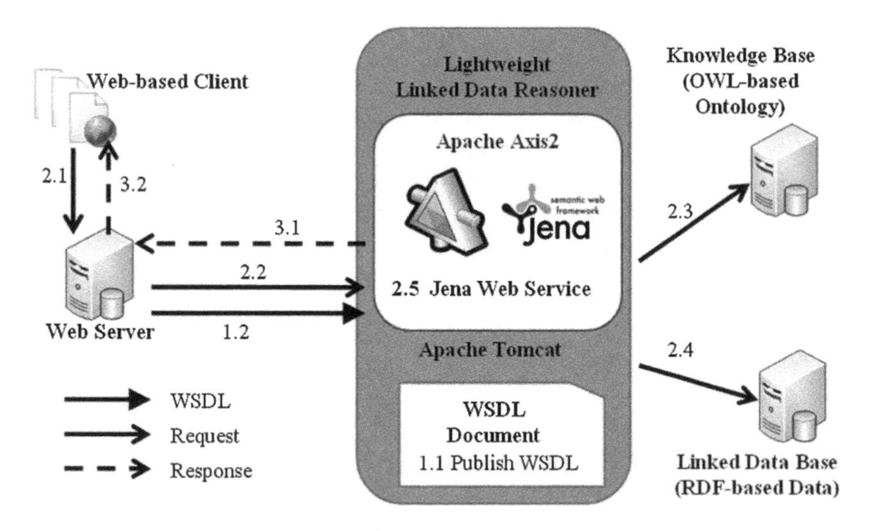

Fig. 2. The flow-oriented LLDR architecture

LLDR is a Jena-based reasoner and is implemented based on Apache Axis2. There are two implementations for the Apache Axis2 Web services engine, including Apache Axis2/Java and Apache Axis2/C. This work adopted the former to develop the LLDR. The partial code of LLDR is shown in Fig. 2.

Knowledge Base is composed of OWL-based ontologies that provide semantic reasoning, and plays the same role as the knowledge base in a traditional expert system.

Linked Data Base is an annotation repository composed of RDF-based documents, which plays the same role as the fact base in a traditional expert system.

Web Server listens to client's request and invokes the remote LLDR. It serves as a broker that receives and filters the information from the LLDR.

The information flow of the LLDR occurs as follows.

1. Steps for finding WSDL.
   1.1 The WSDL of LLDR is automatically generated and published by the Apache Axis2.
   1.2 Web Server parses the WSDL of LLDR to call the LLDR API.
2. Steps for request.
   2.1 The client sends a request with URL to Web Server.
   2.2 The Web Server processes the request. The Web Server then invokes the LLDR with two URL parameters, including ontology and RDF parameters, to assign the ontology and RDF files location, respectively.

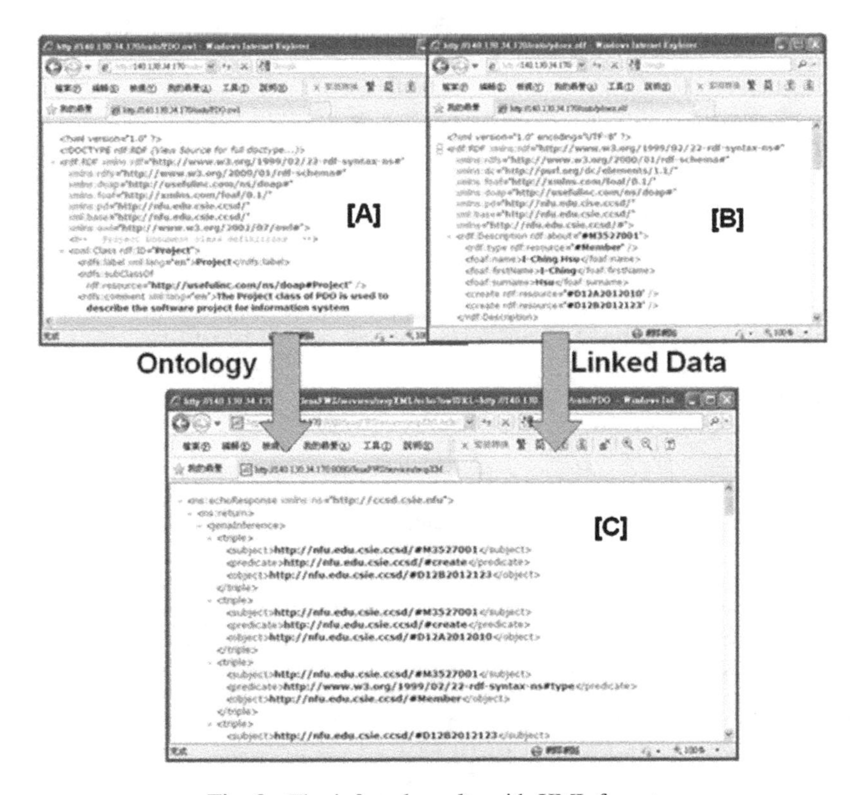

Fig. 3. The inferred results with XML format

2.3 The LLDR accesses the remote domain ontology (as shown in Fig. 3 (A)) based on the ontology parameter.
2.4 The LLDR accesses the remote Linked Data (as shown in Fig. 3 (B)) based on the RDF parameter.
2.5 The LDDR derives new RDF facts from these existing Linked Data and domain ontology.
3. Steps for response.
3.1 The LLDR passes the inferred results with XML format (as shown in Fig. 3 (C)), mentioned in Step 2.5, to Web Server.
3.2 The Web Server parses the XML document to filter the available information, and then responses to the Client.

5 Conclusion

Linked Data contains a structured information source which is written in RDF data model to facilitate the machine-readable. It can facilitate to infer new data resources at runtime through the RDF links, and then provide more complete answers as new data resources appear on the Internet. At present, Linked Data being used only in the promotion of information sharing or exchange is not a semantic inference due to the lack of an easily shared inference engine. This study implements a Lightweight Linked Data Reasoner (LLDR) serves as a Web service that adopts the Jena-based inference engine and develops based on Apache Axis2.

Further research will be to extend the LLDR with advance logic descriptions to support additional intelligence by deducing new adaptation rules. By integrating logic rules into LLDR, this approach can describe additional semantics of Linked Data [9, 10]. Semantic Web Rule language (SWRL) [11] seems to be the most appropriate language to further study, because it currently is the main language for representing logic rules in the Semantic Web.

References

1. Bizer, C., Heath, T., Berners-Lee, T.: Linked data - the story so far. Int. J. Semant. Web Inf. Syst. **5**(3), 1–22 (2009)
2. Bizer, C., Cyganiak, R.: D2R Server. http://www4.wiwiss.fu-berlin.de/bizer/d2r-server/. Accessed 26 Feb 2012
3. Subashini, S., Kavitha, V.: A survey on security issues in service delivery models of cloud computing. J. Netw. Comput. Appl. **34**(1), 1–11 (2011)
4. O'Reilly, T.: What is Web 2.0 (2005). http://www.oreillynet.com/pub/a/oreilly/tim/news/2005/09/30/what-is-web-20.html. Accessed 12 July 2010
5. Hsu, I.-C.: Semantic tag-based profile framework for social tagging systems. Comput. J. **55** (9), 1118–1129 (2012)
6. Xu, G.-X., et al.: Semantic classification method for network Tibetan corpus. Cluster Comput. **20**(1), 155–165 (2017)

7. Hausenblas, M.: Exploiting linked data to build web applications. Internet Comput. IEEE **13** (4), 68–73 (2009)
8. Murugesan, S.: Understanding Web 2.0. IEEE IT Prof. **9**(4), 34–41 (2007)
9. Zhang, X., Lin, E., Lv, Y.: Multi-target search on semantic associations in linked data. Int. J. Seman. Web Inf. Syst. **14**(1), 71–97 (2018)
10. Zhao, Y., Fan, B.: Exploring open government data capacity of government agency: based on the resource-based theory. Govern. Inf. Quart. **35**(1), 1–12 (2018)
11. Horrocks, I., et al.: SWRL: a semantic web rule language combining OWL and RuleML (2003). http://www.daml.org/2003/11/swrl/. Accessed 31 Aug 2011

A System Using Tag Cloud for Recalling Personal Memories

Harumi Murakami(✉) and Ryutaro Murakami

Osaka City University, Sugimoto, Sumiyoshi, Osaka 558-8585, Japan
harumi@osaka-cu.ac.jp
http://murakami.media.osaka-cu.ac.jp/

Abstract. The research presented here extends a previous prototype that supported human recollection with tag clouds created from the use of a personal calendar and Twitter. That system weighted keywords by combining term frequency and the number of photos taken by users to recall memorable events. The aim in this paper is to improve upon our previous work and present a full system that uses tag clouds for recalling personal memories. The main differences from our previous work are as follows. (1) Multiple information sources such as SNSs or instant messengers can be used. (2) To handle multiple information sources, we present a new unified keyword-weighting algorithm. (3) We implemented new functions, such as keyword search, tag search, and photo display, to form a complete system. Preliminary experiments reveal the usefulness of our system in recalling personal memories.

Keywords: Personal memory support · Tag cloud · Calendar · Twitter · LINE · Number of photos

1 Introduction

Memory is crucial for various activities in our daily lives. For example, we may have to write progress reports about what we have done on particular days or weeks. We may have to plan anniversaries and thus recall what we did last year or in previous years. Or we might simply want to reminisce about the day we saw our spouse for the first time. In addition, the amount of information that we manage is increasing. Consequently, we aim to support human memory.

We have presented a prototype that supports human recollection with tag clouds created from the use of calendar and Twitter [1]. Since we assumed that we could identify memorable events based on the days when a user took many photos, we weighted keywords by combining term frequency and the number of photos taken by users to generate tag clouds.

However, we found the following problems in the prototype. (1) Information sources are fixed in a calendar and Twitter. The trends of SNSs are changing, and their preferences are different. We need to easily cope with other information sources. (2) The keyword weighting algorithms for calendar and Twitter are

F. Wotawa et al. (Eds.): IEA/AIE 2019, LNAI 11606, pp. 398–405, 2019.
https://doi.org/10.1007/978-3-030-22999-3_35

different because they are customized to each source. To cope with new information sources, we need to develop an effective simplified weighting algorithm. (3) There is no function of keyword search, which is obviously useful for recalling memory. (4) Photos were used for keyword-weighing algorithms, but the photos themselves were not displayed. Photos are also obviously effective for recalling the past.

The aim of this research is to improve the system developed in our previous work to cope with the above problems and to present a new complete system using tag cloud for recalling personal memories. The main differences from our previous work are as follows. (1) Multiple information sources such as SNSs or instant messengers can be used. (2) To handle multiple information sources, we presented a new unified keyword-weighting algorithm. (3) We implemented new functions, such as keyword search, tag search, and photo display, to build a complete system.

2 Tag Browser

Our research generates a tag cloud by extracting keywords from various kinds of information usage and weighting them using term frequency and the number of photos. First, we obtain the data written by users and generate history structures [2]. Next, we generate tag clouds from these history structures. We call our new system a tag browser. LINE, which is classified as an SNS but is actually the most frequently used instant messenger in Japan, is adopted as well as calendar and Twitter. The user can set the period and the number of tags (default number is 30) and the system displays a tag cloud, information logs, and photos.

Several new functions have been developed for the tag browser. (1) Multiple information sources such as SNSs or instant messengers can be used. (2) Log windows (display position of information sources) are defined automatically based on the amount of log data. (3) The same keyword-weighting algorithm is used for all information sources. (4) The user can change the weighting algorithm using a slider interface. (5) Tag and keyword search functions can be made. (6) Photos are displayed according to date. In what follows, we translated the examples in this paper from Japanese into English for publication.

2.1 Generating History Structure

The basic components constructing the history structure include time, keywords, and log (original text) sets[1]. Nouns and noun phrases are extracted from information logs as keywords. For example, from a LINE message "··· Yes! In Kyoto every day is like a festival!" The system extracts "Kyoto," "every day," and "festival" as keywords. See Fig. 1 for example of history structures.

[1] We have changed from using URI [2] to log in this research. Other attribute information such as URI can be stored optionally. For LINE, the receiver and sender of the message are stored as To and From, respectively.

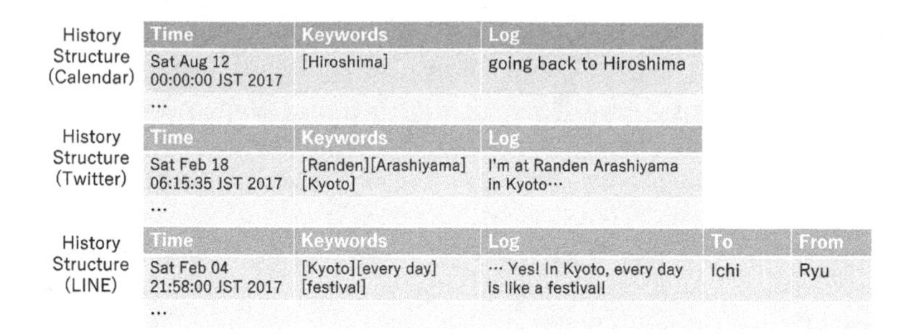

History Structure (Calendar)
Time
Keywords
Log
Sat Aug 12 00:00:00 JST 2017
[Hiroshima]
going back to Hiroshima
...
History Structure (Twitter)
Time
Keywords
Log
Sat Feb 18 06:15:35 JST 2017
[Randen][Arashiyama] [Kyoto]
I'm at Randen Arashiyama in Kyoto···
...
History Structure (LINE)
Time
Keywords
Log
To
From
Sat Feb 04 21:58:00 JST 2017
[Kyoto][every day] [festival]
··· Yes! In Kyoto, every day is like a festival!
Ichi
Ryu
...

Fig. 1. Example of history structures

2.2 Displaying Tag Clouds

Based on the results of experiments, it was decided to display tag cloud to the left and information logs to the right. First, we describe how to display a tag cloud. According to the input (period with or without a keyword), weights of keywords are calculated using Eqs. (1) and (2).

First, W_{i,t_k} defines the weight of term t_k in each information source i:

$$W_{i,t_k} = \sum \frac{1 + C_{photo}(G_{date}(t_k))}{C_{keyword}(G_{date}(t_k))}, \tag{1}$$

where the addition occurs for each term t_k appearing in each history structure. Here, $G_{date}(t_k)$ is a function that obtains the date of term t_k, $C_{photo}(R)$ is a function that obtains the number of photos in range R, and $C_{keyword}(R)$ is a function that obtains the number of keywords (terms) in range R. The denominator works as a normalizing factor for the number of information logs.

Next, W_{t_k} defines the weight of term t_k:

$$W_{t_k} = \sum_{i=1}^{n} \alpha_i W_{i,t_k}, \tag{2}$$

where n is the number of information sources and the default value for α_i is 0.5. In this research, we assigned the following values to i: calendar: 1, Twitter: 2, and LINE: 3.

The size of tags (keywords) is based on the weights determined by Eq. (2). Font colors are designed according to the information sources in which they appear. They reflect the original color image of the applications except for the calendar. If term t_k only appears in the calendar, it is orange, while it is blue in Twitter and green in LINE. If term t_k appears in more than one log, it is red. Tags are sorted by the time of first appearance.

2.3 Displaying Information Logs and Photos

According to the amount of data, information logs are displayed from left to right. In each information source, logs are displayed according to time. When only a period is input, all logs during the period are displayed. In tag or keyword searches, logs that contain the tag or keyword are displayed. Photos are displayed in ascending order according to date. When only a period is input, all photos during that period are displayed. In tag or keyword search, the photos taken on the date matching the date of logs containing the searched tag or keyword are displayed.

2.4 Example

Figure 2 shows an example of the system's basic usage. The user can set the period and number of tags, and a tag cloud is displayed on the left while information logs are displayed on the right. In this case, LINE, Twitter, and calendar logs are displayed from left to right according to the amount of data. Logs and photos are displayed according to time order, and the user can scroll them. The user can change parameter α_i of Eq. (2) ($0 \leq \alpha_i \leq 1$) by using the slider.

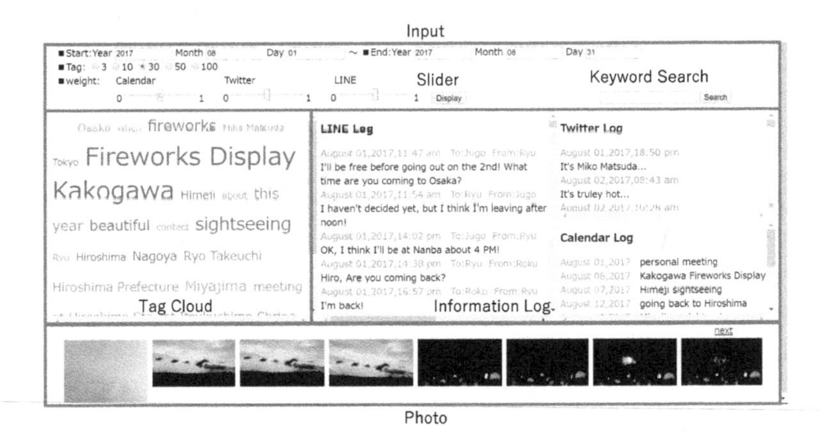

Fig. 2. Screenshot of display for one-month period

Figure 2 shows an example screen of one month (Aug. 2017) for a user. "Fireworks Display" is the largest tag and is shown in red. 10 out of 30 (33%) tags are related to places. In this month' case, many tags work as clues to remember salient events.

When the user clicks a tag in the tag cloud, history structures that contain the keyword are extracted, a new tag cloud is generated, and the information logs of these history structures are displayed. For example, when the user clicks "fireworks," information logs containing "fireworks" are searched and displayed. Photos taken on the dates of the retrieved logs are displayed, and thus the user can recall pleasant memories of "Fireworks Display at Kakogawa."

Figure 3 shows an example of keyword search. When the user inputs "Kyoto," information logs containing "Kyoto" are retrieved. In this case, the period is set to about half a year. Although the user went to Kyoto just once in February as in calendar log, many message exchanges using LINE related to Kyoto are retrieved, and various types of information (plan, memory, etc.) about Kyoto can be recalled. Photos related to Kyoto are also displayed.

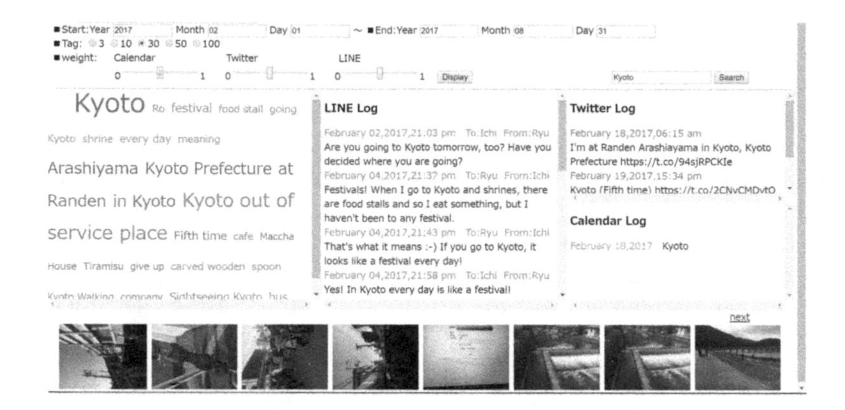

Fig. 3. Screenshot of display for keyword search

The tag browser can be used for business as well. For example, the user can look for photos of certain presentation slides taken at conferences. Although there is no manual tag for the photos, they can be found using a date or information logs searched via tag or keyword search.

3 Experiment

We recruited subjects who satisfied the following conditions: (a) those who take photos and (b) those using more than one of the following: writing a calendar schedule on any device, using Twitter, or using LINE. We gathered data for more than three months, and each experiment used the data of one month.

3.1 Experiment 1: Generating Tag Cloud

We evaluated the usefulness of our algorithm for creating tag clouds. Our subjects were ten males and one female, with an average age of 22.8. We prepared four tag clouds from four weighting algorithms (our algorithm, raw frequency (number of t_k in each history structure), relative frequency (number of t_k divided by number of keywords in each history structure), and a previous algorithm [1]) that display 30 keywords for comparison. The subjects performed the following task: "Rank the four tag clouds by the ease with which they helped you recall your memories." Five subjects selected our algorithm as 1st rank among the four algorithms. Three subjects selected raw frequency, two selected the previous algorithm, and one selected relative frequency. No subject selected our algorithm as 4th (worst) rank. Although the difference was slight, our algorithm was best among the four algorithms, and we found that it was possible to unify a keyword-weighting algorithm.

3.2 Experiment 2: Tag Browser

We evaluated the usefulness of our system. Our subjects were five males and one female, with average age of 23.2. After recalling a one-month-old memory using the system, the subjects answered questions at five levels (5 to 1; 5 is best). Here, we extract some results of the questions in Table 1.

Table 1. Some results of questionnaire

Questions		Mean	SD
Q1	Was the system useful for recalling your memories?	4.50	0.50
Q2	Do you want to use this system in the future?	4.33	0.75
Q3	Did you feel fun while using this system?	4.33	0.75

For the question "Which part was the most useful for recalling your memories? - A: Tag cloud, B: Logs, C: Photos," five subjects answered A: Tag cloud and one subject answered C: Photos. Overall, the results show the usefulness of our system for recalling personal memories.

4 Related Work

This research is a part of our work on personal memory support. Murakami [2] presented the concept of information structure called history structure. The history structure integrates various kinds of information usage. Murakami et al. [3] developed a knowledge-space browser that displays a network rather than a tag cloud. Matsumoto et al. [1] developed the initial prototype of the tag browser displaying a tag cloud. This research improved upon the previous work [1] by further refining the tag browser to its current form.

Our research uses tag clouds for human memory recall, even though little research has used them for this purpose. Chen and Jones [4] developed a prototype system called iCLIPS that searched through personal lifelogs for memory support. In iCLIPS, computer activities and the names of locations and people were displayed in term clouds, which resemble tag clouds. No detailed algorithms for generating term clouds and user studies of the prototype have been reported. We focus on generating better tag clouds rather than accumulating all human activities. We also conducted preliminary experiments that demonstrated our system's usefulness.

Mathur et al. [5] presented a prototype system of a tool called LifeView, which visualizes textual lifelogs for Sentimental Recall and Sharing. In this system, events are manually created by users who manually annotate tags. A tag cloud (these tags) for one event was displayed. On the other hand, our approach automatically extracts keywords from information sources and generates tag clouds.

Aires and Goncalves [6] presented Personal Information Dashboard, a web application that allows users to see, at a glance, various facets of their lives. In this system, Keywords Cloud is a tag cloud-like visualization that shows the most important words from a set of emails, posts, and/or tweets. To calculate the important words, they used tf-idf. The presentation (layout) is spacial (not sorting). Keywords Cloud can be configured to show data from a specific time period. Our tag cloud's algorithm, presentation, and information sources are different from Keywords Cloud.

Many systems and researches have generated tag clouds. The two main purposes are summarizing and navigating certain content. Rivadeneira et al. [7] classified tasks that tag clouds can support into four: (a) search, (b) browsing, (c) impression formation or gisting, and (d) recognition/matching. In general, (a) tag selection algorithms are based on the frequency of the terms or objects, (b) important tags are represented by size or color, and (c) tags are ordered alphabetically.

Venetis et al. [8] evaluated existing algorithms for exploring and understanding a set of objects against tf-idf-based algorithms and presented a maximum covering algorithm that seems a very good choice for most scenarios in their experiment and a popularity algorithm, which is easier to implement and performs well in specific contexts. Both algorithms are based on the number of objects associated with tags. Our research presented a unique algorithm based on the frequency of terms and photos.

Much research has presented ideas for integrating information in the light of Personal Information Management (PIM) [9], to overcome information overload (e.g. [10]). History structure is simply generated from existing information sources. Our research resembles text-based lifelog research. Our approach is simple and does not need special devices to capture information from the world.

5 Summary

We improved upon what we developed in our previous work and presented a complete new system using tag clouds for recalling personal memories. The main differences from our previous work are as follows. (1) Multiple information sources such as SNSs or instant messengers can be used. (2) To handle multiple information sources, we presented a new unified keyword-weighting algorithm. (3) We implemented new functions, such as keyword search, tag search, and photo display, to build a complete system. Preliminary experiments reveal the usefulness of our system for recalling personal memories.

References

1. Matsumoto, M., Matsuura, S., Mitsuhashi, K., Murakami, H.: Supporting human recollection of the impressive events using the number of photos. In: Proceedings of the 6th International Conference on Agents and Artificial Intelligence, ICAART, vol. 1, pp. 538–543 (2014)
2. Murakami, H.: History structure for exploring desktop data. In: Proceedings of the SIGIR 2010 Workshop on Desktop Search (Understanding, Supporting and Evaluating Personal Data Search), pp. 25–26 (2010)
3. Murakami, H., Mitsuhashi, K., Senba, K.: Creating user's knowledge space from various information usages to support human recollection. In: Jiang, H., Ding, W., Ali, M., Wu, X. (eds.) IEA/AIE 2012. LNCS (LNAI), vol. 7345, pp. 596–605. Springer, Heidelberg (2012). https://doi.org/10.1007/978-3-642-31087-4_61
4. Chen, Y., Jones, G.F.: Augmenting human memory using personal lifelogs. In: Proceedings of the 1st Augmented Human International Conference. ACM, New York (2010)
5. Mathur, A., Majumder, A., Datta, S., Menon, S., Malhorta, S., Dahiya, A.: Life-View: a lifelog visualization tool for supporting sentimental recall and sharing. In: Proceedings of the 24th Australian Computer-Human Interaction Conference, OzCHI 2012, pp. 371–380. ACM, New York (2012)
6. Aires, J., Goncalves, D.: Personal information Dashboard - Me, at a Glance. In: PIM 2012 (2012). http://pimworkshop.org/2012/
7. Rivadeneira, A.W., Gruen, D.M., Muller, M.J., Millen, D.R.: Getting our head in the clouds: toward evaluation studies of tagclouds. In: Proceedings of the SIGCHI Conference on Human Factors in Computing Systems, CHI 2007, pp. 995–998. ACM, New York (2007)
8. Venetis, P., Koutrika, G., Garcia-Molina, H.: On the selection of tags for tag clouds. In: Proceedings of the Fourth ACM International Conference on Web Search and Data Mining, WSDM 2011, pp. 835–844. ACM, New York (2011)
9. Jones, W.: Personal information management. Ann. Rev. Inf. Sci. Technol. **41**, 453–504 (2007). ARIST 41
10. Dumais, S., Cutrell, D., Cadiz, J.J., Jancke, G., Sarin, R., Robins, D.C.: Stuff I've seen: a system for personal information retrieval and re-use. In: Proceedings of the 26th Annual International ACM SIGIR Conference on Research and Development in Information Retrieval, SIGIR 2003, pp. 72–79. ACM, New York (2003)

Compressing and Querying Skypattern Cubes

Willy Ugarte[1], Samir Loudni[2], Patrice Boizumault[2], Bruno Crémilleux[2](✉), and Alexandre Termier[3]

[1] Peruvian University of Applied Sciences, Lima, Peru
willyugarte@gmail.com
[2] Normandie Univ., UNICAEN, ENSICAEN, CNRS – UMR GREYC, Caen, France
{samir.loudni,patrice.boizumault,bruno.cremilleux}@unicaen.fr
[3] Univ. Rennes, Inria, CNRS, IRISA, Rennes, France
alexandre.termier@irisa.fr

Abstract. Skypatterns are important since they enable to take into account user preference through Pareto-dominance. Given a set of measures, a skypattern query finds the patterns that are not dominated by others. In practice, different users may be interested in different measures, and issue queries on any subset of measures (a.k.a subspace). This issue was recently addressed by introducing the concept of skypattern cubes. However, such a structure presents high redundancy and is not well adapted for updating operations like adding or removing measures, due to the high costs of subspace computations in retrieving skypatterns. In this paper, we propose a new structure called Compressed Skypattern Cube (abbreviated CSKYC), which concisely represents a skypattern cube, and gives an efficient algorithm to compute it. We thoroughly explore its properties and provide an efficient query processing algorithm. Experimental results show that our proposal allows to construct and to query a CSKYC very efficiently.

Keywords: Skypatterns · Pareto-dominance relation · Skypattern cubes

1 Introduction

The notion of skyline queries [2] has been quite recently integrated into the pattern discovery paradigm to mine skyline patterns (henceforth called *skypatterns*) [11,15]. Given a set of measures, skypatterns are based on a Pareto-dominance relation, which means that no measure can be improved without degrading the others. As an example, a user may prefer patterns with a high frequency, large size and a high confidence. Then a pattern $\mathtt{x_i}$ dominates another pattern $\mathtt{x_j}$ if $\mathtt{freq(x_j)} \geq \mathtt{freq(x_i)}$, $\mathtt{size(x_j)} \geq \mathtt{size(x_i)}$, $\mathtt{conf(x_j)} \geq \mathtt{conf(x_i)}$ where at least one strict inequality holds. The skypattern set contains the patterns that are not dominated by any other pattern. Skypatterns are highly interesting since they do not require thresholds for the measures and the dominance

F. Wotawa et al. (Eds.): IEA/AIE 2019, LNAI 11606, pp. 406–421, 2019.
https://doi.org/10.1007/978-3-030-22999-3_36

relation gives them global interestingness with a semantics easily understood by the user.

In practice, users do not know the exact role of each measure and it is difficult to select beforehand the most appropriate subset of measures. Users would like to keep all potentially useful measures, look at what happens on skypattern sets when removing or adding a measure, thus evaluating the impact of measures, and then converge to a convenient skypattern set.

This issue has been first addressed with the notion of a Skypattern Cube [13], which is the lattice of all possible subsets of measures associated with their skypattern sets. More formally, given a set M of n measures, the $2^n - 1$ possible non-empty skypattern subsets should be precomputed to efficiently handle various queries of users. By comparing two neighboring nodes (differentiated by only one measure), users can observe new skypatterns and the ones which disappear, greatly helping to better understand the role of the measures. To sum up, the cube is a structure that enables to discover the most interesting skypattern sets. The skypattern cube has been exploited in various domains such as bioinformatics [10] and mutagenicity [13]. However there are $2^n - 1$ possible non-empty skypattern sets with high redundancy coming from derivations of skypatterns among subspaces of the cube [13].

In this paper, we propose a new structure called the *Compressed Skypattern Cube* (denoted CSKYC). Each subspace stores skypatterns (called *proper skypatterns*) that do not appear in its descendant ones and the compressed skypattern cube contains only non-empty subspaces. Compared to the original skypattern cube [13], the CSKYC has fewer duplicates among subspaces, and does not need to store all of them. Moreover the cube includes unbalanced skypatterns. For instance, let $M = \{\texttt{freq}, \texttt{size}\}$ and three patterns $\texttt{x}_\texttt{i}$, $\texttt{x}_\texttt{j}$ and $\texttt{x}_\texttt{k}$ such that $\texttt{freq}(\texttt{x}_\texttt{i}) = 10$, $\texttt{size}(\texttt{x}_\texttt{i}) = 1$, $\texttt{freq}(\texttt{x}_\texttt{j}) = 2$, $\texttt{size}(\texttt{x}_\texttt{j}) = 8$, $\texttt{freq}(\texttt{x}_\texttt{k}) = 4$ and $\texttt{size}(\texttt{x}_\texttt{k}) = 5$. Clearly, $\texttt{x}_\texttt{i}$ (resp. $\texttt{x}_\texttt{j}$) is a skypattern for $\{\texttt{freq}\}$ (resp. $\{\texttt{size}\}$), thus $\texttt{x}_\texttt{i}$ (resp. $\texttt{x}_\texttt{j}$) will be instantly a skypattern for M being derived from $\{\texttt{freq}\}$ (resp. $\{\texttt{size}\}$). However, $\texttt{x}_\texttt{k}$ is also a skypattern for M, being more balanced over measures than $\texttt{x}_\texttt{i}$ (resp. $\texttt{x}_\texttt{j}$) which only has an extreme value for $\{\texttt{freq}\}$ (resp. $\{\texttt{size}\}$). Proper skypatterns are often well-balanced skypatterns.

Contributions Overview. We thoroughly explore interesting properties of the compressed skypattern cube and provide an efficient query processing algorithm. Our contributions can be summarized as follows: (i) we provide the summarization structure CSKYC which concisely represents the whole skypattern cube and preserves its essential information. (ii) We propose a bottom-up approach to construct the CSKYC. (iii) We show how this structure can be used efficiently for query processing. Finally, (iv) we present an extensive set of experiments showing the advantages of our proposals.

Paper Organization. The rest of this paper is organized as follows. Section 2 recalls the definitions of the notions used in this paper. Section 3 first introduces the CSKYC, provides algorithms to build it, and shows how the CSKYC can

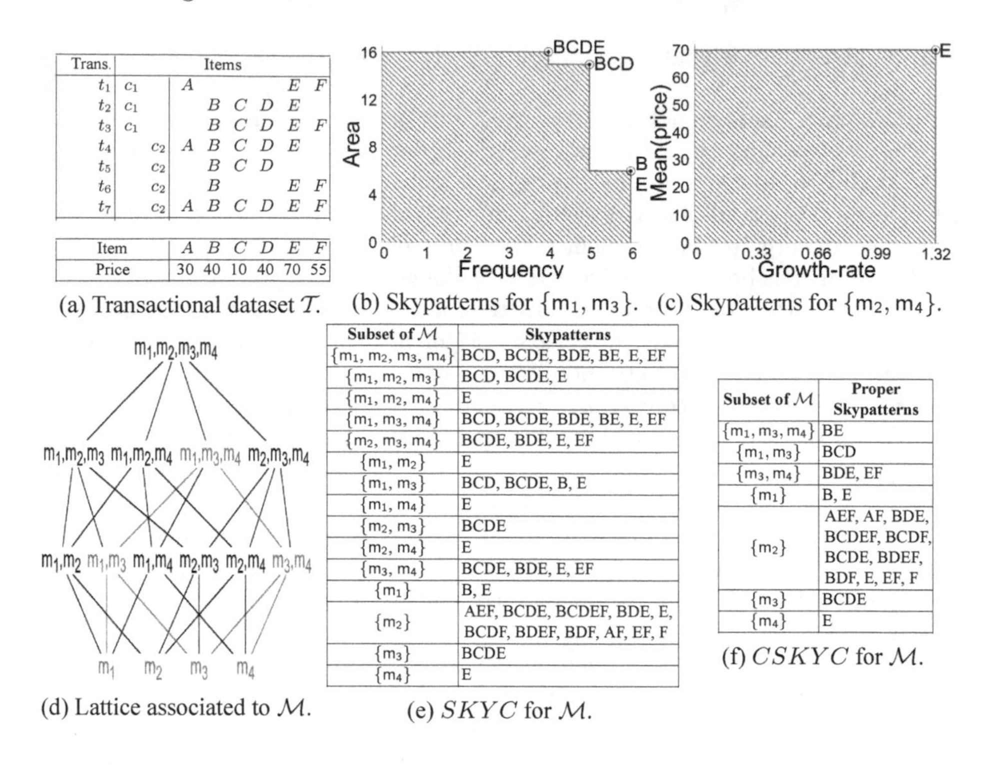

Trans.			Items
t_1	c_1		A E F
t_2	c_1		B C D E
t_3	c_1		B C D E F
t_4		c_2	A B C D E
t_5		c_2	B C D
t_6		c_2	B E F
t_7		c_2	A B C D E F

Item	A	B	C	D	E	F
Price	30	40	10	40	70	55

Subset of $\mathcal{M}$	Skypatterns
$\{m_1, m_2, m_3, m_4\}$	BCD, BCDE, BDE, BE, E, EF
$\{m_1, m_2, m_3\}$	BCD, BCDE, E
$\{m_1, m_2, m_4\}$	E
$\{m_1, m_3, m_4\}$	BCD, BCDE, BDE, BE, E, EF
$\{m_2, m_3, m_4\}$	BCDE, BDE, E, EF
$\{m_1, m_2\}$	E
$\{m_1, m_3\}$	BCD, BCDE, B, E
$\{m_1, m_4\}$	E
$\{m_2, m_3\}$	BCDE
$\{m_2, m_4\}$	E
$\{m_3, m_4\}$	BCDE, BDE, E, EF
$\{m_1\}$	B, E
$\{m_2\}$	AEF, BCDE, BCDEF, BDE, E, BCDF, BDEF, BDF, AF, EF, F
$\{m_3\}$	BCDE
$\{m_4\}$	E

Subset of $\mathcal{M}$	Proper Skypatterns
$\{m_1, m_3, m_4\}$	BE
$\{m_1, m_3\}$	BCD
$\{m_3, m_4\}$	BDE, EF
$\{m_1\}$	B, E
$\{m_2\}$	AEF, AF, BDE, BCDEF, BCDF, BCDE, BDEF, BDF, E, EF, F
$\{m_3\}$	BCDE
$\{m_4\}$	E

(a) Transactional dataset $\mathcal{T}$. (b) Skypatterns for $\{m_1, m_3\}$. (c) Skypatterns for $\{m_2, m_4\}$. (d) Lattice associated to $\mathcal{M}$. (e) $SKYC$ for $\mathcal{M}$. (f) $CSKYC$ for $\mathcal{M}$.

Fig. 1. Running example: $\mathcal{M} = \{m_1{:}\mathtt{freq}(x), m_2{:}\mathtt{gr}_1(x), m_3{:}\mathtt{area}(x), m_4{:}\mathtt{mean}(x.\mathsf{price})\}$. (Color figure online)

handle various skypattern queries. Section 4 is devoted to related works. Finally, Sect. 5 shows our experimental results and Sect. 6 concludes.

2 Preliminaries

Let $\mathcal{I}$ be a set of distinct literals called *items*. A pattern (or itemset) is a non-empty subset of $\mathcal{I}$. The language of patterns corresponds to $\mathcal{L}_\mathcal{I} = 2^\mathcal{I} \setminus \emptyset$. A transactional dataset $\mathcal{T}$ is a multiset of patterns in $\mathcal{L}_\mathcal{I}$. The traditional example is a supermarket database in which, for each transaction t_i, every item in a transaction is a product bought by the customer i. Table 1 summarizes the different notations used throughout the paper.

Example 1. Figure 1a depicts a transactional dataset $\mathcal{T}$ where items in a transaction t_i are denoted $A, \ldots, F$. It serves as example throughout the paper. An attribute (*price*) is associated to each item. For instance, the *Price* of A is \$30. The dataset is partitioned into two classes, class c_1 for clients with loyalty programs and class c_2 for other clients.

Constraint-based pattern mining aims at extracting all patterns $x \in \mathcal{L}_\mathcal{I}$ satisfying a query $q(x)$ which is usually called *theory* [7]: $Th(\mathcal{L}_\mathcal{I}, q) =$

Table 1. Notations.

Symbol	Definition
$\mathcal{T}$	Transactional dataset
$\mathcal{I}$	Set of items
$\mathcal{L}_\mathcal{I} = 2^\mathcal{I} \setminus \emptyset$	Language of patterns
$\mathcal{M}$	Set of measures
$\mathrm{U} \subseteq \mathcal{M}$	Subspace U (i.e. subset of measures of $\mathcal{M}$)
$Sky(\mathcal{L}_\mathcal{I}, \mathrm{U})$	Skypatterns set over $\mathcal{L}_\mathcal{I}$ for U
$\boldsymbol{P}$-$Sky(\mathcal{L}_\mathcal{I}, \mathrm{U})$	Proper Skypattern set over $\mathcal{L}_\mathcal{I}$ for U
ℓ-$Sky(\mathcal{L}_\mathcal{I}, \mathrm{U})$	Large Skypattern set over $\mathcal{L}_\mathcal{I}$ for U
$Desc(\mathcal{L}_\mathcal{I}, \mathrm{U})$	Union of all proper skypattern sets for all descendant subspaces $\mathrm{V} \subset \mathrm{U}$
$SKYC(\mathcal{L}_\mathcal{I}, \mathcal{M})$	Skypattern Cube over $\mathcal{L}_\mathcal{I}$ for $\mathcal{M}$
$CSKYC(\mathcal{L}_\mathcal{I}, \mathcal{M})$	Compressed Skypattern Cube over $\mathcal{L}_\mathcal{I}$ for $\mathcal{M}$
$P \subseteq \mathcal{L}_\mathcal{I}$	Set of patterns
$\mathcal{O}(P)$	Set of data points associated to P
$Skyline(\mathcal{O}(P), \mathrm{U})$	Set of skyline points on $\mathcal{O}(P)$ for U

$\{\mathtt{x} \in \mathcal{L}_\mathcal{I} \mid q(\mathtt{x})\ \textit{is true}\}$. A common example is the minimal frequency constraint ($\mathtt{freq}(\mathtt{x}) \geq \theta$) which provides patterns having a number of occurrences exceeding a given minimal threshold θ. Many other measures for patterns can be considered such as:

- $\mathtt{size}(\mathtt{x}) = |\mathtt{x}|$ is the number of items that x contains.
- $\mathtt{gr}_1(\mathtt{x}) = \frac{(|\mathcal{T}|-|\mathcal{T}_1|)\times \mathtt{freq}_1(\mathtt{x})}{|\mathcal{T}_1| \times (\mathtt{freq}(\mathtt{x}) - \mathtt{freq}_1(\mathtt{x}))}$ where $\mathcal{T}_1$ is a sub-dataset (i.e a class partition) on $\mathcal{T}$.
- $\mathtt{min}(\mathtt{x}.att) = \min_{i \in \mathtt{x}} \{i.att\}$ (resp. $\mathtt{max}(\mathtt{x}.att) = \max_{i \in \mathtt{x}} \{i.att\}$) is the lowest (resp. highest) among item values of x for attribute att.
- $\mathtt{mean}(\mathtt{x}.att) = (\mathtt{min}(\mathtt{x}.att) + \mathtt{max}(\mathtt{x}.att))/2$.

Example 2. $\mathtt{freq}(BC) = 5$, $\mathtt{mean}(BCD.price) = 25$, ...

Skypatterns allow to express a user-preference according to a dominance relation [11].

Definition 1 (Pareto-dominance). *A pattern* x *dominates another pattern* y *w.r.t a measure subset (a.k.a subspace) U, noted by* $\mathtt{x} \succ_U \mathtt{y}$*, iff* $\forall \mathsf{m_i} \in U, \mathsf{m_i}(\mathtt{x}) \geq \mathsf{m_i}(\mathtt{y})$ *and* $\exists \mathsf{m_j} \in U, \mathsf{m_j}(\mathtt{x}) > \mathsf{m_j}(\mathtt{y})$.

Example 3. For $\mathrm{U} = \{\mathsf{m_1}{:}\mathtt{freq}(\mathtt{x}), \mathsf{m_3}{:}\mathtt{area}(\mathtt{x})\}$, pattern BCD dominates pattern BC since $\mathtt{freq}(BCD) = \mathtt{freq}(BC) = 5$ and $\mathtt{area}(BCD) > \mathtt{area}(BC)$.

The Skypattern Operator [11] extracts the skypattern set w.r.t a subspace U.

Definition 2 (Skypattern Operator). *A pattern is a skypattern w.r.t a subspace U iff it is not dominated by any other pattern w.r.t U. The Skypattern Operator returns all the skypatterns w.r.t U:* $Sky(\mathcal{L}_\mathcal{I}, U) = \{x \in \mathcal{L}_\mathcal{I} \mid \nexists\, y \in \mathcal{L}_\mathcal{I}, y \succ_U x\}$.

Example 4. Consider the dataset of Fig. 1a, users may ask for the skypatterns for every combination of the measures $\{m_1, m_2, m_3, m_4\}$. Figures 1b and c depict skypatterns for $\{m_1, m_3\}$ and $\{m_2, m_4\}$ respectively.

As said above, users may query multiple skypattern sets for different subspaces. Furthermore, for a space $\mathcal{M}$ there are $2^{|\mathcal{M}|} - 1$ different skypattern sets. The Skypattern Cube [13] retrieves all skypattern sets for any subspace.

Definition 3 (Skypattern Cube). *Given a set of measures* $\mathcal{M}$*, the Skypattern Cube of* $\mathcal{M}$ *is defined as* $SKYC(\mathcal{L}_\mathcal{I}, \mathcal{M}) = \{(U, Sky(\mathcal{L}_\mathcal{I}, U)) \mid U\} \subseteq \mathcal{M}, U \neq \emptyset$

Example 5. Figure 1d depicts the lattice associated to $\mathcal{M}$ (power set of $\mathcal{M}$: $2^{\mathcal{M}} \setminus \emptyset$). Figure 1e associates to each non-empty subset of $\mathcal{M}$ its skypattern set.

For computing the skypattern cube, [13] has proposed a bottom-up approach using two derivation rules that provide an easy way to automatically infer a large proportion of the skypatterns of a parent node from the skypattern sets of its child nodes without any dominance test (if k measures are associated to a parent node, its child nodes are the nodes defined by the $\binom{k}{k-1}$ subsets of $k-1$ measures).

3 Contributions

This section introduces the CSKYC (Compressed Skypattern Cubes) which concisely represents the entire skypattern cube. The main idea is, for every subspace, to only store its *proper skypatterns*. More precisely, a skypattern x for a subspace U is stored iff $x \in Sky(\mathcal{L}_\mathcal{I}, U)$ and there exists no $V \subset U$ s.t. $x \in Sky(\mathcal{L}_\mathcal{I}, V)$. First, we introduce the CSKYC of a set of measures. Then, we propose a bottom-up approach for building such a CSKYC. Finally, we show how to efficiently query the whole skypattern set for U from the CSKYC.

3.1 The Compressed Skypattern Cube

Definition 4 (Proper Skypattern). *The set of proper skypatterns for a subspace U is the subset of skypatterns on U which are not skypatterns in any subset of U:*

$$\boldsymbol{P}\text{-}Sky(\mathcal{L}_\mathcal{I}, U) = \{x \in Sky(\mathcal{L}_\mathcal{I}, U) \mid \nexists V \subset U, x \in Sky(\mathcal{L}_\mathcal{I}, V)\}$$

Example 6. Consider $U = \{m_1, m_3, m_4\}$ and $U' = \{m_3, m_4\}$. Table 2a shows the measure values for skypatterns for U and U′. Figure 2b illustrates how proper skypattern *BE* (in red) (resp. *BDE* (in light-blue)) is more balanced than other skypatterns for U (resp. U′), having fewer extreme values than the other skypatterns.

Pattern	freq	area	mean(x.*price*)
BCD	5	15	25.00
BCDE	4	16	40.00
E	6	6	70.00
EF	4	8	62.50
BDE	4	12	55.00
BE	5	10	55.00

$Sky(\mathcal{L}_\mathcal{I}, \text{U})$ $Sky(\mathcal{L}_\mathcal{I}, \text{U}')$

(a) Skypatterns for U and U′.

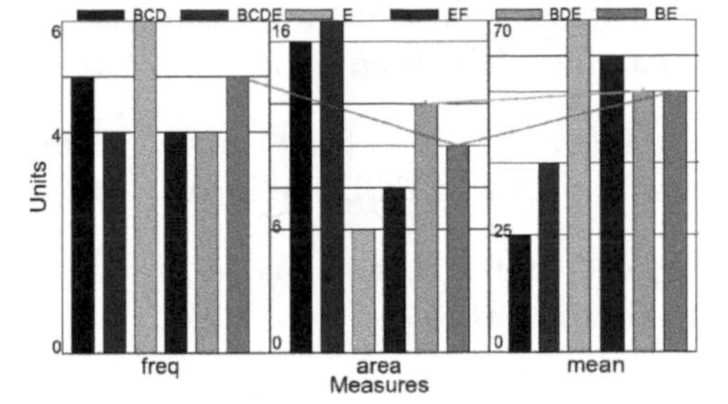

(b) Measure values of skypatterns.

Fig. 2. Example of (Proper) Skypatterns for $\text{U} = \{m_1, m_3, m_4\}$ and $\text{U}' = \{m_3, m_4\}$. (Color figure online)

Based on this notion, we define the compressed skypattern cube.

Definition 5 (Compressed Skypattern Cube). *Given a set of measures $\mathcal{M}$, the compressed skypattern cube of $\mathcal{M}$ is defined as*

$$CSKYC(\mathcal{L}_\mathcal{I}, \mathcal{M}) = \{(U, \boldsymbol{P}\text{-}Sky(\mathcal{L}_\mathcal{I}, U)) \mid U \subseteq \mathcal{M}, U \neq \emptyset, \boldsymbol{P}\text{-}Sky(\mathcal{L}_\mathcal{I}, U) \neq \emptyset\}$$

Example 7. For the dataset shown in Fig. 1a, the $CSKYC(\mathcal{L}_\mathcal{I}, \mathcal{M})$ is depicted in Fig. 1f and its sub-lattice (in red) in Fig. 1d. It contains only 6 non-empty subsets compared to 15 subsets in $SKYC(\mathcal{L}_\mathcal{I}, \mathcal{M})$. Clearly, the CSKYC is much more compact.

3.2 Computing the CSKYC

A first and naive way to get the CSKYC consists in first computing the skypattern cube, and then deriving the CSKYC by removing duplicates from their subspaces. Such an approach is inefficient as the number of subspaces to process is exponential. In this section, we provide a bottom-up algorithm (`CSKYC-BUC`) for building the CSKYC. Given a set of measures $\mathcal{M}$ of size d, the subspaces are organized into d levels, such that the subspaces of size i are in level i. We only keep *non-empty* subspaces (i.e. those containing proper skypatterns). All descendant skypatterns of a subspace are collected to form a large skypattern set ($\boldsymbol{\ell}$-Sky) which are then used as filters, and if no new skypattern is found, the subspace is discarded from the CSKYC.
Let us first give some preliminary definitions in order to compute the CSKYC.

Definition 6 (Indistinct/Incomparable Skypatterns). *Let* x, y *be two skypatterns w.r.t a subspace U: (i)* x, y *are indistinct, noted* $x =_U y$*, iff* $\forall m_i \in U, m_i(x) = m_i(y)$*; (ii)* x, y *are incomparable, noted* $x \prec\succ_U y$*, iff* $x \not\succ_U y$, $y \not\succ_U x$ *and* $x \neq_U y$.

Incomparable skypatterns and indistinct ones for U constitute partitions of $Sky(\mathcal{L}_\mathcal{I}, \text{U})$.

Definition 7 (Indistinct Subspace (IS)). *A subspace U is an* ***Indistinct Subspace (IS)*** *iff all patterns in* $Sky(\mathcal{L}_{\mathcal{I}}, U)$ *are indistinct from each other.*

Example 8. Let $U = \{m_1\}$, and $V = \{m_1, m_3\}$. B and E are indistinct w.r.t. U, while BCDE and BCD are incomparable w.r.t. V.

Lemma 1 states that skypatterns that are common to two different subspaces remain skypatterns in their union.

Lemma 1. $Sky(\mathcal{L}_{\mathcal{I}}, U) \cap Sky(\mathcal{L}_{\mathcal{I}}, V) \subseteq Sky(\mathcal{L}_{\mathcal{I}}, U \cup V)$

Proof (By contradiction). Assume that, for two subspaces U, V s.t. $W = U \cup V$, $\underbrace{\exists x \in Sky(\mathcal{L}_{\mathcal{I}}, U) \cap Sky(\mathcal{L}_{\mathcal{I}}, V)}_{(1)}$, but $\underbrace{x \notin Sky(\mathcal{L}_{\mathcal{I}}, W)}_{(2)}$. From (1): $\underbrace{x \in Sky(\mathcal{L}_{\mathcal{I}}, U)}_{(3)}$ and $\underbrace{x \in Sky(\mathcal{L}_{\mathcal{I}}, V)}_{(4)}$. From (2): $\exists y \in Sky(\mathcal{L}_{\mathcal{I}}, W), y \succ_W x \Rightarrow \underbrace{\forall m_i \in W, m_i(y) \geq m_i(x)}_{(5)}$.

$$\left.\begin{array}{l}\text{From (3): } y \not\succ_U x \Rightarrow \forall m_i \in U, m_i(y) \leq m_i(x). \text{ From (5): } x =_U y \\ \text{From (4): } y \not\succ_V x \Rightarrow \forall m_i \in V, m_i(y) \leq m_i(x). \text{ From (5): } x =_V y\end{array}\right\} x =_W y.$$

Thus, $x \in Sky(\mathcal{L}_{\mathcal{I}}, W)$ leading to a contradiction.

Based on Lemma 1, the following theorem enables us to characterize empty subspaces in the CSKYC, i.e. those without proper skypatterns.

Theorem 1 (Empty subspaces in CSKYC). *Given two subspaces U and V that are IS, if* $Sky(\mathcal{L}_{\mathcal{I}}, U) \cap Sky(\mathcal{L}_{\mathcal{I}}, V) \neq \emptyset$*, then* $U \cup V$ *is an IS and* $\boldsymbol{P}\text{-}Sky(\mathcal{L}_{\mathcal{I}}, U \cup V) = \emptyset$.

Proof (By contradiction). Let U, V two IS and $W = U \cup V$.

- Assume that W is not an IS and $\exists x, y \in Sky(\mathcal{T}, U) \cap Sky(\mathcal{T}, V)$. From Lemma 1: $x, y \in Sky(\mathcal{T}, W)$. Since W is not an IS, $x \prec\succ_W y$. As $x, y \in Sky(\mathcal{T}, U)$, $x, y \in Sky(\mathcal{T}, V)$ and U and V are IS, thus, $x =_U y$ and $x =_V y$. Thus, $x =_W y$ leading to a contradiction.
- Assume that $\exists x \in Sky(\mathcal{T}, U) \cap Sky(\mathcal{T}, V)$ and $\exists y \in \boldsymbol{P}\text{-}Sky(\mathcal{T}, W)$. From Lemma 1: $x \in Sky(\mathcal{T}, W)$. Since W is an IS, $x =_W y$. Thus, $x =_U y$ and $x =_V y$. So, $y \in Sky(\mathcal{T}, U)$ and $y \in Sky(\mathcal{T}, V)$. Thus, $y \notin \boldsymbol{P}\text{-}Sky(\mathcal{T}, W)$ leading to a contradiction.

Example 9. In Fig. 1f, $\boldsymbol{P}\text{-}Sky(\mathcal{L}_{\mathcal{I}}, \{m_1, m_4\}) = \emptyset$ as $Sky(\mathcal{L}_{\mathcal{I}}, \{m_1\}) \cap Sky(\mathcal{L}_{\mathcal{I}}, \{m_4\}) = \{E\}$.

The authors in [13] showed that incomparable skypatterns and some indistinct skypatterns of a child subspace remain also skypatterns in its parent subspace (they are referred to as derivable skypatterns). They also showed that a parent subspace can include non-derivable skypatterns (i.e., those that are not skypatterns in any of its child subspaces). Thus, one can collect the non-empty sets of descendants (which are proper skypatterns) of a subspace to form a large skypattern set and to use them as filters to detect a priori that no proper skypattern exist (see Corollary 1).

Definition 8 (Large Skypattern Set). *The Large Skypattern Set for a subspace U is the union of proper skypattern set of U with all proper skypattern sets of its descendant subspaces* $V \subset U$: $\boldsymbol{\ell}\text{-}Sky(\mathcal{L}_\mathcal{I}, U) = \boldsymbol{P}\text{-}Sky(\mathcal{L}_\mathcal{I}, U) \cup Desc(\mathcal{L}_\mathcal{I}, U)$, *where* $Desc(\mathcal{L}_\mathcal{I}, U) = \bigcup_{V \subset U} \boldsymbol{P}\text{-}Sky(\mathcal{L}_\mathcal{I}, V) = \bigcup_{V \subset U \wedge |V| = |U|-1} \boldsymbol{\ell}\text{-}Sky(\mathcal{L}_\mathcal{I}, V)$.

Example 10. For U = {m_1:**freq**(x), m_3:**area**(x), m_4:**mean**(x.price)}:

$\boldsymbol{\ell}\text{-}Sky(\mathcal{L}_\mathcal{I},)U = \boldsymbol{P}\text{-}Sky(\mathcal{L}_\mathcal{I},)U \cup \boldsymbol{\ell}\text{-}Sky(\mathcal{L}_\mathcal{I}, \{m_1, m_3\}) \cup \boldsymbol{\ell}\text{-}Sky(\mathcal{L}_\mathcal{I}, \{m_1, m_4\}) \cup \boldsymbol{\ell}\text{-}Sky(\mathcal{L}_\mathcal{I}, \{m_3, m_4\})$

$\boldsymbol{P}\text{-}Sky(\mathcal{L}_\mathcal{I},)U$	$\boldsymbol{\ell}\text{-}Sky(\mathcal{L}_\mathcal{I}, \{m_1, m_3\})$	$\boldsymbol{\ell}\text{-}Sky(\mathcal{L}_\mathcal{I}, \{m_1, m_4\})$	$\boldsymbol{\ell}\text{-}Sky(\mathcal{L}_\mathcal{I}, \{m_3, m_4\})$
$B, E, EF, BCD, BE, BDE, BCDE$	$\boldsymbol{P}\text{-}Sky(\mathcal{L}_\mathcal{I}, \{m_1, m_3\}) \cup \boldsymbol{P}\text{-}Sky(\mathcal{L}_\mathcal{I}, \{m_1\}) \cup \boldsymbol{P}\text{-}Sky(\mathcal{L}_\mathcal{I}, \{m_3\})$	$\boldsymbol{P}\text{-}Sky(\mathcal{L}_\mathcal{I}, \{m_1, m_4\}) \cup \boldsymbol{P}\text{-}Sky(\mathcal{L}_\mathcal{I}, \{m_1\}) \cup \boldsymbol{P}\text{-}Sky(\mathcal{L}_\mathcal{I}, \{m_4\})$	$\boldsymbol{P}\text{-}Sky(\mathcal{L}_\mathcal{I}, \{m_3, m_4\}) \cup \boldsymbol{P}\text{-}Sky(\mathcal{L}_\mathcal{I}, \{m_3\}) \cup \boldsymbol{P}\text{-}Sky(\mathcal{L}_\mathcal{I}, \{m_4\})$

Based on Definition 4, the proper skypatterns of any parent subspace can be computed thanks to the following corollary.

Corollary 1. $\boldsymbol{P}\text{-}Sky(\mathcal{L}_\mathcal{I}, U) = Sky(\mathcal{L}_\mathcal{I}, U) \setminus Desc(\mathcal{L}_\mathcal{I}, U)$.

To compute $\boldsymbol{P}\text{-}Sky(\mathcal{L}_\mathcal{I}, \mathrm{U})$, we first retrieve its descendants (which are proper skypatterns), then we seek for skypatterns that are not in $Desc(\mathcal{L}_\mathcal{I}, \mathrm{U})$. Algorithm 1 gives the pseudo-code of our bottom-up approach. It starts by computing $\boldsymbol{P}\text{-}Sky(\mathcal{L}_\mathcal{I}, m_i)$ for every $m_i \in \mathcal{M}$ (level 1) and then follows a level-wise strategy: from the lower level, each level of the lattice is constructed by applying Theorem 1 and, if needed, computing non-derivable skypatterns (cf. line 21). Two data structures, IS and $\boldsymbol{\ell}\text{-}Sky$ are also maintained during the construction process, storing for each subspace its large pattern set and its status. They allow an incremental computation of $\boldsymbol{\ell}\text{-}Sky$.

3.3 Querying $Sky(\mathcal{L}_\mathcal{I}, \mathrm{U})$ from $CSKYC(\mathcal{L}_\mathcal{I}, \mathcal{M})$

When a skypattern set for a given subspace U is queried, the CSKYC may not have a record for U; even if it does, the skypattern set that is stored for U is not complete. We propose a straightforward approach to query the complete skypattern set for U from $CSKYC(\mathcal{L}_\mathcal{I}, \mathcal{M})$.

Our approach is based on the fact that $Sky(\mathcal{L}_\mathcal{I}, \mathrm{U}) \subseteq \boldsymbol{\ell}\text{-}Sky(\mathcal{L}_\mathcal{I}, \mathrm{U})$ and proceeds in two steps (see Algorithm 2): first, approximating $Sky(\mathcal{L}_\mathcal{I}, \mathrm{U})$ by $\boldsymbol{\ell}\text{-}Sky(\mathcal{L}_\mathcal{I}, \mathrm{U})$, and then, performing domination tests to filter dominated patterns.

Algorithm 1. `CSKYC-BUC`: Bottom-up approach for computing CSKYC.

```
   Input: T: a dataset, M: a set of measures.
   Output: The Compressed Skypattern Cube w.r.t M.
 1 CSKYC ← ∅; IS[ ] ← ∅; ℓ-Sky[ ] ← ∅;                    // Initialization step
 2 foreach m_i ∈ M do
 3   | P-Sky ← CP+SKY(L_I, {m_i}) ;                        // Compute P-Sky(L_I, m_i)
 4   | ℓ-Sky[{m_i}] ← P-Sky;
 5   | CSKYC ← CSKYC ∪ {({m_i}, P-Sky)};
 6   | IS[{m_i}] ← true;
 7 for i ← 2 to |M| do
 8   | foreach U ⊆ M s.t. |U| = i do
 9   |   | P-Sky ← ComputeProperSky(U);                    // Compute P-Sky(L_I, U)
10   |   | if P-Sky ≠ ∅ then
11   |   |   | CSKYC ← CSKYC ∪ {(U, P-Sky)};
12 return CSKYC;
13 Function ComputeProperSky(U):
14   | children ← {V ⊂ U | |V| = |U| − 1};                  // Children of U
15   | childrenIS ← {W ∈ children | IS[W] = true};
16   | IS[U] ← false;
17   | if ∃ V, W ∈ childrenIS s.t. Sky[V] ∩ Sky[W] ≠ ∅ then  // Apply theorem 1
18   |   | P-Sky ← ∅;
19   |   | IS[U] ← true;
20   | Desc ← ⋃_{V∈children} ℓ-Sky[V];             // Generate the filter skypatterns
21   | if ¬IS[U] then
22   |   | P-Sky ← CP+SKY(L_I \ Desc, U);                   // Apply corollary 1
23   | ℓ-Sky[U] ← P-Sky ∪ Desc;                            // Update ℓ-Sky for U
24   | return P-Sky
```

(i) *Approximating* $Sky(\mathcal{L}_\mathcal{I}, \mathrm{U})$.
Based on Definition 8, we have that: $\forall\ \mathrm{U} \subseteq \mathcal{M}$, $Sky(\mathcal{L}_\mathcal{I}, \mathrm{U}) \subseteq \boldsymbol{\ell}\text{-}Sky(\mathcal{L}_\mathcal{I}, \mathrm{U})$. The proof is straightforward: $\forall\ \mathtt{x} \in Sky(\mathcal{L}_\mathcal{I}, \mathrm{U})$, either $\mathtt{x} \in \boldsymbol{P}\text{-}Sky(\mathcal{L}_\mathcal{I}, \mathrm{U})$, or $\exists \mathrm{V} \subset \mathrm{U}$ s.t. $\mathtt{x} \in \boldsymbol{P}\text{-}Sky(\mathcal{L}_\mathcal{I}, \mathrm{V})$.

(ii) *Filtering dominated skypatterns.*
To remove dominated skypatterns, we convert the problem into skyline mining operation in $|\mathrm{U}|$ dimensions to process it more efficiently. Let f be a mapping function from a set of patterns $P \subseteq \mathcal{L}_\mathcal{I}$ to $\mathbb{R}^n$ that associates, to each pattern $\mathtt{x}_i \in P$, a data point $f(\mathtt{x}_i) \in \mathbb{R}^n$ with coordinates $(\mathsf{m}_1(\mathtt{x}_i) = v_{i,1}, \ldots, \mathsf{m}_\mathsf{n}(\mathtt{x}_i) = v_{i,n})$. Let us note by $\mathcal{O}(P) = \{f(\mathtt{x}) \mid \mathtt{x} \in P\}$ the set of data points associated to P (see Table 2) and $Skyline(\mathcal{O}(P), \mathrm{U})$ be the set of skyline points of $\mathcal{O}(P)$ w.r.t. U. Thus, $\forall \mathrm{U} \subseteq \mathcal{M}$, $Sky(P, \mathrm{U}) = Skyline(\mathcal{O}(P), \mathrm{U})$. So, applying the skyline operator on $\mathcal{O}(P)$ provides the skypattern set.

Table 2. The multidimensional view for a set of patterns $P \subseteq \mathcal{L}_\mathcal{I}$ w.r.t. a subspace U ($|\text{U}| = \text{n}$).

	Pattern	m_1	m_2	$\cdots$	m_n
$P \subseteq \mathcal{L}_\mathcal{I}$	x_1	$v_{1,1}$	$v_{1,2}$	$\cdots$	$v_{1,n}$
	x_2	$v_{2,1}$	$v_{2,2}$	$\cdots$	$v_{2,n}$
	$\vdots$	$\vdots$	$\vdots$	$\vdots$	$\vdots$
	x_{p-1}	$v_{p-1,1}$	$v_{p-1,2}$	$\cdots$	$v_{p-1,n}$
	x_p	$v_{p,1}$	$v_{p,2}$	$\cdots$	$v_{p,n}$

Algorithm 2. Querying $Sky(\mathcal{L}_\mathcal{I}, \text{U})$ from $CSKYC(\mathcal{L}_\mathcal{I}, \mathcal{M})$

Input: U: a subspace and $CSKYC$: the compressed skyppatern cube w.r.t. $\mathcal{M}$.
Output: $Sky(\mathcal{L}_\mathcal{I}, \text{U})$

1 $\ell\text{-}Sky \leftarrow \bigcup_{\text{V} \subseteq \text{U}} \boldsymbol{P}\text{-}Sky(\mathcal{L}_\mathcal{I}, \text{V})$; // Approximating $Sky(\mathcal{L}_\mathcal{I}, \text{U})$
2 $Sky \leftarrow \texttt{BNL}(\mathcal{O}(\ell\text{-}Sky), \text{U})$; // Filtering dominated skypatterns
3 **return** Sky, $\ell\text{-}Sky$

Theorem 2. $Sky(\mathcal{L}_\mathcal{I}, U) = Skyline(\mathcal{O}(\ell\text{-}Sky(\mathcal{L}_\mathcal{I}, U)), U)$.

Proof. Given a subspace U, we prove the two implications:
($\Rightarrow$) Assume that $\exists \text{x} \in Sky(\mathcal{L}_\mathcal{I}, \text{U})$: $\forall \text{y} \in \mathcal{L}_\mathcal{I}, \text{y} \not\succ_\text{U} \text{x}$. So, $\forall \text{y} \in \ell\text{-}Sky(\mathcal{L}_\mathcal{I}, \text{U}), \text{y} \not\succ_\text{U} \text{x}$ therefore $\text{x} \in Sky(\ell\text{-}Sky(\mathcal{L}_\mathcal{I}, \text{U}), \text{U})$. Thus, $f(\text{x}) \in Skyline(\mathcal{O}(\ell\text{-}Sky(\mathcal{L}_\mathcal{I}, \text{U})), \text{U})$.
($\Leftarrow$) Assume that $\exists f(\text{x}) \in Skyline(\mathcal{O}(\ell\text{-}Sky(\mathcal{L}_\mathcal{I}, \text{U})), \text{U})$. So, $\forall \text{y}_1 \in \ell\text{-}Sky(\mathcal{L}_\mathcal{I}, \text{U})$, $\text{y}_1 \not\succ_\text{U} \text{x}$. From Definition 8: $\forall \text{y}_2 \in \mathcal{L}_\mathcal{I} \setminus \ell\text{-}Sky(\mathcal{L}_\mathcal{I}, \text{U}), \exists \text{y}_3 \in Sky(\mathcal{L}_\mathcal{I}, \text{U}), \text{y}_3 \succ_\text{U} \text{y}_2$. Thus, $\text{y}_2 \not\succ_\text{U} \text{x}$. Therefore, $\text{x} \in Sky(\mathcal{L}_\mathcal{I}, \text{U})$.

The second step is performed using a skyline algorithm based on the BNL approach [2].

4 Related Work

Skylines vs Skypatterns. The notion of skyline queries [2,3] has been recently integrated into pattern discovery to mine skypatterns [11]. Even if these notions seem similar, they correspond to very different extraction tasks. Skyline queries focus on the extraction of dominant tuples of a (point) database (T). The points (objects) are known in advance and then dominance test are applied. The skypattern mining task requires to mine patterns from a dataset (T) that must be Pareto-dominant for a given set of measures. Therefore, the latter problem is much harder since the search space for skypatterns is much larger than the search space for skylines: $O(\mathcal{L}_\mathcal{I} = 2^{|\mathcal{I}|})$ instead of $O(|T|)$. Two methods have been designed for mining skypatterns: Aetheris [11] is a two-step method that benefits from theoretical relationships between condensed representations and skypatterns, while CP+Sky [15] mines skypatterns using dynamic CSPs. Finally, [12] provides a point-to-point comparison between these two approaches.

Skyline Cube vs Skypattern Cube. To offer the best possible response time for a subspace skyline query, skyline cubes (a.k.a SkyCube) were introduced independently by [9,17]. They proposed several strategies to share skyline computation in different subspaces. Pei et al. proposed in [8] *Stellar*, which computes seed skylines groups in the full space, then extend it to build the final set of skyline groups and thus avoid the computation of skylines in all the subspaces. Similarly to the notions of skyline/skypattern, the skyline cube differs from the skypattern cube. A SkyCube tackles a point database looking for skyline point sets for a given set of dimensions. The skypattern cube computation has to deal with all the skypattern sets for a given set of measures. As seen in the previous paragraph, even if these notions are close, computing the skypattern cubes is much harder due to the huge search space. Two methods have been proposed to compute the skypattern cube. The first method, called as CP+SKY+CUBE [13], is based on a bottom-up approach and derivation rules exploiting the relation between the nodes in the lattice. The second method [14] proposes an approximation of the skypattern cube and then applies skyline cube mining in $|\mathcal{M}|$ dimensions on that approximation.

The Compressed Skyline Cube. Probably the closest previous work to our proposal is the so called *compressed skycube* (CSC) [16]. Its compression technique consists in storing for every subspace its partial skyline. It also supports concurrent subspace skyline queries in frequent updated databases. Our CSKYC can be seen as a reshaping of the CSC. However, the compressed skypattern cube computation is much harder due to the huge search space. Indeed, as shown previously, we need to extract patterns from a transactional dataset ($\mathcal{T}$) in order to determinate *proper* skypatterns for a given subspace. Other skycube summarization techniques have also been introduced. For instance, [1] proposed Hashcube, a structure based on bit-strings for storing the whole skycube. The work described in [6] proposed the negative skycube that returns subspaces where objects are not skylines.

5 Experimental Evaluation

This section evaluates constructing and querying the CSKYC on a real-life dataset and benchmarks. We compare the performances of the CSKYC with those of the original SKYC in terms of running-time and space storage, followed by query performance using CSKYC. The implementation of the different algorithms were carried out in C++. All experiments were conducted on a PC running Linux with a core i3 processor at 2.13 GHz.

5.1 Compressed Skypattern Cubes for Mutagenicity Dataset

We performed experiments on a real-life dataset of large size extracted from mutagenicity data [5] (a major problem in risk assessment of chemicals). This dataset has $|\mathcal{T}|$=6,512 transactions encoding chemicals and $|\mathcal{I}| = 1,073$ items encoding frequent closed subgraphs previously extracted from $\mathcal{T}$ with a 2% relative frequency threshold. Chemists use up to $|\mathcal{M}| = 11$ measures, five of them are typically used in contrast mining (e.g. growth rate) and allow to express different kinds of background knowledge. The other six measures are related to topological and chemical properties of the chemicals.

Space Analysis. Figure 3a shows the storage comparison of CSKYC to skypattern cubes of different dimensionality. Column 1 corresponds to the number of measures. Columns 2 and 3 report the total number of skypatterns for SKYC and CSKYC respectively. Column 4 gives their ratio. Columns 5 and 6 report the total number of subspaces for SKYC and CSKYC respectively. Column 7 gives their ratio. For each $|\mathcal{M}| = k$, reported values in columns (2), (3), (5) and (6) represent the averages over all $\binom{11}{k}$ possible skypattern cubes. Overall, CSKYC achieves the best compression of the skypattern sets. The effect of duplicate elimination is greatly amplified for $|\mathcal{M}| \geq 6$. CSKYCachieves up to 20.6$\times$ compression (in number of skypatterns) and permits using $4-7\times$ fewer subspaces. For $|\mathcal{M}| = 11$, the total number of proper skypatterns is 3,853, while for SKYC the total number of skypatterns is 87,374. This lead to a substantial gain greater than 95%.

CPU-Time Analysis. We compare our approach (`CSKYC-BUC`) with two methods: (i) a base-line method (`BL-CSKYC`) for computing CSKYC, and (ii) `CP+SKY+CUBE` proposed in [13] for computing SKYC. `BL-CSKYC` follows a bottom-up strategy: from the lower level, for each level and each subspace of the lattice, we compute its skypatterns, collect the skypatterns of its descendant subspaces, and then we remove all the duplicates. Figure 3b shows the performance of the three methods according to the number of measures $|M|$. The scale is logarithmic. For `CSKYC-BUC` (resp. `CP+SKY+CUBE`)) and for $|M| = k$, the reported CPU-time is the average of CPU-times over all $\binom{11}{k}$ possible CSKYC (resp. SKYC). As we can see, `CSKYC-BUC` clearly outperforms `BL-CSKYC` by several orders of magnitude. This is particularly obvious for higher values of $|M|$ due to the reduced number of skypatterns involved in the construction. For $(2 \leq |\mathcal{M}| \leq 5)$, the average speed-up is 37.3. For $|\mathcal{M}| = 8$, there is an order of magnitude (speed-up value 213.15). For $|\mathcal{M}| = 11$, the speed-up value reaches 949. Finally, `CSKYC-BUC` is an average 2x faster than `CP+SKY+CUBE` for building the CSKYC.

Querying CSKYC. Evaluating the query performance of Algorithm 2, for $|M| = k$, is performed by dividing the total time to sequentially query every subspace by 2^k-1. Each query is extracted from the CSKYC. The reported CPU-time in Fig. 3c are the averages of CPU-times over all $\binom{11}{k}$ possible CSKYCs. By comparing the CPU-times for the two steps of Algorithm 2, overall, the `BNL`

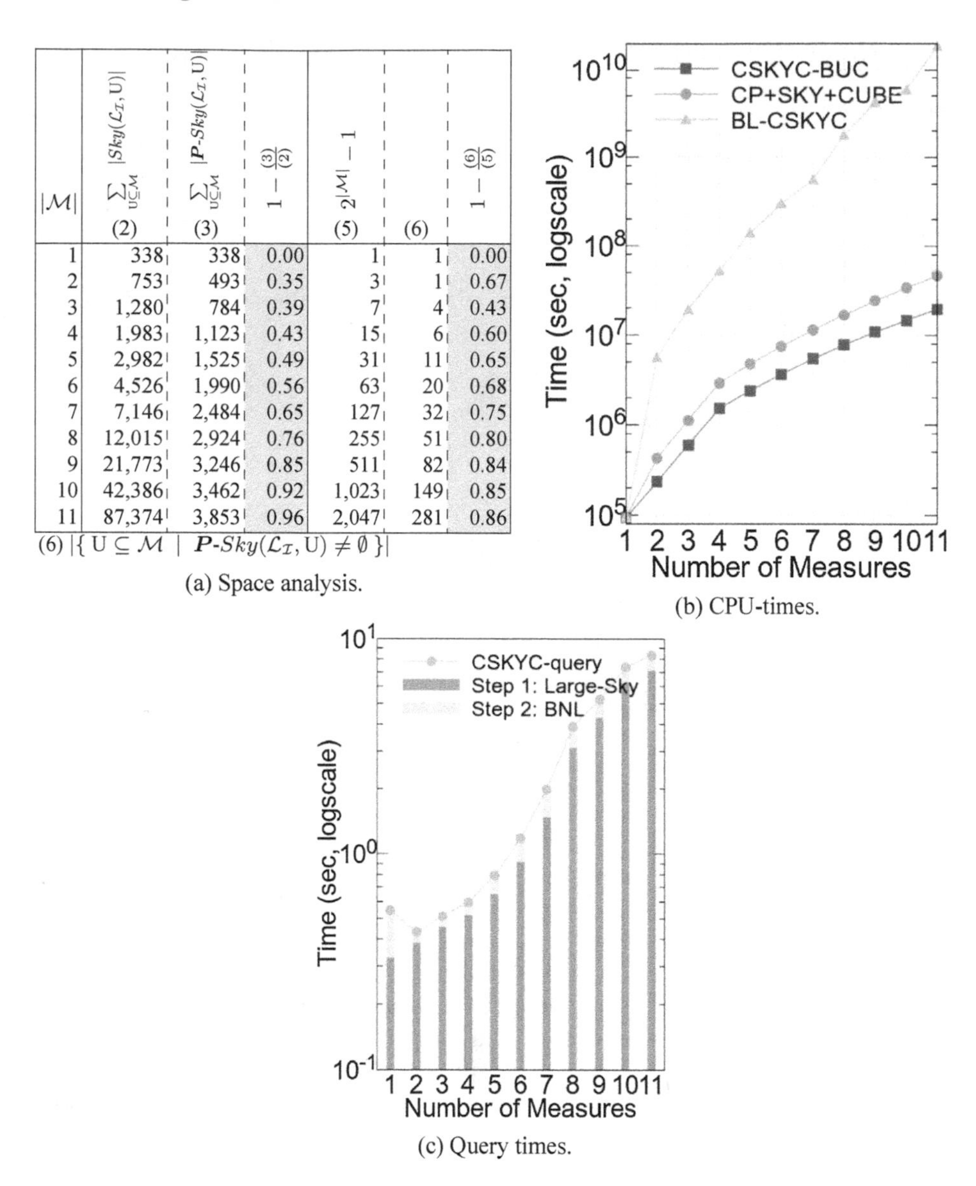

$\lvert\mathcal{M}\rvert$	$\sum_{U\subseteq\mathcal{M}} \lvert Sky(\mathcal{L}_\mathcal{I}, U)\rvert$ (2)	$\sum_{U\subseteq\mathcal{M}} \lvert \boldsymbol{P}\text{-}Sky(\mathcal{L}_\mathcal{I}, U)\rvert$ (3)	$1-\frac{(3)}{(2)}$	$2^{\lvert\mathcal{M}\rvert}-1$ (5)	(6)	$1-\frac{(6)}{(5)}$
1	338	338	0.00	1	1	0.00
2	753	493	0.35	3	1	0.67
3	1,280	784	0.39	7	4	0.43
4	1,983	1,123	0.43	15	6	0.60
5	2,982	1,525	0.49	31	11	0.65
6	4,526	1,990	0.56	63	20	0.68
7	7,146	2,484	0.65	127	32	0.75
8	12,015	2,924	0.76	255	51	0.80
9	21,773	3,246	0.85	511	82	0.84
10	42,386	3,462	0.92	1,023	149	0.85
11	87,374	3,853	0.96	2,047	281	0.86

(6) $\lvert\{\, U \subseteq \mathcal{M} \mid \boldsymbol{P}\text{-}Sky(\mathcal{L}_\mathcal{I}, U) \neq \emptyset \,\}\rvert$

(a) Space analysis.

(b) CPU-times.

(c) Query times.

Fig. 3. Results on Mutagenicity Dataset with $\lvert\mathcal{M}\rvert = 11$.

step is negligible as compared to the fist step. The scale is logarithmic. Experimental results show that query processing of the CSKYC is fast (less than 10 s for $|M| = 11$).

5.2 Compressed Skypattern Cubes for UCI Datasets

Experiments were carried out on 15 datasets from UCI benchmarks [4]. We considered 5 measures $M = \{\texttt{freq}, \texttt{max}, \texttt{area}, \texttt{mean}, \texttt{gr}_1\}$. In order to use mea-

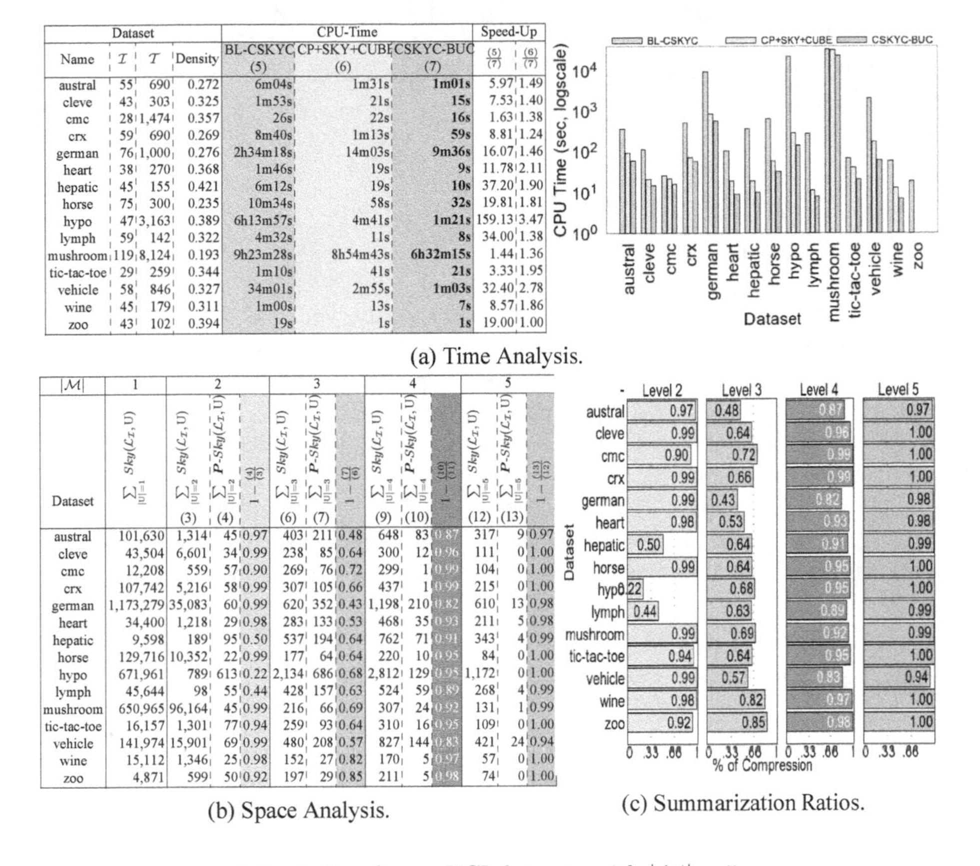

Dataset				CPU-Time			Speed-Up	
Name	$\mathcal{I}$	$\mathcal{T}$	Density	BL-CSKYC (5)	CP+SKY+CUBE (6)	CSKYC-BUC (7)	$\frac{(5)}{(7)}$	$\frac{(6)}{(7)}$
austral	55	690	0.272	6m04s	1m31s	**1m01s**	5.97	1.49
cleve	43	303	0.325	1m53s	21s	**15s**	7.53	1.40
cmc	28	1,474	0.357	26s	22s	**16s**	1.63	1.38
crx	59	690	0.269	8m40s	1m13s	**59s**	8.81	1.24
german	76	1,000	0.276	2h34m18s	14m03s	**9m36s**	16.07	1.46
heart	38	270	0.368	1m46s	19s	**9s**	11.78	2.11
hepatic	45	155	0.421	6m12s	19s	**10s**	37.20	1.90
horse	75	300	0.235	10m34s	58s	**32s**	19.81	1.81
hypo	47	3,163	0.389	6h13m57s	4m41s	**1m21s**	159.13	3.47
lymph	59	142	0.322	4m32s	11s	**8s**	34.00	1.38
mushroom	119	8,124	0.193	9h23m28s	8h54m43s	**6h32m15s**	1.44	1.36
tic-tac-toe	29	259	0.344	1m10s	41s	**21s**	3.33	1.95
vehicle	58	846	0.327	34m01s	2m55s	**1m03s**	32.40	2.78
wine	45	179	0.311	1m00s	13s	**7s**	8.57	1.86
zoo	43	102	0.394	19s	1s	**1s**	19.00	1.00

(a) Time Analysis.

$\|\mathcal{M}\|$	1	2			3			4			5		
Dataset	$\sum_{\|U\|=1} Sky(\mathcal{L}_\mathcal{I}, U)$	$\sum_{\|U\|=2} Sky(\mathcal{L}_\mathcal{I}, U)$ (3)	$\sum_{\|U\|=2} P\text{-}Sky(\mathcal{L}_\mathcal{I}, U)$ (4)	$1-\frac{(4)}{(3)}$	$\sum_{\|U\|=3} Sky(\mathcal{L}_\mathcal{I}, U)$ (6)	$\sum_{\|U\|=3} P\text{-}Sky(\mathcal{L}_\mathcal{I}, U)$ (7)	$1-\frac{(7)}{(6)}$	$\sum_{\|U\|=4} Sky(\mathcal{L}_\mathcal{I}, U)$ (9)	$\sum_{\|U\|=4} P\text{-}Sky(\mathcal{L}_\mathcal{I}, U)$ (10)	$1-\frac{(10)}{(11)}$	$\sum_{\|U\|=5} Sky(\mathcal{L}_\mathcal{I}, U)$ (12)	$\sum_{\|U\|=5} P\text{-}Sky(\mathcal{L}_\mathcal{I}, U)$ (13)	$1-\frac{(13)}{(12)}$
austral	101,630	1,314	45	0.97	403	211	0.48	648	83	0.87	317	9	0.97
cleve	43,504	6,601	34	0.99	238	85	0.64	300	12	0.96	111	0	1.00
cmc	12,208	559	57	0.90	269	76	0.72	299	1	0.99	104	0	1.00
crx	107,742	5,216	58	0.99	307	105	0.66	437	1	0.99	215	0	1.00
german	1,173,279	35,083	60	0.99	620	352	0.43	1,198	210	0.82	610	13	0.98
heart	34,400	1,218	29	0.98	283	133	0.53	468	35	0.93	211	5	0.98
hepatic	9,598	189	95	0.50	537	194	0.64	762	71	0.91	343	4	0.99
horse	129,716	10,352	22	0.99	177	64	0.64	220	10	0.95	84	0	1.00
hypo	671,961	789	613	0.22	2,134	686	0.68	2,812	129	0.95	1,172	0	1.00
lymph	45,644	98	55	0.44	428	157	0.63	524	59	0.89	268	4	0.99
mushroom	650,965	96,164	45	0.99	216	66	0.69	307	24	0.92	131	1	0.99
tic-tac-toe	16,157	1,301	77	0.94	259	93	0.64	310	16	0.95	109	0	1.00
vehicle	141,974	15,901	69	0.99	480	208	0.57	827	144	0.83	421	24	0.94
wine	15,112	1,346	25	0.98	152	27	0.82	170	5	0.97	57	0	1.00
zoo	4,871	599	50	0.92	197	29	0.85	211	5	0.98	74	0	1.00

(b) Space Analysis.

(c) Summarization Ratios.

Fig. 4. Results on UCI datasets with $|\mathcal{M}| = 5$.

sures using numeric values, like `mean`, we generated random values associated to attributes, each value being within the range [0..1]. Figure 4 summarizes the results we obtained.

CPU-Time Analysis. Figure 4a compares the performance of the three methods (with a graphical view). Cols. 1–4 give the characteristics of each dataset (name, number of items ($\mathcal{I}$), number of transactions ($\mathcal{T}$) and density). `CSKYC-BUC` clearly dominates the base-line method. On half of the datasets, there is an order of magnitude (speed-up value at least 11.78) (Col. 8). `CSKYC-BUC` is an average 2 times faster than `CP+SKY+CUBE`.

Space Analysis. Figure 4b compares, for each dataset, the number of proper skypatterns vs. the total number of skypatterns at each level of the cube. For each level i ($2 \leq i \leq 5$), the corresponding summarization ratio is also depicted. Figure 4c shows the graphical view of these ratios. For level 2, on most of the datasets, CSKYC achieves a very high summarization ratios (up to 99%). For level 3, these ratios mostly decrease since both levels (2 and 3) share most subspaces ($\binom{5}{2} + \binom{5}{3} = 20$ in total against 31 for the cube). Finally, for level 4 (resp. 5), CSKYC uses less storage than SKYC by at least 82% (resp. 94%) in size for all datasets we considered. Within these levels, there are almost no *proper* skypatterns since they have few subspaces ($\binom{5}{4} = 5$ for level 4 and $\binom{5}{5} = 1$) for level 5 and each one of these subspaces has a lot of descendant subspaces (14 for level 4 and 30 for level 5).

6 Conclusion

We have presented the compressed skypattern cube which concisely represents the skypattern cube and preserves its essential information. Compared to the original skypattern cube, the compressed skypattern cube has much less duplicates among susbpaces. We have provided an efficient algorithm to compute it and to query the skypattern set for any subspace. Our experimental study shows that CSKYC is particularly efficient in terms of build time and space usage compared to the original skypattern cube. Another interesting property is its ability to efficiently provide the skypattern set of any subspace. As future work, we plan to investigate the incremental maintenance of the CSKYC by allowing to add and/or remove any measure.

References

1. Bøgh, K.S., Chester, S., Sidlauskas, D., Assent, I.: Hashcube: a data structure for space- and query-efficient skycube compression. In: CIKM, pp. 1767–1770 (2014)
2. Börzsönyi, S., Kossmann, D., Stocker, K.: The skyline operator. In: ICDE, pp. 421–430 (2001)
3. Chomicki, J., Godfrey, P., Gryz, J., Liang, D.: Skyline with presorting. In: ICDE, pp. 717–719 (2003)
4. Dua, D., Graff, C.: UCI machine learning repository (2017). http://archive.ics.uci.edu/ml
5. Hansen, K., Mika, S., Schroeter, T., Sutter, A., ter Laak, A., Steger-Hartmann, T., Heinrich, N., Müller, K.: Benchmark data set for in silico prediction of Ames mutagenicity. JCIM **49**(9), 2077–2081 (2009)
6. Hanusse, N., Kamnang Wanko, K., Maabout, S.: Computing and summarizing the negative skycube. In: CIKM, pp. 1733–1742 (2016)
7. Mannila, H., Toivonen, H.: Levelwise search and borders of theories in knowledge discovery. Data Min. Knowl. Discov. **1**(3), 241–258 (1997)
8. Pei, J., Fu, A.W., Lin, X., Wang, H.: Computing compressed multidimensional skyline cubes efficiently. In: ICDE, pp. 96–105 (2007)
9. Pei, J., Jin, W., Ester, M., Tao, Y.: Catching the best views of skyline: a semantic approach based on decisive subspaces. In: VLDB, pp. 253–264 (2005)
10. Pham, H., Lavenier, D., Termier, A.: Identifying genetic variant combinations using skypatterns. In: DEXA Workshops, pp. 44–48. IEEE Computer Society (2016)
11. Soulet, A., Raïssi, C., Plantevit, M., Crémilleux, B.: Mining dominant patterns in the sky. In: ICDM, pp. 655–664 (2011)
12. Ugarte, W., et al.: Skypattern mining: from pattern condensed representations to dynamic constraint satisfaction problems. Artif. Intell. **244**, 48–69 (2017)

13. Ugarte, W., Boizumault, P., Loudni, S., Crémilleux, B.: Computing skypattern cubes. In: ECAI, pp. 903–908 (2014)
14. Ugarte, W., Boizumault, P., Loudni, S., Crémilleux, B.: Computing skypattern cubes using relaxation. In: ICTAI, pp. 859–866 (2014)
15. Ugarte Rojas, W., Boizumault, P., Loudni, S., Crémilleux, B., Lepailleur, A.: Mining (Soft-) Skypatterns using dynamic CSP. In: Simonis, H. (ed.) CPAIOR 2014. LNCS, vol. 8451, pp. 71–87. Springer, Cham (2014). https://doi.org/10.1007/978-3-319-07046-9_6
16. Xia, T., Zhang, D.: Refreshing the sky: the compressed skycube with efficient support for frequent updates. In: SIGMOD Conference, pp. 491–502 (2006)
17. Yuan, Y., Lin, X., Liu, Q., Wang, W., Yu, J.X., Zhang, Q.: Efficient computation of the skyline cube. In: VLDB, pp. 241–252 (2005)

Context-Aware Instance Matching Through Graph Embedding in Lexical Semantic Space

Ali Assi[1(✉)], Hamid Mcheick[2], and Wajdi Dhifli[3]

[1] University of Quebec At Montreal, Montreal H2X3Y7, Canada
assi.ali@courrier.uqam.ca
[2] University of Quebec At Chicoutimi,
555, University Boulevard, Chicoutimi, Canada
hamid_mcheick@uqac.ca
[3] Univ. Lille, EA2694, 3 rue du Professeur Laguesse,
BP83, 59006 Lille Cedex, France
wajdi.dhifli@univ-lille.fr

Abstract. Instance matching is one of the processes that facilitate the integration of independently designed knowledge bases. It aims to link co-referent instances with an `owl:sameAs` connection to allow knowledge bases to complement each other. In this work, we present VDLS, an approach for automatic alignment of instances in RDF knowledge base graphs. VDLS generates for each instance a virtual document from its local description (*i.e.*, data-type properties) and instances related to it through object-type properties (*i.e.*, neighbors). We transform the instance matching problem into a document matching problem and solve it by a vector space embedding technique. We consider the pre-trained word embeddings to assess words similarities at both the lexical and semantic levels. We evaluate our approach on multiple knowledge bases from the instance track of OAEI. The experiments show that VDLS gets prominent results compared to several state-of-the-art existing approaches.

Keywords: Data linking · Instance matching · RDF graph · Semantic web

1 Introduction

Linked Open Data (LOD) includes several Knowledge Bases (KBs) expressed by ontologies in the form of RDF graphs from various domains of applications such as geography, biology, *etc.* These KBs are often created independently from each other. They may contain resources (with distinct descriptions) that are co-referring, but not explicitly defined. *Instance Matching* (IM) is the process of matching instances across these different knowledge bases, that refer to the

F. Wotawa et al. (Eds.): IEA/AIE 2019, LNAI 11606, pp. 422–433, 2019.
https://doi.org/10.1007/978-3-030-22999-3_37

same entity in the real-world (*e.g.*, the same person in two different knowledge bases).

The existing IM approaches can be categorized as domain-dependent [19,20] and domain-independent approaches [7,12,16,21]. An IM approach is called domain-dependent when it deals with KBs related to a specific domain (*e.g.*, music domain). Otherwise, it is called domain-independent. For more details about IM approaches, we refer the reader to [6,15]. In fact, most of the existing approaches mainly depend on the result of the instances' property alignments. The property alignment process aims to match properties from different KBs, that have similar semantics. This process is not trivial since KBs are usually expressed by their specific ontologies in order to describe the RDF graph instances and relations. For example, the information included in the "address" property in a source KB can be represented by several properties (*e.g.*, street, zip-code, *etc.*) in another KB. Thus, the property alignment will not find its corresponding property in the target KB. As a result, such information will be ignored even if it may be worthy to consider it for IM. Indeed, in some cases the description of instances does not carry properties with similar semantics. However, they can contain information with some expressive relationships. For example, given a KB that describes the instance *"Arthur Purdy Stout"* with two properties: "`f:profession`" and "`f:death_Place`", the former property says *"Researcher in Surgical pathology and pathologists"* as his profession. The latter one indicates *"New York City"* where he died. Now given another KB that describes his "`d:place_of_birth`" as *"New York City"* and the information about his career can be deduced from the text given by the property "`d:description`" which states *"Arthur Purdy Stout, M.D., (1885–1967) was a noted American surgeon and diagnostician."*. As you notice, the two descriptions have the same meaning. However, they cannot be inferred by any of the existing property alignment-based approaches.

In this paper, we propose an approach that tackles these drawbacks. Our approach represents instances as virtual documents where each of the latter is represented by a collection of words, and is generated for each instance declared in the KB. It consists of a "bag of words" extracted from the identifier of the instance, predicates, as well as the ones from all of its neighbors. To capture the semantic string similarity between words as "pathologist" and "diagnostician" in the above example, we use the distributed representation of words (*e.g.*, FastText [3], GloVe [18]) also known as word embedding. This latter represents words by a low dimensional dense vector, such that the vectors for similar words (*i.e.*, "pathologist", "diagnostician") are close to each others in their semantic embedding space.

Our major contributions are summarized as follows: (1) We propose a new idea for building virtual documents for instances. (2) We include words embeddings for capturing their semantics (*i.e.*, synonym and terminological variants). (3) We transform the instance matching problem into a document similarity problem and we solve it using lexical semantic similarity technique [13].

We experimentally validate our approach, Virtual Document Lexical Similarity (termed VDLS), on four KBs from the benchmark instance matching track of OAEI 2009 and 2010. The obtained results show that our approach gets highly competitive results compared to state-of-the-art approaches.

2 Preliminaries

In the semantic web, the meaning of a "concept" is similar to the notion of a `Class` in the Object-Oriented Programming (OOP) view. Thus, resources created according to the structure of a class are known as instances of that class [1].

The Resource Description Framework (RDF) data model [9] represents the descriptions of the entities (*i.e.*, concepts) and the instances by RDF expressions, called triples, in the form <subject, predicate, object>. A `subject` can be a URI or a blank node. The latter represents an anonymous entity. An `object` can be a URI, a blank node, or a basic value (*e.g.*, a string, a date, an integer, *etc*). A `predicate` allows to model a relationship between the `subject` and the `object`.

Definition 1 ***(RDF knowledge base graph).*** *An RDF Knowledge Base (KB) graph is a set of facts in the form* <subject, predicate, object> $\in (\mathcal{E} \cup \mathcal{B}) \times \mathcal{P} \times (\mathcal{E} \cup \mathcal{L} \cup \mathcal{B})$, *where* $\mathcal{E}$ *is the set of instances,* $\mathcal{B}$ *is the set of blank nodes,* $\mathcal{P}$ *is the set of predicates and* $\mathcal{L}$ *is the set of literals (basic values).*

An RDF KB graph can adhere or not to an ontology. In the rest of the paper, we write KB shortly to refer to RDF KB graph.

Instance Matching (IM) is the problem of identifying instances that co-refer to the same object of the real world. It can be seen as a process of building the predicate `owl:sameAs` between the co-referent instances belonging to different KBs. Formally:

Definition 2 ***(Instance matching).*** *Given two input sets of instances* $\mathcal{S}$ *and* $\mathcal{T}$ *belonging to two different KBs, the aim of IM is to compute the set* $\mathcal{M} = \{(i_1, i_2) \mid (i_1, i_2) \in \mathcal{S} \times \mathcal{T}, <i_1, \textit{owl:sameAs}, i_2>\}$.

Instance matching is a difficult task [5] mainly due to textual variation of the property values, incompleteness, presence of erroneous information, multilingualism, *etc.*

3 Instance Matching Through Graph Embedding and Lexical Semantic Similarity

3.1 Virtual Document

An instance e is described by a set of triples: $T(e) = \{t_1, \ldots, t_n\}$. All these triples share the same subject e denoted by a URI. For every instance in the KB, a "Virtual Document" (VD) is created. It is represented as a collection of words extracted from different parts of its triples (i.e., from the URI's subject and its properties' values). We consider to treat the URIs as literals as in [17]. This approach is based on the assumption that many URIs encompass valuable information. It detects a pattern from the characters of the URIs. The pattern has the form (prefix-infix-(suffix)). The prefix (P) is the URI domain. The (optional) suffix (S) contains details about the format of the data or named anchor. The infix (I) represents the local identifier of the target instance.

Example 1. In the URI: http://people.csail.mit.edu/lalana_kagal/foaf#me, the prefix is "http://people.csail.mit.edu", the infix is "/lalana_kagal", and the suffix is "/foaf#me".

Definition 3 ***(Statement).*** *A statement t is the smallest irreducible representation for linking one object s to another object o or a literal l via a predicate p. Formally:* $t = <s, p, o>$ *where* $s \in \mathcal{E} \cup \mathcal{B}$, $p \in \mathcal{P}$ *and* $o \in \mathcal{E} \cup \mathcal{B} \cup \mathcal{L}$.

In the following, we refer to the infix of a given URI e by $I(e)$. Indeed, we refer to the different parts of a statement t by $subject(t)$ to designate its subject (*i.e.*, s), by $object(t)$ to designate its object (*i.e.*, o) when this object is a URI or by $value(t)$ when the object is a basic value. We have only included in $T(e)$ the forward statements (triples) from the instance e. We suppose that such triples allow to describe the main features of the instance e. Thus, we define the neighbors of e as follows:

Definition 4 ***(Forward neighbors).*** *Let e be an instance,* $\mathcal{B}$ *be the set of blank nodes and t be any statement with e as its subject. The instance e has a set of forward neighbors denoted by FN and defined as:*

$$FN(e) = \bigcup_{e=subject(t)} \{object(t)\} \tag{1}$$

Definition 5 ***(Local name of an instance).*** *Let e be an instance in* $\mathcal{E}$. *The local name of e, denoted by* $LN(e)$, *is equal to the infix of e, i.e.,* $LN(e) = I(e)$.

Definition 6 ***(Local name of a blank node).*** *Let* $\_{:}b$ *be a blank node* ($\_{:}b \in \mathcal{B}$) *and t be any statement with* $\_{:}b$ *as its subject. The local name of* $\_{:}b$, *denoted by* $LN(\_{:}b)$, *is defined as equal to the local names of its direct 1-hop forward neighbors in the graph:*

$$LN(\_{:}b) = \sum_{o \in FN(\_{:}b)} N(o) \tag{2}$$

where N is a function that returns the name of the object $o \in \mathcal{E} \cup B$. The function N is equal to I when the object $o \in \mathcal{E}$. Whereas, it leads to a recursive extended definition of the local name of a blank node when $o \in \mathcal{B}$.

Definition 7 ***(Recursive local name of a blank node).*** *Let* $\_{:}b$ *be a blank node* ($\_{:}b \in \mathcal{B}$) *and t be any statement with* $\_{:}b$ *as its subject. The recursive local* *name of $\_:b$, denoted by $LN_k(\_:b)$, is defined as the local names of leaf nodes of at most k-hops paths starting form $\_:b$ and ending in object nodes $o \in \mathcal{E}$:*

$$\forall k \geq 1, LN_k(\_:b) = LN(\_:b) + \sum_{\substack{\_:b=subject(t)\\ o=object(t)\\ o\in\mathcal{B}}} LN_{k-1}(o) \tag{3}$$

Note that the recursive definition could terminate in less then k-hops in the case where all the neighbor nodes are leaf nodes. Otherwise, the function terminates in k-hops (in the worst case), if there exists at least a path of length greater or equal to k composed of only blank nodes and that starts from the root blank node of the recursive function. The benefits of this limitation are twofold. Indeed, this allows to alleviate the computation in cases of big KB graph datasets or in cases where the datasets contain long paths of consecutive blank nodes. In addition, this avoids the trap of infinite loops in cases where a cycle of blank nodes exists.

Definition 8 ***(Local description of an instance).*** *Let e be an instance denoted by a Uniform Resource Identifier (URI) and t be any statement with e as its subject. The local description of e, denoted by $Des(e)$, is a collection of words defined by:*

$$Des(e) = \alpha_1 \times LN(e) + \alpha_2 \times Data(e) + \alpha_3 \times \sum_{\substack{e=subject(t)\\ o=object(t)\\ o\in FN(e)}} LN(o) \tag{4}$$

where $Data$ is the set of *values* extracted from $T(e)$ (*i.e.*, basic values) and the coefficients α_1, α_2 and α_3 are three fixed constants in $\{0, 1\}$. LN depends on the strategy used when creating the URIs (*i.e.*, $\alpha_1 = 1$ when the URIs information is meaningful).

Definition 9 ***(Local description of a blank node).*** *Let $\_:b$ be a blank node ($\_:b \in \mathcal{B}$) and t be any statement with $\_:b$ as its subject. The local description of $\_:b$, denoted by $LD(\_:b)$, is defined as equal to its local data values ($Data(\_:b)$) and the local descriptions of its direct 1-hop forward neighbors in the graph:*

$$LD(\_:b) = Data(\_:b) + \sum_{\substack{\_:b=subject(t)\\ o\in FN(\_:b)}} D(o) \tag{5}$$

where D is a function that returns the description of the object $o \in \mathcal{E} \cup \mathcal{B}$. If $o \in \mathcal{E}$, then the function D will be equal to Des. Whereas, in the case where $o \in \mathcal{B}$, then the local description of a blank node will be defined in a recursive way as follows.

Definition 10 ***(Recursive local description of a blank node).*** *Let $\_:b$ be a blank node ($\_:b \in \mathcal{B}$) and t be any statement with $\_:b$ as its subject. The recursive* *local description of* $\_{:}b$*, denoted by* $LD_k(\_{:}b)$*, is defined as the local descriptions of leaf nodes of the k-hops paths starting form* $\_{:}b$ *and ending in object nodes* $o \in \mathcal{E}$*:*

$$\forall k \geq 1, LD_k(\_{:}b) = LD(\_{:}b) + \sum_{\substack{o=object(t)\\ \_{:}b=subject(t)\\ o\in\mathcal{B}}} LD_{k-1}(o) \tag{6}$$

Note also that here the iterations of the recursive definition will terminate in at most k-hops for the same arguments discussed in Definition 7.

Definition 11 ***(Virtual document of an instance).*** *The virtual document of an instance e (denoted by* $VD(e)$*) is defined as:*

$$VD(e) = Des(e) + \alpha \times \sum_{e' \in FN(e)} Des(e') \tag{7}$$

where α is a parameter defined in $\{0, 1\}$. If the node has a rich description, then we can limit VD(e) to the local description of e provided in the set of 1-hop neighbors by setting the parameter α to 0. However, in some applications, the local description of e in the KB graph could be poor and thus it would be judicious to incorporate additional information on e from farther neighbors in the KB by performing walks in the graph. This could be performed by setting α to 1.

3.2 Lexical Semantic Vector

The lexical semantic vector method was introduced in [13] then modified in [10]. Given two VDs (VD_1, VD_2), we create a combined list of vocabulary, denoted by L, that consists of all the unique words in VD_1 and VD_2, *i.e.*, $L = VD_1 \cup VD_2$. Then, we compute the pairwise similarity of each word v_L in L with every word v_1 in VD_1. This leads to create a lexical semantic vector $\boldsymbol{V_1}$ that contains the maximum similarities between each word in L and all words in VD_1 (respectively for VD_2). Formally, each element in the lexical semantic vectors $\boldsymbol{V_1}$ and $\boldsymbol{V_2}$ is defined as:

$$\boldsymbol{V_{1j}} = \max_{1 \leq i \leq |VD_1|} Sim(v_{1i}, v_{Lj}) : \forall v_{Lj} \in L \tag{8}$$

$$\boldsymbol{V_{2j}} = \max_{1 \leq i \leq |VD_2|} Sim(v_{2i}, v_{Lj}) : \forall v_{Lj} \in L \tag{9}$$

where Sim is a similarity measure (cosine in this paper) that computes the similarity between pairs of words based on their embedding vectors.

Definition 12 ***(Lexical semantic word embedding).*** *A word embedding (also called dense distributed representation [2]) is a learned representation where each word is mapped to a real-valued dense vector in a semantic vector space. This is based on the assumption that words with similar contexts will have similar meanings and also will have similar representations (i.e., close vectors in the semantic vector space).*

3.3 Indexing

After determining the VDs that correspond to the source KB, we build an inverted-index, *i.e.*, a binary vector representation (1 the presence or 0 the absence) from the words of these VDs.

Infrequent-Words: One of the common problems encountered while building the inverted-index representation is that multiple used keywords are highly frequent and thus they do not sufficiently provide specificity for the instances. In order to alleviate this drawback, we define *Infrequency* as a measure that quantifies the infrequency of a given word in a dataset:

$$Infrequency(token) = \frac{1}{\log_2 \left(WF(token) + 1\right)} \tag{10}$$

where WF is the number of VDs containing the token. *Infrequency* is theoretically defined in $[0, 1]$. A word appearing only once in a KB has an infrequency of 1. In contrast, the more a word appears in the KB, the more its infrequency is close to 0. In fact, 0 represents a theoretical lower bound for *infrequency* where *WF* leans toward $+\infty$. In practical cases, the maximum word frequency *WF* is bounded by the size of the dataset. Thus, we propose to normalize (MinMax normalization) the infrequency values to make them fall within the range of [0, 1]. Note that a token is considered as an infrequent-word when its infrequency (after normalization) is higher or equal to a predefined threshold γ.

Common Infrequent-Words: Let I_1 and I_2 be two instances for a source and a target KB, respectively. I_2 is considered as a potential matching candidate for I_1 (*i.e.*, *Candidate*(I_1, I_2) = *True*), if both instances share a number of "common infrequent-words" that is higher or equal to a predefined threshold β. Formally:

$$Candidate(I_1, I_2) = \mid I_1 \cap I_2 \mid \geq \beta \tag{11}$$

3.4 Approach Overview

Our approach VDLS (see Fig. 1) starts by parsing the given source and target KBs then it builds the VDs for each instance in them according to the Definition 11. Once the VDs corresponding to the source KB are determined, an inverted-index is set up from their words. Note that infrequency takes effect only over the words of the source KB. By using this index, each source instance gets its candidates from the target KB. A target instance is a candidate for a source instance if both share at least β common infrequent-words in their descriptions as described in Eq. 11. The similarity between the instances is computed between their corresponding Lexical Semantic embedding vectors as detailed in Sect. 3.2. Once this computation step is done, we select for each source instance its best candidate (*i.e.*, top similarity score) as a co-referent.

4 Experimental Evaluation

Datasets. We evaluate our proposed method on two benchmark datasets (Table 1): PR (synthetic) and A-R-S (real) used in OAEI 2010 and 2009, respectively. PR is a small benchmark including two persons data and one restaurants RDF data. A-R-S includes three RDF files named *eprints*, *rexa* and *dblp* related to scientific publications RDF data. For each set, the OAEI provides a mapping file ("gold standard") including the co-referent pairs between the source and target RDF files.

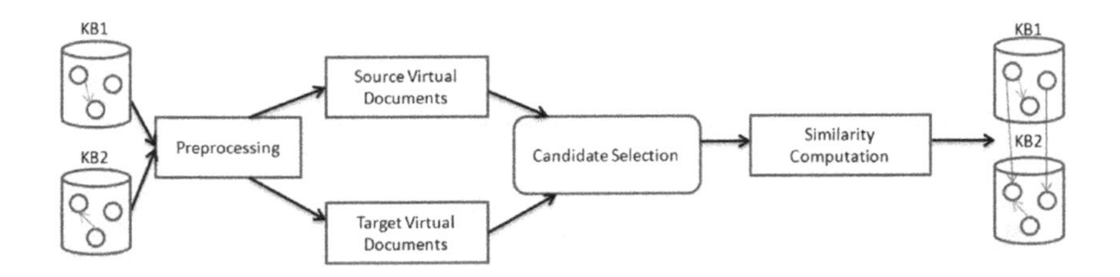

Fig. 1. An overview of our instance matching approach.

Table 1. Benchmarks statistics

Benchmarks	Datasets	Source	Target	Gold standard
PR	Person1	2000 URIs	1000 URIs	500
	Person2	2400 URIs	800 URIs	400
	Restaurant	399 URIS	2256 URIs	112
A-R-S	eprints-rexa	1130 URIS	18,492 URIs	777
	eprints-dblp	1130 URIs	2, 650, 832 URIs	554

Experimental Setup. For word embedding vectors, we use a dataset of pre-trained word vectors embedding developed by Facebook research team using FastText [3]. This dataset is trained on Wikipedia corpus and the generated vectors are of dimensionality equal to 300. If a word does not exist in the pre-trained embeddings, a unique vector will be created at test time. We run the experiments with different infrequency thresholds γ between [0,1] with a step size of 0.1. In Table 2, we report the optimal obtained results. We set $\beta = 2$ in Eq. 11 to represent the minimum required number of words between two instances to be considered as a candidate pair. In addition, we allow the paths to be composed at most of 2 blank nodes (*i.e.*, $k = 2$ in Eq. 3). Our approach is implemented in Java and the KBs are parsed using the RDF4J library. We execute the experiments on a Linux server with 128 GB RAM and two Intel E5-2683 v4 "Broadwell" CPUs of 2.1 Ghz.

4.1 Results and Discussion

Analysis of the Effect of the Context on the IM. We first analyze the effect of including (or not) the context information on the accuracy of the IM process. Table 2 reports the obtained results for VDLS with the context information ($\alpha = 1$) and without it ($\alpha = 0$). We notice that when the instances contain a sufficient local description information, the neighboring information either does not bring any benefit to the matching process (the case of Person1 and Person2) or it leads to an information overload and thus may hinder the IM results (the case of Restaurant). In the case of largely and noisy datasets as in eprints-rexa, the neighboring information permits to relax the effect of the noisy data and thus to enhance the accuracy of the matching process.

Table 2. F-measure results of VDLS with ($\alpha = 1$) and without ($\alpha = 0$) the context information.

Datasets	Person1	Person2	Restaurant	eprints-rexa	eprints-dblp
VDLS ($\alpha = 0$)	**1**	**1**	**1**	0.85	0.89
VDLS ($\alpha = 1$)	**1**	**1**	0.98	**0.87**	**0.9**

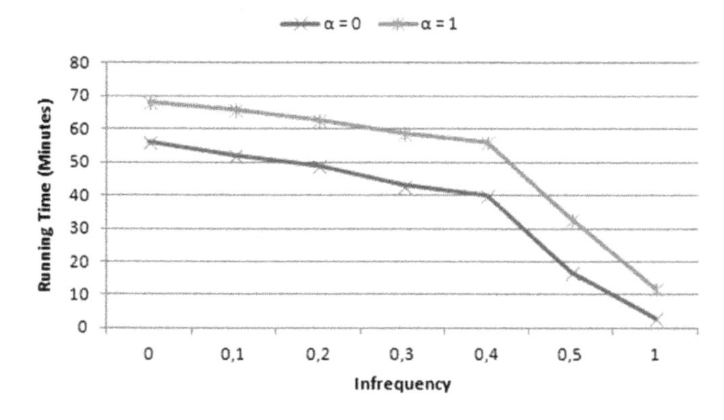

Fig. 2. The effect of Infrequency on the running time of VDLS (using the eprints-rexa datasets).

Analysis of the Effect of Word Infrequency on the Running Time. A higher F-measure requires a higher number of words and thus lower values of infrequency. This requires more running time. Figure 2 shows a clear example of the evolution of running time of VDLS with respect to different infrequency thresholds. Indeed, the required computation time increases when we add more words (lower infrequency) in the VDs. However, such a consideration makes VDLS subject to the "no free lunch" principle [22], where the gain in accuracy comes with an offset of computational cost. Hence, a trade-off between running time and accuracy is essential especially with large scale datasets.

Comparative Analysis. In Table 3, we report the results of multiple state-of-the-art IM approaches on the PR and A-R-S benchmarks and we compare our approach against them. On the PR benchmark, VDLS performed overall better than the other approaches. It was able to correctly retrieve all the co-referent instance, except for Restaurant where VDLS ($\alpha = 1$) where the F-measure was 0.98. This was due to the effect of information overload, where the context (neighbor) information did slightly hinder the accuracy of the IM process. As for the A-R-S benchmark, the IM was more difficult than with the PR benchmark. Indeed, VDLS outperformed all the other approaches in terms of F-measure yet the best results did not exceed 0.87 and 0.9 respectively for eprints-rexa and eprints-dblp. By analyzing the false matchings, we noticed that several of these instances were isolated nodes in the RDF graph and thus their VDs did lack context information.

Table 3. Comparative analyses of F-measure results on PR and A-R-S benchmarks.

Datasets	Person1	Person2	Restaurant	eprints-rexa	eprints-dblp
VDLS ($\alpha = 0$)	**1**	**1**	**1**	0.85	0.89
VDLS ($\alpha = 1$)	**1**	**1**	0.98	**0.87**	**0.90**
PARIS [21]	**1**	**1**	0.91	-	-
ObjectCoref [7]	**1**	0.95	0.90	-	-
ASMOV-D [8]	0.87	0.24	0.70	-	-
CODI [16]	0.91	0.36	0.72	-	-
RIMOM [11]	**1**	0.97	0.81	0.80	0.73
DSSIM [14]	-	-	-	0.38	0.13
HMATCH [4]	-	-	-	0.62	0.65
VMI [12]	-	-	-	0.85	0.66

5 Conclusion and Future Works

In this paper, we proposed a property-independent approach for IM and a new method for building VDs corresponding to the instances. We also transformed the IM problem into a document matching problem and we created lexical semantic vectors to measure the similarity between two VDs. We have compared our approach to state-of-the-art methods on benchmark datasets, and we achieved very promising results.

As future work, we will include more pruning heuristics to reduce the number of candidates for each query instance. It will also be interesting to propose an extension for VDLS that leverages parallel and distributed computation to efficiently handle big data scenarios with large-scale KBs.

References

1. Antoniou, G., Van Harmelen, F.: A Semantic Web Primer. MIT Press, Cambridge (2008)
2. Bengio, Y., Lamblin, P., Popovici, P., Larochelle, H.: Greedy layer-wise training of deep networks. In: Advances in Neural Information Processing Systems, vol. 19. MIT Press, Cambridge, MA (2007)
3. Bojanowski, P., Grave, E., Joulin, A., Mikolov, T.: Enriching word vectors with subword information. arXiv preprint arXiv:1607.04606 (2016)
4. Castano, S., Ferrara, A., Montanelli, S., Lorusso, D.: Instance matching for ontology population. In: Italian Symposium on Advanced Database Systems (SEBD), pp. 121–132. ICAR-CNR (2008)
5. Ferrara, A., Lorusso, D., Montanelli, S., Varese, G.: Towards a benchmark for instance matching. In: Proceedings of the 3rd International Conference on Ontology Matching, vol. 431, pp. 37–48. CEUR-WS. org (2008)
6. Ferraram, A., Nikolov, A., Scharffe, F.: Data linking for the semantic web. Semant. Web Ontol. Knowl. Base Enabled Tools Serv. Appl. **169**, 326 (2013)
7. Hu, W., Chen, J., Qu, Y.: A self-training approach for resolving object coreference on the semantic web. In: Proceedings of the 20th International Conference on World Wide Web, pp. 87–96. ACM (2011)
8. Jean-Mary, Y.R., Shironoshita, E.P., Kabuka, M.R.: Ontology matching with semantic verification. Web Semant. Sci. Serv. Agents World Wide Web **7**(3), 235–251 (2009)
9. Klyne, G., Carroll, J.J.: Resource description framework (RDF): concepts and abstract syntax (2004). http://www.w3.org/TR/2004/REC-rdf-concepts-20040210/
10. Konopik, M., Prazák, O., Steinberger, D., Brychcín, T.: UWB at SemEval-2016 task 2: interpretable semantic textual similarity with distributional semantics for chunks. In: Proceedings of the 10th International Workshop on Semantic Evaluation (SemEval-2016), pp. 803–808 (2016)
11. Li, J., Tang, J., Li, Y., Luo, Q.: RiMOM: a dynamic multistrategy ontology alignment framework. IEEE Trans. Knowl. Data Eng. **21**(8), 1218–1232 (2009)
12. Li, J., Wang, Z., Zhang, X., Tang, J.: Large scale instance matching via multiple indexes and candidate selection. Knowl. Based Syst. **50**, 112–120 (2013)
13. Li, Y., McLean, D., Bandar, Z., O'Shea, J., Crockett, K.A.: Sentence similarity based on semantic nets and corpus statistics. IEEE Trans. Knowl. Data Eng. **18**(8), 1138–1150 (2006)
14. Nagy, M., Vargas-Vera, M., Motta, E.: DSSim - managing uncertainty on the semantic web. In: CEUR Workshop Proceedings, OM, vol. 304. CEUR-WS.org (2007)
15. Nentwig, M., Hartung, M., Ngonga Ngomo, A.C., Rahm, E.: A survey of current link discovery frameworks. Semant. Web **8**(3), 419–436 (2017)
16. Noessner, J., Niepert, M., Meilicke, C., Stuckenschmidt, H.: Leveraging terminological structure for object reconciliation. In: Aroyo, L., et al. (eds.) ESWC 2010. LNCS, vol. 6089, pp. 334–348. Springer, Heidelberg (2010). https://doi.org/10.1007/978-3-642-13489-0_23
17. Papadakis, G., Demartini, G., Fankhauser, P., Kärger, P.: The missing links: discovering hidden same-as links among a billion of triples. In: Proceedings of the 12th International Conference on Information Integration and Web-based Applications & Services, pp. 453–460. ACM (2010)

18. Pennington, J., Socher, R., Manning, C.D.: Glove: global vectors for word representation. In: In EMNLP (2014)
19. Raimond, Y., Sutton, C., Sandler, M.B.: Automatic interlinking of music datasets on the semantic web. In: LDOW, vol. 369 (2008)
20. Rowe, M., Group, O.: Interlinking distributed social graphs. In: In Linked Data on the Web Workshop, WWW 2009, April 2009 (2009)
21. Suchanek, F.M., Abiteboul, S., Senellart, P.: PARIS: probabilistic alignment of relations, instances, and schema. Proc. VLDB Endow. **5**(3), 157–168 (2011)
22. Wolpert, D.H., Macready, W.G.: No free lunch theorems for optimization. IEEE Trans. Evol. Comput. **1**(1), 67–82 (1997)

Identifying Similarities Between Musical Files Using Association Rules

Louis Rompré[1(✉)], Ismaïl Biskri[1(✉)], and Jean-Guy Meunier[2]

[1] Université du Québec à Trois-Rivières, C.P. 500, Trois-Rivières, (QC) G9A 5H7, Canada
rompre.louis@courrier.uqam.ca, ismail.biskri@uqtr.ca
[2] UQAM, C.P. 8888, Succ. Centre-Ville, Montréal, (QC) H3C 3P8, Canada
meunier.jean-guy@uqam.ca

Abstract. The number of music in digital format increases years after years. The amount of data available allows streaming services to offer wide variety of music. This makes these services attractive. Automatic classification process is required to manage and structure all files available. Automatic music classification is an active field of research. Researches rely on machine learning techniques such as deep neural network. These techniques used features extracted from raw data to generate classification. Features have an important impact on results. Selecting the right descriptors is one of the main difficulties associated with automatic music classification. Opacity of current methods makes it difficult to evaluate the contribution of descriptors in the classification process. In this paper, we propose to use association rules to add more transparency and interpretability.

Keywords: Information retrieval · Classification · Music · Association rules · Descriptor · Transparency

1 Introduction

Today, music streaming platforms share millions of subscribers. These platforms give access to large music collections. In addition, they recommend new songs based on user preferences. This particularity contributes to their success. However, finding interesting similar songs hidden in large music databases is not a trivial task.

Robust recognition mechanisms already exist. Audio fingerprint is one of these mechanisms. It allows finding songs from short audio excerpts. Researches made in this area led to the development of industrial products such as Shazam [19]. An audio fingerprint is a unique spectral signature. When an unknown audio excerpt is used as query, its fingerprint is calculated and matched against those stored. A song can be recognized using an excerpt of only few seconds. Audio fingerprinting is a robust recognition tool, but this mechanism is not designed to deal with different versions of a same song nor similar songs [8]. Because of this limitation, audio fingerprint needs to be combined with classification mechanisms.

Automatic music classification is an active domain of research. Many approaches rely on comparison of spectral descriptors. Instead conducting spectral analysis to

F. Wotawa et al. (Eds.): IEA/AIE 2019, LNAI 11606, pp. 434–441, 2019.
https://doi.org/10.1007/978-3-030-22999-3_38

create fingerprint, it's performed to find descriptors. Desired descriptors must exhibit the main characteristics of songs. Even though same descriptors are found in multiple songs, those songs can vary greatly from one to another. Therefore, usefulness of such descriptors is limited. Classification are generated using different machine learning algorithms such as KNN [18] or SVM [6, 10]. Both, descriptors and classifier, impact classification results. Still today, different combinations of descriptors and classifiers are experimented. Recent research has shown promising results with Deep Convolutional Recurrent Neural Network [7]. However, opacity of current methods makes it difficult to verify the contribution of descriptors in the process. We propose to use association rules to visualize how descriptors are distributed into songs. The proposed approach is a tool to better understand the impact of descriptors on classifications.

2 Music Classification

Musical genres are used naturally to describe music. Humans can recognize music easily, but they may have difficulty to classify it. Musicological indicators such as instrumentation, meter, rhythm, tempo, harmony, melody or playing style can be considered to identify genres [17]. Despite this, how these indicators are arranged to formed genres is not so obvious. The main reason is musical genres are ill-defined categories. For instance, consider the GTZAN dataset [18]. This dataset was built by Tzanetakis for the purpose of his research. It contains 1000 excerpts classified into 10 genres: blues, classical, country, disco, hip-hop, jazz, metal, pop, reggae and rock. Each genre is represented by 100 excerpts of the same duration. Since its creation in 2002, GTZAN was widely used as reference dataset. Figure 1 illustrates how Shazam and iTunes classify the classical excerpts of GTZAN dataset while Fig. 2 illustrates how they classify the Hip-Hop excerpts. If you look at these figures, it's obvious that GTZAN, iTunes and Shazam define genres differently, especially hip-hop.

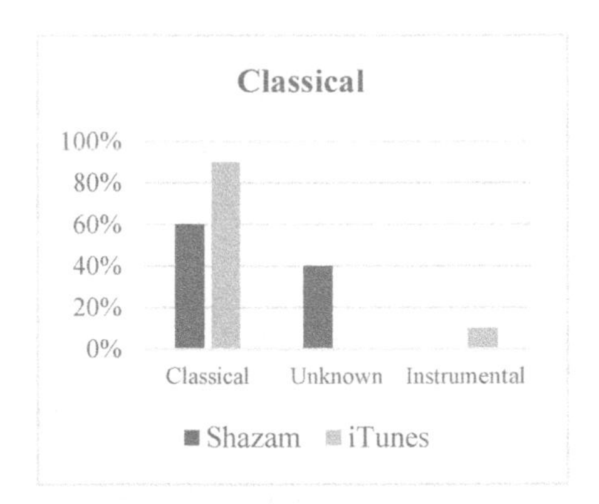

Fig. 1. Classification of GTZAN Classical excerpts according to Shazam and iTunes.

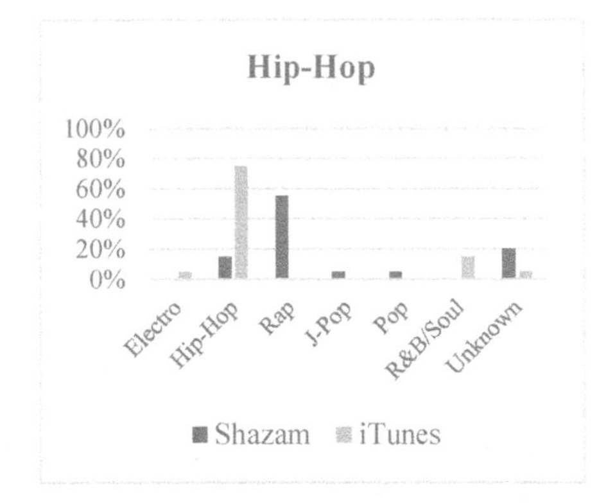

Fig. 2. Classification of GTZAN Hip-Hop excerpts according to Shazam and iTunes.

Genre associated to some music files might seem representative for one individual and aberrant for another. Even worse, our definition of genres evolves over time. Often, more than one genre is associated to a song. Dealing with this complexity and subjectivity is complicated for computers. Selection of relevant descriptors is challenging. The ability of one descriptor to help identifying a specific genre does not guarantee its value for another genre. Still today, selection relies on a trial and error approach. Therefore, there is a need for a tool to help identifying how descriptors are distributed through classes. That's exactly what association rules offer.

3 Association Rules Mining

Association rules mining is an important subfield of data mining. Interest regarding this method started from the research conducted by Agrawal [1]. He showed that association rules can be used to discover interesting patterns in a transactional database. This section presents the notions introduced by Agrawal.

In a general form, a transaction is simply a subset of data. Let $T = t_1, t_2, t_3, \ldots, t_n$ be a set of n transactions, each item i contained in transactions may be a music file content-based descriptor for instance. Let $I = i_1, i_2, i_3, \ldots, i_d$ be a set of d distinct items then a subset of items from I is called an itemset. In order words, an itemset represent a possible combination of items. An itemset is denoted by a capitalized letter such as X or Y. The goal of association rules is to highlight frequent relations between items. This task is performed considering frequent itemsets namely items that appear frequently together in transactions. Measures such as support count or support are used to evaluate how frequent an itemset is. An itemset whose support count or support is greater than a minimum threshold is considered as frequent itemset. Let $T = t_1, t_2, t_3,\ldots, t_n$ be a set of n transactions containing items from I, then the support count of an itemset X represents the number of transactions that contain X. The support count is given by Eq. (1).

$$\sigma(X) = \left|\{t_i.|X \subseteq t_{i,} \, t_i \in T\}\right| \tag{1}$$

The support is a measure like support count, but the overall number of transactions is considered. The support is given by Eq. (2).

$$S(X) = \sigma(X)/n \tag{2}$$

Let X and Y be two frequent itemsets such as $X \cap Y = \emptyset$ an association rule is expressed as $X \rightarrow Y$. It represents the co-occurrence regularities between itemsets X and Y. The quality of an association rule is determined by a measure *m* and a threshold. Thus, an association rule $X \rightarrow Y$ is considered relevant if $m(X \rightarrow Y) \geq$ threshold. Support and confidence are the main measures used to estimate the relevance of an association rule. The support of an association rule is the number of transactions that contain both X and Y. The support is given by Eq. (3).

$$S(X \rightarrow Y) = \sigma(X \cup Y)/n \quad (3)$$

The confidence of a rule is the number of transactions that contain both X and Y among transactions that contain X. The confidence is given by Eq. (4). The support gives the frequency of an association rule while the confidence gives its precision.

$$C(X \rightarrow Y) = \sigma(X \cup Y)/\sigma(X) \quad (4)$$

Extraction of association rules may be divided into two main tasks which are detection of frequent itemsets and generation of rules. Finding frequent itemsets is the most demanding task. From d items, 2^d itemsets can be generated. Many algorithms such as Apriori, FP-Growth, H-Mine and LCM have been proposed to extract association rules.

Association rules were originally designed for market basket analysis, but they are not limited to transactional database. They can be applied to any field if the concept of transaction can be adapted. Recently, association rules have led to interesting results in music information retrieval. They have been applied to obtain relevant feedback during music retrieval process [15], to predict the musical genre [4] and to index audio files [13].

4 Experimentations

Several researches dedicated to automatic music classification try to replicate existing classifications. This approach is somehow an error because targeted classifications represent a point of view that is not necessarily shared by everyone. For instance, many researches use the GTZAN dataset as a reference [3, 5, 12, 14]. This dataset is known to contain faults. These faults were investigated in detail by Sturm [16]. According to him, 10,6% of the dataset is mislabeled. Therefore, researchers able to replicate the GTZAN classification scheme fail in reproducing a universal model of classification but rather the point of view of Tzanetakis. The presence of faults has motivated the choice of GTZAN as dataset. Instead of trying to replicate it, we have decided to use association rules to illustrate how descriptors may impact the classification result.

4.1 Methodology

The excerpts need to be split into transactions to apply association rules mining. So, each excerpt is represented by a set of transactions containing spectrum bins. Spectrum bins are compact descriptors. They can be computed using a straight forward approach. First, the frequencies are obtained using a fast Fourier transform (FFT). The frequencies are then grouped into bins. Firsts bins group low frequencies while lasts bins group high frequencies. Finally, the magnitude of each bin is computed. Few low frequencies are grouped together in firsts bins, but many high frequencies are grouped together in lasts bins because human ear is more sensitive to low frequencies compared to high frequencies. The main advantage of spectrum bins is they allow controlling the number of items inside the transactions. The maximum number of items per transaction is limited to the number of bins considered.

Association rules help identify recurring patterns. In other words, they highlight existing relations between the frequencies which characterize the excerpts. The level of similarity between excerpts is established by comparing the number of association rules they share. The complete procedure is the following:

1. Split audio data: The raw data of excerpts are divided in subsets to further be converted into transactions. This operation aims to represent the signal evolution over time.
2. Compute spectrum bin: The spectrum bins are used to describe the musical content. They are computed for each subset of raw data created in step 1. 31 bins are considered. Only the spectrum bins with the highest magnitude are kept.
3. Extract association rules: Extraction of association rules is performed using the Apriori algorithm [2]. Rules like Bin 3, Bin 4 $\rightarrow$ Bin 1 are extracted. The previous rule can be interpreted as "when the spectrum bins 3 and 4 contain significant frequencies then the spectrum bin 1 also contains significant frequencies".
4. Compare association rules: 80% of excerpts from a genre are used to build a prototype of this genre. Only the association rules shared by all these excerpts are used to build the prototype. The other 20% of the excerpts are kept for evaluation purpose. The association rules of evaluation excerpts are compared with those of all prototypes.
5. Illustrate similarity levels: A similarity matrix is generated to help the visualization of level of similarity between excerpts. Rows represent the evaluation excerpts while columns represent the genre prototypes. Each cell gives the percentage of association rules shared between a given evaluation excerpt and a given genre prototype. A third dimension, the brightness, is added to facilitate the visualization of the data. A dark cell indicates a strong similarity while a white cell means little similarity.

After having been applied the proposed procedure, the visual representation generated can be used to measure the impact of the selected descriptor on the classification result.

4.2 Results

The proposed approach is not a supervised approach. It is not built to reproduce the GTZAN classification. As discussed previously, the musical genres are ill-defined categories. Experiments aim to highlight similarities and differences between the excerpts and genres.

Table 1 shows the classical similarity matrix. Looking at this matrix, it's obvious that the classical evaluation excerpts shared a significant number of association rules with the prototype of classical genre. In fact, 90% of the time, classical evaluation excerpts shared more association rules with classical prototype than any other prototype. Table 1 also illustrates that Classical and Jazz excerpts share many rules.

Table 1. Classical similarity matrix

SONG ID	ROCK	REGGAE	POP	METAL	JAZZ	HIP-HOP	DISCO	COUNTRY	CLASSICAL	BLUES
C95	20%	5%	1%	30%	9%	7%	13%	2%	4%	8%
C90	16%	3%	6%	20%	5%	6%	20%	3%	17%	4%
C85	9%	3%	9%	2%	12%	5%	14%	8%	30%	8%
C80	8%	4%	9%	0%	19%	3%	7%	18%	22%	11%
C75	5%	4%	6%	1%	23%	4%	3%	12%	35%	7%
C70	6%	4%	5%	3%	15%	5%	6%	12%	34%	9%
C65	6%	5%	9%	1%	21%	6%	7%	12%	24%	9%
C60	8%	3%	8%	2%	16%	4%	9%	10%	32%	7%
C55	6%	4%	9%	1%	21%	5%	4%	14%	29%	8%
C50	6%	4%	7%	1%	22%	5%	4%	14%	29%	8%
C45	7%	3%	11%	2%	17%	4%	8%	12%	26%	9%
C40	7%	4%	13%	2%	17%	5%	9%	12%	23%	8%
C35	9%	5%	7%	1%	18%	4%	5%	18%	21%	12%
C30	5%	4%	8%	1%	21%	6%	6%	13%	29%	8%
C25	4%	3%	4%	0%	18%	6%	4%	15%	40%	7%
C20	4%	3%	7%	1%	21%	5%	3%	12%	37%	7%
C15	14%	3%	6%	3%	7%	1%	18%	5%	28%	13%
C10	6%	3%	6%	2%	15%	6%	7%	13%	36%	7%
C05	6%	3%	9%	1%	19%	5%	7%	13%	30%	7%
C00	8%	5%	7%	3%	14%	4%	12%	6%	35%	7%

Table 2 shows the hip-hop similarity matrix. In contrast to the results obtained for classical music, only 10% of the hip-hop evaluation excerpts are more similar with their prototype than any other prototype. Nevertheless, the similarity matrix indicates that there are indeed similarities between hip-hop evaluation excerpts and hip-hop prototype. More surprisingly, the similarity matrix suggests that there is a possible relation between reggae, popular and hip-hop excerpts. After having listening the excerpts, this relation seems effectively existing at least for this dataset. The low level of similarity between hip-hop evaluation files and the classical prototype is also highlighted in the similarity matrix. Thus, considering the model generated from spectrum bins, it would be appropriate to recommend popular or reggae music to hip-hop fan and not appropriate to recommend to him classical music.

Visualizing the distribution of descriptors helps to understand why automatic processes may group songs belonging to different genres into a single category. This can also be useful for measuring the descriptors quality.

Table 2. Hip-Hop similarity matrix

SONG ID	ROCK	REGGAE	POP	METAL	JAZZ	HIP-HOP	DISCO	COUNTRY	CLASSICAL	BLUES
H95	8%	16%	13%	12%	8%	16%	9%	7%	1%	10%
H90	5%	19%	22%	6%	6%	18%	9%	8%	1%	7%
H85	5%	17%	18%	12%	8%	16%	9%	7%	1%	8%
H80	7%	11%	12%	9%	12%	11%	8%	14%	8%	8%
H75	5%	21%	18%	6%	7%	17%	9%	8%	1%	9%
H70	11%	13%	8%	9%	9%	14%	11%	10%	1%	14%
H65	7%	17%	13%	8%	9%	16%	11%	7%	1%	11%
H60	6%	19%	15%	7%	9%	18%	9%	7%	1%	10%
H55	7%	20%	9%	9%	8%	14%	10%	9%	1%	12%
H50	7%	18%	14%	8%	8%	17%	10%	8%	1%	10%
H45	5%	18%	19%	6%	9%	16%	8%	7%	3%	9%
H40	6%	15%	11%	7%	9%	13%	9%	11%	12%	7%
H35	6%	19%	17%	7%	8%	18%	9%	7%	1%	9%
H30	9%	15%	10%	9%	10%	17%	11%	6%	1%	12%
H25	9%	19%	19%	8%	4%	16%	11%	7%	3%	6%
H20	6%	14%	20%	24%	2%	19%	6%	4%	0%	4%
H15	4%	16%	18%	4%	11%	14%	7%	8%	9%	8%
H10	8%	16%	13%	7%	9%	16%	12%	6%	2%	10%
H05	4%	21%	20%	6%	7%	17%	8%	7%	1%	8%
H00	16%	5%	7%	22%	6%	9%	20%	3%	9%	5%

5 Conclusion

Automatic music classification relies on evaluation of descriptors extracted from audio signal. Descriptors have an important impact on result. Selecting the right descriptors is one of the main difficulties associated with automatic music classification. Opacity of current methods makes it difficult to evaluate the contribution of descriptors in the classification process. In this paper, we have proposed to use association rules to add more interpretability. Association rules highlight recurring patterns hidden in data. The same recurring patterns may be observed in several songs. Therefore, association rules can be used to evaluate similarities and dissimilarities between songs. The proposed approach is explanatory in contrast with many other which are black boxes.

Experiments were conducted considering only one kind of descriptors but it easy to imagine that other descriptors can be considerate. This would help to discover which descriptors are meaningful and which aren't.

References

1. Agrawal, R., Imielinski T., Swami, A.: Minning association rules between sets of items in large databases, In: Proceedings of the SIGMOD, pp. 207–216 (1993)
2. Agrawal, R., Srikant, R.: Fast algorithms for mining association rules. In: Proceedings of the 20th International Conference on Very Large Database, pp. 487–499 (1994)
3. Ali, M.A., Siddiqui, Z.A.: Automatic music genres classification using machine learning. Int. J. Adv. Comput. Sci. Appl. **8**, 337–344 (2017)
4. Arjannikov, T., Zhang, J.Z.: An association-based approach to genre classification in music. In: ISMIR, pp. 95–100 (2014)
5. Chang, K., Jang, J.-S.R., Iliopoulos, C.S.: Music genre classification via compressive sampling. In: Proceedings of the International Society Music Information Retrieval, pp. 387–392 (2010)
6. Chang, L., Yu, X., Wan, W., Yao, J.: Research on fast music classification based on SVM in compressed domain. In: Proceedings ICALIP, pp. 638–642 (2008)
7. Choi, K., et al.: Convolutional recurrent neural networks for music classification. In: 2017 IEEE International Conference on Acoustics, Speech and Signal Processing (ICASSP). IEEE (2017)
8. Grosche, P., Müller, M., Serrà, J.: Audio content-based music retrieval. Multimodal Music Process. **3**, 157–174 (2012)
9. Han, J., Pei, J., Yin, Y., Mao, R.: Mining frequent patterns without candidate generation: a frequent-pattern tree approach. Data Min. Knowl. Disc. **8**, 53–87 (2004)
10. Mandel, M.I., Ellis, D.P.W.: Song-level features and support vector machines for music classification, pp. 594–599 (2005)
11. McKay, C., Fujinaga, I.: Automatic genre classification using large high-level musical feature sets. In: Proceedings of ISMIR 2004 (2004)
12. Rajan, R., Murthy, H.A.: Music genre classification by fusion of modified group delay and melodic features. In: Communications (NCC), pp. 1–6 (2017)
13. Rompré, L., Biskri, I., Meunier, J.-G.: Using association rules mining for retrieving genre-specific music files. In: Proceedings of FLAIRS 2017, pp. 706–711 (2017)
14. Schindler, A., Rauber, A.: Capturing the temporal domain in echonest features for improved classification effectiveness. In: Nürnberger, A., Stober, S., Larsen, B., Detyniecki, M. (eds.) AMR 2012. LNCS, vol. 8382, pp. 214–227. Springer, Cham (2014). https://doi.org/10.1007/978-3-319-12093-5_13
15. Shan, M.-K., Chiang, M.-F., Kuo, F.-F.: Relevance feedback for category search in music retrieval based on semantic concept learning. Multimedia Tools Appl. **39**(2), 243–262 (2008)
16. Sturm, B.L.: An analysis of the GTZAN music genre dataset. In: Proceedings of the Second International ACM Workshop on Music Information Retrieval with User-Centered and Multimodal Strategies, pp. 7–12 (2012)
17. Sturm, B.L.: A survey of evaluation in music genre recognition. In: Nürnberger, A., Stober, S., Larsen, B., Detyniecki, M. (eds.) AMR 2012. LNCS, vol. 8382, pp. 29–66. Springer, Cham (2014). https://doi.org/10.1007/978-3-319-12093-5_2
18. Tzanetakis, G., Cook, P.: Musical genre classification of audio signals. IEEE Trans. Speech Audio Process. **10**(5), 293–302 (2002)
19. Wang, A.: An industrial-strength audio algorithm. In: Proceedings of ISMIR 2003, pp. 7–13 (2003)

Intelligent Systems in Real-Life Applications

A Wearable Fall Detection System Using Deep Learning

Eduardo Casilari(✉), Raúl Lora-Rivera, and Francisco García-Lagos

Dpto. Tecnología Electrónica, ETSI Telecomunicación,
Universidad de Málaga, 29071 Málaga, Spain
ecasilari@uma.es

Abstract. Due to the growing aging of the population and the impact of falls on the health and autonomy of the older people, the development of cost-effective non-invasive automatic fall detection systems (FDS) has gained much attention. This work proposes and analyzes the capability of convolutional deep neural networks to detect fall events based on the measurements captured by wearable tri-axial accelerometers that are transported by the user to characterize the mobility of the body. The study is performed on a long public data repository containing the traces obtained from a wide group of experimental users during the execution of a predetermined set of Activities of the Daily Living (ADLs) and mimicked falls. The system is evaluated in term of accuracy, sensitivity and specificity when the network is alternatively fed with the module of the acceleration and the with the tri-axial components of the acceleration.

Keywords: Fall detection systems · Deep learning · Convolutional Neural Networks · Accelerometer · Wearables

1 Introduction

During the last decades, the number of older people that face the risks of living on their own has rocketed because of social changes and, especially, due to the noteworthy increase of life expectancy. In this regard, falls are one of the main factors that may degrade the quality of life and the sense of autonomy of the elderly.

According to the World Health Organization [1], 28–35% of people aged over 65 years of age fall at least once per year while this percentage reaches 32–42% for those over 70, as the frequency of falls are strongly related to age and frailty. In this regard, these rates are even higher (up to 50%, with 40% of them experiencing recurrent falls [2]) for people residing in long-term care institutions or nursing homes. In addition, fall-related injuries and hospitalization are expected to annually increase by 2% until 2030. Consequently, the development of systems for the automatic detection of falls could be in the near future a key element to support both the self-sufficiency of the elderly and the economic sustainability of national health and welfare systems.

Fall Detection Systems (FDSs) are designed to track the movements of the user so that an alerting message (SMS, phone call, etc.) can be automatically emitted as soon as a fall accident is presumed to have occurred. In this sense, FDSs are expected to maximize the probability of identifying falls, while minimizing the false positives, i.e.

F. Wotawa et al. (Eds.): IEA/AIE 2019, LNAI 11606, pp. 445–456, 2019.
https://doi.org/10.1007/978-3-030-22999-3_39

the cases in which a conventional movement or ADL (Activity of Daily Living) is mistaken as a fall and an unnecessary alarm is triggered.

FDSs can be grouped into two general categories [3–5]: context-aware (ambient and/or vision based) and wearable systems. In the first group, the identification of the fall relies on the analysis of the signals measured by a set of environmental sensors (such as vibration sensors, audiovisual equipment, etc.) placed in a pre-determined zone around the subject to be supervised.

Context-aware solutions imply several disadvantages. Firstly, they are only effective in a very restricted monitoring area. They may also entail high installation and maintenance costs and a certain privacy invasion if video-cameras or microphones are permanently operative. In addition, the detection decision can be easily altered by spurious elements [6] such as variations of the lighting, external sounds, movements of another individuals or pets, changes in the furniture, falling objects, etc.

Conversely, wearable systems providing ubiquitous tracking of the patients by leveraging sensing nodes that are seamlessly transported and integrated into the user's clothing or garments. Therefore, these sensors (normally inertial measurement units – IMUs - embedding an accelerometer and a gyroscope) only track variables unmistakably associated to the movements of the user. Owing to the plummeting costs and increasing computing power of wearable electronics, the design of wearable FDSs has gained much research attention during the last decade in the field of mHealth (or application of mobile technologies to Health).

The core of an FDS is the detection algorithm, which is in charge of constantly classifying any movement of the subject as a fall or as an ADL. Coarsely, there are two main categories of detection algorithms [7]: Threshold-Based Approaches try to infer the occurrence of falls by comparing the magnitudes captured by the inertial sensors with one or several reference thresholds. However, the nature and variety of ADLs and falls is quite complex. In fact, the definition itself of the term 'fall' (a loss of balance or an accident which results in an individual coming to rest unintentionally on the ground or other lower level) is still ambiguous and controversial [8]. As a consequence, deterministic 'thresholding' fall detection methods are too inflexible and normally yield poor results.

On the other hand, pattern recognition methods based on machine learning and artificial intelligence strategies have been shown to clearly outperform thresholding strategies [9]. Under pattern recognition architectures with supervised learning (such as k-Nearest Neighbor or Support Vector Machine), the classification algorithms can be automatically parametrized and tuned to optimize the results for a wide set of training samples without heuristically defining arbitrary decision thresholds.

In the field of machine learning, Convolutional Neural Networks (CNNs) are one of the most promising AI techniques. CNNs, consisting of a set of sequenced processing layers interconnected through nodes or 'neurons', are able to autonomously identify underlying structures in big datasets by learning the representation of the data with different layers of abstraction [10].

CNNs were especially conceived and oriented for image processing applications but they have been successfully applied in other domains such as speech, text or audio analysis. More recently, CNNs and deep learning have been adopted in HAR (Human Activity Recognition) systems to substitute traditional strategies that require some kind of 'hand-crafted' feature extraction from the data series [11]. In [12], for example, authors compare the results of applying deep, convolutional, and recurrent approaches to three different datasets containing inertial information captured with wearable sensors.

In this paper, we employ CNNs to assess the capability of these deep learning architectures to perform as the detection algorithm in a wearable FDS scheme. In order to train and test the networks we utilize one of the largest public data repository of accelerometry signals of falls and ADLs.

2 Election and Description of the Dataset

One of the key aspect in the deployment of an FDS is the experimental procedure that must be followed to assess the effectiveness of the detection algorithm.

Given the obvious difficulties of using real-world falls (in particular those experienced by older people), most studies on fall detection systems utilize movements traces captured by monitoring the activity of a group of volunteers (specifically recruited for that objective) while they execute a preconfigured set of ADLs and emulated falls (normally on a mattress or any other cushioned surface) in a controlled test scenario (e.g. a laboratory). In most cases, the generated datasets used for the evaluation are not published, so they cannot be used as the basis for the cross-comparison with alternative detection methods proposed by other authors.

To address this shortcoming and to provide a common benchmarking framework for fall detection algorithms, during the last five years, several data bases have been released by different research groups working on this topic (see [13] for a revision of most existing datasets). All these samples consist of a repository of files containing the numerical series with the measurements captured by one (or several) accelerometers (and in some cases, gyroscopes and magnetometers) located at one or several positions of the body of the volunteers that performed the corresponding movements (ADLs or falls).

After revising the characteristics of the available datasets, we chose the SisFall repository [14] for our analysis, as it combines: a wide group of experimental subjects (38 volunteers comprising 19 males and 19 females) with the highest age range (19–75 years), a numerous amount of samples (4505, including 2707 ADLs and 1798 emulated falls) with a considerably long duration (between 10 and 180 s per sample, with a mean value of 15 s). The typology of the executed activities also presents a remarkable variety of movements: 19 types of ADLs (ranging from sitting down to stumbling or jogging) and 15 types of falls (depending on the direction of the fall and the initial position). As a general rule, every volunteer replicated every considered movement 5 times, except the activities implying jogging and walking, which were executed just once but for a longer period of time (100 s).

During each experiment, the volunteers transported a sensing node that incorporated two accelerometers (ADXL345, MMA8451Q), an ITG3200 gyroscope and SD card to record the measurements. The sensing mote was attached to the waist, which is considered to be good point to capture the human mobility [15] as it is relatively close to the gravity center of the human body and not linked to the individual movements of a particular limb (as the wrist, the ankle or the knee). The sampling rate was set to 200 Hz, which is far above the minimum frequency of 20–40 Hz, that is recommended for a proper characterization of the fall dynamics [16].

In any case, we have to point out that the suitability of evaluating a FDS with falls mimicked by young healthy volunteers on a padded surface is still a controversial issue out of the scope of this paper. Some works, such as that by Klenk in [17], have detected noteworthy differences between the dynamics of real-life falls and those emulated on a controlled testbed. Conversely, after investigating the mobility patterns of actual falls suffered by older people, Jämsa et al. state in [18] that real-life and intentional falls present similar characteristics.

3 Discussion on the Input Features

The detection decision in a wearable FDS is supposed to be produced in real-time by an embedded system with strong limitations in terms of computing and processing power. In addition, one of the main advantages of CNNs is that they can learn autonomously from the internal structure of the input data, without requiring an initial and manual feature extraction.

Consequently, in order to define the input features that will feed the CNN, we propose to directly employ the measurements captured by the sensors. In this manner, we avoid the real-time computation of an arbitrarily selected set of complex statistics derived from the time series of the measurements (statistical moments, autocorrelation, discrete Fourier transform, primary and secondary peaks and 'valleys' of the acceleration module, etc.).

As in most works in the related literature, we consider the accelerometry signals as the basis for the detection algorithm. To simplify the structure and the dimension of the network, we used as inputs the measurements of the ADXL345 accelerometer (a future work should contemplate if the data provided by the gyroscope and the other accelerometer introduce any improvements in the effectiveness of the movement classification).

Falls are always linked to one or several sudden increases of the acceleration values provoked by the impact (or impacts) of the body against the floor [19]. Accordingly, we focus our analysis on a fixed time window around the peak or maximum value) of the acceleration module of every sample, as it is the interval where a fall is most likely to have occurred. This acceleration module or Signal Magnitude Vector (SMV_i) for the i-th sample can be calculated as:

$$SMV_i = \sqrt{A_{x_i}^2 + A_{y_i}^2 + A_{z_i}^2} \tag{1}$$

where A_{x_i}, A_{y_i} and A_{z_i} define the x, y, and z components of the measured acceleration vector (being A_{Y_i} the acceleration measured in the direction perpendicular to the floor plane when the user is standing).

Thus, the maximum of the SMV is straightforwardly computed as:

$$SMV_{max} = SMV_{k_o} = \max(\{SMV_i : i \in [1, N]\}) \tag{2}$$

where N is the number of samples of the trace while k_o is the index of the sample at which the maximum value of the module is found.

The impact of a fall is preceded by a certain phase of 'free fall' (in which the acceleration module tends to zero) and is accompanied (both before and after the contact with the ground) by sudden alterations of the values of the three coordinates of the acceleration, caused by the successive and abrupt changes of the orientation of the body.

As a typical fall lasts between 1 s and 3 s [20], by observing a time window of 5 s around the peak (2.5 s before and after the peak) the most relevant aspects of the dynamics of the fall will be most probably included in analysis patterns that are delivered as input data to the CNN.

By way of example, Fig. 1 shows the evolution of the SMV and the acceleration components for a certain ADL (during which the subject quickly sits on a chair, remains seated some seconds, and then gets up quickly) and an emulated fall (a forward collapse caused by a slip while walking). Figure 2 depicts in turn the same variables for the same movements when we just consider the 5-s observation window around the detected maximum.

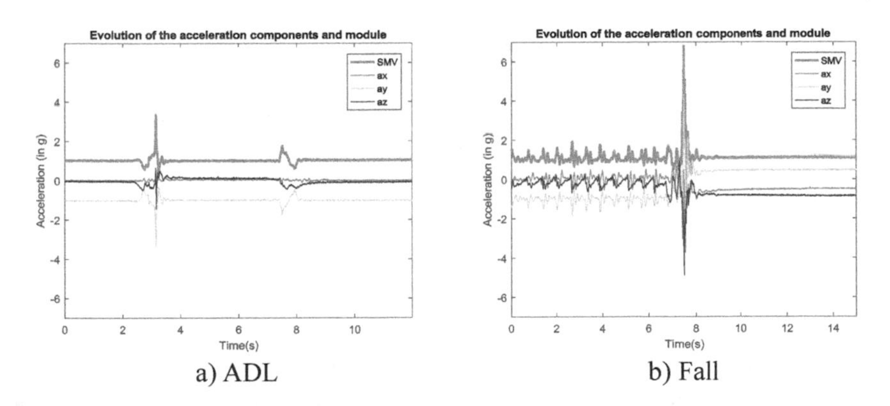

Fig. 1. Example of the evolution of the SMV and the acceleration components during the execution of an ADL and a fall.

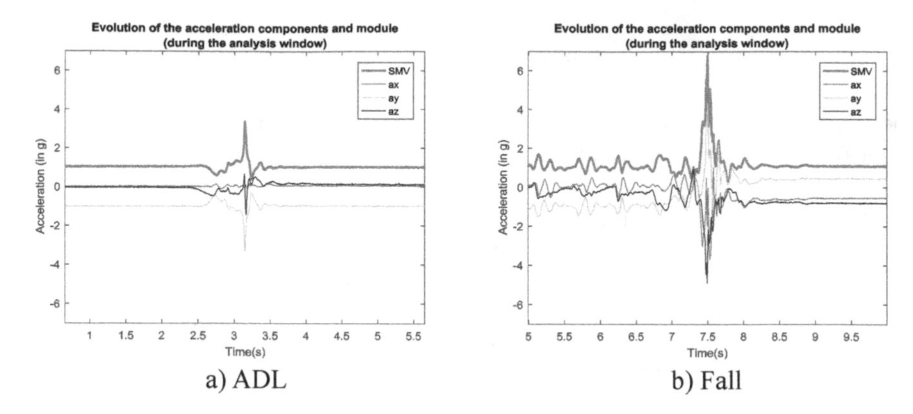

Fig. 2. Example of the evolution of the SMV and the acceleration components during the execution of an ADL and a fall for an observation window of 5 s around the detected peak of the SMV.

The example clearly illustrates that a window of 5 s is enough to characterize the sudden ups and downs in the acceleration signals caused by a fall.

In this regard, we consider two possible cases for the network inputs of the CNN:

- The sequence of the acceleration modules (SMV_j) around the peak:

$$SMV_j \quad \forall j \in \left[k_o - \frac{T}{2}f_s,\, k_o + \frac{T}{2}f_s\right] \tag{3}$$

where f_s represents the sampling rate while T is the duration of the observation window. As the sampling rate of the employed traces is 200 Hz, a window of 5-s requires 1001 values.

- The sequence of three triaxial components of the acceleration around the same peak (which imply 3001 input features at the CNN):

$$A_{x_j}, A_{y_j}, A_{z_j}, \quad \forall j \in \left[k_o - \frac{T}{2}f_s,\, k_o + \frac{T}{2}f_s\right] \tag{4}$$

4 Configuration of the CNN

The general goal of a feedforward neural network is to map a fixed-size input (in our problem, the accelerometer data) into a fixed-size output (a binary probability of 0 or 1, depending on the nature of the movements, ADL or fall). CNNs differ from traditional multilayer perceptrons in some key aspects. Firstly, a CNN consists of several stage or neuron layers (such as convolutional layers, pooling layers, and fully-connected layers) that are in charge of different functions. Secondly, neurons in a layer may be uniquely connected to other neurons in a particular 'zone' or subregion of the following layer, unlike the fully-connected scheme of other categories of neural networks for which the number of parameters (weights) can increase enormously as the number of inputs rises. This enables the neurons in a CNN to be only responsive to a certain selection of the inputs (e.g. a certain phase or period during the fall in our input dataset). As these subregions can overlap, the neurons of a CNN may yield correlated outputs (a correlation that does not exist in other types of neural networks, where neurons are not grouped and produce independent individual results).

During training, the neurons in each layer act as 'feature extractor' by learning nonlinear combinations of the inputs. These learned features from one layer become the inputs for the next layer. For that purpose, every set of neurons pass the weighted sum of their inputs (received from the precedent layer) through a non-linear activation function. Apart from the 'classical' sigmoids, one of the most popular activation functions (used in this work) is the Rectified Linear Unit (ReLU), defined as $f(z) = max(z, 0)$ [10].

The architecture of a CNN is defined through the typology and number of layers that it includes. If a small basic structure with only one or two convolutional layers might be adequate to learn a small set of data, more layers can be needed for complex datasets, as that used in our proposal.

An equivalent 'image' of (1 $\times$ width) 'pixels' is utilized as the network input, where the parameter 'width' indicates the size of the samples (input features of the CNN) in the observation window around the acceleration peak (1001 or 3001).

After the initial input layer that directly receives the accelerometer data, the convolutional layers are responsible for detecting the local conjunctions of features from the inputs. Pooling layers are in turn placed immediately after these convolutional layers. The goal of pooling layers is to down-sample and condense the features obtained by the convolutional stages into a summarized feature map. Each neuron in the pooling layer may condense or represent a region of neurons in the previous layer. The common scheme employed in our architecture is known as max-pooling. Under this paradigm, a pooling unit simply outputs the maximum activation in the input region of neurons.

The previous layers are followed by one fully connected layer, which combines all the features (local information) learned by the previous layer across the input signal (processed as input 'images'), aiming at identifying the largest patterns. Thus, the system is enabled to classify the 'images' between different output types (ADL and falls). For classification problems, as the one considered in this project, a softmax layer is placed after the final fully connected layer.

The global learned features become the inputs to the end layer, which can be a classifier or a regression function that yields an output in the form of a continuous variable. In our FDS, for which a binary response (fall or ADL) is required, we select an output classification function. In this final layer, the output values from the softmax function are assigned to one of the two mutually exclusive classes using the cross-entropy function.

4.1 Cross-Validation Method

To avoid overfitting, we apply the cross-validation method. To this end, the whole repository is split into three independent groups: training, validation and test sets, so that 60%, 20% and 20% of the samples in the original dataset are respectively assigned to these three sets. The division is performed randomly but guaranteeing the same proportional distribution of the falls and ADLs within the three groups.

The validation dataset is employed to evaluate the network after being trained (with the training dataset) for a certain number of epochs. This process is iterated until the error for the validation patterns increases, which is an indication that the network is beginning to overfit the data. Once the training phase is finished, the performance of the CNN is assessed with the test data, which are used as independent input to compute an error that characterizes the network efficacy.

In addition, during training, dropout layers and L2 Regularization techniques are also considered to prevent overfitting. To decide the moment where the training procedure should stop, we set 20 maximum training epochs and a validation patience (or number of times that the loss computed for the validation set can increase) to 3. Figure 3 represents the evolution of the accuracy and loss of the training and validation sets during the learning process.

To sum up, Table 1 recapitulates the structure and basic characteristics of the employed CNN and the parameters employed for the training phase. As it can be observed, the learning architecture includes four feature extraction layers and one final classifying layer. Each feature extractor contains one convolutional layer, one batch

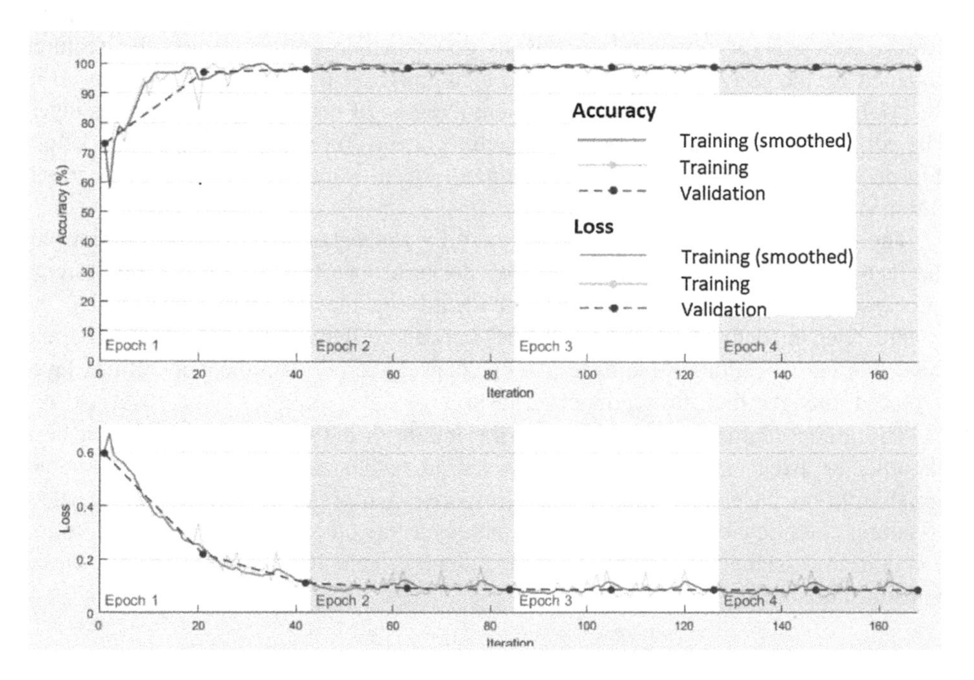

Fig. 3. Training progress: Training and validation accuracy curves

Table 1. Characteristics of the employed CNN.

Training algorithm	Stochastic gradient descent momentum
Layers activation functions	ReLU (hidden layers) and softmax (output layer)
Maximum number of training epochs	20
Iterations per epoch	42
Validation patience	3
Mini-batch size (to estimate the gradient of the loss in every iteration)	64 training instances
Techniques to prevent overfitting	L2Regularization facto and dropout layers
Initial learning rate:	0.0001
Number of feature extraction layers	4
Sub-layers for every extraction layer	4 (1 convolutional, 1 normalization, 1 ReLU and 1 max pooling layers)
Number of filters for each convolutional layer	16 (1st layer), 32 (2nd), 64 (3rd), 128 (4th)
Filter size (for all convolutional layers)	1 × 5
Size of zero-padding	2 samples
Stride	1 × 1 ("non-strided")
Pool size of the max-pooling layer	1 × 5
Classification layers	1 fully-connected layer, 1 softmax layer and 1 final classifier

normalization layer, one ReLU layer (with a nonlinear ReLU activation function) and one max pooling down-sampling layer (except for the final feature extraction layer, which does not contain this final max pooling element). The final classifying layer comprises in turn one fully-connected input layer, one softmax output layer (which applies the normalized exponential function) and one final classifier that bases the decision on a cross-entropy function.

5 Numerical Results

The CNN was implemented through Matlab [21] scripts that employed the so-called Deep Learning Toolbox™ [22].

To characterize the behavior of the CNN we computed three typical performance metrics that are commonly considered in the evaluation of binary classification systems: sensitivity or recall (which is related to the capability of detecting falls), specificity (which describes the efficacy of the detector to avoid false alarms – ADLs misclassified as falls-), and accuracy (which offers a global measurement of the effectiveness of the system). These metrics (defined as a decimal) can be calculated as the following ratios:

$$Sensitivity = \frac{TP}{FN + TP} \tag{5}$$

$$Specificity = \frac{TN}{FP + TN} \tag{6}$$

$$Accuracy = \frac{TP + TN}{TP + FP + TN + FN} \tag{7}$$

where TP and TN respectively represent the amounts of 'True Positives' and 'True Negatives' (i.e. the numbers of falls and ADLs that have been properly identified), while FP and FN indicate the numbers of 'False Positive' and 'False Negatives' (ADLs and falls that have been wrongly recognized).

The results (expressed as percentages) obtained with the test data are presented in Table 2. The table shows that the effectiveness of the FDS (especially the sensitivity) clearly improves when the 3-axis signals are considered. This can be explained by the fact that the information related to the abrupt changes in the direction of the acceleration provoked by the falls most presumably helps the CNN to discriminate the fall patterns.

Table 2. Performance results of the detection system.

Input feature	Performance metric		
	Specificity	Sensitivity	Accuracy
SMV	98.19%	92.84%	96.12%
3-axis signals	99.25%	98.92%	99.1%

These results (with both specificity and sensitivity around 99%) are better than those obtained by other studies that propose fall detection algorithms and employ the same SisFall dataset to assess their efficacy, such as the works in [14, 16, 23–25], where a specificity or/and a sensitivity higher than 0.98 are not achieved. Similarly, our system also shows a better behavior than the Recurrent Neural Networks (RNNs) presented in [26], which is also used as the core of a FDS and tested with three different datasets.

6 Conclusions

This work has presented a detection algorithm aimed at wearable systems for fall alerting and personal tracking. The algorithm, which classifies every movement as a fall or an ADL (Activity of Daily Living) is based on a convolutional neural network fed with the signals captured by an accelerometer. To simplify the required preprocessing of the signals, the system directly bases its decision on the acceleration samples measured during a time window centered around the instant where an acceleration peak is detected and a fall event can be suspected.

The performance of the detection process is assessed by using one of the largest existing public repository of movements containing emulated falls. Results (with specificity and sensitivity around 99%) indicate that the efficacy of the detector notably increases when the three components of the acceleration (instead of the acceleration module) are input into the convolutional neural network. Further studies should analyze the benefits of complementing the characterization of the movements with the measurements of other sensors (e.g. the gyroscope). Additionally, the system should be deployed on a wearable system to evaluate its performance (at least in terms of the false alarms that it provokes) when it is applied to real-life movements during a long monitoring period.

Acknowledgements. This work was supported by Universidad de Málaga, Campus de Excelencia Internacional Andalucia Tech.

References

1. World Health Organization. Ageing & Life Course Unit: WHO global report on falls prevention in older age. World Health Organization, Geneva, Switzerland (2008)
2. Orces, C.H., Alamgir, H.: Trends in fall-related injuries among older adults treated in emergency departments in the USA. Inj. Prev. **20**, 421–423 (2014)
3. Mubashir, M., Shao, L., Seed, L.: A survey on fall detection: principles and approaches. Neurocomputing **100**, 144–152 (2013)
4. Igual, R., Medrano, C., Plaza, I.: Challenges, issues and trends in fall detection systems. Biomed. Eng. Online **12**, 66 (2013)
5. Chaccour, K., Darazi, R., El Hassani, A.H., Andres, E.: From fall detection to fall prevention: a generic classification of fall-related systems. IEEE Sens. J. **17**, 812–822 (2017)
6. Zhang, D., Wang, H., Wang, Y., Ma, J.: Anti-fall: a non-intrusive and real-time fall detector leveraging CSI from commodity WiFi devices. In: Geissbühler, A., Demongeot, J., Mokhtari, M., Abdulrazak, B., Aloulou, H. (eds.) ICOST 2015. LNCS, vol. 9102, pp. 181–193. Springer, Cham (2015). https://doi.org/10.1007/978-3-319-19312-0_15
7. Casilari, E., Luque, R., Morón, M.: Analysis of android device-based solutions for fall detection. Sensors **15**, 17827–17894 (2015)
8. Yoshida, S.: A global report on falls prevention epidemiology of falls. World Health Organization (2007)
9. Aziz, O., Musngi, M., Park, E.J., Mori, G., Robinovitch, S.N.: A comparison of accuracy of fall detection algorithms (threshold-based vs. machine learning) using waist-mounted tri-axial accelerometer signals from a comprehensive set of falls and non-fall trials. Med. Biol. Eng. Comput. **55**, 45–55 (2017)
10. LeCun, Y., Bengio, Y., Hinton, G.: Deep learning. Nature **521**, 436–444 (2015)
11. Ordóñez, F., Roggen, D., Ordóñez, F.J., Roggen, D.: Deep convolutional and LSTM recurrent neural networks for multimodal wearable activity recognition. Sensors **16**, 115 (2016)
12. Hammerla, N.Y., Halloran, S., Plötz, T.: Deep, convolutional, and recurrent models for human activity recognition using wearable. In: Proceedings of the Twenty-Fifth International Joint Conference on Artificial Intelligence, pp. 1533–1540. AAAI, New York, 09–15 July 2016 (2017)

13. Casilari, E., Santoyo-Ramón, J.A., Cano-García, J.M.: Analysis of public datasets for wearable fall detection systems. Sensors **17**, 1513 (2017)
14. Sucerquia, A., López, J.D., Vargas-bonilla, J.F.: SisFall: a fall and movement dataset. Sensors **198**, 1–14 (2017)
15. Ntanasis, P., Pippa, E., Özdemir, A.T., Barshan, B., Megalooikonomou, V.: Investigation of sensor placement for accurate fall detection. In: Perego, P., Andreoni, G., Rizzo, G. (eds.) MobiHealth 2016. LNICST, vol. 192, pp. 225–232. Springer, Cham (2017). https://doi.org/10.1007/978-3-319-58877-3_30
16. Nguyen, L.P., Saleh, M., Le Bouquin Jeannès, R.: An efficient design of a machine learning-based elderly fall detector. In: Ahmed, M.U., Begum, S., Bastel, J.-B. (eds.) HealthyIoT 2017. LNICST, vol. 225, pp. 34–41. Springer, Cham (2018). https://doi.org/10.1007/978-3-319-76213-5_5
17. Klenk, J., et al.: Comparison of acceleration signals of simulated and real-world backward falls. Med. Eng. Phys. **33**, 368–373 (2011)
18. Jämsä, T., Kangas, M., Vikman, I., Nyberg, L., Korpelainen, R.: Fall detection in the older people: from laboratory to real-life. Proc. Est. Acad. Sci. **63**, 341–345 (2014)
19. Hsieh, C.-Y., Liu, K.-C., Huang, C.-N., Chu, W.-C., Chan, C.-T.: Novel hierarchical fall detection algorithm using a multiphase fall model. Sensors (Basel) **17**, 307 (2017)
20. Yu, X.: Approaches and principles of fall detection for elderly and patient. In: Proceedings of the 10th International Conference on e-Health Networking, Applications and Services (HealthCom 2008), pp. 42–47. IEEE, Singapore (2008)
21. Davis, T., Sigmon, K.: MATLAB Primer, 7th edn. http://www.mathworks.com/products/matlab/
22. Deep Learning Toolbox Documentation – MathWorks. https://es.mathworks.com/help/deeplearning/index.html?searchHighlight=DeepLearningNetworkToolbox&s_tid=doc_srchtitle
23. Carletti, V., Greco, A., Saggese, A., Vento, M.: A smartphone-based system for detecting falls using anomaly detection. In: Battiato, S., Gallo, G., Schettini, R., Stanco, F. (eds.) ICIAP 2017. LNCS, vol. 10485, pp. 490–499. Springer, Cham (2017). https://doi.org/10.1007/978-3-319-68548-9_45
24. Mastorakis, G.: Human fall detection methodologies: from machine learning using acted data to fall modelling using myoskeletal simulation (2018). https://eprints.kingston.ac.uk/42275/1/Mastorakis-G.pdf
25. Putra, I.P.E.S., Brusey, J., Gaura, E., Vesilo, R.: An event-triggered machine learning approach for accelerometer-based fall detection. Sensors. **18**, 20 (2017)
26. Mauldin, T.R., Canby, M.E., Metsis, V., Ngu, A.H.H., Rivera, C.C.: SmartFall: a smartwatch-based fall detection system using deep learning. Sensors (Basel) **18**, 3363 (2018)

Analysing the Performance of Fingerprinting-Based Indoor Positioning: The Non-trivial Case of Testing Data Selection

Maciej Grzenda(✉)

Faculty of Mathematics and Information Science, Warsaw University of Technology, ul. Koszykowa 75, 00-662 Warszawa, Poland
M.Grzenda@mini.pw.edu.pl
http://www.mini.pw.edu.pl/~grzendam/

Abstract. Indoor positioning methods make it possible to estimate the location of a mobile object in a building. Many of these methods rely on fingerprinting approach. First, signal strength data is collected in a number of reference indoor locations. Frequently, the vectors of the strength of the signals emitted by WiFi access points acquired in this way are used to train machine learning models, including instance-based models.

In this study, we address the problem of signal strength data acquisition to verify whether different strategies of selecting signal strength data for model testing are equivalent. In the analysed case, the content of a testing data set can be created in a variety of ways. First of all, leave-one-out approach can be adopted. Alternatively, data from randomly selected points or same grid points can be used to estimate method accuracy. We show which of these and other approaches yield different accuracy estimates and in which cases these differences are statistically significant. Our study extends previous studies on analysing the performance of indoor positioning systems. At the same time, it illustrates an interesting problem of testing data acquisition and balancing the conflicting needs of collecting testing data in similar, yet different conditions compared to how training data was acquired.

Keywords: Performance evaluation · Indoor positioning · Regression · Data preprocessing

1 Introduction

Indoor positioning methods (IPM) [1,7,13] also referred to as localisation algorithms [10] are used to estimate the location of mobile objects inside of single or multi-floor buildings. Many IPM solutions rely on *fingerprinting* [2,6,8,10] paradigm. In terminal-centric version of fingerprinting, first measurements of received signal strengths (RSS) observed in a number of reference points (RPs) of a building are made. As a result, a database of labelled RSS vectors linked to

F. Wotawa et al. (Eds.): IEA/AIE 2019, LNAI 11606, pp. 457–469, 2019.
https://doi.org/10.1007/978-3-030-22999-3_40

known $(x, y, floor)$ locations is created. This is referred to as a *radio map*. Next, in on-line phase, the location of an object is estimated using RSS measurements, acquired in an unknown location, as an input for IPM. By comparing the newly acquired RSS vector to the vectors contained in a radio map, the location of the object can be estimated. It is important to note that due to major cost of labelled RSS data acquisition, attempts to reduce the number of records and RPs in which these records are collected are needed.

Most frequently, a positioning algorithm involves the use of machine learning (ML) techniques. The positioning task can be treated as a *regression* (in the case of x and y estimation) or a *classification* (in the case of floor detection) task. Thus, for indoor positioning techniques such as Nearest Neighbour (NN) algorithm [1,13,16], (weighted) kNN [9,13], or random forest [4] are frequently used. Empirical evaluation of the accuracy of individual methods is based on tests made in a number of indoor locations. It relies on comparing true coordinates with the coordinates estimated by the IPM of interest. As different indoor testbeds are used to evaluate individual techniques, the areas used for the tests, the number of RPs, and their granularity differ, which makes the comparison of reported accuracies difficult [7]. Furthermore, comparison of localisation techniques has been also reported to be hindered by lack of standardised representative data [7]. In this study we aim to contribute to the development of IPM performance evaluation practices by investigating the methods of selecting testing data used to analyse the performance of IPMs.

The evaluation of the accuracy of indoor positioning techniques is done by referring to the approaches used in the field of machine learning. A fundamental assumption is to perform the evaluation of the algorithm using the data not included in the training data set. However, the way evaluation guidelines known from ML domain should be adapted to fingerprinting-based algorithms remains largely an open issue.

The key aspect of the evaluation, which is addressed in this study, is the selection of testing data sets used to evaluate individual IPMs. Since the number of records present in a radio map should be relatively large, typically available RSS data is divided into one training and one testing data set. This follows the idea of *holdout* approach [5] frequently present in ML studies i.e. using a separate set of instances to assess model performance. One of the advantages of the holdout approach is that error estimates are obtained on a separate test set i.e. the advantage of these estimates is their independence from training set [5]. Hence, given enough test data is available and the data are representative of the domain of interest, we can obtain reliable error estimates. If the data set is limited, we can use techniques such as cross-validation [3,5], which relies on dividing the data into multiple parts and repeating the evaluation process with one part playing the role of testing data in every iteration. Importantly, also in this case the assumption that the use of a separate testing set is necessary to evaluate ML model is fundamental for the evaluation.

However, the content of the testing data sets can be created in a variety of ways that satisfy the condition of splitting available data into disjoint training part(s) and testing part(s). In the case of fingerprinting-based IPS, typically a separate set of testing points is established, the data from which are used as testing data. In different studies, the testing points are equally distributed in a testing area of a building [10], randomly placed [14] or manually placed in different locations, which are not within the same fixed distance from each other [10,11,14,15]. Last but not least, in some studies how they were placed is not clarified [7,9,12]. Furthermore, leave-one-out approach can be adopted. More precisely, one of the training RSS vectors can be removed from the training data and treated as a testing data set containing a single element. Similarly, the data from a single RP can be moved in every iteration to a testing data set. Obviously, this process has to be repeated to get reliable estimates. Alternatively, an entire RSS data set can be randomly divided into a training and a testing data set.

Moreover, in many studies [2,10,15] measurements are repeated in each location several times. Hence, when random division of a data set is used, some RSS vectors from one location can be potentially placed both in a training and a testing data set. In particular, in [2] out of 100 samples made in each location, 90 are used for training and 10 for testing. Similarly, Dawes and Chin in [1] to create testing data selected 5000 of all RSS vectors, which were sampled across all 56 locations in which RSS measurements were made. It is important to note that similar decisions can be made in other domains. As an example data from the same or different data sources such as sensors or patients can be placed in testing data set. Furthermore, multiple records from the same source can be collected and split into training and testing data. Hence, the problem of testing RSS data selection illustrates a wider class of problems encountered when ML methods are used to develop classification or regression models.

The question arises whether different ways of splitting available RSS data intro training and testing data are equivalent, and if not which of them could be recommended. This can be treated as in intuitive example of the investigation of how the selection of testing data in view of under which conditions (here: in which locations) the data were collected should be performed. To address these open issues and promote progress in the development of indoor positioning methods, we propose guidelines on the selection of testing data set. We use a training data set composed of the data collected in a number of indoor locations as a reference set. Moreover, a reference method of creating testing data sets is proposed. Next, based on machine learning practices and the aforementioned approaches used in the past, a number of other methods of constructing testing data sets are analysed and compared in terms of the equivalence of performance estimates they provide. Finally, statistical significance of accuracy differences caused by varied testing data sets is verified. The remainder of this work is organised as follows:

- First, the data used for the experiments are documented in Sect. 2.
- Next, different methods of creating testing data sets are summarised in Sect. 3. This includes both existing and newly proposed approaches.

- This is followed by the proposal for a method evaluating the impact of testing data selection on IPM performance estimates, which is made in Sect. 4, which is followed by results summarised in Sect. 5.
- Finally, conclusions and suggestions for future research are made.

2 The Reference Data

Similarly to other studies on fingerprinting-based IPMs, real RSS data collected in a reference building is used also in this study. First, extensive RSS data have been collected in a number of Indoor Locations, described by $(x, y, floor)$ coordinates. These indoor locations will be referred to as Points of Interest (POIs) in the remainder of this study. The POIs were densely located in a reference building of the Warsaw University of Technology within a grid of 0.25 m resolution on 3 different floors. Similarly to [7,14], we placed points in the corridors of these floors. In every POI, up to 10 repetitions of measurements were made in each of 4 directions. This yields up to 40 RSS vectors per a point, which is equal to the number of times measurements were made in every point *inter alia* in [10]. A decision was made to use WiFi RSS vectors for the study in terminal-centric approach. Thus, the RSS vectors collected in each POI are composed of the measurements of WiFi signal strengths of the signals emitted by WiFi Access Points.

The objective of the analysis is to verify whether different testing data sets can be used interchangeably i.e. whether there is no statistically significant difference between mean positioning errors attained when testing the same IMP with different testing data sets. For this reason, we propose to develop a complex training data set and vary the content of testing data set under the same positioning method. Should a statistically significant difference between mean error rates observed for different strategies of selecting test data be identified, it could be fully attributed to the fact that the way individual testing data sets were developed was different.

For the scenario described above to be used, a reference training data set, identical or virtually identical for all the runs should be developed. We propose to follow a common practice to develop such a data set. The reference training data set R is created by placing in it the RSS data from a rectangular grid of RPs of a 1.5 m resolution. Furthermore, for every POI used as a RP, and every direction of a terminal, approx. 50% of records available for this combination of RP and direction was placed in the reference training set. In this way, the rest of the data collected in every RP can be used for testing purposes. As a consequence, the entire data set Ω of RSS vectors was split into the reference training set R and other data O. The summary of both data sets is provided in Table 1.

Furthermore, let us denote the set of unique POIs that A data set was collected in by $\xi(A)$. What should be emphasised is that the RSS records have been collected in a large number of locations $|\xi(\Omega)| = 7112$ to make the investigation of various testing grids possible.

Table 1. Summary of the data used in this study

Data set	Role	No. of RSS records	No. of POIs
Ω	All available data	284435	7112
R	Reference training data	4267	225
O	Other data, providing testing data sets	280168	7112

3 Development of Testing Data Sets

3.1 Key Assumptions

As already observed, different approaches are applied to develop testing data sets for fingerprinting-based IPM. These approaches will be referred to as Test Data Acquisition Methods (TDAMs). Frequently, a monitored object moving in an indoor area is allowed to move to any location in a continuous (x, y) space, constrained by the architecture of the building. In addition, in some buildings a user is allowed to move to different floors out of all or allowed floors. This suggests that a reference testing data set should include RSS data from a number of randomly selected testing points (TPs) located in accessible area covered by IPM i.e. having possibly continuous (x, y) coordinates and located possibly on different floors. We will refer to this area as a *reference area*.

While for the purpose of this study RSS measurements in a very dense measurement grid of multiple POIs located in the reference area were made, due to budget constraints such extensive measurements are not likely to be used in production deployments of IPM. It is worth noting here that RSS measurements frequently are made manually and can be only partly automated. This is because of limited access areas such as hospitals or lecture theatres and staircases in which semi-automated movable devices can not be used. Hence, for this study to be representative of a real IPM use, we limit the number of records present in all testing data sets to be no larger than the cardinality of a training data set i.e. $card(T) \leq card(R)$.

3.2 Test Data Acquisition Methods

First of all, we propose the method of developing a testing data set by random sampling of $card(R)$ of RSS records from O set to be treated as a reference TDAM. This is because for IPM evaluation to be representative, ideally data from many locations from reference area, most likely not the same as training locations should be used. In this way, we reflect the fact that a monitored object can appear anywhere in a reference area. This method allows possibly continuous (x, y) coordinates i.e. matches behaviour of a monitored object. Moreover, it allows the object to be present in the RPs, but pays particular attention to other locations. In the case of the data used for this study, the majority of the RSS vectors, the testing RSS vectors are sampled from, come from $7112 - 225$

= 6887 locations not present in the training data. This is because the reference TDAM is applied to O data set described above. This TDAM will be referred to as RANDOM_POIS.

Apart from the reference TDAM, other methods are used in research community or can be proposed by adapting ML rules to the needs of fingerprinting-based evaluation. All methods considered in this study are described in Table 2. In the next stage, individual TDAMs will be compared against RANDOM_POIS to determine whether accuracy estimates they provide are the same.

Table 2. Summary of Test Data Acquisition Methods (TDAMs) considered in this study

TDAM	Description	Dependencies
RANDOM_POIS	**Reference method.** RSS records are sampled from RSS records collected in other locations than the locations of RPs and collected in RP locations, but at different time	$\|T\| = \|R\|$ and $T \subset O$
ONE_RSS_OUT	Leave one RSS record out i.e. remove one RSS record from training data R to use it as testing data T	$\|T\| = 1$ and $\xi(R) \cap \xi(T) \neq \phi$ and $T \subset R$
ONE_POI_OUT	All RSS records from one of RPs are moved to a testing data set	$\|\xi(T)\| = 1$ and $\xi(R) \cap \xi(T) = \phi$ and $T \subset R$
SAME_POIS	All RSS records are sampled from remaining RSS records collected in the RPs	$\xi(T) \subset \xi(R)$ and $\|T\| = \|R\|$ and $T \subset O$
OTHER_ES_POIS	RSS records are sampled from the locations shifted by $[\pm 0.75, \pm 0.75]$m compared to RP locations	$\|T\| = \|R\|$ and $T \subset O$
SAME_LINE_POIS	RSS records are sampled from $(x, y, floor)$ locations sharing either x or y with one of the RPs	$\|T\| = \|R\|$ and $T \subset O$

Importantly, all TDAMs satisfy the condition that RSS data collected for training purposes are not used for model testing. However, individual methods vary in terms of the size of the testing data sets they create and the way the location of TPs is selected. In particular, ONE_RSS_OUT and SAME_POIS generate testing data sets, which include data collected only in the same locations as the data of a training data set.

Figure 1 provides an overview of the testing grids produced by individual TDAMs, illustrated on a schematic plan illustrating the conditions under which IPM systems are deployed. It should be noted here that apart from walls, elevators and other building elements, also obstacles such as printing devices and furniture limit the accessibility and the area of feasible locations of both monitored objects and POIs. It is worth noting here that ONE_RSS_OUT and SAME_POIS

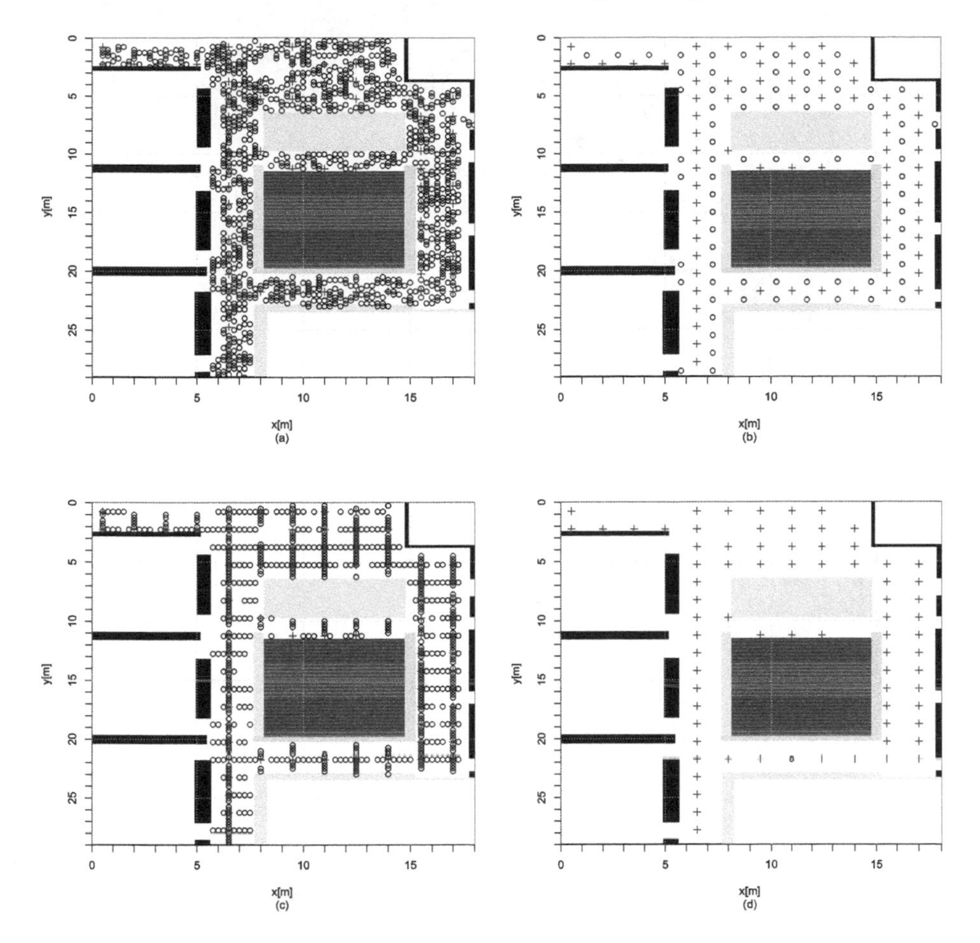

Fig. 1. Sample layout of reference points (cross symbols) and testing points (circles) under: (a) RANDOM_POIS, (b) OTHER_ES_POIS, (c) SAME_LINE_POIS, and (d) ONE_POI_OUT strategy, respectively.

were skipped in Fig. 1, as under these strategies testing locations overlap with training locations.

4 Performance Investigation Method

The key objective of this work is to analyse whether testing data sets developed with individual TDAMs can be used interchangeably. More precisely, a proposal is made to develop for every TDAM multiple positioning models and test them with multiple testing data sets sampled in each run independently from the available data with each TDAM. Next, an average horizontal error is calculated for each of the models on its testing data set. The process we propose is documented in detail in Algorithm 1.

Input: R - original training data, O - remaining RSS data to be used for the development of testing data T, N - the number of algorithm runs for one TDAM, S - TDAM i.e. one of the methods described in Table 2.
Data: $\tilde{T}$ - testing data set generated by TDAM, $r_j \in \tilde{T}$ - j-th record of testing data, $s(r_j)$ - WiFi signal strength vector, $x(r_j), y(r_j)$ - x and y coordinates of the POI where this record has been acquired, respectively, $d : \mathbb{R}^2 \times \mathbb{R}^2 \longrightarrow \mathbb{R}_{\geq 0}$ - Euclidean horizontal distance between true and predicted object location
Result: E_{h} - a vector of the positioning errors of models evaluated on testing data sets developed with S method

```
begin
  for i = 1,...,N do
    if S ∈ {ONE_POI_OUT, ONE_RSS_OUT} then
      (R̃_i, T̃_i) = S(R);
    end
    else
      /* remaining strategies extract testing data from O,
         possibly based on the location of RPs              */
      T̃_i = S(ξ(R), O);
      R̃_i = R;
    end
    /* Train positioning models for x and y coordinates      */
    M_x = train_x(R̃_i);
    M_y = train_y(R̃_i);
    E_h(i) = Σ_{j=1}^{|T̃_i|} d([x(r_j), y(r_j)], [M_x(s(r_j)), M_y(s(r_j))]) / |T̃_i|;
  end
end
```

$$E_{\mathrm{h}}(i) = \frac{\sum_{j=1}^{|\tilde{T}_i|} d\Big([x(r_j), y(r_j)], [M_{\mathrm{x}}(s(r_j)), M_{\mathrm{y}}(s(r_j))]\Big)}{|\tilde{T}_i|};$$

Algorithm 1. The evaluation of the impact of testing data selection on model performance estimates

Algorithm 1 distinguishes between ONE_RSS_OUT and ONE_POI_OUT and the remaining strategies, as the two former strategies do not use O data, but move a subset of the original training data R to the actual training data $\tilde{R}$ used to develop the positioning models. Irrespective of TDAM used, once M_{x} and M_{y} regression models are trained, they are evaluated on testing data $\tilde{T}$ sampled with S method independently in every run out of N runs of the algorithm. This makes it possible to observe the distribution of errors caused by both multiple training sessions of the models and varied content of individual testing data sets developed under the same method S.

5 Results

To analyse the possible impact of testing data selection on performance estimates, Algorithm 1 was executed with $N = 100$ runs. Furthermore, random forest was used to develop positioning models. In every session a random forest composed of 50 trees was built on training data and applied to testing data. This has been repeated for x and y coordinates. We selected random forest method for the evaluation, as the method is resilient to noisy signals. This made it suitable for multiple runs of the method and eliminated the impact of manual feature selection on the evaluation process.

Table 3. Summary of data sets developed with individual Test Data Acquisition Methods (TDAM)

TDAM	$\min(\xi(\tilde{T}))$	$avg(\xi(\tilde{T}))$	$max(\xi(\tilde{T}))$	$max(\|\tilde{T}\|)$
RANDOM_POIS	3250.00	3304.14	3348.00	4267.00
SAME_LINE_POIS	2229.00	2280.83	2317.00	4267.00
ONE_RSS_OUT	1.00	1.00	1.00	1.00
ONE_POI_OUT	1.00	1.00	1.00	19.00
SAME_POIS	225.00	225.00	225.00	4267.00
OTHER_ES_POIS	205.00	205.00	205.00	4267.00

First, the summary data on 100 testing data sets developed with each TDAM is worth investigating. Minimum, mean, and maximum number of locations present in individual testing data sets $\tilde{T}_i$ is provided in Table 3. The table includes also the maximum number of records placed in $\tilde{T}_i$ in one run of a single strategy. What should be noted here is that the size and content of the reference training data set is kept possibly identical in all model development sessions. The only allowed difference is to extract a single RSS vector or RSS vectors from a single RP into a testing data set. The latter divergence from the fixed content of R data set is necessary in the case of ONE_RSS_OUT and ONE_POI_OUT techniques. In the case of remaining TDAM methods, the number of training locations $|\xi(\tilde{R}_i)| = 225$ and the number of RSS vectors $|\tilde{R}_i| = 4267$ in all model development sessions. Importantly, RANDOM_POIS and SAME_LINE_POIS use data from a much larger number of testing locations than other TDAMs.

Figure 2 shows the distribution of E_{h} errors for individual method runs. First of all, ONE_POI_OUT and to a lower extent ONE_RSS_OUT can yield major errors when the vector or vectors placed in a testing data set turn out to be particularly difficult to locate in (x, y) space. Moreover, what follows from Fig. 2(b) is that SAME_POIS strategy i.e. the strategy in which a part of the data collected in RPs is used also for testing purposes yields by far lower error estimates than the reference RANDOM_POIS method.

To verify whether individual differences of E_{h} values are statistically significant, for every TDAM the hypothesis that the mean error rate $mean(E_{\mathrm{h}})$ developed for this TDAM is equal to the mean error developed for reference TDAM has been verified. This has been done based on two vectors of $N = 100$ mean error values E_{h} developed with Algorithm 1. Unpaired t test was used to verify the null hypothesis. Confidence level of the interval of mean difference was set to 0.95. Variances of both vectors were not assumed to be equal. Table 4 contains mean absolute error rates and the results of hypotheses testing, including

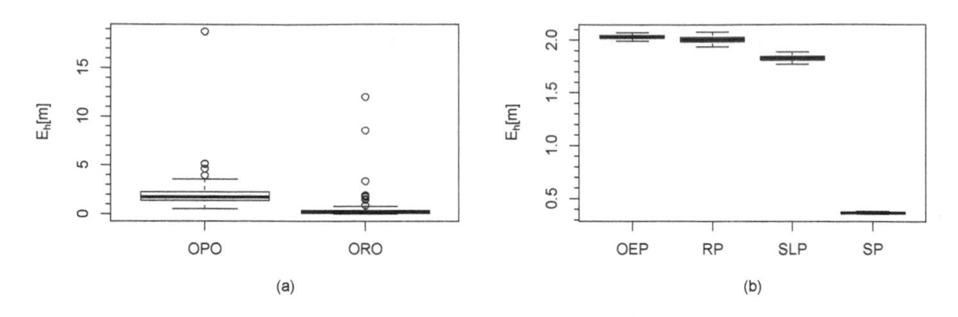

Fig. 2. Distribution of $E_h()$ errors for (a) ONE_POI_OUT (OPO), ONE_RSS_OUT (ORO), and (b) OTHER_ES_POIS (OEP), RANDOM_POIS (RP), SAME_LINE_POIS (SLP) and SAME_POIS (SP) strategy, respectively.

both p values for the hypothesis of mean error equality and confidence intervals of differences in means for every TDAM compaired to the reference TDAM.

First of all, let us observe that the reference method yields an average error of 2.01 m. The two methods using the data from the same locations as training locations $\xi(R)$ i.e. ONE_RSS_OUT and SAME_POIS yield significantly lower error estimates of 0.534 m and 0.369 m, respectively. The confidence interval $[c_{\min}, c_{\max}]$ showing the difference of mean values with negligibly small p value confirms that the hypothesis of the equality of mean error reported on the data sampled from training locations with the mean error reported under reference TDAM can be rejected. This shows that random division of data from the same locations into training and testing data sets results in IPM accuracy estimates largely different from these encountered in unconstrained use of indoor positioning.

As far as ONE_POI_OUT method is concerned, there is no basis to reject the null hypothesis, as the dispersion of errors is too large for statistically significant reasoning. However, this shows that relying on moving the data from one location to testing data in a leave-one-out manner does not provide reliable assessment of IPM performance in a continuous reference area, when the number of runs of the method is $N = 100$ i.e. in spite of major calculation overhead considering the number of training instances involved. It is worth noting here that a complete leave-one-out process would require even more runs i.e. testing IPM based on the number of data divisions equal to the number of training locations.

In the case of SAME_LINE_POIS, error rates are observed to be smaller than in the case of RANDOM_POIS and this difference is confirmed to be statistically significant. Hence, by placing testing locations along the same lines as training locations we get reduced error estimates compared to these attained when testing locations are placed in any point of the reference area. This difference is observed in spite of using testing data acquired in other locations than training data.

Finally, OTHER_ES_POIS yields slightly higher i.e. more pessimistic error estimates than the reference TDAM. Importantly, this difference, while not large is statistically significant. This phenomenon can be explained by the fact that in the case of OTHER_ES_POIS technique the average distance between a TP and the closest RP is larger than in the case of the reference TDAM. Hence, indoor positioning can be more difficult compared to RANDOM_POIS test grids, in which some of testing data may come from even the same locations as training data.

To sum up, extensive calculations made in this study show that individual techniques of selecting testing data out of all available RSS data can largely influence performance estimates of the same positioning technique. This suggests the use of the same standard for selecting testing data used to evaluate the performance of indoor positioning methods. RANDOM_POIS provides a valid choice when a monitored object can move in a continuous space of reference area. OTHER_ES_POIS can be used to provide similar estimates while relying on data collected in a significantly lower number of POIs.

Table 4. Results of testing the hypothesis that mean errors attained for individual TDAMs are equal to mean errors for reference TDAM i.e. RANDOM_POIS method

TDAM	$mean(E_h(T))$ [m]	p value	c_{min}	c_{max}
RANDOM_POIS	2.006	0.000	0.000	0.000
SAME_LINE_POIS	1.828	0.000	0.171	0.186
ONE_RSS_OUT	0.534	0.000	1.175	1.769
ONE_POI_OUT	2.045	0.839	−0.412	0.335
SAME_POIS	0.369	0.000	1.631	1.643
OTHER_ES_POIS	2.033	0.000	-0.033	-0.019

6 Conclusions

The analysis of various testing data acquisition methods made in this study shows that all methods yield horizontal accuracy estimates which are different from the estimates developed for the reference TDAM. Importantly, except for one method the differences that have been observed are statistically significant. This is even though all the methods satisfy the key condition that once RSS records are collected, they are split into the training and testing part and no data is shared between the two data sets. This corresponds to making separate RSS measurements to develop testing data sets. This result contributes to increased understanding of indoor positioning studies that report diverse accuracies of the methods they analyse. More precisely, one of the reasons is the fact these studies frequently follow different strategies of selecting testing data sets and the locations the data in these sets come from.

In the analysed case of indoor positioning, from a deployment point of view, OTHER_ES_POIS strategy is of particular interest. Even though it yields pessimistic error estimates, it relies on a limited number of TPs, which reduces the cost of collecting testing data. In the future, we are planning to analyse further techniques aiming to automate the development of representative testing data sets, while reducing both the cardinality of these data sets and the number of locations in which measurements have to be made.

The results of this study highlight a wider problem that just randomly splitting available records into training and testing data may be not sufficient to get performance estimates representative of real use of machine learning methods. As an example the evaluation of IPS models on testing data coming from the locations already present in training data yields performance estimates substantially different from the evaluation on testing data from any neighbouring location. This illustrates a wider problem that for example testing ML models on the data from the objects already represented in training data may be not sufficient to estimate the ability of ML methods to model behaviour of new objects. Importantly, differences in accuracy estimates observed in this study are statistically significant and in some cases are by far larger than differences typically arising from the use of different regression methods for model development. This emphasises the role of carefully planned evaluation process in selecting the best ML method for a problem of interest.

Acknowledgements. This research was partly supported by the National Centre for Research and Development, grant No. PBS2/B3/24/2014, app. no. 208921.

References

1. Dawes, B., Chin, K.W.: A comparison of deterministic and probabilistic methods for indoor localization. J. Syst. Softw. **84**, 442–451 (2011)
2. Fang, S.H., Lin, T.N., Lin, P.: Location fingerprinting in a decorrelated space. IEEE Trans. Knowl. Data Eng. **20**(5), 685–691 (2008)
3. Flach, P.: Machine Learning: The Art and Science of Algorithms That Make Sense of Data. Cambridge University Press, New York (2012)
4. Grzenda, M.: Reduction of signal strength data for fingerprinting-based indoor positioning. In: Jackowski, K., Burduk, R., Walkowiak, K., Woźniak, M., Yin, H. (eds.) IDEAL 2015. LNCS, vol. 9375, pp. 387–394. Springer, Cham (2015). https://doi.org/10.1007/978-3-319-24834-9_45
5. Japkowicz, N., Shah, M.: Evaluating Learning Algorithms: A Classification Perspective. Cambridge University Press, New York (2011)
6. Jin, Y., Soh, W.S., Wong, W.C.: Error analysis for fingerprint-based localisation. IEEE Commun. Lett. **14**(5), 393–395 (2010)
7. Khalajmehrabadi, A., Gatsis, N., Akopian, D.: Modern WLAN fingerprinting indoor positioning methods and deployment challenges. IEEE Commun. Surv. Tutor. **19**(3), 1974–2002 (2017). https://doi.org/10.1109/COMST.2017.2671454. Thirdquarter
8. Kjargaard, M.B.: Indoor location fingerprinting with heterogeneous clients. Pervasive Mobile Comput. **7**, 31–43 (2011)
9. Lee, M.K., Han, D.S.: Dimensionality reduction of radio map with nonlinear autoencoder. Electron. Lett. **48**(11), 655–657 (2012)
10. Lemic, F., Behboodi, A., Handziski, V., Wolisz, A.: Experimental decomposition of the performance of fingerprinting-based localization algorithms. In: 2014 International Conference on Indoor Positioning and Indoor Navigation (IPIN), pp. 355–364, October 2014. https://doi.org/10.1109/IPIN.2014.7275503

11. Álvarez López, Y., de Cos Gómez, M.E., Álvarez, J.L., Andrés, F.L.H.: Evaluation of an RSS-based indoor location system. Sens. Actuators A Phys. **167**(1), 110–116 (2017). https://doi.org/10.1016/j.sna.2011.02.037. http://www.sciencedirect.com/science/article/pii/S0924424711000999
12. Luo, X., O'Brien, W.J., Julien, C.L.: Comparative evaluation of Received Signal-Strength Index (RSSI) based indoor localization techniques for construction jobsites. Adv. Eng. Inf. **25**, 355–363 (2011)
13. Moghtadaiee, V., Dempster, A.G.: Design protocol and performance analysis of indoor fingerprinting positioning systems. Phys. Commun. **13**(PA), 17–30 (2014). https://doi.org/10.1016/j.phycom.2014.02.004
14. Tao, Y., Zhao, L.: A novel system for WiFi radio map automatic adaptation and indoor positioning. IEEE Trans. Veh. Technol. **67**, 10683–10692 (2018). https://doi.org/10.1109/TVT.2018.2867065
15. Visbrot, R., Kozinsky, A., Freedman, A., Reichman, A., Blaunstein, N.: Measurement campaign to determine and validate outdoor to indoor penetration models for GSM signals in various environments. In: 2011 IEEE International Conference on Microwaves, Communications, Antennas and Electronic Systems (COMCAS 2011), pp. 1–5, November 2011. https://doi.org/10.1109/COMCAS.2011.6105823
16. Yang, Q., Pan, S.J., Zheng, V.W.: Estimating location using Wi-Fi. IEEE Intell. Syst. **23**(1), 8–13 (2008)

GPU-Based Bat Algorithm for Discovering Cultural Coalitions

Amine Kechid(✉) and Habiba Drias

Laboratory for Research in Artificial Intelligence, Computer Science Department, USTHB, BP 32 El Alia Bab Ezzouar, 16111 Algiers, Algeria
{akechid,hdrias}@usthb.dz

Abstract. Nowadays, artificial intelligence makes a great success in our modern social life. The human should be prepared to be able to live with social robots that can provide him comfort and help in solving complex processes. Extending the use of robot's technology is certainly desirable, but preventing certain catastrophes from the misdeeds of artificial intelligence is crucial. One of these troubles could be for instance the creation of robots' coalitions to impose pernicious decisions. As a contribution to cope with such issue, we propose a parallel approach for the detection of cultural coalitions based on Bat Algorithm. This GPU-based bat algorithm approach can treat very large datasets due to the possibility of launching several artificial bats simultaneously, which contribute to reducing the runtime without affecting the performance. To proof the effectiveness of the parallel detection coalition method, we conducted several experiments on datasets of different sizes. These datasets represent the result of cultural artificial agents playing the colored trails (CT) game. For the creation of agents' profiles, we use real cultural datasets generated based on the WV survey. The experimental analysis demonstrates that the use of the proposed method will considerably reduce the runtime.

Keywords: Culture · Coalitions · Social agents · Detecting coalitions

1 Introduction

Few years ago, robots were just a grain of imagination, but with the great advancement in the technological domain, this imagination becomes a reality. Robots were created to facilitate difficult tasks such that we find in industry, military, and health domain. Due to their great success, researchers seek to improve their physical appearance until getting robots made from a special batter that are difficult to distinguish from humans. As a result, robots crawl into human social life. For the best interaction between robots and humans, researchers aim to increase the social intelligence of robots by cloning the human cultural intelligence and transplanting it into the artificial intelligence of the robot.

F. Wotawa et al. (Eds.): IEA/AIE 2019, LNAI 11606, pp. 470–482, 2019.
https://doi.org/10.1007/978-3-030-22999-3_41

As each technology, robot engineering has advantages and drawbacks. While its positives and favor over humanity are great, its drawbacks are also great and could lead to a major scientific disaster. One of these weaknesses could be the creation of coalitions of robots to impose pernicious decisions. The issue is that a subset of users prefer to cooperate within the group and avoid cooperation outside the group [2], such that the elements of the same coalition share the same interests and objectives [19]. As knowing the causes of a problem is the most effective solution to avoid it, detecting coalition becomes nowadays an essential solution to avoid an inevitable Disaster.

The issue of finding coalitions is a complex problem and classified in the NP-hard category [22]. It means that this problem lacks from a resolution algorithm of polynomial complexity, which makes finding solutions needs exponential running time. Techniques of artificial intelligence have proved their effectiveness in solving such problems with approximate methods. Unlike the exact methods that find all the coalitions, intelligent tools find the most efficient ones in a significantly reduced amount of time. Swarm algorithms are one of these methods that are nature-inspired, mimicking some successful characteristics of animal swarms. It is a promising field based on the simulation of the collective intelligence in social insects and some animals such as fishes and birds. The most known algorithms in this category are Ant Colony Optimization (ACO) [4], Particle Swarm Optimization (PSO) [11], Firefly Algorithm (FA) [24], and Bat Algorithm (BA), a meta-heuristic developed in 2010 [25]. This approach simulates the movements of bats when searching for preys. It manages a balance between a local search and PSO optimization, and is appreciated for its good convergence to the optimal solutions.

Although these methods are basically efficient, however with the use of very large datasets the time complexity issue reappears. One way to tackle this hard question is to parallelized the independent parts. Since 2007 and due to the advancement in the graphics hardware, it was possible to exploit this idea with the emergence of the Graphics Processor Units (GPU). This type of processor differs with the old processor in fundamental design philosophy, where multiple threads cooperate to achieve the goal. For example, in the case of matrix multiplications, instead of filing the matrix element by element, all the elements will be filed in a parallel manner. Each thread is associated with one element, where it calculates the sum of the product of all the elements from the row i with all the elements from the column j. This kind of programming can achieve more than 100 times speedup over sequential execution.

In this paper, we propose a parallel approach for the detection of cultural coalitions based on the Bat Algorithm. It aims to perform more efficiently than the existing state-of-the-art algorithms. This approach needs a GPU architecture and some synchronizations with the CPU, following the master/workers model. Technically, the designed method consists in running the master on the CPU, which generates a population of n bats, where each one represents a potential solution. Then, the master launches n threads offloaded on the GPU, where each one performs the tasks of one bat to improve its solution by sharing its best solution with its congeners. The master attempts to receive the result of each bat to rank the bats on the best solutions in the next iteration.

The remainder of the paper is structured as follows. In the next section, we present the related works. Then, in Sect. 3, we present our first contribution on the perception of culture and its integration in the BDI model. In Sect. 4, we present our second contribution, which deals with the adaptation of Bat Algorithm to the coalition detection problem. After that, we present the parallel bat algorithm for the detection of coalitions. Finally, we end this paper with a conclusion after presenting the experimental results.

2 Literature Review

Coalition creation may cause serious problems such as competitive harm, the trust and the reputation abuse. To tackle this issue several approaches were proposed using methods based on Markov chain [3], clustering [23], Rule based, graph based [9], machine learning, convolutional neural networks [1]. In 2007, Metwally and his co-authors proposed a method for detecting coalition in advertising networks named detectives [15]. This method used a sampling approach to detect all pairs of publishers having similarity exceeding a specific threshold. In order to detect coalition members, the authors used these pairs of publishers to construct a similarity graph. The graph is composed of vertices which represent publishers. Two vertices are related with edge if the similarity of visitors between the two vertices is greater than or equal to a specific threshold. After creating the similarity graph, they try to detect the maximal clique. Finding a clique in this graph amounts to finding a set of vertices that are related to each other.

After that, other researchers proposed CATCH Algorithm [14], which can detect coalition groups based on the ratio to gain of cost. A coalition is defined as a group that has its gain per resource (GPR) exceeds a specific threshold and every subgroup has its gain per resource lower than the gain per resource of the group.

After one year, Kerr and Cohen proposed a statistical approach [12], which consists of two steps, clustering and characterizing clusters which contain coalition. In the first step, they use the k-means algorithm [6] to regroup agent interest in several clusters. Agent interest is a vector of n entities which represents the interest of the agent with each other. Coalition members are seen as elements that have a greater interest with coalition members than other members. Therefore, in the second step, the average benefit of each candidate cluster is calculated. After that, for each candidate cluster, they take 100 samples of the same size and calculate the benefit between this cluster and each sample to generate large data. From this data, they use the normal distribution to detect whether the probability of the average benefit calculated above is greater than the threshold.

Another method was also proposed to detect the stock market colluding groups based on spectral clustering algorithm [18]. The authors run several times the k-means algorithm until maximizing a certain modularity function. In each iteration, the number of clusters is incremented. On the other hand, the colluders similarity measure (CSM) [16] uses Colluders Detection Algorithm (CDA) on the list of suspicious nodes to cluster the nodes with similarity more than a specific threshold. Another based clustering method was proposed to detect fraud in internet advertising [21]. It consists of 3 steps: constructing, clustering and filtering. At the first step, they organize the data, then, they transform the coalition detection to a clustering problem and finally, they remove false alarm clusters.

Hybridizing the similarity [15] and GPR [14] methods is the subject of a new approach used for detecting coalition in online advertising [27]. This approach consists of 3 phases: initialization, inductive and finalization. The initialization phase generates the required information in the next step to each metric. In the inductive phase, the method uses the apriori style to generate the candidate coalitions, which are transmitted to each method. The result intersection for each method is used to calculate the candidate coalition for the next iteration. The final coalition is identified in the last step as the intersection of the identified coalition in each method.

In 2017, Zhai and his co-authors proposed an approach to detect potential collusive clique with their activities [26]. This method starts by estimating the probability of being fraud to each potential candidate. Then, if confirmed fraudulent, their potential wealth is calculated.

3 Culture and Its Integration in the BDI Model

In this section, we present the needed background to understand the remainder of the paper. We start by presenting our modeling of culture. Then, we explain how to integrate the concept of culture in the BDI model.

3.1 Our Perception of Culture and Modeling

In previous works [10], we define culture as a multivariate mathematical function, that assigns each tuple $(x_1, x_2, ..., x_n)$ in domain D a class C $= f(x_1, x_2, ..., x_n)$:

$$\begin{array}{llll} f: & D & \rightarrow & R \\ & (x_1, x_2, ..., x_n) & \mapsto & f(x_1, x_2, ..., x_n) \end{array} \tag{1}$$

The tuple is viewed as a vector of socio-cultural factors which is made up of the following elements presented in Table 1. For each factor, we select one or more attributes from the WV survey. This survey is started in 1981 and accessible from the www.worldvaluessurvey.org website. For each country, a dataset is generated, where each instance represents the answers of one person from this country.

After that, we move to the creation of cultural datasets based on the data generated from the survey. From each of the six following countries: Algeria, Germany, China, Japan, Spain and the United States, we create a cultural dataset, where rows represent the individual answers on the selected attributes. These six datasets are analyzed with the Apriori algorithm to generate the frequent cultural characteristics in each region and extract the cultural association rules.

3.2 Culture Integration in the Belief-Desire-Intention (BDI) Model

After we have introduced the culture modeling, we present the integration of this concept in the BDI model [17] to implement multi-agent systems [20]. Cultural attributes are seen as a knowledge base, which can be modified through the encountered events.

The BDI model is used in the creation of rational agents. It consists of 3 concepts: Beliefs which represent the agent knowledge in his environment, desires are agent motivations that represent states of the world that agent wants to reach, and intentions are states of the world that agent undertakes to realize at a given moment. When an agent detects a new event, he updates his beliefs according to the perceived event and his culture to generate an options' list, which represents the objectives that can be instantiated. From these desires, he selects the best choice which represents his intention.

4 Bat Algorithm for Coalitions' Detection

BA is a generic algorithm that can find solutions for several complex issues [8,13]. In this section, we present our previous work, which adapts the BA algorithm to the problem of finding coalitions. We present the formulation of the coalition problem, the solution representation, and the fitness function for the performance evaluation of the artificial bats that encapsulate solutions.

4.1 Problem Formulation

We modeled the problem of finding coalitions as a simple undirected graph G = (V; E), where each vertex of V represents a specific entity (individual) in the system, and there is an edge between two vertices i and j of V if the similarity between the profile interest of the two nodes is less than a specific threshold.

The profile interest of an entity i is a vector of size k, where each dimension j represents the amount of interest between the entity i and j, and k represents the number of entities in the system.

For this problem, we can find one or more solutions. A solution is a set of entities that each one has an edge with each other in the same coalition.

Table 1. Culture dataset attributes

Contribution name	Attribute	Possible values
Reading	Newspaper use	Weekly, Monthly, Less than monthly, Never, No answer
	The use of magazines	
Tradition	Tradition importance	Very Important, Important, Somewhat Important, A little Important, Not Important, Not at all Important, No answer
Age	Age category	Child, Teen, Adult, Old person
Work	Work importance	Very important, Rather important, Not very important, Not at all important, No answer, Don't know
Educational	Highest educational level	From Incomplete primary school to University - level education, with degree, No answer
The interaction with nearby environment	Friends importance	Very important, Rather important, Not very important, Not at all important, No answer, Don't know
Family	Family importance	Very important, Rather important, Not very important, Not at all important, No answer, Don't know
	Teach independence to children	Very important, not very important
	Teach hard work to children	
	Teach feeling of responsibility to children	
	Teach imagination to children	
	Teach tolerance and respect for other people to children	
	Teach thrift to children	
	Teach determination and perseverance to children	
	Teach religious faith to children	
	Teach unselfishness to children	
	Teach obedience to children	
	Teach self-expression to children	
Individual	think up new ideas	Very Important, Important, Somewhat Important, A little Important, Not Important, Not at all Important, No answer
	be rich	
	Living in secure surroundings	
	Luxury and comfort	
	do something for the good of society	
	help people living nearby	
	Being very successful	
	Adventure and taking risks	
	behave properly	
	Looking after the environment	
Universal	Internet use	Daily, Weekly, Monthly, Less than monthly, Never, No answer
	TV use	

4.2 Solution Representation

In this problem, we aim to find the sets of agents that participate in coalitions. As with the BA algorithm each bat encapsulates a single solution, the most appropriate data structure that represents each bat (solution) is a vector. The elements of this vector can take values from 0 to the number of entities in the system. If the value of the element i is equal to 0, it means that this element is not used. Otherwise, it shows the number of the entity that participates in the coalition. For example, if there are 10 entities in the system, the coalition that contains the entities 1, 9, 10, is represented by: 1 9 10 0 0 0 0 0 0 0

4.3 Fitness Function

As BA can generate a lot of solutions, it is necessary to have an effective fitness function to evaluate the quality of the solution and to guide the future solutions. For this purpose, we propose to use the following function.

$$f(x) = \begin{cases} \text{the size of coalition,} & \text{if all the elements are connected.} \\ 0 & \text{Otherwise} \end{cases}$$

4.4 New Solution Generation

Concerning the generation of the new position, we use Algorithm 1 and Eqs. 2 and 3 [7]. This algorithm aims to update some elements from the solution indicated by the actual frequency. This modification starts at a specific element indicated by the actual velocity. So, for each element, it compares the loudness and a random value. If this value is greater, we increment the value of the bit v_i in the solution and save its modulo (k+1). Otherwise, we decrement the value of the bit v_i and save its modulo. To avoid the redundancy of values in the same solution, we update the content of bit v_i to the calculated value if it does not exist in the generated solution. After that, it remains just to increment the value of v_i for passing to the next iteration. This process is repeated until v_i achieve the actual frequency.

5 Parallel Bat Algorithm for the Detection of Coalitions

As described in the previous section, BA is a very effective algorithm and can find solutions to several problems. However, when dealing with very large datasets, we are obliged to increase the number of bats to obtain the desired performance, which increase considerably the runtime of the algorithm. One way to tackle this issue is to launch all the bats in parallel. It means, that all the bats start at the same time. Exploiting this idea needs a new type of processor named Graphics Processor Units (GPU).

In this section, we propose the parallel coalition detection method based on a master/workers paradigm. Whereas the master is executed on the CPU, the

Algorithm 1. Generate new solution

Require: Coalition x_i^{t-1}, Velocity v_i, Loudness A_i ?
Ensure: New coalition x_i^t,
  Begin
  while $v_i \prec f_i^t$ **do**
    if $rand > A_i^t$ **then**
      1- new_entity=((entity at v_i) +1) mod (k+1)
    else
      2- new_entity=((entity at v_i) -1) mod (k+1)
    end if
    if new_entity does not exist in x_i^t **then**
      3- entity at v_i =new_entity
    end if
    4- increment v_i
  end while

workers are offloaded to GPU. Unlike BA, all the bats perform a local search simultaneously. The master starts by initializing randomly one solution to each bat, and initialize the velocity v_i^0 and the frequency f_i^0. After initializing the value of pulse rate r_i^0 and loudness A_i^0, the population is evaluated to extract the best solution x*.

After that, since the maximum number of iterations is not reached again, it copies the input data to the device (GPU) to launch the necessaries number of threads. This process is explained in Algorithm 2. Afterwards, each thread calculates its real index in the set of data by contribution to its index in the block, the index of the block and the number of threads in the block. Each launched thread performs the process of one bat that is, each bat performs a virtual movement using Algorithm 1 to generate a new solution by adjusting frequency f_i, velocity v_i as shown in Eqs. 2 and 3. After evaluating the new solutions, each bat in the population generates a new solution through a random walk using Eq. 4.

$$f_i^t = 1 + (f_{max}) * \beta \tag{2}$$

$$v_i^t = f_{max} - f_i^t - v_i^{t-1} \tag{3}$$

$$x_{new} = x^{'} + \epsilon * A \tag{4}$$

- Where,
  - $x*$ is the current best solution.
  - β is a random vector in the range [0,1].
  - f_{min} and f_{max} are respectively the specified lower and upper bounds for the frequency parameter f.
  - x' is the best solution at the actual iteration.
  - ϵ is a random value between 0 and 1.
  - A is the average loudness of all the bats at the actual iteration.

Finally, the pulse emission rate r_i and the loudness A_i are updated using Eqs. 5 and 6. This principle is explained in Algorithm 3. After every bat ends its tasks and the master receives the solutions, it sorts the solutions according to its quality and then it ranks the bats of the best solutions

$$A_i^t = \alpha * A_i^{t-1} \tag{5}$$

$$r_i^t = r_i^0(1 - e^{\gamma t}) \tag{6}$$

- Where,
  - α and γ are a constants

Algorithm 2. Pseudo code of the parallel Bat algorithm (master)

Require: The graph G.
Ensure: The sets of coalitions.

```
Begin
1- generate at random a population of n bats (n solutions);
for each bat i do
   2- define its loudness A_i, its pulse frequency f_i and its velocity v_i;
   3- set its pulse rate to r_i;
   4- select the best solution x*;
end for
while (Max-Iter not reached) do
   5- cudaMemcpy(Population, cudaMemcpyHostToDevice)
   6- cudaMemcpy(loudness A, cudaMemcpyHostToDevice)
   7- cudaMemcpy(frequency f, cudaMemcpyHostToDevice)
   8- cudaMemcpy(velocity v, cudaMemcpyHostToDevice)
   9- cudaMemcpy(pulse rate r, cudaMemcpyHostToDevice)
   10- Launch n threads
   11- cudaMemcpy(Population, cudaMemcpyDeviceToHost)
   12- cudaMemcpy(loudness A, cudaMemcpyHostToDevice)
   13- cudaMemcpy(frequency f, cudaMemcpyHostToDevice)
   14- cudaMemcpy(velocity v, cudaMemcpyHostToDevice)
   15- cudaMemcpy(pulse rate r, cudaMemcpyHostToDevice)
   16- Rank the bats and find the current best solution x*;
end while
```

---

6 Experiments

To appreciate the performance of the proposed method, we conducted several experiments on datasets of different sizes. These datasets are generated in previous work, where we developed a simulation environment that contains intelligent cultural agents playing the colored trails game (CT) [5]. These datasets consist of several instances, where each one represents the benefit between each agent

Algorithm 3. Pseudo code of the parallel Bat algorithm (Kernel))

Require: Population, loudness A, frequency f, velocity v, pulse rate r.
Ensure: Population, loudness A, frequency f, velocity v, pulse rate r.
  1 i ←blockIdx.x ∗ blockDim.x + threadIdx.x
  2- compute a new solution (f_i, v_i, x_i) using Algorithm 1 and equations 2 and 3.
  if $rand \succ r_i$ **then**
    3- select a solution x' among the best solutions;
    4- improve the solution using equation 4;
  end if
  5- generate at random a new solution (f_i, v_i, x_i);
  if rand ≺ f(x*) **then**
    6- accept the new solution;
    7- increase r_i and reduce A_i using formulas 5 and 6
  end if

and each other in the system. The agents' profiles are collected from the culture dataset based on the WV survey www.worldvaluessurvey.org. We implemented the proposed method with the C-CUDA 4.0 language using a CPU host coupled with a GPU device. The CPU is a quad-core Intel.

In these experiments, we used four datasets. The size of the datasets varies from the 1000 to 4000 instances in increment of 1000. For each dataset, we varied the size of the population from 200 to 500 bats. Figure 1 shows the runtime of the parallel and the sequential method on the first dataset. It is easy to notice that the curve of the proposed method falls under that of the bat algorithm based method, which means that the runtime of the GPU-based bat algorithm method is much reduced by contribution to the bat algorithm based method.

Figure 2 shows the same information as the previous figure, but on the second dataset. We can see that even if we increase the size of the dataset, the parallel method is faster than the sequential one.

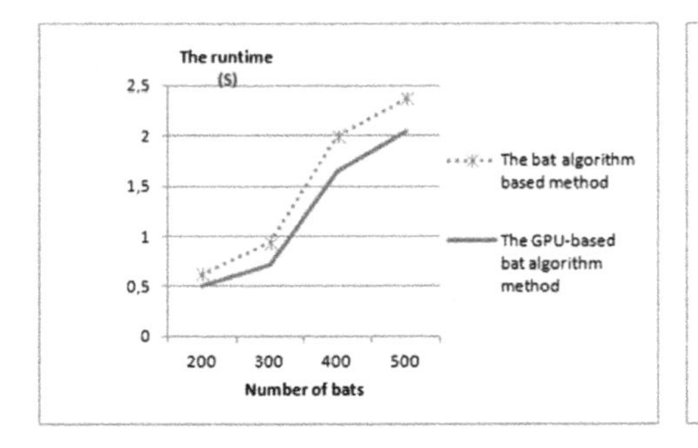

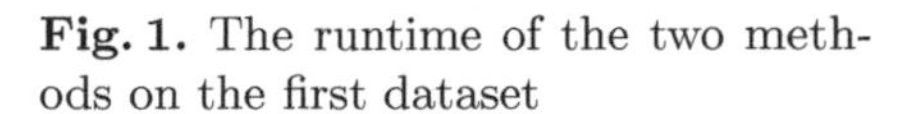
Fig. 1. The runtime of the two methods on the first dataset

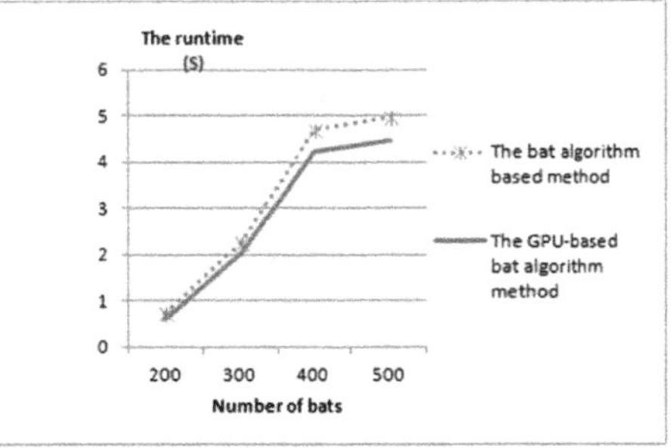

Fig. 2. The runtime of the two methods on the second dataset

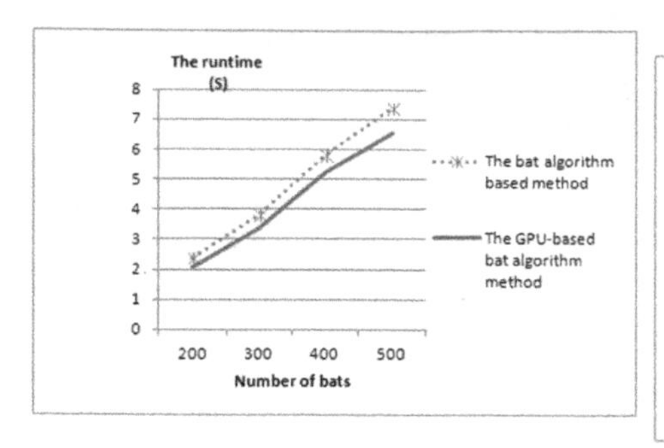

Fig. 3. The runtime of the two methods on the third dataset

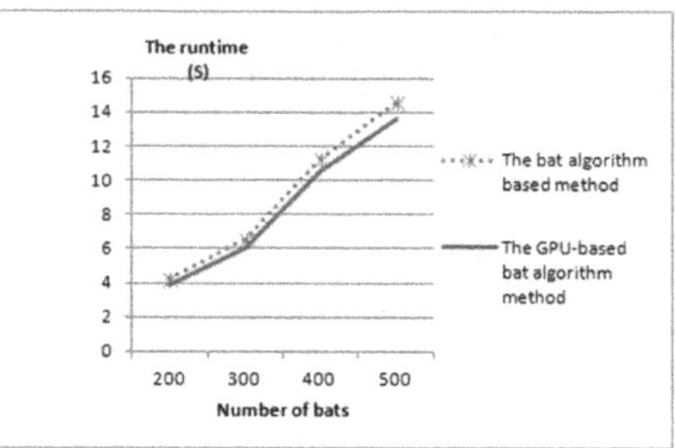

Fig. 4. The runtime of the two methods on the fourth dataset

Figures 3 and 4 show the runtime of the two methods on the third and the fourth dataset, respectively. From these figures, we can confirm the conclusion drawn from the previous experiments, which is the parallel method is faster than the sequential one. So, all the realized experiments share the same conclusion, which is the use of the proposed method reduces the runtime.

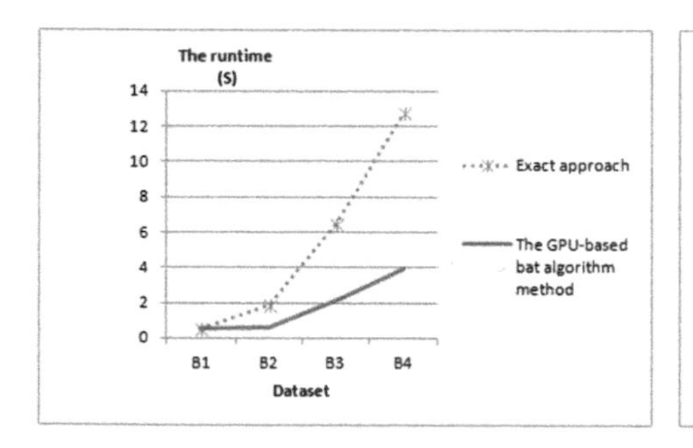

Fig. 5. Comparing the runtime of the proposed method with an exact approach

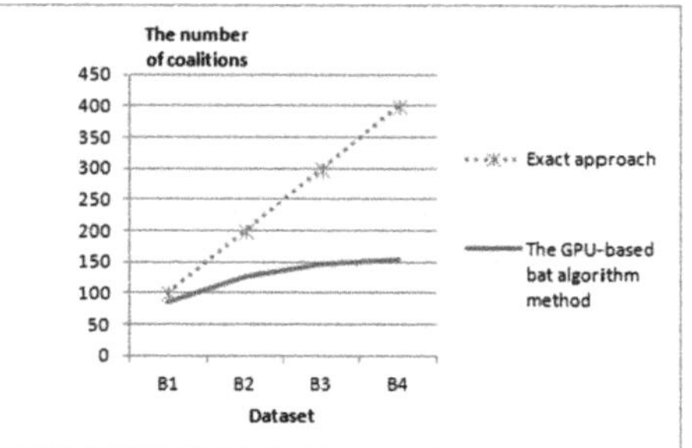

Fig. 6. Comparing the performance of the proposed method with an exact approach

After comparing the parallel and the sequential methods, we pass to compare the proposed method with an exact approach named Similarity-based method [15,27]. Figures 6 and 5 show respectively the performance and the runtime for the two methods on the four datasets. From Fig. 5, we see that the curve of the proposed method is situated bottom the curve of the exact approach, which means that the runtime of the proposed method is well reduced.

On the other hand, from Fig. 6, we see that the curve of the proposed method is situated bottom the curve of the exact approach. It is clear that the performance of the exact approach is higher than the performance of the proposed method, but the performance of the proposed method remains also good.

7 Conclusion

In this paper, we propose a parallel detection coalition method based on Bat Algorithm. This approach allows reducing the runtime while obtaining good results. Unlike the state-of-art methods, our proposal finds the most effective solutions in a significantly reduced amount of time.

When dealing with very large datasets, traditional techniques need to increase the population size, otherwise, we lose in the performance. However, increasing the population's size leads to an increase in the running time. The proposed method allows launching simultaneously several bats on the GPU, which gives good results without affecting the runtime.

To validate the detection coalition method, we implemented our algorithms with C-CUDA 4.0 language on a CPU coupled with a GPU architecture. The obtained result from all the experiments shows the importance of the method.

References

1. Abdallah, A., Maarof, M.A., Zainal, A.: Fraud detection system: a survey. J. Netw. Comput. Appl. **68**, 90–113 (2016)
2. Belmonte, M.V., Conejo, R., Pérez-de-la-Cruz, J.L., Triguero, F.: A stable and feasible payoff division for coalition formation in a class of task oriented domains. In: Meyer, J.-J.C., Tambe, M. (eds.) ATAL 2001. LNCS (LNAI), vol. 2333, pp. 324–334. Springer, Heidelberg (2002). https://doi.org/10.1007/3-540-45448-9_24
3. Davis, B., Conwell, W.: Methods and systems to help detect identity fraud, 12 July 2007, uS Patent App. 11/613,891
4. Dorigo, M., Di Caro, G.: Ant colony optimization: a new meta-heuristic. In: Proceedings of the 1999 Congress on Evolutionary Computation-CEC99 (Cat. No. 99TH8406), vol. 2, pp. 1470–1477. IEEE (1999)
5. Haim, G., An, B., Kraus, S., et al.: Human-computer negotiation in a three player market setting. Artif. Intell. **246**, 34–52 (2017)
6. Han, J., Pei, J., Kamber, M.: Data Mining: Concepts and Techniques. Elsevier, Amsterdam (2011)
7. Heraguemi, K.E., Kamel, N., Drias, H.: Association rule mining based on bat algorithm. In: Pan, L., Păun, G., Pérez-Jiménez, M.J., Song, T. (eds.) BIC-TA 2014. CCIS, vol. 472, pp. 182–186. Springer, Heidelberg (2014). https://doi.org/10.1007/978-3-662-45049-9_29
8. Heraguemi, K.E., Kamel, N., Drias, H.: Multi-swarm bat algorithm for association rule mining using multiple cooperative strategies. Appl. Intell. **45**(4), 1021–1033 (2016)
9. Jiang, M., Cui, P., Beutel, A., Faloutsos, C., Yang, S.: CatchSync: catching synchronized behavior in large directed graphs. In: Proceedings of the 20th ACM SIGKDD International Conference on Knowledge Discovery and Data Mining, pp. 941–950. ACM (2014)
10. Kechid, A., Drias, H.: Association rules mining for culture modeling. In: Rocha, Á., Adeli, H., Reis, L.P., Costanzo, S. (eds.) WorldCIST'18 2018. AISC, vol. 746, pp. 378–387. Springer, Cham (2018). https://doi.org/10.1007/978-3-319-77712-2_36
11. Kennedy, J.: Particle swarm optimization. In: Sammut, C., Webb, G.I. (eds.) Encyclopedia of Machine Learning, pp. 760–766. Springer, Heidelberg (2011). https://doi.org/10.1007/978-0-387-30164-8_630

12. Kerr, R., Cohen, R.: Detecting and identifying coalitions. In: Proceedings of the 11th International Conference on Autonomous Agents and Multiagent Systems, vol. 3, pp. 1363–1364. International Foundation for Autonomous Agents and Multiagent Systems (2012)
13. Khennak, I., Drias, H.: Bat-inspired algorithm based query expansion for medical web information retrieval. J. Med. Syst. **41**(2), 34 (2017)
14. Kim, C., Miao, H., Shim, K.: Catch: a detecting algorithm for coalition attacks of hit inflation in internet advertising. Inf. Syst. **36**(8), 1105–1123 (2011)
15. Metwally, A., Agrawal, D., El Abbadi, A.: Detectives: detecting coalition hit inflation attacks in advertising networks streams. In: Proceedings of the 16th International Conference on World Wide Web, pp. 241–250. ACM (2007)
16. Niknafs, M., Dorri Nogoorani, S., Jalili, R.: A collusion mitigation scheme for reputation systems. ISC Int. J. Inf. Secur. **7**(2), 151–166 (2015)
17. Rao, A.S., Georgeff, M.P., et al.: BDI agents: from theory to practice. In: ICMAS, vol. 95, pp. 312–319 (1995)
18. Sarswat, S., Abraham, K.M., Ghosh, S.K.: Identifying collusion groups using spectral clustering. arXiv preprint arXiv:1509.06457 (2015)
19. Shehory, O., Kraus, S.: Feasible formation of coalitions among autonomous agents in nonsuperadditive environments. Comput. Intell. **15**(3), 218–251 (1999)
20. Shoham, Y., Leyton-Brown, K.: Multiagent Systems: Algorithmic, Game-Theoretic, and Logical Foundations. Cambridge University Press, Cambridge (2008)
21. Tian, T., Zhu, J., Xia, F., Zhuang, X., Zhang, T.: Crowd fraud detection in internet advertising. In: Proceedings of the 24th International Conference on World Wide Web, pp. 1100–1110. International World Wide Web Conferences Steering Committee (2015)
22. Tomita, E., Tanaka, A., Takahashi, H.: The worst-case time complexity for generating all maximal cliques and computational experiments. Theoret. Comput. Sci. **363**(1), 28–42 (2006)
23. Wang, G., Zhang, X., Tang, S., Zheng, H., Zhao, B.Y.: Unsupervised clickstream clustering for user behavior analysis. In: Proceedings of the 2016 CHI Conference on Human Factors in Computing Systems, pp. 225–236. ACM (2016)
24. Yang, X.S.: Nature-Inspired Metaheuristic Algorithms. Luniver Press (2010)
25. Yang, X.S.: A new metaheuristic bat-inspired algorithm. In: NICSO 2010, vol. 284, pp. 65–74. Springer, Heidelberg (2010)
26. Zhai, J., Cao, Y., Yao, Y., Ding, X., Li, Y.: Coarse and fine identification of collusive clique in financial market. Expert Syst. Appl. **69**, 225–238 (2017)
27. Zhang, Q., Feng, W.: Detecting coalition attacks in online advertising: a hybrid data mining approach. Big Data Inf. Anal. **1**(2/3), 227–245 (2016)

Learning Explainable Control Strategies Demonstrated on the Pole-and-Cart System

Domen Šoberl(✉) and Ivan Bratko

Faculty of Computer and Information Science, University of Ljubljana,
Večna pot 113, 1000 Ljubljana, Slovenia
{domen.soberl,ivan.bratko}@fri.uni-lj.si

Abstract. The classical problem of balancing an inverted pendulum is commonly used to evaluate control learning techniques. Traditional learning methods aim to improve the performance of the learned controller, often disregarding comprehensibility of the learned control policies. Recently, Explainable AI (XAI) has become of great interest in the areas where humans can benefit from insights discovered by AI, or need to check whether AI's decisions make sense. Learning qualitative models allows formulation of learned hypotheses in a comprehensible way, closer to human intuition than traditional numerical learning. In this paper, we use a qualitative approach to learning control strategies, which we demonstrate on the problem of balancing an inverted pendulum. We use qualitative induction to learn a qualitative model from experimentally collected numerical traces, and qualitative simulation to search for possible qualitative control strategies, which are tested through reactive execution. Successful behaviors provide a clear explanation of the learned control strategy.

Keywords: Inverted pendulum · Learning qualitative models · Qualitative simulation · Explainable models · Explainable control

1 Introduction

To automatically control and stabilize the behavior of a mechanical system, a controller with a corrective mechanism is required. The controller compares the output from the actual system with the desired output, and applies an appropriate correction based on its mathematical model. When the mathematical model is not known, reinforcement learning is often used to learn a controller through interaction with the environment.

The problem of balancing an inverted pendulum is a clear example of a non-linear system which has become a popular benchmark problem for many control learning methods. The most common implementation of inverted pendulum is the *pole-and-cart* composition, where the pendulum is controlled indirectly by applying forces to the cart. Michie and Chambers [10] were among the first ones

F. Wotawa et al. (Eds.): IEA/AIE 2019, LNAI 11606, pp. 483–494, 2019.
https://doi.org/10.1007/978-3-030-22999-3_42

to study adaptive control on the pole-balancing problem. They implemented a reinforcement learning algorithm called BOXES, which discretized the continuous domain into 'boxes' and kept a record on how actions performed within each 'box'. Later experiments involved various types of neural networks [1,2,6], policy gradient learning [14], and Q-learning [6,9,12]. Ramamoorthy and Kuipers [13] used qualitative modeling to design a controller for the pole-and-cart system. Their control policy, which was derived manually, was robust enough to accommodate a large amount of abuse from the user. It should be noted that in contrast to this, in our work a qualitative model is learned through experimentation, and a control policy is derived automatically from the learned model.

In the area of learning autonomous control, as well as in other areas of reinforcement learning, principal emphasis is on improving the performance of the system over time. This has been achieved to a level that matches and even surpasses human abilities of learning, even in many tasks that were considered computationally unattainable only a few years ago [11,15]. However, performance is not necessarily a sole purpose of machine learning. In many areas, human expertise could greatly benefit from insights gained by artificial intelligence, or when explanation is needed for collaboration and trust between a human and an AI system. The inability of many modern AI learning techniques to explain the newly discovered concepts can thus be argued as a serious limitation [3].

In this paper we discuss a possible approach to learning explainable control strategies, which we demonstrate on the pole-and-cart system. Our methods are based on the principles of qualitative physics—a theory that studies dynamics of physical systems in a human intuitive way. Only a small amount of numerical traces, collected in a matter of seconds, is needed to induce a qualitative control model. Possible control strategies are then found offline by qualitative simulation and tested through execution. Qualitative formulation of both, the model and the found strategies, provides a clear explanation on how to control an inverted pendulum.

The rest of the paper is organized as follows. The following section describes the pole-and-cart system used throughout the paper, and gives an outline of the proposed method. Section 3 demonstrates the use of qualitative induction to learn explainable control models. Section 4 discusses adaptations of qualitative simulation to search for possible control strategies. The method to execute control strategies is proposed in Sect. 5, where the results of execution on the pole-and-cart system are also presented. Section 6 summarizes and concludes the paper.

2 Learning Control Strategies of a Pole-and-Cart System

2.1 The Pole-and-Cart System

A common implementation of inverted pendulum is the *pole-and-cart* system shown in Fig. 1. A pole is freely hinged on top of a wheeled cart that moves along a one-dimensional track. The pole can move vertically in the same direction as the cart. It is assumed that there is no friction between the cart and the track

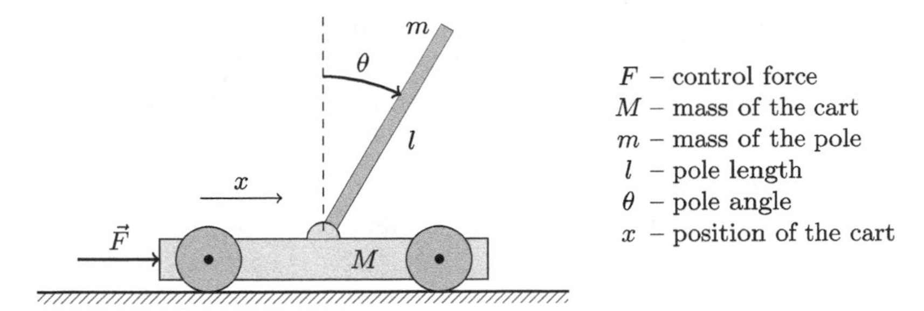

Fig. 1. Forces working on the cart-pole composition while pushing (pulling) the cart with force $\vec{F}$.

or the pole and the cart. The controller can apply force F of a fixed magnitude at discrete time intervals in either direction. This is also known as *bang-bang* control.

The inverted pendulum was simulated using the following parameters: cart mass $M = 1\,\text{kg}$, pole mass $m = 0.1\,\text{kg}$, pole length $l = 1\,\text{m}$, gravitational acceleration $g = 9.81\,\text{m/s}^2$, control force $F = \pm 10\,\text{N}$. Center of mass m was assumed at the center of the pole. Dynamics of the system were modeled by the following differential equations [1]:

$$\ddot{\theta} = \frac{g \sin\theta\,(M+m) - \left(F + ml\dot{\theta}^2 \sin\theta\right)\cos\theta}{\frac{4}{3}l\,(M+m) - ml\cos^2\theta}, \tag{1}$$

$$\ddot{x} = \frac{F - ml\left(\ddot{\theta}\cos\theta + \dot{\theta}^2 \sin\theta\right)}{M+m}. \tag{2}$$

Simulation time step as well as the rate at which the external force was applied was $0.02\,\text{s}$. Motion was bounded to $|x| < 3\,\text{m}$ and $-\frac{\pi}{2} < \theta < \frac{\pi}{2}$, outside which the *failure* signal was raised and the system reset to its initial state $x = 0$ and $\theta = 0$. The task is defined as follows:

1. Move the cart from the initial position x_0 to the goal position x_{goal} and stop there.
2. Keep the pole near its vertical position $\theta = 0$ at all times. However, notice that this goal is sometimes in conflict with the goal of moving the cart towards x_{goal}.

In this paper we do not consider the problem of swing-up control, i.e. finding a strategy to lift the pole from a hanging state.

2.2 Method Overview

Our method of learning explainable control strategies is outlined in Fig. 2. First, experimentation is conducted by the method known as *motor babbling*. Randomly chosen actions are executed while the resulting behaviors are numerically

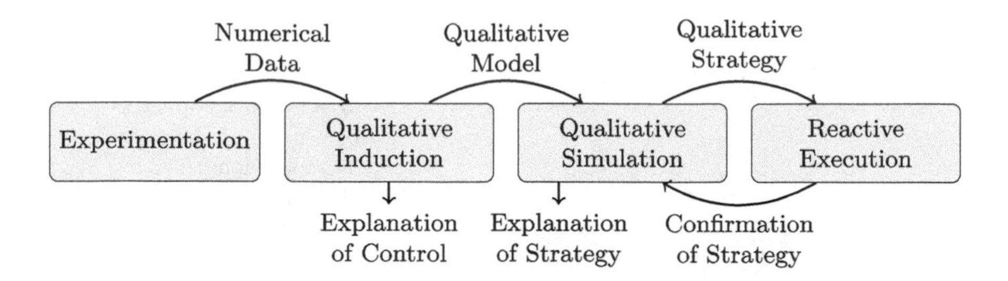

Fig. 2. The method of learning explainable control strategies.

sampled and stored. Relatively small amount of data is typically required, in our case, only a few seconds of experimentation suffices. The collected data is then used with a method known as *qualitative induction*, which is a type of machine learning that looks for qualitative patterns in numerical data. The output is a model that describes dependencies between attributes in the form of monotonic relations. Such a model allows a comprehensible insight into how the system behaves under the defined actions.

Qualitative model is a qualitative approximation of the system's mathematical model, and can be used to simulate possible behaviors of the system. We use an adaptation of QSIM [7], which is an algorithm that simulates qualitative physics of a qualitatively modeled system. Unlike a classical physics simulator that operates in discrete time steps, QSIM captures only qualitatively significant changes at symbolic times, e.g. transition of the pole from the left to the right side. Such sequences of qualitatively formulated transitions are close to human type of reasoning and therefore provide a clear explanation of found control strategies.

Because a qualitative model is an abstraction of the exact mathematical model, certain qualitative solutions may be found which are impossible to execute on our actual system. Each strategy is therefore executed to verify its correctness, until a feasible solution is found. We explain how such execution is possible with reactive guidance presented in [17].

In the following sections we describe each step in more detail and demonstrate how the method applies to the pole-and-cart problem.

3 Learning a Qualitative Control Model

3.1 Qualitative Induction

Our motivation to use qualitative modeling as a means to provide explainable models is closely related to the concept of *qualitative abstraction*. A mathematical model, such as (1, 2), can be viewed as a mathematical abstraction of a physical system. When trying to understand such formulation intuitively, one would usually mentally abstract away certain numerical details and focus on the most significant elements. A possible insight from studying such an equation could be e.g. 'acceleration $\ddot{x}$ can be increased by increasing force F'. This can be seen as yet another level of abstraction—qualitative abstraction.

When qualitative models are learned directly from numerical data, we speak of qualitative induction. Several algorithms exist; two most often used with this type of experiments are QUIN (QUalitative INduction) [5] and Padé (PArtial DErivatives) [18]. In this paper we use the former, although both produce comparable results. QUIN learns so-called qualitative trees, which are similar to the well-known decision trees. The difference is that instead of class values, leaves are labeled with constraints known as *multivariate monotonicity constraints* [16], which are qualitative abstractions of multivariate monotonic functions. Such constraints indicate monotonically increasing and decreasing regions, e.g. $z = M^{+,-}(x, y)$ indicates the existence of a continuously differentiable function $z = f(x, y)$, where $\partial f/\partial x > 0$ and $\partial f/\partial y < 0$. Notation with zero index, e.g. $z = M_0^{+,-}(x, y)$, also indicates $f(0, 0) = 0$. QUIN will form a branch at any point where monotonicity of f changes.

3.2 Qualitative Model of the Pole-and-Cart System

To collect numerical data needed to induce a qualitative model of inverted pendulum, we acquired a sample $(F, x, \dot{x}, \theta, \dot{\theta})$ every $\Delta t = 0.02$ s of the running experimentation. At every Δt, a random force $F \in \{-10, 0, 10\}$[1] was applied to the cart, and the remaining values x, $\dot{x}$, θ, $\dot{\theta}$ recorded at the next Δt, right before executing the next random action. Experimentation started in state $(x, \dot{x}, \ddot{x}, \theta, \dot{\theta}, \ddot{\theta}) = (0, 0, 0, 0, 0, 0)$, and was reset to that state as soon as θ fell out of interval $(-\frac{\pi}{2}, \frac{\pi}{2})$. After 3 s of experimentation, two failed attempts at balancing the pole were observed and 150 samples collected. From the collected data, QUIN learned the following two qualitative constraints in about 2 s:

$$\ddot{\theta} = M^{-,+}(F, \theta) \tag{3}$$

$$\ddot{x} = M^{+}(F) \tag{4}$$

Technically, these are two qualitative trees with a single node—the root only, which means that they apply to all values in the range, within which experiments were conducted. Constraints (3) and (4) are qualitative approximations of differential equations (1) and (2), and offer a simple explanation on how a pole-and-cart system can be controlled:

- Angular acceleration $\ddot{\theta}$ increases when: (i) F decreases and θ increases, or (ii) F decreases and θ is steady, or (iii) F is steady and θ increases. The reverse applies for decreasing $\ddot{\theta}$. If F and θ both increase or both decrease, $\ddot{\theta}$ can either increase, decrease, or remain unchanged.
- Acceleration $\ddot{x}$ increases (decreases) when F increases (decreases).

[1] Values $F \in [-10, 10]$ gave the same results.

The model is approximate not only in terms of quantities, but also functional dependencies. The entropy provided by some attributes is high enough to be treated as noise in the sampled data. We should therefore interpret the induced constraints in terms of confidence. For example, it is very likely that acceleration $\ddot{x}$ will increase with force F, while the effect of θ on $\ddot{x}$ is either non-existent or uncertain. By examining (2), we can see that the true relation between θ and $\ddot{x}$ is a complex one, involving magnitudes of F, $\dot{\theta}$, and $\ddot{\theta}$. On the other hand, with significant gravitational and small centrifugal force being observed during experimentation, it is very likely that angular acceleration $\ddot{\theta}$ will increase with θ.

4 Finding Control Strategies

The induced model alone does not suffice for a successful control of the cart and the pendulum. Suppose the goal is to move the cart to the right while not dropping the pole. The question is, which actions to execute under what conditions. In order to answer this question, long-term behaviors under different actions need to be computed. One way to do that is through qualitative simulation.

4.1 Qualitative Simulation

Qualitative simulation was introduced by Kuipers [7,8], who defined it as a constraint satisfaction problem, and implemented it as the QSIM algorithm. A more recent implementation of QSIM by Bratko can be found in [4]. The main difference between a conventional physics simulator and QSIM is that conventional simulator performs numerical computations in discrete time, whereas QSIM operates with symbolic quantities in symbolic time. Qualitative simulation therefore does not answer the question of precise numerical values at specific times, but provides insights into possible sequences of critical system's transitions called *qualitative behaviors.*

Consider a sinusoidal oscillation $y = A \cdot \sin(t \cdot 2\pi/P)$ in time t, of amplitude A and period P. To perform a numerical simulation, particular values of parameters A and P are needed. On the other hand, qualitative simulation can explain the behavior of such oscillators in general, by computing the following qualitative behavior:

$$\begin{aligned} y : 0/\text{std} &\rightarrow 0..\text{max}/\text{inc} \rightarrow \text{max}/\text{std} \rightarrow 0..\text{max}/\text{dec} \rightarrow \\ &\rightarrow 0/\text{dec} \rightarrow \text{min}..0/\text{dec} \rightarrow \text{min}/\text{std} \rightarrow \text{min}..0/\text{inc} \rightarrow 0/\text{inc} \rightarrow \cdots \end{aligned}$$

Landmarks *min* and *max* symbolize the minimum and the maximum value of variable y. Qualitative state is described by qualitative magnitude and direction of change; e.g. 0/std states that the variable is steady at value 0, and 0..max/inc that it is increasing somewhere within the open interval $(0, \text{max})$.

The algorithm works by satisfying two types of constraints: (i) assumption of smoothness, and (ii) the qualitative model. Assumption of smoothness disallows transitions that exhibit discontinuity of magnitudes, e.g. min..0/inc →

0..max/inc (here intermediate state 0/inc is missing), or discontinuity of time derivatives, e.g. 0..max/inc → 0..max/dec. In this paper we are concerned with two types of constraints provided by model: the monotonicity constraints already discussed in the previous section, and constraints imposed by time derivatives. Here, we presume time derivatives to be defined implicitly by notation, i.e. $\ddot{x}$ is time derivative of $\dot{x}$, $\dot{x}$ of x, etc.

Monotonicity constraint says that y is a function $f(x_1, x_2, \ldots)$ with a property that y *monotonically depends* on its arguments as follows. For all $s_i, (1 \leq i \leq n)$: if $s_i = +$ then y increases with x_i (provided that all other x_i stay constant). More formally this can be written as: if $s_i = +$ then $\partial f / \partial x_i > 0$, and analogously for $s_i = -$.

The criterion imposed by time derivatives is the following. Let x and its time derivative $\dot{x}$ be a part of the qualitative state. State $(x/dir_x, \dot{x}/dir_{\dot{x}}, \ldots)$ is valid if $(\dot{x} > 0 \Leftrightarrow dir_x = \text{inc})$ and $(\dot{x} < 0 \Leftrightarrow dir_x = \text{dec})$ and $(\dot{x} = 0 \Leftrightarrow dir_x = \text{std})$.

4.2 Control Strategies of the Pole-and-Cart System

We define the problem of controlling the pole-and-cart system in the following way:

1. Variables $\theta, \dot{\theta}, \ddot{\theta}, \dot{x}, \ddot{x}$ are qualitatively abstracted to domain $\{\text{neg}, 0, \text{pos}\}$, which corresponds to discretization $\{(-\infty, 0), 0, (0, \infty)\}$. Domain of variable x contains additional landmark x_1 that represents the cart's goal position. Presuming $x_1 > 0$, possible qualitative magnitudes of x are {neg, 0..x_1, x_1, x_1..inf}.
2. The initial position of the cart is always $x = 0$.
3. From the initial position, find a strategy to increase x to x_1 and finish with $\theta = 0$.

A strategy is abstracted from qualitative behavior, found by searching through the qualitative state space generated by QSIM. We implement the search algorithm in the following ways:

- Favor short solutions. Short strategies offer simpler explanations than long ones and hopefully take less time to find. They are therefore first to be tested. This is assured by iterative deepening.
- Favor solutions with effective actions. Consider the constraint $\ddot{\theta} = M^{-,+}(F, \theta)$ with directions F : inc, θ : inc. All three outcomes $\ddot{\theta}$: {inc, std, dec} are valid, where $\ddot{\theta}$: inc indicates that the effect of θ outweighed the effect of F, rendering action F : inc qualitatively ineffective. Solutions with effective actions are more likely to succeed.

The initial condition is $x = 0, \theta = 0$. Table 1 shows a shortest found qualitative behavior that increases position x from an initial state $x = 0, \theta = 0$. This strategy was found by our implementation of QSIM on a typical laptop in about 2 s. Time is discretized by symbolic landmarks t_i, where the sequence of qualitative states alternates between time-point states $S(t_i)$ and time-interval states $S(t_i, t_{i+1})$.

We define explainable strategy as the following reinterpretation of qualitative behavior:

Table 1. Qualitative behavior of an executable strategy to increase position x.

Depth	Time	F	x	$\dot{x}$	$\ddot{x}$	θ	$\dot{\theta}$	$\ddot{\theta}$
0	t_0	0/std	0/std	0/std	0/std	0/std	0/std	0/std
1	(t_0, t_1)	neg/dec	neg/dec	neg/dec	neg/dec	pos/inc	pos/inc	pos/inc
2	t_1	neg/std	neg/dec	neg/dec	neg/std	pos/inc	pos/inc	pos/std
3	(t_1, t_2)	neg/inc	neg/dec	neg/dec	neg/inc	pos/inc	pos/inc	pos/dec
4	t_2	0/inc	neg/dec	neg/std	0/inc	pos/inc	pos/std	0/std
5	(t_2, t_3)	pos/inc	neg/dec	neg/inc	pos/inc	pos/inc	pos/dec	neg/dec
6	t_3	pos/std	neg/std	0/inc	pos/std	pos/std	0/dec	neg/std
7	(t_3, t_4)	pos/dec	neg/inc	pos/inc	pos/dec	pos/dec	neg/dec	neg/inc
8	t_4	pos/dec	0/inc	pos/std	0/dec	pos/dec	neg/std	0/inc
9	(t_4, t_5)	pos/dec	0..x_1/inc	pos/dec	neg/dec	pos/dec	neg/inc	pos/inc
10	t_5	neg/std	x_1/std	0/dec	neg/std	0/std	0/inc	pos/std

- Simplify the behavior to the values of interest; in our case F (magnitude) and x, θ (magnitude and direction). The simplified behavior can be shortened by removing equivalent adjacent states, e.g. states 2 and 3 in Table 1.
- Time-points t_i coincide with important changes in qualitative behavior of the system.
- Time-intervals (t_i, t_{i+1}) explain the type of actions to be taken.

Figure 3 shows a visual representation of two different control strategies. Strategies 1 and 2 are shortest found strategies to increase position x. Suppose Strategy 1 is found first. Its interpretation is the following: (a) in the initial state apply positive force F until (b) a positive velocity of the cart and a negative velocity of the pole is observed. Then apply negative force F to (c) stop the motion of the pole while the cart continues moving forward. (d) Eventually, the pole will gain a certain upward momentum, at which point apply negative force F, to bring (e) the cart and the pole to a full stop exactly at the goal point.

Strategy 2 in Fig. 3 is abstracted from behavior shown in Table 1. Its interpretation is the following: (a) start by applying the negative force F, until (b) negative velocity of the cart and positive velocity of the pole is observed. Then apply positive F to eventually bring (c) the cart and the pole to a stop. By continuing with the positive F, (d) the cart will gain positive velocity, while the pole will start to lift, until (e) the goal state is reached.

5 Execution

Control strategy as defined in the previous section can be broken into control policy that consists of stages of the form:

$$\underset{\text{preconditions}}{\text{State}(t_i)} \longrightarrow \underset{\text{action}}{\text{State}(t_i, t_{i+1})} \longrightarrow \underset{\text{postconditions}}{\text{State}(t_{i+1})}\ .$$

Strategy 1: increase x (unsuccessful)

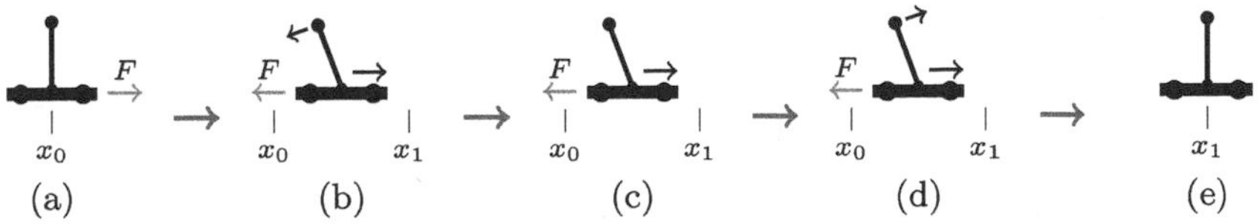

Strategy 2: increase x (successful):

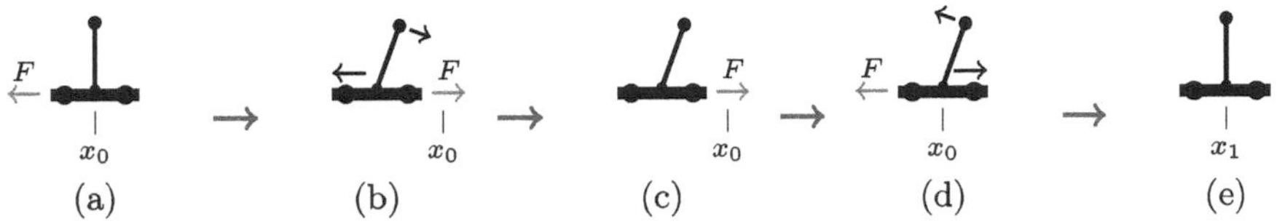

Fig. 3. Two shortest control strategies found by QSIM to increase cart position x. Position x is depicted relative to landmarks x_0 and x_1. Arrows parallel to the cart and at the top of the pole depict directions of motion of the cart and the pole respectively. Actions F : pos and F : neg are respectively symbolized by left and right arrows labeled with symbol F.

However, without incorporating numerical velocities $\dot{x}$ and $\dot{\theta}$ into action decision policy, the execution results in an increasing oscillation of the pole, until the system is thrown out of balance. We were able to achieve successful execution of presented control strategies using the reactive method proposed in [17]. Given a qualitative model (i.e. a set of qualitative constraints), this method is able to implement a continuous transition between two consecutive qualitative states. This is done by executing the following procedure multiple times per second:

1. Observe the current numerical state of the system (plots are shown in Fig. 4).
2. If goal conditions are met, finish the execution.
3. Use given qualitative constraints to determine the effect of each possible action.
4. Estimate the time of arrival $T(x_i)$ of each variable x_i to its goal value. Variables with no goal value have $T(x_i) = 0$.
5. Execute the action that minimizes the total estimation $\sum_{\forall i} T(x_i)$.

We use this method to execute a qualitative control strategy in the following way:
For each consecutive qualitative state S_i in strategy $\mathcal{S}$ do:

1. If state S_i is not the final qualitative state of strategy $\mathcal{S}$, let S_{i+1} be the executor's next goal state.
2. Execute state transition until state S_{i+1} is reached or the given pole-and-cart constraints ($|x| < 3$ m, $|\theta| < \frac{\pi}{2}$) are violated.
3. If the pole-and-cart constraints are violated, execution of strategy $\mathcal{S}$ is unsuccessful.

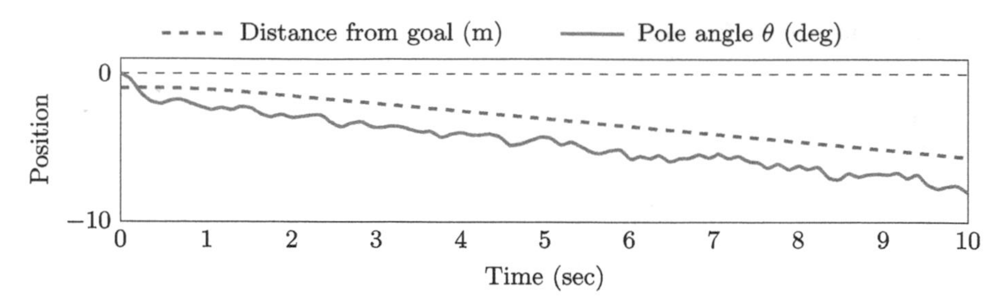

(a) Strategy 1 results in moving the cart away from the goal.

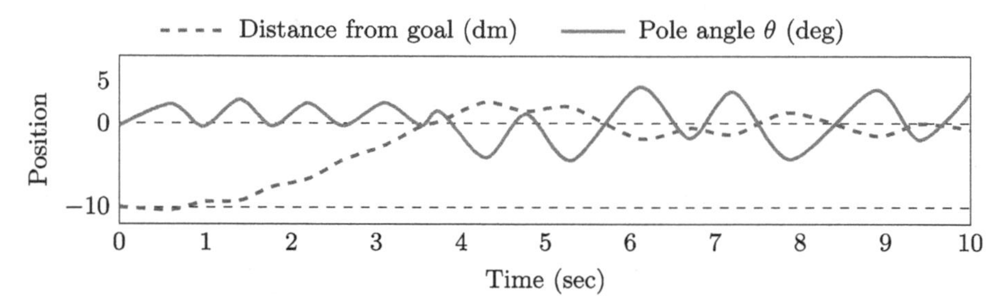

(b) Strategy 2 brings the cart to the goal position. A slight offshoot is observed, which is corrected by the inverted version of this same strategy.

Fig. 4. Execution of control strategies 1 and 2 shown in Fig. 3. The task is to move from the initial position $x_0 = -1$ m to goal position $x_1 = 0$ m, while balancing the pole. Plots show the first 10 s of execution.

Given a goal condition, e.g. $x = x_1/\text{std}, \theta = 0/\text{std}$, the executor takes into account all four goal quantities: $x = x_1$, $\dot{x} = 0$, $\theta = 0$, and $\dot{\theta} = 0$. During the execution, the four numerical values are being read from the sensory input every 0.02 s, and an appropriate action decided at the same rate, with the aim to simultaneously bring all four variables to their goal values. In order to lift the forward inclined pole, the executor will at first issue only pushing actions, but when a certain momentum is build in the pole, the executor would issue occasional pulling action to compensate the momentum. This way, the pole is brought to a near stop before vertical position is achieved. The plot of such execution is shown in Fig. 4b. The slight offshoot made at the goal position $x = x_1$ is corrected by the inverted version of Strategy 2.

This reactive execution procedure sometimes has to balance between the urgency of various goals that occasionally conflict each other (for example, the goal of preventing the pole from falling would require action $F = -10$, whereas the goal of moving the cart to the right would require $F = +10$). Such conflicts are resolved by the current urgency of the goals at various stages of the process. This is also the reason why Strategy 1 results in slowly lowering the pole while moving the cart away from the goal, as seen from plot in Fig. 4a. As the distance from the goal $x = x_1$ increases, the executor tends to increase the frequency of $F = +10$ actions. Details of this can be found in [17].

6 Conclusion

The main advantage of using qualitative methods to learn control is the ability to learn control models and strategies that are easy to interpret, while still informative enough to allow execution. The learning process is fast in comparison to typical reinforcement learning; we showed that a successful control of pole-and-cart composition can be achieved in a matter of seconds. This is mostly due to the fact that little numerical data is needed to induce a qualitative model, and the search for strategies is done offline. The actual trial-and-error learning is left for testing the possible strategies, which is a considerably shorter procedure than learning on the level of individual actions.

The main drawbacks of this approach are the random exploration strategy, which may or may not sufficiently sample the attribute space, and computational complexity of QSIM, which is known to be exponential in the length of the behavior. With more complex domains the time to find a strategy could increase beyond reasonable. This problem could in the future be tackled by testing ambiguous state transitions before the search is finished, and thus limit the state space.

References

1. Anderson, C.W.: Learning to control an inverted pendulum using neural networks. IEEE Control Syst. Mag. **9**(3), 31–37 (1989)
2. Barto, A.G., Sutton, R.S., Anderson, C.W.: Neuronlike adaptive elements that can solve difficult learning control problems. IEEE Trans. Syst. Man Cybern. **SMC-13**(5), 834–846 (1983)
3. Bratko, I.: Alphazero - what's missing? Informatica (Slovenia) **42**, 7–11 (2018)
4. Bratko, I.: Prolog: Programming for Artificial Intelligence, 4th edn. Addison-Wesley, Boston (2011)
5. Bratko, I., Šuc, D.: Learning qualitative models. AI Mag. **24**(4), 107–119 (2003)
6. Hosokawa, S., Kato, J., Nakano, K.: A reward allocation method for reinforcement learning in stabilizing control tasks. Artif. Life Robot. **19**(2), 109–114 (2014)
7. Kuipers, B.: Qualitative simulation. Artif. Intell. **29**(3), 289–338 (1986)
8. Kuipers, B.: Qualitative Reasoning: Modeling and Simulation with Incomplete Knowledge. MIT Press, Cambridge (1994)
9. Linglin, W., Yongxin, L., Xiaoke, Z.: Design of reinforce learning control algorithm and verified in inverted pendulum. In: 2015 34th Chinese Control Conference (CCC), pp. 3164–3168 (2015)
10. Michie, D., Chambers, R.A.: BOXES: an experiment in adaptive control. In: Machine Intelligence, vol. 2, pp. 125–133. Elsevier/North-Holland (1968)
11. Mnih, V., et al.: Human-level control through deep reinforcement learning. Nature **518**, 529–33 (2015)
12. Puriel-Gil, G., Yu, W., Sossa, H.: Reinforcement learning compensation based PD control for inverted pendulum. In: 15th International Conference on Electrical Engineering, Computing Science and Automatic Control (CCE), pp. 1–6 (2018)
13. Ramamoorthy, S., Kuipers, B.: Qualitative heterogeneous control of higher order systems. In: Hybrid Systems: Computation and Control, pp. 417–434. Springer, Heidelberg (2003)

14. Riedmiller, M., Peters, J., Schaal, S.: Evaluation of policy gradient methods and variants on the cart-pole benchmark. In: 2007 IEEE International Symposium on Approximate Dynamic Programming and Reinforcement Learning, pp. 254–261 (2007)
15. Silver, D., et al.: Mastering chess and shogi by self-play with a general reinforcement learning algorithm. CoRR abs/1712.01815 (2017)
16. Wellman, M.P.: Qualitative simulation with multivariate constraints. In: Second International Conference on Principles of Knowledge Representation and Reasoning, pp. 547–557. Morgan Kaufmann (1991)
17. Šoberl, D., Bratko, I.: Reactive motion planning with qualitative constraints. In: Benferhat, S., Tabia, K., Ali, M. (eds.) IEA/AIE 2017. LNCS (LNAI), vol. 10350, pp. 41–50. Springer, Cham (2017). https://doi.org/10.1007/978-3-319-60042-0_5
18. Žabkar, J., Možina, M., Bratko, I., Demšar, J.: Learning qualitative models from numerical data. Artif. Intell. **175**(9–10), 1604–1619 (2011)

Mapping Infected Crops Through UAV Inspection: The Sunflower Downy Mildew Parasite Case

Juan Pablo Rodríguez-Gómez[1(✉)], Maurilio Di Cicco[2], Sandro Nardi[3], and Daniele Nardi[2]

[1] Robotic, Vision and Control group (GRVC), ETS Ingeniera, University of Seville, Seville, Spain
jrodriguezg@us.es

[2] Department of Computer, Control, and Management Engineering "Antonio Ruberti", Sapienza University of Rome, Rome, Italy
nardi@diag.uniroma1.it

[3] Agency for Agro-food Sector Services of the Marche Region, Osimo Stazione, Italy
nardi_sandro@assam.marche.it

Abstract. In agriculture, the detection of parasites on the crops is required to protect the growth of the plants, increase the yield, and reduce the farming costs. A suitable solution includes the use of mobile robotic platforms to inspect the fields and collect information about the status of the crop. Then, by using machine learning techniques the classification of infected and healthy samples can be performed. Such approach requires a large amount of data to train the classifiers, which in most of the cases is not available given constraints such as weather conditions in the inspection area and the hardware limitations of robotic platforms. In this work, we propose a solution to detect the downy mildew parasite in sunflowers fields. A classification pipeline detects infected sunflowers by using a UAV that overflies the field and captures crop images. Our method uses visual information and morphological features to perform infected crop classification. Additionally, we design a simulation environment for precision agriculture able to generate synthetic data to face the lack of training samples due to the limitations to perform the collection of real crop information. Such simulator allows to test and tune the data acquisition procedures thus making the field operations more effective and less failure prone.

Keywords: Point clouds · Support vector machines · Machine learning

1 Introduction

Detecting the presence of parasites and diseases in crops is a common problem in agriculture. The crop inspection task is mainly performed by operators, and

J. P. Rodríguez-Gómez conducted the presented research during his stay at the RoCoCo laboratory of Sapienza University of Rome.

F. Wotawa et al. (Eds.): IEA/AIE 2019, LNAI 11606, pp. 495–503, 2019.
https://doi.org/10.1007/978-3-030-22999-3_43

it requires time and effort, as the task typically involves phytosanitary field evaluation. Sunflower crops require manual visual inspection to avoid the presence at harvest of symptomatic dwarfed plants by downy mildew. Aerial robots are able to inspect the crops in a short time by overflying the terrain and collecting information about the crops. The onboard sensor selection mainly depends on the type of information required for detection. Lidar sensors provide morphological features of the vegetation for the detection of anomalies during the growing stages of the crops [1]. Multispectral imagery provides information to determine parameters such as the normalized difference vegetation index (NDVI) [2], and perform weed and crop classification [3]. Optical RGB sensors provide information to determine the health status of the crops from parameters such as the soil adjusted vegetation index (SAVI), the Triangular Greenness Index (TGI), the plant biomass [4], and the leaf area index (LAI)[5]. In this work, we propose a classification pipeline to detect the downy mildew parasite on sunflower fields, using point clouds retrieved from crop images captured by a UAV.

The data collected on the field can be analyzed using machine learning (ML) as well as the reconstructed point clouds to analyze crop status. For instance, image-based deep learning is used to classify land cover and crops types [6], and to perform crop and weed classification [7]. These applications require a transformation of the point cloud to obtain an ordered and structured input suitable for the CNNs. Support vector machines (SVMs) allow to perform crop classification using directly the point clouds. SVMs are used in precision agriculture to detect infected crops using spectral signatures [8] and soil classification [9]. Our solution uses SVMs to detect infected crops using point clouds instead of images thus reducing the classification complexity. Our aim is to show a viable solution by using SVMs and point clouds rather than finding the optimal solution among different ML approaches. The main limitation to train ML classifiers is the lack of available samples. The problem can be solved by extending the training dataset with synthetic samples. The synthetic information can be retrieved by using simulators able to render photorealistic models of the environment. Game engines are preferred for this task given their rendering capabilities and their successful use in robotic applications [10,11]. Our solution includes the implementation of a simulation tool to generate images of synthetic crops to solve the lack of training samples problem. Summarizing, the contributions of our work are three-fold:

1. A novel pipeline to detect the sunflower downy mildew parasite using only images of the crops.
2. The integration of visual information with the morphological features of the vegetation to perform the detection.
3. The use of a realistic crop simulator to retrieve synthetic data for training and evaluate the best set of parameters to perform the inspection task.

2 Problem Description

The downy mildew is a common problem in sunflower crops, the parasite can reduce the yield up to 100% [12]. The main symptoms are the change of color of the leaves and the size reduction of the plant by the dwarfism phenomenon. These symptoms can be used as classification features to detect the presence of the parasite. In this work, the images of sunflower crops are used to retrieve these features. First, the variation of color in the leaves is analyzed from the pixels intensity of the images. Second, the height information is retrieved from a point cloud of the field computed by using the crop images. Our data show that the infected crops have a height reduction of at least a 60% with respect to the healthy sunflowers. Figure 1 shows the height differences between infected and healthy sunflowers during the last part of the growing stage.

Fig. 1. Infected and healthy sunflowers during the last stage of the growing. A real crop of sunflowers (left), synthetic field rendered from our simulator (right).

3 Simulation Environment

The main goal of using a simulation environment for precision agriculture is to generate synthetic images to extend the dataset to train the classifiers for infected crop detection. Our simulation environment is inspired by [13], however, our solution enlarges the tool capabilities by simulating the behavior of the robot that performs the dataset collection. This extension allows for the evaluation and tuning of critical mission parameters before deploying the real robot for the data collection mission. Parameters such as flying altitude, trajectory, and variable lighting conditions can be evaluated. The simulator includes the model of a UAV to take images of the crops from a top view of the field. The simulation environment is developed using Epic Unreal Engine 4 [14]. It consists of a unique scene including the lighting, dynamics, and rendering properties for a realistic simulation of the data collection mission. The scene includes three types of objects: the environment objects (landscape, mixed vegetation, soil, rocks, etc), the model of the crops, and the UAV. The models of the crops were built from the base model of sunflower in Blender. The material of the crops includes realistic textures to enhance the realism of the crop representation. The models of the infected vegetation include the dwarfism effect and the change of color of the leaves. The models of both infected and healthy sunflowers are shown in Fig. 1. The UAV follows a trajectory defined by a customizable spline that controls the vehicle's trajectory and velocity. During the data collection, the robot flies parallel to the surface to capture images from a top view of the field. The parameters of the camera such as the field of view, the aspect ratio, and light exposure can be set by the user.

4 Classification Pipeline

The process of detecting infected crop samples from the point cloud of the field is described by the sequence of steps that implement the classification pipeline. These steps are described as follows: First, a point cloud of the terrain is computed by using the images captured with the aerial vehicle. Second, the soil component is separated from the vegetation samples using a classifier. Third, the samples are aligned with respect to a non-tilted global frame to retrieve the correct height information. Finally, the infected crops are detected for color and morphological features. Figure 2 shows the block diagram of the classification pipeline.

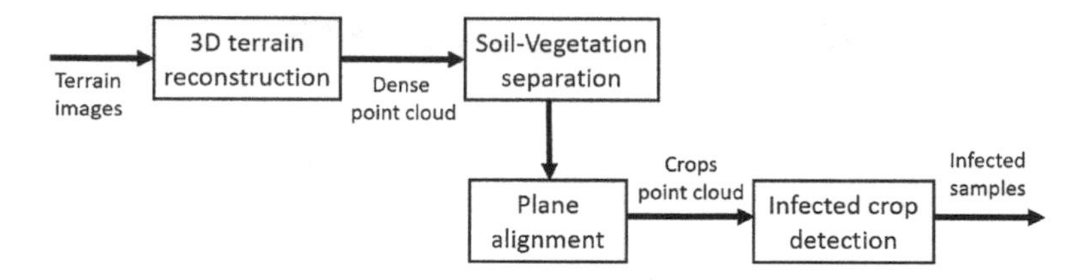

Fig. 2. Block diagram of the classification pipeline.

4.1 3D Terrain Reconstruction

The input of the pipeline is the set of images of the crops. The dense point cloud of the crops is generated by using the Agisoft Photoscan software [15]. The images have an overlapping factor of at least 70% to ensure the reconstruction of the majority of the scene. The point cloud dataset consists of point clouds obtained from synthetic frames and images of real crops. The point cloud samples are labeled using Photoscan. These annotations are used to train the classifiers of the pipeline. For each classifier, we generate a set of annotations. (i) Soil and vegetation labeling for the Soil-Crop SVM, and (ii) infected and healthy crop annotations for the Healthy-Infected crop SVM.

4.2 Soil-Crop Separation and Reference Plane Alignment

This section describes the classification of the field point cloud into vegetation and soil samples and the transformation to rotate the point clouds to a non-tilted plane. Elimination of the soil is performed by a SVM with radial basis kernel using the color of the points as classification features. The training point clouds are obtained from images from both real and synthetic crops under different terrain configurations (e.g. lighting exposure, soil, etc). The synthetic images are used to extend the size of the training dataset. For the soil-vegetation separation, the RGB data of the point clouds were considered as classification features. The output vegetation samples are the input for the next step, while the soil output is used to find the rotation matrix to perform the reference alignment. In a real scenario, the crops may be located on hills or uneven surfaces, therefore, the vegetation's height varies depending on the surface inclination. The alignment rotates the point clouds to a plane in which the z component is referred to a non-tilted surface. After the rotation, the height is normalized between $[0, 1]$. The normalization handles the height mismatches of the point clouds obtained from images captured at different flying altitudes.

4.3 Infected and Healthy Crop Classification

The last step of the pipeline detects the samples with a high probability of being infected with the downy mildew. As in the soil-vegetation classification, a binary SVM with radial basis kernel is used to classify the infected and healthy crop samples. The classification features are divided into: (i) the RGB components of the point cloud to evaluate the change of color on the leaves, (ii) crop normalized height to include the presence of the dwarfism effect.

5 Experiments and Evaluation

The experiments for the crop classification are divided into two parts. First, the soil-vegetation classifier was tested. Second, the performance of the infected-healthy classifier is evaluated by using samples from real crops. Table 1 summarizes the dataset for training, validating and testing the pipeline. The synthetic samples consist of point clouds generated using crop images from the simulator. The simulated crops are distributed on fields of 15×15 m. There are two types of point cloud datasets computed using images of real crops: (i) Frames captured during the middle growing stage of the sunflowers. At this stage, the crop presented few infected samples, therefore, the dataset includes only a few images of crops. (ii) Infected crops images taken during the last growing stage of the sunflowers. The frames of the dataset are available online[1]. The data collection mission was performed by a UAV that flew over the terrain while capturing crop images varying parameters such as the day time (morning, afternoon), flying altitude (5 m, 10 m and 15 m), and weather conditions (clear and partially cloud).

[1] https://mega.nz#F!HiAhRYSS!LruZLdj4-VcqZzSDy63dxQ.

Table 1. Dataset for training, validating and testing the pipeline.

Dataset type	Images	Point clouds (PCs)
Synthetic crops (1)	577	6
Middle growing stage (2)	199	3
Last growing stage (3)	811	14

5.1 Soil-Vegetation Classification

The test evaluated the soil-vegetation SVM performance. The SVM was trained using the synthetic samples (1) and the real crop samples (2). The training point clouds included only a few samples of real crops to evaluate the SVM performance by extending the dataset with synthetic samples. The dataset was divided by following a 5-fold approach to perform cross-validation. One fold was used for validating while the other 4 represent the training samples. For training, we used a distribution of 50% of soil samples (negative) and 50% of vegetation samples (positive). The validation reports an accuracy of 97.40%. For the tests, 5 point cloud samples of dataset 3 were considered. These samples are used to test the complete pipeline. The classifier returns promising results regarding the Accuracy (97.40%), Recall (0.98), FPR(0.06) and Precision(0.96) parameters. An example of the test is shown in Fig. 3, the classifier separates successfully the soil and vegetation samples.

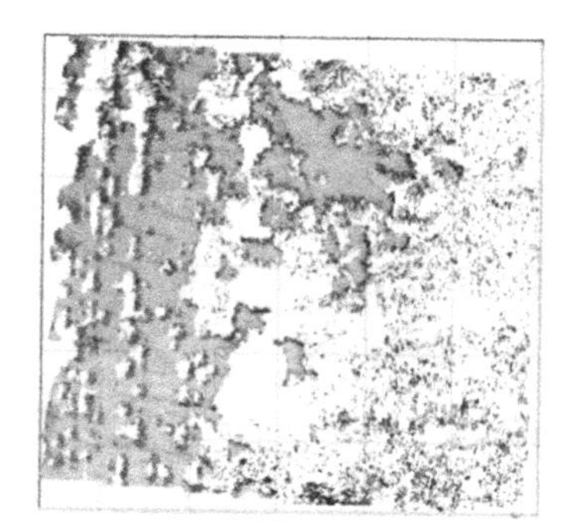

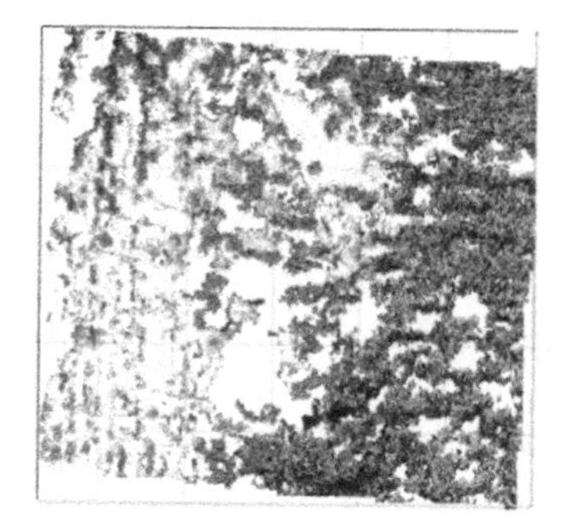

Fig. 3. Classification outcome from the soil-vegetation separation using real crop samples. Point cloud of the soil (left), and point cloud of the vegetation (right).

5.2 Infected Crop Detection

The experiment evaluates the performance of the pipeline. The SVM was trained using 10 point clouds of dataset 3. Synthetic samples were not included for training as the infected 3D model of the sunflower at the final growing stage was not available for simulation. This fact did not affect training as the number of images was enough for training the SVM without the necessity of adding synthetic samples. The training process followed the procedure described in Subsect. 5.1, including cross-validation and grid-search to improve the classification results.

The tests utilized the same 5 testing point clouds of Subsect. 5.1. The samples were divided into four categories based on the weather conditions: (i) clear morning (1 PC), (ii) clear afternoon (2 PCs), (iii) partially cloud morning (1 PC), and (iv) partially cloud evening (1 PC). The day time during the data collection makes an important difference regarding the color appearance of the vegetation. In our experiments, the samples captured during the afternoon present a high bright intensity independently of the weather condition. The testing dataset does not have a 50% class distribution of infected and healthy samples as in the inspected fields the number of infected samples was considerably low with respect to the healthy vegetation. The classification results report a good performance for the four datasets as is shown in Table 2. For each test, the misclassification is below the 11% and the sum of all reports an average classification error of 7.85%. Besides, the low FPR values indicate a low misclassification of the negative samples, which means that our classifier correctly classified the majority of the infected samples. Figure 4 shows two crop maps with detected infected samples. The red points represent the crop areas infected with the parasite.

Table 2. Classification results using the image samples captured from real crops under different lighting conditions and flying altitudes.

Dataset	FPR	Recall	F1	Precision	Accuracy
Cloud M	0.01	0.94	0.97	1.0	94.58%
Cloud A	0.01	0.90	0.95	1.0	91.25%
Clear M	0.12	0.95	0.95	0.96	93.15%
Clear A	0.04	0.82	0.88	0.96	89.24%

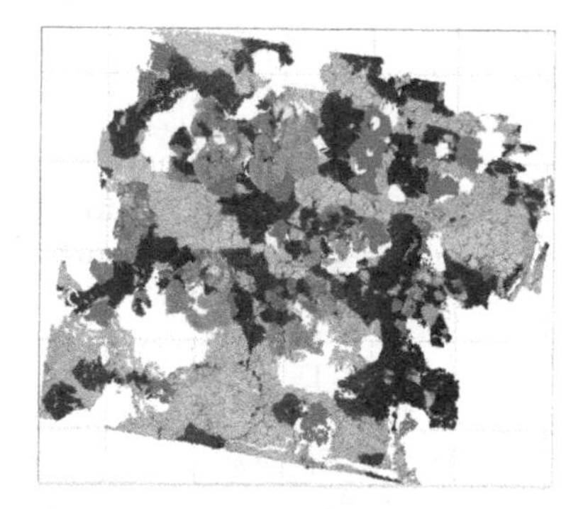

Fig. 4. Map of the terrain highlighting the location of infected crops by red points. Crop map during a morning sunny day (left), map of the crops in a sunny afternoon (right). (Color figure online)

6 Conclusions

In this work, we propose a solution to detect the downy mildew parasite on sunflower fields by using images of the crops and machine learning. The solution includes a classification pipeline to detect infected crops from the point cloud of the terrain and a simulation tool to generate synthetic samples to extend the training dataset. The classification results show the potential for deploying the pipeline in agricultural production. An average classification error below 8% on the whole pipeline shows the effectiveness of the pipeline for finding samples with a high probability of being infected by the downy mildew. The very low rate of negative misclassification guarantees the detection of the majority of the infected samples. Further, the classification results confirm the advantages of using the simulator environment in the agriculture context. Enlarging the training dataset using synthetic images improves the classification in case of lack of training samples. Our future work aims at extending the classification methodology by using other ML architectures [16] that may improve the pipeline's performance.

References

1. Nie, S., Wang, C., Dong, P., Xi, X.: Estimating leaf area index of maize using airborne full-waveform lidar data. Remote Sens. Lett. **7**(2), 111–120 (2016)
2. Candiago, S., Remondino, F., De Giglio, M., Dubbini, M., Gattelli, M.: Evaluating multispectral images and vegetation indices for precision farming applications from UAV images. Remote Sens. **7**(4), 4026–4047 (2015)
3. Sa, I., et al.: weedNet: dense semantic weed classification using multispectral images and MAV for smart farming. IEEE Robot. Autom. Lett. **3**(1), 588–595 (2017)
4. Bendig, J., Bolten, A., Bennertz, S., Broscheit, J., Eichfuss, S., Bareth, G.: Estimating biomass of barley using crop surface models (CSMs) derived from UAV-based RGB imaging. Remote Sens. **6**(11), 10395–10412 (2014)
5. Li, W., Niu, Z., Chen, H., Li, D.: Characterizing canopy structural complexity for the estimation of maize LAI based on ALS data and UAV stereo images. Int. J. Remote Sens. **38**(8–10), 2106–2116 (2017)
6. Kussul, N., Lavreniuk, M., Skakun, S., Shelestov, A.: Deep learning classification of land cover and crop types using remote sensing data. IEEE Geosci. Remote Sens. Lett. **14**(5), 778–782 (2017)
7. Potena, C., Nardi, D., Pretto, A.: Fast and accurate crop and weed identification with summarized train sets for precision agriculture. In: Chen, W., Hosoda, K., Menegatti, E., Shimizu, M., Wang, H. (eds.) IAS 2016. AISC, vol. 531, pp. 105–121. Springer, Cham (2017). https://doi.org/10.1007/978-3-319-48036-7_9
8. Griffel, L.M., Delparte, D., Edwards, J.: Using support vector machines classification to differentiate spectral signatures of potato plants infected with potato virus Y. Comput. Electron. Agric. **153**, 318–324 (2018)
9. Shastry, K.A., Sanjay, H.A., Deexith, G.: Quadratic-radial-basis-function-kernel for classifying multi-class agricultural datasets with continuous attributes. Appl. Soft Comput. **58**, 65–74 (2017)
10. Dosovitskiy, A., Ros, G., Codevilla, F., Lopez, A., Koltun, V.: CARLA: an open urban driving simulator. In: Proceedings of the 1st Annual Conference on Robot Learning, pp. 1–16 (2017)

11. Shah, S., Dey, D., Lovett, C., Kapoor, A.: AirSim: high-fidelity visual and physical simulation for autonomous vehicles. In: Hutter, M., Siegwart, R. (eds.) Field and Service Robotics. SPAR, vol. 5, pp. 621–635. Springer, Cham (2018). https://doi.org/10.1007/978-3-319-67361-5_40
12. Gascuel, Q., Martinez, Y., Boniface, M.-C., Vear, F., Pichon, M., Godiard, L.: The sunflower downy mildew pathogen Plasmopara halstedii. Mol. Plant Pathol. **16**(2), 109–122 (2015)
13. Di Cicco, M., Potena, C., Grisetti, G., Pretto, A.: Automatic model based dataset generation for fast and accurate crop and weeds detection. In: 2017 IEEE/RSJ International Conference on Intelligent Robots and Systems (IROS), pp. 5188–5195. IEEE (2017)
14. Epic Games: Epic Games Unreal 4 (2018). https://www.unrealengine.com/
15. Agisoft: Agisoft photoscan (2016). https://www.unrealengine.com/
16. Qi, C.R., Su, H., Mo, K., Guibas, L.J.: Pointnet: deep learning on point sets for 3D classification and segmentation. In: Proceedings of the IEEE Conference on Computer Vision and Pattern Recognition, pp. 652–660 (2017)

Multipath Routing of Mixed-Critical Traffic in Time Sensitive Networks

Ayman A. Atallah(✉), Ghaith Bany Hamad, and Otmane Ait Mohamed

Department of Electrical and Computer Engineering,
Concordia University, Montreal, Canada
{a_atal,g_banyha,ait}@encs.concordia.ca

Abstract. Distributed embedded systems for safety-critical applications demand reliable communication with real-time characteristics. Switched Ethernet-based network is a practical and scalable solution that allows high-reliability level through path redundancy. The Time-Sensitive Networking (TSN) standard is being developed to support real-time communication which supports a deterministic and low latency communication for safety-critical control applications, namely, Time-Triggered (TT) traffic class. In addition, a bounded-latency traffic class, namely, Audio Video Bridging (AVB) class is introduced. In this paper, we propose a multipath routing technique which tackles both TT and AVB traffic simultaneously. The proposed approach investigates satisfying path redundancy requirements for each message while the imposed interference from TT traffic on AVB traffic is minimized. The considered routing problem is formalized as an Integer Linear Programming (ILP) optimization problem. The Worst Case end-to-end Delay (WCD) is the optimization objective. 50 test cases of various network size and number of messages are solved to evaluate the performance, i.e., interference reduction, and the scalability of the proposed technique. Results demonstrate WCD reduction up to 90% comparing to the typical routing approach that determines TT and AVB routing in separate steps.

Keywords: Time-sensitive network · Routing optimization · Multipath routing · Integer linear programming · Mixed-critical traffic

1 Introduction

The future distributed real-time systems require high bandwidth and cover broad areas. The switched network architecture is a scalable solution for future real-time communication networks. Nonetheless, achieving a deterministic or very low-latency delivery by switched networks is hard due to several inherent limitations. One of these limitations is the non-deterministic queuing delay over the network bridges. Recently, the IEEE Time-Sensitive Networking (TSN) task group has introduced a new standard, namely the TSN standard in order to realize switched networks with real-time capabilities. The TSN network adapts the Ethernet technology since it offers large bandwidth and the low cost. TSN based

F. Wotawa et al. (Eds.): IEA/AIE 2019, LNAI 11606, pp. 504–515, 2019.
https://doi.org/10.1007/978-3-030-22999-3_44

networks support two classes of time-critical traffic, namely, Time-Triggered (TT) and Audion Video Bridging (AVB) classes. The TT messages have deterministic delivery with very low latency, (e.g., <100 μs). Whereas, AVB messages have a guaranteed bounded low Worst-Case Delay (WCD) delivery, (e.g., <2 ms). The messages of non-critical applications are classified as Best Effort (BE) traffic which has the lowest priority and does not guarantee timing requirements.

Each output port in each bridge is equipped with Time-Aware Shaper (TAS) which controls the traffic flow according to the predefined schedule. The TAS ensures that TT messages are transmitted through the network according to a static schedule that guarantees exclusive access to the transmission queue at particular time slots. TT frames have a smooth motion over dedicated links until reaching their destinations. On the other hand, the TAS blocks the non-scheduled traffic (e.g., AVB and BE) from reaching the output ports during the reserved time slots. Such blocking imposes an additional delay, namely, schedule interference on the AVB traffic. Typical routing techniques for TSN networks assume that TT and AVB traffic is routed in separate steps [1–4]. In particular, routing of AVB traffic is optimized to minimize the schedule interference after deciding the TT routing. Such an optimization approach is limited to the fixed TT routing which may concentrate the TT traffic in particular links which hike the maximum WCD of AVB traffic in the network. Therefore, in this paper, we are interested in the problem of routing both TT and AVB traffic simultaneously which allows further interference optimization.

In safety-critical industrial application, redundant communication is required to ensure the high-reliability [5]. Most of the literature that addresses the routing problem considers non-redundant routing, i.e., the fault-tolerance requirements for the critical traffic are not considered [2,4,6]. Existing techniques assume that the network is physically replicated for fault-tolerance. On the other hand, we are interested in the multipath routing to meet the required Redundancy Level (RL) which offers higher efficiency in terms of power, cost, and weight. In particular, multiple replicas of the message are transmitted through disjoint paths, i.e., if one replica is corrupted, delayed or dropped, the message is still received on time.

Contribution: The contribution of this paper can be summarized as the following:

- Multipath routing is proposed to meet the spatial redundancy constraints for the communication of safety-critical applications.
- Joint AVB and TT traffic routing is proposed to improve the design space exploration and allows further optimized solution.
- A new ILP formulation for the routing of mixed-criticality traffic in TSN networks is proposed. The proposed formulation handles various RL constraints for different messages while optimizing the interference imposed on AVB traffic.

The remainder of this paper is structured as follows. The related works are presented in Sect. 2. The definition of the system under consideration is presented in Sect. 3. The proposed ILP formulation is explained in Sect. 4. The experimental results are presented in Sect. 5. Finally, Sect. 6 concludes the work.

2 Related Work

The communication reliability of TSN networks has been a subject of several studies in the recent literature. Authors in [7] provide an excellent overview of the different fault-resilience concepts for IEEE 802.1 TSN networks. In [8], a mathematical analysis of the gain in the transmission reliability due to temporal redundancy is introduced. In [3], a temporal redundancy-based approach to developing a reliability-aware routing algorithm for the TSN network is proposed. This approach utilizes an ILP-based formulation to determine the path and temporal redundancy level for each message such that the required Mean-Time-to-Detect-Error (MTTDE) is satisfied. Despite the resilience efficiency of the temporal redundancy against transient transmission error, it is incapable of tolerating permanent link failure scenario since all replicas are sent through the same route. On the other hand, the spatial redundancy for fault-tolerance is introduced in [9,10]. In [2], an interference-aware heuristic routing algorithm for AVB traffic is presented. In this work, a Greedy Randomized Adaptive Search Procedure (GRASP)-based algorithm to minimize the worst-case delay for the AVB traffic in the presence of TT traffic. In [1], a schedulability-aware routing for TT traffic in TSN network to generate routes that increase the chance of finding a possible schedule is introduced. In this work, the routing problem of multi-hop TT network is formulated as a set of ILP constraints that takes into account an additional parameter, namely, maximum scheduled traffic load. This parameter reduces the chance of getting a conflict between TT messages. Furthermore, several works have addressed routing and scheduling synthesis. In [4,11] ILP formulations to solve the routing and scheduling problems of TT traffic, which offer higher optimization capabilities, are introduced. However, these techniques suffer from scalability limitation, i.e., it is applicable for small size problems. On the other hand, our approach tackles the routing problem to attain higher scalability. To the best of our knowledge, there is currently no work on the multipath routing of TT and AVB traffic.

3 System Model

In this paper, we consider an Ethernet-based multi-hop switched architecture compliant with IEEE TSN standard. An example of a network topology is shown in Fig. 1. This network is composed of 5 bridges (B_1–B_5). Each bridge is connected to several Electronic Control Units (ECUs). A set of full-duplex physical links connects all of these components. We assume one queue is reserved for TT streams while the remaining queues stores non-scheduled traffic. Figure 2 provides a simplified view of the egress port defined in IEEE 802.1Qbv [12]. The mixed-critical traffic is isolated in separate gated buffers controlled by a Gate Control List (GCL) which is configured off-line. We assume that both TT and

AVB messages are transmitted periodically every specific time interval called hyper-period.

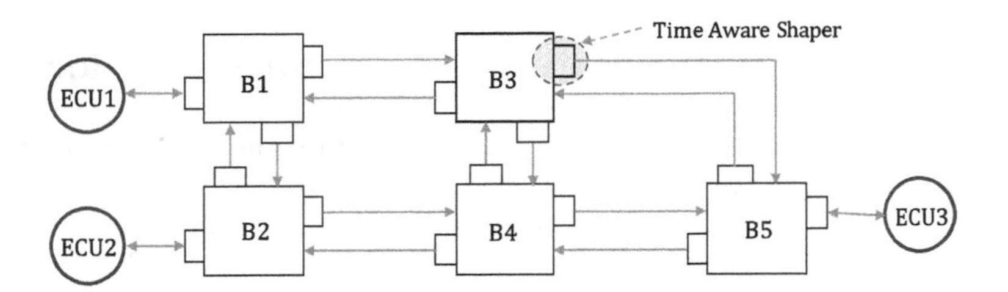

Fig. 1. Example topology of five-bridge TSN network shows the location of time aware shapers in front of each directional link in the network.

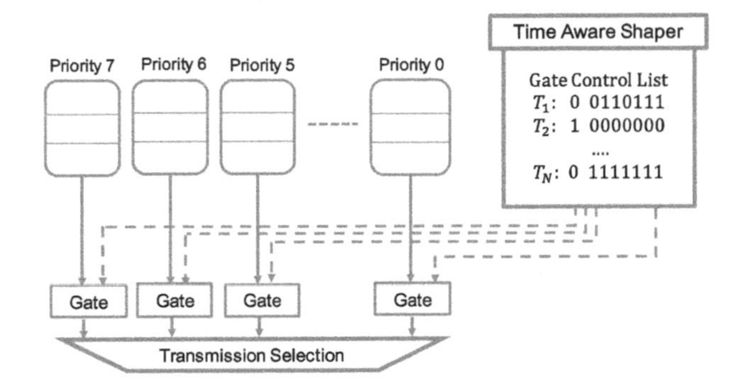

Fig. 2. Simplified view of TSN egress port composes of the TAS including the GCL, 8 egress queues in which the highest priority queue (priority 7) is assigned for scheduled traffic the remaining 7 queues are assigned for non-scheduled traffic, and a priority-based transmission selection.

We consider a set of ECUs $\mathcal{E}$ that exchange a set of TT messages $\mathcal{M}$ through a set of bridges $\mathcal{B}$. The messages in $\mathcal{M}$ should be sent through the network according to an overlap-free schedule such that the output ports of the bridges along the message's path are dedicated for that message before it arrives. The set of ECUs $\mathcal{E}$ as well as the set of TT messages $\mathcal{M}$ are predefined as inputs to the problem. Each message, $m \in \mathcal{M}$, is defined by the following tuple $\langle m.src, m.dest, m.size, m.period, m.rl \rangle$ where $m.src$ and $m.dest$ denote the message source and destination, respectively. $m.size$ and $m.period$ denote the frame size and the period between frames, respectively. Finally, $m.rl$ denotes the RL of the message which specifies the required number of disjoint paths through which the message should be routed. We assume that the design engineers specify RL of the messages according to the criticality of the application that needs these messages. Multiple replicas of each message, m^i where $i \in m.rl$, are transmitted through a specific set of disjoint paths, $r_m^i \in R_m$. The path r_m^i is denoted by an ordered sequence of connected vertices starting from $m.src$ and ending at $m.dest$.

4 Proposed ILP-Based Multipath Routing Technique

This section presents the proposed ILP formulation for the proposed multipath routing of critical traffic in TSN networks. In the following, the formulation for the redundancy and capacity constraints are introduced. Then we present the objective function which minimizes the interference imposed on AVB messages. The ILP problem is defined by the following constants:

1. M_{TT}: The set of TT messages.
2. M_{AVB}: The set of AVB messages.
3. $M = M_{TT} \cup M_{AVB}$: The set of time-sensitive messages to be handle.
4. $m \in M$: Index of messages.
5. O_m: Number of redundant paths that are required for message m.
6. P_m: Period of message m.
7. N_m: The set of possible paths for message m.
8. $n \in N_m$: Index of a possible path for message m.
9. $L_{(m,n)}$: The set of links that compose path n for message m.
10. $Q_{(m,n)}$: The number of links that compose path n for message m.
11. Γ_l: Available bandwidth on link l.
12. γ_m: Total bandwidth occupied by message m during one hyper-period.
13. $T_{(m,l)}$: Transmission time for message m on link l.

The ILP uses the following variables:

1. $X_{(m,n)}$: Binary variable indicates whether message $m \in M$ is transmitted through path n
2. $Xc_{(m,n)}$: Auxiliary binary variable complements $X_{(m,n)}$.
3. $G_{(m,n,l)}$: Binary variable indicates whether message $m \in M$ passes through link l.
4. D: Integer variable represents an upper bound for the WCD of M_{AVB} in μ sec.

4.1 Redundancy Constraints

The following constraints (1)–(5) define the valid routing. Each message should be assigned to a particular number O_m of disjoint paths. This requirement is guaranteed by constraint (1).

$$\forall m \in M : \sum_{n \in N_m} X_{(m,n)} = O_m \tag{1}$$

This constraint is applied to each message to enforce the ILP to select a specific number of redundant paths from the set N_m which represents possible paths between the source and destination nodes of message m. The set N_m is obtained in advance using Yen's algorithm [13]. N_m can include the whole possible paths for small problems. Generating all possible paths is a viable option for small-scale networks. However, for large-scale networks topology such as `Orion` [14] that composes of 31 bridges, it is not practical to consider all possible paths. We address this limitation using a heuristic approach by limiting N_m to the K-shortest paths.

We add constraints (2), (3), and (4) to define the variable $X_{(m,n)}$ as an indication that a certain path n is selected for message m. First, constraint (2) defines when $X_{(m,n)}$ is enforced to be one.

$$\forall m \in M, \forall n \in N_m : X_{(m,n)} + \left(Q_{(m,n)} - \sum_{l \in L_{(m,n)}} G_{(m,n,l)} \right) \geq 1 \qquad (2)$$

This constraint states that when message m is assigned to all links that compose n, then the variable $X_{(m,n)}$ is enforced to be one since the terms inside the brackets are equal to zero. On the other hand, constraints (3) and (4) are introduced to specify when $X_{(m,n)}$ is enforced to be zero.

$$\forall m \in M, \forall n \in N_m : -B \cdot Xc_{(m,n)} + \left(Q_{(m,n)} - \sum_{l \in L_{(m,n)}} G_{(m,n,l)} \right) \leq 0 \qquad (3)$$

where B is a big number able to dominate other terms and deactivate the inequality when the message m is not routed through the path n. This constraint states that, $Xc_{(m,n)}$ is enforced to be one unless all links along path n are allocated for message m, which implies that $X_{(m,n)}$ is equal to one due to the constraint (4).

$$\forall m \in M, \forall n \in N_m : X_{(m,n)} + Xc_{(m,n)} = 1 \qquad (4)$$

In order to ensure that the selected paths for each message are disjoint, we introduce constraint (5).

$$\forall m \in M, \forall l \in L : \sum_{n \in N_m} G_{(m,n,l)} \leq 1 \qquad (5)$$

This constraint states that the redundant paths of message m are not allowed to use any common link.

4.2 Capacity Constraints

Typical link capacity is in range of 100 MB/s to 1 GB/s where a portion of this capacity is dedicated to the high priority traffic (TT and AVB class-A) to prevent network starvation and serve the lower priority traffic. TSN standard allows maximum utilization of 75% of link capacity for high priority traffic [15].

Constraint (6) is added for each link in the network to ensure that the total size of the high priority traffic adheres to this specification.

$$\forall l \in L: \sum_{m \in M} \sum_{n' \in N_m^l} \left(G_{(m,n',l)} \cdot \gamma_m \right) \leq \Gamma_l \tag{6}$$

where $N_m^l \subset N_m$ is the set of possible paths for message m that use link l. The summations in this constraint represent the total size of TT and AVB traffic that passes through link l.

4.3 Interference Objective

As mentioned earlier, we are interested in investigating the optimal routing for TT and AVB messages that minimizes WCD for AVB traffic. Given that variables $\mathbf{X}$ determine the routing of AVB and TT messages, and D_m refers to the maximum interference imposed on message m. Then, the objective is to minimize the maximum D_m for $m \in M_{AVB}$ as depicted in (7).

$$\{X^*\} = \arg\min_{X} \left\{ \max_{m \in M_{AVB}} D_m \right\} \tag{7}$$

In order to define an upper bound for the maximum D_m for $m \in M_{AVB}$, we introduce the following constraint:

$$\forall m \in M_{AVB}, \forall n \in N_m: \\ -B \cdot Xc_{mn} + \sum_{l \in L_{(m,n)}} \sum_{\substack{m' \in M \\ m' \neq m}} \left(G_{(m',n,l)} \cdot T_{(m,l)} \right) \leq D_m \tag{8}$$

Second term represents the total delay due to messages routed through all links along path n. Finally, minimizing the objective variable D_m generates the best routing solution.

5 Experimental Results

In this section, the results of an exhaustive performance and scalability evaluation of the proposed routing technique are presented and discussed. In particular, the runtime, as well as the WCD reduction, are reported for 50 test cases of different sizes. In addition, a full case study from the space area based on the Ethernet-based network in the Orion Crew Exploration Vehicle (CEV) [16] is analyzed. MATLAB 2017a is used to implement the proposed ILP-based algorithm. The reported results have been carried out on a workstation with an Intel Core i7 6820HQ processor running at 3.0 GHz and 16 GB RAM.

5.1 Synthetic Test Cases

The synthetic test cases comprise a different number of messages with $RL \in 1, 2, 3$. A suitable number of messages is selected for each test case with accordance to its RL value to generate 12, 24, 30, 36, or 48 streams. The streams are randomly assigned as AVB or TT streams and routed through a network topology that comprises 12 ports deployed in 5 bridges similar to the topology shown in Fig. 1. The source and destination of each message are randomly assigned from the list of ECUs. The transmission period of the messages is between 0.1 and 10 ms.

The average runtime for a different number of streams and RL is depicted in Fig. 3 for K = 5. Results show good scalability, i.e., less than 1000 sec for the most significant test case (48 streams). On the other hand, Fig. 3 show an increase of the runtime by increasing RL for the same number of streams. This observation can be explained by the fact that routing streams belonging to the same message impose additional constraints to ensure the disjointness between paths.

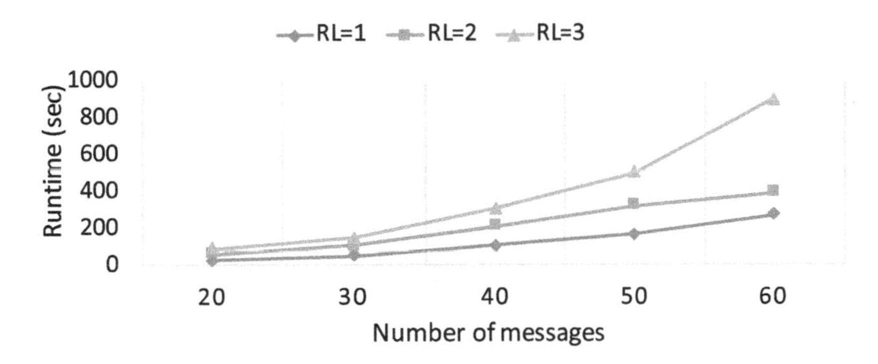

Fig. 3. Runtime of the proposed technique for different number of messages and redundancy level = [1, 2, 3].

To evaluate the performance of the proposed technique, we investigate the resultant WCD reduction comparing to the typical Shortest Path Routing (SPR) technique. Both techniques are implemented in the same framework. All test cases have been solved, and the resulting maximum WCD are shown in Fig. 4 with respect to their sizes. It can be observed from these results that the proposed technique has a significant impact, i.e., 40%, on the WCD comparing to the SPR technique. This impact increases with the number of messages. For instance, in the case of 60 messages, a reduction of 60% is achieved compared to the SPR technique.

5.2 Case Study: Orion Crew Exploration Vehicle

Orion project is intended to be the next-generation Crew Exploration Vehicle (CEV) instead of the ended Space Shuttle Program [16]. Orion has strict relia-

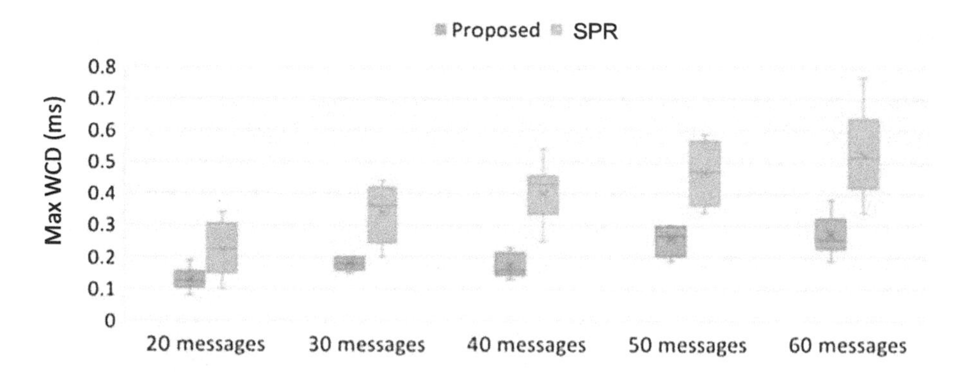

Fig. 4. Worst case delay resulting using the proposed technique and the shortest path routing for different loaded networks

bility requirements, e.g., it has to meet an overall reliability allocation of 0.9999 for up to 5000 h of continuous operation at a time under strict weight and power constraints [14]. Concerning the communication requirements, Orion adapts an Ethernet-based switched communication network [17]. In this section, multipath routing is determined for two setups of 50 and 100 messages with the network topology and the specifications adopted in [17] with RL = 2 (Fig. 5). The relative reduction in WCD of both setups using the proposed technique in comparison with the SPR technique is shown in Fig. 6.

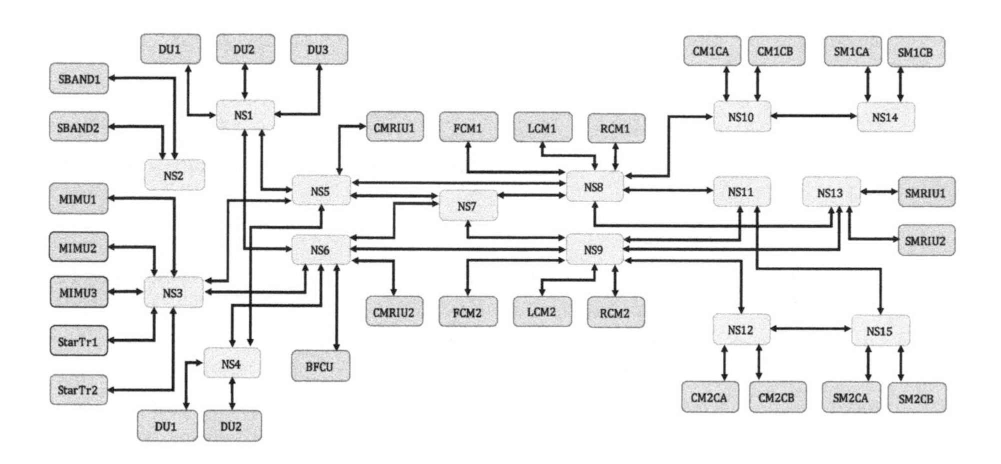

Fig. 5. Network topology in the Orion CEV.

Figure 6 shows the reduction in WCD by considering different values of the shortest path for each message. In particular, these results show that handling AVB and TT messages simultaneously by the proposed technique achieves 30% reduction in the case of 50 messages. Moreover, it shows a further reduction up to 65% for the higher utilized case of 100 messages. On the other hand, results demonstrate that the proposed technique provides further reduction of WCD by increasing the number of considered paths, e.g., by considering the five shortest paths. This observation can be explained by the fact that larger K allows better design space exploration which improves the optimization results.

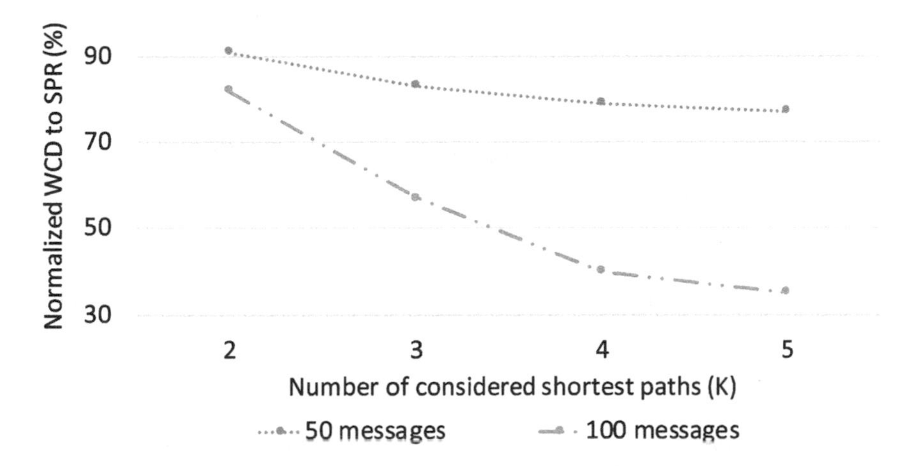

Fig. 6. The relative value of WCD divided on the reference WCD value resulting by shortest path routing approach for different number of messages and different values of K.

In order to investigate the impact of the joint routing of AVB and TT traffic, the reduction in WCD when applying the proposed technique is compared with the 2-step routing technique. In 2-steps routing, the TT traffic is routed in advance. Afterward, the routing of AVB traffic is determined such that WCD is minimized. The relative value of WCD divided on the reference WCD value resulting by 2-step routing approach for a different number of messages, and different values of K are shown in Fig. 7. The proposed technique out-performs the 2-steps routing in both 50- and 100-messages cases, and for all K values. Moreover, it is observed that the advantage and the efficiency of the proposed technique become more evident when the high loaded networks increase the impact of the proposed technique in reducing the AVB traffic delay. For instance, the proposed joint routing reduces the maximum WCD up to 42% for the case of 100 messages with $K = 5$. These results highlight the importance of considering the timing requirements of non-scheduled traffic, e.g., AVB during the routing of TT traffic to ensure optimal traffic planning.

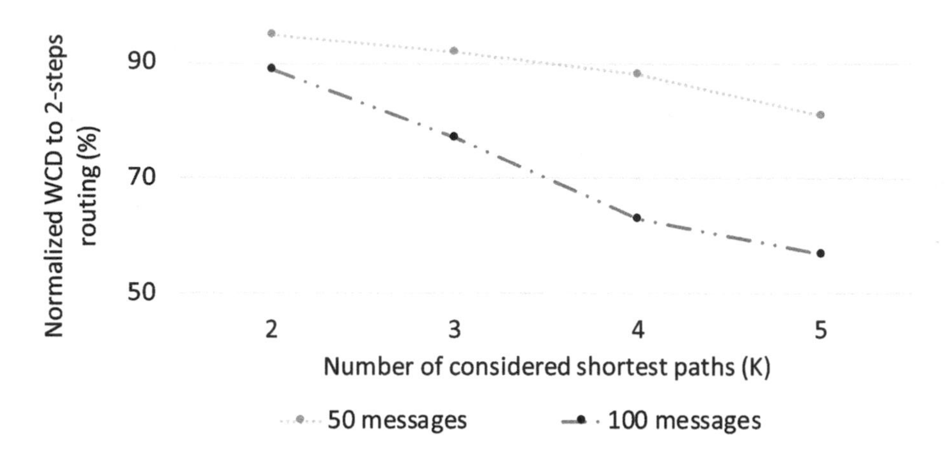

Fig. 7. The relative value of WCD divided on the reference WCD value resulting by 2-steps routing approach for different number of messages and different values of K.

6 Conclusion

In this paper, we described a multipath routing technique for mixed-criticality traffics in TSN networks. The proposed technique determines multipath routing which comprises of a set of disjoint paths (defined as RL) for each AVB or TT message. A new ILP formulation is proposed for the considered routing problem. The scalability of the ILP-based technique is improved by tightening the search space to small subsets of the shortest path for each message. Yen's algorithm is adapted to find these subsets of K shortest path. The proposed approach is evaluated using 50 synthetic test cases in addition to a realistic case study based on Orion CEV. The performance of the proposed technique is compared to the typical SPR technique. Moreover, the gain of the joint routing of AVB and TT message is evaluated compared to typical successive 2-step technique. The results demonstrate out-performance compared to the SPR and 2-step techniques with respect to the maximum WCD imposed on AVB traffic. In particular, the proposed technique reduced the maximum WCD by up to 80% and 65% compared to SPR technique for the considered synthetic test cases, and the Orion case study, respectively. The results show the impact of a larger set of shortest paths on the maximum WCD is settled at K = 5. The scalability of the proposed technique is evaluated on problems with size up to 100 messages. All problems have been solved with runtime less than 1000 s.

References

1. Laursen, S.M., Pop, P., Steiner, W.: Routing optimization of AVB streams in TSN networks. ACM SIGBED Rev. **13**, 43–48 (2016)
2. Nayak, N.G., Durr, F., Rothermel, K.: Routing Algorithms for IEEE802.1Qbv Networks (2017)

3. Atallah, A.A., Bany Hamad, G., Ait Mohamed, O.: Reliability-aware routing of AVB streams in TSN networks. In: Mouhoub, M., Sadaoui, S., Ait Mohamed, O., Ali, M. (eds.) IEA/AIE 2018. LNCS (LNAI), vol. 10868, pp. 697–708. Springer, Cham (2018). https://doi.org/10.1007/978-3-319-92058-0_67
4. Schweissguth, E., Danielis, P., Timmermann, D., Parzyjegla, H., Muhl, G.: ILP-based joint routing and scheduling for time-triggered networks. In: Proceedings of 25th International Conference on Real-Time Networks and Systems, pp. 8–17. ACM (2017)
5. Tanasa, B., Dutta Bordoloi, U., Eles, P., Peng, Z.: Reliability-aware frame packing for the static segment of FlexRay. In: Proceedings of the Ninth ACM International Conference on Embedded Software, pp. 175–184. ACM (2011)
6. Mahfuzi, R., Aminifar, A., Samii, S., Rezine, A., Eles, P., Peng, Z.: Stability-aware integrated routing and scheduling for control applications in Ethernet networks. In: Design, Automation and Test in Europe (DATE) (2018)
7. Kehrer, S., Kleineberg, O., Heffernan, D.: A comparison of fault-tolerance concepts for IEEE 802.1 time-sensitive networks (TSN). In: Emerging Technology and Factory Automation (ETFA), pp. 1–8 (2014)
8. Smirnov, F., Glaß, M., Reimann, F., Teich, J.: Formal reliability analysis of switched Ethernet automotive networks under transient transmission errors. In: Design Automation Conference (DAC), pp. 1–6. IEEE (2016)
9. Gavrilut, V., Zarrin, B., Pop, P., Samii, S.: Fault-tolerant topology and routing synthesis for IEEE time-sensitive networking. In: Proceedings of International Conference on Real-Time Networks and Systems, pp. 267–276. ACM (2017)
10. Atallah, A.A., Hamad, G.B., Ait Mohamed, O.: Fault-resilient topology planning and traffic configuration for IEEE 802.1 Qbv TSN networks. In: 24th International Symposium on On-Line Testing and Robust System Design (IOLTS), pp. 151–156. IEEE (2018)
11. Smirnov, F., Glaß, M., Reimann, F., Teich, J.: Optimizing message routing and scheduling in automotive mixed-criticality time-triggered networks. In: Design Automation Conference (DAC), pp. 1–6. IEEE (2017)
12. IEEE Standard for Local and metropolitan area networks - Bridges and Bridged Networks - Amendment 25: Enhancements for Scheduled Traffic. IEEE Std, March 2015
13. Yen, J.Y.: An algorithm for finding shortest routes from all source nodes to a given destination in general networks. Q. Appl. Math. **27**(4), 526–530 (1970)
14. Paulitsch, M., Schmidt, E., Gstottenbauer B., Scherrer, C., Kantz, H.: Time-triggered communication (industrial applications), pp. 121–152 (2011)
15. ETG.1000.1 EtherCAT Specification. Standard, EtherCAT Technology Group, March 2013
16. McCabe, M., Baggerman, C.: Avionics architecture interface considerations between constellation vehicles. In: IEEE/AIAA Digital Avionics Systems Conference. IEEE (2009)
17. Tamas-Selicean, D., Pop, P., Steiner, W.: Design optimization of TTEthernet-based distributed real-time systems. Real-Time Syst. **51**(1), 1–35 (2015)

On a Clustering-Based Approach for Traffic Sub-area Division

Jiahui Zhu[1], Xinzheng Niu[1](✉), and Chase Q. Wu[2]

[1] Department of Computer Science, University of Electronic Science and Technology of China, Chengdu, Sichuan, China
xinzhengniu@uestc.edu.cn

[2] Department of Computer Science, New Jersey Institute of Technology, Newark, NJ 07102, USA

Abstract. Traffic sub-area division is an important problem in traffic management and control. This paper proposes a clustering-based approach to this problem that takes into account both temporal and spatial information of vehicle trajectories. Considering different orders of magnitude in time and space, we employ a z-score scheme for uniformity and design an improved density peak clustering method based on a new density definition and similarity measure to extract hot regions. We design a distribution-based partitioning method that employs k-means algorithm to split hot regions into a set of traffic sub-areas. For performance evaluation, we develop a traffic sub-area division criterium based on the $S_D bw$ indicator and the classical Davies-Bouldin index in the literature. Experimental results illustrate that the proposed approach improves traffic sub-area division quality over existing methods.

Keywords: Clustering · Density · Hot region · Vehicle trajectory · Traffic sub-area

1 Introduction

As the urbanization process continues to evolve around the globe, vehicle ownership has rapidly increased and traffic congestion becomes a ubiquitous problem in almost every place with a dense population. In most transportation systems, a commonly adopted strategy is to divide a traffic network into small traffic sub-areas to facilitate traffic analysis at a finer-grained level and hence reduce the corresponding complexity. Traffic network division has been the focus of research for decades due to its practical importance and theoretical significance.

In recent years, many methods have been proposed in the literature for traffic sub-area division, including data mining-based region division [13,20,21], and road network-based region division [5,6,13,17,22]. Despite promising results as reported, these methods require to know a priori the number of traffic sub-areas or identify the geographical characteristics of the road network used to divide the traffic network. In addition, some methods are not best suited for non-spherical

F. Wotawa et al. (Eds.): IEA/AIE 2019, LNAI 11606, pp. 516–529, 2019.
https://doi.org/10.1007/978-3-030-22999-3_45

distributed data. Most importantly, one key observation we made about traffic sub-area division is that traffic congestion often occurs in hot regions reflecting the real traffic situation at different times. As such, a direct traffic sub-area division is not sufficient to alleviate traffic congestion. An ideal solution would be to divide the traffic network based on hot regions. Since each hot region has its own lifespan, the analysis of a hot region is similar to that of a transportation network within its lifespan. Based on this observation, to address the limitations of existing methods, we propose a traffic sub-area division approach based on hot regions, referred to as TSAD-HR. This approach first generates a set of hot regions based on spatiotemporal trajectories, each of which is then spilt into a set of traffic sub-areas during the lifespan of each hot region. Finally, all similar traffic sub-areas form the traffic sub-area division of the entire traffic network at different times.

The rest of this paper is organized as follows. Section 2 conducts a survey of related work. The traffic sub-area division problem is defined in Sect. 3. The design of TSAD-HR is detailed in Sect. 4. We provide experimental results in Sect. 5 and conclude our work in Sect. 6.

2 Relation Work

In this section, we first describe existing techniques for trajectory clustering, followed by a survey of related work on traffic sub-area division.

2.1 Trajectory Clustering

Clustering of mobile object trajectories in road networks has attracted a great deal of attention for various purposes such as understanding and exploring potential social connections and common interests of mobile users moving in a road network [7]. As a result, traditional partitioning methods (k-means) and density-based methods (OPTIC [1], DBSCAN [3]) have been extended to cluster large volumes of trajectory datasets generated by sensor networks. In spatial clustering methods, such as k-means and k-medoids methods, clusters are groups of elements characterized by a small distance to the cluster center. An objective function, typically the sum of the distances to a set of putative cluster centers, is optimized [4,8,10] until the best cluster center candidates are found [18]. However, these approaches are not able to detect non-spherical clusters.

Clusters with an arbitrary shape can be easily detected by approaches based on the density of data points. In density-based spatial clustering methods, clusters are regions of high density separated by regions of low density. DBSCAN is one representative density-based clustering algorithm, and OPTICS was devised to reduce the burden of determining parameter values in DBSCAN. Traclus [12], as one state-of-the-art sub-trajectory clustering technique, partitions each trajectory into sub-trajectories using the Minimum Distance Length principle and groups similar sub-trajectories together using a variant of DBSCAN. The work in [15] adapts TraClus to online trajectory clustering and the work in [11] extends

TraClus for trajectory classification. In density-based clustering, one is required to adjust the density threshold dynamically until clustering by density peaks [18] is present. However, the density threshold obtained by [18] is unfavorable when dealing with spatiotemporal trajectories.

2.2 Traffic Sub-area Division

In the literature, most of the related work is based on a traffic network, which is abstracted as a weighted graph where the nodes, links, and weights correspond to the intersections, road segments, and traffic flow parameters, respectively, in a traffic network. Among them, Ma *et al.* [17] designed a traffic sub-area division expert system based on integrated correlation index, which is combined with the correlation index of every two intersections at a normal condition and the particular correlation index at the special traffic condition or request, to trade off dynamic traffic conditions and different division demands. Li *et al.* [13], based on Back Propagation neural network, proposed a method to divide traffic control sub-area dynamically according to traffic flow, distance of intersections and cycle. However, estimating the traffic in a real traffic network itself is a challenging task. Thus, Zhou *et al.* [22] used the traffic states between two adjacent intersections to represent the correlation degree between intersections, and then designed a community detection method for traffic network partition. In addition, to achieve dynamic division of traffic sub-areas, Guo *et al.* [5] redefined the similarity between adjacent junctions according to the adjacent junction's maximum-minimum static traffic sub-area division and the short-term traffic state prediction, and then divided traffic sub-areas quickly in a dynamic manner. Shen *et al.* [6] divided the control area of urban arterial roads into traffic sub-areas on the basis of correlation degrees obtained by a fuzzy computing method applied to a hierarchical structure and optimized the split according to real-time traffic conditions.

All the aforementioned methods consider road networks for traffic sub-area division. Due to time-varying traffic flows, it is difficult to accurately analyse the impact of various factors on the division of traffic sub-areas. In this work, we divide traffic road networks by mining and analyzing data with spatial attributes.

3 Problem Formulation

We provide the following definitions to facilitate the formulation of the traffic sub-area division problem under study:

Definition 1. ***Trajectory (TR):*** *A trajectory TR_i of any object is represented by a list of spatiotemporal points sampled at equal time intervals, denoted as $TR_i = <p_{i,1}, p_{i,2} \cdots p_{i,j}, \ldots, p_{i,n}>$ and $p_{i,j} = (l_{i,j}, t_{i,j})$ denotes the location information $l_{i,j}$ (longitude and latitude) of the object at time $t_{i,j}$.*

Definition 2. ***Hot Region (HR):*** *A hot region is a set of spatiotemporal trajectories or trajectories points in a spatial area A, where the moving objects* *appear during a time interval from start time S to end time E. The trajectories or trajectory points that belong to the same hot region are close to each other according to the similarity measure.*

As illustrated in Fig. 1, x, y, and t are the longitude, latitude, and time of the 3-dimensional space, respectively, and L_1, L_2 and L_3 represent the real traffic road network layers at different times. Eight vehicle trajectories form two hot regions HR_1 and HR_2 marked in different colors. Hot region HR_1 corresponds to area A_1 in the traffic road network, which exists during time interval $[t_1, t_2]$.

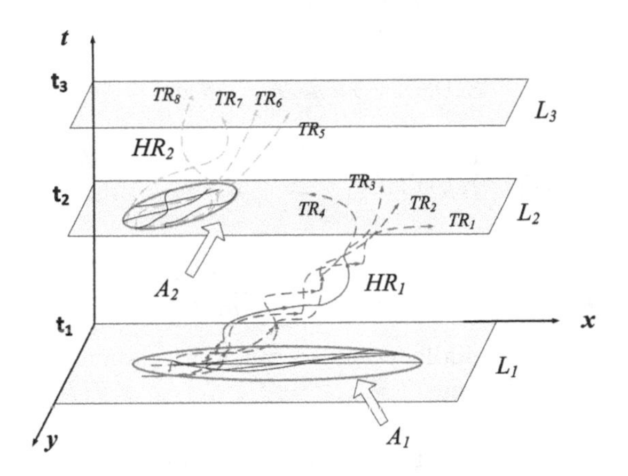

Fig. 1. Hot regions.

Definition 3. ***Traffic Sub-Area (TSA):*** *A traffic sub-area is a part of the entire transportation network and consists of a group of points extracted from different trajectories, denoted as (R_{SA}, N_{SA}), where R_{SA} is a topologically closed polygon representing the location and shape of a place and N_{SA} is the number of objects at the place.*

Definition 4. ***Hot Region Division (HRD):*** *For a given hot region HR_i, the projection of HR_i onto the two-dimensional XY-plane is a traffic network during time interval $[S_i, E_i]$ where each data point represents a location of some moving object. The transportation network formed by the projection of HR_i is split into num different traffic sub-areas $\{TSA_{i_1}, TSA_{i_2}, \ldots, TSA_{i_{num}}\}$, referred to as hot region division, denoted as HRD_i.*

According to the above definitions, we formulate the problem of traffic sub-area division as follows.

We consider a set of trajectories $T = \{TR_1, TR_2, \ldots, TR_{numtrajs}\}$ to be processed to obtain a number of hot region divisions $D = \{HRD_1, HRD_2, \ldots, HRD_{num_H}\}$. For the evaluation of division quality, we define a traffic sub-area division indicator, referred to as TI, based on the S_Dbw index, which combines both cluster compactness (in terms of intra-cluster variance) and density between clusters (in terms of inter-cluster density), as well as the classical Davies-Bouldin index [2].

The TI indicator evaluates the division of hot regions from two aspects: one is the S_Dbw indicator of each hot region projected onto the XY-plane, which takes into account density variation and average scattering for the traffic sub-areas of a hot region. The smaller the indicator S_Dbw is, the better the hot region is divided. The other is the time interval between the hot regions. The larger the time interval is, the smaller impact it has between the hot regions.

We first calculate the S_Dbw indicator S_Dbw_i [16] for each hot region HR_i in the two-dimensional XY-plane as follows:

$$S\_Dbw_i = Dens\_bw_i + Scat_i, \tag{1}$$

$$Dens\_bw_i = \frac{1}{|HRD_i| \cdot (|HRD_i| - 1)} \sum_{m=1}^{|HRD_i|} \sum_{\substack{n=1 \\ m \neq n}}^{|HRD_i|} \frac{ds_i(u_{mn})}{\max\{ds_i(v_m), ds_i(v_n)\}}, \tag{2}$$

$$Scat_i = \frac{1}{|HRD_i|} \sum_{m=1}^{|HRD_i|} \frac{\|\sigma(v_m)\|}{\|\sigma(HRD_i)\|}, \tag{3}$$

where u_{mn} denotes the middle point of the line segment defined by centres v_m and v_n of traffic sub-areas $TSA_{i,m}$ and $TSA_{i,n}$ of HRD_i, $|HRD_i|$ is the number of traffic sub-areas of HRD_i, $\sigma(v_m)$ and $\sigma(HR_i)$ are two-dimensional vectors denoting the variance of trajectory points included in $TSA_{i,m}$ and HR_i respectively, $\|\mathbf{x}\|$ is defined as $\|\mathbf{x}\| = (\mathbf{x}^{\mathrm{T}}\mathbf{x})^{\frac{1}{2}}$ and $ds(u_{mn})$ is defined as:

$$ds_i(u_{mn}) = \sum_{k=1}^{num_{mn}} f_i(x_k, u_{mn}), \tag{4}$$

$$f_i(x, u) = \begin{cases} 0, & if\ dist_S(x, u) > stdev_i, \\ 1, & otherwise, \end{cases} \tag{5}$$

$$stdev_i = \frac{1}{|HRD_i|} \sqrt{\sum_{m=1}^{|HRD_i|} \|\sigma(v_m)\|}, \tag{6}$$

where num_{mn} denotes the number of points that belong to the union of traffic sub-areas $TSA_{i,m}$ and $TSA_{i,n}$, $stdev_i$ is the average deviation of traffic sub-areas of HRD_i, and $dist_S(\cdot, \cdot)$ denotes the spatial distance in the two-dimensional XY-plane, i.e., the Euclidean distance between each other. We define TI as

$$TI = \frac{1}{num_H} \sum_{i=1}^{num_H} \max_{j=1 \sim num_H} \frac{S\_Dbw_i + S\_Dbw_j}{\frac{1}{N(HR_i) \cdot N(HR_j)} \sum_{p_m \in HR_i} \sum_{p_n \in HR_j} dist_T(p_m, p_n)}, \tag{7}$$

where num_H denotes the number of hot regions, $N(HR_i)$ and $N(HR_j)$ are the number of trajectory points included in hot regions HR_i and HR_j, respectively, and $dist_T(\cdot, \cdot)$ denotes the time distance, i.e., the absolute value of the time difference between each other. In general, a smaller TI indicates a larger time interval between hot regions or a better division of a hot region. We define the problem of traffic sub-area division as follows:

Definition 5. ***Traffic Sub-Area Division (TSAD):*** *Given a set of trajectories* $T = \{TR_1, TR_2, \ldots, TR_{numtrajs}\}$*, we wish to compute a set of hot region divisions* $\{HRD_1, HRD_2, \ldots, HRD_{numarea}\}$ *with the minimum* TI*.*

In the real world, moving objects are usually distributed in hot regions, so these areas may consist of numerous trajectories. Conventional methods consider regions where a set of trajectories appear or road networks to be divided into a set of areas. However, these methods can only find a collection of hot regions. Since a hot region often spans across space or time, traffic congestion cannot be effectively mitigated, which motivates the design of a new traffic sub-area division approach.

4 An Approach to Traffic Sub-Area Division: TSAD-HR

We propose a traffic sub-area division approach, which consists of three steps, namely, trajectory processing, hot regions generation and hot regions division. Firstly, all points of trajectories are normalized in each dimension by z-score to eliminate the errors caused by the difference in the dimensions. Then, a series of hot regions are extracted from the trajectories according to similarity measure. Finally, the distribution characteristics of each hot region are analyzed to divide the hot regions into multiple traffic sub-areas. The flowchart of TSAD-HR is shown in Fig. 2. Note that trajectory processing is relatively more straightforward. Hence, we focus our explanation on hot regions generation and hot region division.

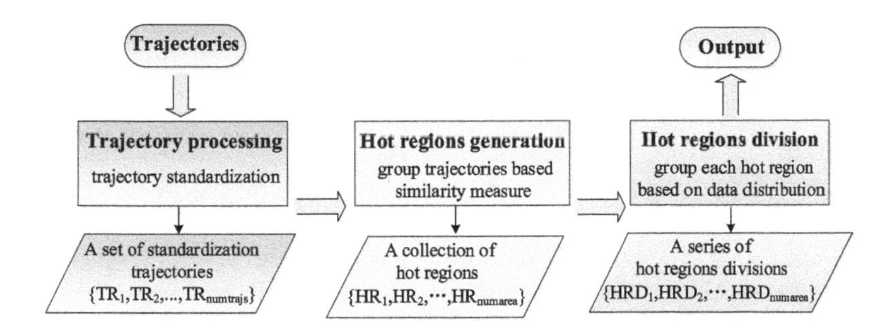

Fig. 2. The flowchart of TSAD-HR.

4.1 Hot Regions Generation

Hot regions generation is mainly based on the similarity measure, which plays a decisive role in the quality of the extracted hot regions. Conventional similarity metric considers a simple Euclidean distance between trajectory samples. However, temporal similarity and spatial similarity are two different aspects of measurement, and the Euclidean distance alone can not reflect their difference. Therefore, similarity is defined in [19] with weight Sim_w as

$$Sim_w = w \cdot Dist_T + (1 - w) \cdot Dist_S, \tag{8}$$

where $Dist_T$ and $Dist_S$ are spatial similarity and temporal similarity, respectively, and w $(0 \leq w \leq 1)$ is the weight of spatial similarity. One problem with this similarity measure is how to comprehensively consider temporal similarity and spatial similarity. It is rather difficult to decide the value of w. In this work, considering the practical traffic significance, it is supposed that temporal similarity has a higher impact on a hot region than spatial similarity. Therefore, we define temporal similarity as the main similarity measure and propose an alternative clustering method based on [4] to extract hot regions according to temporal similarity, which has potential to specify a series of hierarchical hot regions without user intervention. The hot regions identified by this method have a hierarchical structure in time, which can well reflect the traffic situation in different time periods, such as short-term (working days or holidays) traffic congestion period detection, long-term (month, season, year) congested highway identification.

In this method, we consider the sampling points of all trajectories. Given a set of trajectories $T = \{TR_1, TR_2, \ldots, TR_i, \ldots, TR_{numtrajs}\}$, we obtain the trajectory point set $S = \{p_1, p_2, \ldots, p_i, \ldots, p_{num}\}$, where p_i belongs to some trajectory of T. We compute two characteristics for each trajectory point p_i, i.e., density ρ_i and distance δ_i from points of higher density. These two characteristics are on the basis of the temporal distance (similarity) $dist_T(p_i, p_j)$ between two points, where $dist_T(p_i, p_j)$ is calculated in the same way as $dist_T(p_m, p_n)$ in Eq.(7). In [9], the density ρ_i of sample point p_i is define as

$$\rho_i = \sum_j \chi(dist_T(p_i, p_j) - d_c), \tag{9}$$

where $\chi(x) = 1$ if $x < 0$, and $\chi(x) = 0$, otherwise, and d_c is a distance threshold. However, the result is largely affected by threshold d_c, and it is difficult to provide an optimal threshold for different datasets empirically. In this work, inspired by the field in physical space [14], a concept of field is introduced to compute the density of trajectory points. For dataset S, all samples are treated as particles with an equal mass, which have their own fields. The contribution of each sample to others is related to the distance between them: a smaller distance indicates a greater contribution. According to their contributions to a given object, the field strength of each sample point is different, which uncovers different density. Finally, for each point p_i of any trajectory, the new density ξ_i is defined as Eq.(10).

$$\xi_i = \sum_{j}^{num} (1 - \frac{dist_T(p_i, p_j)}{d_{max}}), \tag{10}$$

where d_{max} represents the maximum time distance among all the points of trajectories and num denotes the number of points in S. We calculate σ_i through the minimum $dist_T(p_i, p_j)$ between point p_i and another point p_j with a higher density than p_i. If point p_i has the highest density, its σ_i is the maximum $dist_T(p_i, p_j)$. We define σ_i as follows:

$$\sigma_i = \begin{cases} \min_{j:\xi_j > \xi_i} \ dist_T(p_i, p_j), \ \exists \xi_j > \xi_i, \\ \max_{j=1,2,\ldots,num} \ dist_T(p_i, p_j), \ otherwise. \end{cases} \tag{11}$$

It is assumed in [18] that the cluster center should be surrounded by neighbor objects with lower local density and away from other objects with higher local density. According to this assumption, the points with both high ξ_i and σ_i are recognized as the cluster centers, and the remaining points are assigned to the same cluster as their nearest neighbour of higher density.

4.2 Hot Regions Division

For the collection $H = \{HR_1, HR_2, \ldots, HR_k\}$ of generated hot regions, each hot region HR_i implies traffic conditions of the entire transportation network at the corresponding lifespan, which is comprised of different traffic sub-areas. Therefore, we need to capture those areas of each hot region. In this work, we analyze the distribution characteristics of hot region in order to perform traffic sub-area division in each hot region. Each traffic sub-area can be regarded as a cluster in which the objects are close to each other. Hence, the problem of hot region division is equivalent to the clustering problem. Considering that hot regions are mainly considered in the temporal dimension, we focus on the spatial dimension during the hot region division process. Specifically, we project a hot region to the 2-dimensional XY-plane to create a projected area as the area A_1 in Fig. 1. The following analysis of distribution characteristics is based on the data contained in this projection area. Many methods (in particular, k-means and Gaussian Mixture Model) assume that a cluster adheres to a unimodal distribution. Based on this assumption, we use a statistical method to decide how to split the hot region without a priori setting the number of traffic sub-areas. Given a hot region and a confidence level α, the data from the hot region is mapped to a line between the two points farthest apart. Then, we check if all the mapped points in the line follow a Gaussian distribution at the confidence level α. If not, we split the original data into two parts and iteratively check them in the same way until all traffic sub-areas satisfy Gaussian distribution at the given confidence level.

The splitting rule is illustrated in Fig. 3. We first find two points p_i and p_j in the dataset that are the farthest from each other and take points C_1 and C_2 closest to them as the center, respectively, which can alleviate the influence of outliers on the final result to a certain extent. Next, traditional k-means method

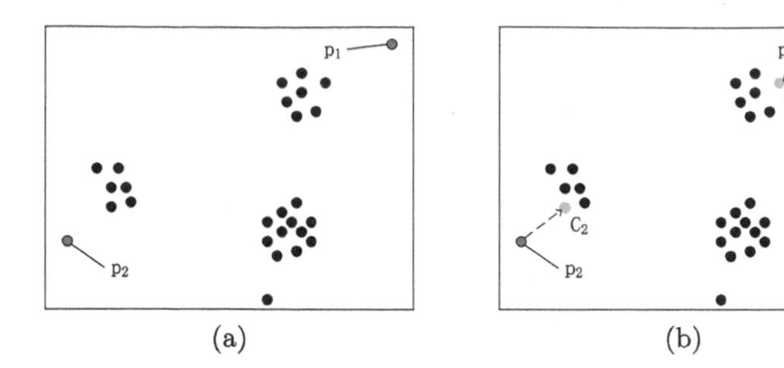

Fig. 3. The selection of initial centers.

is applied based on the chosen initial centers C_1 and C_2 to produce two new subsets.

5 Experimental Evaluation

We implement all the algorithms in Python 2.7, and conduct experiments on a macOS workstation equipped with Intel Core i5 CPU@2.3 GHz and 8 GB of memory. We demonstrate the effectiveness of the proposed algorithm using two datasets. One dataset (Taxi) is the GPS data collected by Microsoft from 10,357 taxis in Beijing. Each taxi's location is sampled every 177 s on average over a time span of a week. The other dataset (Truck) contains 1,100 trajectories from 50 different trucks delivering concrete around Athens, Greece, in which there are 94,098 position records consisting of the truck identifiers, dates and times, and geographical coordinates in GGRS87 reference system. The datasets are publicly available at http://chorochronos.datastories.org/?q=node/10.

For performance comparison, we divide our experiment into two parts. In the first part, we rebuild traffic sub-area division based on hot regions obtained through two other similarity functions: spatiotemporal Euclidean distance and Sim_w. These implementations are referred to as ED-TSAD-HR and Sim_w-TSAD-HR, respectively. We then perform two sets of experiments to analyze the effectiveness of TSAD-HR. In the second part, we extend two existing clustering approaches (TraClus and HDBSCAN) to solve the traffic sub-area division problem, and add the time factor into the parallel traffic sub-area division framework Par3PKM proposed in [21]. These three extended algorithms are named as $TraClus$-TSAD, $HDBSCAN$-TSAD and $Par3PKM$-TSAD, respectively. We evaluate our proposed TSAD-HR in terms of the effectiveness of traffic sub-area division in comparison with $TraClus$-TSAD, $HDBSCAN$-TSAD and $Par3PKM$-TSAD.

Using traffic sub-area division indicator (TI), we first conduct an evaluation of ED-TSAD-HR, Sim_w-TSAD-HR and TSAD-HR. We run these three algorithms on different numbers of trajectories in the Truck dataset ranging from 20 to 70 at an interval of 10. For each size, we execute ten times and then calculate the corresponding TI measurements. Figure 4a plots the mean and standard

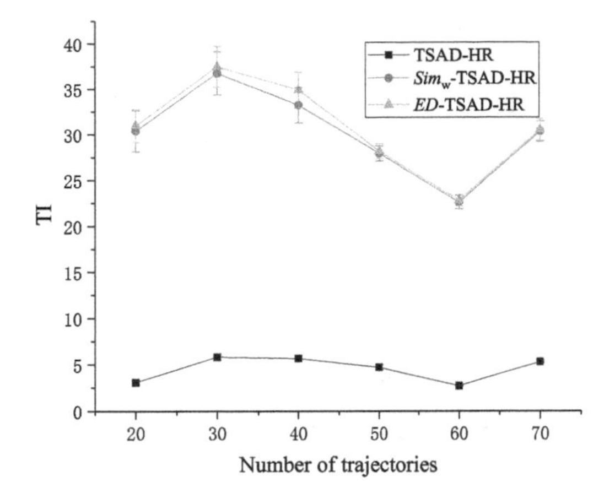

(a) Truck

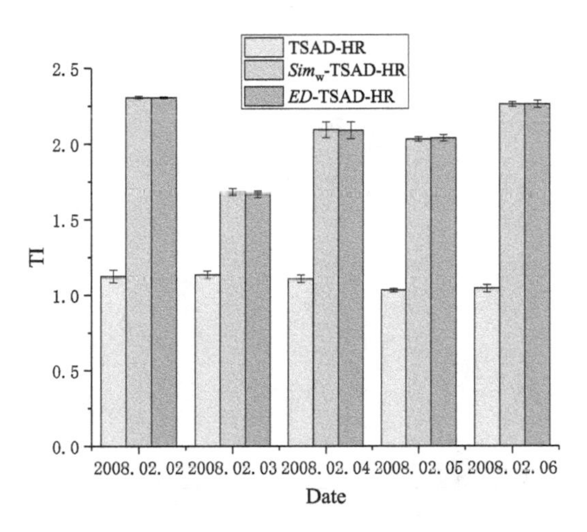

(b) Taxi

Fig. 4. Comparison of traffic sub-area division based on hot regions obtained by different similarity metrics.

deviation of the TI indices across 10 times for each number of trajectories. As illustrated in Fig. (4a), The TI values of ED-TSAD-HR and Sim_w-TSAD-HR are not significantly different, but are much higher than TSAD-HR. It means that it is more beneficial to split the traffic sub-area based on the temporal similarity than considering the spatial similarity and temporal similarity together. To further validate this finding, we conduct experiments on different dates of

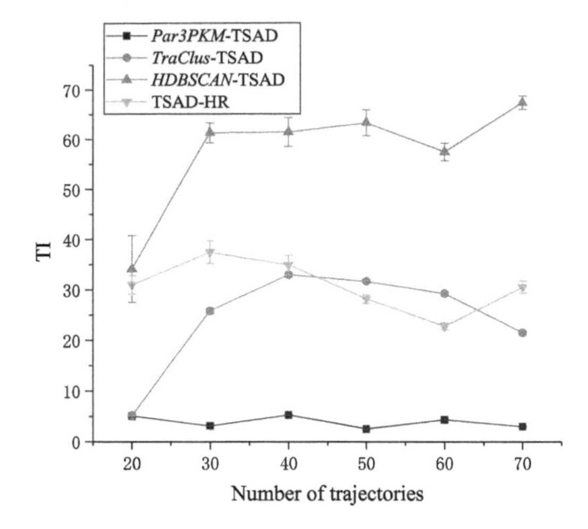

(a) Truck

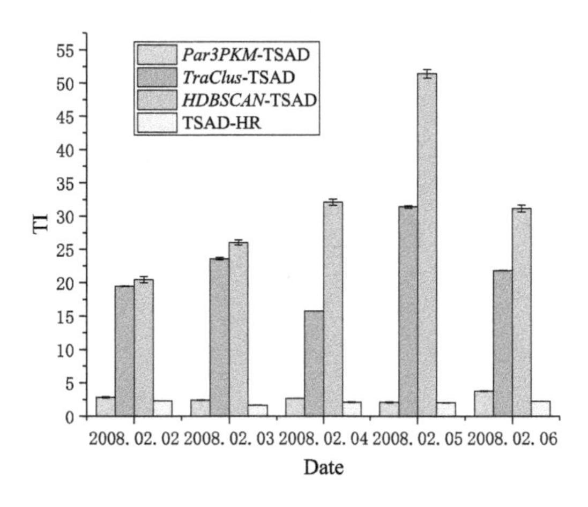

(b) Taxi

Fig. 5. Comparison of traffic sub-area division using TI index.

the Taxi dataset. Similarly, for daily data, we run these algorithms 10 times and calculate the corresponding TI values. Figure (4b) plots the mean and standard deviation of the TI indices across 10 times for each day. We observe that TSAD-HR produces a better traffic division than *ED*-TSAD-HR and Sim_w-TSAD-HR on the daily traffic data. Together with Fig. (4a), we illustrate the efficacy of our proposed method.

We further conduct an evaluation of our proposed method in terms of traffic sub-area quality in comparison with three extended approaches, namely, $TraClus$-TSAD, $HDBSCAN$-TSAD and $Par3PKM$-TSAD. We run these four algorithms on different numbers of truck trajectories ranging from 20 to 70 at an interval of 10. For each size, we execute ten times and then calculate the corresponding TI each time. The mean and standard deviation of the TI indices across 10 times for each number of trajectories are plotted in Fig. (5a). We observe that $HDBSCAN$-TSAD has the highest TI index under all the numbers of trajectories while TSAD-HR has the minimum TI index. These results indicate that TSAD-HR is more effective than the other algorithms in traffic sub-area division. Figure (5b) shows the mean and standard deviation of the TI indices obtained by these four algorithms across 10 times for every date on the Taxi dataset. We observe that $HDBSCAN$-TSAD still has the highest TI index, while $Par3PKM$-TSAD is close to $Dist_T$-TSAD-HR on the Taxi dataset, but still has a TI index slightly higher than TSAD-HR. Therefore, the results from these two datasets show that TSAD-HR effectively splits the traffic network in different datasets with a better quality of division than the other algorithms in comparison.

6 Conclusion

We proposed a traffic sub-area division evaluation index based on the similarity between traffic sub-areas and clustering, and designed a novel traffic sub-area division approach, refer to as TSAD-HR, which takes into account both the time and space factors of trajectories. Extensive experiments demonstrated that TSAD-HR significantly improves traffic sub-area division quality over the existing methods. The proposed traffic sub-area division approach has a wide range of applications in various traffic and location service systems, including urban planning, traffic navigation, logistics and distribution, service recommendation, military scheduling and traffic control.

Acknowledgments. This research is sponsored by the Scientific Research Project of Sichuan Provincial Public Security Department under Grant No. 2015SCYYCX06, the Scientific Research Project of State Grid Sichuan Electric Power Company Information and Communication Company under Grant No. SGSCXT00XGJS1800219, the Science and Technology Planning Project of Sichuan Province under Grant No. 2017FZ0094, the Science and Technology Project of Chengdu under Grant No. 2017-RK00-00021-ZF, and the Joint Funds of the Ministry of Education of China.

References

1. Ankerst, M., Breunig, M.M., Kriegel, H.P., Sander, J.: Optics: ordering points to identify the clustering structure. In: ACM Sigmod Record, vol. 28, pp. 49–60. ACM (1999)
2. Davies, D.L., Bouldin, D.W.: A cluster separation measure. IEEE Trans. Pattern Anal. Mach. Intell. **2**, 224–227 (1979)

3. Ester, M., Kriegel, H.P., Sander, J., Xu, X., et al.: A density-based algorithm for discovering clusters in large spatial databases with noise. In: KDD, vol. 96, pp. 226–231 (1996)
4. Frey, B.J., Dueck, D.: Clustering by passing messages between data points. Science **315**(5814), 972–976 (2007)
5. Guo, H., Cheng, J., Peng, Q., Zhu, C., Mu, Y.: Dynamic division of traffic control sub-area methods based on the similarity of adjacent intersections. In: 2014 IEEE 17th International Conference on Intelligent Transportation Systems (ITSC), pp. 2208–2213. IEEE (2014)
6. Shen, G., Yang, Y.: A dynamic signal coordination control method for urban arterial roads and its application. Front. Inf. Technol. Electron. Eng. **17**(9), 907–918 (2016)
7. Han, B., Liu, L., Omiecinski, E.: Neat: road network aware trajectory clustering. In: IEEE International Conference on Distributed Computing Systems, pp. 142–151 (2012)
8. Höppner, F., Klawonn, F., Kruse, R., Runkler, T.: Fuzzy cluster analysis: methods for classication, data analysis and image recognition. J. Oper. Res. Soc. **51**(6), 769–770 (1999)
9. Hung, C.C., Peng, W.C., Lee, W.C.: Clustering and aggregating clues of trajectories for mining trajectory patterns and routes. VLDB J. Int. J. Very Large Data Bases **24**(2), 169–192 (2015)
10. Kaufman, L., Rousseeuw, P.J.: Finding groups in data: an introduction to cluster analysis. In: DBLP (1990)
11. Lee, J.G., Han, J., Li, X., Gonzalez, H.: Traclass: trajectory classification using hierarchical region-based and trajectory-based clustering. Proc. VLDB Endow. **1**(1), 1081–1094 (2008)
12. Lee, J.G., Han, J., Whang, K.Y.: Trajectory clustering: a partition-and-group framework. In: Proceedings of the 2007 ACM SIGMOD International Conference on Management of Data, pp. 593–604. ACM (2007)
13. Li, C., Xie, Y., Zhang, H., Yan, X.L.: Dynamic division about traffic control subarea based on back propagation neural network. In: 2010 2nd International Conference on Intelligent Human-Machine Systems and Cybernetics (IHMSC), vol. 2, pp. 22–25. IEEE (2010)
14. Li, D., Wang, S., Gan, W., Li, D.: Data field for hierarchical clustering. Int. J. Data Warehousing Mining **7**(4), 43–63 (2011)
15. Li, Z., Lee, J.-G., Li, X., Han, J.: Incremental clustering for trajectories. In: Kitagawa, H., Ishikawa, Y., Li, Q., Watanabe, C. (eds.) DASFAA 2010. LNCS, vol. 5982, pp. 32–46. Springer, Heidelberg (2010). https://doi.org/10.1007/978-3-642-12098-5_3
16. Liu, Y., Li, Z., Xiong, H., Gao, X., Wu, J.: Understanding of internal clustering validation measures, pp. 911–916 (2010)
17. Ma, Y.Y., Yang, X.G.: Traffic sub-area division expert system for urban traffic control. In: 2008 International Conference on Intelligent Computation Technology and Automation (ICICTA), vol. 2, pp. 589–593. IEEE (2008)
18. Rodriguez, A., Laio, A.: Clustering by fast search and find of density peaks. Science **344**(6191), 1492 (2014)
19. Shi, L., Zhang, Y., Zhang, X.: Trajectory data clustering algorithm based on spatio-temporal pattern. J. Comput. Appl. (2017)

20. Wang, B., et al.: Dividing traffic sub-areas based on a parallel K-means algorithm. In: Buchmann, R., Kifor, C.V., Yu, J. (eds.) KSEM 2014. LNCS (LNAI), vol. 8793, pp. 127–137. Springer, Cham (2014). https://doi.org/10.1007/978-3-319-12096-6_12
21. Xia, D., Wang, B., Li, Y., Rong, Z., Zhang, Z.: An efficient mapreduce-based parallel clustering algorithm for distributed traffic subarea division. Discrete Dyn. Nat. Soc. **2015**, 18 (2015)
22. Zhou, Z., Lin, S., Xi, Y.: A dynamic network partition method for heterogenous urban traffic networks. In: International IEEE Conference on Intelligent Transportation Systems, pp. 820–825 (2012)

User-Adaptive Preparation of Mathematical Puzzles Using Item Response Theory and Deep Learning

Ryota Sekiya[1(✉)], Satoshi Oyama[1,2], and Masahito Kurihara[1]

[1] Hokkaido University, Sapporo, Japan
r_sekiya@complex.ist.hokudai.ac.jp,
{oyama,kurihara}@ist.hokudai.ac.jp
[2] RIKEN AIP, Tokyo, Japan

Abstract. The growing use of computer-like tablets and PCs in educational settings is enabling more students to study online courses featuring computer-aided tests. Preparing these tests imposes a large burden on teachers who have to prepare a large number of questions because they cannot reuse the same questions many times as students can easily memorize their solutions and share them with other students, which degrades test reliability. Another burden is appropriately setting the level of question difficulty to ensure test discriminability. Using magic square puzzles as examples of mathematical questions, we developed a method for automatically preparing puzzles with appropriate levels of difficulty. We used crowdsourcing to collect answers to sample questions to evaluate their difficulty. Item response theory was used to evaluate the difficulty of the questions from crowdworkers' answers. Deep learning was then used to build a model for predicting the difficulty of new questions.

Keywords: Computer-aided test · Item response theory · Crowdsourcing · Deep learning · Magic square

1 Introduction

Recent advances in information technology have led to the introduction of computer-like tablets and PCs in educational settings, enabling the worldwide spread of educational systems that leverage the reach of the Internet and the power of computers.

BBC News reported that almost 70% of primary and secondary schools in the United Kingdom use tablet PCs and that 45% of the schools not currently using tablets have plans to introduce them in the future. Their introduction should be especially beneficial to students who have trouble studying using traditional methods [1]. Research by the Ministry of Education, Culture, Sports, Science and Technology in Japan revealed that here were about 1.5 million computers in elementary and junior high schools in Japan in 2018 and that whereas the number of students per computer was 8.4 in 2008, it was 6.4 in 2018 [2], so the introduction of computers into schools in Japan is steadily progressing.

In a computer-supported environment, students can study online courses featuring computer-aided tests(CATs) that are personalized to the student's skill level and

F. Wotawa et al. (Eds.): IEA/AIE 2019, LNAI 11606, pp. 530–537, 2019.
https://doi.org/10.1007/978-3-030-22999-3_46

learning style. Most CAT systems have an item pool for storing many questions from which questions suitable for each student are taken. Various patterns of questions with different levels of difficulty are required, especially in the mathematical area, because the skills of students tend to diverge, so the tests become meaningless if the students remember the questions. System creators must therefore prepare a large number of questions for each level of difficulty, and system administrators must frequently update the questions in the item pool so that students cannot answer the questions without thinking. Both tasks are burdensome.

Another problem with CATs is setting the level of question difficulty. The difficulty of a question is normally determined by considering the accuracy rate for the test, the number of hints or types that will be given, and experiences on this area. But these measures depend on various factors such as the skill and experience of the test creator and the overall skill level of the target group of students. Therefore, if the test creator or student group change, the scores on different tests cannot be directly compared. In such cases, a large number of students is required for each test to avoid degrading test reliability.

Using magic square puzzles as examples of mathematical questions, we developed a user-adaptive method for automatically selecting puzzles with appropriate levels of difficulty. We used crowdsourcing to collect answers for sample questions and used the answers to estimate the difficulty of the questions on the basis of item response theory. We then used machine learning to build a model for predicting the difficulty of new questions. Finally, by using these difficulties and students' skills as measured using a computer adaptive test, we developed a system for recommending questions that can improve the skill of students.

2 Related Work

Many researchers have considered automatic generation of educational questions in various subject areas. Hoshino and Nakagawa [3] focused on the English 4 choice question and used machine learning to identify places in sentences that could be blanked in order to make fill-in-the-blank questions. Hill and Simba [4] generated blank locations and distractors in multiple-choice fill-in-blank questions by calculating word co-occurrence likelihood using n-grams. Sakaguchi et al. [5] used a large number of sets of English sentences given by English learners and journals and found the error correction pairs and the misuse probabilities of words. They then used them and a support vector machine method to generate challenging distractors to confuse the student when choosing an answer. Liu et al. [6] generated questions in Chinese about given Chinese sentences by extracting and ranking five elements from the sentences: when, where, what, which, and who. Rocha and Faron-Zucker [7] and Papasalouros et al. [8] selected questions automatically from a database ora web by using strategy estimation based on domain ontology. Takano and Hashimoto [9] and Furudate et al. [10] created questions automatically by using a knowledge base. Takano and Hashimoto used a knowledge base built in advance to store specific questions while Furudate et al. built the knowledge base itself by using morphological analysis and knowledge patterns to extract previous questions.

Most of these studies used databases, knowledge bases, and ontologies created by relevant experts, so burdens are placed on the experts to build the system and on administrators to update the system. While many related studies used natural language and knowledge obtained from the Web or journals, few studies have used automatic generation of mathematical questions with a large number of various patterns.

3 Item Response Theory (IRT)

Conventional test methods measure the skill of students on the basis of the total points or standard deviation values for a test. The results of such measures are often affected by such factors as uneven test difficulties among test creators or uneven skill levels among different groups of students. This makes is hard to compare the skills of students who take different tests and to compare the difficulties of questions answered by different groups of students.

Item response theory (IRT) is a method for probabilistically estimating latent parameters like the skill levels of students and the difficulties of questions simultaneously from the discrete responses of students to questions. Using IRT makes it possible to estimate parameters universally without the effects of the above-mentioned dependences. This enables comparison of parameter values estimated from different tests and groups of students on the same scale. IRT was derived from Lord's Theory of Test Scores [11].

IRT considers the probability p that student i with skill level θ can answer question j with difficulty b. This process is assumed to follow a logistic regression model with item discrimination a and approximation constant D:

$$p_j(\theta_i) = \frac{1}{1 + \exp(-Da(\theta_i - b_j))}.$$

The larger the θ, the higher the skill level of the student; and the larger the b, the more difficult the question. The likelihood of a student correctly answering a question was calculated using function L,

$$L(u_i|\theta_i) = \prod\nolimits_{j=1}^{n} p_j(\theta_i)^{u_{ij}} (1 - p_j(\theta_i))^{1-u_{ij}}.$$

where u_{ij} represents the response of student i to question j: $u_{ij} = 0$ means that the response was incorrect, and $u_{ij} = 1$ means that it was correct. These are the only observable responses in the model; the other parameters must be estimated. There are similar IRT models, but these models contain too many parameters and are too complex for our purpose. Therefore, we decided to use the above simple model.

We cannot calculate likelihood function L because we cannot observe θ as in a normal logistic regression. To estimate the parameters, we use a two-step algorithm proposed by Mislevy [12] that marginalizes θ and uses Bayesian inference to maximize the likelihood function, which is calculated in a manner similar to that for an expectation-maximization (EM) algorithm.

4 Estimating the Difficulties of Magic Square Puzzles

We used supervised machine learning to prepare many magic square puzzles. Magic square puzzles are a kind of the mathematical puzzles that ask users to guess appropriate integers for some blank cells. In this section, we explain how we calculated the difficulties of these puzzles by using IRT and EM-type IRT models [13] to handle incomplete answer matrices, where students may not have answered all questions and questions may not have been answered by all students. To gather answer data, we used the Lancers crowdsourcing service [14], a service for obtaining input from many people through the Internet. Along with gathering the answer data, we gathered the time it took for the crowdsourcing worker to input an answer, his or her confidence that the answer was correct, and his or her subjective evaluation of the question's difficulty.

We created 330 magic square puzzles and obtained 9059 answers from 448 crowdsourcing workers. We estimated the difficulty of each puzzle by using the IRT model and EM-type IRT method. In addition to solving the puzzles, we also asked the workers to report other factors, as mentioned above: time needed to solve each puzzle, confidence of solution correctness, and subjective difficulty of each puzzle. In addition to the IRT-based difficulty, we also considered the time needed to answer a question by using a computer algorithm and the simple correct rate (rate of correct answers by workers) as measures of puzzle difficulty.

Table 1. Correlations between difficulty measures and other factors

	Time to complete cells	Confidence	Difficulty (subjective)	No. of blank cells
Time with computer algorithm	0.25	−0.24	0.20	0.22
Simple corrective rate	0.87	−0.98	0.95	0.90
Difficulty (IRT)	0.88	−0.97	0.96	0.90

Table 1 shows the correlations between the difficulty measures and the other factors. The top row shows the correlation for the time needed to solve a puzzle by computer using a backtracking algorithm. The middle row shows the correlation for the rate of correct answers by the workers. The bottom row shows the correlation for the difficulty estimated using the IRT. The correlations between the difficulty estimated using the IRT and the other factors were higher than those between the time needed by computer and those factors. This indicates that puzzle difficulty estimated by computer differs greatly from that estimated by people. The finding that the difficulty estimated using the IRT and the simple correct rate show similar correlations with other factors indicates that the differences in skill levels among the workers were relatively small.

From these results, we concluded that the difficulties of the puzzles estimated using the IRT were reliable and thus used them as the true difficulties in the supervised machine learning described in the next section.

5 Difficulty Estimation Using Machine Learning

Although the cost of crowdsourcing is relatively low, asking workers to answer questions to be used in a test is not realistic. Therefore, we used puzzles with various IRT-estimated levels of difficulty as training data and used supervised machine learning to build a model for predicting the difficulty of new puzzles. To evaluate the effectiveness of our approach, we compared the difficulties of puzzles as estimated by machine learning with those estimated by the IRT.

5.1 Model Overview

In a preliminary study, we were unable to accurately estimate the difficulties of puzzles by using a simple neural network model, so we first trained three base models and then trained a combined model using the outputs of the three models. Figure 1 shows the architecture of our combined model.

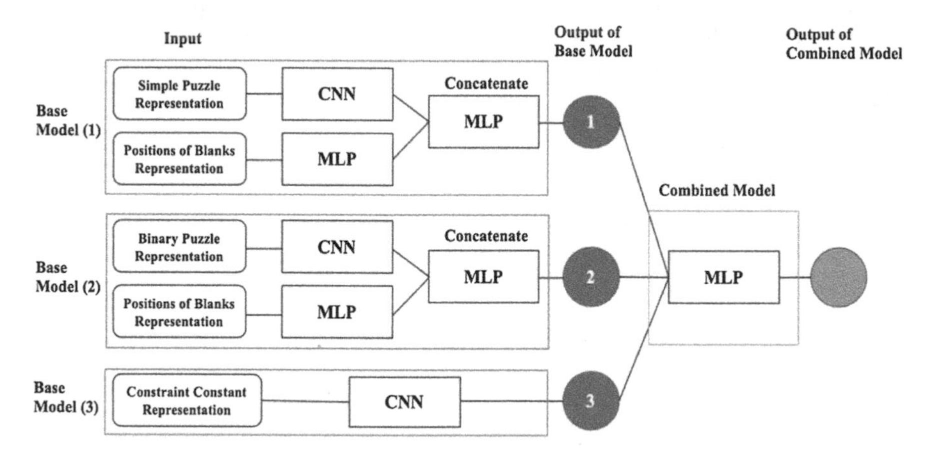

Fig. 1. Architecture of combined model

As inputs to the base models, we used different feature representations of a puzzle: simple puzzle, binary puzzle, positions of blank cells, and constraint condition. We represent the cell in $N \times N$ puzzle X at row i and column j as x_{ij} and the cell in feature matrix Y at row u and column v as y_{uv}. The simple puzzle representation represents a puzzle as an $N \times N$ two-dimensional array, with the puzzle being represented as it is except that when x_{ij} is blank, y_{uv} is filled with a zero. In binary puzzle representation, Y is an $N \times N \times N^2$ three-dimensional array. If the $x_{ij} = a$, a^{th} element of the array of the third dimension is set to one, all the other elements of that array are set to zero. If x_{ij} is blank, all the elements in the corresponding array are set to zero. For example, if $x_{ij} = 11$, the array for the element is $[0,0,0,0,0,0,0,0,0,0,1,0,0,0,0,0]$. The representation of positions of blank cells are represented using a one-dimensional array with N elements. When x_{ij} is a number, the corresponding element in the array is set to zero, and when x_{ij} is blank, the element is set to one. The constraint constant representation uses a $(2N+2) \times N$ two-dimensional array in which each element represents a row, column, diagonal constraint in the original puzzle.

We trained three base models: (1) a combination of a convolutional neural network (CNN) for which the input is the simple puzzle representation and a multi-layer perceptron (MLP) for which the input is the representation of positions of the blank cells, (2) a combination of a CNN for which the input is the binary puzzle representation and a MLP for which the input is the representation of positions of the blank cells, and (3) a single MLP for which the input is the constraint constant representation.

5.2 Model Evaluation

Figure 2 shows the relationship between the ML-based difficulty and the IRT-based difficulty. Table 2 shows correlations between the two difficulties and other factors.

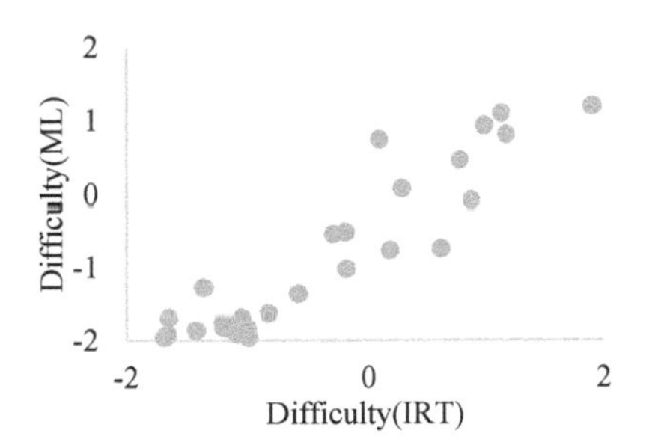

Fig. 2. ML-based difficulty against IRT-based difficulty

Table 2. Correlations between estimated difficulty and other factors

	Time for complete cells	Confidence	Difficulty (subjective)	No. of blank cells
IRT-based difficulty	0.90	−0.96	0.95	0.92
ML-based difficulty	0.88	−0.92	0.92	0.91

In Fig. 2, we can see the strong positive correlation (0.93) between the ML-based difficulty and IRT-based difficulty. In Table 2, we can see that the ML-based difficulty has as strong correlation with other factors as the IRT-based difficulty has. This result shows the learned model can accurately predict the IRT-based difficulty.

Table 3 shows the metrics for the combined model and the three base models. Max and Min are the maximum and minimum values between the true and estimated values; Max represents the positive difference, and Min represents the negative difference. RMSE is the root mean square error of each model; the smaller the value the better.

RCC is Spearman's rank correlation coefficient, which is used to evaluate the similarity of the ranking orders of two values. The larger its value, the more accurate the prediction of which of two puzzles X and $X^{'}$ is more difficult. RCC (blank cells) is the average of RCCs for each number of blank cells. The larger this value, the more accurate the model can distinguish the difficulty of puzzles with the same number of blank cells.

Table 3. Metrics for each model.

	Combined model	Base model (1)	Base model (2)	Base model (3)
Max	0.65	1.14	1.35	1.14
Min	−1.35	−1.62	−1.32	−1.79
RMSE	0.55	0.65	0.55	0.67
RCC	0.89	0.79	0.84	0.84
RCC (blank cells)	0.30	0.50	0.00	0.25

Base model (1) achieved a large RCC (blank cells), but its RCC was small and its RMSE was large. This model can distinguish the questions with the same number of blank cells, but overall accuracy is not good. Base model (2) achieved a good RMSE, but its RCC (blank cells) was small and the difference between Max and Min was very large. This model cannot distinguish the difference in difficulty among questions with the same number of blank cells, and sometime outputs very large outliers despite the overall high accuracy. Compared with the base models, the combined model inherits some strong points of the base models and can accurately rank puzzles based on their difficulties.

6 Conclusion and Future Work

We have developed a method for automatically selecting puzzles with appropriate levels of difficulty as a basis for the automatic preparation of mathematical questions, which will reduce the burden on question creators and system administrators. First, we calculated the numeric difficulties of mathematical questions by using item response theory and EM-type IRT, and then compared the values to those of other measures by using crowdsourced data. Next, we used machine learning to create a combined neural network model for estimating the difficulty of new questions. Testing showed that our combined model outperformed the three base models.

Future work includes developing a recommendation system that provides suitable questions to students on the basis of their skill levels as estimated using the IRT. It also includes building a machine learning model for predicting other values, like the time needed for a specific student to answer a specific question. Although we focused on only magic square puzzles, this method can be applied to other mathematical questions. We will thus work on expanding it for such applications and explore methods that support user-adaptive recommendation of questions.

Acknowledgements. This work was partially supported by JSPS KAKENHI Grant Numbers JP15H02782 and JP18H03337, and by the Telecommunications Advancement Foundation.

References

1. Coughlan, S.: Tablet computers in '70% of schools', BBC NEWS, 3 December 2014. https://www.bbc.com/news/education-30216408
2. Ministry of Education, Culture, Sports, Science and Technology: "The research for situation of educational informatization in school in 2017". https://www.e-stat.go.jp/stat-search/files?page=1&cycle_facet=cycle
3. Hoshino, A., Nakagawa, H.: A real-time multiple-choice question generation for language testing – a preliminary study. In: Proceedings of the 2nd Workshop on Building Educational Applications, Using NLP, pp. 17–20 (2005)
4. Hill, J., Simba, R.: Automatic generation of context-based fill-in-the-blank exercises using co-occurrence likelihoods and Google n-grams. In: Proceedings of the 11th Workshop on Innovative Use of NLP for Building Educational Applications, pp. 23–30 (2016)
5. Sakaguchi, K., Arase, Y., Komachi, M.: Discriminative approach to fill-in-the-blank quiz generation for language learners. In: Proceedings of the 51st Annual Meeting of the Association for Computational Linguistics, pp. 238–242 (2013)
6. Liu, M., Rus, V., Liu, L.: Automatic Chinese factual question generation. IEEE Trans. Learn. Technol. **10**(2), 194–204 (2017)
7. Rocha, O.R., Zucker, C.E.: Automatic generation of educational quizzes from domain ontologies. In: EDULEARN 2017-9th International Conference on Education and New Learning Technologies, pp. 4024–4030 (2017)
8. Papasalouros, A., Kanaris, K., Kotis, K.: Automatic generation of multiple choice questions from domain ontologies. In: Proceeding of the 8th International Conference on Web Intelligence, Mining and Semantics, Article No. 32 (2018)
9. Takano, A., Hashimoto, J.: Drill exercise generation based on the knowledge base. The Special Interest Group Technical Reports of Information Processing Society of Japan, NL-160, pp. 23–28 (2003)
10. Hurudate, M., Takagi, M., Takagi, T.: A proposal and evaluation on a method of automatic construction of knowledge base for automatic generation of exam questions. The Special Interest Group Technical Reports of Information Processing Society of Japan, vol. 128, no. 14 (2015)
11. Lord, F.M.: A Theory of Test Scores (Psychometric Monograph). Psychometric Society, Iowa City (1952)
12. Mislevy, R.J.: Bayes model estimation in item response theory. Psychometrika **51**(2), 177–195 (1986)
13. Sakumura, T., Tokunaga, M., Hirose, H.: Making up the complete matrix from the incomplete matrix using the EM-type IRT and its application. Inf. Process. Soc. Japan J. Trans. Math. Modeling Appl. **7**(2), 17–26 (2014)
14. Lancers Homepage. https://www.lancers.jp

Knowledge Representation and Reasoning

A Formal-Concept-Lattice Driven Approach for Skyline Refinement

Mohamed Haddache[1](✉), Allel Hadjali[2](✉), and Hamid Azzoune[3](✉)

[1] DIF-FS/UMBB, Boumerdes, Algeria
mohamed.haddache@ensma.fr
[2] LIAS/ENSMA, Poitiers, France
allel.hadjali@ensma.fr
[3] LRIA/USTHB, Algers, Algeria
azzoune@yahoo.fr

Abstract. Skyline queries constitute an appropriate tool that can help users to make intelligent decisions in the presence of multidimensional data when different, and often contradictory criteria are to be taken into account. Based on the concept of Pareto dominance, the skyline process extracts the most interesting (not dominated in sense of Pareto) objects from a set of data. However, this process often leads to a huge skyline, which is less informative for the end-users. In this paper, we propose an efficient approach to refine the skyline and reduce its size, using the principle of the formal concepts analysis. The basic idea is to build a formal concept lattice for skyline objects based on the minimal distance between each concept and the target concept. We show that the refined skyline is given by the concept that contains k objects (where k is a user-defined parameter) and has the minimal distance to the target concept. A set of experiments are conducted to demonstrate the effectiveness and efficiency of our approach.

Keywords: Skyline queries · Skyline refinement · Pareto dominance · Lattice of formal concepts

1 Introduction

The skyline queries are introduced by Borzsönyi in [4] to formulate multi-criteria searches. Recently, this concept, has gained much attention in the database community. It has been integrated in many database applications that require decision making and personalized services. Skyline process attempts to identify the most interesting (not dominated in sense of Pareto) objects from a set of data. Skyline queries are based on Pareto dominance relationship. This means that, given a set D of d-dimensional points (objects), a skyline query returns, the skyline S, set of points of D that are not dominated by any other point (object) of D. A point p dominates another point q iff p is better than or equal to q in all dimensions and strictly better than q in at least one dimension.

F. Wotawa et al. (Eds.): IEA/AIE 2019, LNAI 11606, pp. 541–554, 2019.
https://doi.org/10.1007/978-3-030-22999-3_47

A great research effort has been devoted to develop efficient algorithms to skyline computation [12,14,17,20,24,29]. The skyline computation often leads to a huge number of skyline objects which is less informative for the user and does not bring any insight to decision making. In order, to solve this problem and reduce the size of skyline, several algorithms have been developed [2,6,7,9, 13,18,21,23,26]. In this paper, we consider this problem, but with another novel vision. In particular, the idea of the solution advocated is borrowed from the formal concept analysis field. This idea consists in building a formal concept lattice for skyline objects based on the minimal distance between each concept and the target concept (i.e., the ideal object w.r.t the user query). The refined skyline S_{ref} is given by the concept that has the minimal distance to the target concept and contains k objects (k is a parameter given by the user). Starting from this idea, we develop an algorithm to compute the refined skyline, called FLCMD. In summary, our main contributions cover the following points:

- We define an efficient approach to refine the skyline S based on the minimal distance between the concepts lattice and the target concept.
- We develop and implement an algorithm to compute S_{ref} efficiently.
- We conduct a set of thorough experiments to study, analyze and compare the relevance and effectiveness of proposed approach and the naive method.

This paper is structured as follows. In the Sect. 2, we define some necessary notions about the skyline queries, fuzzy set theory, the formal concept analysis and lattice then, we report some works related to the skyline refinement and at the end of this section, we explain the naive approach. In Sect. 3, we present our approach and we give the FLCMD algorithm that compute the refined skyline S_{ref}. Section 4 is dedicated to the experimental study and Sect. 5 concludes this paper and points out some future work.

2 Background and Related Work

2.1 Skyline Queries

Skyline queries [4] represent a very popular and powerful paradigm to extract objects from a multidimensional dataset. They are based on Pareto dominance principle which can be defined as follows:

Definition 1. *Let D be a set of d-dimensional data points and u_i and u_j two points of D. u_i is said to dominate, in Pareto sense, u_j (denoted $u_i \succ u_j$) iff u_i is better than or equal to u_j in all dimensions and strictly better than u_j in at least one dimension. [25]*

Formally, we write:

$$u_i \succ u_j \Leftrightarrow (\forall k \in \{1,..,d\}, u_i[k] \leq u_j[k]) \wedge (\exists l \in \{1,..,d\}, u_i[l] < u_j[l]) \quad (1)$$

where each tuple $u_i = (u_i[1], u_i[2], \cdots, u_i[d])$ with $u_i[k]$ stands for the value of the tuple u_i for the attribute A_k.

In Eq. (1), without loss of generality, we assume that the minimal value, the better.

Definition 2. *The skyline of D, denoted by S, is the set of points which are not dominated by any other point.*

$$u \in S \Leftrightarrow \nexists u' \in D, u' \succ u \tag{2}$$

Example 1. To illustrate the concept of the Skyline, let us consider a database containing information about apartments as shown in Table 1. The list of apartments includes the following information: code apartment, area of apartment (m^2), price in (€) and distance between work and home (apartment) ($dist\_wh$ in km). Ideally, a person is looking to rent an apartment with a minimal price and having a minimal distance to his/her work (price and $dist\_wh$), ignoring the other pieces of information. Applying the traditional skyline on the apartments list of Table 1, returns the following apartments: $\{A_1, A_3, A_5, A_7\}$, see Fig. 1.

Table 1. List of apartments

Code	Area (m^2)	Price (€)	$dist\_wh$ (km)
A_1	60	525	20
A_2	40	400	120
A_3	25	360	85
A_4	30	380	100
A_5	25	340	90
A_6	60	550	95
A_7	65	540	10

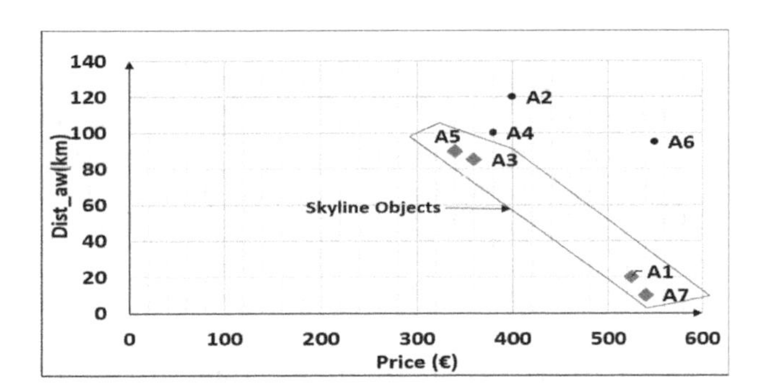

Fig. 1. Skyline of apartments

2.2 Fuzzy Set Theory

The concept of fuzzy sets has been developed by Zadeh [30] in 1965 to represent classes or sets whose limits are imprecise. They can describe gradual transitions between total belonging and rejection. Formally, a fuzzy set F on the universe X is described by a membership function $\mu_F : X \rightarrow [0,1]$, where $\mu_F(x)$ represents the degree of membership of x in F. By definition if $\mu_F(x) = 0$ then the element x does not belong to F, $\mu_F(x) = 1$ then x completely belongs to F, these elements form the core of F denoted by $Cor(F) = \{x \in F \backslash \mu_F(x) = 1\}$. When $0 < \mu_F(x) < 1$ we talk about a partial membership, these elements form the support of F denoted by $supp(F) = \{x \in F \backslash \mu_F(x) > 0\}$. Moreover, $\mu_F(x)$ is closed to 1, more x belongs to F. Let $x, y \in F$, we say that x is preferred to y iff $\mu_F(x) > \mu_F(y)$. If $\mu_F(x) = \mu_F(y)$ then x and y have the same preference. In practice, F can be represented by a trapezoid membership function $(t.m.f)$ $(\alpha, \beta, \varphi, \psi)$ where $[\beta, \varphi]$ is the core and $]\alpha, \psi[$ is its support see Fig. 2.

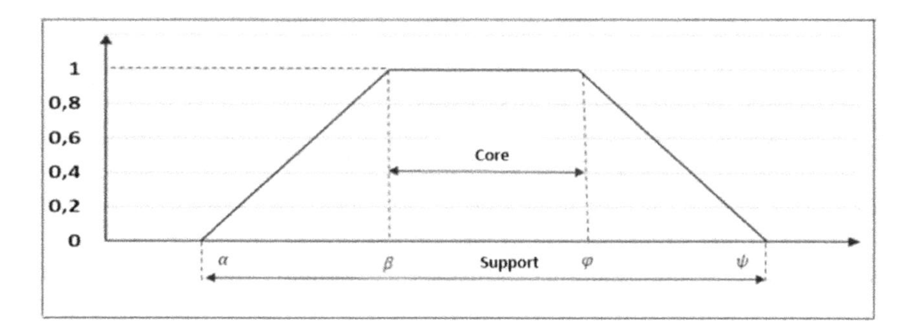

Fig. 2. Trapezoidal fuzzy set.

2.3 Formal Concept Analysis

The theory of formal concept analysis (FCA), proposed by Wille in 1982 [28]. It is based on a formal context $\mathcal{K} = (O, P, R)$, where O is a set of objects, P is a set of properties (attributes) and R a binary relation between O and P. Wille defined a correspondence between sets O and P. These correspondences are called a Galois derivation operator (or sufficiency operator) noted by Δ. Given $A \subset O$, $B \subset P$, A^Δ express all the properties satisfied by all the objects of A and dually B^Δ express the set of objects satisfying all the properties of B (see [28]). The dual pair of operators $((.)^\Delta,(.)^\Delta)$ constitutes a Galois connection which allows to introduce formal concepts. A formal concept of a formal context $\mathcal{K}$ is a pair (A, B) with $A \subset O$, $B \subset P$, $A^\Delta = B$ and $B^\Delta = A$. A and B are respectively called extent and intent of the formal concept (A, B). The set of all formal concepts is equipped with a partial order denoted $\preceq$ defined by: $(A_1, B_1) \preceq (A_2, B_2)$ iff $A_1 \subseteq A_2$ or $B_2 \subseteq B_1$. Ganter and Wille have proved in [10] that, the set of all formal concepts ordered by $\preceq$ forms a complete lattice of the formal context $\mathcal{K}$ denoted by $\mathcal{L}(\mathcal{K})$. In most applications, like in our case the attributes are defined a fuzzy way. In order to take into account relations allowing a gradual satisfaction of a property by an object, a fuzzy FCA was proposed by Burusco and Fuentes-Gonzales [5] and belohlávek et al. in [3]. In this case, the notion of satisfaction can be expressed by a degree $\in [0,1]$. A fuzzy context formal is a tuple (L, O, P, R), where a fuzzy relation $R \in L^{O \times P}$ is a function that is defined $O \times P \longrightarrow L$ which assigns to each object $o \in O$ and for each property $p \in P$ a degree $R(o,p)$ for which the object o has the property p. In general $L = [0,1]$. The generalization of the Galois derivation operator, to fuzzy settings is based on the fuzzy implication defined by belohlávek in [3]. It is defined for an subset $A \in L^O$ (and similarly defined for an subset $B \in L^P$) as follows:

$$A^{\Delta}(p) = \bigwedge_{o \in O} (A(o) \rightarrow R(o,p)) \quad (3)$$

$$B^{\Delta}(o) = \bigwedge_{p \in P} (B(p) \rightarrow R(o,p)) \quad (4)$$

$\rightarrow$: is a fuzzy implication that verify ($0 \rightarrow 0 = 0 \rightarrow 1 = 1 \rightarrow 1 = 1$ and $1 \rightarrow 0 = 0$). We distinguish three type of fuzzy formal concepts. Concept with crisp extent and fuzzy intent, crisp extent and fuzzy intent the third type fuzzy extent and fuzzy intent.

In this paper, we use concept with crisp extent and fuzzy intent, i.e., the set of objects is crisp and the set of properties is fuzzy.

Example 2. To illustrate the computation of formal concepts in our case, let us consider a database containing information about hotels as shown in Table 2. The set of objects O is composed by different hotels $\{h_1, h_2, h_3\}$, the set of properties P contains the properties *cheap* (denoted Ch) and *Near the beach* (denoted Nb), i.e., $P = \{Ch, Nb\}$. $R(o_i, p_j)$ represents the degree for witch the object o_i satisfies the property p_j, for example $R(h_2, ch) = 0.5$ means that the hotel h_2 satisfies the property *cheap* with degree 0.5. Let us consider the sets of objects $A_1 = \{h_2, h_3\}$, $A_2 = \{h_2\}$ and the set of properties $B_1 = \{Ch^{0.5}, Nb^{0.5}\}$. Now, let us describe how to compute $(A_1)^{\Delta}$, $(A_2)^{\Delta}$ and $(B_1)^{\Delta}$. For $(A_1)^{\Delta}$ and $(A_2)^{\Delta}$, we use Eq. (3) and the implication of Gödel defined by

$$p \longrightarrow q = \begin{cases} 1 & \text{if } p \leq q \\ q & \text{else} \end{cases} \quad (5)$$

Table 2. List of hotels

Hotel	Cheap (Ch)	Near the beach (Nb)
h_1	0.0	0.8
h_2	0.5	0.5
h_3	0.5	0.6

$A_1 = \{h_2, h_3\} = \{h_1^0, h_2^1, h_3^1\}$
$(A_1)^\Delta(Ch) = \wedge(0 \to 0,\ 1 \to 0.5,\ 1 \to 0.5) = \wedge(1,\ 0.5,\ 0.5) = 0.5$
$(A_1)^\Delta(Nb) = \wedge(0 \to 0.8,\ 1 \to 0.5,\ 1 \to 0.6) = \wedge(1,\ 0.5,\ 0.6) = 0.5$
$(A_1)^\Delta = \{Ch^{0.5}, Nb^{0.5}\} = B_1$
Similarly, we obtain $(A_2)^\Delta = \{Ch^{0.5}, Nb^{0.5}\} = B_1$

To compute $(B_1)^\Delta$, we use Eq. (4) and the implication of Rescher Gaines defined by

$$p \longrightarrow q = \begin{cases} 1 & \text{if} p \leq q \\ 0 & \text{else} \end{cases} \tag{6}$$

$(B_1)^\Delta(h1) = \wedge(0.5 \to 0,\ 0.5 \to 0.8) = \wedge(0,\ 0.5) = 0$
$(B_1)^\Delta(h2) = \wedge(0.5 \to 0.5,\ 0.5 \to 0.5) = \wedge(1,\ 1) = 1$
$(B_1)^\Delta(h3) = \wedge(0.5 \to 0.5,\ 0.5 \to 0.6) = \wedge(1,\ 1) = 1$
$(B_1)^\Delta = \{h_1^0, h_2^1, h_3^1\} = \{h_2, h_3\} = A_1$.

- $(A_1)^\Delta = B_1$ and $(B_1)^\Delta = A_1$, this means that (A_1, B_1) forms a fuzzy formal concept, A_1 is its extent and B_1 its intent.
- $(A_2)^\Delta = B_1$ but $(B_1)^\Delta = \{h_2, h_3\} \neq A_2$ then, (A_2, B_1) is not a fuzzy formal concept.

2.4 Related Work

The work proposed by Börzsönyi and al. in [4] is the first work that addresses the issue of skyline queries in the database field. They have proposed two different algorithms to process skyline queries in complete database, namely, Block Nested Loop (BNL) and Divide and Conquer (D& C). Later, many algorithms have been developed which are inspired from BNL and D&C [4,8,19,23,27,27]. Several authors have been interested in the problem of huge skyline and have proposed additional mechanisms to refine the skyline and reduce its size.

In [2,7,13,18,21,23,26] ranking functions are used to refine the skyline. The idea of these approaches is to combine the skyline operator with the top-K operator. For each tuple in the skyline, one joins a related score, which is computed by the means of ranking function F. We note that F must be monotonic on all its arguments. Skyline tuples are ordered according to their scores, and the top-K tuples will be returned.

In [11] authors, propose the notion of fuzzy skyline queries, which replaces the standard comparison operators $(=, <, >, \leq, \geq)$ with fuzzy comparison operators defined by user. While in [15], Hadjali and al. have proposed some ideas to introduce an order between the skyline points in order to single out the most interesting ones. In [1], a new definition of dominance relationship based on the fuzzy quantifier "almost all" is introduced to refine the skyline, while in [16] authors, introduce a strong dominance relationship that relies on the relation called "much preferred". This leads to a new extension of skyline, called MPS (Must Preferred Skyline), to find the most interesting skyline tuples. In [22] authors, propose a flexible approach called "$\theta - skyline$" to categorize and refine the skyline set by applying successive relaxations of the dominance conditions with respect to the user's preferences. This approach is based on the ranking method which deals with decision-making in the presence of conflicting choices. Furthermore, they define a global ranking method over the skyline set. In [13], Haddache et al. have proposed an approach based on ELECTRE method borrowed from the outranking domain to refine the skyline.

Furthermore, several researchers have worked on skyline's refinement for the evidential data. In [9] authors, have developed efficient algorithms to retrieve the best evidential skyline objects over uncertain data.

2.5 Naive Method

This approach [6] is based on two steps: (*i*) first compute for each skyline point p, the number of points dominated by p denoted by $num(p)$. (*ii*) The skyline points are sorted according to $num(p)$ in order to choose the $Top-k$.

3 Our Approach

In this section, we will present the main steps of our approach. First, we assume that, we have

- A database formed by a set of m objects (tuples), $O = \{o_1, o_2, \cdots, o_m\}$.
- A set P of n properties (or dimensions or attributes), $P = \{p_1, p_2, \cdots, p_n\}$.
- Each object o_j from the set O is evaluated for every property p_i.
- S the skyline of O, $S = \{o_1, o_2, \cdots, o_t\}$, $t <= m$, t is the size of skyline.
- S_{ref} the refined skyline returned by our approach.
- In our approach we use the implication($\longrightarrow$) of Rescher Gaines defined by Eq. (6).

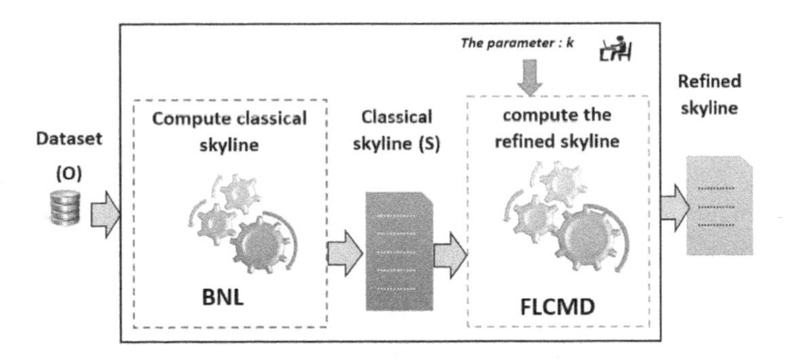

Fig. 3. Steps of our approach

The principle of our approach is to build the fuzzy concept lattice of the skyline points based on the minimal distance between each new concept and the target concept. In summary, our approach is based on the following steps (see Fig. 3).

Algorithm 1. FLCMD

```
   Input: A Skyline S, K: the number of objects chosen by the user
   Output: A refined skyline S_ref
 1 S_ref ← ∅; dist ← 100; stop ← false;
 2 Compute_degree(S); /*compute the degrees of objects*/;
 3 Intent_target ← Compute_target_intent();
 4 for i := 1 to n do
 5 |   Intent_min(i) ← S(1, i);
 6 |   for j := 2 to m do
 7 |   |   if S(j, i) < Intent_min(i) then
 8 |   |   |   Intent_min(i) ← S(j, i);
 9 |   |   end
10 |   end
11 end
12 while stop = false do
13 |   for i := 1 to nb_d do
14 |   |   New_Intent ← Next_intent(Intent_min, i);;
15 |   |   d ← Compute_distance(New_Intent, Intent_target);
16 |   |   if (d < dist) then
17 |   |   |   dist = d; save_Intent ← New_Intent;
18 |   |   |   extent ← Compute_Extent(New_Intent);
19 |   |   |   if (size(extent) = k) then
20 |   |   |   |   S_ref ← extent; stop ← true;
21 |   |   |   end
22 |   |   end
23 |   end
24 |   Intent_min ← save_Intent;
25 end
26 return S_ref;
```

1. First, we calculate the skyline using the Basic Nested Loop algorithm (BNL) for more details see [4].
2. Second, we compute the refined skyline using Algorithm 1 (FLCMD). This algorithm, starts by computing for each object o_i the degree $R(o_i, p_j)$ for witch the o_i minimizes the property p_j chosen by the user. Then, it computes the formal concept whose intent minimizes the properties chosen by the user, i.e., maximizes the degrees $R(o_i, p_j)$ for these properties (this concept is called target concept).
3. FLCMD builds the fuzzy lattice for skyline objects. It starts by computing the formal concept whose intent minimizes the degrees $R(o_i, p_j)$ for the properties chosen by user,
4. The algorithm FLCMD computes all the following concepts of this concept.
5. For each new concept, FLCMD, computes the size of its extent and the distance between its intent and the intent of target concept.
6. If the size of the extent equals k (where k is a user-defined parameter), the process stopped. The refined skyline is given by the objects of this extent

(when the number of extents having a size equals k is greater than 1, FLCMD chooses the extent whose intent has the minimal distance).

7. If the size of the extent is greater than k, FLCMD selects the intent that has the minimal distance and it starts from step 4.

FLCMD algorithm uses the following functions:

- $Next\_intent(Intent\_min, i)$: gives the following intent of $Intent\_min$ on the dimension i.
- $Compute\_Extent(New\_Intent)$: computes the extent of New_Intent, using the equation (4) and the implication given by the Eq. (6).
- $Compute\_distance(New\_Intent, Intent\_target)$: computes the Euclidean distance between New_Intent and $Intent\_target$.

Example 3. To illustrate our approach, let us come back to the skyline calculated in Example 1 presented in Sect. 2.1. As a reminder, we use two properties, namely price and *dist_wh*. Furthermore, we assume that the minimal value, the better. **BNL** algorithm returns as skyline the following apartments: $\{A_1, A_3, A_5, A_7\}$, see Table 3.

Remark *In the following, we note the intent (price$^{\alpha}$, dist_wh$^{\beta}$)* by (α, β).

Table 3. Classic skyline and objects degrees

Classic skyline				Object degrees $R(A_i, P_i)$	
Code	Area (m2)	Price (€)	$dist\_wh(km)$	Price	$dist\_wh$
A_1	60	525	20	0.075	0.875
A_3	25	360	85	0.9	0.0625
A_5	25	340	90	1	0
A_7	65	540	10	0	1

First, we compute for each object skyline A_i the degree $R(A_i, P_i)$ for which A_i minimizes the property P_i. These degrees are given by (see Fig. 4).

- $R(A_i, price) = 1 - (x_1 - 340)/200$, x_1 is the value of A_i w.r.t property *price*.
- $R(A_i, dis\_wh) = 1 - (x_2 - 10)/80$, x_2 is the value of A_i w.r.t property *dis_wh*.

Second, we compute the target intent and the intent that minimizes the degree $R(A_i, P_i)$ w.r.t cheap price and short distance. Using data from Table 3, and the Algorithm 1, one cane observe that $Intent\_target = (1, 1)$ $Intent\_min = (0, 0)$. Then, we compute the following intents of the intent_min.

For $i = 1$, $New\_Intent = (0.075, 0)$. The distance between this intent and the target intent $d = \sqrt{(1 - 0.075)^2 + (1 - 0)^2} = 1.36$, extent $= (A_1, A_3, A_5)$.

For $i = 2$, $New\_Intent = (0, 0.0625)$. The distance between this intent and the target intent $d = \sqrt{(1 - 0)^2 + (1 - 0.0625)^2} = 1.37$, extent$=(A_1, A_3, A_7)$.

If $k = 3$, the process stopped and $S_{ref} = \{A_1, A_3, A_5\}$.

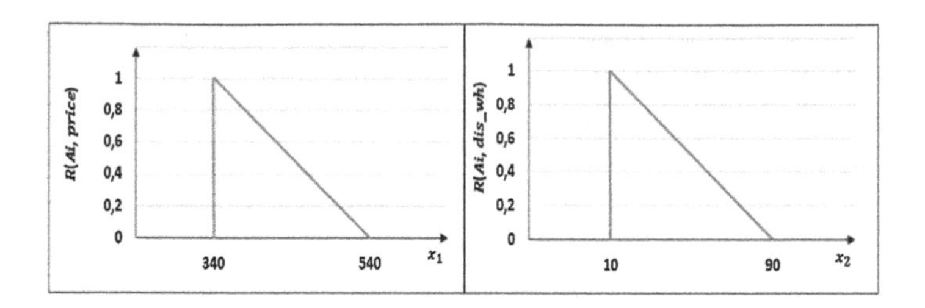

Fig. 4. Objects degrees

If $k < 3$, we select the intent $(0.075, 0)$ (because it has the minimal distance (1.36) with the target intent) then, we compute its following intents and the process continues as shown in Fig. 5. From Fig. 5, we can see that, if $k = 2$, the refined skyline equals $\{A_3, A_5\}$, when $k = 1$ $S_{ref} = \{A_3\}$.

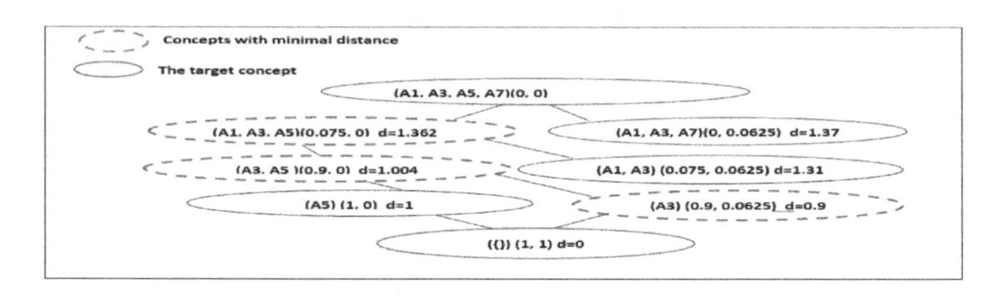

Fig. 5. Lattice of skyline points based on minimal distance

4 Experimental Study

In this section, we present the experimental study that we have conducted. The goal of this study is to prove the effectiveness of our algorithm and its ability to refine huge skyline and compare its relevance to the naive method. All experiments were performed under Windows OS, on a machine with an Intel core i7 2,90 GHz processor, a main memory of 8 GB and 250 GB of disk. All algorithms were implemented with Java. Dataset benchmark is generated using the method described in [4]. The test parameters used are distribution dataset [DIS] (correlated, anti-correlated and independent), the dataset size [D] (100K, 250K, 500K, 1000K, 2000K, 4000K) and the number of dimensions [d] (2, 4, 6, 10, 15). To interpret the results we define the following refinement rate (ref_rate):

$$ref\_rate = \frac{(ntcs - ntrs)}{(ntcs)} \tag{7}$$

where $ntcs$ is the number of tuples of the regular skyline and $ntrs$ is the number of tuples for the refined skyline.

Impact of [DIS]. In this case, we use a dataset with $|D| = 100K$, $d = 6$. Figure 6 shows that the execution time of the two algorithms for anti-correlated data is high compared to the correlated or independent data. This is due to the important number of tuples to refine (14758 tuples for anti-correlated data, 2184 and 89 tuples for independent and correlated data). Figure 6 shows also that our algorithm has the best execution time compared to the naive algorithm (0.004 s for FLCMD, 0.85 s for naive algorithm in the case of correlated data, 10.41 s for FLCMD and 72.32 s for the naive algorithm in the case of anti-correlated data, 0.38 s for FLCMD and 18.2 s for the naive algorithm in the case of independent data). The refinement rate for the two algorithms is very high (for correlated data $= (89-10)/89 = 0.88$, for anti-correlated data $= (14758-10)/14758 = 0.99$ and for independent data $= (2184 - 10)/2184 = 0.995$).

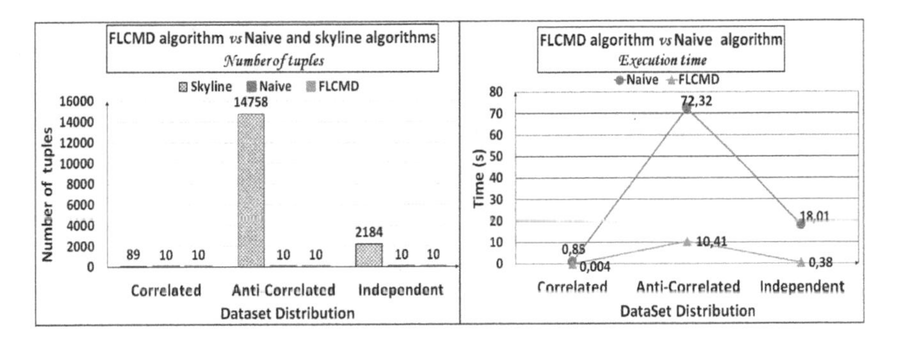

Fig. 6. Impact of [DIS]

Impact of the Size of the Dataset [D]. In this case, we study the impact of the size of the database on the execution time of the refined skyline and the refinement rate for the two algorithms. To do this, we use an anti-correlated database with $d = 4$. Figure 7, shows that, the execution time increases with the increase of the database size. But the execution time of our algorithm remains the best compared to the naive algorithm (the execution time increases from 0.3 s if $|D| = 100K$ to 10.52 s when $|D| = 4000K$ for FLCMD and from 13.48 s if

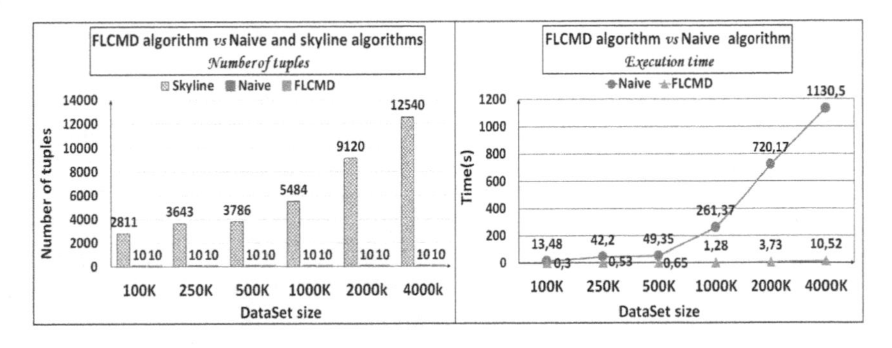

Fig. 7. Impact of [D]

$|D| = 100K$ to 1130.5 s if $|D| = 4000K$ for naive algorithm). The refinement rate for the two algorithms is very high varied from 0.996 ($(2811-10)/2811 = 0.996$) when $|D| = 100K$ to 0.999 ($(12540-10)/12540 = 0.999$) when $|D| = 4000K$.

Impact of the Number of Dimensions [d]. In this case, we study the impact of varying the number of dimensions skyline in the process of computing S_{ref}. We use an anti-correlated distribution data with $|D| = 50K$. Figure 8 shows that the execution time increases with the number of dimensions (from 0.008 s for $d = 2$ to 120.3 s when $d = 15$ for the FLCMD algorithm) and (between 0.5 s and 420 s when d varied from 2 to 15 for naive algorithm). This indicates that our algorithm gives the best execution time compared to naive algorithm. The refinement rate increases from 0.94 ($(187-10)/187 = 0.94$) when $d = 2$ to 0.99 ($(48103-10)/48103 = 0.99$) for $d = 15$.

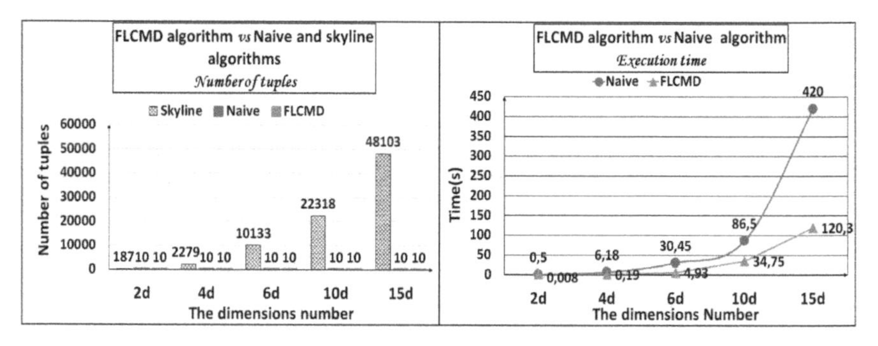

Fig. 8. Impact of [d]

5 Conclusion and Perspectives

In this paper, we addressed the problem of the skyline, especially a huge skyline and we proposed a new approach to reduce its size. The basic idea of this approach is to build a fuzzy concept lattice for skyline objects based on the minimal distance between each concept and the target concept. The process of refinement stopped when we compute the concept that contains k objects (where k is a user-defined parameter) and has the minimal distance with the target concept. The refined skyline is given by the objects of this concept. An algorithm called FLCMD to calculate the refined skyline is proposed. In addition, we implemented the naive algorithm to compare its performance to that of our algorithm. The experimental study we have done showed that, our approach is a good alternative to reduce the size of the classic skyline (the refinement rate reached 99%) and has a reasonable time computation also, the execution time of our algorithm is the best compared to the naive algorithm. As for future work, we will explore, on the one hand the use of semantic distance between concepts to build the refinement lattice and on the other hand, we will use the lattice construction algorithms that gives fuzzy extensions in order to sort the objects of the same concept.

References

1. Abbaci, K., Hadjali, A., Lietard, L., Rocacher, D.: A linguistic quantifier-based approach for skyline refinement. In: Joint IFSA World Congress and NAFIPS Annual Meeting, IFSA/NAFIPS, 24–28 June, Edmonton, Alberta, Canada, pp. 321–326 (2013)
2. Balke, W., Guntzer, U., Lofi, C.: User interaction support for incremental refinement of preference-based queries. In: Proceedings of the First Inter. Conference on Research Challenges in Information Science (RCIS), 23–26 April, Ouarzazate, Morocco, pp. 209–220 (2007)
3. Belohlávek, R.: Fuzzy galois connections. Math. Log. Q. **45**, 497–504 (1999)
4. Börzsönyi, S., Kossmann, D., Stocker, K.: The skyline operator. In: Proceedings of the 17th International Conference on Data Engineering, 2–6 April, Heidelberg, Germany, pp. 421–430 (2001)
5. Burusco, A., Fuentes-González, R.: The study of the l-fuzzy concept lattice. Mathware Soft Comput. **3**(1), 208–209 (1994)
6. Chan, C.Y., Jagadish, H.V., Tan, K., Tung, A.K.H., Zhang, Z.: Finding k-dominant skylines in high dimensional space. In: Proceedings of the International Conference on Management of Data (ACM SIGMOD), 27–29 June, Chicago, Illinois, USA, pp. 503–514 (2006)
7. Chan, C.-Y., Jagadish, H.V., Tan, K.-L., Tung, A.K.H., Zhang, Z.: On high dimensional skylines. In: Ioannidis, Y., et al. (eds.) EDBT 2006. LNCS, vol. 3896, pp. 478–495. Springer, Heidelberg (2006). https://doi.org/10.1007/11687238_30
8. Chomicki, J., Ciaccia, P., Meneghetti, N.: Skyline queries, front and back. SIGMOD Rec. **42**(3), 6–18 (2013)
9. Elmi, S., Tobji, M.A.B., Hadjali, A., Yaghlane, B.B.: Selecting skyline stars over uncertain databases: semantics and refining methods in the evidence theory setting. Appl. Soft Comput. **57**, 88–101 (2017)
10. Ganter, B., Wille, R.: Formal Concept Analysis. Mathematical Foundations. Springer, Heidelberg (1999). https://doi.org/10.1007/978-3-642-59830-2
11. Goncalves, M., Tineo, L.: Fuzzy dominance skyline queries. In: Wagner, R., Revell, N., Pernul, G. (eds.) DEXA 2007. LNCS, vol. 4653, pp. 469–478. Springer, Heidelberg (2007). https://doi.org/10.1007/978-3-540-74469-6_46
12. Gulzar, Y., Alwan, A.A., Salleh, N., Shaikhli, I.F.A.: Processing skyline queries in incomplete database: issues, challenges and future trends. JCS **13**(11), 647–658 (2017)
13. Haddache, M., Belkasmi, D., Hadjali, A., Azzoune, H.: An outranking-based approach for skyline refinement. In: 8th IEEE International Conference on Intelligent Systems, IS 2016, 4–6 September 2016, Sofia, Bulgaria, pp. 333–344 (2016)
14. Hadjali, A., Pivert, O., Prade, H.: Possibilistic contextual skylines with incomplete preferences. In: Second International Conference of Soft Computing and Pattern Recognition, (SoCPaR), 7–10 December, Cergy Pontoise/Paris, France, pp. 57–62 (2010)
15. Hadjali, A., Pivert, O., Prade, H.: On different types of fuzzy skylines. In: Kryszkiewicz, M., Rybinski, H., Skowron, A., Raś, Z.W. (eds.) ISMIS 2011. LNCS (LNAI), vol. 6804, pp. 581–591. Springer, Heidelberg (2011). https://doi.org/10.1007/978-3-642-21916-0_62
16. Hamiche, M., Hadjali, A., Drias, H.: A strong-dominance-based approach for refining the skyline. In: Proceedings of the 12th International Symposium on Programming and Systems (ISPS), 28–30 April, Algiers, Algeria, pp. 1–8 (2015)

17. Khalefa, M.E., Mokbel, M.F., Levandoski, J.J.: Skyline query processing for incomplete data. In: Proceedings of the 24th International Conference on Data Engineering, ICDE 2008, 7–12 April 2008, Cancún, México, pp. 555–565 (2008)
18. Koltun, V., Papadimitriou, C.H.: Approximately dominating representatives. In: Eiter, T., Libkin, L. (eds.) ICDT 2005. LNCS, vol. 3363, pp. 204–214. Springer, Heidelberg (2004). https://doi.org/10.1007/978-3-540-30570-5_14
19. Kossmann, D., Ramsak, F., Rost, S.: Shooting stars in the sky: an online algorithm for skyline queries. In: Proceedings of the 28th International Conference on Very Large Data Bases (VLDB), 20–23 August, Hong Kong, China, pp. 275–286 (2002)
20. Lee, J., Hwang, S.: Scalable skyline computation using a balanced pivot selection technique. Inf. Syst. **39**, 1–21 (2014)
21. Lee, J., You, G., Hwang, S.: Telescope: zooming to interesting skylines. In: Kotagiri, R., Krishna, P.R., Mohania, M., Nantajeewarawat, E. (eds.) DASFAA 2007. LNCS, vol. 4443, pp. 539–550. Springer, Heidelberg (2007). https://doi.org/10.1007/978-3-540-71703-4_46
22. Loyer, Y., Sadoun, I., Zeitouni, K.: Personalized progressive filtering of skyline queries in high dimensional spaces. In: Proceedings of the 17th International Conference on Database Engineering Applications Symposium (IDEAS), 09–11 October, Barcelona, Spain, pp. 186–191 (2013)
23. Papadias, D., Tao, Y., Fu, G., Seeger, B.: An optimal and progressive algorithm for skyline queries. In: Proceedings of the International Conference on Management of Data (ACM SIGMOD), 9–12 June, San Diego, California, USA, pp. 467–478 (2003)
24. Pei, J., Jiang, B., Lin, X., Yuan, Y.: Probabilistic skylines on uncertain data. In: Proceedings of the 33rd International Conference on Very Large Data Bases, University of Vienna, 23–27 September, Austria, pp. 15–6 (2007)
25. Santiago, A., et al.: A survey of decomposition methods for multi-objective optimization. In: Castillo, O., Melin, P., Pedrycz, W., Kacprzyk, J. (eds.) Recent Advances on Hybrid Approaches for Designing Intelligent Systems. SCI, vol. 547, pp. 453–465. Springer, Cham (2014). https://doi.org/10.1007/978-3-319-05170-3_31
26. Sarma, A.D., Lall, A., Nanongkai, D., Lipton, R.J., Xu, J.J.: Representative skylines using threshold-based preference distributions. In: Proceedings of the 27th International Conference on Data Engineering, (ICDE), 11–16 April, Hannover, Germany, pp. 387–398 (2011)
27. Tan, K., Eng, P., Ooi, B.C.: Efficient progressive skyline computation. In: Proceedings of 27th International Conference on Very Large Data Bases (VLDB), 11–14 September, Roma, Italy (2001)
28. Wille, R.: Restructuring lattice theory: an approach based on hierarchies of concepts. In: Rival, I. (ed.) Ordered Sets, pp. 445–470. Springer, Heidelberg (1982). https://doi.org/10.1007/978-94-009-7798-3_15
29. Yiu, M.L., Mamoulis, N.: Efficient processing of top-k dominating queries on multi-dimensional data. In: Proceedings of the 33rd International Conference on Very Large Data Bases (VLDB), 23–27 September, University of Vienna, Austria, pp. 483–494 (2007)
30. Zadeh, L.A.: Fuzzy sets. Inf. Control **8**(3), 338–353 (1965)

A Probabilistic Relational Model for Risk Assessment and Spatial Resources Management

Thierno Kanté[1,2](✉) and Philippe Leray[2](✉)

[1] EDICIA, Nantes, France
[2] LS2N UMR CNRS 6004, DUKe research group, University of Nantes, Nantes, France
thierno-sadou.kante@etu.univ-nantes.fr, philippe.leray@univ-nantes.fr

Abstract. Fault tree (FT) model is one of the most popular techniques for probabilistic risk analysis of large, safety critical systems. Probabilistic graphical models like Bayesian networks (BN) or Probabilistic Relational Models (PRM) provide a robust modeling solution for reasoning under uncertainty. In this paper, we define a general modeling approach using a PRM. This PRM can represent any FT with possible safety barriers, spatial information about localization of events, or resources management. In our proposed approach, we define a direct dependency between the resources allocated to one location and the strength of the barriers related to this same location. We will show how this problem can be fully represented with a PRM by defining its relational schema and its probabilistic dependencies. This model can be used to estimate the probability of some risk scenarios and to assess the presence of resources on each location through barrier's efficiency on risk reduction.

Keywords: Risk assessment · Probabilistic relational models

1 Introduction

Risk is usually defined as the potentially undesirable outcome resulting from an incident, event or event, as determined by its probability and the consequences arising from it. Risk assessment is a process to determine the probability of losses by analyzing potential hazards and evaluating existing conditions of vulnerability that could pose a threat or harm to property, people, livelihoods and the environment on which they depend [1]. There are different approaches to probabilistic risk assessment (PRA), including probabilistic approaches, fault tree (FT) [2], event tree or reliability block diagrams.

A FT is a logical and diagrammatic model that allows to estimate the probability of an event to occur based on the occurrence or non-occurrence of other event. However, FTs cannot handle situations involved complex or uncertain dependencies between the system components. In order to accurately calculate

F. Wotawa et al. (Eds.): IEA/AIE 2019, LNAI 11606, pp. 555–563, 2019.
https://doi.org/10.1007/978-3-030-22999-3_48

the reliability of a fault tree in the presence of statistical dependencies between events, Bobbio et al. [2] present a conversion of FTs into Bayesian Networks (BNs) [3]. However, when the system under study involves a complex structure, the BNs also reach their limits. Therefore, several extensions to BN such as object oriented Bayesian networks or probabilistic relational models (PRMs) [4] have been to solve this lack of modeling for probabilistic risk assessment. A PRM associated to a relational schema $\mathcal{R}$ is composed of a dependency structure $\mathcal{S}$ and a set of parameters Θ. The relational schema $\mathcal{R}$ describes a set of classes with associated descriptive attributes. $\mathcal{S}$ describes the probabilistic dependencies between each attribute and some potential other attributes reachable in all the classes. Θ is a set of conditional distributions, one for each attribute, given tis parents in $\mathcal{S}$, and depicts the strength of the corresponding probabilistic dependency. This PRM is defined for any instance of database corresponding to this relational schema. A Ground Bayesian Network (GBN) can be generated by unrolling the dependencies described in the PRM for one given database. Probabilistic inference can be naively performed on this GBN, but some optimizations have been proposed, taking into account the repetitive structure of the model.

The aim of this work is to propose a methodology for probabilistic risk assessment of socio-technical systems that consider complex structure of the system with uncertainty about causal relationships between risk factors and where some resources allocated in specific areas can influence the system.

The paper is organized as follows. Section 2 reviews some related works about probabilistic risk assessment in an uncertain context. In Sect. 3, we describe how to extend the approach initially proposed in [5] about modeling a fault tree with a Probabilistic Relational Model, in order to take into account spatial information and spatial allocation of resources. Section 4 proposes a toy example that illustrates the interest of our approach in urban area for road safety. In Sect. 5 conclusions and future works are mentioned.

2 Related Work

FT models have been used for instance to examine the factors contributing to the occurrence of road accidents at several urban district [6] or terrorist attack [9]. BNs are mostly used to modeling the propagation of failure events and the degradation of the system [7]. Many papers demonstrate the equivalence between BN and static fault tree [2] and multi-state system through multi-state fault tree (MSFT) [8]. The `ISO 31000:2009` standard suggests evaluating risks in a global view. Bayesian network-based integrated risk analysis approach for industrial systems through bow tie formalism have been also proposed. This methodology is dedicated to modeling human safety barrier (HSB). A BN model for spatial event monitoring have been developed in [10].

PRMs have also proved their applicability to analyze risks related to safety [11]. PRMs take the advantages of both Bayesian networks to model probabilistic dependencies between attributes, and the decomposability of such models into multiple instantiable components depending on the context. Kanté et al. [5] present a conversion of fault tree with possible safety barrier into a probabilistic relational model, where all these information are described in a relational database.

3 Contribution

We propose here to improve the approach proposed in [5], based on a probabilistic relational model for risk assessment of complex systems, by adding spatial information and spatial allocation and management of resources as safety functions. We suppose that each event of the fault tree model have a specific location and in each location, there is the absence or presence of some resources. We assume that these resources affect the safety barriers performance for risk reduction. We also assume that these resources are limited.

Section 3.1 gives some notations related to our previous fault tree model associated to safety barrier. In Sect. 3.2, we formalize our contribution by considering spatial information about event localisation, spatial allocation of resources and probabilistic relationships between barriers and resources. In Sect. 3.3, we finally describe the corresponding Probabilistic Relational Model.

3.1 Fault Tree with Safety Barrier

In [5], we consider a fault tree with safety barriers by a triple $\Psi = \{\mathcal{E}, \mathcal{G}, \mathcal{B}\}$. $\mathcal{E} = \{E_i\}$ is a set of events with a prior probability $priorStrength(E_i)$. $\mathcal{G} = \{G_j\}$ is a set of gates, with $Inputs(G_i) \subset \mathcal{E}$, $Output(G_i) \in \mathcal{E}$ and $Type(G_i) \in \{OR, AND, \ldots\}$. $\mathcal{B} = \{B_k, (E_i, G_j)\}$ is a set of barriers where B_k is a prior barrier associated to one specific event E_i appearing as an input of a given gate G_j, with $BarrierStrength(B_k, E_i, G_j) \in \{absent, low, \ldots, high\}$. One example of a fault tree with safety barriers is given in Fig. 1, with 8 events, 3 OR gates and 2 barriers.

3.2 Modeling Spatial Information and Resources

There exist different way to represent spatial data: points, lines, grids, hierarchies or networks [12]. The nature of our constraint over resources has led us to describe our spatial information by a spatial hierarchical data structure $\mathcal{T}$ where each node $L \in \mathcal{L}$ is a specific location and where the tree structure depicts a set of `belongs-to` relationships (L `belongs-to` $parent(L)$).

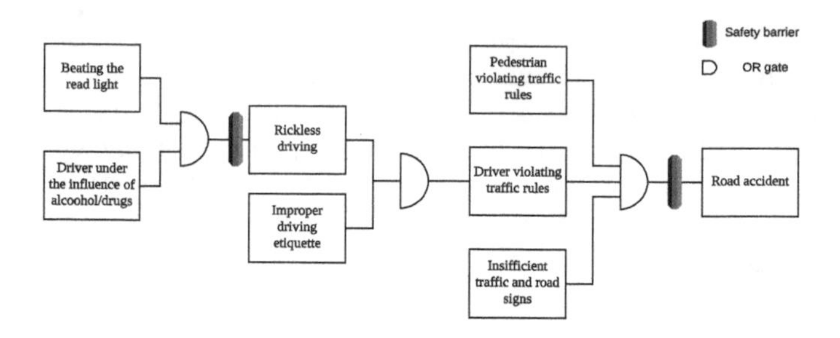

Fig. 1. Example of a Fault tree with safety barriers for road safety

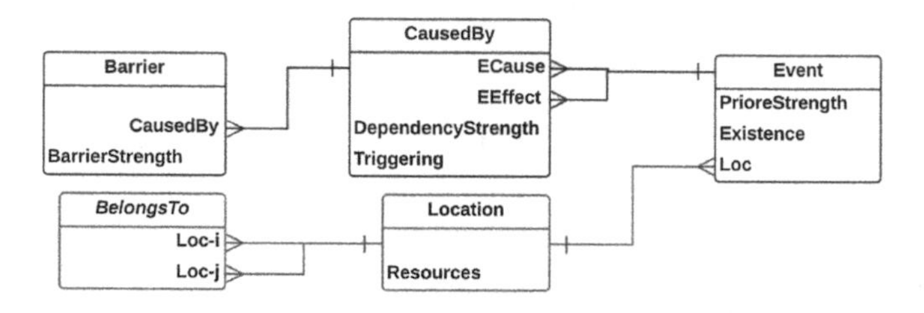

Fig. 2. Relational schema of a fault tree with safety barriers [5] improved for modeling spatial information and resources

As we are interested in spatial resources allocation and management of these resources, we define $Resources(L)$ as the number of resources allocated to the location L, with constraints on the resources. The constraint is a simple recursive formula defined in Eq. 1 describing the fact that the number of resources in a location is the sum of all the resources in the associated sub-locations.

$$Resources(L) = \sum_{L_i \in children(L)} Resources(L_i) \tag{1}$$

By defining these spatial locations, we can now extend our fault tree definition by associating each event E_i to a specific location $Location(E_i) \in \mathcal{L}$.

In order to take into account spatial information, we propose to extend the relational schema defined in [5], as described in Fig. 2. This schema includes the description of the fault tree (with the two entity classes Event and Barrier and the association class CausedBy, depicted in black in the figure) and the description of the spatial data structure (with the entity class Location and the association class BelongsTo, depicted in blue in the same figure). Instances of the classes are defined by fives rules: (i) one instance of Location for each $L_i \in \mathcal{L}$; (ii) one instance of BelongsTo for each edge in $\mathcal{T}$ with $Loc_i = L_i$ and $Loc_j = parent(L_i)$; (iii) one instance of each $E_i \in \mathcal{E}$, and $Loc = Location(E_i)$; (iv) one instance of CausedBy for each $G_j \in \mathcal{G}$ and $E_i \in Inputs(G_j)$ with ECause $= E_i$ and EEffect $= Output(G_j)$; (v) one instance of Barrier for each $\{B_k, (E_i, G_j)\} \in \mathcal{B}$, with CauseById refers to the instance related to gate G_j and input E_i.

3.3 A Spatial PRM for Fault Tree Modeling and Resources Management

We now propose to use the relational schema defined in Sect. 3.2 to define a PRM for fault tree modeling with spatial information and resources management. This PRM, depicted in Fig. 3, includes the probabilistic dependencies defined in [5] for fault tree modeling and extended in this work to take into account the spatial information, the resource management and the dependencies between resources and barriers. *Location.Resources* $\in \{0, 1, \ldots, n\}$ represents the number of resources allocated in a location. The Eq. 1 is transformed into a conditional probability distribution (CPD) between the resources in a location *Location.Resources* and the sum of the resources in the sub-locations. More precisely, the dependency is here a simple equality that can be described by a conditional probability table which is the identity matrix. Finally, we propose to model the fact that a barrier strength depends on the resources allocated in the locations where occur the events related to this barrier.

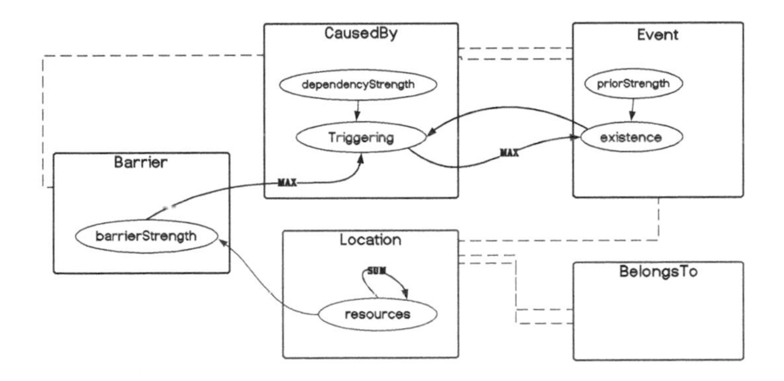

Fig. 3. Spatial PRM for fault tree modeling and resources management

4 Toy Example

In this section, we propose a toy example illustrating the interest of using our approach. In this example, we want to estimate the probability that a particular event is affected in some locations given a certain hazard scenario.

Let's consider the fault tree with safety barriers given in Fig. 1, with 8 events, 3 OR gates and 2 barriers. In terms of spatial information, we will focus on 4 locations: Orvault, Coueron, Nantes and Loire Atlantique, where Orvault, Coueron and Nantes are three cities belonging to Loire Atlantique. We will consider the resources allocated in each location as Police vehicles. We note that increasing the number of resources in one location will increase the barrier strength in the corresponding fault tree. For this running example, we used the PILGRIM Relational C++ Library[1] to implement the PRM model. The ground Bayesian network (GBN) generated from this PRM has been exported, visualized and queried with Genie/Smile software [13].

[1] http://pilgrim.univ-nantes.fr.

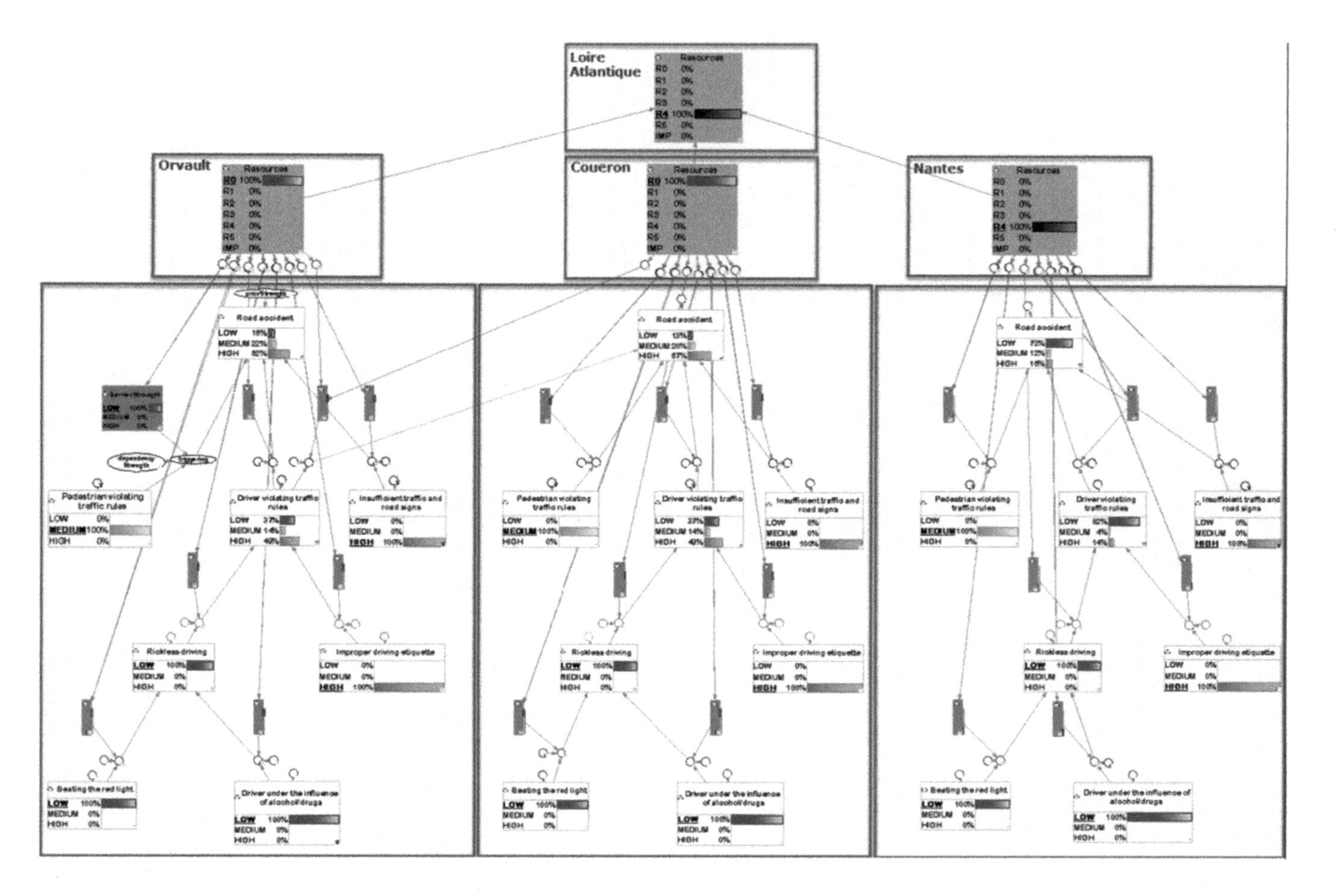

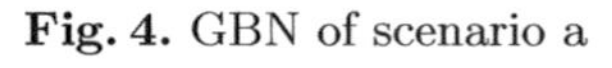

Fig. 4. GBN of scenario a

We present two scenarios in the same context. Given a set of *Event.Existence* and possible *Location.Resources*, the obtained GBN is queried to estimate the probability of *Driver violating traffic rules* and *Road accident* in each city (Orvault, Coueron and Nantes). In both scenarios, we consider that *Improper driving etiquette* is *high*, *Pedestrian violating traffic rules* is *medium* and *Insufficient traffic and road signs* is *high* in the three cities (Orvault, Coueron and Nantes). The difference between both scenarios is the resources allocation (0 in Orvault and Coueron, and 4 in *Nantes* in the first scenario), 0 in Coueron and *Nantes* and 4 in *Orvault* in the second one. Figure 4 shows us the ground Bayesian network and results of probabilistic inference for the first scenario. We can see that, given the observed events, the probability of a *high Driver violating traffic rules* is 49% in *Orvault* and *Coueron* where no resource are present, when it decreases to 14% in *Nantes* where the resources are allocated. For the same reason, the probability of a high *Road accident* is 16% in *Nantes*, and increases to 62% in *Orvault* and even to 67% in *Coueron*, because of the *high Driver violating traffic rules* in *Orvault*. In the example, we see the impact of the resources on the barriers between events: allocating all the resources in *Nantes* helps in blocking raising of the two target events. We can also observe the propagation of the spatial information: the high probability of a source event in *Orvault* increases the probability of a target event in *Coueron*. In the second scenario, the resource allocation has been modified, with all the resources in Orvault. The absence of resources (now in *Coueron* in *Nantes*) gives us a probability of a high *Driver violating traffic rules* equal to 49% and a probability of a *high Road accident* equal to 62%. The presence of the resources in *Orvault* helps in decreasing these probabilities to respectively 14% and 16%. It also increases the barrier strength between the events from *Orvault* to *Coueron*.

We also propose two others scenarios to highlight how the model can be used for simulating effects of resource allocation for resources management. Due to lack of space, we cannot show the associated figures. Firstly, we consider that the number of resources has only be defined for *Loire Atlantique* (4 resources) without any precise information in the three cities. Given the observed events, the probability of a *high Driver violating traffic rules* is 36% in every city. The probability of a *high Road accident* is 44% in *Orvault* and *Nantes*, and increases to 48% in *Coueron*, because of the *high Driver violating traffic rules* in *Orvault*. Secondly, we allocated two resources in *Nantes*, without specifying the number of resources in *Orvault* and *Coueron*. We note that this operation helps in moderately decreasing the probability of a *high Driver violating traffic rules* to 32% and the probability of a *high Road accident* to 40% in *Nantes*. In the same time, the model will directly infer that less resources will be allocated to *Orvault* or *Coueron*, which increases the probability of a *high Driver violating traffic rules* to 50% and the probability of a high *Road accident* to 49% in *Orvault* and 53% in *Coueron*. Then, the decision-maker can use our model as a simulation tool to define the allocated resources.

5 Conclusion and Perspectives

In this paper, we proposed a general framework for probabilistic risk assessment using a probabilistic relational model. Our model is based on a fault tree, with possible safety barriers, where the events are associated to specific locations. The locations are organized into a spatial hierarchical data structure where the resources can be allocated in each location and influence the corresponding barriers strength. We illustrated the characteristics of our model with a toy example. We demonstrated the interest of using a probabilistic relational model to estimate the probability of some target events in a complex system with several locations, events, and dependencies among events in different locations. Finally, we showed that our model can be used as a simulation tool in order to observe the impact of some resource allocation policy. This model has been implemented and is currently being tested by EDICIA, a company specialized in urban safety. As the next step in our research, we also intend to extend this model by also adding a temporal dimension, and by taking into account observations of the events by imprecise probes.

References

1. UNISDR: Terminology for disaster risk reduction. United Nations International Strategy for Disaster Reduction (UNISDR) Geneva, Switzerland (2009)
2. Bobbio, A., Portinale, L., Minichino, M., Ciancamerla, E.: Improving the analysis of dependable systems by mapping fault trees into Bayesian networks. Reliab. Eng. Syst. Saf. **71**(3), 249–260 (2001)
3. Pearl, J.: Probabilistic Reasoning in Intelligent Systems: Networks of Plausible Inference. Elsevier, Amsterdam (2014)
4. Koller, D., et al.: Introduction to Statistical Relational Learning. MIT Press, Cambridge (2007)
5. Kante, T., Leray, P.: A probabilistic relational model approach for fault tree modeling. In: Benferhat, S., Tabia, K., Ali, M. (eds.) IEA/AIE 2017. LNCS, vol. 10351, pp. 154–162. Springer, Cham (2017). https://doi.org/10.1007/978-3-319-60045-1_18
6. Yaghoubpour, Z., Givehchi, S., Tabrizi, M.A., Masoudi, F., Nourian, L.: Public transport risk assessment through fault tree analysis. Int. J. Hum. Capital Urban Manag. **1**(2), 93–102 (2016)
7. Gertsbakh, I.: Reliability Theory: with Applications to Preventive Maintenance. Springer, Heidelberg (2013)
8. Cao, J., Yin, B., Lu, X.: Probabilistic risk assessment of multi-state systems based on Bayesian networks. In: 2016 18th International Conference on Advanced Communication Technology (ICACT), pp. 773–778. IEEE, January 2016
9. Shooman, M.L.: Terrorist risk evaluation using a posteriori fault trees. In: Annual Reliability and Maintainability Symposium, RAMS 2006, pp. 450–455. IEEE, January 2006
10. Jiang, X., Neill, D.B., Cooper, G.F.: A Bayesian network model for spatial event surveillance. Int. J. Approximate Reason. **51**(2), 224–239 (2010)
11. Medina-Oliva, G., Weber, P., Iung, B.: PRM-based patterns for knowledge formalisation of industrial systems to support maintenance strategies assessment. Reliab. Eng. Syst. Saf. **116**, 38–56 (2013)

12. Shekhar, S., Evans, M.R., Kang, J.M., Mohan, P.: Identifying patterns in spatial information: a survey of methods. Wiley Interdisciplinary Rev.: Data Mining Knowl. Discov. **1**(3), 193–214 (2011)
13. Druzdzel, M.J.: SMILE: structural modeling, inference, and learning engine and GeNIe: a development environment for graphical decision-theoretic models. In: AAAI/IAAI, pp. 902–903, July 1999

A Scheme for Continuous Input to the Tsetlin Machine with Applications to Forecasting Disease Outbreaks

Kuruge Darshana Abeyrathna(✉), Ole-Christoffer Granmo, Xuan Zhang, and Morten Goodwin

Centre for Artificial Intelligence Research, University of Agder, Grimstad, Norway
{darshana.abeyrathna,ole.granmo,morten.goodwin}@uia.no,
xuan.z.jiao@gmail.com

Abstract. In this paper, we apply a new promising tool for pattern classification, namely, the *Tsetlin Machine* (TM), to the field of disease forecasting. The TM is interpretable because it is based on manipulating expressions in propositional logic, leveraging a large team of Tsetlin Automata (TA). Apart from being interpretable, this approach is attractive due to its low computational cost and its capacity to handle noise. To attack the problem of forecasting, we introduce a preprocessing method that extends the TM so that it can handle continuous input. Briefly stated, we convert continuous input into a binary representation based on thresholding. The resulting extended TM is evaluated and analyzed using an artificial dataset. The TM is further applied to forecast dengue outbreaks of all the seventeen regions in Philippines using the spatio-temporal properties of the data. Experimental results show that dengue outbreak forecasts made by the TM are more accurate than those obtained by a Support Vector Machine (SVM), Decision Trees (DTs), and several multi-layered Artificial Neural Networks (ANNs), both in terms of forecasting precision and F1-score.

Keywords: Tsetlin Machine · Tsetlin Automata · Learning automata · Pattern recognition with propositional logic · Disease outbreaks forecasting

1 Introduction

The Tsetlin Machine (TM) is a recent pattern classification method that manipulates expressions in propositional logic based on a team of Tsetlin Automata (TAs) [1]. A Tsetlin Automaton (TA) is a fixed structure deterministic automaton that learns the optimal action among the set of actions offered by an environment. Figure 1 shows a two-action TA with 2 N states. The action that the TA performs next is decided by the present state of the TA. States from 1 to N maps to Action 1, while states from N + 1 to 2 N maps to Action 2. The TA interacts with its environment in an iterative way. In each iteration, the TA

F. Wotawa et al. (Eds.): IEA/AIE 2019, LNAI 11606, pp. 564–578, 2019.
https://doi.org/10.1007/978-3-030-22999-3_49

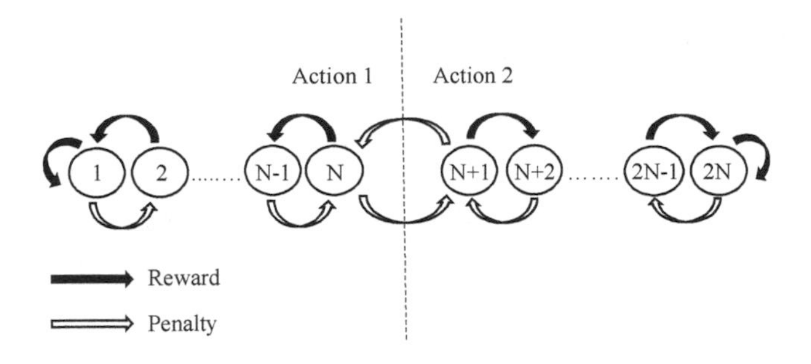

Fig. 1. Transition graph of a two-action Tsetlin Automaton.

performs the action associated with its current state. This, in turn, randomly triggers a reward or a penalty from the environment, according to an unknown probability distribution. If the TA receives a reward, it reinforces the action performed by moving to a "deeper" state, one step closer to one of the ends (left or right side). If the action results in a penalty, the TA moves one step towards the middle state, to weaken the performed action, ultimately jumping to the middle state of the other action. In this manner, with a sufficient number of states, a TA converges to performing the action with the highest probability of producing rewards – the optimal action – with probability arbitrarily close to unity, merely by interacting with the environment [2].

The TM, introduced in 2018 by Granmo [1], uses the TA as a building block to solve complex pattern recognition tasks. The TM operates as follows. Firstly, propositional formulas in disjunctive normal form are used to represent patterns. The TM is thus a general function approximator. The propositional formulas are learned through training on labelled data by employing a collective of TAs organized in a game. The payoff matrix of the game has been designed so that the Nash equilibria (NE) correspond to the optimal configurations of the TM. As a result, the architecture of the TM is relatively simple, facilitating transparency and interpretation of both learning and classification. Additionally, the TM is designed for bit-wise operation. That is, it takes bits as input and uses fast bit manipulation operators for both learning and classification. This gives the TM an inherent computational advantage. Experimental results show that TM outperforms ANNs, Support Vector Machines (SVMs), the Naïve Bayes Classifier (NBC), Random Forests (RF), and Logistic Regression (LR) in diverse benchmarks [1,3]. These promising properties and results make the TM an interesting target for further research.

In this paper, we introduce a novel scheme that improves the accuracy of the TM when features are continuous. In Sect. 2, we provide an overview of related work. Then, in Sect. 3, we present our scheme for handling continuous features. In all brevity, we encode continuous features in binary form based on thresholding. The behavior of the resulting TM is studied in Sect. 4 based on both an artificial dataset and real-life data, focusing on dengue fever forecasting. Section 5 summarizes our research and provides pointers for further work.

2 Related Work

Propositional logic is a well-explored framework for knowledge based pattern classification. In [4], Disjunctive Normal Form (DNF) is used to represent the patterns in clinical and genomic data to find the recurrence of liver cancer. Data is converted to bits by setting thresholds for continuous features. Based on the input features, logical functions for recurrence and non-recurrence are created. Another example is the use of Boolean expressions to capture visual primitives for visual recognition, rather than relying on a data driven approach [5]. In all brevity, the advantage of propositional logic for pattern classification, and knowledge based approaches in general, as opposed to data driven statistical models, is that patterns can be identified even without a single training sample.

Learning propositional formulas to represent patterns in data has a long history [6]. Feldman investigates the hardness of learning DNF [7], Klivans use Polynomial Threshold Functions to build logical expressions [8], while Feldman leverages Fourier analysis [9]. Furthermore, so-called Probably Approximately Correct (PAC) learning has provided fundamental insight into machine learning, as well as providing a framework for learning formulas in DNF [10]. An integer programming approach is applied in [11] to learn disjunctions of conjunctions, providing promising results based on a Bayesian method. In addition to the above techniques, association rule mining models have been extensively applied in [12, 13] to predict sequential events using set of rules. Recent approaches combine Bayesian reasoning with propositional formulas in DNF for robust learning of formulas from data [6]. However, these techniques still suffer when facing noisy non-linear data, which may trap the learning mechanisms in local optima.

An attractive property of TA is that they support online learning in particularly noisy environment. Over several decades the basic TA, shown in Fig. 1, has been extended in several directions. These extensions include the Hierarchy of Twofold Resource Allocation Automata (H-TRAA) for resource allocation [14] and the stochastic searching on the line algorithm by Oommen et al. [15]. Furthermore, teams of Tsetlin Automata have been used to create a distributed coordination system [16], to solve the graph coloring problem [17], and to forecast dengue outbreaks in the Philippines [18]. The TM is a recent addition to the field of TA, addressing complex pattern recognition.

In order to attack the problem of forecasting, this paper introduces a preprocessing method that extends the TM so that it can handle continuous input. To achieve this, we convert continuous input into a binary representation based on thresholding. We use an artificial dataset as well as a real-life dataset to evaluate this approach, namely, forecasting of dengue fever outbreaks in the Philippines.

Different techniques have already been applied to forecast dengue outbreaks in different regions of the world. For instance, Seasonal Autoregressive Integrated Moving Average model is applied to forecast future dengue incidences in Guadeloupe [19] and Bangladesh [20]. In their research, temperature is identified as the best weather parameter to improve the forecasting performances. Here, 1-month ahead forecasting produces the highest accuracy, compared to 3-months and 1-year ahead forecasting. Similarly, dengue incidences in Rio de

Janeiro, Brazil [21], Northeastern Thailand [22], and Southern Thailand [23] are forecasted using Auto-regressive Integrated Moving Average models. Phung et al. investigate the forecasting ability of three regression models on dengue fever incidences in Can Tho city in Vietnam [24]. They find that a Standard Multiple Regression model provides poor forecasting capability. However, the Poisson Distributed Lag model performs well in 12-months ahead forecasting and Seasonal Autoregressive Integrated Moving Average model performs well in 3-months ahead forecasting. The importance of utilizing data from neighboring regions to forecast dengue incidences is identified in [25], using an Artificial Neural Network as the forecasting model with data from the Philippines.

In contrast to the above approaches, we will here investigate whether a rule based approach, based on the Tsetlin Machine, can forecast outbreaks surpassing a decision threshold, across the regions of the Philippines.

3 Methodology

3.1 The Tsetlin Machine Architecture

The TM addresses pattern classification problems where a class can be represented by a collection of sub-patterns, each fixing certain features to distinct values. The TM is designed to uncover these sub-patterns in an effective, yet relatively simple manner. In all brevity, the TM represents a class using a series of clauses. Each clause, in turn, captures a sub-pattern by means of a conjunction of literals, where a literal is a propositional variable or its negation. Each propositional variable takes the value *False* or *True* (in bit form, 0 or 1 respectively).

Let $\boldsymbol{X} = [x_1, x_2, x_3, \ldots, x_n]$ be a feature vector consisting of n propositional variables x_k with domain $\{0, 1\}$. Now suppose the pattern classification problem involves q outputs, and m sub-patterns per output that we need to recognize. Then the resulting pattern classification problem can be captured using $q \times m$ conjunctive clauses C_i^j, $1 \leq j \leq q$, $1 \leq i \leq m$. The output y^j, $1 \leq j \leq q$, of the classifier is given as:

$$C_i^j = 1 \land \left(\bigwedge_{k \in I_i^j} x_k \right) \land \left(\bigwedge_{k \in \bar{I}_i^j} \neg x_k \right). \tag{1}$$

$$y^j = \bigvee\nolimits_{i=1}^{m} C_i^j. \tag{2}$$

Above, I_i^j and $\bar{I}_i^j$ are non-overlapping subsets of the input variable indexes, $I_i^j, \bar{I}_i^j \subseteq \{1, ...n\}$, $I_i^j \cap \bar{I}_i^j = \emptyset$. The subsets decide which of the propositional variables take part in the clause, and whether they are negated or not.

In the TM, the disjunction operator is replaced by a summation operator to increase classification robustness [1]. The structure of the multiclass TM is depicted in Fig. 2b. The sub-figures in Fig. 2 illustrate the three phases of the classification process, i.e., (a) how a team of TAs forms a clause that processes the input features; (b) how a TM is composed by multiple TA teams; and (c)

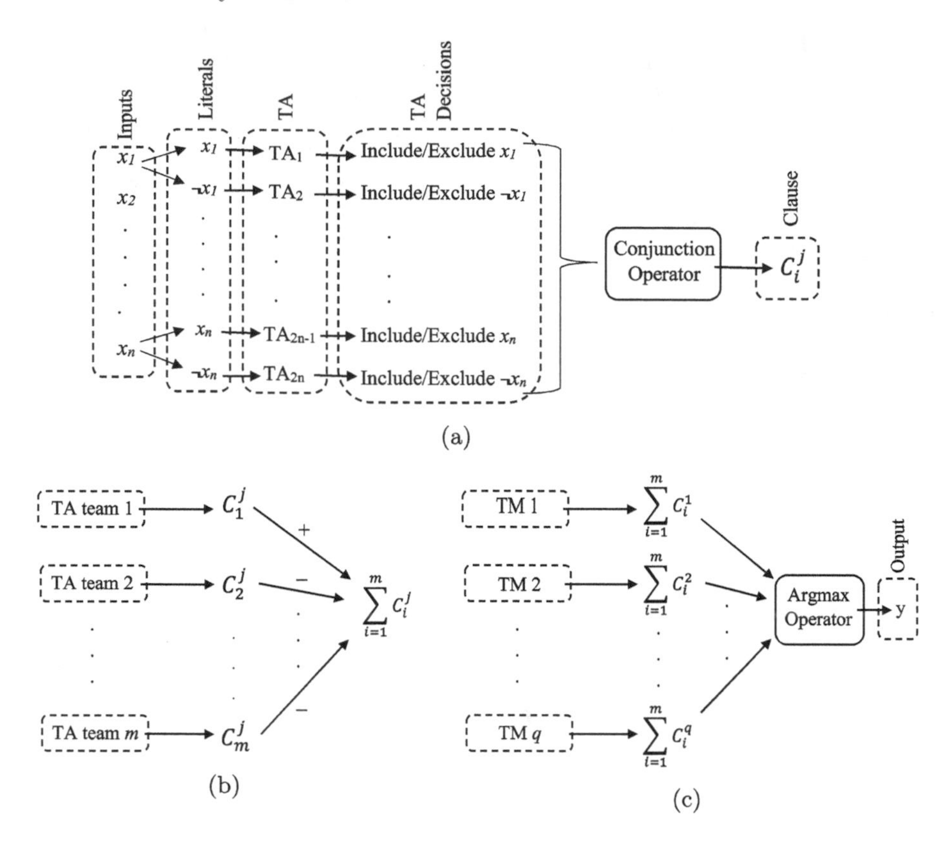

Fig. 2. (a) A TA team forms the clause C_i^j, $1 \leq j \leq q$, $1 \leq i \leq m$. (b) A TM. (c) A multiclass TM.

how a group of TMs are connected to handle multiclass classification problems. We will now detail these phases one by one.

The TA Team

Inputs and Literals. The TM takes n propositional variables $x_1, x_2, x_3, \ldots, x_n$ as input. For each variable x_k, there are two literals, the variable itself and its negation $\neg x_k$.

Tsetlin Automata and Their Decisions. For each clause C_i^j, each literal is assigned a unique TA. This TA decides whether to *include* or *exclude* its assigned literal in the given clause. Thus, for n input variables, we need $2n$ TA. This collective of TA is called a team. The TA team composes a conjunction of the literals that the team has chosen to be included. The conjunction outputs 1 if all of the included literals evaluate to 1, otherwise, the clause outputs 0.

The TM

Clauses and Their Role in a TM. A TM consists of m clauses, each associated with a TA team. The number of clauses needed for a particular class depends on the number of sub-patterns associated with the class. Each clause casts a vote, so that the m clauses jointly decide the output of the TM. Clauses with odd indexes are assigned positive polarity (+) and clauses with even indexes are assigned negative polarity (−). The summation operator aggregates the votes by subtracting the number of negative votes from the number of positive votes.

Note that clauses with positive polarity cast their votes to favor the decision that the input belongs to the class represented by the TM, whereas clauses with negative polarity vote for the input belonging to one of the other classes.

The Multiclass TM

Obtaining the Final Output. With multiple TMs we get a multiclass TM. As shown in Fig. 2c, the final decision is made by the argmax operator to classify the input data to the class that obtained the highest vote sum.

3.2 The TA Game and Orchestration Scheme

We organize learning in the TM as a game being played among the TAs. The Nash Equilibria of the game corresponds to the goal state of the TA, providing the final classifier. In the worst case, the single action of any TA has the power to disrupt the whole game. Therefore, the TAs must be guided carefully towards optimal pattern recognition.

To achieve this, the Tsetlin Machine is built around two kinds of feedback: Type I and Type II feedback. The Reward, Inaction, and Penalty probabilities under these two feedback types are summarized in Table 1, and they are determined based on the clause output (1 or 0), the literal value (1 or 0), and the current action of the TA (include or exclude). Rewards and Penalties are fed to the TA as normal. Inaction means that the state of the TA remains unchanged.

The training process of the TM thus contains several interacting mechanisms. To clarify their roles, we provide a flow chart for the complete procedure, shown in Fig. 3, and explored in the following.

Type I Feedback. As seen in the flowchart, briefly stated, Type I feedback is only activated when the actual output $\hat{y}$ is 1. When the output of the target clause also is 1, Type I Feedback has four roles:

- It reinforces true positive output by assigning a large reward probability $\frac{s-1}{s}$ to the action of including literals that evaluate to 1, and thus contributing to the result of the clause output being 1.
- Conversely, exclude actions are penalized with the same magnitude under these conditions. This is to "tighten" the clause, because it would still output 1 also when the literal considered is included instead.
- Furthermore, if the value of the literal is 0, excluding the literal is the way to go, and exclude actions are thus rewarded with probability $\frac{1}{s}$.

When the output of the clause is 0 (false negative output), Type I feedback has the following effect:

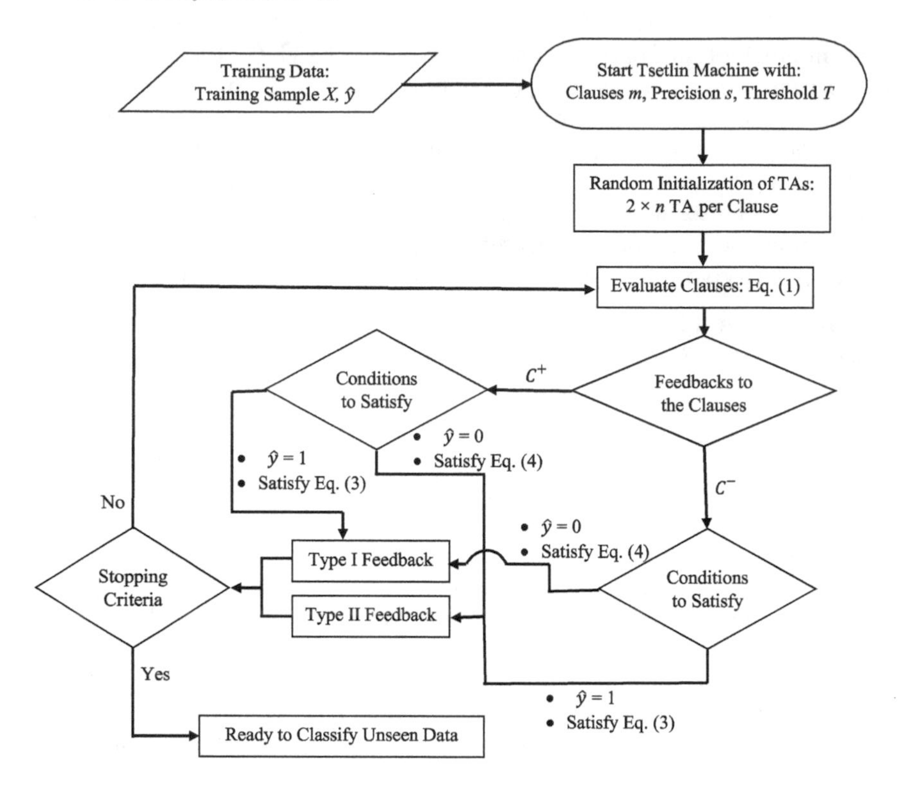

Fig. 3. The training work-flow.

- Type I feedback systematically penalizes include actions with probability $\frac{1}{s}$. Indeed, excluding literals is the only way to invert the output of a clause that outputs 0.
- When the action is exclude, this is rewarded with probability $\frac{1}{s}$, because reinforcing exclude actions will sooner or later invert the output of the clause to 1.

Thus, eventually, Type I feedback combats false negative output, and encourages true positive output.

Type II Feedback. Type II feedback is activated when the actual output $\hat{y}$ is 0, as shown in the flowchart. This type of feedback is designed to eliminate false positive output. That is, when the clause output should be 0, but the clause erroneously evaluates to 1, Type II feedback is triggered. In brief, repeated Type II feedback forces in the end the offending clause to evaluate to 0, simply by including a literal that has the value 0 into the clause (which makes the conjunction of literals evaluate to 0 as well). This is achieved by penalizing, with probability 1, exclude actions for literals that evaluate to 0.

Table 1. Type I and Type II feedback to battle against false negatives and false positives.

Feedback type			I				II			
Clause output			1		0		1		0	
Literal value			1	0	1	0	1	0	1	0
Current state	Include	Reward probability	(s − 1)/s	NA	0	0	0	NA	0	0
		Inaction probability	1/s	NA	(s − 1)/s	(s − 1)/s	1	NA	1	1
		Penalty probability	0	NA	1/s	1/s	0	NA	0	0
	Exclude	Reward probability	0	1/s	1/s	1/s	0	0	0	0
		Inaction probability	1/s	(s − 1)/s	(s − 1)/s	(s − 1)/s	1	0	1	1
		Penalty probability	(s − 1)/s	0	0	0	0	1	0	0

*s is the precision and controls the granularity of the sub-patterns in [1]

To summarize, Type I Feedback reinforces true positive output, while simultaneously reducing false negative output. These dynamics are countered by Type II Feedback, which systematically reduces false positive output.

The Clause Feedback Activation Function. In [1], an additional feedback mechanism is introduced, aiming at allocating the sparse pattern representation resources provided by the clauses as effectively as possible. This is achieved by introducing a target value T for the number of clauses voting from a specific pattern. The idea is to gradually reduce the frequency of feedback for a specific pattern, as the number of votes approaches T. In all brevity, the feedback activation function is basically an activation probability controlled by a Threshold T. The probability of activating Type I Feedback for a specific clause is:

$$\frac{T - max(-T, min(T, \sum_{i=1}^{m} C_i^j))}{2T} \tag{3}$$

For Type II Feedback, the probability is:

$$\frac{T + max(-T, min(T, \sum_{i=1}^{m} C_i^j))}{2T} \tag{4}$$

As seen, for Eq. (3), the activation probability decreases as the number of votes approaches T, and finally when T is reached, the probability becomes 0. Thus ultimately, Type I feedback will not be activated when enough clauses are producing the correct number of votes. This in turn "freezes" the affected clauses since TAs will no longer change state. The crucial point here is that this frees other clauses to seek other sub-patterns, because the "frozen" pattern is no longer attractive for the TA. The same rationale holds for Eq. (4) for Type II feedback. In this way, the pattern representation resources can be allocated more effectively.

3.3 Data Pre-processing

We now come to one of the main contributions of this paper, namely a scheme that allows the Tsetlin Machine to successfully recognize patterns consisting of

Table 2. Conversion of original input features into bits.

Raw data	Thresholds		
	≤3.834	≤5.779	≤10.008
5.779	0	1	1
10.008	0	0	1
5.779	0	1	1
3.834	1	1	1

continuous features, despite being constrained to an internal binary representation. As the TM only takes binary variables as input, we transform continuous features into binary form in a preprocessing step, detailed in the following.

Table 2 illustrates the transformation procedure, using one continuous feature as an example. The same procedure is repeated for each continuous feature in turn. First of all, all the unique values $\{v_1, v_2, \ldots, v_u\}$ of the continuous feature found in the dataset are identified. We consider each unique value v_w to be a potential threshold "$\leq v_w$". Thus each unique value provides a new derived binary feature: is the threshold condition fulfilled or not fulfilled for a particular continuous value v.

As an example, column 1 in Table 2 contains the values of the continues features. As seen, there are three unique values, and these provides three thresholds ≤3.834, ≤5.779, and ≤10.008. Accordingly, three new binary features are introduced, encoding the original raw continuous values, also shown in the table. If the raw continuous value is greater than the threshold, the corresponding bit in the binary form is assigned the value 0; and if the raw continuous value is less than or equal to the threshold, it is given the value 1. For example, in Table 2, for the first value 5.779, it is greater than the first threshold 3.834, so the corresponding binary feature is assigned the value 0 (in column 2). However, being equal to the threshold value of the second threshold 5.779, and less than the third threshold value 10.008, both column 3 and column 4 are assigned the value 1. Therefore, the final binary bits that represent 5.779 becomes 011. Similarly, 10.008 and 3.834 are represented by 001 and 111, respectively.

This new representation becomes particularly powerful due to the capability of the TM to negate features, allowing a clause to specify intervals for continuous features. In the following section, we evaluate this procedure both on an artificial dataset, as well as for the real-life application of forecasting dengue fever outbreaks in the Philippines.

4 Experiments

First, the behavior of the TM is studied using an artificial dataset. Actions chosen by TAs in clauses, clause outputs, and TM outputs, are extensively studied with this dataset. Then, the TM is applied to forecast the dengue outbreaks in the

Philippines. Data and the TM preparation for both tasks are discussed in the following subsections.

4.1 Behavior in Dealing with Artificial Data

4.1.1 Experimental Setup

The dataset consists of two inputs (integers, $0 \leq x_1 \leq 4$ and $0 \leq x_2 \leq 5$). If the sum of the inputs is equal to 9, they are assigned class 1 and the rest is assigned class 0. Since the features in this process are categorical, we use one-hot-encoding to convert them into bits instead of the procedure proposed in the previous section. Input x_1 takes one of five values (0, 1, 2, 3, 4) and input x_2 takes one of six values (0, 1, 2, 3, 4, 5). Therefore, these two features can be expressed using 5 and 6 bits, respectively. An example data sample converted to bits can be found in Table 3.

A Tsetlin Machine with 4 clauses is used to classify the artificial data. Since there are 11 input bits, 22 TAs are needed to form a clause. Each TA is given 100 states per action. Two of those four clauses will vote in favour of class 0. The other two clauses will vote in favour of class 1. The two remaining hyper parameters: Threshold and Precision are set to 1 and 8, respectively.

4.1.2 Behavior Analysis

During the training process, the states of the TAs in each clause are recorded and plotted in Fig. 4. Clause 1 and 3 have positive polarities and clause 2 and 4 have negative polarities. Clause 1 and 4 vote in favour of class 0 while clause 2 and 3 vote in favour of class 1.

The change in states of the TAs in clause 1 and 4 are more dynamic compared to the TAs in clauses 2 and 3. This is due to the much larger number of training samples that belong to class 0. Since more training samples belong to class 0 (~8/9 of the data) than class 1 (~1/9 of the data), clauses 1 and 4 receive feedback more frequently. As seen, the TAs in clauses 2 and 3 move slowly towards the *exclude* action (memory states from 0 to 100) since they receive Type II feedback more often (from class 0 data). However, once the TM is trained, it can classify all the 200 testing samples with 100% accuracy.

Table 3. Converting integer training samples to bits.

Original sample	x_1					x_2						*Out*
	3					5						8
Bit positions	0	1	2	3	4	0	1	2	3	4	5	
Sample in bits	0	0	0	1	0	0	0	0	0	1	0	0

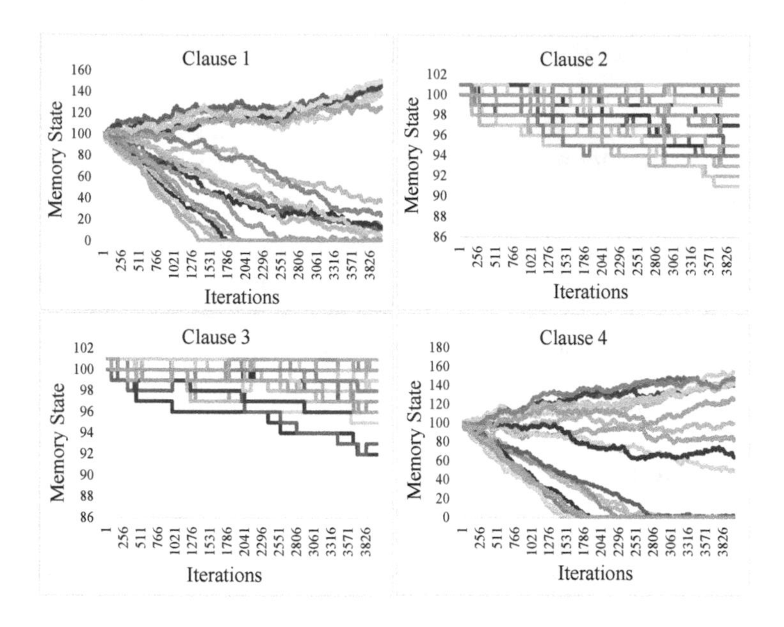

Fig. 4. Variation of actions of TA to classify artificial data.

4.2 Predicting Disease Outbreaks

4.2.1 Experimental Setup

The number of patients who suffer from dengue haemorrhagic fever or dengue shock syndrome has been increasing over the years. Therefore, dengue haemorrhagic fever is considered as an important public health issue, especially in tropical and subtropical countries. To control the mortality rate due to dengue fever, an early warning system, which helps directly on emergency preparedness and resource planning, is called for [24].

The Philippines has 17 administrative regions (from I to XVI with two IV regions; IVA and IVB). Department of Health in the Philippines has collected the number of monthly dengue incidences separately for all these regions from 2008 to 2016.

The number of monthly dengue incidences in most of the regions has been growing from 2008 and peaked in 2013. However, from 2013, the number of incidences has dropped to reach an average value of 9.22 patients per 100,000 population. Considering these trends, we decided to use more than 20 monthly dengue incidences per 100,000 population as an indication of outbreak. Using the data from 2008 to 2015, dengue outbreaks in the months of 2016 are to be predicted for all the regions.

In addition to the dengue incidences in the previous-month and previous-year-same-month of the same region, historical dengue incidences from the neighboring regions and total dengue incidences are considered as input features to forecast the dengue outbreaks. These regions and total dengue incidences are

Table 4. Regions that provide their data to forecast dengue incidences of their neighbors

Target	Selected regions	Target	Selected regions
I	II, III, IVA, XIV, Total	IX	X, XI, XII
II	I, III, IVA, XIV, Total	X	IX, XI, XII, XIV, XV
III	I, II, IVA, XVI	XI	IX, X, XII, XV
IVA	III, IVB, V, XVI, Total	XII	IX, X, XI, XV, Total
IVB	II, IVA, VI, Total	XIII	IX, X, XII, XV, Total
V	IVA, VI, Total	XIV	I, II, III, IVA, IVB, Total
VI	IVB, V, VII, XII, Total	XV	IX, X, XI, XII
XII	IVA, V, VI, XII, Total	XVI	I, III, IVA, V
VIII	V, VI		

selected based on their correlation to the target series. Dengue incidences in the previous-month of the selected regions and total dengue incidences are used as input features. The selected regions to forecast each region are summarized in Table 4.

Once the input features are determined, they are converted to bits using the procedure proposed in Sect. 3. Then they are fed into the TM to predicts dengue outbreaks in each region separately. Each TM has 2000 clauses and the associated TAs are given 100 states per action. The other two hyper parameters, Threshold and Precision, are set to 15 and 8, respectively.

4.2.2 Results

Possible dengue outbreaks in the Philippines for the year 2016 are forecasted by the TM. Results from the TM are compared with results from three other machine learning techniques: ANNs, a SVM, and a Decision Tree (DT). For comprehensiveness, four ANN architectures are used to forecast the dengue outbreaks: ANN-1 – one hidden layer with 5 neurons, ANN-2 – one hidden layer with 20 neurons, ANN-3 – three hidden layers with 20, 150, and 100 neurons, respectively, and ANN-4 – five hidden layers with 20, 200, 150, 100, and 50 neurons. The SVM uses a Radial Basis Function kernel to capture the non-linear patterns in the data. The regularization parameter (C) in this case is fixed at 1.0 with $gamma$ = 1/(the number of input features) to maximize prediction accuracy. The parameters which decide the quality of the DT output, such as maximum tree depth (=max), minimum number of samples required for split (=2), and the minimum number of samples required for a leaf node (=1) are again all adjusted to optimize prediction accuracy. Since there are 17 regions, all of the 204 testing samples are utilized to test the accuracy of each technique. These samples encompass 22 outbreaks to be identified. Each model is executed using 30-fold cross-validation to calculate precision, recall, F1-score, and accuracy. The means and the 95% confidence intervals of these scores can be found in Table 5.

Table 5. Summary of the forecasting outcomes by different models.

	TM	ANN-1	ANN-2	ANN-3	ANN-4	SVM	DT
Precision	0.44 ± 0.02	0.35 ± 0.02	0.39 ± 0.01	0.37 ± 0.02	0.36 ± 0.02	0.43 ± 0.01	0.29 ± 0.02
Recall	0.37 ± 0.02	0.23 ± 0.02	0.31 ± 0.02	0.36 ± 0.02	0.33 ± 0.03	0.14 ± 0.01	0.41 ± 0.02
F1-score	0.40 ± 0.01	0.28 ± 0.02	0.34 ± 0.02	0.36 ± 0.02	0.34 ± 0.03	0.21 ± 0.01	0.34 ± 0.01
Accuracy	0.88 ± 0.01	0.87 ± 0.01	0.87 ± 0.01	0.87 ± 0.02	0.87 ± 0.01	0.89 ± 0.01	0.83 ± 0.01

The TM obtains the highest mean values for precision and F1-score. The second highest precision (0.43) is obtained by the SVM, at the sacrifice of a much lower recall. Conversely, DT produces the highest recall, however, precision suffers. Considering overall performance, captured by the F1 score, the TM obtains the highest mean F1-score (0.40) while ANN-3 obtains the second highest mean F1-score (0.36). Even though mean F1 score peaks at 0.36, as a result of increasing the structural complexity of the ANNs, the score drops again when complexity is increased further. Due to the imbalance of the dataset (182 non-outbreaks and 22 outbreaks), the SVM produces a particularly high accuracy (0.89) by mostly classifying instances as non-outbreaks. Finally, note that both the mean values of precision, recall, F1-score, and accuracy of the TM are higher than what we were able to achieve with the ANN models.

5 Conclusion

In this paper, we proposed a feature pre-processing procedure for the TM so that it can effectively handle continuous input features. This opens up for promising applications in e.g. forecasting, where continuous features are typical. We applied the resulting TM approach to forecast dengue outbreaks in the Philippines, after performing an empirical study on an artificial dataset. While the experiments with the artificial dataset confirmed the desired properties of the new scheme, the results on the real-life dataset further demonstrated competitive performance also with respect to other machine learning approaches. Indeed, it turned out that the TM is more accurate than the evaluated SVMs, Decision Trees, and several multi-layered ANNs, both in terms of forecasting precision and F1-score.

In our further work, we intend to exploit this approach also in other pattern recognition domains where continuous features are dominant. We further intend to investigate how also the output of the TM can be rendered continuous.

References

1. Granmo, O.-C.: The Tsetlin Machine - a game theoretic bandit driven approach to optimal pattern recognition with propositional logic. arXiv:1804.01508
2. Narendra, K.S., Thathachar, M.A.: Learning Automata: An Introduction. Courier Corporation (2012)

3. Berge, G.T., Granmo, O.C., Tveit, T.O., Goodwin, M., Jiao, L., Matheussen, B.V.: Using the Tsetlin Machine to learn human-interpretable rules for high-accuracy text categorization with medical applications. arXiv:1809.04547, September 2018
4. Ogihara, H., Fujita, Y., Hamamoto, Y., Iizuka, N., Oka, M.: Classification based on boolean algebra and its application to the prediction of recurrence of liver cancer. In: 2013 2nd IAPR Asian Conference on Pattern Recognition (ACPR), pp. 838–841. IEEE (2013)
5. Santa Cruz, R., Fernando, B., Cherian, A., Gould, S.: Neural algebra of classifiers. arXiv:1801.08676 (2018)
6. Wang, T., Rudin, C., Doshi-Velez, F., Liu, Y., Klampfl, E., MacNeille, P.: A Bayesian framework for learning rule sets for interpretable classification. J. Mach. Learn. Res. **18**(1), 2357–2393 (2017)
7. Feldman, V.: Hardness of approximate two-level logic minimization and PAC learning with membership queries. J. Comput. Syst. Sci. **75**(1), 13–26 (2009)
8. Klivans, A.R., Servedio, R.A.: Learning DNF in time $2^{O(n^{1/3})}$. J. Comput. Syst. Sci. **68**(2), 303–318 (2004)
9. Feldman, V.: Learning DNF expressions from fourier spectrum. In: Conference on Learning Theory, pp. 17–1 (2012)
10. Valiant, L.G.: A theory of the learnable. Commun. ACM **27**(11), 1134–1142 (1984)
11. Hauser, J.R., Toubia, O., Evgeniou, T., Befurt, R., Dzyabura, D.: Disjunctions of conjunctions, cognitive simplicity, and consideration sets. J. Mark. Res. **47**(3), 485–496 (2010)
12. Rudin, C., Letham, B., Madigan, D.: Learning theory analysis for association rules and sequential event prediction. J. Mach. Learn. Res. **14**(1), 3441–3492 (2013)
13. McCormick, T., Rudin, C., Madigan, D.: A hierarchical model for association rule mining of sequential events: an approach to automated medical symptom prediction. Annals of Applied Statistics (2011)
14. Granmo, O.-C., Oommen, B.J.: Solving stochastic nonlinear resource allocation problems using a hierarchy of twofold resource allocation automata. IEEE Trans. Comput. **59**, 545–560 (2010)
15. Oommen, B.J., Kim, S.-W., Samuel, M.T., Granmo, O.-C.: A solution to the stochastic point location problem in metalevel nonstationary environments. IEEE Trans. Syst. Man, Cybern. Part B (Cybern.) **38**(2), 466–476 (2008)
16. Tung, B., Kleinrock, L.: Using finite state automata to produce self-optimization and self-control. IEEE Trans. Parallel Distrib. Syst. **7**(4), 439–448 (1996)
17. Bouhmala, N., Granmo, O.-C.: Stochastic learning for SAT-encoded graph coloring problems. Int. J. Appl. Metaheuristic Comput. (IJAMC) **1**(3), 1–19 (2010)
18. Abeyrathna, K., Granmo, O.-C., Goodwin, M.: A novel Tsetlin Automata Scheme to Forecast Dengue Outbreaks in the Philippines. In: 2018 IEEE 30th International Conference on Tools with Artificial Intelligence (ICTAI) (2018)
19. Gharbi, M., et al.: Time series analysis of dengue incidence in Guadeloupe, French West Indies: forecasting models using climate variables as predictors. BMC Infect. Dis. **11**(1), 166 (2011)
20. Choudhury, Z.M., Banu, S., Islam, A.M.: Forecasting dengue incidence in Dhaka, Bangladesh: a time series analysis. Dengue Bull. **32**, 29–37 (2008)
21. Luz, P.M., Mendes, B.V., Codeço, C.T., Struchiner, C.J., Galvani, A.P.: Time series analysis of dengue incidence in Rio de Janeiro, Brazil. Am. J. Trop. Med. Hyg. **79**(6), 933–939 (2008)
22. Silawan, T., Singhasivanon, P., Kaewkungwal, J., Nimmanitya, S., Su-wonkerd, W.: Temporal patterns and forecast of dengue infection in Northeastern Thailand. Southeast Asian J. Trop. Med. Public Health **39**(1), 90 (2008)

23. Promprou, S., Jaroensutasinee, M., Jaroensutasinee, K.: Forecasting dengue hemorrhagic fever cases in Southern Thailand Using ARIMA Models. Dengue Bull. **30**, 99–106 (2006)
24. Phung, D., et al.: Identification of the prediction model for dengue incidence in Can Tho City, a Mekong Delta Area in Vietnam. Acta Trop. **141**, 88–96 (2015)
25. Abeyrathna, K., Granmo, O.C., Goodwin, M.: Effect of data from neighbouring regions to forecast dengue incidences in different regions of philippines using artificial neural networks. Norsk Informatikkonferanse (2018)

CEVM: Constrained Evidential Vocabulary Maintenance Policy for CBR Systems

Safa Ben Ayed[1,2](✉), Zied Elouedi[1], and Eric Lefevre[2]

[1] Institut Supérieur de Gestion de Tunis, LARODEC, Université de Tunis, Tunis, Tunisia
safa.ben.ayed@hotmail.fr, zied.elouedi@gmx.fr
[2] Univ. Artois, EA 3926,
Laboratoire de Génie Informatique et d'Automatique de l'Artois (LGI2A), 62400 Béthune, France
eric.lefevre@univ-artois.fr

Abstract. The maintenance of Case-Based Reasoning (CBR) systems has attracted increasing interest within current research since they proved high-quality results in different real-world domains. This kind of systems stores previous experiences, which are described by a vocabulary (e.g., attributes), incrementally in a case base. Actually, the vocabulary presents one among the most important maintenance targets, since it highly contributes in providing accurate solutions and in improving systems' performance, especially within high-dimensional domains. However, there is no policy, in the literature, that offers the ability to exploit prior knowledge (e.g., given by domain-experts) during the maintenance of features describing cases. In this paper, we propose a flexible policy for the most relevant attribute selection based on the attribute clustering concept. This new policy is able, on the one hand, to manage uncertainty using the belief function theory based tools, and on the other hand, to make use of domain-experts knowledge in form of pairwise constraints: If two attributes offer the same information without any added-value, then a *Must-link* constraint between them is generated. Otherwise, if there is no relation between them and they offer different information, then a *Cannot-link* constraint between them is created.

Keywords: Case-Based Reasoning · Vocabulary maintenance · Belief function theory · Uncertainty · Constrained attribute clustering

1 Introduction

Case-Based Reasoning is a methodology of problem-solving that recalls past experiences to solve new problems. It is mainly based on the hypothesis that *similar problems have similar solutions* with offering the possibility to make some adaptations to the provided solution in order to perfectly match the new

F. Wotawa et al. (Eds.): IEA/AIE 2019, LNAI 11606, pp. 579–592, 2019.
https://doi.org/10.1007/978-3-030-22999-3_50

problem's characterizations. After the revision of every provided solution, the problem-solution couple is retained as a new case within the Case Base (CB) [1]. Over the last three decades, CBR systems have been more and more utilized and applied in several areas such as medicine [2], finance [3], and ecology [4]. Obviously, this can only indicate its strength, success and adequacy even with weak-understandable domains. Since CBR systems are now commercially used, the need of their maintenance presents a key issue for overtime success. Hence, more and more research focus on maintaining the different knowledge containers of CBR systems. Actually, there are four knowledge containers, as defined in [5], that may be maintained [6] within a CBR system: *(1)* Case Base, *(2)* Adaptation, *(3)* Similarity, and *(4)* Vocabulary. Obviously, the CB is the elementary container of any CBR system, that several research aimed to maintain [7]. However, the vocabulary container also presents one among the most important maintenance targets, since it can seen as the basis of the different CBR's steps to offer solutions. For Structural CBR systems (SCBR), we can restrict the vocabulary knowledge to the set of attributes describing cases. By this way, to maintain vocabulary for CBR systems, we are faced to two main challenges: First, the elimination of redundant attributes in order to improve CBR systems performance, especially within large-scale domains. Second, the removal of noisy attributes so as to help the CBR system to be conducted to the most accurate solution. To tackle these challenges in the best way possible, a crucial need of uncertainty management within CBR systems knowledge arises. In fact, cases stored in every CB involve real-world experiences which are never exact. Hence, they cause ignorance and overlapping data regions during learning. This uncertainty within knowledge is managed only in some research, in the literature, that focus on maintaining CBR systems vocabulary via the automatic analysis of their content. However, these works suffer from their disability to aid their automatic maintaining mechanism when prior knowledge regarding attributes, which can be provided by domain experts, are available. This limitation can be tackled through *semi-supervised* learning of features, more precisely the *semi-supervised clustering*. It consists, in our settings, at using the pairwise *Must-link* and *Cannot-link* constraints on some instances to help the used unsupervised attributes clustering. Since we intend to learn on features, *Must-link* constraint[1] between two features is generated when a prior knowledge affirms that they offer almost the same information. On the contrary, a *Cannot-link* constraint[2] is created when prior knowledge is available to affirm that there is no relation between them. Based on these ideas, we build our new vocabulary maintenance policy named CEVM, for "Constrained Evidential Vocabulary Maintenance policy for CBR systems", which manages uncertainty within the framework of belief function theory [14,15], and allows the exploitation of experts knowledge, related to attributes relations, using the constrained evidential dissimilarity-based clustering technique called CEVCLUS [16].

[1] Two attributes are surely belonging to the same cluster.

[2] Two attributes cannot belong to the same cluster.

The remaining of this paper is organized as follows. Section 2 is dedicated to define the vocabulary as a maintenance target in CBR systems with explaining our motivation. The related background on the belief function theory is presented in Sect. 3. Throughout Sect. 4, we detail our new established policy aiming at selecting only the most relevant attributes for cases description. We show our experimental study as well as our proposed modes for artificial constraints generation in Sect. 5. Finally, the Sect. 6 is dedicated for the conclusion.

2 Vocabulary Maintenance for CBR Systems

Obviously, CBR systems are made to operate for a long period of time. However, the change of the context along with the incremental learning through experiences give rise to the need of maintaining the vocabulary that describes cases.

2.1 The Vocabulary as a Knowledge Container in a CBR System

The vocabulary knowledge is presented and modeled in [5] as the basis of all the other three knowledge containers. In fact, its definition depends mainly on the knowledge source's nature. In this paper, we focus on attribute-value data. However, in non-structural CBR system, more sophisticated methods may be included within the vocabulary container. For our current purpose, we restrict the vocabulary knowledge container to the set of attributes.

2.2 The Vocabulary as a Maintenance Target

Every encountered experience in our real life can be described with an infinite number of features. However, only some of them are useful to provide the accurate solution for one problem. As already mentioned, there are basically two types of attributes that should be removed to maintain cases' vocabulary. On the one hand, the set of noisy attributes that their removal from the vocabulary conducts to the improvement of the CBR system's decision making. On the other hand, the set of redundant attributes that we define by the ensemble of high correlated features. Actually, we call them redundant since they offer the same information, and the removal of one of them does not affect the whole CBR system's competence in solving new problems, but it may improve its performance in term of response time. Within the same road, some works, such in [8–10], target the vocabulary of CBR systems for maintenance. They are mainly based on selecting the most relevant features, where we cite, for instance, the ReliefF method [11] as one among the baselines of features selection methods. However, existing policies suffer from some weaknesses towards the concepts shown in the following Subsections, where we present our motivation.

2.3 Attribute Clustering and Uncertainty Management During Vocabulary Maintenance

Regrouping attributes according to some proximity data between them can be reached through the attribute clustering concept [12,13]. Actually, applying this concept during maintaining vocabulary leads to preserve relations between features and offers a high amount of flexibility to the CBR framework, where we can substitute every attribute by any other one belonging to the same cluster. Similarly to object clustering, we can consider the set of attributes as the set of objects, and regroup them in such a way that features belong to the same cluster are somehow similar. In contrast, attributes belonging to different clusters are dissimilar. However, uncertainty within attributes clustering has to be managed since attributes-values refer to the description of real-world experiences that are full of uncertainty, vagueness, and imprecision. Further, as mentioned in [18], the vocabulary presents one among the origins of uncertainty in CBR framework. That's why, we make use of one among the most powerful tools for this matter called the belief function theory [14,15], where its basic concepts are shown in Sect. 3.

2.4 Exploiting Prior Knowledge During Vocabulary Maintenance

Usually, research that are interested on knowledge extraction learn via the automatic analysis of available data content without giving the possibility to domain-experts or available prior knowledge to intervene inside this process. Within Case-Based Reasoning framework, systems are generally solving problems within some specific domain, where its experts may provide knowledge that are expensively extracted by machine learning methods. Consequently, it is greatly useful to aid the automatic maintenance process through the exploitation of prior knowledge in form of Must-Link and Cannot-Link constraints. This can be done, for instance, inside a constrained machine learning technique.

3 Belief Function Theory

To handle uncertainty during the decision making process, we use the belief function theory [14,15], called also Dempster-Shafer theory or Evidence theory, which is a powerful mathematical framework used to deal with partial and unreliable information in many fields. We show, during this Section, the fundamental concepts of this theory as well as the evidential clustering and the credal partition concepts.

3.1 Fundamental Concepts

A belief function model is originally defined by a discrete and finite set of elementary events called the frame of discernment Θ of the problem taken into account. The set 2^{Θ} is called the power set and contains all the possible subsets of Θ. The basic belief mass (*bbm*) m^{Θ} is a mapping function from 2^{Θ} to $[0, 1]$ that assigns to every subset A of Θ a degree of belief reflecting the partial knowledge taken by a variable y defined on Θ, and verifies the constraint $\sum_{A \subseteq \Theta} m(A) = 1$. A mass function m is normalized if $m(\emptyset) = 0$. On the opposite case, the interpretation of the mass assigned to the empty set partition consists at measuring the degree of belief towards the hypothesis saying that y does not belong to Θ. This amount of belief can be useful in clustering to identify noises [16]. From a given mass function m, the plausibility function is defined, to measure the maximum amount of belief supporting the different subsets in Θ, as follows:

$$pl(A) = \sum_{B \cap A \neq \emptyset} m(B) \qquad \forall A \subseteq \Theta \tag{1}$$

Given two bbms m_1 and m_2, defined in the same frame of discernment Θ, the following Equation, proposed in [15], presents one among the most known measurements that aim to quantify the degree of conflict between them such that:

$$\kappa = \sum_{A \cap B} m_1(A)\, m_2(B) \tag{2}$$

Authors in [17] proved that if two bbms represent evidence regarding two distinct questions and defined in the same frame Θ, then the plausibility that they acquire the same answer is equal to $1 - \kappa$.

3.2 Evidential Clustering and Credal Partition

We call *Evidential Clustering* the task of regrouping objects[3], according to some attribute-based/dissimilarity-based data, within the frame of belief function theory. In an evidential clustering context, the frame of discernment Θ defines the set of a finite number K of clusters. Besides, the uncertainty regarding the membership of an object o_i to the different clusters is modeled by a bbm m_i on Θ. If we have n objects, the credal partition is, therefore, the n-tuple composed by n mass functions, such that $M = (m_1, ..., m_n)$ [17]. Generally, M is generated after applying an evidential clustering technique to regroup a set of objects according to their similarity while managing the uncertainty in their membership to all the possible partitions of clusters. Since it quantifies uncertainty in a power set space, the credal partition is more general than hard and soft partitions. Nevertheless, it can be converted to any one of these types [16,17]. After generating the credal partition, the decision about the membership may regard the cluster having the highest pignistic probability, which is defined as follows:

$$BetP(\omega) = \sum_{\omega \in A} \frac{m(A)}{|A|} \qquad \forall \omega \in \Theta \tag{3}$$

[3] In our context, these objects represent the set of features that describe cases.

In the case of non normalized mass functions, a preprocessing step of normalization for every bbm should beforehand be applied as follows:

$$m_*(A) = \begin{cases} \dfrac{m(A)}{1 - m(\emptyset)} & if\ A \neq \emptyset \\ 0 & Otherwise \end{cases} \tag{4}$$

After presenting the essential background as well as our motivation, we move on now at detailing our contribution for this paper.

4 Maintaining Vocabulary Through Evidential Constrained Attribute Clustering

At the aim of performing a high-quality attribute selection within a CBR system, our new Constrained Evidential Vocabulary Maintenance policy for CBR systems (CEVM) goes through three main steps, as shown in Fig. 1. It consists, first of all, at generating some dissimilarity data, from the CB, between attributes based on the correlation between their values. Second, CEVM regroups the set of attributes using their dissimilarities and with taking advantage of prior knowledge. After managing uncertainty and generating the credal partition by allowing every attribute to belong to all the partitions of clusters with a degree of belief, we make decision about their membership along with removing noisy and redundant features. More details are given during the three following Subsections.

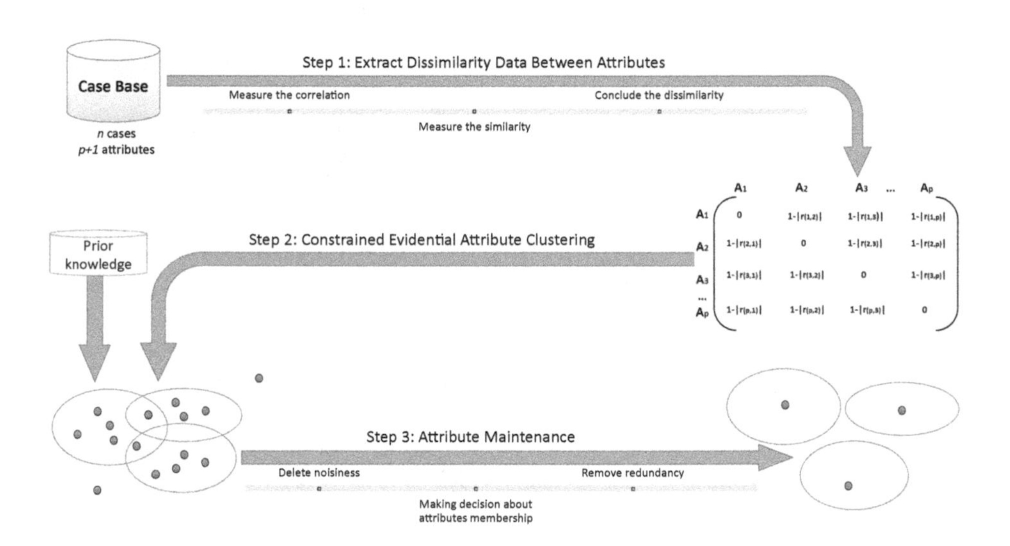

Fig. 1. Steps and substeps of CEVM policy

4.1 Step 1: Extracting Attributes Dissimilarity Data

The notion of dissimilarity between attributes can be defined, according to the context into account, in term of dependency, correlation, etc. To generate the dissimilarity between attributes, three substeps are followed by our new CEVM policy.

1. *Correlation between attributes*: In our context, the origins of dissimilarity data between attributes are generated through measuring the correlation between their values. The idea is that if two attributes are highly correlated, then they offer the same information for solving problems. We use the Pearson's Correlation Coefficient [19] so as to measure the linear association between the different values a_i and b_i of attributes A and B respectively, as follows:

$$r_{AB} = \frac{\sum_{i=1}^{n}(a_i - \overline{a})(b_i - \overline{b})}{\sqrt{\sum_{i=1}^{n}(a_i - \overline{a})^2}\ \sqrt{\sum_{i=1}^{n}(b_i - \overline{b})^2}} \tag{5}$$

   where $\overline{a}$ and $\overline{b}$ are the mean values of features A and B respectively.
   The set of correlations between every two features A and B gives rise to a square relational matrix defined as $R = (r_{AB})$.
2. *Similarity between attributes*: All the correlation values in R are bounded between -1 and 1 [19]. If $r_{AB} \simeq -1$, then there is a high negative correlation and a high similarity between A and B since they offer the same information. Similarly, if $r_{AB} \sim 1$, then there is a high positive correlation and a high similarity between A and B since they offer the same information for learning. However, if $r_{AB} \simeq 0$, then there is no correlation between them, which makes A and B completely dissimilar. As an intuitive consequence, we create the square similarity matrix $S = (s_{AB})$ such as:

$$s_{AB} = |r_{AB}| \tag{6}$$

3. *Attributes dissimilarity data*: After measuring the similarity between features, it is straightforward to compute the square dissimilarity matrix $D = (d_{AB})$ such that:

$$d_{AB} = 1 - s_{AB} \tag{7}$$

   By this way, values in D are also in the interval $[0, 1]$.

4.2 Step 2: Constrained Evidential Attribute Clustering

When we have some background knowledge, it is so gainful to use them throughout learning. Actually, this is the main principle of semi-supervised learning. This step, that aims at regrouping features according to their similarity, is very important to reach two other objectives during vocabulary maintenance. First, managing the uncertainty in attributes membership to clusters from the complete ignorance to the total certainty using the belief function framework. Secondly, exploiting the prior available knowledge supplied, for instance, by the experts

of domain in which the CBR is applied. We used a constrained evidential clustering method based on dissimilarity data between objects[4] called Constrained EVidential CLUStering (CEVCLUS) [16]. It is a variant of EVCLUS [17] that is characterized by its ability to taking into account a prior knowledge in form of two pairwise constraints: The Must-link (*ML*) constraint which concerns two attributes that belong for sure to the same cluster, and the Cannot-Link (*CL*) constraint that concerns the pair of attributes that are known to belong to distinct clusters.
Given m_i and m_j two bbms regarding cluster-membership of attributes A_i and A_j respectively, let $pl_{ij}(\Theta_{ij})$ refers to their plausibility to belong to the same cluster, and $pl_{ij}(\overline{\Theta}_{ij})$ refers to the plausibility of the complementary event. They can be calculated as follows [16]:

$$pl_{ij}(\Theta_{ij}) = 1 - \kappa_{ij} \tag{8a}$$

$$pl_{ij}(\overline{\Theta}_{ij}) = 1 - m_i(\emptyset) - m_j(\emptyset) + m_i(\emptyset)\, m_j(\emptyset) - \sum_{k=1}^{K} m_i(\{\omega_k\}) m_j(\{\omega_k\}) \tag{8b}$$

For the sake of clarity regarding the calculation of this plausibility, let mention that it consists at placing ourselves in the Cartesian product $\Theta^2 = \Theta \times \Theta$ and combining the two vacuous extensions of m_i and m_j [17]. If the resulted combination is denoted by m_{ij}, then pl_{ij} can be computed through m_{ij} using Eq. 1.

To construct the credal partition M, the non-constrained EVCLUS [17] algorithm minimizes a stress function, using a gradient based algorithm, similar to:

$$J(M) = \eta \sum_{i<j} (\kappa_{ij} - \delta_{ij})^2 \tag{9}$$

where $\eta = (\sum_{i<j} \delta_{ij}^2)^{-1}$, and $\delta_{ij} = \varphi(d_{ij})$, with φ is an increasing function such as $\varphi(d) = 1 - exp(-\gamma d^2)$. γ can be calculated as $-log\alpha/d_0^2$, with a recommendation to fix α to 0.05 and d_0, which determines the size of each class, can be set to some quantile of the dissimilarities in D.

The principle of the previous stress function is explained by Eq. 8a. It means that if two attributes are too far in term of distance, then they should have a low plausibility to belong to the same cluster, and a large degree of conflict. In our context, if we have prior knowledge that attributes A_i and A_j surely belong to different clusters, then the constraints $pl_{ij}(\overline{\Theta}_{ij}) = 1$ and $pl_{ij}(\Theta_{ij}) = 0$ are imposed. In contrast, if prior knowledge affirm that they belong to the same cluster, then the constraints $pl_{ij}(\overline{\Theta}_{ij}) = 0$ and $pl_{ij}(\Theta_{ij}) = 1$ are created. By this way, the CEVCLUS algorithm minimizes, using an iterative gradient-based optimization procedure, the following cost function composed by the sum of EVCLUS's stress function [17] and a penalization term:

[4] In our policy, it concerns dissimilarity data between attributes, which are supplied from the previous step.

$$J_C(M) = stress + \frac{\xi}{2(|ML| + |CL|)}(J_{ML} + J_{CL}), \tag{10}$$

with

$$J_{ML} = \sum_{(i,j)\in ML} pl_{ij}(\overline{\Theta}_{ij}) + 1 - pl_{ij}(\Theta_{ij}), \tag{11a}$$

$$J_{CL} = \sum_{(i,j)\in CL} pl_{ij}(\Theta_{ij}) + 1 - pl_{ij}(\overline{\Theta}_{ij}), \tag{11b}$$

where ξ presents the hyper-parameter aiming at arbitrating between the stress function and the constraints.

4.3 Step 3: Attribute Maintenance

Ultimately, we reach our purpose for cases' vocabulary maintenance through removing noisy and redundant features, and keeping only those that are unique and represent the different generated clusters during the previous step. As shown in Fig. 1, this step is composed by the three following substeps:

1. Removing noisy attributes: Since the previous applied clustering method devotes the empty set partition for noisiness allocation, we eliminate attributes characterized by a high belief's assignment to the empty set partition, such that:

$$A_i \in NA \;\; iff \;\; m_i(\emptyset) > \sum_{B_j \subseteq \Theta, B_j \neq \emptyset} m_i(B_j) \tag{12}$$

   where A_i represents the attribute i, and NA presents the set of all the noisy attributes.

2. Making decision about attributes membership to clusters through the highest pignistic probability value, using Eq. 3.
3. Removing redundancy by keeping only one representative attribute for each cluster. This idea gives an amount of flexibility to our policy towards CBR framework: If there is a problem in selecting one representative attribute, then we can re-select and re-flag any other attribute from the same cluster.

Example 1. Let consider some CB's vocabulary described by four attributes A_1, A_2, A_3, and A_4. Let us suppose, now, that the frame of discernment contains two clusters ($\Theta = \{cluster_1, cluster_2\}$), and the values of the credal partition $M = [m_1; m_2; m_3; m_4]$ are given by the previous step as shown in Table 1. First, we note, from Table 1, that $m_2(\emptyset)$ is higher than $m_2(\{cluster_1\})+ m_2(\{cluster_2\})+m_2(\{cluster_1, cluster_2\})$. Then, according to Eq. 12, we flag A_2 as a noisy attribute ($A_2 \in NA$). Consequently, we update the CB's vocabulary by removing the second attribute A_2. Then, we make decision about attributes membership to clusters using BetP defined in Eq. 3. Their corresponding pignistic probability values are shown in Table 2, from which we can conclude that A_1 belongs to $cluster_1$, and A_3 and A_4 belong to $cluster_2$. Finally, we keep only the attribute A_1 as representative of $cluster_1$ and the attribute A_3 as representative of $cluster_2$ to describe the new maintained case base vocabulary.

Table 1. Example of credal partition values

M	$\emptyset$	$\{cluster_1\}$	$\{cluster_2\}$	$\{cluster_1, cluster_2\}$
m_1	0.05	0.75	0.15	0.05
m_2	0.65	0.1	0.1	0.15
m_3	0.1	0.05	0.8	0.05
m_4	0.2	0.1	0.5	0.2

Table 2. Pignistic probability transformation values

	$cluster_1$	$cluster_2$
$\boldsymbol{BetP_1}$	**0.8158**	**0.1842**
~~$BetP_2$~~	~~0.5~~	~~0.5~~
$\boldsymbol{BetP_3}$	**0.0833**	**0.9167**
$BetP_4$	0.25	0.75

5 Experimental Analysis Using Artificial Constraints

Throughout this Section, we establish our experimentation and validate our contribution by developing two variants of our CEVM policy that differ by their way in generating artificial constraints[5].

5.1 Constraints Generation Strategy

Two main modes for *Must-link* and *Cannot-link* constraints generation, such in [7], are build during our experimental analysis:

- Batch mode for constraints generation (CEVM_{bat}): It consists at generating simultaneously a number t of constraints (*Must-link* and *Cannot-link*). For instance, we took t equal to 25% of the total number of attributes. We store the list of these constraints in $listConst$. The activity diagram of CEVM_{bat} is shown in Fig. 2.

[5] Calling domain-experts to generate constraints presents one among our perspectives.

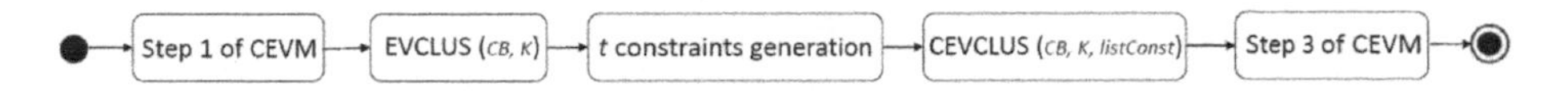

Fig. 2. Activity diagram for batch mode constraints generation

- Alternated mode for constraints generation ($CEVM_{alt}$): It consists at alternating between generating one constraint (*Must-link* or *Cannot-link*) and learning, with storing each one incrementally in *listConst*. Similarly, the number of constraints t is taken equal to $\#attributes \times 25/100$. Its activity diagram is shown in Fig. 3.

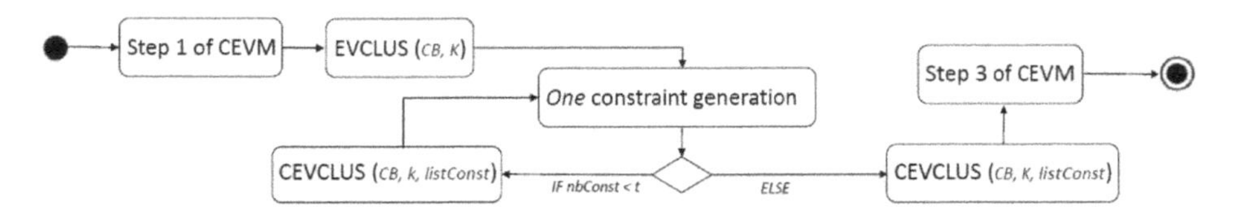

Fig. 3. Activity diagram for alternate mode constraints generation

How we generate a constraint? Actually, we generate artificially a pairwise constraint by handling the uncertainty offered by the credal partition and the pignistic probability transformation (Eq. 3). The idea consists at randomly picking two attributes A_i and A_j and behaving according to the three following situations that may arise:

1. If $\exists$ a cluster $\omega / BetP_i(\omega) > Thresh$ and $BetP_j(\omega) > Thresh$, then generate a *Must-link* constraint between the attributes A_i and A_j.
2. If $\forall$ clusters $\omega_k / |BetP_i(\omega_k) - BetP_j(\omega_k)| > Thresh$, then generate a *Cannot-link* constraint between the attributes A_i and A_j.
3. Else, go back to randomly picking two attributes.

where $Thresh$ is a threshold that aims to answer to the question: *"From which amount of membership certainty in* $[0, 1]$, we consider that the attributes A_i and A_j *belong or not to the same cluster?"*[6].

5.2 Data, Evaluation Criteria, and Experimental Settings

Our new CEVM policy has been implemented using *Matlab R2015a* and the default values for the different CEVCLUS [16] method's parameters have been taken. CEVM with its two variants have been tested on six data sets from U.C.I Repository[7] where the set of attributes are considered as the vocabulary describing cases' problems, and their classes refer to cases' solutions.

[6] During the experimentation, different values to set $Thresh$ have been tested. The best results are offered with $Thresh = 0.55$.

[7] Sonar (SN), Ionosphere (IO), Glass (GL), BreastCancer (BC), German (GR), and Heart (HR): https://archive.ics.uci.edu/ml/.

In order to assess the efficiency of our new vocabulary maintenance policy, we use the two following evaluation criteria:

- The Percentage of Correct Classifications (PCC), which is defined as follows:

$$PCC(\%) = \frac{\#\ Correct\ classifications}{\#\ Total\ classifications} \times 100 \quad (13)$$

  The PCC criterion refers to the competence of CBR systems in solving new problems.
- The Retrieval Time (RT) Criterion, which measures the time spent to offer all the solutions for the different cases instances. It may refer to the performance of CBR systems.

To solve cases' problems, we use the K-Nearest Neighbor (K-NN) since it presents one among the most used machine learning techniques within the CBR framework. We choose to apply *3-NN* so as to avoid the effect of noisy cases during learning. Hence, the RT criterion is exerted around that *3-NN* method. To offer the final results towards the PCC, the 10-fold cross validation technique is used.

5.3 Results and Discussion

In our comparative study, as shown in Table 3, we present results offered by five different sources where two among them present the two variants of our contribution for this paper (CEVM_{bat} and CEVM_{alt}), and the three others present the non maintained case base (Original-CBR), the reliefF method [11] for feature selection (ReliefF-CBR), and the non-constrained vocabulary maintenance policy EvAttClus [8]. Results in Table 3 are offered after varying the number of clusters, or the number of the selected attributes K, from 3 to 9. The most convenient K for every method and every data set is chosen[8].

In term of accuracy, both of our two variants CEVM_{bat} and CEVM_{alt} offer good results comparing to the other methods as well as to the original non-maintained case bases (Original-CBR). We note that the alternate mode for constraints generation is more efficient than the batch mode. We can conclude, furthermore, that better results may be offered if we resort to domain experts. In our context, both of CEVM's variants, that are able to make use of prior knowledge in form of constraints, have been able to maintain all the tested CBs' vocabulary with preserving or even improving their competence in solving problems. For example, they offer PCCs equal to 88.79% and 91.34% for "Glass" data set, where ReliefF-CBR and AttEvClus methods offer PCCs equal to 87.38% and 87.98% respectively. These results can only be explained by CEVM strategy's

[8] ReliefF-CBR: GL and GR ($K = 7$); BC ($K = 8$); SN, IO, and HR ($K = 9$); EvAttClus: IO ($K = 3$); HR ($K = 5$); BC and GR ($K = 8$); SN and GL ($K = 9$); CEVM_{bat}: IO ($K = 3$); HR ($K = 4$); BC and GR ($K = 8$); GL and SN ($K = 9$); CEVM_{alt}: IO ($K = 3$); HR ($K = 4$); GR ($K = 6$); GL ($K = 7$); SN and BC ($K = 8$).

Table 3. Accuracy and retrieval time evaluation

	CB	Original-CBR		ReliefF-CBR		AttEvClus		$CEVM_{bat}$		$CEVM_{alt}$	
		PCC(%)	RT(s)	PCC(%)	RT(s)	PCC(%)	RT(s)	PCC(%)	RT(s)	PCC(%)	RT(s)
1	SN	73.07	0.0642	71.15	0.0078	74.51	0.0082	74.51	0.0081	**75.12**	0.0079
2	IO	86.04	0.0223	84.90	0.0121	**88.03**	0.0085	**88.03**	0.0082	**88.03**	0.0087
3	GL	88.79	0.0141	87.38	0.0089	87.98	0.0093	88.79	0.0092	**91.34**	0.0081
4	BC	96.04	0.0199	96.45	0.0097	**96.63**	0.0097	**96.63**	0.0098	**96.63**	0.0099
5	GR	70.60	0.0319	69.60	0.0119	71.21	0.0122	71.21	0.0123	**73.25**	0.0121
6	HR	56.80	0.0276	59.86	0.0089	60.88	0.0081	**62.78**	0.0078	**62.78**	0.0071

efficiency in detecting noisy and redundant features. In term of retrieval time, we note very competitive results offered by the four vocabulary maintaining policies. However, a slightly higher difference in RT are noted towards Original-CBR. For instance, "Sonar" data set (60 attributes), moved from RT = 0.0642 s to RT = 0.0079 s with $CEVM_{alt}$.

6 Conclusion

In this paper, a new vocabulary maintenance method for CBR systems, called CEVM, with two modes for artificial constraints generation (batch and alternate mode), are proposed. In order to aid its automatic maintaining process, the proposed policies $CEVM_{bat}$ and $CEVM_{alt}$ offer an ability to exploit prior knowledge in form of pairwise constraints within a constrained clustering method. They are also able to manage the uncertainty thanks to the framework of the belief function theory. Finally, the attribute clustering concept for feature selection makes our new CEVM method more flexible for maintaining vocabulary within CBR framework. During experimentation, better results are offered by $CEVM_{alt}$ version than by $CEVM_{bat}$.

References

1. Aamodt, A., Plaza, E.: Case-based reasoning: foundational issues, methodological variations, and system approaches. Artif. Intell. Commun. **7**, 39–52 (1994)
2. Glez-Pea, D., Daz, F., Hernndez, J., Corchado, J., Fdez-Riverola, F.: geneCBR: a translational tool for multiple-microarray analysis and integrative information retrieval for aiding diagnosis in cancer research. BMC Bioinform. **10**, 187 (2009)
3. Chuang, C.L.: Application of hybrid case-based reasoning for enhanced performance in bankruptcy prediction. Inf. Sci. **236**, 174–185 (2013)
4. Lesniak, A., Zima, K.: Cost calculation of construction projects including sustainability factors using the Case Based Reasoning (CBR) method. Sustainability **10**(5), 1608 (2018)
5. Richter, M.M., Michael, M.: Knowledge containers. In: Readings in Case-Based Reasoning. Morgan Kaufmann (2003)

6. Wilson, D.C., Leake, D.B.: Maintaining case-based reasoners: dimensions and directions. Comput. Intell. **17**, 196–213 (2001)
7. Ben Ayed, S., Elouedi, Z., Lefevre, E.: Exploiting domain-experts knowledge within an evidential process for case base maintenance. In: Destercke, S., Denoeux, T., Cuzzolin, F., Martin, A. (eds.) BELIEF 2018. LNCS (LNAI), vol. 11069, pp. 22–30. Springer, Cham (2018). https://doi.org/10.1007/978-3-319-99383-6_4
8. Ben Ayed, S., Elouedi, Z., Lefevre., E.: Maintaining case knowledge vocabulary using a new Evidential Attribute Clustering method. In: 13th International FLINS Conference on Data Science and Knowledge Engineering for Sensing Decision Support, pp. 347–354, Springer, Heidelberg (2018)
9. Arshadi, N., Jurisica, I.: Feature selection for improving case-based classifiers on high-dimensional data sets. In: Florida Artificial Intelligence Research Society Conference (FLAIRS), pp. 99–104 (2005)
10. Leake, D., Schack, B.: Flexible feature deletion: compacting case bases by selectively compressing case contents. In: Hüllermeier, E., Minor, M. (eds.) ICCBR 2015. LNCS (LNAI), vol. 9343, pp. 212–227. Springer, Cham (2015). https://doi.org/10.1007/978-3-319-24586-7_15
11. Kononenko, I.: Estimating attributes: analysis and extensions of RELIEF. In: Bergadano, F., De Raedt, L. (eds.) ECML 1994. LNCS, vol. 784, pp. 171–182. Springer, Heidelberg (1994). https://doi.org/10.1007/3-540-57868-4_57
12. Hong, T.P., Liou, Y.L.: Attribute clustering in high dimensional feature spaces. In: International Conference on Machine Learning and Cybernetics, vol. 4, pp. 2286–2289. IEEE (2007)
13. Maji, P.: Fuzzy-rough supervised attribute clustering algorithm and classification of microarray data. Trans. Syst. Man Cybern. Part B (Cybern.) **41**, 222–233 (2011)
14. Dempster, A.P.: Upper and lower probabilities induced by a multivalued mapping. Ann. Math. Stat. **38**, 325–339 (1967)
15. Shafer, G.: A Mathematical Theory of Evidence. Princeton University Press, Princeton (1976)
16. Antoine, V., Quost, B., Masson, M.H., Denœux, T.: CEVCLUS: evidential clustering with instance-level constraints for relational data. Soft Comput. **18**(7), 1321–1335 (2014)
17. Denœux, T., Masson, M.H.: EVCLUS: evidential clustering of proximity data. IEEE Trans. Syst. Man Cybern. Part B (Cybern.) **34**(1), 95–109 (2004)
18. Weber, R.: Fuzzy set theory and uncertainty in case-based reasoning. Eng. Intell. Syst. Electr. Eng. Commun. **14**, 121–136 (2006)
19. Pearson, K.: Mathematical contributions to the theory of evolution. Philos. Trans. R. Soc. Lond. **187**, 253–318 (1896)

Extended Abduction in Assumption-Based Argumentation

Toshiko Wakaki(✉)

Shibaura Institute of Technology,
307 Fukasaku, Minuma-ku, Saitama-City, Saitama 337–8570, Japan
twakaki@shibaura-it.ac.jp

Abstract. Assumptions are needed to perform non-monotonic reasoning in assumption-based argumentation (ABA), while hypotheses are used to perform abductive or hypothetical reasoning. However not only were hypotheses sometimes confused with assumptions when representing them in ABA frameworks but no work has been done to perform extended abduction in ABA frameworks whose languages contain explicit negation $\neg$. Hence first we define consistency of ABA frameworks w.r.t. $\neg$. Second based on it, we present the framework to perform extended abduction in ABA while treating hypotheses along with assumptions in the framework. Theoretically it is shown that Sakama and Inoue's extended abduction w.r.t. an abductive logic program (ALP) containing classical negation can be captured by our extended abduction in ABA instantiated with the ALP. Finally we provide the method to compute extended abduction in ABA based on answer set programming.

Keywords: Assumption-based argumentation · Extended abduction · Hypotheses · Explicit negation · Consistency

1 Introduction

In logic programming, *negation-as-failure* literals are used to perform non-monotonic and default reasoning, while hypotheses (i.e. abducibles, abducible facts) are used to perform abductive reasoning or hypothetical reasoning [16]. In contrast, as for argumentation, assumptions enable us to perform non-monotonic and default reasoning in assumption-based argumentation (ABA, for short) [1].

However not only were hypotheses sometimes confused with assumptions [6] when representing them in an ABA framework as addressed in [20, Example 1] but no work has been done to perform *extended abduction* [12] in ABA frameworks whose languages contain an explicit negation operator $\neg$, which are the motivation of this study. The following non-monotonic reasoning example illustrates the needs to perform extended abduction in ABA while expressing hypotheses as well as assumptions in the framework.

F. Wotawa et al. (Eds.): IEA/AIE 2019, LNAI 11606, pp. 593–607, 2019.
https://doi.org/10.1007/978-3-030-22999-3_51

Example 1 (Adapted from [13,16]). *"Normally a bird flies unless its wing is broken. Tweety and opus are birds, but tweety broke its wing recently."* The knowledge can be represented by the ABA framework $\langle \mathcal{L}, \mathcal{P}, \mathcal{A}, \overline{} \rangle$ as follows;

$$\begin{aligned} \mathcal{P}: \quad & flies(x) \leftarrow bird(x), normal(x), \qquad ab(x) \leftarrow broken\text{-}w(x), \\ & bird(tweety) \leftarrow, \qquad bird(opus) \leftarrow, \qquad broken\text{-}w(tweety) \leftarrow, \\ \mathcal{A} = & \{normal(tweety), normal(opus)\}, \\ & \overline{normal(x)} = ab(x) \qquad \text{for } x \in \{tweety, opus\}. \end{aligned}$$

We can conclude that *tweety* does not fly but *opus* flies since the argument whose claim is $flies(tweety)$ (i.e. $\{normal(tweety) \vdash flies(tweety)\}$) is not credulously justified, while the argument $\{normal(Opus)\} \vdash flies(opus)$ is skeptically justified under argument-based semantics, e.g. preferred semantics. Now if we observe that *tweety* flies, there is good reason to assume that the wounded has already healed. Then the fact $broken\text{-}w(tweety)$ must be removed from the knowledge base to explain the observation $flies(tweety)$. Hence $broken\text{-}w(tweety)$ that becomes true or not true depending on the situation should be *hypothesis*, and we may consider that $\{broken\text{-}w(tweety), broken\text{-}w(opus)\}$ is the set of hypotheses $\mathcal{H}$. Then in the revised ABA $\langle \mathcal{L}, \mathcal{P} \setminus J, \mathcal{A}, \overline{} \rangle$ where $J = \{broken\text{-}w(tweety)\}$ is erased from $\mathcal{P}$, the argument whose claim is $flies(tweety)$ is skeptically justified successfully. J is called a *negative* explanation of a *positive* observation $flies(tweety)$ [13]. On the other hand, suppose that we later notice that the bird *opus* does not fly any more though the argument whose claim is $flies(opus)$ is skeptically justified in the current ABA $\langle \mathcal{L}, \mathcal{P}, \mathcal{A}, \overline{} \rangle$. Thus we now should revise the knowledge base to block the skeptical justification of the argument by assuming $broken\text{-}w(opus)$, for instance. This situation is characterized by the notion of *anti-explanations* to account for *negative observation* which do not hold [16]. In fact, in the revised ABA $\langle \mathcal{L}, \mathcal{P} \cup I, \mathcal{A}, \overline{} \rangle$ where $I = \{broken\text{-}w(opus)\}$ is added to $\mathcal{P}$, the argument whose claim is $flies(opus)$ is not credulously justified anymore. I is called a *positive* anti-explanation of a *negative* observation $flies(opus)$.

Furthermore when an ABA framework has a language $\mathcal{L}$ containing *explicit negation* $\neg$, it is required to ensure that such a revised ABA framework is consistent since there is the possibility that arguments having contradictory claims each other may exist in the extension of the revised ABA framework due to hypotheses introduction or erasure for explaining an observation. To the best of our knowledge, however, no one has defined the notion of (in)consistency of ABA frameworks though Caminada and Amgoud proposed *rationality postulates* [4] (i.e. the *closure and consistency properties* [9]) to guarantee consistency of rule-based argumentation systems so far. Hence first we define consistency of ABA frameworks w.r.t. $\neg$, which is different from satisfying their postulates, i.e. the so-called *consistency-property* [4,9]. Second based on it, we propose the novel framework which enables us to perform extended abduction w.r.t. (abductive) ABA frameworks whose languages may contain explicit negation $\neg$ while treating hypotheses along with assumptions in the frameworks. Theoretically it is shown that (anti-)explanations of a given observation w.r.t. Sakama and

Inoue's abductive logic program (ALP) [16] containing *classical negation* [11] can be captured by our extended abduction w.r.t. the abductive ABA framework instantiated with the ALP under stable semantics. Finally we provide the method to compute (anti-)explanations of the proposed extended abduction in ABA based on answer set programming (ASP, for short) [11].

The rest of this paper is as follows. After addressing preliminaries in Sect. 2, we present in Sect. 3 our framework to perform extended abduction in ABA. In Sect. 4, we show its computational method in answer set programming. We discuss related work and conclude in Sect. 5.

2 Preliminaries

We briefly review the basic notions used throughout this paper.

Definition 1. An assumption-based argumentation framework (an ABA framework, or an ABA, for short) [1,8] is a tuple $\langle \mathcal{L}, \mathcal{R}, \mathcal{A}, \bar{\ } \rangle$, where $(\mathcal{L}, \mathcal{R})$ is a deductive system, consisting of a language $\mathcal{L}$ and a set $\mathcal{R}$ of inference rules of the form: $b_0 \leftarrow b_1, \ldots, b_m$ $(m \geq 0, b_i \in \mathcal{L})$, $\mathcal{A} \subseteq \mathcal{L}$ is a set of *assumptions*, and $\bar{\ }$ is a total mapping from $\mathcal{A}$ into $\mathcal{L}$, where $\overline{\alpha}$ is referred to as the *contrary* of α. For a rule $r \in \mathcal{R}$ s.t. $b_0 \leftarrow b_1, \ldots, b_m$, let $head(r) = b_0$ (resp. $body(r) = \{b_1, \ldots, b_m\}$). We will sometimes represent $b_0 \leftarrow$ simply as b_0. We enforce that ABA frameworks are *flat*, namely assumptions do not occur in the head of rules.

In ABA, an *argument for (the claim)*$c \in \mathcal{L}$ *supported by* $K \subseteq \mathcal{A}$ ($K \vdash c$ in short) is a (finite) tree with nodes labelled by sentences in $\mathcal{L}$ or by τ, and *attacks* against arguments are directed at the assumptions in their supports as follows.

- An argument $K_1 \vdash c_1$ *attacks* an argument $K_2 \vdash c_2$ iff $c_1 = \bar{\alpha}$ for $\exists \alpha \in K_2$.

Corresponding to a flat ABA $\mathcal{F} = \langle \mathcal{L}, \mathcal{R}, \mathcal{A}, \bar{\ } \rangle$, the abstract argumentation (AA) framework $AF_{\mathcal{F}} = (AR, attacks)$ is constructed based on *arguments* and *attacks* addressed above, and all argumentation semantics [7] can be applied to $AF_{\mathcal{F}}$. For a set $\mathcal{A}rgs$ of arguments, let $\mathcal{A}rgs^{+} = \{A|$ there exists an argument in $\mathcal{A}rgs$ that *attacks* $A\}$. $\mathcal{A}rgs$ is *conflict-free* iff $\mathcal{A}rgs \cap \mathcal{A}rgs^{+} = \emptyset$. $\mathcal{A}rgs$ *defends* an argument A iff each argument that attacks A is attacked by an argument in $\mathcal{A}rgs$.

Definition 2. [1,7,8] Let $\langle \mathcal{L}, \mathcal{R}, \mathcal{A}, \bar{\ } \rangle$ be an ABA framework, and AR the associated set of arguments. Then $\mathcal{A}rgs \subseteq AR$ is: admissible iff $\mathcal{A}rgs$ is conflict-free and *defends* all its elements; a complete argument extension iff $\mathcal{A}rgs$ is admissible and contains all arguments it *defends*; a preferred (resp. grounded) argument extension iff it is a (subset-)maximal (resp. (subset-)minimal) complete argument extension; a stable argument extension iff it is conflict-free and $\mathcal{A}rgs \cup \mathcal{A}rgs^{+} = AR$.

Let *claim*(Ag) be the claim of an argument Ag. The *conclusion* of an argument extension $\mathcal{E}$ is defined by $\texttt{Concs}(\mathcal{E}) = \{c \in \mathcal{L} | c = claim(Ag)$ for $Ag \in \mathcal{E}\}$. Let *Sname* $\in \{complete, preferred, grounded, stable\}$. Originally the various ABA semantics [1] is given by sets of assumptions called assumption extensions. However a flat ABA framework $\mathcal{F}$ can be identified with the corresponding AA framework $AF_{\mathcal{F}}$ since there is a one-to-one correspondence between *Sname* assumption extensions and *Sname* argument extensions [5]. Caminada and Amgoud's rationality postulates [4], i.e. satisfying the *closure* and *consistency* properties [9] are stated as follows.

Definition 3. [9] Given a flat ABA framework $\langle \mathcal{L}, \mathcal{R}, \mathcal{A}, \overline{\ \ } \rangle$, a set $X \subseteq \mathcal{L}$ is said to be contradictory iff X is contradictory w.r.t. $\overline{\ \ }$, i.e. there exists an assumption $\alpha \in \mathcal{A}$ such that $\{\alpha, \overline{\alpha}\} \subseteq X$; or X is contradictory w.r.t. $\neg$, i.e. there exists $\sigma \in \mathcal{L}$ such that $\{\sigma, \neg\sigma\} \subseteq X$ if $\mathcal{L}$ contains an explicit negation operator $\neg$.

Let $CN_{\mathcal{R}} : 2^{\mathcal{L}} \rightarrow 2^{\mathcal{L}}$ be a consequence operator. $CN_{\mathcal{R}}(X)$ is the smallest set s.t. $X \subseteq CN_{\mathcal{R}}(X)$, and for each $r \in \mathcal{R}$, if $body(r) \subseteq CN_{\mathcal{R}}(X)$ then $head(r) \in CN_{\mathcal{R}}(X)$. A set X is said to be inconsistent iff $CN_{\mathcal{R}}(X)$ is contradictory. X is said to be consistent iff it is not inconsistent. A flat ABA $\mathcal{F} = \langle \mathcal{L}, \mathcal{R}, \mathcal{A}, \overline{\ \ } \rangle$ is said to satisfy the *consistency-property* (resp. the *closure-property*) if for each complete extension $\mathcal{E}$ of $AF_{\mathcal{F}}$ generated from it, $\texttt{Concs}(\mathcal{E})$ is consistent (resp. $\texttt{Concs}(\mathcal{E}) = CN_{\mathcal{R}}(\texttt{Concs}(\mathcal{E}))$).

Definition 4. An extended logic program (ELP) is a set of rules of the form:

$$L_0 \leftarrow L_1, \ldots, L_m, not\ L_{m+1}, \ldots, not\ L_n, \qquad (n \geq m \geq 0) \tag{1}$$

where each L_i is a literal, i.e. either an atom or an atom preceded by classical negation $\neg$. Each atom, say A, is called a positive literal, while $\neg A$ is called a negative literal. *not* represents *negation as failure* (NAF). A literal preceded by *not* is called a NAF-literal. A rule $L_0 \leftarrow$ is identified with L_0. Let Lit_P be the set of all ground literals in the language of an ELP P. The semantics of an ELP is given by *answer set semantics* [10,11] (resp. *paraconsistent stable model semantics* [15]) defined as follows.

First, let P be a *not*-free ELP (i.e., for each rule $m = n$). Then, $S \subseteq Lit_P$ is an *answer set* of P if S is a minimal set satisfying the following two conditions: (i) For each ground instance of a rule $L_0 \leftarrow L_1, \ldots, L_m$ in P, if $\{L_1, \ldots, L_m\} \subseteq S$, then $L_0 \in S$. (ii) If S contains a pair of complementary literals, then $S = Lit_P$. Second, let P be any ELP and $S \subseteq Lit_P$. The *reduct* of P by S is a *not*-free ELP P^S which contains $L_0 \leftarrow L_1, \ldots, L_m$ iff there is a ground rule of the form (1) in P such that $\{L_{m+1}, \ldots, L_n\} \cap S = \emptyset$. Then S is an answer set of P if S is an answer set of P^S. In contrast, *paraconsistent stable models* (*p-stable models*, for short) [15] are regarded as answer sets defined without the condition (ii).

An answer set is *consistent* if it is not Lit_P; otherwise it is *inconsistent*. An ELP P is *consistent* if it has a consistent answer set; otherwise P is *inconsistent* under answer set semantics. On the other hand, a p-stable model is *inconsistent* if it contains a pair of complementary literals; otherwise it is *consistent*. For an ELP P, P is *consistent* if it has a consistent p-stable model; otherwise it is *inconsistent* under paraconsistent stable model semantics. An ELP P is called a *normal logic program* (NLP) [10] if every literal L_i in the program is an atom, i.e. $L_i \in HB_P$ for a Herbrand base HB_P of P. Answer sets as well as p-stable models of P coincide with *stable models* [10] when P is an NLP.

3 Extended Abduction in ABA

3.1 Consistency of ABA Frameworks

The following theorems hold w.r.t. the *closure and consistency properties* [4,9].

Theorem 1. *A flat ABA framework satisfies the closure property.*

Proof: Based on [9, Theorems 6 and 2], *this theorem is proved.* □

Theorem 2. *A flat ABA framework* $\mathcal{F} = \langle \mathcal{L}, \mathcal{R}, \mathcal{A}, \overline{\ \ } \rangle$ *satisfies the consistency property if and only if for each complete argument extension* $\mathcal{E}$ *of the AA framework* $AF_{\mathcal{F}}$ *generated from* $\mathcal{F}$, $\texttt{Concs}(\mathcal{E})$ *is not contradictory w.r.t.* $\neg$.

Proof: This is easily proved based on Theorem 1. □

In contrast, we define (in)consistency of a flat ABA framework in a way similar to the way to define (in)consistency of an ELP as follows.

Definition 5. A complete argument extension $\mathcal{E}$ of $AF_{\mathcal{F}}$ is said to be consistent if $\texttt{Concs}(\mathcal{E})$ is consistent, i.e. not contradictory w.r.t. $\neg$; otherwise it is inconsistent. A flat ABA $\mathcal{F}$ is *consistent* under *Sname* semantics if $AF_{\mathcal{F}}$ generated from $\mathcal{F}$ has a consistent *Sname* argument extension; otherwise it is inconsistent.

The following example illustrates the difference between our definition of (in)consistency of a flat ABA framework and the rationality postulates [4,9].

Example 2. Consider the following ELP P.

$$P: \qquad \neg p \leftarrow not\ a, \quad a \leftarrow p, not\ b, \quad p \leftarrow, \quad b \leftarrow not\ a$$

P is consistent under answer set semantics since it has the unique consistent answer set $\{a, p\}$. On the other hand, P has two p-stable models, $S_1 = \{a, p\}$ and $S_2 = \{\neg p, p, b\}$, where S_1 is consistent and S_2 is inconsistent. Hence P is also consistent under paraconsistent stable model semantics.

Next consider the ABA framework $\mathcal{F}_P = \langle \mathcal{L}, P, \mathcal{A}, \overline{\ \ } \rangle$ instantiated with P where $\mathcal{A} = \{not\ a, not\ b\}$, $\overline{not\ a} = a$, $\overline{not\ b} = b$. $\mathcal{F}_P$ has arguments as follows:

$A_1 : \{not\ a\} \vdash \neg p$, $A_2 : \{not\ b\} \vdash a$, $A_3 : \{\ \} \vdash p$,
$A_4 : \{not\ a\} \vdash b$, $A_5 : \{not\ a\} \vdash not\ a$, $A_6 : \{not\ b\} \vdash not\ b$,

and $attacks = \{(A_2, A_1), (A_2, A_5), (A_2, A_4), (A_4, A_2), (A_4, A_6)\}$. There are three complete argument extensions $\mathcal{E}_1$, $\mathcal{E}_2$, $\mathcal{E}_3$ in $\mathcal{F}_P$ as follows,

$\mathcal{E}_1 = \{A_2, A_3, A_6\}$, $concs(\mathcal{E}_1) = \{a, p, not\ b\}$
$\mathcal{E}_2 = \{A_1, A_3, A_4, A_5\}$, $concs(\mathcal{E}_2) = \{\neg p, p, b, not\ a\}$.
$\mathcal{E}_3 = \{A_3\}$, $concs(\mathcal{E}_3) = \{p\}$.

where $\mathcal{E}_1$, $\mathcal{E}_2$ are stable extensions such that $concs(\mathcal{E}_i) \cap Lit_P = S_i$ for $i = 1, 2$. Then regarding $\neg$ in P as explicit negation in $\mathcal{F}_P$, $concs(\mathcal{E}_1)$ and $concs(\mathcal{E}_3)$ are consistent, while $concs(\mathcal{E}_2)$ is inconsistent. Thus $\mathcal{F}_P$ is consistent under stable (resp. complete) semantics since it has the consistent stable extension $\mathcal{E}_1$, while $\mathcal{F}_P$ does not satisfy the *consistency*-property [9] since it has the inconsistent $\mathcal{E}_2$.

3.2 Frameworks for Performing Extended Abduction in ABA

We propose the framework to establish extended abduction in ABA as follows.

Definition 6 ***(ABAs equipped with hypotheses).*** Let $\mathcal{H} \subseteq \mathcal{L}$ be a finite set of *hypotheses. An assumption-based argumentation framework equipped with hypotheses* (or *a h_ABA* for short) is a tuple as follows:

$$\langle \mathcal{L}, (\mathcal{R}, \mathcal{H}), \mathcal{A}, \overline{\ } \rangle,$$

where $\mathcal{H} \cap \mathcal{A} = \emptyset$ and $\langle \mathcal{L}, \mathcal{R}, \mathcal{A}, \overline{\ } \rangle$ is a flat ABA framework. We assume $b_0 \notin \mathcal{H}$ for $\forall (b_0 \leftarrow b_1, \ldots, b_m) \in \mathcal{R}$.

Abductive ABA frameworks w.r.t. a given h_ABA are defined as follows.

Definition 7 ***(Abductive ABA frameworks).*** Given a h_ABA $\langle \mathcal{L}, (\mathcal{R}, \mathcal{H}), \mathcal{A}, \overline{\ } \rangle$, an ABA $\langle \mathcal{L}, \mathcal{P}, \mathcal{A}, \overline{\ } \rangle$ where $\mathcal{P} = \mathcal{R} \cup E = \mathcal{R} \cup \{h \leftarrow | h \in E \subseteq \mathcal{H}\}$ is called an *abductive ABA framework with* (the set of hypotheses) $E \subseteq \mathcal{H}$ from the h_ABA (or an AABA, for short). An AABA with $\mathcal{H}$, i.e. $\langle \mathcal{L}, \mathcal{R} \cup \mathcal{H}, \mathcal{A}, \overline{\ } \rangle$ is called a *universal* AABA. There are $2^{|\mathcal{H}|}$ AABAs w.r.t. a given h_ABA, where $|\mathcal{H}|$ is the cardinality of $\mathcal{H}$.

For an abductive ABA framework, arguments and *attacks* are defined as follows. $\tau \notin \mathcal{L}$ stands for an empty set of premises as used in [8, Definition 10.1].

Definition 8 ***(Arguments and attacks in abductive ABA frameworks).*** Given a h_ABA $\langle \mathcal{L}, (\mathcal{R}, \mathcal{H}), \mathcal{A}, \overline{\ } \rangle$, let $\langle \mathcal{L}, \mathcal{P}, \mathcal{A}, \overline{\ } \rangle$ be an AABA with $E = \mathcal{H} \cap \mathcal{P}$ from the h_ABA, where $\mathcal{P} = \mathcal{R} \cup E$. Then an *argument for (the claim)* $c \in \mathcal{L}$ *supported by assumptions* $K \subseteq \mathcal{A}$ *and hypotheses* $\Delta \subseteq E$ ($K \vdash_\Delta c$, in short) is a (finite) tree with nodes labelled by sentences in $\mathcal{L}$, or by the symbol τ such that

- the root is labelled by c,
- leaves are labelled either by an assumption in $\mathcal{A}$ or by τ,
- for every non-leaf node N whose label is ℓ_N,
  - if $\ell_N \in E \subseteq \mathcal{H}$, the child of N is labelled by τ,
  - otherwise $\ell_N \notin \mathcal{H}$, then there is an inference rule $\ell_N \leftarrow b_1, \ldots, b_m (m \geq 0)$ and either $m = 0$ and the child of N is labelled by τ or $m > 0$ and N has m children, labelled by $b_1, \ldots, b_m$ respectively,
- $K \subseteq \mathcal{A}$ is the set of all assumptions labelling the leaves,
- $\Delta \subseteq E$ is the set of all hypotheses labelling non-leaves.

In an AABA, *attacks* against arguments are directed at the assumptions in their supports as usual as follows.

- An argument $K_1 \vdash_{\Delta_1} c_1$ *attacks* an argument $K_2 \vdash_{\Delta_2} c_2$ iff $\overline{\alpha} = c_1$ for $\exists \alpha \in K_2$.

Like a standard ABA, the AA framework associated with an AABA is generated from the AABA based on arguments and *attacks* defined above.

Definition 9 ***(Abductive AFs associated with abductive ABAs).*** Given a h_ABA $\langle \mathcal{L}, (\mathcal{R}, \mathcal{H}), \mathcal{A}, \overline{} \rangle$, let $AF_\mathcal{P} = (Args_\mathcal{P}, attacks_\mathcal{P})$ be the AA framework associated with an AABA $\mathcal{F}_\mathcal{P} = \langle \mathcal{L}, \mathcal{P}, \mathcal{A}, \overline{} \rangle$ with $E = \mathcal{H} \cap \mathcal{P} = \{h \in \mathcal{H} | \ h \leftarrow \in \mathcal{P}\}$ where $\mathcal{P} = \mathcal{R} \cup E$. $AF_\mathcal{P}$ is called the abductive AF (or the AAF, for short) associated with an AABA $\mathcal{F}_\mathcal{P}$. Hereafter AF_U stands for the AAF $AF_{\mathcal{R} \cup \mathcal{H}} = (Args_{\mathcal{R} \cup \mathcal{H}}, attacks_{\mathcal{R} \cup \mathcal{H}})$ associated with the universal AABA $\langle \mathcal{L}, \mathcal{R} \cup \mathcal{H}, \mathcal{A}, \overline{} \rangle$. W.r.t. $AF_\mathcal{P} = (Args_\mathcal{P}, attacks_\mathcal{P})$, it holds that

$$Args_\mathcal{P} = \{K \vdash_\Delta c \in Args_{\mathcal{R} \cup \mathcal{H}} | \Delta \subseteq E \subseteq \mathcal{H} \text{ for } E = \mathcal{H} \cap \mathcal{P}\},$$
$$attacks_\mathcal{P} = attacks_{\mathcal{R} \cup \mathcal{H}} \cap (Args_\mathcal{P} \times Args_\mathcal{P}).$$

Argument extensions as well as assumption extensions are defined for AAFs and AABAs under *Sname* semantics in a way similar to the way for standard ABAs. We define notations for naming arguments as follows.

⟨**Notation**⟩
When $\mathcal{H} = \emptyset$, an argument $K \vdash_\Delta c$ defined in Definition 8 reduces to an argument $K \vdash c$ in an ABA framework defined in [8, Definition 10.1]. Then when $\Delta = \emptyset$, we sometimes write $K \vdash c$ instead of $K \vdash_\emptyset c$. We often use a unique name to denote an argument, e.g. Ag: $K \vdash_\Delta c$ that is an argument $K \vdash_\Delta c$ with name Ag. With an abuse of notation, the name of an argument sometimes stands for the whole argument, for example, Ag denotes the argument Ag: $K \vdash_\Delta c$.

Example 3. The ABA framework $\langle \mathcal{L}, \mathcal{P}, \mathcal{A}, \overline{} \rangle$ shown in Example 1 is the AABA $\langle \mathcal{L}, \mathcal{P}, \mathcal{A}, \overline{} \rangle$ with the set of hypotheses $\{broken\text{-}w(tweety)\}$ from the h_ABA $\langle \mathcal{L}, (\mathcal{R}, \mathcal{H}), \mathcal{A}, \overline{} \rangle$ where $\mathcal{R} = \mathcal{P} \setminus \{broken\text{-}w(tweety) \leftarrow\}$ and $\mathcal{H} = \{broken\text{-}w(tweety), broken\text{-}w(opus)\}$.

Example 4. Consider the h_ABA $\langle \mathcal{L}, (\mathcal{R}, \mathcal{H}), \mathcal{A}, \overline{} \rangle$, where
$\mathcal{R} = \{\neg p \leftarrow \alpha, a \leftarrow e, p \leftarrow \neg h\}$, $\mathcal{H} = \{e, \neg h\}$, $\mathcal{A} = \{\alpha\}$, $\overline{\alpha} = a$.

There are four AABAs $\mathcal{F}_i = \langle \mathcal{L}, \mathcal{R} \cup E_i, \mathcal{A}, \overline{} \rangle$ $(1 \le i \le 4)$ such that $E_1 = \emptyset$, $E_2 = \{e\}$, $E_3 = \{\neg h\}$, $E_4 = \mathcal{H} = \{e, \neg h\}$. The universal AABA $\mathcal{F}_4$ or AF_U has arguments: A_1:$\{\alpha\} \vdash \neg p$, A_2:$\{\alpha\} \vdash \alpha$, A_3:$\emptyset \vdash_{\{e\}} a$, A_4:$\emptyset \vdash_{\{\neg h\}} p$, A_5:$\emptyset \vdash_{\{e\}} e$, A_6:$\emptyset \vdash_{\{\neg h\}} \neg h$ and $attacks_{\mathcal{R} \cup \mathcal{H}} = \{(A_3, A_1), (A_3, A_2)\}$.

Then each $\mathcal{F}_i$ $(1 \le i < 4)$ has the unique complete argument extension $\mathcal{E}_i$ such that $\mathcal{E}_1 = \{A_1, A_2\}$ with $\texttt{Concs}(\mathcal{E}_1) = \{\neg p, \alpha\}$, $\mathcal{E}_2 = \{A_3, A_5\}$ with $\texttt{Concs}(\mathcal{E}_2) = \{a, e\}$, $\mathcal{E}_3 = \{A_1, A_2, A_4, A_6\}$ with $\texttt{Concs}(\mathcal{E}_3) = \{\neg p, \alpha, p, \neg h\}$ which is contradictory w.r.t. $\neg$, and $\mathcal{E}_4 = \{A_3, A_4, A_5, A_6\}$ with $\texttt{Concs}(\mathcal{E}_4) = \{a, p, e, \neg h\}$. Hence $\mathcal{F}_1$, $\mathcal{F}_2$ and $\mathcal{F}_4$ are consistent, while $\mathcal{F}_3$ is inconsistent under complete semantics.

We are ready to show our framework to perform extended abduction in ABA.

Definition 10 ***(Positive/negative observations).*** A sentence G is called a *positive* observation if it is observed, namely it holds; otherwise called a *negative* observation.

Definition 11 ***(Explanations/anti-explanations).*** Let $\langle \mathcal{L}, (\mathcal{R}, \mathcal{H}), \mathcal{A}, \overline{} \rangle$ be a h_ABA, $\langle \mathcal{L}, \mathcal{P}, \mathcal{A}, \overline{} \rangle$ be an AABA with $\mathcal{H} \cap \mathcal{P}$ from the h_ABA and (I, J)

$\in 2^{\mathcal{H}} \times 2^{\mathcal{H}}$. Let E_{cur} be $\mathcal{H} \cap \mathcal{P}$. Then given $G^+ \in \mathcal{L}$ a sentence representing a positive observation, a pair (I, J) (or $E = (E_{cur} \setminus J) \cup I$) is called a *skeptical* (resp. *credulous*) explanation of G^+ w.r.t. an AABA $\langle \mathcal{L}, \mathcal{P}, \mathcal{A}, ^- \rangle$ with E_{cur} under *Sname* semantics if

1. every (resp. some) consistent *Sname* argument extension of the AABA $\langle \mathcal{L}, (\mathcal{P} \setminus J) \cup I, \mathcal{A}, ^- \rangle$ with $E = (E_{cur} \setminus J) \cup I$ contains an argument with claim G^+,
2. $I \subseteq \mathcal{H} \setminus \mathcal{P}$ and $J \subseteq \mathcal{H} \cap \mathcal{P}$, (in other word, $I \subseteq \mathcal{H} \setminus E_{cur}$ and $J \subseteq E_{cur}$),
3. the AABA $\langle \mathcal{L}, (\mathcal{P} \setminus J) \cup I, \mathcal{A}, ^- \rangle$ with E is consistent under *Sname* semantics.

On the other hand, given $G^- \in \mathcal{L}$ a sentence representing a negative observation, a pair (I, J) (or $E = (E_{cur} \setminus J) \cup I$) is called a *credulous* (resp. *skeptical*) anti-explanation of G^- w.r.t. $\langle \mathcal{L}, \mathcal{P}, \mathcal{A}, ^- \rangle$ with E_{cur} under *Sname* semantics if

1. some (resp. every) consistent *Sname* argument extension of the AABA $\langle \mathcal{L}, (\mathcal{P} \setminus J) \cup I, \mathcal{A}, ^- \rangle$ with $E = (E_{cur} \setminus J) \cup I$ does not contain any argument with claim G^-,
2. $I \subseteq \mathcal{H} \setminus \mathcal{P}$ and $J \subseteq \mathcal{H} \cap \mathcal{P}$, (in other word, $I \subseteq \mathcal{H} \setminus E_{cur}$ and $J \subseteq E_{cur}$),
3. the AABA $\langle \mathcal{L}, (\mathcal{P} \setminus J) \cup I, \mathcal{A}, ^- \rangle$ with E is consistent under *Sname* semantics.

Remarks: Note that $\mathcal{P} = \mathcal{R} \cup E_{cur}$, while $(\mathcal{P} \setminus J) \cup I = \mathcal{R} \cup E$ for $E = (E_{cur} \setminus J) \cup I$.

An (anti-)explanation (I, J) of an observation G is called *minimal* if for any (anti-)explanation (I', J') of G, $I' \subseteq I$ and $J' \subseteq J$ imply $I' = I$ and $J' = J$.

Example 5 (Example 3 Cont.). Both $(\emptyset, \{broken\text{-}w(tweety)\})$ and $(\{broken\text{-}w(opus)\}, \{broken\text{-}w(tweety)\})$ are skeptical explanations of $flies(tweety)$ w.r.t. the AABA with $\{broken\text{-}w(tweety)\}$, while both $(\{broken\text{-}w(opus)\}, \emptyset)$ and $(\{broken\text{-}w(opus)\}, \{broken\text{-}w(tweety)\})$ are credulous anti-explanations of $flies\ (opus)$ w.r.t. the AABA with $\{broken\text{-}w(tweety)\}$ under complete semantics.

Example 6 (Example 4 Cont.). $(\{\neg h\}, \emptyset)$ is the unique skeptical explanation of the positive observation p w.r.t. the AABA $\mathcal{F}_2$ with $E_2 = \{e\}$ under complete semantics due to $\mathcal{F}_4$, whereas both $(\emptyset, \{e\})$ and $(\emptyset, \emptyset))$ are the credulous anti-explanations of the negative observation p w.r.t. the AABA $\mathcal{F}_2$ with $E_2 = \{e\}$ due to $\mathcal{F}_1$ (resp. $\mathcal{F}_2$).

Though Sakama and Inoue [13,16] defined extended abduction[1] w.r.t. an abductive logic program (ALP, for short) under answer set semantics [10,11], it can be also defined under paraconsistent stable model semantics [15] as follows.

[1] Normal abduction [14] is considered as a special case of extended abduction where only hypotheses introduction is considered for explaining positive observations.

Definition 12. Let $\langle P, H\rangle$ be an ALP where P is an ELP and $H \subseteq Lit_P$. It is assumed that for any rule $L \leftarrow \Gamma$ from P, $L \in H$ implies $\Gamma = \emptyset$. Let $(I, J) \in 2^{Lit_P} \times 2^{Lit_P}$ and $LPmodel \in$ {answer set, paraconsistent stable model}.

1. Given $G^+ \in Lit_P$ representing a positive observation, a pair (I, J) is a *skeptical* (resp. *credulous*) explanation of G^+ w.r.t. $\langle P, H\rangle$ under *LPmodel* semantics if G^+ is included in *every* (resp. *some*) consistent *LPmodel* of $(P \setminus J \cup I)$, where $I \subseteq H \setminus P$, $J \subseteq H \cap P$ and $(P \setminus J) \cup I$ is consistent under *LPmodel* semantics.
2. Given $G^- \in Lit_P$ representing a negative observation, (I, J) is a *credulous* (resp. skeptical) ant-explanation of G^- w.r.t. $\langle P, H\rangle$ under *LPmodel* semantics if G^- is not included in *some* (resp. *every*) consistent *LPmodel* of $(P \setminus J \cup I)$, where $I \subseteq H \setminus P$, $J \subseteq H \cap P$ and $(P \setminus J) \cup I$ is consistent under *LPmodel* semantics.

The following theorem states that Sakama and Inoue's extended abduction w.r.t. an ALP $\langle P, H\rangle$ where P is an ELP and $H \subseteq Lit_P$ under paraconsistent stable model semantics (resp. P is an NLP and $H \subseteq HB_P$ under answer set semantics) can be captured by our extended abduction in ABA instantiated with the ALP under stable semantics.

Theorem 3. *Let $\langle P, H\rangle$ be an ALP where P is an ELP and $H \subseteq Lit_P$, and $\langle \mathcal{L}_P, (R, H), Lit_{not}, \overline{}\rangle$ be the h_ABA instantiated with the ALP such that $R = P \setminus H$, $Lit_{not} = \{not\ L \mid L \in Lit_P\}$, $\mathcal{L}_P = Lit_P \cup Lit_{not}$ and $\overline{not\ L} = L$ for $L \in Lit_P$. Then given an observation $G \in Lit_P$, a pair (I, J) is a skeptical (resp. credulous) (anti-)explanation of G w.r.t. an ALP $\langle P, H\rangle$ under paraconsistent stable model semantics iff (I, J) is a skeptical (resp. credulous) (anti-) explanation of G w.r.t. the AABA $\langle \mathcal{L}_P, P, Lit_{not}, \overline{}\rangle$ with $H \cap P$ from the h_ABA $\langle \mathcal{L}_P, (R, H), Lit_{not}, \overline{}\rangle$ under stable semantics.*

Proof: See appendix. □

4 Computing Extended Abduction in ABA Using ASP

We show the method to compute extended abduction in ABA under complete and stable semantics by making use of answer set programming in [18,19].

Based on the idea proposed in [18,19], the AA framework (i.e. the AAF) $AF_{\mathcal{P}}$ associated with an AABA $\mathcal{F}_{\mathcal{P}} = \langle \mathcal{L}, \mathcal{P}, \mathcal{A}, \overline{}\rangle$ with $\mathcal{H} \cap \mathcal{P}$ is translated into the ELP for the respective semantics whose answer set (if exists), say S, embeds Caminada's reinstatement labelling $\mathcal{L}$ [3] such that an atom $in(a)$ is in S if and only if the argument a is labelled `in` by $\mathcal{L}$, (that is, $in(a) \in S$ iff $a \in$ `in`$(\mathcal{L})$) where $\mathcal{L}$ satisfies the conditions for the respective semantics. Furthermore in order to check consistency of the conclusion of each complete (resp. stable) extension, not only is an explicit negation operator $\neg$ contained in the language $\mathcal{L}$ of an AABA (resp. an ABA) expressed by the classical negation sign $\neg$ in the translated ELP but each answer set S of the ELP also embeds the conclusion of the corresponding extension which is consistent. Then each sentence $s \in \mathcal{L}$ in an AABA (resp. an ABA) whose form is $a \in \mathcal{L}$ or $\neg a \in \mathcal{L}$ is expressed by a literal $\hat{s}$ in the translated ELP whose form is a positive literal (i.e. an atom) $\hat{a}$ or a negative literal $\neg\hat{a}$ preceded by classical negation $\neg$. Hereafter let $\hat{V} = \{\hat{v}\ |\hat{v}$ is a literal expressing $v \in V\}$ for a set $V \subseteq \mathcal{L}$.

The following ELP Π which is translated from AF_U (or a universal AABA) is the slightly modified one for the ELP given in [19, Definition 20].

Definition 13. Given a h_ABA $\langle \mathcal{L}, (\mathcal{R}, \mathcal{H}), \mathcal{A}, \overline{\ } \rangle$, AF_U denoting the AAF $AF_{\mathcal{R}\cup\mathcal{H}} = (Args_{\mathcal{R}\cup\mathcal{H}},\ attacks_{\mathcal{R}\cup\mathcal{H}})$ associated with a universal AABA $\langle \mathcal{L}, \mathcal{R} \cup \mathcal{H}, \mathcal{A}, \overline{\ } \rangle$ is translated into the ELP $\Pi = \Pi_{AF} \cup \Pi_{Lab}$, where Π_{AF} is a set of domain-dependent rules as follows:

1. for any argument $(a{:}K \vdash_{\Delta} c) \in Args_{\mathcal{R}\cup\mathcal{H}}$,
   $ag(a) \leftarrow \hat{\Delta}$, where $\hat{\Delta} = \{\hat{h} \mid \hat{h}$ is a literal expressing $h \in \Delta \subseteq \mathcal{H}\} \subseteq \hat{\mathcal{H}}$
2. for any $(a_1, a_2) \in attacks_{\mathcal{R}\cup\mathcal{H}}$,
   $def(a_1, a_2) \leftarrow ag(a_1),\ ag(a_2)$,
3. for any argument $(a{:}K \vdash_{\Delta} c) \in Args_{\mathcal{R}\cup\mathcal{H}}$,
   $\hat{c} \leftarrow in(a)$, where $\hat{c}$ is a literal expressing the claim $c \in \mathcal{L}$,

and Π_{Lab} is the set of domain-independent rules [18, Definition 9] as follows:

4. $in(X) \leftarrow ag(X), not\ ng(X)$,
   $ng(X) \leftarrow ag(X), ag(Y), in(Y), def(Y, X)$,
   $ng(X) \leftarrow ag(X), ag(Y), undec(Y), def(Y, X)$,
5. $out(X) \leftarrow ag(X), ag(Y), in(Y), def(Y, X)$,
6. $undec(X) \leftarrow ag(X), not\ in(X), not\ out(X)$,

where $in(a)$ (resp. $out(a)$, $undec(a)$) denotes that the argument $a{:}K \vdash_{\Delta} c$ whose name is a is labelled `in` (resp. `out`, `undec`) [3].

The rule no.3 plays a role to generate the consistent complete argument extensions along with their conclusions except inconsistent ones.

The following lemma states that the ELP $\Pi \cup \hat{E}$ yields answer sets which embed consistent complete extensions of an AABA $\langle \mathcal{L}, \mathcal{R} \cup E, \mathcal{A}, \overline{\ } \rangle$ with E.

Lemma 1 (Soundness and Completeness Theorems). *Let $\langle \mathcal{L}, (\mathcal{R}, \mathcal{H}), \mathcal{A}, \overline{\ } \rangle$ be a h_ABA. Then $\mathcal{E}$ is a consistent complete (resp. stable) argument extension of the AAF $AF_{\mathcal{R}\cup E} = (Args_{\mathcal{R}\cup E},\ attacks_{\mathcal{R}\cup E})$ associated with an AABA $\langle \mathcal{L}, \mathcal{R} \cup E, \mathcal{A}, \overline{\ } \rangle$ with $E \subseteq \mathcal{H}$ iff there is a consistent answer set S of $\Pi \cup \hat{E}$ (resp. $\Pi \cup \{\leftarrow undec(X)\} \cup \hat{E}$) s.t. $\mathcal{E} = \{a | in(a) \in S\}$ for $\hat{E} \subseteq \hat{\mathcal{H}}$.*

Proof (Sketch). This is easily proved based on [18, Lemma 1]. □

Notice that we can make use of Lemma 1 along with Definition 13 to obtain only the consistent complete (resp. stable) extensions in a standard ABA framework.

Example 7. Consider Example 2. The ABA framework $\mathcal{F}_P$ has no hypotheses, i.e. $\mathcal{H} = \emptyset$. Then the AA framework for $\mathcal{F}_P$ is translated into $\Pi_{AF} \cup \Pi_{Lab}$ where $\Pi_{AF} = \{ag(A_i) | 1 \leq i \leq 6\} \cup \{def(A_2, A_1), def(A_2, A_5), def(A_2, A_4), def(A_4, A_2), def(A_4, A_6)\} \cup \{\neg p \leftarrow in(A_1), a \leftarrow in(A_2), p \leftarrow in(A_3), b \leftarrow in(A_4), not\_a \leftarrow in(A_5), not\_b \leftarrow in(A_6)\}$. Note that not_a and not_b are atoms expressing sentences in $\mathcal{L}$ such as $claim(A_5) = not\ a$ and $claim(A_6) = not\ b$. As a result, $\Pi_{AF} \cup \Pi_{Lab} \cup \{\leftarrow undec(X)\}$ has the unique consistent answer set S such that $\{a | in(a) \in S)\} = \{A_2, A_3, A_6\} = \mathcal{E}_1$ and the conclusion $\{a, p, not\_b\} \subseteq S$ w.r.t. the consistent stable extension $\mathcal{E}_1$.

To obtain every explanation/anti-explanation $\langle I, J \rangle$ (or $E = (E_{cur} \setminus J) \cup I \subseteq \mathcal{H}$) under complete (resp. stable) semantics, we should take into account $2^{|\mathcal{H}|}$ AABAs for a given h_ABA, in other words, $\Pi \cup \hat{E}$ (resp. $\Pi \cup \{\leftarrow undec(X)\} \cup \hat{E}$) w.r.t. any $E \in 2^{\mathcal{H}}$. In order to establish this, Lemma 1 for $\Pi \cup \hat{E}$ for example is extended into Theorem 4 for the ELP $\Pi \cup \Gamma(\mathcal{H})$ where $\Gamma(\mathcal{H})$ is the set of rules as follows. For any hypothesis $h \in \mathcal{H}$, the following pair of rules are in $\Gamma(\mathcal{H})$,

$$h' \leftarrow not\ \hat{h}, \qquad \hat{h} \leftarrow not\ h'$$

where $\hat{h}$ is the literal expressing a hypothesis h, while h' is a newly introduced atom uniquely associated with h representing the complement of h.

Theorem 4 (Soundness and Completeness Theorems). *Let $\langle \mathcal{L}, (\mathcal{R}, \mathcal{H}), \mathcal{A}, \bar{\ } \rangle$ be a h_ABA. $\mathcal{E}$ is a consistent complete (resp. stable) argument extension of the AAF $AF_{\mathcal{R} \cup E}$ associated with an AABA $\langle \mathcal{L}, \mathcal{R} \cup E, \mathcal{A}, \bar{\ } \rangle$ with $E \subseteq \mathcal{H}$ iff there is a consistent answer set S of $\Pi \cup \Gamma(\mathcal{H})$ (resp. $\Pi \cup \{\leftarrow undec(X)\} \cup \Gamma(\mathcal{H})$) s.t. $\mathcal{E} = \{a | in(a) \in S\}$ and $S \cap \hat{\mathcal{H}} = \hat{E}$ for $\hat{E} \subseteq \hat{\mathcal{H}}$.*

Proof: This is proved in a similar way to the proof of [19, Theorem 1]. □

Moreover we provide the new notations, symbols and sets as follows.

- $\Gamma^{+}(\mathcal{H}, \mathtt{E})$ where $\mathtt{E} \subseteq \mathcal{H}$ is the set of rules defined as follows:
  - for the literal $\hat{h}$ expressing a hypothesis $h \in \mathcal{H} \setminus \mathtt{E}$,

$$h^{+} \leftarrow \hat{h},$$

  - for the literal $\hat{h}$ expressing a hypothesis $h \in \mathtt{E}$,

$$h^{-} \leftarrow not\ \hat{h},$$

  where h^{+} (resp. h^{-})[2] is a newly introduced literal associated with h denoting that $h \in \mathcal{H} \setminus \mathtt{E}$ (resp. $h \in \mathtt{E}$) is added (resp. removed) to explain an observation.
- Given a set of literals S, let $S^{+} = \{s^{+} | s \in S\}$ and $S^{-} = \{s^{-} | s \in S\}$. Conversely given a set S^{+} (resp. S^{-}), let $S = \{s | s^{+} \in S^{+}\}$ (resp. $S = \{s | s^{-} \in S^{-}\}$).

[2] h^{+} and h^{-} are literals which correspond to *update atoms* in [16, Definition 3.1].

- $M|_X \stackrel{\text{def}}{=} M \cap X$ for an answer set M and a set X.
- $\mathcal{I} \stackrel{\text{def}}{=} \{in(a) \mid (a : K \vdash_\Delta c) \in Args_{\mathcal{R}\cup\mathcal{H}}\}$.

Hereby the following propositions indicate that credulous explanations (resp. anti-explanations) of our approach can be computed in answer set programming.

Proposition 1. *Let $\langle \mathcal{L}, (\mathcal{R}, \mathcal{H}), \mathcal{A}, \overline{} \rangle$ be a h_ABA. Given a positive observation G, a pair (I, J) is a credulous explanation of G w.r.t. an AABA $\langle \mathcal{L}, \mathcal{P}, \mathcal{A}, \overline{} \rangle$ with $\mathcal{H} \cap \mathcal{P}$ from the h_ABA under complete semantics iff there is a consistent answer set S of the ELP $\Pi \cup \Gamma(\mathcal{H}) \cup \Gamma^+(\mathcal{H}, \mathcal{H} \cap \mathcal{P}) \cup \{\leftarrow not\ G\}$ such that $G \in S$, $\hat{I}^+ = S|_{\hat{\mathcal{H}}^+}$, $\hat{J}^- = S|_{\hat{\mathcal{H}}^-}$, $\hat{E} = S|_{\hat{\mathcal{H}}}$ and $S|_{\mathcal{I}} = \{in(a)|a \in \mathcal{E}$ s.t. $G \in \mathtt{concs}(\mathcal{E})\}$, where $\mathcal{E}$ is a consistent complete argument extension of the AAF associated with the AABA $\langle \mathcal{L}, (\mathcal{P} \setminus J) \cup I, \mathcal{A}, \overline{} \rangle$ with $E = ((\mathcal{H} \cap \mathcal{P}) \setminus J) \cup I$.*

Proof (Sketch). This is proved based on Theorem 4 and [19, Proposition 1]. □

Proposition 2. *Let $\langle \mathcal{L}, (\mathcal{R}, \mathcal{H}), \mathcal{A}, \overline{} \rangle$ be a h_ABA. Given a negative observation G, a pair (I, J) is a credulous anti-explanation of G w.r.t. an AABA $\langle \mathcal{L}, \mathcal{P}, \mathcal{A}, \overline{} \rangle$ with $\mathcal{H} \cap \mathcal{P}$ from the h_ABA under complete semantics iff there is a consistent answer set S of the ELP $\Pi \cup \Gamma(\mathcal{H}) \cup \Gamma^+(\mathcal{H}, \mathcal{H} \cap \mathcal{P}) \cup \{g \leftarrow not\ G\} \cup \{\leftarrow not\ g\}$ for a newly introduced atom g such that $G \notin S$, $\hat{I}^+ = S|_{\hat{\mathcal{H}}^+}$, $\hat{J}^- = S|_{\hat{\mathcal{H}}^-}$, $\hat{E} = S|_{\hat{\mathcal{H}}}$ and $S|_{\mathcal{I}} = \{in(a)|a \in \mathcal{E}$ s.t. $G \notin \mathtt{concs}(\mathcal{E})\}$, where $\mathcal{E}$ is a consistent complete argument extension of the AAF associated with the AABA $\langle \mathcal{L}, (\mathcal{P} \setminus J) \cup I, \mathcal{A}, \overline{} \rangle$ with $E = ((\mathcal{H} \cap \mathcal{P}) \setminus J) \cup I$.*

Proof: This is easily proved like Proposition 1. □

Example 8 (Cont. Examples 6 and 4). W.r.t. AF_U, i.e. the AABA $\mathcal{F}_4$, Π_{AF} is:
$\Pi_{AF} = \{ag(A_1) \leftarrow, ag(A_2) \leftarrow, ag(A_3) \leftarrow e, ag(A_4) \leftarrow \neg h, ag(A_5) \leftarrow e, ag(A_6) \leftarrow \neg h, def(A_3, A_1) \leftarrow ag(A_1), ag(A_3), def(A_3, A_2) \leftarrow ag(A_1), ag(A_3), \neg p \leftarrow in(A_1), a \leftarrow in(A_2), a \leftarrow in(A_3), p \leftarrow in(A_4), e \leftarrow in(A_5), \neg h \leftarrow in(A_6)\}$.

For $\mathcal{H} = \{e, \neg h\}$, let $\hat{\mathcal{H}} = \{e, \neg h\}$, $\hat{\mathcal{H}}^+ = \{e^+, \neg h^+\}$ and $\hat{\mathcal{H}}^- = \{e^-, \neg h^-\}$. Then $\Gamma(\mathcal{H}) = \{e' \leftarrow not\ e,\ e \leftarrow not\ e',\ h' \leftarrow not\ \neg h,\ \neg h \leftarrow not\ h'\}$ and $\Gamma^+(\mathcal{H}, \{e\}) = \{\neg h^+ \leftarrow \neg h, e^- \leftarrow not\ e\}$ for $E_2 = \{e\}$ are obtained. As a result, $\Pi \cup \Gamma(\mathcal{H}) \cup \Gamma^+(\mathcal{H}, \{e\})$ has three answer sets S_i $(1 \leq i \leq 3)$ as follows.

- $p \notin S_1$, $\neg p \in S_1$ where $S_1|_{\hat{\mathcal{H}}^+} = \{\}$, $S_1|_{\hat{\mathcal{H}}^-} = \{e^-\}$, $S_1|_{\hat{\mathcal{H}}} = \{\}$, $S_1|_{\mathcal{I}} = \{in(A_1), in(A_2)\} = \{in(a)|a \in \mathcal{E}_1\}$ for consistent $\mathcal{E}_1$,
- $p \notin S_2$, $\neg p \notin S_2$ where $S_2|_{\hat{\mathcal{H}}^+} = \{\ \}$, $S_2|_{\hat{\mathcal{H}}^-} = \{\}$, $S_2|_{\hat{\mathcal{H}}} = \{e\}$, $S_2|_{\mathcal{I}} = \{in(A_3), in(A_5)\} = \{in(a)|a \in \mathcal{E}_2\}$ for consistent $\mathcal{E}_2$,
- $p \in S_3$, $\neg p \notin S_3$ where $S_3|_{\hat{\mathcal{H}}^+} = \{\neg h^+\}$, $S_3|_{\hat{\mathcal{H}}^-} = \{\}$, $S_3|_{\hat{\mathcal{H}}} = \{e, \neg h\}$, $S_3|_{\mathcal{I}} = \{in(A_3), in(A_4), in(A_5), in(A_6)\} = \{in(a)|a \in \mathcal{E}_4\}$ for consistent $\mathcal{E}_4$.

Then given the observation p, $\Pi \cup \Gamma(\mathcal{H}) \cup \Gamma^+(\mathcal{H}, E_2) \cup \{\leftarrow not\ p\}$ has the unique answer set S_3. Thus due to $\hat{I}^+ = S_3|_{\hat{\mathcal{H}}^+} = \{\neg h^+\}$ and $\hat{J}^- = S_3|_{\hat{\mathcal{H}}^-} = \{\}$, we obtain $(I, J) = (\{\neg h\}, \emptyset)$ is a credulous explanation of p w.r.t. the AABA $\mathcal{F}_2$ with $E_2 = \{e\}$ under complete semantics.

On the other hand, whether a credulous (anti-)explanation (I, J) of an observation G w.r.t. a given AABA is skeptical can be decided based on the following proposition.

Proposition 3. *Let (I, J) be a credulous explanation (resp. anti-explanation) of an observation G w.r.t. an AABA $\langle \mathcal{L}, \mathcal{P}, \mathcal{A}, \overline{} \rangle$ with $\mathcal{H} \cap \mathcal{P}$ from the h_ABA $\langle \mathcal{L}, (\mathcal{R}, \mathcal{H}), \mathcal{A}, \overline{} \rangle$ under complete semantics. Then (I, J) is a skeptical explanation (resp. anti-explanation) of an observation G w.r.t. the AABA under complete semantics iff the ELP $\Pi \cup \hat{E} \cup \{\leftarrow G\}$ (resp. $\Pi \cup \hat{E} \cup \{g \leftarrow not\ G\} \cup \{\leftarrow g\}$ for a new atom g) has no answer sets for $\hat{E}$ expressing $E = ((\mathcal{H} \cap \mathcal{P}) \setminus J) \cup I$.*

Example 9 (Example 8 Cont.). For $(I, J) = (\{\neg h\}, \emptyset)$ and $E_2 = \{e\}$, $E = (E_2 \setminus J) \cup I = \{e, \neg h\}$. Then since $\Pi \cup \{e \leftarrow, \neg h \leftarrow\} \cup \{\leftarrow p\}$ has no answer sets, $(I, J) = (\{\neg h\}, \emptyset)$ is a skeptical explanation of p w.r.t. the AABA $\mathcal{F}_2$ with $\{e\}$.

Propositions 1, 2 and 3 also hold for stable semantics by replacing Π with $\Pi \cup \{\leftarrow undec(X)\}$ in the respective ELP.

5 Related Work and Conclusion

We defined consistency of ABA frameworks whose languages have explicit negation $\neg$. Then based on it, we formulated extended abduction w.r.t. such (abductive) ABA frameworks which can treat hypotheses along with assumptions. Finally we provided its computational method in answer set programming.

Booth et al. [2] studied to perform extended abduction in AA frameworks. They showed in [2, Theorem 4] that their AAF instantiated with an ALP $\langle P, H \rangle$ where P is an NLP and $H \subseteq HB_P$ can capture Sakama and Inoue's extended abduction w.r.t. the ALP under the partial stable semantics. Hence their theorem, i.e. [2, Theorem 4] may be regarded as the special case of our Theorem 3 since an ALP $\langle P, H \rangle$ where P is ELP and $H \subseteq Lit_P$ is considered to instantiate the AABA in Theorem 3. In other words, classical negation $\neg$ may occur in our ALPs, whereas it is not allowed in their ALPs when instantiating the respective abductive argumentation. Thus not only did they not need to take account of ensuring consistency of their instantiated AAFs but their applicable class is also far smaller than ours due to less expressive power of their ALPs.

Sakama formulated extended abduction in AA frameworks [17]. However he did not show how to instantiate his AA frameworks for abduction with ALPs. Hence he took account of neither the rationality postulates [4,9] nor consistency of his framework since every argument is abstract in his approach.

Wakaki et al. [19] proposed abductive argumentation whose underlying language is an ELP. However in their approach, hypotheses erasure is not considered (i.e. $J = \emptyset$) like the traditional normal abduction in logic programming [14].

Our future work is not only to implement the computational method shown in Sect. 4 by using the ASP solver (e.g. DLV) but also to apply our approach to various practical problems such as legal reasoning or dishonest reasoning where abductive explanations are required based on argumentation.

Appendix: Proof of Theorem 3

1. (I, J) is a skeptical explanation (resp. anti-explanation) of G w.r.t. $\langle P, H\rangle$ under paraconsistent stable model semantics
   iff $G \in M$ (resp. $G \notin M$) for *every* consistent p-stable model M of $(P \setminus J) \cup I$, where $(P \setminus J) \cup I$ is consistent under paraconsistent stable model semantics
   iff $G \in \texttt{Concs}(\mathcal{E}) \cap Lit_P$ (resp. $G \notin \texttt{Concs}(\mathcal{E}) \cap Lit_P$) for *every* consistent $\texttt{Concs}(\mathcal{E})$ where $\mathcal{E}$ is the consistent extension of the ABA $\langle \mathcal{L}_P, (P \setminus J) \cup I, Lit_{not}, \overline{}\rangle$ under stable semantics due to [21, Theorem 3]
   iff *every* consistent stable extension of the AABA $\langle \mathcal{L}_P, (P \setminus J) \cup I, Lit_{not}, \overline{}\rangle$ with $((H \cap P) \setminus J) \cup I$ from the h_ABA $\langle \mathcal{L}_P, (P \setminus H, H), Lit_{not}, \overline{}\rangle$ contains (resp. does not contain) an argument with claim G, where the AABA with $((H \cap P) \setminus J) \cup I$ is consistent under stable semantics
   iff (I, J) is a skeptical explanation (resp. anti-explanation) of G w.r.t. AABA $\langle \mathcal{L}_P, P, Lit_{not}, \overline{}\rangle$ with $H \cap P$ from the h_ABA under stable semantics.
2. (I, J) is a credulous explanation (resp. anti-explanation) of G w.r.t. $\langle P, H\rangle$ under paraconsistent stable model semantics
   iff $G \in M$ (resp. $G \notin M$) for *some* consistent p-stable model M of $(P \setminus J) \cup I$, where $(P \setminus J) \cup I$ is consistent under paraconsistent stable model semantics
   iff $G \in \texttt{Concs}(\mathcal{E}) \cap Lit_P$ (resp. $G \notin \texttt{Concs}(\mathcal{E}) \cap Lit_P$) for *some* consistent $\texttt{Concs}(\mathcal{E})$ where $\mathcal{E}$ is the consistent extension of the ABA $\langle \mathcal{L}_P, (P \setminus J) \cup I, Lit_{not}, \overline{}\rangle$ under stable semantics due to [21, Theorem 3]
   iff *some* consistent stable extension of the AABA $\langle \mathcal{L}_P, (P \setminus J) \cup I, Lit_{not}, \overline{}\rangle$ with $((H \cap P) \setminus J) \cup I$ from the h_ABA $\langle \mathcal{L}_P, (P \setminus H, H), Lit_{not}, \overline{}\rangle$ contains (resp. does not contain) an argument with claim G, where the AABA with $((H \cap P) \setminus J) \cup I$ is consistent under stable semantics
   iff (I, J) is a credulous explanation (resp. anti-explanation) of G w.r.t. AABA $\langle \mathcal{L}_P, P, Lit_{not}, \overline{}\rangle$ with $H \cap P$ from the h_ABA under stable semantics. □

References

1. Bondarenko, A., Dung, P.M., Kowalski, R.A., Toni, F.: An abstract, argumentation-theoretic approach to default reasoning. Artif. Intell. **93**, 63–101 (1997)
2. Booth, R., Gabbay, D.M., Kaci, S., Rienstra, T., Van Der Torre, L.W.: Abduction and dialogical proof in argumentation and logic programming. In: Proceedings of ECAI 2014, pp. 117–122 (2014)
3. Caminada, M.: On the issue of reinstatement in argumentation. In: Fisher, M., van der Hoek, W., Konev, B., Lisitsa, A. (eds.) JELIA 2006. LNCS (LNAI), vol. 4160, pp. 111–123. Springer, Heidelberg (2006). https://doi.org/10.1007/11853886_11
4. Caminada, M., Amgoud, L.: On the evaluation of argumentation formalisms. Artif. Intell. **171**(5–6), 286–310 (2007)
5. Caminada, M., Sá, S., Alcântara, J., Dvořák, W.: On the difference between assumption-based argumentation and abstract argumentation. In: BNAIC-2013, pp. 25–32 (2013). IFCoLog J. Log. Appl. **2**(1), 15–34 (2015)
6. Čyras, K., Toni, F.: ABA$^+$: assumption-based argumentation with preferences. In: Proceedings of KR 2016, pp. 553–556 (2016)

7. Dung, P.M.: On the acceptability of arguments and its fundamental role in nonmonotonic reasoning, logic programming and n-person games. Artif. Intell. **77**, 321–357 (1995)
8. Dung, P.M., Kowalski, R.A., Toni, F.: Assumption-based argumentation. In: Simari, G., Rahwan, I. (eds.) Argumentation in Artificial Intelligence, pp. 199–218. Springer, Boston (2009). https://doi.org/10.1007/978-0-387-98197-0_10
9. Dung, P.M., Thang, P.M.: Closure and consistency in logic-associated argumentation. J. Artif. Intell. Res. **49**, 79–109 (2014)
10. Gelfond, M., Lifschitz, V.: The stable model semantics for logic programming. In: Proceedings of ICLP/SLP-1988, pp. 1070–1080. MIT Press (1988)
11. Gelfond, M., Lifschitz, V.: Classical negation in logic programs and disjunctive databases. New Gener. Comput. **9**, 365–385 (1991)
12. Inoue, K., Sakama, C.: Abductive framework for nonmonotonic theory change. In: Proceedings of IJCAI 1995, pp. 204–210 (1995)
13. Inoue, K., Sakama, C.: Specifying transactions for extended abduction. In: Proceedings of KR 1998, pp. 394–405 (1998)
14. Kakas, A.C., Kowalski, R.A., Toni, F.: The role of abduction in logic programming. In: Handbook of Logic in Artificial Intelligence and Logic Programming, vol. 5, pp. 235–324. Oxford University Press (1998)
15. Sakama, C., Inoue, K.: Paraconsistent stable semantics for extended disjunctive programs. J. Log. Comput. **5**(3), 265–285 (1995)
16. Sakama, C., Inoue, K.: An abductive framework for computing knowledge base updates. Theory Pract. Log. Program. **3**(6), 671–713 (2003)
17. Sakama, C.: Abduction in argumentation frameworks and its use in debate games. In: Nakano, Y., Satoh, K., Bekki, D. (eds.) JSAI-isAI 2013. LNCS (LNAI), vol. 8417, pp. 285–303. Springer, Cham (2014). https://doi.org/10.1007/978-3-319-10061-6_19
18. Wakaki, T., Nitta, K.: Computing argumentation semantics in answer set programming. In: Hattori, H., Kawamura, T., Idé, T., Yokoo, M., Murakami, Y. (eds.) JSAI 2008. LNCS (LNAI), vol. 5447, pp. 254–269. Springer, Heidelberg (2009). https://doi.org/10.1007/978-3-642-00609-8_22
19. Wakaki, T., Nitta, K., Sawamura, H.: Computing abductive argumentation in answer set programming. In: McBurney, P., Rahwan, I., Parsons, S., Maudet, N. (eds.) ArgMAS 2009. LNCS (LNAI), vol. 6057, pp. 195–215. Springer, Heidelberg (2010). https://doi.org/10.1007/978-3-642-12805-9_12
20. Wakaki, T.: Assumption-based argumentation equipped with preferences and constraints. In: Moral, S., Pivert, O., Sánchez, D., Marín, N. (eds.) SUM 2017. LNCS (LNAI), vol. 10564, pp. 178–193. Springer, Cham (2017). https://doi.org/10.1007/978-3-319-67582-4_13
21. Wakaki, T.: Assumption-based argumentation equipped with preferences and its application to decision-making, practical reasoning, and epistemic reasoning. J. Comput. Intell. **33**(4), 706–736 (2017)

Integrative Cognitive and Affective Modeling of Deep Brain Stimulation

Seyed Sahand Mohammadi Ziabari(✉)

Social AI Group, Vrije Universiteit Amsterdam, De Boelelaan 1105, Amsterdam, The Netherlands
sahandmohammadiziabari@gmail.com

Abstract. In this paper a computational model of Deep Brain Stimulation (DBS) therapy for post-traumatic stress disorder is presented. The considered therapy has as a goal to decrease the stress level of a stressed individual by using electrode which placed in a specific area in brain. Several areas in brain have been used to decrease the stress level, one of them is Amygdala. The presented temporal-causal network model aims at integrative modeling a Deep Brain Stimulation therapy where the relevant brain areas are modeled in a dynamic manner.

Keywords: Deep Brain Stimulation · Amygdala · Network-Oriented Modeling · PTSD

1 Introduction

Post-Traumatic Stress Disorder (briefly PTSD) [15] is a severe psychiatric mental problem that might happen in an individual after serious trauma or extreme stress. PTSD can generate intense apprehension, impuissance in persons [1]. It has been known that electrical stimulation of the amygdala has a remarkable effect and changes on emotional, perceptual, and behavioral functioning.

What is deep brain stimulation is completely explained in [6, p. 406, 12]:

> 'The implementation of DBS electrodes is a neurosurgical procedure, often performed under local anesthesia with patients fully awake to facilitate precise stimulation mapping. After placement of the electrodes in the desired target below the convexity of the brain, they are connected to a programmable pulse generator, similar to modern cardiac pacemakers, that is implanted under the skin below the collarbone.'

There are variety of indications for PTSD such as insomnia, hypervigilance, and stress [14]. Some brain components are involved in having reaction to the stress like cerebral Cortex, Hippocampus, Hypothalamus, Amygdala [12]. The Network-Oriented Modeling approach based on temporal-causal network models described in [16] is such a modeling approach, and has been used here.

The paper is organized as follows. In Sect. 2 the underlying biological and neurological principles concerning the parts of the brain involved in stress and in the suppression of stress are addressed. In Sect. 3 the integrative temporal-causal network model is introduced. In Sect. 4 simulation results of the model are discussed, and eventually in the last section a conclusion is presented.

F. Wotawa et al. (Eds.): IEA/AIE 2019, LNAI 11606, pp. 608–615, 2019.
https://doi.org/10.1007/978-3-030-22999-3_52

2 Underlying Biological and Neurological Principles

As have been discussed in the literature a deep brain stimulation (DBS) has an important role in treatment with PTSD; e.g., [1]. There are many other therapies which have been explained in [18–27]. For instance, the cognitive models of using Ritalin and doing yoga to decrease the stress level presented in [24] and [25], respectively. The efficiency of deep brain stimulation and also the method is introduced precisely in [1, p. 1]:

> 'Deep Brain Stimulation (DBS) has shown promise in refractory movement disorders, depression and obsessive-compulsive disorder, with deep brain targets chosen by integration of clinical and neuroimaging literature. The basolateral amygdala (BLn) is an optimal target for high-frequency DBS in PTSD based on neurocircuitry findings from a variety of perspective.'
> 'An electroencephalographic [EEG] telemetry session will test safety of stimulation before randomization to staggered-onset, double-blind sham versus active stimulation for two months.'
> 'Deep Brain Stimulation (DBS) refers to the process of delivering an electrical current to a precise location in the brain. DBS is now a common clinical practice. DBS is an invasive treatment, and the potential for benefits must clearly outweigh the risks.'

It notable that only stimulation of the BLn [11] has remarkable effect on curing PTSD as stated in [1, p. 7]:

> 'DBS of gray matter, such as the BLn, would likely have effects opposite those seen from DBS of the white matters in the case of the amygdala.
>
> 1. It supports the model that amygdala overactivity is responsible for the symptoms of PTSD. DBS of the stria terminals physiologically equates to amygdala overactivity. In this patient with no psychiatric history, an increase in amygdala output activity has led to symptoms commonly seen in combat PTSD (for example, irritability, helplessness, depression and suicidality.)
> 2. The correction of the amygdala activity led to resolution of the symptoms. As opposed to DBS of the stria terminals, BLn DBS is expected to reduce amygdala.
> 3. The patient did not suffer from a seizure or deterioration in neuropsychological status during chronic DBS for nine months. This suggests an acceptable safety profile of DBS of the amygdala.'
>
> 'Notably, only stimulation of the BLn, but not the central nucleus, the amygdala outflow tract, or neighboring regions affected by other contacts of the stimulating electrodes, led to benefit. Also important was the absence of significant side effects, including seizures.
> While we thus agree that the BLn is not the only potentially valuable target for DBS in PTSD, we believe, based on the rationale described above, that it is the optimal target and it can be modulated safely with current technology.'

Targeting the exact part of the amygdala named Basolateral nucleus (BLn) is more complicated than it is assumed the exact region is defined in [12];

> 'Targeting the BLn is complicated due to anatomical variations in this region. Using a stereotactic atlas, the inferior limit of the BLn is located 16 mm lateral to the AC, 4 mm posterior to the AC and 18 mm inferior to the AC-PC plane.'

Changing in functionality in noradrenergic neurons is believed to be involved in hyperarousal and reexperiencing symptoms of PTSD [2].

PTSD is also defined clearly in [1, p. 2];

'PTSD was characterized by three clusters of psychiatric symptoms. The first, re-experiencing, involves the emotional and perceptual reliving of traumatic events either spontaneously or in response to 'triggers' that remind one of the events because they bear some similarity to the original circumstance.'

There are many therapies for PTSD such as Exposure therapy (PE) [3], Cognitive Processing Therapy [4], and Eye Movement Desensitization and reprocessing (EMDR) [5].

The connection between cerebellum and striatum, amygdala and striatum has been explained in [7, p. 489], [9].

'Inputs from the cortex primarily project to the striatum.' 'The striatum glutamatergic inputs from several areas, including the cortex, hippocampus, amygdala, and thalamus.'

In [10] they claimed that direct stimulation of the vmPFC can prevent the responsiveness of the amygdala. In [10] it has been claimed that amygdalotomy may regulate and moderate stress level.

3 The Temporal-Causal Network Model

First the Network-Oriented Modelling approach used to model the integrative overall process is briefly explained. As discussed in detail in [16, Chap. 2] this approach is based on temporal-causal network models which can be represented at two levels: by a conceptual representation and by a numerical representation (Fig. 1 and Table 1).

- **Strength of a connection $\omega_{X,Y}$.** Each connection from a state X to a state Y has a *connection weight value* $\omega_{X,Y}$ representing the strength of the connection, often between 0 and 1, but sometimes also below 0 (negative effect) or above 1.
- **Combining multiple impacts on a state $\mathbf{c}_Y(..)$.** For each state a *combination function* $\mathbf{c}_Y(..)$ is chosen to combine the causal impacts of other states on state Y.
- **Speed of change of a state η_Y.** For each state Y a *speed factor* η_Y is used to represent how fast a state is changing upon causal impact.

Table 1. Explanation of some states and their relation to neurological principles

States	Neurological principles	Quotation, References
ws_c	External stressor	External stress-inducing event [16, 17]
ws_{ee}	World (body) state of extreme emotion *ee*	External stress-inducing event [16]
ss_c	Sensor state for perception of the stressor	'Human states can refer, for example, to states of body parts to see (Eyes), hear (ears) and fee (skin).' In [16, p. 51]
srs_{ee}	Sensory and Feeling representation of stressful event	'The dACC was activated during the observe condition. The dACC is associated with attention and the ability to accurately detect emotional signals.' [2, p. 12] and [13]

(continued)

Table 1. *(continued)*

States	Neurological principles	Quotation, References
goal (electrode)	Executive function of stimulation	'Using electrode on basolateral nucleus of Amygdala for Deep brain stimulation'
Hippocampus	Brain parts	Brain parts
Thalamus	Brain parts	Brain parts
Lateral nucleus	component of Amygdala	Lateral part of the Amygdala in brain part
Basolateral nucleus	component of Amygdala (influential for stimulation)	Middle part of the Amygdala in brain part
mPFC	medial Prefrontal Cortex (Reasoning part of brain)	Brain parts
Cerebellum	Brain part	'*The cerebellum receives information from the sensory systems, the spinal cord, and other parts of the brain and then regulates motor movements. The cerebellum coordinates voluntary movements such as posture, balance, coordination, and speech, resulting in smooth and balanced muscular activity.*' [8]
Striatum, mPFC, Hippocampus, Thalamus	Brain part	Brain parts

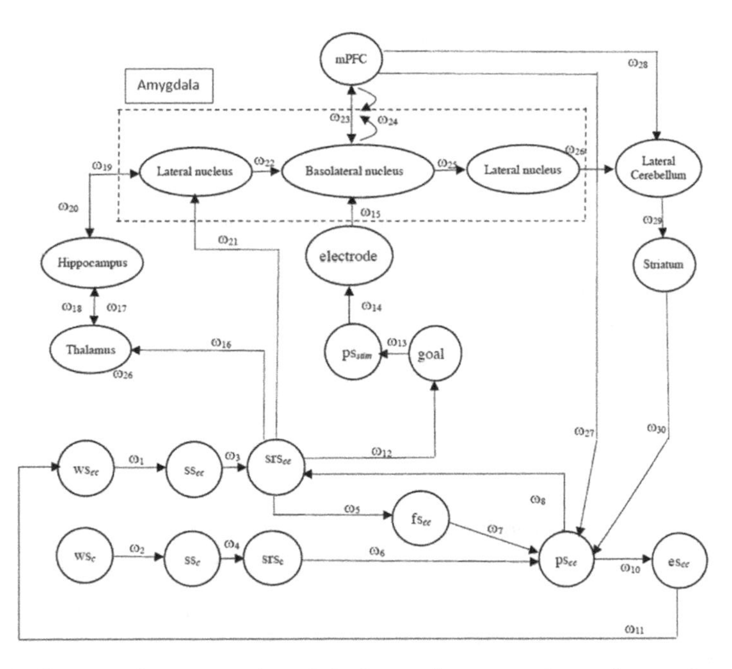

Fig. 1. Conceptual representation of the integrative temporal-causal network model

4 Example Simulation

An example simulation of this process is shown in Fig. 2. Table 2 shows the connection weights used, where the value for ω_{24} is initial value as these weights are adapted over time. The model implemented in the Matlab template provided in [22].

Table 2. Connection weights and scaling factors for the example simulation

Connection weight	ω_1	ω_2	ω_3	ω_4	ω_5	ω_6	ω_7	ω_8	ω_9	ω_{10}	ω_{11}	ω_{12}
Value	1	1	1	1	1	1	1	1	1	1	1	1
Connection weight	ω_{13}	ω_{14}	ω_{15}	ω_{16}	ω_{17}	ω_{18}	ω_{19}	ω_{20}	ω_{21}	ω_{22}	ω_{23}	ω_{24}
Value	1	1	1	1	1	1	1	1	1	1	0.1	0.1
Connection weight	ω_{25}	ω_{26}	ω_{25}	ω_{26}	ω_{27}	ω_{28}	ω_{29}	ω_{30}				
Value	1	1	1	1	−0.9	1	1	−0.5				

Table 3 shows the states which use scale sum function as their combination function. The other states except X_3 (world state of context) and X_{10} (goal) all contains identity function as they have only one incoming connection weight.

Table 3. Scaling factors for the example simulation of states with scale sum function

State	X_5	X_8	X_{13}	X_{14}	X_{15}	X_{16}	X_{17}	X_{19}
λ_i	2	0.8	2	2	2	1	2.1	4

As a biological process, the electrode triggers the activation of Basolateral nucleus in Amygdala at the first state and this impacts other brain parts such as the remarkable one mPFC, left and right lateral nucleus.

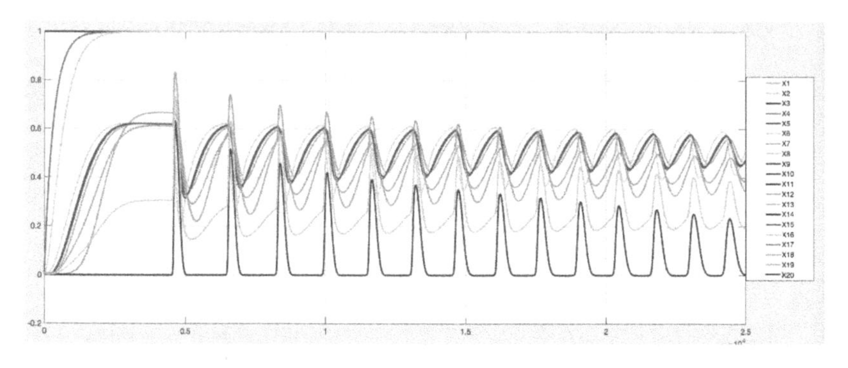

Fig. 2. Simulation results for Deep Brain Stimulation (DBS)

Figure 3 shows a simulation result of using Hebbian learning principle between states basolateral nucleus in amygdala and medial Prefrontal Cortex in one period of using deep brain stimulation.

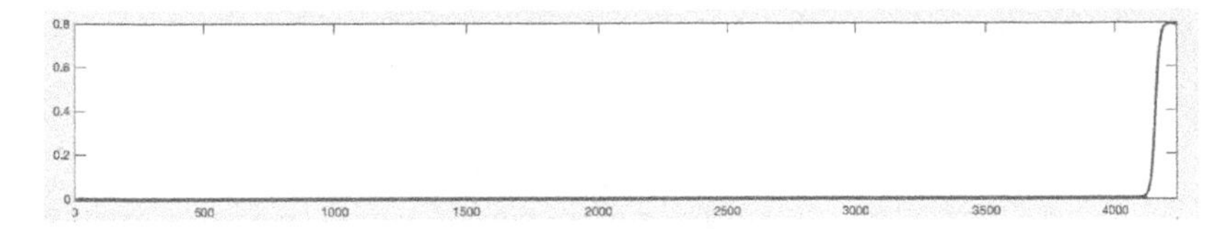

Fig. 3. Simulation results for Deep Brain Stimulation (DBS), Adaptive connections

5 Mathematical Verification

To verify the model that has been proposed mathematical analysis used. For some states in the model the mathematical analysis has been done. Table 4 shows this result. By using the WIMS Linear Solver[1], the following (unique) algebraic solution was obtained for the general case of these equations:

$$X1 = X9$$
$$X2 = X1$$
$$X3 = 1$$
$$X4 = X3$$
$$2 * X5 = 1 * X2 + 1 * X8$$
$$X6 = X4$$
$$X7 = X5$$
$$2 * X8 = 1 * X6 + 1 * X7 - 0.99 * X16 - 0.5 * X20$$
$$X9 = X8$$
$$X10 = 0.2$$
$$X11 = X10$$
$$X12 = X11$$
$$2 * X13 = 1 * X5 + 1 * X14$$
$$2 * X14 = X13 + X15$$
$$X15 = 1 * X14$$
$$X16 = 0.1^{*} X17$$
$$X17 = 0.8 * X14$$
$$X18 = 1 * X17$$
$$2 * X19 = 0.95 * X15$$
$$X20 = X19$$

[1] https://wims.unice.fr/wims/wims.cgi?session=K06C12840B.2&+lang=nl&+module=tool%2Flinear%2Flinsolver.en.

Table 4. Scaling factors for the example simulation of states with scale sum function

State	ws_{ee} X_1	ss_{ee} X_2	ws_c X_3	ss_c X_4	srs_{ee} X_5	srs_c X_6	fs_{ee} X_7	ps_{ee} X_8
Simulation	0.7056	0.7047	1,0000	1,0000	0.7029	1,0000	0.6913	0.7209
Analysis	0.7594	0.7594	1,0000	1,0000	0.7594	0.9987	0.7594	0.7594
Deviation	0.0538	0.0547	0,0000	0,0000	0.0565	0.0013	0.0681	0,0385

6 Conclusion

In this paper an integrative cognitive and affective model of therapy by using deep brain stimulation (DBS) for Post-Traumatic Stress Disorder (PTSD) is introduced in which electrode is used. This affects the preparation state of stress and enables to release the stressed person from a chronic stress. This model can be used as the basis of a virtual agent model to get insight in such processes and to bring up a certain cure or treatment of individuals to perform the therapies of extreme emotions for post-traumatic disorder persons.

References

1. Koek, R.J., et al.: Deep brain stimulation of the basolateral amygdala for treatment-refractory combat post-traumatic stress disorder (PTSD): study protocol for a pilot randomized controlled trial with blinded, staggered onset of stimulation. Trials **15**, 356 (2014). https://doi.org/10.1186/1745-6215-15-356
2. Southwick, S.M., et al.: Noradrenergic and serotonergic function in posttraumatic stress disorder. Arch. Gen. Psychiatry **54**, 749–758 (1997)
3. Foe, E., et al.: Randomized trial of prolonged exposure for posttraumatic stress disorder with and without cognitive restricting: outcome at academic and community clinics. J. Consult. Clin. Psychol. **73**, 953–962 (2005)
4. Monson, C.M., Schnurr, P.P., Resick, P.A., Friedman, M.J., Young-Xu, Y., Stevens, S.P.: Cognitive processing therapy for veterans with military-related posttraumatic stress disorder. J. Consult. Clin. Psychol. **74**, 898–907 (2006)
5. Lee, C., Gavriel, H., Drummond, P., Richards, J., Greenwald, R.: Treatment of PTSD: stress inoculation training with prolonged exposure compared to EMDR. J. Clin. Psychol. **58**, 1071–1089 (2002)
6. Andres, M.L., Lipsman, N.: Probing and regulating dysfunctional circuits using deep brain stimulation. Neuron **77**(3), 406–424 (2013). https://doi.org/10.1016/j.neuron.2013.01.020
7. Gazzaniga, S.G., Ivry, R.B., Mangun, G.R.: Cognitive Neuroscience, The Biology of Mind, 2nd edn (2002)
8. https://www.healthline.com/human-body-maps/cerebellum#1
9. Britt, J.P., Benaliouad, F., McDevitt, R.A., Stuber, G.D., Wise, R.A., Bonci, A.: Synaptic and behavioral profile of multiple glutamatergic inputs to the nucleus Accumbens. Neuron **76**, 790–803 (2012). https://doi.org/10.1016/j.neuron.2012.09.040
10. Quirk, G.J., Likhtik, E., Pelletier, J.G., Pare, D.: Stimulation of medial prefrontal cortex decreases the responsiveness of central amygdala output neurons. J. Neurosci. **23**, 8800–8807 (2003)
11. Lee, G.P., et al.: Clinical and physiological effects of stereotaxic bilateral amygdalotomy for intractable aggression. J. Neuropsychiatry Clin. Neurosci. **10**, 413–420 (1998)

12. Langevin, J.P., et al.: Deep brain stimulation of the basolateral amygdala: targeting technique and electrodiagnostic findings. Brain Sci. **6**, 28 (2016)
13. Foote, S.L., Aston-Jones, G., Bloom, E.F.: Impulse activity of locus coeruleus neurons in awake rats and monkeys is a function of sensory stimulation and arousal. Proc. Natl. Acad. Sci. USA **77**, 3033–3037 (1980)
14. Kozarić-Kovačić, D.: Psychopharmacotherapy of posttraumatic stress disorder. Croat. Med. J. **49**(4), 459–475 (2008). https://doi.org/10.3325/cmj.2008.4.459. PMID:18716993
15. Coupland, N.J.: Brain mechanisms and neurotransmitters. In: Post-traumatic Stress Disorder-Diagnosis, Management and Treatment, pp. 69–100. Martin Dunitz, London (2000)
16. Treur, J.: Network-Oriented Modeling: Addressing Complexity of Cognitive, Affective and Social Interactions. Springer, Cham (2016). https://doi.org/10.1007/978-3-319-45213-5
17. Koelsch, S.: Towards a neural basis of music-evoked emotions. Trends Cogn. Sci. **14**, 131–137 (2010). https://doi.org/10.1016/j.tics.2010.01.002
18. Mohammadi Ziabari, S.S., Treur, J.: Cognitive modeling of mindfulness therapy by autogenic training. In: Satapathy, S.C., Bhateja, V., Somanah, R., Yang, X.-S., Senkerik, R. (eds.) Information Systems Design and Intelligent Applications. AISC, vol. 863, pp. 53–66. Springer, Singapore (2019). https://doi.org/10.1007/978-981-13-3338-5_6
19. Mohammadi Ziabari, S.S., Treur, J.: Computational analysis of gender differences in coping with extreme stressful emotions. In: Proceedings of the 9th International Conference on Biologically Inspired Cognitive Architecture (BICA 2018). Elsevier, Amsterdam (2018)
20. Mohammadi Ziabari, S.S., Treur, J.: An adaptive cognitive temporal-causal network model of a mindfulness therapy based on music. In: Tiwary, U.S. (ed.) IHCI 2018. LNCS, vol. 11278, pp. 180–193. Springer, Cham (2018). https://doi.org/10.1007/978-3-030-04021-5_17
21. Treur, J., Mohammadi Ziabari, S.S.: An adaptive temporal-causal network model for decision making under acute stress. In: Nguyen, N.T., Pimenidis, E., Khan, Z., Trawiński, B. (eds.) ICCCI 2018. LNCS (LNAI), vol. 11056, pp. 13–25. Springer, Cham (2018). https://doi.org/10.1007/978-3-319-98446-9_2
22. Mohammadi Ziabari, S.S., Treur, J.: A modeling environment for dynamic and adaptive network models implemented in Matlab. In: Proceedings of the 4th International Congress on Information and Communication Technology (ICICT 2019), 25–26 February 2019. Springer, London (2019)
23. Mohammadi-Ziabari, S.S., Treur, J.: Integrative biological, cognitive and affective modeling of a drug-therapy for a post-traumatic stress disorder. In: Fagan, D., Martín-Vide, C., O'Neill, M., Vega-Rodríguez, M.A. (eds.) TPNC 2018. LNCS, vol. 11324, pp. 292–304. Springer, Cham (2018). https://doi.org/10.1007/978-3-030-04070-3_23
24. Lelieveld, I., Storre, G., Mohammadi Ziabari, S.S.: A temporal cognitive model of the influence of methylphenidate (Ritalin) on test anxiety. In: Proceedings of the 4th International Congress on Information and Communication Technology (ICICT 2019), 25–26 February 2019. Springer, London (2019)
25. Andrianov, A., Guerriero, E., Mohammadi Ziabari, S.S.: Cognitive modeling of mindfulness therapy: effects of yoga on overcoming stress. In: Proceedings of the 16th International Conference on Distributed Computing and Artificial Intelligence (DCAI 2019), 26-28 June. Avila, Spain (2019)
26. E de Haan, R., Blanker, M., Mohammadi Ziabari, S.S.: Integrative biological, cognitive and affective modeling of caffeine use on stress. In: Proceedings of the 16th International conference on Distributed Computing and Artificial Intelligence (DCAI 2019), 26-28 June. Avila, Spain (2019)
27. Mohammadi Ziabari, S.S.: An adaptive temporal-causal network model for stress extinction using fluoxetine. In: Proceedings of the 15th International Conference on Artificial Intelligence Applications and Innovations (AIAI 2019), 24–26 May 2019, Crete, Greece (2019)

Intelligent Online Configuration for DVFS Multiprocessor Architecture: Fuzzy Approach

Najar Yousra[1(✉)] and Ben Ahmed Samir[2]

[1] High Institute of Computing (ISI), University of Tunis el Manar, Tunis, Tunisia
Yousra_najar@yahoo.co.in, yousra.najar@isi.utm.tn
[2] Faculty of Science, Mathematics, Physics and Natural (FST), University of Tunis el Manar, Tunis, Tunisia

Abstract. The use of fuzzy logic to generate optimal actions for hardware architecture reconfiguration offers flexible and efficient solutions. In this paper, a new fuzzy approach is proposed in order to guarantee the balance between real time periodic application schedulability and energy consumption optimization under multi-core architecture. Dynamic voltage/frequency scaling (DVFS) has been a key technique in exploiting the processors configurable characteristics. However, for large class of applications in embedded real time systems, the variable operating frequency interferes with tasks deadline respect. The problem is seen as multi-criteria multi-objective decision making issue with dependent criteria. The approach calculates, in offline mode and in online mode, the optimal number of activated homogenous cores and their frequency. Simulated and tested on periodic task sets generated with different system charges, the proposed intelligent technique is support decision system that shows significant results.

Keywords: Fuzzy Decision Making (FDM) · Dynamic Voltage/Frequency Scaling (DVFS) · CPU energy consumption · Periodic real time application · Multiprocessor architecture · Schedulability

1 Introduction

Decision Making is the act of choosing between two or more courses of action. However, it must always be remembered that there may not always be a 'correct' decision among the available choices [28]. The theory of decision making formed a basis for more systematic and rational decision making especially in the situation where multiple criteria need to be accounted. This decision theory does not take so much time to fully recognized with the four terms consolidated to be known as multi criteria decision making (MCDM). Fuzzy logic approach is suitable to use when the modeling of human knowledge is necessary and when human knowledge, reasoning and evaluation are needed. This idea was first introduced in the paper [1], it was followed by many researches having the same direction. Applications of fuzzy logic, once thought to be an obscure mathematical curiosity, can be found in many engineering and scientific works. Fuzzy logic has been used in numerous applications such as facial pattern

F. Wotawa et al. (Eds.): IEA/AIE 2019, LNAI 11606, pp. 616–627, 2019.
https://doi.org/10.1007/978-3-030-22999-3_53

recognition, air conditioners, washing machines, vacuum cleaners, antiskid braking systems, transmission systems, control of subway systems and unmanned helicopters, knowledge-based systems for multi-objective optimization of power systems, weather forecasting systems, models for new product pricing or project risk assessment, medical diagnosis and treatment plans, and stock trading [18, 19]. Using fuzzy logic algorithms could enable machines to understand and respond to vague human concepts without having an exhaustive mathematical model. Almost every application, including embedded control, could reap some benefits from fuzzy logic. The use of fuzzy logic to optimize the configuration of hardware architecture was adopted because it offers some advantages [27]. Its incorporation in embedded systems could lead to enhanced performance, increased simplicity and productivity. First, detailed complex modeling of the system is not needed and human logic knowledge is captured. Second, it offers easily adaptive and light weight design support system [4, 5, 27].

2 Research Issue

The usages of energy are growing rapidly with the increase of portable devices, embedded system, automation and much real time devices. Research is going on to provide better power efficiency both at hardware and software level [2]. Dynamic voltage and frequency scaling (DVFS) is a well-known method for reducing power consumption in modern real time applications [3]. Dynamic power, in its cost major parts, is consumed because of the gates switching that dissipated in complementary metal-oxide-semiconductor (CMOS) circuits. DVFS has been easily implemented in real-time system under timing constraints where tasks can be executed with lower CPU frequency. Hence, it could reduce power dissipation especially when the energy leakage is not important. However, it also, significantly, affects real time application performance because lower CPU frequency increases tasks execution time making time contrasts respect a significant issue. Therefore, the chosen scaling factor must give best possible trade-off between energy reduction and time performance. The main challenge is to guarantee that the deadlines will not be violated when frequency is reduced to decrease energy consumption. Besides, multiprocessor architectures are becoming one of the most used solutions in order to meet growing computation requirements of modern applications. Hardware configuration, in this work, concerns design parameters: number of cores and their frequencies. Hence, we treated this issue as multiple criteria multiple objective decision making problem.

In this paper, we present fuzzy decision making system (FDMS) that configures hardware multiprocessor architecture using two design parameters: the number of activated cores and their frequencies. The main objective of the FDMS is to respect the application time constraints and to optimize CPU energy consumption. Therefore, our approach calculates offline parameters before runtime then it updates them during runtime when application temporal parameters change. This paper describes the contribution of the approach in three main parts: The first paragraph is a state of art about dynamic frequency scaling in multiprocessing architecture and about fuzzy logic in embedded systems. The second part is dedicated to the proposed fuzzy decision approach details. Finally, simulations on generated periodic sets with different system charges are detailed to evaluate the approach performance.

3 State of Art: Intelligent DVFS Approaches

Methods used to reduce energy consumption with DVFS are classified into offline scaling factor selection methods and online scaling factor selection methods. While the number of cores is fixed, offline methods return static scaling factor values to identify the speed of processors participating in the execution of tasks. Different methods select the appropriate scaling factor values to eliminate slack times before runtime. Slack time is the amount of time left after a job is finished. Heuristic or an exact method uses the retrieved information to compute the values of scaling factor for processors. However, online scaling factor selection methods are executed during the runtime of the program. They are usually integrated when the instances of tasks are finished. Using information about actual execution time (AET), slack time and energy consumption, scaling factor is updated for each processor. Significant research and development efforts have been made to study voltage and frequency scaling in MPSOC architecture [8]. Many DVFS algorithms are used to raise architecture performances. Typically, they use simple feedback mechanism, such as detecting the amount of idle time on processors over a period of time and then adjust the frequency and voltage to just optimize computational load [10, 11]. The main idea of these studies is about avoiding wasting cycles by reducing the operating frequency which optimizes energy consumption. However, they cannot provide any timeliness guarantees and real time tasks may miss their deadlines. In fact, these algorithms are tightly-coupled with operating system's task management services, since the dynamic selection of DVFS must be coupled to task computation time [10]. Intelligent DVFS approaches are essentially based on Tree decision method, neural networks, deep learning [24, 25], fuzzy logic [4, 5, 16, 17, 23, 26] …

In [23] the author developed an on-line adaptive fuzzy logic controller for DVFS that is able to accurately and robustly predict and track the workload variations even when those variations are highly non stationary or soft. The fuzzy logic controller decides about changing the supply voltage of the circuit under control by observing and predicting the supply-current variations.

The use of heuristic based methodologies to design space optimization for MPSoC involves the knowledge of processor characteristics in order to converge toward a final solution. Qadri in [4, 5] presented a fuzzy based reconfiguration for MPSoc that takes into consideration energy consumption and throughout of real time system. This work argues the advantages of integration fuzzy inference system that calculates the number of cores and the operating frequency. A book is published by springer [27] collected fuzzy logic based approaches that managed multicore architecture configuration for Intel and Xeon. In [26] a FIS is proposed with power and throughputs as criteria and frequency as decision alternative.

4 Energy Model and Schedulability Test Under DVFS for Homogenous Multiprocessor Platform

Since DVFS lowers core frequency, execution time of a program running over that scaled down processor may increase, especially if the program is compute bound. The frequency reduction process can be expressed by the scaling factor S which is the ratio between the highest available frequency (f_{max}) for the CPU and the new frequency (f_{new}) applied to the CPU, as in Eq. (1).

$$S = f_{max}/f_{new} \tag{1}$$

4.1 Periodic Application Schedulability

In this paper, we assume a periodic task model in which each task Γ_i is defined by (r_{i1}, $WCET_i$, AET_i, T_i, D_i): r_i: release time, T_i: period, D_i: deadline and computation time that could be $WCET_i$ (worst case computation time) or AET_i (Actual Execution Time). Furthermore, we focus our attention on implicit deadline task system ($C_i < T_i = D_i$). As in classical treatments of the real-time scheduling problem [14], the relative deadline is assumed to be equal to period so a task must complete its execution before its next release. Given that the worst case computation time is supposed to be the time needed by the task to complete its charge in maximum operating frequency, a real time scheduler allocates processor time to the task in such way that it respects its deadline. In this section, we try to establish a relationship between operating frequency and system schedulability. A system is schedulable when all its tasks respect their deadline under designed architecture and specific real time scheduler. Many schedulability tests where associated with several schedulers such as global EDF (earliest deadline first). For a set of n periodic tasks, when we assume that the platform is composed of m identical processors and that U_n is system utilization, system schedulability is assured if this sufficient and necessary condition is satisfied [13–15]:

$$U_n = \sum_{i=1}^{n} \frac{C_i}{T_i} \leq \frac{m}{S} \tag{2}$$

4.2 Energy Model

The energy consumption of a program is given by formula Eq. (3) [9]. Many searches divide the power consumed by a processor into two power metrics: static power and dynamic power [4, 6, 7, 10, 12]. The first one is consumed as long as the computing unit is on; the latter is only consumed during computation times.

$$E = \int_{0}^{tmax} P(t).dt \; with \, P = P_{dyn} + P_{stat} \tag{3}$$

The dynamic power P_{dyn} is related to the switching activity α, load capacitance C_L, the supply voltage V and operational frequency f, as shown in Eq. (4) [22].

$$P_{dyn} = \alpha \cdot C_L \cdot V^2 \cdot f \tag{4}$$

For earlier processors static power was considered to be negligible [22]:

$$P_{static} = V.N.k_{design} \cdot l_{leak} \tag{5}$$

With N: Number of transistor, k_{design}: design dependent parameter, l_{leak}: technology dependent parameter. The dynamic power is reduced by a factor of S when reducing the frequency of a processor by a factor of S. The energy consumption is measured in Joule, and can be calculated by multiplying the power consumption, measured in watts, by the execution time of the program as follows: Energy = Power.T.

5 Fuzzy Approach to Optimize Energy Consumption and Schedulability

A decision making process involves: Identifying the objective/goal of the decision making process, selection of the criteria, selection of alternatives and aggregation Method. There are two main objectives in this case: minimize energy consumption and guarantee system schedulability under global EDF. Total system charge and initial frequency are approach criterion. In fact, mathematical relation between system charge and the frequency make it a decision-making with interdependent multiple criteria. Decision-making with interdependent multiple criteria is a surprisingly difficult task. If we have clearly conflicting objectives there normally is no optimal solution which would simultaneously satisfy all the criteria [28]. In fact in this case, minimizing CPU speed reduces energy (realizes first objective) but rises computation time and system load (contraries second objective).

Figure 1 represents the proposed approach. Fuzzy inference system (FIS) has two inputs U_n and actual CPU frequency f_o and calculates the number of processor m and the new CPU frequency f_n. This module is in closed loop with energy aware algorithm module that takes (m,f_n) to calculate the approximated consumed energy and the new U_n. Until stabilization, that's when FIS generates the same solution, the loop continues to iterate.

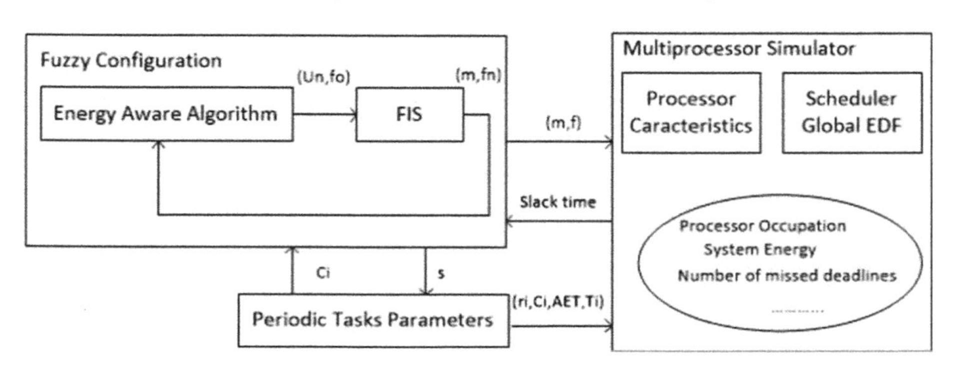

Fig. 1. Fuzzy configuration energy/feasibility aware approach

The main idea of the proposed architecture is to take advantage of fuzzy logic to avoid the establishment of a model describing the exact relations between input and output parameters. The fuzzy system involves: input and output variables identification, associating variables with appropriate membership functions, formulating rule base and defuzzification.

In this problem, both static and dynamic energy for each processor are dependent on frequency and scaling factor. Figure 2 shows the variation of estimated normalized energy depending on frequency values which confirms their proportional relationship. Thus, FIS inputs are U_n and f_o respectively system load and old frequency that it is the frequency adopted in $(t - 1)$. It is obvious in that case that the two criteria are dependent and they represent system state at the instant $(t - 1)$. The FIS outputs are proposed actions $a_i = (m, f_n)$ for the instant t. Tables 1 and 2 present the membership functions associated to input linguistic terms. U_n is presented by four terms: {Verylow, Low, Good and High}. System load U_n is calculated over m processors then to guarantee system schedulability it should not exceed the limit one. This concept explains the limits of trapezoidal membership functions. The universe of discourse varies in positive real values but the interference between sets is centered on [0-1]. It is explained by the fact that when system charge exceeds the limit the application is not feasible under Global EDF.

Table 1. Input1 U_n membership and fuzzy sets

Terms	Fuzzy membership
Verylow	Trapezoid((−1,0), (0,1), (0.3,1), (0.635,0))
Low	Trapezoid((0.284,0), (0.62,1), (0.84,1), (0.986,0))
Good	Trapezoid((0.9,0), (0.954,1), (1.02,1), (1.04,0))
High	Trapezoid((0.098,0), (1.23,1), (30,1), (32,0))

The second input involved in the FIS is old frequency for which is represented by four linguistic terms (f0, f1, f2, f3) corresponding to the four functional points of our study case CPU. For generic purpose, the number of membership terms and functions are generated easily depending on the CPU characteristics. Membership functions are triangular with no interference between sets because the information is almost exact about the frequency value.

Table 2. Input2: f_n membership and fuzzy sets

Terms	Fuzzy membership
f_0	Triangular((f_1 − 1,0), (f_1,1), (f_1 + 1,0))
f_1	Triangular((f_2 − 1,0), (f_2,1), (f_2 + 1,1))
f_2	Triangular((f_3 − 1,0), (f_3,1), (f_3 + 1,0))
f_3	Triangular((f_4 − 1,0), (f_4,1), (f_4 + 1,0))

FIS outputs are the proposed frequency which membership functions are detailed in Table 2 and the number of processor. Table 3 presents three trapezoidal membership functions associated to linguistic terms: inf, same and sup. The functions are centered around m_p which is equal to $m_p = E(Un) + 1$ so the number of iterations to stabilization will be minimized.

Table 3. Output1: m membership and fuzzy sets

Terms	Fuzzy membership
inf	Trapezoid((m_p − 1.5,0), (m_p − 1.25,1), (m_p − 0.75,1), (m_p − 0.65,0))
same	Trapezoid((m_p − 0.45,0), (m_p − 0.35,1), (m_p + 0.35,1), (m_p + 0.45))
sup	Trapezoid((m_p + 0.45,0), (m_p + 0.55,1), (m_p + 1.45,1), (m_p + 1.55))

To establish relationship between the input variables and output parameters, fuzzy logic rules were defined. There are standard criteria for the definition of the rule base. Figure 2 represents the variation of normalized energy ($E_n = E_{new}/E_{max}$) and system charge U_n when changing the scaling factor s for two sets of periodic tasks with different load under Intel x 86 multiprocessor architecture. From some logic facts, mathematical variations and experimental results, the rule base is constructed: When the system utilization U_n is inferior to one then we could decrease the operating frequency trying to minimize energy consumption and idle time. Working with lower frequencies must respect schedulabiliy test. When U_n is superior to one operating frequency is increased until it reaches a feasible scenario. If the system charge doesn't allow it the feed-back module will augment the number of cores until stabilization (Table 4).

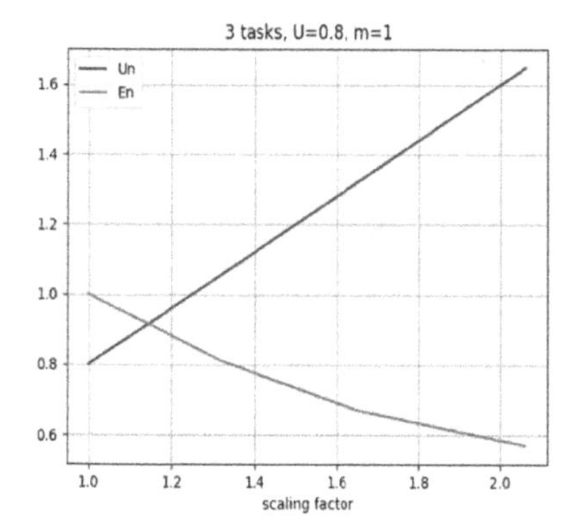

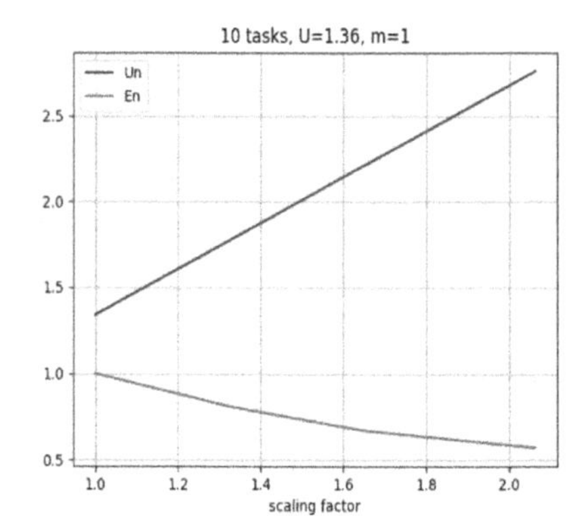

Fig. 2. Normalized energy/system load variation with different scaling factor – CPU Intel x 86 (Set1: 3 tasks U = 0.8 m = 1/Set 2: 10 tasks U = 1.36 m = 1)

Table 4. Rule base of FIS

Rule Base
rule1 = 'if U is verylow and fo is f_1 then fn is f_1 and m is same'
rule2 = 'if U is verylow and fo is f_2 then fn is f_1 and m is same'
rule3 = 'if U is verylow and fo is f_3 then fn is f_2 and m is same'
rule4 = 'if U is verylow and fo is f_4 then fn is f_3 and m is same'
rule5 = 'if U is low and fo is f_1 then fn is f_1 and m is same'
rule6 = 'if U is low and fo is f_2 then fn is f_1 and m is same'
rule7 = 'if U is low and fo is f_3 then fn is f_2 and m is same'
rule8 = 'if U is low and fo is f_4 then fn is f_3 and m is same'
rule9 = 'if U is good and fo is f_1 then fn is f_1 and m is same'
rule10 = 'if U is good and fo is f_1 then fn is f_2 and m is sup '
rule11 = 'if U is good and fo is f_3 then fn is f_3 and m is sup '
rule12 = 'if U is good and fo is f_4 then fn is f_4 and m is sup '
rule13 ='if U is high and fo is f_1 then fn is f_2 and m is sup '
rule14 = 'if U is high and fo is f_2 then fn is f_3 and m is sup '
rule15 = 'if U is high and fo is f_3 then fn is f_4 and m is sup '
rule16 = 'if U is high and fo is f_4 then fn is f_4 and m is sup'

We use an online method DSR (Dynamic Slack Reclamation) for on line power optimization. It uses the slack between WCET and AET to slow down tasks in order to reduce the consumed energy of the processor. Slacks are consumed over processors equally. In fact the same FIS is activated when task's instance is achieved with computation time less than its WCET. New action is proposed by the system with the new recalculated U_n (Fig. 3).

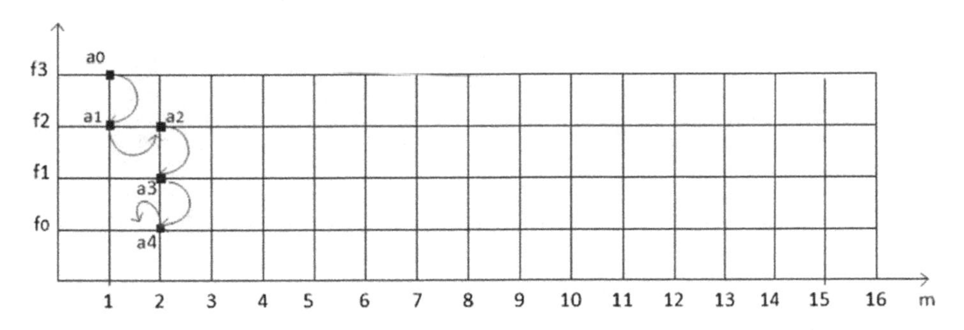

Fig. 3. Example of system response stabilization depending on U_n and E_n values

6 Experimental Results

Our experiments are executed on multiprocessor scheduling simulator that we develop with python. The fuzzy logic engine was implemented conforming to TCE 61131-7 standard [20]. The free Fuzzy logic module (Fuzzython) is an open source class library which is optimized for faster response. The target architectures are multiprocessor systems composed of identical processors supporting DVFS mechanism (Intel 486 GX). Table 5 gives the power consumption of processors for all possible operating points. The scheduler is global EDF, which is dynamic priority based.

Table 5. Periodic generated sets and power consumption and Intel X86 embedded processor datasheet [21]

Simulation system			
Task Set	*Number of tasks*	*Un*	*Random generation method*
Set 1	3	0.8	Rand fix sum
Set 2	7	2.79	Rand fix sum
Set 3	10	1.34	Rand fix sum
Set 4	30	24.77	Rand fix sum

Intel x 86	
Parameter	*Value*
Processor types	Intel 86
Number of cores	[1..16]
Processing frequency	[16 20 25 33] Mhz
Operating voltage	[2.0 2.2 2.4 2.7] V
Dynamic Energy Cons.	[13.1 15.4 18.7 22.9] nJ

The proposed approach was applied to four periodic sets generated using random generation method with different load and task number. The results showed that the algorithm selected different scaling factors for each set depending on U_n. The plots of figure show the normalized energy and the table gives results.

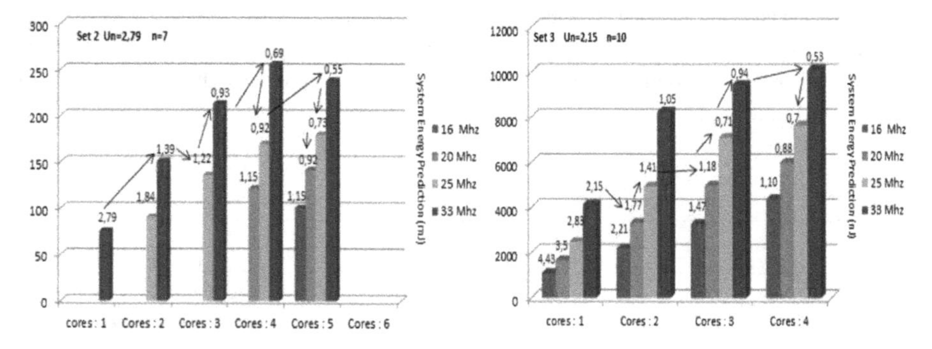

Fig. 4. FIS online decision response for Set 2 (U_n = 2.79 and it number = 8) – D_i = (20,5)

The main emphasis of the proposed fuzzy reconfiguration technique is to assure balance between energy consumption and system schedulability. However, mainly it represents Decision Support System (DSS) for periodic real time applications configuration. In order to fulfill our aims, the proposed algorithm starts the iteration 0 with initial parameters which are the number of cores and the operating frequency ($f_i = f_{max}$ = 33 MHz). And it calculates normalized energy consumption and total system utilization U_n calculated from the task model.

It begins executing the FIS and it completes the offline system configuration when the decision or the action a_i = (m,f) stays invariant. The number of iterations, needed to stabilize the system, depends on real time application utilization value U_n. We could notice that the offline configuration of the most important set (*Set 3* with U_n = 16.33) needed 17 iterations which proved that it stabilizes quickly. The system utilization affects proportionally the number of iterations. The impact of these iterations configuration on individual parameters is discussed as bellow: It is obvious that fuzzy algorithm gives importance to system feasibility rather than optimizing energy consumption. Figure 4 indicates the close loop response in online mode (Table 6).

Table 6. Offline response

Offline configuration				
	Final Di	Energy saving %	Iterations number	Schedulable
Set 1(0.8)	(2,16)	45%	5	Yes
Set 2 (2.97)	(5,20)	33%	8	Yes
Set 3 (2.15)	(3,25)	33%	8	Yes
Set 4 (24.5)	(16,33)	0%	17	No

With every value of m the fuzzy inference looks for the most appropriate operating frequency which influences directly the energy consumption. It gives an obvious conclusion about the correlation of the mean energy consumption with the number of used cores, in one hand, and with the importance of system utilization, in another one. Initiated at 1, the number of required cores is directly affected with system utilization. When the operating frequency calculated by the fuzzy system is invariant and the application stills non schedulable the algorithm raises the number of cores. It could be noticed that when set of periodic tasks has important utilization factor (Set 4) all cores are used with maximum operating frequency scaling. The system is then stabilized with (m = 16 and f_i = 33 MHz). Our system is therefore not feasible under this architecture. The decision support system is mainly responsible of system feasibility guarantee then it takes care of energy consumption. In another hand, when considering task periodic Set number 1 with n = 10 and U_n = 2.15, the fuzzy reconfiguration system is stabilized after 8 iterations: In this case the RT-application is schedulable with two processor m = 4 and energy consumption is optimized when all processors are executed with fi = 25 MHz.

7 Conclusion

This paper addresses the problem of determining the clock frequency and the number of active cores that execute periodic applications with time constraints. The solution is a novel fuzzy inference system (FIS) responsible of the online reconfiguration of homogenous multiprocessor architecture. We proposed power-aware and schedulability-aware design methodology to define the FIS parameters and rule base.

Respecting to the state of art, the obtained power models and GEDF schedulability conditions are plugged in a simulation tool that we developed in python. Our intelligent methodology was illustrated with study cases that were randomly generated under different system charges. The supported architecture is Intel X86 embedded processor with 16 cores that support DVFS mechanism. The results discussed below showed that the FIS responded efficiently in offline and online mode. Therefore, it can be concluded that the proposed reconfigurable fuzzy logic can produce an overall good performance where there is no need for exhaustive mathematical model. In fact, generic aspect of FIS is important to simulate different CPU architectures. In fact, the number of membership functions associated to frequency value can be parameterized depending on the hardware characteristics. In another side, the proposed configurations optimized energy consumption and generated no missed deadlines under global EDF when U_n is appropriate. Some configurations were rejected although energy was lower than the chosen solution because of schedulability issue. However, it will be interesting to study the effect of cache size and associability on energy consumption. In future work, we intend to study the effect of offline FIS activations on computation time of the whole system.

References

1. Pomerol, J.: Artificial intelligence and human decision making. Eur. J. Oper. Res. **2**(99), 3–25 (1997)
2. Nazmul, H., Alam Hossain, Md., Fayezul, I., Priyanka, B., Tahira, Y.: Research on energy efficiency in cloud computing. Int. J. Sci. Eng. Res. **7**(8), 358–367 (2016)
3. Char, J.C., Fakhfakh, A., Couterrier, R., Glerch, A.: Dynamic frequency scaling for energy consumption reduction in synchronous distributed applications. In: 13th IEEE International Symposium on Parallel and Distributed Processing with Applications. IEEE (2016)
4. Qadri, M.Y., Qadri, N.N., McDonald-Maier, K.D.: Fuzzy logic based energy and throughput aware design space exploration for MPSoCs. Microprocess. Microsyst. **3**(2), 68–73 (2015)
5. Qadri, M.Y., McDonald-Maier, K.D., Qadri, N.N.: Energy and throughputs aware fuzzy logic based reconfiguration for MPSoC. J. Intell. Fuzzy Syst. **3**(2), 68–73 (2014)
6. Fakhfakh, M.M.: Energy consumption optimization of parallel applications with iterations using CPU frequency scaling. Thesis (2016)
7. Rauber, T., Runger, G., Schwind, M., Xu, M., Melzner, S.: Energy measurement, modeling, and prediction for processors with frequency scaling. J. Supercomput. **70**(3), 1451–1476 (2014)
8. Rountree, B., Lowenthal, D., Funk, S., Freeh, V.W., De Supinski, B., Schulz, M.: Bounding energy consumption in large-scale MPI programs. In: Proceedings of the 2007 ACM/IEEE Conference on, pp. 1–9 (2007)
9. Cochran, R., Hankendi, C., Coskun, A., Reda, S.,: Identifying the optimal energy-efficient operating points of parallel workloads. In: Proceedings of the International Conference on Computer-Aided Design, ICCAD 2011, IEEE Press, NJ, pp. 608–615 (2011)
10. Henkel, J., Parameswaran, S.: Designing Embedded Processors: A lower Power Perspective. Springer, Heidelberg (2007). https://doi.org/10.1007/978-1-4020-5869-1
11. Da Rosa, T.D., Larrea, V., Calazans, N., Gehm-Moraes, F.: Power consumption reduction in MPSoCs through DFS. In: SBCCI, pp. 1–6 (2012)
12. Parain, F., Banâtre, M., Cabillic, G., Higuera-Toledano, T., Issarny, V.: Lesot: Techniques de réduction de la consommation dans un système embarqué temps réel. Technique et Science Informatiques **20**(10), 1247–1278 (2001)
13. Navet, N., Grajar, B.: Systemés temps réel, Hermes (2006)

14. Baker, T.P., Cirinei, M.: A necessary and sometimes sufficient condition for the feasibility of sets of sporadic hard-deadline tasks. In: Proceedings of IEEE Real-Time Systems Symposium (RTSS), pp 178–190. IEEE Press (2006)
15. Jing, L., Luo, Z., Ferry, D., Agrawal, K., Gill, C., Lu, C.: Global EDF scheduling for parallel real time tasks. Real-Time Syst. **51**(4), 395–439 (2015)
16. Ibrahim, A.: Fuzzy Logic for Embedded Systems Applications. Butterworth, Heinemann, Newton (2003)
17. Najar, Y., Ben Ahmed, S.: Fuzzy multiprocessor architecture reconfiguration based on dynamic frequency scaling. In: Proceedings of ISKE, pp. 761–767. IEEE Press (2017)
18. Bellman, R., Zadeh, L.A.: Decision making in a fuzzy environment. Manag. Sci. **17**, 141–164 (1970)
19. Kickert, W.: Fuzzy Theories on Decision Making: A Critical Review. Frontiers in System Research. Springer, Heidelberg (1979)
20. IEC, International Standard: Programmable Controllers – Part 7: Fuzzy Control Programming, International Electrotechnical Commission, Geneva, Switzerland, iEC Standard (2000)
21. Intel. Embedded Ultra-Low Power Intel486 GX Processor. Datasheet, Intel
22. Chetto, M.: Ordonnancement dans les systèmes temps réel: optimisation de la consommation énergétique. ISTE éditions (2014)
23. Reza P.H., Echeverri, E.J., Pineda, G.: Synthesis and VHDL implementation of fuzzy logic controller for dynamic voltage and frequency scaling (DVFS) goals in digital processors. In: Fuzzy Logic - Controls, Concepts, Theories and Applications (2012)
24. Shen, H., Lu, H., Qiu, Q.: Learning based DVFS for simultaneous temperature, performance and energy management. In: ISQED (2012)
25. Gaurav, D., Tajana, S.R.: Dynamic voltage frequency scaling for multi-tasking systems using online learning. In: ISLPED 2007, Portland, Oregon, USA, pp. 207–212 (2007)
26. Shaheryar, N., Jameel, A.: Real-time implementation of fuzzy logic based DVFS for Leon3 architecture. Asian J. Eng. Sci. Technol. **8**(1) (2018)
27. Ahmed, J., Siyal, M.Y., Najam, S., Najam, Z.: Fuzzy Logic Based Power-Efficient Real-Time Multi-Core System. SAST. Springer, Singapore (2017). https://doi.org/10.1007/978-981-10-3120-5
28. Carlsson, C., Fullér, R.: Fuzzy multiple criteria decision making: recent developments. Fuzzy Sets Syst. **78**, 139–153 (2001)

Introducing the Theory of Probabilistic Hierarchical Learning for Classification

Ziauddin Ursani(✉) and Jo Dicks

Quadram Institute Bioscience, Norwich, UK
ziauddin.ursani@quadram.ac.uk

Abstract. This is the 5th paper in our series of papers on hierarchical learning for classification. Hierarchical learning for classification is an automated method of creating hierarchy list of learnt models that are on the one hand capable of partitioning the training set into equal number of subsets and on the other hand are also capable of classifying elements of each corresponding subset into classes of the problem. In this paper, the probabilistic hierarchical learning for classification has been formalized and presented as a theory. The theory asserts that the accurate models of complex datasets can be produced through hierarchical application of low complexity models. The theory is validated through experiments on five popular real-world datasets. Generalizing ability of the theory is also tested. Comparison with the contemporary literature points towards promising future for this theory. The theory is covered by four postulates, which are carved out elegantly through mathematical formalisms.

Keywords: Hierarchical learning · Probabilistic learning · Set-partitioning

1 Introduction

We have set this introduction to differentiate between hierarchical learning for classification and hierarchical classification itself. The word hierarchical classification has been used in numerous contexts therefore, we have designed this introduction in a way to exclude the irrelevant contexts hierarchically one by one to mark the constrained field of theory of probabilistic hierarchical leaning for classification.

The theory of probabilistic hierarchical leaning for classification is not about hierarchical classification analytically done by human beings. The most profound example of this is the classification of all biological organisms on earth e.g. [1]. The biological organisms are now classified into eight levels i.e., domains, kingdoms, phyla, classes, orders, families, genera and lastly into species in that hierarchical order. This hierarchical classification can be represented through a directed acyclic graph (DAG) e.g. [2]. While considering DAG representation of biological classes, domains can be placed at the root node while species can be placed at the leaf node.

The theory of probabilistic hierarchical leaning for classification is not about hierarchical classification using computational learning methods, where hierarchies are decided meticulously by humans themselves. The classical example of this is a

F. Wotawa et al. (Eds.): IEA/AIE 2019, LNAI 11606, pp. 628–641, 2019.
https://doi.org/10.1007/978-3-030-22999-3_54

Hierarchical Support Vector Machines (H-SVM) [3]. In these methods hierarchies are not set by computers but decided prior to start of a computer program. This is done by merging elements of several classes into one meta-class then applying SVM as a binary classifier between one class against a meta-class. Then again in the next hierarchy another class is extracted from the meta-class and SVM is applied to classify between the newly extracted class against a remaining meta-class. This procedure continues until meta-class retains elements from only one class of the dataset.

The theory of probabilistic hierarchical leaning for classification is not about automated generation of meta-classes either. The automated generation of meta-classes was proposed in 2008 for a handwriting character recognition system [4]. However, to our understanding creation of a meta-class is an artificial creation of class hierarchy where the actual classes are flat not hierarchical.

The theory of probabilistic hierarchical leaning for classification is not about hierarchical classes at all. This theory proposes a model of hierarchical learning even though classes of the dataset are flat. The model of hierarchical learning consists of hierarchy of learnt models rather than hierarchy of classes. The model in each hierarchy is applicable to a subset of the training set created during training in the corresponding hierarchy. Please note that subset created in a hierarchy doesn't represent a single class or a meta-class containing several classes. This is just a subset of the training set containing some of the elements from various classes. Therefore, this subset doesn't represent class hierarchy. This only represents hierarchy of learning where both the model and its area of influence are learnt altogether [5–8].

One might argue that theory of probabilistic hierarchical learning is similar to ensemble learning [9] because both contain multiple models. However, this is an inaccurate assessment. This is because unlike theory of probabilistic hierarchical learning, models in the ensemble learning are not hierarchically learnt. Furthermore, in ensemble learning domain of each of the models covers whole training set, whereas in the theory of probabilistic hierarchical learning sum of domains of all the models is equal to one training set. Therefore, method based on theory of probabilistic hierarchical learning can be much quicker than ensemble learning. Additionally, ensemble learning consists of averaging error of all the models over the whole training set in contravention of the theory of probabilistic hierarchical learning where learnt models are error free in their constrained domains. Finally, models in the ensemble learning are independent of each other therefore they can be applied simultaneously in parallel. However, this is not the case with the theory of probabilistic hierarchical learning where models are actually sub-models of a supermodel in a way that each sub-model is placed at one hierarchy of the supermodel. Therefore, these models can only be applied sequentially or hierarchically on their turn to the rest of the unclassified training set but not applied in parallel to the whole training set.

This paper is structured as follows. In Sect. 2 the theory addresses the question of why hierarchical learning in first place? The theory proposes the model of probabilistic hierarchical learning in Sect. 3. In Sect. 4 the theory sets out the probability of class membership as a corner stone of hierarchical learning. The alternative methodology for class membership under specific circumstances is discussed in Sect. 5. The litmus test of the theory is designed and experimented in Sect. 6. In Sect. 7, generalization ability of theory is experimented. Comparison of results with the literature is done in Sect. 8. Finally, in Sect. 9, conclusions are made, and future work is set out.

2 Hierarchical Learning-Why?

This section sets out very premise of the theory i.e., the "Hierarchical Learning". Why the hierarchical learning in first place? Response to this question is not very difficult to formulate. Complex problems require complex solution methodologies. Since we are classifying the complex datasets, so they require complex solution methodologies. Such complex methodologies are already present in the literature such as Deep Learning e.g. [10, 11] and Recurrent Neural Network e.g. [12]. In these methods we have several hidden layers for training the network. This is because one hidden layer is not enough to grasp the complexity of the problem. In pursuit of our search we have two objectives at hand. One objective is reducing complexity of our model and another objective is increasing its accuracy. Both the objectives are conflicting to each other. As we try to improve on accuracy, we make our model more complex. Therefore, any low complexity discriminant model is impossible to classify any meaningful real-world datasets of practical size. If we try to improve on accuracy with a discriminant model then we will be introducing more and more mathematical operators, which may end up in a very complex discriminant containing several mathematical operators and set of those operators would be very difficult or even impossible to generalize over wide spectrum of datasets. So, what if we keep our discriminant simply restricted to only four elementary mathematical operators $+$, $-$, $\times$ & $\div$? We should not expect from such a low complexity discriminant to classify the whole dataset. However, we can expect from such a discriminant that it may classify only a subset of the training set. If our expectation is reasonable, then this generates an idea. The idea is why not create multiple low complexity discriminants each for specific subset of the training set? In this paper, we have explored this idea. After a rigorous experimentation, we came to conclusion that the only way to materialize such an idea is the development of hierarchical learning procedure where in each hierarchy a model and its corresponding area of influence (subset) be learnt simultaneously. This scenario is depicted in Fig. 1.

It can be seen in Fig. 1 that the model M_1 divides the training set into subsets S_1 and S_1^u, then S_1^u is further divided by model M_2 into subsets S_2 and S_2^u and so on and finally the model M_{n-1} divides the remaining training set into subsets S_{n-1} and S_{n-1}^u. The subset S_{n-1}^u turns out to be equal to subset S_n, as no further division of this subset is needed. This is because it contains the elements belonging to only one class therefore there is no need of another trained model M_n. The scenario in the Fig. 1 can be generalised as a hierarchical model as shown in the Eq. 1.

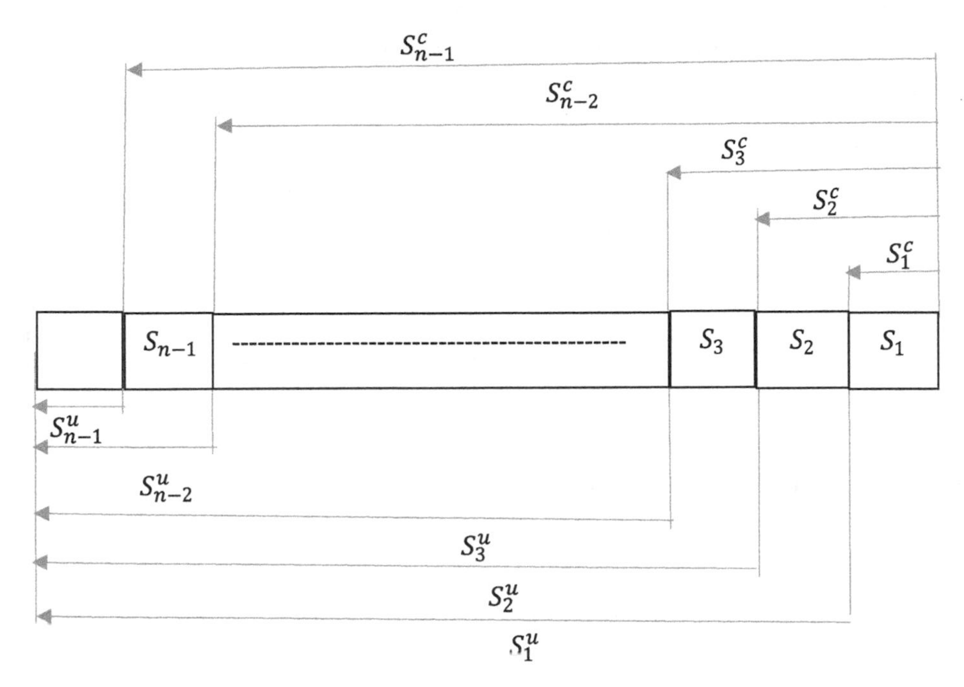

Fig. 1. Successive bifurcation of training set through hierarchical training of low complexity nonlinear discriminants.

$$\forall_{i \in U} H_i : \begin{cases} M_{i-1} \prec M_i : S_i \rightarrow C \\ U = S_i^c \cup S_i^u \\ S_i^c = \cup_{1 \leq j \leq i} S_j \\ S_i^u = \left\{ \cup_{1 \leq j \leq i} S_j \right\}' \end{cases} \tag{1}$$

Where

H_i = Hierarchy level i
M_i = Trained Model at hierarchy level i
S_i = Subset of training set at hierarchy level i
C = Class set
U = Training Set
S_i^c = Set of classified samples at hierarchy level i
S_i^u = Set of unclassified samples at hierarchy level i

The Eq. 1, says that

- at any hierarchy level i, model M_{i-1} precedes model M_i, whose domain is subset S_i and codomain is class set C
- at any hierarchy level i, the training set U is the union of classified S_i^c and unclassified S_i^u samples

- at any hierarchy level i, the classified set S_i^c is the union of all classified subsets preceding and including subset i
- at any hierarchy level i, the unclassified set S_i^u is the complement of the classified set S_i^c

It is emphasized that set of classified samples at any hierarchy level can contain data points from any number of available classes. From the above discussion following postulate can be formulated.

Postulate 1
High complexity model can be replaced with several low complexity models with constrained domains of the training set that could be trained one by one hierarchically until union of all constrained domains covers the whole training set.

3 Hierarchical Learning-How?

Now the question arises how the philosophy of postulate 1 could be materialized? If we have a close look at the model of hierarchical learning in the Fig. 1, we may start doubting the applicability of whole theory in first place. This is because it can be seen in the model in Eq. 1, the trained model in each hierarchy needs to achieve two-pronged classification in parallel i.e., categorization of elements within the corresponding subset into their original classes and also partitioning the remaining training set of unclassified elements into two subsets i.e. subset within its domain and subset outside its domain. There is no doubt as far as ability of model M_i to classify the elements within the subset S_i is concerned. This can normally be achieved using probability of class membership as shown in relation 2 below.

$$P(j) > \forall_{k \neq j}\ P(k) \Rightarrow j \in \{C_k\} \tag{2}$$

The relation 2 says that if the probability of the class membership of element j for class k is greater than its probability of class membership for each of the classes other than class k then the element j is the member of class k. This is the fundamental principle which most of the linear e.g. [13] or nonlinear e.g. [14] discriminants use for the classification. Now the question arises how the second part of classification could be achieved in parallel. If we look at objective of second part of classification carefully then we can make sense of it. Since the second part of classification involves partitioning of elements of training set into two subsets, one within and another outside the domain of the model M_i, therefore we need to decide which elements are within its domain. Naturally those elements which obey the relation (2) are within the domain of the model M_i. Therefore, we slightly modify the probabilistic model of relation 2 for hierarchical learning as shown in relation (3).

$$P(i,j,k) > \forall_{h \neq k}\ P(i,j,h) \Rightarrow j \in \{C_k, S_i\} \tag{3}$$

The relation 3 says that for model M_i if the probability of the class membership of element j for class k, i.e. $P(i,j,k)$ is greater than its probability of class membership for each of the classes other than class k then the element j is the member of class k (C_k) and it is also the member of subset i (S_i), which is domain of model M_i. The elements which do not obey the probabilistic model of relation (3) are outside the domain of model M_i, as shown in relation (4).

$$P(i,j,k) > \forall_{h \neq k} P(i,j,h) \Rightarrow j \in \{C_h, S_i\} \quad (4)$$

The relation 4 says that for model M_i even though the probability of the class membership of element j for class k is greater than its probability of class membership for each of the classes other than class k but the element j is not the member of class k, it is the member of class h instead which is any class other than class k but it is still the member of subset S_i as a misclassified element, which is not desirable. Now the question arises through which mechanism the hierarchical learning could push element j out of subset S_i or domain of model M_i to avoid its misclassification. To understand this, we need to define a technical term 'Highest Misclassifying Margin' (HMM). The HMM is the greatest margin by which the model M_i, could misclassify a sample. The HMM can be calculated through Eq. 5.

$$\Delta_{max} = max\Big(\forall j \in \{C_h, S_i^u\} P(i,j,k) - max(\forall_{h \neq k} P(i,j,h))\Big) \quad (5)$$

From Eq. 5, it can be seen that Δ_{max} (HMM) represents the maximum difference between the probabilities of wrongly assigned class and the maximum of probabilities from rest of the classes. Technically, we can incorporate HMM as computed in Eq. 5, in Eq. 4, to separate element j from subset S_i and thus prevent model M_i from misclassifying it, as shown in relation 6.

$$P(i,j,k) \not> \forall_{h \neq k} P(i,j,h) + \Delta_{max} \Rightarrow j \in S_i^u \quad (6)$$

It can be seen from relation 6, that probability of element j for the class k could not surpass the value on the right-hand side of the relation where Δ_{max} has been added to probabilities of the rest of the classes. This means element j is pushed out to subset S_i^u, which is not within the domain of model M_i and thus remains unclassified and should wait for next round of model training for the classification. However, it should be noted that element j could also be the member of right class C_k if the Δ_{max} was not introduced in the equation. This means that the Δ_{max}, not only pushes all the potentially misclassifying elements out of domain of model M_i but it does also push some of the potentially correctly classifying elements out of the domain of model M_i. However, with the introduction of Δ_{max} in the model, it is now confirmed that there will be no misclassification, either the element j will remain unclassified as in relation 6 or it will be classified correctly as in relation 7 below.

$$P(i,j,k) > \forall_{h \neq k} P(i,j,h) + \Delta_{max} \Rightarrow j \in \{C_k, S_i\} \quad (7)$$

By generalizing the relations (6–7) into one model we get

$$\begin{cases} j \in \{C_k, S_i\} & P(i,j,k) > \forall_{h \neq k} P(i,j,h) + \Delta_{max} \\ j \in S_i^u & otherwise \end{cases} \tag{8}$$

It should be noted that hierarchy i is the last hierarchy iff either $S_i^u = \{\emptyset\}$ or contains members belonging to one class only. We call it a remainder class. In any case whole training set is classified accurately. It should also be noted that if $S_i^u = \{\emptyset\}$ then $\Delta_{max} = 0$ else $\Delta_{max} > 0$. This means that there must always be some misclassifying margin if unclassified set is non-empty. It is emphasized that number of hierarchy levels is not fixed and entirely depends on the structure of the dataset and domain size of the models evolved.

Postulate 2
Misclassification of elements can be eliminated completely during hierarchical training with the incorporation of Highest Misclassifying Margin (HMM) in the fundamental model of probabilistic class membership, thus rendering the hierarchical training model error free.

4 Relative Closeness as Measure of Probability of Class Membership

From the probabilistic model (expression 8) presented in Sect. 3, it can be seen, that hierarchical learning is largely based on probability of class membership therefore it is better to call it probabilistic hierarchical learning. The probabilistic hierarchical learning could only be useful when computation of probability of class membership is easy, helpful and relevant. Now to understand this we need to think about how can we compute probability of element j for the membership of class C_k, with respect to model M_i, i.e., $P(i,j,k)$? It should be computed in a way that supports the probabilistic model (expression 8) in the objective of classification. In doing so, constraints associated with the notion of probability as a quantity could also be avoided and we will learn in a moment what we mean by that. Since the model is very simple therefore, the most natural way of computation of probability of class membership should be based on the relative closeness of element/sample to the mean of the class, i.e., closer the sample to the mean of the class greater should be the probability of its membership of the class. This can be understood from Eq. 9.

$$P_{(i,j,k)} = \begin{cases} \frac{\mu_{(i,j)} - \gamma_{(i,k,min)}}{\gamma_{(i,k,mean)} - \gamma_{(i,k,min)}}, & \mu_{(i,j)} \leq \gamma_{(i,k,mean)} \\ \frac{\gamma_{(i,k,max)} - \mu_{(i,j)}}{\gamma_{(i,k,max)} - \gamma_{(i,k,mean)}}, & otherwise \end{cases} \tag{9}$$

where

$\mu_{(i,j)}$ = value of sample j according to model M_i

$\gamma_{(i,k,min)}$ = estimated minimum value of samples of class C_k according to model M_i

$\gamma_{(i,k,max)}$ = estimated maximum value of samples of class C_k according to model M_i

$\gamma_{(j,k,mean)}$ = estimated mean value of samples of class C_k according to model M_i

The mean value of model M_i of class C_k can be estimated as follows.

$$\gamma_{(i,k,mean)} = \frac{\sum_{j \in C_k} \mu_{(i,j)}}{n_k} \tag{10}$$

where

n_k = Number of samples in the training set of class C_k

The maximum and minimum value among samples of class C_k according to model M_i can be estimated as,

$$\gamma_{\left(i,k,\begin{smallmatrix} max \\ min \end{smallmatrix}\right)} = \gamma_{(i,k,mean)} \pm 3.0^{*}\delta_{(i,k,sd)} \tag{11}$$

where

$\delta_{(i,k,sd)}$ = estimated standard deviation of samples of class C_k according to model M_i

The standard deviation of samples of class C_k according to model M_i can be estimated as,

$$\delta_{(i,k,sd)} = \frac{\sum_{j \in C_k} \left(\mu_{(i,j)} - \gamma_{(i,k,mean)} \right)^2}{n_k} \tag{12}$$

Now the model from Eqs. 9–12 suggests that the class C_k might have the class mean farther than the other classes from the sample j to whom it is assigned. This can be understood from the Fig. 2.

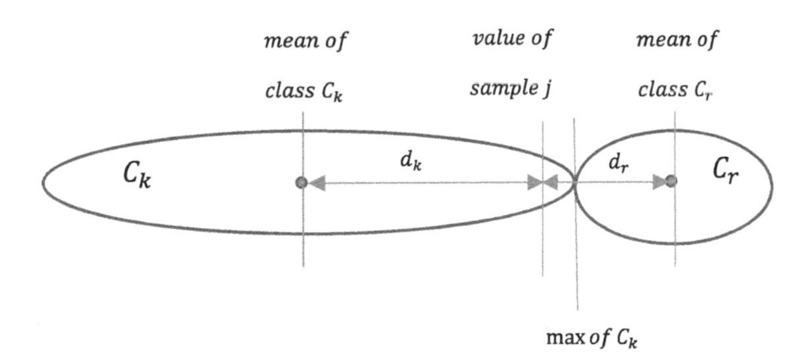

Fig. 2. Principle of relative closeness

It can be seen from Fig. 2, that $d_k > d_r$, but according to model in Eqs. 9–12 the sample will be assigned to class C_k instead of class C_r because value of sample j lies outside the boundary of class C_r but within the boundaries of class C_k. Therefore, even though value of sample j according to model M_i is closer to mean of the class C_r but it is assigned to class C_k because it is closer to class C_k in relative terms or in other words relatively closer to class C_k. However, the measure of relative closeness will only work well when we have good number of training samples belonging to each class, this is because it can be seen from Eq. 11 of the model that estimation of minimum and maximum of the class entirely depend on mean and standard deviation of the class sample values. These quantities are only meaningful when good number of samples are present in the training set.

Postulate 3
Relative closeness of sample to the mean of the class can be a useful measure for computation of probability of class membership when we have good representation of number of samples for each class in the training set.

5 Distance Inverse as a Measure of Probability of Class Membership

Now, since this is a hierarchical learning model, which bifurcates training set in each hierarchy into classified and unclassified samples therefore size of the training set continues to decrease with each hierarchy. In such a situation in the last hierarchy the training set may end up with very few samples such that number of samples of some or all the classes become less than 3. In such a case computation of standard deviation becomes meaningless and so estimation of minimum and maximum. To deal with this scenario, the measure of relative closeness is replaced with the measure of distance inverse. Therefore, Eq. 9 can be modified as Eq. 13 below.

$$P_{(i,j,k)} = \begin{cases} \frac{\mu_{(i,j)} - \gamma_{(i,k,min)}}{\gamma_{(i,k,mean)} - \gamma_{(i,k,min)}}, & \mu_{(i,j)} \leq \gamma_{(i,k,mean)}, \quad n_k \geq 3 \\ \frac{\gamma_{(i,k,max)} - \mu_{(i,j)}}{\gamma_{(i,k,max)} - \gamma_{(i,k,mean)}}, & otherwise, \quad n_k \geq 3 \\ \frac{1}{d_k}, & not\ applicable, \quad otherwise \end{cases} \tag{13}$$

where

n_k = number of samples in the training set for class C_k

It can be seen from the Eq. 13 that distance inverse measure is introduced when number of training samples of the class are less than 3. This changes measure of relative closeness to measure of closeness only or in popular terms measure of nearest neighbor.

Postulate 4
Closeness of sample to the mean of the class can be a useful measure for computation of probability of class membership when we have inadequate representation of number of samples for each class in the training set.

6 Litmus Test of the Theory

What is a litmus test that could validate the theory presented above? Expression 8 presents core model of the theory which is linked to postulate 2, which states that the misclassification can be eliminated completely. This means that learnt models should be able to accurately classify the training set. Therefore, if we can classify some of the popular datasets accurately through hierarchical learning of low complexity models, then this would mean that the basic idea behind the theory is valid. To see that the theory passes this litmus test we chose some of the popular real-world datasets from the UCI repository. The details of those datasets are tabulated in Tables 1 and 2. Table 1 gives feature description and Table 2 gives class description of each dataset.

Table 1. Feature description for each dataset

S. Nr.	Dataset	Nr. of features	Feature names
(1)	(2)	(3)	(4)
1	Iris flower	4	f_1: sepal length, f_2: sepal width, f_3: petal length, f_4: petal width
2	Balance scale	4	f_1: left weight, f_2: left distance, f_3: right weight, f_4: right distance
3	Car evaluation	6	f_1: buying cost, f_2: maintenance cost, f_3: number of doors, f_4: number of seats, f_5: size of lug-boot, f_6: level of safety
4	Banknote authentication	4	f_1: variance of wavelet transformed image (WTI), f_2: skewness of WTI, f_3: curtosis of WTI, f_4: entropy of image
5	Seeds	7	f_1: area, f_2: perimeter, f_3: compactness, f_4: length of kernel (k), f_5: width of k, f_6: asymmetry coeff:, f_7: length of k groove

We devised the training method [5–8] based on hierarchical learning theory above and coded in Microsoft Visual Studio C/C++. Please see details of the parameters and models learnt during training in our earlier works [5–8]. The program was applied on the five datasets described in Tables 1 and 2. The training method was applied for 30 simulations on each dataset on different random seeds. The trained models were then tested back on the same dataset. All the datasets were classified accurately in each simulation.

Table 2. Class description for each dataset

S. Nr.	Dataset	Nr. of classes	c_1	c_2	c_3	c_4	Total Nr. of samples
(1)	(2)	(3)	(4)	(5)	(6)	(7)	(8)
1	Iris flower	3	Setosa 50	Verginica 50	Versicolour 50	–	150
2	Balance scale	3	Balanced 49	Left tipped 288	Right tipped 288	–	625
3	Car evaluation	4	Unacceptable 1210	Acceptable 384	Good 69	Very good 65	1728
4	Banknote authentication	2	True 610	False 762	–	–	1372
5	Seeds	3	Kama 70	Rosa 70	Canadian 70	–	210

7 Generalizing Ability of the Theory

Retrieving accurate models of the complex datasets is an achievement but generalizing ability of such models should also be investigated. Generalizing ability means how such models perform on the unseen data or the data on which the model is not trained. We can devise experiments to test this ability of hierarchical learning. Let us develop hierarchical models on training sets containing only 50% randomly chosen samples of the original dataset and then test the model on the rest of the 50% samples on which they are not trained. The test results on this unseen data will show generalizing ability of the hierarchical model. To cross validate the models we reverse the roles of the training set and test set. Such an approach will reduce any statistical bias towards or against the hierarchical models. Furthermore, repeating this procedure for 30 independent runs will show close to average performance of the hierarchical learning theory. So, these experiments were performed on the same five datasets described in Sect. 6 and the results are reported in Table 3. In Table 3, column 1 shows serial number of the dataset, name of the dataset is given in column 2. Average results of 30 simulations are stated in column 3. Column 4 mentions best result in 30 simulations, column 5 provides percentage of accurate results in 30 simulations. This means the percentage of number of simulations out of 30 where 100% samples are correctly classified. Finally, column 6 just informs that whether data normalization procedure has been applied on the dataset. It can be seen from the results that in all the datasets more than 90% correct results have been obtained on average.

Table 3. Classification results

S. Nr.	Dataset	Average results	Best results	%age of accurate results	Data normalization
(1)	(2)	(3)	(4)	(5)	(6)
1	Iris	92.87%	94%	0.00%	No
2	Balance scale	99.11%	100%	3.33%	No
3	Car evaluation	93.09%	95.08%	0.00%	No
4	Banknote authentication	99.63%	99.93%	0.00%	No
5	Seeds	90.40%	93.33	0.00%	Yes

8 Comparison with State of Art

Now let us see how the results presented in Table 3 compare with the literature. For fair comparisons we need to compare this scheme with recently published methods that are applied on all the above datasets. We have chosen three recently published methods namely Support Vector Machines [15], Decision Trees [16] and random forest [17] that are applied on all the above five datasets. In Table 4 we compare the results of proposed approach with those methods. In Table 4, column 1 gives bibliographical reference, columns 2–6 state average results of five datasets. Number of simulations are mentioned in column 7, column 8 informs about x-validation type and finally size of the training set is revealed in column 9.

Table 4. Comparison with literature

Ref.	Iris	Bal. Scale	Car Evln.	Bn. Auth.	Seeds	Nr. of Sim.	x-valid	Size of Tr. Set
ID	%age	%age	%age	%age	%age	Int	type	%age
(1)	(2)	(3)	(4)	(5)	(6)	(7)	(8)	(9)
[15]	98.00	92.00	78.26	99.12	94.29	5	–	80.00
[16]	92.40	67.10	73.70	90.10	88.70	5	–	75.00
[17]	94.53	80.30	94.70	99.34	93.57	10	10-fld	90.00
Hierarchical learning	92.87	99.11	93.09	99.63	90.40	30	2-fld	50.00

It can be seen from the results that the proposed technique has outsmarted the three methods on the balance scale dataset with a very wide margin i.e. (99.11%/92.00%/80.30%/67.10%). The Hierarchical learning has also beaten the other three methods on the banknote authentication dataset, with smaller margins <1.00%. On the car evaluation dataset, the hierarchical learning has produced much better results than two methods (93.09%/78.26%/73.70) but little worse than the third method. On the rest of the two datasets hierarchical learning has performed better than one of the techniques but worse than other two. However, it should be noted that hierarchical learning has used only 50% of the training set while the other three techniques have used 90%, 80% and 75% of training sets. Keeping this in mind the results produced by hierarchical learning can be regarded as respectable.

9 Conclusion and Future Work

This paper is fifth in our series of papers on hierarchical learning. This paper proposes the theory of probabilistic hierarchical learning covering four postulates. The first postulate says that multiple low complexity models can emulate the effect of high complexity model, when put together hierarchically. The second postulate says that chance of misclassification of the sample can be eliminated with smart use of fundamental model of probabilistic class membership. The third postulate proposes relative closeness rather than absolute nearness of sample to the mean of the class as basis for the probability of class membership. The fourth postulate proposes absolute nearness of sample to the mean of the class as basis for the probability of class membership in case of inadequate class representation in the training set. The theory is not only supported through mathematical analysis but also through experimentation on five popular classification datasets taken from UCI repository. In doing so, generalization ability of theory is also tested and compared with state of art showing satisfactory results. For this theory to work with large spectrum of datasets further theoretical enhancements are still needed which are currently under investigation.

Acknowledgement. Parts of this work were supported by the Biotechnology and Biological Sciences Research Council, through a Responsive Mode award (grant number BB/P022030/1) to J.D.

References

1. Ruggiero, M.A., et al.: A higher-level classification of all living organisms. PLoS ONE **10** (4), e0119248 (2015). https://doi.org/10.1371/journal.pone.0119248
2. Silla Jr., C.N., Freitas, A.A.: A survey of hierarchical classification across different application domains. Data Min. Knowl. Disc. **22**(1–2), 31–72 (2011)
3. Chen, Y., Crawford, M.M., Ghosh, J.: Integrating support vector machines in a hierarchical output space decomposition framework. In: Proceedings of the IEEE International Symposium on Geoscience and Remote Sensing, vol. 2, pp. 949–952 (2004)
4. Freitas, C.O.A., Oliveira, L.S., Aires, S.B.K., Bortolozzi, F.: Metaclasses and zoning mechanism applied to handwriting recognition. J. Univ. Comput. Sci. **14**(2), 211–223 (2008)
5. Ursani, Z., Corne, D.W.: Use of reliability engineering concepts in machine learning for classification. In: 4th International Conference on Soft Computing & Machine Intelligence (IEEE), (ISCMI 2017), Mauritius (2017)
6. Ursani, Z., Corne, D.W.: A novel nonlinear discriminant classifier trained by an evolutionary algorithm. In: 10th International Conference on Machine Learning and Computing (ICMLC 2018), 26–28 February 2018, University of Macau, China, ACM Conference Proceedings (2018). ISBN 978-1-4503-6353-2

7. Ursani, Z., Corne, D.W.: A hierarchical nonlinear discriminant classifier trained through an evolutionary algorithm. In: Tabii, Y., Lazaar, M., Al Achhab, M., Enneya, N. (eds.) BDCA 2018. CCIS, vol. 872, pp. 273–288. Springer, Cham (2018). https://doi.org/10.1007/978-3-319-96292-4_22
8. Ursani, Z., Corne, D.W.: A hierarchical set-partitioning nonlinear discriminant classifier trained by an evolutionary algorithm. In: 2018 International Conference on Artificial Intelligence and Big Data (ICAIBD 2018), 26–28 May 2018, Chengdu, China. IEEE (2018)
9. Opitz, D., Maclin, R.: Popular ensemble methods: an empirical study. J. Artif. Intell. Res. **11** (1999), 169–198 (1999)
10. Goodfellow, I., Bengio, Y., Courville, A.: Deep Learning. MIT Press (2016, in preparation). http://www.deeplearningbook.org, https://github.com/janishar/mit-deep-learning-book-pdf. Accessed 20 Mar 2018
11. Noda, K., Yamaguchi, Y., Nakadai, K., Okuno, H.G., Ogata, T.: Audio-visual speech recognition using deep learning. Appl. Intell. **42**, 722–737 (2015)
12. Chen, Y.C., Wang, J.S.: A Hammerstein-Wiener recurrent neural network with frequency-domain eigensystem realization algorithm for unknown system identification. J. Univ. Comput. Sci. **15**(13), 2547–2565 (2009)
13. Fisher, R.A.: The utilization of multiple measurements in taxonomic problems. Ann. Eugenics **7**, 179–188 (1936)
14. Raymer, M.L., Doom, T.E., Kuhn, L.A., Punch, W.F.: Knowledge discovery in medical and biological datasets using a hybrid Bayes classifier/evolutionary algorithm. IEEE Trans. Syst. Man Cybern. Part B (Cybern.) **33**(5), 802–813 (2003)
15. Bertsimas, D., Dunn, J., Pawlowski, C., Zhuo, Y.D.: Robust classification. INFORMS J. Optim. **1**(1), 2–34 (2019). https://doi.org/10.1287/ijoo.2018.0001
16. Bertsimas, D., Dunn, J.: Optimal classification trees. Mach. Learn. **2017**(106), 1039–1082 (2017)
17. Abellán, J., Mantas, C.J., Castellano, J.G., Moral-García, S.: Increasing diversity in random forest learning algorithm via imprecise probabilities. Expert Syst. Appl. **97**, 228–243 (2018)

Spammers Detection Based on Reviewers' Behaviors Under Belief Function Theory

Malika Ben Khalifa[1,2](✉), Zied Elouedi[1], and Eric Lefèvre[2]

[1] Institut Supérieur de Gestion de Tunis, LARODEC, Université de Tunis, Tunis, Tunisia
malikabenkhalifa2@gmail.com, zied.elouedi@gmx.fr

[2] Univ. Artois, EA 3926, Laboratoire de Génie Informatique et d"Automatique de l"Artois (LGI2A), F-62400 Béthune, France
eric.lefevre@univ-artois.fr

Abstract. Nowadays, we note the dominance of the online reviews which become an essential factor in customers' decision to purchase a product or service. Driven by the immense financial profits from reviews, some corrupt individuals or organizations deliberately post fake reviews to promote their products or to demote their competitors' products, trying to mislead or influence customers. Therefore, it is crucial to spot these spammers in order to detect the deceptive reviews, to protect companies from this harmful action and to ensure the readers confidence. In this way, we propose a novel approach able to detect spammers and to accord a spamicity degree to each reviewer relying on some spammers indicators while handling the uncertainty in the different inputs through the strength of the belief function theory. Tests are conducted on a real database from Tripadvisor to evaluate our method performance.

Keywords: Online reviews · Spammers · Fake reviews · Uncertainty · Belief function theory

1 Introduction

Online reviews are becoming more prevalent nowadays due to the huge use of social media, opinion-sharing websites, blogs, forms and merchant websites. Consumers rely heavily up on reviews posted on these websites when making decisions about which products or services to purchase online. However, reviews are more than just a way for customers to gather information, but also a powerful source information for companies since positive opinions bring significant financial gains for business and individuals. Moreover, negative reviews not only cause financial loss, but also damage the companies' e-reputation. Unfortunately, all this gives an important incentive for fake reviews.

So driven by the desire of profit, spammers create fake reviews and posted them everywhere in order to mislead readers, to influence their decisions and to manipulate their opinion mining. Opinions spam may be positive to promote some companies or negative, to their competitive companies, in order to

F. Wotawa et al. (Eds.): IEA/AIE 2019, LNAI 11606, pp. 642–653, 2019.
https://doi.org/10.1007/978-3-030-22999-3_55

demote them. These review spamming activities make the products and the services identification confusing and complicated. We believe also that more online reviews are used, more spammers will increase and will post more and more deceptive reviews. The spammer detection becomes an essential task since it allows us to stop the appearance of fake opinions. Several methods addressed this problem [5,9], most of them are graph based approaches. The first study [14] proposes a heterogeneous graph model with three types of nodes to define relations between reviewers, reviews, and store. This method used the interrelationship between three based concepts namely; the trustworthiness of reviewers, the honesty of reviews, and the reliability of stores to generate a ranking list of suspicious reviews and reviewers. However, its level of precision amounts to 49% in the fake reviews detection. A similar approach elaborated by authors in [3] which also used a review graph. This method calculated a suspicion score for each node in the review graph and then used an iterative algorithm in order to update these scores based on the graph connectivity. This method has higher precision with respect of the conformity along the human judgments. The third graph based approach was elaborated by Akoglu et al. [1], introduced through a bipartite network. The authors proposed a signed inference algorithm for extending loopy belief propagation (LBP). The output of this algorithm is a list of users ranked by score to get clusters with k reviewers and products. This method was compared to two iterative classifiers, where it succeeded in detecting fraudulent users and spot their fake product ratings. Lim et al. [6] were the first use behavioral indicators of deceptive reviews to spot spammers. Their proposed method is based on the behavior scoring technique for ranking reviewers by measuring the spamming behaviors. The human judgment is used for the evaluation. As a result, the rating of the target products alternated adequately by removing the most suspicious reviews. Since then, behavioral indicators have become an important basis for spammer detection task. In this way, researchers in [8] proposed a method to exploit observed reviewing behaviors in order to detect opinion spammers using a Bayesian inference framework. Moreover, authors in [4] developed an algorithm in order to detect burst patterns in reviews for a specific product. It generated five new spammer behavior features as indicators to used them in review spammer detection. Two types of evaluation are performed: supervised classification and human evaluation. These techniques achieve significant results thanks to the spammers behavior features. Most of these approaches rely on different human evaluators and experts to annotate their data evaluation. Moreover, each method is based on various inputs and aspects. All this won't allow for a safe comparison in this field. In addition, these techniques exhibit some weaknesses fundamentally related to their inability to manage the uncertainty of different reviewers and in reviews information which are often imperfect and imprecise. Ignoring such uncertainty may deeply affect the detection. That is why, treating the uncertainty when dealing with the fake reviewers detection task becomes a widespread interest.

In this paper, we propose a novel method that aims to detect spammers based on the reviewer behavior characteristics under the belief function framework. It is known as a rich tool able to manage several pieces of imperfect information, to combine them, besides taking into account the reliability in the different sources providing them, and making decision under uncertainty. Hence, our approach involves imperfections in the different inputs to spot the spammers and offers also an uncertain output. This latter represents the spamicity degree according to each reviewer in order to identify its reliability.

The remainder of the paper is structured as follows: We firstly present the belief function theory basic fundamentals in Sect. 2. Then, Sect. 3 elucidates our proposed approach. After that, we discuss the experimental results in Sect. 4. Finally, a conclusion and some future works are described in Sect. 5.

2 Belief Function Theory

The belief function theory is one of the useful theories that handles uncertain knowledge. It was introduced by Shafer [10] as a model to represent beliefs. It is considered as a powerful tool able to deal with uncertainty in different levels and to manage various types of imperfection.

2.1 Basic Concepts

The frame of discernment Ω is a finite and exhaustive set of different events associated with a given problem, such set Ω is also called the universe of discourse, defined by:

$$\Omega = \{\omega_1, \omega_2 ..., \omega_n\} \tag{1}$$

The power set 2^Ω contains all possible hypotheses that formed the union of events, and the empty set $\emptyset$ which represents the conflict, defined by:

$$2^\Omega = \{A : A \subseteq \Omega\} \tag{2}$$

A basic belief assignment (*bba*) or a belief mass defined as a function from 2^Ω to $[0, 1]$ that represents the degree of belief given to an element A such that:

$$\sum_{A \subseteq \Omega} m^\Omega(A) = 1 \tag{3}$$

A focal element A is a set of hypotheses with positive mass value $m^\Omega(A) > 0$.

Several kinds of *bba*'s have been proposed [12] in order to express special situations of uncertainty. Here, we underline some special cases of *bba*'s:

- The certain *bba* represents the state of total certainty and it is defined as follows: $m^\Omega(\{\omega_i\}) = 1$ and $\omega_i \in \Omega$.
- The categorical *bba* has a unique focal element A different from the frame of discernment defined by: $m^\Omega(A) = 1$, $\forall A \subset \Omega$ and $m^\Omega(B) = 0$, $\forall B \subseteq \Omega$ $B \neq A$.

- Simple support function: In this case, the *bba* focal elements are $\{A, \Omega\}$. A simple support function is defined as the following equation:

$$m^{\Omega}(X) = \begin{cases} w & \text{if } X = \Omega \\ 1-w & \text{if } X = A \text{ for some } A \subset \Omega \\ 0 & \text{otherwise} \end{cases} \tag{4}$$

Where A is the focus and $w \in [0, 1]$.

2.2 Discounting

The discounting operation [7] allows us to update experts beliefs by taking into consideration their reliability through the degree of trust $(1-\alpha)$ given to each expert with $\alpha \in [0, 1]$ is the discount rate.

When, the *bba* is defined on the set $\{reliable,\ not\ reliable\}$ such that [11]:

$$m(reliable) = 1 - \alpha \quad and \quad m(not\ reliable) = \alpha \tag{5}$$

Accordingly, the discounted *bba*, noted ${}^{\alpha}m^{\Omega}$, m^{Ω} becomes:

$$\begin{cases} {}^{\alpha}m^{\Omega}(A) = (1-\alpha)m^{\Omega}(A) & \forall A \subset \Omega, \\ {}^{\alpha}m^{\Omega}(\Omega) = \alpha + (1-\alpha)m^{\Omega}(\Omega). \end{cases} \tag{6}$$

2.3 Combination Rules

Let m_1^{Ω} and m_2^{Ω} two *bba*'s representing two distinct sources of information defined on the same frame of discernment Ω. Various numbers of combination rules have been proposed in the framework of belief function. They were intended to aggregate a set of *bba*'s in order to get the fused information represented by one *bba*. In what follows, we elucidate those related to our approach.

1. *Conjunctive rule*
   It was settled in [13], denoted by $\bigcirc\!\!\!\!\cap$ and defined as:

$$m_1^{\Omega} \bigcirc\!\!\!\!\cap\, m_2^{\Omega}(A) = \sum_{B \cap C = A} m_1^{\Omega}(B) m_2^{\Omega}(C) \tag{7}$$

2. *Dempster's rule of combination*
   This combination rule is a normalized version of the conjunctive rule [2]. This rule is characterized by a normalization factor denoted by K and it is defined as:

$$(m_1^{\Omega} \oplus m_2^{\Omega})(A) = K.(m_1 \bigcirc\!\!\!\!\cap\, m_2^{\Omega}(A)) \tag{8}$$

   Where

$$K^{-1} = 1 - (m_1^{\Omega}.\bigcirc\!\!\!\!\cap\, m_2^{\Omega}(\emptyset)) \text{ and } (m_1^{\Omega} \oplus m_2^{\Omega})(\emptyset) = 0 \tag{9}$$

2.4 Decision Process

Various solutions have been proposed to choose the most suitable decision for a given problem under the belief function framework. In this work, we adopt the pignistic probability proposed by the Transferable Belief Model [13]. Therefore, it is composed by two level models:

- The credal level where beliefs are defined by *bba*'s then combined.
- The pignistic level where *bba*'s are transformed into pignistic probabilities denoted by $BetP$ and defined as follows:

$$BetP(B) = \sum_{A \subseteq \Omega} \frac{|A \cap B|}{|A|} \frac{m^{\Omega}(A)}{(1 - m^{\Omega}(\emptyset))} \quad \forall\ B \in \Omega \tag{10}$$

3 Spammers Detection Based on Reviewers' Behaviors Under Belief Function Theory

In this section, we elucidate our novel proposed method which deals with different important spammer indicators in an uncertain context through the belief function theory in order to distinguish between fake reviewers and genuine ones.

Our method relies on the four most important spammer behaviors indicators namely; the reviewer's average proliferation, the brust spamicity degree, the reviews helpfulness and the extreme rating providing by each reviewer.

Besides, we adopt the belief function theory to model uncertainty within those indicators. Each reviewer R_i will be represented by two mass functions (*bba*'s), the first one is to model the reviewer reputation $m^{\Omega}_{RR_i}$ and the second one is to represent the reviewer helpfulness m_{RH_i} with $\Omega = \{S, \overline{S}\}$ where S is spammer and $\overline{S}$ is non spammer. Our method follows four main steps detailed in-depth.

3.1 Step 1: Reviewer Reputation

In the spammer review detection field, it has been proved that ordinary reviewers usually write their comments on several products in almost consistent patterns during different periods [5]. Generally, the genuine reviewers post their opinion when they have actually bought new products or used new services. It means that their reviews depend on the number of tested products or services and are also steadily given over time interval. However, spammers are excepted to post a huge number of reviews to limited intended products or services in short time span, say in two or three days. Consequently, these two indicators can construct the reviewer reputation.

In this way, we propose to examine the reviewing history for each reviewer $Hist_{R_i}$ defined as the set of all past reviews written by the reviewer R_i for n discrete products.

The average number of reviews per product is measured through the sum of different reviews given by each reviewer R_i and divided by the total number of reviewed products n. The reviewers average proliferation is calculated through the following equation:

$$AvgP(R_i) = \frac{Hist_{R_i}}{n} \tag{11}$$

If the $AvgP(R_i) > 3$, we can assume that the reviewer is suspicious to be a potential spammer since generally ordinary reviewers do not give more than three reviews per product. The reviewer reputation is then represented by a certain *bba* as follows:

$$m^{\Omega}_{RR_i}(\{S\}) = 1 \tag{12}$$

Else

$$m^{\Omega}_{RR_i}(\{\overline{S}\}) = 1 \tag{13}$$

Example 1. Let us consider the case of five reviewers, for which we have some information about their reviewing history, given an overall rating review for a hotel detailed in the Table 1.
We deal with the $Reviewer_{id} = 1$
So, we calculate the reviewer's average proliferation:
$AvgP(R_1) = \frac{Hist_{R_1}}{n} = \frac{258}{30} = 8.6$
Then, we generate the corresponding *bba*:
$AvgP(R_1) > 3 \Rightarrow m^{\Omega}_{RR_1}(\{S\}) = 1$

Table 1. Hotel reviews and reviewers information

Review	Reviewer_id	Total number of reviews	Total number of product or services	Number of Extreme rating	Number of helpful votes	Number of reviews given in less than 3 days
5*	1	258	30	208	100	200
4*	2	30	10	8	25	4
3*	3	20	12	0	18	2
5*	4	30	16	22	0	15
4*	8	100	92	10	88	10

Moreover, we propose to verify if the reviews are given in a short time of interval or are scattered during the reviewing history.

In our method, we fix the time interval to three days and we measure the brust spamicity degree α_i through the sum of the reviews' number given in less than three days divided by the total number of reviews by each reviewer denoted by TNR_i as follows:

$$\alpha_i = \frac{Number\ of\ reviews\ given\ by\ R_i\ in\ less\ than\ 3\ days}{TNR_i}. \tag{14}$$

Then, we weaken the reviewer reputation *bba* by each corresponding reliability degree (i.e., $(1-\alpha_i)$ or α_i) using the discounting operation (Eq. 6) in order to take into consideration the brust spamicity degree.

This discounted *bba* ${}^{\alpha}m^{\Omega}_{RR_i}$ represented the reviewer reputation using the reviewer's average proliferation and the brust spamicity which are two important spammer indicators.

Example 2. We continue with the previous Example 1, we calculate the brust spamicity degree:
$\alpha_1 = \frac{200}{258} = 0.775$
α_1 is the reliability degree for S, hence we apply the discounting operation as follows:
${}^{\alpha}m^{\Omega}_{RR_1}(\{S\}) = 1 * \alpha_1 = 1 * 0.775 = 0.775$
${}^{\alpha}m^{\Omega}_{RR_1}(\Omega) = (1-\alpha_1) + \alpha_1 * 0 = 0.225$

3.2 Step 2: Reviewer Helpfulness

The reviewer helpfulness is an important indicator to spot spammers. For this reason, we propose to verify if the reviewer post helpful reviews or unhelpful ones in order to mislead readers. Accordingly, we propose to use the Number of Helpful Reviews (NHR) to indicate the helpful ones associated to each reviewer.

Therefore, if ($NHR_i = 0$), the reviewer is suspicious to be spammer, thus we model the reviewer helpfulness by a certain *bba*:

$$m^{\Omega}_{RH_i}(\{S\}) = 1 \tag{15}$$

Else

$$m^{\Omega}_{RH_i}(\{\overline{S}\}) = 1 \tag{16}$$

We propose to penalize the reviewer helpfulness mass by considering the non helpfulness degree for each reviewer R_i denoted by β_i.

So, we propose this discounting factor as follows:

$$\beta_i = \frac{TNR_i - NHR_i}{TNR_i} \tag{17}$$

Then, we use the discounting operation in order to update the *bba* into a simple support function ${}^{\beta}m^{\Omega}_{RH_i}$. Thus, we take into consideration the helpfulness degree.

Generally, customers are not totally satisfied by their consumed products or tested services. Therefore, the innocent reviewer will not usually post extreme rating. However, most spammers perpetually resort to extreme ratings [8], either highest (5*) or lowest (1*), in order to achieve their goal of rapidly raising or bringing down, respectively, the mean score of a product.

When the reviewer had a lot of helpful reviews but they are full of extreme rating, his chances of being genuine reviewer certainly decrease.

In order to take this fact into account, we calculate the extreme rating degree denoted γ_i, corresponding to each reviewer R_i, which is considered as the discounting factor calculated by the number of the extreme rating divided by the total number of reviews given by each reviewer TNR_i as the following equation:

$$\gamma_i = \frac{NER_i}{TNR_i} \tag{18}$$

Where, NER_i is the extreme reviews' number (i.e., $NER_i \in \{1, 5\}$) given by each reviewer R_i.

Then, each simple support function represented the reviewer helpfulness ${}^{\beta}m^{\Omega}_{RH_i}$ is weakened again by its relative reliability degree (i.e., $(1 - \gamma_i)$ or γ_i) through the discounting operation.

Thus, this discounted ${}^{\beta\gamma}m^{\Omega}_{RH_i}$ modeled the reviewer helpfulness based on both the reviewer helpfulness degree and extreme ranting.

Example 3. Let us consider the same Example 1:

- The reviewer helpfulness *bba* corresponding to R_1 is generated as follows: Number of helpful reviews $= 100 > 0 \Rightarrow m^{\Omega}_{RH_1}(\{\overline{S}\}) = 1$
- Then, we calculate the corresponding helpfulness degree: $\beta_1 = \frac{258-100}{258} = 0.612$
- β_1 is the discounting factor $\overline{S}$ and its reliability degree is $(1 - \beta_1)$. So, we apply the discounting operation as follows: ${}^{\beta}m^{\Omega}_{RH_1}(\{\overline{S}\}) = 1 * (1 - \beta_1) = 0.388$ ${}^{\beta}m^{\Omega}_{RH_1}(\Omega) = \beta_1 + (1 - \beta_1) * 0 = 0.612$
- After that, we calculate the extreme rating degree for R_1: $\gamma_1 = \frac{208}{258} = 0.806$
- γ_1 is the discounting factor $\overline{S}$ and its reliability degree is $(1 - \gamma_1)$. So, we reapply the discounting operation as follows: ${}^{\gamma\beta}m^{\Omega}_{RH_1}(\{\overline{S}\}) = 0.388 * (1 - \gamma_1) = 0.388 * (1 - 0.806) = 0.075.$ ${}^{\gamma\beta}m^{\Omega}_{RH_1}(\Omega) = \gamma_1 + (1 - \gamma_1) * 0.612 = 0.925.$

3.3 Step 3: Modeling the Whole Reviewer Trustworthiness

In the interest of representing the whole trustworthiness for each reviewer, we aggregate the reviewer *bba*'s reputation ${}^{\alpha}m^{\Omega}_{RR_i}$ with his helpfulness *bba* ${}^{\beta\gamma}m^{\Omega}_{RH_i}$ using the Dempster combination rule (i.e, $m^{\Omega}_{RT_i} =^{\alpha} m^{\Omega}_{RR_i} \oplus^{\beta\,\gamma} m^{\Omega}_{RH_i}$).

The output of this aggregation is a combined *bba* $m^{\Omega}_{RT_i}$ that represents the whole trustworthiness for each reviewer.

Example 4. Once the *bba*'s representing both the R_1 reputation and helpfulness, calculated in the previous example, are combined we obtain the following *bba*:
$m^{\Omega}_{RT_1}(\{S\}) = 0.761$
$m^{\Omega}_{RT_1}(\{\overline{S}\}) = 0.018$
$m^{\Omega}_{RT_1}(\Omega) = 0.221$

3.4 Step 4: According Spamicity Degree and Making Decision

In order to accord a spamicity degree to each reviewer, we resort the pignistic probability $BetP$. Then, the decision is made either the author is a spammer or innocent as we select the $BetP$ with the greater value as the final decision.

Example 5. After applying the pignistic probability on the *bba* calculated in the previous Example 4 and, we found:
$BetP(\{S\}) = 0.872$
$BetP(\{\overline{S}\}) = 0.128$
The reviewer R_1 is a spammer with a spamicity degree equal to 0.872.

4 Experimentation and Results

The evaluation in spam reviews detection problem has been always a significant barrier, due to the absence of true real world growth data. A common alternative, used by various previous works, is using human evaluators and experts in order to label the dataset. However, the human judgement may provide varying verdicts due to the variability in perception and tolerance without forgetting the human subjectivity.

In this paper, we conducted experiments on real dataset then we propose to validate our method behavior by analyzing some results.

4.1 Evaluation Protocol

Dataset Description
In order to evaluate our method, we used a real world dataset extracted from Tripadvisor which is composed by 6200 reviews given by 1420 reviewers. The dataset contains; the reviews, the reviewed restaurants or hotels and the reviewing historic corresponding to each reviewer which is detailed in the Table 2.

Table 2. Example of reviewer history

The reviewer_id
Total number of reviewed restaurants and hotels
Total number of reviews
The review rating
The review time
Number of helpful ratings

We propose to label our database through one of the most used clustering method K-means where $K = 2$ in order to divide it into two classes; spammer and non spammer, relying on some important features used in the literature [4] such as:

- Duplicate/Near Duplicate Reviews
- Extreme Rating
- Reviewing Burstiness
- The helpfulness degree
- The average mean rating given by each reviewer

Evaluation Criteria
We evaluate our method according to the three following criteria: Accuracy, precision and recall and they can be defined as Eqs. 19, 20, 21 respectively where TP, TN, FP, FN denote True Positive, True Negative, False Positive and False Negative respectively.

$$Accuracy = \frac{(TP + TN)}{(TP + TN + FP + FN)} \tag{19}$$

$$Precision = \frac{TP}{(TP + FN)} \tag{20}$$

$$Recall = \frac{TP}{(TP + FN)} \tag{21}$$

Experimental Results
Our method distinguishes between 229 spammers and 1266 genuine reviewers. We propose to compare it with state-of-art baselines classifier; the Support Vector Machine (SVM) and the Naive Bayes (NB) [5,8,14]. The results are reported in the Table 3.

Table 3. Comparative results

Methods	Accuracy	Precision	Recall
SVM	0.72	0.71	0.70
NB	0.67	0.64	0.59
Our method	**0.98**	**0.96**	**0.94**

Our approach accomplishes the best performance according to accuracy, precision and recall over-passing state-of-art methods. It records at best an accuracy improvement over 30% compared to NB and over 26% compared to SVM.

4.2 Method Behavior Validation

In order to analyze our results, we randomly pick a set of ten reviewers from our Tripadvisor dataset. Table 4 details the reviewers information and presents the results generated by our approach. Our method classifies each reviewer as spammer or innocent by according a spamicity degree to each one.

The $reviewer_{id}$ = 21012Z is detected as a spammer with a high spamicity degree (i.e., 0.91) since he gives various non helpful reviews to some target products in short time interval including a lot of extreme rating in order to over-qualify or to damage them. However, the $reviewer_{id} = 10001E$ is classified as innocent with a very low spamicity degree as his reviews contain various helpful ones and few extreme rating. Moreover, they are spread along the reviewing time interval and each one is given to only one product. Taking also the $reviewer_{id} = 10012B$, almost this one is judged as innocent, he has a high spamicity degree (i.e., 0.47) because most of his reviews are given in less than three days including also some extreme rating, however we can not classified as spammer since he also has several helpful vote and he gives average less than two reviews per product.

Our method can be used in several fields by different reviews websites. In fact, these websites must block the detected spammers in order to stop the appearance of the fake reviews. Moreover and thanks to our uncertain output, they can control the behavior of the innocent ones with a high spamicity degree to prevent their tendency to turn into spammers.

Table 4. Reviewers information and results

Reviewer_id	Total number of reviews	Total number of product or services	Number of Extreme rating	Number of helpful vote	Number of reviews given in less than 3 days	**Decision**	**Spamicity degree**
10012D	258	30	208	100	100	**Spammer**	**0.87**
10013D	30	10	8	25	4	*Innocent*	*0.02*
10021D	20	12	0	18	2	*Innocent*	*0.11*
10010A	30	16	22	0	15	**Spammer**	**0.68**
10012B	16	12	6	10	9	*Innocent*	**0.47**
20012D	40	30	5	32	5	*Innocent*	*0.02*
18012B	30	3	25	0	28	**Spammer**	**0.99**
21012Z	60	5	20	2	50	**Spammer**	**0.91**
10412E	100	92	10	88	10	*Innocent*	*0.01*
10001E	150	150	10	120	15	*Innocent*	*0.01*

5 Conclusion

In this work, we addressed the spammer review detection problem and proposed a novel approach that manages the uncertainty while using the spammer behavior indicators. Our method shows its ability in distinguishing between fake and innocent reviewers while tuning a spamicity degree for each one. As future work, we aim to improve even more our detection by taking into account the semantic aspects through the analysis of the reviews contents.

References

1. Akoglu, L., Chandy, R., Faloutsos, C.: Opinion fraud detection in online reviews by network effects. In: Proceedings of the Seventh International Conference on Weblogs and Social Media, ICWSM, vol. 13, pp. 2–11 (2013)

2. Dempster, A.P.: Upper and lower probabilities induced by a multivalued mapping. Ann. Math. Stat. **38**, 325–339 (1967)
3. Fayazbakhsh, S., Sinha, J.: Review spam detection: a network-based approach. Final Project Report: CSE 590 (Data Mining and Networks) (2012)
4. Fei, G., Mukherjee, A., Liu, B., Hsu, M., Castellanos, M., Ghosh, R.: Exploiting burstiness in reviews for review spammer detection. In: Seventh International AAAI Conference on Weblogs and Social Media, vol. 13, pp. 175–184 (2013)
5. Heydari, A., Tavakoli, M., Ismail, Z., Salim, N.: Leveraging quality metrics in voting model based thread retrieval. World Acad. Sci. Eng. Technol. Int. J. Comput. Elect. Autom. Control Inf. Eng. **10**(1), 117–123 (2016)
6. Lim, P., Nguyen, V., Jindal, N., Liu, B., Lauw, H.: Detecting product review spammers using rating behaviors. In: Proceedings of the 19th ACM International Conference on Information and Knowledge Management, pp. 939–948 (2010)
7. Ling, X., Rudd, W.: Combining opinions from several experts. Appl. Artif. Intell. Int. J. **3**(4), 439–452 (1989)
8. Mukherjee, A., Kumar, A., Liu, B., Wang, J., Hsu, M., Castellanos, M.: Spotting opinion spammers using behavioral footprints. In: Proceedings of the ACM International Conference on Knowledge Discovery and Data Mining, pp. 632–640 (2013)
9. Pan, L., Zhenning, X., Jun, A., Fei, W.: Identifying indicators of fake reviews based on spammer's behavior features. In: Proceedings of the IEEE International Conference on Software Quality, Reliability and Security Companion (QRS-C), pp. 396–403 (2017)
10. Shafer, G.: A Mathematical Theory of Evidence, vol. 1. Princeton University Press, Princeton (1976)
11. Smets, P.: The transferable belief model for expert judgement and reliability problem. Reliab. Eng. Syst. Saf. **38**, 59–66 (1992)
12. Smets, P.: The canonical decomposition of a weighted belief. In: Proceedings of the Fourteenth International Joint Conference on Artificial Intelligence, pp. 1896–1901 (1995)
13. Smets, P.: The transferable belief model for quantified belief representation. In: Smets, P. (ed.) Quantified Representation of Uncertainty and Imprecision. HDRUMS, vol. 1, pp. 267–301. Springer, Dordrecht (1998). https://doi.org/10.1007/978-94-017-1735-9_9
14. Wang, G., Xie, S., Liu, B., Yu, P.S.: Review graph based online store review spammer detection. In: Proceedings of 11th International Conference on Data Mining (ICDM), pp. 1242–1247 (2011)

Mobile and Autonomous Robotics

A Model-Based Reinforcement Learning Approach to Time-Optimal Control Problems

Hsuan-Cheng Liao(✉) and Jing-Sin Liu(✉)

Institute of Information Science, Academia Sinica, Nangang, Taipei 115, Taiwan, ROC
{brianhcliao, liu}@iis.sinica.edu.tw

Abstract. Reinforcement Learning has achieved an exceptional performance in the last decade, yet its application to robotics and control remains a field for deeper investigation due to potential challenges. These include high-dimensional continuous state and action spaces, as well as complicated system dynamics and constraints in robotic settings. In this paper, we demonstrate a pioneering experiment in applying an existing model-based RL framework, PILCO, to the problem of time-optimal control. At first, the algorithm models the system dynamics with Gaussian Processes, successfully reducing the effect of model biases. Then, policy evaluation is done through iterated prediction with Gaussian posteriors and deterministic approximate inference. Finally, analytic gradients are used for policy improvement. A simulation and an experiment of an autonomous car completing a rest-to-rest linear locomotion is documented. Time-optimality and data efficiency of the task are shown in the simulation results, and learning under real-world circumstances is proved possible with our methodology.

Keywords: Model-based reinforcement learning · Time-optimal control · Robotics

1 Introduction

Reinforcement Learning (RL) has become one of the promising approaches to Optimal Control Problems. With the robot modelled as a reward-maximizing agent and the desired behavior expressed as a utility function, it is possible to train the robot for an optimal sequence of actions through its interactions with the environment. Nonetheless, factors such as high-dimensional continuous state and action spaces, as well as complicated system dynamics and nonlinear constraints in robotic settings have increased the difficulty of the problems.

In this paper, we demonstrate applying an existing model-based policy search algorithm, Probabilistic Inference for Learning Control (PILCO) [1], to a Time-Optimal Control Problem, and confront the aforementioned challenges to a certain extent. The method employs non-parametric Gaussian Processes (GP) for probabilistic dynamics modelling. It then uses approximate inference for system predictions and policy evaluation. Finally, policy improvement is made with analytic policy gradients. A simulation and an experiment of an autonomous car driving along a linear path are performed. The task of the vehicle is to complete a rest-to-rest linear locomotion in the shortest time.

F. Wotawa et al. (Eds.): IEA/AIE 2019, LNAI 11606, pp. 657–665, 2019.
https://doi.org/10.1007/978-3-030-22999-3_56

It is shown in the results that the car successfully accomplishes the trajectory with a single switching velocity profile under control constraints, while keeping the advantage of data efficiency.

The remainder of the article is structured as follows: In Sect. 2 we give an brief overview of the related work. The key elements of the PILCO algorithm is elaborated in Sect. 3, including dynamics modelling, trajectory prediction, policy evaluation and policy improvement. This is followed by the simulation and experiment results in Sect. 4. Lastly, we conclude with final remarks in Sect. 5.

2 Related Work

The conventional approaches to TOCP, including Dynamic Programing [2], Convex Optimization [3], and Numerical Integration [4, 5], have been found in applications including manipulators, nonholonomic vehicles, bipedal humanoids. Nonetheless, in most cases, the system dynamics are derived through non-trivial mathematics and physics equations, often still compromised with assumptions which are too simplistic.

The issue of oversimplicity, however, is likely to be alleviated with the advents of RL frameworks, especially the model-based family. These approaches reformulate the problem as a Markov Decision Process for the autonomous agent, which maximizes the long-term rewards and needs no pre-programmed transition dynamics beforehand. The employment of a model of the agent-environment interactions creates an internal simulation during learning process and reduces physical engagement substantially, decreasing potential hazards and mechanical wear of robots.

There have been a surging number of researches related to model-based RL over the past decade. Despite its faster convergence over the model-free frameworks, a severe issue is that system-modelling biases greatly affect the learning performance. Among the modelling techniques such as Receptive Field Weighted Regression (RFWR) and Expectation Maximization (EM), GP is the state-of-the-art practice that achieves the highest accuracy and data efficiency [6]. In contrast to other probabilistic models that maintain a distribution over random variables, GP builds one over functions. Therefore, it has no prior assumption on the function mapping current states and actions towards future states. The fact makes it an effective tool, and is also the reason why we have implemented it. Eventually, having attained the dynamics model, the solution to a RL problem is typically derived from two classes of approaches, namely Value Function and Policy Search [6].

3 Methodology

In this paper, we conveniently adopt the PILCO framework, summarized in Algorithm 1, and apply it to the simulation task with some adjustments [1]. It is proposed by Deisenroth et al., and has reached unprecedented performance in benchmark tasks such as the inverted pendulum and the cart-pole swing-up.

Algorithm 1 PILCO

1: *Define* parametrized policy: $\pi: z_t \times \theta \rightarrow u_t$
2: *Initialize* policy parameters θ randomly
3: *Execute* system and record data
4: **repeat**
5: *Learn* system dynamics model with GP
6: *Predict* system trajectories
7: *Evaluate* policy: $J^{\pi}(\theta) = \sum_{t=0}^{T} \gamma^t \mathbb{E}_X[cost(X_t)|\theta]$
8: *Update* policy parameters by gradients $\mathrm{d}J^{\pi}(\theta)/\mathrm{d}\theta$
9: *Execute* system and record data
10: **until** task completed

Inherent from the characteristics of model-based methods, PILCO exploits the advantage of extracting useful information from observations more efficiently than the model-free approaches. Specifically, PILCO adopts GP probabilistic modelling and inferencing to learn the transition dynamics. Therefore, it effectively handles the input uncertainties and reduces the effect of model errors, eliminating the common drawback of model-based frameworks. It then employs policy search for planning and uses analytic gradients of closed form solutions for optimization. The four core elements of the framework, namely dynamics modelling, trajectory prediction, policy evaluation and policy optimization, are illustrated in this section.

3.1 Dynamics Modelling

GP is chosen to model the latent dynamics function in PILCO considering its resistance against overfitting bias and ability to learn complex systems (Algorithm 1, line 5). The training inputs are the state-action tuples, $\tilde{x}_t = [x_t \ u_t]^{\mathrm{T}} \in \mathbb{R}^{D+F}$, and the targets are the differences between consecutive states, $\Delta_t = x_{t+1} - x_t \in \mathbb{R}^D$.

In this paper, a zero-mean function $m \equiv 0$ is employed and a squared exponential covariance function is defined as:

$$k(\tilde{x}_i, \tilde{x}_j) = \sigma_f^2 \exp\left(-\frac{1}{2}(\tilde{x}_i - \tilde{x}_j)^{\mathrm{T}} \Lambda^{-1} (\tilde{x}_i - \tilde{x}_j)\right), \tag{1}$$

with variance of the function σ_f^2 and $\Lambda := \mathrm{diag}([l_1^2, l_2^2, \ldots, l_{D+F}^2])$ depending on the length scales.

With n training samples $\tilde{X} = [\tilde{x}_1, \ldots \tilde{x}_n]$ and $\mathrm{y} = [\Delta_1, \ldots, \Delta_n]$, the posterior GP hyper-parameters are learned through evidence maximization and describes a one-step prediction model:

$$p(X_{t+1}|X_t, U_t) = \mathcal{N}(X_{t+1}|\mu_{t+1}, \textstyle\sum_{t+1}), \tag{2}$$

$$\mu_{t+1} = X_t + \mathbb{E}_f[\Delta_t], \quad (3)$$

$$\sum\nolimits_{t+1} = \mathrm{var}_f[\Delta_t], \quad (4)$$

where capitals represent random variables.

3.2 Trajectory Prediction

For policy evaluation, PILCO first predicts long-term system trajectories with the learnt transition dynamics (Algorithm 1, line 6). To such end, the one-step prediction process is cascaded from X_0 to X_1, X_1 to X_2, and up to X_T, forming a distribution over the system trajectories. This distribution is assumed Gaussian, $p(X_{t+1}) \sim \mathcal{N}(\mu_{t+1}, \Sigma_{t+1})$, and subsequently approximated by moment matching or linearization of the posterior mean function for further computation in policy evaluation.

$$\mu_{t+1} = \mu_t + \mu_\Delta, \quad (5)$$

$$\sum\nolimits_{t+1} = \Sigma_t + \Sigma_\Delta + cov[x_t, \Delta_{t+1}] + cov[\Delta_{t+1}, x_t], \quad (6)$$

$$cov[x_t, \Delta_{t+1}] = cov[x_t, u_t]\Sigma_u^{-1} cov[u_t, \Delta_{t+1}]. \quad (7)$$

3.3 Policy Evaluation

Having retrieved the predictive trajectories, it remains computing the expected long-term cost. PILCO applies the cost function (8) to state distributions at each time step.

$$J(\theta) = \sum\nolimits_{t=0}^{T} \gamma^t \mathbb{E}_X[\mathrm{cost}(X_t)|\theta] \quad (8)$$

$$\mathbb{E}[\mathrm{cost}(X_t)|\theta] = \int \mathrm{cost}(X_t)\mathcal{N}(X_t|\mu_t, \sum\nolimits_t)\mathrm{d}X_t \quad (9)$$

T is the entire experience time in each episode and γ is the decaying ratio. A saturating cost function (10) is applied for its integrability in (9). It is roughly quadratic around the target state yet smooths out at unity in distant states, which avoids the pitfall of quadratic functions penalizing too heavily the far-off states. We utilize the Euclidean distance from the current state to the target state, and tune the cost width with σ_c.

$$\mathrm{cost}(X) = 1 - \exp\left(-\frac{1}{2\sigma_c^2}\mathrm{dist}(X, x_{target})^2\right) \in [0, 1] \quad (10)$$

With the combination of (8) and (10), a goal-reaching time t' is to be minimized throughout the learning task. This is guaranteed as the algorithm converges to the global optimum because the time-accumulated sum in (8) forces the immediate cost in (10), the distance between the target and current state, to reduce as fast as possible.

3.4 Policy Optimization

The policy is improved episodically through the gradient information of $J(\theta)$ (Algorithm 1, line 8). For analytical tractability, it is required that the expected cost in (9) is differentiable about the state distribution moments, and that the moments of the control distribution are differentiable about the policy parameters θ. Thorough computation of the gradients $\mathrm{d}J^{\pi}(\theta)/\mathrm{d}\theta$, which involves several applications of chain rule, is documented in [1]. Finally, thanks to the analytic expressions of the gradients, any standard gradient-based optimization method such as CG or BFGS can be implemented to search for the set of optimal parameters θ, which minimizes the total cost $J^{\pi}(\theta)$.

4 Results and Discussions

The task of interest in this paper is to drive an autonomous car along a linear path and produce a rest-to-rest time-optimal trajectory. We first run a simulation on a laptop with Intel i7 core and 16 GB RAM, and then have a real-world experiment with an inexpensive robotic car.

4.1 Simulation

Settings. The simulated car has a horizontal length of 30 cm and a mass of 0.5 kg. It starts from the origin in the beginning of every episode, and its goal is to reach the destination without overshooting in the minimum time. The straight-line distance from the origin to the destination is 5 m, and a friction coefficient of 0.1 N/m/s is assumed between the surface and the car. In addition, in order to mimic a real vehicle, there is a boundary of ± 4 m/s^2 on the acceleration control (Fig. 1).

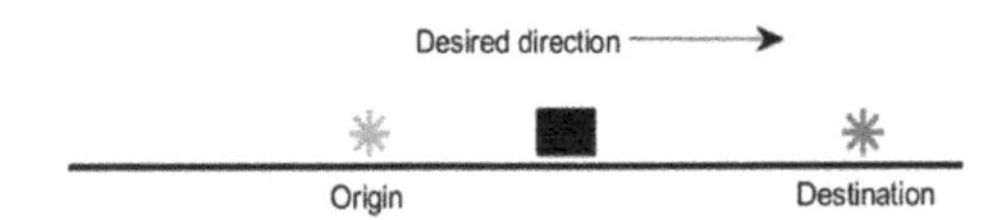

Fig. 1. A side-view of the experiment setup. The black box depicting the car starts from the origin (the green star) and targets at the destination (the red star). (Color figure online)

The agent has no prior knowledge of the domain, and is allowed to interact with the environment for 4 s during each episode (Algorithm 1, line 3/9). We adopt a nonlinear RBF controller as suggested in [1], given as follows:

$$\pi(x, \theta) = \sum\nolimits_{i=1}^{n} w_i \phi_i(x) \tag{11}$$

$$\phi_i(x) = \exp(-\frac{1}{2}(x - \mu_i)^{\mathrm{T}} \Lambda^{-1} (x - \mu_i)) \tag{12}$$

where n is chosen to be 100, and $\theta = \{w_i, \Lambda, \mu_i\} \in \mathbb{R}^{403}$.

A total of 16 episodes (including the initial random policy) are executed, each of which is sampled at a frequency of 40 Hz. The entire process is governed by the given model in (13) for simulation.

$$\dot{x}(t) = \begin{bmatrix} 0 & 1 \\ 0 & \frac{-b}{m} \end{bmatrix} x(t) + \begin{bmatrix} 0 \\ \frac{1}{m} \end{bmatrix} u(t) \tag{13}$$

where b is the friction coefficient, m is the mass of car, $x(t) = [position\ velocity]^{\mathrm{T}}$ and $u(t) = acceleration.$

Results. The simulation data is recorded and explained below.

Figure 2 shows the learning outcome at the final episode, illustrating a time-optimal rest-to-rest linear locomotion. Under the control constraints, the car accelerates and decelerates at its maximum. The velocity profile forms a triangle with one switching point as expected from the classical NI approach.

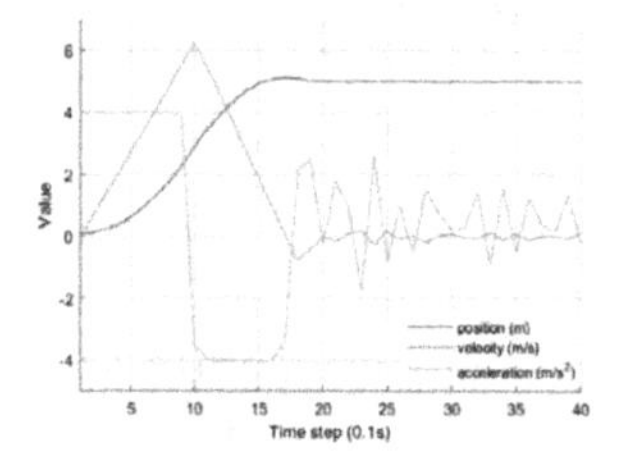

Fig. 2. Outcome of the simulation.

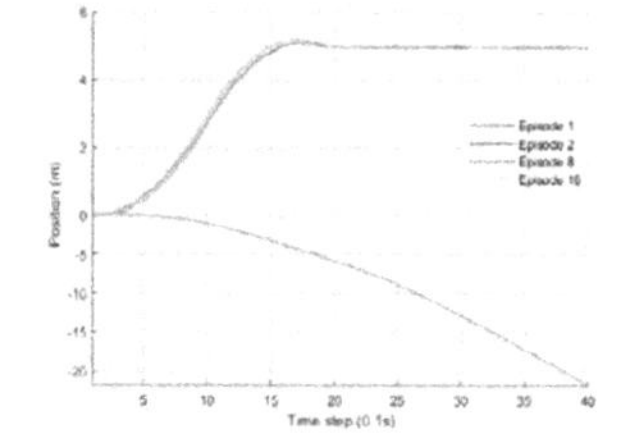

Fig. 3. Position state at various episode.

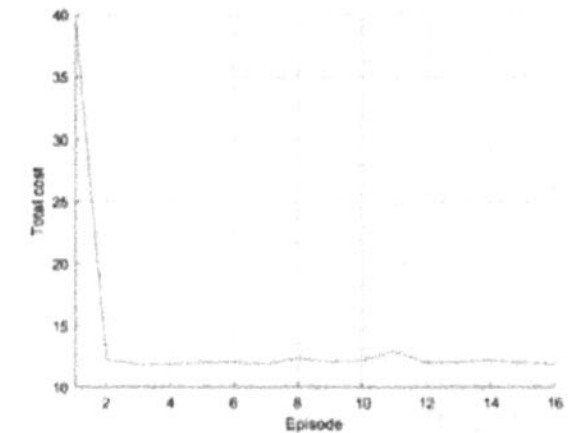

Fig. 4. Simulated total cost against episodes.

We further investigate the efficiency of the algorithm. As shown in Fig. 3, the agent reaches the destination within 2 episodes, producing practically a time-optimal trajectory. The first trajectory is randomly generated, and runs towards the opposite direction of the destination. The rest of the trajectories are relatively identical across the second episode to the last one. The same implication has been drawn from Fig. 4. In the first episode, the total cost is 40 on account of the unity saturating function, yet it is soon reduced down at the second episode and remains at its lowest throughout the task.

4.2 Experiment

Aside from simulating on the laptop, we implement the algorithm onto an inexpensive Raspberry Pi-controlled car, AlphaBot, as shown in Fig. 5.

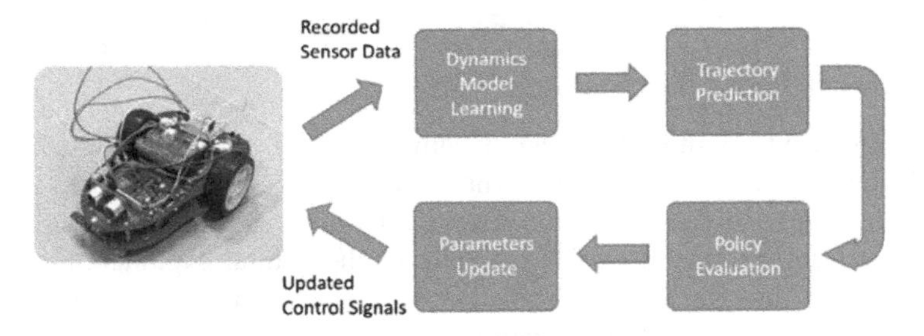

Fig. 5. Block diagram of the experiment.

Configuration. We use the same RBF controller as that in the simulation, and we need not to specify the vehicle parameters nor the system ODE. Instead, these are learnt through the Gaussian Processes.

The small car is designated to begin its route at a distance of 180 cm from the wall and end at 50 cm. It is equipped with photo interrupters and ultrasonic sensors for state measurements. The control signal generated from the RL algorithm ranges from –2 to 2, and is translated into change rate of duty cycles of the motor PWM signal on board.

Results. Figure 6 is the learning outcome of the real-world experiment. It is observed that the vehicle accomplishes nearly a time-optimal path as desired. It accelerates at the maximum over the first half of the path, and decelerates over the second half. The velocity limit stems from the electronic voltage constraints on the robot, and therefore is not directly handled by the robot control signal. The end position, despite having no overshooting and fluctuation, is slightly off the targeted 50 cm because of model errors and sensor inaccuracies. The convergence process is illustrated in Figs. 7 and 8. In Fig. 7, immediate cost at every time step in various episodes are plotted, showing the learning process and the destination arrival time being minimized. In Fig. 8, the total cost starts from 40 in the beginning of the task for the saturating cost function. It reaches its lowest at the seventh episode with a total experience time of 28 s, which is considered still a very efficient learning process.

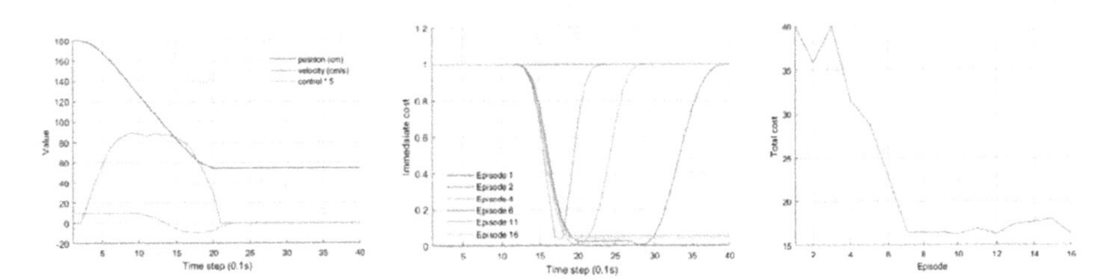

Fig. 6. Outcome of the experiment.

Fig. 7. Immediate cost in different episodes.

Fig. 8. Experimental total cost against episode

4.3 Discussions

We analyze and discuss the results in this section.

PILCO Algorithm. With probabilistic dynamics modelling, the system uncertainty is explicitly handled in policy planning. The benefit makes the algorithm perform efficiently in learning and control tasks. However, it is also observed that sometimes the algorithm is stuck in local optimum. In spite of the natural exploitation-exploration characteristics of the saturating cost function, the resulting policy is not guaranteed to be globally optimal since the optimization problem is not always convex [7].

Differences Between Simulation and Experiment. Successful trials have been produced in both the simulation and experiment, yet there are two noteworthy differences. First, the convergence in the simulation is faster. This is believed to be caused by the relatively complicated circumstances in the real world, where factors such as sensor noises, motor disturbances and environment uncertainties densely exist. Second, although the system is described by simple Newtonian dynamics in our simulation case, this is not the same in the experiment. The mentioned factors and the electronic voltage bounds pose some extent of nonlinearities to the actual system and affect the complexity of the task.

Time Optimality. The time optimality of the task is inherent from the fact that PILCO evaluates the policy over the entire planning horizon. Such property together with the Euclidean distance used in the cost function implicitly forces the agent to arrive at its goal as fast as possible so as to deliver a lower total cost. It is believed that the feature can be generalized to tasks with similar objectives and settings.

Constraints. PILCO enforces control constraints directly through a squashing function applied on the policy. In general, the method serves well for hard constraint boundaries on controls, which is implemented in our case. However, such handling might cause inaccurate predictions around the constraint boundaries [8]. Moreover, apart from hard control boundaries, robotic systems often confront nonlinear state and control constraints. PILCO is not able to meet this type of system requirements also because it looks at the full horizon for policy evaluation. It tends to compromise and balance out the costs generated from the target state and the penalties given for violating the constraints. A few trials are carried out to implement a maximum velocity on the autonomous vehicle through some tuning on the cost function, but in vain. The results indicate that the framework fails to manage such constraints neatly.

5 Conclusion

We have demonstrated a primitive attempt to utilize a model-based policy search framework, PILCO, to a time-optimal control problem in the presence of hard control constraints. A simulation and a real-world experiment of an autonomous car driving along a linear path is illustrated. Time-optimality and data efficiency of the task have been shown in the results.

There are several directions open for future research. First, the task in this paper has a low dimension of state and control spaces. The framework is to be examined in projects with larger scales. Second, the algorithm now performs policy evaluation and optimization offline. It is crucial for future approaches to take these online since most applications encourage real-time operation. Third, time-optimality is forced by the long-term saturating cost in our experiment. It is suggested that future studies inspect the possibility of incorporating Pontryagin Maximum Principle into policy learning, thereby delivering a principled method with more theoretical supports. Lastly, non-linear state and control constraints are to be considered in future work. One potential framework is to extend the predicted system trajectory in PILCO into Model Predictive Control and utilize Sequential Quadratic Programming to satisfy the constraints.

Acknowledgement. We acknowledge Dr. Marc Peter Deisenroth for his kind help when implementing the PILCO algorithm in our project. His advice on system constraints handling was very useful to us.

References

1. Deisenroth, M., Rasmussen, C.: PILCO: a model-based and data-efficient approach to policy search. In: Proceedings of the International Conference on Machine Learning (2011).
2. Shin, K., McKay, N.: A dynamic programming approach to trajectory planning of robotic manipulators. IEEE Trans. Autom. Control **31**(6), 491–500 (1986)
3. Verscheure, D., Demeulenaere, B., Swevers, J., DeSchutter, J., Diehl, M.: Time-optimal path tracking for robots: a convex optimization approach. IEEE Trans. Autom. Control **54**(10), 2318–2327 (2009)
4. Shin, K., McKay, N.: Minimum-time control of robotic manipulators with geometric path constraints. IEEE Trans. Autom. Control **30**(6), 531–541 (1985)
5. Lamiraux, F., Laumond, J.: From paths to trajectories for multibody mobile robots. In: Proceedings of the 5th International Symposium on Experimental Robotics, pp. 301–309 (1998).
6. Polydoros, A.S., Nalpantidis, L.: Survey of model-based reinforcement learning: applications on robotics. J. Intell. Rob. Syst. **86**(2), 153–173 (2017)
7. Rasmussen, C., Kuss, M.: Gaussian processes in reinforcement learning. NIPS **7**, 51–759 (2004)
8. Mayne, D.Q., Rawlings, J.B., Rao, C.V., Scokaert, P.O.: Constrained model predictive control: stability and optimality. Automatica **36**, 789–814 (2000)

Low-Cost Sensor Integration for Robust Grasping with Flexible Robotic Fingers

Padmaja Kulkarni, Sven Schneider(✉), and Paul G. Ploeger

Hochschule Bonn-Rhein-Sieg, Grantham-Allee 20, 53757 Sankt Augustin, Germany
padmaja.kulkarni@smail.inf.h-brs.de,
{sven.schneider,paul.ploeger}@h-brs.de

Abstract. Flexible gripping mechanisms are advantageous for robots when dealing with dynamic environments due to their compliance. However, a major obstacle to using commercially-available flexible fingers is the lack of appropriate feedback sensors. In this paper, we propose a *novel integration* of flexible fingers with commercial off-the-shelf proximity sensors. This integrated system enables us to perform non-interfering measurements of even minor deformations in the flexible fingers and consequently deduce information about grasped objects without the need of advanced fabrication methods. Our experiments have demonstrated that the sensor is capable of robustly detecting grasps on most test objects with an accuracy of 100% without false positives by relying on simple, yet powerful signal processing and can detect deformations of less than 0.03 mm. In addition, the sensor detects objects that are slipping through the flexible fingers.

Keywords: Robotics · Force and tactile sensing · Flexible robots

1 Introduction

The recent trend in rapid prototyping due to new and low-cost manufacturing methods like 3D printing or various methods of molding has fostered research in soft and flexible robotics. Due to their intrinsic ability to adapt to their environment, soft robot fingers promise better results in grasping. For the same reason, they can better handle uncertainties during grasping tasks. Consequently, robots require less elaborate models of their environment or the grasping process as such [1].

However, it is still an open research question how the state of such flexible fingers can be estimated with appropriate sensors. In this paper, we address this question in the context of robotics competitions, in particular RoboCup@Work and RoCKIn@Work. Here, frequently a KUKA youBot robot is employed where we have added off-the-shelf, soft parallel *adaptive gripper fingers* by Festo (see Fig. 1) in various scenarios, involving the grasping and transportation of different industrial objects. In our "naïve" setup we observed the following failure cases related to grasping: (i) During grasping, the fingers collided with and exerted

F. Wotawa et al. (Eds.): IEA/AIE 2019, LNAI 11606, pp. 666–673, 2019.
https://doi.org/10.1007/978-3-030-22999-3_57

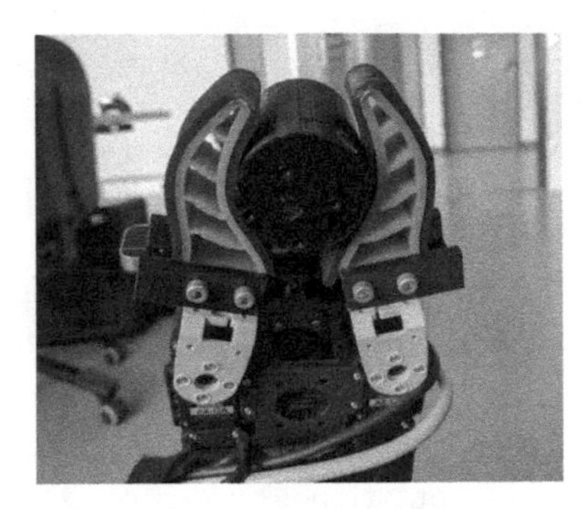

Fig. 1. Flexible fingers without sensors while grasping an object. The fingers' deformation is clearly visible. Each finger consists of two flexible *bands* that meet at the top to form a triangular shape and four horizontal *ribs*.

too much force on the environment. This prevented the fingers from closing properly. (ii) While reaching towards the object, the gripper completely missed the object. (iii) In the third case, the object slipped from the robot's fingers while it was transporting the object to the goal location. Due to the lack of appropriate feedback the robot was unable to detect any of those failures and failed to perform the overall process of transporting and placing the object. Therefore, to improve the robustness of manipulation it is important to detect at least the state of a grasp.

In this paper, we identify and evaluate a sensor (see Fig. 2) that is able to detect failures as those outlined above in our manipulator setup with flexible fingers. To summarize, the main contributions of this paper is the identification of an appropriate sensor (see Fig. 2) that is **low-cost**, both, in terms of hardware and integration process and the **integration** of that sensor into a real-world robot and its application to grasp detection and slip detection.

2 Related Work

The sensor integration into *soft* and *flexible* robotic hands aims at improving the grasping process via the estimation of, for example, (i) the finger state, including finger-object contact points; (ii) forces applied by the fingers; or (iii) the detection of objects slipping from the fingers. Many approaches rely on the attachment of external sensors or the design and fabrication of custom soft hands with embedded sensors [2].

Tactile sensing for soft robotic hands is quite versatile, as almost all approaches involve the design of custom sensors. Frequently, those approaches use custom molding processes and measure the force in the fingers [1,3]. In non-contact situations proximity sensors are employed in addition to the tactile sensor [4,5]. Along with predicting the position of the object, vision sensors can also detect slip or provide haptic feedback, for example, by deriving a force from the deformation observed at the fingertip [6]. Some approaches use mold bend sensors in the flexible fingers for estimating the state of the finger [2,7].

3 Approach

We take the related work as the basis for selecting an appropriate sensor for our use case. Moreover, the following non-functional requirements are considered for the sensor selection.

R1: Low-cost. The sensor as such should be inexpensive and should only impose little additional computational burden on the robot's host computer.
R2: Non-interference. The sensor should not exert forces on the fingers and interfere with the grasping process.
R3: Easy integration. We want to avoid special fabrication techniques which may be expensive or require expert knowledge of the involved procedures.

Afterwards, we provide further details about the sensor and, finally, we describe how this sensor is integrated into both, the flexible fingers and the manipulation architecture.

3.1 Sensor Selection

Tactile sensors always must be molded into flexible material such as polymers and need custom fabrication and integration of sensors into the production of flexible hands, we exclude those types of sensors (**R3**). *Bend sensors* have to be rigidly attached either to the palmal or dorsal surfaces of the fingers. This exerts additional forces on the fingers, impeding the grasping process. Hence, we exclude those sensors (**R2**). Robots already feature different *vision sensors* for object recognition or pose estimation, which may be re-purposed for gathering information about grasped objects. This may interfere with the cameras' primary objectives (**R2**). Moreover, additional vision sensors require more computational power to process RGB or RGB-D data and are therefore excluded.

On the other hand, *proximity sensors* are low cost (**R1**), can be mounted such that they do not interfere with the grasping process of the robot (**R2**) and without requiring any special fabrication technique for the mounting (**R3**). A proximity sensor measures the gripper deformation due to an external force from which we can infer information about the grasp state and the grasped object. Hence, we decide to choose a proximity sensor. Given the setup in our use case, we identify the VCNL4010 [8] (see Fig. 2) sensor as the most suitable since, according, to the specifications in the datasheet its measurement range aligns with the distance between the fingers' base and the lowest rib.

As shown in the circuit diagram in Fig. 3 the VCNL4010 integrates a proximity sensor consisting of an infrared LED and a photo PIN diode with an ambient light sensor (another photo PIN diode). According to the manual the measurement range reaches from 0.1 mm up to 200 mm, whereas the sensitivity peaks between 1 mm and 2 mm. Further details are given in [8].

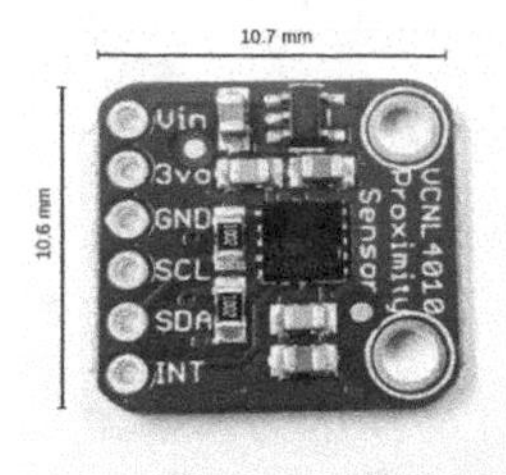

Fig. 2. The breakout board with VCNL4010 distance sensor in center

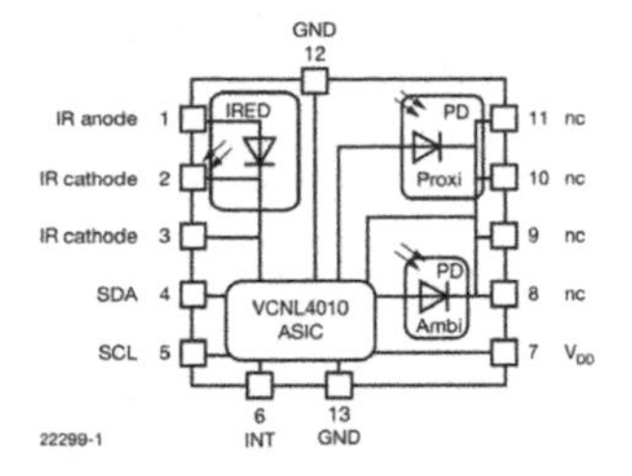

Fig. 3. Circuit diagram of the VCNL4010 sensor [8]

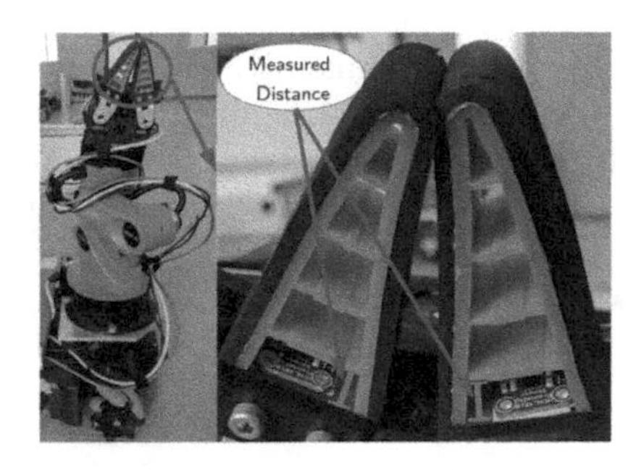

Fig. 4. The overall setup showing the distance measured by the sensors

3.2 Integration into Hardware Architecture

We add one sensor to the lowest compartment of each finger as depicted in Fig. 4. The choice of this location is twofold. Firstly, the gripper's base is a rigid surface, therefore enabling simple mounting of the sensor without interfering with the finger's flexibility. Secondly, this mounting location ensures the sensor's safety by preventing it from getting jammed between the flexible finger elements. A grasped object causes the ribs in the fingers to bend inwards reducing their distances from the mounted sensors (cf. Fig. 1) and, consequently, results in a change of the measured signal.

3.3 Integration into Manipulation Architecture

Our grasping strategy is realized by the following, fairly common, four-step procedure: (i) move the arm to a pre-grasp configuration; (ii) reach towards the object; (iii)close the gripper; (iv) lift the object by moving the arm to a post-grasp configuration.

We exploit the grasping procedure as prior knowledge for the clever evaluation of the sensor signal. This allows us to detect grasps by *simple*, yet *powerful* signal processing approaches, consisting of a moving average filter, an automatic online calibration procedure and signal thresholding.

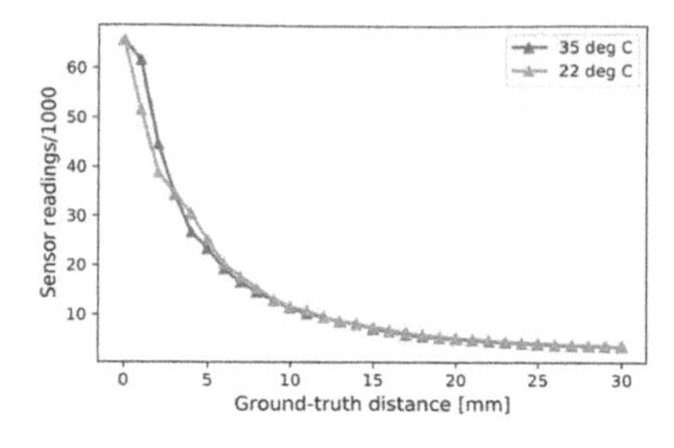

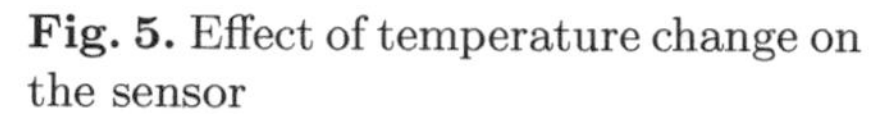
Fig. 5. Effect of temperature change on the sensor

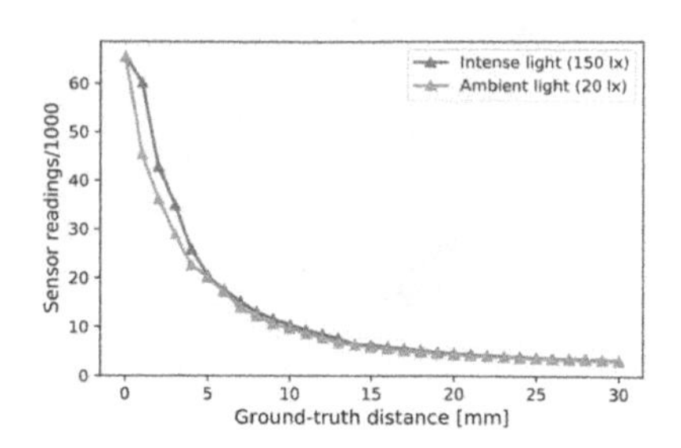

Fig. 6. Effect of light intensity change on the sensor

For the online calibration, we command the robot to close its gripper during the motion to the pre-grasp configuration. Since we know that the gripper is empty, this allows us to store the current sensor readings associated with an empty gripper as a distance reference value. Then, after the arm has moved to the post-grasp configuration, we compare the reference value for the empty gripper against the sensor readings with the (potentially) grasped object. If those two readings deviate by more than a pre-defined threshold, we classify the object as successfully grasped, else the grasp failed.

4 Experiments and Evaluation

We have decided on the following two-step evaluation procedure. Firstly, we evaluate the performance of the stand-alone sensor in various experiments. Secondly, we investigate how the integrated sensor performs in our use case of grasp detection.

4.1 Stand-Alone Sensor Evaluation

To validate the sensor's robustness, we compare the distance values provided by the sensor against ground-truth measurements. In a controlled manner the following parameters are varied to test if the readings are reproducible: (i) the temperature; (ii) the ambient light intensity; (iii) the type of reflecting material; and (iv) different samples of the same sensor. We evaluate the sensor's precision, sensitivity and the amount of noise in the readings. The latter is of special importance for thresholding the readings to detect successful grasps, especially, when small objects only cause minor deformations in the fingers.

To compare the results, the mean squared error is used, as it gives the distance between two readings. Here, the mean squared error is the average of the squares of the difference between the sensor output at two different, controlled conditions (either environmental conditions or different sensors) but at the same distances. Suppose $r_{1,\{1,2,\ldots,n\}}$ is the sequence of first readings and $r_{2,\{1,2,\ldots,n\}}$ is the sequence of second readings, then the mean squared error e is defined as:

$$e = \frac{1}{n} \sum_{i=1}^{i=n} (r_{1,i} - r_{2,i})^2 \tag{1}$$

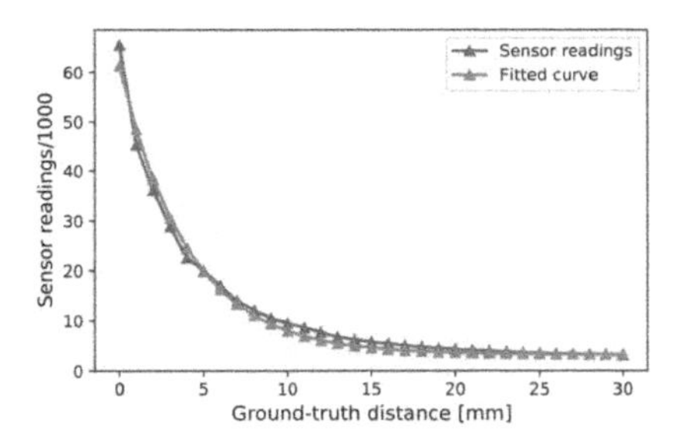

Fig. 7. Curve fitting for one sensor

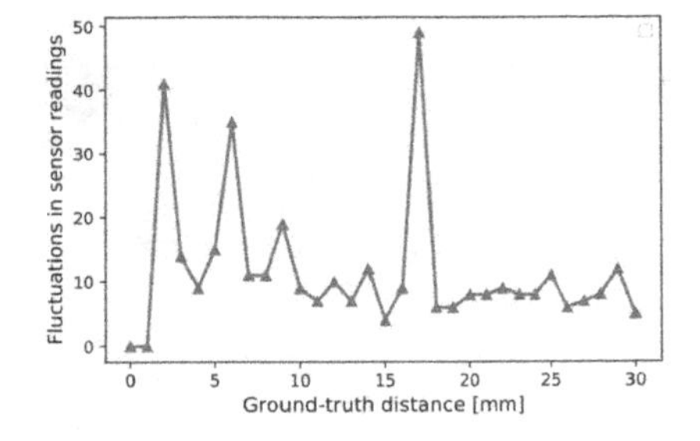

Fig. 8. Fluctuations in the sensor readings for varying distances.

Observations from Figs. 5 and 6 show little effect of temperature and ambient light conditions on the sensor readings, especially in the desired working range from 5 to 10 mm. However, they do depend on the selected sensor sample.

In addition, we can see (Fig. 7, sensor readings curve) that the relationship between sensor readings and ground-truth measurements is non-linear. Hence, to estimate distances from sensor readings, we need to identify (i) the model which maps from sensor readings to distances; and (ii) for each sensor sample a parameterization of that model.

Neither the exact measurement principle nor the conversion to the 16-bit digital representation is specified in the sensor's manual. Thus, to identify a simple model, we opt for a trial-and-error approach by fitting different types of hyperbolas and logarithmic curves to the sensor data. The best-fitting function is the natural logarithm, that, given the sensor output r, estimates the distance d in millimeters by:

$$d = a \log_e \frac{r + b}{c} \tag{2}$$

where a, b and c are the sensor-specific parameters of the model. Figure 7 shows the fitted curve for one particular sensor sample. The parameters are $a = -4.0$, $b = -3201.25$ and $c = 58077.10$ which leads to a mean squared error of 0.15. For a different sensor we identified the parameters as $a = -4.96$, $b = -3397.94$ and $c = 59071.2$ with a mean squared error of 0.12.

It is also important to know the fluctuation in the sensor readings, as it is a deciding factor for thresholding which we will discuss in the context of evaluating the sensor integration. To compute the fluctuation we apply the following procedure: (i) record ten sensor readings at a distance of d mm; (ii) compute the integer average r_i of those ten readings; (iii) repeat the step (i) and (ii) ten times, building a sequence $(r_1, r_2, r_3, \ldots, r_{10})$. This sequence contains a minimum reading r_{min} and a maximum reading r_{max} which define the fluctuation f_d at a distance of d mm according to the formula $f_d = r_{max} - r_{min}$. Figure 8 depicts the sensor reading fluctuations for a ground-truth distance ranging from 0 mm to 30 mm. We can see from the graph that those fluctuations are not correlated with the object's distance from the sensor.

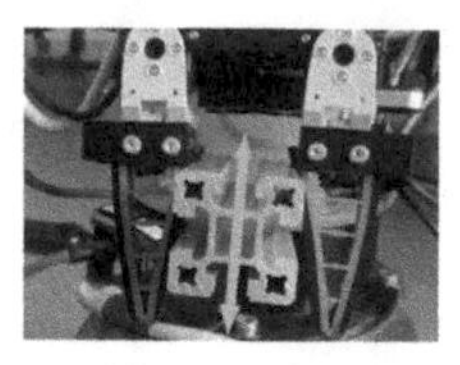

(a) Varying gripper contact points

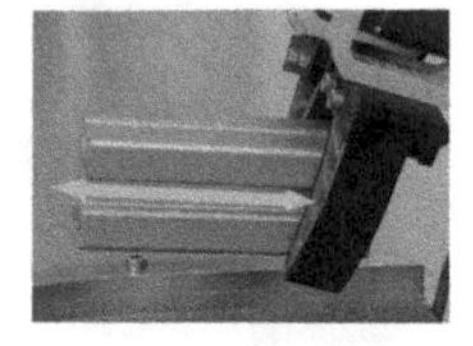

(b) Varying object contact points

Fig. 9. Variation in the grasp contact points

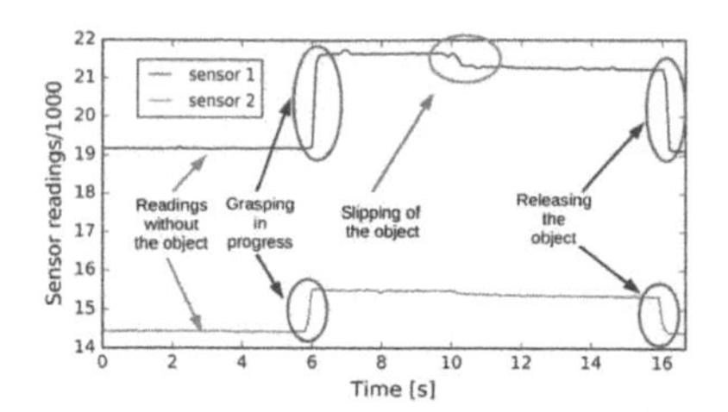

Fig. 10. Time series data of the sensor readings showing the object slipping between the fingers

4.2 Evaluation of Sensor Integration

To investigate the sensor's performance in the overall, integrated setup for grasp detection, we select various objects of different rigidity, size and shape. This selection includes industrial objects, for example, from the RoboCup@Work league, a water bottle, a deformable cloth and t-shirt, as well as a set of keys.

For each of the tested objects, four distinct grasp variations are defined as combinations of (i) grasping with the fingertips vs. grasping with fingers' base (see Fig. 9a); (ii) grasping the object in the middle vs. grasping the object at its end (see Fig. 9b). Additionally, the objects are placed both, in a laying down and standing upright configuration. In each of those variations, the robot is tasked to pick up the object from its back platform, lift it and then verify if the grasp was successful using the proximity sensors. This procedure is repeated ten times.

It is observed that the deformation in the fingers is proportional to the object size. For 16 out of the 19 objects, our integrated setup is able to verify the grasps with 100% accuracy. Exceptions are the distance tube, the challenging thin plate with a thickness of only 3 mm and the impossible-to-detect sheet of paper. Still we achieve a grasp detection rate of 80% for the thin plate. Based on the parameters for the sensor model in Eq. 2, we estimate the minimal deformation that the proximity sensor can detect to be 0.03 mm for the thin plate. No false positives have been observed during the experiments.

Also monitoring the dynamic behaviour of a grasping process reveals interesting insights. Figure 10 shows the time series of the two mounted sensors while grasping an industrial object having $20 \times 20 \times 40$ mm dimensions. We can see that both sensors provide different readings. This is caused, for instance, by the differences in the sensors' mounting, but also by asymmetrically grasping the object. However, in both sensors the signals clearly cross the threshold to trigger the successful grasp detection. Of greater interest, however, are the ripples and the drop that are clearly visible in the first sensor's readings at a time of about 10 s. Those effects are caused by the object slipping through the fingers. Shortly afterwards, the object is stabilized in the fingers again, but at a slightly different finger deformation.

5 Conclusions and Future Work

In this paper we proposed a low-cost (less than 10 Euros) proximity sensor for grasp and slip detection with the flexible robotic hand shown in Fig. 1. This sensor has proven to be easily integrated into the flexible fingers of a real-world robot without the need for advanced fabrication processes. Despite its low price, the sensor is highly sensitive and can detect rib deformations of less than 0.03 mm. For the majority of our test objects the sensor achieves a grasp detection accuracy of 100% with no false positives.

For the future work we would like to derive semantic information, such as a classification or identification label, of a grasped object with the addition of further sensors and using advanced signal processing.

References

1. Ho, V., Hirai, S.: Design and analysis of a soft-fingered hand with contact feedback. IEEE Robot. Autom. Lett. (RA-L) **2**(2), 491–498 (2017)
2. Homberg, B.S., Katzschmann, R.K., Dogar, M.R., Rus, D.: Haptic identification of objects using a modular soft robotic gripper. In: IEEE/RSJ International Conference on Intelligent Robots and Systems (IROS) (2015)
3. Dollar, A.M., Jentoft, L.P., Gao, J.H., Howe, R.D.: Contact sensing and grasping performance of compliant hands. Auton. Robots **28**(1), 65–75 (2010)
4. Ho, V.A., Imai, S., Hirai, S.: Multimodal flexible sensor for healthcare systems. In: 36th Annual International Conference of the IEEE Engineering in Medicine and Biology Society (EMBC) (2014)
5. Liu, Y., Xie, H., Wang, H., Chen, W., Wang, J.: Distance control of soft robot using proximity sensor for beating heart surgery. In: IEEE/SICE International Symposium on System Integration (SII), pp. 403–408 (2016)
6. Mkhitaryan, A., Burschka, D.: Vision based haptic multisensor for manipulation of soft, fragile objects. In: IEEE Sensors, pp. 1–4 (2012)
7. Ozel, S., et al.: A composite soft bending actuation module with integrated curvature sensing. In: IEEE International Conference on Robotics and Automation (ICRA), pp. 4963–4968 (2016)
8. Vishay Semiconductors: Fully Integrated Proximity and Ambient Light Sensor with Infrared Emitter, I2C Interface, and Interrupt Function (2014)

SMT-based Planning for Robots in Smart Factories

Arthur Bit-Monnot[1(✉)], Francesco Leofante[1,2,3], Luca Pulina[1], and Armando Tacchella[3]

[1] University of Sassari, Sassari, Italy
{afbit,lpulina}@uniss.it
[2] RWTH Aachen University, Aachen, Germany
leofante@cs.rwth-aachen.de
[3] University of Genoa, Genoa, Italy
armando.tacchella@unige.it

Abstract. Smart factories are on the verge of becoming the new industrial paradigm, wherein optimization permeates all aspects of production, from concept generation to sales. To fully pursue this paradigm, flexibility in the production means as well as in their timely organization is of paramount importance. AI planning can play a major role in this transition, but the scenarios encountered in practice might be challenging for current tools. We explore the use of SMT at the core of planning techniques to deal with real-world scenarios in the emerging smart factory paradigm. We present special-purpose and general-purpose algorithms, based on current automated reasoning technology and designed to tackle complex application domains. We evaluate their effectiveness and respective merits on a logistic scenario, also extending the comparison to other state-of-the-art task planners.

Keywords: Temporal planning · SMT · Smart factories

1 Introduction

In recent years manufacturing is experiencing a major paradigm shift due to a variety of drivers. From the market side, the push towards high product customization led to a higher proliferation of product variants. From the organizational side, the need to take into account customer feedback led to shorter product cycles. From the technological side, the convergence between traditional industrial automation and information technology brought additional opportunities by combining fields such as, *e.g.*, intelligent Cyber-Physical Systems, additive manufacturing and cloud computing – see, *e.g.*, [27] for a recent survey.

The term *smart factory* is often used to refer to the combination of industrial automation and information technology that should respond to the change drivers, and become the prevalent industrial paradigm, wherein optimization permeates all aspects of production, from concept generation to sales. To fully

F. Wotawa et al. (Eds.): IEA/AIE 2019, LNAI 11606, pp. 674–686, 2019.
https://doi.org/10.1007/978-3-030-22999-3_58

pursue this paradigm, flexibility in production processes as well as in their timely organization is of paramount importance. AI planning has the potential to play a major role in this transition, allowing for more efficient and flexible operation through an on-line automated adaptation and rescheduling of the activities to cope with new operational constraints and demands.

With the above considerations in mind, in this paper we present and evaluate RCLLPlan, a special purpose task planner, and LCP, a general-purpose task planner, specifically conceived to deal with real-world scenarios in the emerging smart factory paradigm. Both RCLLPlan and LCP are based on current automated reasoning technology, namely Satisfiability Modulo Theories (SMT) and Optimization Modulo Theories (OMT) solving — see, *e.g.*, [1,7] and [5,23] for related solvers. SMT solvers developed into a crucial technology in many areas of computer-aided verification — see, *e.g.*, [8]; their application has also been explored in task planning [4,6]. OMT solvers have been introduced more recently, but they already showed promise in some tasks, including task planning [20,21]. The combination of leading-edge SMT and OMT decision procedures with effective encodings, is our recipe to tackle relevant application domains.

In particular, we show the effectiveness of both tools on a logistic scenario based on the RoboCup Logistics League (RCLL) [25], wherein two teams of autonomous robots compete to handle the logistics of materials through several dynamic stages to manufacture products in a smart factory scenario. Using the RCLL as a testbed, we aim to *(i)* compare the performances with other general purpose planners and *(ii)* compare the performances of our specialized version with the general purpose one to highlight strengths and weaknesses of each. The main contribution of our paper is to show that, while considerable investment in research and tech-transfer is required to cope with smart factory needs as a whole, task planners can be made fit to leverage either domain knowledge or suitable encodings towards state-of-the-art decision procedures.

2 Background

2.1 Task Planning

Task planning is a field of Artificial Intelligence concerned with finding a set of actions that would result in desirable state. It is traditionally formulated as a state transition system in which actions allow to transition from one state to another. Solving a planning problem means finding a sequence of actions, i.e. a path, from an initial state to a goal state.

Most research in automated planning has focused in heuristic search techniques in which forward search planners explore the set of states that are reachable from the initial one. Such techniques have proved to very efficiently handle large problems in classical planning, through development of targeted heuristic functions. Such techniques have however proved to be difficult to extend to richer problems such as those involving a rich temporal representation or continuous variables, leading to a renewed interest in constraint-based approach (e.g. [3,6]).

2.2 Planning as Satisfiability

As first shown in [17], classical planning problems can be naturally formulated as propositional satisfiability problems and solved efficiently by SAT solvers. The idea is to encode the existence of a plan of a fixed length p as the satisfiability of a propositional logic formula: the formula for a given p is satisfiable if and only if there is a plan of length p leading from the initial state to the goal state, and a model for the formula represents such plan.

Classical planning abstracts away from time and assumes actions and state transitions to be instantaneous. In contrast, *temporal planning* considers action durations and temporal relations between (possibly concurrently executed) actions. In logistics for instance, plans might need to meet deadlines in order to satisfy some production requirements. The natural encoding of temporal planning problems requires an extension of propositional logic with arithmetic theories, such as the theory of reals or integers. Recent advances in satisfiability checking [1] led to powerful *Satisfiability Modulo Theories* (*SMT*) solvers such as [9,23], which can be used to check the satisfiability of first-order logic formulas over arithmetic theories and thus to solve temporal planning problems.

2.3 SMT and Optimization

Satisfiability Modulo Theories is the problem of deciding the satisfiability of a first-order formula with respect to some decidable theory $\mathcal{T}$. In particular, SMT generalizes the boolean satisfiability problem (SAT) by adding background theories such as the theory of real numbers, the theory of integers, and the theories of data structures (*e.g.*, lists, arrays and bit vectors).

To decide the satisfiability of an input formula φ in Conjunctive Normal Form (CNF), SMT solvers typically first build a *Boolean abstraction* $abs(\varphi)$ of φ where each theory-constraint is replaced by a fresh Boolean variable (proposition). A SAT solver is then called to search for a satisfying assignment S for $abs(\varphi)$. If no such assignment exists then the input formula φ is unsatisfiable. Otherwise, the consistency of the assignment in the underlying theory is checked by a *theory solver*. In this case, if the constraints are consistent then a satisfying solution (*model*) is found for φ. Otherwise, the theory solver returns a theory lemma φ_E giving an *explanation* for the conflict, *e.g.*, the negated conjunction of some inconsistent input constraints. The explanation is used to refine the Boolean abstraction $abs(\varphi)$ to $abs(\varphi) \wedge abs(\varphi_E)$. These steps are iteratively executed until either a theory-consistent Boolean assignment is found, or no more Boolean satisfying assignments exist.

Standard decision procedures for SMT have been extended with optimization capabilities, leading to Optimization Modulo Theories (OMT). OMT extends SMT solving with optimization procedures to find a variable assignment that defines an optimal value for an objective function f (or a combination of multiple objective functions) under all models of a formula φ.

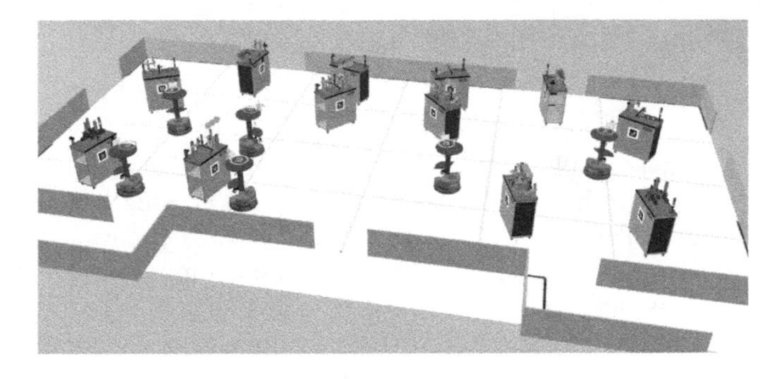

Fig. 1. Simulated RCLL factory environment [28].

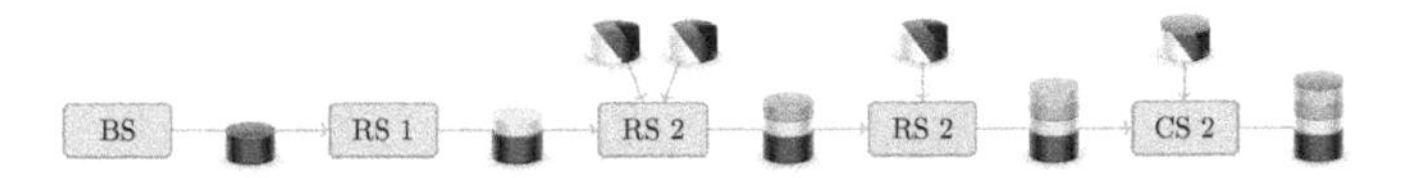

Fig. 2. Example of order configuration for the competition [25]. The order here depicted consists of a red base, three colored rings and a gray cap. (Color figure online)

3 Motivating Case Study: The RoboCup Logistics League

The RoboCup Logistics League (RCLL) models a smart factory scenario where two teams of three autonomous robots compete to handle the logistics of materials to accommodate orders known only at run-time. Competitions take place yearly using a real robotic setup, however, for our experiments we made use of the simulated environment shown in Fig. 1, developed for the Planning and Execution Competition for Logistics Robots in Simulation[1] [24].

Products to be assembled have different complexities and usually require a base, mounting 0 to 3 rings, and a cap as a finishing touch. Bases are available in three different colors, four colors are admissible for rings and two for caps, leading to about 250 different possible combinations. Each order defines which colors are to be used, together with an ordering – see, *e.g.*, Fig. 2.

Several machines are scattered around the factory shop floor (positions are different in each scenarios and announced to the robots at runtime). Each machine completes a particular production step such as providing bases, mounting colored rings or caps.

The objective for a team of autonomous robots is to transport intermediate products between processing machines and optimize a multistage production cycle of different product variants until delivery of final products. Orders that denote the products which must be assembled are posted at run-time by an automated referee box and come with a delivery time window, therefore posing several challenges to state-of-the-art planners [22].

[1] http://www.robocup-logistics.org/sim-comp.

4 SMT-based Planners for Smart Factories

We here present two SMT-based planners: RCLLPlan and LCP.

- RCLLPlan is specifically tailored for the RCLL. This planner generates domain specific encodings for both SMT and OMT using hard-coded rules. RCLLPlan participated and won the Planning and Execution Competition for Logistics Robots in Simulation, the simulated counterpart of the RCLL;
- LCP is a domain-independent planner that accepts as input arbitrary PDDL domain and problem files. LCP internally generates time-oriented SMT encodings which are solved using an off-the-shelf SMT solver. The resulting plan is extracted from the SMT/OMT encoding and validated against VAL [16].

4.1 RCLLPlan: Special Purpose Solution

RCLLPlan implements domain-specific encodings of the *state-based* planning problem defined over the scenario we target here. It builds on the encodings presented in [20, 21] and extends them with the ability to generate different types of encodings specifically tailored to cope with the complexity of the domain. In particular, we evaluate here *(i)* a fine-grained encoding where single actions are encoded separately and *(ii)* an encoding which leverages domain-specific knowledge to build more compact encodings that enable macro planning by grouping action constraints together. In both cases, RCLLPlan can leverage SMT or OMT technology to produce either *feasible* or *optimal* plans, solving a planning as satisfiability problem defined as follows.

To encode the existence of plans of length up to n, we encode a sequence of n ground actions as well as their execution semantics. This requires n copies $A_0, \ldots, A_{n-1}$, of the variable set A defining admissible actions in the RCLL domain, and also $n+1$ copies $\mathcal{V}_0, \ldots, \mathcal{V}_n$ of the propositional and numeric variable sets used to model the RCLL domain, i.e. $\mathcal{V}_i = \{v_i | v \in \mathcal{V}_p \cup \mathcal{V}_n\}$. A *state* s assigns a value to each variable $x_i \in X = \mathcal{V} \cup A$ from their respective domains.

We build a formula defining the *initial* states $I(\mathcal{X})$, one that encodes how actions *affect* states $T(\mathcal{X}, \mathcal{X}')$ and one that defines *goal* states $G(\mathcal{X})$. A *plan* of length p is a sequence of actions $a_0, \ldots, a_p$ such that the implied state sequence $s_0, \ldots, s_p$ satisfies the formula $I(x_0) \wedge \left(\bigwedge_{0 \leq i < p} T(x_i, x_{i+1})\right) \wedge \left(\bigvee_{0 \leq i \leq p} G(x_i)\right)$.

For more details on the specific encodings used by RCLLPlan we refer the reader to [21]. In general the length of a plan is not known a priori and has to be determined empirically by increasing p until a satisfying assignment for the planning formula is found. However, RCLLPlan, being tailored for the domain considered, exploits domain specific knowledge to determine p and simplify plan search.

To support generation of optimal plans with OMT, we introduce additional variables $c \in \mathcal{X}$ to encode the *cost* of executing actions $a \in A$ at time t. We define the total cost associated to a plan as $c_{tot} = \sum_{0 \leq i < p} c_i$ and leverage OMT to minimize it under the side condition that the planning formula holds.[2]

When generating encodings for macro actions we encode domain-specific knowledge explicitly at the logical level, so as to produce encodings that are more compact. Macro actions are encoded based on the following observation. Plans for production in the RCLL often involve action sequences that yield better performance if performed by the same agent. For instance, if a robot is instructed to prepare a base station to provide a base, then it makes sense that the same robot also retrieves the base (under the realistic assumption that providing a base is less expensive than motion planning and navigation for another robot). Following such observations, a logical encoding is built in RCLLPlan where the transition relation contains constraints for macro actions only.

4.2 LCP: General Purpose Algorithm

This domain-dependent planner is complemented by LCP (Lifted Constraint Planner) [4] a domain-independent planner that aims at providing good performance over a large set of problem. LCP supports both temporal PDDL [13] and a subset of ANML [26] to define planning problems.

Like RCLLPlan, it uses a constraint-based encoding where multi-valued variables are related through a set of constraints that can be exploited by SMT solvers. Unlike RCLLPlan's state-oriented representation, LCP relies on time-oriented encoding where the effects and conditions of actions are placed on temporal intervals which are related through temporal constraints ensuring the consistency of a plan.

From PDDL and ANML to Chronicles. LCP uses chronicles [15] as a building block for its internal representation. Chronicles were first introduced in the IxTeT planner [14]. They allow an expressive representation of temporal actions that leverages a constraint-based representation close to the one found in state-of-the-art scheduling solvers such as CP Optimizer [18].

The environment is represented by a finite set of state variables, that describe the evolution of a particular state feature overtime, e.g., the location of a robot over the course of the plan.

A chronicle is composed of a set of decision variables, a set of constraints on these variables as well as some condition and effect statements. Condition statements require a particular state variable to have a given value over a temporal interval while effect statements change the value of a state variable at given point in time. In essence, a chronicle is thus a Constraint Satisfaction Problem (CSP) extended with additional constructs to represent the conditions and effects that are at the core of AI planning.

---

[2] RCLLPlan exploits a simple cost definition in its current state, *i.e.*, minimize time to delivery for each product. However, richer goal structures could be specified.

Chronicles offer a natural representation for the action models present in both the PDDL and ANML languages that are supported by our tool. In practice, one simply needs to map an action's parameters and timepoints into the variables of the chronicles. The action's condition and effect can be straightforwardly encoded into the corresponding condition and effect statements. Additional requirements, e.g. on the duration of an action, are encoded as constraints on the chronicles variables. For this translation, we follow the process demonstrated by other planners such as IxTeT [14] and FAPE [3].

The planning problem itself is also encoded as a chronicle, with effect statements defining the initial state and condition statements representing the goals of the problem.

From Chronicles to Constraint Satisfaction Problems. Planning differs from scheduling in that the actions that will be part of the solution plan are not known beforehand. We escape this problem by generating bounded problems in which a finite set of actions are allowed to be part of a solution plan. More precisely, a bounded planning problem has a finite set of action chronicles, each representing a possible action in the solution plan. Action chronicles are optional: each is associated to a boolean decision variable that is true if the action is part of the solution plan and false otherwise.

Finding a solution to a bounded planning problem means finding an assignment to decision variables that represent action parameters (*i.e.* variables inside the chronicles) and action presence (the boolean variables) such that the set of actions that are present form a consistent plan.

Plan consistency is defined through a set of constraints over the chronicles. Those constraints are detailed in [4] and are sketched below:

- *consistency constraints* enforce that no two effect statements are overlapping. Namely, if two effect statements are part of actions present in the solution, they must either affect different state variables or be active on non-overlapping temporal intervals.
- *support constraints* ensure that any condition in an action is supported by an effect statement. More precisely, it means that there exist an effect statement that changes the state variable to the value required by the condition. In addition, the effect statement must be active before the condition, and there must be no other effect affecting the same state variable active between the start of the effect and the end of the condition.
- *internal constraints* ensure that all constraints defined inside a chronicle hold if the corresponding action is part of the solution.

For a given bounded planning problem, a CSP is built by taking the conjunction of all consistency, support and internal constraints.

This formulation has some important similarities with lifted plan-space planning [3,12,14]. It differs from the former in that the CSPs in LCP contain no dynamic part since possible actions are fixed beforehand. This facilitates the use of off-the-shelf solvers while plan-space planner typically require ad-hoc constraint solving engines. The representation is also closely related to the one of recent constraint programming engines for scheduling that support optional temporal intervals but lack a notion of condition and effects [18,19].

Planning. Planning is done by generating increasingly large bounded planning problems. At each step, an SMT solver is used to prove the existence or absence of a plan for a bounded planning problem, *i.e.*, whether a plan exists with the limited number of actions allowed. The SMT solver is used to find a satisfying assignment (model), which can then be translated into a solution plan. When the solver proves the inconsistency of the CSP, a new bounded planning problem allowing more actions is generated and the process restarts.

In addition to this iterative deepening setting, LCP also supports a configuration that defines the size of the planning problem to solve: specifying for each action in the planning domain its maximal number of occurrences. In this setting, LCP will generate a single bounded planning problem and attempt to find a solution for it, allowing a simple domain-dependent configuration of the planner.

5 Experimental Evaluation

5.1 Experimental Setup

We evaluate different approaches using the RCLL set of benchmarks. Unlike the setting for the RCLL competition that features two competing teams of robots, we use a single team which is closer to the real-world setting in which robots in the same factory are expected to collaborate in the production steps.

The PDDL domain is taken, unmodified, from the referee box of RCLL competition.[3] Problems are generated for different game settings and vary on the number of robots available (from 1 to 3 robots). The complexity of the product to manufacture varies between complexities C0 and C1, respectively corresponding to no-ring and one-ring configurations of the final product.

5.2 Tested Planners

RCLLPlan is evaluated in its two versions, both declined in two subversions depending on whether they seek feasible or optimal plans:

- RCLLPlan-Sat and RCLLPlan-Opt represent the basic RCLL-specific planning components, respectively configured to seek feasible and optimal plans. While this encoding is domain specific, it closely mimics the actions available in the executive system.
- RCLLPlan-Macros-Sat and RCLLPlan-Macros-Opt are the RCLL-specific planning components with macro actions. They are respectively configured to seek feasible and optimal plans. Unlike, RCLLPlan-Sat/RCLLPlan-Opt, the use of macro actions allows for much smaller encodings at the expense of divergence with the underlying execution system and more complexity in its development.

[3] https://github.com/timn/ros-rcll_ros/tree/master/pddl.

The LCP planner is evaluated in two configurations:

- LCP-Gen is our domain-independent planning component running with default options.
- LCP-Spe is our domain-independent planning component with a configuration tailored to the RCLL domains that restricts the number of occurrences of each action. Namely, the bounded planning problem is limited to the same set of actions as RCLLPlan-Sat. This configuration is purely external and does not require touching the internals of the planner.

Finally, we consider two state-of-the-art temporal planners:

- Optic [2] is a forward-search heuristic planner evolved from the POPF [10] planner which was a runner-up the penultimate International Planning Competition (IPC). Optic is a forward search heuristic planner that uses a temporal extension of the h^{FF} heuristic. TFD and YAHSP, the other top competitors of the latest IPCs, were not considered because *(i)* they are not complete with respect to the semantics of PDDL 2.1 [11], and *(ii)* they do not support the PDDL encoding of the RCLL domain that mixes instantaneous and durative actions.
- SMTPlan+ is a recent SMT-based planner for the PDDL+ language [6]. SMTPlan+ is more expressive than the other planners considered as it supports the full range of PDDL+ features, including continuous processes inducing non-linear changes over numeric state variables. It uses a state-oriented SMT encoding, over plans of increasing length.

All SMT-based planners use Z3 [23] in version 4.6.3 as an off-the-shelf solver.

5.3 Results

All benchmarks were run on an Intel i7-3770 @ 3.40GHz with a timeout of 60 s. The timeout is low compared to the usual setting of 30 min of the International Planning Competition but is more adequate to an online planning setting such as the one of RCLL.

C0 Configuration. Results for the C0 configuration of the RCLL problem are given in Table 1. For each planner, the table provides the number of problems solved and the average runtime for successful runs.

For the C0 configuration, all components of RCLLPlan and LCP are able to solve all 60 problems. The only exception is the base configuration of the domain-dependent approach that fails to prove optimality for a number of problems (RCLLPlan-Opt). Runtimes are largely dominated by the domain-specific encoding with macro actions: runtimes are well below a second for both feasibility and optimality (RCLLPlan-Macros-Sat/RCLLPlan-Macros-Opt). Given their generality, both versions of LCP provide good performance solving all problems in handful of seconds. Notably, the performance of the configured version of LCP (LCP-Spe) is on par with that of RCLLPlan-Sat.

Optic has overall poor performance on those domains, solving only 13 problems involving 1 robot and 2 problems involving 2 robots. With the timeout set to 60 s, SMTPlan+ fails to solve any attempted problem.

Table 1. Results for the C0 configuration. For problems with 1 to 3 robots (R1–R3), it indicates for each planner the number of problems solved and the average runtime when a plan was found before the timeout of 60 s.

	C0–R1		C0–R2		C0–R3	
Planner	Solved	Runtime (s)	Solved	Runtime (s)	Solved	Runtime (s)
RCLLPlan-Sat	**20**	0.98	**20**	2.52	**20**	2.73
RCLLPlan-Opt	**20**	15.22	17	42.18	4	54.26
RCLLPlan-Macros-Sat	**20**	**0.13**	**20**	**0.14**	**20**	**0.14**
RCLLPlan-Macros-Opt	**20**	0.40	**20**	0.53	**20**	0.60
LCP-Gen	**20**	7.25	**20**	9.72	**20**	11.52
LCP-Spe	**20**	1.61	**20**	1.93	**20**	2.74
Optic	13	24.39	2	57.60	0	–
SMTPlan+	0	–	0	–	0	–

Table 2. Results for the C1 configuration. For problems with 1 to 3 robots (R1–R3), it indicates for each planner the number of problems solved and the average runtime when a plan was found before the timeout of 60 s.

	C1–R1		C1–R2		C1–R3	
Planner	Solved	Runtime (s)	Solved	Runtime (s)	Solved	Runtime (s)
RCLLPlan-Sat	**20**	11.55	16	10.35	18	24.11
RCLLPlan-Opt	0	–	0	–	0	–
RCLLPlan-Macros-Sat	**20**	**0.37**	**20**	**0.43**	**20**	**0.60**
RCLLPlan-Macros-Opt	**20**	1.82	**20**	9.57	17	6.48
LCP-Gen	2	30.07	1	56.12	1	48.67
LCP-Spe	**20**	7.03	15	10.59	12	12.45
Optic	12	23.46	0	–	0	–
SMTPlan+	0	–	0	–	0	–

C1 Configuration. Problems for configuration C1, where an additional ring must be mounted on the product, show more differentiated results (Table 2). The benchmark is still dominated by the domain specific macro-encoding, that only fails to prove the optimality of three of the most difficult problems.

RCLLPlan-Sat and LCP-Spe continue to show comparable performance both in runtime and number of problems solved in R1 and R2. Both RCLLPlan-Opt and LCP-Gen show their limits, solving respectively none and 4 problems.

Optic almost maintains its performance from the C0 configuration, solving 12 of the 20 problems involving a single robot.

6 Discussion and Conclusion

In this paper we have presented and evaluated two approaches to robot planning for smart factories, leveraging the latest SMT technology for solving typical problems that arise in smart factories. Evaluation of RCLLPlan and LCP on the RCLL benchmarks highlight different trade-offs.

The most involved domain-specific solver, shows unchallenged performance on the RCLL benchmarks highlighting the work that remains to be done on fully automated task planners. Of course the development of such a domain-specific planner induces many difficulties as it is more error prone and time consuming. Perhaps most importantly, adapting it to new operational constraints is challenging as the process must be restarted to account for violated assumptions.

On the other hand, fully domain-independent planners come with a great promise of reusability and adaptability to a wide variety of contexts. The development of LCP is meant to leverage those benefits. While the gap with domain specific solvers remains important, experiments show that LCP does reduce this gap with respect to existing domain-independent planners. In its current state, it would stand as a viable solution when provided with minimal configuration.

The performance gap indicates two directions for future work on LCP. First there is the never-ending quest for performance improvement, the performance of RCLLPlan defining a challenging target to reach. Second, we believe the performance gain from injecting domain-specific knowledge (being in specific solvers or in the configuration of LCP) raises the question of how to inject such knowledge in a principled and solver-independent way. The optional specification of Hierarchical Task Networks (HTN) in the ANML language provide a good opportunity to do this in an incremental and non-intrusive manner. The ANML subset for HTN is currently not supported by LCP and will be the subject of future work.

Acknowledgements. The research of Arthur Bit-Monnot and Luca Pulina has been funded by the EU Commission's H2020 Program under grant agreement N.732105 (CERBERO project). The research of Luca Pulina has been also partially funded by the Sardinian Regional Project PROSSIMO (POR FESR 2014/20-ASSE I) and the FitOptiVis (ID: 783162) project.

References

1. Barrett, C.W., Sebastiani, R., Seshia, S.A., Tinelli, C.: Satisfiability modulo theories. In: Handbook of Satisfiability (2009)
2. Benton, J., Coles, A., Coles, A.: Temporal planning with preferences and time-dependent continuous costs. In: ICAPS (2012)
3. Bit-Monnot, A.: Temporal and hierarchical models for planning and acting in robotics. Ph.D. thesis, Université de Toulouse (2016)

4. Bit-Monnot, A.: A constraint-based encoding for domain-independent temporal planning. In: Hooker, J. (ed.) CP 2018. LNCS, vol. 11008, pp. 30–46. Springer, Cham (2018). https://doi.org/10.1007/978-3-319-98334-9_3
5. Bjørner, N., Phan, A.-D., Fleckenstein, L.: νZ - an optimizing SMT solver. In: Baier, C., Tinelli, C. (eds.) TACAS 2015. LNCS, vol. 9035, pp. 194–199. Springer, Heidelberg (2015). https://doi.org/10.1007/978-3-662-46681-0_14
6. Cashmore, M., Fox, M., Long, D., Magazzeni, D.: A compilation of the full PDDL+ language into SMT. In: ICAPS (2016)
7. Cimatti, A., Franzén, A., Griggio, A., Sebastiani, R., Stenico, C.: Satisfiability modulo the theory of costs: foundations and applications. In: Esparza, J., Majumdar, R. (eds.) TACAS 2010. LNCS, vol. 6015, pp. 99–113. Springer, Heidelberg (2010). https://doi.org/10.1007/978-3-642-12002-2_8
8. Cimatti, A., Griggio, A.: Software model checking via IC3. In: Madhusudan, P., Seshia, S.A. (eds.) CAV 2012. LNCS, vol. 7358, pp. 277–293. Springer, Heidelberg (2012). https://doi.org/10.1007/978-3-642-31424-7_23
9. Cimatti, A., Griggio, A., Schaafsma, B.J., Sebastiani, R.: The MathSAT5 SMT solver. In: Piterman, N., Smolka, S.A. (eds.) TACAS 2013. LNCS, vol. 7795, pp. 93–107. Springer, Heidelberg (2013). https://doi.org/10.1007/978-3-642-36742-7_7
10. Coles, A., Coles, A., Fox, M., Long, D.: Forward-chaining partial-order planning. In: ICAPS (2010)
11. Cushing, W., Kambhampati, S., Mausam, Weld, D.S.: When is temporal planning really temporal? In: IJCAI (2007)
12. Dvorák, F., Barták, R., Bit-Monnot, A., Ingrand, F., Ghallab, M.: Planning and acting with temporal and hierarchical decomposition models. In: ICTAI (2014)
13. Fox, M., Long, D.: PDDL2.1: an extension to PDDL for expressing temporal planning domains. JAIR **20**, 61–124 (2003)
14. Ghallab, M., Laruelle, H.: Representation and control in IxTeT, a temporal planner. In: AIPS (1994)
15. Ghallab, M., Nau, D.S., Traverso, P.: Automated Planning: Theory and Practice (2004)
16. Howey, R., Long, D., Fox, M.: VAL: automatic plan validation, continuous effects and mixed initiative planning using PDDL. In: ICTAI (2004)
17. Kautz, H.A., Selman, B.: Planning as satisfiability. In: ECAI (1992)
18. Laborie, P., Rogerie, J.: Reasoning with conditional time-intervals. In: FLAIRS (2008)
19. Laborie, P., Rogeric, J., Shaw, P., Vilím, P.: Reasoning with conditional time-intervals. Part II: an algebraical model for resources. In: FLAIRS (2009)
20. Leofante, F., Ábrahám, E., Niemueller, T., Lakemeyer, G., Tacchella, A.: On the synthesis of guaranteed-quality plans for robot fleets in logistics scenarios via optimization modulo theories. In: IRI (2017)
21. Leofante, F., Ábrahám, E., Niemueller, T., Lakemeyer, G., Tacchella, A.: Integrated synthesis and execution of optimal plans for multi-robot systems in logistics. Inf. Syst. Front. **21**, 87–107 (2018)
22. Leofante, F., Ábrahám, E., Tacchella, A.: Task planning with OMT: an application to production logistics. In: Furia, C.A., Winter, K. (eds.) IFM 2018. LNCS, vol. 11023, pp. 316–325. Springer, Cham (2018). https://doi.org/10.1007/978-3-319-98938-9_18
23. de Moura, L., Bjørner, N.: Z3: an efficient SMT solver. In: Ramakrishnan, C.R., Rehof, J. (eds.) TACAS 2008. LNCS, vol. 4963, pp. 337–340. Springer, Heidelberg (2008). https://doi.org/10.1007/978-3-540-78800-3_24

24. Niemueller, T., Karpas, E., Vaquero, T., Timmons, E.: Planning competition for logistics robots in simulation. In: PlanRob (2016)
25. Niemueller, T., Lakemeyer, G., Ferrein, A.: The RoboCup logistics league as a benchmark for planning in robotics. In: PlanRob (2015)
26. Smith, D.E., Frank, J., Cushing, W.: The ANML Language. In: ICAPS (2008)
27. Zhong, R.Y., Xu, X., Klotz, E., Newman, S.T.: Intelligent manufacturing in the context of industry 4.0: a review. Engineering **3**(5), 616–630 (2017)
28. Zwilling, F., Niemueller, T., Lakemeyer, G.: Simulation for the RoboCup logistics league with real-world environment agency and multi-level abstraction. In: Bianchi, R.A.C., Akin, H.L., Ramamoorthy, S., Sugiura, K. (eds.) RoboCup 2014. LNCS (LNAI), vol. 8992, pp. 220–232. Springer, Cham (2015). https://doi.org/10.1007/978-3-319-18615-3_18

Soft Biometrics for Social Adaptive Robots

Berardina De Carolis[1(✉)], Nicola Macchiarulo[1], and Giuseppe Palestra[2]

[1] Department of Computer Science, University of Bari, Bari, Italy
berardina.decarolis@uniba.it, macc.nicola@gmail.com
[2] Hero srl, Bari, Italy
giuseppepalestra@gmail.com

Abstract. Soft biometric analysis aims at recognizing personal traits that provide some information about an individual. In this paper, we present a real-time system able to automatically recognize soft biometric traits to enhance the capability of a social robot, Pepper, in this case, to understand characteristics of people present in the environment and to properly interact with them. In particular, the proposed system is able to estimate several traits simultaneously, such as gender, age, the presence of eyeglasses and beard, of people in the field of view of the robot. Our hypothesis is that adding these capabilities to a social robot improves and makes more believable its social behavior. Results of the preliminary evaluation seems to support this hypothesis.

Keywords: Social robots · Soft biometrics · Adaptive behavior

1 Introduction

Soft biometric traits analysis aims at recognizing some characteristics that provide some information about the individual, but do not allow to sufficiently differentiate any two individuals [1]. Soft biometric traits are human characteristics providing categorical information about people such as age, beard, gender, eyeglasses, ethnicity, eye/hair color, length of arms and legs, height, weight, skin/hair color, etc. [2]. For instance, "the young girl with eyeglasses" is a way to identify a person starting from some characteristics. In contrast to "hard" biometrics, soft biometrics provide some vague physical or behavioral information which is not necessarily permanent or distinctive. Such soft biometric traits are usually easier to capture from a distance and do not require cooperation from the subjects.

Soft biometrics can be used effectively in the interaction with a Social Robot for improving its awareness of the surrounding environment and its perception of the humans around. This capability is a key factor for increasing the success of the interaction since it contributes to improving the so-called social believability [3]. For instance, being aware of the observed characteristics of a person could be used to adapt the robot social behavior accordingly.

A way to perform estimation of soft biometric traits is to analyze people's face. Then, using the camera on board of the robot, it is possible to implement a real-time

F. Wotawa et al. (Eds.): IEA/AIE 2019, LNAI 11606, pp. 687–699, 2019.
https://doi.org/10.1007/978-3-030-22999-3_59

system that detects human faces and processes data coming from the robot camera to estimate soft biometrics traits of subjects in the scene.

Several studies have investigated soft biometric and facial expression recognition during the interaction with a Social Robot [4, 5]. They mostly consider one or two soft biometrics traits at the same time (typically gender and age). In this paper, we implemented soft biometric analysis to estimate more than one trait at the same time: age, gender, the presence of eyeglasses and beard. Looking at results of the state of the art and after a comparative evaluation of existing approaches, we decided to implement a system based on deep learning, Convolutional Neural Networks (CNNs) in particular. CNN is a type of feed-forward artificial neural networks in which the connectivity pattern between its neurons is inspired by the organization of the animal visual cortex [19].

The system has been tested on Pepper robot in a real environment: the reception of our Department. The goal of this experiment was twofold:

(i) to test the accuracy of our system in real-time and in a typical environment in which the robot could be used;
(ii) to test the hypothesis that endowing the robot with awareness about people present in the environment, so as to adapt its social behavior, improves the user's perception of its believability.

Results of a preliminary experiment show that users perceived the robot as being more believable when it was aware of the characteristics of people present in the environment and adapted its behavior to them.

The paper is organized as follows: in Sect. 2 an overview on Social Robotics is provided. Then, Sect. 3 presents our approach for estimating soft biometrics traits. Section 4 shows the adaptation process and experimental results. Finally, Sect. 5 is aimed at a final discussion and proposal for future development.

2 Social Robotics

A Social Robot is a physically embodied, autonomous agent that communicates and interacts with humans on a social and emotional level. Social robots represent an emerging field of research focused on developing a "social intelligence" in order to maintain the illusion of dealing with a human being [6, 7]. Social robots are being applied in several domains such as elderly care [8], autism therapy [9], education [10, 11], public places [12], domestic and work environments [12].

To be believable, social robots have to exhibit social intelligence and adapt their behavior to the situation, therefore they should be endowed with a model of the environment and of the user that may include their profile, emotions, personality and past interactions. Then a social robot should be capable of observing and understanding the changes in the environment so as to behave in a proper manner. Based on the acquired information, it makes decisions to react appropriately according to different social situations and according to its role [13]. For instance, when the robot acts as a reception clerk in a public space, a first level of adaptation can be implemented by making the robot aware of which kind of people is present in its sight and adapt the communication to their "visible" characteristics. For instance the robot could adapt the level of formality and the lexicon to the user's age and gender. In this case, soft-biometric traits can be used for this purpose.

In this regard, robots capable of exhibiting sociability and achieving widespread societal acceptance are being used more and more often in human-centered environments. This was the idea behind the development of the Pepper robot by SoftBank Robotics (https://www.softbankrobotics.com/en). It was initially designed for business-to-business applications but, after, the robot became a platform of interest for various other applications, including in the business-to-consumer. In this work, we employ Pepper as a Social Robot.

3 Soft Biometrics Analysis

This section describes the software modules developed for soft biometrics analysis. Soft biometric traits refer to physical and behavioral traits, such as gender, age, height and weight, which are not unique to a specific subject, but are useful for identification, and description of human subjects. Soft biometric traits also can be classified according to permanence and distinctiveness [14]. The permanence of a trait is related to the fact that it doesn't change over time (i.e. gender and ethnicity). Distinctiveness refers to the ability of a trait to differentiate between individuals. For instance, gender classification is a useful preprocessing step also for face recognition since it can be used to reduce the search space for recognizing the person.

Among the most robust and accurate approaches to gender and age classification, we can find some based on the analysis of texture patterns. Many texture features have been used like LBP, Histogram of Oriented Gradients (HOG) and they usually employ Support Vector Machine (SVM) or k-nearest neighbor as classifiers [15, 16]. For instance, Zang and Lu [17] gave a comparison of 6 types of features using three classifiers and showed that for FERET database the best accuracy (99.1%) was obtained with features based on local Gabor binary pattern and LAD (LGBP-LDA) and SVM with automatic confidence (SVMAC). Gunay and Nabiyev [18] used LBP feature as an efficient face descriptor. They divided the faces into small regions from which the LBP histograms are extracted and concatenated into a feature vector. They got 80% age classification rates in FERET database. Although such local descriptors achieve higher results than holistic methods, their performance is affected by variations in expression, pose, illumination and occlusion.

Recently deep learning approaches are being used successfully in this domain. They mainly use the CNN which is a type of feed-forward artificial neural networks in which the connectivity pattern between its neurons is inspired by the organization of the animal visual cortex [19]. Yan et al. [20] proposed an approach which uses the CNN to extract the facial features. Their network has 7 layers and it gives as output 4096 features. For the classification part, they use SVM to classify the face into one of thirteen age groups. Levi and Hassner [21] proposed a network architecture for both age and gender classification. Rothe et al. [22] won the ChaLearn LAP 2015 challenge on apparent age estimation, their proposed CNN uses the VGG-16 architecture [23]. They proposed an approach in which first the face is detected from the input image and then extracts the CNN predictions from an ensemble of 20 networks on the cropped face.

The proposed soft-biometrics module is able to recognize age, gender, eyeglasses and beard presence. Moreover, it also recognizes the color of eyes and hairs. For each soft biometric trait, a specific software module has been implemented as described below.

3.1 Age and Gender Recognition

For age and gender estimation, our system has been based on the work of [22]. Their work focuses on automatic gender and age classification using deep CNNs. The authors use a fine tuned version of the VGG-16 neural network. This work provides age and gender classification outperforming state-of-the-art on both tasks using unconstrained image dataset[1].

3.2 Eyeglasses Detection

Eyeglass detection is a two-class classification problem. For classification, we compared three different approaches present it the literature: one non-learning based and two learning based [24]. The first approach is able to detect the presence or absence of eyeglasses by image processing elaboration while the second approach uses an SVM classifier. The third approach is based on deep learning classification. We implemented and compared the three approaches on the Color FERET v2 dataset [25] and according to results (see [26] for more details) we selected the one based on deep learning to be used by the social robot. CNNs are machine learning algorithms that use neural networks that the received images as input, each of them expressed in 3 dimensions [width, height, channels].

The first step is the identification of the Region Of Interest (ROI). In this case, the ROI represents a region where eyeglasses are certainly present.

The ROI identified is where the eyeglass bridge is located, as shown in Fig. 1a. Using the Dlib machine-learning library [27], 68 face landmark points are located [28]. The ROI is comprised in the rectangle of width equal to the distance between points 21 and 22 and height equal to the distance between points 21 and 28, as shown in Fig. 1b.

For this particular task, we selected 2712 images from the FERET v2 dataset: 2366 people without eyeglasses and 346 people with eyeglasses.

The proposed CNN architecture, depicted in Fig. 2, is based on reduced number of layers to reduce overfitting problem that occurs when training data are small, as in this case. The inputs of the network were ROI images, detected and resized as above described. The dimension of each input was $32 \times 48 \times 3$. This dimension is used to define the first layer of the network, which is the input layer. Then, two convolutional layers were defined as follows.

[1] https://data.vision.ee.ethz.ch/cvl/rrothe/imdb-wiki/.

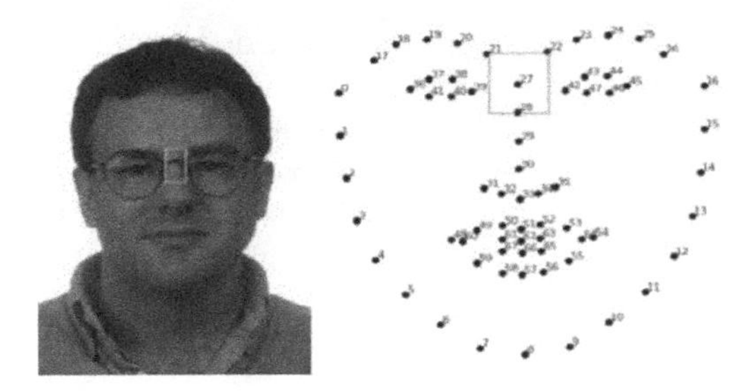

Fig. 1. (a) Eyeglasses detection ROI; (b) ROI based on Dlib 68 face landmarks.

Convolutional layer: divides the image into various overlapping fragments, which are then analyzed to identify their characteristics.

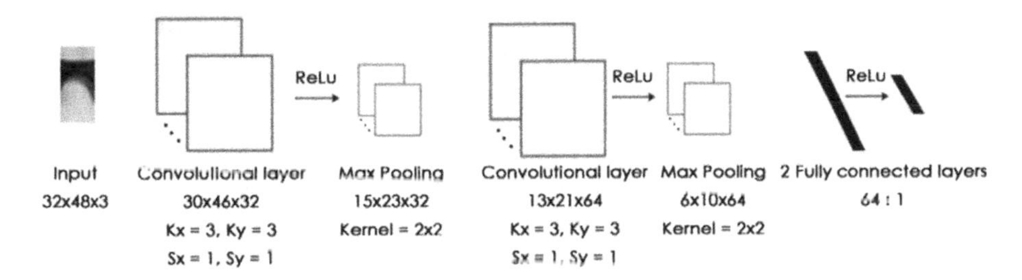

Fig. 2. Proposed eyeglasses detection CNN architecture.

The parameters of this layer are the number of filters (F), the size of the kernel (K_x; K_y), and the displacement factor along the width and height of the matrix (S_x; S_y). Giving an image of size [$W_1H_1D_1$] as input, the output obtained will have dimensions:

$$W_2 = \frac{W_1 - K_x}{S_x + 1} \tag{1}$$

$$H_2 = \frac{H_1 - K_y}{S_y + 1} \tag{2}$$

$$D_2 = F \tag{3}$$

32 filters of size 3 × 3 × 3 pixels are directly used to the input image in the first convolutional layer. A Rectified Linear Unit (ReLU) and a max-pooling layer follow this first convolutional layer. Pooling layer reduces the number of input received through generalization, useful for speeding up the analysis without losing too much precision. The most common pooling algorithms are max-pooling and average-pooling. In this approach, max-pooling was used with a 2 × 2 size kernel in order to select the highest value among 4 neighboring elements of the input matrix.

The second convolutional layer processes the output of the previous layer. The second layer contained 64 filters of size 3 × 3 × 3. As the first one, ReLU and max pooling layer defined with the same parameters as before follow it.

A first fully connected layer works on the received output of the second convolutional layer and contains 64 neurons. A ReLU and a dropout layer follow it. A last fully connected layer maps to the final classes (with or without eyeglasses).

3.3 Beard Detection

Beard detection is also a two-class classification problem. As in the eyeglasses detection we implemented and compared three different approaches non learning based, SVM classifier, and CNNs [25]. Also in this case, according to the accuracy results, we selected the one based on CNNs.

In the case of the beard detection, the ROI to detect the beard is based on the rectangle of width equal to the distance of points 7 and 9 and height equal to the distance between points 6 and 7, as shown in Fig. 3. This region has been chosen because it is assumed that if the beard is present on a person's face, it will surely be present in this region.

Fig. 3. Beard detection ROI based on Dlib 68 face landmark.

The used dataset for training and test was again the Color FERET v2 dataset. For beard detection task, 2712 images have been selected: 2554 people without the beard and 158 bearded people. In order to improve the performance of the beard detection, we performed a data augmentation step that increased the number of pictures to 4292 in which 2554 with the beard and 1738 without it.

So, to get more data, we performed a data augmentation step by making minor alterations to the existing images of people with beard. In particular, we applied minor changes such as horizontal flip, image scaling, random rotation and random variation of brightness and contrast. Then we trained our neural network with additional synthetically modified data.

The CNN architecture described for the eyeglass detection task was also used for beard detection. The only change occurred in the parameters of the input layer in order to accept input images of different size 64 × 16 × 3.

Table 1. Eyeglasses and beard detection accuracy on the dataset

	Accuracy	Precision	Recall	F1
Eyeglasses detection	99.558%	99.246%	98.759%	99.001%
Beard detection	98.578%	98.574%	98.475%	98.524%

3.4 Off-Line Biometric Prediction Accuracy Estimation

We formulated beard and eyeglasses detection as two-class classification problem. For classification, in a previous work [25] we compared three different methods: non-learning method, Support Vector Machines, and Deep Learning.

The evaluation of the accuracy of both beard and eyeglasses detection has been performed by the k-fold cross validation strategy with k = 10. The k folds have been built using a strategy to maintain, in each of them, the same ratio of elements for the classes encountered in the considered dataset. In particular, the gender recognition has been performed with an accuracy of 85% while age estimation reached an accuracy (±1 year) of 84% on the previously mentioned dataset.

In Table 1, we present classification results of the eyeglasses and beard detection using CNNs.

4 Real-Time Evaluation

We performed an experiment in real-time with the aim of evaluating the feasibility and performance of the system usage in real scenarios. Furthermore, we observed the perception of the users interacting with the Robot in terms of social believability introduced by adapting the communication of the robot to the recognized soft biometric traits of the subjects.

A concierge scenario was created in our Department entrance space to test the real time performance of the soft biometric module. As far as age estimation is concerned, we split in intervals the age of the population typical of the environment in which the experiment took place and show the adaptation of the greeting accordingly. Table 2 shows the age intervals.

Table 2. Age intervals

Age group	Age interval
A (young students)	Age $\leq$ 25
B (mature students, young staff)	25 < Age $\leq$ 40
C (middle age staff)	40 < Age $\leq$ 60
D (mature age people)	Age > 60

The system has been tested for two days using the Pepper robot acting as reception clerk and providing information about the Department logistic. In this scenario, Pepper was running in two different modalities. In the first condition Pepper, when a person entered in its field of view, greeted saying the same sentence to everyone *"Hello! How are you today?"* accompanied by a waving greeting gesture. In the second condition Pepper, after having detected the face of the person entering in its field of view, used the functions of the soft-biometrics module, running in the Pepper Department Cloud, for estimating gender, age group, presence of beard and eyeglasses. The robot has two RGB cameras at the forehead and mouth positions. In our system we used the one in the forehead with a resolution of 1280 × 960. In this second case, Pepper adapted the greetings by changing the level of formality/friendliness of the employed language as a consequence of the recognized age. Moreover, some additional sentences were pronounced by the robot to show its awareness of the subject's appearance.

Table 3 shows the adaptation rules implemented as simple behaviors.

Table 3. Adaptation table

Age group	Gender	
	M	F
A	Hi! + fist bump gesture + <if eyeglasses> Nice goggles Bro! + <if beard> Nowadays boys like to have beard … isn't it?	Hi! + fist bump gesture + <if eyeglasses> Nice goggles Sister!
B	Hello! + greeting gesture + <if beard> Nowadays it's on fashion to have beard! <if eyeglasses> Nice frame, did you change your eyeglasses?	Hello! + greeting gesture + <if eyeglasses> Nice frame, did you change eyeglasses? They look great on you
C and D	Good morning/evening! + handshake gesture +<if eyeglasses> Nice frame, I would like to have eyeglasses like yours! + <if beard> Also your beard is very nice	Good morning/evening madame! + handshake gesture + <if eyeglasses> Nice frame, I would like to have eyeglasses like yours

In both cases, after the greetings, Pepper invited the person to give a grade (in a scale from 1 to 5) about the perceived social believability during the interactive experience. In particular, Pepper asked to the user to rate by selecting a grade on its tablet after having explained that to the subject by saying: "We are conducting a study about social robotics, please answer to a simple question on my tablet." On the tablet the following question with the scale was shown: *In a scale from 1 (not at all) to 5 (a lot), how much did you find believable Pepper's behavior in greeting you?*

In total 32 people interacted with Pepper, 16 for each condition. They were distributed as follows: 12 young undergraduate students (6 females and 6 males belonging to the A age group), 10 young researchers (5 females and 5 males belonging to the B age group), 6 middle age professors and technicians (3 females and males belonging to the C age group) and 4 mature professors (2 females and 2 males belonging to the D group). These persons were unknown to the robot and their faces were not present in the training set. Among them 16 were wearing eyeglasses and 14 had the beard (5 had a light beard and 9 had a thick one). They were equally distributed in the two groups.

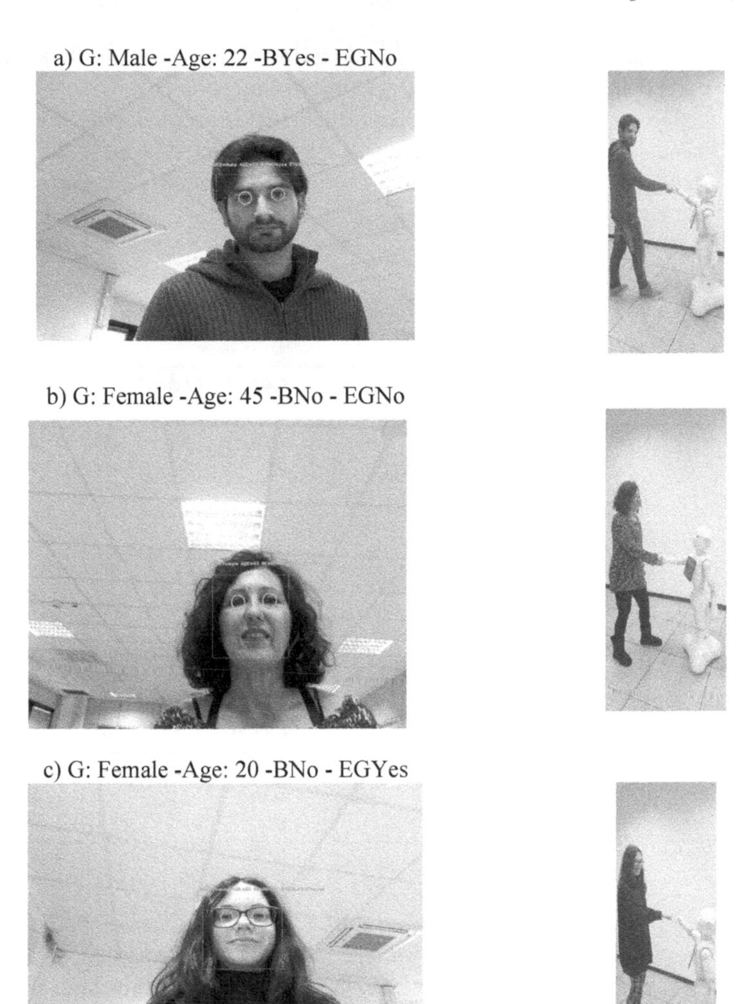

Fig. 4. Three examples of interaction with the Pepper robot. The robot recognizes soft biometrics traits of the interacting subject and greets with a customized behavior.

Figure 4 shows three examples of interaction. On the left-side column the pictures taken from the Pepper camera are shown together with the annotation of the soft-biometrics traits. As you can notice in the right side picture, Pepper adapts the greeting gesture to the estimated age of the subject.

In this real-time test, the system was able to well recognize soft biometric traits (age, gender, beard and eyeglasses presence) of each participant. Results are shown in Table 4. Besides the accuracy we show the confusion matrix for each trait.

About the gender and age estimation, the classification was higher then the one on the dataset maybe because in this case the interaction was mainly frontal and age estimation was performed on intervals. About the eyeglasses, analyzing the images we discovered that only in two cases the frames were not detected because they were thin and almost transparent. The cases in which beard was not detected were related to the fact that in that cases it wasn't very thick.

Table 4. Soft-biometric traits detection accuracy real-time.

	Accuracy	*Confusion Matrix*				
Gender			**M**	**F**		
	87.50%	**M**	16	0		
		F	4	12		
Age	62.50%		**A**	**B**	**C**	**D**
		A	10	2	0	0
		B	2	6	2	0
		C	0	4	2	0
		D	0	1	1	2
EyeGlasses detection	90%		**EGYes**	**EGNo**		
		EGYes	13	3		
		EGNo	0	16		
Beard detection			**BYes**	**BNo**		
	72%	**BYes**	11	3		
		BNo	0	18		

About the evaluation of the social believability of the robot we compared the results of the two groups. In the first condition the rating was lower than in the second condition, in which greetings were adapted to the soft-biometric traits and, according to the t-test result (see Table 5), this difference is significant thus showing that the robot is perceived as more believable when it is aware of the characteristics of the type of persons present in the environment.

Table 5. Comparison of believability rating in the two conditions.

	First condition no adaptation	Second condition greetings adaptation
Mean	2.45	3.8
Stdev	1.21	1.01
T-test p (a = 0.01)	0.000241	

5 Conclusions and Future Work

In this work we endow a social robot, Pepper in this case, with the capability of analyzing soft-biometrics features of people and use these awareness to adapt its behavior for improving the so called social believability.

In our approach we use CNNs to estimate gender, age, and the presence of beard and eyeglasses. Moreover, it considers more than one or two soft biometrics traits at the same time.

To test the real time performance of the system we set an experiment in a real-world scenario in which Pepper was behaving as a receptionist clerk of the Department. Even if performed on a small number of subjects, the experiment was carried out in the wild and the results show that the soft-biometrics module was quite reliable except for age estimation. As far as adaptation is concerned results show that when the greetings were adapted to the characteristics of the person the robot was perceived as more socially believable.

Even if the proposed approach may detect more than one person at time and perform soft-biometric traits recognition for each of them, we did not test it in this experiment. This will be part of our future work that will also aim at improving the soft-biometrics module especially as age-estimation is concerned by improving the dataset and the classification. Moreover we plan to add new traits recognition concerning the body. We also plan to perform another experiment in the wild with a larger number of subjects to have a better evaluation of the performance of our approach.

Acknowledgments. We wish to thanks all the people who participated in the experiments. Funding for this work was provided by the Fondazione Puglia that supported the Italian project "Programmazione Avanzata di Robot Sociali Intelligenti".

References

1. Jain, A.K., Dass, S.C., Nandakumar, K.: Soft biometric traits for personal recognition systems. In: Zhang, D., Jain, Anil K. (eds.) ICBA 2004. LNCS, vol. 3072, pp. 731–738. Springer, Heidelberg (2004). https://doi.org/10.1007/978-3-540-25948-0_99
2. Dantcheva, A., Elia, P., Ross, A.: What else does your biometrics data reveal? A survey on soft biometrics. IEEE Trans. Inf. Forensics Secur. (TIFS) **11**, 441–467 (2015)
3. Dautenhahn, K.: Socially intelligent robots: dimensions of human–robot interaction. Philos. Trans. B Biol. Sci. **362**, 679–704 (2007)
4. Lazzeri, N., Mazzei, D., Zaraki, A., De Rossi, D.: Towards a believable social robot. In: Lepora, N.F., Mura, A., Krapp, H.G., Verschure, P.F.M.J., Prescott, T.J. (eds.) Living Machines 2013. LNCS (LNAI), vol. 8064, pp. 393–395. Springer, Heidelberg (2013). https://doi.org/10.1007/978-3-642-39802-5_45
5. Carcagnì, P., Cazzato, D., Del Coco, M., Mazzeo, P.L., Leo, M., Distante, C.: Soft biometrics for a socially assistive robotic platform. Paladyn J. Behav. Robot. **6**(1) (2015)
6. Alqaderi, M., Rad, A.B.: A multi-modal person recognition system for social robots. Appl. Sci. **8**(3), 387 (2018)
7. Faria, D.R., Vieira, M., Faria, F.C., Premebida, C.: Affective facial expressions recognition for human-robot interaction. In: RO-MAN 2017: IEEE International Symposium on Robot and Human Interactive Communication (2017)

8. Broekens, J., Heerink, M., Rosendal, H.: Assistive social robots in elderly care: review. Gerontechnology **8**, 94–103 (2009)
9. Palestra, G., De Carolis, B., Esposito, F.: Proceedings of the Workshop on Artificial Intelligence with Application in Health co-located with the 16th International Conference of the Italian Association for Artificial Intelligence (AI*IA 2017), Bari, Italy, 14 November 2017 (2017)
10. Mubin, O., Stevens, C.J., Shahid, S., Al Mahmud, A., Dong, J.J.: A review of the applicability of robots in education. J. Technol. Educ. Learn. **1** (2013). 209–0015
11. Saerbeck, M., Schut, T., Bartneck, C., Janse, M.D.: Expressive robots in education: varying the degree of social supportive behavior of a robotic tutor. In: Proceedings of the SIGCHI Conference on Human Factors in Computing Systems (ACM 2010), Atlanta, GA, USA, 10–15 April 2010, pp. 1613–1622 (2010)
12. Leite, I., Martinho, C., Paiva, A.: Social robots for long-term interaction: a survey. Int. J. Soc. Robot. **5**, 291–308 (2013)
13. Ahmad, M.I., Mubin, O., Orlando, J.: A systematic review of adaptivity in human-robot interaction. Multimodal Technol. Interact. **3**(14), 1–25 (2017)
14. Dantcheva, A., Velardo, C., D'angelo, A., Dugelay, J.L.: Bag of soft biometrics for person identification: new trends and challenges. Multimedia Tools Appl. **51**(2), 739–777 (2011)
15. Sun, N., Zheng, W., Sun, C., Zou, C., Zhao, L.: Gender classification based on boosting local binary pattern. In: Wang, J., Yi, Z., Zurada, J.M., Lu, B.-L., Yin, H. (eds.) ISNN 2006. LNCS, vol. 3972, pp. 194–201. Springer, Heidelberg (2006). https://doi.org/10.1007/11760023_29
16. Lian, H.-C., Lu, B.-L.: Multi-view gender classification using local binary patterns and support vector machines. In: Wang, J., Yi, Z., Zurada, J.M., Lu, B.-L., Yin, H. (eds.) ISNN 2006. LNCS, vol. 3972, pp. 202–209. Springer, Heidelberg (2006). https://doi.org/10.1007/11760023_30
17. Zang, J., Lu, B.L.: A support vector machine classifier with automatic confidence and its application to gender classification. Neurocomputing **74**, 1926–1935 (2011)
18. Günay, A., Nablyev, V.V.: Automatic age classification with LBP. In: Proceedings of the 23rd International Symposium on Computer and Information Sciences (ISCIS 2008), October 2008, Istanbul, Turkey, pp. 1–4 (2008)
19. Huo, Z., et al.: Deep age distribution learning for apparent age estimation. In: Proceedings of IEEE Conference on Computer Vision and Pattern Recognition IEEE Las Vegas 2016, pp. 722–729 (2016)
20. Yan, C., Lang, C., Wang, T., Du, X., Zhang, C.: Age estimation based on convolutional neural network. In: Ooi, W.T., Snoek, C.G.M., Tan, H.K., Ho, C.-K., Huet, B., Ngo, C.-W. (eds.) PCM 2014. LNCS, vol. 8879, pp. 211–220. Springer, Cham (2014). https://doi.org/10.1007/978-3-319-13168-9_22
21. Levi, G., Hassner, T.: Age and gender classification using convolutional neural networks. In: The IEEE Conference on Computer Vision and Pattern Recognition (CVPR) Workshops, June 2015
22. Rothe, R., Timofte, R., Gool, L.V.: Dex: deep expectation of apparent age from a single image. In: 2015 IEEE International Conference on Computer Vision Workshop (ICCVW), pp. 252–257, December 2015. BIBLIOGRAPHY 87
23. Simonyan, K., Zisserman, A.: Very deep convolutional networks for large-scale image recognition. arXiv preprint arXiv:1409.1556 (2014)
24. Mohammad, A.S., Rattani, A., Derahkshani, R.: Eyeglasses detection based on learning and non-learning based classi cation schemes. In: 2017 IEEE International Symposium on Technologies for Home-land Security (HST), pp. 1–5. IEEE (2017)

25. Phillips, P.J., Wechsler, H., Huang, J., Rauss, P.J.: The feret database and evaluation procedure for face-recognition algorithms. Image Vis. Comput. **16**(5), 295–306 (1998)
26. De Carolis, B., Macchiarulo, N., Palestra, G.: A comparative study on soft biometric approaches to be used in retail stores. In: ISMIS 2018, pp. 120–129 (2018)
27. King, D.E.: Dlib-ml: a machine learning toolkit. J. Mach. Learn. Res. **10**(Jun), 1755–1758 (2009)
28. Kazemi, V., Josephine, S.: One millisecond face alignment with an ensemble of regression trees. In: 27th IEEE Conference on Computer Vision and Pattern Recognition, CVPR 2014, Columbus, United States, 23 June 2014–28 June 2014 (2014)

Using Particle Filter and Machine Learning for Accuracy Estimation of Robot Localization

Matthias Eder[1,3](✉), Michael Reip[1], and Gerald Steinbauer[2]

[1] IncubedIT Gmbh, Hart bei Graz, Austria
[2] Institute for Software Technology, Graz University of Technology, Graz, Austria
[3] Institute of Innovation and Industrial Management, Graz University of Technology, Graz, Austria
matthias.eder@tugraz.at

Abstract. Robot localization is a fundamental capability of all mobile robots. Because of uncertainties in acting and sensing and environmental factors such as people flocking around robots there is always the risk that a robot loses its localization. Very often behaviors of robots rely on a valid position estimation. Thus, for dependability of robot systems it is of great interest for the system to know the state of its localization component. In this paper we present an approach that allows a robot to asses if the localization is still valid. The approach assumes that the underlying localization approach is based on a particle filter. We use deep learning to identify temporal patterns in the particles in the case of losing/lost localization in combination with weak classifiers from the particle set and perception for boosted learning of a localization monitor. The approach is evaluated in a simulated transport robot environment where a degraded localization is provoked by disturbances cased by dynamic obstacles.

Keywords: Localization · Autonomous robots · Dependability · Machine learning

1 Introduction

Robot localization is a fundamental capability of all mobile intelligent robots. There is a huge corpus of research that has been conducted to allow reliable estimation of the current position using uncertain sensor data and action execution. Typically, these approaches perform some kind of state estimation to determine the current robots pose, using sensors like 2D lasers [15] and Monte Carlo methods based on particle filters [16].

Because of uncertainties and environmental factors such as people flocking around robots there is always the risk that the error of the position estimation increases significantly or that it diverges completely. Very often navigation and

F. Wotawa et al. (Eds.): IEA/AIE 2019, LNAI 11606, pp. 700–713, 2019.
https://doi.org/10.1007/978-3-030-22999-3_60

other behaviors of robots rely on a valid position estimation. Thus, for performance as well as dependability of robot systems it is of great interest for the system to know the state of its localization component.

In this paper we present an approach that allows a robot to asses if the localization is still valid. The approach assumes that the underlying localization approach is based on a particle filter. Following the assumption that the particle set representing a probability distribution of the robot's location bears information about the localization state, we propose to use deep learning to identify temporal patterns in the particle set in the case of losing or lost localization. Thus, we investigated different deep neural networks (non-recurrent and recurrent) for their ability to learn identifying such patterns. Moreover, we propose to combine these networks with weak classifiers obtained from statistical information in the particle set and the actual robot perception for boosted learning of a more reliable localization monitor.

For the training and evaluation of the approach we use a realistic simulation of an industrial transport robot environment where due to access to ground truth a large amount of labeled data can be generated. Moreover, a simulation allows to provoke a degraded localization easily by disturbing the robot's localization system by randomly adding dynamic obstacles.

The remainder of the paper is organized as follows. In the next section we discuss briefly related research. In Sect. 3 we introduce the proposed approach for a localization monitor. A experimental evaluation of the proposed approach is presented in the succeeding section. Finally, in Sect. 5 we draw some conclusions and provide some ideas for future work.

2 Related Research

In this section we briefly discuss related research concerning robot localization and machine learning.

In [11] Röwekämper et al. evaluates the position accuracy of a mobile robot localization method that is based on particle filtering and laser scan matching. For evaluation they used a motion capture system that tracks the position of a robot within its environment with high accuracy. The authors are more interested in evaluating and improving localization approaches rather than providing an online localization monitor.

Convolutional neural networks are commonly used to detect features in images for classifying images. A simple and modern example is presented by Shamov and Shelest [14]. They present the main features of convolutional neural networks and show how they are applied for feature detection. For this purpose they created the task of detecting tower lighthouses from a video stream.

Long-Short Term Memory networks [6] are good at classifying, predicting and processing time series. It is well suited for training to remember longer time series and also focuses on the current input.

Pattern recognition is an important task in visual computing. It is about detecting specific patterns within an given input. In visual computing these inputs are images which are used to detect certain content like humans. Modern pattern recognition methods are based on convolutional neural networks which aim to identify patterns within the input data by training the network on a big set of data. This method can be used to detect licence plates on a car [1] and is also applied in vineyards to detect birds which needs to be scared away [2].

Since particle filters are among the most popular methods for state estimation, a lot of research is done to improve their accuracy and performance. In general the state of a complex system can be estimated correctly by using (infinite) many particles. But with an increasing number of particles the efficiency of the particle filter decreases. Thus, researcher aim at decreasing the number of particles while offering the same performance. One attempt is called adaptive particle filtering [7].

Morales et al. presented a method for object tracking by using a 3D occupancy grid as environment representation and a particle filter based approach for detecting and tracking obstacles [8].

3 Localization Accuracy Estimation Approach

In order to judge if the estimation of the robot's position is accurate we follow the approach depicted in Fig. 1. In our application we need to solve a binary classification problem: either the robot is well localized or delocalized. We declare the robot delocalized if the estimated position and orientation diverged form the true position and orientation more than a given threshold.

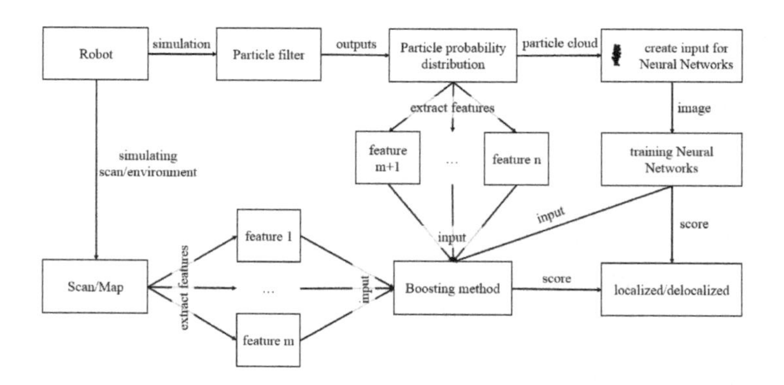

Fig. 1. The overall approach for estimating the accuracy of robot localization.

In order to have access to a ground truth of the position and orientation and to be able to provoke delocalization we simulate the robot's motion and sensing (i.e. laser scans) as well as the environment. The data from the perception as well as the data from the particle-based pose estimation (i.e. particles) are recorded and converted to labeled training data for the following classification steps. For the classification we follow two approaches which we also use in combination.

The first approach follows the assumption that the probability distribution of the robot's pose modelled by the particle set and its temporal change bears information about the localization accuracy. Thus, we train non-recurrent and recurrent neural networks for the classification based on the particle set. The second approach follows the assumption that statistical features (weak classifiers) derived from the perception and the particle set can be used to learn a strong classifier using boosting. Finally, the classification from the trained neural network can be used as an additional classifier in the boosting step.

In the remainder of the section we introduce the following steps of the proposed process in more detail: (1) generation of training data, (2) selection and validation of neural network architectures, (3) validation and selection of features, and (4) learning classifiers.

3.1 Generation of Training Data

This section describes how the training data for the neural networks is generated.

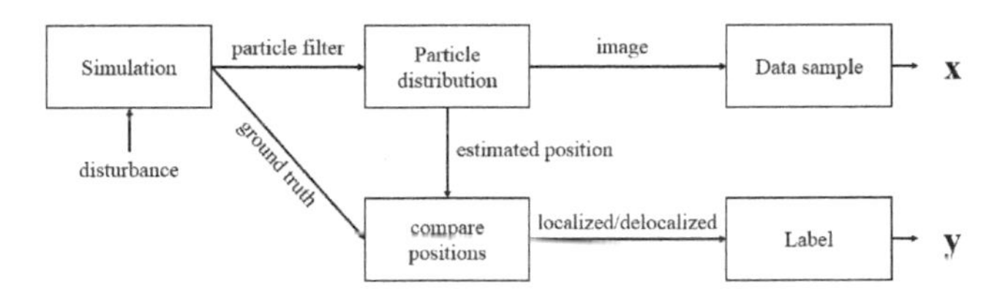

Fig. 2. The procedure of generating a data sample and labelling for training.

Figure 2 shows the overall procedure of generating a training sample and finding the correct label. To create a sample one needs the particle set from the particle filter which is used to create a training image x. A particle represents a potential pose (position and orientation) along with a weight representing the importance of the particle $\langle x, y, \theta, w \rangle$. We convert the particle set to an image because it is able to represent different situation and is suitable for state-of-the-art machine learning tools. This image is labelled by comparing the exact pose $\langle x_{GT}, y_{GT}, \theta_{GT} \rangle$ (retrieved from the simulation) and the estimated pose $\langle x_{PF}, y_{PF}, \theta_{PF} \rangle$ (retrieved by the distribution of the particles). Using the distance between them one can classify the robots localization state and thus also create a label y for the training sample.

$$y = \begin{cases} 0 \text{ , if } \sqrt{(x_{GT} - x_{PF})^2 + y_{GT} - y_{PF})^2} < \alpha \wedge \|\theta_{GT} - \theta_{PF}\| < \beta \\ 1 \text{ , else.} \end{cases} \quad (1)$$

Usually the particles are distributed over a large state space (i.e. the entire environment map). In order to focus on interesting areas the image is centered around the mean of the positions represented in the particle set. Moreover, the image is oriented along the mean of the orientations represented in the particles. We use a size of 36×36 pixel in this work. This values had been manually selected based on run-time and accuracy. Using only a focused representation has the advantage that the trained network can be used for environments of different size and shape.

Figure 3 shows some examples of generated images, stored at different time steps. The first three images illustrate a particle set of a well localized robot while the fourth image is an example of a delocalized robot.

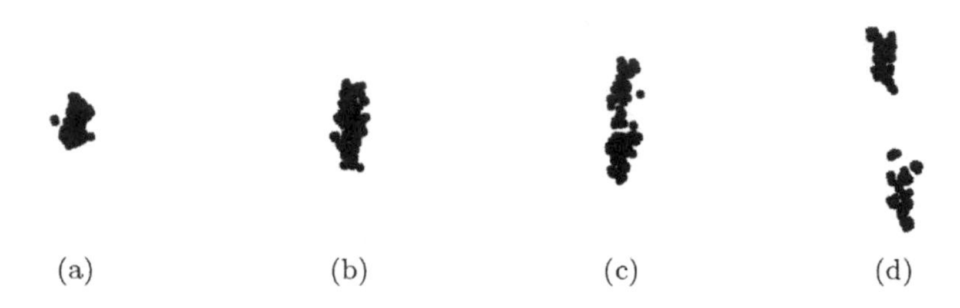

Fig. 3. Example images generated for training the neural network. (a) Localized particle set at time step t. (b) Localized set at time $t + 40$ s. (c) Localized set at time step $t + 90$ s. (d) Delocalized image at time step $t + 110$ s.

The transformation and inflation of the particle cluster are indicators for the localization state of the robot. Thus a neural network can be trained to detect relevant patterns within this transformation.

To generate enough distinctive data samples the robot has to drive randomly within a predefined environment. This leads to different robot locations within the environment and samples representing different localization situation in the environment. Figure 5a shows an example environment. Due the facts that methods based on particle filters are quite robust and we need positive as well as negative examples we need to provoke delocalization. One of the simplest and most realistic attempts is to disturb the lidar sensor using dynamic obstacles that are not represented in the map. For this purpose we randomly generate additional obstacles around the robot in the simulation. Such obstacles cause unexpected measurements and may lead to delocalization.

3.2 Selecting and Validating the Neural Networks

Having created a training set for neural networks one can start defining an appropriate network structure (type, layout) which can be used for localization classification. Thus, several possible network structures need to be trained and validated.

When using data from a particle filters as input a priori it is hard to specify the best network type and structure since it is rather unclear which information is extracted by a neural network for estimating the robot's localization accuracy.

Table 1. Overview of the tested network types and their complexity.

NN type	Complexity	Conv layers	LSTM layers	Other layers	Total layers
CNN	Simple	1	-	2	3
CNN	Mid-complex	2	-	4	6
CNN	Complex	21	-	11	32
LSTM	Simple	-	1	3	3
LSTM	Mid-complex	-	2	3	5
LSTM	Complex	-	3	4	7
LRCN	Simple	2	1	2	5
LRCN	Mid-complex	3	2	5	10
LRCN	Complex	5	2	7	14

A possible network structure is a convolutional neural network (CNN) which is used for pattern recognition. Another possibility is a recurrent network that may learn the transformation of the particle set over a certain time period.

To evaluate a wide range of possibilities, both network types are considered. A CNN for feature extraction and a recurrent network structure for learning the transformation over time. For learning a recurrent network a long-short term memory (LSTM) structure is used since a better performance is expected [6]. Additionally a combination of both, CNN and LSTM, is trained. This is called a Long-Term Recurrent Convolutional Network (LRCN). By using this network structure it is evaluated if the advantages of both former types can be combined.

For each different network types (CNN, LSTM and LRCN) three different network layouts are created and trained. We generate a simple, a moderate, and a complex layout. Those three layouts are chosen to evaluate whether a simple network structure leads to underfitting or a complex structure leads to overfitting. This results in a total number of nine networks to be trained, which are summarized in Table 1.

To validate the usefulness of a network structure one needs to define a metric that allows to compare different network structures in respect to localization performance. A suitable metric is accuracy. Accuracy is a value that describes how well a given data set is classified. If a classification task with two classes is considered it can be easily computed by comparing for each example the result with the desired output.

Having two classes, one can assign to a class either the term *positive* or *negative*. To validate a data sample it is run through the network structure and the output is recorded. The output is then assigned to one of the two classes, depending on the result. By comparing the assigned class with the expected label class one can determine whether a class was correctly assigned. If both, the result and the label match, then the outcome is *true*. When a data sample is classified as positive and the expected class is also positive the data sample is *true positive*. If the two classifications do not match the outcome is *false*. If the data sample is classified as negative and the expected result is positive the outcome is *false negative.* By counting the results on the examples in the data set one can determine the accuracy as

$$acc = \frac{t_p + t_n}{t_p + t_n + f_p + f_n} \tag{2}$$

where t_x indicates the number of correct classification and f_x the number of the incorrect ones with x either p (positive) or n negative.

3.3 Feature Selection

Once an appropriate network structure has been selected useful features for the boosting step needs to be identified. Using boosting it is tried to improve the accuracy of the localization scoring algorithm. Therefore features are considered that are already used to calculate the robot localization, but also novel features are investigated.

In general it is hard to specify which features may be important for localization scoring. Using a simple evaluation one can search for various possible features that might be useful as weak classifiers in scoring the accuracy of a robots localization. By using the thresholds α and β as above the robots localization status can be easily determined and every feature can be inspected on its own. We identified 33 possible features that are based on information from the robot's perception and data from the particle filter. Table 2 summarizes a selection of these features. For a complete list we refer to [3]. During the recording of the training data feature estimations are recorded as well and labelled with the correct class.

Table 2. Examples for features as weak classifiers. Grey rows indicate discriminating features while white rows indicate non-discriminating features.

Feature	unit	$\#bin$	$D(P\|\|Q)$
(1) maximum distance - particle to center of particle set	meters	350	0.265
(3) mean distance - particle to center of particle set	meters	100	0.314
(5) median distance - particle to center of particle set	meters	50	0.321
(6) minum distance - particle to center of particle set	meters	100	0.153
(15) mean distance - laser scan points to closest map point	meters	400	0.086
(16) match ratio - lines in laser scan to lines in map	%	100	0.157
(18) match quality - points in laser scan to lines in map	%	100	0.035
(22) percentage - points in laser scan behind an obstacle	%	100	0.045
(27) average distance - lines in laser scan to closest map line	meters	400	0.042
(31) variance in x direction of the particle set	meters	100	0.403

These recordings are then split up into a localized X_Q and delocalized X_P data set. These sets are then represented as a discrete distribution over k bins. Each bin contains a number of samples which lie within that bin. A discrete probability distribution P is then represented as the number of samples within that bin divided by the total number of samples n

$$P(i) = \frac{|bin(i)|}{n}. \tag{3}$$

The same holds for the distribution Q. Having two discrete distributions P, Q for the localized and delocalized set one can calculate the Kullback-Leibler divergence [9] as

$$D(P||Q) = KL(P, Q) = \sum_{i}^{k} P(i) \times \log \frac{P(i)}{Q(i)} \tag{4}$$

where k is the number of bins. The KL-divergence is only defined if $\forall i : Q(i) = 0 \rightarrow P(i) = 0$ applies. If $P(i) = 0$ the contribution of the i-th bin is also 0. It is defined that $D(P||Q) \geq 0$ for all distributions and $D(P||Q) = 0$ if $P = Q$.

Using the KL-divergence between the probability distribution of the feature values for the positive and negative cases allows to select well discriminating features. The higher the KL-divergence the more different the distributions are and the more informative is the underlying feature.

Table 2 shows the KL-divergence of some example features calculated using the same simulated disturbed navigation as discussed above. Figure 4 depicts the probability distribution of feature values for a discriminating and a non-discriminating feature.

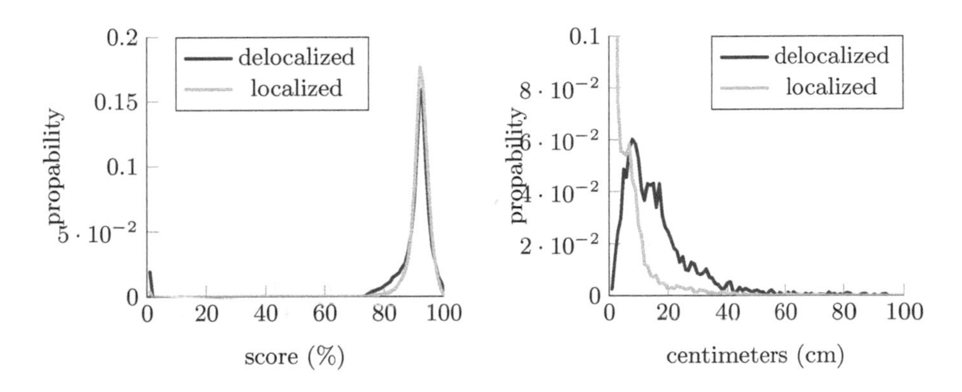

Fig. 4. Probability distributions of feature values for localized and delocalized cases for feature 18 (left, non-discriminating) and feature 31 (right, discriminating).

3.4 Ada Boosting

Boosting is a machine learning approach which uses supervised data for classification. The idea is to combine multiple weak classifier which does not provide enough information about a certain class itself and combine it to one strong classifier that can be used for identifying classes [12, 13]. In general an adaptive boost algorithm takes N training samples $(x_i, y_i), 1 \leq i \leq N$ with $x_i \in \mathbb{R}^K$ and $y_i \in \{-1, +1\}$. x_i is the input vector which contains K different components (features) that are used for training. y_i is the desired label which is either -1 or $+1$. There exist several variants of boosting algorithm which all have a similar structure [4]. In this work the standard discrete Ada Boosting algorithm which uses two classes is used. It uses its N-sized input set and initializes the weights for each input sample with $w_i = 1/N$. Then a weak classifier $f_m(x)$, the weighted training error ϵ_m, and the scaling factor c_m is computed. Then the weights are increased for input samples that have been wrongly classified. After this step the weights are normalized and the steps for finding a new weak classifier are repeated form M times. At the end a final classifier $F(x)$ is found which uses the sign of the weighted sum of the input set.

In the proposed approach we use an appropriate selection of features from the previous section and the same examples from training trajectory used in training the neural network. Moreover, we treat the binary classification of the trained neural networks as additional weak classifier.

4 Experimental Evaluation

In the experimental evaluation we are interested to investigate how well the proposed approaches are able to identify a lost localization of a robot. Moreover, we are interested which network structure preform best and if the approach is able to generalize to different environments. In the remainder of the section we introduce the experimental setup, give details on the preparation of the training data, and discuss the results in respect of the used network structure, and the used features as well as the boosting step.

4.1 Experimental Setup

The training and evaluation is based on a simulation of a team of transport robots in industrial environments based on Stage [5] and provided by an industrial partner. The used navigation software is the same as embedded into the real transport robots and uses laser scans, odometry, a gyro, and a 2D gridmap for the localization based on a particle filter. The simulation is based on ROS Indigo [10] and runs on a standard PC equipped with an Intel i5 (2.5 GHz), 4 GB RAM, and Ubuntu 14.04.

The training and classification is based on the libraries Caffe 1.0.0, OpenCV 3.3, and CUDA 7.5.17 and runs on a standard PC equipped with an Intel i7 (2.1 GHz), 8 GB RAM, Nvida GPU, and Ubuntu 16.04.

Figure 5 depicts the environment map used where the left map was used to collect data for training and the basic evaluation and the right map was used to evaluate if the trained classifiers generalize to other environments.

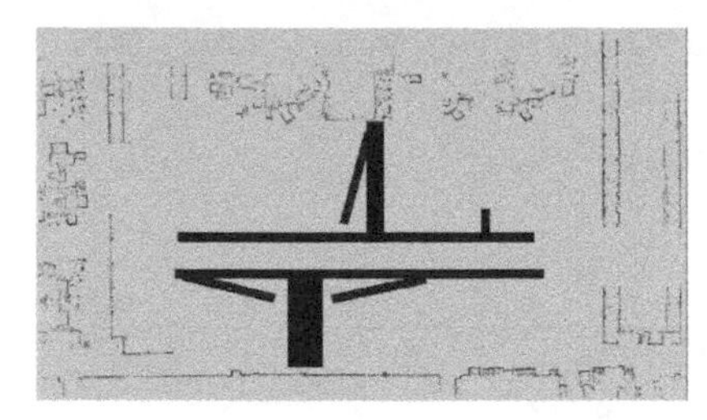

(a) Map for training and basic evaluation.

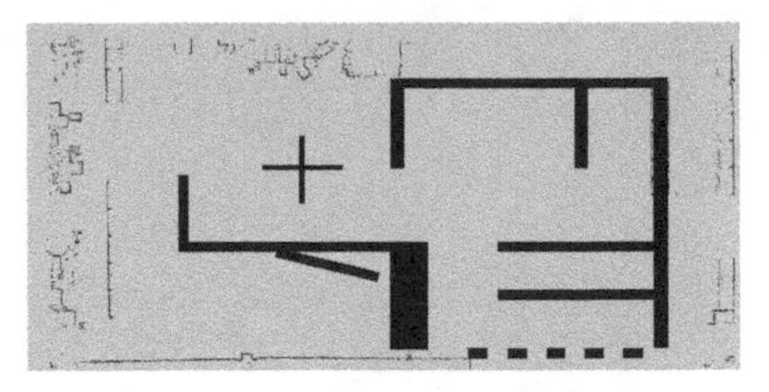

(b) Map for generalization evaluation.

Fig. 5. Environment maps used for training and evaluation. Based on an real industrial environment enriched with additional clutter.

4.2 Training Data

As described in Sect. 3 the robot is randomly driven trough the training environment while randomly dynamic obstacles are generated to provoke delocalization. During this run all relevant output from the particle filter and the perception were recorded. Using the thresholds $\alpha = 1.2\,\mathrm{m}$ and $\beta = 1.2\,\mathrm{rad}$ all data points were labeled. The thresholds were selected in the way that they maximize the KL-distance of an hand-crafted localization score.

This leads to a sequence of 150.000 data points were 50% are labeled delocalized. For the training of the non-recurrent networks this data are equally split into a training and evaluation set and shuffled randomly. For the training of the recurrent networks individual data points are useless. Thus, we split the data points in positive and negative sequences of the length of 2×7. The negative sequences are organized around a time point t where the robot become delocalized and 7 data points from before and after t are added. This lead to 10.714 sequences. Moreover, the same number of positive sequences were generated where the robot stayed localized.

The same data sets were also generated for the second environment which are used to evaluate the generalization of the approach and not used in training.

4.3 Validation of the Neural Networks

In order to evaluate the most appropriate network structure for the estimation of localization accuracy we trained each network structure with the proper (individual data points or sequence of data points) training data and evaluated the results using the evaluation set of the first environment as well as the data set of the second environment. As criterion we use the metric defined in Sect. 3.2.

Table 3 summarizes the results of this evaluation. The number of iterations in training was 100.000 for all networks. In general it can be seen that all networks perform quite well on the training data were we have equally many positive and negative examples. The mid-complex CNN performs best. For the validation set in the same environment the performance drops by about 2–13%. The mid-complex CNN is still in the lead but shows the largest performance loss showing

Table 3. Result for all network structures: simple CNN (A), mid-complex CNN (B), complex CNN (C), simple LSTM (D), mid-complex LSTM (E), complex LSTM (F), simple LRCN (G), mid-complex LRCN (H), complex LRCN (I).

	Accuracy/%								
Environment	A	B	C	D	E	F	G	H	I
Training set	84.52	96.47	91.22	84.42	86.95	87.42	88.73	86.12	86.09
Same environment	79.62	83.25	81.38	82.16	81.99	83.10	82.55	81.34	79.30
New environment	67.63	66.27	62.06	67.79	67.36	66.33	69.55	67.19	58.88

a tendency for overfitting. For the data set from novel environment the performance drops for all networks to around 67%. The leader is here the combined simple network showing that this network structure is able to generalize well.

Table 4. Result of the mid-complex CNN

Environment	tp	tn	fp	fn	Accuracy
Training set	70 050	72 656	2 343	2 949	96.47%
Same environment	59 655	65 220	9 770	15 344	83.25%
New environment	48 284	51 113	23 886	26 715	66.27%

The accuracy is obviously a compressed metric. In order to analyze the performance of the mid-complex CNN, Table 4 shows the detailed results of its evaluation. It can be seen that the true and false results for the positive and negative cases are almost balanced. Thus, in contract to other networks this network is not biased in a sense of being too optimistic (many false negatives - missing delocalizations) or too pessimistic (many false positives - raising a lot of false alarms). We refer to [3] for more detailed results.

4.4 Result of Ada Boosting

First we evaluate how well Ada Boosting using the features introduced in Sect. 3.3 performs on the classification task. The features represent information extracted from the particle filter and the actual perception. Using the discrimination metric 26 promising features out of 33 were selected for boosting. In this evaluation no output of the neural networks had been used.

In order to find the optimal number of weak classifiers learned in boosting we varied the maximum number between 26 and 300. Figure 6a shows the performance of the learned classification on the training and evaluation data set. We selected 276 weak classifiers as an optimal number. Table 5a shows the performance of the so learned classifier on the training and the evaluation sets. The performance for the unknown environment is with 81.83% significantly better

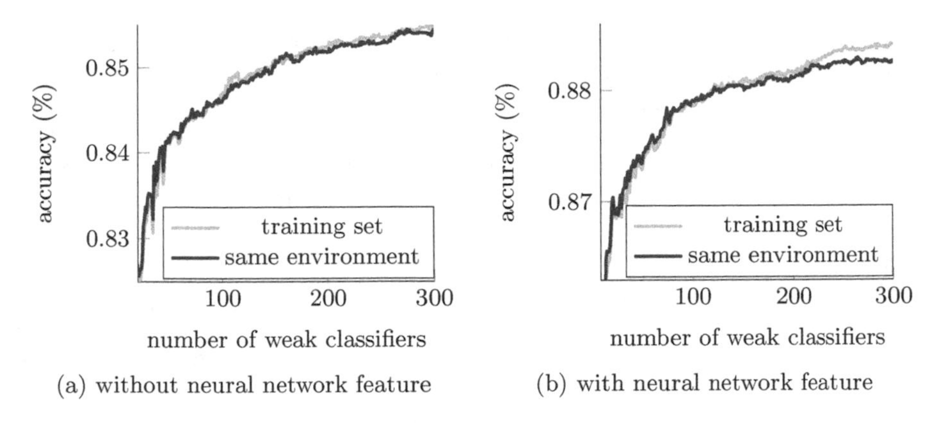

(a) without neural network feature

(b) with neural network feature

Fig. 6. Training and test accuracy of AdaBoost with and without neural network feature on a different number of weak classifiers.

than the trained neural networks. But this comes at a price as the false negative rate is significantly higher than the false positive rate. This can be a problem for robots acting in an open environment.

Figure 6b shows the same learning and evaluation process where additionally the output of the mid-complex LRCN was used as a feature. Here the optimal number of weak classifiers is 229 as the accuracy on the validation set does not significantly increase with further classifiers. Table 5b shows the performance of the so learned classifier on the training and the evaluation sets. It is clearly visible that the additional feature allows only a marginal performance improvement of about 1% on the data from the novel environment.

Table 5. Boosting result using AdaBoost (a) without the neural network feature and 276 weak classifiers (b) with the neural network feature and 229 weak classifiers

Feature	Set	tp	tn	fp	fn	Accuracy
(a) Without neural network feature	Training set	64 258	63 883	11 106	10 731	85.45%
	Same environment	64 423	63 726	11 263	10 566	85.44%
	New environment	48 935	61 540	5 898	18 626	81.83%
(b) with neural network feature	Training set	33 633	32 563	4 886	3 888	88.30%
	Same environment	33 675	32 453	5 097	3 745	88.21%
	New environment	51 550	59 988	7 450	12 011	82.62%

5 Conclusion and Future Work

Estimating its own position in an environment is a crucial capability of intelligent mobile robots. Knowing reliably the state of the localization process is important for the dependability of such robot systems.

Following the observation that most robots use distance sensors, environment maps, and particle filter based approaches we investigate if information generated from these components (patterns and statistical features of the particle set) can be used to train a reliable localization monitor. Using training data from test runs in an simulation environment where delocalization was provoked we trained several deep networks (non-recurrent and recurrent) and a boosting approach. While the trained deep networks showed moderate classification rates in particular in an environment not used for training the boosting approach using the network output as an additional feature showed detection rates of more than 80%. In conclusion, the evaluation of different training settings showed that it is possible to use information obtained from the particle set for scoring the localization accuracy.

For future work it needs to be investigate if delocalization cased by other uncertainties such as slipping can be detected as well. Moreover, a more detailed investigation of networks structures needs do be done. Finally, so far only the position of particles is used for training. It would be interesting if their orientation bears additional useful information.

References

1. Carata, S.V., Neagoe, V.E.: A pulse-coupled neural network approach for image segmentation and its pattern recognition application. In: 2016 International Conference on Communications (COMM), pp. 61–64, June 2016
2. Dolezel, P., Skrabanek, P., Gago, L.: Pattern recognition neural network as a tool for pest birds detection. In: 2016 IEEE Symposium Series on Computational Intelligence (SSCI), pp. 1–6, December 2016
3. Eder, M.: Using particle filters and machine learning approaches for state estimation on robot localization scoring. Master's thesis, Faculty for Computer Science and Biomedical Engineering, Graz University of Technology (2017)
4. Friedman, J., Hastie, T., Tibshirani, R.: Additive logistic regression: a statistical view of boosting. Ann. Stat. **38**(2), 337–407 (2000)
5. Gerkey, B., Vaughan, R.T., Howard, A.: The player/stage project: tools for multi-robot and distributed sensor systems. In: Proceedings of the 11th International Conference on Advanced Robotics, pp. 317–323, June 2003
6. Hochreiter, S., Schmidhuber, J.: Long short-term memory. Neural Comput. **9**(8), 1735–1780 (1997)
7. Sanguino, T.D.J.M., Gomez, F.P.: Toward simple strategy for optimal tracking and localization of robots with adaptive particle filtering. IEEE/ASME Trans. Mechatron. **21**(6), 2793–2804 (2016)
8. Morales, N., Toledo, J., Acosta, L., Sánchez-Medina, J.: A combined voxel and particle filter-based approach for fast obstacle detection and tracking in automotive applications. IEEE Trans. Intell. Transp. Syst. **18**(7), 1824–1834 (2017)
9. Polani, D.: Kullback-Leibler Divergence, pp. 1087–1088. Springer, New York (2013)
10. Quigley, M., et al.: ROS: an open-source robot operating system. In: ICRA Workshop on Open Source Software (2009)
11. Röwekämper, J., Sprunk, C., Tipaldi, G.D., Stachniss, C., Pfaff, P., Burgard, W.: On the position accuracy of mobile robot localization based on particle filters combined with scan matching. In: 2012 IEEE/RSJ International Conference on Intelligent Robots and Systems, pp. 3158–3164, October 2012

12. Schapire, R.E.: A brief introduction to boosting. In: Proceedings of the 16th International Joint Conference on Artificial Intelligence, IJCAI 1999, vol. 2, pp. 1401–1406. Morgan Kaufmann Publishers Inc., San Francisco (1999)
13. Schapire, R.E.: Explaining AdaBoost. In: Schölkopf, B., Luo, Z., Vovk, V. (eds.) Empirical Inference, pp. 37–52. Springer, Heidelberg (2013). https://doi.org/10.1007/978-3-642-41136-6_5
14. Shamov, I.A., Shelest, P.S.: Application of the convolutional neural network to design an algorithm for recognition of tower lighthouses. In: 2017 24th Saint Petersburg International Conference on Integrated Navigation Systems (ICINS), pp. 1–2, May 2017
15. Thrun, S., Burgard, W., Fox, D.: Probabilistic Robotics (Intelligent Robotics and Autonomous Agents). The MIT Press, Cambridge (2005)
16. Thrun, S., Fox, D., Burgard, W., Dellaert, F.: Robust Monte Carlo localization for mobile robots. Artif. Intell. **128**(1), 99–141 (2001)

Natural Language Processing and Sentiment Analysis

Chatbots Assisting German Business Management Applications

Florian Steinbauer[1], Roman Kern[1,2], and Mark Kröll[2](✉)

[1] Institute of Interactive Systems and Data Science, Inffeldgasse 13, Graz, Austria
fsteinbauer@student.tugraz.at
[2] Know-Center GmbH, Inffeldgasse 13, Graz, Austria
{rkern,mkroell}@know-center.at

Abstract. In most companies Business-Management software has become omnipresent in recent years. These systems have been introduced to streamline productivity and handle data in a more centralized fashion. Yet, these systems are often modelled by complex processes which makes navigating through them a challenging task. In this work, we introduce a text-based chatbot to a mid-sized Austrian company to facilitate the interaction with these software solutions and thus support them in managing their customer relationships. To learn more about positive as well as negative effects the chatbot has on the employees, we conduct a technical as well as an empirical evaluation. We, for instance, hypothesize that younger staff, who grew up with computers and smart phones, are more open to the conversational system than older employees. We implement a customized solution which integrates seamlessly into the company's system. The implementation process is informed by related work on Customer-Relationship-Management software, structures of a conversation as well as the typical architecture of a chatbot. In the technical evaluation, objective metrics such as average response time were measured. For the empirical evaluation, a questionnaire was sent out to 15 participating employees asking about subjective metrics such as task ease or user experience. In sum, employees were satisfied with Usage, Interaction Pace and System Response.

Keywords: Business Management application · Chatbot · German language · Natural Language Processing

1 Introduction

Since the early 2000s, Business-Management (BM) software has experienced a massive growth - modern Enterprise-Resource-Planing (ERP) and Customer-Relationship-Management (CRM) systems now cover a wide variety of business processes. Companies using this software can plan every corporate resource from capital, personnel, customers to utilities, communication- and IT-applications.

In the last decade also the number of affected staff has increased. Whereas in the beginnings, only specially skilled employees had to work with these systems,

F. Wotawa et al. (Eds.): IEA/AIE 2019, LNAI 11606, pp. 717–729, 2019.
https://doi.org/10.1007/978-3-030-22999-3_61

now almost every employee gets into contact with this software: from jobs like entering goods receipts to jobs like applying for holidays. Especially senior staff might have difficulties in adapting to this new technology. We thus integrate a text-based chatbot component into an existing CRM system of a mid-sized Austrian company. To learn more about positive as well as negative effects the chatbot has on the employees, we conduct a technical as well as an empirical evaluation. We, for instance, hypothesize that younger staff, who grew up with computers and smart phones, are more open to the conversational system than older employees.

The chatbot component is capable of performing several tasks to support sales personnel in their daily business including *finding a customer*, *ordering goods*, or *reporting a visit*. In the implementation phase we faced several challenges, i.e. (i) to correctly identify the user's intent, (ii) to ask the user for relevant information which is missing or (iii) to detect context switches.

We decided not to use proprietary solutions as, for example, offered by the company *Kore.ai*. The first reason is that by implementing our own customized solution, we are free to extend and adapt according to (our) the company's needs at any time. In this way, we also gain a more seamless integration into the existing system. The second reason is that the chatbot is to support interaction in German language. By self-implementation, we are free to review and select suitable Natural Language Processing tools for the German language.

In the technical evaluation, we measured objective metrics such as average response time - with 809 ms the chatbot's response was quite quick. For the empirical evaluation, a questionnaire was sent out to 15 participating employees asking about subjective metrics such as task ease or user experience. The averaged results from the evaluation are as follows: Usage (4.00), Task Ease (3.00), Interaction pace (4.45), User Experience (3.73), System Response (4.27), Expected Behavior (2.91), and Future Use (3.09)[1]. In sum, employees were satisfied with Usage, Interaction Pace and System Response.

2 Related Work

In this section, we give an overview of the current situation of Customer-Relationship-Management (CRM) today including challenges and prospects as well as some insights into recent developments in chatbot research.

Customer-Relationship-Management

In recent years modern enterprises have focused on deeper customer relations due to the fact that new customer acquisition can be five times more expensive than retention (cf. [2]). While a decade ago only the largest companies deployed computer based strategic sales and marketing instruments, the CRM market has witnessed a growth in recent years. Correia et al. [6] have found that worldwide CRM spendings by companies have risen by 12.3% from 2014 to 2015 and

[1] Rating was between 1 and 5; the higher the better.

reaching $26.3B. These spendings are mainly attributed to the "Big 4" vendors: Salesforce, SAP, Oracle, and Microsoft.

In his work, Kowalke [10] illustrates the effects chatbots can have on CRM systems corroborating and informing our integration efforts. (1) Chatbots can act directly as sales staff by deciding whom to contact when and mimic a sales representative. (2) Chatbots can be used for automatic social media interaction. Adding to the traditional means of conversation like phone or email, a wide variety of social media channels like Facebook, Twitter or WhatsApp have been established, over which customers expect to be able to contact companies. (3) Chatbots can be used for the routing and handling of customer interaction. To give an example, when a customer first interacts with an automatic conversational system, relevant information can be gathered and frequently asked questions can be answered. (4) Chatbots can facilitate access to the CRM systems. Sales personnel spend a lot of time in their car so that provision of a voice-based conversation with the system can save time and money. (5) Chatbots can make CRM systems more efficient. In recent years more functionality was added to the sales and marketing platforms, often having the users switch applications to achieve their goal. When sales personnel can interact with the system in natural language, an automated conversational agent can connect data and perform different sub-tasks to achieve a goal more efficiently.

Most work in this area has been dedicated to using chatbots for interaction purposes: Cui et al. [7], Thomas [14], Voth [15], Xu et al. [18], for instance, created customer service chatbots for e-commerce and banking sector applications respectively. Choi et al. [3] used a conversational system to interactively walk users through product manuals. Bloomberg deployed a chatbot in an application for mobile phones for employees to report if they are unavailable for work due to sickness (cf. [8]).

Chatbots

Early software as ELIZA written by Joseph Weizenbaum (cf. [17]) attempted to simulate a psychotherapeutic conversation. However, ELIZA never attempted to and did not pass the test although it appeared human to several users. Turing estimated that in the year 2000, state-of-the-art machines will be able to fool users in 30% of times. Since 1991 the *Loebner Prize* for the most human-like computer program is yearly awarded[2]. Up to now no software has been able to pass the *Turing Test*. The chatbot *Eugene* mimicking a 13-year old boy from the Ukraine supposedly passed the test by tricking 33% human judges into believing it actually is a child. However, this achievement is disputed. Marcus [13] argues, judges would overlook grammatical errors more easily since a non-native English speaker is simulated.

The work on conversational systems can be grouped into two categories: (i) goal-oriented like *Siri*, *Alexa*, *Google Assistant* or *Cortana* and (ii) non goal-oriented general purpose dialogue systems like *Cleverbot* or *Microsoft Tay*. Microsoft released their Chatbot *Tay* on the 23rd of March 2016, being accessible

[2] http://www.aisb.org.uk/events/loebner-prize, last accessed 2019-03-29.

over Twitter. Microsoft's goal was to see how intelligent systems could learn on their own. Tay was set up to learn more information from Twitter conversations over time and synthesize novel responses. In the background user profiles were created to personalize dialogues.

Since the first advent of conversational systems, a multitude of systems has been created. From initial efforts in mimicking psychotherapists and patients (cf. [5,17]) which can be approached with any topic, the focus has shifted more to assistants which help achieve a narrow goal in a single domain including business-management (cf. [3,8]).

In order to design a chatbot that provides a meaningful experience, we must first understand what expectations people have for this technology, and what opportunities there are for chatbots based on user needs. Zamora [19] has conducted research on 54 participants from the United States and India on their expectations and experiences towards chatbots. The author found that users expect four traits: In order to be accepted and used, the software should be *high-performing* (fast, efficient, and reliable), *smart* (knowledgeable, accurate, and foreseeing), *seamless* (easy, and flexible) and *personable* ("understands me", and likable).

Furthermore, Zamora [19] showed participants were happy to be assisted by bots in personal routine tasks, but as topics like social media or finance were brought up, participants voiced privacy concerns.

3 Concepts & Implementation

At first, we outline the current state of the company's system before the integration of the chatbot component. We then describe the development of the chatbot component. Therefore, we adhere to the general architecture which was introduced by Jurafsky and Martin [9] providing a simplified overview of a conversational system shown in Fig. 1. The architecture itself illustrates several tasks which need to be iteratively completed during a conversation with a chatbot, for instance, (i) to understand what information the user conveys, for instance, by

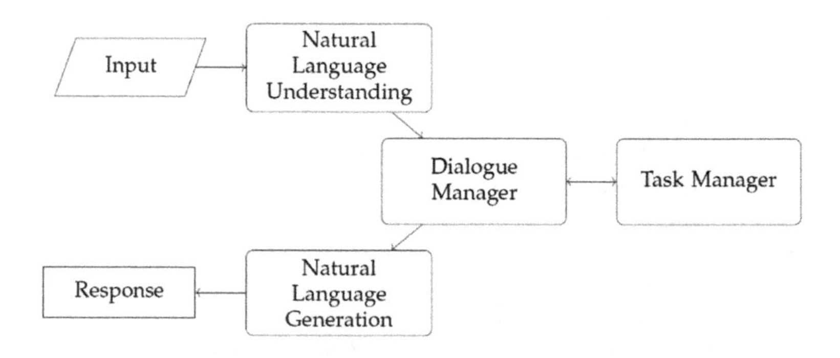

Fig. 1. Generalized architecture of a conversational system (cf. [9]).

identifying the user's intent, (ii) to structure the content, to identify missing, relevant information (to ask for it in case), to memorize the context and recognize context switches, (iii) to select or dynamically generate an adequate response. We integrate our chatbot component into an existing Customer-Relationship-Management (CRM) system[3], containing 30,384 customers and 31 active users at the time of writing. The CRM system provides structured means for managing all customer relations and allows users to interact with existing and potential clients efficiently over many different channels. The system allows for directed customer care, no matter if dealing with a sales- or service customer or a partner. Sales staff and brand ambassadors use the system to educate themselves about the products the customer has in stock, the relevant contact persons, and therefore which products need to be promoted. Later each contact with a client is documented by this user group and orders are added. Sales- and brand managers review the visits and set strategic goals. The key-account-manager uses the data to acquire new strategic customers.

3.1 The Chatbot Component

A main focus when designing the interface for the conversational system was accessibility. The chatbot should always be available to the user no matter the module. Therefore it was placed as an overlay on top of the current user interface, floating in the bottom right corner as shown in Fig. 2. When the system is not

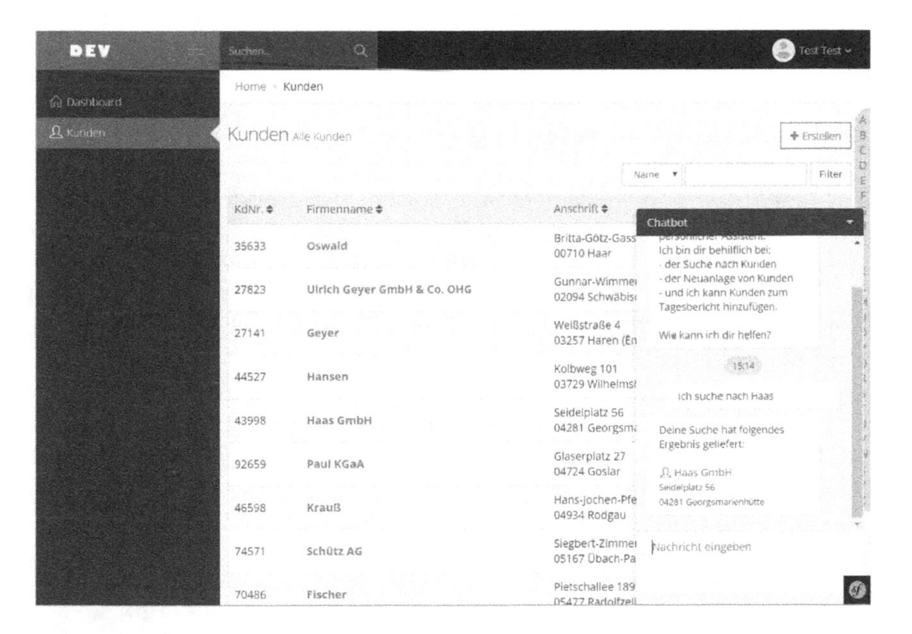

Fig. 2. The CRM's user-interface with the chat-window overlayed in the the right part.

[3] Please note: company details are omitted due to privacy reasons.

needed, by clicking on the header row the chat window can be minimized to save screen space.

The CRM system is implemented based on the Model-View-Controller (MVC) design-pattern. The *Model* is responsible for data storage, the *View* component presents the data and user-interface. The application logic is stored in the *Controller*. Since the whole system has a code-base of more than 2.7 million lines, different modules are grouped into *bundles* for better maintainability. For the chatbot component, we created a new bundle to have all application logic, user-interfaces and software dependencies in a single place.

The chat-window is included after the system is loaded completely to not impair performance of the original system. In the chat-box a history of recent conversations is displayed. When the user sends a message to the chatbot, an asynchronous AJAX request with the input and further context (user-id, location) is sent to the chatbot controller. From there, the received message is first handled by a natural language understanding component, implemented by an input processor pipeline. New input processors can be added dynamically to this pipeline. The extracted intent is then handed to the Intent Processor which manages a set of loaded handlers. The processor tracks the change of intent over the course of the conversation and decides when to switch from one handler to another. Each handler represents a frame and contains relevant slots.

We point out that our chatbot is meant to assist users to operate a German CRM system in a more efficient manner. In the development process we have thus examined existing Natural Language Processing (NLP) components for the German language and have evaluated their availability as well as applicability. The overall architecture is depicted in Fig. 1 and described in detail in the following subsections.

Natural Language Understanding

Intent Classification. At the moment the chatbot performs five tasks: (1) *Searching for an entity in the database*, (2) *Placing orders*, (3) *Logging customers in the daily report of the sales staff*, (4) *Creating new customers* and (5) *Offering help*.

To gain data on how these tasks are normally expressed, we sent out a questionnaire and collected altogether 64 expressions from eight employees between 22 and 27 years of age. To achieve a greater variety in responses, employees were asked to provide alternatives in their expressions. The following list shows an excerpt of the questions.

- You want to search for an entry in the database. How would you ask somebody to perform this search for you?
- You want to add a new record. What would you say to a customer service employee?
- You want to make an order. What would you say to a customer-service employee?
- You want to add a customer or contact to your daily report. How would you formulate this request for a customer-service employee?

Based on the labeled samples gained from the questionnaires, we trained a Support Vector Classifier using a Gaussian radial basis function (RBF) with default parameter settings. We tested the intent classification task via 10-fold cross validation where the RBF model obtained a precision of 0.93 and a recall of 0.90. For classification, we used the machine-learning library PHP-ML[4], since the whole system is based on a PHP framework.

Named Entity Recognition. To structure information and establish a context, we needed to correctly identify and extract named entity categories such as person and location. Two existing libraries qualify for the task of German Named Entity Extraction, i.e. Stanford's CoreNLP (cf. [12]) and GermaNER (cf. [1]). We compared the two libraries in a small evaluation with 23 random entity names (7 persons, 13 companies, 3 locations) which were selected from the database and inserted into sentences of running text. Both libraries were able to consistently identify persons and locations; they had more problems in detecting companies, for example, *Gasthaus zur Linde.* In addition, GermaNER had problems with identifying smaller towns such as *"Serfaus"* or *"Passail"*; we therefore decided to use Stanford's CoreNLP[5].

Dialogue Manager

After extracting information from the input, the next module to process the request is the Dialogue Manager. We decided to model this module as a frame-based agent which was favored over a corpus-based approach, since there are very few conversational data-sets for the German language. The dialogue manager is divided into three sub-modules: (i) the context manager, (ii) the intent selector, and (iii) an array of registered handling classes for the intents.

Context Manager. While interacting with the system, the user provides information to improve the accuracy of the generated responses. For example, this could be information such as the last entities viewed, the coarse location of the user or recent intents. Since this information needs to be kept over multiple sessions, it gets stored in a first-in-first-out fashion and is persisted in a MySQL database. Each entry consists of a user identification, a time-stamp, and a key-value pair. The values are encoded into the data format JSON and therefore any data type, from simple numerical values to complex objects, can be stored.

To ensure timely relevance, a frame is moving over the entries, rendering all entries older than eight minutes invalid. This solution can be improved by allowing each entry to have an individual timeout. For example the location of the user is relevant a longer period of time and should remain available to the system, even if the last location fix is older than five minutes. On the other hand, eight minutes of validity of the current intent can be too long and the system keeps asking about information which is already outdated.

Intent Selector. Let's assume the user wants to search for a customer and sends the following message: *I am searching for the Cafe Central.* This request allows

[4] https://github.com/php-ai/php-ml, last accessed 2019-03-29.

[5] https://stanfordnlp.github.io/CoreNLP/, last accessed 2019-03-29.

the system to identify the search intent. Since most of the times the chatbot cannot fulfill the request after the first interaction, the intent is added to the context of the user. Each further message is analyzed again by the intent selector for a change in the request. If the score of a different intent exceeds a certain threshold (empirically set to 0.4), the system will stop pursuing the last request and switch to the new one. The following example shows a successfully recognized intent switch, before the chatbot has marked the previous one as completed. After searching for an entity, the user wants to add this record to her daily report. Here the message in statement three is classified as intent *Tagesbericht (daily report)* with a confidence score of 0.65. Therefore an intent switch is being performed. The entity from the search before is used in the generation of the daily report, since it was added to the context.

(1) *message* Ich suche nach Maria Huber.
(I am searching for Maria Huber.)

(2) *response* Ergebnisse für "Maria Huber"
(Results for "Maria Huber")
(...)
Wo befindet sich der Kunde?
(Where is the client located?)
(...)

(3) *message* Füge diesen Datensatz zum Tagesbericht hinzu.
(Add this record to the daily report.)
Intent switch: Search → Daily report

(4) *response* Wie wurde der Kunde kontaktiert?
(How was the client contacted?)

The example shows a correct switch from search to daily report.

When the correct intent is identified, the dialogue manager will invoke the corresponding registered intent handler.

Intent Handling. Each handler consists of a set of mandatory and non-mandatory slots which have to be filled in, before the task can be performed. With respect to the supported tasks, the chatbot can handle five intents: *Searching for an entity in the database*, *Placing orders*, *Logging customers in the daily report of the sales staff*, *Creating new customers* and finally *Receiving help*. For each of these intents a different number of slots has to be filled, displayed in Table 1. Except for the search task, regular HTML-forms already existed. Required and optional form fields were transformed into required and optional slots respectively.

When taking again the *Cafe Central* example into consideration, the invoked (search) handler contains the mandatory slot *query* and the optional slots *location* and *type*. For each slot a number of data-transformers can be added. They convert data according to rules into the desired format or can perform extra checks (e.g The start date has always have to be earlier than the ending date). In the case of our search example, as soon as a query is identified (Cafe Central) a list of results, ordered by relevance is returned. Since there are still optional slots unfilled, the system finds the slot to fill to narrow the search down the

Table 1. Required and optional slots for "search", "ordering" and "daily report".

Intent	Required slots	Optional slots
Search	query	type, location
Place order	items	supplier
	quantities	delivery_address_street
	ordertype	delivery_address_zip
	assigned_customer	delivery_address_city
Add to daily report	person	notes
	contacttype	
	presented_products	
	finished_todos	
	result	

most and ask the user to provide information. Since all found establishments are tagged as 'Cafes', asking the user to provide the type of entity searched won't have a big influence to the results. However, the coffeehouses are distributed over Austria and therefore when the user provides a location, like a city name or zip code, the result set can be reduced. This is already partly done in the Natural Language Generation module.

Task Manager & Natural Language Generation
When the necessary information is gathered by the Dialogue Manager, the responsibility is handed over to the Task Manager. Each task which can be invoked by the user usually has an unique entry point. This module forms the link between the chatbot and the application logic. If the task can be performed quickly, the task manager will wait for the result before returning it back to the Dialogue Manager. The task manager simply posts a message to the CRM system and the next available worker will proceed to work off the request.

The candidate response generator is doing all the domain-specific calculations to process the user request. The result of these calculations is a list of response candidates. The response generator uses the context of the conversation as well as intent and entities extracted from the last user message. The response selector scores all response candidates and selects a response which most likely works best for the user.

4 Evaluation

Chatbot evaluation is mostly a human task. While some researchers use the BLEU metric, which was created to evaluate the performance of machine-learning algorithms, Liu et al. [11] have found the metric correlating very poorly with the human perception. To find the *user satisfaction score* a number of methods can be used. The most exhaustive is letting the users fill out a questionnaire after their interaction with the chatbot. We evaluated the chatbot's effects over a two week period. In this time span, 15 company employees provided 465 communication turns. The distribution of usage is shown in the left part of Fig. 3 (age information is given by color). The first five users tested the system most intensively and were all under 40. The majority of the remaining 11 users were 40+. A visual inspection of the distribution suggests that younger users are more open for trying out the chatbot and its features. This observation is in accordance with Chung et al. [4] that age is negatively associated with behavioral intention to participate in online communities.

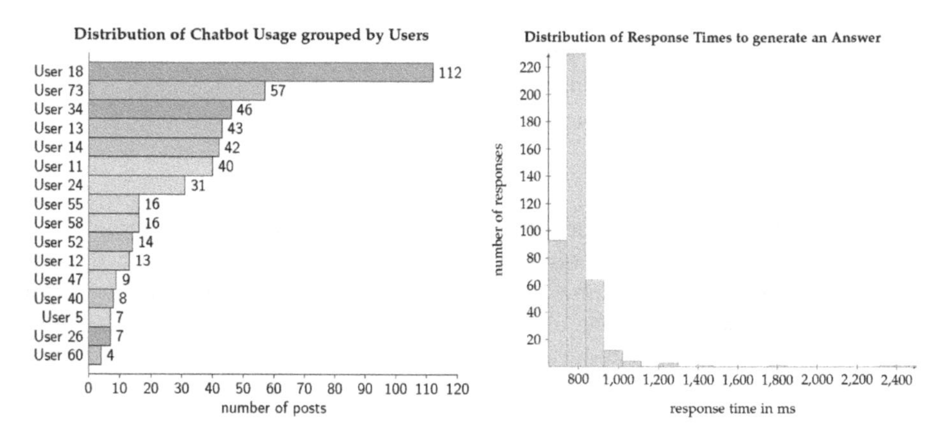

Fig. 3. Left: Distribution of Chatbot usage in the two week evaluation period. Red represents the age group 20–30, blue the age group 30–40 and green marks employees over 40. Right: Distribution of response times for the chatbot to answer. (Color figure online)

Technical Evaluation. For each request, the response time was logged; the corresponding histogram is shown in the right part of Fig. 3. The timer was started as soon as the web-server was invoked and stopped on the instance before the response was returned. The message transmission times are not included. The average response time is 809 ms and the median is 779 ms with a standard deviation of 164 ms and fits a right-skewed normal distribution.

The average length of turns submitted to the system is 13.1 characters long. For each intent the user sent an average of 2.6 turns before the conversation was terminated or an intent switch was performed. A distribution by intents is given in Fig. 4. This distribution is in line with the number of slots which have to be filled to perform each task, i.e. more slots to be filled require longer conversations.

Empirical Evaluation. Informed by Walker et al. [16], we compiled a questionnaire and sent out seven questions to employees with access to the chatbot. Table 2 provides an overview of the empirical evaluation including questions, metric, answering modality as well as average score and standard deviation.

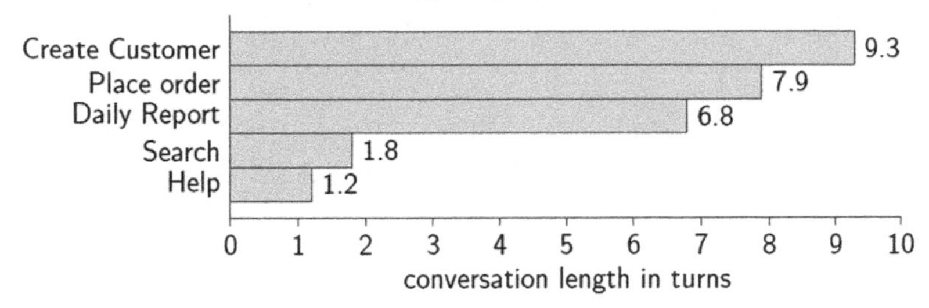

Fig. 4. Conversation length according the required slots to be filled for a task.

Table 2. Overview of the empirical evaluation.

Metric	Question	Modality	Score	Std.D.
Usage	To what extent have you used the system?	1 (*not at all*) to 5 (*in great detail*)	4.00	0.77
Task Ease	Was it difficult or easy to have the chatbot perform a certain task, e.g. searching?	1 (*very difficult*) to 5 (*very easy*)	3.00	1.18
Interaction pace	Was the interaction pace appropriate?	1 (*not appr.*) to 5 (*very appr.*)	4.45	0.68
User Expertise	Was it clear what to do next?	1 (*not clear*) to 5 (*very clear*)	3.73	1.00
System Response	How often did the system respond slowly?	1 (*always*) to 5 (*never*)	4.27	1.00
Expected Behavior	Did the system behave as expected?	1 (*always*) to 5 (*never*)	2.91	0.70
Future Use	Do you plan to use the system in the future?	1 (*most certainly*) to 5 (*cert. not*)	3.09	1.04

The metric *Task Ease* has an average of 3.00 but also the highest statistical deviation of 1.18. So, some users appeared to be satisfied with how tasks could be invoked, others to be unhappy. One user added the remark "The chatbot does not understand anything!!" and another one added "The bar (Name removed) could not be found.". After looking up the respective conversations, we found that for some requests intent classification had failed.

Interaction Pace yielded the highest score with 4.45, although Named Entity Recognition (NER) did not work as reliable as anticipated. Sales staff tend to shorten city names to the corresponding zip codes. For example, entities located in the 10th Viennese city district "Favoriten" are assigned the zip code 1100. This is a common way of quickly targeting entities by their name and rough location. However, the NER-System does not recognize Austrian ZIP-codes. After failing to identify the ZIP-code, the system in many cases falls back to asking directly for the location.

Judging from the score of *System Response*, users were satisfied with the speed of the answer generation. As described in the section above, the average response time was 809 ms. There was one remark added: "The system is getting more and more slowly near the end of the month!". This appears to a subjective experience, since we could not verify this claim by checking the response time logs.

5 Conclusions

In this work we have implemented and integrated a chatbot component into an existing Customer-Relationship-Management (CRM) system of a mid-sized company in Austria. The advantages of (i) a seamless integration, (ii) a flexible extension and adaption to novel requirements and (iii) the focus on the German language led to the decision of developing a customized solution. The provision of a chatbot component in combination with a CRM system for the German speaking market represents itself an added value.

From the evaluation we can learn that the employees were in sum satisfied, yet there is of room for improvement. Employees, for instance, could often not explain the chatbot's behavior and this often could be traced back to a lack of understanding. Gathering more conversational data (as is done during chatbot usage) will contribute to alleviate this issue. It will allow us to improve the context management and one day even to use corpus-based approaches. More data will also enable Natural Language Processing tools such as named entity recognition to retrain the models. Improvement ideas also include having a speech-based chatbot for sales personnel to be briefed during their car rides with all the important information.

Acknowledgements. The Know-Center GmbH Graz is funded within the Austrian COMET Program - Competence Centers for Excellent Technologies - under the auspices of the Austrian Federal Ministry of Transport, Innovation and Technology, the Austrian Federal Ministry of Economy, Family and Youth and by the State of Styria. COMET is managed by the Austrian Research Promotion Agency FFG.

References

1. Benikova, D., Yimam, S.M., Santhanam, P., Biemann, C.: GermaNER: free open German named entity recognition tool. In: German Society for Computational Linguistics and Language Technology (2015)
2. Bergmann, K.: Angewandtes Kundenbindungs-Management. Frankfurt am Main [u.a.]: Lang. (1998)
3. Choi, H., Hamanaka, T., Matsui, K.: Design and implementation of interactive product manual system using chatbot and sensed data. In: 6th Global Conference on Consumer Electronics (GCCE) (2017)

4. Chung, J.E., Park, N., Wang, H., Fulk, J., McLaughlin, M.: Age differences in perceptions of online community participation among non-users: an extension of the technology acceptance model. Comput. Hum. Behav. **26**(6), 1674–1684 (2010)
5. Colby, K.M., Hilf, F.D., Weber, S., Kraemer, H.C.: Turing-like indistinguishability tests for the validation of a computer simulation of paranoid processes. Artif. Intell. **3**, 199–221 (1972)
6. Correia, J.M., Dharmasthira, Y., Poulter, J.: Market share analysis: customer relationship management software. Technical report, Gartner Inc. (2016)
7. Cui, L., Huang, S., Wei, F., Tan, C., Duan, C., Zhou, M.: SuperAgent: a customer service chatbot for e-commerce websites. In: Proceedings of ACL 2017, System Demonstrations (2017)
8. Greenfield, R.: Chatbots are your newest, dumbest co-workers. Bloomberg (2016)
9. Jurafsky, D., Martin, J.H.: Speech and Language Processing: An Introduction to Natural Language Processing, Computational Linguistics, and Speech Recognition, 2nd edn. Pearson Education, London (2009)
10. Kowalke, P.: How chatbots will change Customer Relationship Management (2017)
11. Liu, C., Lowe, R., Serban, I.V., Noseworthy, M., Charlin, L., Pineau, J.: How NOT to evaluate your dialogue system: an empirical study of unsupervised evaluation metrics for dialogue response generation. CoRR (2016)
12. Manning, C.D., Surdeanu, M., Bauer, J., Finkel, J., Bethard, S.J., McClosky, D.: The Stanford CoreNLP natural language processing toolkit. In: Association for Computational Linguistics (ACL) System Demonstrations (2014)
13. Marcus, G.: What comes after the turing test? The New Yorker (2014)
14. Thomas, N.T.: An e-business chatbot using AIML and LSA. In: Proceedings of the International Conference on Advances in Computing, Communications and Informatics (2016)
15. Voth, D.: Practical agents help out [virtual agent system]. IEEE Intell. Syst. **20**(2), 4–6 (2005)
16. Walker, M.A., Passonneau, R., Boland, J.E.: Quantitative and qualitative evaluation of DARPA communicator spoken dialogue systems. In: Association for Computational Linguistics, ACL (2001)
17. Weizenbaum, J.: ELIZA - a computer program for the study of natural language communication between man and machine. Commun. ACM **9**(1), 36–45 (1966)
18. Xu, A., Liu, Z., Guo, Y., Sinha, V., Akkiraju, R.: A new chatbot for customer service on social media. In: Proceedings of the Conference on Human Factors in Computing Systems, CHI (2017)
19. Zamora, J.: I'm sorry, Dave, I'm afraid I can't do that: chatbot perception and expectations. In: Proceedings of the 5th International Conference on Human Agent Interaction, HAI (2017)

An Intelligent Platform with Automatic Assessment and Engagement Features for Active Online Discussions

Michelle L. F. Cheong(✉), Jean Y.-C. Chen, and Bing Tian Dai

School of Information Systems,
Singapore Management University, Singapore, Singapore
{michcheong, jeanchen, btdai}@smu.edu.sg

Abstract. In a university context, discussion forums are mostly available in Learning and Management Systems (LMS) but are often ineffective in encouraging participation due to poorly designed user interface and the lack of motivating factors to participate. Our integrated platform with the Telegram mobile app and a web-based forum, is capable of automatic thoughtfulness assessment of questions and answers posted, using text mining and Natural Language Processing (NLP) methodologies. We trained and applied the Random Forest algorithm to provide instant thoughtfulness score prediction for the new posts contributed by the students, and prompted the students to improve on their posts, thereby invoking deeper thinking resulting in better quality contributions. In addition, the platform is designed with six features to ensure that students remain actively engaged on the platform. We report the performance of our platform based on our implementations for a university course in two runs, and compare with existing systems to show that by using our platform, students' participation and engagement are highly improved, and the quality of posts will increase. Most importantly, our students' performance in the course was shown to be positively correlated with their participation in the system.

Keywords: Discussion forum · Natural Language Processing · Student engagement

1 Online Discussion Forums

Asynchronous online discussion (AOD) forums have been used in many learning contexts to enable students to learn from their peers as a community, and it was mentioned in [1] that "perhaps the most important form of active learning is discussion." Despite the popularity of using threaded forums in Learning Management Systems (LMS), it was argued that they "might not be the best technology to support the interactive and collaborative processes essential to a conversational model of learning" [2]. Some researchers noticed that in threaded forums, students tend to post condensed versions of the explanations of their own ideas rather than responding to the ideas of others [3, 4]. Also, the discussions are often not deep and remain at the surface level such as sharing and comparing information [5]. In [6], they summarized other constraints imposed by threaded discussion forums which include students' tendency to

F. Wotawa et al. (Eds.): IEA/AIE 2019, LNAI 11606, pp. 730–743, 2019.
https://doi.org/10.1007/978-3-030-22999-3_62

attend to unread posts and most recent posts rather than posts with important content, difficulty in promoting interactive dialogues, provides little support for convergent processes, and the lack of timely feedback. Thus, threaded forum discussions found in LMS do not foster productive online discussions naturally, and developing alternative discussion environments is needed to offer better support in asynchronous online discussions. Several suggestions that future discussion environments should improve upon was provided by [6]. First, it is to foster an online community, provide timely feedback, encourage information sharing and support collaborative problem solving. To achieve these objectives, some incentive mechanism may be designed into the environment. Second, the lack of convergence requires a system which can go beyond knowledge sharing to include active processing and synthesizing of the information provided by the community. Third, the use of multi-functional environments or systems that can integrate new media technology to facilitate learning at different phases, such as asynchronous and synchronous discussions. Fourth, to design appropriate instructional activities or strategies that can improve the quality of the discussion.

In [7], the focus is the Starburst system, which uses visualization techniques to present discussion posts as dynamic hyperbolic tree. Their results found that students were more purposeful in selecting which discussion threads to read and they did it in a more connected fashion. In [8], the authors developed and tested a mobile interactive system to compare social knowledge construction behavior of problem-based asynchronous discussion in e-learning and m-learning environments and found that using additional environments led to more options for students, and that using mobile devices positively influenced students' learning performance. In a study by [9], they compared how students provided online interactive feedback using two different systems, Blackboard being the normal threaded forum and the Annotation System developed by [10]. They found that students showed fewer evaluative feedback but more feedback with suggestions for revisions when using Annotation System than Blackboard. Thus, all three pieces of work [8–10] show that software design and interface can indeed influence learner engagement positively.

In this paper, we will discuss about our intelligent Q&A platform, called CAT-IT, which is an integrated platform with the Telegram mobile app and a web-based forum which are synced in real time. The platform is capable of automatic thoughtfulness assessment of the questions and answers posted, using text mining and Natural Language Processing (NLP) methodologies, and provide instant thoughtfulness score prediction to encourage the students to improve their posts, thereby invoking deeper thinking resulting in better quality contributions. In [11], the author proposed four dimensions in assessing text quality including contextual, intrinsic, representation and accessibility, of which, we focused only on contextual in terms of relevancy, and representation in terms of ease of understanding and interpretability. Thus, we will assess posts using a thoughtfulness score, to measure if a statement contains insightful reasoning and relevance to the issues discussed [12]. We present the effectiveness of our newly developed Q&A platform in engaging students to conduct active online discussion, and how the thoughtfulness score prediction encouraged higher quality posts, leading to improved students' performance in the course.

2 System Design and Architecture

2.1 System Design and Features

CAT-IT is a student-centric platform where the interactions in the form of asking questions and providing answers, are all student-driven. We have chosen this design as we believe that while the traditional tutor-student delivery is important, student-student interaction is also important for academic engagement as all participants are equals and there will be no power relationship issues which may hinder active discussions [13]. We have designed six special features to improve and maintain student engagement and discuss the details of each feature below.

Avatar Identity for Anonymity. This is to overcome one main challenge in Q&A platforms, where poorly formed questions and inaccurate answers provided will be ridiculed [14], thus students tend to be wary about how their posts may be viewed by their peers. Our students participate in the platform using their individual avatar identity to provide them a "safe" environment to participate without the fear of being ridiculed. By understanding the "real deal" when students ask questions and provide answers according to their own understanding and perceptions, which could be repeated questions or inaccurate answers, it will allow course instructors to know exactly what the learning challenges are in a more timely fashion and address them during in-class session immediately. In addition, as our data collection involved human subjects, using the avatar identities will allow us to collect data according to IRB requirements.

Gamification and Leaderboard. Gamification is the process of making something more game-like, but it does not involve creating a complete game. It works with something pre-existing that is not a game, and should engage people with something for a purpose that is not solely to entertain or engage them [15]. Several recent works, both online [16] and classroom [17] have utilized gamification as an attractive strategy to enhance student engagement in learning the course content. Points are usually earned and accumulated when a certain desired learning behavior has been attained by the participant, and a leaderboard is used to instill a sense of competition among the participants by displaying the ranks of the participants' performance. In our system, we display the real-time ranking of the avatars on a leaderboard based on their cumulative thoughtfulness score earned, to incite students' desire to stay at the top of the leaderboard.

QA Coins and Bounty. Earning in-game coins is part of gamification. The amount of QA coins earned for a new post is based on a multiplication factor of the thoughtfulness score earned for that particular post. The multiplication factor f(x) follows $\boldsymbol{f}(\boldsymbol{x}) = \boldsymbol{e}^{-\boldsymbol{x}}$, where x is the cumulative thoughtfulness score of that student. The purposeful design of using such a function is to reward students who have low cumulative thoughtfulness score (x) to earn more QA coins, as a form of encouragement to participate. In many gamified platforms, in-game coins are usually used for making in-game purchases such as buying capes and swords to don their avatar characters. However, in our system, each student only has an avatar identity but does not have an avatar character. Thus, the QA coins earned can only be used to "buy" quick response for the best answer provided within a time limit. This further gamifies the process of participation for both the asker and the answerer.

Auto-routing. One of the key success factors of QA platforms is timely feedback. Posted questions which are not answered will lead to loss in interest and engagement, which may result in the eventual failure of the system [18, 19]. Our system will automatically route an unanswered question to the top five and bottom five cumulative thoughtfulness score students, and also to students who have zero participation, once the time limit is reached. Time limit is default to 24 h if the question does not have a bounty and its associated time limit. By routing to the top five students, we hope that students who have more expertise will provide assistance, while by routing to the bottom five and zero participation students, we hope that the less active students can be nudged to participate.

Need Improvement and Up-Vote Buttons. This is a new feature which was added only in our second run, in response to students' feedback that some students asked questions for the sake of asking to earn thoughtfulness score which will be part of their final grade for the course, as reported in our earlier work [20]. Thus the "Need Improvement" button will allow students to suggest that a question or answer needs further improvement. This is similar to a down-vote in other platforms. However, from the education point of view, we believe that there are no stupid questions or stupid answers, and we do not encourage irresponsible down-voting. Thus, for every click of the "Need Improvement" button, one QA coin will be deducted from the student who clicked the button, so that students can be more mindful when doing so. Conversely, we hope to encourage students to acknowledge good contributions from their peers, and use the "Up-Vote" button as a form of positive recognition. By using these two buttons, we hope to capture some emotional cues in student-student interactions.

Periodic Questions Posted by Automatic Chat Bot. This is also a new feature added in the second run. In our first run, we reported that there were about five answers for every question asked [20]. This showed that students found it easier to provide answers than to ask questions. Thus, we allowed our chat bot to automatically ask questions periodically from a question bank, to drive participation from students in answering questions.

2.2 System Architecture

Figure 1 shows the system architecture of our integrated Telegram and web-based forum. The back-end includes databases and machine learning algorithm, automated by Python scripts, Telegram API and Google API. The system has three main databases, where two of them are Excel files, namely the Q&A corpus for the training of the machine learning algorithm for the prediction of thoughtfulness score, and the question bank where the chat bot will draw its periodic question from. The main database is based on MySQL Workbench with tables to store the user database and the user's associated posts, thoughtfulness scores and QA coins earned. There are two front-ends available, the web-based forum (Fig. 2) and the Telegram mobile app (Fig. 3). Students and instructors must register separately on both applications, using their university email address and Telegram account respectively. The platform will link both registrations to be the same user. Using the Telegram app, students can post questions, provide answers, and view posts contributed by other students and himself/herself. The web-based forum has full functionality that lists all conversations by threads, and in addition to what the Telegram app can provide, students can search for posts and view user information and leaderboard results. The real-time synchronization between the two interfaces provides students the option of participating using their computers or their mobile phones on-the-go, improving engagement as noted in [8].

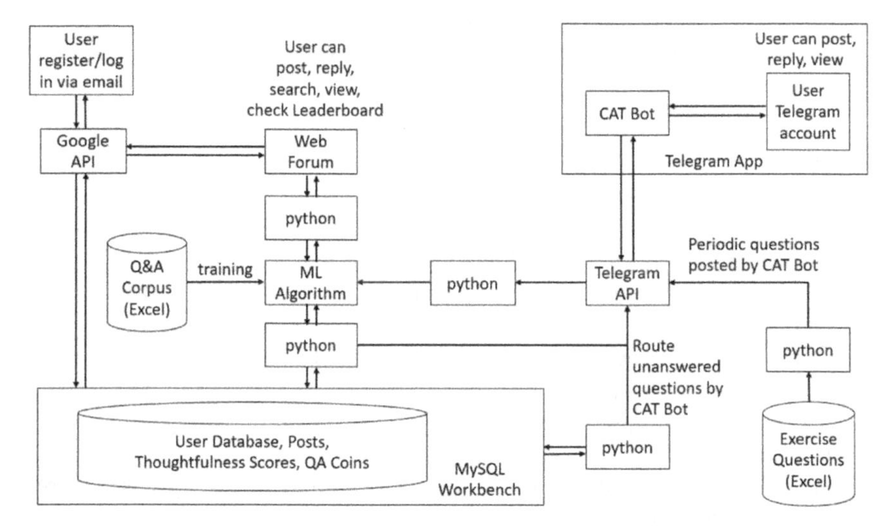

Fig. 1. Schematic diagram of system architecture

Fig. 2. Web-based forum to view, post and search for posts, and to view user info and ranking

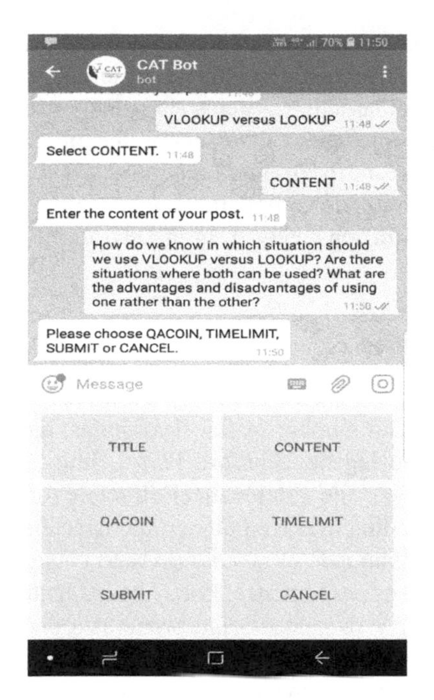

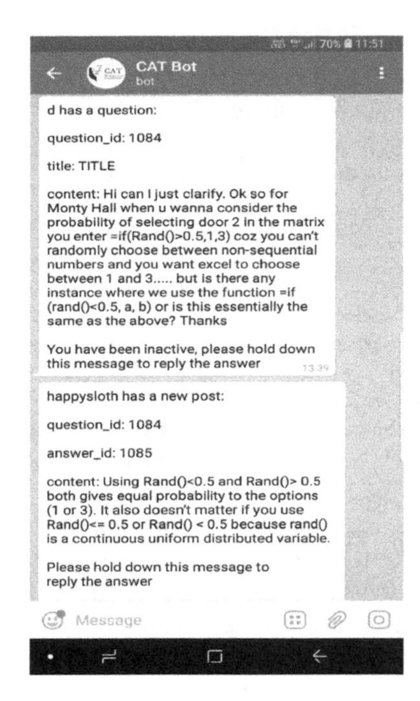

Fig. 3. Telegram app interface for posting question and answering question

3 Data Analytics and Applied Intelligence

3.1 Data Preparation

In order to train a machine learning algorithm to perform the prediction, the team required an initial data set which contains questions and answers representing similar content. An initial data set was crawled from a public forum on Excel (www.excelforum.com) over a four-month period, from June to September 2017. After careful selection based on relevance to the course content, a total of 2377 posts, containing 340 questions, were used to form the training data set. The two course instructors then manually labelled the posts independently on a scale of 0 to 5, based on agreed labeling rules where 0 means that the post is not thoughtful at all; 1 means that the post is a short sentence without details; 2 means that the post contains two to three sentences with some explanations using one or two Excel formulas; 3 means good explanation with example and formula; 4 means good explanation with example, formula and comparison; and 5 means clear explanation, examples, references, suggested formulas and interpretations. We tested inter-coder reliability using consistency index [21] and Cohen's Kappa statistics [22] and achieved 0.906 and 0885 respectively, representing high reliability and deemed that the labeling were consistent and reliable, and thus the labelled data can be used for training the machine learning algorithms.

To implement our system for other courses, the Q&A corpus and chat bot question bank related to the course will need to be prepared. A best performing machine learning model will need to be trained to perform the thoughtfulness prediction. Such work will take about three months at most and the system will be ready for deployment. In the case of a programming course where posts are computer programs, an online judge system will be more suitable. However, such online judge systems are usually used in competitive programming or used by course instructors to assess if the programs submitted by students are correct, and are not meant for discussion purposes. Thus, an integration with open source online judge API for code compilation and execution can be done in future to assess the computer program separately from the text discussion posts.

3.2 Data Pre-processing and Feature Selection

We pre-processed the text using tokenization; stop words removal; word stemming; trim and count the number of URL references and number of Excel formulas; part of speech tagging; and natural language parsing using the Stanford Parser (https://nlp.stanford.edu/software/lex-parser.shtml) to analyze the grammatical structure of the sentences. We generated structural features including average number of characters per word, average number of words per sentence, number of words, average parse tree height, and average number of subordinate clauses per sentence. For syntactic features, we generated average number of noun phrases, verb phrases, and pronounce phrases per sentence. In [23], it was found that discourse relations were correlated with the text quality. Thus, we included "expansion" relation where the second argument expands the first argument or moves its narrative forward; "comparison" relation which highlights the difference between two arguments; and "contingency" relation where an argument causally influences another argument. As discussed in [24], questions can be classified into factoid, list, definition, complex and target. We have adopted the similar concept and proposed an ordinal scale to take care of increasing complexity and discourse features in the different question types according to "where" = 1, "what" = 2, "how" = 3, "when" = 4, and "why' = 5. In addition, we also included whether the post is a question or an answer, and the number of reference links and Excel formulas.

3.3 Machine Learning Models and Selection

With the features, we trained four families of machine learning models to predict the thoughtfulness score. We tested Linear Regression (LR), Neural Network Regression (NN), Support Vector Machine (SVM) and Random Forest (RF). For NN, we used multi-layer perceptron with hyperbolic tangent (tanh) activation function. There are two hidden layers, where the first layer has 1000 neurons and the second layer has 100 neurons. We used Adam, a stochastic gradient descent optimization, with learning rate set at 0.001. For SVM, we have tested several kernels including linear, polynomial, sigmoid and radial basis function (RBF) and found that RBF performed the best. In RF, 10 trees with a maximum of six-feature split used, has the best performance. We used stratified 10-fold cross-validation to split the dataset and the results are given in Table 1. RF outperformed the other three models with the lowest mean squared error (MSE) and the highest R Square value, so we used the RF model to perform automatic thoughtfulness score prediction for the future posts contributed by the students.

Table 1. Machine learning models comparison

	LR	NN	SVM	RF
Mean squared error	0.590	0.276	0.319	0.080
R square	0.188	0.622	0.563	0.892

4 System Process Flow, Implementation and Results

4.1 System Process Flow

The process flow is depicted in Fig. 4. Student starts by asking the first question (Q). A question contains a title, content, optional QA coins as bounty and its associated time limit. Once a question is posted, students can provide answer (A) to the question or to an earlier answer with no depth limit. For every question or answer post, the machine learning (ML) model will predict the thoughtfulness score and prompt the student to improve the post. If the student chooses to improve the post, both posts and their respective thoughtfulness scores will be recorded. In our first pilot run, students were not shown the thoughtfulness scores, so the final post, which could be the first or second post (if student chose to improve the first post) will be posted. While in the second run, the thoughtfulness scores were shown and student can choose either the first or second post to be the final post. For questions with bounty and time limit, the best answer will earn the QA coins posted within the time limit. Once time limit is reached, any unanswered questions will be routed based on our auto-routing rule. For questions with no bounty, the time limit will be automatically set to 24 h.

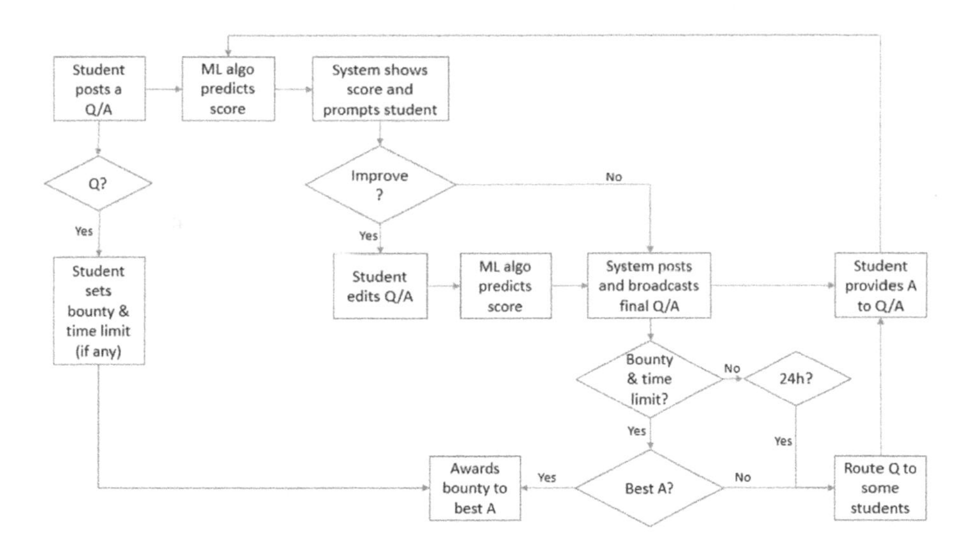

Fig. 4. Process flow

4.2 Implementation

The platform was implemented for a university course on spreadsheets modeling over two runs, with different number of sections, different number of students, different allocation of marks (0%, 5% or 10%) as a percentage of the overall assessment, and with the two additional features (Need-Improvement and Up-Vote buttons, and Automatic Chat bot) added in the second run. For Run 1, participation was mandatory for all sections (5% or 10%). For Run 2, only section G13 was allocation 0% which means voluntary participation (Table 2).

Table 2. Two implementation runs

	# of sections	# students	Settings for different sections
Run 1	3	128	G1 (10%), G15 & G16 (5%)
Run 2	4	147	G1 (10%), G10 & G12 (5%), G13 (0%)

4.3 Comparing Participation Performance with Other Platforms

Table 3 below compares the participation rate and average number of post per active student of our platform with three other similar discussion platforms which were implemented in the past, and their performance results were reported in the literature.

BlikBooks [25] was founded in 2010 and used in more than one-third of UK HE institutions. Its impact on student-tutor and student-student interactions in an International Strategy Development module with 440 students were evaluated [26]. In terms of student-student interaction, only 71 comments were made by 53 students, giving it a low participation rate of 12.1% and 1.34 posts per active student. CaMILE is an anchored forum which was first developed in 1994 and was implemented for students in 17 classes for courses in computer science; chemical engineering; English; history; and literature, culture and communication [27]. Being an anchored forum, discussion topics were provided by the course instructors and thus a higher participation rate would be expected. However, the platform only achieved an average participation rate of 60% and each student only contributed an average of 5.2 posts. On the other hand, SpeakEasy [28] which was also implemented as an anchored forum with discussion

Table 3. Participation performance comparison

Platforms	# of participants	# of active participants	Participation rate	# of meaningful posts	Average # of post per active participant
1. CAT-IT – Run 1	128	101	78.9%	1025	10.15
2. CAT-IT – Run 2	147	123	83.7%	1128*	9.17
3. BlikBooks	440	53	12.1%	71	1.34
4. CaMILE	-	-	60.0%	-	5.2
5. SpeakEasy	180	173	96.1%	-	5.3

* excludes questions posted by chat bot

topics related to Science provided, involved 180 eighth grader, was able to achieve a high participation rate of 96.1% and each student contributed an average of 5.3 posts. Such a high participation rate was achieved due to two main reasons. Firstly, eighth graders would be expected to have more time available for discussions as compared to university students, and secondly, SpeakEasy was implemented as an anchored forum which was part of an assignment.

Comparing to these platforms, our CAT-IT has achieved higher participation rate (at least 78.9%) and higher average number of post per active student in both runs (almost double that of other systems), except when compared to SpeakEasy in terms of participation rate. It is important to note that CAT-IT is not an anchored forum, but a free-form Q&A platform, where all questions and answers are student-driven. In the case of university students who have countless number of time-competing tasks in hand, our platform was able to achieve such high participation rate and high average number of posts shows that it is an effective platform to encourage active student-driven discussion. Both performances will be discussed in the following sections.

4.4 Participation Improvement Due to Specific Features

For both runs, our platform has shown to achieve high participation rate. Of the six special features, we were able to track the improvements in participation due to three of them, specifically QA coins as bounty, auto-routing and automatic chat bot. In applying the QA coins as bounty, the intention is to improve the response time, as unanswered questions will lead to loss in interest and engagement as reported in [18, 19]. Results from Run 1 showed that the average response time for questions with QA coins was 55.6 s, while the average response time for questions without QA coins was 122.5 s. We tested the null hypothesis (H_0) that the average time to response to questions with QA coins is not significantly different than those without QA coins, and obtained a p-value of 0.0283 (<0.05, reject H_0), and t-statistic of −1.996. Thus, we can conclude that by using QA coins as bounty, questions will get faster response.

Our auto-routing feature is also used to reduce unanswered questions. When the time limit is reached, the system will automatically route the unanswered questions to the top five and bottom five cumulative thoughtfulness score students, and also to students who have zero participation. In Run 1, 13 unanswered questions beyond the time limit were auto-routed and 7 were answered, while in Run 2,110 unanswered questions were auto-routed and 20 were answered. This shows that with automatic routing feature, some unanswered questions will receive responses which would otherwise not be forthcoming. Finally, in Run 2, our CAT bot was able to ask questions from a question bank over a 6-week period from Week 2 to Week 7. A total of 140 questions (35 questions each for four sections) was asked by the CAT bot, and 127 of them received answers from the students. The average number of answers received per question for a CAT bot question is 1.96, which is slightly higher than 1.88 for a student question. While the difference is insignificant, it shows that students continue to participate and contribute answers regardless of whether it is a CAT bot question or student question, keeping the discussion active.

4.5 Higher Quality Posts

Once a post is created, our system will prompt the students to improve their posts. Table 4 below shows that in Run 1, where students were not shown the thoughtfulness scores for both attempts, 76.7% of the second attempts were indeed improved. For Run 2, students were shown the thoughtfulness scores for both attempts, and students have the choice to choose which attempt to be the final post, 69.8% of the second attempts were improved. We were surprised to see that a lower percentage of second attempts were improved in Run 2 as compared to Run 1. We offer a couple of plausible explanations why some students still chose the lower thoughtfulness score post as their final post. One is that students did not fully understand how to use the system to choose the higher thoughtfulness score post, and two is that students may have made a mistake in their selections. Nevertheless, our results show that when students chose to improve their posts when prompted, a high percentage of them managed to do so, thus leading to higher quality posts.

Table 4. Improvement in thoughtfulness scores on second attempts

	Number of posts with second attempt	% improved	% did not improve
Run 1	56	76.7%	23.3%
Run 2	53	69.8%	30.2%

4.6 Improved Student Performance in Assessments

We have analyzed and reported in our earlier paper [20] on the positive correlation between students' performance in the course with the thoughtfulness scores they earned, based on the students in Run 1. For Run 2, the same analysis was done and the results are shown in Table 5. It can be seen that for all four sections, the correlations between the average thoughtfulness score and final course performance all displayed positive correlations. The p-values for all are significantly lower than 0.05, except for Section G12 where the p-value is 0.0519, slightly more than 0.05. It is interesting to note that the average thoughtfulness score was higher for sections with allocation of 5% (G10 and G12), as compared to section G1 with allocation of 10%. This is again consistent with what we have reported earlier in our paper [20] which suggested that with a higher stake (10% versus 5%), the quality of posts may not be higher. In totality, we can conclude that students with higher thoughtfulness scores tend to perform better in assessments with appropriate extrinsic motivation, and 5% is better than 10% when higher quality posts are desired.

Table 5. Correlation between average thoughtfulness score and assessment performance

Section	Average thoughtfulness score	Final course performance	
		Correlation	p-value
G1 (10%)	2.03	0.5100	0.0008
G10 (5%)	2.42	0.5029	0.0013
G12 (5%)	2.26	0.3362	0.0519
G13 (0%)	2.16	0.3667	0.0200

5 Conclusions and Future Enhancements

Our intelligent platform with automatic assessment of post contributions and six engagement features has shown to be effective in encouraging active student-driven online discussion, and resulted in better student performance in the course. As compared to other platforms, our CAT-IT can achieve higher participation rate, higher average number of post, and higher quality posts. Most importantly, there is positive correlation between students' performance in the course and their participation in the system. We have successfully applied NLP to automatically assess posts' quality, and show that the thoughtfulness scores of posts increased when students were given the chance to improve their posts, encouraging mindful and purposeful attempts to ask better questions and provide better answers. Our system can be extended to other courses by replacing the Q&A corpus to train the ML model for prediction of thoughtfulness, and the Chat bot question bank. By integrating with open source online judge API for code compilation and execution in future, it will be possible to assess the correctness of computer programs which are contained in the posts, for programming related courses. In our future work, we will improve the NLP analysis to provide guidance to students on how and in what areas to make improvements to their posts, resulting in even higher quality posts.

References

1. Kember, D., McNaught, C.: Enhancing University Teaching: Lessons from Research into Award Winning Teachers. Routledge, Abingdon (2007)
2. Thomas, M.J.W.: Learning with incoherent structures: the space of online discussion forums. J. Comput. Assist. Learn. **18**, 351–366 (2002)
3. Hara, N.M., Bonk, C.J.M., Angeli, C.M.: Content analysis of online discussion in an applied educational psychology course. Instr. Sci. **28**, 115–152 (2000)
4. Larson, B.E., Keiper, T.A.: Classroom discussion and threaded electronic discussion: learning in two arenas. Contemp. Issues Technol. Teach. Educ. **2**(1), 45–62 (2002)
5. Gunawardena, C.N., Lowe, C.A., Anderson, T.: Analysis of a global on-line debate and the development of an interaction analysis model for examining social construction of knowledge in computer conferencing. J. Educ. Comput. Res. **17**(4), 397–431 (1997)

6. Gao, F., Zhang, T., Franklin, T.: Designing asynchronous online discussion environments: recent progress and possible future directions. Br. J. Educ. Technol. **44**(3), 469–483 (2013)
7. Marbouti, F., Wise, A.F.: Starburst: a new graphical interface to support purposeful attention to others' posts in online discussions. Educ. Technol. Res. Dev. **64**, 87–113 (2016)
8. Lan, F.L., Tsai, P.W., Yang, S.H., Hung, C.L.: Comparing the social knowledge construction behavioural patterns of problem-based online asynchronous discussion in e/m-learning environments. Comput. Educ. **59**(4), 1122–1135 (2012)
9. van der Pol, J., van den Berg, B.A.M., Admiraal, W.F., Simons, P.R.J.: Exploration of an e-learning model to foster critical thinking on basic science concepts during work placements. Comput. Educ. **53**, 1–13 (2008)
10. vander Pol, J., Admiraal, W.F., Simons, P.R.J.: The affordance of anchored discussion for the collaborative processing of academic texts. Int. J. Comput. Support. Collab. Learn. **1**(3), 339–357 (2006)
11. Sonntag, D.: Assessing the quality of natural language text data. https://pdfs.semanticscholar.org/8715/1450d6fd7b3f77e83e0b5e66b9fb0b2d40e6.pdf. Accessed 27 Mar 2019
12. Gottipati, S., Jiang, J.: Finding thoughtful comments from social media. In: Proceedings of COLING 2012: Technical Papers, pp. 995–1010 (2012)
13. Kassens-Noor, R.: Twitter as a teaching practice to enhance active and informal learning in higher education: the case of sustainable tweets. Act. Learn. High Educ. **13**, 9–21 (2012)
14. Raphael, R.J.: The 20 dumbest questions on Yahoo Answers. PC World (2009). http://www.pcworld.com/article/184999/strange_questions_yahoo_answers.html. Accessed 27 Mar 2019
15. Andrews, D.: Gamification Systems Development: A practical Guide. The Advanced Services Group, Aston Business School, Birmingham, UK (2017)
16. Barata, G., Gama, S., Jorge, J., Gonçalves, D.: Improving participation and learning with gamification. In: Proceedings of the First International Conference on Gameful Design, Research, and Applications, pp. 10–17. ACM, New York (2013)
17. de Freitas, A.A., de Freitas, M.M.: Classroom Live: a software-assisted gamification tool. Comput. Sci. Educ. **23**(2), 186–206 (2013)
18. Chua, A.Y.K., Banerjee, S.: So fast so good: an analysis of answer quality and answer speed in community question-answering sites. J. Am. Soc. Inf. Sci. Technol. **64**(10), 2058–2068 (2013)
19. Saha, R.K., Saha, A.K., Perry, D.E.: Toward understanding the causes of unanswered questions in software information sites: a case study of stack overflow. In: ESEC/FSE (2013)
20. Cheong, M.L.F., Chen, J.Y.-C., Dai, B.T.: Integrated telegram and web-based forum with automatic assessment of questions and answers for collaborative learning. In: Proceedings of IEEE International Conference on Teaching, Assessment and Learning for Engineering, TALE 2018, Wollongong, Australia (2018)
21. Bennett, E.M., Alpert, R., Goldstein, A.C.: Communications through limited-response questions. Public Opin. Q. **18**(3), 303–308 (1954)
22. Hallgreen, K.A.: Computing inter-rater reliability for observational data: an overview and tutorial. Tutor. Quant. Methods Psychol. **8**(1), 23–34 (2012)
23. Shah, C., Pomerantz, J.: Evaluating and predicting answer quality in community QA. In: SIGIR (2010)
24. Chua, A.Y.K., Banerjee, S.: Answers or no answers: studying question answerability in stack overflow. J. Inf. Sci. **41**(5), 720–731 (2015)
25. Outsell: Blikbooks: Addressing the UK market for higher Ed E-textbooks. https://web.archive.org/web/20131213053912/, http://outsellinc.net/Headline.aspx?ID=455070. Accessed 27 Mar 2019

26. Crowder, M., Antoniadou, M., Stewart, J.: To BlikBook or not to BlikBook: exploring student engagement of an online discussion platform. Innov. Educ. Teach. Int. **56**, 295–306 (2018)
27. Guzdial, M., Turns, J.: Effective discussion through a computer-mediated anchored forum. J. Learn. Sci. **9**(4), 437–469 (2000)
28. Hoadley, C.M., Linn, M.C.: Teaching science through online, peer discussions: SpeakEasy in the knowledge integration environment. Int. J. Sci. Educ. **22**(8), 839–857 (2000)

Automatic Generation of Dictionaries: The Journalistic Lexicon Case

Matteo Cristani, Claudio Tomazzoli, and Margherita Zorzi(✉)

Department of Computer Science, University of Verona, Verona, Italy
margherita.zorzi@univr.it

Abstract. Text normalisation is an important task in the context of Natural Language Processing. By normalisation, free text is mapped into *dictionaries*, i.e. indexed collections of locutions recognised as typical of a particular jaergon. In general, technical dictionaries are difficult to build and validate. They are typically constructed by hand on the basis of everyday human work and they are agreement-based. This is indubitably time consuming and the approach requires a strong human supervision and does not provide a general methodology. In this paper, we perform the first steps towards the to automatic building of a dictionary for Italian journalistic lexicon, called NewsDict, based on sub dictionaries able to characterise main topics occurring in newspaper articles. We exploit a dataset of annotated documents from some Italian newspapers and a statistical techniques based on the Mutual Information Principle. Documents contains information such as the release date and the topic of the article and has been directly annotated by the author. To check the accuracy of the dictionary we built, we develop an initial test. We normalise a control set of journal article into NewsDict. Crossing results presented in this paper against the human annotation, we provide a fist measure of performances of the described methodology.

Keywords: Text normalization · Statistical natural language processing · Automatic generation of dictionaries

1 Introduction

In Statististic natural language processing *normalization* is the activity of transforming a corpus of texts into a list of words with annotations, in turn named an *annotated dictionary*. This is a challenging task that plays now a crucial role in data mining and knowledge acquisition [1–4]. In particular, this holds in scientific/technical setting or in general when narrative styles belong to a jargon that is *discipline-specific* of one that characterizes a specific cultural reality (see e.g. [5], where Baroni et al. provide a method to extract the lexicon related to the nautical universe from a document corpus).

The idea to isolate a part of a common language that is specifically expressive has a long tradition in medicine, for technical dictionaries exist since the thirteen

F. Wotawa et al. (Eds.): IEA/AIE 2019, LNAI 11606, pp. 744–752, 2019.
https://doi.org/10.1007/978-3-030-22999-3_63

century (see, for an initial example of this sensitivity the work of Henry [6]). Nowadays, clinical dictionary represent a crucial resource for physicians (see, e.g., MedDRA or ICD-10), and a number of softwares has been developed to support experts in text classification [7,8]. In general, normalization into a specific lexicon could be useful in all situations characterised by an huge amount of documents of interest. The case of social media is paradigmatic: a reachable sources of (a huge amount of) written text, but these narrative data ranges to a number of slang variations of formal written language. A dictionary could be manually created ex novo. Widely used dictionaries are constructed by hand on the basis of everyday human work and they are agreement-based. This is a good and sound praxis, but it indubitably time consuming. Moreover is not so for dynamic contexts where the lexicon changes rapidly (think, e.g. to the grown of neologisms in linguistic contexts as politics or information technology). The above are all supportive motivations for investigating the following research question: *how can we extract a dictionary from a corpus?* In the recent past, some significant effort has been carried out by the research community in order to identify correct methods for the construction of dictionaries from text [9–11].

In our opinion the above mentioned techniques suffer of a common drawback: they are not completely scalable and incremental; this is why we suggest to adopt the approach that we propose here. We propose a completely automatic technique that generate a dictionary capturing the Italian journalistic lexicon, called NewsDict. The technique is inherently language-independent. We adapt and use a *co-occurrence* based technique inspired to the Mutual Information Principle and generate a set of Italian locutions typical to the journalistic jargon. We employ a dataset of newspaper articles from a set of Italian newspapers as a knowledge base. The data set is annotated by reporters, that manually assigned to each article a topic in front, i.e. national news, local news, sport news, economy, culture and so on. Statistical methods are able to retrieve and filter large and refined sets of *locutions* or *terms*. The construction of the new terminology is completely automatic. Candidate terms are automatically generated from the corpora of narrative documents. Then, new terms are further filtered by frequency and classified by topics, in order to create *sub-dictionaries* for each argument of interest.

We describe here the designed methodology and illustrate results from a set of preliminary experiments. As a As a demonstration test of the accuracy of NewsDict, we normalize a control set of journal articles into NewsDict, retrieving a set of locutions. To extract NewsDict terminology, we use the software DICTOMAT [8]). Then we check if the retrieved terminology belongs to the sub-dictionary of to the category the article belongs to. The paper is organized as follows. Related Work and a quick presentation of the algorithm DICTOMAT are in Sect. 2. Methods are in Sect. 3. In Sect. 4 we described the generation of NewsDict and a first test of its accuracy in task classification. Discussions and Conclusions are in Sect. 5.

2 Background and Related Work

Related Work. In [5,12] some techniques based on Pointwise Mutual Information (PMI) and Information Retrieval (IR) are proposed for recognizing synonyms in specific linguistic domains. In [13,14], authors use co-occurrence based techniques to build a knowledge base and mining algorithms for the clinical linguistic domain.

Similarly, consideration can be applied to the important work done in the community of *Ontology generation*, where the extraction of formal ontologies, often in form of formal definitions in description logic, or in RDF, has been performed [15–19]. There is a long line of investigations involving specifically medical ontologies [20,21] and some studies with specific focus on e-learning problems [22,23]. An interesting aspect dealt with in a recent paper is the evaluation of an ontology extraction method [24], and a recent trend consists in the usage of multi-agent approaches combined with traditional pipeline formed by text mining and rdf [25]. Another ontological based approach can be found in [26]. Machine learning based approaches for lexicon generation are a major trend in NLP. A good survey on this theme can be found in [27] and we list a comparison with the methodology presented here as a high priority future work.

***The NLP Algorithm* DICTOMAT.** In this paper we exploit the normalization software DICTOMAT, an evolution of the NLP algorithm `MagiCoder`, introduced and studied in [8,28–30]. `MagiCoder` has been designed to be robust to dictionary and language changes and performs effectively in normlization task: in [8] we measure, on a data set of 4500 manually revised text, an average recall and precision of 86.9% and 91.8%, respectively.

The main idea of DICTOMAT is that a single linear scan of the free-text is sufficient for recognizing terms from a given dictionary $\mathcal{D}$. From an abstract point of view, it recognizes in the narrative description, by a perfect string matching, *single words* belonging to dictionary locution, where recognised words do not necessarily occupy consecutive positions in the text, (optinally) respecting anyway the order in the dictionary entry[1]. Recognised terms (terms voted by at least one word of the text) are weighted according to information about the syntactical matching. Then they are multisorted and finally a subset of "winning terms" is released as a solution. Differently from other NLP normalization softwares, DICTOMAT does not involve computationally expensive tasks, such as subroutines for permutations and combinations of words. See [8] for a detailed explanation of the algorithm and for a gold standard based test of performances.

3 Methods

The goal of this study is to generate a set of locutions, in order to *built the dictionary* NewsDict, that captures Italian journalistic terminology. The construction

[1] These option can be set by the user. In the case of the journalistic lexicon we decide to respect the order, in order to capture typical expressions and figures of speech of the slang.

of the dictionary is *completely automatic*. Given a corpus of documents, *a set of lexicon driven locutions is generated by statistical methods.*

To generate NewsDict, we adapt some techniques used in linguistics for quantitative analysis and in IR methods [5,12,31–33]. In particular, the cited papers focus on the detection of synonyms (and/or antynomys), i.e. on new locutions semantically correlated to a basis, given dictionary. Here, we address a bit different (and more "venturesome") problem: to generate a dictionary *ex novo*, without the guide of a stable terminology to extend and to use as a support. For this paper, we will focus on bigrams: the locutions of NewsDict are built upon pair of words. We discuss in Sect. 5 how this can be easily generalized in future work. The main idea is to collect "frequent pairs" of words, called *co-occurrence*, appearing in the same contexts (documents).

We show here how co-occurrence methods perform well in a different language (Italian) and for the jornalistic terminology generation. In the paper, we use a general definition of **co-occurrence** [32].

Definition 1 (Co-occurrence). *We say that two different words w_i and w_j are* co-occurring, *or, equivalently, form a* co-occurrence *(pair), when they appear in the same document $d \in \mathsf{D}$ (where D is a collection of documents).*

In the definition of co-occurrence we are considering, *the order of words matter.* The co-occurrence of two words w_i and w_j in the same document within a span of maximally k words is represented by w_i near^k w_j. If $k = 0$, w_i and w_j are contiguous. The co-occurrence of two words w_i and w_j in the same document without regard to their distance is represented by w_i near^∞ w_j.

To discover frequent co-occurrences we adopt a version of the technique used in [5,12]. It is based on the (pointwise) Mutual Information principle (MI) introduced in [12,34]. We compute co-occurring pairs with the following score, closely related to the ones proposed in [12]:

$$\mathsf{SC} : \frac{\mathsf{tfidf}(\mathsf{w}_i \ \mathsf{near}^k \mathsf{w}_j)}{\mathsf{tfidf}(\mathsf{w}_i) \cdot \mathsf{tfidf}(\mathsf{w}_j)}$$

where: (i) $\mathsf{tfidf}(E)$, with E expression, represents the well-known metric TFIDF (Term Frequency, Inverse Document Frequency)[35]; (ii) the co-occurrence of two words w_i and w_j in the same document within a span of maximally k words is represented by w_i near^k w_j. If $k = 0$, w_i and w_j are contiguous;

Score values range from 0 to 1: pairs $(\mathsf{w}_1, \mathsf{w}_2)$ evaluated as maximally co-occurring are scored with 1. Informally, SC computes the score of the co-occurrence of the pair $(\mathsf{w}, \mathsf{w}_i)$ up to a given proximity threshold (or window) k. In this paper, we set k to 0.

4 Experiments

We exploit the following datasets, coming from a corpus D_{news} of articles from the Italian newspapers `L'Arena di Verona` and `il Giornale di Vicenza`:

- D_{gen}: is the dataset we use to generate cooccurence pairs, about 1400 articles from the newspaper `L'Arena di Verona`. Cooccurence pairs, suitably filtered on the basis of the assigned score and the absolute frequency, generate the dictionary NewsDict.
- D_{test}: is the testing data set, of about 3000 articles from the newspaper `Il Giornale di Vicenza`. Documents in D_{test} have been normalized into News-Dictusing the software DICTOMAT, to perform a first test about NewsDict's performances in normalization tasks.

We remark that D_{news} is an annotated corpus: each article has been manually annotated with information as the date and topic by the author. We extract this information and we consider it as a reference gold standard we use to test the results of this study. Topics or categories we extracted are: *National News, Local News, Economics, Market, Culture, Shows, Sport.* Topics play an important role, since we think that the journalistic lexicon can be viewed as a collection of sub-lexicons, able to characterize each category. Following this perspective, we generate different sub-dictionaries, according to different topics we focused on.

4.1 Generation of the Dictionary NewsDict

We describe now the generation of dictionary entries. First, texts from newspaper dataset D_{gen} have been preprocessed. Preprocessing complained standard operations such as *tokenization* (where a token corresponds to a word) and *stop-words removal*. For this paper, we decided to do not apply a *stemming* algorithm (see Sect. 5.).

Second, we generated dictionary entries by applying the score SC. We done this for each topics of interest. For each run we obtain a set of pairs labeled with a value according to the score we computed. As in most IR problems, a crucial point is to discriminate, among the whole set of outputs, a "small" subset of potentially good solutions. The outputs of the SC runs are filtered taking into account only pairs that obtained a score value greater than the threshold 0.48 (heuristically chosen to avoid irrelevant bigrams) and that occur with absolute frequency $f \geq 2$. Threshold f rejects well-scored pairs having a negligible frequency in the whole set of documents.

In Table 1 below we report complete data about retrieved locutions, divided into different topic.

Table 1. Statistics about co-occurrences retrieved by using MI score SC

Topics	Retrieved	Survived
National News	8526	797
Local News	84936	1151
Economy	20797	157
Market	1404	703
Shows	28676	142
Sport	65618	622

4.2 Testing NewsDict Efficiency: A First Experiment

We consider a the data set D_{test}, a corpus of about 3000 of annotated documents from Il giornale di Vicenza. We choose a set of article from a different source (w.r.t. D_{gen}), since we want to test NewsDict in a more challenging setting. We normalized articles from D_{test} by DICTOMAT and, from each document d, we extract a set of locution $\{l_1, \ldots, l_k\}$ of NewsDict and the topic that generates each l_i. Since we are working with bi-grams, in launching DICTOMAT we set that a NewsDict locution has to be completely recognized (both the words of an entry $\langle \mathsf{w}_1, \mathsf{w}_2 \rangle$ appears in the text)[2]. We will call *native* a retrieved locution whose topic match with the one of the processed document. We will call *alien* a retrieved locution whose topic does not match with the one of the processed document. We focus on some topics on which NewsDict provided interesting performances. Articles manually classified as related to Sport have been has seen a hit count of 1756 locutions coming from the dictionary generated by the union of the topic subdictionaries. Once pruned from duplicates, the percentage of native locution is $62,4\%$. For articles manually classified as related to Market 2913 locutions of the automatically generated dictionary have been found. Once pruned from duplicates, the percentage of native locution is $99,4\%$. Also the case of local news is interesting. Articles manually classified as related to Local News have been normalized into 2837 locutions. The percentage of native locution is ("only") $46,5\%$. Notwithstanding, we observe that 521 alien locutions ($18,36\%$) comes from National News subdictionary: this is an awaited results, since it is reasonable that the two sub-lexicons, being the topics strongly related, can also be strongly be overlapped.

5 Discussion and Conclusions

The method described in this paper would be fruitful for the automatic generation of journalistic lexicon based on written language daily used by reporters. Even if we are at the first stage of the investigations, some useful consideration can be yet done. As expected, the automatic generation of locutions performs better on specific jargons and worse on generic linguistic contexts. A national news is reasonably written in a plain Italian language, with a less number of technical expressions and figures of speech. In a news about Market, the reporter reasonably uses a more technical language and a set of standardised linguistic forms.

We have shown, through a preliminary experiment, that the statistical technique we adopted performs effectively in the generation of significative locutions. This is an ongoing project. First, we plain to filter the set of cooccurent words by their POS classification, and probably irrelevant pairs such as pairs of verbs or adjectives. Since we are still working of pairs of words and with a limited distance windows, we claim that the absence of POS classification and filter is not

[2] This selection criterium could be clearly relaxed when we will extend the dictionary to n-grams ($n > 2$).

harmful for the current result, nut will be useful when we will move from pairs to n-uples of words. Generalization to n-uples is a a further short-time task we will address. Moreover, the SC-criterion we used in this papers to compute MI-based co-occurrences could be refined. MI scores could take into account negations in the narrative descriptions and *contexts* (i.e. other words we require to appear in the document in joint with the co-occurring pair) [12].

Acknowledgement. Matteo Cristani and Claudio Tomazzoli gratefully thank Athesis s.p.a. for the financial support to this research through the Google Grant "Premium Semantic Communities", Google DNI Application Number: r4-DPZ3Ln5mmV3O.

References

1. Cristani, M., Tomazzoli, C.: A multimodal approach to relevance and pertinence of documents. In: Fujita, H., Ali, M., Selamat, A., Sasaki, J., Kurematsu, M. (eds.) IEA/AIE 2016. LNCS (LNAI), vol. 9799, pp. 157–168. Springer, Cham (2016). https://doi.org/10.1007/978-3-319-42007-3_14
2. Cristani, M., Tomazzoli, C.: A multimodal approach to exploit similarity in documents. In: Ali, M., Pan, J.-S., Chen, S.-M., Horng, M.-F. (eds.) IEA/AIE 2014. LNCS (LNAI), vol. 8481, pp. 490–499. Springer, Cham (2014). https://doi.org/10.1007/978-3-319-07455-9_51
3. Cristani, M., Fogoroasi, D., Tomazzoli, C.: Measuring homophily. In: CEUR Workshop Proceedings, vol. 1748 (2016)
4. Cristani, M., Cristani, M., Pesarin, A., Tomazzoli, C., Zorzi, M.: Making sentiment analysis algorithms scalable. In: Pautasso, C., Sánchez-Figueroa, F., Systä, K., Murillo Rodríguez, J.M. (eds.) ICWE 2018. LNCS, vol. 11153, pp. 136–147. Springer, Cham (2018). https://doi.org/10.1007/978-3-030-03056-8_12
5. Baroni, M., Bisi, S.: Using cooccurrence statistics and the web to discover synonyms in a technical language. In: Proceedings of the of LREC (2004)
6. Henry, F.P.: A review of the first book on the diseases of the eye, by benvenutus grassus, 1474: exhibition of three other fifteenth century monographs (a) the first medical dictionary, synonyma simonis genuensis, 1473; (b) the first book on diet, by isaac, 1487, (c) the second edition of the first book on diseases of children, by paulus bagellardus, 1487. Med. Libr. Hist. J. **3**(1), 27–40 (1905)
7. Meystre, S., Haug, P.J.: Natural language processing to extract medical problems from electronic clinical documents: performance evaluation. J. Biomed. Inform. **39**(6), 589–599 (2006)
8. Combi, C., Zorzi, M., Pozzani, G., Moretti, U., Arzenton, E.: From narrative descriptions to MedDRA: automagically encoding adverse drug reactions. J. Biomed. Inform. **84**, 184–199 (2018)
9. Forgac, R., Krakovsky, R.: Text processing by using projective art neural networks (2016)
10. Abel, M., Chung, S.: Computing preset dictionaries from text corpora for the compression of messages (2014)
11. Quan, C., Ren, F., He, T., Hu, P.: Automatic construction of biomedical abbreviations dictionary from text (2008)
12. Turney, P.D.: Mining the web for synonyms: PMI-IR versus LSA on TOEFL. In: De Raedt, L., Flach, P. (eds.) ECML 2001. LNCS (LNAI), vol. 2167, pp. 491–502. Springer, Heidelberg (2001). https://doi.org/10.1007/3-540-44795-4_42

13. Schulz, S., Costa, C.M., Kreuzthaler, M., et al.: Semantic relation discovery by using co-occurrence information. In: Proceedings of BioTxtM (2014)
14. Yang, C.C., Yang, H., Jiang, L., Zhang, M.: Social media mining for drug safety signal detection. In: Proceedings of SHB, pp. 33–40. ACM (2012)
15. Buitelaar, P., Olejnik, D., Sintek, M.: A protégé plug-in for ontology extraction from text based on linguistic analysis. In: Bussler, C.J., Davies, J., Fensel, D., Studer, R. (eds.) ESWS 2004. LNCS, vol. 3053, pp. 31–44. Springer, Heidelberg (2004). https://doi.org/10.1007/978-3-540-25956-5_3
16. Wang, W., Barnaghi, P., Bargiela, A.: Probabilistic topic models for learning terminological ontologies. IEEE Trans. Knowl. Data Eng. **22**(7), 1028–1040 (2010)
17. Aussenac-Gilles, N., Sorgel, D.: Text analysis for ontology and terminology engineering. Appl. Ontol. **1**, 35–46 (2005)
18. Faria, C., Serra, I., Girardi, R.: A domain-independent process for automatic ontology population from text. Sci. Comput. Program. **95**(P1), 26–43 (2014)
19. Benafia, A., Mazouzi, S., Maamri, R., Sahnoun, Z., Benafia, S.: From linguistic to conceptual: a framework based on a pipeline for building ontologies from texts. J. Adv. Comput. Intell. Intell. Inform. **20**(6), 941–960 (2016)
20. Milian, K., Hoekstra, R., Bucur, A., ten Teije, A., van Harmelen, F., Paulissen, J.: Enhancing reuse of structured eligibility criteria and supporting their relaxation. J. Biomed. Inform. **56**, 205–219 (2015)
21. Reimer, U., Maier, E., Streit, S., Diggelmann, T., Hoffleisch, M.: Learning a lightweight ontology for semantic retrieval in patient-centered information systems. Int. J. Knowl. Manag. **7**(3), 11–26 (2011)
22. Zouaq, A., Nkambou, R.: Building domain ontologies from text for educational purposes. IEEE Trans. Learn. Technol. **1**(1), 49–62 (2008)
23. Suresh, R., Dinakaran, K., Amulya, R.: Automating ontologies for e-learning. Int. J. Metadata Semant. Ontol. **9**(3), 227–232 (2014)
24. Muresan, S., Klavans, J.: A method for automatically building and evaluating dictionary resources. In: Proceedings of the Third International Conference on Language Resources and Evaluation (LREC 2002), Las Palmas, Canary Islands - Spain, European Language Resources Association (ELRA), May 2002
25. Sellami, Z., Camps, V., Aussenac-Gilles, N.: DYNAMO-MAS: a multi-agent system for ontology evolution from text. J. Data Semant. **2**(2–3), 145–161 (2013)
26. Souvignet, J., Declerck, G., Asfari, H., Jaulent, M.C., Bousquet, C.: Ontoadr a semantic resource describing adverse drug reactions to support searching, coding, and information retrieval. J. Biomed. Inform. **63**, 100–107 (2016)
27. Rahul, M., Shine, S.: A survey of morphosyntactic lexicon generation. In: Proceedings of the International Conference in Emerging Trends in Engineering, Science and Technology, ICETEST 2018, pp. 773–778 (2018)
28. Zorzi, M., Combi, C., Lora, R., Pagliarini, M., Moretti, U.: Automagically encoding adverse drug reactions in MedDRA. In: 2015 International Conference on Healthcare Informatics, ICHI 2015, Dallas, TX, USA, 21–23 October 2015, pp. 90–99. IEEE Computer Society (2015)
29. Zorzi, M., Combi, C., Pozzani, G., Moretti, U.: Mapping free text into MedDRA by natural language processing: a modular approach in designing and evaluating software extensions. In: Proceedings of the 8th ACM International Conference on Bioinformatics, Computational Biology, and Health Informatics, BCB 2017, Boston, MA, USA, 20–23 August 2017, pp. 27–35. ACM (2017)
30. Combi, C., Zorzi, M., Pozzani, G., Arzenton, E., Moretti, U.: Normalizing spontaneous reports into MedDRA: some experiments with MagiCoder. IEEE J. Biomed. Health Inform. **23**(1), 95–102 (2019)

31. Schütze, H., Pedersen, J.O.: A cooccurrence-based thesaurus and two applications to information retrieval. Inform. Process. Manag. **33**(3), 307–318 (1997)
32. Manning, C.D., Schütze, H.: Foundations of Statistical Natural Language Processing. MIT Press, Cambridge (1999)
33. Zorzi, M., Combi, C., Pozzani, G., Arzenton, E., Moretti, U.: A co-occurrence based MedDRA terminology generation: some preliminary results. In: ten Teije, A., Popow, C., Holmes, J.H., Sacchi, L. (eds.) AIME 2017. LNCS (LNAI), vol. 10259, pp. 215–220. Springer, Cham (2017). https://doi.org/10.1007/978-3-319-59758-4_24
34. Church, K.W., Hanks, P.: Word association norms, mutual information, and lexicography. In: Proceedings of ACL 1989, Stroudsburg, PA, USA, pp. 76–83 (1989)
35. Manning, C.D., Raghavan, P., Schütze, H.: Introduction to Information Retrieval. Cambridge University Press, New York (2008)

Decision-Making Support Method Based on Sentiment Analysis of Objects and Binary Decision Tree Mining

Huyen Trang Phan[1(✉)], Van Cuong Tran[2], Ngoc Thanh Nguyen[3], and Dosam Hwang[1(✉)]

[1] Department of Computer Engineering, Yeungnam University, Gyeongsan, Republic of Korea
huyentrangtin@gmail.com, dosamhwang@gmail.com

[2] Faculty of Engineering and Information Technology, Quang Binh University, Dong Hoi, Vietnam
vancuongqbuni@gmail.com

[3] Faculty of Computer Science and Management, Wroclaw University of Science and Technology, Wroclaw, Poland
Ngoc-Thanh.Nguyen@pwr.edu.pl

Abstract. As more and more users express their opinions on many topics on Twitter, the sentiments contained in these opinions are becoming a valuable source of data for politicians, researchers, producers, and celebrities. These sentiments significantly affect the decision-making process for users when they assess policies, plan events, design products, etc. Therefore, users need a method that can aid them in making decisions based on the sentiments contained in tweets. Many studies have attempted to address this problem with a variety of methods. However, these methods have not mined the level of users' satisfaction with objects related to specific topics, nor have they analyzed the level of users' satisfaction with that topic as a whole. This paper proposes a decision-making support method to deal with the aforementioned limitations by combining object sentiment analysis with data mining on a binary decision tree. The results prove the efficacy of the proposed approach in terms of the error ratio and received information.

Keywords: Decision-making · Sentiment analysis · Binary decision tree

1 Introduction

As society rapidly develops day by day, people are increasingly interested in public opinion and there is an increasingly large amount of information available to share. This information has become an important source of data that can satisfy the needs of users. For example, governments may wish to use this data to enact policies that improve the lives of citizens; companies may seek to use this

F. Wotawa et al. (Eds.): IEA/AIE 2019, LNAI 11606, pp. 753–767, 2019.
https://doi.org/10.1007/978-3-030-22999-3_64

information to enhance the quality of their products and attract customers; and celebrities may hope to use this data to gain new fans. Therefore, these organizations and individuals need effective tools to help them make decisions based on information published by users on relevant social networking sites. Decision-making is usually defined as a process of selection, synthesis, and evaluation of multiple opinions or alternatives to yield one that best achieves the aims of the decision maker [2]. In the modern world, the efficacy of the decision-making process depends on the available information and the accuracy of analysis. Users need an automated method to support the decision-making process based on knowledge extracted from data analysis.

As the use and presence of social networking sites has grown, so have the amount of information available on such sites. Twitter is one of the most popular social networking sites. Twitter's userbase has grown rapidly, and approximately 500 million tweets are published every day[1]. According to available statistics[2], the monthly number of active users on Twitter worldwide is 326 million. According to eMarketer, nearly 66% of the businesses who have 100 or more employees have a Twitter account and expect it to rise into 2018. This is the source of available information which if we know a way to exploit, they will bring a lot of benefits. Therefore, the main issue is how to use these tweets for aiding users decision-making with a decision support system.

Sentiment analysis (SA) has become a popular application of natural language processing. SA focuses on characterizing expressions that reflect users' opinion-based sentiments toward entities or facets of entities [6]. Sentiment analysis can be used to support decision making by extracting, analyzing, and predicting the orientation of opinions. Tweet SA is the best way to obtain information on user opinions from Twitter.

A decision tree is a graphical representation of specific decision-making situations when complex branching occurs in a structured decision process. The decision tree is a predictive model based on a branching series of boolean tests that use specific facts to make more generalized conclusions[3]. Data mining with decision tree is used to extract a hidden set of rules, a process that has enabled people to make better decisions in many different areas.

Sentiment analysis and data mining are very helpful tools for analyzing the behavior of users with regard to products, movies, events, and other things [8]. However, separating the two methods and performing each independently reduces their efficiency. Therefore, combining sentiment analysis and data mining can yield a more powerful tool.

Various methods have been developed to automatically analyze personal reviews of products, services, events and policies based on tweets, the result of which aims to support users in decision-making. However, these existing methods support decision-making without considering the combination of data mining and sentiment analysis. Therefore, information on how objects influence users'

[1] http://www.internetlivestats.com/twitter-statistics/.
[2] https://zephoria.com/twitter-statistics-top-ten/.
[3] https://www.techopedia.com/definition/28634/decision-tree.

sentiments is not considered. Users do not know the level of satisfaction other users feel with regard to specific objects that belong to a given topic, nor do they know how satisfied other users are with the topic as a whole. These drawbacks motivated us to propose a method to support decision-making by combining sentiment analysis and data mining. In the proposed method, first, a set of features related to syntactic, lexical, semantic, and sentiment of the words is extracted. Second, objects and their sentiments are determined. Third, the result of objects' sentiment analysis is converted to the form of boolean value. Finally, a binary decision tree is built, and data mining is performed on this tree to estimate the significance of the objects in each topic.

The remainder of the paper is organized as follows. In Sect. 2, we summarize existing work related to approaches for sentiment analysis. The research problem is described in Sect. 3 and the proposed method is presented in Sect. 4. The experimental results and evaluations are shown in Sect. 5. Finally, the conclusion and a discussion of future work are presented in Sect. 6.

2 Related Work

Applying tools for text mining and sentiment analysis to analyze opinions published by users on social networking sites has been the focus of many researchers' work. There are many studies on models and methods for data collection, sentiment analysis, and information extraction to support decision-making. Recent studies have demonstrated the use of acceptably accurate methods for sentiment classification and data mining of tweets.

De Albornoz et al. 2011 [3] predicted overall ratings of a product based on users' opinions of different product features. This system first identifies the features that are relevant to consumers for a particular type of product, as well as the relative importance or salience of said features. The system then extracts from the review the user's opinions on the different features of the product and quantifies these opinions by constructing a vector of feature intensities that represents the review. This vector serves as the input to a machine learning model that classifies the review into different rating categories. The method was applied to over 1000 hotel reviews from booking.com, and it achieved results better than other systems. The authors of the paper [13] built a novel domain-independent decision support model for customer satisfaction research. This model was based on an in-depth analysis of consumer reviews posted on the Internet in natural language. Artificial intelligence techniques, such as web data extraction, sentiment analysis, aspect extraction, aspect-based sentiment analysis, and data mining were used to analyze consumers' reviews. This method evaluated customer satisfaction both qualitatively and quantitatively. The efficacy of the approach was assessed on two datasets related to hotels and banks. The results prove the efficacy of this approach for quantitative research on customer satisfaction. In the paper [12], the authors focus on automatically identifying essential aspects of products from online consumer reviews. Their method consists of the following steps. First, the relevant aspects of the product are recognized by a shallow dependency parser, and consumers' opinions on these aspects are determined via a sentiment classifier. Second, an aspect ranking algorithm is developed to identify the important aspects by simultaneously considering the aspect frequency and the influence of each aspect on consumers' overall opinions. Finally, the aspect ranking results are applied to a document-level sentiment classification, which significantly improves its performance. The experimental results on 11 popular products in four domains demonstrate the effectiveness of the approach.

Generally, the aforementioned methods applied data mining and sentiment analysis to analyze the opinions of users on social networking sites, with the results of the studies demonstrating acceptable accuracy. However, these methods have several disadvantages, as they do not combine data mining with sentiment analysis. Rather, the techniques are applied separately. With regard to supporting decision-making, these methods do not evaluate user satisfaction rates for both entire topics and individual objects within said topics based on the opinions of other users. Unlike previous research, our work focuses on identifying the primary objects and sentiments of those objects from opinions on Twitter, and using these results to construct a binary decision tree. Then, user satisfaction with the topic as well as user satisfaction with objects belonging to that topic are determined, from their tweets, to reasonably aid users in decision-making.

3 Research Problem

This section presents the basic concepts and definitions related to determining user satisfaction for a specific topic from tweets, including objects, sentiments of objects, the user's satisfaction for an object belonging to a specific topic, and the user's satisfaction for the entire topic. The research problems are stated at the end of this section.

3.1 Definitions

Assume that we have a finite set of tweets, T, representing the opinions of users about a specific topic, with T being represented by $T = \{t_1, t_2, ..., t_n\}$, where n is the number of gathered tweets. Let Pt be a set of positive words and Nt be a set of negative words. To determine necessary elements in tweets, each tweet has to be separated into a set of tokens. Let $N_i = \{w_1, w_2, ..., w_g\}$ be a set of tokens of tweet t_i from T.

Definition 1. *A sentiment relation between token w_k and w_h ($w_h \in Pt \cup Nt$) is defined by function Θ given as follows:*

$$\Theta(w_h, w_k) = \begin{cases} 1, & if\ w_k\ referring\ to\ w_h \\ 0, & otherwise. \end{cases} \tag{1}$$

A tweet can have many objects. Let $O_i = \{o_1, o_2, ..., o_m\}$ be a set of objects belonging to t_i where o_j is an object that is assigned a sentiment and is related to the chosen topic.

Definition 2. *An object o_j of sentiments in a tweet is a token w_k that satisfies two conditions simultaneously such as w_k must be a noun or noun phrase and must be related to at least one sentiment word existing in that tweet. o_j is expressed as:*

$$o_j = \{w_k | tag(w_k) = \text{‘}NOUN\text{‘}, \exists w_h \in t_i : \Theta(w_k, w_h) = 1\}. \tag{2}$$

Definition 3. *A sentiment of an object o_j in each tweet is denoted by e_{o_j}. e_{o_j} is positive if the object's description has at least one token w_k belonging to Pt and has the sentiment relation to o_j. e_{o_j} is negative if the object's description has at least one token w_k belonging to Nt and has the sentiment relation to o_j. e_{o_j} is expressed as:*

$$e_{o_j} = \begin{cases} positive, & if\ (\exists w_k \in Pt) \wedge (\Theta(w_k, o_j) = 1) \\ negative, & if\ (\exists w_k \in Nt) \wedge (\Theta(w_k, o_j) = 1). \end{cases} \tag{3}$$

Let o_j^+ and o_j^- be the positive and negative sentiment components of the object o_j, respectively.

Definition 4. *The user's satisfaction for an object of the specific topic is calculated based on the frequency of sentiment components of that object (denoted by ω_{o_j}) and computed by:*

$$\omega_{o_j} = \frac{frequency(o_j^+) - frequency(o_j^-)}{frequency(o_j^+) + frequency(o_j^-)}. \tag{4}$$

Definition 5. *The user's satisfaction for each object o_j in a tweet t_i (denoted by $\omega_{o_{ij}}$) is computed based on the user's satisfaction for this object and the frequency of this object in all tweets. $\omega_{o_{ij}}$ is defined as follows:*

$$\omega_{o_{ij}} = \frac{\omega_{o_j}}{frequency(o_j)}. \tag{5}$$

Definition 6. *The user's satisfaction for the topic (denoted by ω_{t_i}) is measured by the sum of the user's satisfaction for all objects in the tweet. ω_{t_i} is represented by a double $\langle \upsilon_{t_i}, ratio_{\omega_{t_i}} \rangle$ in which υ_{t_i} and $ratio_{\omega_{t_i}}$ are assessed as follows:*

$$\upsilon_{t_i} = \begin{cases} yes, & if\ user\ is\ satisfied\ (ratio_{\omega_{t_i}} \geq 0) \\ no, & if\ user\ is\ not\ satisfied\ (ratio_{\omega_{t_i}} < 0). \end{cases} \tag{6}$$

$$ratio_{\omega_{t_i}} = \sum_{j=1}^{m} \omega_{o_{ij}}. \tag{7}$$

3.2 Problems

In this study, we focus on finding a method to answer the main question as follows: *How can we support users in decision-making for an issue that related to a specific topic from tweets based on the user's satisfaction for each object of this topic as well as for this entire topic?* This question is partitioned in the two following sub-questions:

1. From the existing tweets describing a specific topic, how can we determine the user's satisfaction for objects of a specific topic based on the other users' sentiment?
2. How can we determine the user's satisfaction for the topic based on other users' sentiments for this topic?

4 The Proposed Method

This section presents the method to solve problems identified in Sect. 3.2. The proposed method consists of three steps: determining objects and sentiments of objects in each tweet, converting objects' sentiment in tweets into the form of boolean values, and finally building and mining on the binary decision tree. The workflow of method is shown in Fig. 1. The following subsections explain details of the proposed method.

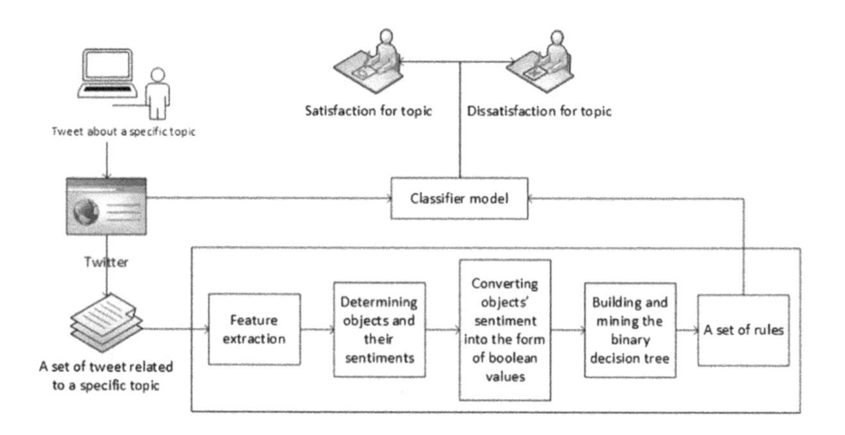

Fig. 1. The workflow of the proposed method.

4.1 Feature Extraction

To determine objects related to sentiments, the features are extracted from in each tweet. In this study, information related to the lexical, syntactic, semantic, and polarity sentiment of words are employed as features [11].

n-grams: the n-grams used in this work include 1-gram, 2-grams, and 3-grams. Each n-gram appearing in a tweet becomes an entry in the feature vector with a feature value corresponding to the term frequency inverse document frequency (tf-idf).

Part-Of-Speech (POS) tags of words: The NLTK toolkit [1] is used to annotate the POS tags. POS tags with their corresponding tf-idf values are the syntactic features and feature values, respectively.

Word embeddings: The 300-dimensional pre-trained word embeddings from Glove[4] are used to compute a tweet embedding as the average of the embeddings of words in the tweet [11].

Special words: Special words include negation, intensifier, and diminishes words. This feature is extracted using a window of 1 to 3 words before a sentiment word and search for these kinds of words [7]. The appearance of special words in the tweet and their tf-idf values become features and feature values, respectively.

Sentiment words: The number of sentiment words in each tweet are used as a feature. The sentiment dictionaries provided by Hu and Liu [4] are employed for determining the positive and negative words in a tweet.

4.2 Determining Objects and Sentiments of Objects in Each Tweet

A combination of Bidirectional Long Short Term Memory (BiLSTM) and Conditional Random Field (CRF) models [5] is used to identify objects and their sentiments in each tweet[5]. This combination leverages the advantages of both models: the creative ability to extract features of the LSTM model and the steady predictability of the CRF model [10]. This model operates via the following steps: a word embedding of each word is put into the BiLSTM layer to extract features discussed in Sect. 4.1. Next, the CRF layer utilizes the aforementioned features to predict labels for each word. In addition to information received from the BiLSTM layer, the CRF also relies on information from previously anticipated labels.

Example 1. Given a set of tweets, $T = \{t_1, t_2, t_3, t_4, t_5, t_6, t_7, t_8, t_9\}$,

t_1: The color of the phone is not lovely, and I also do not like its style.
t_2: The screen is so bright but relatively small.
t_3: The battery is good, the screen is good, and the color is also lovely.
t_4: I do not like the screen on this phone.
t_5: I do not like the screen and battery, but the camera, style, and color are excellent.
t_6: For this phone, I like the battery and the camera.
t_7: The screen is big, but the camera is not clear.
t_8: The phone has a beautiful color, but the style is not lovely.
t_9: The phone is not lovely about color, camera but the style is good.

[4] http://nlp.stanford.edu/projects/glove/.
[5] https://github.com/UKPLab/emnlp2017-bilstm-cnn-crf.

From T, objects and their sentiments are determined based on the combination of BiLSTM and CRF model as follows:

t_1: o_1 = color, e_{o_1} = negative; o_2 = style, e_{o_2} = negative.
t_2: o_1 = screen, e_{o_1} = positive; o_2 = screen, e_{o_2} = negative.
t_3: o_1 = battery,e_{o_1} = positive; o_2 = screen, e_{o_2} = positive; o_3 = color, e_{o_3} = positive.
t_4: o_1 = screen, e_{o_1} = negative.
t_5: o_1 = screen, e_{o_1} = negative, o_2 = style, e_{o_2} = positive, o_3 = color, e_{o_3}=positive, o_4 = battery, e_{o_4} = negative, o_5 = camera, e_{o_5} = positive.
t_6: o_1 = battery, e_{o_1} = positive; o_2 = camera, e_{o_2} = positive.
t_7: o_1 = screen, e_{o_1} = positive; o_2 = camera, e_{o_2} = negative.
t_8: o_1 = style, e_{o_1} = negative; o_2 = color, e_{o_2} = positive.
t_9: o_1 = style, e_{o_1} = positive; o_2 = camera, e_{o_2} = negative, o_3 = color, e_{o_3} = negative.

4.3 Converting Object's Sentiment into a Boolean Value

After the objects and their sentiment in each tweet are assigned labels, the object's sentiment in each tweet will be converted into a boolean value [13] (denoted by e_{w_k}) for simplicity. If the object's sentiment in the tweet is positive, then the value of e_{w_k} is 1. Otherwise, the value of e_{w_k} is 0. The steps to convert the object's sentiment in tweets into a boolean value are explained below in Example 2.

Example 2. From the result of Example 1, we have boolean representations of objects' sentiment as follows:

l_{t_1}: $e_{color} = 0$, $e_{style} = 0$. l_{t_2}: $e_{screen} = 1$, $e_{screen} = 0$.
l_{t_3}: $e_{batterry} = 1$, $e_{screen} = 1$, $e_{color} = 1$. l_{t_4}: $e_{screen} = 0$.
l_{t_5}: $e_{screen} = 0$, $e_{batterry} = 0$, $e_{style} = 1$, $e_{color} = 1$, $e_{camera} = 1$.
l_{t_6}: $e_{battery} = 1$, $e_{camera} = 1$. l_{t_7}: $e_{screen} = 1$, $e_{camera} = 0$.
l_{t_8}: $e_{color} = 1$, $e_{style} = 0$. l_{t_9}: $e_{style} = 1$,$e_{camera} = 0$,$e_{color} = 0$.

4.4 Building and Mining on the Binary Decision Tree

A binary decision tree is a tree used to represent the objects' sentiment in the form of boolean values. It consists of internal nodes and leaves. The internal nodes show the objects' sentiment, e.g., color or ¬color (called attributes). The internal nodes of the tree are selected based on the information gained from the attributes. Each node has the two branches: the right branch represents the absence of an object's sentiment in the tweet (denoted by 0), and the left branch represents the presence of an object's sentiment in the tweet (denoted by 1). Leaves represent the user's satisfaction with the topic and possess one of the two values "Yes" or "No." The leaf's value is "Yes" if the user is satisfied with the topic. The leaf's value is "No" if the user is dissatisfied with the topic.

This binary decision tree is built based on the Iterative Dichotomiser 3 (ID3) algorithm [9].

The constructed binary decision tree allows us to consider the satisfaction of users with not only some objects present in the tweet but also with different objects not present in the tweet, based on certain rules. These rules will support users in making decisions, e.g., when a user wants to buy a phone, the user provides an opinion containing sentiments related to objects of this phone. The user will then be informed whether they should or should not buy it, based on the rules resulting from the binary decision tree.

Example 3. Based on the set of objects and the set of sentiments of those objects extracted in Example 1 and Example 2, there are five objects in T including: *color, style, battery, screen, camera,* and each object has two types of sentiment that are positive (denoted by object) and negative (denoted by $\neg object$) (e.g., color and $\neg color$). Because of the limitation of space, the objects are denoted as o_j $(j = 1, ..., 5)$. The values of ω_{o_j} and $\omega_{o_{ij}}$ are calculated as Table 1.

Table 1. The measure of ω_{o_j} and $\omega_{o_{ij}}$.

	o_1^+	o_1^-	o_2^+	o_2^-	o_3^+	o_3^-	o_4^+	o_4^-	o_5^+	o_5^-
Frequency	2	3	2	2	2	1	3	3	2	2
ω_{o_j}	−0.2		0		0.34		0		0	
$\omega_{o_{ij}}$	−0.04		0		0.11		0		0	

The values of $ratio_{\omega_{t_i}}$ and ω_{t_i} are then calculated by applying Eqs. 6 and 7 in Sect. 3.1 as shown in Table 2.

Table 2. The relationship between objects' sentiments and users' satisfaction in tweets.

	o_1	o_2	o_3	o_4	o_5	$\neg o_1$	$\neg o_2$	$\neg o_3$	$\neg o_4$	$\neg o_5$	$ratio_{\omega_{t_i}}$	v_{t_i}
t_1	0	0	0	0	0	1	1	0	0	0	−0.04	No
t_2	0	0	0	1	0	0	0	0	1	0	0	Yes
t_3	1	0	1	1	0	0	0	0	0	0	0.07	Yes
t_4	0	0	0	0	0	0	0	0	1	0	0	Yes
t_5	1	1	0	0	1	0	0	1	1	0	0.07	Yes
t_6	0	0	1	0	1	0	0	0	0	0	0.11	Yes
t_7	0	0	0	1	0	0	0	0	0	1	0	Yes
t_8	1	0	0	0	0	0	1	0	0	0	−0.04	No
t_9	0	1	0	0	0	1	0	0	0	1	−0.04	No

From Table 2, we can see that there are 3 tweets expressing dissatisfaction with Phone (classified “No”) and 6 tweets expressing satisfaction (classified “Yes”). Hence, the entropy and the information gain of each attribute for Phone topic are computed as in Table 3.

Table 3. The information to choose the root node.

	o_1		o_2		o_3		o_4		o_5		$\neg o_1$		$\neg o_2$		$\neg o_3$		$\neg o_4$		$\neg o_5$	
	1	0	1	0	1	0	1	0	1	0	1	0	1	0	1	0	1	0	1	0
Yes	2	4	1	5	2	4	3	3	2	4	0	6	0	6	1	5	3	3	1	5
No	1	2	1	2	0	3	0	3	0	3	2	1	2	1	0	3	0	3	1	2
Entropy(o_j)	0.92	0.92	1	0.86	0	0.99	0	1	0	0.99	0	0.59	0	0.59	0	0.95	0	1	1	0.86
Gain(topic,o_j)	0		0.025		0.152		0.252		0.152		0.458		0.458		0.07		0.252		0.025	

From Table 3, because $\neg o_1$ and $\neg o_2$ yield the maximum information gain, $\neg o_1$ or $\neg o_2$ is chosen as the root node of the tree. Assuming $\neg o_1$ is selected, the same procedure is repeated for the remaining attributes (see Tables 4 and 5) until we obtain a tree in which the nodes are classified completely.

In Table 5, $\neg o_2$ is chosen as the second node of tree. At node $\neg o_2$, we do not need for further division because its two child nodes are classified completely.

Table 4. The relationship between objects' sentiments and users' satisfaction in each tweet when $\neg o_1 = 0$.

	o_1	o_2	o_3	o_4	o_5	$\neg o_1$	$\neg o_2$	$\neg o_3$	$\neg o_4$	$\neg o_5$	$ratio_{\omega_{t_i}}$	v_{t_i}
t_2	0	0	0	1	0	0	0	0	1	0	0	Yes
t_3	1	0	1	1	0	0	0	0	0	0	0.07	Yes
t_4	0	0	0	0	0	0	0	0	1	0	0	Yes
t_5	1	1	0	0	1	0	0	1	1	0	0.07	Yes
t_6	0	0	1	0	1	0	0	0	0	0	0.11	Yes
t_7	0	0	0	1	0	0	0	0	0	1	0	Yes
t_8	1	0	0	0	0	0	1	0	0	0	−0.04	No

Table 5. The information to choose the second node.

	o_1		o_2		o_3		o_4		o_5		$\neg o_2$		$\neg o_3$		$\neg o_4$		$\neg o_5$	
	1	0	1	0	1	0	1	0	1	0	1	0	1	0	1	0	1	0
Yes	2	4	1	5	2	3	3	3	2	4	1	5	1	5	3	3	1	5
No	1	0	0	1	0	2	0	1	0	1	0	1	0	1	0	1	0	1
Entropy(o_j)	0.92	0	0	0.65	0	0.97	0	0.81	0	0.72	0	0	0	0.65	0	0.81	0	0.65
Gain($o_1 = 0, o_j$)	0.065		-0.099		−0.235		−0.005		−0.058		0.458		−0.099		−0.005		−0.099	

The final binary decision tree is shown in Fig. 2. Looking at Fig. 2, we can see that, from the decision tree, three rules are created. Users can utilize these rules to make a decision when they need to choose a phone. Each rule (X→Y) is characterized by confidence and support measures, in which the support is the percentage of tweets containing both X, Y, and the confidence is the ratio of the number of tweets that contain both X, Y to the number of tweets containing X.

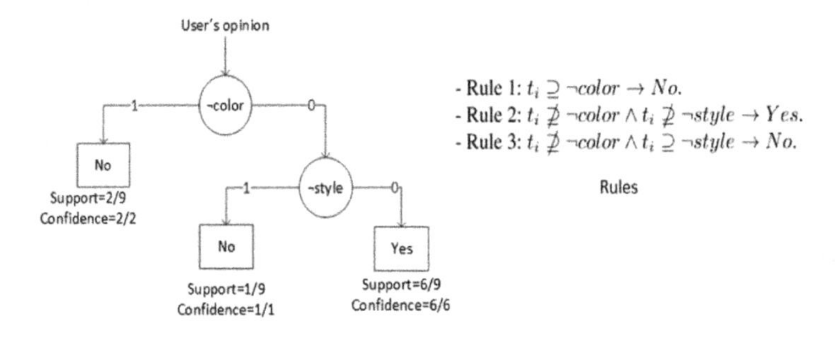

Fig. 2. Example of the binary decision tree.

The problem lies in knowing how to use this binary decision tree to support users in making a decision. Assume there is a user who wants to buy a new phone and this user has an opinion as follows *"This phone has a beautiful color, but the camera and battery are not good."* Should the user buy it or not? This user can be assisted by the three rules created from the decision tree. In this example, there are three objects present in the user's opinion ($color$, $\neg camera$, and $\neg battery$). Thus, this opinion does not contain both $\neg color$ and $\neg style$ and rule 2 will be applied, the result of which is "Yes" with a support value of 6/9 and a confidence value of 6/6, which means that the user should buy this phone. We can see that, although based on the user's opinion, the user does not seem to be satisfied with this phone, according to the binary decision tree, most other users are satisfied with similarly described objects. Therefore, the user should buy this phone.

5 Experiment

5.1 Data Acquisition

The Python package Tweepy[6] was used to collect 5350 tweets which are related to Phone topic and contain at least one sentiment word. In this work, we only deal with English tweets. The non-essential elements in tweets such as punctuation marks, re-tweet symbols, URLs, hashtags, and query term were removed. Each emoji in the tweet was then replaced by descriptive text based on the

[6] https://pypi.org/project/tweepy/.

Python emoji package[7]. It is important to note that tweets are informal and, consequently, users sometimes use acronyms and make spelling errors, which can aect the accuracy of the results. Therefore, the Python-based Aspell library[8] was employed to implement spelling correction. Then, the tweets were divided into two separate database les for use in the training and testing steps. The training set contained 3745 tweets and the testing set included 1605 tweets. All tweets were annotated with three labels (Positive, Negative, and Object) by five manual annotators. The training data file contained two tab-separated columns, with each token on a separate line. The first item on each line is a token; the second item is a label. The tokens that not belonging to factors of interest were annotated as "Other". We annotated the test set as the gold standard to assess the performance. The testing set consisted of 985 tweets classified "Yes" and 620 tweets classified "No". There were 1758 tokens indicating the object, 918 tokens indicating positive sentiment, 840 tokens indicating negative sentiment in the testing set.

5.2 Evaluation Results

Metrics used to assess the proposed method include *precision (P)*, *recall (R)*, and *F-score* (F_1). The values of P, R, and F_1 are computed as follows:

$$P = \frac{TP}{TP + FP} \tag{8}$$

$$R = \frac{TP}{TP + FN} \tag{9}$$

$$F_1 = 2 \times \frac{P \times R}{P + R} \tag{10}$$

where, assuming we have a given class C, TP (True Positive) refers to elements belonging to C and identified as belonging to C, FP (False Positive) refers to elements not belonging to C but classified as C, FN (False Negative) is the number of elements belonging to C but not classified as C, and TN (True Negative) is the number of elements not classified C and not belong to C.

5.3 Result and Discussion

In the testing set, using the combination of BiLSTM and CRF model (Sect. 4.2), nouns related to Phone topic are extracted. The ten nouns with appearing high frequency are then chosen as ten main objects of the topic that are *screen, battery, ram, rom, camera, sound, style, color, application, software.* The result of the object detection process is shown in Table 6.

[7] https://pypi.org/project/emoji/.
[8] https://pypi.org/project/aspell-python-py2/.

Table 6. The performance of objects determination.

	Screen	Battery	Ram	Rom	Camera	Sound	Style	Color	Application	Software
TP	85	146	103	127	178	149	175	209	79	116
FP	31	31	31	35	41	52	25	94	31	20
FN	84	40	57	63	41	34	11	23	23	15
P	0.73	0.83	0.77	0.78	0.81	0.74	0.88	0.69	0.72	0.85
R	0.50	0.79	0.64	0.67	0.81	0.81	0.94	0.90	0.77	0.89
F_1	0.60	0.80	0.70	0.72	0.81	0.78	0.91	0.78	0.75	0.87

Table 6 shows the performance of object detection. According to our assessment, this performance was able to be achieved because the combination of the BiLSTM and CRF models promoted the advantages of each model in detecting objects and their sentiments. In addition, the features extracted also help the training model to more accurately determine the location of words that indicate objects and words define sentiments of objects.

Table 7. The performance of objects' sentiments detection.

Actual	Predicted as Po	Predicted as Ne	P = 0.85
Po = 918	TP = 769	FN = 149	R = 0.84
Ne = 840	FP = 137	TN = 703	$F_1 - 0.84$

From Table 7, it can be seen that the positive (Po) class has performed better the negative (Ne) class. Intuitively, one of the main reasons for the low performance is that the training data contains fewer words indicating a negative sentiment. We believe that, with the construction of a large data warehouse and a better balance between words indicating relevant factors, this result can be significantly improved.

Using the results from determination of the objects and their sentiments, the binary decision tree for the topic is built as in Fig. 3, with six rules being created, shown in Table 8.

Table 8. Extracted rules.

#	Rules
1	$t_i \not\supseteq \neg battery \rightarrow Yes$
2	$t_i \supseteq \neg battery \wedge t_i \not\supseteq style \rightarrow No$
3	$t_i \supseteq \neg battery \wedge t_i \supseteq style \wedge t_i \not\supseteq camera \rightarrow No$
4	$t_i \supseteq \neg battery \wedge t_i \supseteq style \wedge t_i \supseteq camera \wedge t_i \not\supseteq \neg rom \rightarrow Yes$
5	$t_i \supseteq \neg battery \wedge t_i \supseteq style \wedge t_i \supseteq camera \wedge t_i \supseteq \neg rom \wedge t_i \not\supseteq software \rightarrow No$
6	$t_i \supseteq \neg battery \wedge t_i \supseteq style \wedge t_i \supseteq camera \wedge t_i \supseteq \neg rom \wedge t_i \supseteq software \rightarrow Yes$

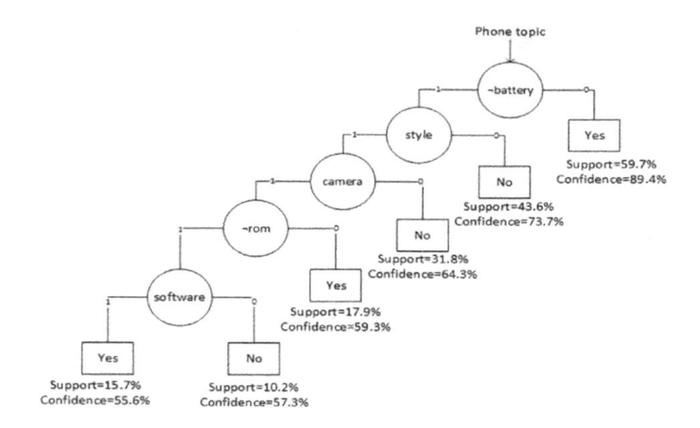

Fig. 3. Example of the binary decision tree.

Table 9. The performance of decision-making.

Actual	Predicted as Yes	Predicted as No	P= 0.83
Yes = 985	TP = 728	FN = 257	R = 0.74
No = 620	FP = 153	TN = 467	F_1 = 0.78

Looking at the Table 9, by applying the binary decision tree and six rules are created from training data for testing data, there are 1195 user's opinions are right supported decision-making, in which, "Yes" class is 728 tweets and "No" class is 467 tweets. This means the proposed method can support users in decision-making the problem related to phone topic with the accuracy of 75%. Thus, the error ratio of the method is about 25%. Therefore, this method is relatively good in supporting decision-making based on mining objects' sentiments from entire tweets. However, the method generally performed with the "No" class better than with the "Yes" class. The cause may be due to uneven distribution between factors in the data. From the above analysis, we find that the proposed method can give the result of decision-making quite good in term of the error ratio and achieved information.

6 Conclusion and Future Work

This paper proposed a method for supporting decision-making by combining sentiment analysis for objects with data mining on a binary decision tree. The proposed method consists of three main steps. First, objects and sentiments of those objects are identified in each tweet. Second, converting objects' sentiment in tweets into the form of boolean values. Finally, a binary decision tree is built and mined. The experimental results demonstrate the efficacy of the proposed approach in terms of the error ratio and received information. Under certain conditions, however, the performance is somewhat poor, which is mainly because of imbalance between labels in the data. Therefore, our future work must focus on increasing the stability of the proposed method.

Acknowledgment. This research was supported by Basic Science Research Program through the National Research Foundation of Korea (NRF) funded by the Ministry of Science, ICT & Future Planning (2017R1A2B4009410). And this work has supported by the National Research Foundation of Korea (NRF) grant funded by the BK21PLUS Program (22A20130012009).

References

1. Baccianella, S., Esuli, A., Sebastiani, F.: SENTIWORDNET 3.0: an enhanced lexical resource for sentiment analysis and opinion mining. In: LREC, vol. 10, pp. 2200–2204 (2010)
2. Bohanec, M.: Decision making: a computer-science and information-technology viewpoint. Interdisc. Description Complex Syst. INDECS **7**(2), 22–37 (2009)
3. de Albornoz, J.C., Plaza, L., Gervás, P., Díaz, A.: A joint model of feature mining and sentiment analysis for product review rating. In: Clough, P., et al. (eds.) ECIR 2011. LNCS, vol. 6611, pp. 55–66. Springer, Heidelberg (2011). https://doi.org/10.1007/978-3-642-20161-5_8
4. Hu, M., Liu, B.: Mining and summarizing customer reviews. In: Proceedings of the Tenth ACM SIGKDD International Conference on Knowledge Discovery and Data Mining, pp. 168–177. ACM (2004)
5. Huang, Z., Xu, W., Yu, K.: Bidirectional LSTM-CRF models for sequence tagging. arXiv preprint arXiv:1508.01991 (2015)
6. Li, J., Hovy, E.: Reflections on sentiment/opinion analysis. In: Cambria, E., Das, D., Bandyopadhyay, S., Feraco, A. (eds.) A Practical Guide to Sentiment Analysis. Socio-Affective Computing, vol. 5, pp. 41–59. Springer, Cham (2017). https://doi.org/10.1007/978-3-319-55394-8_3
7. Narayanan, R., Liu, B., Choudhary, A.: Sentiment analysis of conditional sentences. In: Proceedings of the 2009 Conference on Empirical Methods in Natural Language Processing: Volume 1, vol. 1, pp. 180–189. Association for Computational Linguistics (2009)
8. Pang, B., Lee, L., et al.: Opinion mining and sentiment analysis. Found. Trends Inf. Retrieval **2**(1–2), 1–135 (2008)
9. Quinlan, J.R.: Induction of decision trees. Mach. Learn. **1**(1), 81–106 (1986)
10. Tang, D., Zhang, M.: Deep learning in sentiment analysis. In: Deng, L., Liu, Y. (eds.) Deep Learning in Natural Language Processing, pp. 219–253. Springer, Singapore (2018). https://doi.org/10.1007/978-981-10-5209-5_8
11. Vu, T., Nguyen, D.Q., Vu, X., Nguyen, D.Q., Catt, M., Trenell, M.: NIHRIO at SemEval-2018 task 3: a simple and accurate neural network model for irony detection in twitter. In: Proceedings of The 12th International Workshop on Semantic Evaluation, SemEval@NAACL-HLT, New Orleans, Louisiana, 5–6 June 2018, pp. 525–530 (2018). https://aclanthology.info/papers/S181085/s18-1085
12. Yu, J., Zha, Z.J., Wang, M., Chua, T.S.: Aspect ranking: identifying important product aspects from online consumer reviews. In: Proceedings of the 49th Annual Meeting of the Association for Computational Linguistics: Human Language Technologies, vol. 1, pp. 1496–1505. Association for Computational Linguistics (2011)
13. Yussupova, N., Boyko, M., Bogdanova, D., Hilbert, A.: A decision support approach based on sentiment analysis combined with data mining for customer satisfaction research. Int. J. Adv. Intell. Syst. **8**(1&2) (2015)

Named Entity Recognition Using Gazetteer of Hierarchical Entities

Miha Štravs[1,2] and Jernej Zupančič[3,4](✉)

[1] Faculty of Computer and Information Science, Ljubljana, Slovenia
[2] Faculty of Mathematics and Physics, Ljubljana, Slovenia
miha.stravs996@gmail.com
[3] International Postgraduate School Jožef Stefan, Ljubljana, Slovenia
[4] Jozef Stefan Institute, Ljubljana, Slovenia
jernej.zupancic@ijs.si

Abstract. This paper presents a named entity recognition method which finds predetermined entities in an unstructured text. The method uses word similarities based on typical word transformations (lemmatization and stemming), word embeddings and character level based similarity to map those entities onto words in the text. The approach is language independent, though language-dependent components are used for lemmatization, stemming and word embedding, and works on any given set of entities. Special attention is given to the entities which are represented in a hierarchical form with the hypernymy-hyponymy relation. The proposed method has the following advantages: it finds the normalized form of the recognized entity name; it is easy to adjust to a new domain; it respects the hierarchical organization of entities; and due to the modular approach can be constantly improved just by updating components for lemmatization, stemming or word embedding. The proposed entity recognition method was tested on a test set of tourist queries and hierarchical entities collected from Slovenia.info tourist portal.

Keywords: Natural Language Processing (NLP) · Natural Language Understanding (NLU) · Named Entity Recognition (NER) · Word vectors · Word similarity · n-grams · Bag of words

1 Introduction

Natural language processing (NLP) is an artificial intelligence (AI) research area that addresses the extraction of information out of text or speech in natural language. One of the subtasks in NLP is natural language understanding (NLU). Smart chatbot, an AI that understands natural language and can form an answer in natural language based on information it gathers from the conversation, is

Partially supported by Joint cooperation programme V-A Interreg Slovenia-Austria, project AS-IT-IC.

F. Wotawa et al. (Eds.): IEA/AIE 2019, LNAI 11606, pp. 768–776, 2019.
https://doi.org/10.1007/978-3-030-22999-3_65

one of the NLU applications that has received significant attention in the recent years. Chatbot needs to extract information from the text it receives from a conversation. It needs to understand the intent of the text and important entities mentioned in the text.

Named entity recognition (NER) is about finding the names of the entities in unstructured text. Usually, NER systems are used for tagging the entities in the text with entity tags, such as *"person"*, *"location"*, *"organization"*, *"time"*, and others [11]. While general tags are useful for some applications, accuracy of NER system can be improved by focusing on a target domain [5]. The method discussed in this paper also limits its domain by specifying the entities that will be searched for. This is done by supplying the entities in a form of a gazetteer or entity dictionaries like in [2,3,7]. Other methods aim for independent systems which don't rely on these entity dictionaries but use trained models such as neural networks [4].

The proposed method can extract the entities the user is interested in, even when the entity appears in the text in a different form (different tense, wrong spelling, synonym, added suffix, etc.) as opposed to the majority of systems that detect only names with the exact match.[1] What additionally sets it apart from other domain-based NER methods are the following properties. First, the method is easy to adapt to a new domain by simply changing the entity gazetteers (no large labelled corpora are needed), which can also contain hierarchically organized entities in order to also match a more general entity. Second, this method labels the words in a text using exact information about the detected entity (entity name in its normalized form) instead of just its entity tag. And third, the proposed method is modular in the sense that its parts can easily be swapped when better tools for stemming, lemmatization or word vector embedding are identified, giving it an additional advantage over methods that use end to end training for the NER task such as neural networks.

The rest of the paper goes as follows. In Sect. 2 an overview of the proposed method is given. In Sect. 3 the experimental setup used for testing the proposed method is described, while in Sect. 4 the results are presented and discussed. Section 5 concludes the paper.

2 Overview of the NER Method

The general idea is to map words from the given gazetteer onto words in the input text. An algorithm preprocesses the gazetteer in order to obtain a normalized representation of each present entity. When entities are given in a hierarchical form, each node name in the hierarchical tree is considered an entity. When a sentence is inputted for the NER task it is first preprocessed using the same processing pipeline as for the given entities. Then the entities are ranked according to the similarity to the input sentence representation. If the highest ranked entity is given a score above a predefined threshold, the entity is deemed as recognized and the corresponding words are hidden from the input sentence in

[1] https://spacy.io/api/phrasematcher.

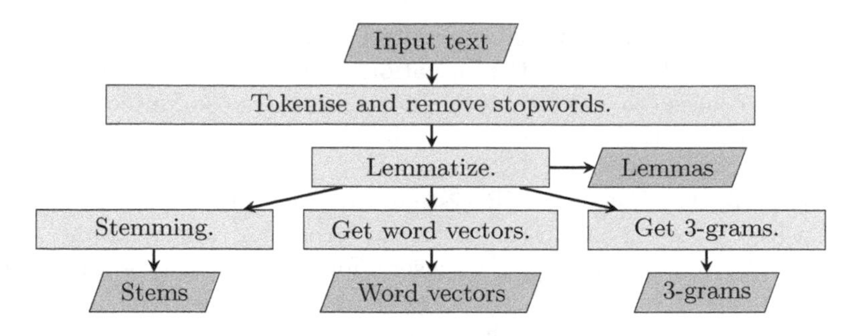

Fig. 1. Entity preprocessing

order to label each word with at most one entity. The ranking of the entities and entity recognition is repeated until the highest score falls below the predefined threshold. The recognized entities are returned as a result.

For instance in the text *"Is there a natural heritage attraction near Maribor I could go see?"* two entities can be recognized: *"natural heritage attraction"* – of an *"attraction"* category and *"Maribor"* – of a *"location"* category.

2.1 Preprocessing

Entities and input text go through the same pipeline (Fig. 1) for data preprocessing. The words are extracted using a tokeniser and non-informative words (stopwords) are deleted. Next, lemmatizer and stemmer are used for word normalization. Normalization is needed for a more accurate comparison of words with different suffixes. Tokenisation and lemmatization is performed using *Spacy*[2] for English. For Slovenian own implementations based on the *reldi-tagger*[3] and *reldi-tokeniser*[4] were used for lemmatization and tokenisation, respectively. Snowball [9] was used for stemming in both languages. Word vectors and 3-grams are then obtained from the lemmas. All obtained 3-grams are merged into one set with no repetition. Word vectors are used to match different words with similar meaning like *"accommodation"* and *"room"*. FastText [1] word vector embeddings converted to a *pymagnitude* [8] format were used.

2.2 Entity Ranking

In order to label the words from the input with an entity, similarity indicator or score for each entity is computed, which indicates the likelihood that an entity really appears in the input. Scoring is performed based on distance metrics and their combination.

[2] https://spacy.io/.
[3] https://github.com/clarinsi/reldi-tagger.
[4] https://github.com/clarinsi/reldi-tokeniser.

Jaccard Distance. Using Jaccard distance the method computes the size of the intersection of multisets (elements can appear in the multiset multiple times) of stems from the input and an entity using Eq. 1.

$$\frac{|\text{inputStems} \cap \text{entityStems}|}{|\text{numberOfEntityWords}|^p}. \tag{1}$$

3-gram Similarity. 3-grams (a set containing all combinations of 3 consecutive characters appearing in the processed input) are obtained from entity and input text lemmas. Jaccard distance is then calculated between the two sets of 3-grams as in Eq. 2.

$$\frac{|\text{input3\_grams} \cap \text{entity3\_grams}|}{|\text{entity3\_grams}|} \tag{2}$$

Common Prefix Similarity. This similarity indicator includes computing the sum of the longest common prefix matching for each word from the entity from all the words from the input, and dividing it by the number of words in the entity to the power of p as in Eq. 3, where common_prefix(iL,eL) is the character length of the common prefix of iL and eL and inputLemmas and entityLemmas are multisets containing lemmas from input and entity, respectively.

$$\frac{\sum_{\text{eL} \in \text{entityLemmas}} \left(\max_{\text{iL} \in \text{inputLemmas}} \frac{|\text{common\_prefix(iL,eL)}|}{|\text{eL}|} \right)}{|\text{numberOfEntityWords}|^p} \tag{3}$$

Word Vector Similarity. For each word in an entity a word from the input text with the largest word vector cosine similarity is chosen. As in common prefix similarity, those maximum similarities are then summed up and divided by the number of words in the entity to some power of p as in Eq. 4, where inputWordVectors and entityWordVectors are lists containing word vectors from input and entity, respectively.

$$\frac{\sum_{\text{eWV} \in \text{entityWordVectors}} \left(\max_{\text{iWV} \in \text{inputWordVectors}} \frac{\text{iWV} \cdot \text{eWV}}{\|\text{iWV}\| \cdot \|\text{eWV}\|} \right)}{|\text{numberOfEntityWords}|^p} \tag{4}$$

Combination 1. Uses a convex combination of the common prefix similarity and word vectors similarity. A convex combination of similarities A and B is defined as $t \cdot A + (1 - t) \cdot B$, where $0 \leq t \leq 1$.

Combination 2. Uses a convex combination of 3-gram similarity and word vectors similarity.

In each similarity score a denominator of $|\text{numberOfEntityWords}|^p$, where $p \in (0, 1)$ is used to prioritize entities with more words. One case where this is useful is when we have input text *"Will Apartment Pine be free next week?"* and in the gazetteer both *"Apartment"* and *"Apartment Pine"* exist. Both entities match perfectly, however, the second one is a more specific choice (entity *"Apartment*

Pine" is an instance of a category entity *"Apartment"*) and is more appropriate in this case.

Synonyms are taken into account by specifying a list of synonyms for each entity. Synonyms are considered as a standalone entity in the preprocessing step, however, they reference back to the original entity in the entity labelling step.

Entities with hypernymy-hyponymy relation are also addressed in the method. Hypernymy entities (more general ones) receive a boost score that is computed based on the score of their hyponymy. In sentence *"I want an apartment in the center of Ljubljana."* the *"Apartment"* entity is more likely to be picked because the entity list has a lot of entities corresponding to different apartments containing word *"Apartment"* in them. Hyponymy boost is limited by a predefined value so that the method does not favor more general entities too much.

The entities with the score that exceed the predefined thresholds are returned as a result of the proposed NER system.

3 Experimental Setup

The proposed method was tested on the tourism domain. In order to measure the method performance, tourism attraction entities and questions from tourists that address those attractions were gathered, cleaned and organized into a benchmark. A list of entities organized in the predefined format required by the method is all that is needed in order to apply the method in a new domain.

3.1 Data

Tourism attraction entities were extracted from 8.915 Slovenian attractions from the main Slovenian tourist portal Slovenia.info[5] and 217 municipality names.[6] Attraction entities were organized in a hierarchical relation. Attractions (lower-most level, e.g. *"Adventure park Postojna"*) were labelled by subcategories (middle level, e.g. *"Adrenaline parks"*) and categories (upper-most level, e.g. *"Adventure sports"*). Each attraction was labelled with at least one of the 13 different categories All the municipality names, however, belonged to the *"Municipality"* category.

The names of attractions and municipalities were mostly in Slovenian and some of them already had English translations. Yandex[7] was used for machine translation of the remaining names.

Tourist queries dataset consisted of sentences labelled with a list of entities from the entity dataset. There were sixty-eight sentences in total. In order to thoroughly test the detection of entities, diverse entities were used and the queries were formed in a way that different syntactic forms of words from entities were used. An example of a tourist query is *"I have problems with my eyes*

[5] https://www.slovenia.info/en/map.
[6] https://sl.wikipedia.org/wiki/Seznam_ob%C4%8Din_v_Sloveniji.
[7] https://translate.yandex.com/.

and would like to go to a spa near Murska Sobota". In this case, the query was labelled with the following entities: *"Disease eyes"*, *"Thermal baths an89d spas"* and *"Murska Sobota"*.

3.2 Experiment

The proposed method was tested on a set of 68 sentences containing 125 entities in total. The performance of the method using each similarity indicator from Subsect. 2.2 was then evaluated by precision (P, calculated by dividing the number of all correct entities found with the number of all the entities found), recall (R, calculated by dividing the number of all correct entities found with the number of all the entities searched for) and F1 score (Eq. 5).

$$\mathrm{F1} = 2 \cdot \frac{R \cdot P}{R + P} \tag{5}$$

In order to obtain the best parameters, local hill-climbing optimization with restarts [10] was used to find near-optimal values for parameters for each similarity indicator.

In order to assess the average performance bootstrapping [6] was used. Bootstrapping is commonly used for the calculation of confidence intervals or for hypothesis testing. It is a statistical technique that belongs to a group of resampling methods. A number of instances are picked from the testing set with repetition and used for the testing, which is then repeated several times. In this case, 100 sentences were sampled 10.000 times. P, R and $F1$ scores were calculated for each sample list of tourist queries.

To summarize, the proposed entity recognition method was tested using two languages (Slovenian and English), six different distance measures, and two different vector embeddings from Wikipedia (Wiki WV) and Common Crawl (CC WV). The performance was compared to the ElasticSearch[8] (ES), where the search was performed using the *_search* endpoint and no further optimization of internal ES parameters was performed. The entities were the indexed documents and the queries were the tourist sentences. Optimal threshold was used to extract the relevant entities. ES was used with four preprocessing steps: in first, no preprocessing was done, in the second, the stopwords were removed, in third and fourth stopword removal was followed by lemmatization and stemming, respectively.

4 Results and Discussion

Figure 2 presents the bootstrapped F1 performance metric scores for Slovenian and English datasets in blue (darker) and orange (lighter) box plots, respectively. Any similarity indicator used in the proposed method significantly outperformed any ES based NER. Comparison to other NER systems was either not possible

[8] https://www.elastic.co/products/elasticsearch.

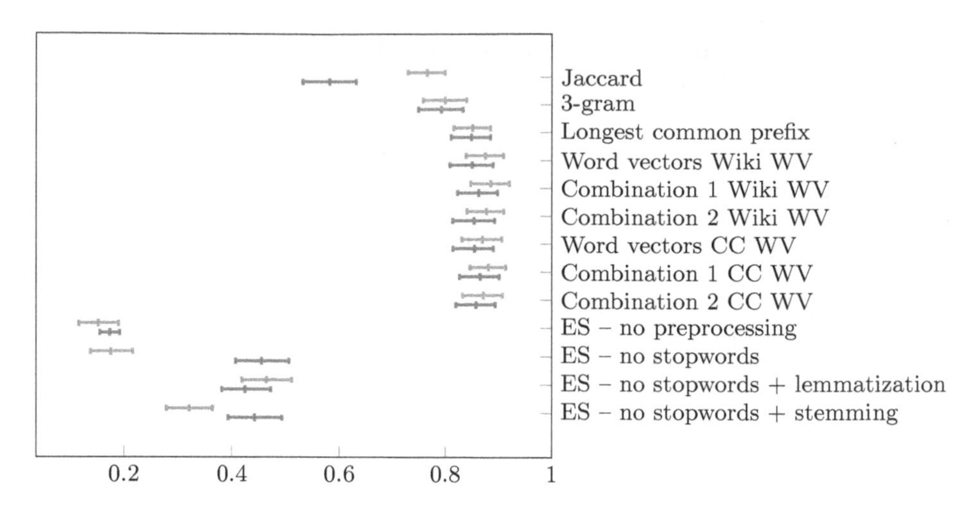

Fig. 2. Box-plots (5th, 50th and 95th percentile) of bootstrapped F1 scores. Blue lines (darker) are used for English and orange (lighter) for Slovenian benchmark. (Color figure online)

(due to unavailable source code or data) or would not provide greater insight (e.g. PhraseMatcher from the Spacy package, which only does exact matching).

Jaccard similarity method was used as a baseline. All other methods, except the 3-gram method, were statistically better according to all performance metrics. All other methods performed somewhat similarly, with the performance of state-of-the-art word vector-based similarity performing only marginally better than the simple longest common prefix method. Since the computation requirement costs of including the word vectors computation into the pipeline greatly increase the results indicate that the added vector embeddings may not always be required. Reason for good performance of the longest common prefix method could be that the English and Slovene language usually have roots of the words at the beginning, giving longest common prefix an advantage, which could be lost in languages such as German, where words are combined in order to form new words.

The only significant difference regarding the language was when using the Jaccard similarity-based method, which performs poorly for the English version, mainly due to the worse recall metric. This could be a result of the poor machine translation.

Different vector embeddings made no noticeable difference indicating that increased computational requirements due to the larger word vector files are not needed.

5 Conclusion and Future Work

A method for NER is presented, which uses word embeddings and character level similarity metrics. What sets it apart from other NER methods are: labelling of words with normalized entity names; ease of adjustment to a new domain with hierarchical entities; and modularity. The possibility to map specific entities onto the text instead of just tagging the entities in the text with tags can be useful in cases when we are interested only in certain entities and would like to know exactly which entities appear in text. One such case would be a tourist support chatbot, which is only interested in attraction entities for which additional can be given. Ease of use, which does not require a large amount of data, is an additional advantage of the proposed method since the majority of users are not in possession of a large corpus of labelled data. The users of the proposed method only need to specify the entities in a hierarchical format they are interested in. An implementation of the method can be found in a code repository in the form of a Python package.[9]

In the future work we plan to test the performance of the method on additional languages and additional test sets, also including domains other than tourism. Due to the modularity of the method, we plan to assess the performance of the method by including other state-of-the-art tools for stemming, lemmatization and word vector embedding.

References

1. Bojanowski, P., Grave, E., Joulin, A., Mikolov, T.: Enriching word vectors with subword information. Trans. Assoc. Comput. Linguist. **5**, 135–146 (2017)
2. Kazama, J., Torisawa, K.: Exploiting Wikipedia as external knowledge for named entity recognition. In: Proceedings of the 2007 Joint Conference on Empirical Methods in Natural Language Processing and Computational Natural Language Learning (EMNLP-CoNLL) (2007)
3. Kozareva, Z.: Bootstrapping named entity recognition with automatically generated gazetteer lists. In: Proceedings of the Eleventh Conference of the European Chapter of the Association for Computational Linguistics: Student Research Workshop, pp. 15–21. Association for Computational Linguistics (2006)
4. Lample, G., Ballesteros, M., Subramanian, S., Kawakami, K., Dyer, C.: Neural architectures for named entity recognition. arXiv preprint arXiv:1603.01360 (2016)
5. Maynard, D., Tablan, V., Ursu, C., Cunningham, H., Wilks, Y.: Named entity recognition from diverse text types. In: Recent Advances in Natural Language Processing 2001 Conference, pp. 257–274 (2001)
6. Mooney, C.Z., Duval, R.D., Duvall, R.: Bootstrapping: A Nonparametric Approach to Statistical Inference. No. 94-95. Sage, Thousand Oaks (1993)
7. Nadeau, D., Turney, P.D., Matwin, S.: Unsupervised named-entity recognition: generating gazetteers and resolving ambiguity. In: Lamontagne, L., Marchand, M. (eds.) AI 2006. LNCS (LNAI), vol. 4013, pp. 266–277. Springer, Heidelberg (2006). https://doi.org/10.1007/11766247_23
8. Patel, A., Sands, A., Callison-Burch, C., Apidianaki, M.: Magnitude: a fast, efficient universal vector embedding utility package. arXiv preprint arXiv:1810.11190 (2018)
9. Porter, M.F.: Snowball: a language for stemming algorithms (2001). http://snowball.tartarus.org/texts/introduction.html

[9] https://repo.ijs.si/DIS-AGENTS/entity-expert.

10. Russell, S.J., Norvig, P.: Artificial Intelligence: A Modern Approach. Pearson Education Limited, Kuala Lumpur (2016)
11. Tjong Kim Sang, E.F., De Meulder, F.: Introduction to the CoNLL-2003 shared task: language-independent named entity recognition. In: Proceedings of the Seventh Conference on Natural Language Learning at HLT-NAACL 2003-Volume 4, pp. 142–147. Association for Computational Linguistics (2003)

The BRAVO: A Framework of Building Reputation Analytics from Voice Online

Bogeun Jo, KyungBae Park, and Sung Ho Ha(✉)

Kyungpook National University, 80 Daehak-ro, Daegu, South Korea
bjo@emich.edu, iamkbpark@gmail.com, hsh@knu.ac.kr

Abstract. This paper provides a framework to efficiently discover production performance and its application plan by analyzing massive amounts of comments from online review data, especially in the field of hospitality. In order to achieve the goal, two stages of text analytics of sentiment analysis and structural topic model estimating are integrated to classify sentimental polarity of each reviews and elicit hidden dimensions of products or services. Based on these dimensions and polarities, this paper verifies key attributes which impact customer satisfaction by adapting logistic regression. This study extends prior research limitation which focused on discovering the product defect by (1) strength detection, (2) time series analysis, and (3) explanation of the relationship between crucial factors and polarity of the review as a proxy of customer satisfaction. By integrating text analytics from computational linguistic and a traditional statistical method, this paper is expected to contribute on both academical and practical implications.

Keywords: Text mining · Customer satisfaction · STM · Sentiment analysis

1 Introduction

Over the past decades, social media grows exponentially with various forms such as social networks and customer reviews. This is mainly attributable to the advancement of information technology in general, and of internet technology in specific [1, 2]. Explosively accumulated textual information, publicly available online, dramatically change the way how customers behave (i.e. shop, travel) and how management reacts accordingly. We have witnessed that consumers are heavily seeking information online and they have no fear of sharing their experiences or opinions regarding products or services. This in turn gives us rich sources of information for both customer and management, triggering the hope of discovering new insights previously unattainable with limited samples of survey or interviewing [3].

From the perspective of the product design and marketing strategies, feedback from customers provides invaluable information to the management. While it plays a critical role in promoting all facets of products, it may also expose the deficiencies or defects of products if not properly retained to improve. Developing and implementing suitable action is time consuming and sensitive. Therefore, it is natural to demand for a mechanism that can effectively identify, categorize, and prioritize any underlying attributes in a timely manner to management.

F. Wotawa et al. (Eds.): IEA/AIE 2019, LNAI 11606, pp. 777–790, 2019.
https://doi.org/10.1007/978-3-030-22999-3_66

In response to such demand, the purpose of the thesis is threefold. First, the study attempts to classify the review documents into different polarity by leveraging the user-comment. Second, it aims to identify buried dimensions from reviews, to track how dimensions have changed over time, and to compare them with the covariate of information, polarity. Third, it aims to evaluate the result of text analytics and further examine the relationships between the dimensions discovered and customer satisfaction.

To achieve this, the thesis utilizes sentiment analysis to classify the polarity of reviews instead of embracing user-rating score originally posted by reviewers. Next, topic model is applied by incorporating covariates of (1) posting date to explore the trend of dimensions, and of (2) polarity to compare the differences in identified topics (dimensions). Eventually, Logistic regression is applied to investigate the impact of discovered dimensions as independent variables on the polarity as a dependent variable.

Before adapting this framework to every single industry, studying the case of hospitality industry would be a good start. Hospitality industry is competitively easy to detect customer satisfaction or dissatisfaction from rich sources of online reviews and to get prompt feedback after service improvement.

In general, the major inspiration of the thesis is the work of Abrahams et al. [4] where specific attention was paid to discover the product defect from user-generated contents with integrated text analytic framework. It is natural to find out weaknesses or shortcomings proactively to provide better customer satisfaction through the improvement of products or service, thereby maximizing the profit. However, in addition to weakness detection, this study also incorporates the analysis of strengths. It will be conceivable to switch from a crisis to an opportunity by identifying weaknesses, and strengthen market dominance by identifying strengths.

2 Literature Review

Before the emergence of internet, there were limited choices for tourism suppliers such as hotels and airlines to distribute their services to consumers, using intermediaries such as travel agents and tour operators. The emergence of electronic commerce has had a massive impact on the tourism and hospitality industry [5]. E-commerce market has already matured and there is nearly no industry untouched by the impact of internet and travel industry is not an exception.

To date, many researches related to social media have begun to receive great attention. Collectively, social media can be defined as online services provided by individual users to create user-generated contents (UGC) and to interact with participants through sharing the generated information. The major sources of UGC come from reviews, blogs, comments, feedback, and so on.

The noticeable evidence from previous studies on review mining reveals that there is a strong relationship between firm's performance and reviews in hospitality industry. Yacouel and Fleischer [6] study the case of the online hotel market. Their study suggests that online travel agents (OTAs) such as Booking.com play a critical role in building hotel reputation and encourage the service provider to place greater effort into service quality. Sparks, So, and Bradley [7] examines how consumers perceive and evaluate potential customers in regard to a negative review and any accompanying hotel response. A timely response yielded favorable customer inference, and this study contribute to the current understandings of effective online reputation management.

Utilizing one of the historical topic models, latent semantic analysis (LSA), Xu and Li [8] analyze online customer reviews of hotels from different types of hotels. The study suggests a clue for hotel service provider to enhance customer satisfaction and ease customer dissatisfaction by improving service and satisfying the customers' needs for the different types of hotels.

Various online medium with a wide variety of methods have been applied by leveraging the textual information to draw more incisive insights, such as sales performance, market reaction, and defect detection. However, only a handful of studies investigate the issues from quality management perspective. To fill such gap and to enhance the literature to a broader perspective, this study proposes an integrated text mining framework to uncover the important latent dimensions that can flexibly extend to examine the relationship with other interests of phenomenon by employing textual analytics and statistical methods.

3 Method

Figure 1 depicts the framework of overall process. Followed by the framework, each component of the framework is discussed more detail.

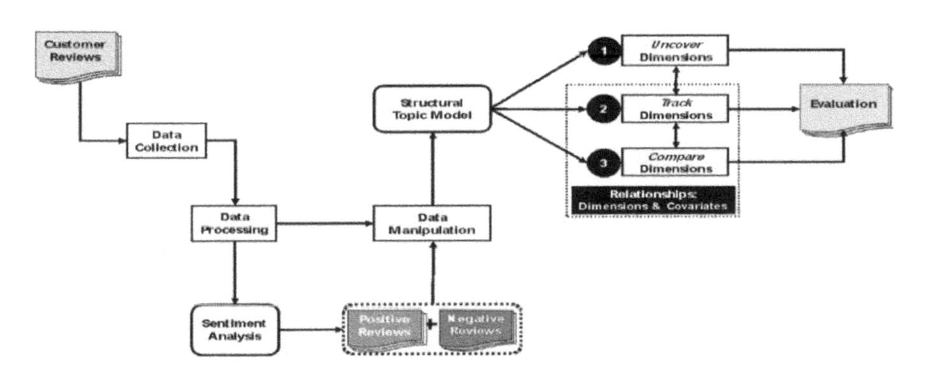

Fig. 1. Research framework

3.1 Data Collection

For this study, TripAdvisor was selected in order to obtain the review data. in that it has been considered to be the most representative online review community of providing and sharing customer experience in the field of hospitality domain. As of 2018, TripAdvisor (https://tripadvisor.mediaroom.com/us-about-us) boasts of having over 730 million reviews and opinions covering approximately 8.1 million accommodations, airlines, experiences, and restaurants around the globe. It provides travelers various sorts of information such as where to stay, how to fly, what to do and where to eat which nearly covers the all the facets of the needs for traveling.

Web scraper was developed to automatically extract all the reviews available at the time of extraction for Wynn Las Vegas hotel located in Las Vegas, NV, USA and produce csv file. After thorough inspection for the structure of information displayed in TripAdvisor and identification of the proper tag information for the contents to be scraped, scraper was built using Python programming. For the purpose of this study, data were collected as having *Posting Date*, *Rating*, *Review Title* and *Review*.

3.2 Data Preprocessing

Prior to employing the appropriate text mining analysis, preprocessing the data for subsequent analysis is a significant step. R 3.4.1 was used for the rest of the study from data preprocessing to the final analysis. In R program, tm package was mainly used in that it offers an excellent text mining framework tool to deal with diverse tasks of handling and preprocessing the textual data [9]. Texts were transformed to lower case, and to filter out stop-words in the review texts the general stop-words (e.g., 'a', 'of', 'the') from the SMART information retrieval system developed by Cornell University were applied. Next, punctuations and numbers were removed.

After processing and manipulating the review data, Part of Speech (POS) tagging was conducted using Stanford CoreNLP with wrapper package, CleanNLP, in R. A POS Tagger is a piece of software that reads text and assigns parts of speech to each word (and other token), such as noun, verb, adjective, etc. Instead of stemming the words, POS was conducted due mainly to extract and utilize lemmatized nouns for the purpose of structural topic model to be analyzed later and in turn, for better cohesiveness as well as interpretability for topics since the form of lemma returns a dictionary form of English word [10].

3.3 Sentiment Analysis

Sentimental analysis or opinion mining is a set of recently developed web mining techniques that performs analysis on sentiment or opinions. Generally, sentiment analysis classifies the state of a polarity for one's emotion or opinion expressed in texts into positive, neutral, or negative state [11].

Apart from the series of preprocessing and manipulating the review data, original review texts, the column of Review, are used for sentiment analysis at document (review) level to classify reviews into different polarity for subsequent analysis.

Sentiment analysis was conducted by neural network-based model called *Recursive Neural Tensor Networks* (RNTN) implemented in Stanford *CoreNLP* 3.7 using the wrapper package, *stansent*, in R to classify reviews. RNTN can update parameters in the neural network based on the various combinations of words and phrases. Thereby, RNTN sentiment analysis is more domain-oriented model in terms of that same words can be assigned with different scores depends on the document [12]. The results from sentiment analysis return decimal sentiment scores from −1 to 1. This study, they are transformed to 2 levels (Negative (0) and Positive (1)) as values of new column, *Polarity*. Table 1 is a sample row of data processed so far.

Table 1. Sample data processed after sentiment analysis

ID	RATING_ DATE	Rating	Sentiment	Polarity	Review Title	Review	LEMMA	LEMMA_ NOUN
41	2016-08-27	5	0.5	1	Mr. Barry	We stayed here for 4 days had a meal at the frank Sinatra restaurant stunning the people there couldn't do enough for us made us feel like one of the family food was lovely please stay here you will love this place it's behind the mgm grand so not on strip lovely and quite clean and beautiful	Stay day meal frank sinatra restaurant stunning people make feel family food lovely stay love place mgm grand lovely clean beautiful	Day meal restaurant people feel family food stay love place mgm

3.4 Structural Topic Model

This research employs Structural Topic Model (STM) [13] to uncover the hidden dimensions in customer review corpora, and to examine the relationships with three covariates: posting date of reviews and the type of polarity of reviews obtained from sentiment analysis.

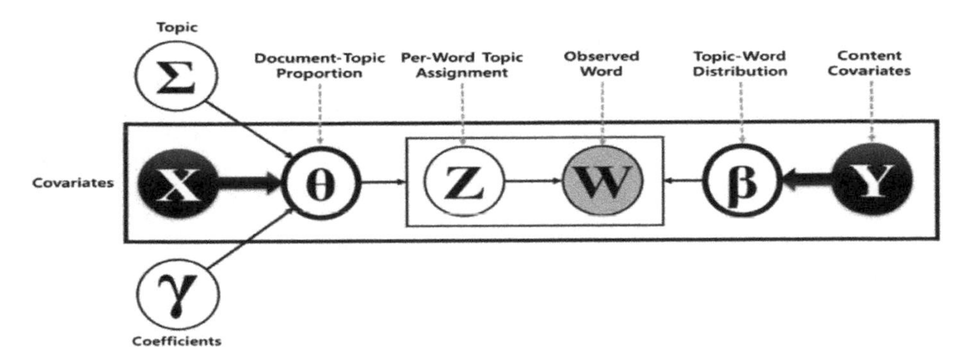

Fig. 2. Plate diagram of structural topic model

Figure 2 depicts the plate diagram for STM. Here, the study identifies underlying dimensions and their relationships with three covariates from customer reviews as follows, and topic refers to a dimension in this study. Each customer review document

d is assumed to be generated such that from a prior distribution, a review (document)-dimension (topic) distribution θ_d is drawn. For each word n in the review, a dimension for that word is pulled from a multinomial distribution from the distribution of dimensions $Z_{d,n} \sim Mult(\theta_d)$. Provisional to the dimensions chosen, the observed word $w_{d,n}$ is selected from a distribution over the vocabulary $w_{d,n} \sim Mult(\beta_{Zd,n})$, and $\beta_{k,v}$ is the probability of pulling the v-th word in the vocabulary for dimension k. Unlike LDA, STM allows researcher to correlate document-topic proportion θ and covariates (X). It further allows to examine the relationships with the prevalence of covariates and topics (dimensions) using a regression model with covariate such that $\theta_d \sim LogisticNormal(X_\gamma, \Sigma)$ [13, 14].

In order to find an optimal topic number, this study applied *searchK* function of the package, *stm*, in R, which computes diagnostic properties for models with different number of topics. Along with the indicators of log-likelihood [15], semantic coherence [16] shown in Fig. 3, a topic number (k) of 14 was drawn after testing different numbers of topics from 5 to 30.

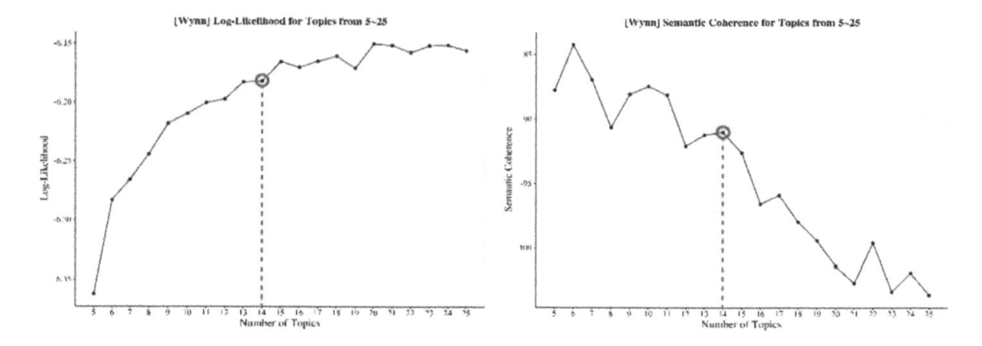

Fig. 3. Plot of log-likelihood and semantic coherence

Now that the optimal number of topics (K) as 14 has discovered, STM is fitting by specifying K = 14. When building the model, covariates of interests, *Posting Date* and *Polarity*, are allowed to be estimated simultaneously in STM.

4 Analysis Results

4.1 Results of Sentiment Analysis

In TripAdvisor, customers can write a review and rate the hotel in 5 star rating scale, ranging from 1 to 5. In order to compare the sentiment-rating scale and user-ratings, unit of ratings was required to transform and match into the same scale. Transformation of user-rating scores into 3 levels of negative, neutral, and positive was performed such that user-ratings of 1 and 2 were collapsed as negative review, and those of 4 and 5 were collapsed as positive review. Neutral reviews required no transformation.

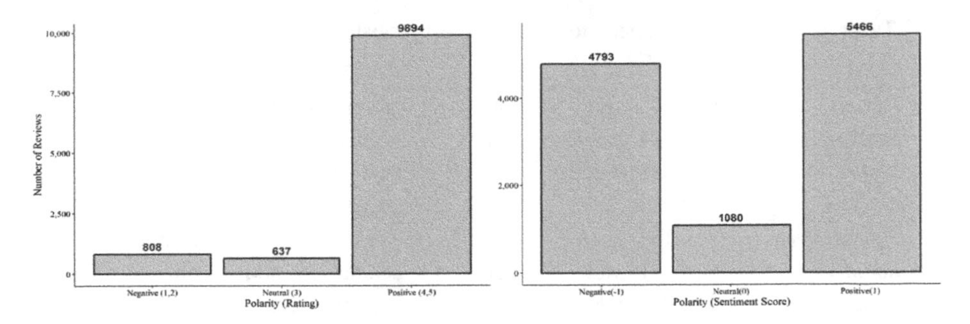

Fig. 4. Polarity distribution of reviews by user-rating and sentiment-rating scale

Sentiment scores ranging from −1 to +1 for each review was first reckoned as a result of sentiment analysis. Sentiment scores between −1 and 0 were collapsed as negative review coded as −1, and scores between 0 and 1 as positive reviews coded as 1. Sentiment score of 0 were collapsed as neutral review. Figure 4 describe the polarity distribution of reviews by user-rating and by sentiment-rating.

It turned out that the reviews are heavily rated as 4 or 5 stars, and those are nearly 90% of the overall reviews. Analyzing textual information to classify the reviews shows us a completely different view of the polarity space. It could be in part because of self-selection bias [17] in that people might rate it differently than they in fact stated. While rating score posted by user provides meaningful insights of overall sentiment about the products or service, it is limited to understand what aspects of products or service people commonly concerned about unless customers reveal on the reviews. Presumably, this is a strong indication to warn the risk of utilizing original user-rating as a measure to classify the polarity of reviews and to detect strengths and weaknesses. Thus, the rescaled rating by data-driven sentiment analysis can provide the results with minimal bias or errors than subjective one-dimensional rating score given by reviewers.

4.2 Results of Structural Topic Model

Descriptive labels are attached to discovered 14 dimensions in reference to the ones suggested in Guo et al. [18]. Two key information, top keywords based on Topic-Word distribution ($\beta_{k,v}$) and the highly associated reviews based on Document-Topic proportions (θ_d), have been examined in order to decide labels for each dimension. Prior to concluding decisive labels for dimensions, cross-validation is often helpful to avoid subjective interpretation of dimensions, thereby increasing the validity of the reliability.

To cross validate, the task of labeling dimensions was conducted by three people from Amazon Mechanical Turk (AMT), credible platform of online labor market. Top 30 keywords for each dimension were provided as a result of structural topic model along with instructions for labeling dimensions. Eventually, labels fulfilled by workers were collected and aggregated for their consistency, in order to finalize naming the dimensions by researcher. Table 2 summarizes the description of dimensions.

Table 2. Summary of fourteen topics with most probable words and proportions(Θ)

Label	#	Θ	word1	word2	word3	word4	word5
Casino	1	0.066	slot	poker	dealer	play	game
Smoking	2	0.034	smoking	king	cigarette	queen	cup
Security Mgt.	3	0.055	security	guy	incident	report	guard
Amenity	4	0.063	fee	wifi	charge	internet	credit
Style & Deco	5	0.157	class	place	world	love	entertainment
Room View	6	0.111	view	mountain	ceiling	window	corner
Bathroom	7	0.069	tub	toilet	shower	sink	bath
Communication	8	0.049	front	bell	desk	stroller	registration
Homeliness	9	0.095	year	trip	time	friend	visit
Wedding	10	0.043	wedding	ceremony	dream	theater	photo
Tower Suite	11	0.049	tower	tableau	parlor	entrance	massage
Location	12	0.084	location	conference	center	mall	facility
Nightclub Noise	13	0.057	noise	nightclub	club	tryst	sleep
Pool	14	0.060	sun	shade	cabana	terrace	bellagio

The dimensions identified by structural topic model reveal both expected and interesting insights. Dimensions of *Security* and *Communication* are the likely dimensions that management constantly deals with customer during the entire stay. By the same token, all dimensions except *Casino*, *Smoking*, *Nightclub Noise* and *Tower Suite* are the expected dimensions of hotels in general. Stated differently, they are expected to be found supporting the topic model successfully identified the key dimensions proposed in previous hospitality literature.

Meanwhile, dimensions of *Casino*, *Smoking*, *Nightclub Noise* and *Tower Suite* are rather location or hotel-specific dimensions. More precisely, considering the hotels under research is located in Las Vegas, *Casino* and *Smoking* seems plausible to be drawn and can be considered as location-specific dimension whereas dimensions of *Nightclub Noise* and *Tower Suite* are deemed as hotel-specific dimension.

However, generally expected dimensions such as *Transportation* in recent study [18] does not appear in our study. It may be explained with the fact that the airport is located in the middle of the city and roughly takes about 10–20 min by taxi to reach the main central strip where the majority of hotels are located. Also, most of nearby hotels or attractions are walking distance and preferable by walk.

4.3 Relations Between Dimensions and Covariates

Topical prevalence was estimate by Posting Date to see how these dimensions vary across time. As can be seen in Fig. 5, we are now able to analyze how particular dimensions' proportions have changed over time beyond the discovery of the dimensions. From the analysis of the topical prevalence by Posting Date from 2005 to 2016, three major patterns (rising, falling, and fluctuating) and one steady pattern (Dimension 11) have emerged. Dimensions of 5, 6, 9, 10, and 12 are rising, while Dimensions of 1, 2, 3, 7, 8, and 14 are falling. Dimensions of 4 and 13 are generally fluctuating upward and then downward. From this trend, we could assume a reasonable order of priority for service improvement.

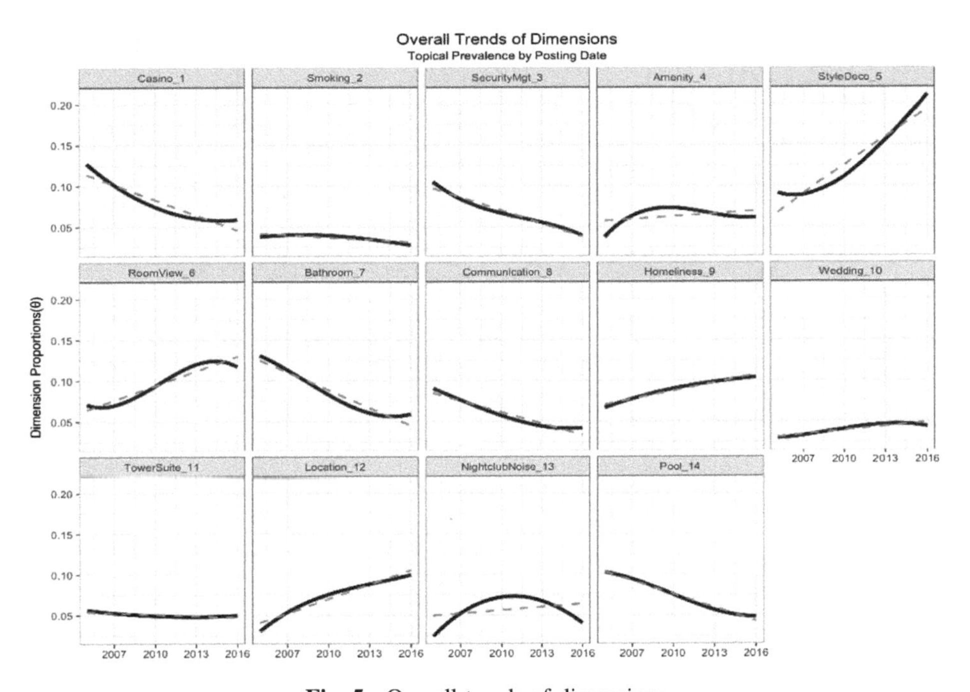

Fig. 5. Overall trends of dimensions

However, it is still limited to understand the dimensions since it only shows the overall trends not considering the effect of polarity. In other words, it is helpful to understand the overarching trends of dimensions in general, but it is still not clear how the positive and negative aspects of the dimension have changed over time with varying degree. By incorporating Polarity variable, this study further expands to examine dimension trends by different polarity as one of the main contributions of the study.

Trends by Polarity splits the trending line into two separate lines and in turn allows us to investigate how dimensions by different valence evolved over time by each dimension. Figure 6 depicts dimension trends with Polarity information over time where a solid line represents the dimensional trend for negativity and a dashed line represents the positivity.

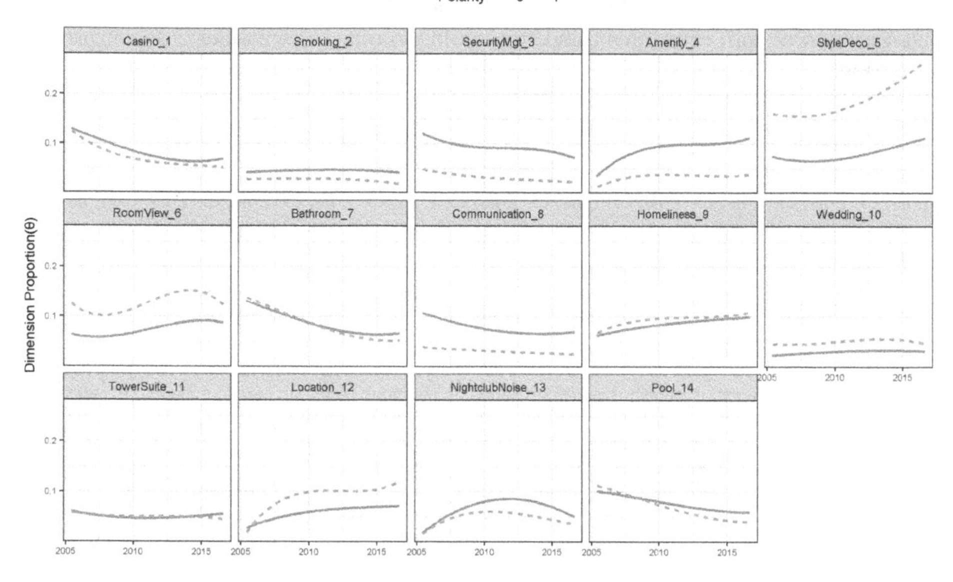

Fig. 6. Topic trends by polarity

From this plot, we can assume that *Style & Decoration, Room View, Homeliness, Wedding, Tower Suite,* and *Location* are positive dimensions, while the others are negative dimensions. For example, dimensions of *Location*, *Style & Decoration*, and *Room View* reveal notable distinction in two lines and have evolved with greater proportions for positivity. The physical advantage of having a good vantage *Location* in Las Vegas seems to naturally lead to positive reviews, and the hotel turned out to have a positive reputation from the guests with a beautiful style and decoration that reminds of the thematic flower decoration near the lobby which is a symbol of the Bellagio Hotel. *Room View* have changed with positive sentiment due mainly to their unique scenery from 'Golf View' which is in the middle of the desert. Stated differently, either side of room view meets a guest satisfaction. Therefore, dimension trends together with dimension by Polarity provides clearer insights to better understand the dimensions.

5 Evaluation

This chapter aims to evaluate the result derived from STM. Since latent topics buried in customer reviews were identified by STM with their review-dimension proportions, the analysis of logistic regression was conducted in order to further assess the impact of 14 discovered topics on *Polarity*. *Polarity* of the review which is a two-class binary variable (0 = Negative, 1 = Positive) is taken as dependent variable, and 14 topic proportions are taken and treated as independent variables to examine how topics influence the *Polarity*. Figure 7 illustrates the process of conducting the evaluation.

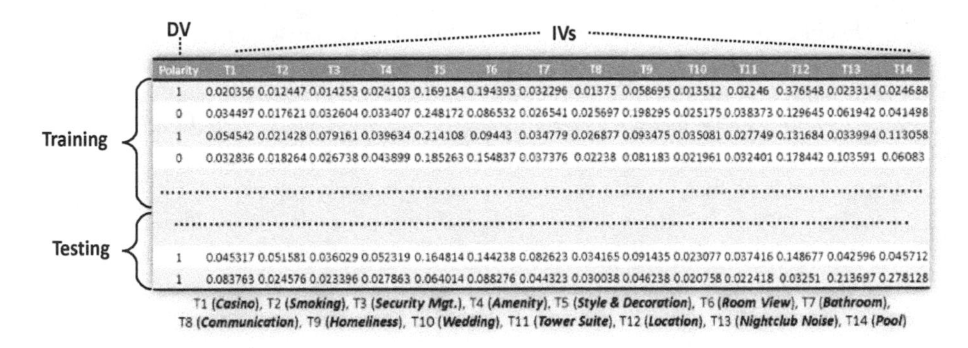

Fig. 7. Illustration of conducting the model evaluation using logistic regression

Prior to fitting the model, independent variables were log-transformed, and correlation and multicollinearity of each variable were assessed. Correlation matrix along with Variance Inflation Factor (VIF) and Tolerance (TOL) suggests that there is no major issue for multicollinearity in our study. Thus, all the 14 variables were used to initially build the model.

Data (10,251 reviews) were split into training set (70% or 7,176 reviews) and testing set (30% or 3,075 reviews). Then, the model was built with training data and tested with testing data for external validity. Initial fit of the full model including all the 14 variables revealed that T1 ($p = .459$), T7 ($p = .892$), and T11 ($p = .299$) were not statistically significant. Though T9 ($p = .035$) found to be statistically significant at .05 level, we decided to keep the variables with the significant level at both .01 and .001 only to be more conservative in building the model. Thus, the revised final model was built without T1, T7, T9, and T11, and the summary is outlined in Table 3.

Table 3. Logistic regression predicting polarity of reviews

	B	SE	Wald	P	Odds ratio	2.5%	97.5%
Smoking	−0.163	0.053	−3.100	0.002**	0.849	0.766	0.942
Security Mgt.	−0.760	0.059	−12.984	0.000***	0.467	0.416	0.524
Amenity	−0.687	0.052	−13.243	0.000***	0.503	0.454	0.557
Style & Decoration	1.180	0.065	18.071	0.000***	3.254	2.866	3.702
Room View	1.122	0.063	17.828	0.000***	3.071	2.718	3.478
Communication	−0.434	0.057	−7.600	0.000***	0.648	0.579	0.724
Wedding	0.528	0.048	10.930	0.000***	1.695	1.543	1.864
Location	0.498	0.058	8.648	0.000***	1.645	1.470	1.842
Nightclub Noise	−0.370	0.049	−7.599	0.000***	0.691	0.627	0.760
Pool	−0.651	0.051	−12.831	0.000***	0.521	0.472	0.576
(Intercept)	−1.586	0.530	−2.992	0.003***	0.205	0.072	0.578

Chi-square of likelihood ratio test is 5400 ($p = .000$) and pseudo R^2 of Cox and Snell and McFadden is .5288 and .5438, respectively

The likelihood ratio test showed that the model was statistically significant at χ^2 (10, N = 7,176) = 5400, p < .001. The model as a whole explained between 52.88% (Cox and Snell) and 54.38% (McFadden) of the variance in *Polarity,* and it shows the accuracy of 85.83%. To test the external validity, the model was tested with test data, and it further shows slightly improved accuracy of 86.73%. Therefore, it is not far-fetching to conclude that logistic regression model is fairly robust.

In logistic regression, the value of odds ratio (exponential of coefficient) demonstrates the magnitude of each independent variable. The strongest positive predictor of determining *Polarity* is *Style & Decoration,* and it records an odds ratio of 3.254. Stated differently, one unit change in *Style & Decoration* will increase the odds by 225% (= [3.254 − 1] * 100). This indicates that the dimension of *Style & Decoration* is one of the major strengths of Wynn hotel that impacts customer satisfaction positively, all other dimensions being equal. Another equivalent strength is *Room View* followed by *Wedding* and *Location.* Overall, the analysis suggests that the physical attractiveness with great location of Wynn hotel heavily influence the guest satisfaction affirmatively.

Security Management appears to be the strongest negative predictor of determining the *Polarity* of the reviews, and it shows the odds ratio of .467. In other words, one unit change in *Security Management* will reduce the odds by 53.3% (= [.467 − 1] * 100). This suggests that the guest satisfaction of Wynn hotel severely suffers from the dimension of *Security Management,* all other dimensions being equal. To comparable extent, the dimensions of *Amenity* (49.7%) and *Pool* (47.9%) found to be one of the major weaknesses hurting the guest satisfaction of Wynn hotel. Though boasting the world-renowned nightclubs in-house, the noise caused from club over the night does harm for guest staying in room.

6 Conclusion

The increasing need for effective mechanisms to turn the influx of scattered data into meaningful information is of great concern. In response to such need, this study proposes an integrated text mining framework in an attempt to discover hidden dimensions with prevalence of time and polarity from online customer reviews. Sentiment analysis was conducted as an objective measure for classifying the polarity of review. Subsequently, topic model was applied by incorporating covariates of *Posting Date* and *Polarity* of reviews into structural topic model in order to capture the relationship between dimensions and covariates. As a result, each customer review is represented as a probabilistic distribution over set of underlying dimensions, where the research interprets them as dimensions of service for Wynn Las Vegas, NV, USA. To evaluate the framework, the effect of each identified dimensions was assessed on polarity as a proxy of customer satisfaction.

Previous researches tend to consider rating information posted by user as given [18]. However, customer satisfaction might not fully reflect in the rating score though one can assume the overall evaluation is already incorporated in the rating score. This study explores customer sentiment analysis sorely from their reviews and demonstrates the significant difference between customer ratings and customer sentimental polarity.

Thus, the rescaled rating by sentiment analysis can provide the results with minimal bias or errors than subjective one-dimensional rating score given by reviewers.

This work offers specific managerial implications. First, it enables managers to effectively identify the managerial strengths and weaknesses at stake. Second, it enables managers to monitor and track the identified dimensions and examine how they vary according to certain sentiment. Third, the review data were from specific property in Las Vegas, and shows property specific dimensions in part. This in turn suggests that manager can target the properties of interest (i.e. competitor) and monitor them to maintain and establish competitive advantage in the market.

There are theoretical implications for researchers. Beyond the discovery of weaknesses and strengths, this study offers a methodological merit and enhancement of how qualitative analysis can be incorporated with quantitative approach to provide richer insights by quantifying massive amounts of publicly available information online. The study offers alternative way of conducting longitudinal study complementing the limitation of traditional approach, which is otherwise costly and time consuming.

This research also has limitations and the findings of this study should be interpreted with caution. Since topic model is generative model, more data are favorable to generate the enhanced results. Another limitation is that customer reviews solely come from TripAdvisor as a single source. In addition, the dataset possibly contains fake reviews which might influence the result. Nonetheless, the potential limitations in size of data and generalizability of the findings does not degrade the value of insights from this study, and therefore does no harm to the purpose of demonstrating the proposed framework.

Also, review data analyzed in this study does not contain any demographic information such as gender, income, and purpose of traveling or staying at the hotel. If complete demographic information is available, the dimensions derived from review can be further incorporated with those demographic variables as additional topical prevalence providing the richer insights.

References

1. Browning, V., So, K.K.F., Sparks, B.: The influence of online reviews on consumers' attributions of service quality and control for service standards in hotels. J. Travel Tour. Mark. **30**(1–2), 23–40 (2013)
2. Litvin, S.W., Goldsmith, R.E., Pan, B.: Electronic word-of-mouth in hospitality and tourism management. Tour. Manag. **29**(3), 458–468 (2008)
3. Tirunillai, S., Tellis, G.J.: Mining marketing meaning from online chatter: strategic brand analysis of big data using latent Dirichlet allocation. J. Mark. Res. **51**(4), 463–479 (2014)
4. Abrahams, A.S., Fan, W., Wang, G.A., Zhang, Z.J., Jiao, J.: An integrated text analytic framework for product defect discovery. Prod. Oper. Manag. **24**(6), 975–990 (2015)
5. Geyskens, I., Gielens, K., Dekimpe, M.G.: The market valuation of internet channel additions. J. Mark. **66**(2), 102–119 (2002)
6. Yacouel, N., Fleischer, A.: The role of cybermediaries in reputation building and price premiums in the online hotel market. J. Travel Res. **51**(2), 219–226 (2012)
7. Sparks, B.A., So, K.K.F., Bradley, G.L.: Responding to negative online reviews: the effects of hotel responses on customer inferences of trust and concern. Tour. Manag. **53**, 74–85 (2016). https://doi.org/10.1016/j.tourman.2015.09.011

8. Xu, X., Li, Y.: The antecedents of customer satisfaction and dissatisfaction toward various types of hotels: a text mining approach. Int. J. Hosp. Manag. **55**, 57–69 (2016)
9. Feinerer, I., Hornik, K.: tm: Text Mining Package. R package version 0.7-3 (2017). https://CRAN.R-project.org/package=tm
10. Martin, F., Johnson, M.: More efficient topic modelling through a noun only approach. Paper presented at the Australasian Language Technology Association workshop 2015 (2015)
11. Liu, B.: Sentiment analysis and opinion mining. In: Synthesis Lectures on Human Language Technologies, vol. 5, no. 1, pp. 1–167 (2012)
12. Socher, R., et al.: Recursive deep models for semantic compositionality over a sentiment treebank. In: Proceedings of the 2013 Conference on Empirical Methods in Natural Language Processing, pp. 1631–1642 (2013)
13. Roberts, M.E., Stewart, B.M., Tingley, D., Airoldi, E.M.: The structural topic model and applied social science. Paper presented at the advances in neural information processing systems workshop on topic models: computation, application, and evaluation (2013)
14. Roberts, M.E., Stewart, B.M., Airoldi, E.M.: A model of text for experimentation in the social sciences. J. Am. Stat. Assoc. **111**(515), 988–1003 (2016)
15. Griffiths, T.L., Steyvers, M.: Finding scientific topics. Proc. Natl. Acad. Sci. **101**(suppl 1), 5228–5235 (2004)
16. Mimno, D., Wallach, H.M., Talley, E., Leenders, M., McCallum, A.: Optimizing semantic coherence in topic models. Paper presented at the proceedings of the conference on empirical methods in natural language processing (2011)
17. Li, X., Hitt, L.M.: Self-selection and information role of online product reviews. Inf. Syst. Res. **19**(4), 456–474 (2008)
18. Guo, Y., Barnes, S.J., Jia, Q.: Mining meaning from online ratings and reviews: tourist satisfaction analysis using latent Dirichlet allocation. Tour. Manag. **59**, 467–483 (2017)

Using Model-Based Reasoning for Enhanced Chatbot Communication

Oliver A. Tazl(✉) and Franz Wotawa

Institute for Software Technology, Graz University of Technology, 8010 Graz, Austria
{oliver.tazl,wotawa}@ist.tugraz.at

Abstract. Chatbots as conversational recommender have gained increasing importance for research and practice with a lot of applications available today. In this paper, we present the methods to support conversational defaults within a human-chatbot conversation that simplifies communication with the purpose of improving the overall recommendation process. In particular, we discuss our model-based reasoning approach for easing user experience during a chat, e.g., in cases where user preferences are mentioned indirectly causing inconsistencies. As a consequence of inconsistencies, it would not be possible for the chatbot to provide answers and recommendations. The presented approach allows for removing inconsistencies during the interactions with the chatbot. Besides the basic foundations, we provide use cases from the intended tourism domain to show the simplification of the conversation process. In particular, we consider recommendations for booking hotels and planning trips.

Keywords: Chatbot · Conversational recommender · Model-based reasoning

1 Introduction

Natural language interfaces (NLI) to communicate with platforms and systems grow in interest and importance in research and industry. Chatbots are one form of AI systems that allows an interaction via NLI. These systems can be based on rule sets or neural networks to provide answers to given requests. Chatbot systems often rely on predefined patterns that lead to a desired behavior (e.g. [16]) and tend to be limited in their interaction space in order to provide correct responses to the user. The user communicates with the recommendation system via natural language. The Natural Language Processing (NLP) extracts the information given by the user and sends it in an abstracted form to the recommender engine. This engine tries to process the user request in the best possible way. Based on our work in [11], we propose the additional possibility of using implicit information and background knowledge along with handling of inconsistencies that are caused by that background knowledge. Also, the recommender

F. Wotawa et al. (Eds.): IEA/AIE 2019, LNAI 11606, pp. 791–798, 2019.
https://doi.org/10.1007/978-3-030-22999-3_67

engine has to provide at least one element in order to create a valid solution. If such a solution could not be provided, the engine has to ask the user to relax and remove some requirements. Requirements provided by the background knowledge are the first to be removed from the set.

The need to handle implicit information within the conversational recommender represents a crucial challenge for a flawless user experience. For example, a user send an inquiry about planning a city trip to a certain destination. Such trips encompass certain requirements, like requiring a hotel near the city center. In fact, these requirements could be presumed by the conversational recommender from background knowledge. This helps the user to find a matching hotel without explicitly asking for these requirements. Parts of these background knowledge will be immutable, e.g. the user might want to book a stay with her or his fiancé, which leads to *adult(2)* but does omit *romantic* as a requirement. This part of the background knowledge could be removed when it leads to an empty result set, which is something to prevent. On the other hand, presuming requirements has also disadvantages. These drawbacks must be addressed by the recommender as well. Also, adding multiple and diverse requirement groups could lead to an inconsistency within these groups. For this reason, the conversational recommender must be also able to deal with this type of problems. The approach of model-based diagnosis helps to detect a possible problem by computing the minimal set of faulty requirements. In the aftermath, these will be used to repair the requirements with the help of the user. Asking the user if a certain background knowledge requirement is crucial, is a way to solve inconsistencies. If negated, the requirement could be removed and the consistency is recovered.

The main contributions of this paper are given below.

1. an approach to simplify the communication between the user and the system by using implicit presumptions, and
2. an algorithm that handles inconsistencies of requirements within the recommendation process.

The remainder of this paper is organized as follows: In the next section we introduce an example domain in the field of tourism and get into more detail of our motivating use case. Next, we introduce our approach of removing inconsistencies within the recommendation process. Finally, we discuss related research and conclude the paper.

2 Use Case

In this section, we depict typical conversations between user and chatbot in the field of tourism. Also, we show the corresponding effects according to the knowledge base of the recommender system.

In Table 1, we show the information contained in the knowledge base along with the recommendation rules. These rules guide the recommender to appropriately react to implicit information that is given by the user. The latter functionality will be shown in a conversation use case. We assume that each *hotel*

Table 1. Item set for $type = hotel_{graz}$

name	familyfriendly	romantic	distance_center	price	stars	spa	distance_pt
Hotel 1	false	false	low	med	1	false	med
Hotel 2	true	false	med	high	1	true	high
Hotel 3	false	true	high	low	1	true	low
Hotel 4	true	false	low	high	2	false	low
Hotel 5	false	true	med	low	2	false	med
Hotel 6	true	false	high	med	2	true	high
Hotel 7	false	true	low	high	3	true	med
Hotel 8	true	false	med	med	3	false	low
Hotel 9	false	true	high	high	3	false	high
Hotel 10	true	false	low	med	4	false	high
Hotel 11	false	true	med	high	4	false	low
Hotel 12	true	false	high	med	4	false	med
Hotel 13	false	true	low	med	5	false	high

consists of a set of attributes, namely: The *name* of the hotel, *family-friendliness* (true or false), *romantic* appearence (true or false), its distance (low, med, high) to the center (*distance_center*) and to the public transport (*distance_pt*). Also, the corresponding *price* is splitted into three categories (low, med, high), the *stars* (1 to 5) and the availability of a *spa* (true or false).

We will demonstrate the chatbot-based conversational recommender in form of a use case. The use case illustrates a recommendation process of a hotel in a specific area. The presented use case will serve as the motivating example for the rest of the paper.

The process starts with the chatbot offering its tourism support. Then, the user requests a well-rated hotel in the city center of Graz, e.g.:

`"I want to book an awesome hotel in Graz near the city center this weekend."`.

The chatbot adds this information to the knowledge base. This information consists of a limitation of four or more stars with medium or high prices in order to ensure a high quality of the selected hotel.

$$awesome \rightarrow (price(med) \vee price(high)) \wedge (stars(4) \vee stars(5)) \\ distance\_center(low) \quad (1)$$

This set of background knowledge is triggered by the implicit information of wanting an "awesome" hotel. It also adds a distance limitation requirement for this hotel in order to adapt the "near the city center" information in the knowledge base. The chatbot acknowledges this addition by appropriately replying to the user. Two of the hotels in the KB, namely *Hotel 10* and *Hotel 13*, fit the requirements of the user.

Next, the chatbot sends an inquiry about the number of people. In turn, the user replies with:

`"The room is for me and my fiancé"`

Here the reply does not contain information about the number of people but the description of an additional person. This implicit detail triggers the chatbot's rule *couple* to add along with $adult(2)$ the need for a romantic hotel with a spa to the knowledge base as well. Once again, the chatbot acknowledges the requirements with a proper response.

$$couple \rightarrow adults(2) \wedge romantic \wedge spa \tag{2}$$

However, adding these requirements might lead to an inconsistency that cannot be resolved by the chatbot. However, such circumstances must be handled by the recommendation system. Here we expect a minimum of at least one hotel in the result set that fits the requirements of the user. As depicted in Table 1, no hotels with spa are available, which was part of the predefined background knowledge for the couple's stay. Now the chatbot requests the user to remove this requirement in order to get a valid set of results to choose from, by asking:

`"There are no hotels with the defined settings. Do you need a spa for your stay?"`.

After the user confirms the removal, a result set of one hotel, namely Hotel 13, will be left, which represents a valid solution for the revised user requirements.

3 Removing Inconsistencies

As explained in the previous section, there might be an inconsistency arising from the communication between a user and a chatbot. The root cause behind such inconsistency represent the user-provided facts, together with background knowledge capturing rule-of-thumbs and other available data. In order to resolve the inconsistencies, we have to retract rules or provided facts. For this purpose, we borrow the idea behind model-based diagnosis (MBD) [1,8,14], which deals with obtaining diagnoses of systems from system models. In MBD, the model of a system comprises several component models and the system structure. Each component model comprises an attached predicate, i.e., AB standing for abnormal, representing the health state of the component. The underlying diagnosis algorithms set AB predicates to true or false in order to remove inconsistencies arising between the provided and the expected (and observed) behavior.

In the following, we describe the application of MBD to the chatbot domain. First, let us have a look at the background knowledge that correspond to the type of the hotel. If a hotel is awesome, we usually expect a higher price and a better classification. However, it might be the case that the price is lower because of the current tourism season or due to comparisons with similar hotels. Hence, we may not always be able to determine that the price will be medium or high. Therefore, we can introduce the predicate AB for this purpose. AB is usually set to false, unless an inconsistency emerges from the underlying assumption. The following logical formulae represent this improved background knowledge:

$$\begin{aligned} awesome \wedge \neg AB(price) \rightarrow price(med) \vee price(high) \\ awesome \rightarrow stars(4) \vee stars(5) \end{aligned} \tag{3}$$

Similar to the hotel, we are able to improve the rules that process inquiries with couples. There we might say that a spa or a romantic hotel may be not always required. Note that in the following logical representation, we distinguish between removing the rule behind the requirement of having a spa from the one requiring a hotel to be classified as romantic. This allows us to retract only the specific part of the background knowledge that causes the inconsistency in the first place.

$$couple \wedge \neg AB(spa) \rightarrow spa, couple \wedge \neg AB(romantic) \rightarrow romantic, \\ couple \rightarrow adults(2) \tag{4}$$

Finally, we may also improve the knowledge behind city trips and introduce AB predicates for the city distance and the distance to public transport, separately:

$$\begin{aligned} citytrip &\rightarrow price(low) \vee price(med) \\ citytrip \wedge \neg AB(distance\_city) &\rightarrow distance\_city(hotel) \\ citytrip \wedge \neg AB(distance\_pt) &\rightarrow distance\_pt(hotel) \end{aligned} \tag{5}$$

Before formalizing the theory behind our approach, let us explain its application on the example in Sect. 2. There, after the first query, the solution contains only two remaining hotels. Neither of them encompasses any spa. However, since we know that the inquiry contains a couple and by taking into account the background knowledge, no solution could be delivered. Hence, when using the new background knowledge with AB predicates, a solution is extracted. When setting $AB(spa)$ to true and all other AB predicates to false, the solution will contain *Hotel 13*. Note that in this approach the rules are not really retracted. Also, they do not provide any effect on the computation of results because they can never be activated. Hence, from a logical point of view, we obtain the same behavior when removing them from the background theory.

The following formalization of the challenge of retracting knowledge in the domain of a chatbot conversation adapts Reiter's formalisms of MBD (see [14]). We start specifying the chatbot conversation formally where we have background knowledge, a set of rules that can be retracted, and a set of facts provided by the user as inputs.

Definition 1 (Chatbot conversation (CC)). *A chatbot conversation (CC) is a tuple (Th, R, U) where Th is a set of logical formulae representing the background knowledge and other information the chatbot offers in order to provide a recommendation, R is a set of rules that can be retracted, and U is a set of facts provided by a user during conversation.*

For the example from Sect. 2 the CC comprises all rules necessary for specifying the hotel table as well as the logical rules (3) to (5) as Th. Th also includes rules for mapping the hotel table to the provided background knowledge, and a rule stating that we want to obtain at least one hotel to be provided to the user as a result of the conversation. R is the set $\{price, spa, romantic, distance\_city, distance\_pt\}$, and the user provided facts U is $\{awesome, couple\}$.

It is worth noting that for all elements r in R there must be a predicate of the form $AB(r)$ in Th used in one or more rules that might be retracted in order to remove inconsistencies. The underlying problem behind is the *chatbot conversation problem* can defined as follows:

Definition 2 (Chatbot conversation problem (CCP)). *Given a chatbot conversation* (Th, R, U)*, the* chatbot conversation problem (CCP) *is the problem of identifying a subset* Δ *of* R *that makes the following logical sentence* $Th \cup U \cup \{\neg AB(c)|c \in R \setminus \Delta\} \cup \{AB(c)|c \in \Delta\}$ *consistent. If this logical sentence is consistent* Δ *is also called a solution of the CCP.*

Obviously, if Δ is the empty set, then all rules comprising AB predicates can be applied and, therefore, all rule-of-thumbs are activated. If this does not lead to an inconsistency, then all such rules can be applied to improve the search for recommendations. Otherwise, some of the rules have to be deactivated by applying AB predicates. Similar to MBD, we define a solution Δ to be minimal or parsimonious if there is no other subset Δ' of Δ that is itself a solution. For our hotel example, there is only one minimal solution $\{spa\}$.

Computing solutions for a CCP is obviously a search problem assigning truth values to the AB predicates and proving the sentence $Th \cup U \cup \{\neg AB(c)|c \in R \setminus \Delta\} \cup \{AB(c)|c \in \Delta\}$ for consistency by using a theorem prover. In the worst case, we have an exponential number of solutions, i.e., all subsets of R. However, in practice, we are interested in minimal solutions and also usually in solutions that include the smallest number of elements from R. In the following, we discuss an algorithm that computes the smallest solutions for a given CCP with respect to cardinality. The algorithm might miss other minimal solutions. However, for practical applications it is usually enough to come up with some smaller but not necessarily all parsimonious solutions. There are plenty of diagnosis algorithms described in literature, e.g., ConDiag [12], which can be easily adapted to provide solutions for the CCP.

4 Related Work and Conclusions

The application of model-based reasoning, and especially model-based diagnosis, in the field of recommender systems is not novel. For example, papers like [3,6,13] compute the minimal sets of faulty requirements. These requirements should be changed in order to find a solution. In these papers, the authors rely on the existence of minimal conflict sets computing the diagnosis for inconsistent requirements. Felfernig et al. [3] present an algorithm that calculates personalized repairs for inconsistent requirements. The algorithm combines concepts of MBD with a collaborative problem solving approach to improve the quality of repairs in terms of prediction accuracy.

Papers that deal with the integration of diagnosis and constraint solving are [2] and [17,18], who proposed a diagnosis algorithm for tree-structured models. The approach is generally applicable due to the fact that all general constraint models can be converted into an equivalent tree-structured model using decomposition methods, e.g., hyper tree decomposition [4,5]. [19] provides more details regarding the coupling of decomposition methods and the diagnosis algorithms for tree-structured models. In addition to that, [15] generalized the algorithms of [2] and [17]. In [10] the authors also propose the use of constraints for diagnosis where conflicts are used to drive the computation. For presenting recommendation tasks as constraint satisfaction problem, we refer to [7].

Human-chatbot communication represents a broad domain. It covers technical aspects as well as psychological and human perspectives. Wallace [21] demonstrates an artificial intelligence robot based on a natural language interface (A.L.I.C.E.) that extends ELIZA [22].

The topic of recommender systems with conversational interfaces is shown in [9], where an adaptive recommendation strategy was shown based on reinforcement learning methods. In the paper [20], the authors proposed a deep reinforcement learning framework to build personalized conversational recommendation agents. In this work, a recommendation model trained from conversational sessions and rankings is also presented.

In this paper, we introduced and discussed an conversational recommender approach, which deals with implicit information given by the user as well as requirements inconsistencies using the techniques of model-based diagnosis.

In the proposed approach, we are handling implicit information provided user by using background knowledge in order to forecast the needs of the user. The background knowledge is gathered by the rule-of-thumb and other available information sources. Along with the facts collected from the user's inputs, the recommender will provide a customed set of e.g. hotels. We also show how handle the case of inconsistent knowledge within the recommendation process that could lead to empty result sets. In a nutshell, we resolve the issue by removing background knowledge in order to provide a valid result set to the user.

We are currently working on an general framework, which allows such recommendations not only within the domain of tourism, but also in other domains where such processes can be used. In the future, we will also use this implementation for performing experiments and user studies in order to evaluate the effectiveness of the proposed approach, when used in practical chatbot settings.

Acknowledgement. The research presented in the paper has been funded in part by the Cooperation Programme Interreg V-A Slovenia-Austria under the project AS-IT-IC (Austrian-Slovenian Intelligent Tourist Information Center).

References

1. Davis, R.: Diagnostic reasoning based on structure and behavior. Artif. Intell. **24**, 347–410 (1984)
2. Fattah, Y.E., Dechter, R.: Diagnosing tree-decomposable circuits. In: Proceedings 14th International Joint Conference on Artificial Intelligence, pp. 1742–1748 (1995)
3. Felfernig, A., Friedrich, G., Schubert, M., Mandl, M., Mairitsch, M., Teppan, E.: Plausible repairs for inconsistent requirements. In: IJCAI International Joint Conference on Artificial Intelligence, pp. 791–796, January 2009

4. Gottlob, G., Leone, N., Scarcello, F.: Hypertree decomposition and tractable queries. In: Proceedings of 18th ACM SIGACT SIGMOD SIGART Symposium on Principles of Database Systems (PODS 1999), pp. 21–32, Philadelphia, PA (1999)
5. Gottlob, G., Leone, N., Scarcello, F.: A comparison of structural CSP decomposition methods. Artif. Intell. **124**(2), 243–282 (2000)
6. Jannach, D.: Finding preferred query relaxations in content-based recommenders. IEEE Intell. Syst. **109**, 81–97 (2008)
7. Jannach, D., Zanker, M., Fuchs, M.: Constraint-based recommendation in tourism: a multiperspective case study. J. IT Tour. **11**, 139–155 (2009)
8. de Kleer, J., Williams, B.C.: Diagnosing multiple faults. Artif. Intell. **32**(1), 97–130 (1987)
9. Mahmood, T., Ricci, F., Venturini, A.: Learning adaptive recommendation strategies for online travel planning. Inf. Commun. Technol. Tour. **2009**, 149–160 (2009)
10. Mauss, J., Sachenbacher, M.: Conflict-driven diagnosis using relational aggregations. In: Proceedings of the 10th International Workshop on Principles of Diagnosis (DX), Loch Awe, Scotland (1999)
11. Nica, I., Tazl, O., Wotawa, F.: Chatbot-based tourist recommendations using model-based reasoning. In: CEUR Workshop Proceedings, vol. 2220, pp. 25–30 (2018)
12. Nica, I., Wotawa, F.: ConDiag - computing minimal diagnoses using a constraint solver. In: Proceedings of the 23rd International Workshop on Principles of Diagnosis (2012)
13. O'Sullivan, B., Papadopoulos, A., Faltings, B., Pu, P.: Representative explanations for over-constrained problems. In: Proceedings of the National Conference on Artificial Intelligence, vol. 1, July 2007
14. Reiter, R.: A theory of diagnosis from first principles. Artif. Intell. **32**(1), 57–95 (1987)
15. Sachenbacher, M., Williams, B.C.: Diagnosis as semiring-based constraint optimization. In: European Conference on Artificial Intelligence, pp. 873–877 (2004)
16. Shawar, B.A., Atwell, E.: Using corpora in machine-learning chatbot systems. Int. J. Corpus Linguist. **10**, 489–516 (2005)
17. Stumptner, M., Wotawa, F.: Diagnosing tree-structured systems. In: Proceedings 15th International Joint Conference on Artificial Intelligence, Nagoya, Japan (1997)
18. Stumptner, M., Wotawa, F.: Diagnosing tree-structured systems. Artif. Intell. **127**(1), 1–29 (2001)
19. Stumptner, M., Wotawa, F.: Coupling CSP decomposition methods and diagnosis algorithms for tree-structured systems. In: Proceedings of the 18th International Joint Conference on Artificial Intelligence (IJCAI), pp. 388–393, Acapulco, Mexico (2003)
20. Sun, Y., Zhang, Y.: Conversational recommender system. In: The 41st International ACM SIGIR Conference on Research & Development in Information Retrieval, SIGIR 2018, pp. 235–244. ACM, New York (2018)
21. Wallace, R.: The anatomy of A.L.I.C.E. In: Epstein, R., Roberts, G., Beber, G. (eds.) Parsing the Turing Test: Philosophical and Methodological Issues in the Quest for the Thinking Computer, pp. 181–210. Springer, Dordrecht (2009). https://doi.org/10.1007/978-1-4020-6710-5_13
22. Weizenbaum, J.: Eliza-a computer program for the study of natural language communication between man and machine. Commun. ACM **9**(1), 36–45 (1966)

Optimization

A Novel Intelligent Technique for Product Acceptance Process Optimization on the Basis of Misclassification Probability in the Case of Log-Location-Scale Distributions

Nicholas Nechval[1(✉)], Gundars Berzins[1], and Konstantin Nechval[2]

[1] BVEF Research Institute, University of Latvia, Raina Blvd 19, Riga 1050, Latvia
nechval@junik.lv
[2] Transport and Telecommunication Institute, Lomonosov Street 1, Riga 1019, Latvia

Abstract. In this paper, to determine the optimal parameters of the product acceptance process under parametric uncertainty of underlying models, a new intelligent technique for optimization of product acceptance process on the basis of misclassification probability is proposed. It allows one to take into account all possible situations that may occur when it is necessary to optimize the product acceptance process. The technique is based on the pivotal quantity averaging approach (PQAA) which allows one to eliminate the unknown parameters from the problem and to use available statistical information as completely as possible. It is conceptually simple and easy to use. One of the most important features of the proposed new intelligent technique for optimization of product acceptance process on the basis of misclassification probability is its great generality, enabling one to optimize diverse problems within one unified framework. To illustrate the proposed technique, the case of log-location-scale distributions is considered under parametric uncertainty.

Keywords: Product acceptance · Misclassification probability · Optimization

1 Introduction

Extensive work has been done on product acceptance processes since their inception. Several text books and papers are available which provide different acceptance processes of product for different probability distribution functions, see, for example, Gupta and Groll [1], Gupta [2], Fertig and Mann [3], Kantam and Rosaiah [4], Kantam et al. [5], Baklizi [6], Wu and Tsai [7], and the references cited therein. Generally, the design of product acceptance processes is based on the population mean or median.

In this paper, a new technique for optimization of product acceptance process under parametric uncertainty (in terms of misclassification probability) is proposed. The mathematical solution to the optimization problem of product acceptance process does not represent any difficulties for the case when the parametric values of the underlying lifetime distributions are known with certainty. In actual practice, such is simply not the

F. Wotawa et al. (Eds.): IEA/AIE 2019, LNAI 11606, pp. 801–818, 2019.
https://doi.org/10.1007/978-3-030-22999-3_68

case. When the model is applied to solve real-world problems, the parameters are estimated and then treated as if they were the true values. The risk associated with using estimates rather than the true parameters is called estimation risk and is often ignored. For efficient optimization of statistical decisions under parametric uncertainty, the classical theory of statistical estimation has little to offer in this type of situation. For this case, a pivotal quantity averaging approach (PQAA) is suggested. It allows one to eliminate unknown parameters from the problem through the pivotal quantity averaging and to find the better decision rules, which have smaller risk than any of the well-known decision rules (Nechval and Vasermanis [8], Nechval et al. [9], Nechval et al. [10]).

The methodology described here can be extended in several different directions to handle various problems that arise in practice. In particular, the proposed methodology can be used (with some modifications) to improve the dynamic pricing in e-business (Nechval et al. [11]), pattern recognition (Nechval et al. [12]), vibration-based diagnostics of fatigued structures (Nechval et al. [13]), airline seat reservation control, etc.

2 Lifetime Distributions from Log-Location-Scale Family

2.1 Gumbel Distribution

This distribution is used in many research fields including, among others, life testing and water resource management. This is the so-called first asymptotic distribution of extreme values, hereafter referred to simply as the extreme value distribution. The distribution is extensively used in a number of areas as a lifetime distribution and sometimes referred to as the Gumbel distribution, after E. J. Gumbel, who had pioneered its use (Gumbel [14]).

Let X_1, …, X_n be the observations of a random sample of size n from the Gumbel distribution with the pdf (probability density function),

$$f_\omega(x) = \frac{1}{\sigma}\exp\left(\frac{x-\lambda}{\sigma}\right)\exp\left[-\exp\left(\frac{x-\lambda}{\sigma}\right)\right], \quad -\infty < x < \infty, \tag{1}$$

and cdf (cumulative distribution function),

$$F_\omega(x) = 1 - \exp\left[-\exp\left(\frac{x-\lambda}{\sigma}\right)\right], \quad -\infty < x < \infty, \tag{2}$$

indexed by location and scale parameters λ and σ, where $\omega = (\lambda, \sigma)$. It is assumed that the parameters $\lambda(-\infty < \lambda < \infty)$ and $\sigma > 0$ are unknown.

The MLE's $\widehat{\lambda}$ and $\widehat{\sigma}$ of the parameters λ and σ, respectively, are solutions of

$$\widehat{\lambda} = \ln\left[n^{-1}\sum_{i=1}^{n}\exp\left(\frac{x_i}{\widehat{\sigma}}\right)\right]^{\widehat{\sigma}}, \tag{3}$$

$$\widehat{\sigma} = \left[\sum_{i=1}^{n} x_i \exp\left(\frac{x_i}{\widehat{\sigma}}\right)\right]\left[\sum_{i=1}^{n} \exp\left(\frac{x_i}{\widehat{\sigma}}\right)\right]^{-1} - \frac{1}{n}\sum_{i=1}^{n} x_i \tag{4}$$

The mean of the Gumbel distribution is given by

$$\begin{aligned} E_{\omega=(\lambda,\sigma)}\{X\} &= \int_{-\infty}^{\infty} x f_{\omega}(x)dx = \int_{-\infty}^{\infty} x\frac{1}{\sigma}\exp\left(\frac{x-\lambda}{\sigma}\right)\exp\left[-\exp\left(\frac{x-\lambda}{\sigma}\right)\right]dx \\ &= \int_{0}^{\infty}[\lambda+\sigma \ln u]\exp(-u)du = \lambda+\sigma\frac{d}{da}\int_{0}^{\infty} u^{a}\exp(-u)du\Big|_{a=0} \\ &= \lambda+\sigma\frac{d}{da}\Gamma(a+1)\Big|_{a=0} = \lambda+\sigma\Gamma'(1) = \lambda-\gamma\sigma, \end{aligned} \tag{5}$$

where

$$u = \exp\left(\frac{x-\lambda}{\sigma}\right), \tag{6}$$

$$\gamma \approx 0.5772 \text{ (Euler's constant)}. \tag{7}$$

In terms of the Gumbel distribution variates, we have that

$$W_1 = \frac{\widehat{\lambda}-\lambda}{\sigma},\ W_2 = \frac{\widehat{\sigma}}{\sigma},\ W_3 = \frac{\widehat{\lambda}-\lambda}{\widehat{\sigma}} \tag{8}$$

are pivotal quantities. The probability density functions of the pivotal quantities do not depend on the parameters.

If the scale parameter σ is known, then a maximum likelihood estimator (MLE) of λ, based on the random sample (X_1, …, X_n) of size n from (1), is given by

$$\widehat{\lambda} = \ln\left[\sum_{i=1}^{n}\exp\left(\frac{X_i}{\sigma}\right)/n\right]^{\sigma}, \tag{9}$$

It can be shown that

$$\begin{aligned} \widehat{\lambda} \sim g_{\overline{\omega}}(\widehat{\lambda}|n) &= \frac{1}{\Gamma(n)(1/n)^{n}}\left[\exp\left(\frac{\widehat{\lambda}-\lambda}{\sigma}\right)\right]^{n} \\ &\times \exp\left[-\exp\left(\frac{\widehat{\lambda}-\lambda}{\sigma}\right)/1/n\right]\frac{1}{\sigma}, \quad -\infty<\widehat{\lambda}<\infty, \end{aligned} \tag{10}$$

where a function of the data $\widehat{\lambda}$ and parameters (λ, σ),

$$\exp\left(\frac{\widehat{\lambda}-\lambda}{\sigma}\right)=W, \tag{11}$$

is said to be a pivot or a pivotal quantity since its probability density function,

$$g(w|n)=\frac{1}{\Gamma(n)(1/n)^n}w^{n-1}\exp\left(-\frac{w}{1/n}\right),\quad w\geq 0, \tag{12}$$

does not depend on the parameters (λ,σ). It follows from (10) and (12) that the cdf of $\widehat{\lambda}$ is given by

$$G_w(h|n)=\int_0^h g_w(\widehat{\lambda}|n)d\widehat{\lambda}=\int_0^h \frac{1}{\Gamma(n)(1/n)^n}\left[\exp\left(\frac{\widehat{\lambda}-\lambda}{\sigma}\right)\right]^n \exp\left[-\exp\left(\frac{\widehat{\lambda}-\lambda}{\sigma}\right)/1/n\right]\frac{1}{\sigma}d\widehat{\lambda}$$
$$=\int_0^{\exp\left(\frac{h-\lambda}{\sigma}\right)} \frac{1}{\Gamma(n)(1/n)^n}w^{n-1}\exp\left(-\frac{w}{1/n}\right)dw=\int_0^{\exp\left(\frac{h-\lambda}{\sigma}\right)} g(w|n)dw=\Pr\left(W\leq\exp\left(\frac{h-\lambda}{\sigma}\right)\right). \tag{13}$$

In this case,

$$\bar{G}_\omega(h|n)=1-G_\omega(h|n)=\Pr\left(W>\exp\left(\frac{h-\lambda}{\sigma}\right)\right). \tag{14}$$

2.2 Weibull Distribution

The Weibull distribution is very flexible, and can, through an appropriate choice of parameters, model many types of failure rate behaviors. It has wide applications in diverse disciplines.

Let $Y_1,\ldots,Y_n$ be the observations of a sample of size n from a two-parameter Weibull distribution with the pdf,

$$f_\theta(y)=\frac{\delta}{\beta}\left(\frac{y}{\beta}\right)^{\delta-1}\exp\left[-\left(\frac{y}{\beta}\right)^\delta\right],\quad y>0,\ \beta>0,\ \delta>0, \tag{15}$$

and cdf,

$$F_\theta(y)=1-\exp\left[-\left(\frac{y}{\beta}\right)^\delta\right],\quad y>0,\ \beta>0,\ \delta>0, \tag{16}$$

indexed by scale and shape parameters β and δ, where $\theta=(\beta,\delta)$. It is assumed that the parameters β and δ are unknown. This distribution is directly related to the extreme-value distribution by the easily shown fact that if Y has a Weibull distribution (15), then

$X = \ln Y$ has a Gumbel distribution with $\lambda = \ln \beta$ and $\sigma = \delta^{-1}$. In analyzing data it is often convenient to work with log times, the Gumbel distribution arises when lifetimes are taken to be Weibull distributed.

The MLE's of the Weibull parameters β and δ are $\widehat{\beta} = \exp \widehat{\lambda}$ and $\widehat{\delta} = \widehat{\sigma}^{-1}$. If desired, the maximum likelihood Eqs. (3) and (4) can be written in Weibull form and solved directly from the start. The equations are

$$\widehat{\beta} = \left(n^{-1}\left(\sum_{i=1}^{n} y_i^{\widehat{\delta}}\right)\right)^{1/\widehat{\delta}}, \tag{17}$$

$$\widehat{\delta} = \left[\left(\sum_{i=1}^{n} y_i^{\widehat{\delta}} \ln y_i\right)\left(\sum_{i=1}^{n} y_i^{\widehat{\delta}}\right)^{-1} - \frac{1}{n}\sum_{i=1}^{n} \ln y_i\right]^{-1}. \tag{18}$$

The mean of the Weibull distribution is given by

$$E_{\theta=(\beta,\delta)}\{Y\} = \int_0^\infty y f_\theta(y)dy = \int_0^\infty y\frac{\delta}{\beta}\left(\frac{y}{\beta}\right)^{\delta-1} \exp\left[-\left(\frac{y}{\beta}\right)^{\delta}\right]dy$$
$$= \beta\int_0^\infty \left[\left(\frac{y}{\beta}\right)^{\delta}\right]^{1/\delta} \exp\left[-\left(\frac{y}{\beta}\right)^{\delta}\right]d\left(\frac{y}{\beta}\right)^{\delta} = \beta\int_0^\infty u^{1/\delta}\exp(-u)du = \beta\Gamma\left(1+\frac{1}{\delta}\right). \tag{19}$$

where $u = (y/\beta)^{\delta}$. In terms of the Weibull variates, we have that

$$V = \left(\frac{\widehat{\beta}}{\beta}\right)^{\delta},\ V_1 = \frac{\delta}{\widehat{\delta}},\ V_2 = \left(\frac{\widehat{\beta}}{\beta}\right)^{\widehat{\delta}} \tag{20}$$

are pivotal quantities. The probability density functions of the pivotal quantities do not depend on the parameters.

If the shape parameter δ is known, then a maximum likelihood estimator (MLE) of β^{δ}, based on the random sample $(Y_1, \ldots, Y_n)$ of size n from (15), is given by

$$\widehat{\beta}^{\delta} = \sum_{i=1}^{n} Y_i^{\delta}/n, \tag{21}$$

It can be shown that

$$\widehat{\beta}^{\delta} \sim g_\theta(\widehat{\beta}^{\delta}|n) = \frac{1}{\Gamma(n)(\beta^{\delta}/n)^n}\left(\widehat{\beta}^{\delta}\right)^{n-1}\exp\left(-\frac{\widehat{\beta}^{\delta}}{\beta^{\delta}/n}\right),\quad \widehat{\beta}^{\delta} \geq 0, \tag{22}$$

where a function of the data $\widehat{\beta}$ and parameters (β, δ),

$$\frac{\widehat{\beta}^{\delta}}{\beta^{\delta}} = V, \tag{23}$$

is said to be a pivot or a pivotal quantity since its probability density function,

$$g(v|n) = \frac{1}{\Gamma(n)(1/n)^n} v^{n-1} \exp\left(-\frac{v}{1/n}\right), \quad v \geq 0, \tag{24}$$

does not depend on the parameters (β, δ). It follows from (22) and (24) that the cdf of $\widehat{\beta}^{\delta}$ is given by

$$\begin{aligned} G_{\theta}(h|n) &= \int_0^h g_{\theta}(\widehat{\beta}^{\delta}|n) d\widehat{\beta}^{\delta} = \int_0^h \frac{1}{\Gamma(n)(\beta^{\delta}/n)^n} \left(\widehat{\beta}^{\delta}\right)^{n-1} \exp\left(-\frac{\widehat{\beta}^{\delta}}{\beta^{\delta}/n}\right) d\widehat{\beta}^{\delta} \\ &= \int_0^{h/\beta^{\delta}} \frac{1}{\Gamma(n)(1/n)^n} v^{n-1} \exp\left(-\frac{v}{1/n}\right) dv = \int_0^{h/\beta^{\delta}} g(v|n) dv = \Pr\left(V \leq \frac{h}{\beta^{\delta}}\right). \end{aligned} \tag{25}$$

In this case,

$$\bar{G}_{\theta}(h|n) = 1 - G_{\theta}(h|n) = \Pr\left(V > \frac{h}{\beta^{\delta}}\right). \tag{26}$$

3 Optimization of Product Acceptance Process

3.1 Problem Statement

Suppose that product lots are submitted for inspection and the lifetimes of individual products have a Weibull distribution with the probability density function (15) and known shape parameter δ. In this case, we consider some situations of optimization of the design parameters of product acceptance process, such as the sample size "n" to test the Weibull MTTF (mean time to failure) "μ" and the separation threshold "h" to satisfy the producer's and consumer's risks. The probability of rejecting a good lot is called the producer's risk, which is denoted by α_p. The probability of accepting a bad lot is called the consumer's risk, which is denoted by α_c. In determining whether or not a product lot is accepted, we have to use the test procedure which satisfies the following conditions:

$$\Pr\left(\text{reject a product lot}\,|\,\text{Weibull MTTF} = \mu_a\right) \le \alpha_p(\text{Producer's risk}), \tag{27}$$

$$\Pr\left(\text{accept a product lot}\,|\,\text{Weibull MTTF} = \mu_r\right) \le \alpha_c(\text{Consumer's risk}), \tag{28}$$

where

$$\begin{aligned}\mu_a &= E_{\theta_a=(\beta_a,\delta)}\{Y\} = \int_0^\infty y f_{\theta_a}(y)dy = \int_0^\infty y \frac{\delta}{\beta_a}\left(\frac{y}{\beta_a}\right)^{\delta-1} \exp\left[-\left(\frac{y}{\beta_a}\right)^{\delta}\right]dy \\ &= \beta_a \Gamma\left(1+\frac{1}{\delta}\right)\end{aligned} \tag{29}$$

is an acceptable Weibull MTTF,

$$\begin{aligned}\mu_r &= E_{\theta_r=(\beta_r,\delta)}\{Y\} = \int_0^\infty y f_{\theta_r}(y)dy = \int_0^\infty y \frac{\delta}{\beta_r}\left(\frac{y}{\beta_r}\right)^{\delta-1} \exp\left[-\left(\frac{y}{\beta_r}\right)^{\delta}\right]dy \\ &= \beta_r \Gamma\left(1+\frac{1}{\delta}\right)\end{aligned} \tag{30}$$

is a rejectable Weibull MTTF, $\mu_a > \mu_r$. It follows from (29) and (30), respectively, that

$$\beta_a^\delta = \left[\mu_a/\Gamma\left(1+\frac{1}{\delta}\right)\right]^\delta \tag{31}$$

and

$$\beta_r^\delta = \left[\mu_r/\Gamma\left(1+\frac{1}{\delta}\right)\right]^\delta. \tag{32}$$

3.2 Situation 1

Consider the situation when the size "n" of a random sample (Y_1, …, Y_n) of product items to test the Weibull MTTF (mean time to failure) "μ" is preassigned, but a product lot misclassification probability,

$$\alpha(h|n) = \alpha_p(h|n) + \alpha_c(h|n), \tag{33}$$

is not preassigned. The problem is to find the separation threshold "h" which minimized the product lot misclassification probability (33).

Let n items be drawn at random from an acceptable two-parameter Weibull lifetime distribution with the probability density function (pdf)

$$f_{\theta_a}(y) = \frac{\delta}{\beta_a}\left(\frac{y}{\beta_a}\right)^{\delta-1} \exp\left[-\left(\frac{y}{\beta_a}\right)^{\delta}\right], \quad y > 0,\, \beta_a > 0,\, \delta > 0, \tag{34}$$

and cumulative distribution function (cdf)

$$F_{\theta_a}(y) = 1 - \exp\left[-\left(\frac{y}{\beta_a}\right)^{\delta}\right], \quad y > 0,\, \beta_a > 0,\, \delta > 0, \tag{35}$$

Then a maximum likelihood estimator (MLE) of β_a^{δ}, based on the random sample $(Y_1, \ldots, Y_n)$ of size n from (34), is given by

$$\widehat{\beta}^{\delta} = \sum_{i=1}^{n} Y_i^{\delta}/n. \tag{36}$$

It follows from (22) that

$$\widehat{\beta}^{\delta} \sim g_{\theta_a}(\widehat{\beta}^{\delta}|n) = \frac{1}{\Gamma(n)(\beta_a^{\delta}/n)^n}\left(\widehat{\beta}^{\delta}\right)^{n-1} \exp\left(-\frac{\widehat{\beta}^{\delta}}{\beta_a^{\delta}/n}\right), \quad \widehat{\beta}^{\delta} \geq 0, \tag{37}$$

where a function of the data $\widehat{\beta}$ and parameters $(\beta_a, \delta) = \theta_a$,

$$\frac{\widehat{\beta}^{\delta}}{\beta_a^{\delta}} = V_a, \tag{38}$$

is said to be a pivot or a pivotal quantity since its probability density function,

$$g(v_a|n) = \frac{1}{\Gamma(n)(1/n)^n} v_a^{n-1} \exp\left(-\frac{v_a}{1/n}\right), \quad v_a \geq 0, \tag{39}$$

does not depend on the parameters (β_a, δ).

It follows from (37) and (39) that the cdf of $\widehat{\beta}^{\delta}$ is given by

$$G_{\theta_a}(h|n) = \Pr\left(\widehat{\beta}^{\delta} \leq h|\beta_a^{\delta}, n\right) = \int_0^h g_{\theta_a}(\widehat{\beta}^{\delta}|n)\,d\hat{\beta}^{\delta}$$
$$= \int_0^{h/\beta_a^{\delta}} \frac{1}{\Gamma(n)(1/n)^n} v_a^{n-1} \exp\left(-\frac{v_a}{1/n}\right) dv_a = \int_0^{h/\beta_a^{\delta}} g(v_a|n)\,dv_a = \Pr\left(V_a \leq \frac{h}{\beta_a^{\delta}}|n\right). \tag{40}$$

Let n items be drawn at random from a rejectable two-parameter Weibull lifetime distribution with the probability density function (pdf)

$$f_{\theta_r}(y) = \frac{\delta}{\beta_r}\left(\frac{y}{\beta_r}\right)^{\delta-1} \exp\left[-\left(\frac{y}{\beta_r}\right)^{\delta}\right], \quad y > 0,\ \beta_r > 0,\ \delta > 0, \tag{41}$$

and cumulative distribution function (cdf)

$$F_{\theta_r}(y) = 1 - \exp\left[-\left(\frac{y}{\beta_r}\right)^{\delta}\right], \quad y > 0,\ \beta_r > 0,\ \delta > 0, \tag{42}$$

Then a maximum likelihood estimator (MLE) of β_r^{δ}, based on the random sample $(Y_1, \ldots, Y_n)$ of size n from (41), is given by

$$\widehat{\beta}^{\delta} = \sum_{i=1}^{n} Y_i^{\delta}/n. \tag{43}$$

It can be shown that

$$\widehat{\beta}^{\delta} \sim g_{\theta_r}(\widehat{\beta}^{\delta}|n) = \frac{1}{\Gamma(n)(\beta_r^{\delta}/n)^n}\left(\widehat{\beta}^{\delta}\right)^{n-1} \exp\left(-\frac{\widehat{\beta}^{\delta}}{\beta_r^{\delta}/n}\right), \quad \widehat{\beta}^{\delta} \geq 0, \tag{44}$$

where a function of the data $\widehat{\beta}$ and parameters (β_r, δ),

$$\frac{\widehat{\beta}^{\delta}}{\beta_r^{\delta}} = V_r, \tag{45}$$

is said to be a pivot or a pivotal quantity since its probability density function,

$$g(v_r|n) = \frac{1}{\Gamma(n)(1/n)^n} v_r^{n-1} \exp\left(-\frac{v_r}{1/n}\right), \quad v_r \geq 0, \tag{46}$$

does not depend on the parameters $(\beta_r, \delta) = \theta_r$. It follows from (44) and (46) that the survival probability of $\widehat{\beta}^{\delta}$ is given by

$$\bar{G}_{\theta_r}(h|n) = \Pr\left(\widehat{\beta}^{\delta} > h|\beta_r^{\delta}, n\right) = \int_h^{\infty} g_{\theta_r}(\widehat{\beta}^{\delta}|n)d\widehat{\beta}^{\delta}$$
$$= \int_{h/\beta_r^{\delta}}^{\infty} \frac{1}{\Gamma(n)(1/n)^n} v_r^{n-1} \exp\left(-\frac{v_r}{1/n}\right) dv_r = \int_{h/\beta_r^{\delta}}^{\infty} g(v_r|n)dv_r = \Pr\left(V_r > \frac{h}{\beta_r^{\delta}}|n\right). \tag{47}$$

It follows from (27) and (40) that a good product lot is rejected if

$$\widehat{\beta}^{\delta} \leq h, \tag{48}$$

where the separation threshold "h" satisfies the following condition:

$$G_{\theta_a}(h|n) = \Pr\left(\widehat{\beta}^{\delta} \leq h|\beta_a^{\delta}, n\right) = \Pr\left(\frac{\widehat{\beta}^{\delta}}{\beta_a^{\delta}} \leq \frac{h}{\beta_a^{\delta}}|n\right) = \Pr\left(V_a \leq \frac{h}{\beta_a^{\delta}}|n\right) = \alpha_p. \tag{49}$$

It follows from (28) and (47) that a bad product lot is accepted if

$$\widehat{\beta}^{\delta} > h, \tag{50}$$

where the separation threshold "h" satisfies the following condition:

$$\bar{G}_{\theta_r}(h|n) = \Pr\left(\widehat{\beta}^{\delta} > h|\beta_r^{\delta}, n\right) = \Pr\left(\frac{\widehat{\beta}^{\delta}}{\beta_r^{\delta}} > \frac{h}{\beta_r^{\delta}}|n\right) = \Pr\left(V_r > \frac{h}{\beta_r^{\delta}}|n\right) = \alpha_c, \tag{51}$$

It follows from (40) and (47) that the product lot misclassification probability (see Fig. 1) is given by

$$\alpha(h|n) = \alpha_p(h|n) + \alpha_c(h|n) = G_{\theta_a}(h|n) + \bar{G}_{\theta_r}(h|n)$$
$$= \int_0^{h/\beta_a^{\delta}} \frac{1}{\Gamma(n)(1/n)^n} v_a^{n-1} \exp\left(-\frac{v_a}{1/n}\right) dv_a + \int_{h/\beta_r^{\delta}}^{\infty} \frac{1}{\Gamma(n)(1/n)^n} v_r^{n-1} \exp\left(-\frac{v_r}{1/n}\right) dv_r$$
$$= \Pr\left(V_a \leq \frac{h}{\beta_a^{\delta}}|n\right) + \Pr\left(V_r > \frac{h}{\beta_r^{\delta}}|n\right). \tag{52}$$

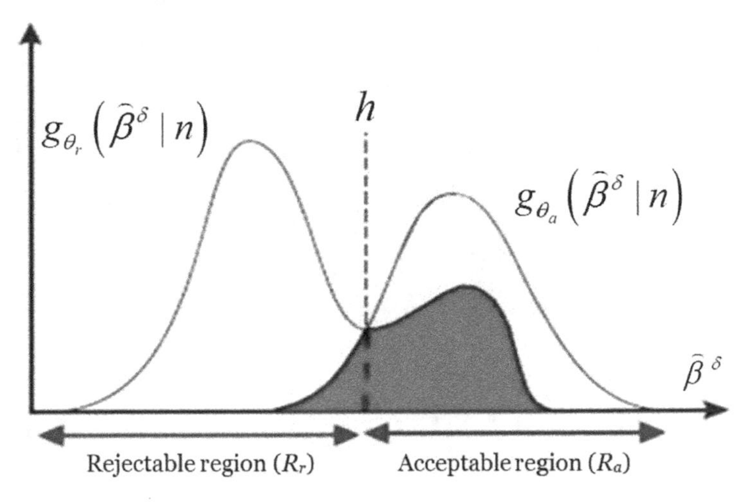

Fig. 1. Product lot misclassification probability $\alpha(h|n)$.

To minimize the probability of product lot misclassification, the separation threshold "h" is determined via (52) as follows:

$$h = \arg\min_{h} \alpha(h|n) = \arg\min_{h}\left(\int_{0}^{h} g_{\theta_a}(\overset{\frown}{\beta}^{\delta}|n)d\overset{\frown}{\beta}^{\delta} + \int_{h}^{\infty} g_{\theta_r}(\overset{\frown}{\beta}^{\delta}|n)d\overset{\frown}{\beta}^{\delta}\right). \tag{53}$$

Thus, the separation threshold h, which is used to obtain the rule of determining whether or not a product lot is accepted, should be "optimal" in the sense of minimizing, on average, the number of incorrect assignments of the product lot into two different regions R_a and R_r.

It follows from the above (see Fig. 1) that classification of the product lot into one of two different regions R_a and R_r (i.e., determining whether or not the product lot is accepted) is carried out through $\overset{\frown}{\beta}^{\delta}$ and h as follows:

$$\overset{\frown}{\beta}^{\delta} \in \begin{cases} R_r = \left\{\overset{\frown}{\beta}^{\delta} : \overset{\frown}{\beta}^{\delta} \leq h\right\} \rightarrow \text{reject the product lot,} \\ R_a = \left\{\overset{\frown}{\beta}^{\delta} : \overset{\frown}{\beta}^{\delta} > h\right\} \rightarrow \text{accept the product lot.} \end{cases} \tag{54}$$

Remark 1. In general, the classification rule (54) can be rewritten as follows: assign the product lot to region R_a (or R_r) for which $g_{\theta_a}(\overset{\frown}{\beta}^{\delta}|n)$ (or $g_{\theta_r}(\overset{\frown}{\beta}^{\delta}|n)$) is largest.

3.3 Situation 2

In the case of Situation 1, a classification scheme is evaluated in terms of its product lot misclassification probability, but this ignores misclassification costs. A rule that ignores costs may cause problems. The costs of product lot misclassification can be defined as follows. The costs are: (1) zero for correct classification, (2) c_p when an estimator $\overset{\frown}{\beta}^{\delta}$ from R_a is incorrectly classified as R_r, and (3) c_c when an estimator $\overset{\frown}{\beta}^{\delta}$ from R_r is incorrectly classified as R_a.

For any classification rule, the average, or *expected cost of product lot misclassification* (ECPLM) is given by

$$\mathrm{ECPLM}(R_r, R_a|n) = c_p \int\limits_{R_r} g_{\theta_a}(\overset{\frown}{\beta}^{\delta}|n)d\overset{\frown}{\beta}^{\delta} + c_c \int\limits_{R_a} g_{\theta_r}(\overset{\frown}{\beta}^{\delta}|n)d\overset{\frown}{\beta}^{\delta}. \tag{55}$$

A reasonable classification rule should have an ECPLM as small, or nearly as small, as possible. The problem is to minimize (55).

Theorem 1. The regions R_r and R_a that minimize the ECPLM are defined by the values of $\overset{\frown}{\beta}^{\delta}$ for which the following inequalities hold:

$$R_r : \frac{g_{\theta_r}(\overset{\frown}{\beta}^{\delta}|n)}{g_{\theta_a}(\overset{\frown}{\beta}^{\delta}|n)} \geq \frac{c_p}{c_c} \left(\begin{pmatrix} \text{density} \\ \text{ratio} \end{pmatrix} \geq \begin{pmatrix} \text{cost} \\ \text{ratio} \end{pmatrix} \right), \tag{56}$$

$$R_a : \frac{g_{\theta_r}(\overset{\frown}{\beta}^{\delta}|n)}{g_{\theta_a}(\overset{\frown}{\beta}^{\delta}|n)} < \frac{c_p}{c_c} \left(\begin{pmatrix} \text{density} \\ \text{ratio} \end{pmatrix} < \begin{pmatrix} \text{cost} \\ \text{ratio} \end{pmatrix} \right), \tag{57}$$

Proof. Noting that

$$\Omega = R_r \cup R_a, \tag{58}$$

so that the total probability

$$1 = \int\limits_{\Omega} g_{\theta_a}(\overset{\frown}{\beta}^{\delta}|n)d\overset{\frown}{\beta}^{\delta} = \int\limits_{R_r} g_{\theta_a}(\overset{\frown}{\beta}^{\delta}|n)d\overset{\frown}{\beta}^{\delta} + \int\limits_{R_a} g_{\theta_a}(\overset{\frown}{\beta}^{\delta}|n)d\overset{\frown}{\beta}^{\delta}, \tag{59}$$

we can write

$$\mathrm{ECPLM}(R_a|n) = c_p \left(1 - \int\limits_{R_a} g_{\theta_a}(\overset{\frown}{\beta}^{\delta}|n)d\overset{\frown}{\beta}^{\delta} \right) + c_c \int\limits_{R_a} g_{\theta_r}(\overset{\frown}{\beta}^{\delta}|n)d\overset{\frown}{\beta}^{\delta}. \tag{60}$$

By the additive property of integrals,

$$\mathrm{ECPLM}(R_a|n) = \int\limits_{R_a} [c_c g_{\theta_r}(\overset{\frown}{\beta}^{\delta}|n) - c_p g_{\theta_a}(\overset{\frown}{\beta}^{\delta}|n)]d\overset{\frown}{\beta}^{\delta} + c_p. \tag{61}$$

Now, c_c and c_p are nonnegative. In addition, $g_{\theta_r}(\overset{\frown}{\beta}^{\delta}|n)$ and $g_{\theta_a}(\overset{\frown}{\beta}^{\delta}|n)$ are nonnegative for all values of $\overset{\frown}{\beta}^{\delta}$ and are the only quantities in ECPLM that depend on $\overset{\frown}{\beta}^{\delta}$. Thus, ECPLM is minimized if R_a includes those values of $\overset{\frown}{\beta}^{\delta}$ for which the integrand

$$[c_c g_{\theta_r}(\overset{\frown}{\beta}^{\delta}|n) - c_p g_{\theta_a}(\overset{\frown}{\beta}^{\delta}|n)] < 0 \tag{62}$$

and excludes those values of $\overset{\frown}{\beta}^{\delta}$ for which this quantity is nonnegative. This completes the proof.

Corollary 1.1. If $c_p/c_c = 1$ (equal product lot misclassification costs), then

$$R_r : g_{\theta_r}(\overset{\frown}{\beta}^{\delta}|n)/g_{\theta_a}(\overset{\frown}{\beta}^{\delta}|n) \geq 1, \quad R_u : g_{\theta_r}(\overset{\frown}{\beta}^{\delta}|n)/g_{\theta_a}(\overset{\frown}{\beta}^{\delta}|n) < 1. \tag{63}$$

It follows from (56) and (57) that classification of the product lot into one of two different regions R_a and R_r (i.e., determining whether or not the product lot is accepted) is carried out through $\overset{\frown}{\beta}^{\delta}$ and h as follows:

$$\overset{\frown}{\beta}^{\delta} \in \begin{cases} R_r = \left\{\overset{\frown}{\beta}^{\delta} : \overset{\frown}{\beta}^{\delta} \leq h\right\} \rightarrow \text{reject the product lot}, \\ R_a = \left\{\overset{\frown}{\beta}^{\delta} : \overset{\frown}{\beta}^{\delta} > h\right\} \rightarrow \text{accept the product lot}, \end{cases} \tag{64}$$

where

$$h = \arg\min_h \left(c_p \int\limits_0^h g_{\theta_a}(\overset{\frown}{\beta}^{\delta}|n)d\overset{\frown}{\beta}^{\delta} + c_c \int\limits_h^\infty g_{\theta_r}(\overset{\frown}{\beta}^{\delta}|n)d\overset{\frown}{\beta}^{\delta} \right). \tag{65}$$

Remark 2. In general, the classification rule (64) can be rewritten as follows: Assign the product lot to region R_a (or R_r) for which $c_p g_{\theta_a}(\overset{\frown}{\beta}^{\delta}|n)$ (or $c_c g_{\theta_r}(\overset{\frown}{\beta}^{\delta}|n)$) is largest.

3.4 Situation 3

Consider Situation 2 when there is the cost c of sampling and putting a product item on test, but the size "n" of a random sample (Y_1, …, Y_n) of product items to test the

Weibull MTTF "μ", the producer's risk, which is denoted by α_p, and the consumer's risk, which is denoted by α_c, are not preassigned.

The problem is to find the optimal separation threshold "h" and integer "n", which minimize the performance index:

$$cn + c_p\alpha_p(h,n) + c_c\alpha_c(h,n) = cn + c_p\int_0^h g_{\theta_a}(\overset{\frown}{\beta}^{\delta}|n)d\overset{\frown}{\beta}^{\delta} + c_c\int_h^\infty g_{\theta_r}(\overset{\frown}{\beta}^{\delta}|n)d\overset{\frown}{\beta}^{\delta}, \quad (66)$$

where the optimal separation threshold "h" for $\overset{\frown}{\beta}^{\delta}$ and integer "n" are given by

$$\begin{aligned}(h,n) &= \arg\min_n\left[cn + \min_h\left(c_p\alpha_p(h|n) + c_c\alpha_c(h|n)\right)\right]\\ &= \arg\min_n\left[cn + \min_h\left(c_p\int_0^h g_{\theta_a}(\overset{\frown}{\beta}^{\delta}|n)d\overset{\frown}{\beta}^{\delta} + c_c\int_h^\infty g_{\theta_r}(\overset{\frown}{\beta}^{\delta}|n)d\overset{\frown}{\beta}^{\delta}\right)\right].\end{aligned} \quad (67)$$

The classification rule of the product lot is given by

$$\overset{\frown}{\beta}^{\delta} \in \begin{cases} R_r = \left\{\overset{\frown}{\beta}^{\delta} : \overset{\frown}{\beta}^{\delta} \le h\right\} \to \text{reject the product lot,}\\ R_a = \left\{\overset{\frown}{\beta}^{\delta} : \overset{\frown}{\beta}^{\delta} > h\right\} \to \text{accept the product lot.}\end{cases} \quad (68)$$

3.5 Situation 4

Consider the situation when the size "n" of a random sample (Y_1, …, Y_n) of product items to test the Weibull MTTF "μ" is not preassigned, but a product lot misclassification probability is preassigned:

$$\alpha(h,n) = \alpha. \quad (69)$$

The problem is to find the optimal separation threshold "h" and the required smallest integer "n", which minimize the performance index:

$$\begin{aligned}[\alpha(h,n)-\alpha]^2 &= \left[\int_0^h g_{\theta_a}(\overset{\frown}{\beta}^{\delta}|n)d\overset{\frown}{\beta}^{\delta} + \int_h^\infty g_{\theta_r}(\overset{\frown}{\beta}^{\delta}|n)d\overset{\frown}{\beta}^{\delta} - \alpha\right]^2\\ &= \left[\int_0^{h/\beta_a^\delta}\frac{1}{\Gamma(n)(1/n)^n}v_a^{n-1}\exp\left(-\frac{v_a}{1/n}\right)dv_a + \int_{h/\beta_r^\delta}^\infty \frac{1}{\Gamma(n)(1/n)^n}v_r^{n-1}\exp\left(-\frac{v_r}{1/n}\right)dv_r - \alpha\right]^2,\end{aligned} \quad (70)$$

where the optimal separation threshold "h" for $\overset{\frown}{\beta}^{\delta}$ and the required smallest integer "n" are given by

$$(h,n) = \arg\min_{n}\left(\min_{h}[\alpha(h|n) - \alpha]^2\right) = \arg\min_{n}\left[\min_{h}\left(\int_{0}^{h} g_{\theta_a}(\overset{\frown}{\beta}^{\delta}|n)d\overset{\frown}{\beta}^{\delta} + \int_{h}^{\infty} g_{\theta_r}(\overset{\frown}{\beta}^{\delta}|n)d\overset{\frown}{\beta}^{\delta} - \alpha\right)^2\right]. \tag{71}$$

The classification rule of the product lot is given by

$$\overset{\frown}{\beta}^{\delta} \in \begin{cases} R_r = \left\{\overset{\frown}{\beta}^{\delta} : \overset{\frown}{\beta}^{\delta} \le h\right\} \to \text{reject the product lot}, \\ R_a = \left\{\overset{\frown}{\beta}^{\delta} : \overset{\frown}{\beta}^{\delta} > h\right\} \to \text{accept the product lot}. \end{cases} \tag{72}$$

3.6 Situation 5

Consider the situation when the size "n" of a random sample (Y_1,..., Y_n) of product items to test the Weibull MTTF "μ" is not preassigned, but the producer's risk, which is denoted by α_p, and the consumer's risk, which is denoted by α_c, are preassigned.

The problem is to find the optimal separation threshold "h" and the required smallest integer "n", which minimize the performance index:

$$(\alpha_p(h,n) - \alpha_p)^2 + (\alpha_c(h,n) - \alpha_c)^2 = \left(\int_{0}^{h} g_{\theta_a}(\overset{\frown}{\beta}^{\delta}|n)d\overset{\frown}{\beta}^{\delta} - \alpha_p\right)^2 + \left(\int_{h}^{\infty} g_{\theta_r}(\overset{\frown}{\beta}^{\delta}|n)d\overset{\frown}{\beta}^{\delta} - \alpha_c\right)^2, \tag{73}$$

where the optimal separation threshold "h" for $\overset{\frown}{\beta}^{\delta}$ and the required smallest integer "n" are given by

$$(h,n) = \arg\min_{n}\left(\min_{h}[(\alpha_p(h|n) - \alpha_p)^2 + (\alpha_c(h|n) - \alpha_c)^2]\right) = \arg\min_{n}\left(\min_{h}\left[\left(\int_{0}^{h} g_{\theta_a}(\overset{\frown}{\beta}^{\delta}|n)d\overset{\frown}{\beta}^{\delta} - \alpha_p\right)^2 + \left(\int_{h}^{\infty} g_{\theta_r}(\overset{\frown}{\beta}^{\delta}|n)d\overset{\frown}{\beta}^{\delta} - \alpha_c\right)^2\right]\right). \tag{74}$$

The classification rule of the product lot is given by

$$\overset{\frown}{\beta}^{\delta} \in \begin{cases} R_r = \left\{ \overset{\frown}{\beta}^{\delta} : \overset{\frown}{\beta}^{\delta} \leq h \right\} \rightarrow \text{reject the product lot,} \\ R_a = \left\{ \overset{\frown}{\beta}^{\delta} : \overset{\frown}{\beta}^{\delta} > h \right\} \rightarrow \text{accept the product lot.} \end{cases} \tag{75}$$

Remark 3. Suppose that product lots are submitted for inspection and the lifetimes of individual products have a Weibull distribution with the probability density function (15) and unknown parameters (β, δ). In this case, the unknown shape parameter δ of the two-parameter Weibull lifetime distribution is replaced by the maximum likelihood estimate $\widehat{\delta}$ given by (18). It follows from (29) and (30) that

$$\beta_a^{\widehat{\delta}} = \left[\mu_a / \Gamma\left(1 + \frac{1}{\widehat{\delta}}\right) \right]^{\widehat{\delta}} \text{ and } \beta_r^{\widehat{\delta}} = \left[\mu_r / \Gamma\left(1 + \frac{1}{\widehat{\delta}}\right) \right]^{\widehat{\delta}}, \tag{76}$$

respectively. It can be shown that

$$\Pr(\overset{\frown}{\beta}^{\widehat{\delta}} \leq h | \beta_a^{\widehat{\delta}}, n, \mathbf{z}^{(n)}) = \frac{1}{\vartheta(\mathbf{z}^{(n)})} \int_0^{\infty} \frac{v_1^{n-2} \prod_{i=1}^{n} z_i^{v_1} G_n \left(\left(\frac{h}{\beta_a^{\widehat{\delta}}} \right)^{v_1} \left[\sum_{i=1}^{n} z_i^{v_1} \right] \right)}{\left(\sum_{i=1}^{n} z_i^{v_1} \right)^n} dv_1, \tag{77}$$

where

$$G_n\left(\left(\frac{h}{\beta_a^{\widehat{\delta}}} \right)^{v_1} \left[\sum_{i=1}^{n} z_i^{v_1} \right] \right) = \int_0^{\left(\frac{h}{\beta_a^{\widehat{\delta}}} \right)^{v_1} \left[\sum_{i=1}^{n} z_i^{v_1} \right]} \frac{1}{\Gamma(n)} \tau^{n-1} \exp(-\tau) d\tau, \tag{78}$$

$$\vartheta(\mathbf{z}^{(n)}) = \int_0^{\infty} v_1^{n-2} \prod_{i=1}^{n} z_i^{v_1} \left(\sum_{i=1}^{n} z_i^{v_1} \right)^{-n} dv_1 \tag{79}$$

is the normalizing constant,

$$\mathbf{Z}^{(n)} = (Z_i, \ldots, Z_n), \quad Z_i = \left(\frac{Y_i}{\overset{\frown}{\beta}} \right)^{\widehat{\delta}}, \; i = 1, \ldots, n, \tag{80}$$

are ancillary statistics, any n–2 of which form a functionally independent set;

$$\Pr(\widehat{\beta^{\delta}} > h|\beta_r^{\widehat{\delta}}, n, \mathbf{z}^{(n)}) = 1 - \Pr(\widehat{\beta^{\delta}} \le h|\beta_r^{\widehat{\delta}}, n, \mathbf{z}^{(n)})$$
$$= 1 - \frac{1}{\vartheta(\mathbf{z}^{(n)})} \int_0^{\infty} \frac{v_1^{n-2} \prod_{i=1}^{n} z_i^{v_1} G_n\left(\left(\frac{h}{\beta_r^{\widehat{\delta}}}\right)^{v_1} \left[\sum_{i=1}^{n} z_i^{v_1}\right]\right)}{\left(\sum_{i=1}^{n} z_i^{v_1}\right)^n} dv_1, \tag{81}$$

where

$$G_n\left(\left(\frac{h}{\beta_r^{\widehat{\delta}}}\right)^{v_1} \left[\sum_{i=1}^{n} z_i^{v_1}\right]\right) = \int_0^{\left(\frac{h}{\beta_r^{\widehat{\delta}}}\right)^{v_1} \left[\sum_{i=1}^{n} z_i^{v_1}\right]} \frac{1}{\Gamma(n)} \tau^{n-1} \exp(-\tau) d\tau. \tag{82}$$

Now, using (77) and (81), the situations discussed above for optimizing the product acceptance process can be viewed in a similar way.

4 Conclusion

The main aim of this paper is to promote the novel ideas of effective optimization (constrained or unconstrained) of product acceptance processes in terms of misclassification probability (under parametric uncertainty of underlying models) for real practical applications. It is expected that these ideas will give interesting and novel contributions to statistical theory and its applications at a good mathematical level, where the theoretical results are obtained through the frequentist (non-Bayesian) statistical approach. It is based on the pivotal quantity averaging (PQA) technique which allows one to eliminate the unknown parameters from the problem and to use available statistical information as completely as possible.

References

1. Gupta, S.S., Groll, P.A.: Gamma distribution in acceptance sampling based on life tests. J. Am. Stat. Assoc. **56**, 942–970 (1961)
2. Gupta, S.S.: Life test sampling plans for normal and lognormal distribution. Technometrics **4**, 151–175 (1962)
3. Fertig, F.W., Mann, N.R.: Life-test sampling plans for two-parameter Weibull populations. Technometrics **22**, 165–177 (1980)
4. Kantam, R.R.L., Rosaiah, K.: Half logistic distribution in acceptance sampling based on life tests. IAPQR Trans. **23**, 117–125 (1998)

5. Kantam, R.R.L., Rosaiah, K., Srinivasa Rao, G.: Acceptance sampling based on life tests: log-logistic model. J. Appl. Stat. **28**, 121–128 (2001)
6. Baklizi, A.: Acceptance sampling based on truncated life tests in the pareto distribution of the second kind. Adv. Appl. Stat. **3**, 33–48 (2003)
7. Wu, C.J., Tsai, T.R.: Acceptance sampling plans for Birnbaum-Saunders distribution under truncated life tests. Int. J. Reliab. Qual. Saf. Eng. **12**, 507–519 (2005)
8. Nechval, N.A., Vasermanis, E.K.: Improved decisions in statistics. Izglitibas soli, Riga (2004)
9. Nechval, N.A., Berzins, G., Purgailis, M., Nechval, K.N.: Improved estimation of state of stochastic systems via invariant embedding technique. WSEAS Trans. Math. **7**, 141–159 (2008)
10. Nechval, N., Purgailis, M., Berzins, G., Cikste, K., Krasts, J., Nechval, K.: Invariant embedding technique and its applications for improvement or optimization of statistical decisions. In: Al-Begain, K., Fiems, D., Knottenbelt, W.J. (eds.) ASMTA 2010. LNCS, vol. 6148, pp. 306–320. Springer, Heidelberg (2010). https://doi.org/10.1007/978-3-642-13568-2_22
11. Nechval, N., Purgailis, M., Nechval, K.: Weibull model for dynamic pricing in e-business. In: Skersys, T., Butleris, R., Nemuraite, L., Suomi, R. (eds.) I3E 2011. IAICT, vol. 353, pp. 292–304. Springer, Heidelberg (2011). https://doi.org/10.1007/978-3-642-27260-8_24
12. Nechval, N.A., Nechval, K.N.: Efficient approach to pattern recognition based on minimization of misclassification probability. Am. J. Theor. Appl. Stat. **5**, 7–11 (2016)
13. Nechval, N.A., Nechval, K.N.: A new technique for vibration-based diagnostics of fatigued structures based on damage pattern recognition via minimization of misclassification probability. In: Ram, M., Davim, J.P. (eds.) Diagnostic Techniques in Industrial Engineering. MIE, pp. 191–206. Springer, Cham (2018). https://doi.org/10.1007/978-3-319-65497-3_7
14. Gumbel, E.J.: Statistics of Extreme. Columbia University Press, New York (1958)

An Investigation of a Bi-level Non-dominated Sorting Algorithm for Production-Distribution Planning System

Malek Abbassi(✉), Abir Chaabani, and Lamjed Ben Said

SMART Lab, ISG, University of Tunis (Université de Tunis), Tunis, Tunisia
malekkabbassi@gmail.com

Abstract. Bi-Level Optimization Problems (BLOPs) belong to a class of challenging problems where one optimization problem acts as a constraint to another optimization level. These problems commonly appear in many real-life applications including: transportation, game-playing, chemical engineering, etc. Indeed, multi-objective BLOP is a natural extension of the single objective BLOP that bring more computational challenges related to the multi-objective hierarchical decision making. In this context, a well-known algorithm called NSGA-II was presented in the literature among the most cited Multi-Objective Evolutionary Algorithm (MOEA) in this research area. The most prominent features of NSGA-II are its simplicity, elitist approach and a non-parametric method for diversity. For this reason, in this work, we propose a bi-level version of NSGA-II, called Bi-NSGA-II, in an attempt to exploit NSGA-II features in tackling problems involving bi-level multiple conflicting criteria. The main motivation of this paper is to investigate the performance of the proposed variant on a bi-level production distribution problem in supply chain management formulated as a Multi-objective Bi-level MDVRP (M-Bi-MDVRP). The paper reveals three Bi-NSGA-II variants for solving the M-Bi-MDVRP basing on different variation operators (M-VMX, VMX, SBX and RBX). The experimental results showed the remarkable ability of our adopted algorithm for solving such NP-hard problem.

Keywords: Bi-level optimization · MOEA · NSGA-II · Bi-level production-distribution problem

1 Introduction

Bi-level programming allows the modeling of situations in which a first decision maker (DM), hereafter the leader, tries to optimize his objective(s) by taking the follower's response to his decisions, explicitly into account. In this way, for each upper solution decision vector, a lower optimization task should be performed firstly to determine the rational response of the follower. This, makes the problem very difficult to solve. Jeroslow has proven that even the simplest case of

F. Wotawa et al. (Eds.): IEA/AIE 2019, LNAI 11606, pp. 819–826, 2019.
https://doi.org/10.1007/978-3-030-22999-3_69

linear bi-level programs, where the lower level problem has a unique optimal solution, is demonstrated as an NP-hard problem [11]. Several real life problems can be cast within this structure and has motivated a number of researchers in different fields. However, there still lack solution methods able to solve efficiently the problem. Resolution methods for BLOPs could be divided into two categories: (1) the classical methods and (2) the evolutionary methods. The latter come to overcome the weakness of the classical methods: (1) the dependence on the mathematical properties of the problem, and (2) the limitation to small scale size problems. In this context, Mathieu et al. proposed the first EA for solving single objective BLOPs (SBLOPs) [12]. Since then, a number of EAs were proposed to solve BLOP, we cite for brevity the following works [1–4,7,8]. Besides, Multi-objective BLOP (MBLOP) which is a multi-objective variant of BLOP presents more difficulties to be addressed. The complexity imposed by the multiple objectives in bi-level optimization, makes the problem more difficult to solve primarily because of the computational and the decision making complexities. For this reason, there exists a significant number of works solving the SBLOPs, however little have been presented to solve the MBLOPs. In this regard, Deb and Sinha (2009b) suggested a bi-level evolutionary multi-objective optimization (BLEMO) algorithm based on NSGA-II applied to both levels. In 2013, Sinha et al., incorporated the DM preferences to the H-BLEMO yielding a new approach called PI-HBLEMO [14]. All the proposed algorithms in this research area are suggested for the continuous MBLOP. We observe the absence of MOEA able to solve the combinatorial case. However, most real life applications could be modeled as combinatorial MBLOPs. Thus, we need to expect a growing interest in solving multi-objective combinatorial BLOPs.

The NSGA-II was presented in the literature as the most cited evolutionary multi-objective optimization (EMO) procedure able to solve efficiently a multi-objective single level optimization problem. The most prominent features of NSGA-II are its simplicity, elitist approach and a method for diversity that does not need any additional parameters. To this end, we aim in this work to exploit the NSGA-II features within the bi-level optimization framework to evaluate the NSGA-II behavior on the combinatorial multi-objective BLOPs. The proposed approach is called Bi-NSGA-II which is referred to "Bi-level Non-dominated Sorting Genetic Algorithm". Indeed, we present in this paper an adaptation of the proposed Bi-NSGA-II on the M-Bi-MDVRP which is a well known problem in supply chain management. This paper should encourage researchers to give more attention to this optimization task of practical importance.

2 Bi-NSGA-II for Production Distribution Planning System

In this work we are exploiting the NSGA-II baseline features with the bi-level framework. We describe with this section the general scheme of Bi-NSGA-II which proceeds as follows: First, a population Pt_{upper} is generated randomly in the upper level decision space. Then, we apply the genetic variation operators in order to generate the offspring population Qt_{upper}. Once the Qt_{upper} is obtained, we perform the evaluation procedure to qualify the solutions in the upper decision space. In this way, we need to lunch the lower procedure to find first the optimal lower solutions. A lower evolution procedure is then executed using a GA procedure to generate the lower reactions regarding each upper solution in Qt_{upper}. At this stage we are able to lunch the sorting algorithm and the crowding distance of NSGA-II to fill the new upper population $Pt + 1_{upper}$. To more understand the Bi-NSGA-II on the M-BI-MDVRP, we describe in the following the step-by-step procedure of the algorithm.

2.1 Upper Level Optimization Procedure

- **Step 1 (Initialization step)** Create an initial upper population P_u with N_u individuals randomly, the encoding solution details are presented in Subsect. 2.3 (cf. Table 1). Next, we perform the lower level task to create a new lower level variables set x_l. We mention here that we combine each upper level variable x_u with the corresponding x_l to obtain the complete solution $x = (x_u, x_l)$.
- **Step 2 (Upper level selection)** Select individuals from the parent population according to the binary tournament selection operator to precise the best individuals that fit into the mating pool.
- **Step 3 (Crossover and Mutation)** Perform the M-VMX crossover operation, described in Sect. 2.4, for each chosen pair of individuals in order to obtain an offspring solution (cf. Fig. 1). Thus, an offspring population Q_u is formed. Then, apply the mutation operation.
- **Step 4 (Offspring upgrading and evaluation)** Appeal the lower level procedure (i.e. solve the inner problem using GA) for each x_u from the Q_u population to get the corresponding optimal vector valued solution x_l^*. To this end, the algorithm is able now to perform the evaluation step on the entire solutions.
- **Step 5 (Elitism)** The parent and offspring populations are then combined together forming a larger population $R_u = P_u \cup Q_u$.
- **Step 6 (Environmental selection)** Choose the N best individuals to the new upper level population based on the non-dominated sorting algorithm and the crowding distance measures of NSGA-II procedure [6].

2.2 Lower Level Optimization Procedure

- **Step 1 (Initialization step)** Generate an initial lower P_l population of N_l individuals randomly.
- **Step 2 (Selection)** Choose lower parent individuals using binary tournament selection operator to perform the variation operators and obtain new offspring.
- **Step 3 (Crossover and Mutation)** The one point crossover operator is applied here for each selected parent to obtain a new offspring generation. Then, apply the mutation operator (permutation).

- **Step 4 (Offspring evaluation)** Evaluate the obtained offspring individuals according to their fitness values.
- **Step5 (Termination step)** The algorithm checks for any new minimum. Finally, we assign the optimal lower solution x_l* to the upper vector valued solution. We note here that the lower level fitness is added to both upper objective fitness to take into consideration the lower behaviour with the upper procedure.

2.3 Solution Encoding

We aim to design a generic bi-level solution for the M-Bi-MDVRP, considering the decision variables and the possible changes in a global visualization of the solution at both upper and lower levels. In this way, an upper level solution is presented by a matrix (a) where $P_{i,j}$ defines the passage order of a vehicle v to a retailer r. This latter takes: (1) $P_{i,j} = 0$ value if v_i does not support r_j. $P_{i,j} = -P_{i,j}$ indicates the starting of a new route for the same vehicle from the corresponding depot to the just assigned costumer. Similarly, we choose a matrix (b) to design the lower level solution, where $Q_{p,d}$ defines the quantity of goods produced by plant p to depot d. We notice that a plant can serve several depots and a depot choose the plant or (the combination of a set of plants) ensuring the minimum possible manufacturing cost. We give in Table 1 an example encoding solution with 9 customers, 4 depots and 4 plants.

Table 1. Example of used solution representations.

(a)

	R_1	R_2	R_3	R_4	R_5	R_6	R_7	R_8	R_9
V_1	0	$P_{i,j}$	-$P_{i,j}$	$P_{i,j}$	0	0	0	0	$P_{i,j}$
V_2	$P_{i,j}$	0	0	0	0	$P_{i,j}$	0	0	0
V_3	0	0	0	0	0	0	0	0	0
V_4	0	0	0	0	$P_{i,j}$	0	$P_{i,j}$	$P_{i,j}$	0

(b)

	P_1	P_2	P_3	P_4	Total
D_1	$Q_{p,d}$	$Q_{p,d}$	0	$Q_{p,d}$	Q_{total}
D_2	0	0	0	$Q_{p,d}$	Q_{total}
D_3	0	0	$Q_{p,d}$	$Q_{p,d}$	Q_{total}
D_4	$Q_{p,d}$	0	$Q_{p,d}$	0	Q_{total}

2.4 A Modified Vehicle Merge Crossover

The Vehicle Merge Crossover (VMX) is a well-known crossover operator customized to the vehicle routing problem proposed by Hosny and Mumford (2009). This variation operator reduces implicitly the number of the vehicles used to deliver the costumers by merging two selected vehicle routes from the parent solutions and inserting them in the offspring one. The strength of this crossover operator is its ability to generate a feasible offspring solution with minimal repairing techniques [10]. Indeed, we performed a modification with this operator aiming to improve its exploration rate to deal with large scale search space that characterize the bi-level optimization framework. The resulting operator which is called Modified Vehicle Merge Crossover (M-VMX), allows a larger covering

Parent$_1$

	C	C	C	C	C	C	C	C	C
V	0	2	0	0	1	0	0	0	0
V	1	0	-3	0	0	2	0	0	0
V	0	0	0	1	0	0	0	2	0
V	0	0	0	0	0	0	1	0	2

Parent$_2$

	C	C	C	C	C	C	C	C	C
V	0	1	0	0	0	0	2	0	0
V	0	0	1	0	0	0	0	0	0
V	1	0	0	-4	0	2	0	3	5
V	0	0	0	0	1	0	0	0	0

Child

	C	C	C	C	C	C	C	C	C
V	0	1	0	0	3	0	2	0	0
V	1	0	-3	0	2	0	0	0	0
V	0	0	0	0	0	0	0	0	0
V	1	0	0	-4	0	2	0	3	5

Fig. 1. M-VMX crossover operator example with 9 costumers and 4 vehicles.

of the search space. This operator enhances the offspring's ability to inherit a promising solution genes from its parents. These modifications can improve both convergence and diversity properties of the algorithm by discovering good search directions at an early stage. Moreover, we tried to improve the procedure efficiently by reducing the number of corrections needed. This fact is guaranteed by using an archive to block both assigned vehicles and visited customers. As shown in Fig. 1 the process proceeds as follows: we select randomly two vehicles and we merge all their routes.

Then we choose a random vehicle that is not assigned yet from the offspring solution to fill it with the new combined routes. The pseudocode of the M-VMX is presented by Algorithm 1.

3 Experimental Study

In this section, we evaluate the performance of our proposed Bi-NSGA-II variant on an NP-hard combinatorial problem which is well studied in the literature. For this reason, three Bi-NSGA-II variants were investigated to determine the best components allowing to the best procedure performance. We conduct a comparison study between the Bi-NSGA-II variants applying the proposed M-VMX, the RBX, and the SBX variation operators, respectively, which are detailed as follows:

RBX: The Route Based Crossover operator copies routes from a parent, and then appends the other routes from the second parent [13].
SBX: The Sequence Based Crossover creates a new route by taking costumers from the starting selected parent route to a random stopping position. Then, it completes the new route with costumers from a random position to the end at the second selected parent route. Finally, it fills the offspring solution with the remaining routes [13].

We conduct a set of experiments on 31 independent runs for each instance from the used set of benchmarks, and we use the Wilcoxon rank-sum test [9]. We

Algorithm 1. M-VMX Pseudocode

Input: *Parent*1, *Parent*2, *Archive*, $Nb_{vehicle}$, *pos*.
Output: Offspring.
1: **Begin**
2: **For** each v in $Nb_{vehicle}$ **do**
3: Select vehicles randomly from *Parent*1, *Parent*2;
4: *Archive* $\longleftarrow$ *Add*(v,r);
5: *pos* $\longleftarrow$ a random position in the offspring solution;
6: $v_{pos} \longleftarrow$ *Insert-from*(*Archive*);
7: *Archive* $\longleftarrow$ *Clear*(*Archive*);
8: **End For**
9: **End**

generate two sets of instances from the 33 published MDVRP data-set (i.e. bipr and bip) provided in [5] for the experiment analysis. Indeed, to ensure a fair comparison we implemented the three versions of the algorithm under the same design of experimentation: Population size = *100*, crossover probability = *0.9*, mutation probability = *0.1*. The termination criteria which is defined by the Function Evaluations is fixed to *25000*.

3.1 Results and Discussion

Table 2 summarizes the HyperVolume results for each of the used benchmarks. It is shown clearly that Bi-NSGA-II with our M-VMX yields better results regarding all used instances which explain that the M-VMX keeps a good covering of the search space in terms of convergence and population diversity. This result can be observed clearly from Fig. 2 which shows the algorithm stability for each couple (algorithm, instance). Regarding the two other used crossover operators:

Table 2. HV values for Bi-NSGA-II algorithms with M-VMX, SBX, and RBX.

Instance	VMX	SBX	RBX	Instance	VMX	SBX	RBX	Instances	VMX	SBX	RBX
Mbip01	**0.516**	0.48	0.462	Mbip12	**0.473**	0.45	0.34	Mbip23	**0.454**	0.442	0.42
Mbip02	**0.44**	0.42	0.359	Mbip13	**0.496**	0.48	0.491	Mbipr01	0.43	**0.5**	0.385
Mbip03	**0.487**	0.47	0.39	Mbip14	**0.469**	0.45	0.41	Mbipr02	**0.498**	0.46	0.495
Mbip04	**0.49**	0.44	0.421	Mbip15	**0.492**	0.38	0.41	Mbipr03	**0.459**	0.41	0.39
Mbip05	**0.495**	0.47	0.432	Mbip16	**0.469**	0.32	0.38	Mbipr04	**0.49**	0.44	0.42
Mbip06	**0.493**	0.48	0.42	Mbip17	**0.484**	0.45	0.425	Mbipr05	0.459	**0.468**	0.445
Mbip07	**0.47**	0.447	0.367	Mbip18	**0.511**	0.51	0.505	Mbipr06	0.519	0.38	**0.67**
Mbip08	**0.488**	0.421	0.449	Mbip19	0.484	**0.49**	0.425	Mbipr07	**0.465**	0.34	0.319
Mbip09	**0.487**	0.438	0.476	Mbip20	**0.57**	0.304	0.5	Mbipr08	**0.507**	0.31	0.314
Mbip10	0.448	**0.46**	0.434	Mbip21	**0.457**	0.43	0.412	Mbipr09	**0.469**	0.383	0.45
Mbip11	**0.506**	0.46	0.423	Mbip22	**0.508**	0.49	0.33	Mbipr10	**0.547**	0.45	0.321

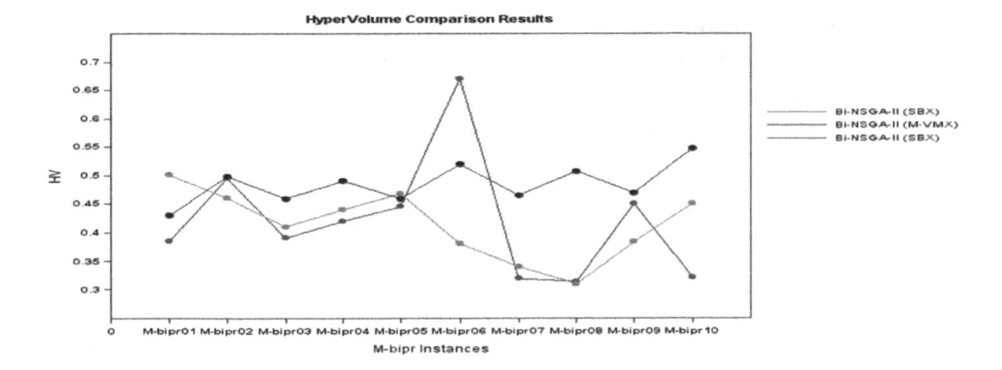

Fig. 2. HV comparison for M-bipr benchmarks.

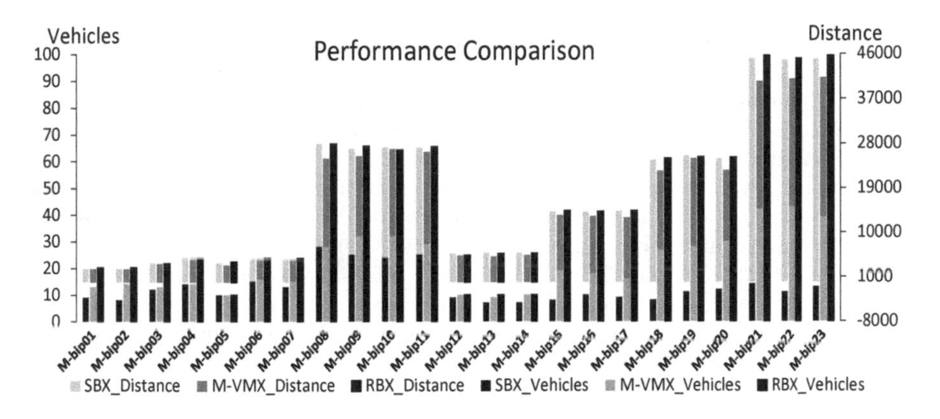

Fig. 3. Results for upper level problem.

a wobbling behaviour is detected among the used benchmarks for both RBX and SBX.

These results are also confirmed by Fig. 3 which reports the quality of solution in terms of generated distance and used vehicles. Such results can be explained by the ability of the M-VMX to more explore the search space which has an impact on the solution quality. We conclude that M-VMX can be successfully adapted with Bi-NSGA-II to solve this NP-hard problem.

4 Conclusion

In this work, we suggested a new bi-level variant of the NSGA-II procedure to solve the multi-objective Bi-MDVRP. We investigated in the experimental study three crossover operators with our Bi-NSGA-II. The results show the outperformance of our M-VMX operator regarding to the other used crossovers. Future works may focus mainly on designing an improved version of Bi-NSGA-II using parallel programming concept. It would be challenging to reduce the handled complexity of a multi-objective bi-level algorithm since it consumes a high computational cost complexity. An other interesting research direction would be the integration of new diversification measures since NSGA-II applies the crowding distance measure only in the case of two identical individuals ranks and with exceeded front size replacement.

References

1. Bell, M.: Transportation networks: recent methodological advances: selected proceedings of the 4th euro transportation meeting. In: Association of European Operational Research Societies (EURO), Working Group on Transportation, Meeting, 4th, 1996, Newcastle, United Kingdom (1998)
2. Chaabani, A., Bechikh, S., Said, L.B.: A co-evolutionary decomposition-based algorithm for bi-level combinatorial optimization. In: 2015 IEEE Congress on Evolutionary Computation(CEC), pp. 1659–1666. IEEE (2015)
3. Chaabani, A., Bechikh, S., Said, L.B.: A memetic evolutionary algorithm for bi-level combinatorial optimization: a realization between Bi-MDVRP and Bi-CVRP. In: 2016 IEEE Congress on Evolutionary Computation (CEC), pp. 1666–1673. IEEE (2016)
4. Chaabani, A., Said, L.B.: Transfer of learning with the co-evolutionary decomposition-based algorithm-II: a realization on the bi-level production-distribution planning system. Appl. Intell. **49**(3), 963–982 (2019)
5. Cordeau, J.F., Gendreau, M., Laporte, G.: A tabu search heuristic for periodic and multi-depot vehicle routing problems. Netw. Int. J. **30**(2), 105–119 (1997)
6. Deb, K., Pratap, A., Agarwal, S., Meyarivan, T.: A fast and elitist multiobjective genetic algorithm: NSGA-II. IEEE Trans. Evol. Comput. **6**(2), 182–197 (2002)
7. Deb, K., Sinha, A.: An evolutionary approach for bilevel multi-objective problems. In: Shi, Y., Wang, S., Peng, Y., Li, J., Zeng, Y. (eds.) MCDM 2009. CCIS, vol. 35, pp. 17–24. Springer, Heidelberg (2009a). https://doi.org/10.1007/978-3-642-02298-2_3
8. Deb, K., Sinha, A.: Solving bilevel multi-objective optimization problems using evolutionary algorithms. In: Ehrgott, M., Fonseca, C.M., Gandibleux, X., Hao, J.-K., Sevaux, M. (eds.) EMO 2009. LNCS, vol. 5467, pp. 110–124. Springer, Heidelberg (2009b). https://doi.org/10.1007/978-3-642-01020-0_13
9. Derrac, J., García, S., Molina, D., Herrera, F.: A practical tutorial on the use of nonparametric statistical tests as a methodology for comparing evolutionary and swarm intelligence algorithms. Swarm Evol. Comput. **1**(1), 3–18 (2011)
10. Hosny, M.I., Mumford, C.L.: Investigating genetic algorithms for solving the multiple vehicle pickup and delivery problem with time windows. In: MIC2009, Metaheuristic International Conference (2009)
11. Jeroslow, R.G.: The polynomial hierarchy and a simple model for competitive analysis. Math. Program. **32**(2), 146–164 (1985)
12. Mathieu, R., Pittard, L., Anandalingam, G.: Genetic algorithm based approach to bi-level linear programming. RAIRO-Oper. Res. **28**(1), 1–21 (1994)
13. Potvin, J.Y., Bengio, S.: The vehicle routing problem with time windows part II: genetic search. INFORMS J. Comput. **8**(2), 165–172 (1996)
14. Sinha, A., Deb, K.: Bilevel multi-objective optimization and decision making. In: Talbi, E.G. (ed.) Metaheuristics for Bi-level Optimization, vol. 482, pp. 247–284. Springer, Heidelberg (2013). https://doi.org/10.1007/978-3-642-37838-6_9

Optimization of Bridges Reinforcement by Conversion to Tied Arch Using an Animal Migration Algorithm

Andrés Morales[1,4], Broderick Crawford[1,4], Ricardo Soto[1,4], José Lemus-Romani[1,4](✉), Gino Astorga[2,4], Agustín Salas-Fernández[1,4], and José-Miguel Rubio[3,4]

[1] Pontificia Universidad Católica de Valparaíso, Valparaíso, Chile
{broderick.crawford,ricardo.soto}@pucv.cl,
{andres.morales.f,jose.lemus.r,juan.salas.f}@mail.pucv.cl
[2] Universidad de Valparaíso, Valparaíso, Chile
gino.astorga@uv.cl
[3] Covenant University, Ota, Nigeria
sanjay.misra@covenantuniversity.edu.ng
[4] Universidad Tecnológica de Chile INACAP, Ñuñoa, Chile
jrubiol@inacap.cl

Abstract. Nowadays there are studies that show that bridges collapse mainly by scour caused by hydraulic action which implies high reconstruction costs. We can avoid the collapse of the bridge, reinforcing it by means of a system of cable-stayed arches, which will make it possible to hold the deck and eliminate the damaged elements. To this end, it is necessary to optimize the order and the magnitudes of adjustment of the tension of the hangers. In this paper, the use of the Animal Migration Optimization algorithm is proposed for the search of an optimal solution in a gradual way, which consists of two processes: the simulation of how groups of animals move from a current position to a new, and how during migration some animals leave the group and join others. Finally, we present experimental results, where the performance of the Animal Migration Optimization algorithm can be observed.

Keywords: Reinforcement of Bridges · Metaheuristics · Animal Migration Optimization · Combinatorial optimization

1 Introduction

At present, there are studies that show that bridges collapse due to erosion, corrosion, but mainly due to the scour caused in the piers by hydraulic action, which implies high reconstruction costs depending on the size, shape and materials. There are several possibilities to avoid the collapse of the bridge, but in this paper the reinforcement technique will be used through a system of cable-stayed steel arches, which will support the deck and eliminate the damaged elements [13].

F. Wotawa et al. (Eds.): IEA/AIE 2019, LNAI 11606, pp. 827–834, 2019.
https://doi.org/10.1007/978-3-030-22999-3_70

For constructive reasons, the tension of the hangers can not be carried out simultaneously, but must be sequential. For this, it is necessary to optimize the order and the magnitudes of adjustment of the tension of the hangers to avoid instability and cause damage to the structure. To optimize the order and the magnitudes of the adjustment of the tension of the hangers, it is proposed to use an algorithm that searches for an optimal solution inspired by the migration behavior of the animals. This is a phenomenon present in the main groups of animals such as mammals, birds, fish, insects. The animals generally migrate because of the local climate, availability of food, season of the year, among other reasons.

The Animal Migration Optimization (AMO) [8] is divided into two processes: simulation of how groups of animals move from the current position to a new and the simulation of how some animals leave the group and others join during the migration. The optimization algorithms that they learn from nature have been applied in various areas of research with great success [2–4, 11]. However, until now there is no algorithm that achieves good performance in all areas.

The problem of the bridges will be presented in Sect. 2, then in Sect. 3 the proposed algorithm for the optimization of sequence and magnitudes of tension, in Sect. 4 show the results obtained after the implementation, then in Sect. 5 we compare the results obtained through statistical techniques, ending with the conclusions of the work.

2 Problem

The most frequent problem that bridges that are in contact with water present are erosion, corrosion and scour [10, 13]. Erosion is defined as the wear and tear on the surface of a body caused by friction with water [12]. Corrosion is the deterioration of the material generated by oxidation. Scour is the removal of support material due to the movement of waves against a cliff and the water swirls [9].

The three factors mentioned above, generate collateral damage on the deck and deterioration of the structure, which is directly related to a high reconstruction cost. Scour, is the main cause of deterioration of the piers that support a bridge, and can cause serious damage, which can trigger the total collapse of the structure.

2.1 Reinforcement System

Build a new bridge has associated high costs of material, labor, in addition to the interruption of traffic for a long time [6]. It is for this reason, that the most appropriate option is the reinforcement of the bridges to prolong their useful life and thus avoid the high costs of a collapse.

Currently, there are precedents that show that different types of reinforcements have been worked on, but the techniques used do not provide solutions to the main problem, which is the action of water against the piers.

According to the problem described above, the solution of a reinforcement has been proposed by means of a arch, which goes over the bridge deck from one end to the other. The arch is composed of a passive hangers that can be configured as a network, crosses, fans, among others, which will lift the deck, to later eliminate the piers that initially support the deck, and thus, directly attack the scour, erosion and corrosion problems caused by hydraulic contact.

For a deck supported by N pier, there are at least N upper hangers, next to the passive hangers, type Network in this case. The proposed method leads to a constructive process of the superstructure, beginning with the installation of the arch, continuing with the upper hangers, which will be tightened, to finally eliminate the lower piers definitively, while the applied forces are distributed and stabilized. The change that occurs from the original bridge, to the final result, where you can see the bridge supported by the hangers. The construction process of the proposed superstructure is divided into the following stages [13], first the arch installed on a bridge to be reinforced, then proceed installation of hangers, continuing with tensioned hangers, ending with the eliminate the piers.

However, it is not possible to tense the three hangers simultaneously, since there are constructive and operational problems, because the combination between the applied force and the tensioning order, must maintain the efforts of the original bridge, that is, the differences of efforts of the modified board, it must be minimal compared to the original board. In addition, it must be considered that the application of excessive forces of tension in the hangers, can have as a consequence damages in the board and even the collapse of the bridge. All calculations to be made regarding the stress state of the bridge will be evaluated by SAP2000 through API [1]. The method consists of evaluating the original bridge to be reinforced and the new structuring that the bridge presents after the reinforcement. It is to know the tensional states of both instances of the bridge to act properly and not to make mistakes during the reinforcement process. To model and evaluate both bridges, SAP2000 will be used, which is a tool aimed at computer-aided structural design.

In addition, the Band Admissible Modified (BAM) can be generated, which describes the admissible tensions for the modified model, that is, any tension in the modified deck that is not inside the BAM described from the original model is discarded, because it will probably cause damage to the deck of the bridge.

2.2 Objective Function

The objective function is defined as the difference of the effort of the original bridge and the modified bridge in each cut, for each beam:

$$min \sum_{i=1}^{2} \sum_{k=1}^{k} |\sigma o_{i,k} - \sigma m_{i,k}| \tag{1}$$

Where:
$\sigma o_{i,k}$ is the tension of the original bridge in the beam i and the cut k.
$\sigma m_{i,k}$ is the tension of the modified bridge in the beam i and the cut k.

2.3 Constraints

There are two constraints that are:

- The hangers cannot be jacking simultaneously.

$$order_1, order_2, ..., order_n \in \{1, 2, ..., n\} \quad (2)$$

$$order_w \neq order_j; \; \forall w, j \quad con \; w \neq j \quad w, j \in \{1, 2, ..., n\} \quad (3)$$

- The effort of the modified bridge deck should not pass the limits of the BAM:

$$\sigma m \geq \sigma o \quad (4)$$

$$\sigma m \geq f_{ct} \quad (5)$$

$$\sigma m \leq f_{cmax2} \qquad (in \; intermediate \; stages) \quad (6)$$

$$\sigma m \leq f_{cmax} \qquad (in \; final \; stage) \quad (7)$$

Where:

σm, is the tension of the modified bridge.
σo, is the tension of the original bridge.
f_{ct}, is the maximum tension to traction admissible for the concrete.
f_{cmax2}, the maximum tension to compression for the concrete, extended. This property allows to make use of the maximum capacities of the concrete in limited time.
f_{cmax}, is the maximum tension to admissible compression for the concrete.

3 Animal Migration Optimization Algorithm

The Animal Migration Algorithm is based on the long-distance movement of animals, usually seasonally by: weather, food availability, reproduction, escape from predators, among other factors.

The process of migration according to mathematical models must respect three rules: move in the same direction as your neighbors; stay close to your neighbors; and avoid collisions with your neighbors. Based on these rules, this new algorithm called animal migration optimization was proposed, which is implemented by means of concentric "zones" around each animal [8].

To simplify the description of this algorithm, two idealized assumptions are used:

- The leader animal with high quality of position will be retained in the next generation.
- The number of animals is fixed, that is, when one animal leaves the group another one joins.

In the animal migration optimization algorithm, there are two processes: migration and updating of the population. In the first one, the movement of groups of animals is simulated, from the current position to a new, respecting three rules:

- A void collisions with their neighbors
- Move in the same direction as your neighbors
- Stay close to your neighbors

In the second process, it simulates how some animals leave the group while others join during the migration.

Of the three rules mentioned above in the migration process, it can be said that: for the first rule, the position of each animal in the group is required to be different; for the second and third rulers, the animal is required to move to a new position based on the position of its neighbors. To define the concept of local neighborhood, we will use a ring type topology, where it is established that the neighborhood has a size of five animals for each dimension of the animal. That is, if the animal has an index i, then the neighborhood will be $i-2$, $i-1$, i, $i+1$, $i+2$.

The algorithm begins with the initialization process with a random population of size *NPopulation* with individuals represented with a vector of D dimensions respecting the minimum and maximum limits of each dimension. This is denoted as shown below:

$$\begin{aligned} \overrightarrow{X}_{min} &= \{x_{1,min}, x_{2,min}, ..., x_{D,min}\} \\ \overrightarrow{X}_{max} &= \{x_{1,max}, x_{2,max}, ..., x_{D,max}\} \end{aligned} \tag{8}$$

Then the neighborhood topology is constructed and the migration process begins by selecting a random neighbor to the animal and updating the animal's position according to the neighbor, as detailed in formula 9, in which $X_{neighborhood,G}$ is the current position of the animal. neighbor and δ is generated from random numbers controlled by a Gaussian distribution.

$$X_{i,G+1} = X_{i,G} + \delta \cdot (X_{neighborhood,G} - X_{i,G}) \tag{9}$$

During the process of updating the population, some animal may leave the group and then another will join to keep the number of animals fixed and the animals will be replaced by some new individual with a probability Pa. The probability is used according to the quality of the fitness. For the best fitness, the probability Pa is 1. For the worst fitness, the probability is $1/N\,Population$. Where r_1, $r_2 \in [1, ..., NPopulation]$ are randomly chosen integers, and $r1 \neq 2 \neq i$. After producing the new solution $X_{i,G+1}$, it will be evaluated and compared with the $X_{i,G}$, If the objective fitness of $X_{i,G+1}$ is smaller than the fitness of $X_{i,G}$, $X_{i,G+1}$ is accepted as a new basic solution; otherwise, $X_{i,G}$ would be obtained.

4 Implementation

The implementation began with the coding and integration of the AMO metaheuristics without modifications in Python. It is important to evaluate the behavior of the AMO algorithm in this state to discover the strengths and weaknesses when trying to solve the bridge problem. This problem was proposed by Valenzuela using Genetic Algorithm (GA) [13] and was compared against Black Hole

(BH) [5] by Matus [10]. BH got better results than GA so our experiments will be compared against BH.

The adjustment of parameters it was performed using parametric sweep technique [7], for to execute the 11 instances and seek solutions with AMO.

4.1 Experimental Results

The Animal Migration Optimization algorithm was implemented in Python 3.6 and executed on a computer with the following hardware features, Operating System: Windows 7 64bits, Intel Core i5-6300U de 2.5 GHz, 8 GB of memory RAM and 256 GB of Solid-State Drive.

It is important to note that the AMO algorithm for each iteration always evaluates the entire population twice, unlike other algorithms that only replace one element of the population for each iteration. That is, if a population of 50 animals is considered, for each iteration a total of 100 fitness evaluations are performed, 50 in the migration process and 50 in the update process.

4.2 Distribution Comparison

With results discussed in the previous section, we now proceed to compare the distribution of the data using violin plots. The best fitness obtained in each of the 15 executions for the 11 evaluated instances are shown (Fig. 1).

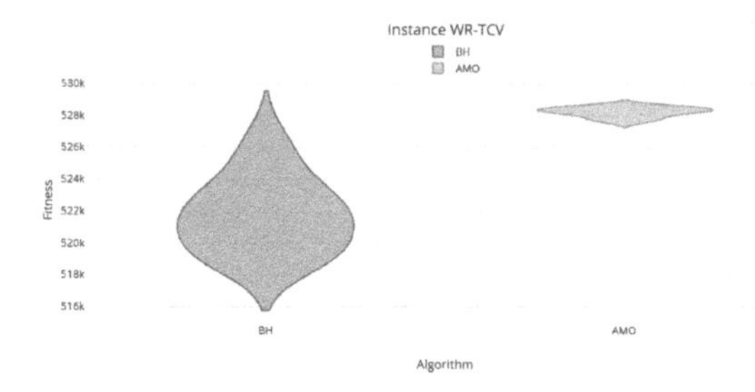

Fig. 1. Instance distribution WR-TCV

From the obtained graphs it can be seen that in all instances, the BH algorithm obtained better results than AMO, which was trapped in a local optimum quickly, due to its large number of fitness evaluations per iteration.

5 Comparison Between AMO and BH

In this subsection we compared the fitness value of the results obtained with AMO and those obtained by Matus et al. with BH [10]. Table 1 shows the minimum value obtained for each instance and the average value after 15 executions. For all instances the minimum value and the average obtained in [10] with BH could not be improved by AMO.

Table 1. Fitness comparison

Instance	Minimum		Average	
	AMO	BH	AMO	BH
PV-TCV	530203.8144	517407.4101	530546.1924	522138.9652
HW-TCV	527345.4042	516990.6841	527906.0618	520214.6409
PT-TCV	529917.4893	517173.9008	530170.6056	521966.7778
AB-TCV	527680.9578	517068.7794	528777.9321	521241.8882
WR-TCV	527673.7935	518685.8652	528259.9454	521745.6059
VC-TCV	526796.5818	515963.2675	527251.9041	520900.6045
CC-TCV	526673.2552	515608.0563	527015.2236	520404.1486
TC-TCV	536247.4154	519571.1341	537050.1275	521994.3783
RD-AA10	521235.3103	508556.0242	521740.3494	514299.2048
RC-AA10	518882.406	510072.7437	519566.1721	514367.665
CR-AA10	518853.1123	511024.8086	519695.0326	514913.0491

5.1 Statistical Test

To compare the results statistically of each one of the instances, the non-parametric Mann-Whitney-Wilcoxon test was used to evaluate the heterogeneity of the samples. To do this, the following hypotheses were proposed, than H_0: AMO is better than BH, and H_1: BH is better than AMO.

The results obtained reflect that for all instances, it is obtained that AMO (2) is better than BH (1) with an error of 99.99%, therefore, H_0 is rejected and H_1 is accepted.

6 Conclusion

In this work, a population-based algorithm called Animal Migration Optimization [8] was used to try to improve the results obtained with the Black Hole algorithm [10] for the problem of Reinforcement of Bridges by Cable-Stayed Arch Conversion.

The first step was the implementation the metaheuristic, which was successfully achieved after understanding the operation of its operators. Regarding the results obtained, it can be said that using BH we obtained better results than AMO in all instances. AMO is trapped in a local optimum quickly, due to the large number of fitness evaluations per iteration.

This allows us to deduce that it is necessary to modify the search strategy of the AMO algorithm, to achieve a balance between intensification and diversification, considering, for example, a certain number of iterations maintaining the same fitness. With this adjustment and others that should be analyzed in more detail, we would expect to obtain better results.

Acknowledgements. Broderick Crawford is supported by Grant CONICYT/ FONDECYT/REGULAR/1171243, Ricardo Soto is supported by Grant CONICYT /FONDECYT /REGULAR/1190129, Gino Astorga is supported by Postgraduate Grant Pontificia Universidad Católica de Valparaíso 2015, José Lemus is Beneficiario Beca Postgrado PUCV 2018. This work was funded by the CONICYT PFCHA/ DOCTORADO BECAS NACIONAL/2019 - 21191692.

References

1. SAP2000, http://www.csiespana.com/software/2/sap2000
2. Crawford, B., Soto, R., Astorga, G., García, J., Castro, C., Paredes, F.: Putting Continuous Metaheuristics to Work in Binary Search Spaces. Complexity **2017**, (2017)
3. J. García, B. Crawford, R. Soto, C. Castro, and F. Paredes. A k-means binarization framework applied to multidimensional knapsack problem. Applied Intelligence, pages 1–24, 2017
4. J. García, B. Crawford, R. Soto, and P. García. A Multi Dynamic Binary Black Hole Algorithm Applied to Set Covering Problem. International Conference on Harmony Search Algorithm, pages 42–51. Springer, 2017
5. Hatamlou, A.: Black hole: A new heuristic optimization approach for data clustering. Information Sciences **222**, 175–184 (2013)
6. J. Jia, M. Ibrahim, M. Hadi, W. Orabi, M. Ali, and Y. Xiao. Estimation of the Total Cost of Bridge Construction for use in Accelerated Bridge Construction Selection Decisions. In Transportation Research Board 95th Annual Meeting, page 17, Washington DC, United States, 2016
7. Lanza-Gutierrez, J., Crawford, B., Soto, R., Berrios, N., Gomez-Pulido, J.A., Paredes, F.: Analyzing the effects of binarization techniques when solving the set covering problem through swarm optimization. Expert Systems with Applications **70**, 67–82 (2017)
8. Li, X., Zhang, J., Yin, M.: Animal migration optimization: an optimization algorithm inspired by animal migration behavior. Neural Computing and Applications **24**(7–8), 1867–1877 (2013)
9. C. Lin, J. Han, C. Bennett, and R. L. Parsons. Case History Analysis of Bridge Failures due to Scour. In Climatic Effects on Pavement and Geotechnical Infrastructure. 2014
10. S. Matus, R. Soto, B. Crawford. Optimización del refuerzo de puentes mediante arco atirantado con black hole algorithm. Master's thesis Escuela de Ingeniería Informática, Pontificia Universidad Católica de Valparaíso, Valparaíso, Chile, 2018
11. R. Soto, B. Crawford, R. Olivares, and N. Pacheco. Resolving the Manufacturing Cell Design Problem via Hunting Search. In International Conference on Industrial, Engineering and Other Applications of Applied Intelligent Systems, pages 414–420. Springer, 2018
12. C. Swann and C. Mullen. Predicting erosion impact on highway and railway bridge substructures. Technical report, 2016
13. M. Valenzuela. Refuerzo de puentes de luces medias por conversión en arco atirantado tipo network. PhD thesis, Universitat Politècnica de Catalunya, Barcelona, España, 2012

Optimization of Monetary Incentive in Ridesharing Systems

Fu-Shiung Hsieh(✉)

Chaoyang University of Technology, Taichung 41349, Taiwan
fshsieh@cyut.edu.tw

Abstract. Although ridesharing is a potential transport model for reducing fuel consumption, green-house gas emissions and improving efficiency, it is still not widely adopted due to the lack of providing monetary incentives for ridesharing participants. Most studies regarding ridesharing focus on travel distance reduction, cost savings and successful matching rate in ridesharing systems, which do not directly provide monetary incentives for the ridesharing participants. In this paper, we address this issue by proposing a performance index for ridesharing based on monetary incentives. We formulate a problem to optimize monetary incentives in ridesharing systems as non-linear integer programming problem. To cope with computational complexity, an evolutionary computation approach based on a variant of PSO is adopted to solve the non-linear integer programming problem for ridesharing systems based on cooperative coevolving particle swarms. The results confirm the effectiveness the proposed algorithm in solving the nonlinear constrained ridesharing optimization problem with binary decision variables and rational objective function.

Keywords: Particle swarm · Coevolution · Ridesharing · Integer programming

1 Introduction

Ridesharing has been identified as a potential transport model in the past years and has attracted many studies in the literature. Several performance indices have been proposed in the literature to measure ridesharing systems. These performance indices include overall travel distance reduction, successful matching rate [1] and cost savings [2, 3] in ridesharing systems. However, these performance indices do not directly provide monetary incentives for ridesharing participants. In this paper, we address this issue by proposing a performance index based on monetary incentives for ridesharing systems and formulate a problem monetary incentive optimization problem. The monetary incentive optimization problem is a non-linear integer programming problem. To cope with computational complexity, an evolutionary computation approach based on a variant of PSO is adopted to solve the non-linear integer programming problem for ridesharing systems based on cooperative coevolving particle swarms optimization (DCCPSO) approach. The DCCPSO algorithm combines the cooperative coevolution of particle swarms, decomposition strategy, random grouping [3–5] and transformation

F. Wotawa et al. (Eds.): IEA/AIE 2019, LNAI 11606, pp. 835–840, 2019.
https://doi.org/10.1007/978-3-030-22999-3_71

of solutions from continuous solution space to discrete solution space to tackle dimensionality problem and discrete solution space. To assess the effectiveness of applying the proposed algorithm to this problem, we conduct experiments for several test cases. The numerical results indicate that the proposed DCCPSO algorithm is still significantly more effective than several algorithms in the literature in solving a constrained optimization problem with nonlinear objective function and binary decision variables.

The remainder of this paper is organized as follows. In Sect. 2, we first we formulate the monetary incentive optimization problem. We briefly introduce the way to handle constraints and the fitness function in Sect. 3 and present the DCCPSO algorithm in Sect. 4, respectively. We present our numerical results in Sect. 5 and conclude this paper in Sect. 6.

2 Problem Formulation for Monetary Incentive Optimization in Ridesharing Systems

As the goal of this paper is to present a decision model to optimize monetary incentive for participants in ridesharing systems, a problem formulation will be presented in this section. To make it clear for readers to compare the differences between the problem formulation presented in this paper and the previous work in [2], the problem formulation of this paper will basically follows most of the notation used in [2]. We will point out the differences between this paper and [2] in this section later.

To formulate the problem, the following notations are defined. Consider a ridesharing system with P passengers and D drivers. A passenger $p \in \{1, 2, 3, \ldots P\}$ submits a bid $b_p^P = (s_{p1}^1, s_{p2}^1, s_{p3}^1, \ldots, s_{pP}^1, s_{p1}^2, s_{p2}^2, s_{p3}^2, \ldots, s_{pP}^2, f_p)$, where s_{pk}^1 and s_{pk}^2 are the number of seats requested by p to be picked up passengers at location L_k and the number of seats released after dropping the passengers at location L_{P+k}, respectively, and f_p is the price of the bid. A driver $d \in \{1, \ldots, D\}$ may submit multiple bids but at most only one bid can be a winning bid for driver d. $b_{dj}^D = (q_{dj1}^1, q_{dj2}^1, \ldots, q_{djk}^1, \ldots, q_{djP}^1, q_{dj1}^2, q_{dj2}^2, \ldots, q_{djk}^2, \ldots, q_{djP}^2, \pi_{dj}, c_{dj}, a_d)$ represents the $j-th$ bid submitted by driver d, which specifies a_d available seats to pick up q_{djk}^1 passengers at location L_k and drop q_{djk}^2 passengers at location L_{P+k} at price c_{dj} by taking the route π_{dj}.

The decision variables of the monetary incentive optimization problem in ridesharing systems include x_{dj} and y_p, where x_{dj} indicates the $j-th$ bid placed by driver d is a winning bid ($x_{dj} = 1$) or not ($x_{dj} = 0$) and y_p indicates the bid placed by passenger p is a winning bid ($y_p = 1$) or not ($y_p = 0$). The carpooling problem is formulated as follows:

$$\max_{x,y} F(x,y) = \frac{\left[\left(\sum_{p=1}^{P} y_p\left(f_p\right)\right) - \left(\sum_{d=1}^{D}\sum_{j=1}^{J_d} x_{dj}(c_{dj} - o_{dj})\right) + \left(\sum_{p=1}^{P} y_p\left(T^P f_p\right)\right) + \left(\sum_{d=1}^{D}\sum_{j=1}^{J_d} x_{dj} T^D (c_{dj} - o_{dj})\right)\right]}{\left(\sum_{p=1}^{P} y_p\left(f_p\right)\right) + \left(\sum_{d=1}^{D}\sum_{j=1}^{J_d} x_{dj} c_{dj}\right)}$$

$$\sum_{d=1}^{D}\sum_{j=1}^{J_d} x_{dj}q_{djk}^1 = y_p s_{pk}^1 \;\; \forall \mathrm{p} \in \{1,2,\ldots,\mathrm{P}\} \; \forall \mathrm{k} \in \{1,2,\ldots,\mathrm{P}\} \tag{1}$$

$$\sum_{d=1}^{D}\sum_{j=1}^{J_d} x_{dj}q_{djk}^2 = y_p s_{pk}^2 \;\; \forall \mathrm{p} \in \{1,2,\ldots,\mathrm{P}\} \; \forall \mathrm{k} \in \{1,2,\ldots,\mathrm{P}\} \tag{2}$$

$$\sum_{p=1}^{P} y_p f_p + \sum_{d=1}^{D}\sum_{j=1}^{J_d} x_{dj}o_{dj} \geq \sum_{d=1}^{D}\sum_{j=1}^{J_d} x_{dj}c_{dj} \tag{3}$$

$$\sum_{j=1}^{J_d} x_{dj} \leq 1 \;\; \forall d \in \{1,\ldots,D\} \tag{4}$$

$$\begin{aligned} & x_{dj} \in \{0,1\} \;\; \forall d \in \{1,\ldots,D\} \;\; \forall j \in \{1,\ldots,J_d\} \\ & y_p \in \{0,1\} \;\; \forall \mathrm{p} \in \{1,2,\ldots,\mathrm{P}\} \end{aligned} \tag{5}$$

There are several constraints that must be satisfied, including: (a) capacity constraints of cars (1) and (2), (b) the nonnegative cost savings constraints (3) and (c) each driver can only have at most one winning bid (4). The value of decision variables must be 1 or zero (5). Note that the objective function $F(x, y)$ for monetary incentive optimization in ridesharing systems is a nonlinear function, which is not a simple linear function as the one used in [2].

3 Fitness Function

The monetary incentive optimization problem in ridesharing systems formulated in the previous section is a nonlinear integer programming problem with binary decision variables and rational objective function. The constraints must be handled properly. In this paper, a method based on biasing feasible over infeasible solutions [6] is adopted in this paper to handle constraints. By using this constraint handling method, the fitness function $F_1(x, y)$ is defined as follows:

$$F_1(x,y) = \begin{cases} F(x,y) & \textit{if } (x,y) \textit{ satisfies constraints } (1)-(5) \\ U_1(x,y) & \textit{otherwise} \end{cases}, \text{ where}$$

$$U_1(x,y) = S_{f\min} + U_2(x,y) + U_3(x,y) + U_4(x,y)$$
$$U_2(x,y) = \left|\left(\sum_{p=1}^{P}\sum_{k=1}^{K}(\sum_{d=1}^{D}\sum_{j=1}^{J_d} x_{dj}q^1_{djk} - y_p s^1_{pk})\right)\right|,\ U_3(x,y) = \left(\sum_{p=1}^{P}\sum_{k=1}^{K}(\sum_{d=1}^{D}\sum_{j=1}^{J_d} x_{dj}q^2_{djk} - y_p s^2_{pk})\right)$$
$$U_4(x,y) = \left|\sum_{d=1}^{D}(1 - \sum_{j=1}^{J_d} x_{dj})\right|,$$

where $S_{f\min} = \min_{(x,y)\in S_f} F(x,y)$, the object function value of the worst feasible solution in the current population and $S_f = \{(x,y)|(x,y)$ is a solution in the current population, (x,y) satisfies constraints (1)–(5).$\}$ is the set of all feasible solutions in the current population.

4 Discrete Cooperative Coevolving Particle Swarm Optimization (DCCPSO) Algorithm

As PSO performs very poorly as the size of a problem increases, the algorithm adopted in this paper to solve the monetary incentive optimization problem is based on modification of the DCCPSO algorithm proposed in [2] in order to test the capability of the DCCPSO algorithm in solving a nonlinear integer programming problem. The DCCPSO algorithm tackles the dimensionality issue by dynamically decomposes a higher dimensional problem into lower dimensional subproblems based on random grouping, solves subproblems individually and cooperatively evolves solutions. The DCCPSO algorithm selects an integer *ds* from a set of integers *DS* to decompose the decision variables into *NS* swarms. Let *MAX_FES* be the maximum number of fitness evaluations. To concisely present the algorithm, we use z to denote the vector obtained by concatenating all the decision variables x and y. The context vector $\hat{z}$ is constructed by concatenating all global best particles from all *NS* swarms. Let $SW_s.z_i$ be the i-th particle in the s-th swarm SW_s. Let $SW_s.z_i^p$ be personal best of the i-th particle in swarm s and let $SW_s.\hat{z}$ be the global best of the component of the swarm s. In each iteration, personal best and velocity of each particle, swarm best and the context vector and are updated. Let ω_1 and ω_2 be the weighting factor and θ be the scaling factor. An outline of the DCCPSO algorithm is as follows.

DCCPSO Algorithm

While the number of fitness evaluation $< MAX\_FES$

Step 1: Select ds from DS and randomly partition the set of decision variables into NS subsets, each with ds decision variables
Initialize swarm SW_s for each $s \in \{1,2,..., NS\}$

Step 2: For each $s \in \{1,2,..., NS\}$
For each particle $i \in SW_s$
Construct the D dimensional vector z_i consisting $\hat{z}$ with its s-th component being replaced by $SW_s.z_i$ to evaluate fitness function
Update personal best $SW_s.z_i^p$ if z_i is better than $SW_s.z_i^p$
Update swarm best $SW_s.\hat{z}$ if $SW_s.z_i^p$ is better than $SW_s.\hat{z}$
End For
Update the context vector ($\hat{z}$)
End For
For each $s \in \{1,2,..., NS\}$
For each particle $i \in SW_s$
For each $d \in \{1,2,..., ds\}$
Update velocity $SW_s.vz_{id}$ with a Gaussian random variable

$$N(\omega_1 SW_s.z_i^p + \omega_2 SW_s.\hat{z}, (\theta |SW_s.z_i^p - SW_s.\hat{z}|)^2)$$

End For
End For
End For

End While

5 Numerical Results

To study the effectiveness of the DCCPSO algorithm, we conduct experiments on several test cases by applying DCCPSO algorithm, the discrete version of PSO algorithm and a recent variant of PSO algorithm called ALPSO [6] to compare the performance. We set the parameters as follows: $\omega_1 = 0.5$, $\omega_2 = 0.5$, $\theta = 1.0$ and $DS = \{2, 5, 10\}$. Table 1 illustrates the results obtained by the proposed DCCPSO algorithm and those obtained by PSO and ALPSO algorithms for population size $NP = 20$. It indicates that the proposed DCCPSO algorithm obtains the same fitness function value as the PSO and ALPSO algorithms for small examples (Case 1 and Case 2). But the DCCPSO algorithm significantly outperforms the PSO and ALPSO algorithms as the size of the examples grow.

Table 1. Results of several test cases for population size: 20.

Case	Drivers	Passengers	PSO	DCCPSO	ALPSO
	D	P	Fitness value/FES	Fitness value/FES	Fitness value/FES
1	3	6	0.2943/8.4	0.2943/8.2	0.2943/10
2	7	10	0.13/191.2	0.13/22.9	0.13/167.2
3	20	20	0.215/19699.6	0.381/5842.2	0.257/23974.6
4	30	30	0.196/30238.3	0.508/5913.6	0.17/27111.6
5	40	40	−0.894/27060.2	0.342/4427.2	−0.898/17938.6

6 Conclusions

Most studies on ridesharing consider performance indices such as overall travel distance reduction, cost savings and successful matching rate in ridesharing systems. These performance indices are not directly linking to the monetary incentives for the ridesharing participants. We address this issue by proposing a monetary incentive performance index and formulate a monetary incentive optimization problem for ridesharing systems. The monetary incentive optimization problem is a non-linear integer programming problem. To cope with computational complexity, an evolutionary computation approach based on a variant of PSO is adopted to solve the non-linear integer programming problem for ridesharing systems based on cooperative coevolving particle swarms optimization approach. To assess the effectiveness of the proposed algorithm, we conduct experiments for several test cases. The numerical results indicate that the proposed algorithm is significantly more effective than several algorithms in the literature in solving a constrained optimization problem with a nonlinear objective function and binary decision variables.

Acknowledgment. This paper was supported in part by Ministry of Science and Technology, Taiwan, under Grant MOST-106-2410-H-324-002-MY2.

References

1. Furuhata, M., Dessouky, M., Ordóñez, F., Brunet, M., Wang, X., Koenig, S.: Ridesharing: the state-of-the-art and future directions. Transp. Res. Part B Methodol. **57**, 28–46 (2013)
2. Hsieh, F.S., Zhan, F.M., Guo, Y.H.: A solution methodology for carpooling systems based on double auctions and cooperative coevolutionary particle swarms. Appl. Intell. **49**(2), 741–763 (2019). https://doi.org/10.1007/s10489-018-1288-x
3. Bergh, F., Engelbrecht, A.P.: A cooperative approach to particle swarm optimization. IEEE Trans. Evol. Comput. **8**(3), 225–239 (2004)
4. Potter, M.A., De Jong, K.A.: A cooperative coevolutionary approach to function optimization. In: Davidor, Y., Schwefel, H.-P., Männer, R. (eds.) PPSN 1994. LNCS, vol. 866, pp. 249–257. Springer, Heidelberg (1994). https://doi.org/10.1007/3-540-58484-6_269
5. Yang, Z., Tang, K., Yao, X.: Large scale evolutionary optimization using cooperative coevolution. Inf. Sci. **178**(15), 2985–2999 (2008)
6. Deb, K.: An efficient constraint handling method for genetic algorithms. Comput. Methods Appl. Mech. Eng. **186**(2–4), 311–338 (2000)

Pareto Optimality for Conditional Preference Networks with Comfort

Sultan Ahmed and Malek Mouhoub(✉)

Department of Computer Science, University of Regina, Regina, SK, Canada
{ahmed28s,mouhoubm}@uregina.ca

Abstract. A *Conditional Preference Network with Comfort* (CPC-net) graphically represents both preference and comfort. Preference and comfort indicate user's habitual behavior and genuine decisions correspondingly. Given that these two concepts might be conflicting, we find it necessary to introduce Pareto optimality when achieving outcome optimization with respect to a given acyclic CPC-net. In this regard, we propose a backtrack search algorithm, that we call Solve-CPC, to return the Pareto optimal outcomes. The formal properties of the algorithm are presented and discussed.

Keywords: Decision theory · CP-net · Preference · Pareto optimality · Backtrack search

1 Introduction

Representation and reasoning about user's choices is critical to the success of many automated decision making applications in Artificial Intelligence. Choices involve habitual behavior and genuine decisions [10]. Habitual behavior is user's expressed desire on options, represented using preferences [11,13]. For example, a user prefers tea to coffee. A *Conditional Preference Network* (CP-net) [6] is a graphical tool to represent and reason about user's conditional *ceteris paribus* preference statements, in a compact manner. On the other hand, genuine decisions [10] require to perceive and gather information on the options, and to choose an option based on that information. For example, given the prices of tea and coffee, the user chooses the one with the lower price. This type of choices depends on the environmental factors such as cost, and is represented using user's comfort [2]. In particular cases, the absolute comfort of an option is computed by taking the inverse of the cost, i.e., the less cost an option has, the more comfort the user feels. In a decision making process, the user is often interested in an outcome with highest preference and highest comfort. This leads to a bi-criteria decision problem [9,12]. To graphically represent both preference and comfort, a *CP-net with Comfort* (CPC-net) [2,3] is introduced as an extension of the CP-net model. In a CPC-net, a Preference-Comfort order (PC order) captures both preference and comfort on the values of a given variable. Both the CP-net and

F. Wotawa et al. (Eds.): IEA/AIE 2019, LNAI 11606, pp. 841–853, 2019.
https://doi.org/10.1007/978-3-030-22999-3_72

the CPC-net share a common directed graph to encode the preferential dependencies. In a CPC-net, every variable is associated with a *Conditional Preference Table with Comfort* (CPTC) that represents a set of PC orders on the values of the variable for each parent combination.

For an acyclic CPC-net, the *preference optimal outcome* has the highest preference, while the *comfort optimal outcome* has the highest comfort [2]. These two outcomes can be obtained using the forward sweep procedure on the CPC-net [2,6]. However, we argue that the above two outcomes are not comparable given that the criteria they are optimizing might be conflicting. For example, the preference optimal outcome has the highest preference but might have a very low comfort. In this case, the user might be interested to have a compromise outcome with a moderate preference and a moderate comfort, by maximizing both as much as possible. In this regard, we define a Pareto optimal outcome that is not dominated by any other outcome using both preference and comfort.

In this paper, we propose a recursive backtrack search algorithm that we call Solve-CPC to obtain the set of Pareto optimal outcomes with respect to an acyclic CPC-net. In every call, Solve-CPC finds a variable with no parent(s), and initiates a branch for every value of the variable such that the value is not locally dominated by any other value. If a value is dominated by another value, the related branch is pruned. We prove that this pruning criterion is sound, i.e., we do not disregard any Pareto optimal outcome. If a branch is not pruned, a sub CPC-net is formed by removing the root variable. Then, the Solve-CPC is recursively called with the sub CPC-net until an empty sub CPC-net is obtained. When all variables are instantiated, a Pareto optimal outcome is produced. The Solve-CPC saves each Pareto optimal outcome, and finally returns the Pareto set. We then discuss the formal properties of Solve-CPC, i.e., its correctness, completeness and complexity.

The paper is organized as follows. In Sect. 2, we outline the necessary background. In Sect. 3, we propose Solve-CPC algorithm. The formal properties of Solve-CPC are presented in Sect. 4. Finally, we conclude the paper in Sect. 5.

2 Background

2.1 Preferences and Comfort

We assume a set of variables $V = \{X_1, X_2, \cdots, X_n\}$ with the finite domains $D(X_1), D(X_2), \cdots, D(X_n)$. We use $D(\cdot)$ to denote the domain of a set of variables as well. The decision maker wants to express habitual behavior using preferences over the complete assignments on V. Each complete assignment can be seen as an outcome of the decision maker's action. The set of all outcomes is denoted by O. A preference order $\succ$ is a binary relation over O. For $o_1, o_2 \in O$, $o_1 \succ o_2$ indicates that o_1 is strictly preferred to o_2. The preference order $\succ$ is a strict partial order, i.e., $\succ$ is irreflexive, asymmetric and transitive. The preference order $\succ$ is a total order, if $\succ$ is also complete, i.e., for every $o_1, o_2 \in O$, we have either $o_1 \succ o_2$ or $o_2 \succ o_1$. Note that indifference is not considered for simplicity.

The size of O is exponential in the number of variables, i.e., the number of outcomes $|O|$ is m^n, where n is the number of variables and m their domain values. Therefore, direct assessment of the preference order is generally impractical. In this case, the notions of *preferential independence* and *conditional preferential independence* play a key role to represent the preference order compactly, at least if the preference order is partial. These notions are standard and well-known notions of independence in Multi-Attribute Utility Theory [6,7,11].

On the other hand, user's genuine decision on an option is represented using comfort [2]. Given the cost of an option s, the comfort $c(s)$ is calculated as: $c(s) = 1/cost(s)$. Therefore, the comfort is inversely proportional to the cost. Given $x_1 \in D(X_1)$ and $x_2 \in D(X_2)$, we naturally get: $cost(x_1x_2) = cost(x_1) + cost(x_2)$ that implies $c(x_1x_2) = c(x_1)c(x_2)/(c(x_1) + c(x_2))$. Generally, for a complete assignment (or an outcome) $x_1x_2 \cdots x_n$ on V, we get

$$c(x_1x_2 \cdots x_n) = \sum_{i=1}^{n} c(x_i) / \sum_{i=1}^{n} (\prod_{j=1, j \neq i}^{n} c(x_j)).$$

Every outcome $o \in O$ has a corresponding comfort $c(o)$. We denote the pair as $(o, c(o))$.

Lemma 1 [2]. $c(x_1y) > c(x_2y) \Leftrightarrow c(x_1) > c(x_2)$.

To represent user's preference with comfort, a Preference-Comfort relation (PC relation) is used. Note that, for $X \subseteq V$ and $x_1, x_2 \in D(X)$, $x_1 \succ x_2$ means that user prefers x_1 to x_2 given X is preferentially independent of $V - X$.

Definition 1 *(PC relation)* [2]. *Let $X \subseteq V$ and $x_1, x_2 \in D(X)$. We say that there is a Preference-Comfort relation (PC relation) ${}_c{\succ}$ on x_1 and x_2, iff $x_1 \succ x_2$ or $c(x_1) > c(x_2)$ holds. It is denoted as $(x_1, c(x_1))\ {}_c{\succ}\ (x_2, c(x_2))$. For brevity, we use $(x_1, c(x_1))\ {}_c{\succ}\ (x_2, c(x_2))$ and $x_1\ {}_c{\succ}\ x_2$ interchangeably. If both $x_1 \succ x_2$ and $c(x_1) > c(x_2)$ hold, the PC relation is harmonic and is denoted as $x_1\ {}_h{\succ}\ x_2$. If $x_1 \succ x_2$ holds and $c(x_1) > c(x_2)$ does not hold, the PC relation is preference-harmonic and is denoted as $x_1\ {}_{ph}{\succ}\ x_2$. If $x_1 \succ x_2$ does not hold and $c(x_1) > c(x_2)$ holds, the PC relation is comfort-harmonic and is denoted as $x_1\ {}_{ch}{\succ}\ x_2$.*

A harmonic PC relation is irreflexive, asymmetric and transitive. A preference-harmonic PC relation is irreflexive, asymmetric and transitive. A comfort-harmonic PC relation is irreflexive, asymmetric and transitive. Also, the preference-harmonic and comfort-harmonic PC relations are converse to each other, i.e., for $x_1, x_2 \in D(X)$, $x_1\ {}_{ph}{\succ}\ x_2 \Leftrightarrow x_2\ {}_{ch}{\succ}\ x_1$. We now define the PC order as an ordering of two or more assignments according to the PC relations between the assignments.

Definition 2 *(PC order)* [2]. *Let $X \subseteq V$. A Preference-Comfort order (PC order) ${}_c{\succ}^X$ for X is an ordering on $D(X)$, such that for every consecutive*

$x_1, x_2 \in D(X)$ in the order ${}_{c}{\succ}^{X}$, there is a PC relation $x_1 \;{}_{c}{\succ}\; x_2$. If every PC relation in the PC order is harmonic, the order is harmonic. If every PC relation in the PC order is preference-harmonic, the order is preference-harmonic. If every PC relation in the PC order is comfort-harmonic, the order is comfort-harmonic. If the PC order is not harmonic, preference-harmonic or comfort-harmonic, it is hybrid.

Given the preference order and the comforts over the values $D(X)$ of a variable X, we can find the possible PC orders by considering each permutation on $D(X)$. For every consecutive values x_1 and x_2 in a permutation, we can determine if $x_1 \succ x_2$ holds, or not, since the preference order is complete. On the other hand, given the comforts, we can determine if $c(x_1) > c(x_2)$ holds or not. Therefore, we can check if the PC relation $x_1 \;{}_{c}{\succ}\; x_2$ holds or not. If there is a PC relation $x_1 \;{}_{c}{\succ}\; x_2$ for every two consecutive values x_1 and x_2 in a permutation, a PC order can be formed by assigning the PC relation between the consecutive values in the permutation.

Example 1. Let us assume that X is a variable such that $D(X) = \{x_1, x_2, x_3\}$. We are given that the preference relations and comforts are $x_1 \succ x_2$, $x_2 \succ x_3$, $x_1 \succ x_3$, $c(x_1) = 1.6$, $c(x_2) = 1.2$ and $c(x_3) = 1.4$. We have $x_1 \succ x_2$ and $c(x_1) > c(x_2)$. Therefore, there is a harmonic PC relation over x_1 and x_2, which is $x_1 \;{}_{h}{\succ}\; x_2$.

On the other hand, to obtain the PC orders over x_1, x_2 and x_3, we check the six permutations $x_1x_2x_3$, $x_1x_3x_2$, $x_2x_1x_3$, $x_2x_3x_1$, $x_3x_1x_2$ and $x_3x_2x_1$. For the first permutation $x_1x_2x_3$, we have that $x_1 \;{}_{h}{\succ}\; x_2$ and $x_2 \;{}_{ph}{\succ}\; x_3$ hold. By using these PC relations, we can obtain the PC order $x_1 \;{}_{h}{\succ}\; x_2 \;{}_{ph}{\succ}\; x_3$, which is a hybrid PC order. For the second permutation $x_1x_3x_2$, $x_1 \;{}_{h}{\succ}\; x_3$ and $x_3 \;{}_{ch}{\succ}\; x_2$ hold. Therefore, we get the PC order $x_1 \;{}_{h}{\succ}\; x_3 \;{}_{ch}{\succ}\; x_2$. For the third permutation $x_2x_1x_3$, let us consider the first two consecutive values x_2 and x_1. Between these consecutive values, $x_2 \;{}_{c}{\succ}\; x_1$ does not hold; and therefore, there is no PC order for the permutation $x_2x_1x_3$. Note that we do not need to check on the second two consecutive values x_1 and x_3. Similarly, we can show that there is no PC order for the rest of the permutations $x_2x_3x_1$, $x_3x_1x_2$ and $x_3x_2x_1$. We conclude that there are two PC orders on $D(X)$, which are $x_1 \;{}_{h}{\succ}\; x_2 \;{}_{ph}{\succ}\; x_3$ and $x_1 \;{}_{h}{\succ}\; x_3 \;{}_{ch}{\succ}\; x_2$. □

2.2 CP-net

A *Conditional Preference Network* (CP-net) [6] graphically represents user's conditional *ceteris paribus* preference statements using the notions of *preferential independence* and *conditional preferential independence*. A CP-net consists of a directed graph, in which, preferential dependencies over the set V of variables are represented using directed arcs. An arc $\overrightarrow{(X_i, X_j)}$ for $X_i, X_j \in V$ indicates that the preference orders over $D(X_j)$ depend on the actual value of X_i. For each variable $X \in V$, there is a *Conditional Preference Table* (CPT) that gives the preference orders over $D(X)$ for each $p \in D(Pa(X))$, where $Pa(X)$ is the set of X's parent(s).

Definition 3 *(CP-net)* [6]. *A Conditional Preference Network (CP-net) N over variables $V = \{X_1, X_2, \cdots, X_n\}$ is a directed graph over $X_1, X_2, \cdots, X_n$ whose nodes are annotated with $CPT(X_i)$ for each $X_i \in V$.*

Example 2. A CP-net with four variables, A, B, C and D, is shown in Fig. 1, where $D(A) = \{a_1, a_2, a_3\}$, $D(B) = \{b_1, b_2\}$, $D(C) = \{c_1, c_2, c_3\}$ and $D(D) = \{d_1, d_2\}$. The preference order over $D(A)$ is $a_1 \succ a_2 \succ a_3$. The preferences over $D(C)$ depend on the actual value of A and B. For the combination $a_1 b_1$ of A and B, the preference order is $c_1 \succ c_2 \succ c_3$. □

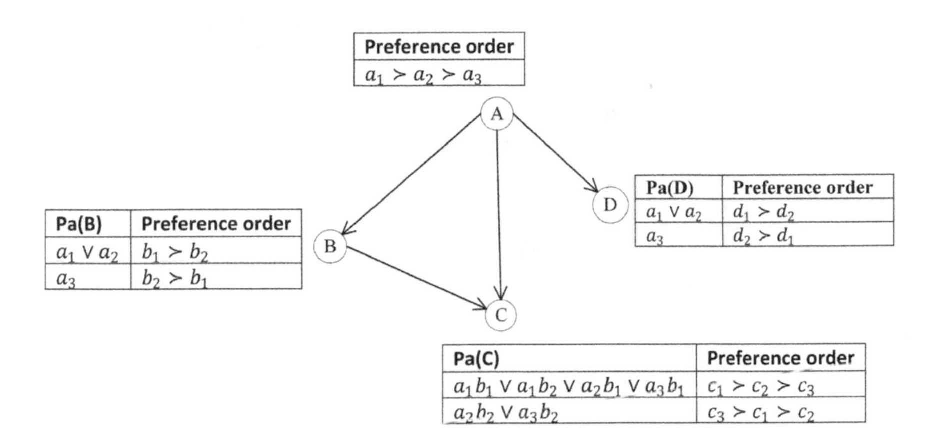

Fig. 1. A CP-net

The semantics of a CP-net is defined in terms of the set of preference orders that are consistent with the set of preference constraints imposed by the CPTs. A preference order $\succ$ on the outcomes of a CP-net N satisfies the CPT of a variable X, iff $\succ$ orders any two outcomes that differ only on the value of X consistently with the preference order on $D(X)$ for each $p \in D(Pa(X))$. The preference order $\succ$ satisfies N iff $\succ$ satisfies each CPT of N. If o_1 and o_2 are two outcomes of N, we say that N entails $o_1 \succ o_2$, denoted as $N \models o_1 \succ o_2$, iff $o_1 \succ o_2$ holds in every preference order that satisfies N. Similarly, $N \not\models o_1 \succ o_2$ indicates that there exists at least a preference order over the outcomes which satisfies N however does not entail $o_1 \succ o_2$.

2.3 CPC-net

The *CP-net with Comfort* (CPC-net) [2,3] is an extension of the CP-net model to incorporate comfort information. The CP-net and the CPC-net have a common structure, i.e., the same directed graph. In both cases, conditional preferential independence is used to obtain this structure. However, in a CPC-net, for each variable $X \in V$, there is a *CPT with Comfort* (CPTC) that gives the PC orders over $D(X)$ for each $p \in D(Pa(X))$.

Definition 4 *(CPC-net)* [2]. *A CP-net with Comfort (CPC-net) N_c over variables $V = \{X_1, X_2, \cdots, X_n\}$ is a directed graph over $X_1, X_2, \cdots, X_n$ whose nodes are annotated with $CPTC(X_i)$ for each $X_i \in V$.*

Example 3. The CPC-net corresponding to the CP-net of Fig. 1 is shown in Fig. 2, where $c(a_1) = 0.3$, $c(a_2) = 0.1$, $c(a_3) = 0.5$, $c(b_1) = 0.05$, $c(b_2) = 0.01$, $c(c_1) = 0.4$, $c(c_2) = 0.7$, $c(c_3) = 0.6$, $c(d_1) = 0.07$ and $c(d_2) = 0.1$. Given the preference order $a_1 \succ a_2 \succ a_3$ in Fig. 1 and the comforts over $D(A)$, we find four PC orders $(a_1, 0.3)\ {}_{h}\!\succ (a_2, 0.1)\ {}_{ph}\!\succ (a_3, 0.5)$, $(a_1, 0.3)\ {}_{ph}\!\succ (a_3, 0.5)\ {}_{ch}\!\succ (a_2, 0.1)$, $(a_2, 0.1)\ {}_{ph}\!\succ (a_3, 0.5)\ {}_{ch}\!\succ (a_1, 0.3)$ and $(a_3, 0.5)\ {}_{ch}\!\succ (a_1, 0.3)\ {}_{h}\!\succ (a_2, 0.1)$. These PC orders are shown with the $CPTC(A)$ in Fig. 2. The PC order $a_1\ {}_{h}\!\succ a_2\ {}_{ph}\!\succ a_3$ indicates that a_1 has more preference and comfort than a_2, while a_2 has more preference but less comfort than a_3. On the other hand, the variable C preferentially depends on the variables A and B. Given a_1b_1, a_1b_2, a_2b_1 or a_3b_1 of A and B, we have $c_1 \succ c_2 \succ c_3$ over $D(C)$ in Fig. 1. For this preference order and the comforts over $D(C)$, we find four PC orders $(c_1, 0.4)\ {}_{ph}\!\succ (c_2, 0.7)\ {}_{h}\!\succ (c_3, 0.6)$, $(c_2, 0.7)\ {}_{ch}\!\succ (c_1, 0.4)\ {}_{ph}\!\succ (c_3, 0.6)$, $(c_2, 0.7)\ {}_{h}\!\succ (c_3, 0.6)\ {}_{ch}\!\succ (c_1, 0.4)$ and $(c_3, 0.6)\ {}_{ch}\!\succ (c_1, 0.4)\ {}_{ph}\!\succ (c_2, 0.7)$. Similarly, we can find the PC-orders for the combinations a_2b_2 and a_3b_2. □

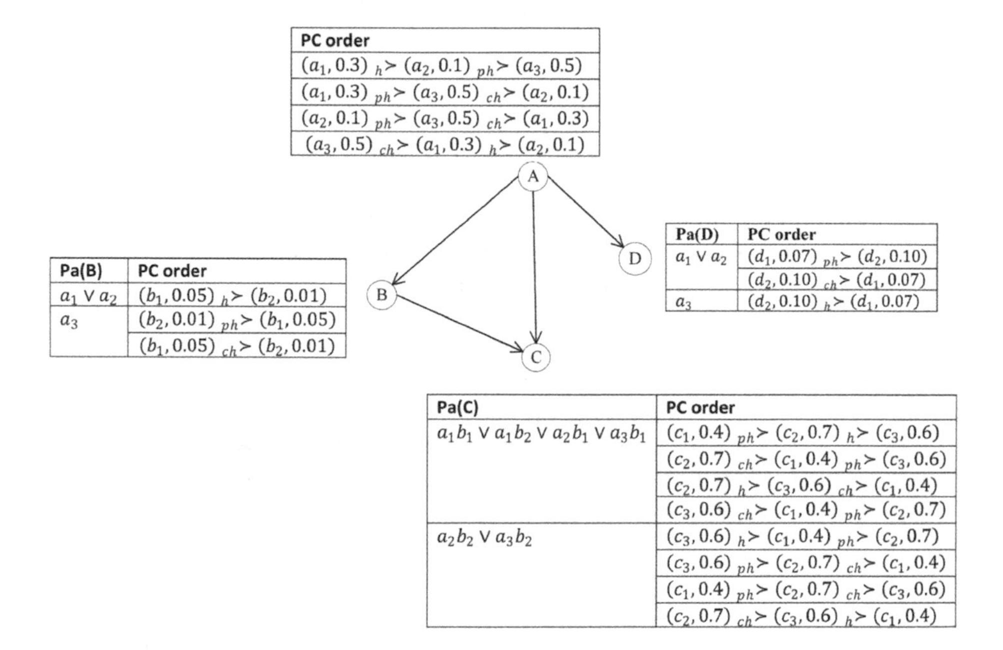

PC order
$(a_1, 0.3)\ {}_{h}\!\succ (a_2, 0.1)\ {}_{ph}\!\succ (a_3, 0.5)$
$(a_1, 0.3)\ {}_{ph}\!\succ (a_3, 0.5)\ {}_{ch}\!\succ (a_2, 0.1)$
$(a_2, 0.1)\ {}_{ph}\!\succ (a_3, 0.5)\ {}_{ch}\!\succ (a_1, 0.3)$
$(a_3, 0.5)\ {}_{ch}\!\succ (a_1, 0.3)\ {}_{h}\!\succ (a_2, 0.1)$

Pa(B)	PC order
$a_1 \vee a_2$	$(b_1, 0.05)\ {}_{h}\!\succ (b_2, 0.01)$
a_3	$(b_2, 0.01)\ {}_{ph}\!\succ (b_1, 0.05)$
	$(b_1, 0.05)\ {}_{ch}\!\succ (b_2, 0.01)$

Pa(D)	PC order
$a_1 \vee a_2$	$(d_1, 0.07)\ {}_{ph}\!\succ (d_2, 0.10)$
	$(d_2, 0.10)\ {}_{ch}\!\succ (d_1, 0.07)$
a_3	$(d_2, 0.10)\ {}_{h}\!\succ (d_1, 0.07)$

Pa(C)	PC order
$a_1b_1 \vee a_1b_2 \vee a_2b_1 \vee a_3b_1$	$(c_1, 0.4)\ {}_{ph}\!\succ (c_2, 0.7)\ {}_{h}\!\succ (c_3, 0.6)$
	$(c_2, 0.7)\ {}_{ch}\!\succ (c_1, 0.4)\ {}_{ph}\!\succ (c_3, 0.6)$
	$(c_2, 0.7)\ {}_{h}\!\succ (c_3, 0.6)\ {}_{ch}\!\succ (c_1, 0.4)$
	$(c_3, 0.6)\ {}_{ch}\!\succ (c_1, 0.4)\ {}_{ph}\!\succ (c_2, 0.7)$
$a_2b_2 \vee a_3b_2$	$(c_3, 0.6)\ {}_{h}\!\succ (c_1, 0.4)\ {}_{ph}\!\succ (c_2, 0.7)$
	$(c_3, 0.6)\ {}_{ph}\!\succ (c_2, 0.7)\ {}_{ch}\!\succ (c_1, 0.4)$
	$(c_1, 0.4)\ {}_{ph}\!\succ (c_2, 0.7)\ {}_{ch}\!\succ (c_3, 0.6)$
	$(c_2, 0.7)\ {}_{ch}\!\succ (c_3, 0.6)\ {}_{h}\!\succ (c_1, 0.4)$

Fig. 2. A CPC-net.

It has been shown that the CP-net is a special instance of the CPC-net [2]. Given that preference and comfort can be conflicting, an acyclic CPC-net has two optimal outcomes, one with the highest preference and the other with the highest comfort. Like for finding the optimal outcome in an acyclic CP-net, each of these two outcomes can be obtained using the forward sweep procedure [6] with linear time in the number of variables. If N_c is the corresponding CPC-net of a CP-net N over V, and o_1 and o_2 are two outcomes, we say that N_c entails $o_1 \;_h\!\succ o_2$, denoted as $N_c \models o_1 \;_h\!\succ o_2$, iff $N \models o_1 \succ o_2$ holds and $c(o_1) > c(o_2)$ holds in N_c. Similarly, $N_c \not\models o_1 \;_h\!\succ o_2$ indicates that N_c does not entail $o_1 \;_h\!\succ o_2$.

3 Pareto Optimality

With respect to a CPC-net, we define a Pareto optimal outcome that is not dominated by any other outcome using both preference and comfort.

Definition 5 *(Pareto optimal outcome). Given a CPC-net N_c over V, an outcome o_1 is Pareto optimal if and only if there is no other outcome o_2 such that $N_c \models o_2 \;_h\!\succ o_1$.*

To get the set of all Pareto optimal outcomes for an acyclic CPC-net, we propose a backtrack search algorithm that we call Solve-CPC. Solve-CPC is a recursive function with two parameters: a sub CPC-net N_c, which is initially the original CPC-net $N_{c_{orig}}$ and a parameter K that is the assignment to all variables in $N_{c_{orig}} - N_c$, which is initially *empty*. Solve-CPC returns the set of Pareto optimal outcomes in R_c.

Solve-CPC begins with a root variable X, i.e., a variable with no parent(s). For every value x_i of X, a new branch is created. We check the pruning criterion of the branch in lines 4–5. If the value is dominated by another value, the branch is terminated. We claim that the pruning criterion is correct, i.e., we do not disregard any Pareto optimal outcome. We formally prove this claim using Theorem 1 in Sect. 4. If the pruning criterion is not met, the branch is extended. In line 7, a sub CPC-net is built by removing X from the given CPC-net and by restricting each $CPTC(Y)$ to $X = x_i$ such that Y is a child of X.

In line 8, the sub CPC-net is divided in components such that, for every two components C_1 and C_2, there is no edge between any variable from C_1 and any variable from C_2. This ensures that every component is an acyclic CPC-net, and that the preference and comfort encoded in the component do not depend on any variable of any other component. Using lines 9–10, Solve-CPC is called for each component individually. The independency guarantees that a component is not needed to pass to the call of Solve-CPC for another component. This factorization might save a substantial amount of computing time in practice. However, we leave this experimental study for a future research. Every outcome generated by Solve-CPC is added to R_c in line 13. We show in Sect. 4 that: if Solve-CPC generates an outcome, then this outcome is not dominated by any other outcome. Therefore, Solve-CPC generates only Pareto optimal outcomes.

Example 4. Let us apply Solve-CPC for the CPC-net N_c of Fig. 2. The corresponding search tree is illustrated in Fig. 3. For the initial call (call 0), variable

Algorithm 1. Solve-CPC(N_c, K)

Input: Acyclic CPC-net N_c; assignment K to the variables of $N_{c_{orig}} - N_c$
Output: Set R_c of all solutions that are Pareto optimal with respect to N_c

1: Choose any variable X with no parents in N_c
2: Set: $R_c = \{\}$ (i.e., initialize the set of local results)
3: **for all** $x_i \in D(X)$ **do**
4:    **if** $x_j \ {}_h\!\succ x_i$ holds for any $x_j \in D(X) - \{x_i\}$ **then**
5:       **continue** with the next iteration
6:    **else**
7:       Construct a sub CPC-net N_{c_i} by removing X from N_c and, for each variable Y that is a child of X, revising $CPTC(Y)$ by restricting each row to $X = x_i$
8:       Let $N_{c_i}^1, N_{c_i}^2, \cdots, N_{c_i}^l$ be the components of N_{c_i} that are connected by edges of N_{c_i}
9:       **for all** $k \in \{1, 2, \cdots, l\}$ **do**
10:          S_i^k=Solve-CPC($N_{c_i}^k, K \cup x_i$)
11:       **end for**
12:       **for all** $o \in x_i \times S_i^1 \times S_i^2 \times \cdots \times S_i^l$ **do**
13:          Add o to R_c
14:       **end for**
15:    **end if**
16: **end for**
17: **return** R_c

A is chosen in line 1. In line 2, we initialize $R_{c_0} = \{\}$. Lines 4–15 are repeated for a_1, a_2 and a_3 as follows.

Since pruning criterion in line 4 is not met for a_1, the branch for a_1 is not terminated (see Fig. 3). In line 7, a sub CPC-net $N_{c_{a_1}}$ is built by restricting the CPTCs for B, C and D to $A = a_1$. The sub CPC-net is shown in Fig. 4. Note that in the sub CPC-net, C does not preferentially depend on B as this dependency is particularly for $A = a_2$ or $A = a_3$, not for $A = a_1$. There is no edges in $N_{c_{a_1}}$, and so Solve-CPC is called individually for the components $N_{c_{a_1}}^B$, $N_{c_{a_1}}^C$ and $N_{c_{a_1}}^D$. Note that we do not need to pass $N_{c_{a_1}}^C$ and $N_{c_{a_1}}^D$ to the call for $N_{c_{a_1}}^B$. As the redundancy is eliminated, it saves computation. Consider the call Solve-CPC ($N_{c_{a_1}}^B, a_1$) (call 1). In line 2, $R_{c_1} = \{\}$. Lines 4–15 repeat for b_1 and b_2. Pruning criterion is not met for b_1. In line 7, after removing B from $N_{c_{a_1}}^B$, we get an empty CPC-net. Lines 8–11 are skipped. Using lines 12–14, b_1 is added to R_{c_1}, i.e., $R_{c_1} = \{b_1\}$. On the other hand, pruning criterion is met for b_2 and the branch for $B = b_2$ is terminated. Solve-CPC ($N_{c_{a_1}}^B, a_1$) returns $\{b_1\}$ and it is stored in $S_{a_1}^B$ for call 0. Similarly, Solve-CPC ($N_{c_{a_1}}^C, a_1$) and Solve-CPC ($N_{c_{a_1}}^D, a_1$) return $\{c_1, c_2\}$ and $\{d_1, d_2\}$ correspondingly. In call 0, we get: $S_{a_1}^C = \{c_1, c_2\}$ and $S_{a_1}^D = \{d_1, d_2\}$. Now, every $o \in \{a_1\} \times S_{a_1}^B \times S_{a_1}^C \times S_{a_1}^D$ is added to R_{c_0} using lines 12–14, i.e., we get: $R_{c_0} = \{a_1b_1c_1d_1, a_1b_1c_1d_2, a_1b_1c_2d_1, a_1b_1c_2d_2\}$. For $A = a_2$, pruning criterion is met as $a_1 \ {}_h\!\succ a_2$ and the branch is terminated. For $A = a_3$, pruning criterion is not met. For the branch $A = a_3$, Solve-CPC similarly produces four Pareto optimal outcomes (see Fig. 3). After adding each of these Pareto optimal outcomes to R_{co}, we get: $R_{co} = \{a_1b_1c_1d_1,\ a_1b_1c_1d_2,\ a_1b_1c_2d_1,\ a_1b_1c_2d_2,\ a_3b_1c_1d_2,\ a_3b_1c_2d_2,\ a_3b_2c_2d_2,\ a_3b_2c_3d_2\}$. □

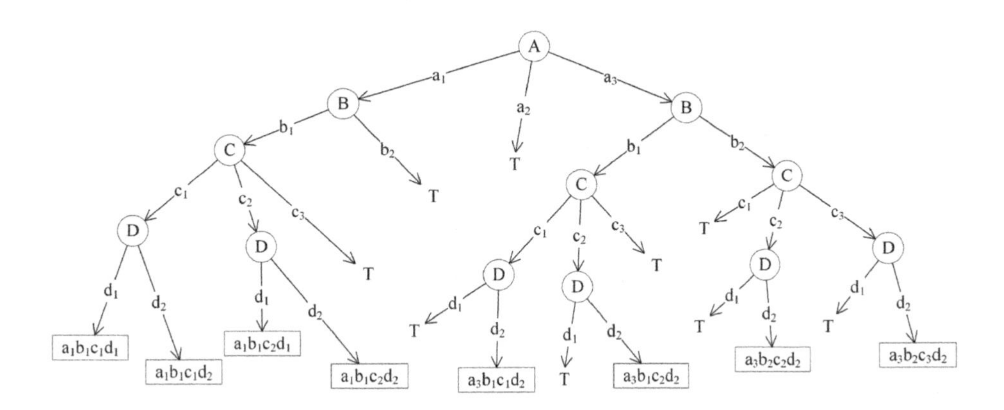

Fig. 3. The search tree of the CPC-net of Fig. 2 using Solve-CPC algorithm.

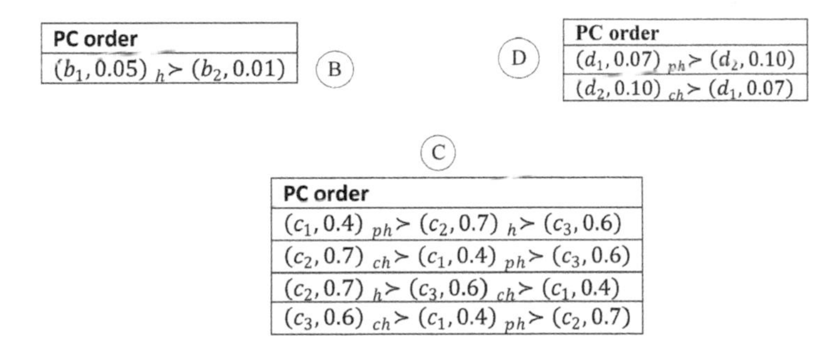

Fig. 4. A sub CPC-net produced during the execution of Solve-CPC for the CPC-net of Fig. 2.

4 Solve-CPC Formal Properties

4.1 Correctness and Completeness

Before we present our main result, we need to prove the following three lemmas.

Lemma 2. $(c(x_1) > c(x_2)) \wedge (c(y_1) > c(y_2)) \Rightarrow c(x_1y_1) > c(x_2y_2)$.

Proof. $(c(x_1) > c(x_2)) \wedge (c(y_1) > c(y_2))$
$\Rightarrow (c(x_1y_1) > c(x_2y_1)) \wedge (c(x_2y_1) > c(x_2y_2))$ [By Lemma 1]
$\Rightarrow c(x_1y_1) > c(x_2y_2)$ □

Lemma 3. *If variables X and Y are preferentially independent, we get:* $(x_1 \succ x_2) \wedge (y_1 \succ y_2) \Rightarrow x_1y_1 \succ x_2y_2$ *for* $x_1, x_2 \in D(X)$ *and* $y_1, y_2 \in D(Y)$.

Proof. $(x_1 \succ x_2) \wedge (y_1 \succ y_2)$
$\Rightarrow (x_1y_1 \succ x_2y_1) \wedge (x_2y_1 \succ x_2y_2)$ [This is a direct consequence of *ceteris paribus* semantics given that X and Y are preferentially independent]
$\Rightarrow x_1y_1 \succ x_2y_2$ [By transitivity principle] □

Lemma 4. *If variables X and Y are preferentially independent, we get:* $\neg(x_1 \ {}_h\!\succ x_2) \wedge \neg(y_1 \ {}_h\!\succ y_2) \Rightarrow \neg(x_1y_1 \ {}_h\!\succ x_2y_2)$ *for* $x_1, x_2 \in D(X)$ *and* $y_1, y_2 \in D(Y)$.

Proof. $\neg(x_1 \ {}_h\!\succ x_2) \wedge \neg(y_1 \ {}_h\!\succ y_2)$
$\Rightarrow \neg((x_1 \succ x_2) \wedge (c(x_1) > c(x_2))) \wedge \neg((y_1 \succ y_2) \wedge (c(y_1) > c(y_2)))$
$\Rightarrow \neg((x_1 \succ x_2) \wedge (c(x_1) > c(x_2)) \wedge (y_1 \succ y_2) \wedge (c(y_1) > c(y_2)))$ [Since, $\neg A \wedge \neg B \Rightarrow \neg(A \wedge B)$]
$\Rightarrow \neg((x_1 \succ x_2) \wedge (y_1 \succ y_2) \wedge (c(x_1) > c(x_2)) \wedge (c(y_1) > c(y_2)))$
$\Rightarrow \neg((x_1y_1 \succ x_2y_2) \wedge (c(x_1y_1) > c(x_2y_2)))$ [By Lemmas 2 and 3]
$\Rightarrow \neg(x_1y_1 \ {}_h\!\succ x_2y_2)$ □

We now provide our main result: the correctness and completeness of Solve-CPC.

Theorem 1. *Given a CPC-net* N_c*, an outcome* o_1 *belongs to the set* R_c *generated by the algorithm Solve-CPC if and only if there is no other outcome* o_2 *such that* $o_2 \ {}_h\!\succ o_1$.

Proof. Let R be the desired set of Pareto optimal outcomes of N_c. To prove the theorem, we have to prove that:

1. Completeness: Every Pareto optimal outcome with respect to N_c is produced by Solve-CPC, i.e., $R \subseteq R_c$; and
2. Correctness: Any outcome produced by Solve-CPC is Pareto optimal with respect to N_c, i.e., $R_c \subseteq R$.

We prove these below:

1. The outcomes of N_c are pruned by Solve-CPC only at the search space pruning using lines 4–5. This is enough to prove the completeness that no Pareto optimal outcome is pruned out. Let us assume that branch $X = x_i$ is pruned for a variable X using lines 4–5. Let us assume that the set of variables Z is already instantiated to assignment K and let the set of remaining variables be W. We have that $Pa(X) \subseteq Z$ and so every variable in $Pa(X)$ is also instantiated. Let $Pa(X)$'s instantiation is p where $p \subseteq K$. Given line 4, there exists $x_j \in$

$D(X) - \{x_i\}$ such that $x_j \;_h\!\succ x_i$. Now, it is easy to see that, for every outcome Kx_iw for $w \in D(W)$ for the branch $X = x_i$, there is a corresponding outcome Kx_jw that is searched by Solve-CPC for the branch $X = x_j$. For every such correspondence, we have that $Kx_jw \;_h\!\succ Kx_iw$, since $x_j \;_h\!\succ x_i$ holds and $Pa(X)$ are instantiated to $p \subseteq K$. Therefore, there is no Pareto optimal outcome for the branch $X = x_i$ that is pruned by Solve-CPC.

2. To prove the correctness, it is enough to prove that $(N_c \not\models o_i \;_h\!\succ o_j) \wedge (N_c \not\models o_j \;_h\!\succ o_i)$ for every two Pareto optimal outcomes o_i and o_j of N_c generated by Solve-CPC. We prove this by induction in the number of variables of N_c. The claim trivially holds if N_c involves one variable X, since Solve-CPC discards every $x_i \in D(X)$ if there exists $x_j \;_h\!\succ x_i$ for any $x_j \in D(X) - \{x_i\}$ (lines 4–5). Assume that the correctness holds for every CPC-net, where the number of variables is less than or equal to $n-1$. Let N_c be a CPC-net with n variables and X be a variable of N_c with no parents. Let the Pareto set produced by Solve-CPC involves two outcomes o_i and o_j of N_c.

First, suppose that o_i and o_j provide the same value to X, that is $o_i = x_l o'_i$ and $o_j = x_l o'_j$, for some $x_l \in D(X)$. In this case, o'_i and o'_j belong to the output of the same recursive call to Solve-CPC with N_{c_l}, i.e., o'_i and o'_j belong to the same branch $X = x_l$. Thus, using our inductive hypothesis, we get that $(N_{c_l} \not\models o'_i \;_h\!\succ o'_j) \wedge (N_{c_l} \not\models o'_j \;_h\!\succ o'_i)$. Since N_{c_l} is independent of X given $X = x_l$, we have that $(N_c \not\models x_l o'_i \;_h\!\succ x_l o'_j) \wedge (N_c \not\models x_l o'_j \;_h\!\succ x_l o'_i)$, that is $(N_c \not\models o_i \;_h\!\succ o_j) \wedge (N_c \not\models o_j \;_h\!\succ o_i)$.

Second, suppose that o_i and o_j provide two different values to X, that is $o_i = x_l o'_i$ and $o_j = x_m o'_j$, for $x_l, x_m \in D(X)$. In this case, o'_i belongs to the output of the recursive call to Solve-CPC with N_{c_l} and o'_j belongs to the output of the recursive call to Solve-CPC with N_{c_m}. That is, o'_i belongs to the branch $X = x_l$ and o'_j belongs to the branch $X = x_m$. None of these branches is pruned; and the lines 4–5 imply that $\neg(x_l \;_h\!\succ x_m)$ and $\neg(x_m \;_h\!\succ x_l)$. By our inductive hypothesis, we get $\neg(o'_i \;_h\!\succ o'_j)$ with respect to N_{c_m} and $\neg(o'_j \;_h\!\succ o'_i)$ with respect to N_{c_l}. By applying Lemma 4 on $\neg(x_l \;_h\!\succ x_m)$ and $\neg(o'_i \;_h\!\succ o'_j)$, we get $\neg(x_l o'_i \;_h\!\succ x_m o'_j)$, i.e., $\neg(o_i \;_h\!\succ o_j)$. Similarly for $\neg(x_m \;_h\!\succ x_l)$ and $\neg(o'_j \;_h\!\succ o'_i)$, we get $\neg(x_m o'_j \;_h\!\succ x_l o'_i)$, i.e., $\neg(o_j \;_h\!\succ o_i)$. Therefore, we find $(N_c \not\models o_i \;_h\!\succ o_j) \wedge (N_c \not\models o_j \;_h\!\succ o_i)$. □

4.2 Complexity Analysis

We now analyze the complexity of Solve-CPC.

Lemma 5. *If m is the number of values of a variable X, then the number of non-dominated values is m in the worst case, while it is 1 in the best case.*

Proof. Let the domain of X be $D(X) = \{x_1, x_2, \cdots, x_m\}$ with preference order $x_1 \succ x_2 \succ \cdots \succ x_m$. In the worst case, let comfort on $D(X)$ be as follows: $c(x_m) > c(x_{m-1}) > \cdots > c(x_1)$. For every $x_i \in D(X)$, we get: $x_i \;_{ch}\!\succ x_j$ and $x_i \;_{ph}\!\succ x_k$ such that $1 \leq j \leq i-1$ and $i+1 \leq k \leq m$. Therefore, there is no

$x \in D(X) - \{x_i\}$ such that $x \underset{h}{\succ} x_i$; and thus no value of X is dominated by any other value.

In the best case, let the preference order and the comfort on $D(X)$ be $x_1 \succ x_2 \succ \cdots \succ x_m$ and $c(x_1) > c(x_2) > \cdots > c(x_m)$ correspondingly. For every $x \in D(X) - \{x_1\}$, we get: $x_1 \underset{h}{\succ} x$. Therefore, x_1 dominates every other value of X. □

Theorem 2. *If n is the number of variables and m is the number of values of each variable in a CPC-net N_c, then Solve-CPC produces m^n outcomes in the worst case and 1 outcome in the best case.*

Proof. We prove this by induction in the number of variables. Using Lemma 5, the claim is trivially true for a single variable CPC-net for both the best and the worst cases. Suppose that the theorem is true for any CPC-net with $n-1$ variables. Let N_c be a CPC-net with n variables and X be a variable of N_c with no parents. X is chosen in line 1 of Solve-CPC in its initial call.

In the best case, using Lemma 5, there is only one value, say x_i, of X that dominates every other value. Therefore, every branch of X is pruned except the branch for $X = x_i$ where a sub CPC-net N_{c_i} is constructed by removing X from N_c and, for each variable Y that is a child of X, revising $CPTC(Y)$ by restricting each row to $X = x_i$. Using our inductive hypothesis, Solve-CPC produces one outcome for N_{c_i} in the best case; therefore Solve-CPC produces one outcome for N_c.

In the worst case, using Lemma 5, no value of X is dominated by any other value. Therefore, none of the m branches is pruned. For every $x_i \in D(X)$ and the corresponding branch for $X = x_i$, a sub CPC-net N_{c_i} is constructed by removing X from N_c and, for each variable Y that is a child of X, revising $CPTC(Y)$ by restricting each row to $X = x_i$. Using the inductive hypothesis, Solve-CPC produces m^{n-1} outcomes for every N_{c_i} in the worst case. Therefore, the total number of outcomes produced by Solve-CPC is $m \cdot m^{n-1}$, which is m^n. □

Theorem 1 indicates that if an outcome is produced by Solve-CPC, then this outcome is never dominated by other outcomes. Therefore, Solve-CPC has the anytime property. In this case, if user has limited time available, Solve-CPC can be stopped at anytime and the already generated solutions can be collected. For example, if one wants to find a single Pareto optimal outcome, the algorithm can be terminated as soon as the first outcome is produced. Producing a single outcome requires traversing through each of the CPC-net variables in a topological ordering; therefore it is in linear time in the number of variables.

5 Conclusion

We have proposed the algorithm Solve-CPC to obtain all Pareto optimal outcomes with respect to an acyclic CPC-net. Solve-CPC is a backtrack search algorithm that prunes branches based on a local criterion. The formal properties of the algorithm indicate that it returns every Pareto optimal outcome. A single Pareto optimal outcome can be produced in linear time in the number of variables, while the number of such outcomes might be exponential. In the near future, we plan to extend the CPC-net by including hard feasibility constraints using the CSP framework [8] as was done for CP-nets [1,4,5]. Here, the size of the Pareto set will be reduced as some outcomes become infeasible due to constraints.

Acknowledgment. This research was funded by Natural Sciences and Engineering Research Council of Canada (RGPIN-2016-05673).

References

1. Ahmed, S., Mouhoub, M.: Constrained optimization with partial CP-nets. In: IEEE International Conference on Systems, Man, and Cybernetics, pp. 3361–3366 (2018)
2. Ahmed, S., Mouhoub, M.: Extending conditional preference network with user's genuine decisions. In: Proceedings of IEEE International Conference on Systems, Man, and Cybernetics, pp. 4216–4223 (2018)
3. Ahmed, S., Mouhoub, M.: Transformation between CP-net and CPC-net. In: Proceedings of International Conference on Industrial, Engineering and Other Applications of Applied Intelligent Systems, pp. 292–300 (2018)
4. Alanazi, E., Mouhoub, M.: Variable ordering and constraint propagation for constrained CP-nets. Appl. Intell. **44**(2), 437–448 (2016)
5. Boutilier, C., Brafman, R., Domshlak, C., Hoos, H., Poole, D.: Preference-based constrained optimization with CP-nets. Comput. Intell. **20**, 137–157 (2004)
6. Boutilier, C., Brafman, R.I., Domshlak, C., Hoos, H.H., Poole, D.: CP-nets: a tool for representing and reasoning with conditional ceteris paribus preference statements. J. Artif. Intell. Res. (JAIR) **21**, 135–191 (2004)
7. Brafman, R.I., Domshlak, C., Shimony, S.E.: On graphical modeling of preference and importance. J. Artif. Intell. Res. **25**, 389–424 (2006)
8. Dechter, R.: Constraint Processing. Morgan Kaufmann, San Francisco (2003)
9. Fishburn, P.C., Lavalle, I.H.: MCDA: theory, practice and the future. J. Multicriteria Decis. Anal. **8**(1), 1–2 (1999)
10. Katona, G.: Psychological Analysis of Economic Behavior. McGraw-Hill, New York (1951)
11. Keeney, R.L., Raiffa, H.: Decisions with Multiple Objectives: Preferences and Value Trade-Offs. Cambridge University Press, New York (1993)
12. Triantaphyllou, E.: Multi-criteria decision making methods. In: Triantaphyllou, E. (ed.) Multi-Criteria Decision Making Methods: A Comparative Study, pp. 5–21. Springer, Boston (2000). https://doi.org/10.1007/978-1-4757-3157-6_2
13. Zajonc, R.B.: Feeling and thinking: preferences need no inferences. Am. Psychol. **35**(2), 151–175 (1980)

Solving the Set Covering Problem Using Spotted Hyena Optimizer and Autonomous Search

Ricardo Soto[1], Broderick Crawford[1], Emanuel Vega[1](✉), Alvaro Gómez[1], and Juan A. Gómez-Pulido[2]

[1] Pontificia Universidad Católica de Valparaíso, Valparaíso, Chile
{ricardo.soto,broderick.crawford}@pucv.cl,
{emanuel.vega.m,alvaro.gomez.r}@mail.pucv.cl

[2] University of Extremadura, 10003 Caceres, Spain
jangomez@unex.es

Abstract. The Set Covering Problem (SCP) is an important combinatorial optimization problem that finds application in a large variety of practical areas, particularly in airline crew scheduling or vehicle routing and facility placement problems. To solve de SCP we employ the Spotted Hyena Optimizer (SHO), which is a metaheuristic inspired by the natural behavior of the spotted hyenas. In this work, in order to improve the performance of our proposed approach we use Autonomous Search (AS), a case of adaptive systems that allows modifications of internals components on the run. We illustrate interesting experimental results where the proposed approach is able to obtain global optimums for a set of well-known set covering problem instances.

Keywords: Set covering problem · Spotted hyena optimizer · Autonomous search

1 Introduction

The set covering problem (SCP) is one of the well-known Karp's 21 NP-complete problems, where the goal is to find a subset of columns in a 1-0 matrix such that they cover all the rows of the matrix at a minimum cost. Several applications of the SCP can be seen in the real world, for instance, bus crew scheduling [1], location of emergency facilities [2], and vehicle routing [3]. Actually, exact methods and metaheuristics techniques have been proposed in the literature for solving the SCP. Among the exact algorithms, Beasley and Jornsten [4] used Lagrangian heuristic, feasible solution exclusion constraints was proposed in [5].

In this paper, we propose a discrete Spotted Hyena Optimizer (SHO) to solve the SCP. The SHO is based on a natural behavior of spotted hyenas. The standard SHO works on continuous search spaces but we adapt this approach in order to handle the binary domains of the SCP by using a transfer function [6] and a discretization method. Additionally, in order to improve the performance of our

F. Wotawa et al. (Eds.): IEA/AIE 2019, LNAI 11606, pp. 854–861, 2019.
https://doi.org/10.1007/978-3-030-22999-3_73

approach, we employ Autonomous Search (AS) [7]. The goal is to improve the performance by modifying the parameters, either by self-adaptation or supervised adaptation [8]. In the literature we can find more recent metaheuristics for solving the SCP, such as binary cat swarm optimization [9], binary firefly algorithm [10], and shuffle frog leaping algorithm [11]. We will perform experimental evaluations on a set of well-known instances from the Beasley's ORlibrary. We illustrate promising results where the proposed approach competes with previous reported techniques for solving the set covering problem. The rest of this paper is organized as follows. In Sect. 2, we describe the mathematical model of the SCP. The Spotted Hyena Optimizer is explained in Sect. 3. Finally, Sect. 4 illustrates the experimental results, followed by conclusions and future work.

2 Set Covering Problem

The Set Covering Problem (SCP) is formally defined as follow: Let's mxn binary matrix $A = (a_{ij})$ and a positive n-dimensional vector $C = (c_j)$, where each element c_j of C gives the cost of selecting the column j of matrix A. If a_{ij} is equal to 1, then it means that row i is covered by column j, otherwise it is not. The goal of the SCP is to find a minimum cost of columns in A such that each row in A is covered by at least one column. A mathematically definition of the SCP can be expressed as follows:

$$Minimize \sum_{j-1}^{n} c_j x_j \tag{1}$$

$$Subject\ to \sum_{j=1}^{n} a_{ij} x_j \geq 1, \quad i = 1, 2, ..., m \tag{2}$$

$$x_j \in \{0, 1\}, \quad j = 1, 2, .., m \tag{3}$$

where x_j is 1 if column j is in the solution, otherwise it is 0. Constraint (2) ensures that each row i is covered by at least one column.

3 Spotted Hyena Optimizer

The main feature of SHO is the cohesive clustering in his population [13]. The mathematical model is described as follows.

1. Encircling prey: SHO takes the current best candidate solution as the target prey, which is close to the optimum in a search space not known a priori. The other agents will try to move towards the best position defined.

$$D_h = |B \cdot P_p(x) - P(x)| \tag{4}$$

$$P(x+1) = P_p(x) - E \cdot D_h \tag{5}$$

where D_h is the distance between the current spotted hyena and the prey, x indicates the current iteration, B and E are coefficient vectors, P_p is the position of the prey, and P is the position of the spotted hyena. The vectors B and E are defined as follows:

$$B = 2 \cdot rd_1 \tag{6}$$

$$E = 2h \cdot rd_2 - h \tag{7}$$

$$h = 5 - (Iteration * (5/Max_{iteration})) \tag{8}$$

where $Iteration = 1, 2, 3, \ldots, Max_{iteration}$, rd_1 and rd_2 are random vectors in [0,1].

2. Hunting: The search agents make a cluster towards the best spotted hyena so far to update their positions. The following equations are proposed in this mechanism:

$$D_h = |B \cdot P_h - P_k| \tag{9}$$

$$P_k = P_h - E \cdot D_h \tag{10}$$

$$C_h = P_k + P_{k+1} + ... + P_{k+N} \tag{11}$$

where P_h is the best spotted hyena, and P_k indicates the position of other spotted hyenas. Here, N is the number of spotted hyenas, which is computed as follows:

$$N = count_{nos}(P_h, P_{h+1}, P_{h+2}, ..., (P_h + M)) \tag{12}$$

Here, M is a random vector [0.5, 1], *nos* defines the number of solutions and count all candidate solutions plus M, and C_h is a cluster of N number of optimal solutions.

3. Attacking Prey: SHO works around the Cluster forcing the spotted hyenas to assault towards the prey. The following equation is proposed:

$$P(x+1) = C_h/N \tag{13}$$

Here, $P(x+1)$ updates the positions of other spotted hyenas according to the position of the best search agent and save the best solution.

4. Search for Prey: Spotted hyenas mostly search the prey based on the position of the cluster of spotted hyenas, which reside in vector C_h. SHO makes use of the coefficient vector E and B with random values to force the search agents to move far away from the prey. This mechanism allows the algorithm to search globally.

3.1 Discretization of Decision Variables

As previously mentioned, SHO works on a continuous space, so in order to tackle the SCP a transformation of domain is needed. In this work, this task is performed by applying a transfer function [6] and a discretization method.

When Eqs. (5), (10), and (13) are calculated, we apply the binarization strategy, the best result for the Spotted Hyena Optimizer is $V_2 + Standard$, described as follows.

1. Standard: If condition is satisfied, standard method returns 1, otherwise returns 0.
$$X_i^d(t+1) = \begin{cases} 1, & if\ rand\ \leq T(x_i^d(t+1)) \\ 0, & otherwise \end{cases} \tag{14}$$
2. V2:
$$V_2 : T(x_i^d(t+1)) = \left|tanh\left(x_i^d(t+1)\right)\right| \tag{15}$$

3.2 Autonomous Search

Autonomous Search (AS) is a modern approach which let the solvers automatically re-configure their own solving parameters in order to improve the process when poor performances are detected [14]. We modify the population of spotted hyenas if any of the following 3 criteria is met:

1. Increase of Population by equal: If the subtract of fitness values between the best and worst spotted hyena is equal to 0, we increase the population by 5 and generate random spotted hyenas.
2. Increase of population by variation: If the coefficient variation is less than 0, 3, high cohesion between the spotted hyenas, we increase the population by 5 and generate random spotted hyenas.
3. Decrease of population: if two or more spotted hyenas have the same fitness, we remove at least 1 spotted hyena and decrease the population by the same amount.

4 Experimental Results

The effectiveness of our proposed approach has been tested using 50 different well-known instances, organized in 8 sets from the Beasley's ORlibrary. In the literature there have been reported different types of methods to improve the work for the SCP [12]. In this work we employ two of them, (1) Column Domination: once a set of rows L_j is covered by another column j' and $c'_j < c_j$, we say that column j is dominated by j', then column j is removed from the solution. (2) Column Inclusion: If a row is covered by only one column after the above domination, this means that there is no better column to cover those rows, therefore this column must be included in the optimal solution. For experimental evaluation, the configuration uses 50 agents as initial population size, and 6,000 iterations for the SHO. We compare our results with the global optimum (S_{opt}) and recent approaches for solving SCP. In this context, we compare with binary cat swarm optimization (BCSO) [9] and binary electromagnetism-like algorithm

(BELA) [15]. Additionally, the results are evaluated using the relative percentage deviation (RPD). The RPD quantifies the deviation of the objective value S_{min} from S_{opt} for each instance and it is calculated as follows:

$$RDP = \frac{(S - S_{opt})}{S_{opt}} \times 100 \tag{16}$$

Tables 1 and 2 illustrates the results from instances set group 4, 5, 6, A, B, C, D, and E respectively, where the difficulty increase between the last and the following group. Regarding optimum reached, SHO is able to get 14 in total in contrast with BCSO who gets 1, while BELA is unable to optimally solve any instance. Now, comparing the overall performance, we highlight that in several instances SHO remains very close to optimal values giving room for improvement in our future work. Finally, the results shows that no great variation exists between the best and average values obtained for SHO in all the instances set. Thus, it can be seen that the algorithm has a considerable robustness and symmetric convergence, illustrated in Fig. 1 for the instance D.1 where SHO gets optimum value.

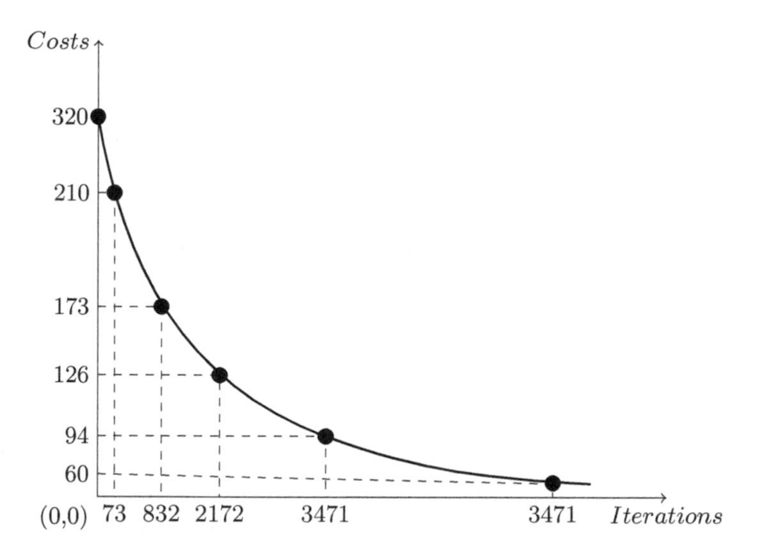

Fig. 1. Convergence chart for instance D.1 solved by SHO.

Table 1. Results for instances set of groups 4, 5, 6, A.

Instance		4.1	4.2	4.3	4.4	4.5	4.6	4.7	4.8	4.9	4.10
S_{opt}		429	512	516	494	512	560	430	492	641	514
SHO	S_{min}	430	528	532	505	514	**560**	**430**	503	669	518
	S_{avg}	432	528	532	506	514	562	430	503	669	518
	RPD	0.23	3.12	3.10	2.22	0.39	0.00	0.00	2.23	4.36	0.77
BCSO	S_{min}	279	339	247	251	230	232	332	320	295	285
	S_{avg}	480	594	607	578	554	650	467	567	725	552
	RPD	7	11.3	14.3	10.7	6.4	13.8	7.4	11	10.9	4.5
BELA	S_{min}	447	559	537	527	527	607	448	509	682	571
	S_{avg}	448	559	539	530	529	608	449	512	682	571
	RPD	4.20	9.18	4.07	6.68	2.93	8.39	4.19	3.46	6.40	11.09
Instance		5.1	5.2	5.3	5.4	5.5	5.6	5.7	5.8	5.9	5.10
S_{opt}		253	302	226	242	211	213	293	288	279	265
SHO	S_{min}	257	312	234	**242**	**211**	216	296	291	280	271
	S_{avg}	257	314	234	242	211	217	296	292	281	271
	RPD	1,58	3,31	3,53	0.00	0.00	1,40	1,02	1,04	0,35	2,26
BCSO	S_{min}	279	339	247	251	230	232	332	320	295	285
	S_{avg}	287	340	251	253	230	243	338	330	297	287
	RPD	10.30	12.30	9.30	3.70	9.00	8.90	13.30	11.10	5.70	7.50
BELA	S_{min}	280	318	242	251	225	247	316	315	314	280
	S_{avg}	281	321	240	252	227	248	317	317	315	282
	RPD	10.67	5.30	7.08	3.72	6.64	15.96	7.85	9.38	12.54	5.66
Instance		6.1	6.2	6.3	6.4	6.5	A.1	A.2	A.3	A.4	A.5
S_{opt}		138	146	145	131	161	253	252	232	234	236
SHO	S_{min}	140	**146**	148	133	165	256	259	239	235	**236**
	S_{avg}	140	146	148	133	166	256	259	239	235	236
	RPD	1,44	0.00	2,06	1,52	2,48	1,18	2,77	3,01	0,42	0.00
BCSO	S_{min}	151	152	160	138	169	286	274	257	248	244
	S_{avg}	160	157	164	142	173	287	276	263	251	244
	RPD	9.40	4.10	10.30	5.30	5.00	13.00	8.70	10.80	6.00	3.00
BELA	S_{min}	152	160	160	140	184	261	279	252	250	241
	S_{avg}	152	161	163	142	187	264	281	253	252	243
	RPD	10.14	9.59	10.34	6.87	14.29	3.16	10.71	8.62	6.84	2.12

Table 2. Results for instances set of groups B, C, D, E.

Instance		B.1	B.2	B.3	B.4	B.5	C.1	C.2	C.3	C.4	C.5
S_{opt}		69	76	80	79	72	227	219	243	219	215
SHO	S_{min}	72	81	81	81	**72**	233	223	251	225	**215**
	S_{avg}	72	81	81	82	72	234	223	251	225	215
	RPD	4,34	6,57	1,25	2,53	0.00	2,64	1,82	3,29	2,73	0.00
BCSO	S_{min}	79	86	85	89	73	242	240	277	250	243
	S_{avg}	79	89	85	89	73	242	241	278	250	244
	RPD	14.50	13.20	6.30	12.70	1.40	6.60	9.60	14.00	12.30	13.00
BELA	S_{min}	86	88	85	84	78	237	237	271	246	224
	S_{avg}	87	88	87	88	81	238	239	271	248	225
	RPD	24.64	15.79	6.25	6.33	8.33	4.41	8.22	11.52	12.33	4.19
Instance		D.1	D.2	D.3	D.4	D.5	E.1	E.2	E.3	E.4	E.5
S_{opt}		60	66	72	62	61	29	30	27	28	28
SHO	S_{min}	**60**	68	76	**62**	**61**	**29**	31	**27**	29	**28**
	S_{avg}	60	68	76	62	61	29	31	27	29	28
	RPD	0.00	3,03	5,55	0.00	0.00	0.00	3,33	0.00	3,57	0.00
BCSO	S_{min}	65	70	79	64	65	**29**	34	31	32	30
	S_{avg}	66	70	81	67	66	30	34	32	33	30
	RPD	8.30	6.10	9.70	3.20	6.60	0.00	13.30	14.80	14.30	7.10
BELA	S_{min}	62	73	79	67	66	30	35	34	33	30
	S_{avg}	62	74	81	69	67	31	35	34	34	31
	RPD	3.33	10.61	9.72	8.06	8.20	3.45	16.67	25.93	17.86	7.14

5 Conclusion and Future Work

In this work, we have proposed a discrete Spotted Hyena Optimizer (SHO) algorithm to solve the Set Covering Problem. After testing 65 non-unicost instances from the Beasley's OR-Library, our approach is able to reach some global optimum values. However, in several instances, SHO remains very close to optimal values, particularly for the hardest instances. Finally, concerning future work, we aim to improve our proposed SHO by adding hybridization with local Search Techniques. The main objective is to improve the exploitation of our approach. Finally, we propose to add smart behaviors considering Machine learning [16].

Acknowledgment. Broderick Crawford was supported by Grant CONICYT/FONDECYT/REGULAR 1171243, Ricardo Soto was supported by Grant CONICYT/FONDECYT/REGULAR 1190129, and Juan A. Gómez-Pulido was supported by Grant IB16002.

References

1. Smith, B.: Impacs-a bus crew scheduling system using integer programming. Math. Program. **42**(1–3), 181–187 (1988)
2. Toregas, C., Swain, R., Revelle, C., Bergman, L.: The location of emergency service facilities. Oper. Res. **19**, 1363–1373 (1971)
3. Foster, B., Ryan, D.: An integer programming approach to the vehicle scheduling problem. Oper. Res. Q. **27**, 367–384 (1976)
4. Beasley, J., JSmsten, K.: Enhancing an algorithm for set covering problems. Eur. J. Oper. Res. **58**, 293–300 (1992)
5. Beasley, J.: A Lagrangian heuristic for set covering problems. Naval Res. Logist. **37**, 151–164 (1990)
6. Mirjalili, S., Lewis, A.: S-shaped versus v-shaped transfer functions for binary particle swarm optimization. Swarm Evol. Comput. **9**, 1–14 (2013)
7. Hamadi, Y., Monfroy, E., Saubion, F.: An Introduction to Autonomous Search. Autonomous Search, pp. 1–11 (2011)
8. Soto, R., et al.: Using autonomous search for solving constraint satisfaction problems via new modern approaches. Swarm Evol. Comput. **30**, 64–77 (2016)
9. Crawford, B., Soto, R., Berríos, N., Johnson, F., Paredes, F.: Solving the set covering problem with binary cat swarm optimization. In: Tan, Y., Shi, Y., Buarque, F., Gelbukh, A., Das, S., Engelbrecht, A. (eds.) ICSI 2015. LNCS, vol. 9140, pp. 41–48. Springer, Cham (2015). https://doi.org/10.1007/978-3-319-20466-6_4
10. Crawford, B., Soto, R., Olivares-Suárez, M., Paredes, F.: A binary firefly algorithm for the set covering problem. In: Silhavy, R., Senkerik, R., Oplatkova, Z.K., Silhavy, P., Prokopova, Z. (eds.) Modern Trends and Techniques in Computer Science. AISC, vol. 285, pp. 65–73. Springer, Cham (2014). https://doi.org/10.1007/978-3-319-06740-7_6
11. Crawford, B., et al.: Binarization methods for shuffled frog leaping algorithms that solve set covering problems. In: Silhavy, R., Senkerik, R., Oplatkova, Z.K., Prokopova, Z., Silhavy, P. (eds.) Software Engineering in Intelligent Systems. AISC, vol. 349, pp. 317–326. Springer, Cham (2015). https://doi.org/10.1007/978-3-319-18473-9_31
12. Fisher, M., Kedia, P.: Optimal solution of set covering/partitioning problems using dual heuristics. Manage. Sci. **36**(6), 674–688 (1990)
13. Dhiman, G., Kumar, V.: Spotted hyena optimizer: a novel bio-inspired based metaheuristic technique for engineering applications. Adv. Eng. Softw. **114**, 48–70 (2017)
14. Hamadi, Y., Monfroy, E., Saubion, F.: What is Autonomous Search? Technical Report MSR-TR-2008-80, Microsoft Research (2008)
15. Soto, R., Crawford, B., Muñoz, A., Johnson, F., Paredes, F.: Preprocessing, repairing and transfer functions can help binary electromagnetism-like algorithms. In: Artificial Intelligence Perspectives and Applications. Advances in Intelligent (2015)
16. Nakib, A., Hilia, M., Heliodore, F., Talbi, E.: Design of metaheuristic based on machine learning: a unified approach. In: IPDPS Workshops, pp. 510–518 (2017)

Author Index

CPSIA information can be obtained
at www.ICGtesting.com
Printed in the USA
LVHW080200180619
621569LV00003B/54/P